A METHOD FOR ANALYSIS OF TRANSMISSION LINES TERMINATED BY NONLINEAR LOADS

MATHEMATICS RESEARCH DEVELOPMENTS

MATHEMATICS RESEARCH DEVELOPMENTS

A METHOD FOR ANALYSIS OF TRANSMISSION LINES TERMINATED BY NONLINEAR LOADS

VASIL G. ANGELOV, PH.D.

Novinka

New York

NOTICE TO THE READER

Library of Congress Cataloging-in-Publication Data

Angelov, Vasil Georgiev.
  A method for analysis of transmission lines terminated by nonlinear loads / Vasil Georgiev Angelov (Department of Mathematics, Faculty of Mine Electromechanics, University of Mining and Geology, Sofia, Bulgaria).
      pages cm
  Includes bibliographical references and index.
  ISBN 978-1-62618-909-6 (soft cover)
  1. Electric lines--Testing. 2. Electric lines--Mathematical models. 3. Telecommunication lines--Mathematical models. 4. Telecommunication lines--Testing. I. Title.
  TK3207.A54 2013
  621.319'2--dc23
                    2013028421

Published by Nova Science Publishers, Inc. † *New York*

CONTENTS

Preface — ix

Introduction — xi

Chapter I **Preliminaries** — 1

Abstract — 1

Introduction — 1

1.1. Transmission Lines and Derivation of Their Equations — 2

1.2. An Equivalence of a Mixed Problem for an Inhomogeneous Lossless Transmission Line System to an Initial Value Problem for a Neutral Equation — 4

1.3. Transformation of Transmission Lines Equations for Uniform Plane Waves — 19

1.4. Tools from Fixed-point theory in Metric and Uniform Spaces — 24

1.5. Tools from the Theory of Differential Equations with Deviating Arguments — 36

1.6. Discussion on the Operator Presentation of Periodic and Oscillatory Problems — 42

Conclusion — 52

Chapter II **Lossless Transmission Lines Terminated by a Nonlinear R-Load** — 53

Abstract — 53

Introduction — 53

2.1. Derivation of the Boundary Conditions, Formulation of the Mixed Problem, and Reducing to an Initial Value Problem for a Neutral Equation — 54

2.2. Periodic Regimes for a Neutral Equation with Polynomial Nonlinearity — 61

2.3. Measurable Periodic Regimes — 79

2.4. Oscillatory Regimes for Neutral Equation with Polynomial Nonlinearity — 87

2.5. Another Way to Reducing the Mixed Problem to a Neutral System and the Existence-Uniqueness of a Periodic Solution — 100

	2.6. Periodic Regimes for Transmission Lines Terminated by a Resistive Element with Exponential V-I Characteristic	**113**
	2.7. Measurable Periodic Solutions for a Neutral Equation with an Exponential Nonlinearity	**123**
	2.8. Oscillatory Regimes in the Exponential Case	**128**
	2.9. Another Way to Reducing the Mixed Problem to a Periodic Problem for a Neutral Equation with an Exponential Nonlinearity	**139**
	Conclusion	**154**
Chapter III	**Lossless Transmission Lines Terminated by in Series Connected *RLC*-Loads**	**157**
	Abstract	**157**
	Introduction	**157**
	3.1. Derivation of the Boundary Conditions and Formulation of the Mixed Problem	**158**
	3.2. Reducing the Mixed Problem to a Periodic Problem on the Boundary	**160**
	3.3. Analysis of the Arising Nonlinearities	**166**
	3.4. An Operator Formulation of the Periodic Problem	**172**
	3.5. The Existence-Uniqueness of a Periodic Solution for a Nonlinear Neutral System	**183**
	3.6. Numerical Example	**206**
	3.7. The Existence-Uniqueness of a Periodic Solution of a System with Linear and Time-Varying in Series Connected RLC-Loads	**214**
	3.8. Applications for Non-Uniform Lossless Transmission Lines	**228**
	3.9. Oscillatory Regimes for the Nonlinear System	**232**
	Conclusion	**241**
Chapter IV	**Lossless Transmission Lines Terminated by Parallel Connected *RLC*-Loads**	**243**
	Abstract	**243**
	Introduction	**243**
	4.1. Derivation of Boundary Conditions, Formulation of The Mixed Problem and Reduction to a Periodic Problem for a Neutral System	**244**
	4.2. Operator Presentation of the Periodic Problem for a Neutral System with Linear RLC-Loads	**250**
	4.3. Periodic Solutions of the Linear Neutral System	**259**
	4.4. Analysis of the Arising Nonlinearities	**274**
	4.5. Periodic Solutions of the Nonlinear Neutral System	**279**
	4.6. A Different Approach to Obtaining a Nonlinear Neutral System	**306**
	4.7. Oscillatory Regimes for the Nonlinear Neutral System	**309**
	Conclusion	**319**

Chapter V   **Lossy Transmission Lines Terminated by a Nonlinear Resistive Element** — 321

Abstract — 321

Introduction — 321

5.1. *Formulation of the Mixed Problem for Lossy Transmission Lines Terminated by a Nonlinear R-Element* — 322

5.2. *Reducing the Mixed Problem to an Initial Value Problem for a Neutral Equation* — 327

5.3. *Exponentially Vanishing Oscillatory Regimes for Neutral Equations with Polynomial Nonlinearities* — 330

5.4. *A Different Manner to Reducing the Mixed Problem to an Oscillatory Problem for a Neutral Equation* — 346

5.5. *Oscillatory Regimes for Neutral Equations with Exponential Nonlinearity* — 363

5.6. *Another Approach to Reducing the Mixed Problem to an Oscillatory One for a Neutral Equation with an Exponential Nonlinearity* — 380

5.7. *Transmission Lines with Time-Varying Specific Parameters* — 396

Conclusion — 401

Chapter VI   **Distortionless Lossy Transmission Lines Terminated by Parallel Connected *RLC*-Loads** — 403

Abstract — 403

Introduction — 403

6.1. *Derivation of the Boundary Conditions and Reducing the Mixed Problem to an Initial Value Problem* — 404

6.2. *Analysis of the Arising Nonlinearities* — 413

6.3. *An Operator Presentation of the Periodic Problem* — 416

6.4. *Existence-Uniqueness of a Periodic Solution for the Nonlinear Neutral System* — 425

6.5. *Numerical Example* — 446

6.6. *Oscillatory Regimes for the Neutral System* — 452

6.7. *Another Manner to Reducing the Mixed Problem to an Oscillatory One* — 464

6.8. *Lossy Transmission Lines with Time-Varying Specific Parameters* — 489

Conclusion — 496

Chapter VII   **Distortionless Lossy Transmission Lines Terminated by in Series Connected *RLC*-Loads** — 497

Abstract — 497

Introduction — 497

7.1. *Derivation of the Boundary Conditions, Formulation of the Mixed Problem and Analysis of the Nonlinearities* — 498

7.2. *Reducing the Mixed Problem to an Initial Value Problem for a Nonlinear Neutral System* — 503

7.3. *Another Manner to Reducing the Mixed Problem to an Oscillatory One* — **510**

7.4. *The Existence-Uniqueness of an Oscillatory Solution for the Nonlinear Neutral System* — **517**

7.5. *Numerical Example* — **543**

7.6. *Applications to Transmission Lines Terminated by Linear and Time-Varying Loads* — **548**

Chapter VIII Lossless and Lossy Transmission Lines Terminated by in Series Connected *RL*-Loads Parallel to the *C*-Load — **557**

Abstract — **557**

Introduction — **557**

8.1. *Derivation of the Boundary Conditions for Lossless Transmission Line Equations and Formulation of the Mixed Problem* — **558**

8.2. *Reducing the Mixed Problem to an Initial Value Problem on the Boundary* — **560**

8.3. *A Different Approach to Reducing the Mixed Problem to a Periodic Problem on the Boundary* — **565**

8.4. *Estimates of the Arising Nonlinearities and Introducing Metrics* — **567**

8.5. *Operator Presentation of the Periodic Problem* — **571**

8.6. *Numerical Example* — **600**

8.7. *Lossy Transmission Lines Terminated by Nonlinear RL-Loads Parallel to C-Load* — **605**

8.8. *Reducing the Mixed Problem to an Initial Value Problem on the Boundary* — **606**

8.9. *Operator Presentation of the Oscillatory Problem* — **611**

Conclusion — **646**

General Conclusion — **647**

Suggestions for Further Studies — **648**

References — **653**

Index — **659**

PREFACE

The subject of transmission lines has grown in importance in the last few decades because of the applications not only in the area of power transmission but in the area of high-speed very large scale integrated (VLSI) interconnects as well. In many cases, transmission lines are connected to nonlinear circuits. For instance, interconnects of high-speed VLSI chips can be modeled as transmission lines loaded with nonlinear elements. The transmission line theory is based on the Telegrapher's equations, which from mathematical point of view present a first-order hyperbolic system of partial differential equations with unknown functions for the voltage and the current. The derivation of this system can be found in many books (cf. for instance [32], [50], [72], [83], [86], [90], [91], [95], [100]-[102], [107], [108]). It turns out that with some additional conditions the Lord Kelvin transmission line theory [100]-[102] is equivalent to the Maxwell theory (cf. [100]-[102]), placing transmission lines theory alongside classical electromagnetic theory. In application, transmission lines are terminated by nonlinear loads at both ends. This yields (by means of Kirchhoff's laws) boundary conditions dependent upon the corresponding characteristics of the nonlinear (or linear) RLC-elements. For such a system, an initial-boundary value problem (or briefly mixed problem) can be formulated. In both cases of lossless and lossy transmission lines, the mixed problem for linear hyperbolic systems can be reduced to an initial value problem for a neutral equation (or system of equations) on the boundary (cf. [1], [41], [28], [29], [104]). Thus we reach systems whose future state depends on the past ones. The main reason for this dependency is the finite velocity of propagation of interaction, and in particular the electromagnetic waves. Therefore we have to draw our attention in the theory of differential equations with retarded arguments. Our exposition is based on earlier results from researchers such as V. Volterra [113], A. D. Myshkis [87], [88], M. A. Krasnoselskii [74], [75], G. A. Kamenskii [69], [70], R. Bellman and K. L. Cooke [23], L. E. Elsgolz and S.B. Norkin [53], J. G. Borisovich [30], [31], C. Corduneanu [42], [43], B. N. Sadovskii [99], A. Halanay [60], A. Halanay and J. Yorke [61], W. Melvin [85], A. M. Zverkin [127], and many others.

The primary purpose of this book is to propose a method for reducing the mixed problem for Telegrapher's equations not only for lossless transmission lines but for lossy ones as well. For the obtained neutral equation (or neutral system) on the boundary we formulate conditions for the existence-uniqueness of periodic and oscillatory solutions. Arising nonlinearities in the neutral systems caused by nonlinear characteristics of the RLC-loads may lead to many new effects such as instability, chaos, generation of higher order harmonics, etc. Our existence-uniqueness theorems guarantee stable solutions. Since the characteristics of the loaded elements are polynomial and exponential functions we obtain neutral equations (or systems) with polynomial and exponential nonlinearities. This makes mathematical treatment very difficult because most results refer to Lipschitzian nonlinearities. Our approach is based on the choice of suitable function spaces, the using extensions of families of A. Bielecki [22] and the discovering of operators whose fixed-points are periodic or oscillatory solutions of the formulating problems. By means of fixed-point theorems for contractive mappings in metric and uniform spaces (cf. I. A. Rus [98], V. G. Angelov [14]) we find conditions for existence-uniqueness of periodic and oscillatory solutions. We obtain also successive approximations of the solution with respect to a suitable family of pseudo-metrics and give an estimate of the rate of convergence. Although the question of finding the initial approximation is not trivial (cf. L. Collatz [40]) we show that one can begin with a harmonic initial approximation. The rate of convergence depends on the parameters of the transmission lines and characteristics of the nonlinear RLC-loads. Our conditions are applicable even in the case of non-uniform transmission lines, frequency depending parameters and time-varying parameters provided a system of inequalities to be satisfied. Using A. Bielecki-type pseudo-metrics, we can essentially improve the rate of convergence. The suitable choice of some constants implies that our method is better for higher frequencies.

Sofia, 2013
V. G. Angelov

INTRODUCTION

Circuits with lumped parameters have negligible influence upon circuits that have retreated a great distance. This means that the radiating energy can be disregarded. Since the absorption of the energy in external fields can be disregarded too, circuits with lumped parameters can generate electromagnetic fields, but cannot radiate and detect signals. For that purpose one can use the circuits with distributed parameters such as transmission lines. In such circuits the basic nonlinear elements − inductors, capacitors and resistors − are not lumped at particular part but to a certain extent are distributed uniformly along the circuit. A transmission line is a system of two or more closely spaced, parallel conductors. Our results refer only two conductor lines. The transmission line is said to be uniform when the size of the conductors, their spacing, and dielectric between the two conductors are the same along the entire transmission line. In other words a uniform transmission line is one in which the cross-sectional dimensions of the line, i.e. conductor cross section as well as dielectric cross sections, are the same at any two points along the line. Many practical lines do not have uniform characteristics but comprise of interconnections of dissimilar sections of transmission lines, each having a given set of R, G, L and C per unit length. Such lines are called piece-wise uniform and can be referred to as composite lines. In other cases the characteristics may vary continuously throughout the length of the line.

Without claim of completeness, we will identify the following books when the theory, examples and applications of the transmission lines are exposed: L. Brillouin and M. Parodi [32], S. G. Burns and P. R. Bond [33], N. Dodov [50], J. Dunlop and D. G. Smith [51], M. P. Flynn and J. J. Kiang [54], Ju. S. Kolesov and D. I. Shvitra [72], P. C. Magnusson, G. C. Alexander and V. K. Tripathi [83], G. Miano and A. Maffucci [86], C. R. Paul [90], D. Pozar [91], S. Rosenstark [95], S. Ramo, J. R.Whinnery, T. van Duzer [97], J. E. Rowe [96], S. A. Schelkunoff [100]-[102], W. A. Smirnov [107], K. Szimony, [108], P. Vizmuller [111], A. H. Zemanian [123] and many others.

As we have already mentioned in the Preface, the basic problem involved in the analysis of all these two-conductor lines is to determine the currents on the conductors and the voltage between the two conductors at all points along the line. The partial differential equations that govern the voltage and current of a transmission line are referred to as the transmission line equations. These equations constitute the model of the transmission line, known as Telegrapher's equations, derived by Lord Kelvin almost simultaneously as the introduction of the famous J. C. Maxwell field's system of equations. The fundamental structure of the electromagnetic fields surrounding the conductors of a transmission line is a transverse electromagnetic field structure in which the electric and magnetic field intensity vectors at each point in space have no component parallel to the line conductors. The field vectors are perpendicular to the line axis and lie on a plane perpendicular to the line. Because of this, waves propagate along the line, and this presents the transverse electromagnetic mode of propagation wherein the field vectors are transverse to the direction of propagation. These waves are called TEM waves.

We would like to mention, however (cf. S. A. Schelkunoff [100]-[102]), that the classical equations can be used at much higher frequencies than one would expect from their conventional derivation, based on the assumption of quasi-stationary fields. This is particularly so in the case of an analog of transmission lines in the study of electronic beams (cf. for instance J. E. Rowe [96], P. Bloom, R. W. Peter [25], E. Kelebekler [71], K. Szimony [108], R. Gould [56] and others). An explanation of this fact is contained in the theory of Schelkunoff that appears equivalent to that of Maxwell (cf. [102]).

The transmission line equations for two conductors taking into account the lossies are

$$\frac{\partial u(x,t)}{\partial x} + L\frac{\partial i(x,t)}{\partial t} + Ri(x,t) = 0,$$

$$\frac{\partial i(x,t)}{\partial x} + C\frac{\partial u(x,t)}{\partial t} + Gu(x,t) = 0$$

where R, L, G, C are respectively the resistance, inductance, conductance and capacitance per unit-length along the line. The unknown functions $i(x,t)$, $u(x,t)$ are the instantaneous values of the current in one conductor, and the transverse voltage from it to the other conductor. Let $\Lambda > 0$ be the length of the line. In order to find $i(x,t)$, $u(x,t)$ one has to prescribe initial conditions

$$i(x,0) = i_0(x), \ u(x,0) = u_0(x), x \in [0,\Lambda].$$

The boundary conditions can be obtained using Kirchhoff's current or voltage laws. Their explicit form depends on the characteristics of nonlinear RLC-loads terminated at both ends. Since the characteristics are in general nonlinear ones, we obtain from mathematical point of view nonlinear boundary conditions, and hence a nonlinear mixed problem for the hyperbolic systems.

Although our results are formulated only for current and voltage, they are applicable to a much wider range of cases. In [100], Schelkunoff derives transmission line equations of transverse plane electric waves. The obtained equations are similar to equations connecting the voltage and the current in a transmission line having Z for its distributed series impedance and Y for the shunt admittance:

$$\frac{\partial F}{\partial x} = -ZU, \qquad \frac{\partial U}{\partial x} = -YF$$

where $Z = j\omega\mu$, $Y = (g + j\omega\varepsilon)$. Here the unknown functions are U: the magnetomotive force between a given point and infinity along a path contained completely in the equiphase surface passing through the given point, and F: the electric flux through a curve drawn from a given point to infinity. Obviously Z and Y are derived for harmonic solutions. The general form of the above system becomes

$$\frac{\partial F}{\partial z} + \mu \frac{\partial U}{\partial t} = 0, \qquad \frac{\partial U}{\partial z} + \varepsilon \frac{\partial F}{\partial t} + gF = 0 \, .$$

In [25] S. Bloom and R. W. Peter consider a transmission line an analog of a modulated electron beam, and derive the following system

$$\frac{\partial V}{\partial x} = jXI, \qquad \frac{\partial I}{\partial x} = jBV$$

where $X = \omega \dfrac{p^2}{\omega^2 \varepsilon_0 \sigma}$ is a series reactance of the beam and $B = \dfrac{\omega \varepsilon_0 \sigma}{p^2}\left(\dfrac{2\pi}{\lambda_p}\right)^2$ is shunt susceptance per unit length. In this case, the above system in general form is:

$$\frac{\partial V}{\partial x} = \frac{p^2}{\omega^2 \varepsilon_0 \sigma} \frac{\partial I}{\partial t}, \qquad \frac{\partial I}{\partial x} = \frac{\omega \varepsilon_0 \sigma}{p^2}\left(\frac{2\pi}{\lambda_p}\right)^2 \frac{\partial V}{\partial t} \, .$$

In order to point out the advantages of our method we will briefly describe other methods for investigation of transmission lines.

We will briefly discuss the various methods of analysis of transmission lines, beginning with the numerical methods developed to solve for time domain voltage and current on a lossy and/or non-uniform transmission line. The lumped element model (LEM) (cf. D. Dhaene and D. D. Zutter [48]) is a method in which a lossy transmission line is presented as a cascade of infinitesimally small cells, each characterized by its distributed resistance, conductance, inductance and capacitance (RLC). In a variation on this method, the lossy transmission line is modelled using a combination of lumped elements and ideal lossless transmission lines (cf. T. Komuro [76]). The major advantage of LEM is that equivalent circuits can be easily and quickly implemented in existing circuit simulators and extended to include coupled transmission lines with an arbitrary termination. This method provides an accurate analysis of the transmission line's transient behaviour when the line is terminated by a nonlinear circuit. A drawback is that computation speed suffers because the time domain response is obtained by repeated calculations carried out over a range of frequencies, followed by a numerical convolution (specifically, an inverse Laplace transformation). In addition, a characteristic of lumped models is that they fail to accurately characterize interconnects at high frequencies.

The method of characteristics, a direct time domain calculation that provides an analytical solution to the wave equation for the lossless case, was first used for transmission lines by F. H. Branin [27]. With this method the transmission line response is modelled as a sum of modes. For this reason it is referred to as a modal extraction technique. To include losses, A. G. Grudis and C. S. Chang [58] suggested a cascaded method of characteristics model. The cascaded method of characteristics performs well for a long line, though for short lines it is not as fast or as accurate as other methods (cf. I. M. Elfadel, H. M. Huang, A. E. Rudehli, A. Dounavis, M. S. Nakhla [52]). A variation of the method of characteristics using lumped matrix rational approximations was proposed by Grivet-Talocia, *at el.* [57]. They analyse a lossy transmission line in the frequency domain, then apply fast Fourier transforms and convolution to recover information about the transient time domain response. This method is useful for short as well as long transmission lines, but for some signals the response is degraded when the frequency points are poorly sampled by the fast Fourier transforms or the convolution spans a large time interval. Xu, *et al* [118] introduced an improved method of characteristics technique that relies on the Laplace transform, rather than on fast Fourier transforms and convolution.

The modified method of characteristics is not subject to instability, and requires less computation time and less memory than previous method of characteristics methods. Because this method of characteristics requires a Taylor series in the frequency domain as well as a Padé approximation for the exponential terms, it is expected that it would be very difficult to apply to non-uniform line cases. Time/frequency characteristics of lossy transmission lines have been obtained using the finite element method (FEM) [78], [80], [121] and finite-difference time-domain (FDTD) [35], [77], [103], [109], [120]. S.Y. Lee, A. Konard and R. Saldanha [78] solved the Telegrapher's equations using FEM, which was shown to give accurate results for lossy structures as well as for non-uniform lines. R. Lucic, *et al.* [80] and S. H. You and E. F. Kuester [121] developed a hybrid LEM-FEM method. The LEM-FEM method takes advantage of the topology of LEM and removes the disadvantage of instability that can arise in FEM. A modified 'on-line FDTD' scheme was introduced by X. Zhong, *et al.* [125] based on a semi-implicit approximation. Although Zhong's method improves computational accuracy, it does not match the broadband accuracy of the standard FDTD method. T. K. Sekine, *et al.* [103] proposed FDTD with a new boundary condition for non-uniform transmission lines. This method has proven to be faster than the method of characteristics, and more accurate than other current FDTD absorbing boundary conditions for an inhomogeneous line. A drawback of FDTD is that it is subject to numerical dispersion [103] when the signal is propagated over a significant number of time steps. In addition, due to the interleaving structure of FDTD, voltage and current are not computed at the same spatial points on the transmission line. This can lead to small errors in the characterization of capacitance, inductance and characteristic impedance calculated from numerical results.

Another attractive numerical technique, the Transmission Line Method (TLM) [63], [34] belonging to the general class of differential equation time-domain numerical modelling methods. With TLM the transmission line is presented by a series of nodes connected by transmission line segments. The relationship between incident and scattered voltages and currents at the nodes is determined using a scattering matrix. Additional elements such as transmission-line stubs can be added to the connecting nodes so that a non-uniform transmission line is fashioned. It has been reported [34] that compared with standard second-order FDTD, the TLM method has much less numerical dispersion.

There are several specialized techniques for analysing non-uniform lossy transmission lines [81] [120]. In [81] the authors concatenate small, linearly-tapered transmission line sections that are each presented by the exact frequency domain ABCD matrix. This allows a piecewise-linear approximation of the characteristic impedance of a general shaped non-uniform transmission line, which is a much more accurate presentation than the piecewise constant profile used in similar methods. Xu, *et al.* have applied the differential quadrature method to the Telegrapher's equation: [119], [120], [92]. In these cases the differential equations are discretized into a set of algebraic equations and solved in the frequency domain. The time domain response is obtained using inverse Laplace transform and recursive convolution. It is claimed [120] that the differential quadrature method leads to higher computational efficiency than conventional FEM and FDTD methods, and can be easily applied to non-uniform and multi-conductor transmission lines.

Finally, we note the interesting results from J. Jeong and R. Nevels [66], Jeong [67], in which they present an analytical solution for the coupled Telegrapher's equations in terms of the voltage and current on a homogeneous lossy transmission line and multi-conductor transmission line. The resulting Telegrapher's equation solution is in the form of an exact time-domain propagator operating on the line voltage and current, showing that these analytical equations lead to a stable numerical method that can be used in the analysis of both homogeneous and inhomogeneous transmission lines. For analytical solutions we also note: [106].

The primary purposes of this book are twofold.

The first purpose is to derive boundary conditions, and to formulate a mixed problem for every configuration of nonlinear elements in which the line is terminated, due to the fact that every circuit type generates specific difficulties. Then we demonstrate a method for reducing the mixed problem for a hyperbolic system of first-order partial differential equations, to an initial value problem on the boundary for a neutral equation or system of equations. From the mathematical point of view these results are well known, but we give a common scheme not only for lossless transmission lines but for lossy ones as well. These results for lossy transmission lines are not generally recognized to date.

The second objective is to provide a general method for analysis of transmission lines terminated by nonlinear elements. Because of the relevance to many applications, we consider widespread configurations of nonlinear (and in particular linear) loads − resistive, capacitive and inductive elements. Reducing the mixed problem for corresponding configurations (in series connected, parallel connected, and RL-loads in series parallel to C-load), we obtain a neutral equation or system of equations on the boundary. We formulate problems for the existence-uniqueness of periodic and oscillatory solutions. Several critical problems arise: 1) Finding appropriate functional spaces and appropriate operators (which act in these spaces) whose fixed-points are solutions of the problems stated; 2) to overcome the difficulties generated by nonlinearities of corresponding nonlinear characteristics of RLC-elements. These nonlinearities are included in the neutral equation or systems. Typically, contractive fixed-point theorems are successfully applied to equations with nonlinearities of the Lipschitz type. But here we encounter worse nonlinearities: polynomial, and exponential; 3) the choice between a metric and a family of pseudo-metrics in uniform spaces. We extend the idea of A. Bielecki, introducing exponential weighted pseudo-metrics and thus control the polynomial and exponential rate of the nonlinearities. In the end we obtain conditions for existence-uniqueness of periodic and oscillatory solutions in the form of simple inequalities between basic quantities, which can easily be checked for design. Specifically, we use a metric fixed-point theorem for the periodic solutions and fixed-point theorems in uniform spaces for oscillating solutions.

First, we study a transmission line loaded by a circuit with a nonlinear negative resistance load, in series connected *RLC*-loads, parallel connected *RLC*-loads, and in series connected *RL*-loads parallel to a nonlinear *C*-load. We use the results of Chapter I as a base, and reduce the mixed problem for a hyperbolic system describing the transmission lines to an initial value problem for neutral functional differential equations on the boundary. Then we present a suitable operator formulation of the problem for periodic or oscillatory solutions. Finally, using the fixed-point method (cf. [14], [98]) we obtain conditions for the existence and uniqueness of a periodic or oscillatory solution for the above-mentioned problem. The main difficulty comes from the arising nonlinearities. They are of polynomial and exponential type. Our method overcomes the arising difficulties, and we obtain approximated solutions with a high level of accuracy. We show that our results can be applied to non-uniform lines when the characteristic impedance depends on the frequency. Such case occurs in the transmission line analog of electron beams. Chapters II, III, and IV are devoted to lossless transmission lines, while V, VI, and VII contain the same case, but for lossy transmission lines. Chapter VIII contains an analysis of another configuration of nonlinear loads for both cases of lossless and lossy transmission lines.

The basic results explained in this book have already been published in [3]-[21]. We would like to point that a lot of papers have been devoted to an existence theorem without theorems showing the uniqueness of the solution (there are exceptions cf. for instance [82]). The question naturally arises: which of the many solutions provided are unique? In contrast to such papers, we consider an existence-uniqueness theorem and obtain a sequence of successive approximations approaching a solution with respect to a suitable metric. The solution obtained, even when approximated, has an explicit form beginning with simple combinations of trigonometric functions.

Chapter I begins with a short introduction to the transmission line theory. **§ 1.1** contains a derivation of the transmission line system of equations based on typical distributed parameter per-unit-length equivalent circuits. In **§ 1.2** the proof of the equivalence of mixed problem for transmission line system to an initial value problem for a neutral equation on the boundary is given. It is shown also that the same reasoning is valid for the uniform plane waves in **§ 1.3.** In **§ 1.4** tools from the fixed-point theory are given following [14] and [98]. These results are a primary method used in this book. Paragraph **§ 1.5** contains a short exposition of the basic results from the theory of differential equations with deviating arguments (cf. [23], [30], [31], [42], [43], [53], [60], [61], [65], [69], [70], [74], [75], [85], [86], [93], [96], [99], [113], [126], [127]). Finally in **§ 1.6**, we discuss various methods (cf. [30], [31], [74], [75], [99]) for the definition of operators whose fixed-points are periodic solutions of neutral equations. We extend these results to solve our periodic and oscillatory problems.

Chapter II is devoted to an analysis of lossless transmission lines terminated by a nonlinear resistive load. **§ 2.1** contains a reduction of the mixed problem for a hyperbolic system to a periodic initial value problem on the boundary with line voltage an unknown function. In **§ 2.2,** the existence-uniqueness of the T_0-periodic solution for the obtained neutral equation is proved. Here, the *V-I* characteristic of the *R*-load is of a polynomial type. The proof does not depend on the degree of the polynomial. In **§ 2.3** we formulate conditions for the existence-uniqueness of a measurable solution. In **§ 2.4** sufficient conditions for the existence-uniqueness of an oscillatory regime of a neutral equation are formulated. In **§ 2.5** a different way of reducing the mixed problem to an initial value problem is given, and conditions for the existence-uniqueness of a periodic solution of the neutral equation is proved. § **2.6** considers the same problem as § **2.2** but using *V-I* characteristics of an exponential type of *R*-load. In § **2.7** the same problem as in § **2.6** is considered, in the class of measurable functions. In § **2.8** conditions for the existence-uniqueness of an oscillatory regime are formulated. Finally, in § **2.9** we propose a different method of reducing the mixed problem to a periodic one on the boundary for a neutral equation with exponential nonlinearity.

The main purpose of **Chapter III** is to analyse the process in a lossless transmission line terminated by a series of connected *RLC*-loads. First in § **3.1** we derive the boundary conditions of the mixed problem for transmission line equations as a consequence of Kirchhoff's law. In § **3.2** we reduce the mixed problem to an initial value problem for a neutral equation. In § **3.3** we analyze arising nonlinearities of nonlinear loads. Particular difficulties are generated by the singularities of the capacitive functions. We also obtain an existence-uniqueness result for periodic solutions of the obtained neutral equations. In § **3.4** we introduce a new operator formulation of the periodic problem. In § **3.5** we prove a theorem for the existence-uniqueness of a periodic solution for the nonlinear neutral system. In § **3.6** we demonstrate conditions of the existence-uniqueness theorem using specific numeric data. In § **3.7** we show the existence-uniqueness of a periodic solution of a system with linear in series connected *RLC*-loads. Our goal is to show a unified approach, namely, the fixed-point iteration method. In § **3.8** we apply our methods to transmission line analogs of electron beams where specific parameters depend on the frequency.

The main purpose of **Chapter IV** is to analyse the process in a lossless transmission line terminated by parallel connected *RLC*-loads. First, in § **4.1** we derive the boundary conditions as a consequence of Kirchhoff's law, formulate the mixed problem for transmission line equations, and propose a manner to reduce the mixed problem to an initial value problem for a neutral equation on the boundary. In § **4.2** we give an operator presentation of the periodic problem for a neutral system of equations. In § **4.3** we prove an existence-uniqueness theorem for periodic regimes of a transmission line terminated by linear *RLC*-loads. In § **4.4** we analyze arising nonlinearities in view of the nonlinear characteristics of the loads. In § **4.5** we prove a theorem for the existence-uniqueness of periodic solutions for the nonlinear neutral system, that is, when the loads are nonlinear. In § **4.6** we propose a different approach to reducing the mixed problem as an initial value problem for a nonlinear neutral system. Finally in § **4.7** we prove the existence-uniqueness theorem for an oscillatory solution of the neutral system.

In **Chapter V** we consider the same problem as in Chapter II, but for lossy transmission lines. We extend our technique of reducing the mixed problem to an initial value one and use the fixed-point method to solve the obtained neutral equation. In § **5.1** we first formulate the mixed problem for a lossy transmission line system, then in § **5.2** propose a reduction of the mixed problem to a periodic initial value problem on the boundary with the line voltage as an unknown function. A theorem for the existence-uniqueness of a T_0-periodic solution of the obtained neutral equation is proved. Here the *V-I* characteristic of the *R*-load is of a polynomial type. The proof does not depend on the degree of the polynomial. In § **5.3** we give conditions for the existence-uniqueness of an oscillatory solution for the neutral equation with polynomial nonlinearity. In § **5.4** we propose another manner for reducing the mixed problem to an oscillatory problem for a neutral equation. In § **5.5** we formulate conditions for the existence-uniqueness of an oscillatory regime for a neutral equation with exponential nonlinearity. In § **5.6** we propose a new approach to reducing the mixed problem of lossy transmission lines to an oscillatory problem for a neutral equation with exponential nonlinearity. In § **5.7** we consider a lossy transmission line with time-varying specific parameters. We introduce a condition (extending the Heaviside condition) that allows us to reduce this case similar to the previous one.

In **Chapter VI** we consider lossy transmission lines terminated by parallel-connected nonlinear RLC-loads. This means $R \neq 0, G \neq 0$ which generates additional difficulties. We assume that the Heaviside condition is fulfilled, that is, $R / L = G / C$. It implies that the line is without distortion. So in § **6.1** we derive boundary conditions corresponding to circuit configuration of the loads. Then we reduce the mixed problem for the lossy transmission line system to a periodic problem on the boundary. In § **6.2** we provide an analysis of the arising nonlinearities. In § **6.3** we propose an operator presentation of the periodic problem. In § **6.4** we prove the existence-uniqueness theorem for a periodic solution for the nonlinear neutral system. In § **6.5** we consider a numerical example. In § **6.6** we give conditions for the existence-uniqueness of an oscillatory solution for the obtained neutral system. In § **6.7** we propose another manner by which to reduce the mixed problem to an oscillatory one on the boundary. Finally in § **6.8** we introduce a generalized Heaviside condition in order to obtain distortionless propagation.

In **Chapter VII** we consider distortionless lossy transmission lines terminated by in series connected RLC-loads. In § **7.1** we derive boundary conditions corresponding to the terminated nonlinear loads, then formulate the mixed problem for transmission line systems. In § **7.2** we reduce the mixed problem for the transmission line system to an initial value problem for a nonlinear neutral system. In § **7.3** we propose a different manner in which to reduce the mixed problem to a periodic problem on the boundary. In § **7.4** an existence-uniqueness theorem for a periodic solution for the neutral system is proved. As in the previous chapters we apply the fixed-point method in suitable metric and uniform spaces. In § **7.5** we show how to apply the method to the specific problems outlined. Finally In § **7.6** we apply the same method to transmission lines terminated by linear and time-varying loads. The generalized Heaviside condition is used again for transmission lines with time-varying specific parameters.

In **Chapter VIII** transmission lines terminated by a different (from previous chapters) configuration of nonlinear loads are considered, namely in series connected RL-loads parallel to C-load. In § **8.1** we derive boundary conditions and formulate the mixed problem for lossless transmission lines. We want to emphasize that in this case the RL current function cannot be excluded and we have to consider four equations instead of two as in the previous chapters. In § **8.2** we reduce the mixed problem to a periodic initial value problem on the boundary. In § **8.3** we provide a different approach to reducing the mixed problem to a periodic one on the boundary. In § **8.4** we analyze the arising nonlinearities. In § **8.5** we give an operator presentation of the periodic problem. In § **8.6** using a numerical example, we demonstrate how to apply our method to specific problems. In § **8.7** we consider lossy transmission lines terminated by the same configuration of nonlinear RLC-loads. In § **8.8** we reduce the mixed problem to an initial value problem on the boundary. Finally we give an

operator presentation of the oscillatory problem in § **8.8** and give a numerical example in §
8.9.

We would like to point out that the present monograph is addressed to a wide range of
readers − not only to researchers, mathematicians, physicists, and engineers, but Ph.D.
students as well. For those who are not interested in mathematical proofs we recommend you
focus only on the inequalities in the numerical examples. They guarantee the existence-
uniqueness of a periodic or oscillatory solution.

Chapter I

PRELIMINARIES

ABSTRACT

Our goal is to provide the necessary tools from the theory of fixed-points (some of which belong to the author) and from the theory of differential equations with deviating arguments. As for methods to reduce the mixed problem for hyperbolic system to the initial value problem for neutral equations on the boundary, it can be said that we generalize some techniques for transmission line equations with delay.

INTRODUCTION

We begin with a short introduction to transmission line theory. § 1.1 contains a derivation of the transmission line system of equations based on the usual distributed parameter per-unit-length equivalent circuits. In § 1.2 we provide proof of the equivalence of the mixed problem for transmission line system to an initial value problem for a neutral equation on the boundary. It is also shown that the same reasoning is valid for uniform plane waves. In § 1.3 it is also shown that the same reasoning is valid for uniform plane waves. In § 1.4 tools from fixed-point theory are given following [14] and [98]. These results are the primary method used in this book. § 1.5 contains short exposition of the basic results from the theory of differential equations with deviating arguments (cf. [23], [30], [31], [42], [43], [53], [60], [61], [65], [69], [70], [74], [75], [85], [86], [93], [96], [99], [113], [126] and [127]). Finally in § 1.6 we discuss various methods (cf. [30], [31], [74], [75], [99]) of defining of operators whose fixed-points are periodic solutions of neutral equations. We extend these results to solve our periodic and oscillatory problems.

The transmission line equations can be derived for a general two-conductor line by three methods: 1) from the integral forms of Maxwell's equations; 2) from the differential forms of Maxwell's equations and; 3) from the usual distributed parameter, per-unit-length equivalent circuit. Here we briefly note the third method following C. R. Paul [90] (cf. also L. Brillouin & M. Parodi [32], J. Dunlop & D. G. Smith [51], P. C. Magnusson, G. C. Alexander & V. K. Tripathi [83], G. Miano & A. Maffucci [86], D. Pozar [91], J. E. Rowe [96], V. A. Smirnov [107], K. Szimony [103], P. Vizmuller [111]).

1.1. TRANSMISSION LINES AND DERIVATION OF THEIR EQUATIONS

First we recall (cf. C. Paul [90]) a basic definition: an electronic circuit or electromagnetic radiating structure is said to be electrically small if its largest dimension Λ is significantly smaller than a wavelength λ, that is, $\Lambda \ll \lambda$. Let us take an x-axis in the direction of the line. The concept stems from the fact that lumped circuit concepts are only valid for structures whose largest dimension is electrically small. If a structural dimension is electrically large, one may break it into the union of electrically small substructures and can then present each substructure with a lumped circuit model. In order to apply this to a transmission line, one may consider breaking it into small, Δx length subsections (as in Fig. 1.1). The per-unit-length inductance L presents the magnetic flux as passing between the conductors due to the current on those conductors. One may lump this in each Δx subsection by multiplying the per-unit-length by parameter Δx. For a uniform line, this can be done for all such subsections as shown in Fig. 1.1. Similarly, the per-unit-length capacitance C represents the displacement current flowing between the two conductors and can be similarly lumped in each subsection. The per-unit-length conductance G presents the transverse conduction current flowing between the two conductors and can be lumped in q in a similar fashion.

Small conductor losses can be handled in this equivalent circuit in an approximate manner by including the per-unit-length resistance R (the total for both conductors) in series with the inductance element. From the per-unit-length equivalent circuit shown in Fig. 1.1 one obtains

$$u(x+\Delta x,t)-u(x,t)=-R\,\Delta x\,i(x,t)-L\,\Delta x\frac{\partial i(x,t)}{\partial t}, \qquad (1.1.1)$$

and similarly

$$i(x+\Delta x,t)-i(x,t)=-G\Delta x u(x+\Delta x,t)-C\Delta x\frac{\partial u(x+\Delta x,t)}{\partial t}. \qquad (1.1.2)$$

Dividing (1.1.1) by Δx and taking the limit as $\Delta x \to 0$ we get the first transmission line equation:

$$\frac{\partial u(x,t)}{\partial x}=-Ri(x,t)-L\frac{\partial i(x,t)}{\partial t}. \qquad (1.1.3)$$

From (1.1.1) one obtains:

$$u(x+\Delta x,t)=u(x,t)-R\,\Delta x\,i(x,t)-L\,\Delta x\frac{\partial i(x,t)}{\partial t}.$$

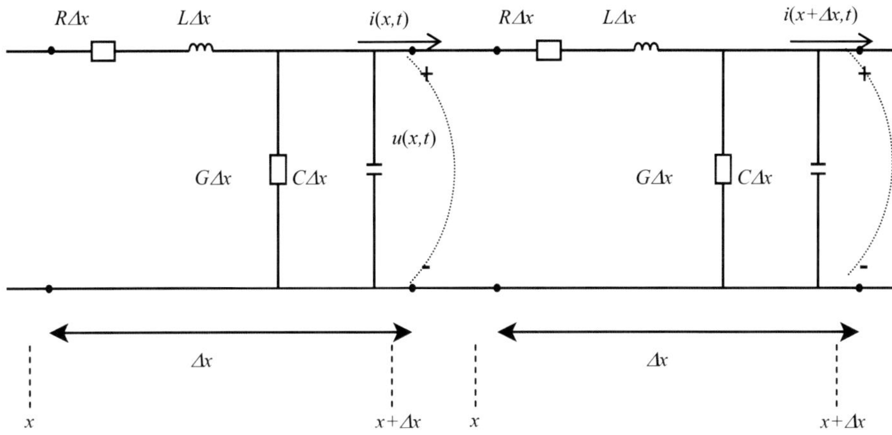

Figure 1.1.

After a differentiation in *t* we get:

$$\frac{\partial u(x+\Delta x,t)}{\partial t} = \frac{\partial u(x,t)}{\partial t} - R\Delta x\frac{\partial i(x,t)}{\partial t} - L\Delta x\frac{\partial^2 i(x,t)}{\partial t^2}.$$

Then replacing in (1.1.2) one has:

$$\frac{i(x+\Delta x,t)-i(x,t)}{\Delta x} = -Gu(x,t) - C\frac{\partial u(x+\Delta x,t)}{\partial t} + \Delta x\left[GRi(x,t) + (GL+RC)\frac{\partial i(x,t)}{\partial t} + LC\frac{\partial^2 i(x,t)}{\partial t^2} \right]$$

and taking the limit as $\Delta x \to 0$ we obtain the second transmission line equation:

$$\frac{\partial i(x,t)}{\partial x} = -Gu(x,t) - C\frac{\partial u(x,t)}{\partial t}. \tag{1.1.4}$$

The system of transmission line equations then becomes:

$$\frac{\partial u(x,t)}{\partial x} + L\frac{\partial i(x,t)}{\partial t} + Ri(x,t) = 0,$$
$$\frac{\partial i(x,t)}{\partial x} + C\frac{\partial u(x,t)}{\partial t} + Gu(x,t) = 0. \tag{1.1.5}$$

This is a first-order hyperbolic system of partial differential equations. For such a system, an initial-boundary value (or mixed) problem can be formulated: to find a solution of the system (1.1.5) in the set

$$\Pi = \{(x,t) : x \in [0,\Lambda], t \in [0,T]\},$$

where $\Lambda > 0, T > 0$ are fixed constants. The solution should satisfy prescribed initial conditions and boundary conditions.

1.2. An Equivalence of a Mixed Problem for an Inhomogeneous Lossless Transmission Line System to an Initial Value Problem for a Neutral Equation

1.2.1. Lossless Transmission Lines with Distributed Sources

If we consider an ideal transmission line (that is, neglecting the losses $R = 0$, $G = 0$) we obtain lossless transmission line equations:

$$C\frac{\partial u(x,t)}{\partial t} + \frac{\partial i(x,t)}{\partial x} = 0,$$

$$L\frac{\partial i(x,t)}{\partial t} + \frac{\partial u(x,t)}{\partial x} = 0.$$

If the transmission line is excited by known incidental electromagnetic fields generated by nearby radiating structures, the effects of these fields may be taken into account by including suitable distributed sources along the line. Such lines are described by the following system of first order partial differential equations:

$$C\frac{\partial u(x,t)}{\partial t} + \frac{\partial i(x,t)}{\partial x} = j(x,t) ,$$

$$L\frac{\partial i(x,t)}{\partial t} + \frac{\partial u(x,t)}{\partial x} = e(x,t) \tag{1.2.1}$$

where the distributed sources $e(x,t)$ and $j(x,t)$ depend on the incident electromagnetic field and on the structure of the guiding system modeled by the lossless transmission line.

The main problem for the system (1.2.1) is to determine the unknown functions, namely voltage $u(x,t)$ and current $i(x,t)$, in the transmission line terminated at each end by linear or nonlinear circuit elements. We suppose the length of the line is Λ and its ends are $x = 0$ and $x = \Lambda$.

The natural problem for the system (1.2.1) from a mathematical point of view is the initial-boundary value problem (a mixed problem), which is to find a solution $(u(x,t), \; i(x,t))$ on the set $\Pi = \{(x,t) : x \in [0,\Lambda], t \in [0,T]\}$. The solution should satisfy the initial condition

$$u(x,0) = u_0(x), \; i(x,0) = i_0(x), x \in [0,\Lambda] \tag{1.2.2}$$

and the boundary conditions (written in general form)

$$f_p\left(u(0,t),\frac{\partial u(0,t)}{\partial t},\int_0^t u(0,\tau)d\tau,u(\Lambda,t),\frac{\partial u(\Lambda,t)}{\partial t},\int_0^t u(\Lambda,\tau)d\tau,\right.$$
$$\left. i(0,t),\frac{\partial i(0,t)}{\partial t},\int_0^t i(0,\tau)d\tau,i(\Lambda,t),\frac{\partial i(\Lambda,t)}{\partial t},\int_0^t i(\Lambda,\tau)d\tau,t\right)=0 \tag{1.2.3}$$

$(p=0,1)$, where f_0 and f_1 are nonlinear functions defined for $R^{12}\times R_+^1$. The type of the functions depends on the loads at the ends of the line.

The mixed problem formulated for the system (1.2.1) can be reduced to an initial value problem for a neutral equation on the boundary. We extend the method given by K. L. Cooke & D. W. Krumme [41]).

First we write the system (1.2.1) in matrix form:

$$A_1\frac{\partial U(x,t)}{\partial t}+A_2\frac{\partial U(x,t)}{\partial x}=A_3(x,t) \tag{1.2.4}$$

where

$$A_1=\begin{bmatrix}C & 0\\ 0 & L\end{bmatrix},\quad A_2=\begin{bmatrix}0 & 1\\ 1 & 0\end{bmatrix},\quad A_3=\begin{bmatrix}j(x,t)\\ e(x,t)\end{bmatrix},\quad U=\begin{bmatrix}u(x,t)\\ i(x,t)\end{bmatrix},$$

$$\frac{\partial U(x,t)}{\partial t}=\begin{bmatrix}\dfrac{\partial u(x,t)}{\partial t}\\[2mm] \dfrac{\partial i(x,t)}{\partial t}\end{bmatrix},\quad \frac{\partial U(x,t)}{\partial x}=\begin{bmatrix}\dfrac{\partial u(x,t)}{\partial x}\\[2mm] \dfrac{\partial i(x,t)}{\partial x}\end{bmatrix}.$$

Since $|A_1|\neq 0$, then the matrix equation (1.2.4) can be written in the form

$$\frac{\partial U}{\partial t}+A_1^{-1}A_2\frac{\partial U}{\partial x}=A_1^{-1}A_3,$$

where $A_1^{-1}=\begin{bmatrix}1/C & 0\\ 0 & 1/L\end{bmatrix}$ or

$$\frac{\partial U(x,t)}{\partial t}+A\frac{\partial U(x,t)}{\partial x}=M(x,t). \tag{1.2.5}$$

Here $A = A_1^{-1}A_2 = \begin{bmatrix} 0 & 1/C \\ 1/L & 0 \end{bmatrix}$ and

$$M = A_1^{-1}A_3 = \begin{bmatrix} 1/C & 0 \\ 0 & 1/L \end{bmatrix}\begin{bmatrix} j(x,t) \\ e(x,t) \end{bmatrix} = \begin{bmatrix} (1/C)j(x,t) \\ (1/L)e(x,t) \end{bmatrix}.$$

In order to transform the matrix A into a diagonal form, we solve the characteristic equation: $\begin{vmatrix} -\lambda & 1/C \\ 1/L & -\lambda \end{vmatrix} = 0$. Its roots are $\lambda_1 = \dfrac{1}{\sqrt{LC}}$, $\lambda_2 = -\dfrac{1}{\sqrt{LC}}$. For the eigenvectors we obtain the following systems:

$$\begin{vmatrix} -\dfrac{1}{\sqrt{LC}}\xi_1 & +\dfrac{1}{L}\xi_2 = 0 \\ \dfrac{1}{C}\xi_1 & -\dfrac{1}{\sqrt{LC}}\xi_2 = 0 \end{vmatrix} \quad \text{and} \quad \begin{vmatrix} \dfrac{1}{\sqrt{LC}}\xi_1 & +\dfrac{1}{L}\xi_2 = 0 \\ \dfrac{1}{C}\xi_1 & +\dfrac{1}{\sqrt{LC}}\xi_2 = 0 \end{vmatrix}.$$

Hence $\left(\xi_1^{(1)},\xi_2^{(1)}\right) = \left(\sqrt{C},\sqrt{L}\right)$, $\left(\xi_1^{(2)},\xi_2^{(2)}\right) = \left(-\sqrt{C},\sqrt{L}\right)$.

H denotes the matrix formed by eigenvectors: $H = \begin{bmatrix} \sqrt{C} & \sqrt{L} \\ -\sqrt{C} & \sqrt{L} \end{bmatrix}$. Its inverse matrix is

$H^{-1} = \begin{bmatrix} \dfrac{1}{2\sqrt{C}} & -\dfrac{1}{2\sqrt{C}} \\ \dfrac{1}{2\sqrt{L}} & \dfrac{1}{2\sqrt{L}} \end{bmatrix}$. If we denote by $A^{can} = \begin{bmatrix} \dfrac{1}{\sqrt{LC}} & 0 \\ 0 & -\dfrac{1}{\sqrt{LC}} \end{bmatrix}$, then it is known

that $A^{can} = HAH^{-1}$.

Introduce new variables $Z=HU$, (or $U = H^1 Z$) where

$$Z = \begin{bmatrix} V(x,t) \\ I(x,t) \end{bmatrix}, \quad H = \begin{bmatrix} \sqrt{C} & \sqrt{L} \\ -\sqrt{C} & \sqrt{L} \end{bmatrix}, \quad U = \begin{bmatrix} u(x,t) \\ i(x,t) \end{bmatrix}.$$

Then

$$\begin{vmatrix} V(x,t) = \sqrt{C}\,u(x,t) + \sqrt{L}\,i(x,t) \\ I(x,t) = -\sqrt{C}\,u(x,t) + \sqrt{L}\,i(x,t) \end{vmatrix}$$

or

$$\left|\begin{array}{l} u(x,t) = \dfrac{1}{2\sqrt{C}} V(x,t) - \dfrac{1}{2\sqrt{C}} I(x,t), \\[3mm] i(x,t) = \dfrac{1}{2\sqrt{L}} V(x,t) + \dfrac{1}{2\sqrt{L}} I(x,t). \end{array}\right.$$

$$(1.2.6)$$

Replacing $U = H^{-1}Z$ in the matrix equation (1.2.5) we obtain

$$\frac{\partial\left(H^{-1}Z(x,t)\right)}{\partial t} + A\frac{\partial\left(H^{-1}Z(x,t)\right)}{\partial x} = M \, ,$$

$$\frac{\partial H^{-1}}{\partial t} Z + H^{-1}\frac{\partial Z}{\partial t} + A\left(\frac{\partial H^{-1}}{\partial x} Z + H^{-1}\frac{\partial Z}{\partial x}\right) = M \, .$$

But $\dfrac{\partial H^{-1}}{\partial t} = \widetilde{0}$ and $\dfrac{\partial H^{-1}}{\partial x} = \widetilde{0}$ where $\widetilde{0} = \begin{bmatrix} 0 & 0 \\ 0 & 0 \end{bmatrix}$. Consequently

$$H^{-1}\frac{\partial Z}{\partial t} + AH^{-1}\frac{\partial Z}{\partial x} = M \, .$$

Multiplying from the left by the matrix H we obtain

$$\frac{\partial Z}{\partial t} + HAH^{-1}\frac{\partial Z}{\partial x} = HM$$

or

$$\frac{\partial Z}{\partial t} + A^{can}\frac{\partial Z}{\partial x} = \Phi \qquad\qquad (1.2.7)$$

where

$$\Phi = HM = \begin{bmatrix} \sqrt{C} & \sqrt{L} \\ -\sqrt{C} & \sqrt{L} \end{bmatrix}\begin{bmatrix} \dfrac{1}{C} j(x,t) \\[3mm] \dfrac{1}{L} e(x,t) \end{bmatrix} = \begin{bmatrix} \dfrac{1}{\sqrt{C}} j(x,t) + \dfrac{1}{\sqrt{L}} e(x,t) \\[3mm] -\dfrac{1}{\sqrt{C}} j(x,t) + \dfrac{1}{\sqrt{L}} e(x,t) \end{bmatrix} = \begin{pmatrix} \Phi_1(x,t) \\ \Phi_2(x,t) \end{pmatrix}.$$

System (1.2.7) might be written in more detail as

$$\frac{\partial V(x,t)}{\partial t}+\frac{1}{\sqrt{LC}}\frac{\partial V(x,t)}{\partial x}=\frac{1}{\sqrt{C}}j(x,t)+\frac{1}{\sqrt{L}}e(x,t),$$

$$\frac{\partial I(x,t)}{\partial t}-\frac{1}{\sqrt{LC}}\frac{\partial I(x,t)}{\partial x}=-\frac{1}{\sqrt{C}}j(x,t)+\frac{1}{\sqrt{L}}e(x,t).$$

$$(1.2.8)$$

Remark 1.2.1. Let us emphasize that the above derivations are valid only when H^1 does not depend on x and t, while the next considerations are valid and for $A^{can}=A^{can}(x,t)$.

Recall that in the matrix equation (1.2.7)

$$Z=\begin{pmatrix}V(x,t)\\I(x,t)\end{pmatrix},\quad A^{can}=\begin{pmatrix}\lambda_1(x,t)&0\\0&\lambda_2(x,t)\end{pmatrix},\quad \Phi=\begin{pmatrix}\Phi_1(x,t)\\\Phi_2(x,t)\end{pmatrix}.$$

We suppose that $\lambda_1(x,t)>0$ and $\lambda_2(x,t)<0$.

The initial condition in the new variables becomes:

$$Z(x,0)=\begin{pmatrix}V(x,0)\\I(x,0)\end{pmatrix}=\begin{pmatrix}\sqrt{C}\,u(x,0)+\sqrt{L}\,i(x,0)\\-\sqrt{C}\,u(x,0)+\sqrt{L}\,i(x,0)\end{pmatrix}=\begin{pmatrix}V_0(x)\\I_0(x)\end{pmatrix}=Z_0(x)\qquad (1.2.9)$$

for $x\in[0,\Lambda]$.

The boundary conditions become:

$$\tilde{f}_p\left(V(0,t),\frac{\partial V(0,t)}{\partial t},\int_0^t V(0,\tau)d\tau,V(\Lambda,t),\frac{\partial V(\Lambda,t)}{\partial t},\int_0^t V(\Lambda,\tau)d\tau,\right.$$

$$\left. I(0,t),\frac{\partial I(0,t)}{\partial t},\int_0^t I(0,\tau)d\tau,I(\Lambda,t),\frac{\partial I(\Lambda,t)}{\partial t},\int_0^t I(\Lambda,\tau)d\tau,t\right)=0\ (p=0,1)$$

$$(1.2.10)$$

The characteristics of the system (1.2.8) form two families of curves with slopes

$$\frac{dx}{dt}=\lambda_1(x,t)\text{ and }\frac{dx}{dt}=\lambda_2(x,t).$$

$$(1.2.11)$$

We assume that through each point $(x,t)\in\Pi=\{(x,t)\in R^2:x\in[0,\Lambda]$ and $t\in[0,\infty)\}$ there are two characteristics, C_1 with positive slope and C_2 with negative slope (cf. Fig. 1.2).

The curve C_1 extends to the right until it intersects the straight line $x=\Lambda$, and to the left until it intersects the straight line $x=0$ or $t=0$, whereas C_2 intersects $x=0$ on the left and either $x=\Lambda$ or $t=0$ on the right.

Introduce directional differentiation along the characteristic C_p:

$$D_p = \frac{\partial}{\partial t} + \frac{\partial}{\partial x}\frac{dx}{dt} = \frac{\partial}{\partial t} + \lambda_p(x,t)\frac{\partial}{\partial x} \quad (p=1,2).$$

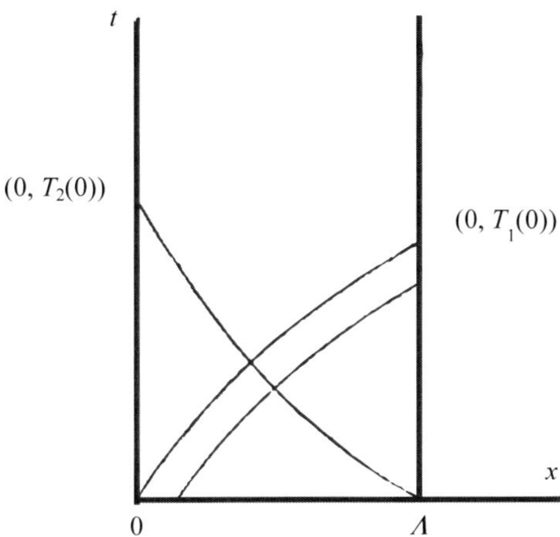

Figure 1.2.

Let $(0,\hat{t})$ be a point on the ray $x = 0$, $t \geq 0$. Let the integral curve C_1 of the first equation from (1.2.11) be

$$g_1(x,t) = c_1 = \text{const.}$$

The curve C_1 goes through the point $(0,\hat{t})$, if the coordinates of the point satisfy the above equation, i.e. the constant c_1 can be defined from the equation $c_1 = g_1(0,\hat{t})$. Therefore the curve C_1 through the point $(0,\hat{t})$ is:

$$g_1(x,t) = g_1(0,\hat{t}).$$

Now we are able to find the intersection point of the last curve with the ray $x = \Lambda$, $t \geq 0$, replacing in the above equation $x = \Lambda \Rightarrow g_1(\Lambda,t) = g_1(0,\hat{t})$. Solving this equation with respect to t we obtain some value t_1, such that $t_1 > \hat{t}$, since $\lambda_1(x,t) > 0$. Denote this by

$$t_1 = \hat{t} + T_1(\hat{t}). \text{ In any event } T_1 = T_1(\hat{t}).$$

In the same way, we choose an arbitrary point $\left(\Lambda,\hat{t}\right)$ from the ray $x=\Lambda,\, t\geq 0$ and define the integral curve C_2 of the second equation from (1.2.11):

$$g_2(x,t)=c_2=\text{const.}$$

We define the constant c_2 in such way that the curve goes through the point $(\Lambda,\hat{t})$. So the characteristic C_2 goes through the point $(\Lambda,\hat{t})$, that is:

$$g_2(x,t)=g_2(\Lambda,\hat{t}).$$

The last curve intersects the ray $x=0, t\geq 0$ in a point defined by the equation $g_2(0,t)=g_2(\Lambda,\hat{t})$. Denote by t_2 its solution where t_2 depends on $\hat{t}$ $\;(\lambda_2(x,t)<0)$ and set

$$t_2=\hat{t}+T_2(\hat{t}).$$

Integrating the equation

$$D_1 V=\frac{\partial V}{\partial t}+\frac{\partial V}{\partial x}\frac{dx}{dt}=\frac{\partial V}{\partial t}+\lambda_1(x,t)\frac{\partial V}{\partial x}=\Phi_1(x,t)$$

along the characteristic C_1 from $(0,t)$ to $(\Lambda,t+T_1(t))$ we obtain

$$V(\Lambda,t+T_1(t))=V(0,t)+\int\limits_{t}^{t+T_1(t)}\Phi_1(x_1(t_1),t_1)dt_1\quad\left(t\geq 0\right).\tag{1.2.12}$$

Then

$$\frac{dV(0,t)}{dt}=\frac{\partial V(\Lambda,t+T_1(t))}{\partial t}-\frac{d}{dt}\left(\int\limits_{t}^{t+T_1(t)}\Phi_1(x_1,t_1)dt_1\right)=$$

$$(1+\dot{T_1}(t))\frac{\partial V(\Lambda,t+T_1(t))}{\partial t}-\frac{d}{dt}\left(\int\limits_{t}^{t+T_1(t)}\Phi_1(x_1,t_1)dt_1\right).$$

It should be supposed that $\dot{T}_p(t)\neq -1, (p=1,2)..$ Otherwise if we take $\Phi_1(x,t)=0$ the above equality yields O $\dfrac{dV(0,t)}{dt}=0\Rightarrow V(0,t)=\text{const.}$

In analogous way we integrate the equation $D_2 I=\Phi_2(x,t)$ along the characteristic C_2 from $(0,t+T_2(t))$ to (Λ,t) and obtain:

$$I(\Lambda,t) = I(0,t+T_2(t)) + \int_{t+T_2(t)}^{t} \Phi_2(x_2,t_1)dt_1, \quad t \geq 0 \tag{1.2.13}$$

and then

$$\frac{dI(\Lambda,t)}{dt} = \frac{\partial I(0,t+T_2(t))}{\partial t} + \frac{d}{dt}\left(\int_{t+T_2(t)}^{t}\Phi_2(x_2,t_1)dt_1\right) =$$

$$(1+\dot{T}_2(t))\frac{\partial I(0,t+T_2(t))}{\partial t} + \frac{d}{dt}\left(\int_{t+T_2(t)}^{t}\Phi_2(x_2,t_1)dt_1\right).$$

Replacing (1.2.12) and (1.2.13) into the boundary condition (1.2.10) we obtain nonlinear equations.

First we put

$$V(t) = V(\Lambda,t), \quad I(t) = I(0,t).$$

Then after differentiation and integration of (1.2.12), written in the form

$$V(0,t) = V(\Lambda,t+T_1(t)) - \int_{t}^{t+T_1(t)} \Phi_1(x_1(t_1),t_1)dt_1$$

we have:

$$\frac{dV(0,t)}{dt} = (1+\dot{T}_1(t))\dot{V}(t+T_1(t)) - \frac{d}{dt}\left(\int_{t}^{t+T_1(t)} \Phi_1(x_1(t_1),t_1)dt_1\right) \tag{1.2.14}$$

and

$$\int_{0}^{t}V(0,\tau)d\tau = \int_{0}^{t}V(\tau+T_1(\tau))d\tau - \int_{0}^{t}\int_{\tau}^{\tau+T_1(\tau)} \Phi_1(x_1(t_1),t_1)dt_1 d\tau.$$

Analogously after differentiation and integration of (1.2.13) we have

$$\frac{dI(\Lambda,t)}{dt} = (1+\dot{T}_2(t))\dot{I}(t+T_2(t)) + \frac{d}{dt}\left(\int_{t+T_2(t)}^{t} \Phi_2(x_2(t_1),t_1)dt_1\right) \tag{1.2.15}$$

and

$$\int_0^t I(\Lambda,\tau)d\tau = \int_0^t I(\tau+T_2(\tau))d\tau + \int_0^t \int_{\tau+T_2(\tau)}^\tau \Phi_2(x_2(t_1),t_1)dt_1 d\tau .$$

Replacing in (1.2.10) we obtain the system:

$$\tilde{f}_p\left(V(t),V(t+T_1(t)),\dot{V}(t),\dot{V}(t+T_1(t)),\int_0^t V(\tau)d\tau,\int_0^t V(\tau+T_1(\tau))d\tau,\right.$$
$$\left. I(t),I(t+T_2(t)),\dot{I}(t),\dot{I}(t+T_2(t)),\int_0^t I(\tau)d\tau,\int_0^t I(\tau+T_2(\tau))d\tau,t\right)=0 \tag{1.2.16}$$

$(p=1,2)$.

We point out that $\tilde{f}_p$ are different functions from f_p.

Obviously the system (1.2.16) is a nonlinear system of integro-differential equations with unknown functions $V(t)$ and $I(t)$.

In the following, we show that the system (1.2.16) is equivalent to the mixed problem (1.2.8) - (1.2.10).

We have already shown that every solution of the problem (1.2.8) - (1.2.10) is a solution of the system (1.2.16). We will prove the converse.

Let us suppose that $V(t)$ and $I(t)$ satisfy the system (1.2.16) for $t \geq 0$. Define $V(\Lambda,t):=V(t)$ and $I(0,t):=I(t), t\geq 0$. Then we define $V(x,t)$ and $I(x,t)$ throughout the strip Π by integration of the system $D_1 V = \Phi_1(x,t), D_2 I = \Phi_2(x,t)$ along the characteristics using the prescribed values of $V(\Lambda,t)$ and $I(0,t)$ on the boundary of Π. It is easy to see that equations (1.2.12) and (1.2.13) are valid and also (1.2.14) and (1.2.15). Therefore the system (1.2.16) implies that the boundary conditions (1.2.10) are satisfied.

Equations (1.2.8) are satisfied since $V(x,t)$ and $I(x,t)$ are solutions for the system $D_1 V = \Phi_1(x,t), D_2 I = \Phi_2(x,t)$. So we have obtained that every solution of (1.2.16) generates a unique solution to the problem (1.2.8) - (1.2.10). We observe that initial condition (1.2.9) (cf. Fig. 1.2) leads by integration along characteristic C_1 emanating from $x \in [0,\Lambda], t=0$ to values of $V(\Lambda,t)$ (or $V(t)$) on $x=\Lambda, 0\leq t \leq T_1(0)$, and by integration along characteristic C_2 emanating from this segment to values of $I(0,t)$ (or $I(t)$) on $x=0$, $0\leq t \leq T_2(0)$.

Remark 1.2.2. Here we consider the particular case $e \equiv 0$, $L=$ const., $C=$ const. We obtain $\lambda_1 = \dfrac{1}{\sqrt{LC}} =$ const, $\lambda_2 = -\dfrac{1}{\sqrt{LC}} =$ const. and $\Phi = 0$. Then the system (1.2.11) becomes

$$\frac{dx}{dt} = \frac{1}{\sqrt{LC}} \quad \text{and} \quad \frac{dx}{dt} = -\frac{1}{\sqrt{LC}} \quad \left(v = \frac{1}{\sqrt{LC}} \right)$$

and the integral curves $C_p\,(p = 0,1)$ of the system (1.2.11) are the families of straight lines

$$x - vt = c_1 \quad \text{and} \quad x + vt = c_2 .$$

The curve C_1 goes through the point $(0,\hat{t})$ if $c_1 = g_1(0,\hat{t})$. The constant c_1 can be defined from the equation $c_1 = g_1(0,\hat{t})$. Therefore the curve C_1 through the point $(0,\hat{t})$ is:

$$x - vt = -v\hat{t}.$$

Now we are able to find the intersection point of the last curve with the ray $x = \Lambda$, $t \geq 0$, replacing $x = \Lambda \Rightarrow g_1(\Lambda,t) = g_1(0,\hat{t}) \Leftrightarrow \Lambda - vt = -v\hat{t}$ in the above equation. Solving this equation with respect to t we obtain some value $t_1 = \hat{t} + \dfrac{\Lambda}{v} \equiv \hat{t} + T$, $t_1 > \hat{t}$. Obviously in this case $T = \text{const.} > 0$.

In the same way, choosing $(\Lambda,\hat{t})$ from the ray $x = \Lambda$, $t \geq 0$, the integral curve C_2 from the system (1.2.11) becomes:

$$x + vt = c_2 = \text{const.}$$

The constant c_2 is defined such that C_2 to go through the point $(\Lambda,\hat{t})$, that is,

$$x + vt = \Lambda + v\hat{t}.$$

The last curve and the ray $x = 0, t \geq 0$ are intersected at the point defined by the equation $vt = \Lambda + v\hat{t}$. Denote by t_2 its solution $t_2 = \dfrac{\Lambda}{v} + \hat{t} \equiv \hat{t} + T$.

Integrating the equation

$$D_1 V = \frac{\partial V}{\partial t} + \frac{\partial V}{\partial x}\frac{dx}{dt} = \frac{\partial V}{\partial t} + v\frac{\partial V}{\partial x} = 0$$

along the characteristic C_1 from $(0,t)$ to $(\Lambda, t+T)$ we obtain

$$V(\Lambda, t+T) = V(0,t) \quad (t \geq 0). \tag{1.2.17}$$

Then

$$\frac{\partial V(0,t)}{\partial t} = \frac{\partial V(\Lambda, t+T)}{\partial t} \equiv \frac{dV(\Lambda, t+T)}{dt}.$$

In an analogous way we integrate the equation $D_2 I = 0$ along the characteristic C_2 from $(0, t+T)$ to (Λ, t) and obtain:

$$I(\Lambda, t) = I(0, t+T) \quad t \geq 0 \tag{1.2.18}$$

and then

$$\frac{\partial I(\Lambda, t)}{\partial t} = \frac{\partial I(0, t+T)}{\partial t} \equiv \frac{dI(0, t+T)}{dt}.$$

Remark 1.2.3. For the particular case $e(x,t) \neq 0, L = \text{const.}, C = \text{const.}$ we have

$$V(\Lambda, t+T) = V(0,t) + \int_{t}^{t+T} \left(\frac{1}{\sqrt{C}} j(x_1(t_1), t_1) + \frac{1}{\sqrt{L}} e(x_1(t_1), t_1) \right) dt_1 \quad (t \geq 0). \tag{1.2.19}$$

Then

$$\frac{dV(0,t)}{dt} = \frac{\partial V(\Lambda, t+T)}{\partial t} - \frac{d}{dt}\left(\int_{t}^{t+T} \left(\frac{1}{\sqrt{C}} j(x_1(t_1), t_1) + \frac{1}{\sqrt{L}} e(x_1(t_1), t_1) \right) dt_1 \right) =$$

$$= (1+\dot{T})\frac{\partial V(\Lambda, t+T)}{\partial t} - \frac{d}{dt}\left(\int_{t}^{t+T} \left(\frac{1}{\sqrt{C}} j(x_1(t_1), t_1) + \frac{1}{\sqrt{L}} e(x_1(t_1), t_1) \right) dt_1 \right) = \tag{1.2.20}$$

$$= \frac{dV(\Lambda, t+T)}{\partial t} - \frac{j(x_1(t+T), t+T) - j(x_1(t), t)}{\sqrt{C}} - \frac{e(x_1(t+T), t+T) - e(x_1(t), t)}{\sqrt{L}}$$

Similarly in view of the equation

$$I(\Lambda, t) = I(0, t+T) + \int_{t+T}^{t} \left(-\frac{1}{\sqrt{C}} j(x_1(t_1), t_1) + \frac{1}{\sqrt{L}} e(x_1(t_1), t_1) \right) dt_1$$

we find

$$\frac{dI(\Lambda,t)}{dt} = \frac{dI(0,t+T)}{dt} + \frac{d}{dt}\left(\int_{t+T}^{t}\left(-\frac{1}{\sqrt{C}}j(x_1(t_1),t_1) + \frac{1}{\sqrt{L}}e(x_1(t_1),t_1)\right)dt_1\right) =$$

$$= \frac{dI(0,t+T)}{dt} - \frac{j(x_2(t),t) - j(x_2(t+T),t+T)}{\sqrt{C}} + \frac{e(x_2(t),t) - e(x_2(t+T),t+T)}{\sqrt{L}}.$$

If $j(t),\ e(t) \neq 0,\ L = \text{const.},\ C = \text{const.}$ the above relations become

$$V(\Lambda,t+T) = V(0,t) + \int_{t}^{t+T}\left(\frac{1}{\sqrt{C}}j(s) + \frac{1}{\sqrt{L}}e(s)\right)ds;$$

$$\frac{dV(0,t)}{dt} = \frac{dV(\Lambda,t+T)}{\partial t} - \frac{j(t+T)-j(t)}{\sqrt{C}} - \frac{e(t+T)-e(t)}{\sqrt{L}};$$

$$I(\Lambda,t) = I(0,t+T) + \int_{t+T}^{t}\left(-\frac{1}{\sqrt{C}}j(s) + \frac{1}{\sqrt{L}}e(s)\right)ds;$$

$$\frac{dI(\Lambda,t)}{dt} = \frac{dI(0,t+T)}{dt} - \frac{j(t)-j(t+T)}{\sqrt{C}} + \frac{e(t)-e(t+T)}{\sqrt{L}}.$$

1.2.2. Lossy Transmission Lines with Distributed Sources

Lossy transmission lines may also be influenced by known incident electromagnetic fields generated by nearby radiating structures. The effect of these incident fields may be taken into account by including suitable distributed sources $j(x,t), e(x,t)$ along the line. They depend on the incident electromagnetic field as well as on the structure of the guiding system modeled by the transmission line.

Let us consider such a system

$$C\frac{\partial u(x,t)}{\partial t} + \frac{\partial i(x,t)}{\partial x} + Gu(x,t) = j(x,t),$$

$$L\frac{\partial i(x,t)}{\partial t} + \frac{\partial u(x,t)}{\partial x} + Ri(x,t) = e(x,t)$$

$$(1.2.21)$$

$$(x,t) \in \Pi = \left\{(x,t) \in \Pi^2 : (x,t) \in [0,\Lambda] \times [0,\infty)\right\}.$$

We suppose that Heaviside condition $R/L = G/C$ is satisfied.
First we present the system (1.2.21) in the form

$$\frac{\partial u(x,t)}{\partial t} + \frac{1}{C}\frac{\partial i(x,t)}{\partial x} + \frac{G}{C}u(x,t) = \frac{1}{C}j(x,t),$$

$$\frac{\partial i(x,t)}{\partial t} + \frac{1}{L}\frac{\partial u(x,t)}{\partial x} + \frac{R}{L}i(x,t) = \frac{1}{L}e(x,t)$$

and then in a matrix form:

$$\frac{\partial Y(x,t)}{\partial t} + A_1\frac{\partial Y(x,t)}{\partial x} + A_2 Y(x,t) = A_3(x,t) \tag{1.2.22}$$

where

$$Y(x,t) = \begin{bmatrix} u(x,t) \\ i(x,t) \end{bmatrix}, \quad \frac{\partial Y(x,t)}{\partial t} = \begin{bmatrix} \dfrac{\partial u(x,t)}{\partial t} \\ \dfrac{\partial i(x,t)}{\partial t} \end{bmatrix}, \quad \frac{\partial Y(x,t)}{\partial x} = \begin{bmatrix} \dfrac{\partial u(x,t)}{\partial x} \\ \dfrac{\partial i(x,t)}{\partial x} \end{bmatrix},$$

$$A_1 = \begin{bmatrix} 0 & \dfrac{1}{C} \\ \dfrac{1}{L} & 0 \end{bmatrix}, \quad A_2 = \begin{bmatrix} \dfrac{G}{C} & 0 \\ 0 & \dfrac{R}{L} \end{bmatrix}, \quad A_3(x,t) = \begin{bmatrix} \dfrac{1}{C}j(x,t) \\ \dfrac{1}{L}e(x,t) \end{bmatrix}.$$

In order to transform the matrix $A_1 = \begin{bmatrix} 0 & 1/C \\ 1/L & 0 \end{bmatrix}$ in a diagonal form we have to solve

the characteristic equation $\begin{vmatrix} -\lambda & 1/C \\ 1/L & -\lambda \end{vmatrix} = 0$. Its roots are $\lambda_1 = 1/\sqrt{LC}, \ \lambda_2 = -1/\sqrt{LC}$.

For the eigenvectors, we obtain the following systems:

$$\begin{vmatrix} -\dfrac{1}{\sqrt{LC}}\xi_1 + \dfrac{1}{L}\xi_2 = 0 \\ \dfrac{1}{C}\xi_1 - \dfrac{1}{\sqrt{LC}}\xi_2 = 0 \end{vmatrix} \quad \text{and} \quad \begin{vmatrix} \dfrac{1}{\sqrt{LC}}\xi_1 + \dfrac{1}{L}\xi_2 = 0 \\ \dfrac{1}{C}\xi_1 + \dfrac{1}{\sqrt{LC}}\xi_2 = 0 \end{vmatrix}.$$

Hence $\left(\xi_1^{(1)}, \xi_2^{(1)}\right) = \left(\sqrt{C}, \sqrt{L}\right), \ \left(\xi_1^{(2)}, \xi_2^{(2)}\right) = \left(-\sqrt{C}, \sqrt{L}\right)$.

Denote by H the matrix formed by the eigenvectors $H = \begin{bmatrix} \sqrt{C} & \sqrt{L} \\ -\sqrt{C} & \sqrt{L} \end{bmatrix}$. Its inverse is

$$H^{-1} = \begin{bmatrix} \dfrac{1}{2\sqrt{C}} & -\dfrac{1}{2\sqrt{C}} \\ \dfrac{1}{2\sqrt{L}} & \dfrac{1}{2\sqrt{L}} \end{bmatrix}. \text{ Denote by } A^{\text{can}} = \begin{bmatrix} \dfrac{1}{\sqrt{LC}} & 0 \\ 0 & -\dfrac{1}{\sqrt{LC}} \end{bmatrix}. \text{ Then } A^{\text{can}} = HA_1 H^{-1}.$$

Next, when we introduce new variables $Z = HY$, (or $H^{-1}Z = Y$) where

$$Z = \begin{bmatrix} U(x,t) \\ I(x,t) \end{bmatrix}, \quad H = \begin{bmatrix} \sqrt{C} & \sqrt{L} \\ -\sqrt{C} & \sqrt{L} \end{bmatrix}, \quad Y = \begin{bmatrix} u(x,t) \\ i(x,t) \end{bmatrix}.$$

Then

$$\left| \begin{array}{l} U(x,t) = \sqrt{C}\, u(x,t) + \sqrt{L}\, i(x,t), \\ I(x,t) = -\sqrt{C}\, u(x,t) + \sqrt{L}\, i(x,t) \end{array} \right. \qquad (1.2.23)$$

or

$$\left| \begin{array}{l} u(x,t) = \dfrac{1}{2\sqrt{C}} U(x,t) - \dfrac{1}{2\sqrt{C}} I(x,t), \\ i(x,t) = \dfrac{1}{2\sqrt{L}} U(x,t) + \dfrac{1}{2\sqrt{L}} I(x,t). \end{array} \right. \qquad (1.2.24)$$

Replacing $H^{-1}Z = Y$ in the matrix equation (1.2.22) we obtain

$$\frac{\partial \left(H^{-1}Z(x,t) \right)}{\partial t} + A_1 \frac{\partial \left(H^{-1}Z(x,t) \right)}{\partial x} + A_2 \left(H^{-1}Z(x,t) \right) = A_3(x,t).$$

Since H^{1} is a constant matrix we have

$$H^{-1} \frac{\partial Z}{\partial t} + A_1 H^{-1} \frac{\partial Z}{\partial x} + A_2 H^{-1} Z = A_3.$$

After multiplication from the left by H we obtain

$$\frac{\partial Z}{\partial t} + HA_1 H^{-1} \frac{\partial Z}{\partial x} + HA_2 H^{-1} Z = HA_3$$

or

$$\frac{\partial Z(x,t)}{\partial t} + A^{can} \frac{\partial Z(x,t)}{\partial x} + \left(HA_2 H^{-1}\right) Z(x,t) = HA_3(x,t) \qquad (1.2.25)$$

where

$$HA_2 H^{-1} = \begin{bmatrix} \dfrac{1}{2}\left(\dfrac{G}{C}+\dfrac{R}{L}\right) & \dfrac{1}{2}\left(-\dfrac{G}{C}+\dfrac{R}{L}\right) \\[2ex] \dfrac{1}{2}\left(-\dfrac{G}{C}+\dfrac{R}{L}\right) & \dfrac{1}{2}\left(\dfrac{G}{C}+\dfrac{R}{L}\right) \end{bmatrix},$$

$$HA_3 = \begin{bmatrix} \sqrt{C} & \sqrt{L} \\ -\sqrt{C} & \sqrt{L} \end{bmatrix} \begin{bmatrix} \dfrac{1}{C} j(x,t) \\[2ex] \dfrac{1}{L} e(x,t) \end{bmatrix} = \begin{bmatrix} \dfrac{1}{\sqrt{C}} j(x,t) + \dfrac{1}{\sqrt{L}} e(x,t) \\[2ex] -\dfrac{1}{\sqrt{C}} j(x,t) + \dfrac{1}{\sqrt{L}} e(x,t) \end{bmatrix}.$$

As we have already mentioned, we consider transmission lines without distortion, so the following Heaviside condition $R/L = G/C$ is fulfilled. Then $HA_2 H^{-1}$ can be simplified

$$HA_2 H^{-1} = \begin{bmatrix} R/L & 0 \\ 0 & R/L \end{bmatrix}$$

and the equation (1.2.25) can be written in the form:

$$\frac{\partial U(x,t)}{\partial t} + \frac{1}{\sqrt{LC}} \frac{\partial U(x,t)}{\partial x} + \frac{R}{L} U(x,t) = \frac{1}{\sqrt{C}} j(x,t) + \frac{1}{\sqrt{L}} e(x,t),$$
$$\frac{\partial I(x,t)}{\partial t} - \frac{1}{\sqrt{LC}} \frac{\partial I(x,t)}{\partial x} + \frac{R}{L} I(x,t) = -\frac{1}{\sqrt{C}} j(x,t) + \frac{1}{\sqrt{L}} e(x,t). \qquad (1.2.26)$$

System (1.2.26) could be further simplified with the substitution:

$$W(x,t) = e^{\frac{R}{L}t} U(x,t), \quad J(x,t) = e^{\frac{R}{L}t} I(x,t),$$

or

$$U(x,t) = e^{-\frac{R}{L}t} W(x,t), \quad I(x,t) = e^{-\frac{R}{L}t} J(x,t). \qquad (1.2.27)$$

Using this system (1.2.24) we obtain

$$\left|u(x,t)=\frac{1}{2\sqrt{C}}\,e^{-\frac{R}{L}t}W(x,t)-\frac{1}{2\sqrt{C}}\,e^{-\frac{R}{L}t}J(x,t),\right.$$

$$\left|i(x,t)=\frac{1}{2\sqrt{L}}\,e^{-\frac{R}{L}t}W(x,t)+\frac{1}{2\sqrt{L}}\,e^{-\frac{R}{L}t}J(x,t).\right.$$

Then putting $U(x,t)$ and $I(x,t)$ from (1.2.27) into (1.2.26) we find

$$\frac{\partial W(x,t)}{\partial t}+\frac{1}{\sqrt{LC}}\frac{\partial W(x,t)}{\partial x}=\frac{1}{\sqrt{C}}\,j(x,t)+\frac{1}{\sqrt{L}}\,e(x,t),$$

$$\frac{\partial J(x,t)}{\partial t}-\frac{1}{\sqrt{LC}}\frac{\partial J(x,t)}{\partial x}=-\frac{1}{\sqrt{C}}\,j(x,t)+\frac{1}{\sqrt{L}}\,e(x,t).$$

1.3. TRANSFORMATION OF TRANSMISSION LINES EQUATIONS FOR UNIFORM PLANE WAVES

Here we extend the transformation technique to uniform plane wave equations. The term "uniform" means that the electric and magnetic field intensity vectors $\vec{E}$ and $\vec{H}$ are independently positioned in each plane. The term "plane" means that at any point in the space these vectors lie in a plane, and the planes at any two points are parallel. The uniform plane wave is an example of TEM propagation in that field vectors of those waves are also orthogonal to the direction of propagation of the wave. Note that TEM waves on transmission lines do not coincide with uniform plane waves. The field vectors of the transmission line fields are not independent of position in the planes perpendicular to the line axis. Consequently, waves on transmission lines are plane waves, but not uniform plane ones.

We derive the system for uniform plane waves in a lossless media from Maxwell system (alternately, Faraday's and Ampere's laws [90], [107]):

$$\mathrm{rot}\,\vec{E}=-\mu\frac{\partial\vec{H}}{\partial t},$$

$$\mathrm{rot}\,\vec{H}=\sigma\vec{E}+\varepsilon\frac{\partial\vec{H}}{\partial t}+\vec{j}_s$$

where $\vec{E}$ is the electric field intensity vector, while $\vec{H}$ – the magnetic field intensity vector, ε – permittivity, μ – permeability of the medium, σ – conductivity, and $\vec{j}_s$ – is the impressed current, which can be viewed as the source of the fields. But for simple media $\vec{j}_s=\vec{0}$. Therefore the above system becomes

$$\text{rot } \vec{E} = -\mu \frac{\partial \vec{H}}{\partial t},$$

$$\text{rot } \vec{H} = \sigma \vec{E} + \varepsilon \frac{\partial \vec{H}}{\partial t}$$

or in detail

$$\frac{\partial E_z}{\partial y} - \frac{\partial E_y}{\partial z} = -\mu \frac{\partial H_x}{\partial t}, \qquad \frac{\partial E_x}{\partial z} - \frac{\partial E_z}{\partial x} = -\mu \frac{\partial H_y}{\partial t}, \qquad \frac{\partial E_y}{\partial x} - \frac{\partial E_x}{\partial y} = -\mu \frac{\partial H_z}{\partial t}, \quad (1.3.1)$$

$$\frac{\partial H_z}{\partial y} - \frac{\partial H_y}{\partial z} = \sigma E_x + \varepsilon \frac{\partial E_x}{\partial t}, \quad \frac{\partial H_x}{\partial z} - \frac{\partial H_z}{\partial x} = \sigma E_y + \varepsilon \frac{\partial E_y}{\partial t}, \quad \frac{\partial H_y}{\partial x} - \frac{\partial H_x}{\partial y} = \sigma E_z + \varepsilon \frac{\partial E_z}{\partial t}. \quad (1.3.2)$$

We may assume the electric and magnetic intensity vectors to be directed in Ox, that is, $E_y(x,y,z,t) = 0$, $E_z(x,y,z,t) = 0$. The uniformity of the field vectors means that they must be independent on x and y, that is:

$$\frac{\partial E_x}{\partial x} = \frac{\partial E_x}{\partial y} = 0.$$

This implies $\vec{E} = E_x(z,t)\vec{e}_x$. Substituting in (1.3.1) we conclude that the magnetic field has only a second component $\vec{H} = H_y(z,t)\vec{e}_y$. It is easy to see that

$$\frac{\partial H_y}{\partial x} = \frac{\partial H_y}{\partial y} = 0.$$

Finally substituting in (1.3.1) and (1.3.2) we obtain

$$\frac{\partial E_x(z,t)}{\partial z} = -\mu \frac{\partial H_y(z,t)}{\partial t},$$

$$-\frac{\partial H_y(z,t)}{\partial z} = \sigma E_x(z,t) + \varepsilon \frac{\partial E_x(z,t)}{\partial t}$$

or

$$\frac{\partial E_x(z,t)}{\partial t} = -\frac{1}{\varepsilon} \frac{\partial H_y(z,t)}{\partial z} - \frac{\sigma}{\varepsilon} E_x(z,t),$$

$$\frac{\partial H_y(z,t)}{\partial t} = -\frac{1}{\mu}\frac{\partial E_x(z,t)}{\partial z}.$$

Omitting indices and changing z by x we have

$$\frac{\partial E(x,t)}{\partial t} + \frac{1}{\varepsilon}\frac{\partial H(x,t)}{\partial x} + \frac{\sigma}{\varepsilon}E(x,t) = 0,$$

$$\frac{\partial H(x,t)}{\partial t} + \frac{1}{\mu}\frac{\partial E(x,t)}{\partial x} = 0. \qquad (1.3.3)$$

First we write (1.3.3) in a matrix form:

$$\begin{bmatrix} \dfrac{\partial E}{\partial t} \\[2mm] \dfrac{\partial H}{\partial t} \end{bmatrix} + \begin{bmatrix} 0 & \dfrac{1}{\varepsilon} \\[2mm] \dfrac{1}{\mu} & 0 \end{bmatrix}\begin{bmatrix} \dfrac{\partial E}{\partial x} \\[2mm] \dfrac{\partial H}{\partial x} \end{bmatrix} + \begin{bmatrix} \dfrac{\sigma}{\varepsilon} & 0 \\[2mm] 0 & 0 \end{bmatrix}\begin{bmatrix} E \\[2mm] H \end{bmatrix} = \begin{bmatrix} 0 \\[2mm] 0 \end{bmatrix}. \qquad (1.3.4)$$

Denoting by

$$\vec{U} = \begin{bmatrix} E \\ H \end{bmatrix}, \quad \frac{\partial \vec{U}}{\partial t} = \begin{bmatrix} \dfrac{\partial E}{\partial t} \\[2mm] \dfrac{\partial H}{\partial t} \end{bmatrix}, \quad \frac{\partial \vec{U}}{\partial x} = \begin{bmatrix} \dfrac{\partial E}{\partial x} \\[2mm] \dfrac{\partial H}{\partial x} \end{bmatrix}, \quad A_1 = \begin{bmatrix} 0 & \dfrac{1}{\varepsilon} \\[2mm] \dfrac{1}{\mu} & 0 \end{bmatrix}, \quad A_2 = \begin{bmatrix} \dfrac{\sigma}{\varepsilon} & 0 \\[2mm] 0 & 0 \end{bmatrix}, \quad \vec{0} = \begin{bmatrix} 0 \\ 0 \end{bmatrix}$$

we have

$$\frac{\partial \vec{U}(x,t)}{\partial t} + A_1 \frac{\partial \vec{U}(x,t)}{\partial x} + A_2 \vec{U}(x,t) = \vec{0} \qquad (1.3.5)$$

In order to transform the matrix $A_1 = \begin{bmatrix} 0 & 1/\varepsilon \\ 1/\mu & 0 \end{bmatrix}$ in a diagonal form we have to solve

the characteristic equation $\begin{vmatrix} -\lambda & 1/\varepsilon \\ 1/\mu & -\lambda \end{vmatrix} = 0.$ Its roots are $\lambda_1 = \dfrac{1}{\sqrt{\varepsilon\mu}}, \; \lambda_2 = -\dfrac{1}{\sqrt{\varepsilon\mu}}.$

For the eigenvectors we obtain the following systems:

$$\left|\begin{array}{l}-\dfrac{1}{\sqrt{\varepsilon\mu}}\xi_1+\dfrac{1}{\varepsilon}\xi_2=0\\[2mm]\dfrac{1}{\mu}\xi_1-\dfrac{1}{\sqrt{\varepsilon\mu}}\xi_2=0\end{array}\right. \quad\text{and}\quad \left|\begin{array}{l}\dfrac{1}{\sqrt{\varepsilon\mu}}\xi_1+\dfrac{1}{\varepsilon}\xi_2=0\\[2mm]\dfrac{1}{\mu}\xi_1+\dfrac{1}{\sqrt{\varepsilon\mu}}\xi_2=0\end{array}\right. .$$

Hence $\left(\xi_1^{(1)},\xi_2^{(1)}\right)=\left(\sqrt{\varepsilon},\sqrt{\mu}\right),\qquad \left(\xi_1^{(2)},\xi_2^{(2)}\right)=\left(-\sqrt{\varepsilon},\sqrt{\mu}\right).$

Denote by H the matrix formed by eigenvectors $H=\begin{bmatrix}\sqrt{\varepsilon}&\sqrt{\mu}\\-\sqrt{\varepsilon}&\sqrt{\mu}\end{bmatrix}$. Its inverse is

$$H^{-1}=\begin{bmatrix}\dfrac{1}{2\sqrt{\varepsilon}}&-\dfrac{1}{2\sqrt{\varepsilon}}\\[3mm]\dfrac{1}{2\sqrt{\mu}}&\dfrac{1}{2\sqrt{\mu}}\end{bmatrix}.$$ If we denote by $A^{can}=\begin{bmatrix}\dfrac{1}{\sqrt{\varepsilon\mu}}&0\\[3mm]0&-\dfrac{1}{\sqrt{\varepsilon\mu}}\end{bmatrix}$, then we know that

$$HA_1H^{-1}=\begin{bmatrix}\sqrt{\varepsilon}&\sqrt{\mu}\\-\sqrt{\varepsilon}&\sqrt{\mu}\end{bmatrix}\begin{bmatrix}0&\dfrac{1}{\varepsilon}\\[2mm]\dfrac{1}{\mu}&0\end{bmatrix}\begin{bmatrix}\dfrac{1}{2\sqrt{\varepsilon}}&-\dfrac{1}{2\sqrt{\varepsilon}}\\[3mm]\dfrac{1}{2\sqrt{\mu}}&\dfrac{1}{2\sqrt{\mu}}\end{bmatrix}=\begin{bmatrix}\dfrac{1}{\sqrt{\mu\varepsilon}}&0\\[3mm]0&-\dfrac{1}{\sqrt{\mu\varepsilon}}\end{bmatrix}=A^{can}.$$

Introduce new variables $\vec{Z}=H\vec{U}$ (or $H^{-1}\vec{Z}=\vec{U}$) where

$$\vec{Z}=\begin{bmatrix}\breve{E}(x,t)\\\breve{H}(x,t)\end{bmatrix},\quad H=\begin{bmatrix}\sqrt{\varepsilon}&\sqrt{\mu}\\-\sqrt{\varepsilon}&\sqrt{\mu}\end{bmatrix},\quad \vec{U}=\begin{bmatrix}E(x,t)\\H(x,t)\end{bmatrix}$$

Then

$$\left|\begin{array}{l}\breve{E}(x,t)=\sqrt{\varepsilon}\,E(x,t)+\sqrt{\mu}\,H(x,t),\\\breve{H}(x,t)=-\sqrt{\varepsilon}\,E(x,t)+\sqrt{\mu}\,H(x,t)\end{array}\right.$$

or

$$\left|\begin{aligned}
E(x,t) &= \frac{1}{2\sqrt{\varepsilon}}\,\breve{E}(x,t) - \frac{1}{2\sqrt{\varepsilon}}\,\breve{H}(x,t),\\[2mm]
H(x,t) &= \frac{1}{2\sqrt{\mu}}\,\breve{E}(x,t) + \frac{1}{2\sqrt{\mu}}\,\breve{H}(x,t).
\end{aligned}\right.$$

(1.3.6)

Substituting $\vec{U} = H^{-1}\vec{Z}$ in the matrix equation (1.3.5) we obtain

$$\frac{\partial\left(H^{-1}\vec{Z}\right)}{\partial t} + A_1\frac{\partial\left(H^{-1}\vec{Z}\right)}{\partial x} + A_2\left(H^{-1}\vec{Z}\right) = \vec{0}\,.$$

Since H is a constant matrix (and consequently H^{-1}), we have:

$$H^{-1}\frac{\partial\vec{Z}}{\partial t} + \left(A_1 H^{-1}\right)\frac{\partial\vec{Z}}{\partial x} + \left(A_2 H^{-1}\right)\vec{Z} = \vec{0}\,.$$

After multiplication from the left by H, we obtain

$$\frac{\partial\vec{Z}}{\partial t} + \left(H A_1 H^{-1}\right)\frac{\partial\vec{Z}}{\partial x} + \left(H A_2 H^{-1}\right)\vec{Z} = \vec{0}$$

or

$$\frac{\partial Z(x,t)}{\partial t} + A^{\mathrm{can}}\,\frac{\partial Z(x,t)}{\partial x} + B\vec{Z}(x,t) = \overline{0}$$

(1.3.7)

where

$$B = H A_2 H^{-1} = \begin{bmatrix} \sqrt{\varepsilon} & \sqrt{\mu} \\ -\sqrt{\varepsilon} & \sqrt{\mu} \end{bmatrix}\begin{bmatrix} \dfrac{\sigma}{\varepsilon} & 0 \\ 0 & 0 \end{bmatrix}\begin{bmatrix} \dfrac{1}{2\sqrt{\varepsilon}} & -\dfrac{1}{2\sqrt{\varepsilon}} \\[2mm] \dfrac{1}{2\sqrt{\mu}} & \dfrac{1}{2\sqrt{\mu}} \end{bmatrix} = \begin{bmatrix} \dfrac{\sigma}{2\varepsilon} & -\dfrac{\sigma}{2\varepsilon} \\[2mm] -\dfrac{\sigma}{2\varepsilon} & \dfrac{\sigma}{2\varepsilon} \end{bmatrix},$$

$$\frac{\partial\breve{E}(x,t)}{\partial t} + \frac{1}{\sqrt{\varepsilon\mu}}\,\frac{\partial\breve{E}(x,t)}{\partial x} + \frac{\sigma}{2\varepsilon}\,\breve{E}(x,t) - \frac{\sigma}{2\varepsilon}\,\breve{H}(x,t) = 0,$$

$$\frac{\partial\breve{H}(x,t)}{\partial t} - \frac{1}{\sqrt{\varepsilon\mu}}\,\frac{\partial\breve{H}(x,t)}{\partial x} - \frac{\sigma}{2\varepsilon}\,\breve{E}(x,t) + \frac{\sigma}{2\varepsilon}\,\breve{H}(x,t) = 0.$$

and finally, we put $\breve{E}(x,t) = e^{-\frac{\sigma}{2\varepsilon}t}\,\hat{E}(x,t)$, $\breve{H}(x,t) = e^{-\frac{\sigma}{2\varepsilon}t}\,\hat{H}(x,t)$. Then

$$-\frac{\sigma}{2\varepsilon}e^{-\frac{\sigma}{2\varepsilon}t}\,\hat{E}(x,t) + e^{-\frac{\sigma}{2\varepsilon}t}\,\frac{\partial \hat{E}(x,t)}{\partial t} + \frac{1}{\sqrt{\varepsilon\mu}}e^{-\frac{\sigma}{2\varepsilon}t}\,\frac{\partial \hat{E}(x,t)}{\partial x} + \frac{\sigma}{2\varepsilon}e^{-\frac{\sigma}{2\varepsilon}t}$$

$$\hat{E}(x,t) - \frac{\sigma}{2\varepsilon}e^{-\frac{\sigma}{2\varepsilon}t}\,\hat{H}(x,t) = 0,$$

$$-\frac{\sigma}{2\varepsilon}e^{-\frac{\sigma}{2\varepsilon}t}\,\hat{H}(x,t) + e^{-\frac{\sigma}{2\varepsilon}t}\,\frac{\partial \hat{H}(x,t)}{\partial t} - \frac{1}{\sqrt{\varepsilon\mu}}e^{-\frac{\sigma}{2\varepsilon}t}\,\frac{\partial \hat{H}(x,t)}{\partial x} + \frac{\sigma}{2\varepsilon}e^{-\frac{\sigma}{2\varepsilon}t}$$

$$\hat{H}(x,t) - \frac{\sigma}{2\varepsilon}e^{-\frac{\sigma}{2\varepsilon}t}\,\hat{E}(x,t) = 0.$$

Consequently,

$$\frac{\partial \hat{E}(x,t)}{\partial t} + \frac{1}{\sqrt{\varepsilon\mu}}\frac{\partial \hat{E}(x,t)}{\partial x} - \frac{\sigma}{2\varepsilon}\hat{H}(x,t) = 0,$$

$$\frac{\partial \hat{H}(x,t)}{\partial t} - \frac{1}{\sqrt{\varepsilon\mu}}\frac{\partial \hat{H}(x,t)}{\partial x} - \frac{\sigma}{2\varepsilon}\hat{E}(x,t) = 0.$$

1.4. Tools from Fixed-point theory in Metric and Uniform Spaces

Here we reveal the mathematical methods for solving the main problems formulated in this book. In order to formulate the basic assertions we need some preliminary results – definitions and theorems from the fixed-point theory, following [14]. The proofs, in general, we omit. For details we refer the reader to [14], a monograph devoted to this subject (cf. also [98]). We provide here the proofs necessary for the exposition and lacking in the monographs.

By X we mean a nonempty set of elements and call them its points. The space X is said to be topological one if a given collection of its sets is defined, called open sets, having the following properties: 1) the empty set $\varnothing$ and the whole space X are open sets; 2) the union of arbitrarily many and intersection of finitely many open sets are again open ones.

The family of all open sets in the space X forms a topology. The complement $\overline{A} = X \setminus A$ of an open set A is called a closed set. It is easy to verify that a union of finitely many closed sets and intersection of arbitrarily many closed sets is again a closed set. The topology can also be defined by the closed sets. This allows us to define a convergence in X. The notion "limit" is uniquely determined provided the topological space to be T_2-separable or Hausdorff separable. For the applications the notion of metric space plays an important role. It extends the distance from the well-known Euclidean space E^3. If $A(A_1, A_2, A_3)$ and $B(B_1, B_2 B_3)$ are different points in E^3, the distance is

$$d(A,B) = \sqrt{(A_1 - B_1)^2 + (A_2 - B_2)^2 + (A_3 - B_3)^2}\,.$$

It has the following properties: 1) if $A \equiv B$, then $d(A,B) = 0$ and conversely: $d(A,B) = 0$ it implies $A \equiv B$; 2) $d(A,B) \geq 0$; 3) $d(A,B) = d(B,A)$; 4) $d(A,B) \leq d(A,C) + d(C,B)$ for every three points. As usually, these properties can be set as the base of the definition of an abstract metric space.

Indeed, for an arbitrarily set X let $\rho : X \times X \to [0,\infty)$ be a map. This means that to every ordered pair $(x,y) \in X$ a nonnegative real number $\rho(x,y)$ is assigned, possessing the following properties:

1) $\rho(x,y) \geq 0$; 2) $\rho(x,y) = 0 \Leftrightarrow x = y$; 3) $\rho(x,y) = \rho(y,x)$; 4) $\rho(x,y) \leq \rho(x,z) + \rho(z,y)$.

In this case, the pair (X,ρ) is called a metric space. The set $S(x_0,\varepsilon) = \{x \in X : \rho(x,x_0) < \varepsilon\}$ is said to be an open ball with center x_0 and radius $\varepsilon > 0$, analogous to the open ball in E^3. The sets of the type $S(x_0,\varepsilon)$ ($x_0 \in X$, $\varepsilon > 0$) define a topology in X.

Later on, we use metric and uniform spaces as well as particular cases − normed linear space and locally convex topological vector spaces (cf. the references of [14], [40], [98]).

From now on we denote by R^1 the set of real numbers, and in general, $R^n = \underbrace{R^1 \times R^1 \times ... \times R^1}_{n\text{-}\mathrm{times}}$ − the set of all real ordered n-tuples $(x_1, x_2, ..., x_n)$.

If in the space X two operations are introduced: the addition of two elements, that is, $x + y$ and the multiplication of an element by a real number $\alpha \in R$, that is, αx and these operations satisfy 8 axioms known from the linear algebra, then X is said to be a linear space. The linear space X is called normed if to every element a non-negative real number is assigned $\| x \|$ (norm of x), satisfying the conditions: 1) $\| x \| = 0$ if and only if $x = \overline{0}$ ($\overline{0}$ − zero element of X); 2) $\| x + y \| \leq \| x \| + \| y \|$ (triangle inequality); 3) $\| \lambda x \| = | \lambda | \| x \|$ for every $\lambda \in R$. The last property metric implies that for $\lambda = -1$, $\| -x \| = \| x \|$. In the linear space X the metric $\rho(x,y) = \| x - y \|$ can be introduced, that is, every normed linear space becomes a metric one, even though the converse is not true.

We recall some basic notions from metric fixed-point theory (cf. [40], [98]). An infinite sequence $\{x_n\}_{n=1}^{\infty}$ of elements of the metric space (X,ρ) is called a convergent to the element $x_0 \in X$, if $\lim_{n \to \infty} \rho(x_n, x_0) = 0$. The sequence $\{x_n\}_{n=1}^{\infty}$ is said to be a Cauchy sequence, if $\lim_{m,n \to \infty} \rho(x_n, x_m) = 0$. The space (X,ρ) is complete, if for every Cauchy sequence $\{x_n\}_{n=1}^{\infty}$ there is an element $x_0 \in (X,\rho)$ such that $\lim_{n \to \infty} \rho(x_n, x_0) = 0$. The last property plays an important role in fixed-point theorems. We apply it in the following sections to solve various problems for nonlinear differential and functional differential equations.

A subset $K \subset X$ of a topological space X is called sequentially compact, if every sequence of points of K contains a subsequence converging to a point of X. A subset of a metric space is compact if and only if (iff) it is closed and sequentially compact. An example of a non-compact set presents the real axis $X \equiv R^1$ with the usual metric $\rho(x, y) = |x - y|$ and the subset $K = \{nx : x \in R, x \neq 0, n = 0,1,2,...\}$. Indeed, K is not a convergent sequence and does not contain a convergent subsequence.

A subset K of a metric space is totally bounded, if for every $\varepsilon > 0$ it is possible to cover K by a finite number of spheres $S(x_i, \varepsilon), (i = 1,2,...n)$ with centers in K. For $K \subset X$ (with X being a metric space) the following statements are equivalent: 1) K is sequentially compact; 2) $\mathrm{cl}\, K$ is compact; 3) K is totally bounded; and $\mathrm{cl}\, K$ is complete. A subset D of a metric space (X, ρ) is said to be relatively compact if its closure is a compact set.

Every linear normed space complete with respect to its norm is said to be a Banach space.

1.4.1. Iteration Methods and Fixed Points

Let (X, ρ) be a complete metric space and let $D \subset X$ be a subset of X. Let $B : D \to D$ be an operator – either linear or nonlinear.

The operator B has a fixed-point in D, if there exists an element $u \in D$, such that $B(u) = u$. One known way to reach the solution of $B(u) = u$ is to form a sequence of successive approximations (iterations) $u_{n+1} = B(u_n)$ $(n = 0,1,2,...)$, beginning with an arbitrarily initial approximation $u_0 \in D$. Under prescribed assumptions the above sequence of iterations tends to the fixed-point of B. The convergence depends on the choice of the initial approximation u_0. For a given fixed-point u the set of all elements u_n converging to u, (that is, $\lim_{n \to \infty} u_n = u$), is called a set of convergence of u.

The main theorem for existence and uniqueness of a fixed-point for contractive operators acting on some complete metric space (X, ρ) is provided by Banach and Cacciopoli. An operator $B : X \to X$ is said to be a contractive one, if for every two elements $x, y \in X$ the following inequality is satisfied

$$\rho(B(x), B(y)) \leq k\rho(x, y)$$

for some constant $0 < k < 1$. In this case the number k is called a contractive constant.

Theorem 1.4.1. If $B : D \to D$ $(D \subset X)$ is a contractive operator then it possesses a unique fixed point ξ in D and $\xi = \lim\limits_{n \leftarrow \infty} x_n$, where $x_{n+1} = B(x_n)$ with arbitrarily chosen $x_0 \in D$. The rate of convergence is given by the inequality

$$\rho(x_n, \xi) \le \frac{k^n}{1-k} \rho(x_1, x_0), \ (0 < k < 1).$$

The natural extension of the Banach contraction mapping principle can be obtained by introducing nonlinear contractions:

$$\rho(B(x), B(y)) \le \Phi(\rho(x, y)).$$

The Banach contraction condition is a particular case obtained by choosing $\Phi(t) = kt$, $(0 < k < 1)$. Various contractive functions generate fixed-point theorems (cf. References in [20] and [98]).

Let the mapping B_y depend on some parameter $y \in Y$, running in some metric space Y. If $B_y(x) : X \to Y$ is continuous at point $y_0 \in Y$, then the following theorem is valid:

Theorem 1.4.2 (continuous dependence on parameter). If for every $y \in Y$ the mapping $B_y(x)$ satisfies the condition

$$\rho(B_y(x), B_y(\overline{x})) \le k\rho(x, \overline{x}),$$

where k does not depend on y and the mapping $B_y(x)$ is continuous at $y_0 \in Y$, then the solution of the equation $x = B_y(x)$ is continuous at y_0, that is, $x = x(y)$ is a continuous mapping at y_0.

The above-mentioned theorems in Banach and metric spaces are extended to uniform spaces, and in particular to locally convex topological vector spaces (cf. [14]).

1.4.2. Uniform Spaces and Locally Convex Topological Vector Spaces

The notion of uniform space is introduced by A. Weil in 1938 (cf. [14]), who considers topological groups. Uniform spaces are a natural extension of metric spaces, and as we see below, many function spaces could not be treated as metric spaces, but at the same time they do form uniform spaces.

For a given set X we consider the subsets of $X \times X$. Their elements are ordered pairs (x, y). For every subset $U \subset X \times X$, define the set U^{-1} in the following way:

$U^{-1} = \{(x,y): \ (y,x) \in U\}$. For $U \subset X \times X$ and $V \subset X \times X$ we define $U \circ V = \{(x,z) \in X \times X : \text{ for some } y \in X \ (x,y) \in U \text{ and } (y,z) \in V\}$. Next, introduce the sets $\Delta(X) = \{(x,x) : x \in X\}$ and $U[A] = \{y : (x,y) \in U \text{ for some } x \in A\}$ for every $A \subset X$. Then $U[x] = U[\{x\}]$.

A uniformity on the set X is said to be a non-empty family $\boldsymbol{U}$ of subsets of $X \times X$, such that:

1. every element $U \in \boldsymbol{U}$ contains $\Delta(X)$,

2. if $U \in \boldsymbol{U}$, then $U^{-1} \in \boldsymbol{U}$,

3. if $U \in \boldsymbol{U}$, then there is $V \in \boldsymbol{U}$, such that $V \circ V \subset U$,

4. if U и $V \in \boldsymbol{U}$, then $U \bigcap V \in \boldsymbol{U}$,

5. and if $U \in \boldsymbol{U}$ and $V \subset U$, then $V \in \boldsymbol{U}$.

The ordered pair $(X, \boldsymbol{U})$ is said to be a uniform space (cf. [14]).

The collection of all subsets $T \subset X$ with the property for every $x \in T$ there is $U \in \boldsymbol{U}$, such that $U[x] \subset T$, forms a topology on X, called a uniform topology. In terms of applications, this statement plays an important role: every uniformity on X is generated by the family of all uniformly continuous pseudo-metrics on $X \times X$.

The metric turns out to be a pseudo-metric if condition 2) from the definition of a metric is replaced by the following one:

2') if $x = y$, then $\rho(x, y) = 0$, but the converse is not true.

Obviously every metric is a pseudo-metric but the converse is not true.

Further on, we consider a family $\boldsymbol{A} = \{\rho_\alpha(x,y) : \alpha \in A\}$, where A is the index set. Let us introduce the set called neighborhoods of the diagonal $\Delta(X)$:

$$V(\varepsilon, \alpha) = \{(x,y) : \rho_\alpha(x, y < \varepsilon)\} \ (\varepsilon > 0 \ \text{и} \ \alpha \in A).$$

All finite intersections of the above sets generate a uniformity $\boldsymbol{U}$ called a uniformity generated by a family $\boldsymbol{A}$. The following is a useful result:

Theorem 1.4.3. A topology τ on X is generated by some uniformity $\boldsymbol{U}$ iff the topological space (X, τ) is completely regular.

Since the locally convex topological vector spaces are completely regular, their topology is uniformizible. Consequently all results in uniform spaces are valid in locally convex topological vector spaces.

Here we provide a typical example of a uniform space that is not metric.

Let us consider the family of all continuous functions $C(R^1) f : R^1 \to R^1$, defined on the whole real axis. They are, in general, unbounded, therefore we cannot define the distance between them since $\sup\{|f(t) - g(t)| : t \in R^1\} = \infty$.

One can introduce, however, a family of pseudo-distances, as follows

$$\rho_K(f,g) = \sup\{|f(t) - g(t)| : t \in K\},$$

where K runs over all compact subsets $K \subset R^1$. Obviously $\rho_K(f,g)$ satisfies all conditions for metrics with the exception of the 2). Indeed, $\rho_K(f,g) = 0$ for some K does not imply $f \equiv g$. Therefore $\rho_K(f,g)$ is a pseudo-metric. Fig. 1.3 is an example of the case when two functions

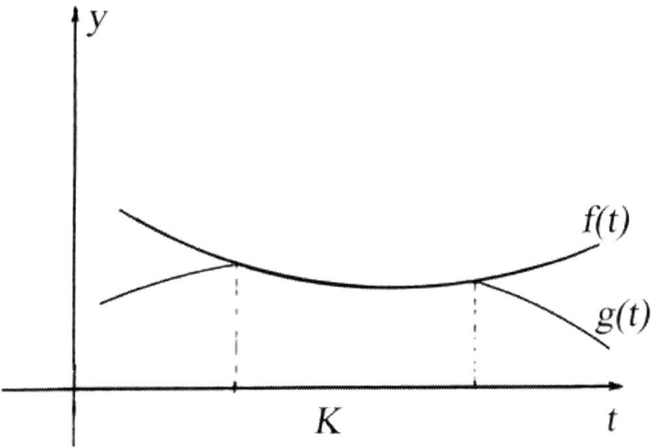

Figure 1.3.

coincide on the interval $K \Rightarrow \rho_K(f,g) = 0$, but outside the interval K they take different values. In this case the family of pseudo-metrics $\mathbf{A} = \{\rho_K(.,.) : K \in A\}$ defines a uniformity on $C(R^1)$. The index set A consists of all compact subsets $K \subset R^1$.

1.4.3. Fixed Points of Φ-Contractive Mappings in Uniform Spaces

Here we define the notion Φ-contractive mapping acting on uniform space. This is introduced by the author in [3], [14], and fixed-point theorems are proved. For references in detail cf. [14]. These theorems make clear there is a problem for the Lipschitz constant of the neutral part in the right-hand side of a nonlinear (and even linear) functional differential equation of a neutral type. Such equations arise in the problems considered in all later chapters.

By $(X, \mathbf{A})$ we mean a Hausdorff sequentially complete uniform space, whose uniformity is generated by a saturated family of pseudo-metrics $\mathbf{A} = \{\rho_\alpha(x,y) : \alpha \in A\}$, where A is an index set. By $(\Phi) = \{\Phi_\alpha(.) : \alpha \in A\}$ we denote a family of functions $\Phi_\alpha(t) = R_+^1 \to R_+^1$ $(R_+^1 = [0, \infty))$. We call them Φ-contractive functions if the following conditions are fulfilled: (cf. [14]):

(Φ1) $\Phi_\alpha(t)$ is monotone increasing and right continuous in t for every fixed $\alpha \in A$;

(Φ2) $0 < \Phi_\alpha(t) < t$ for every $\alpha \in A$ and $t > 0$. (It follows $\Phi_\alpha(0) = 0$).

Let $j : A \to A$ be an arbitrary mapping of the index set into itself. The iterations can be defined inductively as follows:

$$j^n(t) = j(j^{n-1}(\alpha)), \; j^0(\alpha) = \alpha \; (n = 1,2,3,...).$$

The delays or advance arguments define the map j (cf. [14]).

Let M be a subset of X and $B : M \to M$ be a linear or nonlinear mapping. B is called Φ-contractive on M, if for any fixed $\alpha \in A$,

$$\rho_\alpha(Bx, By) \leq \Phi_\alpha(\rho_{j(\alpha)}(x,y))$$

is satisfied for every $x, y \in M$.

A set M is said to be bounded in $(X, \mathbf{A})$ if for every $\alpha \in A$

$$\rho_\alpha^M = \sup\left[\rho_\alpha(x,y) : x, y \in M\right] < \infty.$$

A uniform space $(X, \mathbf{A})$ is called quasi-complete if all bounded and closed subsets are complete in $(X, \mathbf{A})$.

Theorem 1.4.4 [14]. Let M be a bounded and closed subset of a sequentially complete uniform space $(X, \mathbf{A})$. Let $B : X \to X$ be a Φ-contractive operator, and for any fixed $\alpha \in A$ let there exist a function $\overline{\Phi}_\alpha(.)$ with properties **(Φ1)** and **(Φ2)**, such that

$$\sup\left\{\Phi_{j^n(\alpha)}(t) : n = 1,2,....\right\} \leq \overline{\Phi}_\alpha(t) \; \text{и} \; \rho_{j^n(\alpha)}^M \leq \rho_\alpha^M \; (n = 1,2,...).$$

Then B has a unique fixed point $x \in M$ and $\lim_{n\to\infty} B^n(x_0) = x$ for arbitrary chosen initial approximation $x_0 \in M$.

Corollary 1.4.1 [14]. Let conditions of Theorem 1.2.4 be fulfilled and let the operator B^s be Φ-contractive, where $B^s(x) = B(B^{s-1}(x))$, s is a natural number. Then B has a unique fixed point.

Proof: Since the operator B^s is Φ-contractive, it has a unique fixed point $\xi \in X$, that is, $B^s(\xi) = \xi$. But $B(B^s(\xi)) = B^s(B(\xi))$, hence $B(\xi) = B^s(B(\xi))$. This means $B(\xi)$ is a fixed

point of B^s. But the fixed point of B^s is unique and therefore coincides with ξ, that is.
$\xi = B(\xi)$. The last equality shows that ξ is the unique fixed point of B.

If $B: X \to X$ maps the whole space into itself then the following theorem is valid:

Theorem 1.4.5 [14]. If:

1) $B: X \to X$ is Φ-contractive,

2) for every $\alpha \in A$ there is a function $\overline{\Phi}_\alpha(t)$ with properties **(Φ1)** и **(Φ2)** and such that $\sup\{\Phi_{j^n(\alpha)}(t): n = 0,1,2,...\} \le \overline{\Phi}_\alpha(t)$ and $\overline{\Phi}_\alpha(t)/t$ is monotone increasing,

3) there is an element $x_0 \in X$ and and a constant $q(\alpha) > 0$ such that

$$\rho_{j^n(\alpha)}(x_0, Bx_0) \le q(\alpha) \; (n = 0,1,2,...).$$

Then B has at least one fixed point in X.

Theorem 1.4.6 [14]. In addition to conditions of Theorem 1.2.5 suppose for every $x, y \in X$ and $\alpha \in A$ there is a constant $q = q(x, y, \alpha) > 0$ such that

$$\rho_{j^n(\alpha)}(x, y) \le q < \infty \; (n = 0,1,...).$$

Then the fixed-point of B is unique and the following rate of convergence is valid:

$$\rho_\alpha(x_n, x) \le k_\alpha^n / (1 - k_\alpha), \tag{1.4.1}$$

where $k_\alpha = \overline{\Phi}_\alpha(q_\alpha)/q_\alpha$.

In fact, for every two natural numbers n and p and for every $\alpha \in A$:

$$\rho_\alpha(x_{n+p}, x_n) \le$$

$$\le \rho_\alpha(B^{n+1}(x_0), B^n(x_0)) + \rho_\alpha(B^{n+2}(x_0), B^{n+1}(x_0)) + ... + \rho_\alpha(B^{n+p}(x_0), B^{n+p-1}(x_0)) \le$$

$$\le \Phi_\alpha(\Phi_{j(\alpha)}(...\Phi_{j^{n-1}(\alpha)}(\rho_{j^n(\alpha)}(x_1, x_0))...)) + \Phi_\alpha(\Phi_{j(\alpha)}(...\Phi_{j^n(\alpha)}(\rho_{j^{n+1}(\alpha)}(x_1, x_0))...)) + ...$$

$$+ \Phi_\alpha(\Phi_{j(\alpha)}(...\Phi_{j^{n+p-2}(\alpha)}(\rho_{j^{n+p-1}(\alpha)}(x_1, x_0))...)) \le \overline{\Phi}_\alpha^n(q_\alpha) + \overline{\Phi}_\alpha^{n+1}(q_\alpha) + ... + \overline{\Phi}_\alpha^{n+p-1}(q_\alpha). \cdot$$

In view of $k_\alpha = \overline{\Phi}_\alpha(q_\alpha)/q_\alpha < 1$ we have

$$\overline{\Phi}_\alpha^{n+s+1}(q_\alpha) \le \overline{\Phi}_\alpha^{n+s}(q_\alpha).k_\alpha \;\; (s = 0,1,\ldots,p-1; \; n = 1,2,\ldots),$$

since $\overline{\Phi}_\alpha(t)/t$ is monotone increasing. So we obtain

$$\overline{\Phi}_\alpha^{n+s+1}(q_\alpha) \le \overline{\Phi}_\alpha^{n}(q_\alpha).k_\alpha^{s+1} \;\; (s = 0,1,\ldots,p-1; \; n = 1,2,\ldots) \;\; \text{and} \;\; \overline{\Phi}_\alpha^{n}(q_\alpha) \le k_\alpha^{n}.$$

Thus

$$\rho_\alpha(x_{n+p}, x_n) \le k_\alpha^{n} \sum_{s=0}^{p-1} k_\alpha^{s} = k_\alpha^{n}.\frac{1-k_\alpha^{p}}{1-k_\alpha} \xrightarrow[p\to\infty]{} \frac{k_\alpha^{n}}{1-k_\alpha} \;\; (n = 1,2,\ldots)$$

i.e. we obtain (1.4.1).

Corollary 1.4.2 [14]. Let B be from Theorem 1.4.6 and let B^s be Φ-contractive where $B^s(x) = B(B^{s-1}(x)), s \in N$. Then B has a unique fixed point.

The following extension is valid:

Theorem 1.4.7 [14]. Let $B : X \to X$ be Φ-contractive mapping and:

1)   for every $\alpha \in A$ there is $k_\alpha \in (0,1)$, such that

$$\sup\{\Phi_{j^k(\alpha)}(t) : k = 0,1,2,\ldots\} \le k_\alpha t, t \in R_+^1,$$

2)   there is $x_0 \in X$, such that for every $\alpha \in A$

$$\lim_{n\to\infty} \rho_{j^n(\alpha)}(x_0, B(x_0)) = \infty \;\; \text{and} \;\; \lim_{n\to\infty}\left(a_{n+1}^\alpha / a_n^\alpha\right) k_\alpha = q^\alpha < 1, \;\text{where}$$

$$a_n^\alpha = \rho_{j^n(\alpha)}\left(x_0, B(x_0)\right).$$

Then B has a fixed point.

Remark 1.4.2. If some $a_n^\alpha = 0$ then the condition $\lim_{n\to\infty}\left(a_{n+1}^\alpha / a_n^\alpha\right) k_\alpha = q^\alpha < 1$ should be replaced by: $\lim_{n\to\infty}(k_\alpha \sqrt[n]{a_n^\alpha}) = q^\alpha < 1$.

A uniform space is called j-unbounded if for every two elements $x, y \in X$ and $\alpha \in A$ is satisfied $\lim_{n\to\infty} b_n^\alpha = \infty$, where $b_n^\alpha = \rho_{j^n(\alpha)}(x,y)$ and

$$\lim_{n\to\infty}\left(b_{n+1}^\alpha / b_n^\alpha\right) k_\alpha = q^\alpha < 1 \left(\lim_{n\to\infty}(k_\alpha \sqrt[n]{a_n^\alpha}) = q^\alpha < 1\right).$$

Theorem 1.4.8 [14]. In addition to conditions of Theorem 1.4.7 we suppose X is j-unbounded. Then the fixed-point is unique.

1.4.4. Fixed-points of Mappings Acting on a Product of Metric and Uniform Spaces

An operator $B:(X,\mathbf{A})\to(X,\mathbf{A})$ is said to be Φ-contractive if for every $x,y\in X$ and $\alpha\in A$

$$\rho_\alpha(Bx,By)\le\Phi_\alpha(\rho_{j(\alpha)}(x,y))$$

and a uniform space $(X,\mathbf{A})$ is said to be j-bounded if there is a constant $Q=Q(\alpha,x,y)>0$ such that $\rho_{j^k(\alpha)}(x,y)\le Q$ $(k=0,1,2,...)$ for every $x,y\in X$ and $\alpha\in A$.

Then Theorem 1.4.6 can be formulated in the following manner: if $(X,\mathbf{A})$ is j-bounded, then the fixed-point of B is unique.

So the fixed-point results can be extended on the product of uniform spaces. They will be used for solving of systems of functional differential equations.

Let $\{(X_i,\mathbf{A}_i)\}_{i=1}^n$ be a finite family of uniform spaces. We suppose that all index sets of the families $\mathbf{A}_1,...,\mathbf{A}_n$ coincide, that is, $\mathbf{A}_1=\mathbf{A}_2=...=\mathbf{A}_n=\mathbf{A}$. So we obtain $\mathbf{A}_i-\{\rho_\alpha^{(i)}(x,y):\alpha\in\mathbf{A}\}$. The elements of the Cartesian product $X=X_1\times X_2\times...\times X_n$ we mean $(x_1,x_2,...,x_n)\in X$, where $x_i\in X_i$ $(i=1,...,n)$. For the Cartesian product X we introduce a family of pseudo-metrics in the following way: $\mathbf{A}=\{\rho_\alpha(x,y):\alpha\in\mathbf{A}\}$, where

$$\rho_\alpha(x,y)=\max\{\rho_\alpha^{(i)}(x_i,y_i):i=1,2,...,n\}.$$

One can check that the convergence in the product space (X,ρ) is equivalent to the convergence of each coordinate sequence, that is, if $\{x^{(k)}\}_{k=1}^\infty$ is a sequence in X, tending to $x^{(0)}\in X$, i.e. $\lim_{k\to\infty}\rho_\alpha(x^{(k)},x^{(0)})=0$, then every coordinate sequence $x_i^{(k)}$ tends to $x_i^{(0)}$, i.e.

$$\lim_{k\to\infty}\rho_\alpha^{(i)}(x_i^{(k)},x_i^{(0)})=0\quad(i=1,2,...,n).$$

In other words

$$\rho_\alpha(x^{(k)},x^{(0)})\xrightarrow[k\to\infty]{}0\Leftrightarrow\begin{cases}\rho_\alpha^{(1)}(x_1^{(k)},x_1^{(0)})\xrightarrow[k\to\infty]{}0\\ ....\\ \rho_\alpha^{(n)}(x_n^{(k)},x_n^{(0)})\xrightarrow[k\to\infty]{}0.\end{cases}$$

If every space $(X_i, \mathbf{A}_i)$ is sequentially complete, then so is the product space $(X, \mathbf{A})$.

Theorem 1.4.9. Let $B_i : (X_i, \mathbf{A}_i) \to (X_i, \mathbf{A}_i)$ $(i = 1, 2, ..., n)$ be a family of Φ-contractive mappings and $B : (X, \mathbf{A}) \to (X, \mathbf{A})$ be a map defined in the following way: $B(x_1, x_2, ..., x_n) = (B_1(x_1), B_2(x_2), ..., B_n(x_n))$ for every n-tuple $(x_1, x_2, ..., x_n) \in X$. If there is an element $x^{(0)} = (x_1^{(0)}, x_2^{(0)}, ..., x_n^{(0)})$ such that $\rho_{j^k(\alpha)}^{(i)}(x_i^{(0)}, T_i(x_i^{(0)})) \leq Q^{(i)}(\alpha) < \infty$ for $i = 1, 2, ..., n$ $(k = 0, 1, ...)$, then B is a Φ-contractive operator and has a fixed point in $(X, \mathbf{A})$.

The uniqueness theorem is as follows:

Theorem 1.4.10. In addition to conditions of Theorem 1.4.9 we suppose the spaces $(X_i, \mathbf{A}_i)$ are j-bounded. Then $(X, \mathbf{A})$ is j-bounded and B contains a unique fixed point.

Now we formulate an iterative test for a mapping very useful for various applications. We say that the iterative test for a Φ-contractive mapping is conclusive in $(X, \mathbf{A})$, if for every Φ-contractive mapping

$$B : (X, \mathbf{A}) \to (X, \mathbf{A})$$

is valid. If for some $x_0 \in X$ the sequence $\{B^n(x_0)\}_{n=1}^{\infty}$ does not converge to some point of X, then B does not have a fixed point.

1.4.5. Continuous Dependence of Fixed Points on the Operators

Let $\{B_k\}_{k=1}^{\infty}$ be a sequence of operators $B_k : (X, \mathbf{A}) \to (X, \mathbf{A})$, $y_k (k = 1, 2, ...)$ be the fixed point of B_k, and $T_0 : (X, \mathbf{A}) \to (X, \mathbf{A})$ be Φ-a contractive operator with fixed point y_0 (cf. [14]).

To these conditions (Φ) we add

$(\Phi 3)$ $\Phi_{j(\alpha)}(t) \leq \Phi_{\alpha}(t)$;

$(\Phi 4)$ $\Phi_{\alpha}(t_1 + t_2) \leq \Phi_{\alpha}(t_1) + \Phi_{\alpha}(t_2)$ for t_1 and $t_2 \in R_+^1$.

The sequence $\{B_k\}_{k=1}^{\infty}$ tends uniformly to B_0 if for every $\varepsilon > 0$ there is $\nu = \nu(\varepsilon) > 0$ such that $\rho_{\alpha}(B_k(y), B_0(y)) < \varepsilon$ for every $k > \nu$, $y \in X$ and $\alpha \in \mathbf{A}$. The sequence $\{B_k\}_{k=1}^{\infty}$ tends point-wise to B_0 in $(X, \mathbf{A})$ if $\lim_{k \to \infty} \rho_{\alpha}(B_k(x), B_0(x)) = 0$ for every $x \in X$ and $\alpha \in \mathbf{A}$.

Theorem 1.4.11. If $\{B_k\}_{k=1}^{\infty}$ tends uniformly to B_0, then the sequence of fixed points $\{y_k\}_{k=1}^{\infty}$ tends to y_0 in the topology of $(X,\mathbf{A})$.

If the space $(X,\mathbf{A})$ is locally compact it is completely regular and consequently uniformizible.

Theorem 1.4.12. Let $(X,\mathbf{A})$ be a locally compact Hausdorff quasicomplete j-bounded uniform space, every operator $\{B_k\}_{k=1}^{\infty}$, $B_k : (X,\mathbf{A}) \to (X,\mathbf{A})$ be Φ-contractive with fixed point y_k $(k=0,1,2,...)$ and $\{B_k\}_{k=1}^{\infty}$ tends point-wise to B_0. Then $\{y_k\}_{k=1}^{\infty}$ tends to y_0.

Let $\mathbf{A}_1 = \{\rho_{\alpha_1}(x,y) : \alpha_1 \in \mathbf{A}\}$ and $\mathbf{A}_2 = \{\rho_{\alpha_2}(x,y) : \alpha_2 \in \mathbf{A}\}$ be two families of pseudo-metrics. They are said to be equivalent if the identity mapping from $(X, \mathbf{A}_1)$ to $(X, \mathbf{A}_2)$ is a homeomorphism. If we suppose that $\mathbf{A}_1$ and $\mathbf{A}_2$ are of the same cardinality, further on we denote by $\mathbf{A}$ the index set of all families of pseudo-metrics.

A sequence of families of pseudo-metrics $\{\mathbf{A}_n\}_{n=1}^{\infty}$ tends uniformly to the family $\mathbf{A}_0$ if for every $\varepsilon > 0$ there is N, such that for every $n > N$ the following inequality $|\rho_\alpha^{(n)}(x,y) - \rho_\alpha^{(0)}(x,y)| < \varepsilon$ is satisfied for all $x,y \in X$ and $\alpha \in \mathbf{A}$.

Theorem 1.4.13. Let $\{\mathbf{A}_n\}_{n=1}^{\infty}$ be a sequence of families of pseudo-metrics on X tending uniformly to the family $\mathbf{A}_0$. Let $\{B_n\}_{n=1}^{\infty}$ be a sequence of operators tending point-wise to the operator B_0 on X and each B_n is Φ-contractive with respect to $\mathbf{A}_n$. Then $\{B_n\}_{n=1}^{\infty}$ tends $\mathbf{A}_0$-uniformly to B_0 on every compact set $K \subset X$.

Prior to formulating the next theorem we introduce the concept of j-locally compact space (cf. [14]). A uniform space X is said to be j-locally compact if for every point y_0 and for every finite family of index sets $\alpha_1, \alpha_2, ... \alpha_p \in \mathbf{A}$ there exists $\varepsilon = \varepsilon(\alpha_1, \alpha_2, ... \alpha_p) > 0$, such that the set

$$K(\alpha_1,\alpha_2,...\alpha_p)(y_0,\varepsilon) = \{x \in X : \rho_\alpha(y_0,x) < \varepsilon(\alpha_1,\alpha_2,...\alpha_p)\}$$

is compact, $\varepsilon(\alpha_1,\alpha_2,...\alpha_p) \leq \varepsilon(j(\alpha_1), j(\alpha_2),..., j(\alpha_p))$ and

$$K(\alpha_1,\alpha_2,...\alpha_p)(y_0,\varepsilon(\alpha_1,\alpha_2,...\alpha_p)) \subset K(\overline{\alpha}_1,\overline{\alpha}_2,...\overline{\alpha}_p)(y_0,\varepsilon(\overline{\alpha}_1,\overline{\alpha}_2,...\overline{\alpha}_p))$$

for every $\overline{\alpha}_i \in j^{-1}(\alpha_i)$ $(i=1,2,...,p)$.

Theorem 1.4.14. Let $(X,\mathbf{A})$ be j-bounded complete uniform space and let a sequence of operators $\{B_k\}_{k=1}^{\infty}$ and the family of pseudo-metrics $\{\mathbf{A}_k\}_{k=1}^{\infty}$ satisfy the conditions of

Theorem 1.4.13. If B_0 is Φ-contractive mapping with respect to the family A_0 and B_n has a fixed point $y_n (n = 0,1,2,...)$, then the sequence $\{y_n\}_{n=1}^{\infty}$ tends to y_0.

1.5. TOOLS FROM THE THEORY OF DIFFERENTIAL EQUATIONS WITH DEVIATING ARGUMENTS

Here we examine some tools from the theory of differential equations with deviating arguments. They arise in reducing the mixed problems for transmission line equations to differential equations with delays on the boundary.

A differential equation for which the unknown function and its derivatives are involved with different values of the argument, is called a differential equation with deviating arguments.

For instance

$$\dot{x}(t) = a(t)x(t) + b(t)x(t-\tau) + c(t), \tag{1.5.1}$$

where $a(t)$, $b(t)$ and $c(t)$ are prescribed functions, τ is a positive constant and $x(t)$ is the unknown function is such an equation with constant delay. The dot means the derivative of $x(t)$, i.e. $\dot{x}(t)$ is the derivative of $x(t)$. The last equation is linear with respect to the unknown function, while as we see in the applications, nonlinear ones also arise.

The general nonlinear equation is:

$$\dot{x}(t) = f(t,x(t),x(t-\tau)), \tag{1.5.2}$$

where $f(t,u,v)$ is a prescribed nonlinear function.

As far as we know such equations arise for the first time in the studies of Condorcet in the second half of the 18th century – in 1771. In 1909 V. Volterra introduces retardations in the elasticity theory [113]. In 1949 A. D. Myshkis lays a foundation of the modern theory of differential equations with deviating arguments (cf. [87]). Neutral equations are then introduced by G. A. Kamenskii [69]. The theory is further developed after that. Since we are not able to list all the authors, we mention only some of whose results are related to our consideration: E. Wright, R. Bellman, K. Cook, L. E. Elsgolz, S. B. Norkin, M. A. Krasnoselskii, C. Corduneanu, R. D. Driver, J. G. Borisovich, B. N. Sadowskii, J. K. Hale, R. Nussbaum and many others (cf. references of [14]). The theory of differential equations with deviating arguments is developed along the lines of ordinary differential equations. However, the presence of the deviating arguments gives rise to additional difficulties. For instance, the delay in a self-controlled system can generate self-excited oscillations and even non-stability in the system.

Prescribing an initial function to the initial set, the initial value problem can be formulated as follows: to find a function $x(t)$ (belonging to some class, for instance, continuous functions or measurable ones), which satisfies (1.5.2) on some interval

$[t_0, t_0 + T_0]$, where t_0 is the initial point and $T_0 > 0$. In contrast to the initial value problem for an ordinary differential equation

$$\dot{x}(t) = f(t, x(t)), x(t_0) = x_0$$

in this case one should prescribe an initial function $\varphi(t)$ for $t \in [t_0 - \tau, t_0]$ (cf. Fig. 1.4).

The interval $[t_0 - \tau, t_0]$ is called an initial set.

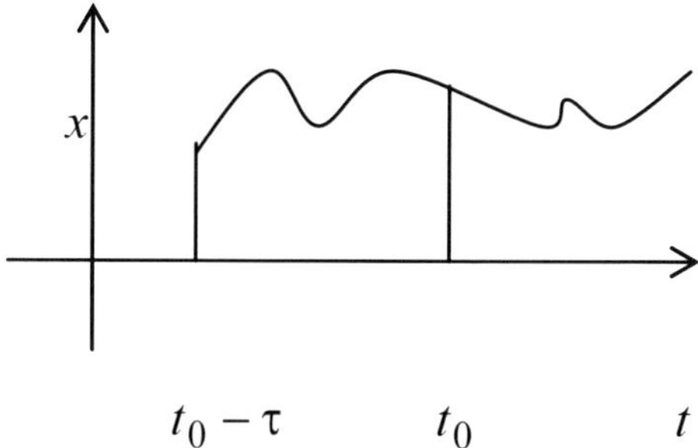

Figure 1.4.

When the delay depends on the argument $\tau = \tau(t)$ and $\tau(t) \geq 0$, the initial set (we denote it by E_{t_0}) consists of the point t_0 and those values of $t - \tau(t)$, which satisfy the inequality $t - \tau(t) \leq t_0$ for $t \geq t_0$. Then $x(t)$ coincides with the initial function $\varphi(t)$ for $t \in E_{t_0}$. So for the equation $\dot{x}(t) = x^2(t) - 3x(t - \sin^2 t)$ (with initial point $t_0 = 0$) the initial function $\varphi(t)$ should be defined on $[-1, 0]$, since the values of $-\sin^2 t$ are between -1 and 0.

If we consider the equation $\dot{x}(t) = x^2(t) - 3x(t/2)$ with an initial point $t_0 = 0$, then the initial set reduces to the point $t_0 = 0$, i.e. $E_{t_0} = \{0\}$. If we choose, however, $t_0 = 1$, then it is easy to see that $E_{t_0} = [1/2 \,; 1]$.

Let us point out that $x(t)$ can belong to various classes of functions. Moreover its derivatives could be seen in the classical or generalized sense ([53], [85]).

A general form of a neutral type differential equation with delay is:

$$\dot{x}(t) = F(t, x(t), x(t - \tau), \dot{x}(t - \tau)).$$

The last equation is of a first-order type.

The formulation of the initial value problem and classification can be extended for the system of differential equations with deviating arguments, written in a vector form.

Remark 1.5.1. We do not use the functional denotation $\dot{x}(t) = F(t, x_t)$ for equations with retarded arguments and therefore we do not use the concept of functional differential equations. Nevertheless we shall sometimes use functional differential equations instead of differential equations with retarded arguments.

It is known that an n-th order ordinary differential equation might be replaced by an equivalent first-order system with n unknown functions, but the system obtained is not always equivalent to the initial equation.

For instance, let us suppose that $\tau = \text{const.} > 0$. Let us assign to the initial value problem for the equation

$$\left|\begin{array}{l} x^{(n)}(t) = F(t, x(t),...,x^{(n-1)}(t), x(t-\tau), \dot{x}(t-\tau),...,x^{(n-1)}(t-\tau)), t > t_0 \\ x^{(k)}(t_0 + 0) = \varphi_k(t_0), \\ x^{(k)}(t-\tau) = \varphi_k(t-\tau), \ t - \tau < t_0 \end{array}\right. \qquad (1.5.3)$$

$(k = 0,1,...,n-1)$ the corresponding first-order system $(x_0(t) \equiv x(t))$

$$\left|\begin{array}{l} \dot{x}_0(t) = x_1(t) \\ \dot{x}_1(t) = x_2(t) \\ \dotfill \\ \dot{x}_{n-2}(t) = x_{n-1}(t) \\ \dot{x}_{n-1}(t) = F(t, x_0(t),...,x_{n-1}(t), x_0(t-\tau),...,x_{n-1}(t-\tau)), \end{array}\right. \quad \text{for } t > t_0 \qquad (1.5.4)$$

$$x_k(t_0 + 0) = \varphi_k(t_0)$$

$$x_k(t-\tau) \equiv \varphi_k(t-\tau) \ \text{ for } \ t-\tau < t_0$$

$(k = 0,1,..., n\text{-}1)$.

The initial functions $\varphi_k(t)$ for differential equations with retarded arguments of n-th order are derivatives of a prescribed function $\varphi(t)$:

$$\varphi_k(t) = \varphi^{(k)}(t) \quad (k = 0,1,...,n-1). \qquad (1.5.5)$$

In this case, the initial value problem (1.5.3) with the additional condition (1.5.5) is not equivalent to the initial value problem (1.5.3). It is equivalent to (1.5.4) with conditions (1.5.5), but they are not well-suited to functional differential systems.

A natural method for solving the functional differential equations with constant delay is the step method. The steps are of length τ, if $\tau = \text{const.} > 0$.

Let us consider the initial value problem:

$$\dot{x}(t) = a\, x(t-\tau), \quad t > t_0 \qquad (1.5.6)$$

$$x(t) = \varphi(t) \equiv C = \text{const.}, \quad t \leq t_0,$$

where a and $\tau > 0$ are constants.

It is easy to ascertain that the solution is

$$x(t) = C \sum_{n=0}^{\left[\frac{t-t_0}{\tau}\right]+1} a^n \frac{(t - t_0 - (n-1)\tau)^n}{n!},$$

where $\left[\dfrac{t - t_0}{\tau}\right]$ is the integer part of $\dfrac{t - t_0}{\tau}$ (for instance $[5,37] = 5$).

We next expose the main features of differential equations with retarded arguments. We show that even when the derivatives of the right-hand side F and of the initial function φ exist up to some order, the solution of the initial value problem has a jump discontinuity in the k-th derivative at the point $t_0 + (k-1)\tau$, while the derivatives of lower order are continuous ones.

Indeed, the first derivative $\dot{x}(t)$ has a jump discontinuity at the initial point t_0, since integrating the equation $\dot{x}(t) = F(t, x(t), \varphi_0(t - \tau))$ for $t \in [t_0, t_0 + \tau]$, one can satisfy the condition $x(t_0) = \varphi(t_0)$. The condition $\dot{x}(t_0 + 0) = \dot{\varphi}(t_0 - 0)$, however, cannot be satisfied in general. Just a suitable choice of the initial function $\varphi_0(t)$ can imply a continuity of the derivative of the solution at t_0. For that purpose $\varphi_0(t)$ should be satisfy the conformity condition

$$\dot{\varphi}_0(t_0 - 0) = F(t_0, \varphi_0(t), \varphi_0(t_0 - \tau)).$$

At the point $t_0 + \tau$ the solution derivative is continuous, since the equation

$$\dot{x}(t) = F(t, x(t), x(t - \tau))$$

is satisfied and its right hand side is continuous at the point $t_0 + \tau$ as a function of t ($x(t)$ is continuous at t_0).

Differentiating the last equation, we get

$$\ddot{x}(t) = \frac{\partial F}{\partial t} + \frac{\partial F}{\partial x}\dot{x}(t) + \frac{\partial F}{\partial x}\dot{x}(t - \tau).$$

In general, the second derivative $\ddot{x}(t)$ is continuous at the point $t_0 + \tau$, since the derivative $\dot{x}(t - \tau)$, as is mentioned above, is discontinuous at $t = t_0 + \tau$. The second derivative $\ddot{x}(t)$ is continuous at the point $t = t_0 + 2\tau$, since $\dot{x}(t-\tau)$ and $x(t-\tau)$ are continuous at the point $t = t_0 + 2\tau$. Similarly we see the derivative $x^{(k)}(t)$ is discontinuous at $t = t_0 + (k-1)\tau$, but derivatives of lower order are continuous, certainly if F is differentiable a sufficient number of times.

Such a smoothing occurs at the interior points too. Indeed, if φ_0 and F are continuous then the solution $x(t)$ of the equation

$$\dot{x}(t) = F(t, x(t), x(t-\tau))$$

will possess a continuous derivative for $t_0 < t < t_0 + \tau$. If F is continuously differentiable then the equation will have twice the differentiable solutions for $t_0 + \tau < t < t_0 + 2\tau$ and so on.

Therefore, for increasing t the, solution $x(t)$ becomes smoother. It follows that periodic solutions of functional differential equations are infinitely differentiable, provided F is infinitely differentiable.

The method of successive integration (or step method) can be applied even in cases where the delay is a function of t, i.e. $\tau = \tau(t)$. In this case, the solution of the initial value problem

$$\begin{aligned} \dot{x}(t) &= F(t, x(t), x(t-\tau(t))) \\ x(t) &= \varphi(t), t \in E_{t_0} = [t_0, \gamma(t_0)] \end{aligned} \tag{1.5.7}$$

($\gamma(t)$ is the inverse function of $t - \tau(t)$, if it exists.) one can define from the equation without delays

$$\begin{aligned} \dot{x}(t) &= F(t, x(t), \varphi_0(t-\tau(t))) \\ x(t_0) &= \varphi(t_0) \end{aligned}$$

(cf. [53]).

Further on, from the equation

$$\begin{aligned} \dot{x}(t) &= F(t, x(t), \varphi_1(t-\tau(t))) \\ x(\gamma(t_0)) &= \varphi_1(\gamma(t_0)) \end{aligned}$$

(where $\varphi_1(t)$ is the extension of $\varphi_0(t)$), one can define the solution on the interval $E_{\gamma(t_0)} = [\gamma(t_0), \gamma(\gamma(t_0))]$ by the solution of (1.5.7) on the interval $[t_0, \gamma(t_0)]$.

In the particular case when $t - \tau(t)$ is monotone increasing, $\varphi_1(t)$ is a solution of (1.5.7) on the interval $[t_0, \gamma(t_0)]$, and proceeding in the same way we can reduce this problem to an integration of equation without retardation.

Finally, we consider the main features of differential equations with retarded arguments of a neutral type. Let us consider neutral equation of first order with a constant delay τ:

$$\dot{x}(t) = F(t, x(t), x(t - \tau), \dot{x}(t - \tau)).$$
(1.5.8)

We would like to point out that in contrast of the differential equations with delays, here the initial function $\varphi(t)$ for (1.5.8) should be differentiable. Indeed, it is easy to see that the derivative at (1.5.8) takes values to the left from the initial point t_0. Using the step method, on the first step we obtain an equation without deviation of the argument

$$\dot{x}(t) = F(t, x(t), \varphi_0(t - \tau), \dot{\varphi}_0(t - \tau)) \text{ for } t_0 \le t < t_0 + \tau.$$
(1.5.9)

Denote by $\varphi_1(t)$ its solution for $[t_0, t_0 + \tau]$. The next step is

$$\dot{x}(t) = F(t, x(t), \varphi_1(t - \tau), \dot{\varphi}_1(t - \tau)) \text{ for } t_0 + \tau \le t < t_0 + 2\tau$$
(1.5.10)

and so on.

There is no smoothing of the solution for neutral equations on every following step. Indeed, we have $\dot{\varphi}(t_0 - 0) \ne \dot{x}(t_0 + 0)$, not only for $t = t_0 + \tau$ but for $\dot{\varphi}(t_0 + \tau - 0) \ne \dot{x}(t_0 + \tau + 0)$ as well. The solution does not become smoother even for the interior points. That is why, when we look for a continuous solution of neutral equations

$$\dot{x}(t) = F(t, x(t), x(t - \tau(t)), \dot{x}(t - \tau(t))), t > t_0$$
$$x(t) = \varphi(t), \quad t < t_0$$
(1.5.11)

we assume that conformity condition at $t = t_0$ is satisfied

$$\dot{\varphi}(t_0) = F(t_0, \varphi(t_0), \varphi(t - \tau(t_0)), \dot{\varphi}(t_0 - \tau(t_0))). \textbf{ (CC)}$$

If the last condition is not satisfied then the derivative of the solution of (1.5.11) will have discontinuities for sufficiently large t, independently of the smoothness of the initial function $\varphi(t)$ and the right-hand side F.

The same is valid for the differential-difference equations (arising mainly in the following chapters):

$$\dot{x}(t) = F(t, x(t), x(t-\tau), \dot{x}(t-\tau)), t > t_0$$
$$x(t) = \varphi(t), \quad t \in [t_0 - \tau, \ t_0] \ (\tau = \text{const.} > 0)$$

$$(1.5.12)$$

and

$$\dot{\varphi}(t_0) = F(t_0, \varphi(t_0), \varphi(t-\tau), \dot{\varphi}(t_0 - \tau)).$$

Many papers are devoted (cf. for instance W. Melvin [85], A. Zverkin [127]) to generalized solutions for neutral functional differential equations. Since the conformity condition is not always satisfied in the applications, the most suitable spaces are that of measurable L^∞ functions with derivatives also from L^∞. We call them Sobolev spaces [105] (cf. also [112]).

Remark 1.5.2. In order to overcome the difficulty caused by the conformity condition we have introduced operator K and we can consider operator functions B to the right of the initial point (cf. § **1.6**).

1.6. DISCUSSION ON THE OPERATOR PRESENTATION OF PERIODIC AND OSCILLATORY PROBLEMS

The main purpose of the present paragraph is to comment on the choice of the functional spaces, and the family of pseudo-metrics (metrics) defining their topologies and operators acting on them.

From Chapters II - VIII we follow the scheme:

(1) Derivation of boundary conditions on the basis of Kirchhoff's law for a given transmission line terminated by some configuration of nonlinear elements;
(2) Formulation of a mixed problem for the transmission line system. It is a first-order hyperbolic system of partial differential equations (Telegrapher's equations for a lossless or lossy transmission line);
(3) Reduction of the mixed problem to an initial value problem for a neutral equation or neutral system of equations on the boundary;
(4) Formulation of the problem for existence-uniqueness of periodic and oscillatory solutions of the obtained neutral equation or system. For that purpose we choose suitable function spaces and introduce the corresponding operators acting on them. These operators are chosen in such a way that their fixed points are periodic or oscillatory solutions of the neutral equations;
(5) Formulation of conditions for the existence-uniqueness theorems which provide that operators defined become contractive ones. These conditions are of two types: 1) conditions that imply the operator maps the chosen set into itself and 2) conditions that assure the operator is contractive.

We have seen that operators with fixed points act on closed spaces or subspaces. Consider the set

$$M_U = \left\{ u(.) \in C^1_{T_0}[T,3T] : |u(t)| \le U_0 e^{\mu(t-T-kT_0)}, \right.$$

$$\left. t \in [T + kT_0, T + (k+1)T_0] \ (k = 0,1,2,...,2m-1) \right\}$$

where U_0, T_0, μ, $\mu T_0 = \mu_0$ are positive constants and $mT_0 = T$ ($m > 1$ is integer). Introduce the following family of metrics:

$$\rho_\mu^{(k)}(u,\overline{u}) = \max\left\{ e^{-\mu(t-T-kT_0)} |u(t) - \overline{u}(t)| : t \in [T + kT_0, T + (k+1)T_0] \right\},$$

$$\rho^{(k)}(u,\overline{u}) = \max\left\{ |u(t) - \overline{u}(t)| : t \in [T + kT_0, T + (k+1)T_0] \right\},$$

$$\hat{\rho}(u,\overline{u}) = \max\left\{ |u(t) - \overline{u}(t)| : t \in [T,3T] \right\},$$

$$\rho_\mu^{(k)}(\dot{u},\dot{\overline{u}}) = \max\left\{ e^{-\mu(t-T-kT_0)} |\dot{u}(t) - \dot{\overline{u}}(t)| : t \in [T + kT_0, T + (k+1)T_0] \right\}$$

$(k = 0,1,2,...,2m-1).$

The space M_U becomes metric with the metric:

$$\hat{\rho}_\mu((u,\dot{u}),(\overline{u},\dot{\overline{u}})) = \max\left\{ \hat{\rho}(u,\overline{u}), \rho_\mu^{(k)}(\dot{u},\dot{\overline{u}}) : k = 0,1,2,...,2m-1 \right\}.$$

It is easy to verify that

$$\rho_\mu^{(k)}(u,\overline{u}) \le \rho^{(k)}(u,\overline{u}) \le e^{\mu_0} \rho_\mu^{(k)}(u,\overline{u}), \ (k = 0,1,2,...)$$

and

$$\hat{\rho}(u,\overline{u}) = \max\left\{ \rho^{(0)}(u,\overline{u}), \rho^{(1)}(u,\overline{u}),..., \rho^{(2m-1)}(u,\overline{u}) \right\} \le$$

$$\le e^{\mu T_0} \max\left\{ \rho_\mu^{(0)}(u,\overline{u}), \rho_\mu^{(1)}(u,\overline{u}),..., \rho_\mu^{(2m-1)}(u,\overline{u}) \right\}.$$

The set M_U turns into a complete metric space (cf. § 1.2) with respect to the metric $\hat{\rho}_\mu((u,\dot{u}),(\overline{u},\dot{\overline{u}}))$.

We mark the completeness of M_U. Let us take the metric $\rho_\mu^{(k)}(u,\overline{u})$ and let $\left\{ u_p(t) \right\}_{p=1}^\infty$ be a Cauchy sequence. This means

$$|u_n(t) - u_m(t)| \le \rho^{(k)}(u_n,u_m) \le e^{\mu_0} \rho_\mu^{(k)}(u_n,u_m) < \varepsilon.$$

In view of the completeness of the set of the real numbers, we conclude that for every fixed $t \in [T + kT_0, T + (k+1)T_0]$ the sequence $\{u_p(t)\}_{p=1}^{\infty}$ is point-wise convergent to a function $\breve{u}(t)$. The last one is continuous and T_0-periodic. Indeed, fix $t_0 \in [T + kT_0, T + (k+1)T_0]$. Then the continuity follows from the inequalities

$$\left|\breve{u}(t) - \breve{u}(t_0)\right| \leq \left|\breve{u}(t) - u_p(t)\right| + \left|u_p(t) - u_p(t_0)\right| + \left|u_p(t_0) - \breve{u}(t_0)\right| < 3\varepsilon.$$

The periodicity follows from

$$\left|\breve{u}(t) - \breve{u}(t + T_0)\right| \leq \left|\breve{u}(t) - u_p(t)\right| + \left|u_p(t) - u_p(t + T_0)\right| + \left|u_p(t + T_0) - \breve{u}(t + T_0)\right| < 2\varepsilon$$

For the derivatives, we take into account that $\{\dot{u}_p(t)\}_{p=1}^{\infty}$ is uniformly convergent and consequently $\lim \dot{u}_p(t) = \dot{\breve{u}}(t)$.

It remains to show that if $u_p(.) \in M_U \Rightarrow \breve{u}(.) \in M_U$. Indeed, since

$$\left|u_p(t)\right| \leq U_0 e^{\mu(t - T - kT_0)} \qquad \text{then}$$

$$\left|\breve{u}(t)\right| - \left|\breve{u}(t)\right| \leq \left|\breve{u}(t) - u_p(t)\right| \leq e^{\mu 0}\rho_\mu^{(k)}(\breve{u}, u_p) < \varepsilon \Rightarrow$$

$$\left|\breve{u}(t)\right| \leq \left|u_p(t)\right| + \varepsilon \leq U_0 + \varepsilon \Rightarrow \left|\breve{u}(t)\right| \leq U_0 \leq U_0 e^{\mu(t - T - kT_0)}.$$

We can similarly prove the same assertion for the derivatives.

To find the oscillatory solution we proceed in the following way.

Introduce a set $S = \{t_k\}_{k=0}^{\infty}$ of zeros of some class of functions with some properties.

We introduce the set $C^1[t_0, \infty)$, $(t_0 \equiv T)$ consisting of all continuous and bounded functions piece-wise differentiable with bounded derivatives on every interval $[t_k, t_{k+1}]$. Let us note that the functions from $C^1[t_0, \infty)$ might not be differentiable at t_k. That is why we introduce a topology of uniform convergence of the functions on $[t_0, t_{k+1}]$ and of the derivatives on every interval $[t_k, t_{k+1}]$, which need the introduction of uniform spaces (cf. [14]).

The operator $K : u(t) \to \bar{u}(t) \equiv u(t - 2T)$ maps the set $C_S^1[t_0, \infty)$, $(t_0 \equiv T)$ of all continuous oscillatory functions on $[t_0, \infty)$, $(t_0 \equiv T)$ into itself (in contrast to the same operator in the space of periodic functions).

This operator might be defined as follows:

$$K : u(t) \rightarrow \bar{u}(t) = \begin{cases} \upsilon_0(t - 2T), \ t \in [T, 3T] \\ u(t - 2T), t \in [3T, \infty) \end{cases},$$

$$\left(\text{resp. } K : u(t) \rightarrow \bar{u}(t) = \begin{cases} U_0(t - T), \ t \in [T, 2T] \\ U(t - T), t \in [2T, \infty) \end{cases} \right)$$

where the initial function $\upsilon_0(t)$ is defined on $[-T, T]$ $\left(\text{resp. } U_0(t) \text{ is defined on } [0, T] \right)$.

In this way one can replace $u(t - 2T)$ (resp. $u(t - T)$) with $\bar{u}(t)$.

We note some previous results from M. A. Krasnoselskii [74], M. A. Krasnoselskii *et al.* [75], J. G. Borisovich [30], [31] and B. N. Sadovskii [99] concerning operators whose fixed points are periodic solutions. Here we extend these results in the following way.

We define an operator $B(u)$ on every interval $[T + kT_0, T + (k+1)T_0], (k = 0, 1, 2, ...)$ of periodicity or on every interval $[t_k, t_{k+1}], (k = 0, 1, 2, ...)$ where t_k are the zeros of the solution. The main reason is to introduce a family of weighted pseudo-metrics:

$$\rho_\mu^{(k)}(u, \bar{u}) = \max \left\{ \rho^{-\mu(t - T - kT_0)} |u(t) - \bar{u}(t)| : t \in [T + kT_0, \ T + (k+1)T_0] \right\},$$
$$(k = 0, 1, 2, ...).$$

We also introduce the set

$$M_U = \left\{ u(.) \in C_{T_0}^1[T, 3T] : |u(t)| \le U_0 e^{\mu(t - T - kT_0)}, \right.$$
$$t \in [T + kT_0, \ T + (k+1)T_0] \ (k = 0, 1, 2, ...) \right\}.$$

where $U_0, T_0, \mu, \mu T_0 = \mu_0 = $ const. are positive constants.

Let us consider a periodic problem for the neutral equation

$$\dot{u}(t) = F(t, u(t), u(t - 2T), \dot{u}(t - 2T)) \equiv U(u)(t), t > T \tag{1.6.1}$$
$$u(t) = \upsilon_0(t), \quad \dot{u}(t) = \dot{\upsilon}_0(t), \ t \in [-T, \ T]$$

with initial point T assuming $F(t, ., ., .)$ is T_0-periodic in t. The fixed-point method is directly related to the appropriate choice of the operator whose fixed points are periodic solutions. The operator B acts on spaces of periodic functions.

Consider a Sadovskii-type operator:

$$B(u)(t) := u(T) + \int_T^t U(u)(s)\,ds - \left(\frac{t - T}{T_0} - \frac{1}{2} \right) \int_T^{T+T_0} U(u)(s)\,ds, \quad t \ge T . \tag{1.6.2}$$

The following lemma plays a key role (in our book):

Key Lemma. Periodic problem (1.6.1) has a periodic solution iff the operator (1.6.2) has a fixed-point.

Proof: Let $u(t)$ be a T_0-periodic solution of (1.6.1). Then, integrating (1.6.1) we obtain

$$u(t) = u(T) + \int_T^t U(u)(s)ds \, .$$

Insert $t = T + T_0$, which yields

$$u(T + T_0) = u(T) + \int_T^{T+T_0} U(u)(s)ds \;\; \Rightarrow \;\; \int_T^{T+T_0} U(u)(s)ds = 0 \, .$$

Therefore

$$u(t) = u(T) + \int_T^t U(u)(s)ds - \left(\frac{t-T}{T_0} - \frac{1}{2} \right) \int_T^{T+T_0} U(u)(s)ds \Leftrightarrow u = B(u),$$

that is, operator B has a fixed point.

Conversely, let B have a fixed point $u(.)$, that is,

$$u(t) = u(T) + \int_T^t U(u)(s)ds - \left(\frac{t-T}{T_0} - \frac{1}{2} \right) \int_T^{T+T_0} U(u)(s)ds \, .$$

Then

$$u(T) = u(T + T_0) = \int_T^{T+T_0} U(u)(s)ds - \left(\frac{T+T_0-T}{T_0} - \frac{1}{2} \right) \int_T^{T+T_0} U(u)(s)ds \Leftrightarrow \frac{1}{2} \int_T^{T+T_0} U(u)(s)ds = 0.$$

Then $u(t) = u(T) + \int_T^t U(u)(s)ds \Leftrightarrow \dot{u}(t) = U(u)(t)$, that is, (1.6.1) has a periodic solution.

The lemma is thus proved.

Clearly $\dfrac{t-T-T_0}{T_0} - \dfrac{1}{2}$ plays a crucial role in the second part of the proof. Specifically, if the operator has a fixed point, then it is a periodic solution of (1.6.1). The availability of the fraction $\dfrac{1}{2}$ implies $\int_T^{T+T_0} U(u)(s)ds = 0 \, .$

If we define the operator B in the following way:

$$B(u)(t) := \begin{cases} u_0(t), t \in [0,T] \\ u(T) + \int\limits_{T}^{t} U(u)(s)ds - \left(\dfrac{t-T}{T_0} - \dfrac{1}{2} \right) \int\limits_{T}^{T+T_0} U(u)(s)ds, \ t \geq T \end{cases}$$

one can see, however, that the function $B(u)(t)$ has a discontinuity at $t = T$. Indeed

$$B(u)(t) := \begin{cases} u(T), \ t = T - 0 \\ u(T) + \int\limits_{T}^{T+T_0} U(u)(s)ds, \ t = T + 0. \end{cases}$$

An analogous difficulty arises in the Borisovich-type operators:

$$B(u)(t) := u(T) + \int\limits_{T}^{t} U(u)(s)ds - \dfrac{t-T-1}{T_0} \int\limits_{T}^{T+T_0} U(u)(s)ds.$$

Here again we have a discontinuity of the operator function at the initial point:

$$B(u)(t) := \begin{cases} u(T), t = T - 0 \\ u(T) + \dfrac{1}{T_0} \int\limits_{T}^{T+T_0} U(u)(s)ds, t = T + 0. \end{cases}$$

We overcome this difficulty in two ways.

1) The first way (for periodic solutions) is to introduce an operator

$$(K_0 u)(t) : u(.) \rightarrow \bar{\upsilon}_0(t) = \upsilon_0(t - 2T), \ t \in [T, 3T] \ ,$$

that is, $\upsilon_0(t)$ for $t \in [-T, T]$ is translated to the right over interval $[T, 3T]$. So in (1.6.1) $u(t-2T)$ and $\dot{u}(t-2T)$ are replaced by the initial function $\bar{\upsilon}_0(t)$ and its derivative $\dot{\bar{\upsilon}}_0(t)$ that were translated to the right. This means that the right-hand side of (1.6.1) becomes

$$U(u)(t) = F(t, u(t), \bar{\upsilon}_0(t), \dot{\bar{\upsilon}}_0(t)), t \geq T \ .$$

Obviously $(K_0 u)(t) = (Ku)(t) = \bar{\upsilon}_0(t)$ for $t \in [T, 3T]$.

In this case the operator function $B(u)(t)$ might be defined only for $t \in [T, 3T]$, because we cannot continue the initial function through the whole axis.

If we consider the periodic problem

$$\dot{u}(t)=F(t,u(t),u(t-T),\dot{u}(t-T))\equiv U(u)(t), t>T \qquad (1.6.1\text{-}1)$$

$$u(t)=\upsilon_0(t), \quad \dot{u}(t)=\dot{\upsilon}_0(t), \quad t\in[0,\ T]$$

then

$$K:u(.)\to\bar{u}(t)=\begin{cases}\upsilon_0(t-T),\, t\in[T,2T]\\ u(t-T),\, t\in[2T,\infty)\end{cases}$$

and $K_0:u(.)\to\bar{\upsilon}_0(t)=\upsilon_0(t-T)$, $t\in[T,2T]$.

We notice that K does not map the space of T_0-periodic functions on $[T,\infty)$ into itself because the obtained function is not T_0-periodic over the whole $[T,\infty)$. That is why from here on, we shall consider a periodic problem only on $[T,2T]$, then on $[2T,3T]$ and so on. With this, the problem for the discontinuity at the initial point disappears.

2)   The second way is to introduce an operator by applying the formula

$$B(u)(t):=\begin{cases}u_0(t),t\in[0,T],\\ u(T)+\displaystyle\int_T^t U(u)(s)ds-\frac{t-T}{T_0}\displaystyle\int_T^{T+T_0}U(u)(s)ds,\, t\geq T.\end{cases} \qquad (1.6.3)$$

Then obviously $B(u)(t):=\begin{cases}u(T),t=T-0\\ u(T),t=T+0\end{cases}$. We can show that $\left|\displaystyle\int_T^{T+T_0}U(u)(s)ds\right|=0$ using the advantages of the introduced weighted pseudo-metrics and the sets of periodic functions.

We notice, however, that the operator function (1.6.3) has a discontinuity of the derivative. Indeed,

$$\dot{B}(u)(t):=\begin{cases}\dot{u}_0(t),t\in[0,T],\\ U(u)(t)-\dfrac{1}{T_0}\displaystyle\int_T^{T+T_0}U(u)(s)ds,\, t\geq T\end{cases}$$

and obviously $\dot{u}_0(T)=U(u)(T)-\dfrac{1}{T_0}\displaystyle\int_T^{T+T_0}U(u)(s)ds$ is not always satisfied. This difficulty can be overcome by defining the operator by the formula

$$B(u)(t):=u(T)+\int_T^t U(u)(s)ds-\frac{t-T}{T_0}\int_T^{T+T_0}U(u)(s)ds, t\geq T$$

where $U(u)(t) = F(t, u(t), \dot{u}(t), \ddot{u}(t))$, $(Ku)(t) = \ddot{u}(t)$, $t \geq T$. In this way we consider the periodic or oscillatory problem only to the right of the initial point.

In Chapters III, IV, VI and VII, new problems arise, generated by the availability of integrals of the unknown functions. We have to introduce a set of such periodic functions such that their integrals become periodic functions again. The condition that ensures this is $\int_{T}^{T+T_0} u(t)\,dt = 0$. This requires introducing a new operator, namely

$$B(u)(t) := \int_{T}^{t} U(u)(s)\,ds - \left(\frac{t-T}{T_0} - \frac{1}{2} \right) \int_{T}^{T+T_0} U(u)(s)\,ds - \frac{1}{T_0} \int_{T}^{T+T_0} \int_{T}^{s} U(u)(\theta)\,d\theta\,ds.$$

It is easy to verify that $\int_{T}^{T+T_0} B(u)(t)\,dt = 0$.

Our considerations are based on the new definition of the operators. We define the operator corresponding on the periodic solutions on every interval

$$[T + kT_0; T + (k+1)T_0], \ (k = 0,1,2,...,m-1)$$

in the following way:

$$B(u)(t) := \int_{T+kT_0}^{t} U(u)(s)\,ds - \frac{t-T-kT_0}{T_0} \int_{T+kT_0}^{T+(k+1)T_0} U(u)(s)\,ds, \ t \in [T + kT_0, T + (k+1)T_0],$$

$(k = 0,1,2,...,m-1)$, where on the right-hand side of (1.6.1) $u(t-T)$ and $\dot{u}(t-T)$ are replaced by the functions $\bar{\upsilon}_0(t)$ and $\dot{\upsilon}_0(t)$.

In the case of periodic solutions we prove existence-uniqueness of such solutions on a finite interval.

When we look for oscillatory solutions we introduce the set $S_T = \{\tau_k\}_{k=1}^{n}, n \in N$ of zeros of the initial function, that is, $\upsilon_0(\tau_k) = 0$ such that $\tau_1 = -T$, $\tau_n = T \equiv t_0$. Besides $\max\{\tau_{k+1} - \tau_k : k = 0,1,...,n\} \leq T_0$. Consider also a strictly increasing sequence of real numbers $S = \{t_k\}_{k=0}^{\infty}$ satisfying the following conditions (C):

(C1) $\lim\limits_{k \to \infty} t_k = \infty$;

(C2) for every k there is $s < k$ such that $t_k - T = t_s$, where $t_s \in S_T \cup S$.

It follows

$$0 \leq \inf\{t_{k+1} - t_k : k = 0,1,2,...\} \leq \sup\{t_{k+1} - t_k : k = 0,1,2,...\} = T_0 < \infty$$

and $t_k - 2T = t_q$ for some $q < k$.

Consider the set

$$M_S = \left\{u(.) \in C^1[t_0, \infty) : u(t_k) = 0 \, (k = 0,1,2,...)\right\}$$

and

$$M_{SU} = \left\{u(.) \in M_S : |u(t)| \leq U_0 e^{\mu(t - t_k)}, t \in [t_k, t_{k=1}]\right\}$$

where U_0, μ are positive constants. Recall that $\mu T_0 = \mu_0 = \text{const.} > 0$.

In the case of oscillatory ones, we define the operator by the formulas

$$B(u)(t) := \int_{t_k}^{t} U(u)(s)ds - \frac{t - t_k}{t_{k+1} - t_k} \int_{t_k}^{t_{k+1}} U(u)(s)ds, \quad t \in [t_k, t_{k+1}], \, (k = 0,1,2,... \,).$$

One can verify that $B(u)(t)$ is continuous function for $t \geq t_0$, but not necessary differentiable at $t = t_k$. That is why we introduce a family of pseudo-metrics which imply a uniform convergence on every interval $[t_0, t_{k+1}]$ for the functions and a uniform convergence on every $[t_k, t_{k+1}]$ for the derivatives. In other words, the topology of the set of the oscillatory functions could not be described by one metric alone, so this requires introducing a countable family of pseudo-metrics.

It is easy to verify that the operators

$$K : u(t) \to \bar{u}(t) = \begin{cases} \upsilon_0(t - 2T), \, t \in [t_0, 3t_0], \, (t_0 = T) \\ u(t - 2T), t \in [3t_0, \infty) \end{cases},$$

$$\left(\text{resp. } K : u(t) \to \bar{u}(t) = \begin{cases} U_0(t - T), \, t \in [t_0, 2t_0] \\ U(t - T), t \in [2t_0, \infty) \end{cases} \right)$$

map the set M_S into itself.

In the chapters following, we formulate conditions denoted by **(IN)**. They are concerned with the initial functions. Indeed, we look for a solution belonging to the set

$$M_U = \left\{u(.) \in C^1_{T_0}[T, 3T] : |u(t)| \leq U_0 e^{\mu(t - T - kT_0)}, \right.$$

$$t \in [T + kT_0, \, T + (k+1)T_0]; \, (k = 0,1,2,..., m-1)\Big\}.$$

In order to ensure that the operator B maps M_U into itself we have to impose suitable conditions on the initial function. Recall $T = mT_0$ and assume

$$\left| \upsilon_0(t) \right| \le U_0 e^{\mu(t-kT_0)}, t \in [kT_0, \ (k+1)T_0], (k = 0,1,...,m-1).$$

If $t \in [T, 2T] \Rightarrow t - T \in [0,T]$ then $\bar{\upsilon}_0(t) = u(t-T) = \upsilon_0(t-T), t \in [T, 2T]$. Consequently

$$\left| \bar{u}(t) \right| \le U_0 e^{\mu(t-T-kT_0)}, t \in [T+kT_0, \ T+(k+1)T_0], \ (k = 0,1,...,m-1).$$

We complete the discussion with the conditions (usually denoted by **(E)**) on the source functions. Since $T = mT_0$ we notice $E_p(t-T) = E_p(t), (p = 0,1)$, and consequently $E_p(t-2T) = E_p(t), (p = 0,1)$.

1.6.1. Some Often Used Inequalities and Statements

We use the following inequalities:

1) $\left| \dfrac{t-T-kT_0}{T_0} \right| \le 1, \quad t \in [T+kT_0, T+(k+1)T_0], \ (k = 0,1,2,...)$;

2) For $t \in [T+kT_0, T+(k+1)T_0]$

$$e^{-\mu(t-T-kT_0)} \left| f(t) - g(t) \right| \le e^{-\mu(t-T-kT_0)} \left| \int_{T+kT_0}^{t} \left(\dot{f}(s) - \dot{g}(s) \right) ds \right|$$

$$\le e^{-\mu(t-T-kT_0)} \int_{T+kT_0}^{t} \left| \dot{f}(s) - \dot{g}(s) \right| ds \le$$

$$\le e^{-\mu(t-T-kT_0)} \rho_\mu^{(k)}\left(\dot{f}, \dot{g} \right) \int_{T+kT_0}^{t} e^{-\mu(s-T-kT_0)} ds \le$$

$$\le e^{-\mu(t-T-kT_0)} \rho_\mu^{(k)}\left(\dot{f}, \dot{g} \right) \frac{e^{-\mu(t-T-kT_0)} - 1}{\mu} \le \frac{\rho_\mu^{(k)}\left(\dot{f}, \dot{g} \right)}{\mu},$$

which imply $\rho_\mu^{(k)}(f,g) \le \dfrac{\rho_\mu^{(k)}(\dot{f}, \dot{g})}{\mu}$.

3) We must make an estimate for the function $e^u - 1$:

$$\left| e^h - 1 \right| \le e^{|h|} - 1 = |h|\,|\xi(h)| \,, \ \ \xi(h) = \frac{e^h - 1}{h}, \ h \in (0;\infty), \xi(0) = 1.$$

We have noticed that $\xi(h) = \dfrac{e^h - 1}{h}$ is well defined at $h = 0$ because $\lim\limits_{h \to 0} \dfrac{e^h - 1}{h} = 1$.

It is easy to verify that $\xi(h)$ is an increasing function. Indeed, $\xi'(h) = \dfrac{he^h - e^h + 1}{h^2}$.

For $h \ge 1$, clearly $\xi'(h) = \dfrac{(h-1)e^h + 1}{h^2}$. For $0 \le h < 1$ we denote by $g(h) = (1-h)e^h$.

Then we have

$$he^h - e^h + 1 > 0 \Leftrightarrow (1-h)e^h < 1 \Leftrightarrow g(h) < g(0).$$

Therefore $\lim\limits_{(t_{k+1}-t_k) \to 0} \dfrac{e^{\mu(t_{k+1}-t_k)} - 1}{\mu(t_{k+1} - t_k)} = 1, \ k \to \infty$, and then $\dfrac{e^{\mu(t_{k+1}-t_k)} - 1}{\mu(t_{k+1} - t_k)} \le \dfrac{e^{\mu T_0} - 1}{\mu T_0}$.

4) $\rho^{(k)}(u,\bar{u}) \le e^{\mu_0} \rho_\mu^{(k)}(u,\bar{u}) \le \dfrac{e^{\mu_0} \rho_\mu^{(k)}(\dot{u},\ddot{u})}{\mu}; \ \ \rho^{(k)}(i,\bar{i}) \le e^{\mu_0} \rho_\mu^{(k)}(i,\bar{i}) \le \dfrac{e^{\mu_0} \rho_\mu^{(k)}(\dot{i},\ddot{i})}{\mu}$.

5) $\dfrac{e^{n\mu_0} - 1}{n} = \dfrac{(e^{\mu_0} - 1)(e^{(n-1)\mu_0} + e^{(n-2)\mu_0} + ... + 1)}{n} \le \dfrac{(e^{\mu_0} - 1)ne^{(n-1)\mu_0}}{n} = (e^{\mu_0} - 1)e^{(n-1)\mu_0}$.

CONCLUSION

Our main goal is to provide results from other areas of mathematics that will be necessary for the following Chapters. Let us point out that first we expose known results from [41] and then extend these transformations to the more general cases. These are cases of lossy transmission line equations and uniform plane equations.

We briefly expose some basic results from fixed-point theory. Our main goal is to show that only a small part of the fixed-point theory is needed for the application to the problems considered in this book.

We also expose specific difficulties arising in the theory of neutral equations with retarded arguments due to the reducing of the mixed problem for transmission line equations leading to a neutral equation (or system) on the boundary.

Chapter II

LOSSLESS TRANSMISSION LINES TERMINATED BY A NONLINEAR R-LOAD

ABSTRACT

In the present chapter we analyze a lossless transmission line terminated by a nonlinear resistive element and parallel to it a linear capacitve element. Two ways of reducing the mixed problem for Telegrapher' equations are presented. The *V-I* characteristics of the R-element can possess polynomial nonlinearities and exponential ones. This leads to the same type of nonlinearities in the neutral equations. Using the fixed-point method and chosing suitable function spaces with weighted family of pseudo-metrics, we obtain existence-uniqueness of periodic and oscillatory solution.

INTRODUCTION

The problem of analysis of lossless transmission lines terminated by a nonlinear element with negative resistance occurs in many investigations, and used in various applications (cf. L. A. Bessonov [24], R. Brayton [28], [29], L. O. Chua, C. A. Desoer & E. S. Kuh [37], L. O. Chua & P. M. Lin [38], L. O. Chua [39], V. Damgov [44], L. V. Danilov [45], S. Darlington [46], A. Deutsch *et al.* [47], [55], D. G. Haigh *et al* [59], L. Jiang *et al.* [68], Ju. Kolesov & D. Shvitra [72], A. Maffucci & G. Miano [82], [86], V. P. Matkhanov [84], J. Nagumo & M. Shimura [89], M. Shimura [104], Wu [116], Wu & Xia [117], A. M. Zaezdnii [122] etc.

As we have already mentioned, the processes in a transmission line are governed by the transmission line equations. They form a first-order hyperbolic system of partial differential equations (Telegrapher's equations) (cf. for instance [115]):

$$C\frac{\partial u(x,t)}{\partial t} + \frac{\partial i(x,t)}{\partial x} = 0, \quad L\frac{\partial i(x,t)}{\partial t} + \frac{\partial u(x,t)}{\partial x} = 0.$$

The commonly accepted approach (proposed by Lord Kelvin in 1884) is to eliminate the one unknown function (for instance $i(x, t)$) and to obtain from the last system a second-order hyperbolic equation for the unknown voltage:

$$\frac{\partial^2 u}{\partial t^2} = v^2 \frac{\partial^2 u}{\partial x^2}$$

where $v = 1/\sqrt{LC}$ is the speed of propagation of the traveling waves.

The last equation contains two families of characteristics, namely straight lines

$$x - vt = \text{const.}, \quad x + vt = \text{const.}$$

The methods for solving the above equation are presented in a number of books (cf. for instance V. S. Vladimirov [112], F.Witham [115]).

In contrast to this approach, we proceed from the above system but avoid increasing the order of the equation. First we formulate the mixed problem and then we reduce it to an initial value problem on the boundary (cf. Chapter I and [41]).

The content of this chapter is as follows: Paragraph § 2.1 contains a reduction of the mixed problem for the hyperbolic system to a periodic initial value problem on the boundary with an unknown function the voltage of the line. In § 2.2 the existence-uniqueness of the T_0-periodic solution for the obtained neutral equation is proved. Here the V-I characteristic of the R-load is of polynomial type. The proof does not depend on the degree of the polynomial. In § 2.3 we formulate conditions for the existence-uniqueness of any measurable solution. In § 2.4 sufficient conditions for existence-uniqueness of an oscillatory regime of a neutral equation are formulated. In § 2.5 a different way of reducing the mixed problem to an initial value problem is given, and conditions for the existence-uniqueness of a periodic solution for the neutral equation is proved. In § 2.6 we consider the same problem as in § 2.2, where V-I characteristics of the R-load are of an exponential type. In § 2.7 the same problem as in § 2.6 is considered, but in the context of a class of measurable functions. In § 2.8 conditions for existence-uniqueness of an oscillatory regime are formulated. Finally, in § 2.9, we propose a different method of reducing of the mixed problem to a periodic one on the boundary, for neutral equations with exponential nonlinearity.

2.1. DERIVATION OF THE BOUNDARY CONDITIONS, FORMULATION OF THE MIXED PROBLEM, AND REDUCING TO AN INITIAL VALUE PROBLEM FOR A NEUTRAL EQUATION

Here we consider a transmission line terminated by a nonlinear with negative resistance R-load with V-I characteristic $i = f(u)$ and parallel connected capacitance C_0, where $E(t)$ the source is a function and R_0 is the resistance of the source (cf. Fig. 2.1).

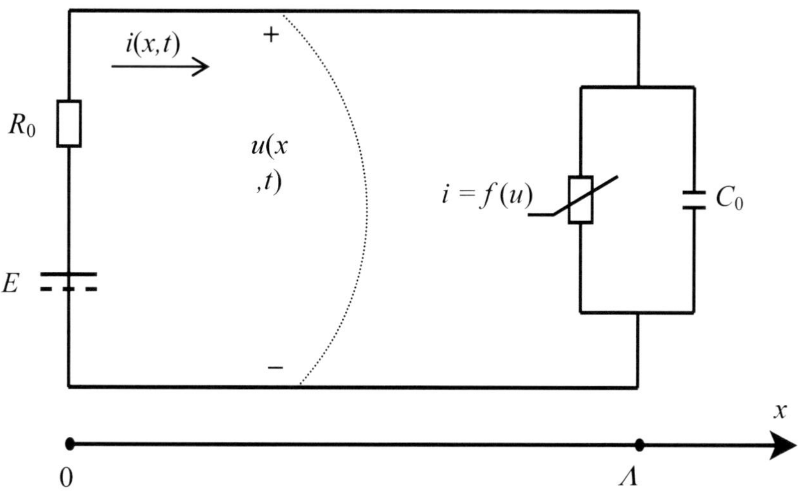

Figure 2.1.

The processes in such line are described by the system (Telegrapher's equations):

$$C\frac{\partial u(x,t)}{\partial t} + \frac{\partial i(x,t)}{\partial x} = 0,$$

$$L\frac{\partial i(x,t)}{\partial t} + \frac{\partial u(x,t)}{\partial x} = 0.$$

(2.1.1)

Our goal is to transform the lossless transmission line equations (2.1.1) in order to reduce the mixed problem to an initial value problem on the boundary. Then we formulate a suitable operator whose fixed-points are periodic solutions of the obtained neutral equation.

We rewrite (2.1.1) in the form

$$\frac{\partial u(x,t)}{\partial t} + \frac{1}{C}\frac{\partial i(x,t)}{\partial x} = 0, \quad \frac{\partial i(x,t)}{\partial t} + \frac{1}{L}\frac{\partial u(x,t)}{\partial x} = 0.$$

Multiplying the second equation by $\sqrt{L/C}$, one obtains:

$$\frac{\partial u(x,t)}{\partial t} + \frac{1}{C}\frac{\partial i(x,t)}{\partial x} = 0, \qquad \sqrt{\frac{L}{C}}\frac{\partial i(x,t)}{\partial t} + \sqrt{\frac{1}{LC}}\frac{\partial u(x,t)}{\partial x} = 0.$$

(2.1.2)

Adding the above equations we get:

$$\frac{\partial u(x,t)}{\partial t} + \sqrt{\frac{L}{C}}\frac{\partial i(x,t)}{\partial t} + \frac{1}{C}\frac{\partial i(x,t)}{\partial x} + \frac{1}{\sqrt{LC}}\frac{\partial u(x,t)}{\partial x} = 0$$

or

$$\frac{\partial}{\partial t}\left(u(x,t)+\sqrt{\frac{L}{C}}\,i(x,t)\right)+\frac{1}{\sqrt{LC}}\frac{\partial}{\partial x}\left(u(x,t)+\sqrt{\frac{L}{C}}\,i(x,t)\right)=0.$$

Subtracting the equations of (2.1.2) we get:

$$\frac{\partial u(x,t)}{\partial t}-\sqrt{\frac{L}{C}}\frac{\partial i(x,t)}{\partial t}+\frac{1}{C}\frac{\partial i(x,t)}{\partial x}-\frac{1}{\sqrt{LC}}\frac{\partial u(x,t)}{\partial x}=0$$

or

$$\frac{\partial}{\partial t}\left(u(x,t)-\sqrt{\frac{L}{C}}\,i(x,t)\right)-\frac{1}{\sqrt{LC}}\frac{\partial}{\partial x}\left(u(x,t)-\sqrt{\frac{L}{C}}\,i(x,t)\right)=0.$$

We recall accepted denotations for the characteristic impedance $Z_0=\sqrt{L/C}$, the speed of propagation of the waves $v=1/\sqrt{LC}$ and the time of propagation of the wave along the whole line $T=\Lambda/(1/\sqrt{LC})=\Lambda\sqrt{LC}$. In this way we obtain:

$$\frac{\partial}{\partial t}\left(u(x,t)+Z_0\,i(x,t)\right)+v\frac{\partial}{\partial x}\left(u(x,t)+Z_0\,i(x,t)\right)=0,$$
$$\frac{\partial}{\partial t}\left(u(x,t)-Z_0\,i(x,t)\right)-v\frac{\partial}{\partial x}\left(u(x,t)-Z_0\,i(x,t)\right)=0.$$

$$(2.1.3)$$

Let us put

$$U(x,t)=u(x,t)+Z_0\,i(x,t),\quad I(x,t)=u(x,t)-Z_0\,i(x,t)$$

and then solve the last system with respect to $u(x,t),\ i(x,t)$:

$$u(x,t)=\frac{1}{2}U(x,t)+\frac{1}{2}I(x,t),$$
$$i(x,t)=\frac{1}{2Z_0}U(x,t)-\frac{1}{2Z_0}I(x,t).$$

$$(2.1.3\text{-}1)$$

This is how, from (2.1.3), we reach the system

$$\frac{\partial U(x,t)}{\partial t} + v\frac{\partial U(x,t)}{\partial x} = 0,$$

$$\frac{\partial I(x,t)}{\partial t} - v\frac{\partial I(x,t)}{\partial x} = 0.$$

Characteristics of this system are the families of straight lines:

$$x + vt = C_1 = \text{const.}, \quad x - vt = C_2 = \text{const.}$$

Thus the solution of the system is the pair

$$U(x,t) = \Phi(x - vt), \quad I(x,t) = \Psi(x + vt),$$

where Φ и Ψ are arbitrarily smooth functions. Then

$$u(x,t) + Z_0 i(x,t) = \Phi(x - vt),$$
$$u(x,t) - Z_0 i(x,t) = \Psi(x + vt).$$

Solving the last system with respect to $u(x,t)$ and $i(x,t)$ we obtain

$$u(x,t) = \frac{1}{2}\left[\Phi(x - vt) + \Psi(x + vt)\right],$$

$$i(x,t) = \frac{1}{2Z_0}\left[\Phi(x - vt) - \Psi(x + vt)\right]. \tag{2.1.4}$$

The mixed problem for the hyperbolic system

$$\frac{\partial u}{\partial t} + \frac{1}{C}\frac{\partial i}{\partial x} = 0, \quad \sqrt{\frac{L}{C}}\frac{\partial i}{\partial t} + \sqrt{\frac{1}{LC}}\frac{\partial u}{\partial x} = 0 \tag{2.1.5}$$

could be formulated by finding a solution of (2.1.5) under the following boundary conditions (yielding by Kirchhoff's law):

$$E(t) - u(0,t) = R_0 i(0,t), \quad t \geq 0 \tag{2.1.6-1}$$

$$C_0 \frac{du(\Lambda,t)}{dt} = i(\Lambda,t) - f(u(\Lambda,t)), \, t \geq 0 \tag{2.1.6-2}$$

and initial conditions

$$u(x,0) = u_0(x),\ i(x,0) = i_0(x),\ x \in [0,\Lambda] \tag{2.1.7}$$

where $u_0(x)$, $i_0(x)$, $E(t)$ and $f(u)$ are prescribed functions, while C_0, R_0 are prescribed constants.

In what follows, we reduce the problem (2.1.5), (2.1.6-1), (2.1.6-2), (2.1.7) to an initial value problem for functional differential equations of neutral type on the boundary.

From (2.1.4) for $x = \Lambda$ we obtain

$$u(\Lambda,t) = \frac{1}{2}\big[\Phi(\Lambda - vt) + \Psi(\Lambda + vt)\big],$$

$$i(\Lambda,t) = \frac{1}{2Z_0}\big[\Phi(\Lambda - vt) - \Psi(\Lambda + vt)\big]$$

We solve the above system with respect to $\Phi(\Lambda - vt)$ and $\Psi(\Lambda + vt)$:

$$\Phi(\Lambda - vt) = u(\Lambda,t) + Z_0 i(\Lambda,t),$$
$$\Psi(\Lambda + vt) = u(\Lambda,t) - Z_0 i(\Lambda,t). \tag{2.1.8}$$

Then we set

$$\Lambda - vt \equiv -vt' \quad \Rightarrow \quad \Lambda + vt' = vt \quad \Rightarrow \quad t = \big(\Lambda + vt'\big)/v \quad \Rightarrow \quad t = t' + T$$

and

$$\Lambda + vt \equiv vt'' \quad \Rightarrow \quad -\Lambda + vt'' = vt \quad \Rightarrow \quad t = \big(-\Lambda + vt''\big)/v \quad \Leftrightarrow \quad t = t'' - T\ .$$

Replacing $\Lambda - vt$ and $\Lambda + vt$ in (2.1.8) we get

$$\Phi(-vt') = u\big(\Lambda, t' + T\big) + Z_0 i\big(\Lambda, t' + T\big),$$
$$\Psi(vt'') = u\big(\Lambda, t'' - T\big) - Z_0 i\big(\Lambda, t'' - T\big).$$

Equalizing the time, or in other words formally replacing t' by t in the first equation and t'' by t in the second one we obtain:

$$\Phi(-vt) = u\big(\Lambda, t + T\big) + Z_0 i\big(\Lambda, t + T\big),$$
$$\Psi(vt) = u\big(\Lambda, t - T\big) - Z_0 i\big(\Lambda, t - T\big). \tag{2.1.9}$$

Then we put $x = 0$ in (2.1.4):

$$u(0,t) = \frac{1}{2}\Phi(-vt) + \frac{1}{2}\Psi(vt),$$

$$i(0,t) = \frac{1}{2Z_0}\Phi(-vt) - \frac{1}{2Z_0}\Psi(vt).$$

Substituting $\Phi(-vt)$ and $\Psi(vt)$ from (2.1.9) in the previous equalities we have:

$$u(0,t) = \frac{u(\Lambda,t+T) + Z_0 i(\Lambda,t+T) + u(\Lambda,t-T) - Z_0 i(\Lambda,t-T)}{2},$$

$$i(0,t) = \frac{u(\Lambda,t+T) + Z_0 i(\Lambda,t+T) - u(\Lambda,t-T) + Z_0 i(\Lambda,t-T)}{2Z_0}.$$

Now substitute $u(0,t)$, $i(0,t)$ in the first boundary condition (2.1.6-1):

$$E(t) - \frac{u(\Lambda,t+T) + Z_0 i(\Lambda,t+T) + u(\Lambda,t-T) - Z_0 i(\Lambda,t-T)}{2} -$$

$$- R_0 \frac{u(\Lambda,t+T) + Z_0 i(\Lambda,t+T) - u(\Lambda,t-T) + Z_0 i(\Lambda,t-T)}{2Z_0} = 0.$$

After obvious transformations (multiplying by 2 and dividing by $Z_0 + R_0$) we obtain:

$$\frac{2E(t)}{Z_0 + R_0} - \frac{1}{Z_0}u(\Lambda,t+T) - i(\Lambda,t+T) + \frac{R_0 - Z_0}{Z_0(Z_0 + R_0)}u(\Lambda,t-T) + \frac{Z_0 - R_0}{Z_0 + R_0}i(\Lambda,t-T) = 0.$$

Let us put $\tau \equiv t + T$. Since $t \geq 0$ (cf. (2.1.6-1), (2.1.6-2)) it follows that the initial point becomes $t_0 = T$. Then

$$\frac{2E(\tau-T)}{Z_0 + R_0} - \frac{1}{Z_0}u(\Lambda,\tau) - i(\Lambda,\tau) + \frac{R_0 - Z_0}{Z_0(Z_0 + R_0)}u(\Lambda,\tau-2T) + \frac{Z_0 - R_0}{Z_0 + R_0}i(\Lambda,\tau-2T) = 0.$$

Replacing τ by t we obtain:

$$i(\Lambda,t) - \frac{Z_0 - R_0}{Z_0 + R_0}i(\Lambda,t-2T) = \frac{2E(t-T)}{Z_0 + R_0} - \frac{1}{Z_0}u(\Lambda,t) - \frac{1}{Z_0}\frac{Z_0 - R_0}{Z_0 + R_0}u(\Lambda,t-2T), t \geq T \qquad (2.1.10)$$

The second boundary condition (2.1.6-2) yields

$$i(\Lambda,t) = C_0 \frac{du(\Lambda,t)}{dt} + f(u(\Lambda,t)),$$

$$i(\Lambda,t-2T) = C_0 \frac{du(\Lambda,t-2T)}{dt} + f(u(\Lambda,t-2T))$$

and replacing $i(\Lambda,t)$ and $i(\Lambda,t-2T)$ in (2.1.10) we obtain

$$C_0 \frac{du(\Lambda,t)}{dt} + f(u(\Lambda,t)) - \frac{Z_0 - R_0}{Z_0 + R_0}\left[C_0 \frac{du(\Lambda,t-2T)}{dt} + f(u(\Lambda,t-2T)) \right] =$$

$$= \frac{2E(t-T)}{Z_0 + R_0} - \frac{1}{Z_0} u(\Lambda,t) - \frac{1}{Z_0}\frac{Z_0 - R_0}{Z_0 + R_0} u(\Lambda,t-2T) \ .$$

Since we have choosen the initial point to be $t_0 = T$ then the initial set should be $[-T,T]$.

We choose the unknown function to be $u(\Lambda,t)$ and put $u(t) = u(\Lambda,t)$ and recall $E(t-T) = E(t)$ (cf. Chapter I). In that case the initial value problem for the above equation could be formulated as follows: find a solution of the following functional differential equation of neutral type

$$\frac{du(t)}{dt} = \frac{1}{C_0}\frac{2E(t)}{Z_0 + R_0} - \frac{1}{C_0} f(u(t)) + \frac{1}{C_0}\frac{Z_0 - R_0}{Z_0 + R_0} f(u(t-2T)) -$$
$$- \frac{1}{C_0 Z_0} u(t) - \frac{1}{C_0 Z_0}\frac{Z_0 - R_0}{Z_0 + R_0} u(t-2T) + \frac{Z_0 - R_0}{Z_0 + R_0}\frac{du(t-2T)}{dt}, \ t \geq T \tag{2.1.11}$$

$$u(t) = \upsilon_0(t), \frac{du(t)}{dt} = \frac{d\upsilon_0(t)}{dt}, t \in [-T,T],$$

where the initial function $\upsilon_0(t)$ is obtained in the following way: we translate the initial function $u_0(x)$ (cf. Fig. 2.2) defined on the interval $t = 0$, $x \in [0,\Lambda]$ along the characteristic $x - vt = \eta$ on the boundary $(x,t) \in \Lambda \times [0,T]$, and so we obtain an initial function $\upsilon_0(t)$ on $[0,T]$. If the initial function $u_0(x)$ is Λ_0-periodic one with $\Lambda_0 = \Lambda/m$ for some $m \in \{2,3,4,...\}$, then $\upsilon_0(t)$ should be T_0-periodic function where $T_0 = T/m$. Since we look for periodic solutions we can extend $\upsilon_0(t)$ periodically from $[0,T]$ on the interval $[-T,T]$.

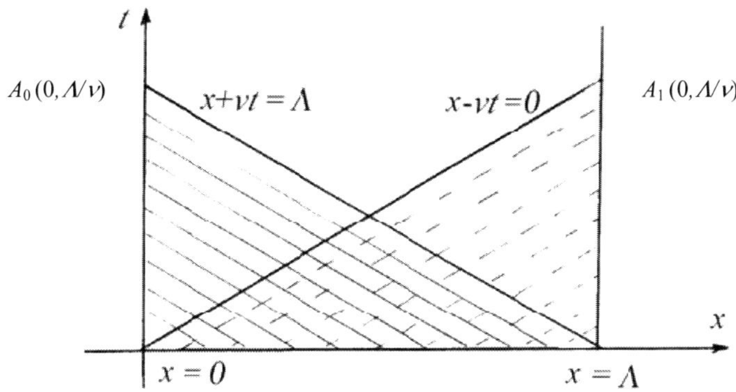

Figure 2.2.

Remark 2.1.1. Let us describe in detail the process of translating the initial functions from $[0,\Lambda]$ to $[0,T]$ (cf Fig. 2.2). Indeed, by the straight line $\eta = v\xi + x$ to every point $(x,0) \in [0,\Lambda] \times \{0\}$ one can assign a point $\left[\Lambda,\ t \equiv (\Lambda - x)/v\right]$ belonging to the straight line $x = \Lambda$. Therefore we can define the initial function $\upsilon_0(t)$ using the formula:

$$x \rightarrow t = \frac{\Lambda - x}{v} \Rightarrow u_0(x) \rightarrow \upsilon_0(t) \equiv \upsilon_0\big((\Lambda - x)/v\big).$$

Remark 2.1.2. Once we have found the voltage $u(\Lambda,t)$ we can find the current of the transmission line by the formula $i(\Lambda,t) = f(u(\Lambda,t)) + C_0 \dfrac{du(\Lambda,t)}{dt}$ since $f(.)$ is a prescribed function while C_0 is a prescribed constant.

2.2. PERIODIC REGIMES FOR A NEUTRAL EQUATION WITH POLYNOMIAL NONLINEARITY

Our goal here is to prove an existence-uniqueness of a continuously differentiable T_0-periodic solution of the problem (2.1.11), where the resistive element has a *V-I* characteristic of polynomial type of arbitrary degree, that is, $i = f(u) = \sum_{n=1}^{p} r_n u^n$. In most applications third-order polynomials $f(u) = r_1 u + r_3 u^3$ (cf. [55], [117]) are most commonly found.

Using the denotation $\dfrac{du(t)}{dt} \equiv \dot{u}(t)$ we can rewrite (2.1.11) in the following form:

$$\dot{u}(t) = \frac{2E(t)}{C_0(Z_0 + R_0)} - \frac{1}{C_0 Z_0} u(t) - \frac{1}{C_0} \sum_{n=1}^{p} r_n [u(t)]^n - \frac{Z_0 - R_0}{Z_0 C_0(Z_0 + R_0)} u(t - 2T) + \qquad (2.2.1)$$

$$+\frac{Z_0-R_0}{C_0(Z_0+R_0)}\sum_{n=1}^{p}r_n\left[u(t-2T)\right]^n+\frac{Z_0-R_0}{Z_0+R_0}\dot{u}(t-2T)\equiv F(u(t),u(t-2T),\dot{u}(t-2T)),\ \ t\in[T,\infty)$$

$$u(t)=\upsilon_0(t),\ \dot{u}(t)=\dot{\upsilon}_0(t),\ t\in[-T,T].$$

Formulation of the main problem: to find a continuously differentiable T_0-periodic solution $u(t)$ of the problem (2.2.1) on the interval $[T,\infty)$ coinciding with prescribed T_0-periodic initial function $\upsilon_0(t)$ on $[-T,T]$, where $\upsilon_0(.)\in C^1_{T_0}[-T,T]$. We look for a solution by applying the step method: first we find a solution for $[T,3T]$ with an initial function of $[-T,T]$, then the obtained solution becomes an initial function and we find a solution for $[3T,5T]$, and so on.

Remark 2.2.1. We proceed in this way because if we consider the problem globally, and taking an arbitrary T_0-periodic function $u(t)$ on the whole semiaxis $[T,\infty)$, in view of the delay $t-2T$ we have to translate the initial function to the right, over $[T,3T]$ and define the operator K (cf. **§ 1.6**). We recall that we have to restrict ourselves to the interval $[T,3T]$ because otherwise the function $(Ku)(t)=\bar{u}(t)=\begin{cases}\upsilon_0(t-2T),\ t\in[T,3T]\\ u(t-2T),\ t\in[3T,\infty)\end{cases}$ is not T_0-periodic for the whole semiaxis $[T,\infty)$.

Remark 2.2.2. Since the above equation is a differential equation of neutral type (cf. Chapter I) we must comment on the conformity condition **(CC)** (cf. [23], [53], [65], [69], [85], [87], [99]) guaranteeing an existence of a smooth derivative of the solution to the above equation. For the last equation **(CC)** appears in the following way:

$$\dot{\upsilon}_0(T)=\frac{2E(T)}{C_0(Z_0+R_0)}-\frac{1}{C_0Z_0}\upsilon_0(T)-\frac{1}{C_0}\sum_{n=1}^{p}r_n\left[\upsilon_0(T)\right]^n-\frac{Z_0-R_0}{Z_0C_0(Z_0+R_0)}\upsilon_0(-T)+$$

$$+\frac{Z_0-R_0}{C_0(Z_0+R_0)}\sum_{n=1}^{p}r_n\left[\upsilon_0(-T)\right]^n+\frac{Z_0-R_0}{Z_0+R_0}\dot{\upsilon}_0(-T). \tag{2.2.2}$$

Obviously **(CC)** is satisfied, provided:

$$E(T)=0,\upsilon_0(-T)=0,\dot{\upsilon}_0(-T)=0,\ \upsilon_0(T)=0,\ \dot{\upsilon}_0(T)=0.$$

With the assumption (v) $\upsilon_0(-T)=0,\ \dot{\upsilon}_0(-T)=0.$

In view of $mT_0=T$ it follows that $\upsilon_0(-T)=0,\ \dot{\upsilon}_0(T)=0$, that is, **(CC)** is satisfied. The periodicity of the initial function $\upsilon_0(t)$

$$\upsilon_0(T)=0 \text{ and } u(T)=u(T+kT_0)=0\ \ (k=0,1,2,...\ ).$$

By $C_{T_0}^1[T,\infty)$ we mean the space of all continuous T_0-periodic functions with continuous derivatives. The main difficulty is to define a suitable operator whose fixed points are T_0-periodic solutions of the periodic problem (2.2.2).

We define an operator $B(u)$ in the following way:

$$B_k(u)(t) := \int\limits_{T+kT_0}^{t} U(u)(s)\,ds - \frac{t-T-kT_0}{T_0}\int\limits_{T+kT_0}^{T+(k+1)T_0} U(u)(s)\,ds,$$

$$t \in [T+kT_0, T+(k+1)T_0]$$

$$(k = 0,1,2,\ldots,2m-1)\,. \tag{2.2.3}$$

Here $U(u)(t)$ is the right-hand side of (2.2.1) where the unknown function with the delay $u(t-2T)$ is replaced by the function $\bar{u}(t)$. On $[T,3T]$ $\bar{u}(t)$ coincides with $\bar{\upsilon}_0(t)$:

$$U(u)(t) \equiv \frac{2E(t)}{C_0(Z_0+R_0)} - \frac{1}{C_0Z_0}u(t) - \frac{1}{C_0}\sum_{n=1}^{p}r_n\big[u(t)\big]^n - \frac{Z_0-R_0}{Z_0C_0(Z_0+R_0)}\bar{u}(t) +$$

$$+ \frac{Z_0-R_0}{C_0(Z_0+R_0)}\sum_{n=1}^{p}r_n\big[\bar{u}(t)\big]^n + \frac{Z_0-R_0}{Z_0+R_0}\frac{d\bar{u}(t)}{dt} \tag{2.2.4}$$

that is, the right hand side of (2.2.1) becomes type

$$U(u)(t) \equiv \frac{2E(t)}{C_0(Z_0+R_0)} + F(u(t),\bar{u}(t),\dot{\bar{u}}(t)) = \frac{2E(t)}{C_0(Z_0+R_0)} + F(u(t),\bar{\upsilon}_0(t),\dot{\bar{\upsilon}}_0(t)),\ \ t \in [T,3T]$$

Thus we do not need the conformity condition **(CC)** and can omit the restrictive condition $\dot{\upsilon}_0(-T) = 0 \Rightarrow \dot{\upsilon}_0(T) = 0.$

Remark 2.2.3. Recall that the function $\bar{u}(t)$ was defined in Chapter I (cf. § 1.6) (cf. T. Jankowski & M. Kwapisz [65]).

Lemma 2.2.1. If $u(.) \in C_{T_0}^1[T,\infty)$, $E(.) \in C_{T_0}^1[0,\infty)$, $\upsilon_0(.) \in C_{T_0}^1[-T,T]$, then $\bar{\upsilon}_0(.) \in C_{T_0}^1[T,3T]$ and consequently $U(u)(.) \in C_{T_0}[T,3T]$.

Remark 2.2.4. It is understood that the frequency of the source $E(t)$ should not be smaller than the generated frequency (cf. Dobrev, Jordanova [49]). Consequently, in view of $\omega_0 = 2\pi/T_0$ the period should not be larger than the period of the generated signal. Here we assume that $E(t)$ has the same frequency $\omega_0 = 2\pi/T_0$.

Below, we show that $B(u)(t)$ is a periodic function on $[T,3T]$. For that purpose we have to establish some auxiliary assertions.

Proposition 2.2.1. The function $B_k(u)(t)$ for $t \in [T + kT_0, T + (k+1)T_0]$ equals $B_{k+1}(u)(t)$ for $t \in [T + (k+1)T_0, T + (k+2)T_0]$, that is,

$$B_k(u)(t - T_0) = B_{k+1}(u)(t), \ t \in [T + (k+1)T_0, T + (k+2)T_0].$$

Proof: Prior to begin the proof we see that $B_{k+1}(u)(t)$ is a T_0-periodic continuation of $B_k(u)(t), \ t \in [T + kT_0, T + (k+1)T_0]$ on the interval $[T + (k+1)T_0, T + (k+2)T_0])$.

In order to compare $B_k(u)(t)$ with $B_{k+1}(u)(t)$, we have to translate $B_k(u)(t)$ on the interval $[T + (k+1)T_0, T + (k+2)T_0]$.

Indeed, let us define

$$\widetilde{B}_k(u)(t) := B_k(u)(t - T_0) \ \text{for} \ t \in [T + (k+1)T_0, T + (k+2)T_0].$$

So we have to prove

$$\widetilde{B}_k(u)(t) := B_{k+1}(u)(t) \ \text{for} \ t \in [T + (k+1)T_0, T + (k+2)T_0]:$$

$$\widetilde{B}_k(u)(t) := B_k(u)(t - T_0) = \int_{T+kT_0}^{t-T_0} U(u)(s)ds - \frac{t - T_0 - T - kT_0}{T_0} \int_{T+kT_0}^{T+(k+1)T_0} U(u)(s)ds.$$

Let us change the variable $\theta = s + T_0$ in the last integral. Then in view of periodicity of $U(u)(t)$ we have

$$\widetilde{B}_k(u)(t) = \int_{T+(k+1)T_0}^{t} U(u)(\theta - T_0)d\theta - \frac{t - T - (k+1)T_0}{T_0} \int_{T+(k+1)T_0}^{T+(k+2)T_0} U(u)(\theta - T_0)d\theta =$$

$$= \int_{T+(k+1)T_0}^{t} U(u)(\theta)d\theta - \frac{t - T - (k+1)T_0}{T_0} \int_{T+(k+1)T_0}^{T+(k+2)T_0} U(u)(\theta)d\theta = B_{k+1}(u)(t) \ .$$

Proposition 2.2.1 is thus proved.

Next, we rewrite the above periodic problem as:

$$\frac{du(t)}{dt} = U(u)(t), \ t \in [T, 3T]; \ u(T) = u(T + kT_0) = 0, \ (k = 0,1,2,...,2m-1). \quad (2.2.5)$$

Introduce the set

$$M_U = \left\{ u(.) \in C_{T_0}^1[T,3T] : |u(t)| \le U_0 e^{\mu(t - T - kT_0)}, t \in [T + kT, \ T + (k+1)T_0]; (k = 0,1,2,...,2m-1) \right\}$$

where U_0, T_0, μ, $\mu T_0 = \mu_0$ are positive constants (chosen below).

Introduce the following family of metrics (cf. A. Bielecki [22]) for $k = 0,1,2,...,2m-1$:

$$\rho_\mu^{(k)}(u,\overline{u}) = \max\left\{ e^{-\mu(t-T-kT_0)}\left|u(t)-\overline{u}(t)\right| : t \in [T+kT_0, T+(k+1)T_0] \right\},$$

$$\rho^{(k)}(u,\overline{u}) = \max\left\{ \left|u(t)-\overline{u}(t)\right| : t \in [T+kT_0, T+(k+1)T_0] \right\},$$

$$\hat{\rho}(u,\overline{u}) = \max\left\{ \left|u(t)-\overline{u}(t)\right| : t \in [T,3T] \right\},$$

$$\rho_\mu^{(k)}(\dot{u},\dot{\overline{u}}) = \max\left\{ e^{-\mu(t-T-kT_0)}\left|\dot{u}(t)-\dot{\overline{u}}(t)\right| : t \in [T+kT_0, T+(k+1)T_0] \right\}.$$

Remark 2.2.5. In fact, the space M_U becomes metric one because we can put

$$\hat{\rho}_\mu((u,\dot{u}),(\overline{u},\dot{\overline{u}})) = \max\left\{ \hat{\rho}(u,\overline{u}), \rho_\mu^{(k)}(\dot{u},\dot{\overline{u}}) : k = 0,1,2,...,2m-1 \right\}. \qquad (2.2.6)$$

Remark 2.2.6. It is easy to verify that (cf. Chapter I)

$$\rho_\mu^{(k)}(u,\overline{u}) \le \rho^{(k)}(u,\overline{u}) \le e^{\mu T_0}\rho_\mu^{(k)}(u,\overline{u}), \quad (k = 0,1,2,...) ;$$

$$\max\left\{\rho^{(0)}(u,\overline{u}),\rho^{(1)}(u,\overline{u}),...,\rho^{(k)}(u,\overline{u})\right\} \le e^{\mu_0} \max\left\{\rho_\mu^{(0)}(u,\overline{u}),\rho_\mu^{(1)}(u,\overline{u}),...,\rho_\mu^{(k)}(u,\overline{u})\right\} \equiv e^{\mu_0}\hat{\rho}_\mu^{(k)}(u,\overline{u});$$

$$\hat{\rho}(u,\overline{u}) = \max\left\{\rho^{(0)}(u,\overline{u}),\rho^{(1)}(u,\overline{u}),...,\rho^{(2m-1)}(u,\overline{u})\right\}.$$

We need the following conditions:

(E)
$$E(.) \in C_{T_0}^1[0,\infty), E(T) = 0, \left|E(t)\right| \le U_E e^{\mu(t-T-nT_0)}(n = 0,1,2,...), t \in [T+nT_0, T+(n+1)T_0] ;$$

(IN) $\upsilon_0(t) \in C_{T_0}^1[-T,T]$ and

$$\left|\upsilon_0(t)\right| \le U_0\, e^{\mu(t+T-nT_0)}\ (n = 0,1,...,2m-1),\ t \in [-T+nT_0, -T+(n+1)T_0]$$

Remark 2.2.7. Condition **(IN)** implies

$$\left|\overline{u}(t)\right| \le U_0\, e^{\mu(t-T-kT_0)}, t \in [T+kT_0, T+(k+1)T_0]\ (k = 0,1,2,...,2m-1) .$$

Lemma 2.2.2. The initial value problem (2.2.5) has a solution $u(.) \in M_U$ iff operator B has a fixed point $u \in M_U$, that is, $u = B(u)$.

Proof: Let $u(.) \in M_U$ be a solution of (2.2.5), that is:

$$\frac{du(t)}{dt} = U(u)(t), \quad u(T) = u(T + kT_0) = 0.$$

Then after integration we have $u(t) = \int_{T+kT_0}^{t} U(u)(t)dt$.

Let us put $t = T + (k+1)T_0$ in the last equation. Since $u(T + (k+1)T_0) = 0$ we obtain

$$0 = u(T + (k+1)T_0) = \int_{T+kT_0}^{T+(k+1)T_0} U(u)(t)dt \Rightarrow \int_{T+kT_0}^{T+(k+1)T_0} U(u)(t)dt = 0 \ .$$

Therefore $u(t) = \int_{T+kT_0}^{t} U(u)(t)dt \iff u(t) = B(u)$, that is, $u(.) \in M_U$ is a fixed point of B.

Conversely, let B have a fixed point $u(.) \in M_U$, i.e. $u = B(u)$. Then, recalling, $\mu T_0 = \mu_0$ we have

$$\beta \equiv \left| \int_{T+kT_0}^{T+(k+1)T_0} U(u)(s)ds \right| \le$$

$$\le \frac{2}{C_0(Z_0 + R_0)} \int_{T+kT_0}^{T+(k+1)T_0} |E(t)|dt + \frac{1}{C_0 Z_0} \int_{T+kT_0}^{T+(k+1)T_0} |u(t)|dt + \frac{1}{C_0} \sum_{n=1}^{p} |r_n| \int_{T+kT_0}^{T+(k+1)T_0} |u(t)|^n dt +$$

$$+ \frac{1}{Z_0 C_0} \frac{|Z_0 - R_0|}{Z_0 + R_0} \int_{T+kT_0}^{T+(k+1)T_0} |\bar{u}(t)|dt + \frac{1}{C_0} \frac{|Z_0 - R_0|}{Z_0 + R_0} \sum_{n=1}^{p} |r_n| \int_{T+kT_0}^{T+(k+1)T_0} |\bar{u}(t)|^n dt + \frac{|Z_0 - R_0|}{Z_0 + R_0} \left| \int_{T+kT_0}^{T+(k+1)T_0} \dot{\bar{u}}(t)dt \right| \le$$

$$\le \frac{2U_E}{C_0(Z_0 + R_0)} \frac{e^{\mu T_0} - 1}{\mu} + \frac{U_0}{C_0 Z_0} \frac{e^{\mu T_0} - 1}{\mu} + \frac{1}{C_0} \sum_{n=1}^{p} |r_n| U_0^n \int_{T+kT_0}^{T+(k+1)T_0} e^{n\mu(t-T-kT_0)}dt +$$

$$+ \frac{U_0}{Z_0 C_0} \frac{|Z_0 - R_0|}{Z_0 + R_0} \frac{e^{\mu T_0} - 1}{\mu} + \frac{1}{C_0} \frac{|Z_0 - R_0|}{Z_0 + R_0} \sum_{n=1}^{p} |r_n| U_0^n \int_{T+kT_0}^{T+(k+1)T_0} e^{n\mu(t-T-kT_0)}dt +$$

$$+\frac{|Z_0-R_0|}{Z_0+R_0}\left|\bar{\upsilon}_0(T+(k+1)T_0-2T)-\bar{\upsilon}_0(T+kT_0-2T)\right|\le$$

$$\le\frac{2U_E}{C_0(Z_0+R_0)}\frac{e^{\mu T_0}-1}{\mu}+\frac{U_0}{C_0 Z_0}\frac{e^{\mu T_0}-1}{\mu}+\frac{1}{C_0}\sum_{n=1}^{p}|r_n|U_0^n\frac{e^{n\mu T_0}-1}{n\mu}+$$

$$+\frac{U_0}{C_0 Z_0}\frac{|Z_0-R_0|}{Z_0+R_0}\frac{e^{\mu T_0}-1}{\mu}+\frac{1}{C_0}\frac{|Z_0-R_0|}{Z_0+R_0}\sum_{n=1}^{p}|r_n|U_0^n\frac{e^{n\mu T_0}-1}{n\mu}\le$$

$$\le\frac{2U_E}{C_0(Z_0+R_0)}\frac{e^{\mu T_0}-1}{\mu}+\frac{U_0}{C_0 Z_0}\frac{e^{\mu T_0}-1}{\mu}+\frac{1}{C_0}\frac{e^{\mu T_0}-1}{\mu}\sum_{n=1}^{p}|r_n|U_0^n e^{(n-1)\mu T_0}+$$

$$+\frac{U_0}{C_0 Z_0}\frac{|Z_0-R_0|}{Z_0+R_0}\frac{e^{\mu T_0}-1}{\mu}+\frac{1}{C_0}\frac{|Z_0-R_0|}{Z_0+R_0}\frac{e^{\mu T_0}-1}{\mu}\sum_{n=1}^{p}|r_n|U_0^n e^{(n-1)\mu T_0}\le$$

$$\le\frac{e^{\mu 0}-1}{\mu C_0}\left[\frac{2U_E}{Z_0+R_0}+\left(1+\frac{|Z_0-R_0|}{Z_0+R_0}\right)\left(\frac{U_0}{Z_0}+\sum_{n=1}^{p}|r_n|U_0^n e^{(n-1)\mu 0}\right)\right]\equiv M(\mu).$$

Let us assume that $\left|\displaystyle\int_{T+kT_0}^{T+(k+1)T_0}U(u)(s)ds\right|=\gamma>0$. But we have just proved that $\gamma\le M(\mu)$.

For sufficiently large $\mu>0$ (and sufficiently small $T_0>0$) one can reach the inequality

$M(\mu)<\gamma$. The obtained contradiction implies $\displaystyle\int_{T+kT_0}^{T+(k+1)T_0}U(u)(s)ds=0$. It follows that

$u(t)=\displaystyle\int_{T+kT_0}^{t}U(u)(s)ds$ and after a differentiation we obtain (2.2.4).

Lemma 2.2.2 is thus proved.

Theorem 2.2.1. Let the assumptions

(E) $E(.)\in C_{T_0}^1[0,\infty),E(0)=0,$

$\left|E(t)\right|\le U_E e^{\mu(t-T-kT_0)},t\in[T+kT_0,T+(n+1)T_0],(k=0,1,...)$;

(IN) $\upsilon_0(t)\in C_{T_0}^1[-T,T],$

$\left|\upsilon_0(t)\right|\le U_0\, e^{\mu(t+T-nT_0))}\ (n=0,1,...,2m-1),\ t\in[-T+nT_0,-T+(n+1)T_0]$;

be fulfilled.

Observe a unique T_0-periodic solution of (2.2.2) on $[T, 3T]$.

Proof: We show that B maps M_U into itself.

We show that for $u \in M_U$ the function $B(u)(t)$ is T_0-periodic on $[T, \infty)$. But this is proved in Proposition 2.2.1.

We prove that $B(u)(t)$ is a continuously differentiable function.

Indeed, since

$$B_k(u)(t) := \int_{T+kT_0}^{t} U(u)(s)ds - \frac{t - T - kT_0}{T_0} \int_{T+kT_0}^{T+(k+1)T_0} U(u)(s)ds,$$

then

$$B_k(u)(T+(k+1)T_0) = \int_{T+kT_0}^{T+(k+1)T_0} U(u)(s)ds - \frac{T+(k+1)T_0 - T - kT_0}{T_0} \int_{T+kT_0}^{T+(k+1)T_0} U(u)(s)ds = 0;$$

$$B_{k+1}(u)(T+(k+1)T_0) := \int_{T+(k+1)T_0}^{T+(k+1)T_0} U(u)(s)ds - \frac{T+(k+1)T_0 - T - (k+1)T_0}{T_0} \int_{T+(k+1)T_0}^{T+(k+2)T_0} U(u)(s)ds = 0.$$

For the derivatives, we obtain

$$\frac{dB_k(u)(t)}{dt} = U(u)(t) - \frac{1}{T_0} \int_{T+kT_0}^{T+(k+1)T_0} U(u)(s)ds,$$

$$\frac{dB_{k+1}(u)(t)}{dt} = U(u)(t) - \frac{1}{T_0} \int_{T+(k+1)T_0}^{T+(k+2)T_0} U(u)(s)ds$$

and therefore since $\displaystyle\int_{T+kT_0}^{T+(k+1)T_0} U(u)(s)ds = \int_{T+(k+1)T_0}^{T+(k+2)T_0} U(u)(s)ds$ (recall that $U(u)(s)$ is T_0-periodic), we obtain

$$\frac{dB_k(u)(T+(k+1)T_0)}{dt} = U(u)(T+(k+1)T_0) - \frac{1}{T_0} \int_{T+kT_0}^{T+(k+1)T_0} U(u)(s)ds =$$

$$= U(u)(T+(k+1)T_0) - \frac{1}{T_0} \int_{T+(k+1)T_0}^{T+(k+2)T_0} U(u)(s)ds = \frac{dB_{k+1}(u)(T+(k+1)T_0)}{dt}.$$

Next, we have to show that

$$|(Bu)(t)| \le U_0 e^{\mu(t-T-kT_0)}, t \in [T+kT_0, T+(k+1)T_0].$$

Recall that in accordance with Remark 2.2.9, condition (**IN**) of Theorem 2.2.1 implies

$$\left|u(t-2T)\right| = \left|\vec{v}_0(t)\right| \le U_0 e^{\mu(t-2T-kT_0)}, t \in [T+kT_0, T+(k+1)T_0].$$

In addition, $\displaystyle\int_{T+kT_0}^{t} \dot{\vec{v}}_0(s)\,ds = \vec{v}_0(t) - \vec{v}_0(T+kT_0) = \vec{v}_0(t)$ and consequently

$$\int_{T+kT_0}^{t} U(u)(s)\,ds \equiv \frac{2}{C_0(Z_0+R_0)} \int_{T+kT_0}^{t} E(s)\,ds - \frac{1}{C_0 Z_0} \int_{T+kT_0}^{t} u(s)\,ds - \frac{1}{C_0} \sum_{n=1}^{p} r_n \int_{T+kT_0}^{t} [u(s)]^n\,ds -$$

$$- \frac{Z_0-R_0}{Z_0 C_0(Z_0+R_0)} \int_{T+kT_0}^{t} \bar{u}(s)\,ds + \frac{Z_0-R_0}{C_0(Z_0+R_0)} \sum_{n=1}^{p} r_n \int_{T+kT_0}^{t} [\bar{u}(s)]^n\,ds + \frac{Z_0-R_0}{Z_0+R_0}\left(\bar{u}(t) - \bar{u}(T+kT_0)\right) =$$

$$= \frac{2}{C_0(Z_0+R_0)} \int_{T+kT_0}^{t} E(s)\,ds - \frac{1}{C_0 Z_0} \int_{T+kT_0}^{t} u(s)\,ds - \frac{1}{C_0} \sum_{n=1}^{p} r_n \int_{T+kT_0}^{t} [u(s)]^n\,ds -$$

$$- \frac{Z_0-R_0}{Z_0 C_0(Z_0+R_0)} \int_{T+kT_0}^{t} \bar{u}(s)\,ds + \frac{Z_0-R_0}{C_0(Z_0+R_0)} \sum_{n=1}^{p} r_n \int_{T+kT_0}^{t} [\bar{u}(s)]^n\,ds + \frac{Z_0-R_0}{Z_0+R_0}\bar{u}(t).$$

For every $u(.) \in M_U$ we obtain

$$\left|(Bu)(t)\right| \le \left|\int_{T+kT_0}^{t} U(u)(s)\,ds\right| + \left|\int_{T+kT_0}^{T+(k+1)T_0} U(u)(s)\,ds\right| \equiv U_1 + U_2.$$

Then

$$U_1 \le \left[\frac{2U_E}{C_0(Z_0+R_0)} \int_{T+kT_0}^{t} e^{\mu(s-T-kT_0)}\,ds + \frac{U_0}{C_0 Z_0} \int_{T+kT_0}^{t} e^{\mu(s-T-kT_0)}\,ds + \frac{U_0^n}{C_0} \sum_{n=1}^{p} |r_n| \int_{T+kT_0}^{t} e^{n\mu(s-T-kT_0)}\,ds + \right.$$

$$\left. + \frac{|Z_0-R_0|U_0}{Z_0 C_0(Z_0+R_0)} \int_{T+kT_0}^{t} e^{\mu(s-T-kT_0)}\,ds + \frac{|Z_0-R_0|U_0^n}{C_0(Z_0+R_0)} \sum_{n=1}^{p} |r_n| \int_{T+kT_0}^{t} e^{n\mu(s-T-kT_0)}\,ds \right] + \frac{|Z_0-R_0|}{Z_0+R_0}|\bar{u}(t)| \le$$

$$\le \frac{2U_E}{C_0(Z_0+R_0)} \frac{e^{\mu(t-T-kT_0)}-1}{\mu} + \frac{U_0}{C_0 Z_0} \frac{e^{\mu(t-T-kT_0)}-1}{\mu} + \frac{1}{C_0} \sum_{n=1}^{p} |r_n| U_0^n \int_{T+kT_0}^{t} e^{n\mu(s-T-kT_0)}\,ds +$$

$$+ \frac{U_0}{C_0 Z_0} \frac{|Z_0-R_0|}{Z_0+R_0} \frac{e^{\mu(t-T-kT_0)}-1}{\mu} + \frac{1}{C_0} \frac{|Z_0-R_0|}{Z_0+R_0} \sum_{n=1}^{p} |r_n| U_0^n \int_{T+kT_0}^{t} e^{n\mu(s-T-kT_0)}\,ds + \frac{|Z_0-R_0|}{Z_0+R_0} U_0 e^{\mu(t-T-kT_0)} + \le$$

$$\leq \frac{2U_E}{C_0(Z_0+R_0)}\frac{e^{\mu(t-T-kT_0)}-1}{\mu}+\frac{U_0}{C_0Z_0}\frac{e^{\mu(t-T-kT_0)}-1}{\mu}+\frac{1}{C_0}\sum_{n=1}^{p}|r_n|U_0^n\frac{e^{n\mu(t-T-kT_0)}-1}{n\mu}+$$

$$+\frac{U_0}{Z_0C_0}\frac{|Z_0-R_0|}{Z_0+R_0}\frac{e^{\mu(t-T-kT_0)}-1}{\mu}+\frac{1}{C_0}\frac{|Z_0-R_0|}{Z_0+R_0}\sum_{n=1}^{p}|r_n|U_0^n\frac{e^{n\mu(t-T-kT_0)}-1}{n\mu}+$$

$$\frac{|Z_0-R_0|}{Z_0+R_0}U_0\,e^{\mu(t-T-kT_0)}\leq$$

$$\leq\frac{1}{C_0}\left[\frac{2U_E}{Z_0+R_0}\frac{e^{\mu(t-T-kT_0)}-1}{\mu}+\frac{U_0}{Z_0}\frac{e^{\mu(t-T-kT_0)}-1}{\mu}+\frac{e^{\mu(t-T-kT_0)}-1}{\mu}\sum_{n=1}^{p}|r_n|U_0^n e^{(n-1)\mu(t-T-kT_0)}+\right.$$

$$\left.+\frac{U_0}{Z_0}\frac{|Z_0-R_0|}{Z_0+R_0}\frac{e^{\mu(t-T-kT_0)}-1}{\mu}+U_0\frac{|Z_0-R_0|}{Z_0+R_0}\frac{e^{\mu(t-T-kT_0)}-1}{\mu}\sum_{n=1}^{p}|r_n|U_0^n e^{(n-1)\mu(t-T-kT_0)}\right]+\frac{|Z_0-R_0|}{Z_0+R_0}U_0\,e^{\mu(t-T-kT_0)}\leq$$

$$\leq e^{\mu(t-T-kT_0)}\left[\frac{1}{\mu C_0}\left(\frac{2U_E}{Z_0+R_0}+\left(1+\frac{|Z_0-R_0|}{Z_0+R_0}\right)\left(\frac{U_0}{Z_0}+\sum_{n=1}^{p}|r_n|\left(U_0 e^{\mu_0}\right)^n\right)\right)+U_0\frac{|Z_0-R_0|}{Z_0+R_0}\right];$$

$$U_2\leq\left[\frac{2U_E}{C_0(Z_0+R_0)}\int_{T+kT_0}^{T+(k+1)T_0}e^{\mu(s-T-kT_0)}ds+\frac{U_0}{C_0Z_0}\int_{T+kT_0}^{T+(k+1)T_0}e^{\mu(s-T-kT_0)}ds+\frac{U_0^n}{C_0}\sum_{n=1}^{p}|r_n|\int_{T+kT_0}^{T+(k+1)T_0}e^{n\mu(s-T-kT_0)}ds+\right.$$

$$\left.+\frac{|Z_0-R_0|U_0}{Z_0C_0(Z_0+R_0)}\int_{T+kT_0}^{T+(k+1)T_0}e^{\mu(s-T-kT_0)}ds+\frac{|Z_0-R_0|U_0^n}{C_0(Z_0+R_0)}\sum_{n=1}^{p}|r_n|\int_{T+kT_0}^{T+(k+1)T_0}e^{n\mu(s-T-kT_0)}ds\right]+$$

$$+\frac{|Z_0-R_0|}{Z_0+R_0}\left|\vec{u}(T+(k+1)T_0)-\vec{u}(T+kT_0)\right|\leq$$

$$\leq e^{\mu(t-T-kT_0)}\frac{e^{\mu_0}-1}{\mu C_0}\left(\frac{2U_E}{Z_0+R_0}+\left(1+\frac{|Z_0-R_0|}{Z_0+R_0}\right)\left(\frac{U_0}{Z_0}+\sum_{n=1}^{p}|r_n|\left(U_0 e^{\mu_0}\right)^n\right)\right).$$

Therefore

$$\left|B_k(u)(t)\right|\leq$$

$$\leq e^{\mu(t-T-kT_0)}\frac{1}{\mu C_0}\left(\frac{2U_E}{Z_0+R_0}+\left(1+\frac{|Z_0-R_0|}{Z_0+R_0}\right)\left(\frac{U_0}{Z_0}+\sum_{n=1}^{p}|r_n|\left(U_0 e^{\mu_0}\right)^n\right)\right)+\frac{|Z_0-R_0|}{Z_0+R_0}U_0\,e^{\mu(t-T-kT_0)}+$$

$$+e^{\mu(t-T-kT_0)}\frac{\left(e^{\mu_0}-1\right)}{\mu C_0}\left(\frac{2U_E}{Z_0+R_0}+\left(1+\frac{|Z_0-R_0|}{Z_0+R_0}\right)\left(\frac{U_0}{Z_0}+\sum_{n=1}^{p}|r_n|\left(U_0 e^{\mu_0}\right)^n\right)\right)\leq$$

$$\leq e^{\mu(t-T-kT_0)}\left\{\frac{e^{\mu 0}}{\mu C_0}\left[\frac{2U_E}{Z_0+R_0}+\left(1+\frac{|Z_0-R_0|}{Z_0+R_0}\right)\left(\frac{U_0}{Z_0}+\sum_{n=1}^{p}|r_n|\left(U_0 e^{\mu 0}\right)^n\right)\right]+U_0\frac{|Z_0-R_0|}{Z_0+R_0}\right\}\leq$$

$$\leq e^{\mu(t-T-kT_0)}U_0.$$

The last inequality is satisfied for sufficiently large $\mu>0$ because $\dfrac{|Z_0-R_0|}{Z_0+R_0}<1$.

Consequently, the operator B maps M_U into itself.

The next step is to show that B is a contractive operator.

For every $u(.),\overline{u}(.)\in M_U$ we get for $t\in[T+kT_0,T+(k+1)T_0]$:

$$\left|B_k(u)(t)-B_k(\overline{u})(t)\right|\leq$$

$$\left|\int_{T+kT_0}^{t}\left(U(u)(s)-U(\overline{u})(s)\right)ds\right|+\left|\int_{T+kT_0}^{T+(k+1)T_0}\left(U(u)(s)-U(\overline{u})(s)\right)ds\right|\equiv V_1+V_2.$$

We have

$$V_1\leq\frac{1}{C_0 Z_0}\int_{T+kT_0}^{t}|u(s)-\overline{u}(s)|ds+\frac{1}{C_0}\sum_{n=1}^{p}|r_n|\int_{T+kT_0}^{t}|u^n(s)-\overline{u}^n(s)|ds+$$

$$\frac{|Z_0-R_0|}{Z_0 C_0(Z_0+R_0)}\int_{T+kT_0}^{t}|\overline{u}(s)-\overline{u}(s)|ds+$$

$$+\frac{|Z_0-R_0|}{C_0(Z_0+R_0)}\sum_{n=1}^{p}|r_n|\int_{T+kT_0}^{t}|\overline{u}^n(s)-\overline{u}^n(s)|ds+\frac{|Z_0-R_0|}{Z_0+R_0}\left|\int_{T+kT_0}^{t}\left(\dot{\overline{u}}(s)-\dot{\overline{u}}(s)\right)ds\right|\leq$$

$$\leq\frac{\rho_\mu^{(k)}(u,\overline{u})}{C_0 Z_0}\int_{T+kT_0}^{t}e^{\mu(s-T-kT_0)}ds+\frac{1}{C_0}\sum_{n=1}^{p}|r_n|n\sup\left\{|u(s)|^{n-1}:s\in s\in[T+kT_0,T+(k+1)T_0\right\}\int_{T+kT_0}^{t}|u(s)-\overline{u}(s)|ds\leq$$

$$\leq\frac{\rho_\mu^{(k)}(u,\overline{u})}{C_0 Z_0}\frac{e^{\mu(t-T-kT_0)}-1}{\mu}+\frac{\rho_\mu^{(k)}(u,\overline{u})}{C_0}\frac{e^{\mu(t-T-kT_0)}-1}{\mu}\sum_{n=1}^{p}|r_n|n U_0^{n-1}e^{(n-1)\mu T_0}\leq$$

$$\leq\frac{\rho_\mu^{(k)}(\dot{u},\dot{\overline{u}})}{\mu C_0 Z_0}\frac{e^{\mu(t-T-kT_0)}-1}{\mu}+\frac{\rho_\mu^{(k)}(\dot{u},\dot{\overline{u}})}{\mu C_0}\frac{e^{\mu(t-T-kT_0)}-1}{\mu}\sum_{n=1}^{p}|r_n|n U_0^{n-1}e^{(n-1)\mu T_0}\leq$$

$$\leq \frac{e^{\mu(t-T-kT_0)}-1}{\mu^2 C_0}\rho_\mu^{(k)}(\dot{u},\dot{\overline{u}})\left(\frac{1}{Z_0}+\sum_{n=1}^{p}|r_n|nU_0^{n-1}e^{(n-1)\mu_0}\right)$$

and

$$V_2 \leq \frac{1}{C_0 Z_0}\int_{T+kT_0}^{T+(k+1)T_0}|u(s)-\overline{u}(s)|ds+\frac{1}{C_0}\sum_{n=1}^{p}|r_n|\int_{T+kT_0}^{T+(k+1)T_0}|u^n(s)-\overline{u}^n(s)|ds \leq$$

$$\frac{\rho_\mu^{(k)}(u,\overline{u})}{C_0 Z_0}\int_{T+kT_0}^{T+(k+1)T_0}e^{\mu(s-T-kT_0)}ds+$$

$$+\frac{1}{C_0}\sum_{n=1}^{p}|r_n|n\sup\left\{|u(s)|^{n-1}:s\in s\in[T+kT_0,T+(k+1)T_0\right\}\int_{T+kT_0}^{T+(k+1)T_0}|u(s)-\overline{u}(s)|ds \leq$$

$$\leq \frac{\rho_\mu^{(k)}(\dot{u},\dot{\overline{u}})}{\mu C_0 Z_0}\frac{e^{\mu T_0}-1}{\mu}+\frac{\rho_\mu^{(k)}(\dot{u},\dot{\overline{u}})}{\mu C_0}\frac{e^{\mu T_0}-1}{\mu}\sum_{n=1}^{p}|r_n|nU_0^{n-1}e^{(n-1)\mu T_0} \leq$$

$$\leq e^{\mu(t-T-kT_0)}\rho_\mu^{(k)}(\dot{u},\dot{\overline{u}})\frac{e^{\mu_0}-1}{\mu^2 C_0}\left(\frac{1}{Z_0}+\sum_{n=1}^{p}|r_n|nU_0^{n-1}e^{(n-1)\mu_0}\right).$$

Therefore, for every fixed $k=0,1,2,\ldots,2m-1$

$$\left|B_k(u)(t)-B_k(\overline{u})(t)\right| \leq$$

$$\leq e^{\mu(t-T-kT_0)}\hat{\rho}_\mu((u,\dot{u}),(\overline{u},\dot{\overline{u}}))\frac{1}{\mu^2 C_0}\left(\frac{1}{Z_0}+\sum_{n=1}^{p}|r_n|nU_0^{n-1}e^{(n-1)\mu T_0}\right)+$$

$$+e^{\mu(t-T-kT_0)}\hat{\rho}_\mu((u,\dot{u}),(\overline{u},\dot{\overline{u}}))\frac{e^{\mu_0}-1}{\mu^2 C_0}\left(\frac{1}{Z_0}+\sum_{n=1}^{p}|r_n|nU_0^{n-1}e^{(n-1)\mu_0}\right) \leq$$

$$\leq e^{\mu(t-T-kT_0)}\hat{\rho}_\mu((u,\dot{u}),(\overline{u},\dot{\overline{u}}))\frac{e^{\mu_0}}{\mu^2 C_0}\left(\frac{1}{Z_0}+\sum_{n=1}^{p}|r_n|n\left(U_0 e^{\mu_0}\right)^{n-1}\right)\equiv e^{\mu(t-T-kT_0)}K_U\hat{\rho}_\mu((u,\dot{u}),(\overline{u},\dot{\overline{u}})) \leq$$

$$\leq e^{\mu T_0}K_U\hat{\rho}_\mu((u,\dot{u}),(\overline{u},\dot{\overline{u}})).$$

Taking the supremum of the left hand side for $t\in[T+kT_0,T+(k+1)T_0$ we obtain

$$\rho^{(k)}(B_k(u),B_k(\overline{u}))\leq e^{\mu T_0}K_U\hat{\rho}_\mu((u,\dot{u}),(\overline{u},\dot{\overline{u}}))\ (k=0,1,2,\ldots,2m-1).$$

But since $e^{\mu T_0} K_U \hat{\rho}_\mu((u,\dot{u}),(\overline{u},\dot{\overline{u}}))$ does not depend on k, it follows

$$\hat{\rho}(B(u),B(\overline{u})) \leq e^{\mu_0} K_U \hat{\rho}_\mu((u,\dot{u}),(\overline{u},\dot{\overline{u}})).$$

It remains to estimate the derivative of B:

$$\left|\dot{B}_k(u)(t) - \dot{B}_k(\overline{u})(t)\right| \leq$$

$$\left|U(u)(t) - U(\overline{u})(t)\right| + \frac{1}{T_0}\left|\int_{T+kT_0}^{T+(k+1)T_0}[U(u)(s) - U(\overline{u})(s)]ds\right| \equiv \dot{U}_1 + \dot{U}_2.$$

We have

$$\dot{U}_1 \leq \frac{1}{C_0 Z_0}|u(t) - \overline{u}(t)| + \frac{1}{C_0}\sum_{n=1}^{p}|r_n|\|u^n(t) - \overline{u}^n(t)\| \leq$$

$$\leq e^{\mu(t-T-kT_0)}\frac{\rho_\mu^{(k)}(u,\overline{u})}{C_0 Z_0} + e^{\mu(t-T-kT_0)}\frac{\rho_\mu^{(k)}(u,\overline{u})}{C_0}\sum_{n=1}^{p}|r_n|nU_0^{n-1}e^{(n-1)\mu(t-T-kT_0)} \leq$$

$$\leq e^{\mu(t-T-kT_0)}\frac{\rho_\mu^{(k)}(\dot{u},\dot{\overline{u}})}{\mu C_0}\left(\frac{1}{Z_0} + \sum_{n=1}^{p}|r_n|nU_0^{n-1}e^{(n-1)\mu T_0}\right) \leq$$

$$\leq e^{\mu(t-T-kT_0)}\frac{1}{\mu C_0}\left(\frac{1}{Z_0} + \sum_{n=1}^{p}|r_n|n\left(U_0 e^{\mu_0}\right)^{n-1}\right)\hat{\rho}_\mu((u,\dot{u}),(\overline{u},\dot{\overline{u}}))$$

and

$$\dot{U}_2 = \frac{1}{T_0}V_2 = \frac{1}{T_0}\left|\int_{T+kT_0}^{T+(k+1)T_0}[U(u)(s) - U(\overline{u})(s)]ds\right| \leq$$

$$\leq e^{\mu(t-T-kT_0)}\rho^{(k)}(\dot{u},\dot{\overline{u}})\frac{e^{\mu_0}-1}{\mu^2 T_0 C_0}\left(\frac{1}{Z_0} + \sum_{n=1}^{p}|r_n|nU_0^{n-1}e^{(n-1)\mu_0}\right) \leq$$

$$\leq e^{\mu(t-T-kT_0)}\frac{e^{\mu_0}-1}{\mu_0}\frac{1}{\mu C_0}\left(\frac{1}{Z_0} + \sum_{n=1}^{p}|r_n|n\left(U_0 e^{\mu_0}\right)^{n-1}\right)\hat{\rho}_\mu((u,\dot{u}),(\overline{u},\dot{\overline{u}})).$$

Consequently

$$\left|\dot{B}_k(u)(t) - \dot{B}_k(\overline{u})(t)\right| \le e^{\mu(t-T-kT_0)} \frac{1}{\mu C_0}\left(\frac{1}{Z_0} + \sum_{n=1}^{p}|r_n|n\left(U_0 e^{\mu_0}\right)^{n-1}\right)\hat{\rho}_\mu((u,\dot{u}),(\overline{u},\dot{\overline{u}})) +$$

$$+ e^{\mu(t-T-kT_0)}\frac{e^{\mu_0}-1}{\mu_0}\frac{1}{\mu C_0}\left(\frac{1}{Z_0} + \sum_{n=1}^{p}|r_n|n\left(U_0 e^{\mu_0}\right)^{n-1}\right)\hat{\rho}_\mu((u,\dot{u}),(\overline{u},\dot{\overline{u}})) \le$$

$$\le e^{\mu(t-T-kT_0)}\left(1+\frac{e^{\mu_0}-1}{\mu_0}\right)\frac{1}{\mu C_0}\left(\frac{1}{Z_0} + \sum_{n=1}^{p}|r_n|n\left(U_0 e^{\mu_0}\right)^{n-1}\right)\hat{\rho}_\mu((u,\dot{u}),(\overline{u},\dot{\overline{u}})) \equiv$$

$$\equiv e^{\mu(t-T-kT_0)}\dot{K}_U\,\hat{\rho}_\mu((u,\dot{u}),(\overline{u},\dot{\overline{u}})).$$

Taking the supremum over $[T+kT_0, T+(k+1)T_0]$

$$e^{-\mu(t-T-kT_0)}\left|\dot{B}_k(u)(t) - \dot{B}_k(\overline{u})(t)\right| \le \dot{K}_U\,\hat{\rho}_\mu((u,\dot{u}),(\overline{u},\dot{\overline{u}}))$$

we obtain

$$\rho_\mu^{(k)}(\dot{B}_k u, \dot{B}_k \overline{u}) \le \dot{K}_U\,\hat{\rho}_\mu((u,\dot{u}),(\overline{u},\dot{\overline{u}})).$$

On the other hand from the above we have

$$\hat{\rho}(B(u), B(\overline{u})) \le e^{\mu T_0}K_U\hat{\rho}_\mu((u,\dot{u}),(\overline{u},\dot{\overline{u}})).$$

Denoting by $K = \max\left\{e^{\mu_0}K_U, \dot{K}_U\right\} < 1$ we conclude

$$\hat{\rho}_\mu((Bu,\dot{B}u),(B\overline{u},\dot{B}\overline{u})) \le K\,\hat{\rho}_\mu((u,\dot{u}),(\overline{u},\dot{\overline{u}})).$$

For sufficiently large μ the operator B is contractive.

The unique fixed point of B is a periodic solution of (2.1.1) on $[T,3T]$.

Theorem 2.2.1 is thus proved.

2.2.1. Numerical Example

To apply the above theorem we have to present all the inequalities from the proof:

$$\frac{e^{\mu_0}}{\mu C_0}\left[\frac{2}{Z_0+R_0}\frac{U_E}{U_0} + \left(1+\frac{|Z_0-R_0|}{Z_0+R_0}\right)\left(\frac{1}{Z_0} + \sum_{n=1}^{p}|r_n|\left(U_0 e^{\mu_0}\right)^{n-1}\right)\right] + \frac{|Z_0-R_0|}{Z_0+R_0} \le 1$$

and

$$K_U = \frac{e^{\mu_0}}{\mu^2 C_0}\left(\frac{1}{Z_0} + \sum_{n=1}^{p}|r_n|n\left(U_0 e^{\mu_0}\right)^{n-1}\right) < 1;$$

$$\dot{K}_U = \left(1 + \frac{e^{\mu_0}-1}{\mu_0}\right)\frac{1}{\mu C_0}\left(\frac{1}{Z_0} + \sum_{n=1}^{p}|r_n|n\left(U_0 e^{\mu_0}\right)^{n-1}\right) < 1.$$

Consider a transmission line with specific parameters:

$$\Lambda = 5m,\ L = 0{,}2\,\mu H/m,\ C = 80\,pF/m,\ v = 1/\sqrt{LC} = 1/\sqrt{0{,}2.10^{-6}.80.10^{-12}} = 1/\left(4.10^{-9}\right) = 2{,}5.10^{8},$$

$$Z_0 = \sqrt{L/C} = \sqrt{\left(0{,}2.10^{-6}\right)/\left(80.10^{-12}\right)} = 50\ \Omega,\ R_0 = 35\Omega,\ C_0 = 8pF = 8.10^{-12}F.$$

In this case, $T = \Lambda\sqrt{LC} = 2.10^{-8}s;\ |Z_0 - R_0|/(Z_0 + R_0) = 1/19$.

Let us check the propagation of waves with $\lambda_0 = (1/4)10^{-3}\,m$. We have

$$f_0 = 1/\left(\lambda_0\sqrt{LC}\right) = 1/\left((1/4)10^{-3}.4.10^{-9}\right) = 10^{12}\ Hz \Rightarrow T_0 = 1/f_0 = 10^{-12}\ \text{sec}.$$

Let us choose $\mu = 10^{12}$, then $\mu T_0 = \mu_0 - 1$ and $T = 2.10^{-8}.10^{12}T_0 - 20000.T_0$. We also have $\mu C_0 = 10^{12}.8.10^{-12} = 8;\ \mu^2 C_0 = 10^{24}.8.10^{-12} = 8.10^{12}$.

If the *V-I* characteristic of the nonlinear resistive element is $f(u) = -0{,}12\,u + 0{,}8u^3$, then for $U_E = U_0 = 0{,}1$ the above inequalities become:

$$\frac{e}{8}\left[\frac{2}{85} + \frac{20}{19}\left(\frac{1}{50} + 0{,}12 + 0{,}8.10^{-2}e^2\right)\right] + \frac{1}{19} \le 1 \Leftrightarrow 0{,}08 + 0{,}053 \le 1;$$

$$eK_U = e\frac{e}{8.10^{12}}\left(\frac{1}{50} + 0{,}12 + 3.0{,}8e^2 U_0^2\right) = \frac{e}{10^{12}}\left(0{,}05 + 6U_0^2\right) = \frac{0{,}11e}{10^{12}};$$

$$\dot{K}_U = \frac{e}{8}\left(0{,}02 + 0{,}12 + 3.0{,}8e^2 U_0^2\right) = 0{,}048 + 6{,}05 U_0^2 = 0{,}1085$$

and therefore $K = \{eK_U, \dot{K}_U\} = 0{,}1085 < 1$.

Let us choose the source function $E(t) = U_E \sin(\omega_0 t),\ U_E \le U_0$. We find the second successive approximation only on the interval $[T, T + T_0]$ beginning with the first:

$$u^{(0)}(t) = \begin{cases} u_0 \sin(\omega_0 t), t \in [-T,T] \\ 0, \qquad t \in [T, T+T_0], \end{cases}$$

$$\dot{u}^{(0)}(t) = \begin{cases} -u_0 \omega_0 \cos(\omega_0 t) = \dot{u}_0 \cos(\omega_0 t), \ t \in [-T,T] \\ 0, \qquad\qquad\qquad\qquad t \in [T, T+T_0]. \end{cases}$$

If we want **(CC)** to be satisfied we must change $u^{(0)}(t)$ on the "small" set $[-T; -T+\Delta]$ and then we continue periodically with the obtained function on $[-T,T]$ such that the conformity condition $\upsilon_0(-T) = 0$, $\dot{\upsilon}_0(-T) = 0$ can be satisfied. Denote the new initial function by $\tilde{u}^{(0)}(t)$. Then $\tilde{u}^{(0)}(-T) = \tilde{u}^{(0)}(T) = \dot{\tilde{u}}^{(0)}(-T) = \dot{\tilde{u}}^{(0)}(T) = 0$. For sufficiently small $\Delta > 0$ we obtain

$$\int_T^t u^{(0)}(s-2T)ds \approx \int_{T+\Delta}^t \tilde{u}^{(0)}(s-2T)ds \ \text{ and } \ \int_T^{T+T_0} u^{(0)}(s-2T)ds \approx \int_{T+\Delta}^{T+T_0-\Delta} \tilde{u}^{(0)}(s-2T)ds .$$

Denote by $\delta = \dfrac{Z_0 - R_0}{Z_0 + R_0}$ and then in view of

$$\omega_0 T_0 = 2\pi \implies \sin(\omega_0 s - \omega_0 T) = \sin(\omega_0 s - \omega_0 T_0.3.10^8) = \sin \omega_0 s$$

we obtain

$$u^{(1)}(t) = \frac{2}{C_0(Z_0+R_0)}\int_T^t U_E \sin(\omega_0(s-T))ds - \frac{1}{C_0 Z_0}\int_T^t \tilde{u}^{(0)}(s)ds + \frac{1}{C_0}\left(0{,}12\int_T^t \tilde{u}^{(0)}(s)ds - 0{,}8\int_T^t \left(\tilde{u}^{(0)}(s)\right)^3 ds\right) -$$

$$-\frac{\delta}{Z_0 C_0}\int_T^t \tilde{u}^{(0)}(s-2T)ds \frac{\delta}{C_0}\left[-0{,}12\int_T^t \tilde{u}^{(0)}(s-2T)ds + 0{,}8\int_T^t \left(\tilde{u}^{(0)}(s-2T)\right)^3 ds\right] + \delta\int_T^t \dot{\tilde{u}}^{(0)}(s-2T)ds -$$

$$-\left(\frac{t-T}{T_0} - \frac{1}{2}\right)\left[\frac{2}{C_0(Z_0+R_0)}\int_T^{T+T_0} E(s)ds - \frac{1}{C_0 Z_0}\int_T^{T+T_0} \tilde{u}^{(0)}(s)ds + \right.$$

$$+\frac{1}{C_0}\left(0{,}12\int_T^{T+T_0}\tilde{u}^{(0)}(s)ds - 0{,}8\int_T^{T+T_0}\left(\tilde{u}^{(0)}(s)\right)^3 ds\right) - \frac{\delta}{Z_0 C_0}\int_T^{T+T_0}\tilde{u}^{(0)}(s-2T)ds -$$

$$\left.-\frac{0{,}12\kappa}{C_0}\int_T^{T+T_0}\tilde{u}^{(0)}(s-2T)ds + \frac{0{,}8\delta}{C_0}\int_T^{T+T_0}\left(\tilde{u}^{(0)}(s-2T)\right)^3 ds + \delta\int_T^{T+T_0}\dot{\tilde{u}}^{(0)}(s-2T)ds\right] =$$

$$= \frac{2U_E}{C_0(Z_0 + R_0)} \int\limits_T^t \sin \omega_0 s\, ds - \frac{\delta}{Z_0 C_0} \int\limits_T^t u_0 \sin(\omega_0 s)\, ds -$$

$$- 0{,}12 \frac{\delta}{C_0} \int\limits_T^t u_0 \sin(\omega_0 s)\, ds + 0{,}8 \frac{\delta}{C_0} \int\limits_T^t (u_0 \sin(\omega_0 s))^3\, ds + \delta\left(u^{(0)}(t-2T) - u^{(0)}(-T)\right) -$$

$$- \left(\frac{t-T}{T_0} - \frac{1}{2}\right) \frac{0{,}8\delta}{C_0} \int\limits_T^{T+T_0} (u_0 \sin(\omega_0 s))^3\, ds =$$

$$= \left(\frac{2U_E}{C_0(Z_0 + R_0)} - \frac{\delta u_0}{Z_0 C_0} - 0{,}12 \frac{\delta u_0}{C_0}\right) \frac{1 - \cos \omega_0 t}{\omega_0} +$$

$$0{,}8 \frac{\delta u_0^3}{C_0}\left(\frac{1 - \cos \omega_0 t}{\omega_0} - \frac{1 - \cos^3 \omega_0 t}{3\omega_0}\right) + \delta u_0 \sin \omega_0 t =$$

$$= \frac{1}{C_0}\left(\frac{2U_E}{Z_0 + R_0} - \frac{\delta u_0}{Z_0} - 0{,}12\delta u_0 + 0{,}8\delta u_0^3\right)\frac{1 - \cos \omega_0 t}{\omega_0} - \frac{0{,}8\delta\, u_0^3}{C_0} \frac{1 - \cos^3 \omega_0 t}{3\omega_0} + \delta u_0 \sin \omega_0 t$$

Further on we have

$$e^{-\mu(t-T)}\left|u^{(0)}(t) - u^{(1)}(t)\right| \le (1 - |\delta|)u_0 + \frac{2}{\omega_0 C_0}\left[\frac{2U_0}{Z_0 + R_0} + |\delta|u_0\left(\frac{1}{Z_0} + 0{,}12\right)\right] + \frac{5{,}6|\delta|u_0^3}{3\omega_0 C_0} \equiv P,$$

that is, $\rho(u^{(1)}, u^{(0)}) \le P$. One can check that $\omega_0 C_0 = \dfrac{5}{\pi}$, and consequently

$$P \le U_0\left[1 - |\delta| + \frac{2}{\omega_0 C_0}\left(\frac{2}{Z_0 + R_0} + \frac{|\delta|}{Z_0} + 0{,}12|\delta| + \frac{5{,}6|\delta|U_0^2}{3}\right)\right] \le U_0\left(1{,}1 + 0{,}032U_0^2\right) \approx 1{,}1.U_0.$$

For the derivative for $t \in [T, T + T_0]$ we have:

$$\dot{u}^{(1)}(t) = \frac{2E(t)}{C_0(Z_0 + R_0)} - \frac{1}{C_0 Z_0} u^{(0)}(t) + \frac{1}{C_0}\left(0{,}12 u^{(0)}(t) - 0{,}8\left(u^{(0)}(t)\right)^3\right) -$$

$$+ \frac{\delta}{C_0}\left[-0{,}12 u^{(0)}(t-2T) + 0{,}8\left(u^{(0)}(t-2T)\right)^3\right] + \delta \dot{u}^{(0)}(t-2T) -$$

$$-\frac{1}{T_0}\left[\frac{2}{C_0(Z_0+R_0)}\int\limits_T^{T+T_0}E(s)ds-\frac{1}{C_0Z_0}\int\limits_T^{T+T_0}u^{(0)}(s)ds+\right.$$

$$\frac{1}{C_0}\left(0,12\int\limits_T^{T+T_0}u^{(0)}(s)ds-0,8\int\limits_T^{T+T_0}\left(u^{(0)}(s)\right)^3ds\right)-$$

$$-\frac{\delta}{Z_0C_0}\int\limits_T^{T+T_0}u^{(0)}(s-2T)ds$$

$$\left.-\frac{0,12\delta}{C_0}\int\limits_T^{T+T_0}u^{(0)}(s-2T)ds+\frac{0,8\delta}{C_0}\int\limits_T^{T+T_0}\left(u^{(0)}(s-2T)\right)^3ds+\delta\int\limits_T^{T+T_0}\dot{u}^{(0)}(s-2T)ds\right]=$$

$$=\frac{2U_E\sin(\omega_0t)}{C_0(Z_0+R_0)}+\frac{\delta}{C_0}\left[-0,12u_0\sin\omega_0t+0,8\left(u_0\sin\omega_0t\right)^3\right]+\delta\dot{u}_0\cos\omega_0t-$$

$$-\frac{1}{T_0}\left[-\frac{\delta}{Z_0C_0}\int\limits_T^{T+T_0}u_0\sin\omega_0s\,ds-\frac{0,12\delta}{C_0}\int\limits_T^{T+T_0}u_0\sin\omega_0s\,ds+\frac{0,8\delta}{C_0}\int\limits_T^{T+T_0}\left(u_0\sin\omega_0s\right)^3ds+\right.$$

$$\left.\int\limits_T^{T+T_0}\dot{u}_0\cos\omega_0s\,ds\right]=$$

$$=\frac{2U_E\sin(\omega_0t)}{C_0(Z_0+R_0)}+\frac{\delta}{C_0}\left[-0,12u_0\sin\omega_0t+0,8\left(u_0\sin\omega_0t\right)^3\right]+\delta\dot{u}_0\cos\omega_0t\,.$$

We notice that

$$e^{-\mu(t-T)}\left|\dot{u}^{(1)}(t)-\dot{u}^{(0)}(t)\right|\leq(1-|\delta|)|u_0|\omega_0+\frac{1}{C_0}\left(\frac{2U_E}{Z_0+R_0}+0,12.|\delta|u_0+0,8.|\delta|u_0^3\right)\leq$$

$$\leq U_0\left[(1-|\delta|)\omega_0+\frac{1}{C_0}\left(\frac{2}{Z_0+R_0}+0,12.|\delta|+0,8.|\delta|U_0^2\right)\right]\leq6,3.10^{11}\equiv\dot{P}\,.$$

Thus $\rho(\dot{u}^{(1)},\dot{u}^{(0)})\leq\dot{P}$ and therefore $\rho_\mu(u^{(1)},u^{(0)})\leq\max\{P,\dot{P}\}\approx6,3.10^{11}$.

In this way we obtain the rate of convergence

$$\rho_\mu^{(k)}(u^{(n+1)\cdot},u^{(n)})\leq\frac{(0,108)^n}{0,9967}\max\{P,\dot{P}\}\approx(0,108)^n.6,3.10^{11}\,.$$

Remark 2.2.8. Note that the above inequalities do not depend on the index of pseudo-metrics k. This means that for the solution on $[3T, 5T]$, $[5T, 7T]$,... the same estimates are valid.

2.3. MEASURABLE PERIODIC REGIMES

Here we approach the same problem, but using the class of measurable functions (cf. P. Halmos [62]). Consider discrete signals as step-wise functions − a particular case with measurable functions. In this approach, one may also avoid the problem with the conformity condition **(CC)** (cf. [85]).

By $L_{T_0}^{\infty,1}[-T,\infty)$ we mean the space consisting of all measurable essentially bounded T_0-periodic functions whose derivatives are $L_{T_0}^{\infty}$-functions (cf. S. L. Sobolev [105]).

We note that in § 2.3 all equalities and inequalities will be satisfied almost everywhere.

Assume that $\upsilon_0(.) \in L_{T_0}^{\infty,1}[-T,T]$ and $T = mT_0$.

Introduce the set

$$M_U^{\infty} = \left\{ u(.) \in L_{T_0}^{\infty}[-T,3T] : u(t) = \upsilon_0(t), t \in [-T,T], |u(t)| \le U_0 e^{\mu(t-T-kT_0)}, t \in [T+kT_0,\ T+(k+1)T_0] \right\}$$

$(k = 0,1,2,...,2m-1$) where U_0, T_0, μ, $\mu T_0 = \mu_0 = $ const. are positive constants.

The set M_U^{∞} turns into a complete metric space (cf. Chapter I, § 1.2) with a metric

$$\hat{\rho}_{\mu}((u,\dot{u}),(\overline{u},\dot{\overline{u}})) = \max\left\{ \hat{\rho}(u,\overline{u}), \rho_{\mu}^{(k)}(\dot{u},\dot{\overline{u}}) : k = 0,1,2,...,2m-1 \right\}$$

where

$$\rho_{\mu}^{(k)}(u,\overline{u}) = \operatorname{ess\,sup}\left\{ e^{-\mu(t-T-kT_0)}|u(t)-\overline{u}(t)| : t \in [T+kT_0, T+(k+1)T_0] \right\}$$

$$\rho^{(k)}(u,\overline{u}) = \operatorname{ess\,sup}\left\{ |u(t)-\overline{u}(t)| : t \in [T+kT_0, T+(k+1)T_0] \right\},$$

$$\hat{\rho}(u,\overline{u}) = \operatorname{ess\,sup}\left\{ |u(t)-\overline{u}(t)| : t \in [T,3T] \right\},$$

$$\rho_{\mu}^{(k)}(\dot{u},\dot{\overline{u}}) = \operatorname{ess\,sup}\left\{ e^{-\mu(t-T-kT_0)}|\dot{u}(t)-\dot{\overline{u}}(t)| : t \in [T+kT_0, T+(k+1)T_0] \right\}$$

Remark 2.3.1. It is easy to verify that the metrics

$$\rho_\mu^{(k)}(u,\overline{u}) = \text{ess sup}\left\{e^{-\mu(t-T-kT_0)}\left|u(t)-\overline{u}(t)\right| : t \in [T+kT_0, T+(k+1)T_0]\right\}$$

and

$$\rho_{L^\infty}^{(k)}(u,\overline{u}) = \text{ess sup}\left\{\left|u(t)-\overline{u}(t)\right| : t \in [T+kT_0, T+(k+1)T_0]\right\}$$

are equivalent. Indeed,

$$e^{-\mu(t-T-kT_0)}\left|u(t)-\overline{u}(t)\right| \le \left|u(t)-\overline{u}(t)\right| \;\Rightarrow\; \rho_\mu^{(k)}(u,\overline{u}) \le \rho_{L^\infty}^{(k)}(u,\overline{u}),$$

that is, $\left|u(t)-\overline{u}(t)\right| \le e^{\mu(t-T-kT_0)}\left|u(t)-\overline{u}(t)\right| \;\Rightarrow\; \rho_{L^\infty}^{(k)}(u,\overline{u}) \le e^{\mu T_0}\rho_\mu^{(k)}(u,\overline{u}).$

Define the operator $B(u)(t)$ on every interval $[T+kT_0, T+(k+1)T_0]$ for $u(.)\in M_U^\infty$ by the formulas:

$$B_k(u)(t) = \upsilon_{0T} + \int_{T+kT_0}^{t} U(u)(s)ds - \frac{t-T-kT_0}{T_0}\int_{T+kT_0}^{T+(k+1)T_0} U(u)(s)ds \;,\, t\in[T+kT_0, T+(k+1)T_0] \qquad (2.3.1)$$

$(k=0,1,2,\ldots,2m-1)$, where $\upsilon_{0T} = \upsilon_0(T)$,

$$U(u)(t) \equiv \frac{2E(t-T)}{C_0(Z_0+R_0)} - \frac{1}{C_0 Z_0}u(t) - \frac{1}{C_0}\sum_{n=1}^{P} r_n[u(t)]^n - \frac{Z_0-R_0}{Z_0 C_0(Z_0+R_0)}\vec{u}(t) +$$

$$+ \frac{Z_0-R_0}{C_0(Z_0+R_0)}\sum_{n=1}^{P} r_n[\vec{u}(t)]^n + \frac{Z_0-R_0}{Z_0+R_0}\dot{\vec{u}}(t),\, t\in[T,3T]$$

and $u(t-2T)$ are replaced by $\vec{\upsilon}_0(t)\equiv\vec{u}(t)$ (cf. Remark 2.2.3).

Remark 2.3.2. We notice that $U(u)(t)$ is a measurable function for every $u(.)\in L_{T_0}^{1,\infty}[-T,\infty)$. Indeed, let us present the right-hand side of (2.2.1) in the form

$$U(u)(t) \equiv \frac{2E(t)}{C_0(Z_0+R_0)} + F(u(t),\vec{u}(t),\dot{\vec{u}}(t)).$$

Function $E(t)$ is measurable by the assumption **(E)** of Theorem 2.2.1. Function $F(u,v,w)$ is continuous. Since $u(t)$, $\vec{u}(t)$, $\dot{\vec{u}}(t)$ are measurable functions then $F(u(t),\vec{u}(t)(t),\dot{\vec{u}}(t-2T))$ is also measurable as a composition of continuous and measurable functions (cf. P. Halmos [62]).

Remark 2.3.3. In this case one can choose the initial function to be a step-wise one. In other words we can consider discrete initial signals.

In the next lemma we assume that the initial function and source function satisfy the condition, as in paragraph § 2.2.

Lemma 2.3.1. The periodic problem

$$\dot{u}(t) = U(u)(t), t \in [T, 3T],$$
$$u(t) = \upsilon_0(t), \ \dot{u}(t) = \dot{\upsilon}_0(t), \ t \in [-T, T] \tag{2.3.2}$$

has a solution $u(.) \in M_U^\infty$ iff the operator B has a fixed point $u \in M_U^\infty$, that is, $u = B(u)$.

Proof: Let $u(.) \in M_U^\infty$ be a solution of the periodic problem, that is, $u(.)$ satisfies (2.3.2). Then after integration on every $[T + kT_0, T + (k+1)T_0]$ we have

$$u(t) = \upsilon_{0T} + \int_{T+kT_0}^{t} U(u)(s)ds \ .$$

For $t = T + (k+1)T_0$ we obtain

$$0 = u(T + (k+1)T_0) - \upsilon_{0T} \equiv u(T + (k+1)T_0) - u(T + kT_0) = \int_{T+kT_0}^{T+(k+1)T_0} U(u)(t)dt \Rightarrow$$

$$\int_{T+kT_0}^{T+(k+1)T_0} U(u)(t)dt = 0.$$

Then (2.3.1) becomes

$$B_k(u)(t) = \upsilon_{0T} + \int_{T+kT_0}^{t} U(u)(s)ds, \ t \in [T + kT_0, T + (k+1)T_0]$$

and consequently

$$u(t) = \upsilon_{0T} + \int_{T+kT_0}^{t} U(u)(s)ds \ t \in [T + kT_0, T + (k+1)T_0], (k = 0,1,2,...,2m-1)$$

that is, $u(.) \in M_U^\infty$ is a fixed point of B.

Conversely, if B has a fixed point $u(.) \in M_U^\infty$, i.e. $u = B(u)$, then repeating the inequalities a.e. from Lemma 2.2.1 we obtain

$$\left| \int_{T+kT_0}^{T+(k+1)T_0} U(u)(s)ds \right| \leq$$

$$\leq \frac{2}{C_0(Z_0+R_0)}\int_{T+kT_0}^{T+(k+1)T_0}|E(t)|dt + \frac{1}{C_0Z_0}\int_{T+kT_0}^{T+(k+1)T_0}|u(t)|dt + \frac{1}{C_0}\sum_{n=1}^{p}|r_n|\int_{T+kT_0}^{T+(k+1)T_0}|u(t)|^n dt +$$

$$+\frac{1}{Z_0C_0}\frac{|Z_0-R_0|}{Z_0+R_0}\int_{T+kT_0}^{T+(k+1)T_0}|\bar{u}(t)|dt + \frac{1}{C_0}\frac{|Z_0-R_0|}{Z_0+R_0}\sum_{n=1}^{p}|r_n|\int_{T+kT_0}^{T+(k+1)T_0}|\bar{u}(t)|^n dt + \frac{|Z_0-R_0|}{Z_0+R_0}\left|\int_{T+kT_0}^{T+(k+1)T_0}\dot{\bar{u}}(t)dt\right| \leq$$

$$\leq \frac{2U_E}{C_0(Z_0+R_0)}\frac{e^{\mu T_0}-1}{\mu} + \frac{U_0}{C_0Z_0}\frac{e^{\mu T_0}-1}{\mu} + \frac{1}{C_0}\sum_{n=1}^{p}|r_n|U_0^n\int_{T+kT_0}^{T+(k+1)T_0}e^{n\mu(t-T-kT_0)}dt +$$

$$+\frac{U_0}{Z_0C_0}\frac{|Z_0-R_0|}{Z_0+R_0}\frac{e^{\mu T_0}-1}{\mu} + \frac{1}{C_0}\frac{|Z_0-R_0|}{Z_0+R_0}\sum_{n=1}^{p}|r_n|U_0^n\int_{T+kT_0}^{T+(k+1)T_0}e^{n\mu(t-T-kT_0)}dt +$$

$$+\frac{|Z_0-R_0|}{Z_0+R_0}|u(T+(k+1)T_0-2T)-u(T+kT_0-2T)| \leq$$

$$\leq \frac{2U_E}{C_0(Z_0+R_0)}\frac{e^{\mu T_0}-1}{\mu} + \frac{U_0}{C_0Z_0}\frac{e^{\mu T_0}-1}{\mu} + \frac{1}{C_0}\sum_{n=1}^{p}|r_n|U_0^n\frac{e^{n\mu T_0}-1}{n\mu} +$$

$$+\frac{U_0}{C_0Z_0}\frac{|Z_0-R_0|}{Z_0+R_0}\frac{e^{\mu T_0}-1}{\mu} + \frac{1}{C_0}\frac{|Z_0-R_0|}{Z_0+R_0}\sum_{n=1}^{p}|r_n|U_0^n\frac{e^{n\mu T_0}-1}{n\mu} \leq$$

$$\leq \frac{2U_E}{C_0(Z_0+R_0)}\frac{e^{\mu T_0}-1}{\mu} + \frac{U_0}{C_0Z_0}\frac{e^{\mu T_0}-1}{\mu} + \frac{1}{C_0}\frac{e^{\mu T_0}-1}{\mu}\sum_{n=1}^{p}|r_n|U_0^n e^{(n-1)\mu T_0} +$$

$$+\frac{U_0}{C_0Z_0}\frac{|Z_0-R_0|}{Z_0+R_0}\frac{e^{\mu T_0}-1}{\mu} + \frac{1}{C_0}\frac{|Z_0-R_0|}{Z_0+R_0}\sum_{n=1}^{p}|r_n|U_0^n\frac{(e^{\mu T_0}-1)e^{(n-1)\mu T_0}}{\mu} \leq$$

$$\leq \frac{e^{\mu T_0}-1}{\mu C_0}\left[\frac{2U_E}{Z_0+R_0}+\left(1+\frac{|Z_0-R_0|}{Z_0+R_0}\right)\left(\frac{U_0}{Z_0}+\sum_{n=1}^{p}|r_n|\left(Ue^{\mu T_0}\right)^n\right)\right] \equiv M(\mu).$$

We have proved that $\gamma \leq M(\mu) \Rightarrow \int_{T+kT_0}^{T+(k+1)T_0}U(u)(s)ds = 0$.

Further on, the proof can be accomplished as the one of Lemma 2.2.1.

Theorem 2.3.1. Let the following conditions be fulfilled:

(IN) The initial function $\upsilon_0(.) \in L_{T_0}^{\infty}[-T,T]$, $|\upsilon_{0T}| < U_0$;

$$\left|\upsilon_0(t)\right| \le U_0 e^{\mu(t+T-kT_0))} \ , t \in [-T+kT_0, -T+(k+1)T_0], (k=0,1,...,2m-1);$$

(E) $E(.) \in L^\infty_{T_0}(0,\infty), \left|E(t)\right| \le U_E e^{\mu(t-T-kT_0)}(k=0,1,2,...), t \in [T+kT_0, T+(k+1)T_0].$

In this case, there is a unique measurable T_0-periodic solution of the initial value problem (2.3.2) belonging to M_U^∞.

Proof: We show that B maps M_U^∞ into itself.

Indeed, first we notice that $B(u)(t)$ is a measurable function as a composition of continuous and measurable functions. Also for $u \in M_U^\infty$ the function $B(u)(t)$ is T_0-periodic on $[T,\infty)$. Indeed, let $t \in [T+kT_0, T+(k+1)T_0]$ then

$t+T_0 \in [T+(k+1)T_0, T+(k+2)T_0].$

Therefore

$$B(u)(t+T_0) = B_{k+1}(u)(t+T_0) = \upsilon_{0T} + \int_{T+(k+1)T_0}^{t+T_0} U(u)(s))ds - \frac{t+T_0-T-(k+1)T_0}{T_0} \int_{T+(k+1)T_0}^{T+(k+2)T_0} U(u)(s))ds \cdot$$

Let change the variable in the intgrals $s-T_0 = \theta.$ Then

$$B_{k+1}(u)(t+T_0) = \upsilon_{0T} + \int_{T+(k+1)T_0}^{t+T_0} U(u)(s))ds - \frac{t+T_0-T-(k+1)T_0}{T_0} \int_{T+(k+1)T_0}^{T+(k+2)T_0} U(u)(s))ds =$$

$$= \upsilon_{0T} + \int_{T+kT_0}^{t} U(u)(\theta-T_0))d\theta - \frac{t-T-kT_0}{T_0} \int_{T+kT_0}^{T+(k+1)T_0} U(u)(\theta-T_0))d\theta =$$

$$= \upsilon_{0T} + \int_{T+kT_0}^{t} U(u)(\theta))d\theta - \frac{t-T-kT_0}{T_0} \int_{T+kT_0}^{T+(k+1)T_0} U(u)(\theta))d\theta = B_k(u)(t)(t).$$

Thus, the equalities are satisfied almost everywhere.

We next have to show that $\left|B_k(u)(t)\right| \le U_0 e^{\mu(t-T-kT_0)}.$

For every $u(.) \in M_U^\infty$ and $t \in [T+kT_0, T+(k+1)T_0]$ proceeding as in Theorem 2.2.1 we obtain

$$\left|B_k(u)(t)\right| \le$$

$$\leq |v_{0T}| + e^{\mu(t-T-kT_0)} \frac{1}{\mu C_0}\left(\frac{2U_E}{Z_0+R_0} + \left(1+\frac{|Z_0-R_0|}{Z_0+R_0}\right)\left(\frac{U_0}{Z_0} + \sum_{n=1}^{p} |r_n| U_0^n e^{(n-1)\mu T_0}\right)\right) +$$

$$+ e^{\mu(t-T-kT_0)} \frac{e^{\mu T_0}-1}{\mu C_0}\left(\frac{2U_E}{Z_0+R_0} + \left(1+\frac{|Z_0-R_0|}{Z_0+R_0}\right)\left(\frac{U_0}{Z_0} + \sum_{n=1}^{p} |r_n| U_0^n e^{(n-1)\mu T_0}\right)\right) +$$

$$+ \frac{|Z_0-R_0|}{Z_0+R_0} U_0 \, e^{\mu(t-T-kT_0)} \leq$$

$$\leq e^{\mu(t-T-kT_0)}\left[|v_{0T}| + \frac{e^{\mu_0}}{\mu C_0}\left(\frac{2U_E}{Z_0+R_0} + \left(1+\frac{|Z_0-R_0|}{Z_0+R_0}\right)\left(\frac{U_0}{Z_0} + \sum_{n=1}^{p} |r_n| U_0^n e^{(n-1)\mu_0}\right)\right) + \frac{|Z_0-R_0|}{Z_0+R_0} U_0 \right] \leq e^{\mu(t-T-kT_0)} U_0$$

for sufficiently large $\mu > 0$.

Consequently the operator B maps M_U^{∞} into itself.

To show that B is a contractive operator we choose $u(.),\overline{u}(.) \in M_U^{\infty}$ and get:

$$\left| B_k(u)(t) - B_k(\overline{u})(t)\right| \leq$$

$$\left| \int_{T+kT_0}^{t} \bigl(U(u)(s) - U(\overline{u})(s)\bigr)ds\right| + \left| \int_{T+kT_0}^{T+(k+1)T_0} \bigl(U(u)(s) - U(\overline{u})(s)\bigr)ds\right| \equiv V_1 + V_2.$$

We have

$$V_1 \leq \frac{1}{C_0 Z_0} \int_{T+kT_0}^{t} |u(s)-\overline{u}(s)|\,ds + \frac{1}{C_0}\sum_{n=1}^{p} |r_n| \int_{T+kT_0}^{t} |u^n(s)-\overline{u}^n(s)|\,ds +$$

$$+ \frac{|Z_0-R_0|}{Z_0 C_0 (Z_0+R_0)} \int_{T+kT_0}^{t} |\overline{u}(s)-\overline{u}(s)|\,ds +$$

$$+ \frac{|Z_0-R_0|}{C_0(Z_0+R_0)}\sum_{n=1}^{p} |r_n| \int_{T+kT_0}^{t} |\overline{u}^n(s)-\overline{u}^n(s)|\,ds + \frac{|Z_0-R_0|}{Z_0+R_0}\left| \int_{T+kT_0}^{t} \bigl(\dot{\overline{u}}(s)-\dot{\overline{u}}(s)\bigr)ds\right| \leq$$

$$\leq \frac{\rho_\mu^{(k)}(u,\overline{u})}{C_0 Z_0} \int_{T+kT_0}^{t} e^{\mu(s-T-kT_0)}\,ds +$$

$$+ \frac{1}{C_0}\sum_{n=1}^{p} |r_n|\, n\, \mathrm{ess}\sup\left\{ |u(s)|^{n-1} : s \in [T+kT_0, T+(k+1)T_0\right\} \int_{T+kT_0}^{t} |u(s)-\overline{u}(s)|\,ds \leq$$

$$\leq \frac{\rho_\mu^{(k)}(u,\overline{u})}{C_0 Z_0} \frac{e^{\mu(t-T-kT_0)}-1}{\mu} + \frac{\rho_\mu^{(k)}(u,\overline{u})}{C_0} \frac{e^{\mu(t-T-kT_0)}-1}{\mu} \sum_{n=1}^{p} |r_n|\, n\, U_0^{n-1} e^{(n-1)\mu T_0} \leq$$

$$\leq \frac{e^{\mu(t-T-kT_0)}}{\mu C_0}\,\frac{\rho_\mu^{(k)}(\dot{u},\dot{\overline{u}})}{\mu}\left(\frac{1}{Z_0}+\sum_{n=1}^{p}\left|r_n\right|n\,U_0^{n-1}e^{(n-1)\mu_0}\right)$$

and

$$V_2 \leq \frac{1}{C_0 Z_0}\int_{T+kT_0}^{T+(k+1)T_0}\left|u(s)-\overline{u}(s)\right|ds+\frac{1}{C_0}\sum_{n=1}^{p}\left|r_n\right|\int_{T+kT_0}^{T+(k+1)T_0}\left|u^n(s)-\overline{u}^n(s)\right|ds \leq$$

$$\leq \frac{\rho_\mu^{(k)}(u,\overline{u})}{C_0 Z_0}\int_{T+kT_0}^{T+(k+1)T_0}e^{\mu(s-TskT_0)}ds+$$

$$+\frac{1}{C_0}\sum_{n=1}^{p}\left|r_n\right|n\,\mathrm{ess\,sup}\left\{\left|u(s)\right|^{n-1}:s\in[T+kT_0,T+(k+1)T_0\right\}\int_{T+kT_0}^{T+(k+1)T_0}\left|u(s)-\overline{u}(s)\right|ds \leq$$

$$\leq \frac{\rho_\mu^{(k)}u,\overline{u})}{C_0 Z_0}\frac{e^{\mu T_0}-1}{\mu}+\frac{\rho_\mu^{(k)}(u,\overline{u})}{C_0}\frac{e^{\mu T_0}-1}{\mu}\sum_{n=1}^{p}\left|r_n\right|n\,U_0^{n-1}e^{(n-1)\mu T_0} \leq$$

$$\leq e^{\mu(t-T-kT_0)}\frac{\rho_\mu^{(k)}(\dot{u},\dot{\overline{u}})}{\mu}\frac{e^{\mu_0}-1}{\mu C_0}\left(\frac{1}{Z_0}+\sum_{n=1}^{p}\left|r_n\right|n\,U_0^{n-1}e^{(n-1)\mu_0}\right).$$

Therefore

$$\left|B_k(u)(t)-B_k(\overline{u})(t)\right| \leq$$

$$\leq e^{\mu(t-T-kT_0)}\frac{\rho_\mu^{(k)}(\dot{u},\dot{\overline{u}})}{\mu}\frac{1}{\mu C_0}\left(\frac{1}{Z_0}+\sum_{n=1}^{p}\left|r_n\right|n\,U_0^{n-1}e^{(n-1)\mu T_0}\right)+$$

$$+e^{\mu(t-T-kT_0)}\frac{\rho_\mu^{(k)}(\dot{u},\dot{\overline{u}})}{\mu}\frac{e^{\mu_0}-1}{\mu C_0}\left(\frac{1}{Z_0}+\sum_{n=1}^{p}\left|r_n\right|n\,U_0^{n-1}e^{(n-1)\mu_0}\right) \leq$$

$$\leq e^{\mu(t-T-kT_0)}\rho_\mu^{(k)}(\dot{u},\dot{\overline{u}})\frac{e^{\mu_0}}{\mu^2 C_0}\left(\frac{1}{Z_0}+\sum_{n=1}^{p}\left|r_n\right|n\,U_0^{n-1}e^{(n-1)\mu_0}\right)\leq e^{\mu(t-T-kT_0)}K_U\,\hat{\rho}_\mu((u,\dot{u}),(\overline{u},\dot{\overline{u}}))\,.$$

Thus it follows that $\hat{\rho}(B(u),B(\overline{u}))\leq e^{\mu_0}K_U\,\hat{\rho}_\mu((u,\dot{u}),(\overline{u},\dot{\overline{u}}))$.

For the derivatives, we have

$$\left|\dot{B}_k(u)(t)-\dot{B}_k(\overline{u})(t)\right|\leq e^{\mu(t-T-kT_0)}\frac{1}{\mu C_0}\left(\frac{1}{Z_0}+\sum_{n=1}^{p}\left|r_n\right|n\left(U_0 e^{\mu_0}\right)^{n-1}\right)\hat{\rho}_\mu((u,\dot{u}),(\overline{u},\dot{\overline{u}}))+$$

$$+ e^{\mu(t-T-kT_0)} \frac{e^{\mu_0}-1}{\mu_0} \frac{1}{\mu C_0} \left(\frac{1}{Z_0} + \sum_{n=1}^{p} |r_n| n \left(U_0 e^{\mu_0}\right)^{n-1} \right) \hat{\rho}_\mu((u,\dot{u}),(\overline{u},\dot{\overline{u}})) \le$$

$$\le e^{\mu(t-T-kT_0)} \left(1 + \frac{e^{\mu_0}-1}{\mu_0} \right) \frac{1}{\mu C_0} \left(\frac{1}{Z_0} + \sum_{n=1}^{p} |r_n| n \left(U_0 e^{\mu_0}\right)^{n-1} \right) \hat{\rho}_\mu((u,\dot{u}),(\overline{u},\dot{\overline{u}})) \equiv$$

$$\equiv e^{\mu(t-T-kT_0)} \dot{K}_U \, \hat{\rho}_\mu((u,\dot{u}),(\overline{u},\dot{\overline{u}}))$$

or $\rho_\mu^{(k)}(\dot{B}_k(u),\dot{B}_k(\overline{u})) \le \dot{K}_U \, \hat{\rho}_\mu((u,\dot{u}),(\overline{u},\dot{\overline{u}}))$.

As in Theorem 2.2.1, we conclude

$$\hat{\rho}_\mu((Bu,\dot{B}u),(B\overline{u},\dot{B}\overline{u})) \le K \, \hat{\rho}_\mu((u,\dot{u}),(\overline{u},\dot{\overline{u}}))$$

where $K = \max\{e^{\mu_0} K_U, \dot{K}_U\} < 1$ and the unique fixed point of B is a measurable T_0-periodic solution of (2.1.11).

Theorem 2.3.1 is thus proved.

Remark 2.3.4. Let us compare the just obtained inequalities with those of the previous paragraph, for the data from the previous numerical example

$$|v_{0T}| + \frac{e^{\mu_0}}{\mu C_0} \left[\frac{2U_E}{Z_0+R_0} + \left(1 + \frac{|Z_0-R_0|}{Z_0+R_0} \right) \left(\frac{U_0}{Z_0} + \sum_{n=1}^{p} |r_n| U_0^n e^{(n-1)\mu_0} \right) \right] + \frac{|Z_0-R_0|}{Z_0+R_0} U_0 \le U_0 ;$$

$$K_U = \frac{e^{\mu_0}}{\mu^2 C_0} \left(\frac{1}{Z_0} + \sum_{n=1}^{3} |r_n| n \left(U_0 e^{\mu_0}\right)^{n-1} \right);$$

$$\dot{K}_U = \left(1 + \frac{e^{\mu_0}-1}{\mu_0} \right) \frac{1}{\mu C_0} \left(\frac{1}{Z_0} + \sum_{n=1}^{p} |r_n| n \left(U_0 e^{\mu_0}\right)^{n-1} \right) < 1$$

with continuously differentiable case from Theorem 2.2.1

$$\frac{e^{\mu_0}}{\mu C_0} \left[\frac{2U_E}{Z_0+R_0} + \left(1 + \frac{|Z_0-R_0|}{Z_0+R_0} \right) \left(\frac{U_0}{Z_0} + \sum_{n=1}^{3} |r_n| (U_0)^n e^{(n-1)\mu_0} \right) \right] + U_0 \frac{|Z_0-R_0|}{Z_0+R_0} \le U_0 ;$$

$$K_U = \frac{e^{\mu_0}}{\mu^2 C_0} \left(\frac{1}{Z_0} + \sum_{n=1}^{p} |r_n| n U_0^{n-1} e^{(n-1)\mu_0} \right);$$

$$\dot{K}_U = \left(1 + \frac{e^{\mu_0}-1}{\mu_0}\right)\frac{1}{\mu C_0}\left(\frac{1}{Z_0} + \sum_{n=1}^{p}|r_n|n\left(U_0 e^{\mu_0}\right)^{n-1}\right) < 1.$$

It is easy to see that the rate of convergence is the same, but in measurable cases the conformity condition **(CC)** is already unnecessary.

2.4. OSCILLATORY REGIMES FOR NEUTRAL EQUATION WITH POLYNOMIAL NONLINEARITY

Here we deal with the problem of oscillating solutions of (2.1.1), that is:

$$\dot{u}(t) = \frac{2E(t)}{C_0(Z_0+R_0)} - \frac{u(t)}{C_0 Z_0} - \frac{1}{C_0}\sum_{n=1}^{p} r_n[u(t)]^n - \frac{Z_0-R_0}{Z_0 C_0(Z_0+R_0)}u(t-2T)+$$

$$+\frac{Z_0-R_0}{C_0(Z_0+R_0)}\sum_{n=1}^{p} r_n[u(t-2T)]^n + \frac{Z_0-R_0}{Z_0+R_0}\dot{u}(t-2T),\ t\in[T,\infty) \qquad (2.4.1)$$

$$u(t) = \upsilon_0(t),\ \dot{u}(t) = \dot{\upsilon}_0(t),\ t\in[-T,T].$$

Let us put $t_0 \equiv T$. Now we are able to formulate the main problem: to find an oscillatory solution for problem (2.4.1) with advanced prescribed zeros on an interval $[t_0,\infty)$, where $\upsilon_0(t)$ is a prescribed initial oscillating function on the interval $[-t_0,t_0]$.

Let $S_T = \{\tau_k\}_{k=0}^{n}, n\in N$ be the set of zeros of the initial function, that is, $\upsilon_0(\tau_k) = 0$ such that $\tau_0 = -T$, $\tau_n = T \equiv t_0$. Besides $\max\{\tau_{k+1} - \tau_k : k = 0,1,...,n\} \le T_0$.

Let $S = \{t_k\}_{k=0}^{\infty}$ be a strictly increasing sequence of real numbers satisfying the following conditions **(C)**:

(C1) $\lim\limits_{k\to\infty} t_k = \infty$;

(C2) for every k $\left(t_k - T \ge \tau_0\right)$ there is $s < k$ such that $t_k - T = t_s$ where $t_s \in S_T \cup S$.

Condition **(C2)** implies

$$0 < \inf\{t_{k+1} - t_k : k = 0,1,2,...\} \le \sup\{t_{k+1} - t_k : k = 0,1,2,...\} = T_0 < \infty.$$

Introduce the set $C^1[t_0,\infty)$ consisting of all continuous functions whose derivatives are bounded and continuous on every interval $[t_k,t_{k+1}]$. We note the right and left derivatives at

t_k of these functions might not coincide. For this reason, we introduce below a topology of uniform convergence on every interval $[t_k, t_{k+1}]$ of the derivatives that needs the introduction of uniform spaces (cf. [14]). In fact, we introduce a countable family of pseudo-metrics, as there are the intervals $[t_k, t_{k+1}]$.

Let us consider the set

$$M_S = \left\{ u(.) \in C^1[t_0, \infty) : u(t_k) = 0 \ (k = 0,1,2,...) \right\}$$

and

$$M_{SU} = \left\{ u(.) \in M_S : |u(t)| \le U_0 e^{\mu(t-t_k)}, t \in [t_k, t_{k+1}] \right\},$$

where U_0, μ are positive constants and $\mu_0 = \mu T_0 = \text{const}.$

Introduce the following family of pseudo-metrics

$$\rho^{(k)}(u, \overline{u}) = \max \left\{ |u(t) - \overline{u}(t)| : t \in [t_k, t_{k+1}] \right\},$$

$$\hat{\rho}^{(k)}(u, \overline{u}) = \max \left\{ |u(t) - \overline{u}(t)| : t \in [t_0, t_{k+1}] \right\},$$

$$\rho_\mu^{(k)}(u, \overline{u}) = \max \left\{ e^{-\mu(t-t_k)} |u(t) - \overline{u}(t)| : t \in [t_k, t_{k+1}] \right\},$$

$$\hat{\rho}_\mu^{(k)}(u, \overline{u}) = \max \left\{ \rho_\mu^{(0)}(u, \overline{u}), \rho_\mu^{(1)}(u, \overline{u}),..., \rho_\mu^{(k)}(u, \overline{u}) \right\},$$

$$\rho_\mu^{(k)}(\dot{u}, \dot{\overline{u}}) = \max \left\{ e^{-\mu(t-t_k)} |\dot{u}(t) - \dot{\overline{u}}(t)| : t \in [t_k, t_{k+1}] \right\},$$

$$\hat{\rho}_\mu^{(k)}(\dot{u}, \dot{\overline{u}}) = \max \left\{ \rho_\mu^{(0)}(\dot{u}, \dot{\overline{u}}), \rho_\mu^{(1)}(\dot{u}, \dot{\overline{u}}),..., \rho_\mu^{(k)}(\dot{u}, \dot{\overline{u}}) \right\}.$$

Remark 2.4.1. The following inequalities imply the equivalence of both families of pseudo-metrics

$$\rho_\mu^{(k)}(u, \overline{u}) \le \rho^{(k)}(u, \overline{u}) \le e^{\mu_0} \rho_\mu^{(k)}(u, \overline{u}), \ (k = 0,1,2,...).$$

It is easy to verify that

$$\hat{\rho}^{(k)}(u, \overline{u}) = \max \left\{ \rho^{(0)}(u, \overline{u}), \rho^{(1)}(u, \overline{u}),..., \rho^{(k)}(u, \overline{u}) \right\} \le$$
$$\le e^{\mu_0} \max \left\{ \rho_\mu^{(0)}(u, \overline{u}), \rho_\mu^{(1)}(u, \overline{u}),..., \rho_\mu^{(k)}(u, \overline{u}) \right\} = e^{\mu_0} \hat{\rho}_\mu^{(l)}(u, \overline{u}),$$

where l is some integer from $\{0,1,2,...,k\}$.

The set M_{SU} turns into a complete uniform space with respect to the saturated family of pseudo-metrics

$$\hat{\rho}_{\mu}^{(k)}((u,\dot{u}),(\overline{u},\dot{\overline{u}})) = \max\{\hat{\rho}^{(k)}(u,\overline{u}),\hat{\rho}_{\mu}^{(k)}(\dot{u},\dot{\overline{u}})\},(k=0,1,2,...).$$

Obviously the index set of the saturated family of pseudo-metrics is the set of ordered pairs $A = \{[t_0,t_{k+1}],[t_k,t_{k+1}]\}_{k=0}^{\infty}$.

Define an operator B by the formulas

$$B(u)(t):= \int_{t_k}^{t} U(u)(s)ds - \frac{t-t_k}{t_{k+1}-t_k} \int_{t_k}^{t_{k+1}} U(u)(s)ds, \ \ t\in[t_k,t_{k+1}], \ (k=0,1,2,...\ ),$$

where

$$(Uu)(t) = \frac{2E(t)}{C_0(Z_0+R_0)} - \frac{u(t)}{C_0 Z_0} - \frac{1}{C_0}\sum_{n=1}^{p} r_n[u(t)]^n - \frac{Z_0-R_0}{Z_0 C_0(Z_0+R_0)}\overline{u}(t)+$$

$$+ \frac{Z_0-R_0}{C_0(Z_0+R_0)}\sum_{n=1}^{p} r_n[\overline{u}(t)]^n + \frac{Z_0-R_0}{Z_0+R_0}\dot{\overline{u}}(t), \ t\in[T,\infty)$$

.

In this manner we consider the problem to the right of the initial point and avoid the difficulty with the conformity condition **(CC)**

$$\frac{du(t_0)}{dt} = \frac{2E(t_0)}{C_0(Z_0+R_0)} - \frac{u(t_0)}{C_0 Z_0} - \frac{1}{C_0}\sum_{n=1}^{p} r_n[u(t_0)]^n - \frac{(Z_0-R_0)u(-t_0)}{Z_0 C_0(Z_0+R_0)}+$$

$$+ \frac{Z_0-R_0}{C_0(Z_0+R_0)}\sum_{n=1}^{p} r_n[u(-t_0)]^n + \frac{Z_0-R_0}{Z_0+R_0}\frac{du(-t_0)}{dt}$$

,

so the restrictive condition $u(-t_0) = \dot{u}(-t_0) = 0 \Rightarrow u(t_0) = \dot{u}(t_0) = 0$ becomes unnecessary.

We have already proved in Chapter I that M_{SU} is a closed subset of $C^1[t_0,\infty)$ with respect to the above family of pseudo-metrics.

Lemma 2.4.1. Let $E(.) \in C^1_{ST \cup S}[-T,\infty), |E(t)| \leq U_E e^{\mu(t-t_k)}, t\in[t_k,t_{k+1}],(k=0,1,...)$. In that case, problem (2.4.1) has a solution $u(.) \in M_{SU}$ iff the operator B has a fixed point in M_{SU}, that is,

$$u(t) = B(u)(t). \tag{2.4.2}$$

Proof: Let $u(.) \in M_{SU}$ be a solution of (2.4.1). Integrating (2.4.1) on every interval $[t_k, t] \subset [t_k, t_{k+1}]$ $(k = 0,1,2 \dots)$ we obtain

$$u(t) - u(t_k) = \int_{t_k}^{t} U(u)(s)ds \;\Leftrightarrow\; u(t) = \int_{t_k}^{t} U(u)(s)ds \,.$$

Consequently

$$u(t) = \int_{t_k}^{t} U(u)(s)ds \;\Rightarrow\; 0 = u(t_{k+1}) = \int_{t_k}^{t_{k+1}} U(u)(s)ds \;\Rightarrow\; \int_{t_k}^{t_{k+1}} U(u)(s)ds = 0 \,. \qquad (2.4.3)$$

Therefore, $u(t)$ satisfies

$$u(t) = \int_{t_k}^{t} U(u)(s)ds - \frac{t - t_k}{t_{k+1} - t_k} \int_{t_k}^{t_{k+1}} U(u)(s)ds, \; t \in [t_k, t_{k+1}] \Leftrightarrow u = B(u),$$

that is, $u(.)$ is a fixed point of B.

Conversely, let $u(.) \in M_{SU}$ be a solution of $u = B(u)$, that is,

$$u(t) = \int_{t_k}^{t} U(u)(s)ds - \frac{t - t_k}{t_{k+1} - t_k} \int_{t_k}^{t_{k+1}} U(u)(s)ds, t \in [t_k, t_{k+1}] \,.$$

Then recall that with $\mu_0 = \mu T_0$, we obtain

$$\left| \int_{t_k}^{t_{k+1}} U(u)(s)ds \right| \leq$$

$$\leq \frac{2}{C_0(Z_0 + R_0)} \int_{t_k}^{t_{k+1}} |E(t)|dt + \frac{1}{C_0 Z_0} \int_{t_k}^{t_{k+1}} |u(t)|dt + \frac{1}{C_0} \sum_{n=1}^{p} |r_n| \int_{t_k}^{t_{k+1}} |u(t)|^n dt +$$

$$+ \frac{1}{Z_0 C_0} \frac{|Z_0 - R_0|}{Z_0 + R_0} \int_{t_k}^{t_{k+1}} |u(t - 2T)|dt +$$

$$+ \frac{1}{C_0} \frac{|Z_0 - R_0|}{Z_0 + R_0} \sum_{n=1}^{p} |r_n| \int_{t_k}^{t_{k+1}} |u(t - 2T)|^n dt + \frac{|Z_0 - R_0|}{Z_0 + R_0} \left| \int_{t_k}^{t_{k+1}} \dot{u}(t - 2T)dt \right| \leq$$

$$\leq \frac{2U_E e^{\mu 0}}{C_0(Z_0+R_0)}\frac{e^{\mu(t_{k+1}-t_k)}-1}{\mu}+\frac{U_0}{C_0 Z_0}\frac{e^{\mu(t_{k+1}-t_k)}-1}{\mu}+\frac{1}{C_0}\sum_{n=1}^{p}|r_n|U_0^n\int_{t_k}^{t_{k+1}}e^{n\mu(t-t_k)}dt+$$

$$+\frac{U_0 e^{\mu 0}}{Z_0 C_0}\frac{|Z_0-R_0|}{Z_0+R_0}\frac{e^{\mu(t_{k+1}-t_k)}-1}{\mu}+\frac{1}{C_0}\frac{|Z_0-R_0|}{Z_0+R_0}\sum_{n=1}^{p}|r_n|U_0^n e^{n\mu 0}\int_{t_k}^{t_{k+1}}e^{\mu(t-t_k)}dt+\frac{|Z_0-R_0|}{Z_0+R_0}\left|u(t_{k+1}-2T)-u(t_k-2T)\right|\leq$$

$$\leq \frac{2U_E e^{\mu 0}}{C_0(Z_0+R_0)}\frac{e^{\mu T_0}-1}{\mu}+\frac{U_0}{C_0 Z_0}\frac{e^{\mu T_0}-1}{\mu}+\frac{1}{C_0}\sum_{n=1}^{p}|r_n|U_0^n\frac{e^{n\mu T_0}-1}{n\mu}+$$

$$+\frac{U_0 e^{\mu 0}}{Z_0 C_0}\frac{|Z_0-R_0|}{Z_0+R_0}\frac{e^{\mu T_0}-1}{\mu}+\frac{1}{C_0}\frac{|Z_0-R_0|}{Z_0+R_0}\sum_{n=1}^{p}|r_n|U_0^n\sum_{n=1}^{p}|r_n|U_0^n e^{n\mu 0}\frac{e^{\mu T_0}-1}{\mu}\leq$$

$$\leq \frac{e^{\mu T_0}-1}{\mu}\left(\frac{2U_E e^{\mu 0}}{C_0(Z_0+R_0)}+\frac{U_0}{C_0 Z_0}+\frac{1}{C_0}\sum_{n=1}^{p}|r_n|U_0^n\frac{ne^{(n-1)\mu T_0}}{n}+\right.$$

$$\left.+\frac{U_0 e^{\mu 0}}{Z_0 C_0}\frac{|Z_0-R_0|}{Z_0+R_0}+\frac{e^{\mu 0}}{C_0}\frac{|Z_0-R_0|}{Z_0+R_0}\sum_{n=1}^{p}|r_n|U_0^n e^{(n-1)\mu 0}\right)\leq$$

$$\leq \frac{e^{\mu 0}-1}{\mu C_0}\left[\frac{2U_E e^{\mu 0}}{Z_0+R_0}+\left(1+e^{\mu 0}\frac{|Z_0-R_0|}{Z_0+R_0}\right)\left(\frac{U_0}{Z_0}+\sum_{n=1}^{p}|r_n|U_0^n e^{(n-1)\mu 0}\right)\right]\equiv M(\mu).$$

Let us assume that $\left|\int_{t_k}^{t_{k+1}}U(u)(t)dt\right|=\gamma>0$. We have just obtained that $\gamma\leq M(\mu)$.

Then for a sufficiently large $\mu>0$ (and sufficiently small $T_0>0$) we can reach the inequality $M(\mu)<\gamma$. The obtained contradiction implies $\int_{t_k}^{t_{k+1}}U(u)(t)dt=0$. It follows that

$$u(t)=\int_{t_k}^{t}U(u)(s)ds,$$ and after a differentiation we obtain (2.4.1).

Lemma 2.4.1 is thus proved.

Remark 2.4.2. We could define the operator B in the following manner:

$$B_k(u)(t)=\int_{t_k}^{t}U(u)(s)ds-\left(\frac{t-t_k}{t_{k+1}-t_k}-\frac{1}{2}\right)\int_{t_k}^{t_{k+1}}U(u)(s)ds-\frac{1}{t_{k+1}-t_k}\int_{t_k}^{t_{k+1}}\int_{t_k}^{t}U(u)(s)dsdt,\ t\in[t_k,t_{k+1}]$$

In that case it becomes easy to verify that $\int_{t_k}^{t_{k+1}} B_k(u)(s)ds = 0$. This means that the

function space might be chosen such that $\int_{t_k}^{t_{k+1}} u(s)ds = 0$ which implies that every function

$u(s)$ changes the sign, at least in $[t_k, t_{k+1}]$. Indeed,

$$\int_{t_k}^{t_{k+1}} B_k(u)(t)dt = \int_{t_k}^{t_{k+1}} \int_{t_k}^{t} U(u)(s)ds - \int_{t_k}^{t_{k+1}} \left(\frac{t-t_k}{t_{k+1}-t_k} - \frac{1}{2} \right) dt \int_{t_k}^{t_{k+1}} U(u)(s)ds - \int_{t_k}^{t_{k+1}} \int_{t_k}^{t} U(u)(s)dsdt =$$

$$= -\frac{t_{k+1}-t_k}{2} \left(\frac{t-t_k}{t_{k+1}-t_k} - \frac{1}{2} \right)^2 \Bigg|_{t_k}^{t_{k+1}} = 0.$$

Theorem 2.4.1. Let the following conditions be fulfilled:

1) The initial function $\upsilon_0(.) \in C^1[-T,T]$ satisfies condition **(IN)** and $\upsilon_0(-t_0) = 0 \Rightarrow \upsilon_0(t_0) = 0$;

2) $E(.) \in C_S^1[-T,\infty), E(t_k - T) = E(t_q); |E(t)| \le U_0 e^{\mu(t-t_k)}, t \in [t_k, t_{k+1}], E(t_k) = 0, (k = 0,1,...)$.

In that case, there exists a unique oscillatory solution of (2.4.1), belonging to M_{SU}.

Proof: We show that B maps M_{SU} into itself, that is, $u \in M_{SU}$ implies $B(u) \in M_{SU}$.

First we notice that $B(u)(t)$ is continuous on $[t_0, \infty)$:

$$B_0(u)(t_0) = \int_{t_0}^{t_0} U(u)(s)ds - \frac{t_0 - t_0}{t_1 - t_0} \int_{t_0}^{t_1} U(u)(s)ds = 0,$$

$$\lim_{t \to t_{k+1}(t<t_{k+1})} B_k(u)(t) = \lim_{t \to t_{k+1}(t<t_{k+1})} \left(\int_{t_k}^{t} U(u)(s)ds - \frac{t-t_k}{t_{k+1}-t_k} \int_{t_k}^{t_{k+1}} U(u)(s)ds \right) = 0,$$

$$\lim_{t \to t_{k+1}(t>t_{k+1})} B_{k+1}(u)(t) = \lim_{t \to t_{k+1}(t>t_{k+1})} \left(\int_{t_{k+1}}^{t} U(u)(s)ds - \frac{t-t_{k+1}}{t_{k+2}-t_{k+1}} \int_{t_{k+1}}^{t_{k+2}} U(u)(s)ds \right) = 0$$

and $B(u)(t)$ is differentiable on every (t_k, t_{k+1}).

It is also easy to verify that $B(u)(t_k) = 0$ and $B(u)(t_{k+1}) = 0$.

We have to establish that $|B_k(u)(t)| \le U_0 e^{\mu(t-t_k)}, t \in [t_k, t_{k+1}]$.

Since $\left|\dfrac{t-t_k}{t_{k+1}-t_k}\right| \le 1$, $t \in [t_k, t_{k+1}]$ for sufficiently large μ we obtain $t \in [t_k, t_{k+1}]$:

$$\left|B_k(u)(t)\right| \le \left|\int\limits_{t_k}^{t} U(u)(s)ds\right| + \left|\int\limits_{t_k}^{t_{k+1}} U(u)(s)ds\right| \equiv B_1 + B_2 .$$

Obviously every function $u(.) \in M_{SU}$ has an upper bound $U_0 e^{\mu T_0}$ including the initial function. Then in view of

$$\int\limits_{t_k}^{t_{k+1}}\left|u(t-2T)\right|dt \le U_0 e^{\mu_0} \int\limits_{t_k}^{t_{k+1}} e^{\mu(t-t_k)}dt = U_0 \frac{e^{\mu(t_{k+1}-t_k)}-1}{\mu}$$

$$\int\limits_{t_k}^{t_{k+1}}\left|E(t)\right|dt \le U_E \int\limits_{t_k}^{t_{k+1}} e^{\mu(t-t_k)}dt \le U_0 \int\limits_{t_k}^{t_{k+1}} e^{\mu(t-t_k)}dt = U_0 \frac{e^{\mu(t_{k+1}-t_k)}-1}{\mu}$$

we obtain

$$B_1 \le \left[\frac{2}{C_0(Z_0+R_0)}\int\limits_{t_k}^{t}\left|E(s)\right|ds + \frac{1}{C_0 Z_0}\int\limits_{t_k}^{t}\left|u(s)\right|ds + \frac{1}{C_0}\sum_{n=1}^{p}\left|r_n\right|\left|\int\limits_{t_k}^{t}\left|u(s)\right|^n ds\right| + \right.$$

$$\left.+\frac{1}{Z_0 C_0}\frac{\left|Z_0-R_0\right|}{Z_0+R_0}\int\limits_{t_k}^{t}\left|u(s-2T)\right|ds + \frac{1}{C_0}\frac{\left|Z_0-R_0\right|}{Z_0+R_0}\sum_{n=1}^{p}\left|r_n\right|\left|\int\limits_{t_k}^{t}\left|u(s-2T)\right|^n ds\right] + \frac{\left|Z_0-R_0\right|}{Z_0+R_0}\left|\int\limits_{t_k}^{t}\dot{u}(s-2T)ds\right| \le\right.$$

$$\le \left[\frac{2U_E}{C_0(Z_0+R_0)}\frac{e^{\mu(t-t_k)}-1}{\mu} + \frac{U_0}{C_0 Z_0}\frac{e^{\mu(t-t_k)}-1}{\mu} + \frac{1}{C_0}\sum_{n=1}^{p}\left|r_n\right|U_0^n\int\limits_{t_k}^{t}e^{n\mu(s-t_k)}ds + \right.$$

$$\left.+\frac{U_0}{Z_0 C_0}\frac{\left|Z_0-R_0\right|}{Z_0+R_0}\frac{e^{\mu(t-t_k)}-1}{\mu} + \frac{1}{C_0}\frac{\left|Z_0-R_0\right|}{Z_0+R_0}\sum_{n=1}^{p}\left|r_n\right|U_0^n e^{n\mu_0}\int\limits_{t_k}^{t}e^{\mu(s-t_k)}ds\right] + \frac{\left|Z_0-R_0\right|}{Z_0+R_0}\left|u(t-2T)\right| \le$$

$$\le \frac{2U_0}{C_0(Z_0+R_0)}\frac{e^{\mu(t-t_k)}-1}{\mu} + \frac{U_0}{C_0 Z_0}\frac{e^{\mu(t-t_k)}-1}{\mu} + \frac{e^{\mu(t-t_k)}-1}{\mu}\frac{1}{C_0}\sum_{n=1}^{p}\left|r_n\right|U_0^n\frac{ne^{(n-1)\mu(t-t_k)}}{n} +$$

$$+\frac{U_0}{Z_0 C_0}\frac{\left|Z_0-R_0\right|}{Z_0+R_0}\frac{e^{\mu(t-t_k)}-1}{\mu} + \frac{e^{\mu(t-t_k)}-1}{\mu}\frac{1}{C_0}\frac{\left|Z_0-R_0\right|}{Z_0+R_0}\sum_{n=1}^{p}\left|r_n\right|U_0^n e^{n\mu_0} + e^{\mu(t-t_k)}\frac{\left|Z_0-R_0\right|}{Z_0+R_0}U_0 \le$$

$$\le e^{\mu(t-t_k)}U_0\left\{\frac{1}{\mu C_0}\left[\frac{2}{Z_0+R_0} + \left(1+\frac{\left|Z_0-R_0\right|}{Z_0+R_0}\right)\left(\frac{1}{Z_0} + \sum_{n=1}^{p}\left|r_n\right|U_0^{n-1}e^{(n-1)\mu_0}\right)\right] + \frac{\left|Z_0-R_0\right|}{Z_0+R_0}\right\}$$

and

$$B_2 \le \left[\frac{2U_0}{C_0(Z_0+R_0)} \frac{e^{\mu(t_{k+1}-t_k)}-1}{\mu} + \frac{U_0}{C_0 Z_0} \frac{e^{\mu(t_{k+1}-t_k)}-1}{\mu} + \frac{1}{C_0} \sum_{n=1}^{p} |r_n| U_0^{\ n} \int_{t_k}^{t_{k+1}} e^{n\mu(s-t_k)} ds + \right.$$

$$\left. + \frac{U_0}{Z_0 C_0} \frac{|Z_0-R_0|}{Z_0+R_0} \frac{e^{\mu(t_{k+1}-t_k)}-1}{\mu} + \frac{1}{C_0} \frac{|Z_0-R_0|}{Z_0+R_0} \sum_{n=1}^{p} |r_n| U_0^{\ n} e^{n\mu 0} \int_{t_k}^{t_{k+1}} e^{\mu(s-t_k)} ds \right] +$$

$$+ \frac{|Z_0-R_0|}{Z_0+R_0} |u(t_{k+1}-2T)-u(t_k-2T)| \le$$

$$\le \frac{2U_0}{C_0(Z_0+R_0)} \frac{e^{\mu(t_{k+1}-t_k)}-1}{\mu} + \frac{U_0}{C_0 Z_0} \frac{e^{\mu(t_{k+1}-t_k)}-1}{\mu} + \frac{1}{C_0} \sum_{n=1}^{p} |r_n| U_0^{\ n} \frac{e^{n\mu(t_{k+1}-t_k)}-1}{n\mu} +$$

$$+ \frac{U_0}{Z_0 C_0} \frac{|Z_0-R_0|}{Z_0+R_0} \frac{e^{\mu(t_{k+1}-t_k)}-1}{\mu} + \frac{1}{C_0} \frac{|Z_0-R_0|}{Z_0+R_0} \sum_{n=1}^{p} |r_n| U_0^{\ n} e^{n\mu 0} \frac{e^{\mu(t_{k+1}-t_k)}-1}{\mu} \le$$

$$\le \frac{2U_0}{C_0(Z_0+R_0)} \frac{e^{\mu(t_{k+1}-t_k)}-1}{\mu} + \frac{U_0}{C_0 Z_0} \frac{e^{\mu(t_{k+1}-t_k)}-1}{\mu} + \frac{1}{C_0} \frac{e^{\mu(t_{k+1}-t_k)}-1}{\mu} \sum_{n=1}^{p} |r_n| U_0^{\ n} e^{(n-1)\mu 0} +$$

$$+ \frac{U_0}{Z_0 C_0} \frac{|Z_0-R_0|}{Z_0+R_0} \frac{e^{\mu(t_{k+1}-t_k)}-1}{\mu} + e^{\mu 0} \frac{1}{C_0} \frac{|Z_0-R_0|}{Z_0+R_0} \frac{e^{\mu(t_{k+1}-t_k)}-1}{\mu} \sum_{n=1}^{p} |r_n| U_0^{\ n} e^{(n-1)\mu 0} \le$$

$$\le \frac{e^{\mu 0}-1}{\mu C_0} U_0 \left[\frac{2}{Z_0+R_0} + \left(1+\frac{|Z_0-R_0|}{Z_0+R_0}\right)\left(\frac{1}{Z_0} + \sum_{n=1}^{p} |r_n| \left(U_0 e^{\mu 0}\right)^{n-1}\right) \right].$$

Therefore for a sufficiently large $\mu > 0$ we obtain

$$|B_k(u)(t)| \le$$

$$\le e^{\mu(t-t_k)} U_0 \left\{ \frac{1}{\mu C_0} \left[\frac{2}{Z_0+R_0} + \left(1+\frac{|Z_0-R_0|}{Z_0+R_0}\right)\left(\frac{1}{Z_0} + \sum_{n=1}^{p} |r_n|\left(U_0 e^{\mu 0}\right)^{n-1}\right) \right] + \frac{|Z_0-R_0|}{Z_0+R_0} \right\} +$$

$$+ \left(e^{\mu 0}-1\right) U_0 \frac{1}{\mu C_0} \left[\frac{2}{Z_0+R_0} + \left(1+\frac{|Z_0-R_0|}{Z_0+R_0}\right)\left(\frac{1}{Z_0} + \sum_{n=1}^{p} |r_n|\left(U_0 e^{\mu 0}\right)^{n-1}\right) \right] \le$$

$$\le e^{\mu(t-t_k)} U_0 \left\{ \frac{e^{\mu 0}}{\mu C_0} \left[\frac{2}{Z_0+R_0} + \left(1+\frac{|Z_0-R_0|}{Z_0+R_0}\right)\left(\frac{1}{Z_0} + \sum_{n=1}^{p} |r_n|\left(U_0 e^{\mu 0}\right)^{n-1}\right) \right] + \frac{|Z_0-R_0|}{Z_0+R_0} \right\} \le e^{\mu(t-t_k)} U_0 .$$

Consequently the operator B maps M_{SU} into itself.

We next show that B is a contractive operator. Indeed,

$$\left|B_k(u)(t) - B_k(\widetilde{u})(t)\right| \le$$

$$\left|\int_{t_k}^{t}[U(u)(s) - U(\widetilde{u})(s)]ds\right| + \left|\int_{t_k}^{t_{k+1}}[U(u)(s) - U(\widetilde{u})(s)]ds\right| \equiv P_1 + P_2, \ t \in [t_k, t_{k+1}].$$

Before making estimates, we notice that for every $[t_k, t_{k+1}]$ there is $[t_q, t_{q+1}]$ such that $t_k - 2T = t_q,\ t_{k+1} - 2T = t_{q+1}$. If $t_{k+1} - 2T = t_{q+1} \le t_0 = T \Rightarrow \rho^{(q)}(u, \widetilde{u}) = 0$.

Then in view of $q < k$ we have

$$P_1 \le \left[\frac{1}{C_0 Z_0}\int_{t_k}^{t}|u(s) - \widetilde{u}(s)|ds + \frac{1}{C_0}\sum_{n=1}^{p}|r_n|\int_{t_k}^{t}|u^n(s) - \widetilde{u}^n(s)|ds + \frac{1}{Z_0 C_0}\frac{|Z_0 - R_0|}{Z_0 + R_0}\int_{t_k}^{t}|u(s - 2T) - \widetilde{u}(s - 2T)|ds + \right.$$

$$\left. + \frac{1}{C_0}\frac{|Z_0 - R_0|}{Z_0 + R_0}\sum_{n=1}^{p}|r_n|\int_{t_k}^{t}|u^n(s - 2T) - \widetilde{u}^n(s - 2T)|ds\right] + \frac{|Z_0 - R_0|}{Z_0 + R_0}\left|\int_{t_k}^{t}\left(\dot{u}(s - 2T) - \dot{\widetilde{u}}(s - 2T)\right)ds\right| \le$$

$$\le \frac{\rho^{(k)}(u, \widetilde{u})}{C_0 Z_0}\frac{e^{\mu(t - t_k)} - 1}{\mu} + \frac{1}{C_0}\sum_{n=1}^{p}n|r_n|\sup\left\{|u(\theta)|^{n-1} : \theta \in [t_k, t_{k+1}]\right\}\int_{t_k}^{t}|u(s) - \widetilde{u}(s)|ds \le$$

$$+ \rho^{(q)}(u, \widetilde{u})\frac{1}{Z_0 C_0}\frac{|Z_0 - R_0|}{Z_0 + R_0}\frac{e^{\mu(t - t_k)} - 1}{\mu} +$$

$$+ \frac{1}{C_0}\frac{|Z_0 - R_0|}{Z_0 + R_0}\sum_{n=1}^{p}n|r_n|\operatorname{ess\,sup}\left\{|u^{n-1}(\theta)| : \theta \in [t_q, t_{q+1}]\right\}\rho^{(q)}(u, \widetilde{u})\frac{e^{\mu(t - t_k)} - 1}{\mu} +$$

$$+ \frac{|Z_0 - R_0|}{Z_0 + R_0}|u(t - 2T) - \widetilde{u}(t - 2T)| \le$$

$$\le \frac{e^{\mu_0}\rho_\mu^{(k)}(u, \widetilde{u})}{C_0 Z_0}\frac{e^{\mu(t - t_k)} - 1}{\mu} + \frac{e^{\mu_0}\rho_\mu^{(k)}(u, \widetilde{u})}{C_0}\frac{e^{\mu(t - t_k)} - 1}{\mu}\sum_{n=1}^{p}n|r_n|U_0^{n-1}e^{(n-1)\mu_0} +$$

$$+ \frac{e^{\mu_0}\rho_\mu^{(q)}(u, \widetilde{u})}{Z_0 C_0}\frac{|Z_0 - R_0|}{Z_0 + R_0}\frac{e^{\mu(t - t_k)} - 1}{\mu} +$$

$$+ \frac{e^{\mu_0}\rho_\mu^{(q)}(u, \widetilde{u})}{C_0}\frac{e^{\mu(t - t_k)} - 1}{\mu}\frac{|Z_0 - R_0|}{Z_0 + R_0}\sum_{n=1}^{p}n|r_n|U_0^{n-1}e^{(n-1)\mu_0} + e^{\mu_0}\rho_\mu^{(q)}(u, \widetilde{u})\frac{|Z_0 - R_0|}{Z_0 + R_0} \le$$

$$\le \frac{e^{\mu(t - t_k)} - 1}{\mu}e^{\mu_0}\left(\frac{\rho_\mu^{(k)}(\dot{u}, \dot{\widetilde{u}})}{\mu C_0 Z_0} + \frac{\rho_\mu^{(k)}(\dot{u}, \dot{\widetilde{u}})}{\mu C_0}\sum_{n=1}^{p}n|r_n|U_0^{n-1}e^{(n-1)\mu_0} +\right.$$

$$\left. + \frac{\rho_\mu^{(q)}(\dot{u}, \dot{\widetilde{u}})}{\mu Z_0 C_0}\frac{|Z_0 - R_0|}{Z_0 + R_0} + \frac{\rho_\mu^{(q)}(\dot{u}, \dot{\widetilde{u}})}{\mu C_0}\frac{|Z_0 - R_0|}{Z_0 + R_0}\sum_{n=1}^{p}n|r_n|U_0^{n-1}e^{(n-1)\mu_0}\right) + e^{\mu_0}\frac{\rho_\mu^{(q)}(\dot{u}, \dot{\widetilde{u}})}{\mu}\frac{|Z_0 - R_0|}{Z_0 + R_0} \le$$

$$\leq e^{\mu(t-t_k)}\hat{\rho}_\mu^{(k)}(\dot{u},\dot{\tilde{u}})\left[\frac{e^{\mu_0}}{\mu^2 C_0}\left(1+\frac{|Z_0-R_0|}{Z_0+R_0}\right)\left(\frac{1}{Z_0}+\sum_{n=1}^{p}n|r_n|U_0^{n-1}e^{(n-1)\mu_0}\right)+\frac{e^{\mu_0}}{\mu}\frac{|Z_0-R_0|}{Z_0+R_0}\right]$$

and

$$P_2\leq\hat{\rho}_\mu^{(k)}(\dot{u},\dot{\tilde{u}})\left(e^{\mu_0}-1\right)\frac{e^{\mu_0}}{\mu^2 C_0}\left[\left(1+\frac{|Z_0-R_0|}{Z_0+R_0}\right)\left(\frac{1}{Z_0}+\sum_{n=1}^{p}n|r_n|\left(U_0 e^{\mu_0}\right)^{n-1}\right)\right].$$

Consequently

$$\left|B_k(u)(t)-B_k(\tilde{u})(t)\right|\leq$$

$$\leq e^{\mu(t-t_k)}\hat{\rho}_\mu^{(k)}(\dot{u},\dot{\tilde{u}})\frac{e^{\mu_0}}{\mu}\left[\frac{1}{\mu C_0}\left(1+\frac{|Z_0-R_0|}{Z_0+R_0}\right)\left(\frac{1}{Z_0}+\sum_{n=1}^{p}n|r_n|\left(U_0 e^{\mu_0}\right)^{n-1}\right)+\frac{|Z_0-R_0|}{Z_0+R_0}\right]+$$

$$+e^{\mu(t-t_k)}\hat{\rho}_\mu^{(k)}(\dot{u},\dot{\tilde{u}})\frac{e^{\mu_0}-1}{\mu}\frac{e^{\mu_0}}{\mu C_0}\left[\left(1+\frac{|Z_0-R_0|}{Z_0+R_0}\right)\left(\frac{1}{Z_0}+\sum_{n=1}^{p}n|r_n|\left(U_0 e^{\mu_0}\right)^{n-1}\right)\right]\leq$$

$$\leq e^{\mu(t-t_k)}\hat{\rho}_\mu^{(k)}((u,\dot{u}),(\overline{u},\dot{\overline{u}}))\left[\frac{e^{2\mu_0}}{\mu^2 C_0}\left(1+\frac{|Z_0-R_0|}{Z_0+R_0}\right)\left(\frac{1}{Z_0}+\sum_{n=1}^{p}n|r_n|\left(U_0 e^{\mu_0}\right)^{n-1}\right)+\frac{e^{\mu_0}}{\mu}\frac{|Z_0-R_0|}{Z_0+R_0}\right]\equiv$$

$$\equiv e^{\mu(t-t_k)}K_u\hat{\rho}_\mu^{(k)}((u,\dot{u}),(\overline{u},\dot{\overline{u}})).$$

Since

$$\left|B_0(u)(t)-B_0(\tilde{u})(t)\right|\leq e^{\mu(t-t_0)}K_u\hat{\rho}_\mu^{(0)}((u,\dot{u}),(\overline{u},\dot{\overline{u}})),t\in[t_0,t_1],$$

$$\left|B_1(u)(t)-B_1(\tilde{u})(t)\right|\leq e^{\mu(t-t_1)}K_u\hat{\rho}_\mu^{(1)}((u,\dot{u}),(\overline{u},\dot{\overline{u}})),t\in[t_1,t_2],$$

$$\left|B_k(u)(t)-B_k(\tilde{u})(t)\right|\leq e^{\mu(t-t_k)}K_u\hat{\rho}_\mu^{(k)}((u,\dot{u}),(\overline{u},\dot{\overline{u}})),t\in[t_k,t_{k+1}]$$

and

$$\rho^{(0)}(B_0(u),B_0(\tilde{u}))\leq e^{\mu T_0}K_u\hat{\rho}_\mu^{(0)}((u,\dot{u}),(\overline{u},\dot{\overline{u}})),t\in[t_0,t_1],$$

$$\rho^{(1)}(B_1(u),B_1(\tilde{u}))\leq e^{\mu T_0}K_u\hat{\rho}_\mu^{(1)}((u,\dot{u}),(\overline{u},\dot{\overline{u}})),t\in[t_1,t_2],$$

$$\rho^{(k)}(B_k(u),B_k(\tilde{u}))\leq e^{\mu T_0}K_u\hat{\rho}_\mu^{(k)}((u,\dot{u}),(\overline{u},\dot{\overline{u}})),t\in[t_k,t_{k+1}]$$

we get

$$\hat{\rho}^{(k)}(B(u),B(\tilde{u})) \le e^{\mu T_0} K_u \hat{\rho}_\mu^{(k)}((u,\dot{u}),(\overline{u},\dot{\overline{u}})).$$

It remains to estimate the derivative of B. Indeed, we have

$$\left| \dot{B}_k(u)(t) - \dot{B}_k(\overline{u})(t) \right| \le$$

$$\left| U(u)(t) - U(\overline{u})(t) \right| + \frac{1}{t_{k+1} - t_k} \left| \int_{t_k}^{t_{k+1}} [U(u)(s) - U(\overline{u})(s)] \, ds \right| \equiv \dot{B}_1 + \dot{B}_2.$$

But $q < k$ and then

$$\dot{B}_1 \le \frac{1}{C_0 Z_0}|u(t) - \overline{u}(t)| + \frac{1}{C_0}\sum_{n=1}^{p}|r_n||u^n(t) - \overline{u}^n(t)| + \frac{1}{C_0 Z_0}\frac{|Z_0 - R_0|}{Z_0 + R_0}|u(t - 2T) - \overline{u}(t - 2T)| +$$

$$+ \frac{1}{C_0}\frac{|Z_0 - R_0|}{Z_0 + R_0}\sum_{n=1}^{p}|r_n||u^n(t - 2T) - \overline{u}^n(t - 2T)| + \frac{|Z_0 - R_0|}{Z_0 + R_0}|\dot{u}(t - 2T) - \dot{\overline{u}}(t - 2T)| \le$$

$$\le \frac{e^{\mu(t-t_k)}\rho_\mu^{(k)}(u,\overline{u})}{C_0 Z_0} + \frac{e^{\mu(t-t_k)}\rho_\mu^{(k)}(u,\overline{u})}{C_0}\sum_{n=1}^{p}|r_n|n\,U_0^{n-1}e^{(n-1)\mu_0} +$$

$$+ e^{\mu(t-t_q)}\rho_\mu^{(q)}(u,\overline{u})\frac{1}{C_0 Z_0}\frac{|Z_0 - R_0|}{Z_0 + R_0} +$$

$$+ e^{\mu(t-t_q)}\frac{\rho_\mu^{(q)}(u,\overline{u})}{C_0}\frac{|Z_0 - R_0|}{Z_0 + R_0}\sum_{n=1}^{p}|r_n|U_0^{n-1}e^{(n-1)\mu_0} + e^{\mu(t-t_q)}\frac{|Z_0 - R_0|}{Z_0 + R_0}\rho_\mu^{(q)}(\dot{u},\dot{\overline{u}}) \le$$

$$\le \frac{e^{\mu(t-t_k)}\rho_\mu^{(k)}(\dot{u},\dot{\overline{u}})}{\mu C_0 Z_0} + \frac{e^{\mu(t-t_k)}\rho_\mu^{(k)}(\dot{u},\dot{\overline{u}})}{\mu C_0}\sum_{n=1}^{p}|r_n|n\,U_0^{n-1}e^{(n-1)\mu_0} +$$

$$+ e^{\mu(t-t_q)}\frac{\rho_\mu^{(q)}(\dot{u},\dot{\overline{u}})}{\mu}\frac{1}{C_0 Z_0}\frac{|Z_0 - R_0|}{Z_0 + R_0} +$$

$$+ e^{\mu(t-t_q)}\frac{\rho_\mu^{(q)}(\dot{u},\dot{\overline{u}})}{\mu C_0}\frac{|Z_0 - R_0|}{Z_0 + R_0}\sum_{n=1}^{p}|r_n|U_0^{n-1}e^{(n-1)\mu_0} + e^{\mu(t-t_q)}\frac{|Z_0 - R_0|}{Z_0 + R_0}\rho_\mu^{(q)}(\dot{u},\dot{\overline{u}}) \le$$

$$\le e^{\mu(t-t_k)}\hat{\rho}_\mu^{(k)}(\dot{u},\dot{\overline{u}})\left[\frac{1}{\mu C_0}\left(1 + e^{\mu_0}\frac{|Z_0 - R_0|}{Z_0 + R_0}\right)\left(\frac{1}{Z_0} + \sum_{n=1}^{p}|r_n|n\,U_0^{n-1}e^{(n-1)\mu_0}\right) + e^{\mu_0}\frac{|Z_0 - R_0|}{Z_0 + R_0}\right].$$

Prior to estimating the next summand, we recall that the function $\xi(h) = \dfrac{e^h - 1}{h}, h \ge 0$

is increasing and $\lim\limits_{h \to 0}\xi(h) = 1$ (cf. 1.6.1).

Then

$$\dot{B}_2 \le \frac{1}{t_{k+1}-t_k}\left|\int_{t_k}^{t_{k+1}}\left(U(\dot{u})(s)-U(\dot{\overline{u}})(s)\right)ds\right| \le$$

$$\le \frac{\rho^{(k)}(\dot{u},\dot{\overline{u}})}{\mu C_0}\frac{e^{\mu(t_{k+1}-t_k)}-1}{\mu(t_{k+1}-t_k)}\left[\frac{1}{Z_0}\left(1+e^{\mu_0}\frac{|Z_0-R_0|}{Z_0+R_0}\right)+\left(1+e^{\mu_0}\frac{|Z_0-R_0|}{Z_0+R_0}\right)\sum_{n=1}^{p}|r_n|\,n\left(U_0 e^{\mu_0}\right)^{n-1}\right] \le$$

$$\le \frac{\rho^{(k)}(\dot{u},\dot{\overline{u}})}{\mu C_0}\frac{e^{\mu_0}-1}{\mu_0}\left(1+e^{\mu_0}\frac{|Z_0-R_0|}{Z_0+R_0}\right)\left(\frac{1}{Z_0}+\sum_{n=1}^{p}|r_n|\,n\left(U_0 e^{\mu_0}\right)^{n-1}\right).$$

Therefore

$$\left|\dot{B}_k(u)(t)-\dot{B}_k(\overline{u})(t)\right| \le$$

$$\le e^{\mu(t-t_k)}\hat{\rho}_\mu^{(k)}(\dot{u},\dot{\overline{u}})\left[\frac{1}{\mu C_0}\left(1+e^{\mu_0}\frac{|Z_0-R_0|}{Z_0+R_0}\right)\left(\frac{1}{Z_0}+\sum_{n=1}^{p}|r_n|\,n\,U_0^{n-1}e^{(n-1)\mu_0}\right)+e^{\mu_0}\frac{|Z_0-R_0|}{Z_0+R_0}\right]+$$

$$+\frac{\hat{\rho}^{(k)}(\dot{u},\dot{\overline{u}})}{\mu C_0}\frac{e^{\mu_0}-1}{\mu_0}\left(1+e^{\mu_0}\frac{|Z_0-R_0|}{Z_0+R_0}\right)\left(\frac{1}{Z_0}+\sum_{n=1}^{p}|r_n|\,n\left(U_0 e^{\mu_0}\right)^{n-1}\right) \le$$

$$\le e^{\mu(t-t_k)}\hat{\rho}_\mu^{(k)}(\dot{u},\dot{\overline{u}})\left[\left(1+\frac{e^{\mu_0}-1}{\mu_0}\right)\frac{1}{\mu C_0}\left(1+e^{\mu_0}\frac{|Z_0-R_0|}{Z_0+R_0}\right)\left(\frac{1}{Z_0}+\sum_{n=1}^{p}|r_n|\,n\,U_0^{n-1}e^{(n-1)\mu_0}\right)+e^{\mu_0}\frac{|Z_0-R_0|}{Z_0+R_0}\right] \le$$

$$\le e^{\mu(t-t_k)}\dot{K}_u\,\hat{\rho}_\mu^{(k)}((u,\dot{u}),(\overline{u},\dot{\overline{u}})).$$

It follows

$$\rho_\mu^{(k)}(\dot{B}(u),\dot{B}(\overline{u})) \le \dot{K}_u\hat{\rho}_\mu^{(k)}((u,\dot{u}),(\overline{u},\dot{\overline{u}})).$$

Therefore

$$\hat{\rho}_\mu^{(k)}\left((B(u),\dot{B}(u)),(B(\overline{u}),\dot{B}(\overline{u}))\right) \le \max\left\{e^{\mu_0}K_u,\dot{K}_u\right\}\hat{\rho}_\mu^{(k)}((u,\dot{u}),(\overline{u},\dot{\overline{u}})).$$

We have to verify that M_{SU} is j-bounded. Indeed, since j is an identity mapping,

$$\hat{\rho}_\mu^{j^n(k)}((u,\dot{u}),(\overline{u},\dot{\overline{u}})) \le \hat{\rho}_\mu^{(k)}((u,\dot{u}),(\overline{u},\dot{\overline{u}})) < \infty \quad (n=0,1,2,\dots).$$

Therefore, in consideration of the fixed-point Theorem 1.2.7 for contractive mappings in uniform spaces (cf. Chapter I), operator B has a unique fixed point and it is an oscillating solution of (2.4.1).

Theorem 2.4.1 is thus proved.

2.4.1. Numerical Example

Finally we collect all inequalities needed for the applications:

$$\frac{e^{\mu_0}}{\mu C_0}\left[\frac{2}{Z_0+R_0}+\left(1+\frac{|Z_0-R_0|}{Z_0+R_0}\right)\left(\frac{1}{Z_0}+\sum_{n=1}^{p}|r_n|\left(U_0 e^{\mu_0}\right)^{n-1}\right)\right]+\frac{|Z_0-R_0|}{Z_0+R_0}\leq 1;$$

$$e^{\mu_0}K_u=e^{\mu_0}\left[\frac{e^{2\mu_0}}{\mu^2 C_0}\left(1+\frac{|Z_0-R_0|}{Z_0+R_0}\right)\left(\frac{1}{Z_0}+\sum_{n=1}^{p}n|r_n|\left(U_0 e^{\mu_0}\right)^{n-1}\right)+\frac{e^{\mu_0}}{\mu}\frac{|Z_0-R_0|}{Z_0+R_0}\right]<1;$$

$$\dot{K}_u=\left(1+\frac{e^{\mu_0}-1}{\mu_0}\right)\frac{1}{\mu C_0}\left(1+e^{\mu_0}\frac{|Z_0-R_0|}{Z_0+R_0}\right)\left(\frac{1}{Z_0}+\sum_{n=1}^{p}|r_n|\,n\,U_0^{n-1}e^{(n-1)\mu_0}\right)+e^{\mu_0}\frac{|Z_0-R_0|}{Z_0+R_0}<1$$

For a transmission line with $\Lambda=1\,m,\ L=0,2\,\mu H/m,\ C=80\,pF/m,$

$$v=1/\sqrt{LC}=1/\sqrt{0,2.10^{-6}.80.10^{-12}}=1/\left(4.10^{-9}\right)=2,5.10^{8},$$

$$Z_0=\sqrt{L/C}=\sqrt{0,2.10^{-6}/80.10^{-12}}=50\ \Omega,\quad R_0=45\Omega,\quad C_0=8\,pF=8.10^{-12}\,F$$

we obtain

$$T=\Lambda\sqrt{LC}=4.10^{-9}s;\ |Z_0-R_0|/(Z_0+R_0)=1/19\approx 0,053.$$

Let us check the propagation of waves with $\lambda_0=10^{-4}\,m$:

$$f_0=1/\left(\lambda_0\sqrt{LC}\right)=1/\left(10^{-4}.4.10^{-9}\right)=(1/4)10^{13}\,Hz\ \Rightarrow\ T_0=1/f_0=4.10^{-13}\,\text{sec.};\ l_0=2.10^{-13}\,\text{sec}\cdot$$

If we choose $\mu=(1/4).10^{13}$, then $\mu T_0=\mu_0=1,\ \mu l_0=1/2,$

$$T=4.10^{-9}.(1/4).10^{13}T_0=10000.T_0,\ \mu T=(1/4)10^{13}.2.10^{-8}=(1/2)10^{5},$$

$$\mu C_0=(1/4)10^{13}.8.10^{-12}=20,\ \mu^2 C_0=(1/16).10^{26}.8.10^{-12}=(1/2)10^{14}.$$

If we take a nonlinear resistive element with a *V-I* characteristic

$f(u) = -0,12u + 0,8u^3$ then for $U_0 = 0,1$ we have

$$\frac{e}{20}\left[\frac{2}{95} + \left(1 + \frac{e}{19}\right)\left(\frac{1}{50} + 0,12 + 0,8(0,1e)^2\right)\right] + \frac{1}{19} \leq 1;$$

$$\dot{K}_u = \frac{e}{20}\left(1 + \frac{e}{19}\right)\left(\frac{1}{50} + 0,12 + 3.0,8(0,1e)^2\right) + \frac{e}{19} = 0,193 < 1.$$

Therefore $K = 0,193$.

Remark 2.4.3. The method provided above is suitable for non-uniform lines

$$C_e \frac{\partial u(x,t)}{\partial t} + \frac{\partial i(x,t)}{\partial x} = 0,$$

$$L_e \frac{\partial i(x,t)}{\partial t} + \frac{\partial u(x,t)}{\partial x} = 0 .$$

The above inequalities could be satisfied for sufficiently large $\mu > 0$ even though $Z_0 = \sqrt{L_e/C_e}$ decreases for increasing frequencies. (cf. Remark 2.2.7).

2.5. ANOTHER WAY TO REDUCING THE MIXED PROBLEM TO A NEUTRAL SYSTEM AND THE EXISTENCE-UNIQUENESS OF A PERIODIC SOLUTION

Here we consider again problems (2.1.5), (2.1.6-1), (2.1.6-2), (2.1.7) but unlike § 2.1 we propose another method for reducing the mixed problem for the hyperbolic system to a periodic problem for a neutral system on the boundary. We proceed from the transformation (2.1.3-1)

$$u(x,t) = \frac{1}{2}U(x,t) + \frac{1}{2}I(x,t),$$

$$i(x,t) = \frac{1}{2Z_0}U(x,t) - \frac{1}{2Z_0}I(x,t)$$

and then for $x = 0$ and $x = \Lambda$, we have

$$u(0,t) = \frac{1}{2}U(0,t) + \frac{1}{2}I(0,t) \qquad u(\Lambda,t) = \frac{1}{2}U(\Lambda,t) + \frac{1}{2}I(\Lambda,t)$$

and

$$i(0,t) = \frac{1}{2Z_0}U(0,t) - \frac{1}{2Z_0}I(0,t) \qquad i(\Lambda,t) = \frac{1}{2Z_0}U(\Lambda,t) - \frac{1}{2Z_0}I(\Lambda,t).$$

respectively.

Replacing them in the boundary conditions, we obtain a system with respect to the new variables $U(x,t), I(x,t)$:

$$E(t) - \frac{1}{2}U(0,t) - \frac{1}{2}I(0,t) = R_0 \frac{U(0,t) - I(0,t)}{2Z_0}, \quad t \geq T$$

$$C_0 \frac{d}{dt}\left(\frac{U(\Lambda,t) + I(\Lambda,t)}{2}\right) = \frac{1}{2Z_0}U(\Lambda,t) - \frac{1}{2Z_0}I(\Lambda,t) - f\left(\frac{U(\Lambda,t) + I(\Lambda,t)}{2}\right), t \geq T$$

An integration along the characteristics (cf. Chapter I) yields:

$$U(0,t) = U(\Lambda,t+T), \quad I(\Lambda,t) = I(0,t+T).$$

Substitute this in the previous system and obtain:

$$E(t) - \frac{1}{2}U(\Lambda,t+T) + \frac{1}{2}I(0,t) = R_0 \frac{U(\Lambda,t+T) - I(0,t)}{2Z_0}, \quad t \geq T$$

$$C_0 \frac{d}{dt}\left(\frac{U(\Lambda,t) + I(0,t+T)}{2}\right) = \frac{1}{2Z_0}U(\Lambda,t) - \frac{1}{2Z_0}I(0,t+T) - f\left(\frac{U(\Lambda,t) + I(0,t+T)}{2}\right), t \geq T$$

Let us replace $t+T$ by t. Then the above system becomes:

$$E(t-T) - \frac{1}{2}U(\Lambda,t) - \frac{1}{2}I(0,t-T) = R_0 \frac{U(\Lambda,t) - I(0,t-T)}{2Z_0}, \quad t \geq T$$

$$C_0 \frac{d}{dt}\left(\frac{U(\Lambda,t-T) + I(0,t)}{2}\right) = \frac{1}{2Z_0}U(\Lambda,t-T) - \frac{1}{2Z_0}I(0,t) - f\left(\frac{U(\Lambda,t-T) + I(0,t)}{2}\right), \quad t \geq T,$$

and the initial conditions become

$$U_0(x) = U(x,0) = u(x,0) + Z_0\, i(x,0) = u_0(x) + Z_0\, i_0(x),$$

$$I_0(x) = I(x,0) = u(x,0) - Z_0\, i(x,0) = u_0(x) - Z_0\, i_0(x), \ x \in [0,\Lambda].$$

As above, one can shift the initial functions along the characteristics on interval $[0,T]$ and obtain initial functions $U_0(t)$, $I_0(t)$ on the initial set $t \in [0,T]$. They are assumed to be $C_{T_0}^1[0,T]$ functions.

If we assume unknown functions to be $U(\Lambda,t) \equiv U(t)$, $I(0,t) \equiv I(t)$ then we can formulate the following periodic problem for the case $f(u) = \sum_{n=1}^{m} r_n u^n$: to find a periodic solution of the system

$$U(t) = 2E(t-T) - I(t-T) - \frac{R_0}{Z_0}U(t) + \frac{R_0}{Z_0}I(t-T), \quad t \in [T,\infty)$$

$$\frac{dI(t)}{dt} = -\frac{dU(t-T)}{dt} + \frac{1}{C_0 Z_0}U(t-T) - \frac{1}{C_0 Z_0}I(t) - \frac{2}{C_0}\sum_{n=1}^{m} r_n \left(\frac{U(t-T)+I(t)}{2}\right)^n, \; t \in [T,\infty)$$

$$U(t) = U_0(t), \; I(t) = I_0(t), \; t \in [0,T]. \tag{2.5.1}$$

By $C_{T_0}^1[T,\infty)$ we mean the space of all continuous T_0-periodic functions with continuous derivatives. Introduce the sets

$$M_U = \left\{u(.) \in C_{T_0}^1[T,\infty) : |U(t)| \leq V_0 e^{\mu(t-T-kT_0)}, t \in [T+kT, \; T+(k+1)T_0]; (k=0,1,2,...)\right\}$$

$$M_I = \left\{u(.) \in C_{T_0}^1[T,\infty) : |I(t)| \leq J_0 e^{\mu(t-T-kT_0)}, t \in [T+kT, \; T+(k+1)T_0]; (k=0,1,2,...)\right\}$$

which are closed subsets of $C_{T_0}^1[T,\infty)$. Here V_0, J_0, T_0, μ are positive constants (chosen below). The set $M_U \times M_I$ turns into a complete uniform space (cf. Chapter I, § 1.2) with respect to the family of pseudo-metrics

$$\hat{\rho}_\mu((U,\dot{U},I,\dot{I}),(\overline{U},\dot{\overline{U}},\overline{I},\dot{\overline{I}})) = \max\left\{\hat{\rho}(U,\overline{U}), \rho^{(k)}(\dot{U},\dot{\overline{U}}), \hat{\rho}(I,\overline{I}), \rho^{(k)}(\dot{I},\dot{\overline{I}}) : k = 0,1,2,...,m-1\right\},$$

where

$$\rho_\mu^{(k)}(U,\overline{U}) = \max\left\{e^{-\mu(t-T-kT_0)}|U(t)-\overline{U}(t)| : t \in [T+kT_0, T+(k+1)T_0]\right\}$$

$$\rho^{(k)}(U,\overline{U}) = \max\left\{|U(t)-\overline{U}(t)| : t \in [T+kT_0, T+(k+1)T_0]\right\},$$

$$\hat{\rho}(U,\overline{U}) = \max\left\{|U(t)-\overline{U}(t)| : t \in [T,2T]\right\}$$

$$\rho_\mu^{(k)}(\dot{U},\dot{\bar{U}}) = \max\left\{ e^{-\mu(t-T-kT_0)}\left|\dot{U}(t)-\dot{\bar{U}}(t)\right| : t \in [T+kT_0, T+(k+1)T_0] \right\}.$$

$$\rho_\mu^{(k)}(I,\bar{I}) = \max\left\{ e^{-\mu(t-T-kT_0)}\left|I(t)-\bar{I}(t)\right| : t \in [T+kT_0, T+(k+1)T_0] \right\},$$

$$\rho^{(k)}(I,\bar{I}) = \max\left\{ \left|I(t)-\bar{I}(t)\right| : t \in [T+kT_0, T+(k+1)T_0] \right\},$$

$$\hat{\rho}(I,\bar{I}) = \max\left\{ \left|I(t)-\bar{I}(t)\right| : t \in [T, 2T] \right\},$$

$$\rho_\mu^{(k)}(\dot{I},\dot{\bar{I}}) = \max\left\{ e^{-\mu(t-T-kT_0)}\left|\dot{I}(t)-\dot{\bar{I}}(t)\right| : t \in [T+kT_0, T+(k+1)T_0] \right\}.$$

Let us write the conformity condition (**CC**):

$$\frac{dI(T)}{dt} = -\frac{dU(0)}{dt} + \frac{1}{C_0 Z_0}U(0) - \frac{1}{C_0 Z_0}I(T) - \frac{2}{C_0}\sum_{n=1}^{p} r_n\left(\frac{U(0)+I(T)}{2}\right)^n.$$

We have to assume $U(0) = 0,\ I(0) = 0, \dot{U}(0) = 0,\ \dot{I}(0) = 0$. In view of the condition $T = mT_0$ it follows that $I(T) = 0,\ \dot{I}(T) = 0$.

We look for a periodic solution, first on $[T, 2T]$, then on $[2T, 3T]$ and so on.

In order to avoid (**CC**) which imposes the restrictive condition $\dot{U}(0) = 0$ and $\dot{I}(0) = 0$ we define the operator $B = \left(B_U(U,I), B_I(U,I)\right)$ on every interval $[T+kT_0, T+(k+1)T_0]$ ($k = 0, 1, 2, \ldots$) with the expressions:

$$B_U^{(k)}(U,I)(t) := W(U,I)(t),\ t \in [T+kT_0, T+(k+1)T_0]$$

$$B_I^{(k)}(U,I)(t) = U(T+kT_0) + \int_{T+kT_0}^{t} J(U,I)(s)\,ds - \left(\frac{t-T-kT_0}{T_0}\right)\int_{T+kT_0}^{T+(k+1)T_0} J(U,I)(s)\,ds,$$

$$t \in [T+kT_0, T+(k+1)T_0], (k = 0,1,2,\ldots,m-1) \text{ where}$$

$$W(U,I)(t) = 2E(t-T) - \bar{I}_0(t) - \frac{R_0}{Z_0}U(t) + \frac{R_0}{Z_0}\bar{I}_0(t),$$

$$J(U,I) = -\frac{d\bar{U}_0(t)}{dt} + \frac{\bar{U}_0(t)}{C_0 Z_0} - \frac{1}{C_0 Z_0}I(t) - \frac{2}{C_0}f\left(\frac{\bar{U}_0(t)+I(t)}{2}\right),\ t \in [T, 2T]$$

and $\vec{U}_0(.), \vec{I}_0(.) \in C^1_{T_0}[T,2T]$ are translated to the right as initial functions $U_0(.), I_0(.)$ on $[T,2T]$, that is, $\vec{U}_0(.) = U_0(t-T), t \in [T,2T]$, $\vec{I}_0(.) = I_0(t-T), t \in [T,2T]$.

Theorem 2.5.1. With the following conditions fulfilled:

1) The initial functions $U_0(t), I_0(t) \in C^1_{T_0}[0,T]$ satisfy the conditions

$$U(0) = 0, \ I(0) = 0,$$
$$\left|U_0(t)\right| \leq V_0 e^{-\beta} e^{\mu(t-T-kT_0)}, \left|I_0(t)\right| \leq J_0 e^{-\beta} e^{\mu(t-T-kT_0)}, t \in [kT_0,(k+1)T_0];$$

$(k = 0,1,2,...,m-1,)$, where $\beta > 0$ is a constant chosen below;

2) $E(.) \in C^1_{T_0}[0,\infty); \quad \left|E(t)\right| \leq V_E e^{\mu(t-T-kT_0)}, t \in [T+kT_0, T+(k+1)T_0]$;

3) $R_0/Z_0 < 1$

there exists a unique T_0-periodic solution of the initial value problem (2.5.1), belonging to $M = M_U \times M_I$.

Proof: We omit the proof that $B_U^{(k)}(U,I)(t)$ and $B_I^{(k)}(U,I)(t)$ are continuously differentiable T_0-periodic functions.

We show that

$$\left|B_U^{(k)}(U,I)(t)\right| \leq V_0 e^{\mu(t-T-kT_0)}, \left|B_I^{(k)}(U,I)(t)\right| \leq J_0 e^{\mu(t-T-kT_0)}.$$

In other words, for $t \in [T+kT_0, T+(k+1)T_0]$ we have

$$\left|B_U^{(k)}(U,I)(t)\right| \leq 2\left|E(t)\right| + \left|\vec{I}_0(t)\right| + \frac{R_0}{Z_0}\left|U(t)\right| + \frac{R_0}{Z_0}\left|\vec{I}_0(t)\right| \leq$$

$$\leq 2e^{\mu(t-T-kT_0)}V_E + e^{\mu(t-T-kT_0)}J_0 e^{-\beta} + \frac{R_0}{Z_0}e^{\mu(t-T-kT_0)}V_0 + \frac{R_0}{Z_0}e^{\mu(t-T-kT_0)}J_0 e^{-\beta} \leq$$

$$\leq e^{\mu(t-T-kT_0)}\left(2V_E + \frac{R_0}{Z_0}V_0 + J_0 e^{-\beta} + \frac{R_0}{Z_0}J_0 e^{-\beta}\right) \leq e^{\mu(t-T-kT_0)}V_0$$

since $\dfrac{R_0}{Z_0} < 1$ and $\beta > 0$ sufficiently large.

For the second component we have

$$\left|B_I^{(k)}(U,I)(t)\right| \le \left|\int_{T+kT_0}^{t} J(U,I)(s)ds\right| + \left|\int_{T+kT_0}^{T+(k+1)T_0} J(U,I)(s)ds\right| \equiv I_1 + I_2;$$

$$I_1 \le \left|\int_{T+kT_0}^{t}\dot{\bar{U}}_0(s)ds\right| + \frac{1}{C_0 Z_0}\int_{T+kT_0}^{t}\left|\bar{U}_0(s)\right|ds + \frac{1}{C_0 Z_0}\int_{T+kT_0}^{t}\left|I(s)\right|ds + \frac{2}{C_0}\sum_{n=1}^{p}\left|r_n\right|\int_{T+kT_0}^{t}\left|\frac{1}{2}\bar{U}_0(s)+\frac{1}{2}I(s)\right|^n ds \le$$

$$\le \left|\bar{U}_0(t)\right| + \frac{V_0 e^{-\beta}}{C_0 Z_0}\frac{e^{\mu(t-T-kT_0)}-1}{\mu} + \frac{J_0}{C_0 Z_0}\frac{e^{\mu(t-T-kT_0)}-1}{\mu} +$$

$$\frac{2}{C_0}\sum_{n=1}^{m}\frac{\left|r_n\right|}{2^n}\int_{T+kT_0}^{t}\left|\bar{U}_0(s)+I(s)\right|^n ds \le$$

$$\le e^{\mu(t-T-kT_0)}V_0 e^{-\beta} + \frac{e^{\mu(t-T-kT_0)}-1}{\mu}\frac{V_0 e^{-\beta}}{C_0 Z_0} + \frac{e^{\mu(t-T-kT_0)}-1}{\mu}\frac{J_0}{C_0 Z_0} +$$

$$+\frac{1}{C_0}\sum_{n=1}^{m}\frac{\left|r_n\right|}{2^{n-1}}\left(V_0 e^{-\beta}+J_0\right)^n\frac{e^{n\mu(t-T-kT_0)}-1}{n\mu} \le$$

$$\le e^{\mu(t-T-kT_0)}V_0 e^{-\beta} + \frac{e^{\mu(t-T-kT_0)}-1}{\mu}\frac{V_0 e^{-\beta}}{C_0 Z_0} + \frac{e^{\mu(t-T-kT_0)}-1}{\mu}\frac{J_0}{C_0 Z_0} +$$

$$+\frac{e^{\mu(t-T-kT_0)}-1}{\mu}\frac{1}{C_0}\sum_{n=1}^{m}\frac{\left|r_n\right|}{2^{n-1}}\left(V_0 e^{-\beta}+J_0\right)^n e^{(n-1)\mu T_0} \le$$

$$\le e^{\mu(t-T-kT_0)}V_0 e^{-\beta} + \frac{e^{\mu(t-T-kT_0)}-1}{\mu C_0}\left(V_0 e^{-\beta}+J_0\right)\left(\frac{1}{Z_0}+\sum_{n=1}^{m}\left|r_n\right|\left(\frac{V_0 e^{-\beta}+J_0}{2}\right)^{n-1}e^{(n-1)\mu_0}\right) \le$$

$$\le e^{\mu(t-T-kT_0)}\left[V_0 e^{-\beta} + \frac{V_0 e^{-\beta}+J_0}{\mu C_0}\left(\frac{1}{Z_0}+\sum_{n=1}^{m}\left|r_n\right|\left(\frac{V_0 e^{-\beta}+J_0}{2}\right)^{n-1}e^{(n-1)\mu_0}\right)\right]$$

and

$$I_2 \le \left|\int_{T+kT_0}^{T+(k+1)T_0}\dot{\bar{U}}_0(s)ds\right| + \frac{1}{C_0 Z_0}\int_{T+kT_0}^{T+(k+1)T_0}\left|\bar{U}_0(s)\right|ds + \frac{1}{C_0 Z_0}\int_{T+kT_0}^{T+(k+1)T_0}\left|I(s)\right|ds +$$

$$+\frac{2}{C_0}\sum_{n=1}^{m}\left|r_n\right|\int_{T+kT_0}^{T+(k+1)T_0}\left|\frac{\bar{U}_0(s)+I(s)}{2}\right|^n ds \le$$

$$\le \frac{V_0 e^{-\beta}}{C_0 Z_0}\frac{e^{\mu T_0}-1}{\mu} + \frac{J_0}{C_0 Z_0}\frac{e^{\mu T_0}-1}{\mu} +$$

$$\frac{2}{C_0}\sum_{n=1}^{m}\frac{|r_n|}{2^n}\int_{T+kT_0}^{T+(k+1)T_0}\left|\vec{U}_0(s)+I(s)\right|^n ds \le$$

$$\le \frac{V_0 e^{-\beta}}{C_0 Z_0}\frac{e^{\mu T_0}-1}{\mu}+\frac{J_0}{C_0 Z_0}\frac{e^{\mu T_0}-1}{\mu}+$$

$$\frac{1}{C_0}\sum_{n=1}^{m}\frac{|r_n|}{2^{n-1}}\left(V_0 e^{-\beta}+J_0\right)^n\int_{T+kT_0}^{T+(k+1)T_0}e^{n\mu(s-T-kT_0)}ds \le$$

$$\le \frac{e^{\mu T_0}-1}{\mu}\frac{V_0 e^{-\beta}+J_0}{C_0 Z_0}+\frac{1}{C_0}\sum_{n=1}^{m}\frac{|r_n|\left(V_0 e^{-\beta}+J_0\right)^n}{2^{n-1}}\frac{e^{n\mu T_0}-1}{n\mu}\le$$

$$\le \frac{e^{\mu_0}-1}{\mu C_0}\left(\frac{V_0 e^{-\beta}+J_0}{Z_0}+\sum_{n=1}^{m}\frac{|r_n|\left(V_0 e^{-\beta}+J_0\right)^n}{2^{n-1}}e^{(n-1)\mu_0}\right).$$

Consequently

$$\left|B_I^{(k)}(U,I)(t)\right|\le$$

$$\le e^{\mu(t-T-kT_0)}\left[V_0 e^{-\beta}+\frac{V_0 e^{-\beta}+J_0}{\mu C_0}\left(\frac{1}{Z_0}+\sum_{n=1}^{m}|r_n|\left(\frac{V_0 e^{-\beta}+J_0}{2}\right)^{n-1}e^{(n-1)\mu_0}\right)\right]+$$

$$+ e^{\mu(t-T-kT_0)}\frac{\left(e^{\mu_0}-1\right)\left(V_0 e^{-\beta}+J_0\right)}{\mu C_0}\left(\frac{1}{Z_0}+\sum_{n=1}^{m}\frac{|r_n|\left(V_0 e^{-\beta}+J_0\right)^{n-1}}{2^{n-1}}e^{(n-1)\mu_0}\right)\le$$

$$\le e^{\mu(t-T-kT_0)}\left\{V_0 e^{-\beta}+\frac{e^{\mu_0}\left(V_0 e^{-\beta}+J_0\right)}{\mu C_0}\left[\frac{1}{Z_0}+\sum_{n=1}^{m}|r_n|\left(\frac{e^{\mu_0}\left(V_0 e^{-\beta}+J_0\right)}{2}\right)^{n-1}\right]\right\}\le J_0 e^{\mu(t-T-kT_0)}.$$

The last inequality is satisfied for a sufficiently large $\mu>0$.

In the following we show that B is contractive operator. Indeed, for every $k=0,1,2,...,m-1$:

$$\left|B_U^{(k)}(U,I)(t)-B_U^{(k)}(\overline{U},\overline{I})(t)\right|\le\left|\overline{I}_0(t)-\overline{I}_0(t)\right|+\frac{R_0}{Z_0}\left|U(t)-\overline{U}(t)\right|+\frac{R_0}{Z_0}\left|\overline{I}_0(t)-\overline{I}_0(t)\right|\le$$

$$\le e^{\mu(t-T-kT_0)}\frac{R_0}{Z_0}\rho_\mu^{(k)}(U,\overline{U})\le e^{\mu(t-T-kT_0)}\frac{R_0}{Z_0}\frac{\rho_\mu^{(k)}(\dot{U},\dot{\overline{U}})}{\mu}\le$$

$$\le e^{\mu(t-T-kT_0)}\hat{\rho}_\mu((U,\dot{U},I,\dot{I}),(\overline{U},\dot{\overline{U}},\overline{I},\dot{\overline{I}}))\frac{R_0}{\mu Z_0}\equiv$$

$$\equiv e^{\mu(t-T-kT_0)}K_U\hat{\rho}_\mu((U,\dot{U},I,\dot{I}),(\overline{U},\dot{\overline{U}},\overline{I},\dot{\overline{I}}))\le e^{\mu_0}K_U\hat{\rho}_\mu((U,\dot{U},I,\dot{I}),(\overline{U},\dot{\overline{U}},\overline{I},\dot{\overline{I}})),$$

therefore

$$\hat{\rho}(B_U(U,I),B_U(\overline{U},\overline{I})) \le e^{\mu_0}K_U\,\hat{\rho}_\mu((U,\dot{U},I,\dot{I}),(\overline{U},\dot{\overline{U}},\overline{I},\dot{\overline{I}})).$$

For the derivatives we have

$$\left|\dot{B}_U^{(k)}(U,I)(t)-\dot{B}_U^{(k)}(\overline{U},\overline{I})(t)\right| \le \left|\dot{\overline{I}}_0(t)-\dot{\overline{I}}_0(t)\right| + \frac{R_0}{Z_0}\left|\dot{U}(t)-\dot{\overline{U}}(t)\right| + \frac{R_0}{Z_0}\left|\dot{\overline{I}}_0(t)-\dot{\overline{I}}_0(t)\right| \le$$

$$\le e^{\mu(t-T-kT_0)}\frac{R_0}{Z_0}\rho_\mu^{(k)}(\dot{U},\dot{\overline{U}}) \le$$

$$\le e^{\mu(t-T-kT_0)}\hat{\rho}_\mu((U,\dot{U},I,\dot{I}),(\overline{U},\dot{\overline{U}},\overline{I},\dot{\overline{I}}))\frac{R_0}{Z_0} \equiv e^{\mu(t-T-kT_0)}\dot{K}_U\,\hat{\rho}_\mu((U,\dot{U},I,\dot{I}),(\overline{U},\dot{\overline{U}},\overline{I},\dot{\overline{I}})) \le$$

$$\le e^{\mu(t-T-kT_0)}\dot{K}_U\,\hat{\rho}_\mu((U,I),(\overline{U},\overline{I})).$$

Thus it follows that

$$\rho_\mu^{(k)}(\dot{B}_U^{(k)}(U,I),\dot{B}_U^{(k)}(\overline{U},\overline{I})) \le \dot{K}_U\,\hat{\rho}_\mu((U,\dot{U},I,\dot{I}),(\overline{U},\dot{\overline{U}},\overline{I},\dot{\overline{I}})).$$

Similarly, we obtain

$$\left|B_I^{(k)}(U,I)(t)-B_I^{(k)}(\overline{U},\overline{I})(t)\right| \le$$

$$\le \left|\int_{T+kT_0}^{t}\left(J(U,I)(s)-J(\overline{U},\overline{I})(s)\right)ds\right| + \left|\int_{T+kT_0}^{T+(k+1)T_0}\left(J(U,I)(s)-J(\overline{U},\overline{I})(s)\right)ds\right| \equiv J_1 + J_2.$$

We have

$$J_1 \le \left|\int_{T+kT_0}^{t}\left(\dot{\overline{U}}_0(s)-\dot{\overline{U}}_0(s)\right)ds\right| + \frac{1}{C_0Z_0}\int_{T+kT_0}^{t}\left|\overline{U}_0(s)-\overline{U}_0(s)\right|ds + \frac{1}{C_0Z_0}\int_{T+kT_0}^{t}\left|I(s)-\overline{I}(s)\right|ds +$$

$$+\frac{2}{C_0}\sum_{n=1}^{m}|r_n|\int_{T+kT_0}^{t}\left|\left(\frac{1}{2}\overline{U}_0(T)+\frac{1}{2}I(s)\right)^n-\left(\frac{1}{2}\overline{U}_0(s)+\frac{1}{2}\overline{I}(s)\right)^n\right|ds \le$$

$$\le \frac{\rho^{(k)}(I,\overline{I})}{C_0Z_0}\frac{e^{\mu(t-T-kT_0)}}{\mu} +$$

$$+\frac{1}{C_0}\sum_{n=1}^{m}n|r_n|\sup\left\{\left|\frac{1}{2}\overline{U}_0(s)+\frac{1}{2}I(s)\right|^{n-1}:s\in[T+kT_0,T+(k+1)T_0]\right\}\int_{T+kT_0}^{t}\left|I(s)-\overline{I}(s)\right|ds \le$$

$$\leq \frac{e^{\mu(t-T-kT_0)}-1}{\mu}\frac{\rho_\mu^{(k)}(\dot{I},\dot{\bar{I}})}{\mu C_0 Z_0} + \frac{1}{C_0}\sum_{n=1}^{m} n|r_n|\frac{\left(V_0 e^{-\beta}+J_0\right)^{n-1}}{2^{n-1}} e^{(n-1)\mu_0}\int_{T+kT_0}^{t} e^{\mu(s-T-kT_0)} e^{-\mu(s-T-kT_0)}\left|I(s)-\bar{I}(s)\right|ds \leq$$

$$\leq \frac{e^{\mu(t-T-kT_0)}-1}{\mu}\frac{\rho_\mu^{(k)}(\dot{I},\dot{\bar{I}})}{\mu C_0 Z_0} + \frac{1}{C_0}\sum_{n=1}^{m} n|r_n|\frac{\left(V_0 e^{-\beta}+J_0\right)^{n-1} e^{(n-1)\mu_0}}{2^{n-1}}\frac{\rho_\mu^{(k)}(\dot{I},\dot{\bar{I}})}{\mu}\int_{T+kT_0}^{t} e^{\mu(s-T-kT_0)}ds \leq$$

$$\leq \frac{e^{\mu(t-T-kT_0)}-1}{\mu}\frac{\rho_\mu^{(k)}(\dot{I},\dot{\bar{I}})}{\mu C_0 Z_0} + \frac{1}{C_0}\sum_{n=1}^{m} n|r_n|\frac{\left(V_0 e^{-\beta}+J_0\right)^{n-1} e^{(n-1)\mu_0}}{2^{n-1}}\frac{\rho_\mu^{(k)}(\dot{I},\dot{\bar{I}})}{\mu}\frac{e^{\mu(t-T-kT_0)}-1}{\mu} \leq$$

$$\leq \frac{e^{\mu(t-T-kT_0)}-1}{\mu}\frac{\rho_\mu^{(k)}(\dot{I},\dot{\bar{I}})}{\mu C_0}\left(\frac{1}{Z_0}+\sum_{n=1}^{m} n|r_n|\frac{\left(V_0 e^{-\beta}+J_0\right)^{n-1}}{2^{n-1}} e^{(n-1)\mu T_0}\right) \leq$$

$$\leq e^{\mu(t-T-kT_0)}\hat{\rho}_\mu((U,\dot{U},I,\dot{I}),(\bar{U},\dot{\bar{U}},\bar{I},\dot{\bar{I}}))\frac{1}{\mu^2 C_0}\left(\frac{1}{Z_0}+\sum_{n=1}^{m} n|r_n|\left(e^{\mu_0}\frac{V_0 e^{-\beta}+J_0}{2}\right)^{n-1}\right)$$

and

$$J_2 \leq \left|\int_{T+kT_0}^{T+(k+1)T_0}\left(J(U,I)(s)-J(\bar{U},\bar{I})(s)\right)ds\right| \leq$$

$$\leq \left|\int_{T+kT_0}^{T+(k+1)T_0}\left(\dot{\bar{U}}_0(s)-\dot{\bar{U}}_0(s)\right)ds\right| + \frac{1}{C_0 Z_0}\int_{T+kT_0}^{T+(k+1)T_0}\left|\bar{U}_0(s)-\bar{U}_0(s)\right|ds + \frac{1}{C_0 Z_0}\int_{T+kT_0}^{T+(k+1)T_0}\left|I(s)-\bar{I}(s)\right|ds +$$

$$+\frac{2}{C_0}\sum_{n=1}^{m}|r_n|\int_{T+kT_0}^{T+(k+1)T_0}\left|\left(\frac{1}{2}\bar{U}_0(s)+\frac{1}{2}I(s)\right)^{n}-\left(\frac{1}{2}\bar{U}_0(s)+\frac{1}{2}\bar{I}(s)\right)^{n}\right|ds \leq$$

$$\leq \frac{\rho_\mu^{(k)}(I,\bar{I})}{C_0 Z_0}\frac{e^{\mu T_0}-1}{\mu}+$$

$$+\frac{1}{C_0}\sum_{n=1}^{m} n|r_n|\sup\left\{\left|\frac{1}{2}\bar{U}_0(s)+\frac{1}{2}I(s)\right|^{n-1}: s\in[T+kT_0,T+(k+1)T_0]\right\}\int_{T+kT_0}^{T+(k+1)T_0}\left|I(s)-\bar{I}(s)\right|ds \leq$$

$$\leq \frac{\rho_\mu^{(k)}(\dot{I},\dot{\bar{I}})}{\mu C_0 Z_0}\frac{e^{\mu T_0}-1}{\mu}+\frac{1}{C_0}\sum_{n=1}^{m} n|r_n|\frac{\left(V_0 e^{-\beta}+J_0\right)^{n-1}}{2^{n-1}}\frac{\rho_\mu^{(k)}(\dot{I},\dot{\bar{I}})}{\mu}\int_{T+kT_0}^{T+(k+1)T_0} e^{n\mu(s-T-kT_0)}ds \leq$$

$$\leq \frac{\rho_\mu^{(k)}(\dot{I},\dot{\bar{I}})}{\mu C_0 Z_0}\frac{e^{\mu T_0}-1}{\mu}+\frac{1}{C_0}\sum_{n=1}^{m} n|r_n|\frac{\left(V_0 e^{-\beta}+J_0\right)^{n-1} e^{(n-1)\mu T_0}}{2^{n-1}}\frac{\rho_\mu^{(k)}(\dot{I},\dot{\bar{I}})}{\mu}\frac{e^{\mu T_0}-1}{\mu} \leq$$

$$\leq \frac{\rho_\mu^{(k)}(\dot{I},\dot{\bar{I}})}{\mu C_0 Z_0}\frac{e^{\mu T_0}-1}{\mu}+\frac{1}{C_0}\frac{e^{\mu T_0}-1}{\mu}\sum_{n=1}^{m}n|r_n|\frac{\left(V_0 e^{-\beta}+J_0\right)^{n-1}}{2^{n-1}}e^{(n-1)\mu T_0}\frac{\rho_\mu^{(k)}(\dot{I},\dot{\bar{I}})}{\mu}\leq$$

$$\leq \frac{e^{\mu T_0}-1}{\mu}\frac{\rho_\mu^{(k)}(\dot{I},\dot{\bar{I}})}{\mu C_0}\left(\frac{1}{Z_0}+\sum_{n=1}^{m}n|r_n|\frac{\left(V_0 e^{-\beta}+J_0\right)^{n-1}}{2^{n-1}}e^{(n-1)\mu T_0}\right)\leq$$

$$\leq e^{\mu(t-T-kT_0)}\hat{\rho}_\mu((U,\dot{U},I,\dot{I}),(\bar{U},\dot{\bar{U}},\bar{I},\dot{\bar{I}}))\frac{e^{\mu_0}-1}{\mu^2 C_0}\left(\frac{1}{Z_0}+\sum_{n=1}^{m}n|r_n|\left(e^{\mu_0}\frac{V_0 e^{-\beta}+J_0}{2}\right)^{n-1}\right).$$

Therefore

$$\left|B_I^{(k)}(U,I)(t)-B_I^{(k)}(\bar{U},\bar{I})(t)\right|\leq$$

$$\leq e^{\mu(t-T-kT_0)}\hat{\rho}_\mu((U,\dot{U},I,\dot{I}),(\bar{U},\dot{\bar{U}},\bar{I},\dot{\bar{I}}))\frac{1}{\mu^2 C_0}\left(\frac{1}{Z_0}+\sum_{n=1}^{m}n|r_n|\left(e^{\mu_0}\frac{V_0 e^{-\beta}+J_0}{2}\right)^{n-1}\right)+$$

$$+e^{\mu(t-T-kT_0)}\hat{\rho}_\mu((U,\dot{U},I,\dot{I}),(\bar{U},\dot{\bar{U}},\bar{I},\dot{\bar{I}}))\frac{e^{\mu_0}-1}{\mu^2 C_0}\left(\frac{1}{Z_0}+\sum_{n=1}^{m}n|r_n|\left(e^{\mu_0}\frac{V_0 e^{-\beta}+J_0}{2}\right)^{n-1}\right)\leq$$

$$\leq e^{\mu(t-T-kT_0)}\hat{\rho}_\mu((U,\dot{U},I,\dot{I}),(\bar{U},\dot{\bar{U}},\bar{I},\dot{\bar{I}}))\frac{e^{\mu_0}}{\mu^2 C_0}\left(\frac{1}{Z_0}+\sum_{n=1}^{m}n|r_n|\left(e^{\mu_0}\frac{V_0 e^{-\beta}+J_0}{2}\right)^{n-1}\right)\equiv$$

$$\equiv e^{\mu(t-T-kT_0)}K_I\,\hat{\rho}_\mu((U,\dot{U},I,\dot{I}),(\bar{U},\dot{\bar{U}},\bar{I},\dot{\bar{I}}))\leq e^{\mu T_0}K_I\,\hat{\rho}_\mu((U,\dot{U},I,\dot{I}),(\bar{U},\dot{\bar{U}},\bar{I},\dot{\bar{I}}))$$

It follows that

$$\hat{\rho}(B_I(U,I),B_I(\bar{U},\bar{I}))\leq e^{\mu_0}K_I\,\hat{\rho}_\mu((U,\dot{U},I,\dot{I}),(\bar{U},\dot{\bar{U}},\bar{I},\dot{\bar{I}})).$$

For the derivative we get

$$\left|\dot{B}_I^{(k)}(U,I)(t)-\dot{B}_I^{(k)}(U,I)(t)\right|\leq|J(U,I)(t)-J(U,I)(t)|+\frac{1}{T_0}\left|\int_T^{T+T_0}\left(J(U,I)(s)-J(\bar{U},\bar{I})(s)\right)ds\right|\equiv \dot{J}_1+\dot{J}_2.$$

We have

$$\dot{J}_1\leq\left|\dot{\vec{U}}_0(s)-\dot{\vec{U}}_0(s)\right|+\frac{1}{C_0 Z_0}\left|\vec{U}_0(s)-\vec{U}_0(s)\right|+\frac{1}{C_0 Z_0}\left|I(t)-\bar{I}(t)\right|+$$

$$+ \frac{2}{C_0} \sum_{n=1}^{m} |r_n| \left| \left(\frac{1}{2}\vec{U}_0(t) + \frac{1}{2}I(t) \right)^n - \left(\frac{1}{2}\vec{U}_0(t) + \frac{1}{2}\bar{I}(t) \right)^n \right| \leq$$

$$\leq e^{\mu(t-T-kT_0)} \frac{\rho_\mu^{(k)}(I,\bar{I})}{C_0 Z_0} + \frac{1}{C_0} \sum_{n=1}^{m} n|r_n| \frac{\left(V_0 e^{-\beta} + J_0\right)^{n-1}}{2^{n-1}} e^{(n-1)\mu_0} e^{\mu(t-T-kT_0)} \rho_\mu^{(k)}(I,\bar{I}) \leq$$

$$\leq e^{\mu(t-T-kT_0)} \frac{\rho_\mu^{(k)}(\dot{I},\dot{\bar{I}})}{\mu C_0} \left(\frac{1}{Z_0} + \sum_{n=1}^{m} n|r_n| \frac{\left(V_0 e^{-\beta} + J_0\right)^{n-1}}{2^{n-1}} e^{(n-1)\mu T_0} \right) \leq$$

$$\leq e^{\mu(t-T-kT_0)} \hat{\rho}_\mu((U,\dot{U},I,\dot{I}),(\overline{U},\dot{\overline{U}},\bar{I},\dot{\bar{I}})) \frac{1}{\mu C_0} \left(\frac{1}{Z_0} + \sum_{n=1}^{m} n|r_n| \left(e^{\mu_0} \frac{V_0 e^{-\beta} + J_0}{2} \right)^{n-1} \right)$$

and

$$\dot{J}_2 \leq \frac{1}{T_0} \int_{T+kT_0}^{T+(k+1)T_0} \left| J(U,I)(s) - J(\overline{U},\bar{I})(s) \right| ds \leq$$

$$\leq \frac{1}{T_0} \int_{T+kT_0}^{T+(k+1)T_0} e^{\mu(t-T-kT_0)} dt \; \hat{\rho}_\mu((U,\dot{U},I,\dot{I}),(\overline{U},\dot{\overline{U}},\bar{I},\dot{\bar{I}})) \frac{1}{\mu C_0} \left(\frac{1}{Z_0} + \sum_{n=1}^{m} n|r_n| \left(e^{\mu_0} \frac{V_0 e^{-\beta} + J_0}{2} \right)^{n-1} \right) \leq$$

$$\leq e^{\mu(t-T-kT_0)} \hat{\rho}_\mu((U,\dot{U},I,\dot{I}),(\overline{U},\dot{\overline{U}},\bar{I},\dot{\bar{I}})) \frac{e^{\mu_0}-1}{\mu_0} \frac{1}{\mu C_0} \left(\frac{1}{Z_0} + \sum_{n=1}^{m} n|r_n| \left(e^{\mu_0} \frac{V_0 e^{-\beta} + J_0}{2} \right)^{n-1} \right)$$

Therefore

$$\left| \dot{B}_I^{(k)}(U,I)(t) - \dot{B}_I^{(k)}(\overline{U},\bar{I})(t) \right| \leq$$

$$\leq e^{\mu(t-T-kT_0)} \hat{\rho}_\mu((U,\dot{U},I,\dot{I}),(\overline{U},\dot{\overline{U}},\bar{I},\dot{\bar{I}})) \frac{1}{\mu C_0} \left(\frac{1}{Z_0} + \sum_{n=1}^{m} n|r_n| \left(e^{\mu_0} \frac{V_0 e^{-\beta} + J_0}{2} \right)^{n-1} \right) +$$

$$+ e^{\mu(t-T-kT_0)} \hat{\rho}_\mu((U,\dot{U},I,\dot{I}),(\overline{U},\dot{\overline{U}},\bar{I},\dot{\bar{I}})) \frac{e^{\mu_0}-1}{\mu_0} \frac{1}{\mu C_0} \left(\frac{1}{Z_0} + \sum_{n=1}^{m} n|r_n| \left(e^{\mu_0} \frac{V_0 e^{-\beta} + J_0}{2} \right)^{n-1} \right) \leq$$

$$\leq e^{\mu(t-T-kT_0)} \hat{\rho}_\mu((U,\dot{U},I,\dot{I}),(\overline{U},\dot{\overline{U}},\bar{I},\dot{\bar{I}})) \left(1 + \frac{e^{\mu_0}-1}{\mu_0} \right) \frac{1}{\mu C_0} \left(\frac{1}{Z_0} + \sum_{n=1}^{m} n|r_n| \left(e^{\mu_0} \frac{V_0 e^{-\beta} + J_0}{2} \right)^{n-1} \right) \equiv$$

$$\equiv e^{\mu(t-T-kT_0)}\dot{K}_I\,\hat{\rho}_\mu((U,\dot{U},I,\dot{I}),(\overline{U},\dot{\overline{U}},\overline{I},\dot{\overline{I}}))\,.$$

Therefore

$$\rho_\mu^{(k)}(\dot{B}_I^{(k)}(U,I),\dot{B}_I^{(k)}(\overline{U},\overline{I}))\le \dot{K}_I\,\hat{\rho}_\mu((U,\dot{U},I,\dot{I}),(\overline{U},\dot{\overline{U}},\overline{I},\dot{\overline{I}}))\,.$$

The obtained four inequalities imply

$$\hat{\rho}_\mu((B_U,\dot{B}_U,B_I,\dot{B}_I),((\overline{B}_U,\dot{\overline{B}}_U,\overline{B}_I,\dot{\overline{B}}_I))\le K\hat{\rho}_\mu((U,\dot{U},I,\dot{I}),(\overline{U},\dot{\overline{U}},\overline{I},\dot{\overline{I}}))\ ,$$

where $K=\max\left\{e^{\mu_0}K_U,\dot{K}_U,e^{\mu_0}K_I,\dot{K}_I\right\}<1$.

Then the fixed point of *B* is a periodic solution of the problem stated. Theorem 2.5.1 is thus proved.

2.5.1. Numerical Example

Here we would like to compare the conditions of the last theorem with those from § 2.2.

$$2V_E+\frac{R_0}{Z_0}V_0+J_0e^{-\beta}\left(1+\frac{R_0}{Z_0}\right)\le V_0;$$

$$V_0e^{-\beta}+\frac{e^{\mu_0}\left(V_0e^{-\beta}+J_0\right)}{\mu C_0}\left[\frac{1}{Z_0}+\sum_{n=1}^{m}|r_n|\left(\frac{e^{\mu_0}\left(V_0e^{-\beta}+J_0\right)}{2}\right)^{n-1}\right]\le J_0;$$

$$K_U=\frac{e^{-\beta}}{\mu}+\frac{(1+e^{-\beta})R_0}{\mu Z_0}<1;\ \dot{K}_U=e^{-\beta}+\frac{(1+e^{-\beta})R_0}{Z_0}<1;$$

$$K_I=\frac{e^{\mu_0}}{\mu^2 C_0}\left(\frac{1}{Z_0}+\sum_{n=1}^{m}n|r_n|\frac{\left(V_0e^{-\beta}+J_0\right)^{n-1}}{2^{n-1}}e^{(n-1)\mu_0}\right)<1;$$

$$\dot{K}_I=\left(1+\frac{e^{\mu_0}-1}{\mu_0}\right)\frac{1}{\mu C_0}\left(\frac{1}{Z_0}+\sum_{n=1}^{m}n|r_n|\frac{\left(V_0e^{-\beta}+J_0\right)^{n-1}}{2^{n-1}}e^{(n-1)\mu_0}\right)<1.$$

Here we choose a line with larger characteristics impedance. Let

$$\Lambda = 1m \quad L = 0,45\,\mu H/m,\ C = 80\,pF/m,$$

$$v = \frac{1}{\sqrt{LC}} = \frac{1}{\sqrt{0,45.10^{-6}.80.10^{-12}}} = \frac{1}{6.10^{-9}} = 1,66.10^{8},$$

$$Z_0 = \sqrt{\frac{L}{C}} = \sqrt{\frac{0,45.10^{-6}}{80.10^{-12}}} = 75\,\Omega, T = \Lambda\sqrt{LC} = 1.6.10^{-9} = 6.10^{-7}\ \text{sec};$$

$$R_0 = 15\Omega,\ C_0 = 5pF = 5.10^{-12}F.$$

For waves with $\lambda_0 = \frac{1}{6}10^{-3}\,m$ we have $f_0 = 1/\left(\lambda_0\sqrt{LC}\right) = \frac{6}{10^{-3}6.10^{-9}} = 10^{12}$

$\Rightarrow T_0 = \dfrac{1}{f_0} = 10^{-12}\ \text{sec}.$ If   we   choose $\mu = 10^{12}$,   then $\mu T_0 = \mu_0 = 1$ and

$T = 6.10^{-7}.10^{12}T_0 = 600000.T_0$. We also have

$$\mu C_0 = 10^{12}.5.10^{-12} = 5;\quad \mu^2 C_0 = 10^{24}.5.10^{-12} = 5.10^{12}.$$

If the *V-I* characteristic of the nonlinear resistive element is $f(u) = -0,12u + 0,8u^3$, then the above inequalities for $\beta = 3$, $V_0 = J_0 = 10^{-1}$, $V_E = 10^{-1}V_0$ become :

$$\frac{2}{10}V_0 + \frac{R_0}{Z_0}V_0 + V_0e^{-\beta}\left(1 + \frac{R_0}{Z_0}\right) \le V_0 \Leftrightarrow \frac{2}{10} + \frac{15}{75} + e^{-3}\left(1 + \frac{15}{75}\right) \le 1$$

$$10^{-1}e^{-3} + \frac{e(e^{-\beta} + 1)}{\mu C_0}\left(\frac{1}{Z_0} + \sum_{n=1}^{3}\frac{|r_n|\left(V_0(e^{-\beta} + 1)\right)^{n-1}}{2^{n-1}}e^{(n-1)}\right) \le 1 \Leftrightarrow$$

$$\Leftrightarrow 0,005 + 0,58\left(0,013 + 0,12 + 0,0022e^2\right) \le 1;$$

$$K_U = \frac{R_0}{\mu Z_0} < 1 \Rightarrow \frac{15}{75.10^{12}} < 1;\ \dot{K}_U = e^{-3} + \frac{(1+e^{-3})15}{75} < 1;$$

$$K_I = \frac{e}{\mu^2 C_0}\left(\frac{1}{Z_0} + \sum_{n=1}^{3}n|r_n|\frac{(J_0)^{n-1}}{2^{n-1}}e^{(n-1)}\right) = \frac{e}{5.10^{12}}\left(\frac{1}{75} + \sum_{n=1}^{3}n|r_n|\left(\frac{e}{20}\right)^{n-1}\right) < 1;$$

$$\dot{K}_I = \frac{e}{\mu C_0}\left(\frac{1}{Z_0} + \sum_{n=1}^{3}n|r_n|\left(\frac{1+e^{-3}}{2}J_0e\right)^{n-1}\right) = \frac{e}{5}\left(0,13 + 0,05\right) = \frac{e}{5}0,18 \approx 0,1 < 1;$$

$$K = \dot{K}_U = \frac{1 + 6e^{-3}}{5} = 0,26 < 1.$$

Obviously the ratio $\dfrac{R_0}{Z_0} < 1$ plays an important role for the rate of convergence.

Remark 2.5.1. The just described method is not suitable for non-uniform lines because $\dfrac{R_0}{Z_0} < 1$ cannot be made smaller at the expense of large μ.

2.6. PERIODIC REGIMES FOR TRANSMISSION LINES TERMINATED BY A RESISTIVE ELEMENT WITH EXPONENTIAL V-I CHARACTERISTIC

Our goal here is to demonstrate that our approach is applicable to a wider class of nonlinearities, namely the exponential ones.

Here we consider problem (2.2.1) from § 2.2 where the V-I characteristic $i = f(u)$ contains an exponential nonlinearity. Such type V-I characteristics describe Esaki diode nonlinearity and are encountered in many microwave devices (cf. [37], [39], [55], [56], [68], [122]). In this case, the current is a sum of diffusion current and tunnel current:

$$i = f(u) = i_{diff} + i_{tunn} = \widehat{I}_0(e^{\frac{q}{k\tau}u} - 1) + \frac{I_{\max}u}{U_{\max}} e^{1 - \frac{u}{U_{\max}}}.$$

The usually accepted denotations are $\alpha = q/(k\tau)$, $U_m = U_{\max}$, $I_m = I_{\max}$. Then the last equation becomes

$$i = f(u) = \widehat{I}_0(e^{\alpha u} - 1) + \frac{I_m}{U_m} u e^{1 - \frac{u}{U_m}} \tag{2.6.1}$$

where I_0, I_m, α, U_m are positive constants.

Thus we have to replace the function $f(u)$ by (2.6.1) in the initial value problem (2.1.11) and obtain:

$$\frac{du(t)}{dt} = \frac{1}{C_0}\frac{2E(t)}{Z_0 + R_0} - \frac{1}{C_0 Z_0}u(t) - \frac{\widehat{I}_0(e^{\alpha u(t)} - 1)}{C_0} - \frac{I_m}{C_0 U_m}u(t)e^{1 - \frac{u(t)}{U_m}} -$$

$$- \frac{Z_0 - R_0}{C_0 Z_0(Z_0 + R_0)}u(t - 2T) + \frac{\widehat{I}_0}{C_0}\frac{Z_0 - R_0}{Z_0 + R_0}(e^{\alpha u(t - 2T)} - 1) + \frac{I_m}{C_0 U_m}\frac{Z_0 - R_0}{Z_0 + R_0}u(t - 2T)e^{1 - \frac{u(t - 2T)}{U_m}} +$$

$$+\frac{Z_0 - R_0}{(Z_0 + R_0)}\frac{du(t-2T)}{dt}\equiv U(u)(t),\quad t\in[T,3T]\tag{2.6.2}$$

$$u(t)=\upsilon_0(t),\ \frac{du(t)}{dt}=\frac{d\upsilon_0(t)}{dt},\ t\in[-T,T].$$

Recall that $\delta=\dfrac{Z_0 - R_0}{Z_0 + R_0}<1$.

The conformity condition **(CC)** is

$$\frac{du(T)}{dt}=\frac{1}{C_0}\frac{2E(T)}{Z_0+R_0}-\frac{1}{C_0 Z_0}u(T)-\frac{I_0(e^{\alpha u(T)}-1)}{C_0}-\frac{I_m}{C_0 U_m}u(T)e^{1-\frac{u(T)}{U_m}}-$$

$$-\frac{\delta}{C_0 Z_0}\upsilon_0(-T)+\frac{\delta\widehat{I}_0}{C_0}(e^{\alpha\upsilon_0(-T)}-1)+\frac{I_m\delta}{C_0 U_m}\upsilon_0(-T)e^{1-\frac{\upsilon_0(-T)}{U_m}}+\delta\frac{d\upsilon_0(-T)}{dt}$$

This is satisfied, provided $\upsilon_0(-T)=\dot{\upsilon}_0(-T)=0,\ E(T)=0$.

Next we formulate the main problem: to find a continuously differentiable T_0-periodic solution of (2.6.2) on the interval $[T,\infty)$ coinciding with prescribed T_0-periodic function on the interval $[-T,T]$.

We define an operator whose fixed-points are periodic solutions of (2.6.2) using the expressions:

$$B_k(u)(t):=u(T+kT_0)+\int_{T+kT_0}^{t}U(u)(s)ds-\frac{t-T-kT_0}{T_0}\int_{T+kT_0}^{T+(k+1)T_0}U(u)(s)ds,\tag{2.6.3}$$

$t\in[T+kT_0,T+(k+1)T_0]$ ($k=0,1,2,...,2m-1$), where $U(u)(t)$ is the right-hand side of (2.6.2)

$$U(u)(t)\equiv\frac{1}{C_0}\frac{2E(t)}{Z_0+R_0}-\frac{1}{C_0 Z_0}u(t)-\frac{\widehat{I}_0(e^{\alpha u(t)}-1)}{C_0}-\frac{I_m}{C_0 U_m}u(t)e^{1-\frac{u(s)}{U_m}}-$$

$$-\frac{\delta}{C_0 Z_0}\bar{u}(t)+\delta\frac{\widehat{I}_0}{C_0}(e^{\alpha\bar{u}(t)}-1)+\delta\frac{I_m}{C_0 U_m}\bar{u}(t)e^{1-\frac{\bar{u}(t)}{U_m}}+\delta\frac{d\bar{u}(t)}{dt},\quad t\in[T,3T].$$

As in § 2.2 $\bar{\upsilon}_0(t)$ the initial function is translated to the right on the interval $[T,3T]$.

In this manner the difficulty caused by **(CC)** is avoided.

One can verify also that $U(u)(t)$ is a continuously differentiable and T_0-periodic function.

We rewrite the above initial value problem as:

$$\frac{du(t)}{dt} = U(u)(t),\ t \in [T, 3T],$$

$$u(T) = u(T + kT_0) = 0. \tag{2.6.4}$$

Recall that $C_{T_0}^1[T, \infty)$ is the space of all continuous T_0-periodic functions with continuous derivatives and

$$M_U = \left\{ u(.) \in C_{T_0}^1[T, \infty) : |u(t)| \le U_0 e^{\mu(t-T-kT_0)},\ t \in [T + kT,\ T + (k+1)T_0], (k = 0,1,2,...,2m-1) \right\}$$

where U_0, T_0, μ are positive constants (chosen below). The set M_U turns out to be a complete metric space with a metric

$$\hat{\rho}((u,\dot{u}),(\overline{u},\dot{\overline{u}})) = \max\left\{ \hat{\rho}(u,\overline{u}), \rho_\mu^{(k)}(\dot{u},\dot{\overline{u}}) : k = 0,1,2,...,2m-1 \right\},$$

where

$$\rho^{(k)}(u,\overline{u}) = \max\left\{ |u(t) - \overline{u}(t)| : t \in [T + kT_0, T + (k+1)T_0] \right\},$$

$$\rho_\mu^{(k)}(u,\overline{u}) = \max\left\{ e^{-\mu(t-T-kT_0)} |u(t) - \overline{u}(t)| : t \in [T + kT_0, T + (k+1)T_0] \right\},$$

$$\hat{\rho}(u,\overline{u}) = \max\left\{ |u(t) - \overline{u}(t)| : t \in [T, 3T] \right\},$$

$$\rho_\mu^{(k)}(\dot{u},\dot{\overline{u}}) = \max\left\{ e^{-\mu(t-T-kT_0)} |\dot{u}(t) - \dot{\overline{u}}(t)| : t \in [T + kT_0, T + (k+1)T_0] \right\}.$$

Prior to formulating the next lemma we note that Remark 2.2.5 and 2.2.6 are valid.

Lemma 2.6.1. The initial value problem (2.6.4) has a solution $u(.) \in M_U$ iff the operator B has a fixed point $u \in M_U$, that is, $u = B(u)$.

The proof is analogous to the one of Lemma 2.2.1.

Further on we assume:

(U) $0 < U_0 e^{\mu T_0} \le U_m.$

Besides this, we notice

$$g(u) = ue^{1-\frac{u}{U_m}}, \qquad\qquad\qquad (2.6.5)$$

$$g'(u) = e^{1-\frac{u}{U_m}}\left(1-\frac{u}{U_m}\right) \Rightarrow |g'(u)| \leq \left(1+\frac{U_0 e^{\mu_0}}{U_m}\right)e^{1+\frac{U_0 e^{\mu_0}}{U_m}} \leq \left(1+\frac{U_m}{U_m}\right)e^{1+\frac{U_m}{U_m}} \leq 2e^2$$

Theorem 2.6.1. Let the following conditions be fulfilled:

(E) $E(.) \in C^1_{T_0}[0,\infty), E(0) = 0,$

$$|E(t)| \leq U_E e^{\mu(t-T-kT_0)}, t \in [T + kT_0, T + (n+1)T_0], (k = 0,1,2,...);$$

(IN) $\upsilon_0(t) \in C^1_{T_0}[-T,T],$

$$|\upsilon_0(t)| \leq U_0\ e^{\mu(t+T-nT_0))}, t \in [-T + nT_0, -T + (n+1)T_0], (n = 0,1,...2m-1);$$

(V) $\upsilon_0(-T) = 0.$

In this case there exists a unique T_0-periodic solution of (2.6.4), belonging to M_U.

Proof: First we show that the operator B maps the set M_U into itself. As with the proof of Theorem 2.2.1, one can show that $(Bu)(.) \in C^1_{T_0}[T,\infty)$.

We must first show that $|B_k(u)(t)| \leq U_0 e^{\mu(t-T-kT_0)}, t \in [T + kT_0, T + (k+1)T_0]$.
Indeed,

$$|B_k(u)(t)| \leq \left|\int_{T+kT_0}^{t} U(u)(s)ds\right| + \left|\int_{T+kT_0}^{T+(k+1)T_0} U(u)(s)ds\right| \equiv B_1 + B_2.$$

Bearing in mind that $\xi(h) = \dfrac{e^h - 1}{h}$ is non-decreasing function (cf. 1.6.1, 3)) we have

$$B_1 \leq \left[\frac{2}{C_0(Z_0+R_0)}\int_{T+kT_0}^{t}|E(s)|ds + \frac{1}{C_0 Z_0}\int_{T+kT_0}^{t}|u(s)|ds + \frac{\hat{I}_0}{C_0}\int_{T+kT_0}^{t}\left|e^{\alpha u(s)}-1\right|ds + \frac{I_m}{C_0 U_m}\int_{T+kT_0}^{t}|u(s)|e^{1-\frac{u(s)}{U_m}}ds + \right.$$

$$+\frac{1}{C_0 Z_0}\frac{|Z_0-R_0|}{Z_0+R_0}\int_{T+kT_0}^{t}\bar{u}(s)ds + \frac{\hat{I}_0}{C_0}\frac{|Z_0-R_0|}{Z_0+R_0}\int_{T+kT_0}^{t}\left|e^{\alpha\bar{u}(s)}-1\right|ds +$$

$$\left.+\frac{I_m}{C_0 U_m}\frac{|Z_0-R_0|}{Z_0+R_0}\int_{T+kT_0}^{t}|\bar{u}(s)|e^{1-\frac{\bar{u}(s)}{U_m}}ds + \frac{|Z_0-R_0|}{Z_0+R_0}\left|\int_{T+kT_0}^{t}\dot{\bar{u}}(s)ds\right|\right] \leq$$

$$\leq \frac{e^{\mu(t-T-kT_0)}-1}{\mu C_0}\left[\frac{2U_E}{(Z_0+R_0)}+\frac{U_0}{Z_0}+\alpha\widehat{I}_0 U_0\xi(\alpha U_m)+\frac{I_m U_0 e^2}{U_m}+\frac{U_0}{Z_0}\frac{|Z_0-R_0|}{Z_0+R_0}+\right.$$

$$\left.+\frac{|Z_0-R_0|}{Z_0+R_0}\alpha\widehat{I}_0 U_0\xi(\alpha U_m)+\frac{I_m U_0 e^2}{U_m}\frac{|Z_0-R_0|}{Z_0+R_0}\right]+\frac{|Z_0-R_0|}{Z_0+R_0}U_0 e^{\mu(t-T-kT_0)}\leq$$

$$\leq e^{\mu(t-T-kT_0)}\left\{\frac{1}{\mu C_0}\left[\frac{2U_E}{Z_0+R_0}+U_0\left(1+\frac{|Z_0-R_0|}{Z_0+R_0}\right)\left(\frac{1}{Z_0}+\alpha\widehat{I}_0\xi(\alpha U_m)+\frac{I_m e^2}{U_m}\right)\right]+U_0\frac{|Z_0-R_0|}{Z_0+R_0}\right\}$$

and

$$B_2\leq\frac{1}{C_0(Z_0+R_0)}\int_{T+kT_0}^{T+(k+1)T_0}|E(s)|ds+\frac{1}{2C_0 Z_0}\int_{T+kT_0}^{T+(k+1)T_0}|u(s)|ds+\frac{\widehat{I}_0}{2C_0}\int_{T+kT_0}^{T+(k+1)T_0}\left|e^{\alpha u(s)}-1\right|ds+$$

$$+\frac{I_m}{2C_0 U_m}\int_{T+kT_0}^{T+(k+1)T_0}|u(s)|e^{1-\frac{u(s)}{U_m}}ds+\frac{1}{2C_0 Z_0}\frac{|Z_0-R_0|}{Z_0+R_0}\int_{T+kT_0}^{T+(k+1)T_0}|\bar{u}(s)|ds+\frac{\widehat{I}_0}{2C_0}\frac{|Z_0-R_0|}{Z_0+R_0}\int_{T+kT_0}^{T+(k+1)T_0}\left|e^{\alpha\bar{u}(s)}-1\right|ds+$$

$$+\frac{I_m}{2C_0 U_m}\frac{|Z_0-R_0|}{Z_0+R_0}\int_{T+kT_0}^{T+(k+1)T_0}|\bar{u}(s)|e^{1-\frac{\bar{u}(s)}{U_m}}ds+\frac{1}{2}\frac{|Z_0-R_0|}{Z_0+R_0}\left|\int_{T+kT_0}^{T+(k+1)T_0}\dot{\bar{u}}(s)ds\right|\leq$$

$$\leq\frac{e^{\mu T_0}-1}{\mu C_0}\left[\frac{2U_E}{(Z_0+R_0)}+\frac{U_0}{Z_0}+\alpha\widehat{I}_0 U_0\xi(\alpha U_m)+\frac{I_m U_0 e^2}{U_m}+\frac{U_0}{Z_0}\frac{|Z_0-R_0|}{Z_0+R_0}+\frac{|Z_0-R_0|}{Z_0+R_0}\alpha I_0 U_0\xi(\alpha U_m)+\right.$$

$$\left.+\frac{I_m U_0 e^2}{U_m}\frac{|Z_0-R_0|}{Z_0+R_0}\right]\leq$$

$$\leq e^{\mu(t-T-kT_0)}\frac{e^{\mu T_0}-1}{\mu C_0}\left[\frac{2U_E}{Z_0+R_0}+U_0\left(1+\frac{|Z_0-R_0|}{Z_0+R_0}\right)\left(\frac{1}{Z_0}+\alpha\widehat{I}_0\xi(\alpha U_m)+\frac{I_m e^2}{U_m}\right)\right].$$

Therefore, for $t\in[T+kT_0;\,T+(k+1)T_0]$ we have

$$\left|(B_k u)(t)\right|\leq B_1+B_2\leq$$

$$\leq e^{\mu(t-T-kT_0)}\left\{\frac{1}{\mu C_0}\left[\frac{2U_E}{Z_0+R_0}+U_0\left(1+\frac{|Z_0-R_0|}{Z_0+R_0}\right)\left(\frac{1}{Z_0}+\alpha\widehat{I}_0\xi(\alpha U_m)+\frac{I_m e^2}{U_m}\right)\right]+U_0\frac{|Z_0-R_0|}{Z_0+R_0}\right\}+$$

$$+e^{\mu(t-T-kT_0)}\frac{e^{\mu T_0}-1}{\mu C_0}\left[\frac{2U_E}{Z_0+R_0}+U_0\left(1+\frac{|Z_0-R_0|}{Z_0+R_0}\right)\left(\frac{1}{Z_0}+\alpha\widehat{I}_0\xi(\alpha U_m)+\frac{I_m e^2}{U_m}\right)\right]\leq$$

$$\leq e^{\mu(t-T-kT_0)}\left\{\frac{e^{\mu_0}}{\mu C_0}\left[\frac{2U_E}{Z_0+R_0}+U_0\left(1+\frac{|Z_0-R_0|}{Z_0+R_0}\right)\left(\frac{1}{Z_0}+\alpha\widehat{I}_0\xi(\alpha U_m)+\frac{I_m e^2}{U_m}\right)\right]+U_0\frac{|Z_0-R_0|}{Z_0+R_0}\right\}\leq$$

$$\leq e^{\mu(t-T-kT_0)}U_0$$

which is satisfied for a sufficiently large $\mu > 0$ since $|\delta| = \dfrac{|Z_0 - R_0|}{Z_0 + R_0} < 1$.

Consequently the operator B maps M_U into itself.

It remains to show that B is a contractive operator.

Recalling inequalities 1) and 2) from Remark 2.2.7, and $|g'(u)| \le 2e^2$, and the definition of $\vec{\upsilon}_0(t)$, for every $u(.), \overline{u}(.) \in M_U$ we get:

$$\left| B_k(u)(t) - B_k(\overline{u})(t) \right| \le$$

$$\le \left| \int_{T+kT_0}^{t} [U(u)(s) - U(\overline{u})(s)] ds \right| + \left| \int_{T+kT_0}^{T+(k+1)T_0} [U(u)(s) - U(\overline{u})(s)] ds \right| \equiv U_1 + U_2.$$

We have

$$U_1 \le \frac{1}{C_0 Z_0} \int_{T+kT_0}^{t} |u(s) - \overline{u}(s)| ds + \frac{\widehat{I}_0}{C_0} \int_{T+kT_0}^{t} \left| e^{\alpha u(s)} - e^{\alpha \overline{u}(s)} \right| ds +$$

$$\frac{I_m}{C_0 U_m} \int_{T+kT_0}^{t} \left| u(s) e^{1 - \frac{u(s)}{U_m}} - \overline{u}(s) e^{1 - \frac{\overline{u}(s)}{U_m}} \right| ds +$$

$$+ \frac{1}{C_0 Z_0} \frac{|Z_0 - R_0|}{Z_0 + R_0} \int_{T+kT_0}^{t} |\vec{u}(t) - \vec{u}(t)| ds + \frac{\widehat{I}_0}{C_0} \frac{|Z_0 - R_0|}{Z_0 + R_0} \int_{T+kT_0}^{t} \left| e^{\alpha \vec{u}(t)} - e^{\alpha \vec{u}(t)} \right| ds +$$

$$+ \frac{I_m}{C_0 U_m} \frac{|Z_0 - R_0|}{Z_0 + R_0} \int_{T+kT_0}^{t} \left| \vec{u}(t) e^{1 - \frac{\vec{u}(t)}{U_m}} - \vec{u}(t) e^{1 - \frac{\vec{u}(t)}{U_m}} \right| ds + \frac{|Z_0 - R_0|}{Z_0 + R_0} \left| \int_{T+kT_0}^{t} (\dot{\vec{u}}_0(t) - \dot{\vec{u}}(t)) ds \right| \le$$

$$\le \frac{\rho_\mu^{(k)}(u, \overline{u})}{C_0 Z_0} \int_{T+kT_0}^{t} e^{\mu(s - T - kT_0)} ds + \frac{\alpha \widehat{I}_0 e^{\alpha U_m}}{C_0} \int_{T+kT_0}^{t} |u(s) - \overline{u}(s)| ds + \frac{2e^2 I_m}{C_0 U_m} \int_{T+kT_0}^{t} |u(s) - \overline{u}(s)| ds \le$$

$$\le \frac{\rho_\mu^{(k)}(u, \overline{u})}{C_0 Z_0} \frac{e^{\mu(t - T - kT_0)} - 1}{\mu} + \frac{\alpha \widehat{I}_0 e^{\alpha U_m} \rho_\mu^{(k)}(u, \overline{u})}{C_0} \frac{e^{\mu(t - T - kT_0)} - 1}{\mu} + \frac{2e^2 I_m \rho_\mu^{(k)}(u, \overline{u})}{C_0 U_m} \frac{e^{\mu(t - T - kT_0)} - 1}{\mu} \le$$

$$\le \frac{\rho_\mu^{(k)}(\dot{u}, \dot{\overline{u}})}{\mu C_0 Z_0} \frac{e^{\mu(t - T - kT_0)} - 1}{\mu} + \frac{\alpha \widehat{I}_0 e^{\alpha U_m} \rho_\mu^{(k)}(\dot{u}, \dot{\overline{u}})}{\mu C_0} \frac{e^{\mu(t - T - kT_0)} - 1}{\mu} + \frac{2e^2 I_m \rho_\mu^{(k)}(\dot{u}, \dot{\overline{u}})}{\mu C_0 U_m} \frac{e^{\mu(t - T - kT_0)} - 1}{\mu} \le$$

$$\le e^{\mu(t - T - kT_0)} \left(\frac{\rho_\mu^{(k)}(\dot{u}, \dot{\overline{u}})}{\mu^2 C_0 Z_0} + \frac{\alpha \widehat{I}_0 e^{\alpha U_m} \rho_\mu^{(k)}(\dot{u}, \dot{\overline{u}})}{\mu^2 C_0} + \frac{2e^2 I_m \rho_\mu^{(k)}(\dot{u}, \dot{\overline{u}})}{\mu^2 C_0 U_m} \right) \le$$

$$\leq e^{\mu(t-T-kT_0)}\frac{1}{\mu^2 C_0}\left(\frac{1}{Z_0}+\alpha\widehat{I}_0 e^{\alpha U_m}+\frac{2e^2 I_m}{U_m}\right)\rho_\mu^{(k)}(\dot{u},\dot{\overline{u}})$$

and

$$U_2=\left|\int\limits_{T+kT_0}^{T+(k+1)T_0}\left[U(u)(s)-U(\overline{u})(s)\right]ds\right|\leq$$

$$\leq\frac{1}{C_0 Z_0}\int\limits_{T+kT_0}^{T+(k+1)T_0}\left|u(s)-\overline{u}(s)\right|ds+\frac{\widehat{I}_0}{C_0}\int\limits_{T+kT_0}^{T+(k+1)T_0}\left|e^{\alpha u(s)}-e^{\alpha\overline{u}(s)}\right|ds+\frac{I_m}{C_0 U_m}\int\limits_{T+kT_0}^{T+(k+1)T_0}\left|u(s)e^{1-\frac{u(s)}{U_m}}-\overline{u}(s)e^{1-\frac{\overline{u}(s)}{U_m}}\right|ds+$$

$$+\frac{1}{C_0 Z_0}\frac{\left|Z_0-R_0\right|}{Z_0+R_0}\int\limits_{T+kT_0}^{T+(k+1)T_0}\left|\overline{u}(t)-\overline{u}(t)\right|ds+\frac{\widehat{I}_0}{C_0}\frac{\left|Z_0-R_0\right|}{Z_0+R_0}\int\limits_{T+kT_0}^{T+(k+1)T_0}\left|e^{\alpha\overline{u}(s)}-e^{\alpha\overline{u}(s)}\right|ds+$$

$$+\frac{I_m}{C_0 U_m}\frac{\left|Z_0-R_0\right|}{Z_0+R_0}\int\limits_{T+kT_0}^{T+(k+1)T_0}\left|\overline{u}(s)e^{1-\frac{\overline{u}(s)}{U_m}}-\overline{u}(s)e^{1-\frac{\overline{u}(s)}{U_m}}\right|ds+\frac{\left|Z_0-R_0\right|}{Z_0+R_0}\left|\int\limits_{T+kT_0}^{T+(k+1)T_0}\left(\dot{\overline{u}}(s)-\dot{\overline{u}}(s)\right)ds\right|\leq$$

$$\leq\frac{\rho_\mu^{(k)}(u,\overline{u})}{C_0 Z_0}\int\limits_{T+kT_0}^{T+(k+1)T_0}e^{\mu(s-T-kT_0)}ds+\frac{\alpha\widehat{I}_0 e^{\alpha U_m}\rho_\mu^{(k)}(u,\overline{u})}{C_0}\int\limits_{T+kT_0}^{T+(k+1)T_0}e^{\mu(s-T-kT_0)}ds+$$

$$+\frac{2e^2 I_m\rho_\mu^{(k)}(u,\overline{u})}{C_0 U_m}\int\limits_{T+kT_0}^{T+(k+1)T_0}e^{\mu(s-T-kT_0)}ds\leq$$

$$\leq\rho_\mu^{(k)}(\dot{u},\dot{\overline{u}})\frac{e^{\mu T_0}-1}{\mu^2 C_0}\left(\frac{1}{Z_0}+\alpha\widehat{I}_0 e^{\alpha U_m}+\frac{2e^2 I_m}{U_m}\right)\leq e^{\mu(t-T-kT_0)}\frac{e^{\mu_0}-1}{\mu^2 C_0}\left(\frac{1}{Z_0}+\alpha\widehat{I}_0 e^{\alpha U_m}+\frac{2e^2 I_m}{U_m}\right)\rho_\mu^{(k)}(\dot{u},\dot{\overline{u}}).$$

Therefore, for $t\in[T+kT_0;\,T+(k+1)T_0]$

$$\left|B_k(u)(t)-B_k(\overline{u})(t)\right|\leq$$

$$\leq e^{\mu(t-T-kT_0)}\frac{1}{\mu^2 C_0}\left(\frac{1}{Z_0}+\alpha\widehat{I}_0 e^{\alpha U_m}+\frac{2e^2 I_m}{U_m}\right)\rho_\mu^{(k)}(\dot{u},\dot{\overline{u}})+$$

$$+e^{\mu(t-T-kT_0)}\frac{e^{\mu_0}-1}{\mu^2 C_0}\left(\frac{1}{Z_0}+\alpha\widehat{I}_0 e^{\alpha U_m}+\frac{2e^2 I_m}{U_m}\right)\rho_\mu^{(k)}(\dot{u},\dot{\overline{u}})\leq$$

$$\leq e^{\mu(t-T-kT_0)}\frac{e^{\mu 0}}{\mu^2 C_0}\left(\frac{1}{Z_0}+\alpha\widehat{I}_0 e^{\alpha U_m}+\frac{2e^2 I_m}{U_m}\right)\rho_\mu^{(k)}(\dot{u},\dot{\overline{u}})\equiv e^{\mu(t-T-kT_0)}K_U\,\rho_\mu^{(k)}(\dot{u},\dot{\overline{u}})\leq$$

$$\leq e^{\mu T_0}K_U\hat{\rho}_\mu((u,\dot{u}),(\overline{u},\dot{\overline{u}})).$$

it follows

$$\hat{\rho}(B(u),B(\overline{u}))\leq e^{\mu 0}K_U\hat{\rho}_\mu((u,\dot{u}),(\overline{u},\dot{\overline{u}})).$$

The remaining task here is to estimate the derivative of B. We have

$$\left|\dot{B}_k(u)(t)-\dot{B}_k(\overline{u})(t)\right|\leq$$

$$\leq\left|U(u)(s)-U(\overline{u})(s)\right|+\frac{1}{T_0}\left|\int_{T+kT_0}^{T+(k+1)T_0}[U(u)(s)-U(\overline{u})(s)]ds\right|\equiv\dot{U}_1+\dot{U}_2.$$

But

$$\dot{U}_1\leq\frac{1}{C_0 Z_0}\left|u(t)-\overline{u}(t)\right|+\frac{\widehat{I}_0}{C_0}\left|e^{\alpha u(t)}-e^{\alpha\overline{u}(t)}\right|+\frac{I_m}{C_0 U_m}\left|u(t)e^{1-\frac{u(s)}{U_m}}-\overline{u}(t)e^{1-\frac{\overline{u}(s)}{U_m}}\right|+$$

$$+\frac{1}{C_0 Z_0}\frac{|Z_0-R_0|}{Z_0+R_0}\left|\vec{u}(t)-\vec{u}(t)\right|+\frac{\widehat{I}_0}{C_0}\frac{|Z_0-R_0|}{Z_0+R_0}\left|e^{\alpha\vec{u}(t)}-e^{\alpha\vec{u}(t)}\right|+$$

$$+\frac{I_m}{C_0 U_m}\frac{|Z_0-R_0|}{Z_0+R_0}\left|\vec{u}(t)e^{1-\frac{\vec{u}(t)}{U_m}}-\vec{u}(t)e^{1-\frac{\vec{u}(t)}{U_m}}\right|+\frac{|Z_0-R_0|}{Z_0+R_0}\left|\dot{\vec{u}}(t)-\dot{\vec{u}}(t)\right|\leq$$

$$\leq\frac{1}{C_0 Z_0}\left|u(t)-\overline{u}(t)\right|+\frac{\alpha e^{\alpha U_m}\widehat{I}_0}{C_0}\left|u(t)-\overline{u}(t)\right|+\frac{2e^2 I_m}{C_0 U_m}\left|u(t)-\overline{u}(t)\right|\leq$$

$$\leq e^{\mu(t-T-kT_0)}\rho_\mu^{(k)}(u,\overline{u})\left(\frac{1}{C_0 Z_0}+\frac{\alpha e^{\alpha U_m}\widehat{I}_0}{C_0}+\frac{2e^2 I_m}{C_0 U_m}\right)\leq e^{\mu(t-T-kT_0)}\frac{\rho_\mu^{(k)}(\dot{u},\dot{\overline{u}})}{\mu}\left(\frac{1}{C_0 Z_0}+\frac{\alpha e^{\alpha U_m}\widehat{I}_0}{C_0}+\frac{2e^2 I_m}{C_0 U_m}\right)\leq$$

$$\leq e^{\mu(t-T-kT_0)}\frac{1}{\mu C_0}\left(\frac{1}{Z_0}+\alpha e^{\alpha U_m}\widehat{I}_0+\frac{2e^2 I_m}{U_m}\right)\rho_\mu^{(k)}(\dot{u},\dot{\overline{u}}).$$

We also have

$$\dot{U}_2=\frac{1}{T_0}U_2=\frac{1}{T_0}\left|\int_{T+kT_0}^{T+(k+1)T_0}[U(u)(s)-U(\overline{u})(s)]ds\right|\leq$$

$$\leq e^{\mu(t-T-kT_0)}\frac{1}{T_0}\frac{e^{\mu_0}-1}{\mu^2 C_0}\left(\frac{1}{Z_0}+\alpha\widehat{I}_0 e^{\alpha U_m}+\frac{2e^2 I_m}{U_m}\right)\rho_\mu^{(k)}(\dot{u},\dot{\overline{u}})\leq$$

$$\leq e^{\mu(t-T-kT_0)}\frac{e^{\mu_0}-1}{\mu_0}\frac{1}{\mu C_0}\left(\frac{1}{Z_0}+\alpha\widehat{I}_0 e^{\alpha U_m}+\frac{2e^2 I_m}{U_m}\right)\rho_\mu^{(k)}(\dot{u},\dot{\overline{u}}).$$

Therefore

$$\left|\dot{B}_k(u)(t)-\dot{B}_k(\overline{u})(t)\right|\leq$$

$$\leq e^{\mu(t-T-kT_0)}\frac{1}{\mu C_0}\left(\frac{1}{Z_0}+\alpha e^{\alpha U_m}\widehat{I}_0+\frac{2e^2 I_m}{U_m}\right)\rho_\mu^{(k)}(\dot{u},\dot{\overline{u}})+$$

$$+e^{\mu(t-T-kT_0)}\frac{e^{\mu_0}-1}{\mu_0}\frac{1}{\mu C_0}\left(\frac{1}{Z_0}+\alpha\widehat{I}_0 e^{\alpha U_m}+\frac{2e^2 I_m}{U_m}\right)\rho_\mu^{(k)}(\dot{u},\dot{\overline{u}})\leq$$

$$\leq e^{\mu(t-T-kT_0)}\left(1+\frac{e^{\mu_0}-1}{\mu_0}\right)\frac{1}{\mu C_0}\left(\frac{1}{Z_0}+\alpha\widehat{I}_0 e^{\alpha U_m}+\frac{2e^2 I_m}{U_m}\right)\rho_\mu^{(k)}(\dot{u},\dot{\overline{u}})\equiv\dot{K}_U\rho_\mu^{(k)}(\dot{u},\dot{\overline{u}})\leq\dot{K}_U\hat{\rho}_\mu((u,\dot{u}),(\overline{u},\dot{\overline{u}})).$$

and it follows that

$$\rho_\mu^{(k)}(\dot{B}_k(u),\dot{B}_k(\overline{u}))\leq\dot{K}_U\hat{\rho}_\mu((u,\dot{u}),(\overline{u},\dot{\overline{u}})).$$

Consequently

$$\hat{\rho}_\mu((B(u),\dot{B}(u)),(B\overline{u},\dot{B}(\overline{u})))\leq K\hat{\rho}_\mu((u,\dot{u}),(\overline{u},\dot{\overline{u}}))$$

where $K=\max\left\{e^{\mu_0}K_U,\dot{K}_U\right\}<1$, for sufficiently large $\mu>0$. Therefore operator B has a unique fixed point in M_U. This fixed point is a T_0-periodic solution of (2.6.4).

Theorem 2.6.1 is thus proved.

2.6.1. Numerical Example

We next consider all inequalities of the previous theorem.

$$\frac{e^{\mu_0}}{\mu C_0}\left[\frac{2U_E}{(Z_0+R_0)U_0}+\left(1+|\delta|\right)\left(\frac{1}{Z_0}+\alpha\widehat{I}_0\xi(\alpha U_m)+\frac{I_m e^2}{U_m}\right)\right]+|\delta|\leq 1;$$

$$K_U = \frac{e^{\mu_0}}{\mu^2 C_0}\left(\frac{1}{Z_0} + \alpha \hat{I}_0 e^{\alpha U_m} + \frac{2e^2 I_m}{U_m}\right) < 1;$$

$$\dot{K}_U = \left(\frac{e^{\mu_0}-1}{\mu_0}+1\right)\frac{1}{\mu C_0}\left(\frac{1}{Z_0} + \alpha \hat{I}_0 e^{\alpha U_m} + \frac{2e^2 I_m}{U_m}\right) < 1.$$

For a transmission line with specific parameters: $\Lambda = 3,5\,m$, $L = 0,45\,\mu H/m$, $C = 80\,pF/m$ we obtain $v = \dfrac{1}{\sqrt{LC}} = \dfrac{1}{\sqrt{0,45.10^{-6}.80.10^{-12}}} = \dfrac{1}{6.10^{-9}} = 1,66.10^8$,

$Z_0 = \sqrt{L/C} = 75\,\Omega$, $R_0 = 45\,\Omega$, $C_0 = 10\,pF = 10^{-11}\,F$.

Then $T = \Lambda\sqrt{LC} = 2.10^{-8}s$; $\left|Z_0 - R_0\right|/\left(Z_0 + R_0\right) = 0,25$; $2U_E = U_0$.

Let us now check the propagation of millimeter waves $\lambda_0 = (1/3)10^{-3}\,m$. We have

$$f_0 = \frac{1}{\lambda_0\sqrt{LC}} = \frac{1}{(1/3)10^{-3}.6.10^{-9}} = \frac{1}{2}10^{12}\,Hz \Rightarrow T_0 = \frac{1}{f_0} = 2.10^{-12}\,\text{sec.}$$

If we choose $\mu = \dfrac{1}{2}10^{12}$, then $\mu T_0 = \mu_0 = 1$ and $T = 2.10^{-8}.\dfrac{1}{2}10^{12}T_0 = 10000.T_0$.

We also have

$$\mu C_0 = (1/2)10^{12}.10.10^{-12} = 5;\quad \mu^2 C_0 = (1/4)10^{24}.10.10^{-12} = 1,25.10^{12},$$

$$E(t) = E_0 \sin(\omega_0 t) = 0,5\sin(\omega_0 t)$$

$$(\omega_0 = 2\pi/T_0 = 2\pi f_0 = 3,14.10^{12}).$$

Here $I_0 = 10^{-8}\,A$, $\alpha = q/k\tau = 1/(52.10^{-3}) = 19,23$; $I_m = 1,9.10^{-3}\,A$;

$$\frac{I_m}{U_m} = \frac{1,9.10^{-3}}{0,15} = 12,67.10^{-3},$$

$U_m = 0,15$; $1/U_m = 6,66$; $\alpha U_m = 2,8845$; $e^{\alpha U_m} \le e^3 = 20,124$.

Then

$$\frac{e}{5}\left[\frac{2}{120} + \frac{5}{4}\left(\frac{1}{75} + 10^{-8}\frac{e^{2,8845}-1}{0,15} + e^2 12,67.10^{-3}\right)\right] + \frac{1}{4} \le 1;$$

$$K_U = \frac{e}{1,25.10^{12}}\left(\frac{1}{75}+19,23.10^{-8}20,124+14,8.12,67.10^{-3}\right)<1\,;$$

$$\dot{K}_U = \frac{e}{5}\left(\frac{1}{75}+19,23.10^{-8}.20,124+14,8.12,67.10^{-3}\right)\approx 0,11<1.$$

Obviously $K = \dot{K}_U = 0,11<1$.

2.7. MEASURABLE PERIODIC SOLUTIONS FOR A NEUTRAL EQUATION WITH AN EXPONENTIAL NONLINEARITY

Let us recall that in order to obtain a solution in the class of step-wise functions arising in the discretization of the signals, we have to utilize the space of measurable functions.
Introduce the set

$$M_U^{\infty} = \left\{u(.)\in L_{T_0}^{\infty}[T,\infty): |u(t)|\le U_0 e^{\mu(t-T-kT_0)}, t\in[T+kT_0,T+(k+1)T_0]\right\}.$$

The set M_U^{∞} turns out to be a complete uniform space (cf. Chapter I, § 1.2) with a saturated family of pseudo-metrics

$$\hat{\rho}_{\mu}(u,\overline{u}) = \max\left\{\hat{\rho}(u,\overline{u}),\rho_{\mu}^{(k)}(\dot{u},\dot{\overline{u}}):k=0,1,2,,2m-1\right\}$$

(see § 2.3) where

$$\rho_{\mu}^{(k)}(u,\overline{u}) = \text{ess sup}\left\{e^{-\mu(t-T-kT_0)}|u(t)-\overline{u}(t)|:t\in[T+kT_0,T+(k+1)T_0]\right\},\ (k=0,1,2,...,2m-1),$$

$$\hat{\rho}(u,\overline{u}) = \text{ess sup}\left\{|u(t)-\overline{u}(t)|:t\in[T,3T]\right\}.$$

Define the operator $B(u)(t)$ on every interval $[T+kT_0,T+(k+1)T_0]$ for $u(.)\in M_U^{\infty}$ using the formulas:

$$B_k(u)(t) = \upsilon_{0T} + \int_{T+kT_0}^{t}U(u)(s)ds - \frac{t-T-kT_0}{T_0}\int_{T+kT_0}^{T+(k+1)T_0}U(u)(s)ds\ ,\ t\in[T+kT_0,T+(k+1)T_0] \qquad (2.7.1)$$

$(k=0,1,2,...,2m-1)$, where $\upsilon_{0T}=\upsilon_0(T)$,

$$U(u)(t)\equiv \upsilon_{0T} + \frac{1}{C_0}\frac{2E(t)}{Z_0+R_0} - \frac{1}{C_0Z_0}u(t) - \frac{\widehat{I}_0(e^{\alpha u(t)}-1)}{C_0} - \frac{I_m}{C_0U_m}u(t)e^{1-\frac{u(s)}{U_m}} -$$

$$-\frac{\delta}{C_0 Z_0}\bar{u}(t)+\delta\frac{\widehat{I_0}}{C_0}(e^{\alpha\bar{u}(t)}-1)+\delta\frac{I_m}{C_0 U_m}\bar{u}(t)e^{1-\frac{\bar{u}(t)}{U_m}}+\delta\frac{d\bar{u}(t)}{dt}\ ,\quad t\in[T,3T]$$

and $\bar{u}(t)=\begin{cases}v_0(t-2T),\, t\in[T,3T]\\ u(t-2T),\, t\in[3T,\infty)\end{cases}$, $\bar{v}_0(t)=v_0(t-2T),\, t\in[T,3T]$.

In fact we use the restriction of $\bar{u}(t)$ on $[T,3T]$, which coincides with $v_0(t)$.

Lemma 2.7.1. The problem (2.6.2) has the solution $u(.)\in M_U^\infty$ iff the operator B has a fixed point $u\in M_U^\infty$, that is, $u=B(u)$.

The proof is analogous to that of Lemma 2.6.1.

Theorem 2.7.1. Let the following conditions be fulfilled:

(IN) The initial function $v_0(.)\in L_{T_0}^\infty[-T,T]$, $|v_{0T}|<U_0$;

$$|v_0(t)|\le U_0 e^{\mu(t+T-kT_0)}\ ,\, t\in[-T+kT_0,-T+(k+1)T_0],\,(k=0,1,...,2m-1);$$

(E) $E(.)\in L_{T_0}^\infty(0,\infty)$,

$$|E(t)|\le U_E e^{\mu(t-T-kT_0)}\,(n=0,1,2,...),t\in[T+kT_0,T+(n+1)T_0].$$

In that case there exists a unique T_0-periodic solution of (2.6.2), belonging to M_U^∞.

Proof: The proof is analogous to that of Theorem 2.6.1.

The operator B maps the set M_U^∞ into itself. Indeed, for $u\in M_U^\infty$ the function $B(u)(t)$ is T_0-periodic on $[T,\infty)$ and $B(u)(t)=B(u)(t+T_0)$ a.e.

We show that $|u(t)|\le U_0 e^{\mu(t-T-kT_0)}\ \Rightarrow\ |B_k(u)(t)|\le U_0 e^{\mu(t-T-kT_0)}$.

Indeed, for every $u(.)\in M_U^\infty$ and sufficiently large μ we obtain (recall (2.6.5))

$$|(B_k u)(t)|\le|v_{0T}|+\left|\int_{T+kT_0}^{t}U(u)(s)ds\right|+\frac{1}{2}\left|\int_{T+kT_0}^{T+(k+1)T_0}U(u)(s)ds\right|\equiv|v_{0T}|+P_1+P_2.$$

We have

$$P_1\le\frac{2U_E}{C_0(Z_0+R_0)}\frac{e^{\mu(t-T-kT_0)}-1}{\mu}+\frac{U_0}{C_0 Z_0}\frac{e^{\mu(t-T-kT_0)}-1}{\mu}+\frac{\alpha U_0\widehat{I_0}}{C_0}\xi(\alpha U_m)\frac{e^{\mu(t-T-kT_0)}-1}{\mu}+$$

$$+\frac{I_m U_0}{C_0 U_m}e^2\frac{e^{\mu(t-T-kT_0)}-1}{\mu}+\frac{U_0}{C_0 Z_0}\frac{|Z_0-R_0|}{Z_0+R_0}\frac{e^{\mu(t-T-kT_0)}-1}{\mu}+$$

$$+\frac{|Z_0-R_0|}{Z_0+R_0}\frac{\alpha U_0\widehat{I_0}}{C_0}\xi(\alpha U_m)\frac{e^{\mu(t-T-kT_0)}-1}{\mu}+\frac{|Z_0-R_0|}{Z_0+R_0}\frac{I_m U_0}{C_0 U_m}e^2\frac{e^{\mu(t-T-kT_0)}-1}{\mu}+$$

$$+ e^{\mu(t-T-kT_0)} \frac{|Z_0 - R_0|}{Z_0 + R_0} U_0 \le$$

$$\le e^{\mu(t-T-kT_0)} \left\{ \frac{1}{\mu C_0} \left[\frac{2U_E}{Z_0 + R_0} + U_0 \left(1 + \frac{|Z_0 - R_0|}{Z_0 + R_0} \right) \left(\frac{1}{Z_0} + \alpha \widehat{I}_0 \xi(\alpha U_m) + \frac{I_m e^2}{U_m} \right) \right] + U_0 \frac{|Z_0 - R_0|}{Z_0 + R_0} \right\}$$

and

$$P_2 \le \frac{2}{C_0(Z_0 + R_0)} \int\limits_{T+kT_0}^{T+(k+1)T_0} |E(s-T)| ds + \frac{1}{C_0 Z_0} \int\limits_{T+kT_0}^{T+(k+1)T_0} |u(s)| ds + \frac{\widehat{I}_0}{C_0} \int\limits_{T+kT_0}^{T+(k+1)T_0} \left| e^{\alpha u(s)} - 1 \right| ds +$$

$$+ \frac{I_m}{C_0 U_m} \int\limits_{T+kT_0}^{T+(k+1)T_0} |u(s)| e^{1-\frac{u(s)}{U_m}} ds + \frac{1}{C_0 Z_0} \frac{|Z_0 - R_0|}{Z_0 + R_0} \int\limits_{T+kT_0}^{T+(k+1)T_0} |\vec{v}_0(s)| ds + \frac{\widehat{I}_0}{C_0} \frac{|Z_0 - R_0|}{Z_0 + R_0} \int\limits_{T+kT_0}^{T+(k+1)T_0} \left| e^{\alpha \vec{v}_0(s)} - 1 \right| ds +$$

$$+ \frac{I_m}{C_0 U_m} \frac{|Z_0 - R_0|}{Z_0 + R_0} \int\limits_{T+kT_0}^{T+(k+1)T_0} |\vec{v}_0(s)| e^{1-\frac{\vec{v}_0(s)}{U_m}} ds + \frac{|Z_0 - R_0|}{Z_0 + R_0} \left| \int\limits_{T+kT_0}^{T+(k+1)T_0} \dot{\vec{v}}_0(s) ds \right| \le$$

$$\le e^{\mu(t-T-kT_0)} \frac{e^{\mu_0} - 1}{\mu C_0} \left[\frac{2U_E}{Z_0 + R_0} + U_0 \left(1 + |\delta| \right) \left(\frac{1}{Z_0} + \alpha \widehat{I}_0 \xi(\alpha U_m) + \frac{I_m e^2}{U_m} \right) \right].$$

Therefore for sufficiently large $\mu > 0$

$$\left| B_k(u)(t) \right| \le |\upsilon_{0T}| + P_1 + P_2 \le e^{\mu(t-T-kT_0)} |\upsilon_{0T}| +$$

$$+ e^{\mu(t-T-kT_0)} \left\{ \frac{e^{\mu_0}}{\mu C_0} \left[\frac{2U_E}{Z_0 + R_0} + U_0 \left(1 + |\delta| \right) \left(\frac{1}{Z_0} + \alpha \widehat{I}_0 \xi(\alpha U_m) + \frac{I_m e^2}{U_m} \right) \right] + U_0 |\delta| \right\} \le e^{\mu(t-T-kT_0)} U_0 .$$

Consequently, operator B maps M_U into itself.

It remains to show that B is a contractive operator.

For every $u(.), \overline{u}(.) \in M_U^\infty$ we get:

$$\left| B_k(u)(t) - B_k(\widetilde{u})(t) \right| \le$$

$$\left| \int\limits_{T+kT_0}^{t} [U(u)(s) - U(\widetilde{u})(s)] ds \right| + \frac{1}{2} \left| \int\limits_{T+kT_0}^{T+(k+1)T_0} [U(u)(s) - U(\widetilde{u})(s)] ds \right| \equiv M_1 + M_2 .$$

We have

$$M_1 \le \frac{1}{C_0 Z_0} \int\limits_{T+kT_0}^{t} |u(s) - \widetilde{u}(s)| ds + \frac{\widehat{I}_0}{C_0} \int\limits_{T+kT_0}^{t} \left| e^{\alpha u(s)} - e^{\alpha \widetilde{u}(s)} \right| ds + \frac{I_m}{C_0 U_m} \int\limits_{T+kT_0}^{t} \left| u(s) e^{1-\frac{u(s)}{U_m}} - \widetilde{u}(s) e^{1-\frac{\widetilde{u}(s)}{U_m}} \right| ds +$$

$$+\frac{1}{C_0 Z_0}\frac{|Z_0-R_0|}{Z_0+R_0}\int_{T+kT_0}^{t}|\bar{u}(s)-\vec{u}(s)|ds+\frac{\widehat{I}_0}{C_0}\frac{|Z_0-R_0|}{Z_0+R_0}\int_{T+kT_0}^{t}\left|e^{\alpha\bar{u}(s)}-e^{\alpha\vec{u}(s)}\right|ds+$$

$$+\frac{I_m}{C_0 U_m}\frac{|Z_0-R_0|}{Z_0+R_0}\int_{T+kT_0}^{t}\left|\bar{u}(s)e^{1-\frac{\bar{u}(s)}{U_m}}-\bar{\upsilon}_0(s)e^{1-\frac{\vec{u}(s)}{U_m}}\right|ds+\frac{|Z_0-R_0|}{Z_0+R_0}\left|\int_{T+kT_0}^{t}\left(\dot{\bar{u}}(s)-\dot{\vec{u}}(s)\right)ds\right|\le$$

$$\le e^{\mu(t-T-kT_0)}\left[\frac{1}{\mu C_0}\left(\frac{1}{Z_0}+\alpha\widehat{I}_0 e^{\alpha U_m}+\frac{2e^2 I_m}{U_m}\right)\right]\rho_\mu^{(k)}(u,\widetilde{u})$$

and

$$M_2=\left|\int_{T+kT_0}^{T+(k+1)T_0}\left[U(u)(s)-U(\widetilde{u})(s)\right]ds\right|\le$$

$$\le\frac{1}{C_0 Z_0}\int_{T+kT_0}^{T+(k+1)T_0}|u(s)-\widetilde{u}(s)|ds+\frac{\widehat{I}_0}{C_0}\int_{T+kT_0}^{T+(k+1)T_0}\left|e^{\alpha u(s)}-e^{\alpha\widetilde{u}(s)}\right|ds+\frac{I_m}{C_0 U_m}\int_{T+kT_0}^{T+(k+1)T_0}\left|u(s)e^{1-\frac{u(s)}{U_m}}-\widetilde{u}(s)e^{1-\frac{\widetilde{u}(s)}{U_m}}\right|ds+$$

$$+\frac{1}{C_0 Z_0}\frac{|Z_0-R_0|}{Z_0+R_0}\int_{T+kT_0}^{T+(k+1)T_0}|\bar{u}(s)-\vec{u}(s)|ds+\frac{\widehat{I}_0}{C_0}\frac{|Z_0-R_0|}{Z_0+R_0}\int_{T+kT_0}^{T+(k+1)T_0}\left|e^{\alpha\bar{u}(s)}-e^{\alpha\vec{u}(s)}\right|ds+$$

$$+\frac{I_m}{C_0 U_m}\frac{|Z_0-R_0|}{Z_0+R_0}\int_{T+kT_0}^{T+(k+1)T_0}\left|\bar{u}(s)e^{1-\frac{\bar{u}(s)}{U_m}}-\bar{\upsilon}_0(s)e^{1-\frac{\bar{u}(s)}{U_m}}\right|ds+\frac{|Z_0-R_0|}{Z_0+R_0}\left|\int_{T+kT_0}^{T+(k+1)T_0}\left(\dot{\bar{u}}(s)-\dot{\vec{u}}(s)\right)ds\right|\le$$

$$\le e^{\mu(t-T-kT_0)}\frac{e^{\mu_0}-1}{\mu C_0}\left(\frac{1}{Z_0}+\alpha\widehat{I}_0 e^{\alpha U_m}+\frac{2e^2 I_m}{U_m}\right)\rho_\mu^{(k)}(u,\widetilde{u}).$$

Therefore

$$\left|B_k(u)(t)-B_k(\widetilde{u})(t)\right|\le e^{\mu(t-T-kT_0)}\frac{e^{\mu_0}}{\mu C_0}\left(\frac{1}{Z_0}+\alpha\widehat{I}_0 e^{\alpha U_m}+\frac{2e^2 I_m}{U_m}\right)\hat{\rho}_\mu((u,\dot{u}),(\widetilde{u},\dot{\widetilde{u}}))\equiv$$

$$\equiv e^{\mu(t-T-kT_0)}K_U\,\hat{\rho}_\mu((u,\dot{u}),(\widetilde{u},\dot{\widetilde{u}})).$$

It follows $\hat{\rho}(B(u),B(\bar{u}))\le e^{\mu_0}K_U\,\hat{\rho}_\mu((u,\dot{u}),(\widetilde{u},\dot{\widetilde{u}}))$.

For the derivatives we have

$$\left|\dot{B}_k(u)(t)-\dot{B}_k(\widetilde{u})(t)\right|\le$$

$$\leq \left| U(u)(t) - U(\widetilde{u})(t) \right| + \frac{1}{T_0} \left| \int_{T+kT_0}^{T+(k+1)T_0} [U(u)(s) - U(\widetilde{u})(s)] ds \right| \equiv \dot{M}_1 + \dot{M}_2 .$$

But

$$\dot{M}_1 \leq \frac{1}{C_0 Z_0} \left| u(t) - \widetilde{u}(t) \right| + \frac{\hat{I}_0}{C_0} \left| e^{\alpha u(t)} - e^{\alpha \widetilde{u}(t)} \right| + \frac{I_m}{C_0 U_m} \left| u(t) e^{1-\frac{u(t)}{U_m}} - \widetilde{u}(t) e^{1-\frac{\widetilde{u}(t)}{U_m}} \right| \leq$$

$$\leq e^{\mu(t-T-kT_0)} \left[\frac{1}{\mu C_0} \left(\frac{1}{Z_0} + \alpha \hat{I}_0 e^{\alpha U_m} + \frac{2e^2 I_m}{U_m} \right) \right] \rho_\mu^{(k)}(\dot{u}, \dot{\widetilde{u}})$$

and

$$\dot{M}_2 \leq \frac{1}{T_0} \left| \int_{T+kT_0}^{T+(k+1)T_0} [U(u)(s) - U(\widetilde{u})(s)] ds \right| \leq$$

$$\leq \frac{1}{C_0 Z_0} \int_{T+kT_0}^{T+(k+1)T_0} |u(s) - \widetilde{u}(s)| ds + \frac{\hat{I}_0}{C_0} \int_{T+kT_0}^{T+(k+1)T_0} \left| e^{\alpha u(s)} - e^{\alpha \widetilde{u}(s)} \right| ds + \frac{I_m}{C_0 U_m} \int_{T+kT_0}^{T+(k+1)T_0} \left| u(s) e^{1-\frac{u(s)}{U_m}} - \widetilde{u}(s) e^{1-\frac{\widetilde{u}(s)}{U_m}} \right| ds +$$

$$+ \frac{1}{C_0 Z_0} \frac{|Z_0 - R_0|}{Z_0 + R_0} \left| \int_{T+kT_0}^{t} (\dot{\widetilde{u}}(s) - \dot{u}(s)) ds \right| \leq \int_{T+kT_0}^{T+(k+1)T_0} |\dot{\widetilde{u}}(s) - \dot{u}(s)| ds + \frac{\hat{I}_0}{C_0} \frac{|Z_0 - R_0|}{Z_0 + R_0} \int_{T+kT_0}^{T+(k+1)T_0} \left| e^{\alpha \widetilde{u}(s)} - e^{\alpha u(s)} \right| ds +$$

$$+ \frac{I_m}{C_0 U_m} \frac{|Z_0 - R_0|}{Z_0 + R_0} \int_{T+kT_0}^{T+(k+1)T_0} \left| \widetilde{u}(s) e^{1-\frac{\widetilde{u}(s)}{U_m}} - u(s) e^{1-\frac{u(s)}{U_m}} \right| ds + \frac{|Z_0 - R_0|}{Z_0 + R_0} \left| \int_{T+kT_0}^{T+(k+1)T_0} (\dot{\widetilde{u}}(s) - \dot{u}(s)) ds \right| \leq$$

$$\leq e^{\mu(t-T-kT_0)} \frac{e^{\mu_0} - 1}{\mu_0} \frac{1}{\mu C_0} \left[\frac{1}{Z_0} + \alpha \hat{I}_0 e^{\alpha U_m} + \frac{2e^2 I_m}{U_m} \right] \rho_\mu^{(k)}(\dot{u}, \dot{\widetilde{u}}) .$$

Then

$$\hat{\rho}_\mu((B(u), \dot{B}(u)), (B(u), \dot{B}(\widetilde{u}))) \leq K \hat{\rho}_\mu((u, \dot{u}), (\widetilde{u}, \dot{\widetilde{u}}))$$

where $K = \max \left\{ e^{\mu_0} K_U, \dot{K}_U \right\} < 1$, for sufficiently large $\mu > 0$. Therefore B has a unique fixed point, which is a measurable periodic solution of (2.6.2).

Theorem 2.7.1 is thus proved.

2.8. OSCILLATORY REGIMES IN THE EXPONENTIAL CASE

Here we deal with the problem of the existence of oscillating solutions in the following neutral equation with exponential nonlinearity:

$$\frac{du(t)}{dt} = \frac{1}{C_0}\frac{2E(t)}{Z_0 + R_0} - \frac{1}{C_0 Z_0}u(t) - \frac{\widehat{I}_0(e^{\alpha u(t)} - 1)}{C_0} - \frac{I_m}{C_0 U_m}u(t)e^{1-\frac{u(s)}{U_m}} - \frac{\delta}{C_0 Z_0}u(t - 2T) +$$

$$+ \frac{\delta \widehat{I}_0}{C_0}(e^{\alpha u(t-2T)} - 1) + \frac{\delta . I_m}{C_0 U_m}u(t-2T)e^{1-\frac{u(t-2T)}{U_m}} + \delta\frac{du(t-2T)}{dt}, \quad t \geq T \qquad (2.8.1)$$

$$u(t) = \upsilon_0(t), \frac{du(t)}{dt} = \frac{d\upsilon_0(t)}{dt}, t \in [-T, T]$$

where $(x,t) \in \Pi = \left\{(x,t) \in \Pi^2 : (x,t) \in [0,\Lambda] \times [0,\infty)\right\}$, $u(t) = u(\Lambda, t)$, $\delta = \dfrac{Z_0 - R_0}{Z_0 + R_0}$.

Note that the initial function $\upsilon_0(t)$ can be obtained by translating the initial function $u_0(x)$, as in § 2.1.

Now we are able to formulate the main problem: to find a solution of (2.8.1) with advanced prescribed zeros on an interval $[t_0, \infty)$, $T = t_0$, where $\upsilon_0(t)$ is a prescribed oscillating function on the interval $[-T, T]$.

Let $S_T = \{\tau_k\}_{k=1}^n, n \in N$ be the set of zeros of the initial function, that is, $\upsilon_0(\tau_k) = 0$ such that $\tau_1 = -T$, $\tau_n = T \equiv t_0$. Besides $\max\{\tau_{k+1} - \tau_k : k = 0,1,...,n\} \leq T_0$.

Let $S = \{t_k\}_{k=0}^\infty$ be a strictly increasing sequence of real numbers satisfying the following conditions (C):

(C1) $\lim\limits_{k \to \infty} t_k = \infty$;

(C2) for every k there is $s < k$ such that $t_k - T = t_s$ where $t_s \in S_T \cup S$.

It follows

$$0 \leq \inf\{t_{k+1} - t_k : k = 0,1,2,...\} \leq \sup\{t_{k+1} - t_k : k = 0,1,2,...\} = T_0 < \infty$$

and $t_k - 2T = t_q$ for some $q < k$.

Consider the sets

$$M_S = \left\{u(.) \in C^1[t_0, \infty) : u(t_k) = 0 \, (k = 0,1,2,...)\right\},$$

$$M_{SU} = \left\{ u(.) \in M_S : |u(t)| \leq U_0 e^{\mu(t-t_k)}, t \in [t_k, t_{k=1}] \right\}$$

where U_0, μ are positive constants. Recall that $\mu T_0 = \mu_0 = \text{const.} > 0$.

Introduce the following family of pseudo-metrics

$$\rho^{(k)}(u,\overline{u}) = \max\left\{ |u(t) - \overline{u}(t)| : t \in [t_k, t_{k+1}] \right\},$$

$$\hat{\rho}^{(k)}(u,\overline{u}) = \max\left\{ |u(t) - \overline{u}(t)| : t \in [t_0, t_{k+1}] \right\},$$

$$\rho_\mu^{(k)}(u,\overline{u}) = \max\left\{ e^{-\mu(t-t_k)} |u(t) - \overline{u}(t)| : t \in [t_k, t_{k+1}] \right\},$$

$$\hat{\rho}_\mu^{(k)}(u,\overline{u}) = \max\left\{ \rho_\mu^{(0)}(u,\overline{u}), \rho_\mu^{(1)}(u,\overline{u}),..., \rho_\mu^{(k)}(u,\overline{u}) \right\},$$

$$\rho_\mu^{(k)}(\dot{u},\dot{\overline{u}}) = \max\left\{ e^{-\mu(t-t_k)} |\dot{u}(t) - \dot{\overline{u}}(t)| : t \in [t_k, t_{k+1}] \right\},$$

$$\hat{\rho}_\mu^{(k)}(\dot{u},\dot{\overline{u}}) = \max\left\{ \rho_\mu^{(0)}(\dot{u},\dot{\overline{u}}), \rho_\mu^{(1)}(\dot{u},\dot{\overline{u}}),..., \rho_\mu^{(k)}(\dot{u},\dot{\overline{u}}) \right\}.$$

We use the inequalities from Remark 2.4.1. The set M_{SU} turns into a complete uniform space with respect to the saturated family of pseudo-metrics

$$\hat{\rho}_\mu^{(k)}((u,\dot{u}),(\overline{u},\dot{\overline{u}})) = \max\left\{ \hat{\rho}^{(k)}(u,\overline{u}), \hat{\rho}_\mu^{(k)}(\dot{u},\dot{\overline{u}}) \right\} (k = 0,1,2,...).$$

Define an operator *B* by the formulas

$$B(u)(t) := \int_{t_k}^{t} U(u)(s)ds - \frac{t-t_k}{t_{k+1}-t_k} \int_{t_k}^{t_{k+1}} U(u)(s)ds, \quad t \in [t_k, t_{k+1}], \ (k = 0,1,2,...)$$

where

$$U(u)(t) = \frac{1}{C_0}\frac{2E(t)}{Z_0 + R_0} - \frac{1}{C_0 Z_0}u(t) - \frac{\widehat{I}_0(e^{\alpha u(t)} - 1)}{C_0} - \frac{I_m}{C_0 U_m}u(t)e^{1-\frac{u(s)}{U_m}} - \frac{\delta}{C_0 Z_0}\overline{u}(t) +$$

$$+ \frac{\delta \widehat{I}_0}{C_0}(e^{\alpha \overline{u}(t)} - 1) + \frac{\delta . I_m}{C_0 U_m}\overline{u}(t)e^{1-\frac{\overline{u}(t)}{U_m}} + \delta\frac{d\overline{u}(t)}{dt}, \quad t \geq T.$$

In this manner, considering the problem to the right from *T*, we do not need the conformity condition **(CC)**.

Remark 2.8.1. Prior to formulating the next results we recall

$$0 < U_0 e^{\mu(t_{k+1}-t_k)} \le U_0 e^{\mu T_0} \le U_m$$

and

$$g(u) = u e^{1-\frac{u}{U_m}},$$

$$g'(u) = e^{1-\frac{u}{U_m}}\left(1-\frac{u}{U_m}\right) \Rightarrow |g'(u)| \le \left(1+\frac{U_0 e^{\mu T_0}}{U_m}\right) e^{1+\frac{U_0 e^{\mu T_0}}{U_m}} \le 2 e^{1+\frac{U_0 e^{\mu 0}}{U_m}} \le 2e^2.$$

The following lemma is true:

Lemma 2.8.1. Let $E(.) \in C^1_{S_T \cup S}[0,\infty), |E(t)| \le U_0 e^{\mu(t-t_k)}, t \in [t_k,t_{k+1}]$. Problem (2.8.1) has a solution $u(.) \in M_{SU}$ iff the operator B has a fixed-point in M_{SU}, that is, $u(t) = B(u)(t)$.

Proof: Let $u(.) \in M_{SU}$ be a solution of (2.8.1). Then, integrating (2.8.1) on the interval $[t_k,t] \subset [t_k,t_{k+1}]$ $(k = 0,1,2 \ldots)$ we obtain

$$u(t) - u(t_k) = \int_{t_k}^{t} U(u)(s)ds \iff u(t) = \int_{t_k}^{t} U(u)(s)ds$$

and then

$$u(t) = \int_{t_k}^{t} U(u)(s)ds \Rightarrow 0 = u(t_{k+1}) = \int_{t_k}^{t_{k+1}} U(u)(s)ds \Rightarrow \int_{t_k}^{t_{k+1}} U(u)(s)ds = 0 \; (2.8.2)$$

Therefore $u(t)$ satisfies

$$u(t) = \int_{t_k}^{t} U(u)(s)ds - \frac{t-t_k}{t_{k+1}-t_k}\int_{t_k}^{t_{k+1}} U(u)(s)ds, \; t \in [t_k,t_{k+1}] \iff u = B(u),$$

that is, $u(.)$ is a fixed point of B.

Conversely, let $u(.) \in M_{SU}$ be a solution of $u = B(u)$, that is,

$$u(t) = \int_{t_k}^{t} U(u)(s)ds - \frac{t-t_k}{t_{k+1}-t_k}\int_{t_k}^{t_{k+1}} U(u)(s)ds.$$

Then

$$\left|\int_{t_k}^{t_{k+1}} U(u)(s)ds\right| \le$$

$$\le \frac{1}{C_0}\frac{2}{Z_0+R_0}\int_{t_k}^{t_{k+1}} E(t-T)dt + \frac{1}{C_0 Z_0}\int_{t_k}^{t_{k+1}}|u(t)|dt + \frac{\widehat{I}_0}{C_0}\int_{t_k}^{t_{k+1}}\left|e^{\alpha u(t)}-1\right|dt + \frac{I_m}{C_0 U_m}\int_{t_k}^{t_{k+1}}\left|u(t)e^{1-\frac{u(t)}{U_m}}\right|dt +$$

$$+\frac{1}{C_0 Z_0}\frac{|Z_0-R_0|}{Z_0+R_0}\int_{t_k}^{t_{k+1}}|u(t-2T)|dt + \frac{\widehat{I}_0}{C_0}\frac{|Z_0-R_0|}{Z_0+R_0}\int_{t_k}^{t_{k+1}}\left|e^{\alpha u(t-2T)}-1\right|dt +$$

$$+\frac{I_m}{C_0 U_m}\frac{|Z_0-R_0|}{Z_0+R_0}\int_{t_k}^{t_{k+1}}\left|u(t-2T)e^{1-\frac{u(t-2T)}{U_m}}\right|dt + \frac{|Z_0-R_0|}{Z_0+R_0}\int_{t_k}^{t_{k+1}}\frac{du(t-2T)}{dt}dt \le$$

$$\le \frac{2U_0 e^{\mu_0}}{C_0(Z_0+R_0)}\frac{e^{\mu(t_{k+1}-t_k)}-1}{\mu} + \frac{U_0}{C_0 Z_0}\frac{e^{\mu(t_{k+1}-t_k)}-1}{\mu} + \frac{\alpha U_0 \widehat{I}_0}{C_0}\xi(\alpha U_m)\frac{e^{\mu(t_{k+1}-t_k)}-1}{\mu} +$$

$$+\frac{I_m U_0 e^2}{C_0 U_m}\frac{e^{\mu(t_{k+1}-t_k)}-1}{\mu} + \frac{U_0 e^{\mu_0}}{C_0 Z_0}\frac{|Z_0-R_0|}{Z_0+R_0}\frac{e^{\mu(t_{k+1}-t_k)}-1}{\mu} + \frac{\alpha U_0 e^{\mu_0}\widehat{I}_0}{C_0}\frac{|Z_0-R_0|}{Z_0+R_0}\xi(\alpha U_m)\frac{e^{\mu(t_{k+1}-t_k)}-1}{\mu} +$$

$$+\frac{|Z_0-R_0|}{Z_0+R_0}\frac{I_m e^{\mu_0}U_0 e^2}{C_0 U_m}\frac{e^{\mu(t_{k+1}-t_k)}-1}{\mu} + \frac{|Z_0-R_0|}{Z_0+R_0}|u(t_{k+1}-2T)-u(t_k-2T)| \le$$

$$\le \frac{e^{\mu_0}-1}{\mu C_0}U_0\left[\frac{2e^{\mu_0}}{Z_0+R_0}+\left(1+e^{\mu_0}|\delta|\right)\left(\frac{1}{Z_0}+\alpha\widehat{I}_0\xi(\alpha U_m)+\frac{I_m e^2}{U_m}\right)\right]\equiv M(\mu) \quad.$$

Let us assume that $\left|\int_{t_k}^{t_{k+1}} U(u)(t)dt\right| = \gamma > 0$. On the other hand, we have obtained

$\gamma \le M(\mu)$. Recall $\mu T_0 = \mu_0 = \text{const.}$ and for sufficiently large $\mu > 0$ (and sufficiently small $T_0 > 0$) one can reach the inequality $M(\mu) < \gamma$. The obtained contradiction implies

$\int_{t_k}^{t_{k+1}} U(u)(t)dt = 0$. It follows $u(t) = \int_{t_k}^{t} U(u)(s)ds$ and after a differentiation we obtain

(2.8.1).

Lemma 2.8.1 is thus proved.

Theorem 2.8.1. Let the following conditions be fulfilled:

1) The initial function $\upsilon_0(.) \in C^1[-T,T]$ satisfies $\left|\upsilon_0(t)\right| \le U_0 e^{\mu(t-\tau_k)}, t \in [\tau_k,\tau_{k+1}]$, $\upsilon_0(t_0) = 0 \ (\Rightarrow \upsilon_0(0)=0)$;

2) $E(.) \in C_S^1[0,\infty), |E(t)| \le U_E e^{\mu(t-t_k)}, t \in [t_k,t_{k+1}], E(t_k-T) = E(t_q), q < k$.

In this case, there exists a unique oscillatory solution of (2.8.1), belonging to M_{SU}.

Proof: We show that B maps M_{SU} into itself, that is, $u \in M_{SU}$ implies $B(u) \in M_{SU}$.

We know that $B(u)(t)$ is continuous on $[t_0, \infty)$ and differentiable on every $[t_k, t_{k+1}]$. This can be verified in the proof of Theorem 2.4.1. We also have $B(u)(t_k) = 0$ and $B(u)(t_{k+1}) = 0$.

We show that $\left| B_k(u)(t) \right| \le U_0 e^{\mu(t-t_k)}, t \in [t_k, t_{k+1}]$.

For every $u(.) \in M_{SU}$ the function $B(u)(t)$ is continuous and differentiable on every interval $[t_k, t_{k+1}]$. We see that $\left| \dfrac{t - t_k}{t_{k+1} - t_k} \right| \le 1$, $t \in [t_k, t_{k+1}]$. For sufficiently large μ and $t \in [t_k, t_{k+1}]$ we obtain:

$$\left| B_k(u)(t) \right| \le \left| \int_{t_k}^{t} U(u)(s)ds \right| + \left| \int_{t_k}^{t_{k+1}} U(u)(s)ds \right| \equiv P_1 + P_2.$$

We have

$$P_1 \le \left[\frac{2}{C_0(Z_0 + R_0)} \int_{t_k}^{t} \left| E(s) \right| ds + \frac{1}{C_0 Z_0} \int_{t_k}^{t} \left| u(s) \right| ds + \frac{\widehat{I}_0}{C_0} \int_{t_k}^{t} \left| e^{\alpha u(s)} - 1 \right| ds + \right.$$

$$+ \frac{I_m}{C_0 U_m} \int_{t_k}^{t} \left| u(s) \right| e^{1 - \frac{u(s)}{U_m}} ds +$$

$$+ \frac{1}{C_0 Z_0} \frac{\left| Z_0 - R_0 \right|}{Z_0 + R_0} \int_{t_k}^{t} \left| u(s - 2T) \right| ds + \frac{\widehat{I}_0}{C_0} \frac{\left| Z_0 - R_0 \right|}{Z_0 + R_0} \int_{t_k}^{t} \left| e^{\alpha u(s - 2T)} - 1 \right| ds +$$

$$+ \frac{I_m}{C_0 U_m} \frac{\left| Z_0 - R_0 \right|}{Z_0 + R_0} \int_{t_k}^{t} \left| u(s - 2T) \right| e^{1 - \frac{u(s - 2T)}{U_m}} ds + \frac{\left| Z_0 - R_0 \right|}{Z_0 + R_0} \left| \int_{t_k}^{t} u(s - 2T) ds \right| \le$$

$$\le \frac{2 U_E e^{\mu_0}}{C_0(Z_0 + R_0)} \frac{e^{\mu(t-t_k)} - 1}{\mu} + \frac{U_0}{C_0 Z_0} \frac{e^{\mu(t-t_k)} - 1}{\mu} + \frac{\alpha U_0 \widehat{I}_0}{C_0} \xi(\alpha U_m) \frac{e^{\mu(t-t_k)} - 1}{\mu} + \frac{I_m U_0 e^2}{C_0 U_m} \frac{e^{\mu(t-t_k)} - 1}{\mu} +$$

$$+ \frac{U_0 e^{\mu_0}}{C_0 Z_0} \frac{\left| Z_0 - R_0 \right|}{Z_0 + R_0} \frac{e^{\mu(t-t_k)} - 1}{\mu} + \frac{\alpha U_0 \widehat{I}_0 e^{\mu_0}}{C_0} \frac{\left| Z_0 - R_0 \right|}{Z_0 + R_0} \xi(\alpha U_m) \frac{e^{\mu(t-t_k)} - 1}{\mu} +$$

$$+ \frac{I_m U_0 e^{\mu_0} e^2}{C_0 U_m} \frac{\left| Z_0 - R_0 \right|}{Z_0 + R_0} \frac{e^{\mu(t-t_k)} - 1}{\mu} + \frac{\left| Z_0 - R_0 \right|}{Z_0 + R_0} \left| u(t - 2T) \right| \le$$

$$\leq e^{\mu(t-t_k)}\left\{\frac{U_0}{\mu C_0}\left[\frac{2e^{\mu_0}}{Z_0+R_0}+\left(1+e^{\mu_0}\frac{|Z_0-R_0|}{Z_0+R_0}\right)\left(\frac{1}{Z_0}+\alpha\widehat{I}_0\xi(\alpha U_m)+\frac{I_m e^2}{U_m}\right)\right]+e^{\mu_0}\frac{|Z_0-R_0|}{Z_0+R_0}U_0\right\}$$

and

$$P_2\leq\left[\frac{2}{C_0(Z_0+R_0)}\int\limits_{t_k}^{t_{k+1}}|E(s-T)|ds+\frac{1}{C_0 Z_0}\int\limits_{t_k}^{t_{k+1}}|u(s)|ds+\frac{\widehat{I}_0}{C_0}\int\limits_{t_k}^{t_{k+1}}\left|e^{\alpha u(s)}-1\right|ds+\right.$$

$$+\frac{I_m}{C_0 U_m}\int\limits_{t_k}^{T+T_0}|u(s)|e^{1-\frac{u(s)}{U_m}}ds+\frac{1}{C_0 Z_0}\frac{|Z_0-R_0|}{Z_0+R_0}\int\limits_{t_k}^{t_{k+1}}|u(s-2T)|ds$$

$$+\frac{\widehat{I}_0}{C_0}\frac{|Z_0-R_0|}{Z_0+R_0}\int\limits_{t_k}^{t_{k+1}}\left|e^{\alpha u(s-2T)}-1\right|ds+$$

$$+\frac{I_m}{C_0 U_m}\frac{|Z_0-R_0|}{Z_0+R_0}\int\limits_{t_k}^{t_{k+1}}|u(s-2T)|e^{1-\frac{u(s-2T)}{U_m}}ds+\left.\frac{|Z_0-R_0|}{Z_0+R_0}\left|\int\limits_{t_k}^{t_{k+1}}\dot{u}(s-2T)ds\right|\right]\leq$$

$$\leq e^{\mu(t-t_k)}\frac{\left(e^{\mu_0}-1\right)U_0}{\mu C_0}\left[\frac{2e^{\mu_0}}{Z_0+R_0}+\left(1+e^{\mu_0}\frac{|Z_0-R_0|}{Z_0+R_0}\right)\left(\frac{1}{Z_0}+\alpha\widehat{I}_0\xi(\alpha U_m)+\frac{I_m e^2}{U_m}\right)\right].$$

Therefore, for sufficiently large $\mu>0$ and $\mu T_0=\text{const.}$

$$|B_k(u)(t)|\leq P_1+P_2\leq$$

$$\leq e^{\mu(t-t_k)}\left\{\frac{U_0}{\mu C_0}\left[\frac{2e^{\mu_0}}{Z_0+R_0}+\left(1+e^{\mu_0}\frac{|Z_0-R_0|}{Z_0+R_0}\right)\left(\frac{1}{Z_0}+\alpha\widehat{I}_0\xi(\alpha U_m)+\frac{I_m e^2}{U_m}\right)\right]+e^{\mu_0}\frac{|Z_0-R_0|}{Z_0+R_0}U_0\right\}+$$

$$+e^{\mu(t-t_k)}\frac{\left(e^{\mu_0}-1\right)U_0}{\mu C_0}\left[\frac{2e^{\mu_0}}{Z_0+R_0}+\left(1+e^{\mu_0}\frac{|Z_0-R_0|}{Z_0+R_0}\right)\left(\frac{1}{Z_0}+\alpha\widehat{I}_0\xi(\alpha U_m)+\frac{I_m e^2}{U_m}\right)\right]\leq$$

$$\leq e^{\mu(t-t_k)}\left\{\frac{e^{\mu_0}U_0}{\mu C_0}\left[\frac{2e^{\mu_0}}{Z_0+R_0}+\left(1+e^{\mu_0}|\delta|\right)\left(\frac{1}{Z_0}+\alpha\widehat{I}_0\xi(\alpha U_m)+\frac{I_m e^2}{U_m}\right)\right]+e^{\mu_0}|\delta|U_0\right\}\leq e^{\mu(t-t_k)}U_0.$$

In what follows we show that B is a contractive operator.
For $t\in[t_k,t_{k+1}]$:

$$|B_k(u)(t)-B_k(\overline{u})(t)|\leq\left|\int\limits_{t_k}^{t}[U(u)(s)-U(\overline{u})(s)]ds\right|+\left|\int\limits_{t_k}^{t_{k+1}}[U(u)(s)-U(\overline{u})(s)]ds\right|\equiv M_1+M_2$$

$$M_1 = \left| \int_{t_k}^{t} \left[U(u)(s) - U(\overline{u})(s) \right] ds \right| \le$$

$$\le \left[\frac{1}{C_0 Z_0} \int_{t_k}^{t} |u(s) - \overline{u}(s)| ds + \frac{\widehat{I}_0}{C_0} \int_{t_k}^{t} \left| e^{\alpha u(s)} - e^{\alpha \overline{u}(s)} \right| ds + \frac{I_m}{C_0 U_m} \int_{t_k}^{t} \left| u(s) e^{1 - \frac{u(s)}{U_m}} - \overline{u}(s) e^{1 - \frac{\overline{u}(s)}{U_m}} \right| ds + \right.$$

$$+ \frac{1}{C_0 Z_0} \frac{|Z_0 - R_0|}{Z_0 + R_0} \int_{t_k}^{t} |u(s) - \overline{u}(s)| ds + \frac{\widehat{I}_0}{C_0} \frac{|Z_0 - R_0|}{Z_0 + R_0} \int_{t_k}^{t} \left| e^{\alpha u(s)} - e^{\alpha \overline{u}(s)} \right| ds +$$

$$+ \frac{I_m}{C_0 U_m} \frac{|Z_0 - R_0|}{Z_0 + R_0} \int_{t_k}^{t} \left| u(s) e^{1 - \frac{u(s)}{U_m}} - \overline{u}(s) e^{1 - \frac{\overline{u}(s)}{U_m}} \right| ds + \frac{|Z_0 - R_0|}{Z_0 + R_0} \left| \int_{t_k}^{t} \left(\frac{du(s)}{ds} - \frac{d\overline{u}(s)}{ds} \right) ds \right| \le$$

$$\le \frac{1}{C_0 Z_0} \rho_\mu^{(k)}(u,\overline{u}) \frac{e^{\mu(t-t_k)} - 1}{\mu} + \frac{\alpha e^{\alpha U_m} \widehat{I}_0}{C_0} \rho_\mu^{(k)}(u,\overline{u}) \frac{e^{\mu(t-t_k)} - 1}{\mu} + \frac{2e^2 I_m}{C_0 U_m} \rho_\mu^{(k)}(u,\overline{u}) \frac{e^{\mu(t-t_k)} - 1}{\mu} +$$

$$+ \frac{1}{C_0 Z_0} \frac{|Z_0 - R_0|}{Z_0 + R_0} \rho_\mu^{(k)}(u,\overline{u}) \frac{e^{\mu(t-t_k)} - 1}{\mu} + \frac{\widehat{I}_0 \alpha e^{\alpha U_m}}{C_0} \frac{|Z_0 - R_0|}{Z_0 + R_0} \rho_\mu^{(k)}(u,\overline{u}) \frac{e^{\mu(t-t_k)} - 1}{\mu} +$$

$$+ \frac{2e^2 I_m}{C_0 U_m} \frac{|Z_0 - R_0|}{Z_0 + R_0} \rho_\mu^{(k)}(u,\overline{u}) \frac{e^{\mu(t-t_k)} - 1}{\mu} + \frac{|Z_0 - R_0|}{Z_0 + R_0} \rho_\mu^{(k)}(u,\overline{u}) \le$$

$$\le \frac{1}{C_0 Z_0} \rho_\mu^{(k)}(u,\overline{u}) \frac{e^{\mu(t-t_k)} - 1}{\mu} + \frac{\alpha e^{\alpha U_m} \widehat{I}_0}{C_0} \rho_\mu^{(k)}(u,\overline{u}) \frac{e^{\mu(t-t_k)} - 1}{\mu} + \frac{2e^2 I_m}{C_0 U_m} \rho_\mu^{(k)}(u,\overline{u}) \frac{e^{\mu(t-t_k)} - 1}{\mu} +$$

$$+ \frac{1}{C_0 Z_0} \frac{|Z_0 - R_0|}{Z_0 + R_0} \rho_\mu^{(k)}(u,\overline{u}) \frac{e^{\mu(t-t_k)} - 1}{\mu} + \frac{\widehat{I}_0 \alpha e^{\alpha U_m}}{C_0} \frac{|Z_0 - R_0|}{Z_0 + R_0} \rho_\mu^{(k)}(u,\overline{u}) \frac{e^{\mu(t-t_k)} - 1}{\mu} +$$

$$+ \frac{2e^2 I_m}{C_0 U_m} \frac{|Z_0 - R_0|}{Z_0 + R_0} \rho_\mu^{(k)}(u,\overline{u}) \frac{e^{\mu(t-t_k)} - 1}{\mu} + \frac{|Z_0 - R_0|}{Z_0 + R_0} \rho_\mu^{(k)}(u,\overline{u}) \le$$

$$\le e^{\mu(t-t_k)} \widehat{\rho}_\mu^{(k)}(u,\overline{u}) \left[\frac{1}{\mu C_0 Z_0} + \frac{\alpha e^{\alpha U_m} \widehat{I}_0}{\mu C_0} + \frac{2e^2 I_m}{\mu C_0 U_m} + \frac{1}{\mu C_0 Z_0} \frac{|Z_0 - R_0|}{Z_0 + R_0} + \frac{\alpha e^{\alpha U_m} I_0}{\mu C_0} \frac{|Z_0 - R_0|}{Z_0 + R_0} + \right.$$

$$\left. + \frac{2e^2 I_m}{\mu C_0 U_m} \frac{|Z_0 - R_0|}{Z_0 + R_0} + \frac{|Z_0 - R_0|}{Z_0 + R_0} \right] \le$$

$$\le e^{\mu(t-t_k)} \frac{\widehat{\rho}_\mu^{(k)}(\dot{u},\dot{\overline{u}})}{\mu} \left[\frac{1}{\mu C_0 Z_0} \left(1 + \frac{|Z_0 - R_0|}{Z_0 + R_0} \right) + \frac{\alpha e^{\alpha U_m} \widehat{I}_0}{\mu C_0} \left(1 + \frac{|Z_0 - R_0|}{Z_0 + R_0} \right) + \frac{2e^2 I_m}{\mu C_0 U_m} \left(1 + \frac{|Z_0 - R_0|}{Z_0 + R_0} \right) + \right.$$

$$\left. + \frac{|Z_0 - R_0|}{Z_0 + R_0} \right] \le$$

$$\le e^{\mu(t-t_k)}\frac{\hat{\rho}_\mu^{(k)}(\dot{u},\dot{\overline{u}})}{\mu}\left[\left(1+\frac{|Z_0-R_0|}{Z_0+R_0}\right)\frac{1}{\mu C_0}\left(\frac{1}{Z_0}+\alpha\,e^{\alpha U_m}\hat{I}_0+\frac{2e^2 I_m}{U_m}\right)+\frac{|Z_0-R_0|}{Z_0+R_0}\right]$$

and

$$M_2=\left|\int_{t_k}^{t_{k+1}}\left[U(u)(s)-U(\overline{u})(s)\right]ds\right|\le$$

$$\le\left[\frac{1}{C_0 Z_0}\int_{t_k}^{t_{k+1}}|u(s)-\widetilde{u}(s)|ds+\frac{\hat{I}_0}{C_0}\int_{t_k}^{t_{k+1}}\left|e^{\alpha u(s)}-e^{\alpha\overline{u}(s)}\right|ds+\frac{I_m}{C_0 U_m}\int_{t_k}^{t_{k+1}}\left|u(s)e^{1-\frac{u(s)}{U_m}}-\widetilde{u}(s)e^{1-\frac{\overline{u}(s)}{U_m}}\right|ds+\right.$$

$$+\frac{1}{C_0 Z_0}\frac{|Z_0-R_0|}{Z_0+R_0}\int_{t_k}^{t_{k+1}}|u-\overline{u}(s)|ds+\frac{\hat{I}_0}{C_0}\frac{|Z_0-R_0|}{Z_0+R_0}\int_{t_k}^{t_{k+1}}\left|e^{\alpha u(s)}-e^{\alpha\overline{u}(s-2T)}\right|ds+$$

$$+\frac{I_m}{C_0 U_m}\frac{|Z_0-R_0|}{Z_0+R_0}\int_{t_k}^{t_{k+1}}\left|\vec{u}(s)e^{1-\frac{u(s)}{U_m}}-\vec{\widetilde{u}}(s)e^{1-\frac{\overline{u}(s)}{U_m}}\right|ds+\frac{|Z_0-R_0|}{Z_0+R_0}\left|\int_{t_k}^{t_{k+1}}(\dot{u}(s)-\dot{\overline{u}}(s))ds\right|\le$$

$$\le\frac{\rho_\mu^{(k)}(u,\overline{u})}{C_0 Z_0}\frac{e^{\mu(t_{k+1}-t_k)}-1}{\mu}+\frac{\alpha\,e^{\alpha U_0 e^{\mu T_0}}\hat{I}_0\rho_\mu^{(k)}(u,\overline{u})}{C_0}\frac{e^{\mu(t_{k+1}-t_k)}-1}{\mu}+\frac{2e^2 I_m\rho_\mu^{(k)}(u,\overline{u})}{C_0 U_m}\frac{e^{\mu(t_{k+1}-t_k)}-1}{\mu}+$$

$$+\frac{\rho_\mu^{(k)}(u,\overline{u})}{C_0 Z_0}\frac{|Z_0-R_0|}{Z_0+R_0}\frac{e^{\mu(t_{k+1}-t_k)}-1}{\mu}+\frac{\rho_\mu^{(k)}(u,\overline{u})\alpha\,e^{\alpha U_0 e^{\mu_0}}\hat{I}_0}{C_0}\frac{|Z_0-R_0|}{Z_0+R_0}\frac{e^{\mu(t_{k+1}-t_k)}-1}{\mu}+$$

$$+\frac{\rho_\mu^{(k)}(u,\overline{u})I_m}{C_0 U_m}\frac{|Z_0-R_0|}{Z_0+R_0}2e^2\frac{e^{\mu(t_{k+1}-t_k)}-1}{\mu}\le$$

$$\le\frac{\rho_\mu^{(k)}(\dot{u},\dot{\overline{u}})}{\mu C_0}\frac{e^{\mu T_0}-1}{\mu}\left[\frac{1}{Z_0}+\alpha\,e^{\alpha U_m}I_0+\frac{2e^2 I_m}{U_m}+\frac{1}{Z_0}\frac{|Z_0-R_0|}{Z_0+R_0}+\alpha\,e^{\alpha U_m}\hat{I}_0\frac{|Z_0-R_0|}{Z_0+R_0}+\right.$$

$$+\frac{I_m}{U_m}\frac{|Z_0-R_0|}{Z_0+R_0}2e^2\Bigg]\le$$

$$\le e^{\mu(t-t_k)}\hat{\rho}_\mu^{(k)}(\dot{u},\dot{\overline{u}})\frac{e^{\mu_0}-1}{\mu}\left[\left(1+\frac{|Z_0-R_0|}{Z_0+R_0}\right)\frac{1}{\mu C_0}\left(\frac{1}{Z_0}+\alpha\,e^{\alpha U_m}\hat{I}_0+\frac{2e^2 I_m}{U_m}\right)\right].$$

Therefore

$$\left|B_k(u)(t)-B_k(\overline{u})(t)\right|\le$$

$$\le e^{\mu(t-t_k)}\hat{\rho}_\mu^{(k)}(\dot{u},\dot{\overline{u}})\frac{1}{\mu}\left[\left(1+\frac{|Z_0-R_0|}{Z_0+R_0}\right)\frac{1}{\mu C_0}\left(\frac{1}{Z_0}+\alpha\,e^{\alpha U_m}\hat{I}_0+\frac{2e^2 I_m}{U_m}\right)+\frac{|Z_0-R_0|}{Z_0+R_0}\right]+$$

$$+ e^{\mu(t-t_k)} \hat{\rho}_\mu^{(k)}(\dot{u},\dot{\overline{u}}) \frac{e^{\mu_0}-1}{\mu}\left[\left(1+\frac{|Z_0-R_0|}{Z_0+R_0}\right)\frac{1}{\mu C_0}\left(\frac{1}{Z_0}+\alpha\, e^{\alpha U_m} I_0 + \frac{2e^2 I_m}{U_m}\right)\right] \le$$

$$\le e^{\mu(t-t_k)} \hat{\rho}_\mu^{(k)}(\dot{u},\dot{\overline{u}})\frac{e^{\mu_0}}{\mu^2 C_0}\left(1+|\delta|\right)\left(\frac{1}{Z_0}+\alpha\, e^{\alpha U_m}\widehat{I}_0 + \frac{2e^2 I_m}{U_m}\right)\equiv$$

$$\equiv e^{\mu(t-t_k)} K \hat{\rho}_\mu^{(k)}((u,\dot{u}),(\overline{u},\dot{\overline{u}}))$$

and then

$$\hat{\rho}^{(k)}(B(u),B(\overline{u})) \le e^{\mu_0} K_u \hat{\rho}_\mu^{(k)}((u,\dot{u}),(\overline{u},\dot{\overline{u}})) \,.$$

Finally, we estimate the derivative of B:

$$\left|\dot{B}_k(u)(t)-\dot{B}_k(\overline{u})(t)\right| \le \left|U(u)(s)-U(\overline{u})(s)\right| + \frac{1}{t_{k+1}-t_k}\left|\int_{t_k}^{t_{k+1}}\left[U(u)(s)-U(\overline{u})(s)\right]ds\right| \equiv \dot{U}_1 + \dot{U}_2 \,.$$

We have

$$\dot{U}_1 \le \frac{1}{C_0 Z_0}\left|u(t)-\overline{u}(t)\right| + \frac{\widehat{I}_0}{C_0}\left|e^{\alpha u(t)}-e^{\alpha \overline{u}(t)}\right| + \frac{I_m}{C_0 U_m}\left|u(t)e^{1-\frac{u(s)}{U_m}} - \overline{u}(t)e^{1-\frac{\overline{u}(s)}{U_m}}\right| +$$

$$+ \frac{1}{C_0 Z_0}\frac{|Z_0-R_0|}{Z_0+R_0}\left|u(t)-\overline{u}(t)\right| + \frac{\widehat{I}_0}{C_0}\frac{|Z_0-R_0|}{Z_0+R_0}\left|e^{\alpha u(t)}-e^{\alpha \overline{u}(t-2T)}\right| +$$

$$+ \frac{I_m}{C_0 U_m}\frac{|Z_0-R_0|}{Z_0+R_0}\left|u(t)e^{1-\frac{u(t)}{U_m}} - \overline{u}(t)e^{1-\frac{\overline{u}(t)}{U_m}}\right| + \frac{|Z_0-R_0|}{Z_0+R_0}\left|\dot{u}(t)-\dot{\overline{u}}(t)\right| \le$$

$$\le \frac{1}{C_0 Z_0}\rho_\mu^{(k)}(u,\overline{u})e^{\mu(t-t_k)} + \frac{\alpha\, e^{\alpha U_m}\widehat{I}_0}{C_0}\rho_\mu^{(k)}(u,\overline{u})e^{\mu(t-t_k)} + \frac{2e^2 I_m}{C_0 U_m}\rho_\mu^{(k)}(u,\overline{u})e^{\mu(t-t_k)} +$$

$$+ \frac{1}{C_0 Z_0}\frac{|Z_0-R_0|}{Z_0+R_0}\rho_\mu^{(k)}(u,\overline{u})e^{\mu(t-t_k)} + \frac{\alpha\, e^{\alpha U_m}\widehat{I}_0}{C_0}\frac{|Z_0-R_0|}{Z_0+R_0}\rho_\mu^{(k)}(u,\overline{u})e^{\mu(t-t_k)} +$$

$$+ \frac{2e^2 I_m}{C_0 U_m}\frac{|Z_0-R_0|}{Z_0+R_0}\rho_\mu^{(k)}(u,\overline{u})e^{\mu(t-t_k)} + \frac{|Z_0-R_0|}{Z_0+R_0}\rho_\mu^{(k)}(\dot{u},\dot{\overline{u}})e^{\mu(t-t_k)} \le$$

$$\le e^{\mu(t-t_k)} \hat{\rho}_\mu^{(k)}(\dot{u},\dot{\overline{u}})\left[\frac{1}{\mu C_0}\left(1+\frac{|Z_0-R_0|}{Z_0+R_0}\right)\left(\frac{1}{Z_0}+\alpha\, e^{\alpha U_m}\widehat{I}_0 + \frac{2e^2 I_m}{U_m}\right) + \frac{|Z_0-R_0|}{Z_0+R_0}\right]$$

and

$$\dot{U}_2 = \frac{1}{t_{k+1}-t_k}\left|\int_{t_k}^{t_{k+1}}\left[U(u)(s)-U(\overline{u})(s)\right]ds\right| \leq$$

$$\leq \frac{1}{t_{k+1}-t_k}\left[\frac{1}{C_0 Z_0}\int_{t_k}^{t_{k+1}}\left|u(s)-\overline{u}(s)\right|ds + \frac{I_0}{C_0}\int_{t_k}^{t_{k+1}}\left|e^{\alpha u(s)}-e^{\alpha\overline{u}(s)}\right|ds + \frac{I_m}{C_0 U_m}\int_{t_k}^{t_{k+1}}\left|u(s)e^{1-\frac{u(s)}{U_m}}-\overline{u}(s)e^{1-\frac{\overline{u}(s)}{U_m}}\right|ds + \right.$$

$$+\frac{1}{C_0 Z_0}\frac{|Z_0-R_0|}{Z_0+R_0}\int_{t_k}^{t_{k+1}}\left|u(s)-\overline{u}(s)\right|ds + \frac{\widehat{I}_0}{C_0}\frac{|Z_0-R_0|}{Z_0+R_0}\int_{t_k}^{t_{k+1}}\left|e^{\alpha u(s)}-e^{\alpha\overline{u}(s)}\right|ds +$$

$$\left.+\frac{I_m}{C_0 U_m}\frac{|Z_0-R_0|}{Z_0+R_0}\int_{t_k}^{t_{k+1}}\left|u(s)e^{1-\frac{u(s)}{U_m}}-\overline{u}(s)e^{1-\frac{\overline{u}(s)}{U_m}}\right|ds + \frac{|Z_0-R_0|}{Z_0+R_0}\left|\int_{t_k}^{t_{k+1}}\left(\dot{u}(s)-\dot{\overline{u}}(s)\right)ds\right|\right] \leq$$

$$\leq \frac{\hat{\rho}_\mu^{(k)}(\dot{u},\dot{\overline{u}})}{\mu C_0}\frac{e^{\mu(t_{k+1}-t_k)}-1}{\mu(t_{k+1}-t_k)}\left(\frac{1}{Z_0}+\alpha\, e^{\alpha U_m}\widehat{I}_0 + \frac{2e^2 I_m}{U_m} + \right.$$

$$\left.+\frac{1}{Z_0}\frac{|Z_0-R_0|}{Z_0+R_0}+\alpha\, e^{\alpha U_m}I_0\frac{|Z_0-R_0|}{Z_0+R_0}+\frac{2e^2 I_m}{U_m}\frac{|Z_0-R_0|}{Z_0+R_0}\right) \leq$$

$$\leq e^{\mu(t-t_k)}\hat{\rho}_\mu^{(k)}(\dot{u},\dot{\overline{u}})\frac{e^{\mu_0}-1}{\mu_0}\frac{1}{\mu C_0}\left(1+\frac{|Z_0-R_0|}{Z_0+R_0}\right)\left(\frac{1}{Z_0}+\alpha\, e^{\alpha U_m}\widehat{I}_0 + \frac{2e^2 I_m}{U_m}\right).$$

Therefore

$$\left|\dot{B}_k(u)(t)-\dot{B}_k(\overline{u})(t)\right| \leq$$

$$\leq e^{\mu(t-t_k)}\hat{\rho}_\mu^{(k)}(\dot{u},\dot{\overline{u}})\left[\frac{1}{\mu C_0}\left(1+\frac{e^{\mu_0}-1}{\mu_0}\right)\left(1+|\delta|\right)\left(\frac{1}{Z_0}+\alpha\, e^{\alpha U_m}\widehat{I}_0 + \frac{2e^2 I_m}{U_m}\right)+|\delta|\right] \equiv$$

$$\equiv e^{\mu(t-t_k)}\dot{K}_u\rho_\mu^{(k)}(\dot{u},\dot{\overline{u}}) \leq e^{\mu(t-t_k)}\dot{K}_u\hat{\rho}_\mu^{(k)}((u,\dot{u}),(\overline{u},\dot{\overline{u}})).$$

It follows

$$\rho_\mu^{(k)}(\dot{B}_k(u),\dot{B}_k(\overline{u})) \leq \dot{K}_u\hat{\rho}_\mu^{(k)}((u,\dot{u}),(\overline{u},\dot{\overline{u}})).$$

Consequently

$$\hat{\rho}_\mu^{(k)}((B(u),\dot{B}(u)),(B(\overline{u}),\dot{B}(\overline{u}))) \leq \hat{K}\hat{\rho}_\mu^{(k)}((u,\dot{u}),(\overline{u},\dot{\overline{u}})) \quad (k=0,1,2,\ldots)$$

where $\hat{K} = \max\{e^{\mu_0} K_u, \dot{K}_u\} < 1$ does not depend on u and k.

We have to verify that M_{SU} is j-bounded. As we know, since j is an identity mapping then

$$\hat{\rho}_{\mu}^{j^n(k)}((u,\dot{u}),(\overline{u},\dot{\overline{u}})) \le \hat{\rho}_{\mu}^{(k)}((u,\dot{u}),(\overline{u},\dot{\overline{u}})) < \infty \quad (n=0,1,2,\ldots)$$

which implies M_{SU} is j-bounded.

Therefore, in view of the fixed-point Theorem 1.2.7 for contractive mappings in uniform spaces (cf. Chapter I), operator B has a unique fixed point and it is an oscillating solution of (2.8.1).

Theorem 2.8.1 is thus proved.

2.8.1. Numerical Example

We collect all inequalities necessary for our applications.

$$\frac{e^{\mu_0}}{\mu C_0}\left[\frac{2}{Z_0 + R_0} + (1+e^{\mu_0}|\delta|)\left(\frac{1}{Z_0} + \widehat{I}_0 \frac{e^{\alpha U_m}-1}{U_m} + \frac{I_m e^2}{U_m}\right)\right] + e^{\mu_0}|\delta| \le 1;$$

$$K = \frac{e^{\mu_0}}{\mu^2 C_0}\left(\frac{1}{Z_0} + \alpha e^{\alpha U_m}\widehat{I}_0 + \frac{2e^2 I_m}{U_m}\right) < 1;$$

$$\dot{K} = \frac{1}{\mu C_0}\left(1 + \frac{e^{\mu_0}-1}{\mu_0}\right)\left(\frac{1}{Z_0} + \alpha e^{\alpha U_m}\widehat{I}_0 + \frac{2e^2 I_m}{U_m}\right) < 1.$$

Consider a line with the following specific parameters

$$\Lambda = 2\,m, \; L = 0{,}2\,\mu H/m, \; C = 80\,pF/m, \; v = 1/\sqrt{LC} = 1/(4.10^{-9}) = 2{,}5.10^8$$

$$Z_0 = \sqrt{L/C} = 50\,\Omega, \quad R_0 = 45\Omega, \; C_0 = 8pF = 8.10^{-12}\,F.$$

Then $T = \Lambda\sqrt{LC} = 8.10^{-9}\,s$; $|\delta| = \dfrac{Z_0 - R_0}{Z_0 + R_0} = \dfrac{1}{19} = 0{,}0526$.

Let us check the propagation of millimeter waves $\lambda_0 = 0{,}25.10^{-3}\,m$. We have

$$f_0 = \frac{1}{\lambda_0\sqrt{LC}} = \frac{1}{0{,}25.10^{-3}.4.10^{-9}} = 10^{12}\,Hz \Rightarrow T_0 = \frac{1}{f_0} = 10^{-12}\,sec.$$

Choose $\mu = 10^{12}$, then $\mu T_0 = \mu_0 = 1$ and

$$\mu C_0 = 10^{12}.8.10^{-12} = 8, \quad \mu^2 C_0 = 8.10^{12}, \quad \widehat{I}_0 = 10^{-8}\,A,$$

$$\alpha = \frac{q}{k\tau} = \frac{1}{2.26.10^{-3}} = 19{,}23\,;\frac{I_m}{U_m} = \frac{1{,}9.10^{-3}\,A}{0{,}15V} = 12{,}67.10^{-3},$$

$$U_m = 0{,}15;\quad \alpha U_m = 2{,}8845.$$

From this, the above inequalities become

$$\frac{e}{8}\left[\frac{2}{95} + \left(1 + \frac{e}{19}\right)\left(\frac{1}{50} + 10^{-8}\frac{e^3 - 1}{3} + 12{,}67.10^{-3}e^2\right)\right] + \frac{e}{19} \le 1,$$

$$K = \frac{e^2}{4.10^{12}}\left(\frac{1}{50} + \alpha e^{2{,}8845}.10^{-8} + 2.12{,}67.10^{-3}e^2\right) < 1,$$

$$\dot{K} = \frac{e}{8}\left(\frac{1}{50} + 19{,}23\,e^3.10^{-8} + 2e^2 12{,}67.10^{-3}\right) = 0{,}07 < 1.$$

If we take a line with $Z_0 = 75\,\Omega$ then K becomes smaller.

Remark 2.8.1. The method described here is suitable for non-uniform lines as

$$C_e\frac{\partial u(x,t)}{\partial t} + \frac{\partial i(x,t)}{\partial x} = 0, \quad L_e\frac{\partial i(x,t)}{\partial t} + \frac{\partial u(x,t)}{\partial x} = 0.$$

The above inequalities could be satisfied for sufficiently large $\mu > 0$ even though $Z_0 = \sqrt{L_e / C_e}$ decreases for increasing frequencies (cf. Remark 2.2.7).

2.9. ANOTHER WAY TO REDUCING THE MIXED PROBLEM TO A PERIODIC PROBLEM FOR A NEUTRAL EQUATION WITH AN EXPONENTIAL NONLINEARITY

We proceed from the transformation

$$u(x,t) = \frac{1}{2}U(x,t) + \frac{1}{2}I(x,t),$$

$$i(x,t) = \frac{1}{2Z_0}U(x,t) - \frac{1}{2Z_0}I(x,t)$$

and then

$$u(0,t) = \frac{1}{2}U(0,t) + \frac{1}{2}I(0,t) \qquad u(\Lambda,t) = \frac{1}{2}U(\Lambda,t) + \frac{1}{2}I(\Lambda,t)$$

$$\text{and}$$

$$i(0,t) = \frac{1}{2Z_0}U(0,t) - \frac{1}{2Z_0}I(0,t) \qquad i(\Lambda,t) = \frac{1}{2Z_0}U(\Lambda,t) - \frac{1}{2Z_0}I(\Lambda,t)$$

Replace into the boundary conditions

$$E(t) - \frac{1}{2}U(0,t) - \frac{1}{2}I(0,t) = R_0 \frac{U(0,t) - I(0,t)}{2Z_0}, \quad t \geq T$$

$$C_0 \frac{d}{dt}\left(\frac{U(\Lambda,t-T) + I(0,t)}{2}\right) = \frac{1}{2Z_0}U(\Lambda,t) - \frac{1}{2Z_0}I(\Lambda,t) -$$

$$f\left(\frac{1}{2}U(\Lambda,t) + \frac{1}{2}I(\Lambda,t)\right), t \geq T .$$

But $U(0,t) = U(\Lambda,t+T), \quad I(\Lambda,t) = I(0,t+T)$ and then

$$E(t) - \frac{1}{2}U(\Lambda,t+T) + \frac{1}{2}I(0,t) = R_0 \frac{U(\Lambda,t+T) - I(0,t)}{2Z_0}, \quad t \geq T$$

$$C_0 \frac{d}{dt}\left(\frac{U(\Lambda,t-T) + I(0,t)}{2}\right) = \frac{1}{2Z_0}U(\Lambda,t) - \frac{1}{2Z_0}I(0,t+T) -$$

$$f\left(\frac{1}{2}U(\Lambda,t) + \frac{1}{2}I(0,t+T)\right), t \geq T .$$

Let us replace $t+T$ by t. Then the system above becomes:

$$E(t-T) - \frac{1}{2}U(\Lambda,t) - \frac{1}{2}I(0,t-T) = R_0 \frac{U(\Lambda,t) - I(0,t-T)}{2Z_0}, \quad t \geq T$$

$$C_0 \frac{d}{dt}\left(\frac{U(\Lambda,t-T)+I(0,t)}{2}\right) = \frac{1}{2Z_0}U(\Lambda,t-T) - \frac{1}{2Z_0}I(0,t) -$$

$$f\left(\frac{1}{2}U(\Lambda,t-T) + \frac{1}{2}I(0,t)\right), t \geq T .$$

The initial conditions are

$$U_0(x) = U(x,0) = u(x,0) + Z_0\, i(x,0) = u_0(x) + Z_0\, i_0(x),$$
$$I_0(x) = I(x,0) = u(x,0) - Z_0\, i(x,0) = u_0(x) - Z_0\, i_0(x) \, , \, x \in [0,\Lambda].$$

As above, one can shift the initial functions along the characteristics on the interval $[0,T]$ and obtain initial functions $U_0(t)$, $I_0(t)$ on the initial set $t \in [0,T]$. They are assumed $C^1_{T_0}[0,T]$ functions.

If we assume unknown functions to be $U(\Lambda,t) \equiv U(t)$, $I(0,t) \equiv I(t)$ then we can formulate the following periodic problem for the system:

$$U(t) = 2E(t) - I(t-T) - \frac{R_0}{Z_0}U(t) + \frac{R_0}{Z_0}I(t-T), \qquad t \in [T,2T]$$

$$\frac{dI(t)}{dt} = -\frac{dU(t-T)}{dt} + \frac{1}{C_0 Z_0}U(t-T) - \frac{1}{C_0 Z_0}I(t) - \qquad (2.9.1)$$

$$-\frac{2}{C_0}\widehat{I}_0(e^{\alpha\frac{U(t-T)+I(t)}{2}} - 1) + \frac{I_m}{U_m}\frac{U(t-T)+I(t)}{2}e^{1-\frac{U(t-T)+I(t)}{2U_m}} \quad , t \in [T,2T]$$

$$U(t) = U_0(t), \;\; I(t) = I_0(t), \;\; \frac{dU(t)}{dt} = \frac{dU_0(t)}{dt}, \frac{dI(t)}{dt} = \frac{dI_0(t)}{dt}, t \in [0,T]$$

The conformity condition is

$$\frac{dI(T)}{dt} = -\frac{dU(0)}{dt} + \frac{1}{C_0 Z_0}U(0) - \frac{1}{C_0 Z_0}I(T) -$$

$$-\frac{2}{C_0}\widehat{I}_0\left(e^{\alpha\frac{U(0)+I(T)}{2}} - 1\right) + \frac{I_m}{U_m}\frac{U(0)+I(T)}{2}e^{1-\frac{U(0)+I(T)}{2U_m}}$$

satisfied for $U(0) = \dot{U}(0) = 0$, $I(0) = \dot{I}(0) = 0 \Rightarrow \dot{I}(T) = 0$ because $T = mT_0$.

By $C^1_{T_0}[T,\infty)$ we mean the space of all continuous T_0-periodic functions with continuous derivatives. Introduce the sets

$$M_U = \left\{ u(.) \in C^1_{T_0}[T,\infty) : |U(t)| \le U_0 e^{\mu(t-T-kT_0)}, t \in [T+kT,\ T+(k+1)T_0];\ (k=0,1,2,...,m-1) \right\}$$

$$M_I = \left\{ u(.) \in C^1_{T_0}[T,\infty) : |I(t)| \le J_0 e^{\mu(t-T-kT_0)}, t \in [T+kT,\ T+(k+1)T_0];\ (k=0,1,2,...,m-1) \right\}$$

These are closed subsets of $C^1_{T_0}[T,\infty)$. Here $V_0, J_0, T_0, \mu > 0$ are constants (chosen below).

The set $M_U \times M_I$ turns into a complete metric space (cf. § 1.2) with respect to the metric

$$\hat{\rho}_\mu((U,\dot{U},I,\dot{I}),(\overline{U},\dot{\overline{U}},\overline{I},\dot{\overline{I}})) = \max\left\{ \hat{\rho}(U,\overline{U}), \rho^{(k)}(\dot{U},\dot{\overline{U}}), \hat{\rho}(I,\overline{I}), \rho^{(k)}(\dot{I},\dot{\overline{I}}) : k = 0,1,2,...,m-1 \right\}$$

where

$$\hat{\rho}(U,\overline{U}) = \max\left\{ |(U(t)-\overline{U}(t)| : t \in [T,2T] \right\}$$

$$\rho_\mu^{(k)}(U,\overline{U}) = \max\left\{ e^{-\mu(t-T-kT_0)}|(U(t)-\overline{U}(t)| : t \in [T+kT_0, T+(k+1)T_0] \right\},$$

$$\rho_\mu^{(k)}(\dot{U},\dot{\overline{U}}) = \max\left\{ e^{-\mu(t-T-kT_0)}|\dot{U}(t)-\dot{\overline{U}}(t)| : t \in [T+kT_0, T+(k+1)T_0] \right\},$$

$$\hat{\rho}(I,\overline{I}) = \max\left\{ |I(t)-\overline{I}(t)| : t \in [T,2T] \right\},$$

$$\rho_\mu^{(k)}(I,\overline{I}) = \max\left\{ e^{-\mu(t-T-kT_0)}|I(t)-\overline{I}(t)| : t \in [T+kT_0, T+(k+1)T_0] \right\},$$

$$\rho_\mu^{(k)}(\dot{I},\dot{\overline{I}}) = \max\left\{ e^{-\mu(t-T-kT_0)}|\dot{I}(t)-\dot{\overline{I}}(t)| : t \in [T+kT_0, T+(k+1)T_0] \right\}.$$

Define an operator $B = \left(B_U(U,I), B_I(U,I) \right)$ by the expressions:

$$B_U^{(k)}(U,I)(t) := W(U,I)(t),\ t \in [T+kT_0, T+(k+1)T_0]$$

$$B_I^{(k)}(U,I)(t) = \int_{T+kT_0}^{t} J(U,I)(s)ds - \frac{t-T-kT_0}{T_0} \int_{T+kT_0}^{T+(k+1)T_0} J(U,I)(s)ds,\ t \in [T+kT_0, T+(k+1)T_0]$$

$(k = 0, 1, 2, \ldots)$ where

$$W(U,I)(t) = 2E(t) - \bar{I}_0(t) - \frac{R_0}{Z_0}U(t) + \frac{R_0}{Z_0}\bar{I}_0(t),$$

$$J(U,I) = -\frac{d\bar{U}_0(t)}{dt} + \frac{\bar{U}_0(t)}{C_0 Z_0} - \frac{1}{C_0 Z_0}I(t) - \frac{2}{C_0}f\left(\frac{\bar{U}_0(t) + I(t)}{2}\right), \ t \in [T, 2T]$$

and $\bar{U}_0(t)$, $\bar{I}_0(t)$ are translated to the right on $[T, 2T]$ initial functions. In this manner, the difficulty with **(CC)** is overcome by considering the problem to the right from *T*.

We omit the proof that $B_U^{(k)}(U,I)(t)$ and $B_I^{(k)}(U,I)(t)$ are continuously differentiable T_0-periodic functions.

The next theorem is valid for "sufficiently small" initial conditions.

Theorem 2.9.1. Let the following conditions be fulfilled:

1) The initial functions $U_0(t), I_0(t) \in C_{T_0}^1[0,T]$;

$$\left|U_0(t)\right| \le U_0 e^{-\beta}e^{\mu(t-T-kT_0)}, \left|I_0(t)\right| \le J_0 e^{-\beta}e^{\mu(t-T-kT_0)}, t \in [kT_0, (k+1)T_0]; (k = 0,1,2,...,m-1)$$

$U_0(0) = 0$, $I_0(0) = 0$;

2) $E(.) \in C_{T_0}^1[T,\infty); \left|E(t)\right| \le U_E e^{\mu(t-T-kT_0)}, t \in [T + kT_0, T + (k+1)T_0]; (k = 0,1,...);$

3) $2U_E + \dfrac{R_0}{Z_0} < 1;\ \dfrac{U_0 e^{-\beta} + J_0}{2}e^{\mu 0} \le U_m$ and $\dfrac{U_0 + J_0 e^{-\beta}}{2}e^{\mu 0} \le U_m.$

Then there exists a unique T_0-periodic solution of the initial value problem (2.9.1), belonging to $M = M_U \times M_I$ but over interval $[T, 2T]$.

Proof: As in the previous theorems, we can prove that functions $B_U(U,I), B_I(U,I)$ are continuously differentiable and periodic ones. Prior to continuing we observe

$$\left|U(t-T)\right| = \left|\bar{U}_0(t)\right| \le U_0 e^{-\beta}e^{\mu(t-T-kT_0)},$$

$$\left|I(t-T)\right| = \left|\bar{I}_0(t)\right| \le J_0 e^{-\beta}e^{\mu(t-T-kT_0)}, t \in [T + kT_0, T + (k+1)T_0]$$

and recall that $\xi(h) = \dfrac{e^h - 1}{h}$ is non-decreasing and

$$\left|e^h - 1\right| \le e^{|h|} - 1 = |h|\xi(|h|),\ h \in (-\infty;\infty)$$

(cf. 1.6.1).

It remains to show that $\left|B_U^{(k)}(U,I)(t)\right| \le U_0 e^{\mu(t-T-kT_0)}$, $\left|B_I^{(k)}(U,I)(t)\right| \le J_0 e^{\mu(t-T-kT_0)}$ for $t \in [T+kT_0, T+(k+1)T_0]$.

Indeed we have,

$$W(U,I)(t) = 2E(t) - \vec{I}_0(t) - \frac{R_0}{Z_0}U(t) + \frac{R_0}{Z_0}\vec{I}_0(t),$$

$$\left|B_U^{(k)}(U,I)(t)\right| \le 2\left|E(t)\right| + \left|\vec{I}_0(t)\right| + \frac{R_0}{Z_0}\left|U(t)\right| + \frac{R_0}{Z_0}\left|\vec{I}_0(t)\right| \le$$

$$\le 2U_E e^{\mu(t-T-kT_0)} + J_0 e^{-\beta}e^{\mu(t-T-kT_0)} + \frac{R_0}{Z_0}U_0 e^{\mu(t-T-kT_0)} + \frac{R_0}{Z_0}J_0 e^{-\beta}e^{\mu(t-T-kT_0)} \le$$

$$\le e^{\mu(t-T-kT_0)}\left[2U_E + e^{-\beta}J_0 + \frac{R_0}{Z_0}U_0 + \frac{R_0}{Z_0}J_0 e^{-\beta}\right] \le U_0 e^{\mu(t-T-kT_0)}$$

since $2U_E + \dfrac{R_0}{Z_0} < 1$.

Then

$$\left|B_I^{(k)}(U,I)(t)\right| \le \left|\int_{T+kT_0}^{t} J(U,I)(s)ds\right| + \left|\int_{T+kT_0}^{T+(k+1)T_0} J(U,I)(s)ds\right| \equiv I_1 + I_2 .$$

But

$$I_1 \le \left|\int_{T+kT_0}^{t} J(U,I)(s)ds\right| \le \left|\int_{T+kT_0}^{t}\dot{\vec{U}}_0(s)ds\right| + \frac{1}{C_0 Z_0}\int_{T+kT_0}^{t}\left|\vec{U}_0(s)\right|ds + \frac{1}{C_0 Z_0}\int_{T+kT_0}^{t}\left|I(s)\right|ds +$$

$$+\frac{2I_0}{C_0}\int_{T+kT_0}^{t}\left|e^{\frac{\alpha(\vec{U}_0(s)+I(s))}{2}} - 1\right|ds + \frac{2I_m}{C_0 U_m}\int_{T+kT_0}^{t}\left(\left|\vec{U}_0(s)\right|+\left|I(s)\right|\right)e^{1+\frac{(\left|\vec{U}_0(s)\right|+\left|I(s)\right|)}{2U_m}}ds \le$$

$$\le \left|\vec{U}_0(t)\right| + \frac{U_0 e^{-\mu T}}{C_0 Z_0}\frac{e^{\mu(t-T-kT_0)}-1}{\mu} + \frac{J_0}{C_0 Z_0}\frac{e^{\mu(t-T-kT_0)}-1}{\mu} +$$

$$+\frac{2I_0}{C_0}\int_{T+kT_0}^{t}\left(e^{\frac{\alpha(\left|\vec{U}_0(s)\right|+\left|I(s)\right|)}{2}} - 1\right)ds + \frac{2I_m e^2}{C_0 U_m}\int_{T+kT_0}^{t}\left(\left|\vec{U}_0(s)\right|+\left|I(s)\right|\right)ds \le$$

$$\leq e^{\mu(t-T-kT_0)}U_0 e^{-\beta} + \frac{e^{\mu(t-T-kT_0)}-1}{\mu}\frac{U_0 e^{-\beta}}{C_0 Z_0} + \frac{e^{\mu(t-T-kT_0)}-1}{\mu}\frac{J_0}{C_0 Z_0}$$

$$+\frac{\alpha I_0}{C_0}\left(e^{\frac{\alpha\left(U_0 e^{-\beta}+J_0\right)}{2}}-1\right)\frac{e^{\mu(t-T-kT_0)}-1}{\mu} + \frac{2e^2 I_m\left(U_0 e^{-\beta}+J_0\right)}{C_0 U_m}\frac{e^{\mu(t-T-kT_0)}-1}{\mu} \leq$$

$$\leq e^{\mu(t-T-kT_0)}U_0 e^{-\beta} + \frac{e^{\mu(t-T-kT_0)}-1}{\mu C_0}\left[\frac{U_0 e^{-\beta}+J_0}{Z_0} + \alpha I_0\left(e^{\frac{\alpha\left(U_0 e^{-\beta}+J_0\right)}{2}}-1\right) + \frac{2e^2 I_m\left(U_0 e^{-\beta}+J_0\right)}{U_m}\right]$$

and

$$I_2 \leq \left|\int_{T+kT_0}^{T+(k+1)T_0} J(U,I)(s)\,ds\right| \leq$$

$$\leq \left|\int_{T+kT_0}^{T+(k+1)T_0}\dot{\bar{U}}_0(s)\,ds\right| + \frac{1}{C_0 Z_0}\int_{T+kT_0}^{T+(k+1)T_0}\left|\bar{U}_0(s)\right|ds + \frac{1}{C_0 Z_0}\int_{T+kT_0}^{T+(k+1)T_0}\left|I(s)\right|ds +$$

$$+\frac{\widehat{I}_0}{C_0}\int_{T+kT_0}^{T+(k+1)T_0}\left(e^{\frac{\alpha\left(\left|\bar{U}_0(s)\right|+\left|I(s)\right|\right)}{2}}-1\right)ds + \frac{I_m}{C_0 U_m}\int_{T+kT_0}^{T+(k+1)T_0}\left(\left|\bar{U}_0(s)\right|+\left|I(s)\right|\right)e^{1+\frac{\left(\left|\bar{U}_0(s)\right|+\left|I(s)\right|\right)}{2U_m}}ds \leq$$

$$\leq \frac{U_0 e^{-\beta}}{C_0 Z_0}\frac{e^{\mu T_0}-1}{\mu} + \frac{J_0}{C_0 Z_0}\frac{e^{\mu T_0}-1}{\mu} +$$

$$+\frac{J_0}{C_0 Z_0}\frac{e^{\mu T_0}-1}{\mu} + \frac{\widehat{I}_0}{C_0}\left(e^{\frac{\alpha\left(U_0 e^{-\beta}+J_0\right)}{2}}-1\right)\frac{e^{\mu T_0}-1}{\mu} + \frac{2I_m e^2}{C_0 U_m}\left(U_0 e^{-\beta}+J_0\right)\frac{e^{\mu T_0}-1}{\mu} \leq$$

$$\leq \frac{e^{\mu T_0}-1}{\mu C_0}\left[\frac{U_0 e^{-\beta}+J_0}{Z_0} + \widehat{I}_0\left(e^{\frac{\alpha\left(U_0 e^{-\beta}+J_0\right)}{2}}-1\right) + \frac{2I_m e^2\left(U_0 e^{-\beta}+J_0\right)}{U_m}\right].$$

Therefore

$$\left|B_I^{(k)}(U,I)(t)\right| \leq I_1 + I_2 \leq$$

$$\leq e^{\mu(t-T-kT_0)}\left\{U_0 e^{-\beta}+\frac{1}{\mu C_0}\left[\frac{U_0 e^{-\beta}+J_0}{Z_0}+\alpha \hat{I}_0\left(e^{\frac{\alpha\left(U_0 e^{-\beta}+J_0\right)}{2}}-1\right)+\frac{2e^2 I_m\left(U_0 e^{-\beta}+J_0\right)}{U_m}\right]\right\}+$$

$$+e^{\mu(t-T-kT_0)}\frac{e^{\mu_0}-1}{\mu C_0}\left[\frac{U_0 e^{-\beta}+J_0}{Z_0}+\hat{I}_0\left(e^{\frac{\alpha\left(U_0 e^{-\beta}+J_0\right)}{2}}-1\right)+\frac{2 I_m e^2\left(U_0 e^{-\beta}+J_0\right)}{U_m}\right]\leq$$

$$\leq e^{\mu(t-T-kT_0)}\left\{U_0 e^{-\beta}+\frac{e^{\mu_0}}{\mu C_0}\left[\frac{U_0 e^{-\beta}+J_0}{Z_0}+\alpha \hat{I}_0\left(e^{\frac{\alpha\left(U_0 e^{-\beta}+J_0\right)}{2}}-1\right)+\frac{2e^2 I_m\left(U_0 e^{-\beta}+J_0\right)}{U_m}\right]\right\}\leq J_0 e^{\mu(t-T-kT_0)}$$

for sufficiently large $\mu>0$ and $\beta>0$.

In the following we show that B is contractive operator.
We have

$$\left|B_U^{(k)}(U,I)(t)-B_U^{(k)}(\overline{U},\overline{I})(t)\right|\leq\left|\vec{I}_0(t)-\vec{I}_0(t)\right|+\frac{R_0}{Z_0}\left|U(t)-\overline{U}(t)\right|+\frac{R_0}{Z_0}\left|\vec{I}_0(t)-\vec{I}_0(t)\right|\leq$$

$$\leq e^{\mu(t-T-kT_0)}\frac{R_0}{Z_0}\rho^{(k)}(U,\overline{U})\leq e^{\mu(t-T-kT_0)}\frac{R_0}{Z_0}\frac{\rho^{(k)}(\dot{U},\dot{\overline{U}})}{\mu}\leq$$

$$\leq e^{\mu(t-T-kT_0)}\hat{\rho}_\mu((U,\dot{U},I,\dot{I}),(\overline{U},\dot{\overline{U}},\overline{I},\dot{\overline{I}}))\frac{R_0}{\mu Z_0}\equiv e^{\mu(t-T-kT_0)}K_U\,\hat{\rho}_\mu((U,\dot{U},I,\dot{I}),(\overline{U},\dot{\overline{U}},\overline{I},\dot{\overline{I}}))$$

Therefore

$$\hat{\rho}(B_U(U,I),B_U(\overline{U},\overline{I}))\leq e^{\mu_0}K_U\,\hat{\rho}_\mu((U,\dot{U},I,\dot{I}),(\overline{U},\dot{\overline{U}},\overline{I},\dot{\overline{I}})).$$

Further on we have

$$\left|\dot{B}_U^{(k)}(U,I)(t)-\dot{B}_U^{(k)}(\overline{U},\overline{I})(t)\right|\leq\left|\dot{\vec{I}}_0(t)-\dot{\vec{I}}_0(t)\right|+\frac{R_0}{Z_0}\left|\dot{U}(t)-\dot{\overline{U}}(t)\right|+\frac{R_0}{Z_0}\left|\dot{\vec{I}}_0(t)-\dot{\vec{I}}_0(t)\right|\leq$$

$$\leq e^{\mu(t-T-kT_0)}\frac{R_0}{Z_0}\rho^{(k)}(\dot{U},\dot{\overline{U}})\leq$$

$$\leq e^{\mu(t-T-kT_0)}\rho_\mu^{(k)}((U,I),(\overline{U},\overline{I}))\frac{R_0}{Z_0}\leq e^{\mu(t-T-kT_0)}\dot{K}_U\,\hat{\rho}_\mu((U,\dot{U},I,\dot{I}),(\overline{U},\dot{\overline{U}},\overline{I},\dot{\overline{I}}))$$

or

$$\rho_\mu^{(k)}(\dot{B}_U^{(k)}(U,I),\dot{B}_U^{(k)}(\overline{U},\overline{I}))\le \dot{K}_U\,\hat{\rho}_\mu((U,\dot{U},I,\dot{I}),(\overline{U},\dot{\overline{U}},\overline{I},\dot{\overline{I}})).$$

For the second component we obtain

$$\left|B_I^{(k)}(U,I)(t)-B_I^{(k)}(\overline{U},\overline{I})(t)\right|\le$$

$$\le\left|\int_{T+kT_0}^{t}\left(J(U,I)(s)-J(\overline{U},\overline{I})(s)\right)ds\right|+\left|\int_{T+kT_0}^{T+(k+1)T_0}\left(J(U,I)(s)-J(\overline{U},\overline{I})(s)\right)ds\right|\equiv J_1+J_2.$$

We have

$$J_1\le\int_{T+kT_0}^{t}\left|\dot{\bar{U}}_0(s)-\dot{\bar{\overline{U}}}_0(s)\right|ds+\frac{1}{C_0Z_0}\int_{T+kT_0}^{t}\left|\bar{U}_0(s)-\bar{U}_0(s)\right|ds+\frac{1}{C_0Z_0}\int_{T+kT_0}^{t}\left|I(s)-\overline{I}(s)\right|ds+$$

$$+\frac{2I_0}{C_0}\cdot\int_{T+kT_0}^{t}\left|e^{\alpha\left(\frac{\bar{U}_0(s)+I(s)}{2}\right)}-e^{\alpha\left(\frac{\bar{U}_0(s)+\overline{I}(s)}{2}\right)}\right|ds+$$

$$+\frac{2I_m}{C_0U_m}\cdot\int_{T+kT_0}^{t}\left|\left(\bar{U}_0(s)+I(s)\right)e^{1-\left(\frac{\bar{U}_0(s)+I(s)}{2U_m}\right)}-\left(\bar{U}_0(s)+\overline{I}(s)\right)e^{1-\left(\frac{\bar{U}_0(s)+\overline{I}(s)}{2U_m}\right)}\right|ds\le$$

$$\le\frac{\rho^{(k)}(I,\overline{I})}{C_0Z_0}\int_{T+kT_0}^{t}e^{\mu(s-T-kT_0)}ds+\frac{2\alpha\hat{I}_0}{C_0}\frac{1}{2}e^{\alpha\frac{U_0e^{-\beta}+J_0}{2}}\int_{T+kT_0}^{t}\left|I(s)-\overline{I}(s)\right|ds+\frac{2e^2I_m}{C_0U_m}\frac{1}{2}\int_{T+kT_0}^{t}\left|I(s)-\overline{I}(s)\right|ds\le$$

$$\le\frac{e^{\mu(t-T-kT_0)}-1}{\mu}\frac{e^{\mu_0}\rho_\mu^{(k)}(\dot{I},\dot{\overline{I}})}{\mu C_0}\left[\frac{1}{Z_0}+\alpha\hat{I}_0e^{\alpha\frac{U_0e^{-\beta}+J_0}{2}}+\frac{e^2I_m}{U_m}\right]\le$$

$$\le e^{\mu(t-T-kT_0)}\hat{\rho}_\mu((U,\dot{U},I,\dot{I}),(\overline{U},\dot{\overline{U}},\overline{I},\dot{\overline{I}}))\frac{e^{\mu_0}}{\mu^2 C_0}\left(\frac{1}{Z_0}+\alpha\hat{I}_0e^{\alpha\frac{U_0e^{-\beta}+J_0}{2}}+\frac{e^2I_m}{U_m}\right)$$

and

$$J_2\le\left|\int_{T+kT_0}^{T+(k+1)T_0}\left(J(U,I)(s)-J(\overline{U},\overline{I})(s)\right)ds\right|\le$$

$$\leq \int_{T+kT_0}^{T+(k+1)T_0} \left|\dot{\vec{U}}_0(s)-\dot{\vec{U}}_0(s)\right|ds + \frac{1}{C_0 Z_0}\int_{T+kT_0}^{T+(k+1)T_0}\left|\vec{U}_0(s)-\vec{U}_0(s)\right|ds + \frac{1}{C_0 Z_0}\int_{T+kT_0}^{T+(k+1)T_0}\left|I(s)-\bar{I}(s)\right|ds +$$

$$+\frac{2\hat{I}_0}{C_0}\cdot \int_{T+kT_0}^{T+(k+1)T_0}\left|e^{\alpha\frac{\vec{U}_0(s)+I(s)}{2}} - e^{\alpha\frac{\vec{U}_0(s)+\bar{I}(s)}{2}}\right|ds +$$

$$+\frac{2e^2 I_m}{C_0 U_m}\int_{T+kT_0}^{T+(k+1)T_0}\left|\left(\vec{U}_0(s)+I(s)\right)e^{1-\frac{\vec{U}_0(s)+I(s)}{2U_m}} - \left(\vec{U}_0(s)+\bar{I}(s)\right)e^{1-\frac{\vec{U}_0(s)+\bar{I}(s)}{2U_m}}\right|ds \leq$$

$$\leq \rho_\mu^{(k)}((U,I),(\overline{U},\bar{I}))\frac{e^{\mu T_0}-1}{\mu^2 C_0}e^{\mu T_0}\left(\frac{1}{Z_0}+\alpha\hat{I}_0 e^{\alpha\frac{U_0 e^{-\beta}+J_0}{2}}+\frac{e^2 I_m}{U_m}\right)\leq$$

$$\leq e^{\mu(t-T-kT_0)}\hat{\rho}_\mu((U,\dot{U},I,\dot{I}),(\overline{U},\dot{\overline{U}},\bar{I},\dot{\bar{I}}))\frac{e^{\mu 0}-1}{\mu^2 C_0}e^{\mu 0}\left(\frac{1}{Z_0}+\alpha\hat{I}_0 e^{\alpha\frac{U_0 e^{-\beta}+J_0}{2}}+\frac{e^2 I_m}{U_m}\right).$$

Then

$$\left|B_I^{(k)}(U,I)(t) - B_I^{(k)}(\overline{U},\bar{I})(t)\right| \leq$$

$$\leq e^{\mu(t-T-kT_0)}\rho_\mu^{(k)}((U,I),(\overline{U},\bar{I}))\frac{1}{\mu^2 C_0}e^{\mu 0}\left(\frac{1}{Z_0}+\alpha\hat{I}_0 e^{\alpha\frac{U_0 e^{-\beta}+J_0}{2}}+\frac{e^2 I_m}{U_m}\right)$$

$$+ e^{\mu(t-T-kT_0)}\rho_\mu^{(k)}((U,I),(\overline{U},\bar{I}))\frac{e^{\mu 0}-1}{\mu^2 C_0}e^{\mu 0}\left(\frac{1}{Z_0}+\alpha\hat{I}_0 e^{\alpha\frac{U_0 e^{-\beta}+J_0}{2}}+\frac{e^2 I_m}{U_m}\right)\leq$$

$$\leq e^{\mu(t-T-kT_0)}\hat{\rho}_\mu((U,\dot{U},I,\dot{I}),(\overline{U},\dot{\overline{U}},\bar{I},\dot{\bar{I}}))\frac{e^{2\mu 0}}{\mu^2 C_0}\left(\frac{1}{Z_0}+\alpha\hat{I}_0 e^{\alpha\frac{U_0 e^{-\beta}+J_0}{2}}+\frac{e^2 I_m}{U_m}\right)\equiv$$

$$\equiv e^{\mu(t-T-kT_0)}K_I\hat{\rho}_\mu((U,\dot{U},I,\dot{I}),(\overline{U},\dot{\overline{U}},\bar{I},\dot{\bar{I}}))$$

and hence

$$\hat{\rho}(B_I(U,I),B_I(\overline{U},\bar{I})) \leq e^{\mu 0}K_I\hat{\rho}_\mu((U,\dot{U},I,\dot{I}),(\overline{U},\dot{\overline{U}},\bar{I},\dot{\bar{I}})).$$

Finally we get

$$\left| \dot{B}_I^{(k)}(U,I)(t) - \dot{B}_I^{(k)}(\overline{U},\overline{I})(t) \right| \le$$

$$\le \left| J(U,I)(t) - J(\overline{U},\overline{I})(t) \right| + \frac{1}{T_0}\left| \int_{T+kT_0}^{T+(k+1)T_0} \left(J(U,I)(s) - J(\overline{U},\overline{I})(s) \right) ds \right| \equiv \dot{J}_1 + \dot{J}_2.$$

We have

$$\dot{J}_1 \le \left| \dot{\vec{U}}_0(t) - \dot{\overline{\vec{U}}}_0(t) \right| + \frac{1}{C_0 Z_0}\left| \vec{U}_0(t) - \overline{\vec{U}}_0(t) \right| + \frac{1}{C_0 Z_0}\left| I(t) - \overline{I}(t) \right| +$$

$$+ \frac{2\widehat{I}_0}{C_0}\left| e^{\alpha\frac{\vec{U}_0(t)+I(t)}{2}} - e^{\alpha\frac{\overline{\vec{U}}_0(t)+\overline{I}(t)}{2}} \right| +$$

$$\frac{2I_m}{C_0 U_m}\left| \left(\vec{U}_0(t)+I(t)\right)e^{1-\frac{\vec{U}_0(t)+I(t)}{2U_m}} - \left(\overline{\vec{U}}_0(t)+\overline{I}(t)\right)e^{1-\frac{\overline{\vec{U}}_0(t)+\overline{I}(t)}{2U_m}} \right| \le$$

$$\le e^{\mu(t-T-kT_0)}\frac{\rho_\mu^{(k)}(I,\overline{I})}{C_0 Z_0} + e^{\mu(t-T-kT_0)}\frac{\alpha\widehat{I}_0}{C_0}e^{\alpha\frac{U_0 e^{-\beta}+J_0}{2}}\rho_\mu^{(k)}(I,\overline{I}) + e^{\mu(t-T-kT_0)}\frac{e^2 I_m}{C_0 U_m}\rho_\mu^{(k)}(I,\overline{I}) \le$$

$$\le e^{\mu(t-T-kT_0)}\left(\frac{\rho_\mu^{(k)}(\dot{I},\dot{\overline{I}})}{\mu C_0 Z_0} + \frac{\alpha\widehat{I}_0}{C_0}e^{\alpha\frac{U_0 e^{-\beta}+J_0}{2}}\frac{\rho_\mu^{(k)}(\dot{I},\dot{\overline{I}})}{\mu} + \frac{e^2 I_m}{C_0 U_m}\frac{\rho_\mu^{(k)}(\dot{I},\dot{\overline{I}})}{\mu} \right) \le$$

$$\le e^{\mu(t-T-kT_0)}\hat{\rho}_\mu((U,\dot{U},I,\dot{I}),(\overline{U},\dot{\overline{U}},\overline{I},\dot{\overline{I}}))\frac{1}{\mu C_0}\left(\frac{1}{Z_0} + \alpha\widehat{I}_0 e^{\alpha\frac{U_0 e^{-\beta}+J_0}{2}} + \frac{e^2 I_m}{U_m} \right)$$

and

$$\dot{J}_2 \le \frac{1}{T_0}\left| \int_{T+kT_0}^{T+(k+1)T_0} \left(J(U,I)(s) - J(\overline{U},\overline{I})(s) \right) ds \right| \le$$

$$\le e^{\mu(t-T-kT_0)}\frac{e^{\mu_0}-1}{\mu_0}\frac{1}{\mu C_0}\left(\frac{1}{Z_0} + \alpha\widehat{I}_0 e^{\alpha\frac{U_0 e^{-\beta}+J_0}{2}} + \frac{e^2 I_m}{U_m} \right)\hat{\rho}_\mu((U,\dot{U},I,\dot{I}),(\overline{U},\dot{\overline{U}},\overline{I},\dot{\overline{I}}))$$

Consequently

$$\left| \dot{B}_I^{(k)}(U,I)(t) - \dot{B}_I^{(k)}(\overline{U},\overline{I})(t) \right| \le$$

$$\le e^{\mu(t-T-kT_0)}\frac{1}{\mu C_0}\left(\frac{1}{Z_0}+\alpha\widehat{I}_0 e^{\alpha\frac{U_0 e^{-\beta}+J_0}{2}}+\frac{e^2 I_m}{U_m}\right)\hat{\rho}_\mu((U,\dot{U},I,\dot{I}),(\overline{U},\dot{\overline{U}},\overline{I},\dot{\overline{I}}))+$$

$$+e^{\mu(t-T-kT_0)}\frac{e^{\mu_0}-1}{\mu_0}\frac{1}{\mu C_0}\left(\frac{1}{Z_0}+\alpha\widehat{I}_0 e^{\alpha\frac{U_0 e^{-\beta}+J_0}{2}}+\frac{e^2 I_m}{U_m}\right)\hat{\rho}_\mu((U,\dot{U},I,\dot{I}),(\overline{U},\dot{\overline{U}},\overline{I},\dot{\overline{I}}))\le$$

$$\le e^{\mu(t-T-kT_0)}\left(1+\frac{e^{\mu_0}-1}{\mu_0}\right)\frac{1}{\mu C_0}\left(\frac{1}{Z_0}+\alpha\widehat{I}_0 e^{\alpha\frac{U_0 e^{-\beta}+J_0}{2}}+\frac{e^2 I_m}{U_m}\right)\hat{\rho}_\mu((U,\dot{U},I,\dot{I}),(\overline{U},\dot{\overline{U}},\overline{I},\dot{\overline{I}}))\equiv$$

$$\equiv e^{\mu(t-T-kT_0)}\dot{K}_I\,\hat{\rho}_\mu((U,\dot{U},I,\dot{I}),(\overline{U},\dot{\overline{U}},\overline{I},\dot{\overline{I}})).$$

It follows

$$\rho_\mu^{(k)}(\dot{B}_I^{(k)}(U,I),\dot{B}_I^{(k)}(\overline{U},\overline{I}))\le\dot{K}_I\,\hat{\rho}_\mu((U,\dot{U},I,\dot{I}),(\overline{U},\dot{\overline{U}},\overline{I},\dot{\overline{I}})).$$

Therefore

$$\hat{\rho}_\mu((B_U(U,I),\dot{B}_U(U,I)),(B_I(U,I),\dot{B}_I(U,I)),(B_U(\overline{U},\overline{I}),\dot{B}_U(\overline{U},\overline{I})),(B_I(\overline{U},\overline{I}),\dot{B}_I(\overline{U},\overline{I})))\le$$

$$\le K\,\hat{\rho}_\mu((U,\dot{U},I,\dot{I}),(\overline{U},\dot{\overline{U}},\overline{I},\dot{\overline{I}}))$$

where $K=\max\left\{e^{\mu_0}K_U,\dot{K}_U,e^{\mu_0}K_I,\dot{K}_I\right\}<1$.

Consequently, the operator B is contractive and its fixed point is a periodic solution of (2.9.1).

Theorem 2.9.1 is thus proved.

2.9.1. Numerical Example

All inequalities implying an existence-uniqueness theorem are for $\mu_0=1$

$$\frac{R_0}{Z_0}\le 1;\ 2U_E+\frac{R_0}{Z_0}U_0+e^{-\beta}J_0\left(1+\frac{R_0}{Z_0}\right)\le U_0;$$

$$U_0 e^{-\beta}+\frac{e^{\mu_0}\left(U_0 e^{-\beta}+J_0\right)}{\mu C_0}\left[\frac{1}{Z_0}+\frac{2e^2 I_m}{U_m}+\frac{\alpha^2\widehat{I}_0}{2}\xi(\alpha U_m)\right]\le J_0;$$

$$eK_U = \frac{eR_0}{\mu Z_0} < 1 \, ; \; eK_I = \frac{e^2}{\mu^2 C_0} \left(\frac{1}{Z_0} + \alpha \hat{I}_0 e^{\alpha \frac{U_0 e^{-\beta} + J_0}{2}} + \frac{e^2 I_m}{U_m} \right) < 1 \, ;$$

$$\dot{K}_U = \frac{R_0}{Z_0} < 1 \, ; \; \dot{K}_I = \frac{e}{\mu C_0} \left(\frac{1}{Z_0} + \alpha \hat{I}_0 e^{\alpha \frac{U_0 e^{-\beta} + J_0}{2}} + \frac{e^2 I_m}{U_m} \right) < 1 \, .$$

Clearly, the ratio $\dfrac{R_0}{Z_0} < 1$ plays an important role for the rate of convergence. It is better for larger Z_0 and smaller R_0.

2.9.2. Applications to Electron Wave Interaction Phenomena

Here we demonstrate briefly applications of our method to transmission line analogous to thous considered by Rowe [96]. We use denotations given in [96].

1) Electromagnetic wave propagation on helical conductors is described by the following system:

$$\frac{\partial I_x}{\partial x} + \left(\frac{\beta}{\gamma} \right)^2 \frac{2\pi \varepsilon_0}{I_0(\gamma a) K_0(\gamma a)} \frac{\partial V}{\partial t} = 0,$$

$$\frac{\partial V}{\partial x} + \frac{\mu_0 \cot^2 \psi}{2\pi} I_1(\gamma a) K_1(\gamma a) \frac{\partial I_x}{\partial x} = 0.$$

Then obviously

$$L_e = \frac{\mu_0 \cot^2 \psi}{2\pi} I_1(\gamma a) K_1(\gamma a) \; [H/m] \text{ and}$$

$$C_e = \left(\frac{\beta}{\gamma} \right)^2 \frac{2\pi \varepsilon_0}{I_0(\gamma a) K_0(\gamma a)} \approx \frac{2\pi \varepsilon_0}{I_0(\gamma a) K_0(\gamma a)} \; [F/m].$$

The impedance of the line is

$$Z_0 = \sqrt{\frac{L_e}{C_e}} = \sqrt{\frac{\mu_0}{\varepsilon_0}} \, \frac{\cot \psi}{2\pi} \sqrt{I_0(\gamma a) K_0(\gamma a) I_1(\gamma a) K_1(\gamma a)}$$

and the characteristic phase velocity

$$v_0 = \frac{1}{\sqrt{L_e C_e}} = \frac{c}{\cot\psi} \sqrt{\frac{I_0(\gamma a)K_0(\gamma a)}{I_1(\gamma a)K_1(\gamma a)}} \,.$$

Let us note that in this case μ_0 is the magnetic permeability of vacuum, whereas in the previous calculations we have denoted a constant $\mu_0 = \mu T_0$.

2) A transmission line analog of an electron beam leads to the system

$$\frac{\partial V}{\partial x} = L_e \frac{\partial I}{\partial t}, \qquad \frac{\partial I}{\partial x} = C_e \frac{\partial V}{\partial t},$$

where $L_e = \dfrac{p^2}{\omega^2 \varepsilon_0 \sigma}[H/m]$, $C_e = \dfrac{\omega \sigma \varepsilon_0}{p^2}\left(\dfrac{2\pi}{\lambda_p}\right)^2 [F/m]$

(cf. S. Bloom & R.W. Peter [25]).

3) In [100], Schelkunoff derives transmission line equations of transverse plane electric waves. The obtained equations are similar to the equations connecting voltage and current in a transmission line having Z for its distributed series impedance and Y for the shunt admittance

$$\frac{\partial F}{\partial z} + \mu \frac{\partial U}{\partial t} = 0, \qquad \frac{\partial U}{\partial z} + \varepsilon \frac{\partial F}{\partial t} + gF = 0$$

where U is the magnetomotive force between a given point and infinity along a path contained completely in the equiphase surface passing through the given point, and F is the electric flux through a curve drawn from a given point to infinity.

4) The study of plasma interaction with travelling waves leads to a system

$$\frac{\partial i_x}{\partial x} + C_e \frac{\partial V}{\partial t} = 0, \qquad \frac{\partial V}{\partial x} + L_e \frac{\partial i_x}{\partial x} = 0 \,,$$

where

$$L_e = \frac{\tau}{2\pi b \omega_p^2 \varepsilon_0}\frac{I_0(\tau b)}{I_1(\tau b)}[H/m], \quad C_e = \frac{2\pi b \beta^2 \varepsilon_0}{\gamma\left[1-(\omega/\omega_p)^2\right]}\frac{K_1(\gamma b)}{K_0(\gamma b)}[F/m].$$

So the mixed problem for a transmission line analog terminated by a nonlinear negative resistance load (with V-I characteristic $i = f(u)$)

$$\frac{\partial u}{\partial t} + \frac{1}{C_e}\frac{\partial i}{\partial x} = 0, \quad \sqrt{\frac{L_e}{C_e}}\frac{\partial i}{\partial t} + v\frac{\partial u}{\partial x} = 0 \ E(t) - u(0,t) = R_0 i(0,t), \ t \geq 0$$

$$C_0 \frac{du(\Lambda,t)}{dt} = i(\Lambda,t) - f(u(\Lambda,t)), t \geq 0$$

$$u(x,0) = u_0(x), i(x,0) = i_0(x), x \in [0,\Lambda]$$

might be reduced to an initial value problem for the neutral equation on the boundary

$$\frac{du(t)}{dt} = \frac{1}{C_0}\frac{2E(t)}{\sqrt{L_e/C_e}+R_0} - \frac{1}{C_0}f(u(t)) + \frac{1}{C_0}\frac{\sqrt{L_e/C_e}-R_0}{\sqrt{L_e/C_e}+R_0}f(u(t-2T)) -$$

$$- \frac{\sqrt{L_e/C_e}}{C_0}u(t) - \frac{\sqrt{L_e/C_e}}{C_0}\frac{\sqrt{L_e/C_e}-R_0}{\sqrt{L_e/C_e}+R_0}u(t-2T) + \frac{\sqrt{L_e/C_e}-R_0}{\sqrt{L_e/C_e}+R_0}\frac{du(t-2T)}{dt}, t \geq T$$

$$u(t) = \upsilon_0(t), \frac{du(t)}{dt} = \frac{d\upsilon_0(t)}{dt}, t \in [-T,T]$$

In the case of exponential characteristics the inequalities guaranteeing an existence-uniqueness of a periodic solution become:

$$U_0 e^{-\beta} + \frac{e^{\mu_0}}{\mu C_0}\left[\left(U_0 e^{-\beta} + J_0\right)\sqrt{L_e/C_e} + \alpha I_0\left(e^{\frac{\alpha\left(U_0 e^{-\beta}+J_0\right)}{2}} - 1\right) + \frac{2e^2 I_m\left(U_0 e^{-\beta} + J_0\right)}{U_m}\right] \leq J_0;$$

$$K_U = \frac{R_0}{\mu}\sqrt{L_e/C_e} < 1; \ K_I = \frac{e^{\mu_0}}{\mu^2 C_0}\left(\sqrt{L_e/C_e} + \alpha I_0 e^{\alpha\frac{U_0 e^{-\beta}+J_0}{2}} + \frac{e^2 I_m}{U_m}\right) < 1;$$

$$\dot{K}_U = \sqrt{L_e/C_e} < 1; \ \dot{K}_I = \left(1 + \frac{e^{\mu_0}-1}{\mu_0}\right)\frac{1}{\mu C_0}\left(\sqrt{L_e/C_e} + \alpha I_0 e^{\alpha\frac{U_0 e^{-\beta}+J_0}{2}} + \frac{e^2 I_m}{U_m}\right) < 1.$$

We have

$$\frac{C_e}{L_e} = \frac{1}{1-(\omega/\omega_p)^2}\frac{(2\pi\beta b\omega_p\varepsilon_0)^2}{\gamma\tau}\frac{K_1(\gamma b)}{K_0(\gamma b)}\frac{I_1(\tau b)}{I_0(\tau b)}; \ \frac{eJ_0}{2} \leq U_m;$$

and the remaining inequalities can be written substituting the calculated quantities.

It is known that Z_0 decreases as the frequency γa increases [96]. This can be compensated for by choosing sufficiently large $\mu > 0$. Consequently, the above existence-uniqueness result is valid even in the case of non-uniform lines, provided the above inequalities are satisfied. Finally we conclude that, provided the above inequalities hold, the periodic solution does exist. Indeed, one can even check the propagation of ultraviolet waves $\lambda_0 = 2{,}5.10^{-8} m$.

CONCLUSION

We compare the inequalities from numerical example 2.2.1 with those from numerical example 2.5.1 for the case of polynomial nonlinearities:

$$\Lambda = 5m,\ L = 0{,}2\,\mu H/m,\ C = 80\,pF/m,\ v = 1/\left(4.10^{-9}\right) = 2{,}5.10^{8},\ Z_0 = 50\ \Omega,$$

$$R_0 = 35\Omega,\ \beta = 4,\ C_0 = 8pF = 8.10^{-12}\,F,\ T = \Lambda\sqrt{LC} = 2.10^{-8}s;$$

$$\left|Z_0 - R_0\right|/\left(Z_0 + R_0\right) = 1/19,\ \lambda_0 = \left(1/4\right)10^{-3}m,\ f_0 = 1/\left(\lambda_0\sqrt{LC}\right) = 10^{12}\,Hz,$$

$$T_0 = 1/f_0 = 10^{-12}\ \text{sec}.$$

For $\mu = 10^{12}$ we have $\mu T_0 = \mu_0 = 1, T = 2.10^{-8}.10^{12}\,T_0 = 20000.T_0,$

$$\mu C_0 = 10^{12}.8.10^{-12} = 8;\ \mu^2 C_0 = 10^{24}.8.10^{-12} = 8.10^{12},\ f(u) = -0{,}12\,u + 0{,}8u^3.$$

Then for $U_0 = J_0 = 0{,}1$ we obtain

$$\frac{e}{8}\left(1 + \frac{e}{19}\right)\left(\frac{1}{50} + 0{,}12 + 0{,}8\left(10^{-1}e\right)^2\right) + \frac{1}{19} \le 1;$$

$$K_U = \frac{e}{8.10^{12}}\left(\frac{1}{Z_0} + \left|r_1\right| + 3\left|r_3\right|\left(U_0 e\right)^2\right) < 1;\ \dot{K}_U = \frac{e}{8}\left(\frac{1}{50} + 0{,}12 + 3.0{,}8\left(10^{-1}e\right)^2\right) < 1$$

and therefore $K = \left\{eK_U, \dot{K}_U\right\} = 0{,}1085 < 1$.

For 2.5.1 we have

$$\frac{R_0}{Z_0} \le 1;\ \frac{e}{\mu C_0}\left(\frac{1}{Z_0} + \left|r_1\right| + \frac{\left|r_3\right|\left(J_0 e\right)^2}{2^2}\right) \le 1;\ K_U = \frac{R_0}{\mu Z_0} < 1;$$

$$K_I = \frac{e}{\mu^2 C_0}\left(\frac{1}{Z_0} + \left|r_1\right| + 3\left|r_3\right|\frac{\left(J_0 e\right)^2}{2^2}\right) < 1;$$

$$\dot{K}_U = \frac{R_0}{Z_0} < 1 \, ; \; \dot{K}_I = \frac{e}{\mu C_0}\left(\frac{1}{Z_0} + |r_1| + 3|r_3|\frac{(J_0 e)^2}{2^2}\right) < 1 \, ; \; \dot{K}_I = 0{,}063 < 1 .$$

Then $K = \max\{\dot{K}_U, \dot{K}_I\} = \left\{\dfrac{35}{50} \, ; \, 0{,}063\right\} = \dfrac{7}{10} < 1 .$

Obviously the ratio $R_0 / Z_0 < 1$ plays an important role in the rate of convergence. Therefore for R_0 / Z_0 close to 1 it is better to apply the first method.

In the case of exponential nonlinearity 2.8.1 we have $\delta = \dfrac{Z_0 - R_0}{Z_0 + R_0}$;

$$\frac{e}{\mu C_0}\left(\frac{1}{Z_0} + \alpha I_0 \xi(\alpha U_m) + \frac{I_m e^2}{U_m}\right) \leq 1 \, ; \; K = \frac{e}{\mu^2 C_0}\left(\frac{1}{Z_0} + \alpha e^{\alpha U_m}\widehat{I}_0 + \frac{2e^2 I_m}{U_m}\right) < 1 ;$$

$$\dot{K} = \frac{e}{\mu C_0}\left(\frac{1}{Z_0} + \alpha\, e^{\alpha U_m}\widehat{I}_0 + \frac{2e^2 I_m}{U_m}\right) < 1$$

while for 2.9.1: $\dfrac{R_0}{Z_0} \leq 1 \, ; \; eK_U = \dfrac{eR_0}{\mu Z_0} < 1 \, ; \; \dot{K}_U = \dfrac{R_0}{Z_0} < 1 \, ;$

$$\frac{e}{\mu C_0}\left[\frac{J_0}{Z_0} + 2\alpha I_0 \sinh\left(\frac{\alpha J_0}{2}\right) + \frac{2e^2 I_m J_0}{U_m}\right] \leq J_0 ;$$

$$e^{\mu_0}K_I = \frac{e^2}{\mu^2 C_0}\left(\frac{1}{Z_0} + \alpha\widehat{I}_0 e^{\alpha\frac{J_0}{2}} + \frac{e^2 I_m}{U_m}\right) < 1 ;$$

$$\dot{K}_I = \frac{e}{\mu C_0}\left(\frac{1}{Z_0} + \alpha\widehat{I}_0 e^{\alpha\frac{J_0}{2}} + \frac{e^2 I_m}{U_m}\right) = 0{,}0002 < 1$$

and then

$$K = \max\{\dot{K}_U, \dot{K}_I\} = \left\{\frac{35}{50} \, ; \, 0{,}0002\right\} = \frac{35}{50} < 1 ;$$

$$I_0 = 10^{-8} A; \; \alpha = \frac{q}{k\tau} = \frac{1}{2.26.10^{-3}} = 19{,}23 \, ; \; \frac{I_m}{U_m} = \frac{1{,}9.10^{-3} A}{0{,}15 V} = 12{,}67.10^{-3}$$

that is, the method from 2.8.1 is the better one to use.

Chapter III

Lossless Transmission Lines Terminated by in Series Connected *RLC*-Loads

Abstract

The main purpose of the present chapter is to analyse the processes in a lossless transmission line terminated by in series connected nonlinear *RLC*-loads at both ends (cf. Fig. 3.1). First we formulate boundary conditions using Kirchhoff's law, then we show that the first manner (from Chapter II) of reducing the mixed problem to neutral systems leads to equations whose neutral part contain Lipschitz constants larger than 1. We then use the second manner and obtain another neutral system. With suitable operators and function spaces we prove an existence-uniqueness of periodic solutions in both nonlinear and linear elements. Finally we show that even oscillatory solutions of the nonlinear neutral system exist.

Introduction

Lossless transmission lines terminated by circuits with in series connected *RLC* elements have many applications. In contrast to the case from Chapter II, we have more complicated nonlinear elements that lead to more complicated nonlinear neutral equations. In particular, the capacitive functions have singularities that require refining of the definition domains and the various estimates of the capacitive functions and their derivatives.

In § 3.1 we derive the boundary conditions of the mixed problem for transmission line equations as a consequence of Kirchhoff's law. In § 3.2 we reduce the mixed problem to an initial value problem for a neutral equation. In § 3.3 we analyze arising nonlinearities of the nonlinear loads. Particular difficulties generate the singularities of the capacitive functions. We also obtain an existence-uniqueness result for a periodic solution of the neutral equations obtained. In § 3.4 we introduce a new operator formulation of the periodic problem. In § 3.5 we prove a theorem for existence-uniqueness of a periodic solution of the nonlinear neutral system. In § 3.6 we demonstrate the conditions of the existence-uniqueness theorem using specific numerical data. In § 3.7 existence-uniqueness of a periodic solution of a system with linear in series connected *RLC*-loads is obtained. Our goal is to give a unified approach via the fixed-point method. In § 3.8 we apply our methods to transmission line analog of electron beams where specific parameters depend on the frequency. Finally in § 3.9 we give conditions that imply an existence-uniqueness of an oscillatory solution.

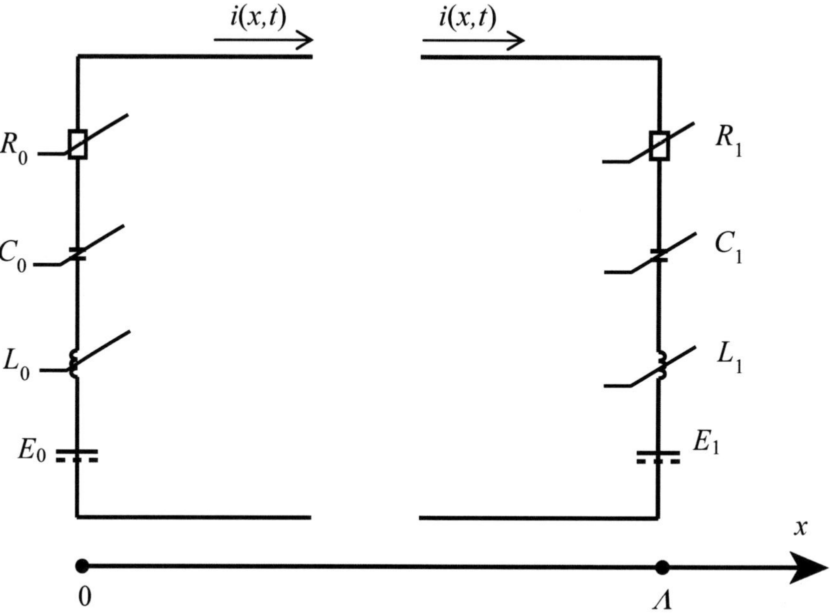

Figure 3.1.

3.1. DERIVATION OF THE BOUNDARY CONDITIONS AND FORMULATION OF THE MIXED PROBLEM

In accordance with Fig. 3.1 for the voltages at the left end $x = 0$ we obtain (cf. [16]):

$$-u(0,t) = u_{R_0} + u_{C_0} + u_{L_0} - E_0(t),\qquad(3.1.1)$$

where $E_0(t)$ is the source voltage. Then

$$-u(0,t) = R_0(i(0,t)) + u_{C_0} + \frac{d\Psi_0}{dt} - E_0(t).$$

Here we derive an explicit form of the characteristics.

First we introduce denotations $\widetilde{L}_p(i) = L_p(i).i$; $\widetilde{C}_p(u) = C_p(u).u$ $(p = 0,1)$.

To calculate the voltage of the condensers C_p we proceed from the relation (assuming $u_{C_p}(T) \equiv u(T) = 0$):

$$i_{C_p} = \frac{dq_{C_p}}{dt} = \frac{d(\widetilde{C}_p(u))}{dt} \Rightarrow \int_T^t i(x,\tau)d\tau = \widetilde{C}_p(u),$$

$$u_{C_p} = u(0,t) = \widetilde{C}_p^{-1}\left(\int_T^t i(0,\tau)d\tau \right).$$

In order to calculate the voltage of the inductor we use the relation

$$u_{\Psi_p} = \frac{d\Psi_p}{dt} = \frac{d\widetilde{L}_p(i)}{dt} = \frac{d(L_p(i).i)}{dt} = \left[i\frac{dL_p(i)}{di} + L_p(i) \right]\frac{di}{dt}, \quad (p=0,1)$$
.

Therefore, the first boundary condition for $x = 0$ is:

$$u(0,t) = E_0(t) - R_0(i(0,t)) - \widetilde{C}_0^{-1}\left(\int_T^t i(0,\tau)d\tau \right) -$$

$$-\left[i(0,t)\frac{dL_0(i(0,t))}{di} + L_0(i(0,t)) \right]\frac{di(0,t)}{dt}. \tag{3.1.2}$$

For the right end, that is, $x = \Lambda$ (cf. Fig. 3.1) we obtain

$$u(\Lambda,t) = E_1(t) + R_1(i(\Lambda,t)) + \widetilde{C}_1^{-1}\left(\int_T^t i(\Lambda,\tau)d\tau \right) +$$

$$+\left[i(\Lambda,t)\frac{dL_1(i(\Lambda,t))}{di} + L_1(i(\Lambda,t)) \right]\frac{di(\Lambda,t)}{dt} \tag{3.1.3}$$

where $R_p(.), C_p(.)$ and $L_p(.), (p=0,1)$ are characteristics of the nonlinear loads. They are, in general, nonlinear functions.

Now we are able to formulate the initial-boundary value problem (or mixed problem) for the hyperbolic transmission line equations, which lead us to find a solution $(u(x,t),i(x,t))$ of the hyperbolic system

$$\frac{\partial u(x,t)}{\partial t} + L\frac{\partial i(x,t)}{\partial x} = 0$$

$$\frac{\partial i(x,t)}{\partial t} + C\frac{\partial u(x,t)}{\partial x} = 0 \tag{3.1.4}$$

for $(x,t) \in \Pi = \{(x,t) \in R^2 : 0 \le x \le \Lambda,\ t \ge 0\}$, satisfying the initial conditions

$$u(x,0) = u_0(x),\ i(x,0) = i_0(x) \text{ for } x \in [0,\Lambda] \tag{3.1.5}$$

and the boundary conditions

$$u(0,t) = E_0(t) - R_0(i(0,t)) - \tilde{C}_0^{-1}\left(\int_T^t i(0,\tau)d\tau\right) -$$

$$\left[i(0,t)\frac{dL_0(i(0,t))}{di} + L_0(i(0,t))\right]\frac{di(0,t)}{dt} \tag{3.1.6}$$

for $x = 0$ and

$$u(\Lambda,t) = E_1(t) + R_1(i(\Lambda,t)) + \tilde{C}_1^{-1}\left(\int_T^t i(\Lambda,\tau)d\tau\right) +$$

$$\left[i(\Lambda,t)\frac{dL_1(i(\Lambda,t))}{di} + L_1(i(\Lambda,t))\right]\frac{di(\Lambda,t)}{dt} \tag{3.1.7}$$

for $x = \Lambda$.

3.2. REDUCING THE MIXED PROBLEM TO A PERIODIC PROBLEM ON THE BOUNDARY

Proceeding exactly as in Chapter II we transform the lossless transmission line system

$$\frac{\partial u(x,t)}{\partial x} + L\frac{\partial i(x,t)}{\partial t} = 0,\quad \frac{\partial i(x,t)}{\partial x} + C\frac{\partial u(x,t)}{dt} = 0 \tag{3.2.1}$$

as follows: rewrite (3.2.1) in the form

$$\frac{\partial u(x,t)}{\partial t} + \frac{1}{C}\frac{\partial i(x,t)}{\partial x} = 0,\quad \frac{\partial i(x,t)}{\partial t} + \frac{1}{L}\frac{\partial u(x,t)}{\partial x} = 0 \tag{3.2.2}$$

and repeat the reasonings from § 2.4. The first and the second boundary condition (resp. (3.1.6) and (3.1.7)), in view of

$$u(0,t) = \frac{u(\Lambda,t+T) + Z_0 i(\Lambda,t+T) + u(\Lambda,t-T) - Z_0 i(\Lambda,t-T)}{2},$$

$$i(0,t) = \frac{u(\Lambda,t+T) + Z_0 i(\Lambda,t+T) - u(\Lambda,t-T) + Z_0 i(\Lambda,t-T)}{2Z_0}.$$

yield the system

$$\frac{du(t)}{dt} = -Z_0 \frac{di(t)}{dt} + \frac{du(t-2T)}{dt} - Z_0 \frac{di(t-2T)}{dt} + \frac{2Z_0 E_0(t-T)}{\widetilde{L}_0(i)} - \tag{3.2.3}$$

$$- Z_0 \frac{u(t) + Z_0 i(t) + u(t-2T) - Z_0 i(t-2T)}{\widetilde{L}_0(i)} - \frac{2Z_0 R_0\big(i(0,t-T)\big)}{\widetilde{L}_0(i)} - \frac{2Z_0}{\widetilde{L}_0(i)} \widetilde{C}_0^{-1}\left(\int_T^t \big(i(0,s-T)\big)ds\right)$$

$$\frac{di(t)}{dt} = \frac{u(t) - E_1(t) - R_1(i(t))}{\widetilde{L}_1(i)} - \frac{1}{\widetilde{L}_1(i)} \widetilde{C}_1^{-1}\left(\int_T^t i(\tau)d\tau\right) \tag{3.2.4}$$

where

$$\widetilde{L}_0(i) = i(0,t-T)\frac{dL_0\big(i(0,t-T)\big)}{di} + L_0\big(i(0,t-T)\big) > 0,$$

$$\widetilde{L}_1(i) = i(t)\frac{dL_1(i(t))}{di} + L_1(i(t)) > 0$$

should be strictly positive.

Since the Lipschitz constant of the neutral part should be smaller than 1 (cf. R. Bellman & K. L. Cooke [23]; T. Jankowski & M. Kwapisz [65], G. A. Kamenskii [69], A. D. Myshkis [85], [86]) we have to check the coefficient before $\dfrac{di(t)}{dt}$ in (3.2.3). It is $-Z_0$. But in the applications $\big|-Z_0\big| > 1$. This means we have to proceed in a different manner to reduce the mixed problem to a neutral system.

We proceed again from the system

$$\frac{\partial u(x,t)}{\partial t} + \frac{1}{C}\frac{\partial i(x,t)}{\partial x} = 0, \qquad \frac{\partial i(x,t)}{\partial t} + \frac{1}{L}\frac{\partial u(x,t)}{\partial x} = 0.$$

As in § 2.1, from the last system we obtain

$$\frac{\partial}{\partial t}\big(u(x,t)+Z_0\,i(x,t)\big)+v\frac{\partial}{\partial x}\big(u(x,t)+Z_0\,i(x,t)\big)=0,$$

$$\frac{\partial}{\partial t}\big(u(x,t)-Z_0\,i(x,t)\big)-v\frac{\partial}{\partial x}\big(u(x,t)-Z_0\,i(x,t)\big)=0$$

and then introduce new variables (U,I):

$$U(x,t)=u(x,t)+Z_0\,i(x,t),\ \ I(x,t)=u(x,t)-Z_0\,i(x,t)$$

for which

$$\frac{\partial U(x,t)}{\partial t}+v\frac{\partial U(x,t)}{\partial x}=0\ ,\ \frac{\partial I(x,t)}{\partial t}-v\frac{\partial I(x,t)}{\partial x}=0\ .$$

Then the initial conditions can be obtained from (3.1.5) for $t=0$:

$$U(x,0)=u(x,0)+Z_0\,i(x,0)=u_0(x)+Z_0\,i_0(x)\equiv U_0(x),$$

$$I(x,0)=u(x,0)-Z_0\,i(x,0)=u_0(x)-Z_0\,i_0(x)\equiv I_0(x).$$

(3.2.5)

In order to obtain new boundary conditions, we proceed from the inverse transformation

$$u(x,t)=\frac{1}{2}U(x,t)+\frac{1}{2}I(x,t),$$

$$i(x,t)=\frac{1}{2Z_0}U(x,t)-\frac{1}{2Z_0}I(x,t).$$

For $x=0$

$$u(0,t)=\frac{1}{2}U(0,t)+\frac{1}{2}I(0,t),\ i(0,t)=\frac{1}{2Z_0}U(0,t)-\frac{1}{2Z_0}I(0,t)$$

and for $x=\Lambda$

$$u(\Lambda,t)=\frac{1}{2}U(\Lambda,t)+\frac{1}{2}I(\Lambda,t),\ \ i(\Lambda,t)=\frac{1}{2Z_0}U(\Lambda,t)-\frac{1}{2Z_0}I(\Lambda,t).$$

Replacing the last expressions into (3.1.6) and (3.1.7) we obtain

$$\frac{U(0,t)+I(0,t)}{2}=E_0(t)-R_0\left(\frac{U(0,t)-I(0,t)}{2Z_0}\right)-\tilde{C}_0^{-1}\left(\frac{1}{2Z_0}\int_T^t U(0,\tau)d\tau-\frac{1}{2Z_0}\int_T^t I(0,\tau)d\tau\right)-$$

$$-\left[\left(\frac{U(0,t)-I(0,t)}{2Z_0}\right)\frac{dL_0}{di}+L_0\left(\frac{U(0,t)-I(0,t)}{2Z_0}\right)\right]\times\left(\frac{1}{2Z_0}\frac{dU(0,t)}{dt}-\frac{1}{2Z_0}\frac{dI(0,t)}{dt}\right); \qquad (3.2.6)$$

$$\frac{U(\Lambda,t)+I(\Lambda,t)}{2}=E_1(t)+R_1\left(\frac{U(\Lambda,t)-I(\Lambda,t)}{2Z_0}\right)+\widetilde{C}_1^{-1}\left(\frac{1}{2Z_0}\int_T^t U(\Lambda,\tau)d\tau-\frac{1}{2Z_0}\int_T^t I(\Lambda,\tau)d\tau\right)+$$

$$+\left[\left(\frac{U(\Lambda,t)-I(\Lambda,t)}{2Z_0}\right)\frac{dL_1}{di}+L_1\left(\frac{U(\Lambda,t)-I(\Lambda,t)}{2Z_0}\right)\right]\times\left(\frac{1}{2Z_0}\frac{dU(\Lambda,t)}{dt}-\frac{1}{2Z_0}\frac{dI(\Lambda,t)}{dt}\right). \qquad (3.2.7)$$

So we have to solve the homogenous system

$$\begin{pmatrix}\dfrac{\partial U}{\partial t}\\[2ex]\dfrac{\partial I}{\partial t}\end{pmatrix}+\begin{pmatrix}\dfrac{1}{\sqrt{CL}} & 0\\[2ex] 0 & -\dfrac{1}{\sqrt{CL}}\end{pmatrix}\begin{pmatrix}\dfrac{\partial U}{\partial x}\\[2ex]\dfrac{\partial I}{\partial x}\end{pmatrix}=0$$

under initial conditions (3.2.5) and boundary conditions (3.2.6), (3.2.7) for (x,t) belonging to

$$\Pi=\left\{(x,t)\in R^2:x\in[0,\Lambda],\,t\in[0,\infty)\right\}.$$

Integration along the characteristics yields (cf. Chapter I):

$$U(\Lambda,t+T)=U(0,t),\ I(\Lambda,t)=I(0,t+T).$$

Then the last mixed problem is reduced to an initial value problem for the system

$$\frac{1}{2}U(\Lambda,t+T)+\frac{1}{2}I(0,t)=E_0(t)-R_0\left(\frac{U(\Lambda,t+T)-I(0,t)}{2Z_0}\right)-$$

$$-\widetilde{C}_0^{-1}\left(\frac{1}{2Z_0}\int_T^t U(\Lambda,\tau+T)d\tau-\frac{1}{2Z_0}\int_T^t I(0,\tau)d\tau\right)- \qquad (3.2.8)$$

$$-\left[\left(\frac{U(\Lambda,t+T)-I(0,t)}{2Z_0}\right)\frac{dL_0}{di}+L_0\left(\frac{U(\Lambda,t+T)-I(0,t)}{2Z_0}\right)\right]\times$$

$$\times\left(\frac{1}{2Z_0}\frac{dU(\Lambda,t+T)}{dt}-\frac{1}{2Z_0}\frac{dI(0,t)}{dt}\right);$$

$$\frac{1}{2}U(\Lambda,t)+\frac{1}{2}I(0,t+T)=E_1(t)+R_1\left(\frac{U(\Lambda,t)-I(0,t+T)}{2Z_0}\right)+$$

$$+\widetilde{C}_1^{-1}\left(\frac{1}{2Z_0}\int\limits_T^t U(\Lambda,\tau)d\tau+\frac{1}{2Z_0}\int\limits_T^t I(0,\tau+T)d\tau\right)+ \qquad (3.2.9)$$

$$+\left[\left(\frac{U(\Lambda,t)-I(0,t+T)}{2Z_0}\right)\frac{dL_1}{di}+L_1\left(\frac{U(\Lambda,t)-I(0,t+T)}{2Z_0}\right)\right]\times\left(\frac{1}{2Z_0}\frac{dU(\Lambda,t)}{dt}-\frac{1}{2Z_0}\frac{dI(0,t+T)}{dt}\right).$$

We choose unknown functions to be $U(t)=U(\Lambda,t),\ I(t)=I(0,t)$ and then (3.2.8), (3.2.9) become:

$$\frac{1}{2}U(t+T)+\frac{1}{2}I(t)=E_0(t)-R_0\left(\frac{U(t+T)-I(t)}{2Z_0}\right)-$$

$$-\widetilde{C}_0^{-1}\left(\frac{1}{2Z_0}\int\limits_T^t U(\tau+T)d\tau-\frac{1}{2Z_0}\int\limits_T^t I(\tau)d\tau\right)-$$

$$-\left[\left(\frac{U(t+T)-I(t)}{2Z_0}\right)\frac{dL_0}{di}+L_0\left(\frac{U(t+T)-I(t)}{2Z_0}\right)\right]\left(\frac{1}{2Z_0}\frac{dU(t+T)}{dt}-\frac{1}{2Z_0}\frac{dI(t)}{dt}\right);$$

$$\frac{1}{2}U(t)+\frac{1}{2}I(t+T)=E_1(t)+R_1\left(\frac{U(t)-I(t+T)}{2Z_0}\right)+$$

$$+\widetilde{C}_1^{-1}\left(\frac{1}{2Z_0}\int\limits_T^t U(\tau)d\tau-\frac{1}{2Z_0}\int\limits_T^t I(\tau+T)d\tau\right)+$$

$$+\left[\left(\frac{U(t)-I(t+T)}{2Z_0}\right)\frac{dL_1}{di}+L_1\left(\frac{U(t)-I(t+T)}{2Z_0}\right)\right]\left(\frac{1}{2Z_0}\frac{dU(t)}{dt}-\frac{1}{2Z_0}\frac{dI(t+T)}{dt}\right).$$

Change the variables in the integrals $\tau+T=s\Rightarrow d\tau=ds$. For $\tau=T\Rightarrow s=2T$ and for $\tau=t\Rightarrow s=t+T$. In view of

$$\int\limits_T^t U(\tau+T)d\tau=\int\limits_{2T}^{t+T} U(s)ds \text{ and } \int\limits_T^t I(\tau)d\tau=\int\limits_{2T}^{t+T} I(s-T)ds$$

we obtain

$$\frac{1}{2}U(t+T)+\frac{1}{2}I(t)=E_0(t)-R_0\left(\frac{U(t+T)-I(t)}{2Z_0}\right)-\widetilde{C}_0^{-1}\left(\frac{1}{2Z_0}\int\limits_{2T}^{t+T} U(s)ds-\frac{1}{2Z_0}\int\limits_{2T}^{t+T} I(s-T)ds\right)-$$

$$-\left[\frac{U(t+T)-I(t)}{2Z_0}\frac{dL_0}{di}+L_0\left(\frac{U(t+T)-I(t)}{2Z_0}\right)\right]\left(\frac{1}{2Z_0}\frac{dU(t+T)}{dt}-\frac{1}{2Z_0}\frac{dI(t)}{dt}\right)$$

and

$$\frac{1}{2}U(t)+\frac{1}{2}I(t+T)= E_1(t)+R_1\left(\frac{U(t)-I(t+T)}{2Z_0}\right)+\widetilde{C}_1^{-1}\left(\frac{1}{2Z_0}\int\limits_{2T}^{t+T}U(s)ds-\frac{1}{2Z_0}\int\limits_{2T}^{t+T}I(s-T)ds\right)+$$

$$+\left[\frac{U(t)-I(t+T)}{2Z_0}\frac{dL_1}{di}+L_1\left(\frac{U(t)-I(t+T)}{2Z_0}\right)\right]\left(\frac{1}{2Z_0}\frac{dU(t)}{dt}-\frac{1}{2Z_0}\frac{dI(t+T)}{dt}\right).$$

Replacing $t+T$ by t we finally obtain the following neutral system with respect to the unknown functions $U(t), I(t)$:

$$\frac{U(t)}{2}+\frac{I(t-T)}{2}=E_0(t-T)-R_0\left(\frac{U(t)-I(t-T)}{2Z_0}\right)-$$

$$-\widetilde{C}_0^{-1}\left(\frac{1}{2Z_0}\int\limits_{T}^{t}U(s)ds-\frac{1}{2Z_0}\int\limits_{T}^{t}I(s-T)ds\right)- \qquad (3.2.10)$$

$$-\left[\frac{U(t)-I(t-T)}{2Z_0}\frac{dL_0}{di}+L_0\left(\frac{U(t)-I(t-T)}{2Z_0}\right)\right]\left(\frac{1}{2Z_0}\frac{dU(t)}{dt}-\frac{1}{2Z_0}\frac{dI(t-T)}{dt}\right);$$

$$\frac{U(t-T)}{2}+\frac{I(t)}{2}= E_1(t-T)+R_1\left(\frac{U(t-T)-I(t)}{2Z_0}\right)+$$

$$+\widetilde{C}_1^{-1}\left(\frac{1}{2Z_0}\int\limits_{T}^{t}U(s-T)ds-\frac{1}{2Z_0}\int\limits_{T}^{t}I(s)ds\right)+ \qquad (3.2.11)$$

$$+\left[\frac{U(t-T)-I(t)}{2Z_0}\frac{dL_1}{di}+L_1\left(\frac{U(t-T)-I(t)}{2Z_0}\right)\right]\left(\frac{1}{2Z_0}\frac{dU(t-T)}{dt}-\frac{1}{2Z_0}\frac{dI(t)}{dt}\right).$$

System (3.2.10) - (3.2.11) is a differential system of a neutral type because the derivatives take a part at different instants $\dfrac{dU(t)}{dt},\dfrac{dU(t-T)}{dt},\dfrac{dI(t)}{dt},\dfrac{dI(t-T)}{dt}.$

Recalling that $\widetilde{L}_p(i)=i.L_p(i)\ (p=0,1)$ and $\dfrac{d\widetilde{L}_p(i)}{di}=i.\dfrac{L_p(i)}{di}+L_p(i)$ and introducing denotations:

$$i_0(U,I)(t)=\frac{U(t)-I(t-T)}{2Z_0},\qquad i_1(U,I)(t)=\frac{U(t-T)-I(t)}{2Z_0}$$

the above system becomes

$$\left[i_0(U,I)(t)\frac{dL_0(i_0(U,I)(t))}{di} + L_0\big(i_0(U,I)(t)\big)\right]\left(\frac{1}{2Z_0}\frac{dU(t)}{dt} - \frac{1}{2Z_0}\frac{dI(t-T)}{dt}\right) =$$

$$= -\frac{1}{2}U(t) - \frac{1}{2}I(t-T) + E_0(t-T) - R_0(i_0(U,I)(t)) - \widetilde{C}_0^{-1}\left(\int_T^t i_0(U,I)(s)ds\right);$$

$$\left[i_1(U,I)(t)\frac{dL_1(i_1(U,I)(t))}{di} + L_1\big(i_1(U,I)(t)\big)\right]\left(\frac{1}{2Z_0}\frac{dU(t-T)}{dt} - \frac{1}{2Z_0}\frac{dI(t)}{dt}\right) =$$

$$= \frac{1}{2}U(t-T) + \frac{1}{2}I(t) - E_1(t-T) - R_1(i_1(U,I)(t)) - \widetilde{C}_1^{-1}\left(\int_T^t i_1(U,I)(s)ds\right).$$

If

$$\widetilde{L}_0\left(\frac{U(t)-I(t-T)}{2Z_0}\right) = \frac{U(t)-I(t-T)}{2Z_0}\frac{dL_0(i_0(U,I)(t))}{di} + L_0\left(\frac{U(t)-I(t-T)}{2Z_0}\right) > 0,$$

$$\widetilde{L}_1\left(\frac{U(t-T)-I(t)}{2Z_0}\right) = \frac{U(t-T)-I(t)}{2Z_0}\frac{dL_1(i_1(U,I)(t))}{di} + L_1\left(\frac{U(t-T)-I(t)}{2Z_0}\right) > 0$$

then we rewrite the above neutral system as:

$$\frac{dU(t)}{dt} = \frac{dI(t-T)}{dt} - Z_0\frac{U(t)+I(t-T)-2E_0(t-T)+2R_0(i_0(U,I)(t))}{\widetilde{L}_0(i_0(U,I)(t))} -$$

$$- \frac{2Z_0}{\widetilde{L}_0(i_0(U,I)(t))}\widetilde{C}_0^{-1}\left(\int_T^t i_0(U,I)(s)ds\right) \tag{3.2.12}$$

$$\frac{dI(t)}{dt} = \frac{dU(t-T)}{dt} - Z_0\frac{U(t-T)+I(t)-2E_1(t-T)-2R_1(i_1(U,I)(t))}{\widetilde{L}_1(i_1(U,I)(t))} +$$

$$+ \frac{2Z_0}{\widetilde{L}_1(i_1(U,I)(t))}\widetilde{C}_1^{-1}\left(\int_T^t i_1(U,I)(s)ds\right).$$

3.3. ANALYSIS OF THE ARISING NONLINEARITIES

Proceeding from the known relation, we have

$$i = \frac{dq}{dt} = \frac{d\widetilde{C}_p(u)}{dt} \Rightarrow \int_T^t i(x,\tau)d\tau = \widetilde{C}_p(u) \ \ (p=0,1)$$

and $u_{C_p} = u(0,t) = \widetilde{C}_p^{-1}\left(\int_T^t i(0,\tau)d\tau\right)$, where

$$C_p(u) = \frac{c_p}{\sqrt[h]{1-(u/\Phi_p)}} = \frac{c_p\sqrt[h]{\Phi_p}}{\sqrt[h]{\Phi_p - u}} \ (p=0,1) \ c_p, \Phi_p, \ \ h \in [2,3]$$

are positive constants. We choose an interval $|u| \le \phi_0 < \Phi = \min\{\Phi_0, \Phi_1\}$ so as to isolate the singularity of $C_p(u)$. We have to consider the problem for the existence of the inverse function of

$$\widetilde{C}_p(u) = uC_p(u) = c_p\sqrt[h]{\Phi_p}\,\frac{u}{\sqrt[h]{\Phi_p - u}}.$$

Calculate the derivatives

$$\frac{d\widetilde{C}_p(u)}{du} = C_p(u) + u\frac{dC_p(u)}{du} =$$

$$= c_p\sqrt[h]{\Phi_p}\left(\frac{1}{\sqrt[h]{\Phi_p - u}} + \frac{u}{h\sqrt[h]{(\Phi_p - u)^{h+1}}}\right) = \frac{c_p\sqrt[h]{\Phi_p}}{\sqrt[h]{\Phi_p - u}}\,\frac{h(\Phi_p - u) + u}{h(\Phi_p - u)}.$$

Then $\dfrac{d\widetilde{C}_p(u)}{du} > 0 \ \Leftrightarrow \ \dfrac{h(\Phi_p - u) + u}{h(\Phi_p - u)} > 0 \Leftrightarrow h(\Phi_p - u) + u > 0 \Leftrightarrow u < \dfrac{h}{h-1}\Phi_p$.

Obviously $u \in [-\phi_0, \phi_0] \Rightarrow -\phi_0 \le u \le \phi_0 < \Phi_p < \dfrac{h}{h-1}\Phi_p$.

Let $U_0 \le \phi_0$. Then

$$\widetilde{C}_p(u) = \frac{u\,c_p\sqrt[h]{\Phi_p}}{\sqrt[h]{\Phi_p - u}} : [-U_0; U_0] \to \left[-c_p\sqrt[h]{\Phi_p}\,U_0\,\frac{1}{\sqrt[h]{\Phi_p + U_0}}; \ c_p\sqrt[h]{\Phi_p}\,U_0\,\frac{1}{\sqrt[h]{\Phi_p - U_0}}\right].$$

Since $\dfrac{d\widetilde{C}_p(u)}{du} > 0$, the inverse function exists:

$$\widetilde{C}_p^{-1}(\lambda):\left[\frac{-c_p\sqrt[h]{\Phi_p}\,U_0}{\sqrt[h]{\Phi_p+U_0}},\ \frac{c_p\sqrt[h]{\Phi_p}\,U_0}{\sqrt[h]{\Phi_p-U_0}}\right]\rightarrow[-U_0,U_0]$$

and

$$\left|\widetilde{C}_p^{-1}(\lambda)\right|\le U_0\le\phi_0. \tag{3.3.1}$$

The explicit form of the inverse function for $h=2$ is:

$$\frac{u\,c_p\sqrt{\Phi_p}}{\sqrt{\Phi_p-u}}=\lambda \Rightarrow c_p\Phi_p u^2=\Phi_p\lambda^2-\lambda^2 u \Rightarrow c_p^2\Phi_p u^2+\lambda^2 u-\Phi_p\lambda^2=0,$$

$$\widetilde{C}_p^{-1}(\lambda)=\frac{1}{2c_p^2\Phi_p}\left(\sqrt{\lambda^4+4c_p^2\Phi_p^2\lambda^2}-\lambda^2\right)=|\lambda|\frac{\sqrt{\lambda^2+4c_p^2\Phi_p^2}-|\lambda|}{2c_p^2\Phi_p}\ (p=0,1) \tag{3.3.2}$$

and

$$\frac{d\widetilde{C}_p^{-1}(\lambda)}{d\lambda}=\frac{1}{2c_p^2\Phi_p}\left(\frac{4\lambda^3+8c_p^2\Phi_p^2\lambda}{2\sqrt{\lambda^4+4c_p^2\Phi_p^2\lambda^2}}-2\lambda\right)=\lambda\frac{\lambda^2+2c_p^2\Phi_p^2-\sqrt{\lambda^4+4c_p^2\Phi_p^2\lambda^2}}{c_p^2\Phi_p\sqrt{\lambda^4+4c_p^2\Phi_p^2\lambda^2}}.$$

It is easy to verify that:

$$\left|\widetilde{C}_p^{-1}(\lambda)\right|\le|\lambda|\frac{\sqrt{\lambda^2+4c_p^2\Phi_p^2}-|\lambda|}{2c_p^2\Phi_p}\le|\lambda|\frac{\sqrt{\left(\dfrac{c_p\sqrt{\Phi_p}\,\phi_0}{\sqrt{\Phi_p-\phi_0}}\right)^2+4c_p^2\Phi_p^2}}{2c_p^2\Phi_p}=$$

$$|\lambda|\frac{2\Phi_p-\phi_0}{2c_p\sqrt{\Phi_p}\sqrt{\Phi_p-\phi_0}}\equiv H_p|\lambda|;$$

$$|\lambda|\le\frac{c_p\sqrt{\Phi_p}\,\phi_0}{\sqrt{\Phi_p+\phi_0}}=\min\left\{\left|\frac{-c_p\sqrt{\Phi_p}\,\phi_0}{\sqrt{\Phi_p+\phi_0}}\right|,\left|\frac{c_p\sqrt{\Phi_p}\,\phi_0}{\sqrt{\Phi_p-\phi_0}}\right|\right\}$$

and

$$\left|\frac{d\widetilde{C}_p^{-1}(\lambda)}{d\lambda}\right| \le \frac{|\lambda|}{c_p^2\Phi_p}\left(\frac{|\lambda|^2 + 2c_p^2\Phi_p^2}{\sqrt{\lambda^4 + 4c_p^2\Phi_p^2\lambda^2}} + 1\right) \le \frac{|\lambda|}{c_p^2\Phi_p}\left(\frac{|\lambda|^2 + 2c_p^2\Phi_p^2}{|\lambda|\sqrt{\lambda^2 + 4c_p^2\Phi_p^2}} + 1\right) \le$$

$$\le \frac{1}{c_p^2\Phi_p}\left(\sqrt{\lambda^2 + 4c_p^2\Phi_p^2} + |\lambda|\right) \le \frac{1}{c_p^2\Phi_p}\left(\sqrt{\left(\frac{c_p\sqrt{\Phi_p}\,\phi_0}{\sqrt{\Phi_p - \phi_0}}\right)^2 + 4c_p^2\Phi_p^2} + \frac{c_p\sqrt{\Phi_p}\,\phi_0}{\sqrt{\Phi_p - \phi_0}}\right) =$$

$$= \frac{c_p\sqrt{\Phi_p}}{c_p^2\Phi_p}\left(\frac{\sqrt{\phi_0^2 + 4\Phi_p^2 - 4\Phi_p\phi_0}}{\sqrt{\Phi_p - \phi_0}} + \frac{\phi_0}{\sqrt{\Phi_p - \phi_0}}\right) = \frac{2\sqrt{\Phi_p}}{c_p\sqrt{\Phi_p - \phi_0}}\ .$$

Therefore

$$\left|\frac{d\widetilde{C}_p^{-1}(\lambda)}{d\lambda}\right| \le \frac{2\sqrt{\Phi_p}}{c_p\sqrt{\Phi_p - \phi_0}}\ ,\ (p = 0,1). \tag{3.3.3}$$

Remark 3.3.1. An explicit form of the inverse function may be obtained for $h = 3$.

Indeed, $\dfrac{u\,c_p\sqrt[3]{\Phi_p}}{\sqrt[3]{\Phi_p - u}} = \lambda \Leftrightarrow u^3 + \dfrac{\lambda^3}{c_p^3\Phi_p}u - \dfrac{\lambda^3}{c_p^3} = 0$, and then in view of Cardano

formula we have

$$\widetilde{C}_p^{-1}(\lambda) == \sqrt[3]{\frac{\lambda^3}{2c_p^3} + \sqrt{\frac{\lambda^6}{4c_p^6} + \frac{\lambda^9}{27c_p^9\Phi_p^3}}} + \sqrt[3]{\frac{\lambda^3}{2c_p^3} - \sqrt{\frac{\lambda^6}{4c_p^6} + \frac{\lambda^9}{27c_p^9\Phi_p^3}}}\ .$$

Remark 3.3.2. In the applications very often one uses a polynomial form of $C_p(u)$ obtained in the following manner:

$$C_p(u) = c_p u\left(1 - \frac{u}{\Phi_p}\right)^{-\frac{1}{2}} = c_p u\left(1 - \frac{1}{2}\frac{u}{\Phi_p} + \frac{3}{8}\left(\frac{u}{\Phi_p}\right)^2 - ...\right) \approx c_p u\left(1 - \frac{1}{2}\frac{u}{\Phi_p} + \frac{3}{8}\left(\frac{u}{\Phi_p}\right)^2\right)$$

under condition $\left|u/\Phi_p\right| \le U_0/\Phi_p < 1$. Here one can proceed as in the case of $R_p(i)$ and $L_p(i)$.

We consider *I-V* and *I-L* as characteristics of a polynomial type. Namely we assume that

$$R_p(i) = \sum_{n=1}^{m} r_n^{(p)} i^n \text{ and } L_p(i) = \sum_{n=0}^{m} l_n^{(p)} i^n \quad (p = 0,1).$$

Since (cf. L. O. Chua, C. A. Desoer & E. S. Kuh [37], P. N. Matkhanov [84], A. M. Zaezdnii [122])

$$u_{\Psi_p} = \frac{d\Psi_p}{dt} = \frac{d(i.L_p(i))}{dt} = \left(i \frac{dL_p(i)}{di} + L_p(i) \right) \frac{di}{dt} \quad (p = 0,1)$$

then $\Psi_p(i) = i \cdot L_p(i) = i \cdot \sum_{n=0}^{m} l_n^{(p)} i^n$ and we get

$$\widetilde{L}_p = i \frac{dL_p(i)}{di} + L_p(i) = i \sum_{n=0}^{m} n l_n^{(p)} i^{n-1} + \sum_{n=0}^{m} l_n^{(p)} i^n = \sum_{n=0}^{m} (n+1) l_n^{(p)} i^n, \quad (p = 0,1).$$

We have to estimate $\dfrac{d\widetilde{L}_p(i)}{di}$:

$$\frac{d\widetilde{L}_p}{di} = i \frac{dL_p(i)}{di} + i \frac{d^2 L_p(i)}{di^2} + \frac{dL_p(i)}{di} = (i+1) \frac{dL_p(i)}{di} + i \frac{d^2 L_p(i)}{di^2} = \sum_{n=1}^{m} \widetilde{l}_n^{(p)} i^n .$$

Recalling that

$$i_0(U,I)(t) = \frac{U(t) - I(t-T)}{2Z_0}, \qquad i_1(U,I)(t) = \frac{U(t-T) - I(t)}{2Z_0}$$

we write

$$\widetilde{L}_0(i_0(U,I)) = i_0(U,I) \frac{dL_0(i_0(U,I))}{di} + L_0\big(i_0(U,I)\big) =$$

$$= \sum_{n=0}^{m} (n+1) l_n^{(0)} \big(i_0(U,I)\big)^n = \sum_{n=0}^{m} (n+1) l_n^{(0)} \left(\frac{U(t) - I(t-T)}{2Z_0} \right)^n ;$$

$$\widetilde{L}_1(i_1(U,I)) = i_1(U,I) \frac{dL_1(i_1(U,I))}{di} + L_1\big(i_1(U,I)\big) =$$

$$= \sum_{n=0}^{m}(n+1)l_n^{(1)}\left(i_1(U,I)\right)^n = \sum_{n=0}^{m}(n+1)l_n^{(1)}\left(\frac{U(t-T)-I(t)}{2Z_0}\right)^n.$$

From here, the following assumptions will be fulfilled:

(E): $E_p(.) \in C_{T_0}^1[0,\infty),\ (p=0,1),\ \int_{T+kT_0}^{T+(k+1)T_0} E_p(t)dt = 0,\ E_p(0)=0,$

$$\left|E_p(t)\right| \le U_{E_p} e^{\mu(t-T-kT_0)},\ t \in [T+kT_0, T+(k+1)T_0]\ (k=0,1,2,3,...),\ U_{E_p} \le U_0$$

(IN): $U_0(.), I_0(.) \in C_{T_0}^1[0,T],\ \int_{-T+kT_0}^{-T+(k+1)T_0} U_0(t)dt = 0,\ \int_{-T+kT_0}^{-T+(k+1)T_0} I_0(t)dt = 0;$

$$\left|U_0(t)\right| \le e^{-\beta} e^{\mu(t-kT_0)},\ \left|I_0(t)\right| \le e^{-\beta} e^{\mu(t-kT_0)}\ (k=0,1,2,...,m-1);\ \beta = \text{const.} > 0.$$

Remark 3.3.3. It is easy to see that $U(t-T)=\bar{U}_0(t), I(t-T)=\bar{I}_0(t)$ for $t \in [T,2T]$, where $\bar{U}_0(t), \bar{I}_0(t)$ are translated to the right in initial functions over $[T,2T]$.

Remark 3.3.4. It follows $\left|U(t-T)\right| = \left|\bar{U}_0(t)\right| \le e^{-\beta} U_0 e^{\mu(t-T-kT_0)},$

$$\left|I(t-T)\right| = \left|\bar{I}_0(t)\right| \le e^{-\beta} I_0 e^{\mu(t-T-kT_0)};\quad (k=0,1,2,3,...,m-1).$$

then

$$\left|i_0(U,I)(t)\right| = \left|\frac{U(t)-I(t-T)}{2Z_0}\right| \le \frac{e^{\mu_0}(U_0 + I_0 e^{-\beta})}{2Z_0},$$

$$\left|i_1(U,I)(t)\right| = \left|\frac{U(t-T)-I(t)}{2Z_0}\right| \le \frac{e^{\mu_0}(U_0 e^{-\beta} + I_0)}{2Z_0}.$$

Let $[-\mathrm{I}_0, \mathrm{I}_0]$ be an interval such that

$$\left|i_0(U,I)(t)\right| \le \mathrm{I}_0,\ \left|i_1(U,I)(t)\right| \le \mathrm{I}_0 \tag{3.3.4}$$

to imply

$$\widetilde{L}_0\left(\frac{U(t)-I(t-T)}{2Z_0}\right)(t)=\sum_{n=0}^{m}(n+1)l_n^{(0)}\left(\frac{U(t)-I(t-T)}{2Z_0}\right)^n\geq\sum_{n=0}^{m}(n+1)l_n^{(0)}\left(\frac{e^{\mu_0}(U_0+I_0e^{-\beta})}{2Z_0}\right)^n\equiv\hat{L}_0>0$$

and

$$\widetilde{L}_1\left(\frac{U(t-T)-I(t)}{2Z_0}\right)(t)=\sum_{n=0}^{m}(n+1)l_n^{(1)}\left(\frac{U(t-T)-I(t)}{2Z_0}\right)^n\geq\sum_{n=0}^{m}(n+1)l_n^{(1)}\left(\frac{e^{\mu_0}(U_0e^{-\beta}+I_0)}{2Z_0}\right)^n\equiv\hat{L}_1>0.$$

One can choose $I_0=\min\left\{\dfrac{e^{\mu_0}(U_0+I_0e^{-\beta})}{2Z_0};\dfrac{e^{\mu_0}(U_0e^{-\beta}+I_0)}{2Z_0}\right\}.$

3.4. AN OPERATOR FORMULATION OF THE PERIODIC PROBLEM

Now we formulate the main problem: Find a continuously differentiable T_0-periodic solution $(U(t),I(t))$ of (3.4.1) on the interval $[T,\infty)$ coinciding with prescribed T_0-periodic initial functions $U_0(t)$, $I_0(t)$ on the interval $[0,T]$:

$$\frac{dU(t)}{dt}=\frac{dI(t-T)}{dt}-Z_0\frac{U(t)+I(t-T)-2E_0(t)+2R_0\big(i_0(U,I)(t)\big)}{\widetilde{L}_0(U,I)(t)}-$$

$$-\frac{2Z_0}{\widetilde{L}_0(U,I)(t)}\widetilde{C}_0^{-1}\left(\int_T^t i_0(U,I)(S)ds\right),\ t\in[T,2T]$$

$$\frac{dI(t)}{dt}=\frac{dU(t-T)}{dt}-Z_0\frac{U(t-T)+I(t)-2E_1(t)-2R_1\big(i_1(U,I)(t)\big)}{\widetilde{L}_1(U,I)(t)}+$$

$$+\frac{2Z_0}{\widetilde{L}_1(U,I)(t)}\widetilde{C}_1^{-1}\left(\int_T^t i_1(U,I)(S)ds\right),\ t\in[T,2T]$$

$$\text{(3.4.1)}$$

$$U(t)=U_0(t),\ I(t)=I_0(t),\ \frac{dU(t)}{dt}=\frac{dU_0(t)}{dt},\ \frac{dI(t)}{dt}=\frac{dI_0(t)}{dt},\qquad t\in[0,T]$$

By $C_{T_0}^1[T,\infty)$ we mean the space of all continuous T_0-periodic functions with continuous derivatives. The main difficulty is to define a suitable operator whose fixed points are the solutions sought.

First we introduce the sets

$$M_U=\left\{U(.)\in C_{T_0}^1[T,2T]:\int_{T+kT_0}^{T+(k+1)T_0}U(t)dt=0\ ;k=0,1,2,3,...,m-1\right\},$$

$$M_I = \left\{ I(.) \in C_{T_0}^1[T,2T]: \int_{T+kT_0}^{T+(k+1)T_0} I(t)dt = 0 \; ; k = 0,1,2,3,...,m-1 \right\},$$

$$M_U^* = \left\{ U(.) \in M_U : |U(t)| \le U_0 e^{\mu(t-T-kT_0)}, t \in [T+kT_0, T+(k+1)T_0], \; k = 0,1,2,3,....m-1 \right\},$$

$$M_I^* = \left\{ I(.) \in M_I : |I(t)| \le I_0 e^{\mu(t-T-kT_0)}, t \in [T+kT_0, T+(k+1)T_0], k = 0,1,2,3,...,m-1 \right\}.$$

Prior to introducing a family of pseudo-metrics we notice that if

$$\rho_\mu^{(k)}(U,\overline{U}) = \max\left\{ e^{-\mu(t-t_k)} |U(t)-\overline{U}(t)| : t \in [T+kT_0, T+(k+1)T_0] \right\},$$

$$\rho^{(k)}(U,\overline{U}) = \max\left\{ |U(t)-\overline{U}(t)| : t \in [T+kT_0, T+(k+1)T_0] \right\}$$

it follows

$$\rho_\mu^{(k)}(U,\overline{U}) \le \rho^{(k)}(U,\overline{U}) \le e^{\mu T_0}\rho_\mu^{(k)}(U,\overline{U}), \; (k=0,1,2,...).$$

Then we define

$$\rho^{(k)}(U,\overline{U}) = \max\left\{ |U(t)-\overline{U}(t)| : t \in [T+kT_0, T+(k+1)T_0] \right\},$$

$$\rho^{(k)}(I,\overline{I}) = \max\left\{ |I(t)-\overline{I}(t)| : t \in [T+kT_0, T+(k+1)T_0] \right\},$$

$$\hat{\rho}(U,\overline{U}) = \max\left\{ |U(t)-\overline{U}(t)| : t \in [T,2T] \right\},$$

$$\hat{\rho}(I,\overline{I}) = \max\left\{ |I(t)-\overline{I}(t)| : t \in [T,2T] \right\},$$

$$\rho_\mu^{(k)}(U,\overline{U}) = \max\left\{ e^{-\mu(t-t_k)} |U(t)-\overline{U}(t)| : t \in [T+kT_0, T+(k+1)T_0] \right\},$$

$$\rho_\mu^{(k)}(I,\overline{I}) = \max\left\{ e^{-\mu(t-t_k)} |I(t)-\overline{I}(t)| : t \in [T+kT_0, T+(k+1)T_0] \right\},$$

$$\rho_\mu^{(k)}(\dot{U},\dot{\overline{U}}) = \max\left\{ e^{-\mu(t-t_k)} |\dot{U}(t)-\dot{\overline{U}}(t)| : t \in [T+kT_0, T+(k+1)T_0] \right\},$$

$$\rho_\mu^{(k)}(\dot{I},\dot{\overline{I}}) = \max\left\{ e^{-\mu(t-t_k)} |\dot{I}(t)-\dot{\overline{I}}(t)| : t \in [T+kT_0, T+(k+1)T_0] \right\},$$

$$\hat{\rho}_\mu((\dot{U},\dot{I}),(\dot{\overline{U}},\dot{\overline{I}})) = \max\left\{ \rho_\mu^{(k)}(\dot{U},\dot{\overline{U}}), \rho_\mu^{(k)}(\dot{I},\dot{\overline{I}}) : k = 0,1,2,...,m-1 \right\}.$$

It is easy to verify that

$$\hat{\rho}(U,\overline{U}) = \max\left\{\rho^{(0)}(U,\overline{U}), \rho^{(1)}(U,\overline{U}),...,\rho^{(k)}(U,\overline{U})\right\} \le$$
$$\le e^{\mu T_0} \max\left\{\rho_\mu^{(0)}(U,\overline{U}), \rho_\mu^{(1)}(U,\overline{U}),...,\rho_\mu^{(k)}(U,\overline{U})\right\},$$
$$\hat{\rho}(I,\overline{I}) = \max\left\{\rho^{(0)}(I,\overline{I}), \rho^{(1)}(I,\overline{I}),...,\rho^{(k)}(I,\overline{I})\right\} \le$$
$$\le e^{\mu T_0} \max\left\{\rho_\mu^{(0)}(I,\overline{I}), \rho_\mu^{(1)}(I,\overline{I}),...,\rho_\mu^{(k)}(I,\overline{I})\right\}. \tag{3.4.2}$$

The set $M_U^* \times M_I^*$ turns into a complete metric space with respect to the metric

$$\hat{\rho}_\mu(((W,\dot{W}),(J,\dot{J})),((\overline{W},\dot{\overline{W}}),(\overline{J},\dot{\overline{J}}))) = \max\left\{\hat{\rho}(U,\overline{U}), \hat{\rho}(I,\overline{I}), \rho_\mu^{(k)}(\dot{U},\dot{\overline{U}}), \rho_\mu^{(k)}(\dot{I},\dot{\overline{I}}) : k = 0,1,2,...,m-1\right\}$$

Since we are looking for a continuously differentiable solution using the above system, let us point out that the conformity condition (CC) is:

$$\widetilde{L}_0 \frac{dU(T)}{dt} = \widetilde{L}_0 \frac{dI(0)}{dt} + 2Z_0 E_0(0) - Z_0 U(T) - Z_0 I(0) - 2Z_0 R_0\left(\frac{U(T)-I(0)}{2Z_0}\right),$$
$$\widetilde{L}_1 \frac{dI(T)}{dt} = \widetilde{L}_1 \frac{dU(0)}{dt} + 2Z_0 E_1(0) - Z_0 U(0) - Z_0 I(T) + 2Z_0 R_1\left(\frac{U(0)-I(T)}{2Z_0}\right).$$

It is satisfied provided we choose the initial functions such that

$$U_0(0) = 0, I_0(0) = 0, \dot{U}_0(0) = 0, \dot{I}_0(0) = 0 \text{ and } E_0(0) = E_1(0) = 0.$$

The periodicity implies

$$U_0(T) = 0, I_0(T) = 0, \dot{U}_0(T) = 0, \dot{I}_0(T) = 0.$$

In order to overcome the restrictive conditions $\dot{U}_0(T) = 0, \dot{I}_0(T) = 0$, we introduce an operator in the following way: $B(U,I)(t) = \left(B_U(U,I)(t), B_I(U,I)(t)\right)$ $t \in [T,2T]$ is defined on every interval $[T + kT_0, T + (k+1)T_0]$ $(k = 0,1,2,...,m-1)$, $T = mT_0$ assuming $U(T) = I(T) = 0 \Rightarrow U(T + kT_0) = I(T + kT_0) = 0$ by the formulas:

$$B_U^{(k)}(U,I)(t) := \int_{T+kT_0}^{t} V(U,I)(s)ds - \left(\frac{t-T-kT_0}{T_0} - \frac{1}{2}\right)\int_{T+kT_0}^{T+(k+1)T_0} V(U,I)(s)ds - \frac{1}{T_0}\int_{T+kT_0}^{T+(k+1)T_0}\int_{T+kT_0}^{t} V(U,I)(s)dsdt$$

$$B_I^{(k)}(U,I)(t) := \int_{T+kT_0}^{t} J(U,I)(s)ds - \left(\frac{t-T-kT_0}{T_0} - \frac{1}{2}\right)\int_{T+kT_0}^{T+(k+1)T_0} J(U,I)(s)ds - \frac{1}{T_0}\int_{T+kT_0}^{T+(k+1)T_0}\int_{T+kT_0}^{t} J(U,I)(s)dsdt$$

where

$$V(U,I)(t) = \frac{d\vec{I}_0(t)}{dt} - Z_0 \frac{U(t) + \vec{I}_0(t) - 2E_0(t) + 2R_0\big(i_0(U,I)(t)\big)}{\widetilde{L}_0(i_0(U,I))(t)} -$$

$$- \frac{2Z_0}{\widetilde{L}_0(i_0(U,I))(t)} \widetilde{C}_0^{-1}\left(\int_T^t i_0(U,I)(s)ds\right),$$

$$J(U,I)(t) = \frac{d\vec{U}_0(t)}{dt} - Z_0 \frac{\vec{U}_0(t) + I(t) - 2E_1(t) - 2R_1\big(i_1(U,I)(t)\big)}{\widetilde{L}_1(i_1(U,I))(t)} +$$

$$+ \frac{2Z_0}{\widetilde{L}_1(i_1(U,I))(t)} \widetilde{C}_1^{-1}\left(\int_T^t i_1(U,I)(s)ds\right),$$

$$i_0(U,I)(t) = \frac{U(t) - \vec{I}_0(t)}{2Z_0}, \quad i_1(U,I)(t) = \frac{\vec{U}_0(t) - I(t)}{2Z_0}$$

and $\vec{U}_0(t)$, $\vec{I}_0(t)$ are initial functions translated to the right on $[T, 2T]$.

Lemma 3.4.1. If $U(.) \in M_U$ $\big(\text{resp.} I(.) \in M_I\big)$ then $F(t) = \int_{T+kT_0}^t U(\tau)d\tau$

$\left(\text{resp. } G(t) = \int_{T+kT_0}^t I(\tau)d\tau\right)$ is a T_0-periodic function.

Proof: Our result is

$$F(t + T_0) = \int_{T+kT_0}^{t+T_0} U(\tau)d\tau = \int_{T+kT_0}^t U(\tau)d\tau + \int_t^{t+T_0} U(\tau)d\tau =$$

$$= F(t) + \int_t^{T+kT_0} U(\tau)d\tau + \int_{T+kT_0}^{T+(k+1)T_0} U(\tau)d\tau + \int_{T+(k+1)T_0}^{t+T_0} U(\tau)d\tau = F(t) + \int_t^{T+kT_0} U(\tau)d\tau + \int_{T+kT_0}^t U(\theta)d\theta = F(t).$$

Lemma 3.4.1 is thus proved.

Lemma 3.4.2. If the initial functions and source functions are T_0-periodic ones and $(U,I) \in M_U \times M_I$ then $V(U,I)(t)$, $J(U,I)(t)$ are T_0-periodic functions.

The proof is straightforward and is based on Lemma 3.4.1.

Lemma 3.4.3. For every $\big(U(.), I(.)\big) \in M_U \times M_I$ it follows

$$\int_{T+kT_0}^{T+(k+1)T_0} \int_{T+kT_0}^s V(U,I)(\theta)d\theta ds = \int_{T+(k+1)T_0}^{T+(k+2)T_0} \int_{T+(k+1)T_0}^s V(U,I)(\theta)d\theta ds \quad (k = 0,1,2,...)$$

$$\int_{T+kT_0}^{T+(k+1)T_0}\int_{T+kT_0}^{s}J(U,I)(\theta)d\theta ds = \int_{T+(k+1)T_0}^{T+(k+2)T_0}\int_{T+(k+1)T_0}^{s}J(U,I)(\theta)d\theta ds \ \ (k=0,1,2,\ldots)$$

Proof:

1st manner. We notice that in view of Lemma 3.4.2 $V(U,I)(t), J(U,I)(t)$ are T_0-periodic functions.

Let us rewrite

$$\Gamma_k(s) = \int_{T+kT_0}^{s}V(U,I)(\theta)d\theta \ \text{ and } \ \Gamma_{k+1}(s) = \int_{T+(k+1)T_0}^{s}V(U,I)(\theta)d\theta$$

in the form

$$\Gamma_k(\eta) = \int_{T+kT_0}^{\eta+T+kT_0}V(U,I)(\theta)d\theta \ \text{ and } \ \Gamma_{k+1}(\eta) = \int_{T+(k+1)T_0}^{\eta+T+(k+1)T_0}V(U,I)(\theta)d\theta .$$

Changing the variable $\tau = \theta - T_0$ we obtain

$$\Gamma_{k+1}(\eta) = \int_{T+(k+1)T_0}^{T+(k+1)T_0+\eta}V(U,I)(\theta)d\theta = \int_{T+kT_0}^{T+kT_0+\eta}V(U,I)(\tau+T_0)d\tau = \int_{T+kT_0}^{T+kT_0+\eta}V(U,I)(\tau)d\tau = \Gamma_k(\eta).$$

Consequently

$$\int_{T+kT_0}^{T+(k+1)T_0}\Gamma_k(\eta)d\eta = \int_{T+(k+1)T_0}^{T+(k+2)T_0}\Gamma_{k+1}(\eta)d\eta \ \ (k=0,1,2,\ldots).$$

The proof is analogous for $J(U,I)(t)$.

2nd manner. Changing the variable $\xi = s - T_0$ we have

$$\int_{T+(k+1)T_0}^{T+(k+2)T_0}\int_{T+(k+1)T_0}^{s}V(U,I)(\theta)d\theta ds = \int_{T+kT_0}^{T+(k+1)T_0}\int_{T+(k+1)T_0}^{\xi+T_0}V(U,I)(\theta)d\theta d\xi =$$

$$= \int_{T+kT_0}^{T+(k+1)T_0}\left(\int_{T+(k+1)T_0}^{\xi}V(U,I)(\theta)d\theta + \int_{\xi}^{\xi+T_0}V(U,I)(\theta)d\theta\right)d\xi = \int_{T+kT_0}^{T+(k+1)T_0}\left(\int_{T+(k+1)T_0}^{\xi}V(U,I)(\theta)d\theta\right)d\xi =$$

$$= \int_{T+kT_0}^{T+(k+1)T_0}\left(\int_{T+(k+1)T_0}^{T+kT_0}V(U,I)(\theta)d\theta + \int_{T+kT_0}^{\xi}V(U,I)(\theta)d\theta\right)d\xi = \int_{T+kT_0}^{T+(k+1)T_0}\left(\int_{T+kT_0}^{\xi}V(U,I)(\theta)d\theta\right)d\xi .$$

Lemma 3.4.3 is thus proved.

Lemma 3.4.4. The function $B_U(U,I)(t)$ belongs to M_U and $B_I(U,I)(t)$ − to M_I.

Proof: We show that if $(U(.),I(.)) \in M_U \times M_I$ then

$$B(U,I)(t) = \big(B_U(U,I)(t),\, B_I(U,I)(t)\big)$$

is also T_0-periodic on $[T,\infty)$. Indeed, since $V(U,I)(t)$ is T_0-periodic we have

$$\int_t^{t-T_0} V(U,I)(s))ds + \int_{T+kT_0}^{T+(k+1)T_0} V(U,I)(s))ds = 0 \text{ and}$$

$$\int_{T+(k+1)T_0}^{T+(k+2)T_0} V(U,I)(s))ds = \int_{T+kT_0}^{T+(k+1)T_0} V(U,I)(s))ds.$$

Besides if $t \in [T+kT_0, T+(k+1)T_0] \Rightarrow t+T_0 \in [T+(k+1)T_0, T+(k+2)T_0]$.

Then we have to show that

$$B_U^{(k)}(U,I)(t-T_0) = B_U^{(k+1)}(U,I)(t),\ t \in [T+(k+1)T_0, T+(k+2)T_0].$$

In view of Lemma 3.4.3 we obtain

$$B_U^{(k)}(U,I)(t-T_0) = \int_{T+kT_0}^{t-T_0} V(U,I)(s)ds - \left(\frac{t-T_0-T-kT_0}{T_0} - \frac{1}{2}\right)\int_{T+kT_0}^{T+(k+1)T_0} V(U,I)(s)ds -$$

$$-\frac{1}{T_0}\int_{T+kT_0}^{T+(k+1)T_0}\int_{T+kT_0}^{t} V(U,I)(s)dsdt =$$

$$= \int_{T+kT_0}^{T+(k+1)T_0} V(U,I)(s)ds + \int_{T+(k+1)T_0}^{t} V(U,I)(s)ds + \int_t^{t-T_0} V(U,I)(s)ds - \left(\frac{t-T-(k+1)T_0}{T_0} - \frac{1}{2}\right)\int_{T+(k+1)T_0}^{T+(k+2)T_0} V(U,I)(s)ds -$$

$$-\frac{1}{T_0}\int_{T+kT_0}^{T+(k+1)T_0}\int_{T+kT_0}^{t} V(U,I)(s)dsdt =$$

$$= \int_{T+(k+1)T_0}^{t} V(U,I)(s)ds - \left(\frac{t-T-(k+1)T_0}{T_0} - \frac{1}{2}\right)\int_{T+(k+1)T_0}^{T+(k+2)T_0} V(U,I)(s)ds -$$

$$-\frac{1}{T_0}\int_{T+(k+1)T_0}^{T+(k+2)T_0}\int_{T+(k+1)T_0}^{t} V(U,I)(s)dsdt = B_U^{(k+1)}(U,I)(t)\ .$$

On can show in the same way that $B_U^{(k)}(U,I)(t-T_0) = B_U^{(k+1)}(U,I)(t)\ .$

Let us show that $\displaystyle\int_{T+kT_0}^{T+(k+1)T_0} B_U^{(k)}(U,I)(t)\,dt = 0$. Indeed, it is easy to verify that

$$\int_{T+kT_0}^{T+(k+1)T_0}\left(\frac{t-T-kT_0}{T_0}-\frac{1}{2}\right)dt = 0.$$

Then

$$\int_{T+kT_0}^{T+(k+1)T_0} B_U^{(k)}(U,I)(t)\,dt = \int_{T+kT_0}^{T+(k+1)T}\int_{T+kT_0}^{t} V(U,I)(s)\,ds\,dt -$$

$$-\int_{T+kT_0}^{T+(k+1)T_0}\left(\frac{t-T-kT_0}{T_0}-\frac{1}{2}\right)dt\int_{T+kT_0}^{T+(k+1)T_0} V(U,I)(s)\,ds - \int_{T+kT_0}^{T+(k+1)T_0}\int_{T+kT_0}^{t} V(U,I)(s)\,ds\,dt = 0$$

and analogously $\displaystyle\int_{T+kT_0}^{T+(k+1)T_0} B_I^{(k)}(U,I)(t)\,dt = 0$.

We have to prove that $B_U(U,I)(t)$, $B_I(U,I)(t)$ are continuously differentiable functions. Indeed,

$$B_U^{(k)}(U,I)(T+(k+1)T_0) = \int_{T+kT_0}^{T+(k+1)T_0} V(U,I)(s)\,ds - \left(\frac{T+(k+1)T_0-T-kT_0}{T_0}-\frac{1}{2}\right)\int_{T+kT_0}^{T+(k+1)T_0} V(U,I)(s)\,ds -$$

$$-\frac{1}{T_0}\int_{T+kT_0}^{T+(k+1)T_0}\int_{T+kT_0}^{s} V(U,I)(\theta)\,d\theta\,ds = \frac{1}{2}\int_{T+kT_0}^{T+(k+1)T_0} V(U,I)(s)\,ds - \frac{1}{T_0}\int_{T+kT_0}^{T+(k+1)T_0}\int_{T+kT_0}^{s} V(U,I)(\theta)\,d\theta\,ds.$$

On the other hand

$$B_U^{(k+1)}(U,I)(T+(k+1)T_0) := \int_{T+(k+1)T_0}^{T+(k+1)T_0} V(U,I)(s)\,ds - \left(\frac{T+(k+1)T_0-T-(k+1)T_0}{T_0}-\frac{1}{2}\right)\int_{T+(k+1)T_0}^{T+(k+2)T_0} V(U,I)(s)\,ds -$$

$$-\frac{1}{T_0}\int_{T+(k+1)T_0}^{T+(k+2)T_0}\int_{T+(k+1)T_0}^{s} V(U,I)(\theta)\,d\theta\,ds = \frac{1}{2}\int_{T+(k+1)T_0}^{T+(k+2)T_0} V(U,I)(s)\,ds - \frac{1}{T_0}\int_{T+(k+1)T_0}^{T+(k+2)T_0}\int_{T+(k+1)T_0}^{s} V(U,I)(\theta)\,d\theta\,ds.$$

Since $V(U,I)(s)$ is a T_0-periodic function then

$$\int_{T+kT_0}^{T+(k+1)T_0} V(U,I)(s)\,ds = \int_{T+(k+1)T_0}^{T+(k+2)T_0} V(U,I)(s)\,ds.$$

In view of Lemma 3.4.3

$$\int\limits_{T+kT_0}^{T+(k+1)T_0} \int\limits_{T+kT_0}^{s} V(U,I)(\theta)d\theta ds = \int\limits_{T+(k+1)T_0}^{T+(k+2)T_0} \int\limits_{T+(k+1)T_0}^{s} V(U,I)(\theta)d\theta ds.$$

Therefore

$$B_U^{(k)}(U,I)(T+(k+1)T_0) = B_U^{(k+1)}(U,I)(T+(k+1)T_0).$$

For the derivatives we obtain

$$\frac{dB_U^{(k)}(U,I)(t)}{dt} = V(U,I)(t) - \frac{1}{T_0}\int\limits_{T+kT_0,\,T+(k+1)T}^{T+(k+1)T_0} V(U,I)(s)ds,\ t \in [T+kT_0,\,T+(k+1)T_0]$$

$$\frac{dB_U^{(k+1)}(U,I)(t)}{dt} = V(U,I)(t) - \frac{1}{T_0}\int\limits_{T+(k+1)T_0}^{T+(k+2)T_0} V(U,I)(s)ds,\ t \in [T+(k+1)T_0,\,T+(k+2)T_0]$$

and therefore in view of

$$\frac{1}{T_0}\int\limits_{T+kT_0}^{T+(k+1)T_0} V(U,I)(s)ds \ = \ \frac{1}{T_0}\int\limits_{T+(k+1)T_0}^{T+(k+2)T_0} V(U,I)(s)ds$$

it follows

$$\frac{dB_U^{(k)}(U,I)(T+(k+1)T_0)}{dt} = V(U,I)(T+(k+1)T_0) - \frac{1}{T_0}\int\limits_{T+kT_0}^{T+(k+1)T_0} V(U,I)(s)ds \ =$$

$$= V(U,I)(T+(k+1)T_0) - \frac{1}{T_0}\int\limits_{T+(k+1)T_0}^{T+(k+2)T_0} V(U,I)(s)ds \ = \ \frac{dB_U^{(k+1)}(U,I)(T+(k+1)T_0)}{dt}$$

Lemma 3.4.4 is thus proved.

Lemma 3.4.5. The periodic problem (3.4.1) has a unique solution $(U(.),I(.)) \in M_U \times M_I$ if the operator B has a fixed point $(U(.),I(.)) \in M_U \times M_I$, that is,

$$U = B_U(U,I), \quad I = B_I(U,I).$$

Proof: Let $(U(.),I(.)) \in M_U \times M_I$ be a solution of the periodic problem:

$$\frac{dU(t)}{dt} = V(U,I)(t), t \in [T,2T]$$

$$U(t) = U_0(t), \frac{dU(t)}{dt} = \frac{dU_0(t)}{dt}, t \in [0,T] \tag{3.4.3}$$

$$\frac{dI(t)}{dt} = J(U,I)(t), t \in [T,2T]$$

$$I(t) = I_0(t), \frac{dI(t)}{dt} = \frac{dI_0(t)}{dt}, t \in [0,T]. \tag{3.4.4}$$

After integration we have ($U(T) = 0, I(T) = 0$):

$$U(t) = \int_{T+kT_0}^{t} V(U,I)(t)dt, \quad I(t) = \int_{T+kT_0}^{t} J(U,I)(t)dt$$

and replacing in the last equalities

$$t = T + (k+1)T_0$$

we obtain

$$U(T + (k+1)T_0) = \int_{T+kT_0}^{T+(k+1)T_0} V(U,I)(t)dt \Rightarrow \int_{T+kT_0}^{T+(k+1)T_0} V(U,I)(t)dt = 0,$$

$$I(T + (k+1)T_0) = \int_{T+kT_0}^{T+(k+1)T_0} J(U,I)(t)dt \Rightarrow \int_{T+kT_0}^{T+(k+1)T_0} J(U,I)(t)dt = 0.$$

For $B_U^{(k)}(U,I)(t)$ and $B_I^{(k)}(U,I)(t)$ remain

$$B_U^{(k)}(U,I)(t) = \int_{T+kT_0}^{t} V(U,I)(s)ds - \frac{1}{T_0} \int_{T+kT_0}^{T+(k+1)T_0} \int_{T+kT_0}^{t} V(U,I)(s)dsdt,$$

$$B_I^{(k)}(U,I)(t) = \int_{T+kT_0}^{t} J(U,I)(s)ds - \frac{1}{T_0} \int_{T+kT_0}^{T+(k+1)T_0} \int_{T+kT_0}^{t} J(U,I)(s)dsdt.$$

Changing the order of integration we obtain

$$\int\limits_{T+kT_0}^{T+(k+1)T_0} \int\limits_{T+kT_0}^{t} V(U,I)(s)\,ds\,dt = (T+(k+1)T_0)\int\limits_{T+kT_0}^{T+(k+1)T_0} V(U,I)(s)\,ds - \int\limits_{T+kT_0}^{T+(k+1)T_0} sV(U,I)(s)\,ds =$$

$$= (T+(k+1)T_0)\int\limits_{T+kT_0}^{T+(k+1)T_0} \frac{dU(s)}{ds}\,ds - \int\limits_{T+kT_0}^{T+(k+1)T_0} sV(U,I)(s)\,ds = -\int\limits_{T+kT_0}^{T+(k+1)T_0} sV(U,I)(s)\,ds.$$

But

$$\int\limits_{T+kT_0}^{T+(k+1)T_0} sV(U,I)(s)\,ds = \int\limits_{T+kT_0}^{T+(k+1)T_0} s\frac{dU(s)}{ds}\,ds = sU(s)\Big|_{T+kT_0}^{T+(k+1)T_0} - \int\limits_{T+kT_0}^{T+(k+1)T_0} U(s)\,ds = 0-0 = 0.$$

Therefore

$$B_U^{(k)}(U,I)(t) = \int\limits_{T+kT_0}^{t} V(U,I)(s)\,ds; \quad B_I^{(k)}(U,I)(t) = \int\limits_{T+kT_0}^{t} J(U,I)(s)\,ds$$

that is, $U = B_U(U,I)$ $I = B_I(U,I)$.

Conversely, let $B = (B_U, B_I)$ have a fixed point $(U(.),I(.)) \in M_U \times M_I$, that is, $U(t) = B_U^{(k)}(U,I)(t)$ and $I(t) = B_I^{(k)}(U,I)(t)$ for $t \in [T+kT_0, T+(k+1)T_0]$. Therefore

$$U(T+kT_0) = B_U^{(k)}(U,I)(T+(k+1)T_0)$$

and

$$I(T+kT_0) = B_I^{(k)}(U,I)(T+(k+1)T_0)$$

or

$$0 = U(T+kT_0) = B_U^{(k)}(U,I)(T+(k+1)T_0) =$$

$$= \int\limits_{T+kT_0}^{T+(k+1)T_0} V(U,I)(s)\,ds - \frac{T+(k+1)T_0 - T - kT_0}{T_0}\int\limits_{T+kT_0}^{T+(k+1)T_0} V(U,I)(s)\,ds -$$

$$-\frac{1}{T_0}\int\limits_{T+kT_0}^{T+(k+1)T_0} \int\limits_{T+kT_0}^{s} V(U,I)(\theta)\,d\theta\,ds + \frac{1}{2}\int\limits_{T+kT_0}^{T+(k+1)T_0} V(U,I)(s)\,ds =$$

$$= -\frac{1}{T_0}\int\limits_{T+kT_0}^{T+(k+1)T_0} \int\limits_{T+kT_0}^{s} V(U,I)(\theta)\,d\theta\,ds + \frac{1}{2}\int\limits_{T+kT_0}^{T+(k+1)T_0} V(U,I)(s)\,ds.$$

It follows that

$$\frac{1}{T_0}\int\limits_{T+kT_0}^{T+(k+1)T_0}\int\limits_{T+kT_0}^{s}V(U,I)(\theta)\,d\theta\,ds=\frac{1}{2}\int\limits_{T+kT_0}^{T+(k+1)T_0}V(U,I)(s)\,ds.$$

We show that $\displaystyle\int\limits_{T+kT_0}^{T+(k+1)T_0}V(U,I)(s)\,ds=0$. Indeed,

$$\left|\int\limits_{T+kT_0}^{T+(k+1)T_0}V(U,I)(s)\,ds\right|\le\int\limits_{T+kT_0}^{T+(k+1)T_0}\left|V(U,I)(s)\right|ds\le$$

$$\le\left|\int\limits_{T+kT_0}^{T+(k+1)T_0}\frac{dI(t-T)}{dt}\,dt\right|+\frac{Z_0}{\hat{L}_0}\int\limits_{T+kT_0}^{T+(k+1)T_0}\left|U(t)\right|dt+\frac{Z_0}{\hat{L}_0}\int\limits_{T+kT_0}^{T+(k+1)T_0}\left|I(t-T)\right|dt+2\frac{Z_0}{\hat{L}_0}\int\limits_{T+kT_0}^{T+(k+1)T_0}\left|E_0(t-T)\right|dt+$$

$$+\frac{2Z_0}{\hat{L}_0}\int\limits_{T+kT_0}^{T+(k+1)T_0}\left|R_0\big(i_0(U,I)(t)\big)\right|dt+\frac{2Z_0}{\hat{L}_0}\int\limits_{T+kT_0}^{T+(k+1)T_0}\left|\widetilde{C}_0^{-1}\left(\int\limits_{T+kT_0}^{t}i_0(U,I)(s)\,ds\right)\right|dt\le$$

$$\le\frac{Z_0}{\hat{L}_0}U_0\int\limits_{T+kT_0}^{T+(k+1)T_0}e^{\mu(t-T-kT_0)}dt+\frac{Z_0}{\hat{L}_0}I_0e^{-\beta}\int\limits_{T+kT_0}^{T+(k+1)T_0}e^{\mu(t-T-kT_0)}dt+\frac{Z_0}{\hat{L}_0}2U_{E0}\int\limits_{T+kT_0}^{T+(k+1)T_0}e^{\mu(t-T-kT_0)}dt+$$

$$+\frac{2Z_0}{\hat{L}_0}\int\limits_{T+kT_0}^{T+(k+1)T_0}\sum_{n=1}^{m}\left|r_n^{(0)}\right|\left|i_0(U,I)(t)\right|^n dt+\frac{2Z_0}{\hat{L}_0}\int\limits_{T+kT_0}^{T+(k+1)T_0}\int\limits_{T+kT_0}^{t}\left|i_0(U,I)(s)\right|ds\,dt\le$$

$$\le\frac{Z_0}{\hat{L}_0}U_0\frac{e^{\mu T_0}-1}{\mu}+\frac{Z_0}{\hat{L}_0}I_0e^{-\beta}\frac{e^{\mu T_0}-1}{\mu}+2\frac{Z_0}{\hat{L}_0}U_{E0}\frac{e^{\mu T_0}-1}{\mu}+2\frac{Z_0}{\hat{L}_0}\sum_{n=1}^{m}\left|r_n^{(0)}\right|\frac{\left(U_0+I_0e^{-\beta}\right)^n}{(2Z_0)^n}\int\limits_{T+kT_0}^{T+(k+1)T_0}e^{n\mu(t-T-kT_0)}dt+$$

$$+\frac{U_0+I_0e^{-\beta}}{\hat{L}_0}H_0\int\limits_{T+kT_0}^{T+(k+1)T_0}\int\limits_{T+kT_0}^{t}e^{\mu(s-T-kT_0)}ds\,dt\le$$

$$\le\frac{Z_0}{\hat{L}_0}\left(U_0+I_0e^{-\beta}\right)\frac{e^{\mu T_0}-1}{\mu}+2\frac{Z_0}{\hat{L}_0}U_{E0}\frac{e^{\mu T_0}-1}{\mu}+2\frac{Z_0}{\hat{L}_0}\sum_{n=1}^{m}\left|r_n^{(0)}\right|\frac{\left(U_0+I_0e^{-\beta}\right)^n}{(2Z_0)^n}\frac{e^{n\mu T_0}-1}{n\mu}+$$

$$+\frac{U_0+I_0e^{-\beta}}{\hat{L}_0}H_0\frac{e^{\mu T_0}-1}{\mu^2}\le\frac{Z_0}{\hat{L}_0}U_0\frac{e^{\mu T_0}-1}{\mu}+\frac{Z_0}{\hat{L}_0}I_0\frac{e^{\mu T_0}-1}{\mu}+$$

$$+2\frac{Z_0}{\hat{L}_0}U_{E0}\frac{e^{\mu T_0}-1}{\mu}+2\frac{Z_0}{\hat{L}_0}\frac{e^{\mu T_0}-1}{\mu}\sum_{n=1}^{m}\left|r_n^{(0)}\right|\frac{\left(U_0+I_0e^{-\beta}\right)^n}{(2Z_0)^n}e^{(n-1)\mu T_0}+\frac{\left(U_0+I_0e^{-\beta}\right)}{\hat{L}_0}H_0\frac{e^{\mu T_0}-1}{\mu^2}\le$$

$$\le\frac{e^{\mu T_0}-1}{\mu\hat{L}_0}\left(2Z_0U_{E0}+Z_0\big(U_0+I_0e^{-\beta}\big)+2Z_0\sum_{n=1}^{m}\left|r_n^{(0)}\right|\left(\frac{U_0+I_0e^{-\beta}}{2Z_0}\right)^n e^{(n-1)\mu_0}+\frac{\left(U_0+I_0e^{-\beta}\right)}{\mu}H_0\right)\equiv M_0(\mu)$$

and

$$\left| \int\limits_{T+kT_0}^{T+(k+1)T_0} J(U,I)(s)ds \right| \le \left| \int\limits_{T+kT_0}^{T+(k+1)T_0} \frac{dU(t-T)}{dt} dt \right| + \frac{Z_0}{\hat{L}_1} \int\limits_{T+kT_0}^{T+(k+1)T_0} |U(t-T)| dt + \frac{Z_0}{\hat{L}_1} \int\limits_{T+kT_0}^{T+(k+1)T_0} |I(t)| dt +$$

$$+2\frac{Z_0}{\hat{L}_1} \int\limits_{T+kT_0}^{T+(k+1)T_0} |E_1(t-T)| dt +2\frac{Z_0}{\hat{L}_1} \int\limits_{T+kT_0}^{T+(k+1)T_0} |R_1(i_1(U,I)(t))| dt + \frac{2Z_0}{\hat{L}_1} \int\limits_{T+kT_0}^{T+(k+1)T_0} \left| \widetilde{C}_1^{-1}\left(\int\limits_T^t i_1(U,I)(s)ds \right) \right| dt \le$$

$$\le \frac{e^{\mu_0}-1}{\mu \hat{L}_1}\left(2Z_0 U_{E_1} + Z_0\left(U_0 e^{-\beta}+I_0\right) + 2Z_0 \sum_{n=1}^{m} |r_n^{(1)}| \left(\frac{U_0 e^{-\beta}+I_0}{2Z_0}\right)^n e^{(n-1)\mu_0} + \frac{\left(U_0 e^{-\beta}+I_0\right)}{\mu} H_1 \right) \equiv M_1(\mu).$$

Since $\mu T_0 = \mu_0 = \text{const.}$ then $\lim\limits_{\mu\to\infty} M_p(\mu) = 0 \ (p=0,1)$ which implies

$$\left| \int\limits_{T+kT_0}^{T+(k+1)T_0} V(U,I)(s)ds \right| = 0, \ \left| \int\limits_{T+kT_0}^{T+(k+1)T_0} J(U,I)(s)ds \right| = 0 . \text{ Consequently}$$

$$B_U^{(k)}(U,I)(t) = \int\limits_{T+kT_0}^{t} V(U,I)(s)ds, \quad B_U^{(k)}(U,I)(t) = \int\limits_{T+kT_0}^{t} V(U,I)(s)ds .$$

Differentiating the last equalities, we ascertain that the fixed point of the operator is a periodic solution of (3.4.2), (3.4.3).

Lemma 3.4.5 is thus proved.

3.5. THE EXISTENCE-UNIQUENESS OF A PERIODIC SOLUTION FOR A NONLINEAR NEUTRAL SYSTEM

Here we formulate the periodic problem: to find a T_0-periodic solution of the following system belonging to $M_U^* \times M_I^*$:

$$\frac{dU(t)}{dt} = \frac{dI(t-T)}{dt} + \frac{2Z_0}{\widetilde{L}_0(i_0)(t)} E_0(t-T) - \frac{Z_0}{\widetilde{L}_0(i_0)(t)} U(t) - \frac{Z_0}{\widetilde{L}_0(i_0)(t)} I(t-T) -$$

$$- \frac{2Z_0}{\widetilde{L}_0(i_0)(t)} R_0(i_0) - \frac{2Z_0}{\widetilde{L}_0(i_0)(t)} \widetilde{C}_0^{-1}\left(\int\limits_T^t i_0 ds \right) \equiv V(U,I)(t) , \ t\in[T,2T],$$

(3.5.1)

$$\frac{dI(t)}{dt} = \frac{dU(t-T)}{dt} + \frac{2Z_0}{\widetilde{L}_1(i_1)(t)} E_1(t-T) - \frac{Z_0}{\widetilde{L}_1(i_1)(t)} U(t-T) - \frac{Z_0}{\widetilde{L}_1(i_1)(t)} I(t) +$$

$$+\frac{2Z_0}{\widetilde{L}_1(i_1)(t)}R_1(i_1)+\frac{2Z_0}{\widetilde{L}_1(i_1)(t)}\widetilde{C}_1^{-1}\left(\frac{1}{2Z_0}\int_T^t i_1 ds\right)\equiv J(U,I)(t),t\in[T,2T],$$

$$U(t)=U_0(t),\ \frac{dU(t)}{dt}=\frac{dU_0(t)}{dt},\ \ t\in[0,T];$$

$$I(t)=I_0(t),\ \frac{dI(t)}{dt}=\frac{dI_0(t)}{dt},\ \ t\in[0,T]$$

where $U_0(t),I_0(t)$ are prescribed T_0-periodic initial functions.

Theorem 3.5.1. Let be the following assumptions be valid:

$$U_0(-T)=0,\ I_0(-T)=0\ ,\ E_p(.)\in C_{T_0}^1[T;\infty),E_p(0)=0,\ \ (p=0,1),$$

$$\left|E_p(t)\right|\le U_{E_p}\,e^{\mu(t-T-kT_0)}\le U_0 e^{\mu(t-T-kT_0)},t\in[T+kT_0,T+(k+1)T_0],(k=0,1,...);$$

$$\left|U_0(t)\right|\le e^{-\beta}U_0 e^{\mu(t-nT_0)},\left|I_0(t)\right|\le e^{-\beta}I_0 e^{\mu(t-nT_0)},t\in[nT_0,(n+1)T_0],(n=0,1,2,...,m-1);\beta=\text{const.}>0.$$

$$\frac{U_0+I_0 e^{-\beta}}{2Z_0}\le I_0;\frac{U_0 e^{-\beta}+I_0}{2Z_0}\le I_0.$$

Then for sufficiently large $\mu>0$ and $\mu T_0=\mu_0=$ const. there exists a unique T_0-periodic solution of (3.5.1).

Remark 3.5.1. For $t\in[T+kT_0,T+(k+1)T_0]$, $(k=0,1,2,...,m-1)$ it follows

$$\left|U_0(t-T)\right|=\left|\vec{U}_0(t)\right|\le e^{-\beta}U_0 e^{\mu(t-T-kT_0)},\left|I_0(t-T)\right|=\left|\vec{I}_0(t)\right|\le e^{-\beta}I_0 e^{\mu(t-T-kT_0)}.$$

Proof: The above lemmas imply that $(B_U,B_I)\in M_U\times M_I$. It remains to show $(B_U,B_I)\in M_U^*\times M_I^*$.

Using the inequalities $\left|\dfrac{t-T-kT_0}{T_0}-\dfrac{1}{2}\right|\le\dfrac{1}{2},t\in[T+kT_0,T+(k+1)T_0]$

and

$$\left|C_p^{-1}(I)\right|\le H_p\left|I\right|\equiv\frac{2\Phi_p-\phi_0}{2c_p\sqrt{\Phi_p}\sqrt{\Phi_p-\phi_0}}\left|I\right|,(p=0,1)$$

we get

$$\left|B_U^{(k)}(U,I)(t)\right| \le \left|\int_{T+kT_0}^{t} V(U,I)(s)ds\right| + \frac{1}{2}\left|\int_{T+kT_0}^{T+(k+1)T_0} V(U,I)(s)ds\right| + \frac{1}{T_0}\left|\int_{T+kT_0}^{T+(k+1)T_0}\int_{T+kT_0}^{t} V(U,I)(s)dsdt\right|$$

$$\le \left|\int_{T+kT_0}^{t} V(U,I)(s)ds\right| + \frac{1}{2}\left|\int_{T+kT_0}^{T+(k+1)T_0} V(U,I)(s)ds\right| + \frac{1}{T_0}\left|\int_{T+kT_0}^{T+(k+1)T_0}\int_{T+kT_0}^{t} V(U,I)(s)dsdt\right| \equiv$$

$$\equiv V_1 + V_2 + V_3.$$

But

$$V_1 = \left|\int_{T+kT_0}^{t} V(U,I)(s)ds\right| \le$$

$$\le \left|\int_{T+kT_0}^{t} \dot{\bar{I}}_0(s)ds\right| + \frac{Z_0}{\hat{L}_0}\int_{T+kT_0}^{t}|U(s)|ds + \frac{Z_0}{\hat{L}_0}\int_{T+kT_0}^{t}|\bar{I}_0(s)|ds + \frac{2Z_0}{\hat{L}_0}\int_{T+kT_0}^{t}\left|R_0\left(\frac{U(s)-\bar{I}_0(s)}{2Z_0}\right)\right|ds +$$

$$+\frac{2Z_0}{\hat{L}_0}\int_{T+kT_0}^{t}\left|\widetilde{C}_0^{-1}\left(\frac{1}{2Z_0}\int_{T+kT_0}^{s}(U(\tau)-\bar{I}_0(\tau))d\tau\right)\right|ds + \frac{2Z_0}{\hat{L}_0}\int_{T+kT_0}^{t}|E_0(s-T)|ds \le$$

$$\le \left|\bar{I}_0(t-T)\right| + \frac{Z_0 U_0}{\hat{L}_0}\int_{T+kT_0}^{t} e^{\mu(s-T-kT_0)}ds + \frac{Z_0 I_0 e^{-\beta}}{\hat{L}_0}\int_{T+kT_0}^{t} e^{\mu(s-T-kT_0)}ds +$$

$$+\frac{2Z_0}{\hat{L}_0}\sum_{n=1}^{m}\int_{T+kT_0}^{t}\left|r_n^{(0)}\left|\left(\frac{U(s)-\bar{I}_0(s)}{2Z_0}\right)\right|^n\right|ds +$$

$$+\frac{2Z_0}{\hat{L}_0}H_0\int_{T+kT_0}^{t}\frac{1}{2Z_0}\int_{T+kT_0}^{s}|U(\tau)-\bar{I}_0(\tau)|d\tau ds + \frac{2Z_0 U_{E0}}{\hat{L}_0}\int_{T+kT_0}^{t} e^{\mu(s-T-kT_0)}ds \le$$

$$\le e^{\mu(t-T-kT_0)}I_0 e^{-\beta} + \frac{Z_0 U_0}{\hat{L}_0}\frac{e^{\mu(t-T-kT_0)}-1}{\mu} + \frac{Z_0 I_0 e^{-\beta}}{\hat{L}_0}\frac{e^{\mu(t-T-kT_0)}-1}{\mu} +$$

$$+\frac{2Z_0}{\hat{L}_0}\sum_{n=1}^{m}\left|r_n^{(0)}\right|\left(\frac{U_0+I_0 e^{-\beta}}{2Z_0}\right)^n\int_{T+kT_0}^{t} e^{n\mu(s-T-kT_0)}ds +$$

$$+\frac{2Z_0}{\hat{L}_0}\frac{2\Phi_0-\phi_0}{c_0\sqrt{\Phi_0}\sqrt{\Phi_0-\phi_0}}\frac{U_0+I_0 e^{-\beta}}{2Z_0}\int_{T+kT_0}^{t}\frac{e^{\mu(s-T-kT_0)}-1}{\mu}ds + \frac{2Z_0 U_{E0}}{\hat{L}_0}\frac{e^{\mu(t-T-kT_0)}-1}{\mu} \le$$

$$\leq e^{\mu(t-T-kT_0)}I_0 e^{-\beta} + \frac{Z_0 U_0}{\hat{L}_0}\frac{e^{\mu(t-T-kT_0)}-1}{\mu} + \frac{Z_0 I_0 e^{-\beta}}{\hat{L}_0}\frac{e^{\mu(t-T-kT_0)}-1}{\mu} +$$

$$+\frac{2Z_0}{\hat{L}_0}\sum_{n=1}^{m}\left|r_n^{(0)}\right|\left(\frac{U_0+I_0 e^{-\beta}}{2Z_0}\right)^n \frac{e^{n\mu(t-T-kT_0)}-1}{n\mu} +$$

$$+\frac{2Z_0}{\hat{L}_0}\frac{2\Phi_0-\phi_0}{c_0\sqrt{\Phi_0}\sqrt{\Phi_0-\phi_0}}\frac{U_0+I_0 e^{-\beta}}{2Z_0}\frac{e^{\mu(t-T-kT_0)}-1}{\mu^2} + \frac{2Z_0 U_{E_0}}{\hat{L}_0}\frac{e^{\mu(t-T-kT_0)}-1}{\mu} \leq$$

$$\leq e^{\mu(t-T-kT_0)}I_0 e^{-\beta} + \frac{Z_0 U_0}{\hat{L}_0}\frac{e^{\mu(t-T-kT_0)}-1}{\mu} + \frac{Z_0 I_0 e^{-\beta}}{\hat{L}_0}\frac{e^{\mu(t-T-kT_0)}-1}{\mu} +$$

$$+\frac{2Z_0}{\hat{L}_0}\frac{e^{\mu(t-T-kT_0)}-1}{\mu}\sum_{n=1}^{m}\left|r_n^{(0)}\right|\left(\frac{U_0+I_0 e^{-\beta}}{2Z_0}\right)^n e^{(n-1)\mu T_0} +$$

$$+\frac{2Z_0}{\hat{L}_0}\frac{2\Phi_0-\phi_0}{c_0\sqrt{\Phi_0}\sqrt{\Phi_0-\phi_0}}\frac{U_0+I_0 e^{-\beta}}{2Z_0}\frac{e^{\mu(t-T-kT_0)}-1}{\mu^2} + \frac{2Z_0 U_{E_0}}{\hat{L}_0}\frac{e^{\mu(t-T-kT_0)}-1}{\mu} \leq$$

$$\leq e^{\mu(t-T-kT_0)}I_0 e^{-\beta} + \frac{e^{\mu(t-T-kT_0)}-1}{\mu\hat{L}_0}\left[\left(Z_0+\frac{H_0}{\mu}\right)\left(U_0+I_0 e^{-\beta}\right)+2Z_0\sum_{n=1}^{m}\left|r_n^{(0)}\right|(I_0)^n e^{(n-1)\mu_0}+2Z_0 U_{E_0}\right];$$

$$V_2 = \left|\left(\frac{t-T-kT_0}{T_0}-\frac{1}{2}\right)\int_{T+kT_0}^{T+(k+1)T_0}V(U,I)(s)ds\right| \leq$$

$$\leq \frac{1}{2}\left|\int_{T+kT_0}^{T+(k+1)T_0}\dot{\bar{I}}_0(s)ds\right| + \frac{Z_0}{2\hat{L}_0}\int_{T+kT_0}^{T+(k+1)T_0}\left|U(s)\right|ds + \frac{Z_0}{2\hat{L}_0}\int_{T+kT_0}^{T+(k+1)T_0}\left|\bar{I}_0(s)\right|ds + \frac{Z_0}{\hat{L}_0}\int_{T+kT_0}^{T+(k+1)T_0}\left|R_0\left(\frac{U(s)-\bar{I}_0(s)}{2Z_0}\right)\right|ds +$$

$$+\frac{Z_0}{\hat{L}_0}\int_{T+kT_0}^{T+(k+1)T_0}\left|\widetilde{C}_0^{-1}\left(\frac{1}{2Z_0}\int_{T+kT_0}^{s}(U(\tau)-\bar{I}_0(\tau))d\tau\right)\right|ds + \frac{Z_0}{\hat{L}_0}\int_{T+kT_0}^{T+(k+1)T_0}\left|E_0(s-T)\right|ds \leq$$

$$\leq e^{\mu(t-T-kT_0)}\frac{e^{\mu T_0}-1}{2\mu\hat{L}_0}\left[\left(Z_0+\frac{H_0}{\mu}\right)\left(U_0+I_0 e^{-\beta}\right)+2Z_0\sum_{n=1}^{m}\left|r_n^{(0)}\right|(I_0)^n e^{(n-1)\mu_0}+2Z_0 U_{E_0}\right]$$

and

$$V_3 = \left|\frac{1}{T_0}\int_{T+kT_0}^{T+(k+1)T_0}\int_{T}^{t}V(U,I)(s)ds\,dt\right| \leq \frac{1}{T_0}\int_{T+kT_0}^{T+(k+1)T_0}\left|\int_{T+kT_0}^{t}V(U,I)(s)ds\right|dt \leq$$

$$\leq e^{\mu(t-T-kT_0)}\frac{e^{\mu_0}-1}{\mu_0}\left[I_0 e^{-\beta}+\frac{1}{\mu\hat{L}_0}\left(\left(Z_0+\frac{H_0}{\mu}\right)\left(U_0+I_0 e^{-\beta}\right)+2Z_0\sum_{n=1}^{m}\left|r_n^{(0)}\right|(I_0)^n e^{(n-1)\mu_0}+2Z_0 U_{E_0}\right)\right].$$

Therefore, for sufficiently large μ we have

$$\left| B_U^{(k)}(U,I)(t) \right| \le$$

$$\le e^{\mu(t-T-kT_0)} I_0 e^{-\beta} + \frac{e^{\mu(t-T-kT_0)}}{\mu \hat{L}_0}\left[\left(Z_0 + \frac{H_0}{\mu}\right)\left(U_0 + I_0 e^{-\beta}\right) + 2Z_0 \sum_{n=1}^{m}\left|r_n^{(0)}\right|\left(I_0\right)^n e^{(n-1)\mu_0} + 2Z_0 U_{E_0}\right] +$$

$$+ e^{\mu(t-T-kT_0)}\frac{1}{\mu \hat{L}_0}\left[\left(Z_0 + \frac{H_0}{\mu}\right)\left(U_0 + I_0 e^{-\beta}\right) + 2Z_0 \sum_{n=1}^{m}\left|r_n^{(0)}\right|\left(I_0\right)^n e^{(n-1)\mu_0} + 2Z_0 U_{E_0}\right] +$$

$$+ e^{\mu(t-T-kT_0)}\frac{e^{\mu_0}-1}{\mu_0}\left[I_0 e^{-\beta} + \frac{1}{\mu \hat{L}_0}\left(\left(Z_0 + \frac{H_0}{\mu}\right)\left(U_0 + I_0 e^{-\beta}\right) + 2Z_0 \sum_{n=1}^{m}\left|r_n^{(0)}\right|\left(I_0\right)^n e^{(n-1)\mu_0} + 2Z_0 U_{E_0}\right)\right] \le$$

$$\le e^{\mu(t-T-kT_0)}\left\{ \left(1 + \frac{e^{\mu_0}-1}{\mu_0}\right) I_0 e^{-\beta} + \left(1 + \frac{e^{\mu_0}-1}{2} + \frac{e^{\mu_0}-1}{\mu_0}\right)\frac{1}{\mu \hat{L}_0} \times \right.$$

$$\left. \times \left[\left(Z_0 + \frac{H_0}{\mu}\right)\left(U_0 + I_0 e^{-\beta}\right) + 2Z_0 \sum_{n=1}^{m}\left|r_n^{(0)}\right|\left(I_0\right)^n e^{(n-1)\mu_0} + 2Z_0 U_{E_0}\right]\right\} \le U_0 e^{\mu(t-T-kT_0)}.$$

For the second component of the operator B we have

$$\left| B_I^{(k)}(U,I)(t) \right| \le \left| \int_{T+kT_0}^{t} J(U,I)(s)ds \right| + \frac{1}{2}\left| \int_{T+kT_0}^{T+(k+1)T_0} J(U,I)(s)ds \right| + \frac{1}{T_0}\left| \int_{T+kT_0}^{T+(k+1)T_0} \int_{T+kT_0}^{t} J(U,I)(s)dsdt \right| \equiv$$

$$\equiv J_1 + J_2 + J_3.$$

But

$$J_1 \le \left| \int_{T+kT_0}^{t} J(U,I)(s)ds \right| \le$$

$$\le \left| \int_{T+kT_0}^{t} \frac{d\vec{U}_0(s)}{ds}ds \right| + \frac{Z_0}{\hat{L}_1}\int_{T+kT_0}^{t}\left|\vec{U}_0(s)\right|ds + \frac{Z_0}{\hat{L}_1}\int_{T+kT_0}^{t}\left|I(s)\right|ds + \frac{2Z_0}{\hat{L}_1}\int_{T+kT_0}^{t}\left|R_1\left(i_1(U,I)(s)\right)\right|ds +$$

$$+ \frac{2Z_0}{\hat{L}_1}\int_{T+kT_0}^{t}\left|\widetilde{C}_1^{-1}\left(\frac{1}{2Z_0}\int_{T+kT_0}^{s} i_1(U,I)(\theta)d\theta\right)\right|dt + 2\frac{Z_0}{\hat{L}_1}\int_{T+kT_0}^{t}\left|E_1(s-T)\right|ds \le$$

$$\le e^{\mu(t-T-kT_0)}U_0 e^{-\beta} + \frac{Z_0}{\hat{L}_1}U_0 e^{-\beta}\frac{e^{\mu(t-T-kT_0)}-1}{\mu} + I_0\frac{Z_0}{\hat{L}_1}\frac{e^{\mu(t-T-kT_0)}-1}{\mu} + 2\frac{Z_0}{\hat{L}_1}\sum_{n=1}^{m}\left|r_n^{(1)}\right|\left(\frac{U_0 e^{-\beta}+I_0}{2Z_0}\right)^n \frac{e^{n\mu(t-T-kT_0)}-1}{n\mu} +$$

$$+ \frac{2Z_0}{\hat{L}_1}H_1\int_{T+kT_0}^{t}\int_{T+kT_0}^{s}\left|i_1(U,I)(\theta)\right|d\theta dt + 2U_{E_1}\frac{Z_0}{\hat{L}_1}\frac{e^{\mu(t-T-kT_0)}-1}{\mu} \le$$

$$\leq e^{\mu(t-T-kT_0)}U_0 e^{-\beta} + \frac{Z_0 U_0 e^{-\beta}}{\hat{L}_1}\frac{e^{\mu(t-T-kT_0)}-1}{\mu} + \frac{I_0 Z_0}{\hat{L}_1}\frac{e^{\mu(t-T-kT_0)}-1}{\mu} + \frac{e^{\mu(t-T-kT_0)}-1}{\mu}\frac{2Z_0}{\hat{L}_1}\sum_{n=1}^{m}\left|r_n^{(1)}\right|\left(\frac{U_0 e^{-\beta}+I_0}{2Z_0}\right)^{n}e^{(n-1)\mu T_0} +$$

$$+\frac{2Z_0}{\hat{L}_1}\frac{2\Phi_1-\phi_0}{2c_1\sqrt{\Phi_1}\sqrt{\Phi_1-\phi_0}}\int_{T+kT_0}^{t}\int_{T+kT_0}^{s}\left|\frac{\vec{U}_0(\theta)-I(\theta)}{2Z_0}\right|d\theta dt + 2U_{E_1}\frac{Z_0}{\hat{L}_1}\frac{e^{\mu(t-T-kT_0)}-1}{\mu}\leq$$

$$\leq e^{\mu(t-T-kT_0)}U_0 e^{-\beta} + \frac{Z_0 U_0 e^{-\beta}}{\hat{L}_1}\frac{e^{\mu(t-T-kT_0)}-1}{\mu} + \frac{I_0 Z_0}{\hat{L}_1}\frac{e^{\mu(t-T-kT_0)}-1}{\mu} + \frac{e^{\mu(t-T-kT_0)}-1}{\mu}\frac{2Z_0}{\hat{L}_1}\sum_{n=1}^{m}\left|r_n^{(1)}\right|\left(\frac{U_0 e^{-\beta}+I_0}{2Z_0}\right)^{n}e^{(n-1)\mu T_0} +$$

$$+\frac{2Z_0}{\hat{L}_1}\frac{2\Phi_1-\phi_0}{2c_1\sqrt{\Phi_1}\sqrt{\Phi_1-\phi_0}}\frac{U_0 e^{-\beta}+I_0}{2Z_0}\frac{e^{\mu(t-T-kT_0)}-1}{\mu^2} + 2U_{E_1}\frac{Z_0}{\hat{L}_1}\frac{e^{\mu(t-T-kT_0)}-1}{\mu}\leq$$

$$\leq e^{\mu(t-T-kT_0)}\left[U_0 e^{-\beta} + \frac{1}{\mu\hat{L}_1}\left(\left(Z_0+\frac{H_1}{\mu}\right)(U_0 e^{-\beta}+I_0)+2Z_0\sum_{n=1}^{m}\left|r_n^{(1)}\right|(I_0)^n e^{(n-1)\mu_0} + 2Z_0 U_{E_1}\right)\right];$$

$$J_2 \leq \frac{1}{2}\left|\int_{T+kT_0}^{T+(k+1)T_0}J(U,I)(s)\,ds\right| \leq$$

$$\leq\left|\int_{T+kT_0}^{T+(k+1)T_0}\frac{d\vec{U}_0(s)}{ds}\,ds\right| + \frac{Z_0}{\hat{L}_1}\int_{T+kT_0}^{T+(k+1)T_0}\left|\vec{U}_0(s)\right|ds + \frac{Z_0}{\hat{L}_1}\int_{T+kT_0}^{T+(k+1)T_0}\left|I(s)\right|ds + \frac{2Z_0}{\hat{L}_1}\int_{T+kT_0}^{T+(k+1)T_0}\left|R_1\left(i_1(U,I)(s)\right)\right|ds +$$

$$+\frac{2Z_0}{\hat{L}_1}\int_{T+kT_0}^{T+(k+1)T_0}\left|\widetilde{C}_1^{-1}\left(\frac{1}{2Z_0}\int_{T+kT_0}^{s}i_1(U,I)(\theta)d\theta\right)\right|dt + 2\frac{Z_0}{\hat{L}_1}\int_{T+kT_0}^{T+(k+1)T_0}\left|E_1(s-T)\right|ds \leq$$

$$\leq e^{\mu(t-T-kT_0)}\frac{e^{\mu_0}-1}{2}\frac{1}{\mu\hat{L}_1}\left(\left(Z_0+\frac{H_1}{\mu}\right)(U_0 e^{-\beta}+I_0)+2Z_0\sum_{n=1}^{m}\left|r_n^{(1)}\right|(I_0)^n e^{(n-1)\mu_0} + 2Z_0 U_{E_1}\right)$$

and

$$J_3 = \frac{1}{T_0}\left|\int_{T+kT_0}^{T+(k+1)T_0}\int_{T+kT_0}^{t}J(U,I)(s)\,ds\,dt\right| \leq$$

$$\leq\frac{1}{T_0}\int_{T+kT_0}^{T+(k+1)T_0}e^{\mu(t-T-kT_0)}dt\left[U_0 e^{-\beta} + \frac{1}{\mu\hat{L}_1}\left(\left(Z_0+\frac{H_1}{\mu}\right)(U_0 e^{-\beta}+I_0)+2Z_0\sum_{n=1}^{m}\left|r_n^{(1)}\right|\left(\frac{U_0 e^{-\beta}+I_0}{2Z_0}\right)^{n}e^{(n-1)\mu_0} + 2Z_0 U_{E_1}\right)\right] \leq$$

$$\leq\frac{e^{\mu_0}-1}{\mu_0}\left[U_0 e^{-\beta} + \frac{1}{\mu\hat{L}_1}\left(\left(Z_0+\frac{H_1}{\mu}\right)(U_0 e^{-\beta}+I_0)+2Z_0\sum_{n=1}^{m}\left|r_n^{(1)}\right|(I_0)^n e^{(n-1)\mu_0} + 2Z_0 U_{E_1}\right)\right].$$

Then

$$\left|B_I^{(k)}(U,I)(t)\right| \leq$$

$$\leq e^{\mu(t-T-kT_0)}\left[\left(1+\frac{e^{\mu_0}-1}{\mu_0}\right)U_0 e^{-\beta}+\left(1+\frac{e^{\mu_0}-1}{2}+\frac{e^{\mu_0}-1}{\mu_0}\right)\frac{1}{\mu\hat{L}_1}\times\right.$$

$$\left.\times\left(\left(Z_0+\frac{H_1}{\mu}\right)\left(U_0 e^{-\beta}+I_0\right)+2Z_0\sum_{n=1}^{m}\left|r_n^{(1)}\right|\left(I_0\right)^n e^{(n-1)\mu_0}+2Z_0 U_{E_1}\right)\right]\leq e^{\mu(t-T-kT_0)}I_0.$$

It remains to show that the operator $B=(B_U,B_I)$ is contractive on $M_U^*\times M_I^*$.

Recall that $\rho_\mu^{(k)}(U,\overline{U})\leq\dfrac{\rho_\mu^{(k)}(\dot{U},\dot{\overline{U}})}{\mu}$. In view of

$$i_0(t)=\frac{U(t)-\vec{I}_0(t)}{2Z_0},\ \bar{i}_0(t)=\frac{\overline{U}(t)-\vec{I}_0(t)}{2Z_0},\ i_1(t)=\frac{\vec{U}_0(t)-I(t)}{2Z_0},\ \bar{i}_1(t)=\frac{\vec{U}_0(t)-\overline{I}(t)}{2Z_0}$$

and

$$\widetilde{L}_0(i_0)(t)=\sum_{n=0}^{m}(n+1)l_n^{(0)}\left(i_0(t)\right)^n,\ \ \widetilde{L}_1(i_1)(t)=\sum_{n=0}^{m}(n+1)l_n^{(1)}\left(i_1(t)\right)^n$$

we obtain

$$\frac{d\widetilde{L}_0(i_0)}{di}=i_0\frac{dL_0(i_0)}{di}+i_0\frac{d^2L_0(i_0)}{di^2}+\frac{dL_0(i_0)}{di}=\sum_{n=1}^{m}\widetilde{l}_n^{(0)}i_0^{\ n}=\sum_{n=1}^{m}\widetilde{l}_n^{(0)}\left(\frac{U(t)-\vec{I}_0(t)}{2Z_0}\right)^n,$$

$$\frac{d\widetilde{L}_1(i_1)}{di}=i\frac{dL_1(i_1)}{di}+i\frac{d^2L_1(i_1)}{di^2}+\frac{dL_1(i_1)}{di}=\sum_{n=1}^{m}\widetilde{l}_n^{(1)}i_1^{\ n}=\sum_{n=1}^{m}\widetilde{l}_n^{(1)}\left(\frac{\vec{U}_0(t)-I(t)}{2Z_0}\right)^n.$$

Preliminary inequalities (0):

$$\left|\frac{1}{\widetilde{L}_0(i_0(s))}-\frac{1}{\widetilde{L}_0(\bar{i}_0(s))}\right|\leq\frac{1}{\hat{L}_0^2}\left|\frac{d\widetilde{L}_0(i_0)}{di}\right|\left|\frac{U(s)-\vec{I}_0(s)}{2Z_0}-\frac{\overline{U}(s)-\vec{I}_0(s)}{2Z_0}\right|\leq$$

$$\leq\frac{1}{\hat{L}_0^2}\sum_{n=1}^{m}\left|\widetilde{l}_n^{(0)}\right|\left(\frac{U_0+I_0 e^{-\beta}}{2Z_0}\right)^n\frac{\left|U(s)-\overline{U}(s)\right|}{2Z_0}\leq\frac{1}{\hat{L}_0^2}\sum_{n=1}^{m}\left|\widetilde{l}_n^{(0)}\right|\left(I_0\right)^n\frac{\left|U(s)-\overline{U}(s)\right|}{2Z_0};$$

$$\left|\frac{R_0(i_0(s))}{\widetilde{L}_0(i_0(s))}-\frac{R_0(\bar{i}_0(s))}{\widetilde{L}_0(i_0(s))}\right|\leq\frac{1}{\hat{L}_0}\left|R_0(i_0(s))-R_0(\bar{i}_0(s))\right|\leq\frac{1}{\hat{L}_0}\sum_{n=1}^{m}\left|r_n^{(0)}\right|n.\sup\left|i_0\right|^{n-1}\left|i_0(s)-\bar{i}_0(s)\right|\leq$$

$$\leq\frac{1}{\hat{L}_0}\sum_{n=1}^{m}\left|r_n^{(0)}\right|n.\left(\frac{U_0+I_0 e^{-\beta}}{2Z_0}\right)^{n-1}\frac{\left|U(s)-\overline{U}(s)\right|}{2Z_0}\leq\frac{1}{\hat{L}_0}\sum_{n=1}^{m}\left|r_n^{(0)}\right|n.\left(I_0\right)^{n-1}\frac{\left|U(s)-\overline{U}(s)\right|}{2Z_0};$$

$$\left|\frac{\widetilde{C}_0^{-1}\left(\int\limits_{T+kT_0}^{s} i_0(\tau)d\tau\right)}{\widetilde{L}_0(i_0(s))}-\frac{\widetilde{C}_0^{-1}\left(\int\limits_{T+kT_0}^{s}\bar{i}_0(\tau)d\tau\right)}{\widetilde{L}_0(i_0(s))}\right|\le\frac{1}{\hat{L}_0}\left|\widetilde{C}_0^{-1}\left(\int\limits_{T+kT_0}^{s} i_0(\tau)d\tau\right)-\widetilde{C}_0^{-1}\left(\int\limits_{T+kT_0}^{s}\bar{i}_0(\tau)d\tau\right)\right|\le$$

$$\le\frac{1}{\hat{L}_0}\left|\frac{d\widetilde{C}_0^{-1}(I)}{dI}\right|\int\limits_{T+kT_0}^{s}\left|i_0(\tau)-\bar{i}_0(\tau)\right|d\tau\le\frac{1}{\hat{L}_0}\frac{2\sqrt{\Phi_0}}{c_0\sqrt{\Phi_0-\phi_0}}\int\limits_{T+kT_0}^{s}\left|i_0(\tau)-\bar{i}_0(\tau)\right|d\tau$$

Using the previous inequalities, we find

$$\left|B_U^{(k)}(U,I)(t)-B_U^{(k)}(\overline{U},\overline{I})(t)\right|\le\left|\int\limits_{T+kT_0}^{t}\left(V(U,I)(s)-V(\overline{U},\overline{I})(s)\right)ds\right|+$$

$$+\left|\left(\frac{t-T-kT_0}{T_0}-\frac{1}{2}\right)\int\limits_{T+kT_0}^{T+(k+1)T_0}\left(V(U,I)(s)-V(\overline{U},\overline{I})(s)\right)ds\right|+\left|\frac{1}{T_0}\int\limits_{T+kT_0}^{T+(k+1)T_0}\int\limits_{T+kT_0}^{t}\left(V(U,I)(s)-V(\overline{U},\overline{I})(s)\right)dsdt\right|\equiv$$

$$\equiv V_1+V_2+V_3.$$

Since

$$V_1\le\left|\int\limits_{T+kT_0}^{t}\left(V(U,I)(s)-V(\overline{U},\overline{I})(s)\right)ds\right|\le$$

$$\le\left|\int\limits_{T+kT_0}^{t}\left(\dot{I}(s-T)-\dot{\overline{I}}(s-T)\right)ds\right|+2Z_0\int\limits_{T+kT_0}^{t}\left|E_0(s)\right|\left|\frac{1}{\widetilde{L}_0(i_0(s))}-\frac{1}{\widetilde{L}_0(\bar{i}_0(s))}\right|ds+$$

$$+Z_0\int\limits_{T+kT_0}^{t}\left|\frac{U(s)}{\widetilde{L}_0(i_0(s))}-\frac{\overline{U}(s)}{\widetilde{L}_0(\bar{i}_0(s))}\right|ds+Z_0\int\limits_{T+kT_0}^{t}\left|\frac{I(s-T)}{\widetilde{L}_0(i_0(s))}-\frac{\overline{I}(s-T)}{\widetilde{L}_0(\bar{i}_0(s))}\right|ds+$$

$$+2Z_0\int\limits_{T+kT_0}^{t}\left|\frac{R_0(i_0(s))}{\widetilde{L}_0(i_0(s))}-\frac{R_0(\bar{i}_0(s))}{\widetilde{L}_0(\bar{i}_0(s))}\right|ds+2Z_0\int\limits_{T+kT_0}^{t}\left|\frac{\widetilde{C}_0^{-1}\left(\int\limits_{T+kT_0}^{s}i_0(\tau)d\tau\right)}{\widetilde{L}_0(i_0(s))}-\frac{\widetilde{C}_0^{-1}\left(\int\limits_{T+kT_0}^{s}\bar{i}_0(\tau)d\tau\right)}{\widetilde{L}_0(\bar{i}_0(s))}\right|ds\le$$

$$\le 2Z_0 U_{E_0}\int\limits_{T+kT_0}^{t}e^{\mu(s-T-kT_0)}\left|\frac{1}{\widetilde{L}_0(i_0(s))}-\frac{1}{\widetilde{L}_0(\bar{i}_0(s))}\right|ds+$$

$$+Z_0\int\limits_{T+kT_0}^{t}\left|\frac{U(s)}{\widetilde{L}_0(i_0(s))}-\frac{\overline{U}(s)}{\widetilde{L}_0(i_0(s))}\right|ds+Z_0\int\limits_{T+kT_0}^{t}\left|\frac{\overline{U}(s)}{\widetilde{L}_0(i_0(s))}-\frac{\overline{U}(s)}{\widetilde{L}_0(\bar{i}_0(s))}\right|ds+$$

$$+ Z_0 \int_{T+kT_0}^{t} \left| \frac{I(s-T)}{\widetilde{L}_0(i_0(s))} - \frac{\overline{I}(s-T)}{\widetilde{L}_0(i_0(s))} \right| ds + Z_0 \int_{T+kT_0}^{t} \left| \frac{\overline{I}(s-T)}{\widetilde{L}_0(i_0(s))} - \frac{\overline{I}(s-T)}{\widetilde{L}_0(\overline{i}_0(s))} \right| ds +$$

$$+ 2Z_0 \int_{T+kT_0}^{t} \left| \frac{R_0(i_0(s))}{\widetilde{L}_0(i_0(s))} - \frac{R_0(\overline{i}_0(s))}{\widetilde{L}_0(i_0(s))} \right| ds + 2Z_0 \int_{T+kT_0}^{t} \left| \frac{R_0(\overline{i}_0(s))}{\widetilde{L}_0(i_0(s))} - \frac{R_0(\overline{i}_0(s))}{\widetilde{L}_0(\overline{i}_0(s))} \right| ds +$$

$$+ 2Z_0 \int_{T+kT_0}^{t} \left| \frac{1}{\widetilde{L}_0(i_0(s))} \widetilde{C}_0^{-1}\left(\int_{T+kT_0}^{s} i_0(\tau)d\tau \right) - \frac{1}{\widetilde{L}_0(i_0(s))} \widetilde{C}_0^{-1}\left(\int_{T+kT_0}^{s} \overline{i}_0(\tau)d\tau \right) \right| ds +$$

$$+ 2Z_0 \int_{T+kT_0}^{t} \left| \frac{1}{\widetilde{L}_0(i_0(s))} \widetilde{C}_0^{-1}\left(\int_{T+kT_0}^{s} \overline{i}_0(\tau)d\tau \right) - \frac{1}{\widetilde{L}_0(\overline{i}_0(s))} \widetilde{C}_0^{-1}\left(\int_{T+kT_0}^{s} \overline{i}_0(\tau)d\tau \right) \right| ds \leq$$

$$\leq 2Z_0 U_{E_0} \int_{T+kT_0}^{t} \left| \frac{1}{\widetilde{L}_0(i_0(s))} - \frac{1}{\widetilde{L}_0(\overline{i}_0(s))} \right| e^{\mu(s-T-kT_0)} ds +$$

$$+ \frac{Z_0}{\hat{L}_0} \rho^{(k)}(U,\overline{U}) \int_{T+kT_0}^{t} e^{\mu(s-T-kT_0)} ds + Z_0 U_0 \int_{T+kT_0}^{t} e^{\mu(s-T-kT_0)} \left| \frac{1}{\widetilde{L}_0(i_0(s))} - \frac{1}{\widetilde{L}_0(\overline{i}_0(s))} \right| ds +$$

$$+ Z_0 I_0 e^{-\beta} \int_{T+kT_0}^{t} e^{\mu(s-T-kT_0)} \left| \frac{1}{\widetilde{L}_0(i_0(s))} - \frac{1}{\widetilde{L}_0(\overline{i}_0(s))} \right| ds + \frac{2Z_0}{\hat{L}_0} \int_{T+kT_0}^{t} \left| R_0(i_0(s)) - R_0(\overline{i}_0(s)) \right| ds +$$

$$+ 2Z_0 \sup\left\{ \sum_{n=1}^{m} \left| r_n^{(0)} \right| \left\| \overline{i}_0(s) \right\|^n \right\} \int_{T+kT_0}^{t} \left| \frac{1}{\widetilde{L}_0(i_0(s))} - \frac{1}{\widetilde{L}_0(\overline{i}_0(s))} \right| ds +$$

$$+ \frac{2Z_0}{\hat{L}_0} \int_{T+kT_0}^{t} \left| \widetilde{C}_0^{-1}\left(\int_{T+kT_0}^{s} i_0(\tau)d\tau \right) - \widetilde{C}_0^{-1}\left(\int_{T+kT_0}^{s} \overline{i}_0(\tau)d\tau \right) \right| ds +$$

$$+ 2Z_0 \sup\left\{ \left| \widetilde{C}_0^{-1}\left(\int_{T+kT_0}^{s} \overline{i}_0(\tau)d\tau \right) \right| \right\} \int_{T+kT_0}^{t} \left| \frac{1}{\widetilde{L}_0(i_0(s))} - \frac{1}{\widetilde{L}_0(\overline{i}_0(s))} \right| ds \leq$$

$$\leq \left[2Z_0 U_{E_0} + Z_0 e^{\mu T_0} U_0 + Z_0 I_0 e^{-\beta} e^{\mu T_0} + 2Z_0 \sum_{n=1}^{m} \left| r_n^{(0)} \right| \left(\frac{U_0 + I_0 e^{-\beta}}{2Z_0} \right)^n + 2Z_0 U_0 \right] \int_{T+kT_0}^{t} \left| \frac{1}{\widetilde{L}_0(i_0(s))} - \frac{1}{\widetilde{L}_0(\overline{i}_0(s))} \right| ds$$

$$+ \frac{Z_0}{\hat{L}_0} \rho^{(k)}(U,\overline{U}) \int_{T+kT_0}^{t} e^{\mu(s-T-kT_0)} ds + Z_0 I_0 e^{-\beta} \int_{T+kT_0}^{t} e^{\mu(s-T-kT_0)} \left| \frac{1}{\widetilde{L}_0(i_0(s))} - \frac{1}{\widetilde{L}_0(\overline{i}_0(s))} \right| ds +$$

$$+ \frac{2Z_0}{\hat{L}_0} \int_{T+kT_0}^{t} \frac{1}{\hat{L}_0} \sum_{n=1}^{m} \left| r_n^{(0)} \right| n \cdot \left(\frac{U_0 + I_0 e^{-\beta}}{2Z_0} \right)^{n-1} \frac{\left| U(s) - \overline{U}(s) \right|}{2Z_0} ds +$$

$$+ \frac{2Z_0}{\hat{L}_0} \int_{T+kT_0}^{t} \frac{1}{\hat{L}_0} \frac{2\sqrt{\Phi_0}}{c_0 \sqrt{\Phi_0 - \phi_0}} \int_{T+kT_0}^{s} \left| i_0(\tau) - \overline{i}_0(\tau) \right| d\tau \, ds \leq$$

$$\leq \left[U_{E_0} + e^{\mu_0}\left(U_0 + I_0 e^{-\beta}\right) + \sum_{n=1}^{m}\left|r_n^{(0)}\right|\left(\frac{U_0 + I_0 e^{-\beta}}{2Z_0}\right)^n + U_0 \right] \times$$

$$\times \frac{1}{\hat{L}_0^2}\sum_{n=1}^{m}\left|\tilde{l}_n^{(0)}\right|\left(\frac{U_0 + I_0 e^{-\beta}}{2Z_0}\right)^n \int_{T+kT_0}^{t}\left|U(s)-\overline{U}(s)\right|ds + \frac{Z_0}{\hat{L}_0}\rho^{(k)}(U,\overline{U})\int_{T+kT_0}^{t}e^{\mu(s-T-kT_0)}ds +$$

$$+ \frac{1}{\hat{L}_0^2}\sum_{n=1}^{m}\left|r_n^{(0)}\right|\left|n.\left(\frac{U_0 + I_0 e^{-\beta}}{2Z_0}\right)^{n-1}\int_{T+kT_0}^{t}\left|U(s)-\overline{U}(s)\right|ds + \frac{1}{\hat{L}_0^2}\frac{2\sqrt{\Phi_0}}{c_0\sqrt{\Phi_0-\phi_0}}\int_{T+kT_0}^{t}\int_{T+kT_0}^{s}\left|U(\tau)-\overline{U}(\tau)\right|d\tau\,ds \leq$$

$$\leq \left[\frac{1}{\hat{L}_0^2}\left(U_{E_0} + e^{\mu_0}\left(U_0 + I_0 e^{-\beta}\right) + \sum_{n=1}^{m}\left|r_n^{(0)}\right|\left(\frac{U_0 + I_0 e^{-\beta}}{2Z_0}\right)^n + U_0\right)\sum_{n=1}^{m}\left|\tilde{l}_n^{(0)}\right|\left(\frac{U_0 + I_0 e^{-\beta}}{2Z_0}\right)^n + \right.$$

$$\left. + \frac{1}{\hat{L}_0}\left(Z_0 + \sum_{n=1}^{m}\left|r_n^{(0)}\right|\left|n.\left(\frac{U_0 + I_0 e^{-\beta}}{2Z_0}\right)^{n-1} + \frac{2\sqrt{\Phi_0}}{\mu\hat{L}_0 c_0\sqrt{\Phi_0-\phi_0}}\right)\right]\rho^{(k)}(U,\overline{U})\int_{T+kT_0}^{t}e^{\mu(s-T-kT_0)}ds \leq$$

$$\leq \frac{e^{\mu(t-T-kT_0)}-1}{\mu^2}e^{\mu_0}\left[\frac{1}{\hat{L}_0^2}\left(U_{E_0} + e^{\mu_0}\left(U_0 + I_0 e^{-\beta}\right) + \sum_{n=1}^{m}\left|r_n^{(0)}\right|\left(I_0\right)^n + U_0\right)\sum_{n=1}^{m}\left|\tilde{l}_n^{(0)}\right|\left(I_0\right)^n + \right.$$

$$\left. + \frac{1}{\hat{L}_0}\left(Z_0 + \sum_{n=1}^{m}\left|r_n^{(0)}\right|\left|n.\left(I_0\right)^{n-1} + \frac{2\sqrt{\Phi_0}}{\mu\hat{L}_0 c_0\sqrt{\Phi_0-\phi_0}}\right)\right]\rho_\mu^{(k)}(\dot{U},\dot{\overline{U}}) ;$$

$$V_2 = \left|\left(\frac{t-T-kT_0}{T_0}-\frac{1}{2}\right)\int_{T+kT_0}^{T+(k+1)T_0}\left(V(U,I)(s)-V(\overline{U},\overline{I})(s)\right)ds\right| \leq$$

$$\leq \frac{1}{2}\left[2Z_0 U_{E_0}\int_{T+kT_0}^{T+(k+1)T_0}\left|\frac{1}{\tilde{L}_0(i_0(s))}-\frac{1}{\tilde{L}_0(\bar{i}_0(s))}\right|ds + \right.$$

$$+ Z_0\int_{T+kT_0}^{T+(k+1)T_0}\left|\frac{U(s)}{\tilde{L}_0(i_0(s))}-\frac{\overline{U}(s)}{\tilde{L}_0(i_0(s))}\right|ds + Z_0\int_{T+kT_0}^{T+(k+1)T_0}\left|\frac{\overline{U}(s)}{\tilde{L}_0(i_0(s))}-\frac{\overline{U}(s)}{\tilde{L}_0(\bar{i}_0(s))}\right|ds +$$

$$+ Z_0\int_{T+kT_0}^{T+(k+1)T_0}\left|\frac{I(s-T)}{\tilde{L}_0(i_0(s))}-\frac{\overline{I}(s-T)}{\tilde{L}_0(i_0(s))}\right|ds + Z_0\int_{T+kT_0}^{T+(k+1)T_0}\left|\frac{\overline{I}(s-T)}{\tilde{L}_0(i_0(s))}-\frac{\overline{I}(s-T)}{\tilde{L}_0(\bar{i}_0(s))}\right|ds +$$

$$+ 2Z_0\int_{T+kT_0}^{T+(k+1)T_0}\left|\frac{R_0(i_0(s))}{\tilde{L}_0(i_0(s))}-\frac{R_0(\bar{i}_0(s))}{\tilde{L}_0(i_0(s))}\right|ds + 2Z_0\int_{T+kT_0}^{T+(k+1)T_0}\left|\frac{R_0(\bar{i}_0(s))}{\tilde{L}_0(i_0(s))}-\frac{R_0(\bar{i}_0(s))}{\tilde{L}_0(\bar{i}_0(s))}\right|ds +$$

$$+ 2Z_0\int_{T+kT_0}^{T+(k+1)T_0}\left|\frac{1}{\tilde{L}_0(i_0(s))}\tilde{C}_0^{-1}\left(\int_{T+kT_0}^{s}i_0(\tau)d\tau\right)-\frac{1}{\tilde{L}_0(i_0(s))}\tilde{C}_0^{-1}\left(\int_{T+kT_0}^{s}\bar{i}_0(\tau)d\tau\right)\right|ds +$$

$$+ 2Z_0\int_{T+kT_0}^{T+(k+1)T_0}\left|\frac{1}{\tilde{L}_0(i_0(s))}\tilde{C}_0^{-1}\left(\int_{T+kT_0}^{s}\bar{i}_0(\tau)d\tau\right)-\frac{1}{\tilde{L}_0(\bar{i}_0(s))}\tilde{C}_0^{-1}\left(\int_{T+kT_0}^{s}\bar{i}_0(\tau)d\tau\right)\right|ds \right] \leq$$

$$\leq \left[2Z_0 U_{E_0} + Z_0 e^{\mu T_0} U_0 + Z_0 I_0 e^{-\beta} e^{\mu T_0} + 2Z_0 \sum_{n=1}^{m} \left|r_n^{(0)}\right| \left(\frac{U_0 + I_0 e^{-\beta}}{2Z_0}\right)^n + 2Z_0 U_0 \right] \int\limits_{T+kT_0}^{T+(k+1)T_0} \left| \frac{1}{\widetilde{L}_0(i_0(s))} - \frac{1}{\widetilde{L}_0(\bar{i}_0(s))} \right| ds$$

$$+ \frac{Z_0}{\hat{L}_0} \rho^{(k)}(U,\overline{U}) \int\limits_{T+kT_0}^{T+(k+1)T_0} e^{\mu(s-T-kT_0)} ds + Z_0 I_0 e^{-\beta} \int\limits_{T+kT_0}^{T+(k+1)T_0} e^{\mu(s-T-kT_0)} \left| \frac{1}{\widetilde{L}_0(i_0(s))} - \frac{1}{\widetilde{L}_0(\bar{i}_0(s))} \right| ds +$$

$$+ \frac{2Z_0}{\hat{L}_0} \int\limits_{T+kT_0}^{T+(k+1)T_0} \frac{1}{\hat{L}_0} \sum_{n=1}^{m} \left|r_n^{(0)}\right| n.\left(\frac{U_0 + I_0 e^{-\beta}}{2Z_0}\right)^{n-1} \frac{\left|U(s) - \overline{U}(s)\right|}{2Z_0} ds +$$

$$+ \frac{2Z_0}{\hat{L}_0} \int\limits_{T+kT_0}^{T+(k+1)T_0} \frac{1}{\hat{L}_0} \frac{2\sqrt{\Phi_0}}{c_0 \sqrt{\Phi_0 - \phi_0}} \int\limits_{T+kT_0}^{s} \left|i_0(\tau) - \bar{i}_0(\tau)\right| d\tau \, ds \leq$$

$$\leq \frac{e^{\mu T_0} - 1}{2\mu^2} e^{\mu_0} \left[\frac{1}{\hat{L}_0^{\,2}} \left(U_{E_0} + e^{\mu_0}\left(U_0 + I_0 e^{-\beta}\right) + \sum_{n=1}^{m}\left|r_n^{(0)}\right|(I_0)^n + U_0 \sum_{n=1}^{m}\left|\widetilde{l}_n^{(0)}\right|(I_0)^n + \right.\right.$$

$$+ \frac{1}{\hat{L}_0}\left(Z_0 + \sum_{n=1}^{m}\left|r_n^{(0)}\right| n.(I_0)^{n-1} + \frac{2\sqrt{\Phi_0}}{\mu\hat{L}_0 c_0 \sqrt{\Phi_0 - \phi_0}}\right) \Bigg] \rho_\mu^{(k)}(\dot{U},\dot{\overline{U}})$$

and

$$V_3 \leq \frac{1}{T_0} \int\limits_{T+kT_0}^{T+(k+1)T_0} \left| \int\limits_{T+kT_0}^{t} \left(V(U,I)(s) - V(\overline{U},\bar{I})(s)\right) ds \right| dt \leq$$

$$\leq \rho_\mu^{(k)}((U,I),(\overline{U},\bar{I})) \frac{1}{T_0} \int\limits_{T+kT_0}^{T+(k+1)T_0} e^{\mu(t-T-kT_0)} dt \times$$

$$\times \frac{1}{\mu^2}\left[\frac{1}{\hat{L}_0^{\,2}}\left(U_{E_0} + e^{\mu_0}\left(U_0 + I_0 e^{-\beta}\right) + \sum_{n=1}^{m} r_n^{(0)}\left(\frac{U_0 + I_0 e^{-\beta}}{2Z_0}\right)^n + U_0 \sum_{n=1}^{m}\widetilde{l}_n^{(0)}\left(\frac{U_0 + I_0 e^{-\beta}}{2Z_0}\right)^n + \right.\right.$$

$$+ \frac{1}{\hat{L}_0}\left(Z_0 + \sum_{n=1}^{m}\left|r_n^{(0)}\right| n.\left(\frac{U_0 + I_0 e^{-\beta}}{2Z_0}\right)^{n-1} + \frac{2\sqrt{\Phi_0}}{\mu\hat{L}_0 c_0 \sqrt{\Phi_0 - \phi_0}}\right) \Bigg] \leq$$

$$\leq \rho_\mu^{(k)}(\dot{U},\dot{\overline{U}}) \frac{e^{\mu_0} - 1}{\mu_0} \frac{1}{\mu^2} e^{\mu_0}\left[\frac{1}{\hat{L}_0^{\,2}}\left(U_{E_0} + e^{\mu_0}\left(U_0 + I_0 e^{-\beta}\right) + \sum_{n=1}^{m}\left|r_n^{(0)}\right|(I_0)^n + U_0 \sum_{n=1}^{m}\left|\widetilde{l}_n^{(0)}\right|(I_0)^n + \right.\right.$$

$$+ \frac{1}{\hat{L}_0}\left(Z_0 + \sum_{n=1}^{m}\left|r_n^{(0)}\right| n.(I_0)^{n-1} + \frac{2\sqrt{\Phi_0}}{\mu\hat{L}_0 c_0 \sqrt{\Phi_0 - \phi_0}}\right) \Bigg],$$

it follows

$$\left|B_U^{(k)}(U,I)(t)-B_U^{(k)}(\overline{U},\overline{I})(t)\right|\le e^{\mu(t-T-kT_0)}\hat{\rho}_\mu(((U,\dot{U}),(I,\dot{I})),((\overline{U},\dot{\overline{U}}),(\overline{I},\dot{\overline{I}})))\times$$

$$\times e^{\mu_0}\frac{1}{\mu^2}\left[\frac{1}{\hat{L}_0^{\,2}}\left(U_{E_0}+e^{\mu_0}\left(U_0+I_0e^{-\beta}\right)+\sum_{n=1}^m\left|r_n^{(0)}\right|(\mathrm{I}_0)^n+U_0\sum_{n=1}^m\left|\tilde{l}_n^{(0)}\right|(\mathrm{I}_0)^n+\right.\right.$$

$$\left.\left.+\frac{1}{\hat{L}_0}\left(Z_0+\sum_{n=1}^m\left|r_n^{(0)}\right|n.(\mathrm{I}_0)^{n-1}+\frac{2\sqrt{\Phi_0}}{\mu\hat{L}_0c_0\sqrt{\Phi_0-\phi_0}}\right)\right)\right]+$$

$$+e^{\mu(t-T-kT_0)}\hat{\rho}_\mu(((U,\dot{U}),(I,\dot{I})),((\overline{U},\dot{\overline{U}}),(\overline{I},\dot{\overline{I}})))\frac{e^{\mu_0}-1}{2}\frac{1}{\mu^2}e^{\mu_0}\times$$

$$\times\left[\frac{1}{\hat{L}_0^{\,2}}\left(U_{E_0}+e^{\mu_0}\left(U_0+I_0e^{-\beta}\right)+\sum_{n=1}^m\left|r_n^{(0)}\right|(\mathrm{I}_0)^n+U_0\sum_{n=1}^m\left|\tilde{l}_n^{(0)}\right|(\mathrm{I}_0)^n+\right.\right.$$

$$\left.\left.+\frac{1}{\hat{L}_0}\left(Z_0+\sum_{n=1}^m\left|r_n^{(0)}\right|n.(\mathrm{I}_0)^{n-1}+\frac{2\sqrt{\Phi_0}}{\mu\hat{L}_0c_0\sqrt{\Phi_0-\phi_0}}\right)\right)\right]+$$

$$+e^{\mu(t-T-kT_0)}\hat{\rho}_\mu(((U,\dot{U}),(I,\dot{I})),((\overline{U},\dot{\overline{U}}),(\overline{I},\dot{\overline{I}})))\frac{e^{\mu_0}-1}{\mu_0}\frac{1}{\mu^2}e^{\mu_0}\left[\frac{1}{\hat{L}_0^{\,2}}\left(U_{E_0}+e^{\mu_0}\left(U_0+I_0e^{-\beta}\right)+\sum_{n=1}^m\left|r_n^{(0)}\right|(\mathrm{I}_0)^n+U_0\times\right.\right.$$

$$\times\sum_{n=1}^m\left|\tilde{l}_n^{(0)}\right|(\mathrm{I}_0)^n+\frac{1}{\hat{L}_0}\left(Z_0+\sum_{n=1}^m\left|r_n^{(0)}\right|n.(\mathrm{I}_0)^{n-1}+\frac{2\sqrt{\Phi_0}}{\mu\hat{L}_0c_0\sqrt{\Phi_0-\phi_0}}\right)\right]\le$$

$$\le e^{\mu(t-T-kT_0)}\hat{\rho}_\mu(((U,\dot{U}),(I,\dot{I})),((\overline{U},\dot{\overline{U}}),(\overline{I},\dot{\overline{I}})))\left(1+\frac{e^{\mu_0}-1}{2}+\frac{e^{\mu_0}-1}{\mu_0}\right)\frac{e^{\mu_0}}{\mu^2}\left[\frac{1}{\hat{L}_0^{\,2}}\left(U_{E_0}+e^{\mu_0}\left(U_0+I_0e^{-\beta}\right)+\sum_{n=1}^m\left|r_n^{(0)}\right|(\mathrm{I}_0)^n+U_0\times\right.\right.$$

$$\times\sum_{n=1}^m\left|\tilde{l}_n^{(0)}\right|(\mathrm{I}_0)^n+\frac{1}{\hat{L}_0}\left(Z_0+\sum_{n=1}^m\left|r_n^{(0)}\right|n.(\mathrm{I}_0)^{n-1}+\frac{2\sqrt{\Phi_0}}{\mu\hat{L}_0c_0\sqrt{\Phi_0-\phi_0}}\right)\right]\equiv$$

$$\equiv e^{\mu(t-T-kT_0)}K_U\hat{\rho}_\mu(((U,\dot{U}),(I,\dot{I})),((\overline{U},\dot{\overline{U}}),(\overline{I},\dot{\overline{I}}))).$$

Thus

$$\hat{\rho}\left(B_U(U,I),B_U(\overline{U},\overline{I})\right)\le e^{\mu_0}K_U\hat{\rho}_\mu(((U,\dot{U}),(I,\dot{I})),((\overline{U},\overline{I}),(\dot{U},\dot{\overline{I}}))).$$

For the derivative, we obtain

$$\left|\dot{B}_U^{(k)}(U,I)(t)-\dot{B}_U^{(k)}(\overline{U},\overline{I})(t)\right|\le\left|V(U,I)(t)-V(\overline{U},\overline{I})(t)\right|$$

$$+\left|\frac{1}{T_0}\int_{T+kT_0}^{T+(k+1)T_0}\left(V(U,I)(s)-V(\overline{U},\overline{I})(s)\right)ds\right|\equiv\dot{V}_1+\dot{V}_2.$$

We have

$$\dot{V}_1 \le \left| \dot{I}(t-T) - \dot{\bar{I}}(t-T) \right| + 2Z_0 |E_0(t)| \left| \frac{1}{\widetilde{L}_0(i_0(t))} - \frac{1}{\widetilde{L}_0(\bar{i}_0(t))} \right| +$$

$$+ Z_0 \left| \frac{U(t)}{\widetilde{L}_0(i_0(t))} - \frac{\overline{U}(t)}{\widetilde{L}_0(\bar{i}_0(t))} \right| + Z_0 \left| \frac{U(t-T)}{\widetilde{L}_0(i_0(t))} - \frac{\overline{U}(t-T)}{\widetilde{L}_0(\bar{i}_0(t))} \right| +$$

$$+ 2Z_0 \left| \frac{1}{\widetilde{L}_0(i_0(t))} R_0(i_0(t)) - \frac{1}{\widetilde{L}_0(\bar{i}_0(t))} R_0(\bar{i}_0(t)) \right| +$$

$$+ 2Z_0 \left| \frac{1}{\widetilde{L}_0(i_0(t))} \widetilde{C}_0^{-1}\left(\int_{T+kT_0}^{t} i_0(\tau)d\tau \right) - \frac{1}{\widetilde{L}_0(\bar{i}_0(t))} \widetilde{C}_0^{-1}\left(\int_{T+kT_0}^{t} \bar{i}_0(\tau)d\tau \right) \right| \le$$

$$\le 2Z_0 |E_0(t)| \left| \frac{1}{\widetilde{L}_0(i_0(t))} - \frac{1}{\widetilde{L}_0(\bar{i}_0(t))} \right| +$$

$$Z_0 \left| \frac{U(t)}{\widetilde{L}_0(i_0(t))} - \frac{\overline{U}(t)}{\widetilde{L}_0(i_0(t))} \right| + Z_0 \left| \frac{\overline{U}(t)}{\widetilde{L}_0(i_0(t))} - \frac{\overline{U}(t)}{\widetilde{L}_0(\bar{i}_0(t))} \right| +$$

$$+ Z_0 \left| \frac{I(t-T)}{\widetilde{L}_0(i_0(t))} - \frac{\bar{I}(t-T)}{\widetilde{L}_0(i_0(t))} \right| + Z_0 \left| \frac{\bar{I}(t-T)}{\widetilde{L}_0(i_0(t))} - \frac{\bar{I}(t-T)}{\widetilde{L}_0(\bar{i}_0(t))} \right| +$$

$$+ Z_0 I_0 e^{-\beta} e^{\mu_0} \left| \frac{1}{\widetilde{L}_0(i_0(t))} - \frac{1}{\widetilde{L}_0(\bar{i}_0(t))} \right| + \frac{2Z_0}{\hat{L}_0} \left| R_0(i_0(t)) - R_0(\bar{i}_0(t)) \right| +$$

$$+ 2Z_0 \sup\left\{ R_0(\bar{i}_0(t)) \right\} \left| \frac{1}{\widetilde{L}_0(i_0(t))} - \frac{1}{\widetilde{L}_0(\bar{i}_0(t))} \right| +$$

$$\frac{2Z_0}{\hat{L}_0} \left| \widetilde{C}_0^{-1}\left(\int_{T+kT_0}^{t} i_0(\tau)d\tau \right) - \widetilde{C}_0^{-1}\left(\int_{T+kT_0}^{t} \bar{i}_0(\tau)d\tau \right) \right| +$$

$$+ 2Z_0 \sup\left\{ \left| \widetilde{C}_0^{-1}\left(\int_{T+kT_0}^{t} \bar{i}_0(\tau)d\tau \right) \right| \right\} \left| \frac{1}{\widetilde{L}_0(i_0(t))} - \frac{1}{\widetilde{L}_0(\bar{i}_0(t))} \right| \le$$

$$\le 2Z_0 U_{E_0} e^{\mu(t-T-kT_0)} e^{-\beta} \frac{1}{\hat{L}_0^2} \left| \frac{d\widetilde{L}_0}{di_0} \right| \left| i_0(t) - \bar{i}_0(t) \right| + \frac{Z_0}{\hat{L}_0} \left| U(t) - \overline{U}(t) \right| + Z_0 U_0 e^{\mu(t-T-kT_0)} \frac{1}{\hat{L}_0^2} \left| \frac{d\widetilde{L}_0}{di_0} \right| \left| i_0(t) - \bar{i}_0(t) \right| +$$

$$+ \frac{Z_0}{\hat{L}_0} \left| I(t-T) - \bar{I}(t-T) \right| + Z_0 I_0 e^{\mu(t-T-kT_0)} e^{-\beta} \frac{1}{\hat{L}_0^2} \left| \frac{d\widetilde{L}_0}{di_0} \right| \left| i_0(t) - \bar{i}_0(t) \right| +$$

$$+ Z_0 I_0 e^{-\beta} e^{\mu_0} \frac{1}{\hat{L}_0^2} \left| \frac{d\widetilde{L}_0}{di_0} \right| \left| i_0(t) - \bar{i}_0(t) \right| + \frac{2Z_0}{\hat{L}_0} \left| \sum_{n=1}^{m} R_0 |i_0(t)|^n - \sum_{n=1}^{m} R_0 |\bar{i}_0(t)|^n \right| +$$

$$+ 2Z_0 \sup\left\{ R_0\left(\bar{i}_0(t)\right) : t \in [T + kT_0; T + (k+1)T_0] \right\} \frac{1}{\hat{L}_0^2} \left| \frac{d\widetilde{L}_0}{di_0} \right| \left| i_0(t) - \bar{i}_0(t) \right| +$$

$$+ \frac{2Z_0}{\hat{L}_0} \frac{2\sqrt{\Phi_0}}{c_0\sqrt{\Phi_0 - \phi_0}} \int_{T+kT_0}^{t} \left| i_0(\tau) - \bar{i}_0(\tau) \right| d\tau +$$

$$+ 2Z_0 \sup\left\{ \left\| \widetilde{C}_0^{-1}\left(\int_{T+kT_0}^{t} \bar{i}_0(\tau)d\tau \right) \right\| : t \in [T + kT_0; T + (k+1)T_0] \right\} \frac{1}{\hat{L}_0^2} \left| \frac{d\widetilde{L}_0}{di_0} \right| \left| i_0(t) - \bar{i}_0(t) \right| \le$$

$$\le e^{\mu(t-T-kT_0)}\left(2Z_0 U_{E_0} + Z_0 U_0 e^{\mu_0} + Z_0 I_0 e^{-\beta} e^{\mu_0} + 2Z_0 \sum_{n=1}^{m} \left| r_n^{(0)} \right| \left(\frac{U_0 + I_0 e^{-\beta}}{2Z_0} \right)^n + 2Z_0 U_0 \right) \times$$

$$\times \frac{1}{\hat{L}_0^2} \sum_{n=1}^{m} \left| \widetilde{l}_n^{(0)} \right| \left(\frac{U_0 + I_0 e^{-\beta}}{2Z_0} \right)^n \frac{\left| U(s) - \overline{U}(s) \right|}{2Z_0} +$$

$$+ \frac{Z_0}{\hat{L}_0} \rho^{(k)}(U, \overline{U}) + \frac{2Z_0}{\hat{L}_0} \sum_{n=1}^{m} \left| r_n^{(0)} \right| n \left(\frac{U_0 + I_0 e^{-\beta}}{2Z_0} \right)^{n-1} \left| i_0(t) - \bar{i}_0(t) \right| +$$

$$+ \frac{2\sqrt{\Phi_0}}{c_0\sqrt{\Phi_0 - \phi_0}} \frac{2Z_0}{\hat{L}_0^2} \int_{T+kT_0}^{t} \left| i_0(\tau) - \bar{i}_0(\tau) \right| d\tau \le$$

$$\le e^{\mu(t-T-kT_0)}\left(U_{E_0} + U_0 e^{\mu_0} + I_0 e^{-\beta} e^{\mu_0} + \sum_{n=1}^{m} \left| r_n^{(0)} \right| \left(\frac{U_0 + I_0 e^{-\beta}}{2Z_0} \right)^n + U_0 \right) \times$$

$$\times \frac{1}{\hat{L}_0^2} \sum_{n=1}^{m} \left| \widetilde{l}_n^{(0)} \right| \left(\frac{U_0 + I_0 e^{-\beta}}{2Z_0} \right)^n \rho^{(k)}(U, \overline{U}) +$$

$$+ \frac{Z_0}{\hat{L}_0} \rho^{(k)}(U, \overline{U}) + \frac{1}{\hat{L}_0} \sum_{n=1}^{m} \left| r_n^{(0)} \right| n \left(\frac{U_0 + I_0 e^{-\beta}}{2Z_0} \right)^{n-1} \rho^{(k)}(U, \overline{U}) +$$

$$+ \frac{2\sqrt{\Phi_0}}{c_0\sqrt{\Phi_0 - \phi_0}} \frac{1}{\hat{L}_0^2} \int_{T+kT_0}^{t} \left| U(\tau) - \overline{U}(\tau) \right| d\tau \le$$

$$\le e^{\mu(t-T-kT_0)}\left(U_{E_0} + U_0 e^{\mu_0} + I_0 e^{-\beta} e^{\mu_0} + \sum_{n=1}^{m} \left| r_n^{(0)} \right| \left(\frac{U_0 + I_0 e^{-\beta}}{2Z_0} \right)^n + U_0 \right) \times$$

$$\times \frac{1}{\hat{L}_0^2} \sum_{n=1}^{m} \left| \widetilde{l}_n^{(0)} \right| \left(\frac{U_0 + I_0 e^{-\beta}}{2Z_0} \right)^n \rho^{(k)}(U, \overline{U}) +$$

$$+ e^{\mu(t-T-kT_0)} \frac{Z_0}{\hat{L}_0} \rho^{(k)}(U,\overline{U}) + e^{\mu(t-T-kT_0)} \frac{1}{\hat{L}_0} \sum_{n=1}^{m} \left| r_n^{(0)} \right| n \left(\frac{U_0 + I_0 e^{-\beta}}{2Z_0} \right)^{n-1} \rho^{(k)}(U,\overline{U}) +$$

$$+ \frac{2\sqrt{\Phi_0}}{c_0 \sqrt{\Phi_0 - \phi_0}} \frac{1}{\hat{L}_0^{\,2}} \rho^{(k)}(U,\overline{U}) \frac{e^{\mu(t-T-kT_0)}}{\mu} \le$$

$$\le e^{\mu(t-T-kT_0)} \rho^{(k)}(U,\overline{U}) \left[\frac{1}{\hat{L}_0^{\,2}} \left(U_{E0} + e^{\mu_0}\left(U_0 + I_0 e^{-\beta} \right) + \sum_{n=1}^{m} \left| r_n^{(0)} \right| \left(\frac{U_0 + I_0 e^{-\beta}}{2Z_0} \right)^n + U_0 \right) \sum_{n=1}^{m} \left| \tilde{l}_n^{(0)} \right| \left(\frac{U_0 + I_0 e^{-\beta}}{2Z_0} \right)^n + \right.$$

$$\left. + \frac{1}{\hat{L}_0} \left(Z_0 + \sum_{n=1}^{m} \left| r_n^{(0)} \right| n \left(\frac{U_0 + I_0 e^{-\beta}}{2Z_0} \right)^{n-1} + \frac{2\sqrt{\Phi_0}}{\mu \hat{L}_0 c_0 \sqrt{\Phi_0 - \phi_0}} \right) \right] \le e^{\mu(t-T-kT_0)} \hat{\rho}_\mu(((U,\dot{U}),(I,\dot{I})),((\overline{U},\dot{\overline{U}}),(\overline{I},\dot{\overline{I}}))) \times$$

$$\times \frac{1}{\mu} \left[\frac{1}{\hat{L}_0^{\,2}} \left(U_{E0} + e^{\mu_0}\left(U_0 + I_0 e^{-\beta} \right) + \sum_{n=1}^{m} \left| r_n^{(0)} \right| \left(\frac{U_0 + I_0 e^{-\beta}}{2Z_0} \right)^n + U_0 \right) \sum_{n=1}^{m} \left| \tilde{l}_n^{(0)} \right| \left(\frac{U_0 + I_0 e^{-\beta}}{2Z_0} \right)^n + \right.$$

$$\left. + \frac{1}{\hat{L}_0} \left(Z_0 + \sum_{n=1}^{m} \left| r_n^{(0)} \right| n \left(\frac{U_0 + I_0 e^{-\beta}}{2Z_0} \right)^{n-1} + \frac{2\sqrt{\Phi_0}}{\mu \hat{L}_0 c_0 \sqrt{\Phi_0 - \phi_0}} \right) \right]$$

and

$$\dot{V}_2 \le \frac{1}{T_0} \int_{T+kT_0}^{T+(k+1)T_0} \left| \dot{(V(U,I)(s) - V(U',I)(s))} \right| ds \le e^{\mu(t-l-kt_0)} \hat{\rho}_\mu(((U,\dot{U}),(I,\dot{I})),((U',\dot{U'}),(I,\dot{I}))) \frac{2}{T_0} \frac{c^{\mu T_0}-1}{2\mu^2} \times$$

$$\times \left[\frac{1}{\hat{L}_0^{\,2}} \left(U_{E0} + e^{\mu_0}\left(U_0 + I_0 e^{-\beta} \right) + \sum_{n=1}^{m} \left| r_n^{(0)} \right| \left(\frac{U_0 + I_0 e^{-\beta}}{2Z_0} \right)^n + U_0 \right) \sum_{n=1}^{m} \left| \tilde{l}_n^{(0)} \right| \left(\frac{U_0 + I_0 e^{-\beta}}{2Z_0} \right)^n + \right.$$

$$\left. + \frac{1}{\hat{L}_0} \left(Z_0 + \sum_{n=1}^{m} \left| r_n^{(0)} \right| n. \left(\frac{U_0 + I_0 e^{-\beta}}{2Z_0} \right)^{n-1} + \frac{2\sqrt{\Phi_0}}{\mu \hat{L}_0 c_0 \sqrt{\Phi_0 - \phi_0}} \right) \right]$$

$$\le e^{\mu(t-T-kT_0)} \hat{\rho}_\mu(((U,\dot{U}),(I,\dot{I})),((\overline{U},\dot{\overline{U}}),(\overline{I},\dot{\overline{I}}))) \times$$

$$\times \frac{e^{\mu_0}-1}{\mu_0} \frac{1}{\mu} \left[\frac{1}{\hat{L}_0^{\,2}} \left(U_{E0} + e^{\mu_0}\left(U_0 + I_0 e^{-\beta} \right) + \sum_{n=1}^{m} \left| r_n^0 \right| \left(\frac{U_0 + I_0 e^{-\beta}}{2Z_0} \right)^n + U_0 \right) \sum_{n=1}^{m} \left| \tilde{l}_n^{(0)} \right| \left(\frac{U_0 + I_0 e^{-\beta}}{2Z_0} \right)^n + \right.$$

$$\left. + \frac{1}{\hat{L}_0} \left(Z_0 + \sum_{n=1}^{m} \left| r_n^0 \right| n \left(\frac{U_0 + I_0 e^{-\beta}}{2Z_0} \right)^{n-1} + \frac{2\sqrt{\Phi_0}}{\mu \hat{L}_0 c_0 \sqrt{\Phi_0 - \phi_0}} \right) \right].$$

Then

$$\left| \dot{B}_U^{(k)}(U,I)(t) - \dot{B}_U^{(k)}(\overline{U},\overline{I})(t) \right| \le$$

$$\leq e^{\mu(t-T-kT_0)}\hat{\rho}_\mu(((U,\dot{U}),(I,\dot{I})),((\overline{U},\dot{\overline{U}}),(\overline{I},\dot{\overline{I}})))\times$$

$$\times\frac{1}{\mu}\left\{\frac{1}{\hat{L}_0^2}\left[U_{E_0}+e^{\mu_0}\left(U_0+I_0e^{-\beta}\right)+\sum_{n=1}^m r_n^0\left(\frac{U_0+I_0e^{-\beta}}{2Z_0}\right)^n+U_0\right]\sum_{n=1}^m\widetilde{l}_n^{(0)}\left|\left(\frac{U_0+I_0e^{-\beta}}{2Z_0}\right)^n+\right.\right.$$

$$\left.\left.+\frac{1}{\hat{L}_0}\left[Z_0+\sum_{n=1}^m|r_n^0|n\left(\frac{U_0+I_0e^{-\beta}}{2Z_0}\right)^{n-1}+\frac{2\sqrt{\Phi_0}}{\mu\hat{L}_0c_0\sqrt{\Phi_0-\phi_0}}\right]\right\}+$$

$$+e^{\mu(t-T-kT_0)}\hat{\rho}_\mu(((U,\dot{U}),(I,\dot{I})),((\overline{U},\dot{\overline{U}}),(\overline{I},\dot{\overline{I}})))\frac{e^{\mu_0}-1}{\mu_0}\frac{1}{\mu}\left\{\frac{1}{\hat{L}_0^2}\left[U_{E_0}+e^{\mu_0}\left(U_0+I_0e^{-\beta}\right)+\right.\right.$$

$$+\sum_{n=1}^m|r_n^0|\left(\frac{U_0+I_0e^{-\beta}}{2Z_0}\right)^n+U_0\left]\sum_{n=1}^m|\widetilde{l}_n^{(0)}|\left(\frac{U_0+I_0e^{-\beta}}{2Z_0}\right)^n+\right.$$

$$\left.\left.+\frac{1}{\hat{L}_0}\left[Z_0+\sum_{n=1}^m|r_n^0|n\left(\frac{U_0+I_0e^{-\beta}}{2Z_0}\right)^{n-1}+\frac{2\sqrt{\Phi_0}}{\mu\hat{L}_0c_0\sqrt{\Phi_0-\phi_0}}\right]\right\}\leq$$

$$\leq e^{\mu(t-T-kT_0)}\hat{\rho}_\mu(((U,\dot{U}),(I,\dot{I})),((\overline{U},\overline{I}),(\dot{\overline{U}},\dot{\overline{I}})))\left(1+\frac{e^{\mu_0}-1}{\mu_0}\right)\times$$

$$\times\frac{1}{\mu}\left[\frac{1}{\hat{L}_0^2}\left(U_{E_0}+e^{\mu_0}\left(U_0+I_0e^{-\beta}\right)+\sum_{n=1}^m|r_n^0|(I_0)^n+U_0\right)\sum_{n=1}^m|\widetilde{l}_n^{(0)}|(I_0)^n+\frac{1}{\hat{L}_0}\left(Z_0+\sum_{n=1}^m|r_n^0|n(I_0)^{n-1}+\frac{2\sqrt{\Phi_0}}{\mu\hat{L}_0c_0\sqrt{\Phi_0-\phi_0}}\right)\right]\equiv$$

$$\equiv\dot{K}_U\,\hat{\rho}_\mu(((U,\dot{U}),(I,\dot{I})),((\overline{U},\overline{I}),(\dot{\overline{U}},\dot{\overline{I}}))).$$

Thus

$$\rho_\mu^{(k)}(\dot{B}_U^{(k)}(U,I),\dot{B}_U^{(k)}(\overline{U},\overline{I}))\leq\dot{K}_U\,\hat{\rho}_\mu(((U,\dot{U}),(I,\dot{I})),((\overline{U},\overline{I}),(\dot{\overline{U}},\dot{\overline{I}}))).$$

Preliminary inequalities (1):

$$\left|\frac{1}{\widetilde{L}_1(i_1(s))}-\frac{1}{\widetilde{L}_1(\overline{i}_1(s))}\right|\leq\frac{1}{\hat{L}_1^2}\left|\frac{d\widetilde{L}_1(i_1)}{di}\right|\left|\frac{\overline{U}_0(s)-I(s)}{2Z_0}-\frac{\overline{U}_0(s)-\overline{I}(s)}{2Z_0}\right|\leq$$

$$\leq\frac{1}{\hat{L}_1^2}\sum_{n=1}^m|\widetilde{l}_n^{(1)}|\left(\frac{U_0e^{-\beta}+I_0}{2Z_0}\right)^n\frac{|I(s)-\overline{I}(s)|}{2Z_0}\leq\frac{1}{\hat{L}_1^2}\sum_{n=1}^m|\widetilde{l}_n^{(1)}|(I_0)^n\frac{|I(s)-\overline{I}(s)|}{2Z_0};$$

$$\left|\frac{R_1(i_1(s))}{\widetilde{L}_1(i_1(s))}-\frac{R_1(\overline{i}_1(s))}{\widetilde{L}_1(i_1(s))}\right|\leq\frac{1}{\hat{L}_1}|R_1(i_1(s))-R_1(\overline{i}_1(s))|\leq\frac{1}{\hat{L}_1}\sum_{n=1}^m|r_n^{(1)}|n.\sup\{|i_1|^{n-1}\}|i_1(s)-\overline{i}_1(s)|\leq$$

$$\leq \frac{1}{\hat{L}_1} \sum_{n=1}^{m} \left|r_n^{(1)}\right| n.\left(\frac{U_0 e^{-\beta} + I_0}{2Z_0}\right)^{n-1} \frac{\left|I(s) - \bar{I}(s)\right|}{2Z_0} \leq \frac{1}{\hat{L}_1} \sum_{n=1}^{m} \left|r_n^{(1)}\right| n.(I_0)^{n-1} \frac{\left|I(s) - \bar{I}(s)\right|}{2Z_0};$$

$$\left| \frac{\widetilde{C}_1^{-1}\left(\int_{T+kT_0}^{s} i_1(\tau)d\tau\right)}{\widetilde{L}_1(i_1(s))} - \frac{\widetilde{C}_1^{-1}\left(\int_{T+kT_0}^{s} \bar{i}_1(\tau)d\tau\right)}{\widetilde{L}_1(i_1(s))}\right| \leq \frac{1}{\hat{L}_1}\left|\widetilde{C}_1^{-1}\left(\int_{T+kT_0}^{s} i_1(\tau)d\tau\right) - \widetilde{C}_1^{-1}\left(\int_{T+kT_0}^{s} \bar{i}_1(\tau)d\tau\right)\right| \leq$$

$$\leq \frac{1}{\hat{L}_1}\left|\frac{d\widetilde{C}_1^{-1}(I)}{dI}\right| \int_{T+kT_0}^{s} \left|i_1(\tau) - \bar{i}_1(\tau)\right|d\tau \leq \frac{1}{\hat{L}_1} \frac{2\sqrt{\Phi_1}}{c_1\sqrt{\Phi_1 - \phi_0}} \int_{T+kT_0}^{s} \left|i_1(\tau) - \bar{i}_1(\tau)\right|d\tau .$$

For the second component of B we have

$$\left|B_I^{(k)}(U,I)(t) - B_I^{(k)}(\overline{U},\bar{I})(t)\right| \leq \left|\int_{T+kT_0}^{t} \left(J(U,I)(s) - J(\overline{U},\bar{I})(s)\right)ds\right| +$$

$$+\left|\left(\frac{t-T-kT_0}{T_0} - \frac{1}{2}\right) \int_{T+kT_0}^{T+(k+1)T_0} \left(J(U,I)(s) - J(\overline{U},\bar{I})(s)\right)ds\right| + \left|\frac{1}{T_0}\int_{T+kT_0}^{T+(k+1)T_0}\int_{T+kT_0}^{t}\left(J(U,I)(s) - J(\overline{U},\bar{I})(s)\right)dsdt\right| \equiv$$

$$\equiv J_1 + J_2 + J_3$$

and

$$J_1 \leq \left|\int_{T+kT_0}^{t} \left(J(U,I)(s) - J(\overline{U},\bar{I})(s)\right)ds\right| \leq$$

$$\leq \int_{T+kT_0}^{t} \left|\frac{dU(s-T)}{ds} - \frac{d\overline{U}(s-T)}{ds}\right|ds + 2Z_0 \int_{T+kT_0}^{t} \left|\frac{E_1(s)}{\widetilde{L}_1(i_1(s))} - \frac{E_1(s)}{\widetilde{L}_1(\bar{i}_1(s))}\right|ds +$$

$$+ Z_0 \int_{T+kT_0}^{t} \left|\frac{U(s-T)}{\widetilde{L}_1(i_1(s))} - \frac{\overline{U}(s-T)}{\widetilde{L}_1(\bar{i}_1(s))}\right|ds + Z_0 \int_{T+kT_0}^{t} \left|\frac{I(s)}{\widetilde{L}_1(i_1(s))} - \frac{\bar{I}(s)}{\widetilde{L}_1(\bar{i}_1(s))}\right|ds +$$

$$+ 2Z_0 \int_{T+kT_0}^{t} \left|\frac{1}{\widetilde{L}_1(i_1(s))} R_1(i_1(s)) - \frac{1}{\widetilde{L}_1(\bar{i}_1(s))} R_1(\bar{i}_1(s))\right|ds +$$

$$+ 2Z_0 \int_{T+kT_0}^{t} \left|\frac{\widetilde{C}_1^{-1}(i_1(s))}{\widetilde{L}_1(i_1(s))} - \frac{\widetilde{C}_1^{-1}(\bar{i}_1(s))}{\widetilde{L}_1(\bar{i}_1(s))}\right|ds \leq$$

$$\leq 2Z_0 U_{E_1} \int_{T+kT_0}^{t} \left| \frac{1}{\widetilde{L}_1(i_1(s))} - \frac{1}{\widetilde{L}_1(\bar{i}_1(s))} \right| ds +$$

$$+ Z_0 \int_{T+kT_0}^{t} \left| \frac{U(s-T)}{\widetilde{L}_1(i_1(s))} - \frac{\overline{U}(s-T)}{\widetilde{L}_1(\bar{i}_1(s))} \right| ds + Z_0 \int_{T+kT_0}^{t} \left| \frac{\overline{U}(s-T)}{\widetilde{L}_1(i_1(s))} - \frac{\overline{U}(s-T)}{\widetilde{L}_1(\bar{i}_1(s))} \right| ds +$$

$$+ Z_0 \int_{T+kT_0}^{t} \left| \frac{I(s)}{\widetilde{L}_1(i_1(s))} - \frac{\overline{I}(s)}{\widetilde{L}_1(i_1(s))} \right| ds + Z_0 \int_{T+kT_0}^{t} \left| \frac{\overline{I}(s)}{\widetilde{L}_1(i_1(s))} - \frac{\overline{I}(s)}{\widetilde{L}_1(\bar{i}_1(s))} \right| ds +$$

$$+ 2Z_0 \int_{T+kT_0}^{t} \left| \frac{1}{\widetilde{L}_1(i_1(s))} R_1(i_1(s)) - \frac{1}{\widetilde{L}_1(i_1(s))} R_1(\bar{i}_1(s)) \right| ds +$$

$$+ 2Z_0 \int_{T+kT_0}^{t} \left| \frac{1}{\widetilde{L}_1(i_1(s))} R_1(\bar{i}_1(s)) - \frac{1}{\widetilde{L}_1(\bar{i}_1(s))} R_1(\bar{i}_1(s)) \right| ds +$$

$$+ 2Z_0 \int_{T+kT_0}^{t} \left| \frac{\widetilde{C}_1^{-1}\left(\int_{T+kT_0}^{s} i_1(\tau)d\tau \right)}{\widetilde{L}_1(i_1(s))} - \frac{\widetilde{C}_1^{-1}\left(\int_{T+kT_0}^{s} \bar{i}_1(\tau)d\tau \right)}{\widetilde{L}_1(i_1(s))} \right| ds +$$

$$+ 2Z_0 \int_{T+kT_0}^{t} \left| \frac{\widetilde{C}_1^{-1}\left(\int_{T+kT_0}^{s} \bar{i}_1(\tau)d\tau \right)}{\widetilde{L}_1(i_1(s))} - \frac{\widetilde{C}_1^{-1}\left(\int_{T+kT_0}^{s} \bar{i}_1(\tau)d\tau \right)}{\widetilde{L}_1(\bar{i}_1(s))} \right| ds \leq$$

$$\leq 2Z_0 U_{E_1} \int_{T+kT_0}^{t} \left| \frac{1}{\widetilde{L}_1(i_1(s))} - \frac{1}{\widetilde{L}_1(\bar{i}_1(s))} \right| ds + Z_0 U_0 e^{-\beta} e^{\mu_0} \int_{T+kT_0}^{t} \left| \frac{1}{\widetilde{L}_1(i_1(s))} - \frac{1}{\widetilde{L}_1(\bar{i}_1(s))} \right| ds +$$

$$+ \frac{Z_0}{\hat{L}_1} \int_{T+kT_0}^{t} \left| I(s) - \overline{I}(s) \right| ds + Z_0 I_0 e^{\mu_0} \int_{T+kT_0}^{t} \left| \frac{1}{\widetilde{L}_1(i_1(s))} - \frac{1}{\widetilde{L}_1(\bar{i}_1(s))} \right| ds +$$

$$+ 2Z_0 \frac{1}{\hat{L}_1} \int_{T+kT_0}^{t} \left| R_1(i_1(s)) - R_1(\bar{i}_1(s)) \right| ds +$$

$$+ 2Z_0 \sup\left| R_1(\bar{i}_1(s)) \right| \int_{T+kT_0}^{t} \left| \frac{1}{\widetilde{L}_1(i_1(s))} - \frac{1}{\widetilde{L}_1(\bar{i}_1(s))} \right| ds +$$

$$+ \frac{2Z_0}{\hat{L}_1} \int_{T+kT_0}^{t} \left| \widetilde{C}_1^{-1}\left(\int_{T+kT_0}^{s} i_1(\tau)d\tau \right) - \widetilde{C}_1^{-1}\left(\int_{T+kT_0}^{s} \bar{i}_1(\tau)d\tau \right) \right| ds +$$

$$+ 2Z_0 \sup\left| \widetilde{C}_1^{-1}\left(\int_{T+kT_0}^{s} \bar{i}_1(\tau)d\tau \right) \right| \int_{T+kT_0}^{t} \left| \frac{1}{\widetilde{L}_1(i_1(s))} - \frac{1}{\widetilde{L}_1(\bar{i}_1(s))} \right| ds \leq$$

$$\leq \left(2Z_0 U_{E_1} + Z_0 \left(U_0 e^{-\beta} + I_0 \right) e^{\mu_0} + 2Z_0 \sup \left| R_1 \left(\bar{i}_1(s) \right) \right| + 2Z_0 \sup \left| \overline{C}_1^{-1} \left(\int_{T+kT_0}^{s} \bar{i}_1(\tau) d\tau \right) \right| \right) \times$$

$$\times \int_{T+kT_0}^{t} \left| \frac{1}{\widetilde{L}_1(i_1(s))} - \frac{1}{\widetilde{L}_1(\bar{i}_1(s))} \right| ds + \frac{Z_0}{\hat{L}_1} \int_{T+kT_0}^{t} \left| I(s) - \bar{I}(s) \right| ds +$$

$$2Z_0 \frac{1}{\hat{L}_1} \int_{T+kT_0}^{t} \left| R_1 \left(i_1(s) \right) - R_1 \left(\bar{i}_1(s) \right) \right| ds +$$

$$+ \frac{2Z_0}{\hat{L}_1} \int_{T+kT_0}^{t} \left| \widetilde{C}_1^{-1} \left(\int_{T+kT_0}^{s} i_1(\tau) d\tau \right) - \widetilde{C}_1^{-1} \left(\int_{T+kT_0}^{s} \bar{i}_1(\tau) d\tau \right) \right| ds + 2Z_0 \sup \left| \widetilde{C}_1^{-1} \left(\int_{T+kT_0}^{s} \bar{i}_1(\tau) d\tau \right) \right| \int_{T+kT_0}^{t} \left| \frac{1}{\widetilde{L}_1(i_1(s))} - \frac{1}{\widetilde{L}_1(\bar{i}_1(s))} \right| ds \leq$$

$$\leq \left(2Z_0 U_{E_1} + Z_0 \left(U_0 e^{-\beta} + I_0 \right) e^{\mu_0} + 2Z_0 \sum_{n=1}^{m} \left| r_n^{(1)} \right| \left(\frac{U_0 e^{-\beta} + I_0}{2Z_0} \right)^n + 2Z_0 U_0 \right) \times$$

$$\times \frac{1}{\hat{L}_1^{\,2}} \sum_{n=1}^{m} \left| \widetilde{l}_n^{(1)} \right| \left(\frac{U_0 e^{-\beta} + I_0}{2Z_0} \right)^n \int_{T+kT_0}^{t} \frac{\left| I(s) - \bar{I}(s) \right|}{2Z_0} ds +$$

$$+ \frac{Z_0}{\hat{L}_1} \rho^{(k)}(I, \bar{I}) \int_{T+kT_0}^{t} e^{\mu(s - T - kT_0)} ds +$$

$$+ 2Z_0 \frac{1}{\hat{L}_1} \sum_{n=1}^{m} \left| r_n^{(1)} \right| n \left(\frac{U_0 e^{-\beta} + I_0}{2Z_0} \right)^{n-1} \int_{T+kT_0}^{t} \left| i_1(s) - \bar{i}_1(s) \right| ds +$$

$$+ \frac{2Z_0}{\hat{L}_1^{\,2}} \frac{2\sqrt{\Phi_1}}{c_1 \sqrt{\Phi_1 - \phi_0}} \int_{T+kT_0}^{t} \int_{T+kT_0}^{s} \left| i_1(\tau) - \bar{i}_1(\tau) \right| d\tau ds \leq$$

$$\leq \left(U_{E_1} + \left(U_0 e^{-\beta} + I_0 \right) e^{\mu_0} + \sum_{n=1}^{m} \left| r_n^{(1)} \right| \left(\frac{U_0 e^{-\beta} + I_0}{2Z_0} \right)^n + U_0 \right) \frac{1}{\hat{L}_1^{\,2}} \sum_{n=1}^{m} \left| \widetilde{l}_n^{(1)} \right| \left(\frac{U_0 e^{-\beta} + I_0}{2Z_0} \right)^n \rho^{(k)}(I, \bar{I}) \frac{e^{\mu(t - T - kT_0)} - 1}{\mu} +$$

$$+ \frac{Z_0}{\hat{L}_1} \rho^{(k)}(I, \bar{I}) \frac{e^{\mu(t - T - kT_0)} - 1}{\mu} + \frac{1}{\hat{L}_1} \sum_{n=1}^{m} \left| r_n^{(1)} \right| n \left(\frac{U_0 e^{-\beta} + I_0}{2Z_0} \right)^{n-1} \rho^{(k)}(I, \bar{I}) \frac{e^{\mu(t - T - kT_0)} - 1}{\mu} +$$

$$+ \frac{1}{\hat{L}_1^{\,2}} \frac{2\sqrt{\Phi_1}}{c_1 \sqrt{\Phi_1 - \phi_0}} \rho^{(k)}(I, \bar{I}) \frac{e^{\mu(t - T - kT_0)} - 1}{\mu^2} \leq$$

$$\leq e^{\mu(t - T - kT_0)} \frac{\hat{\rho}_\mu \left(\left((U, \dot{U}), (I, \dot{I}) \right), \left((\overline{U}, \bar{I}), (\dot{\overline{U}}, \dot{\bar{I}}) \right) \right)}{\mu^2} \times$$

$$\times \left[\frac{1}{\hat{L}_1^{\,2}} \left(U_{E_1} + \left(U_0 e^{-\beta} + I_0 \right) e^{\mu_0} + \sum_{n=1}^{m} \left| r_n^{(1)} \right| \left(\frac{U_0 e^{-\beta} + I_0}{2Z_0} \right)^n + U_0 \right) \sum_{n=1}^{m} \left| \widetilde{l}_n^{(1)} \right| \left(\frac{U_0 e^{-\beta} + I_0}{2Z_0} \right)^n + \right.$$

$$+ \frac{1}{\hat{L}_1}\left(Z_0 + \sum_{n=1}^{m}\left|r_n^{(1)}\right| n\left(\frac{U_0 e^{-\beta}+I_0}{2Z_0}\right)^{n-1} + \frac{2\sqrt{\Phi_1}}{\mu \hat{L}_1 c_1 \sqrt{\Phi_1 - \phi_0}}\right)\right];$$

$$J_2 = \left|\left(\frac{t-T-kT_0}{T_0}-\frac{1}{2}\right)\int_{T+kT_0}^{T+(k+1)T_0}\left(J(U,I)(s)-J(\overline{U},\overline{I})(s)\right)ds\right| \le$$

$$\le \left(U_{E_1} + \left(U_0 e^{-\beta}+I_0\right)e^{\mu_0} + \sum_{n=1}^{m}\left|r_n^{(1)}\right|\left(\frac{U_0 e^{-\beta}+I_0}{2Z_0}\right)^n + U_0\right)\frac{1}{\hat{L}_1^{2}}\sum_{n=1}^{m}\left|\tilde{l}_n^{(1)}\right|\left(\frac{U_0 e^{-\beta}+I_0}{2Z_0}\right)^n \rho^{(k)}(I,\overline{I})\frac{e^{\mu T_0}-1}{2\mu} +$$

$$+ \frac{Z_0}{\hat{L}_1}\rho^{(k)}(I,\overline{I})\frac{e^{\mu T_0}-1}{2\mu} + \frac{1}{\hat{L}_1}\sum_{n=1}^{m}\left|r_n^{(1)}\right| n\left(\frac{U_0 e^{-\beta}+I_0}{2Z_0}\right)^{n-1}\rho^{(k)}(I,\overline{I})\frac{e^{\mu T_0}-1}{2\mu} +$$

$$+ \frac{1}{\hat{L}_1^{2}}\frac{2\sqrt{\Phi_1}}{c_1\sqrt{\Phi_1-\phi_0}}\rho^{(k)}(I,\overline{I})\frac{e^{\mu T_0}-1}{2\mu^2} \le$$

$$\le \left(U_{E_1} + \left(U_0 e^{-\beta}+I_0\right)e^{\mu_0} + \sum_{n=1}^{m}\left|r_n^{(1)}\right|\left(\frac{U_0 e^{-\beta}+I_0}{2Z_0}\right)^n + U_0\right)\frac{1}{\hat{L}_1^{2}}\sum_{n=1}^{m}\left|\tilde{l}_n^{(1)}\right|\left(\frac{U_0 e^{-\beta}+I_0}{2Z_0}\right)^n e^{\mu_0}\frac{\rho_\mu^{(k)}(\dot{I},\dot{\overline{I}})}{\mu}\frac{e^{\mu T_0}-1}{2\mu} +$$

$$+ \frac{Z_0}{\hat{L}_1}e^{\mu_0}\frac{\rho_\mu^{(k)}(\dot{I},\dot{\overline{I}})}{\mu}\frac{e^{\mu T_0}-1}{2\mu} + \frac{1}{\hat{L}_1}\sum_{n=1}^{m}\left|r_n^{(1)}\right| n\left(\frac{U_0 e^{-\beta}+I_0}{2Z_0}\right)^{n-1}e^{\mu_0}\frac{\rho_\mu^{(k)}(\dot{I},\dot{\overline{I}})}{\mu}\frac{e^{\mu T_0}-1}{2\mu} +$$

$$+ \frac{1}{\hat{L}_1^{2}}\frac{2\sqrt{\Phi_1}}{c_1\sqrt{\Phi_1-\phi_0}}e^{\mu_0}\frac{\rho_\mu^{(k)}(\dot{I},\dot{\overline{I}})}{\mu}\frac{e^{\mu T_0}-1}{2\mu^2} \le$$

$$\le e^{\mu(t-T-kT_0)}\hat{\rho}_\mu\left(((U,\dot{U}),(I,\dot{I})),((\overline{U},\dot{\overline{U}}),(\overline{I},\dot{\overline{I}}))\right) \times$$

$$\times \frac{e^{\mu_0}-1}{2}\frac{e^{\mu_0}}{\mu^2}\left[\left(U_{E_1} + \left(U_0 e^{-\beta}+I_0\right)e^{\mu_0} + \sum_{n=1}^{m}\left|r_n^{(1)}\right|\left(\frac{U_0 e^{-\beta}+I_0}{2Z_0}\right)^n + U_0\right)\frac{1}{\hat{L}_1^{2}}\sum_{n=1}^{m}\left|\tilde{l}_n^{(1)}\right|\left(\frac{U_0 e^{-\beta}+I_0}{2Z_0}\right)^n +\right.$$

$$\left.+ \frac{Z_0}{\hat{L}_1} + \frac{1}{\hat{L}_1}\sum_{n=1}^{m}\left|r_n^{(1)}\right| n\left(\frac{U_0 e^{-\beta}+I_0}{2Z_0}\right)^{n-1} + \frac{1}{\hat{L}_1^{2}}\frac{2\sqrt{\Phi_1}}{c_1\sqrt{\Phi_1-\phi_0}}\right]$$

and

$$J_3 \le \frac{1}{T_0}\int_{T+kT_0}^{T+(k+1)T_0}\left|\int_{T+kT_0}^{t}\left(J(U,I)(s)-J(\overline{U},\overline{I})(s)\right)ds\right|dt \le$$

$$\leq e^{\mu(t-T-kT_0)}\frac{\hat{\rho}_\mu(((U,\dot{U}),(I,\dot{I})),((\overline{U},\overline{I}),(\dot{\overline{U}},\dot{\overline{I}})))}{\mu^2}\times$$

$$\times\frac{e^{\mu_0}-1}{\mu_0}\left\{\frac{1}{\hat{L}_1^{\,2}}\left[U_{E_1}+e^{\mu_0}(U_0e^{-\beta}+I_0)+\sum_{n=1}^{m}\left|r_n^{(1)}\right|\left(\frac{U_0e^{-\beta}+I_0}{2Z_0}\right)^n\right]+U_0\sum_{n=1}^{m}\left|\widetilde{l}_n^{\,(1)}\right|\left(\frac{U_0e^{-\beta}+I_0}{2Z_0}\right)^n+\right.$$

$$\left.+\frac{1}{\hat{L}_1}\left[Z_0+\sum_{n=1}^{m}\left|r_n^{(1)}\right|n\left(\frac{U_0e^{-\beta}+I_0}{2Z_0}\right)^{n-1}+\frac{2\sqrt{\Phi_1}}{\mu\hat{L}_1c_1\sqrt{\Phi_1-\phi_0}}\right]\right\}.$$

Then

$$\left|B_I^{(k)}(U,I)(t)-B_I^{(k)}(\overline{U},\overline{I})(t)\right|\leq e^{\mu(t-T-kT_0)}\hat{\rho}_\mu(((U,\dot{U}),(I,\dot{I})),((\overline{U},\overline{I}),(\dot{\overline{U}},\dot{\overline{I}})))\times$$

$$\times\frac{1}{\mu^2}\left\{\frac{1}{\hat{L}_1^{\,2}}\left[U_{E_1}+e^{\mu_0}(U_0e^{-\beta}+I_0)+\sum_{n=1}^{m}\left|r_n^1\right|(I_0)^n+U_0\right]\sum_{n=1}^{m}\left|\widetilde{l}_n^{\,(1)}\right|(I_0)^n+\right.$$

$$\left.+\frac{1}{\hat{L}_1}\left[Z_0+\sum_{n=1}^{m}\left|r_n^{(1)}\right|n(I_0)^{n-1}+\frac{2\sqrt{\Phi_1}}{\mu\hat{L}_1c_1\sqrt{\Phi_1-\phi_0}}\right]\right\}+$$

$$+e^{\mu(t-T-kT_0)}\hat{\rho}_\mu(((U,\dot{U}),(I,\dot{I})),((\overline{U},\overline{I}),(\dot{\overline{U}},\dot{\overline{I}})))\frac{e^{\mu_0}-1}{2}\frac{1}{\mu^2}\left\{\frac{1}{\hat{L}_1^{\,2}}\left[U_{E_1}+e^{\mu_0}(U_0e^{-\beta}+I_0)+\right.\right.$$

$$\left.\left.+\sum_{n=1}^{m}\left|r_n^{(1)}\right|(I_0)^n+U_0\right]\sum_{n=1}^{m}\left|\widetilde{l}_n^{\,(1)}\right|(I_0)^n+\frac{1}{\hat{L}_1}\left[Z_0+\sum_{n=1}^{m}\left|r_n^{(1)}\right|n(I_0)^{n-1}+\frac{2\sqrt{\Phi_1}}{\mu\hat{L}_1c_1\sqrt{\Phi_1-\phi_0}}\right]\right\}+$$

$$+e^{\mu(t-T-kT_0)}\hat{\rho}_\mu(((U,\dot{U}),(I,\dot{I})),((\overline{U},\overline{I}),(\dot{\overline{U}},\dot{\overline{I}})))\frac{e^{\mu_0}-1}{\mu_0}\frac{1}{\mu^2}\left\{\frac{1}{\hat{L}_1^{\,2}}\left[U_{E_1}+e^{\mu_0}(U_0e^{-\beta}+I_0)+\right.\right.$$

$$\left.\left.+\sum_{n=1}^{m}\left|r_n^{(1)}\right|(I_0)^n+U_0\right]\sum_{n=1}^{m}\left|\widetilde{l}_n^{\,(1)}\right|(I_0)^n+\frac{1}{\hat{L}_1}\left[Z_0+\sum_{n=1}^{m}\left|r_n^{(1)}\right|n(I_0)^{n-1}+\frac{2\sqrt{\Phi_1}}{\mu\hat{L}_1c_1\sqrt{\Phi_1-\phi_0}}\right]\right\}\leq$$

$$\leq e^{\mu(t-T-kT_0)}\hat{\rho}_\mu(((U,\dot{U}),(I,\dot{I})),((\overline{U},\overline{I}),(\dot{\overline{U}},\dot{\overline{I}})))\times$$

$$\times\left(1+\frac{e^{\mu_0}-1}{2}+\frac{e^{\mu_0}-1}{\mu_0}\right)\frac{e^{\mu_0}}{\mu^2}\left\{\frac{1}{\hat{L}_1^{\,2}}\left[U_{E_1}+e^{\mu_0}(U_0e^{-\beta}+I_0)+\sum_{n=1}^{m}\left|r_n^{(1)}\right|(I_0)^n+U_0\right]\sum_{n=1}^{m}\left|\widetilde{l}_n^{\,(1)}\right|(I_0)^n+\right.$$

$$\left.+\frac{1}{\hat{L}_1}\left[Z_0+\sum_{n=1}^{m}\left|r_n^{(1)}\right|n(I_0)^{n-1}+\frac{2\sqrt{\Phi_1}}{\mu\hat{L}_1c_1\sqrt{\Phi_1-\phi_0}}\right]\right\}\equiv$$

$$\equiv e^{\mu(t-T-kT_0)}K_I\,\hat{\rho}_\mu(((U,\dot{U}),(I,\dot{I})),((\overline{U},\overline{I}),(\dot{\overline{U}},\dot{\overline{I}})))\leq e^{\mu_0}K_I\,\hat{\rho}_\mu(((U,\dot{U}),(I,\dot{I})),((\overline{U},\overline{I}),(\dot{\overline{U}},\dot{\overline{I}})))$$

It follows $\hat{\rho}(B_I(U,I),B_I(\overline{U},\overline{I}))\leq e^{\mu_0}K_I\,\hat{\rho}_\mu(((U,\dot{U}),(I,\dot{I})),((\overline{U},\overline{I}),(\dot{\overline{U}},\dot{\overline{I}}))).$

Finally, for the derivative we obtain

$$\left|\dot{B}_I(U,I)(t)-\dot{B}_I(\overline{U},\overline{I})(t)\right|\le\left|J(U,I)(t)-J(\overline{U},\overline{I})(t)\right|+$$

$$+\left|\frac{1}{T_0}\int_{T+kT_0}^{T+(k+1)T_0}\left(J(U,I)(s)-J(\overline{U},\overline{I})(s)\right)ds\right|\equiv\dot{J}_1+\dot{J}_2.$$

But

$$\dot{J}_1\le\left|\dot{U}(t-T)-\dot{\overline{U}}(t-T)\right|+2Z_0U_{E_1}\left|\frac{1}{\widetilde{L}_1(i_1(t))}-\frac{1}{\widetilde{L}_1(\overline{i}_1(t))}\right|+$$

$$+Z_0\left|\frac{U(t-T)}{\widetilde{L}_1(i_1(t))}-\frac{U(t-T)}{\widetilde{L}_1(\overline{i}_1(t))}\right|+Z_0\left|\frac{U(t-T)}{\widetilde{L}_1(\overline{i}_1(t))}-\frac{\overline{U}(t-T)}{\widetilde{L}_1(\overline{i}_1(t))}\right|+$$

$$+Z_0\left|\frac{I(t)}{\widetilde{L}_1(i_1(t))}-\frac{I(t)}{\widetilde{L}_1(\overline{i}_1(t))}\right|+Z_0\left|\frac{I(t)}{\widetilde{L}_1(\overline{i}_1(t))}-\frac{\overline{I}(t)}{\widetilde{L}_1(\overline{i}_1(t))}\right|+$$

$$+2Z_0\left|\frac{R_1(i_1(t))}{\widetilde{L}_1(i_1(t))}-\frac{R_1(i_1(t))}{\widetilde{L}_1(\overline{i}_1(t))}\right|+2Z_0\left|\frac{R_1(i_1(t))}{\widetilde{L}_1(\overline{i}_1(t))}-\frac{R_1(\overline{i}_1(t))}{\widetilde{L}_1(\overline{i}_1(t))}\right|+$$

$$+2Z_0\left|\frac{\widetilde{C}_1^{-1}\left(\int_{T+kT_0}^t i_1(\tau)d\tau\right)}{\widetilde{L}_1(i_1(t))}-\frac{\widetilde{C}_1^{-1}\left(\int_{T+kT_0}^t i_1(\tau)d\tau\right)}{\widetilde{L}_1(\overline{i}_1(t))}\right|+2Z_0\left|\frac{\widetilde{C}_1^{-1}\left(\int_{T+kT_0}^t i_1(\tau)d\tau\right)}{\widetilde{L}_1(\overline{i}_1(t))}-\frac{\widetilde{C}_1^{-1}\left(\int_{T+kT_0}^t \overline{i}_1(\tau)d\tau\right)}{\widetilde{L}_1(\overline{i}_1(t))}\right|\le$$

$$\le2Z_0U_{E_1}e^{\mu(t-T-kT_0)}\left|\frac{1}{\widetilde{L}_1(i_1(t))}-\frac{1}{\widetilde{L}_1(\overline{i}_1(t))}\right|+$$

$$+Z_0I_0e^{\mu(t-T-kT_0)}\left|\frac{1}{\widetilde{L}_1(i_1(t))}-\frac{1}{\widetilde{L}_1(\overline{i}_1(t))}\right|+\frac{Z_0}{\hat{L}_1}\left|I(t)-\overline{I}(t)\right|+$$

$$+2Z_0\sum_{n=1}^m\left|r_n^{(1)}\right|\left(\frac{U_0e^{-\beta}+I_0}{2Z_0}\right)^n\left|\frac{1}{\widetilde{L}_1(i_1(t))}-\frac{1}{\widetilde{L}_1(\overline{i}_1(t))}\right|+\frac{2Z_0}{\hat{L}_1}\left|R_1(i_1(t))-R_1(\overline{i}_1(t))\right|+$$

$$+2Z_0U_0\left|\frac{1}{\widetilde{L}_1(i_1(t))}-\frac{1}{\widetilde{L}_1(\overline{i}_1(t))}\right|+\frac{2Z_0}{\hat{L}_1^2}\left|\widetilde{C}_1^{-1}\left(\int_{T+kT_0}^t i_1(\tau)d\tau\right)-\widetilde{C}_1^{-1}\left(\int_{T+kT_0}^t \overline{i}_1(\tau)d\tau\right)\right|\le$$

$$\le\left(2Z_0U_{E_1}+Z_0U_0e^{-\beta}e^{\mu(t-T-kT_0)}+Z_0I_0e^{\mu(t-T-kT_0)}+2Z_0\sum_{n=1}^m\left|r_n^{(1)}\right|\left(\frac{U_0e^{-\beta}+I_0}{2Z_0}\right)^n+2Z_0U_0\right)\left|\frac{1}{\widetilde{L}_1(i_1(t))}-\frac{1}{\widetilde{L}_1(\overline{i}_1(t))}\right|+$$

$$+e^{\mu(t-T-kT_0)}\frac{Z_0}{\hat{L}_1}\rho^{(k)}(I,\overline{I})+\frac{2Z_0}{\hat{L}_1}\sum_{n=1}^m\left|r_n^{(1)}\right|n\left(\frac{U_0e^{-\beta}+I_0}{2Z_0}\right)^{n-1}\frac{\left|I(t)-\overline{I}(t)\right|}{2Z_0}+$$

$$+ \frac{2Z_0}{\hat{L}_1^2} \frac{2\sqrt{\Phi_1}}{c_1\sqrt{\Phi_1 - \phi_0}} \int\limits_{T+kT_0}^{t} \frac{\left|I(\tau) - \bar{I}(\tau)\right|}{2Z_0} d\tau \le$$

$$\le \left(2Z_0 U_{E_1} + Z_0 U_0 e^{-\beta} e^{\mu(t-T-kT_0)} + Z_0 I_0 e^{\mu(t-T-kT_0)} + 2Z_0 \sum_{n=1}^{m} \left|r_n^{(1)}\right| \left(\frac{U_0 e^{-\beta} + I_0}{2Z_0} \right)^n + 2Z_0 U_0 \right) \times$$

$$\times \frac{1}{\hat{L}_1^2} \sum_{n=1}^{m} \left|\widetilde{l}_n^{(1)}\right| \left(\frac{U_0 e^{-\beta} + I_0}{2Z_0} \right)^n \frac{\left|I(t) - \bar{I}(t)\right|}{2Z_0} +$$

$$+ e^{\mu(t-T-kT_0)} \frac{Z_0}{\hat{L}_1} \rho^{(k)}(I,\bar{I}) + e^{\mu(t-T-kT_0)} \rho^{(k)}(I,\bar{I}) \frac{1}{\hat{L}_1} \sum_{n=1}^{m} \left|r_n^{(1)}\right| n \left(\frac{U_0 e^{-\beta} + I_0}{2Z_0} \right)^{n-1} +$$

$$+ \frac{1}{\hat{L}_1^2} \frac{2\sqrt{\Phi_1}}{c_1\sqrt{\Phi_1 - \phi_0}} \rho^{(k)}(I,\bar{I}) \int\limits_{T+kT_0}^{t} e^{\mu(\tau-T-kT_0)} d\tau \le$$

$$\le e^{\mu(t-T-kT_0)} \rho^{(k)}(I,\bar{I}) \left[\left(2U_{E_1} + U_0 e^{-\beta} e^{\mu T_0} + I_0 e^{\mu T_0} + \sum_{n=1}^{m} \left|r_n^{(1)}\right| \left(\frac{U_0 e^{-\beta} + I_0}{2Z_0} \right)^n + U_0 \right) \times \right.$$

$$\left. \times \frac{1}{\hat{L}_1^2} \sum_{n=1}^{m} \left|\widetilde{l}_n^{(1)}\right| \left(\frac{U_0 e^{-\beta} + I_0}{2Z_0} \right)^n + \frac{Z_0}{\hat{L}_1} + \frac{1}{\hat{L}_1} \sum_{n=1}^{m} \left|r_n^{(1)}\right| n \left(\frac{U_0 e^{-\beta} + I_0}{2Z_0} \right)^{n-1} + \frac{1}{\hat{L}_1^2} \frac{2\sqrt{\Phi_1}}{c_1\sqrt{\Phi_1 - \phi_0}} \right] \le$$

$$\le e^{\mu(t-T-kT_0)} \frac{\rho_\mu^{(k)}(\dot{I},\dot{\bar{I}})}{\mu} \left[\frac{1}{\hat{L}_1^2} \left(2U_{E_1} + U_0 e^{-\beta} e^{\mu T_0} + I_0 e^{\mu T_0} + \sum_{n=1}^{m} \left|r_n^{(1)}\right| \left(\frac{U_0 e^{-\beta} + I_0}{2Z_0} \right)^n + U_0 \right) \sum_{n=1}^{m} \left|\widetilde{l}_n^{(1)}\right| \left(\frac{U_0 e^{-\beta} + I_0}{2Z_0} \right)^n + \right.$$

$$\left. + \frac{1}{\hat{L}_1} \left(Z_0 + \sum_{n=1}^{m} \left|r_n^{(1)}\right| n \left(\frac{U_0 e^{-\beta} + I_0}{2Z_0} \right)^{n-1} + \frac{2\sqrt{\Phi_1}}{\hat{L}_1 c_1\sqrt{\Phi_1 - \phi_0}} \right) \right] \le$$

$$\le e^{\mu(t-T-kT_0)} \frac{\hat{\rho}_\mu(((U,\dot{U}),(I,\dot{I})),((\overline{U},\bar{I}),(\dot{\overline{U}},\dot{\bar{I}})))}{\mu} \times$$

$$\times \left[\frac{1}{\hat{L}_1^2} \left(2U_{E_1} + U_0 e^{-\beta} e^{\mu T_0} + I_0 e^{\mu T_0} + \sum_{n=1}^{m} \left|r_n^{(1)}\right| \left(\frac{U_0 e^{-\beta} + I_0}{2Z_0} \right)^n + U_0 \right) \sum_{n=1}^{m} \left|\widetilde{l}_n^{(1)}\right| \left(\frac{U_0 e^{-\beta} + I_0}{2Z_0} \right)^n + \right.$$

$$\left. + \frac{1}{\hat{L}_1} \left(Z_0 + \sum_{n=1}^{m} \left|r_n^{(1)}\right| n \left(\frac{U_0 e^{-\beta} + I_0}{2Z_0} \right)^{n-1} + \frac{2\sqrt{\Phi_1}}{\hat{L}_1 c_1\sqrt{\Phi_1 - \phi_0}} \right) \right]$$

and

$$\dot{J}_2 \le \frac{1}{T_0} \int\limits_{T+kT_0}^{T+(k+1)T_0} \left| J(U,I)(s) - J(\overline{U},\bar{I})(s) \right| ds \le$$

$$\le e^{\mu(t-T-kT_0)}\hat{\rho}_\mu(((U,\dot{U}),(I,\dot{I})),((\overline{U},\overline{I}),(\dot{\overline{U}},\dot{\overline{I}})))\frac{e^{\mu_0}-1}{\mu_0}\times$$

$$\times\frac{1}{\mu}\left\{\frac{1}{\hat{L}_1^2}\left[2U_{E_1}+e^{\mu_0}\left(U_0e^{-\beta}+I_0\right)+\sum_{n=1}^{m}\left|r_n^{(1)}\right|\left(\frac{U_0e^{-\beta}+I_0}{2Z_0}\right)^n+U_0\right]\sum_{n=1}^{m}\left|\tilde{l}_n^{(1)}\right|\left(\frac{U_0e^{-\beta}+I_0}{2Z_0}\right)^n+\right.$$

$$\left.+\frac{1}{\hat{L}_1}\left[Z_0+\sum_{n=1}^{m}\left|r_n^{(1)}\right|n\left(\frac{U_0e^{-\beta}+I_0}{2Z_0}\right)^{n-1}+\frac{2\sqrt{\Phi_1}}{\mu\hat{L}_1c_1\sqrt{\Phi_1-\phi_0}}\right]\right\},$$

then

$$\rho_\mu^{(k)}(\dot{B}_I^{(k)}(U,I),\dot{B}_I^{(k)}(\overline{U},\overline{I}))\le$$

$$\le\hat{\rho}_\mu(((U,\dot{U}),(I,\dot{I})),((\overline{U},\dot{\overline{U}}),(\overline{I},\dot{\overline{I}})))\times$$

$$\times\left(1+\frac{e^{\mu_0}-1}{\mu_0}\right)\frac{1}{\mu}\left\{\frac{1}{\hat{L}_1^2}\left[2U_{E_1}+e^{\mu_0}\left(U_0e^{-\beta}+I_0\right)+\sum_{n=1}^{m}\left|r_n^{(1)}\right|\left(I_0\right)^n+U_0\right]\sum_{n=1}^{m}\left|\tilde{l}_n^{(1)}\right|\left(I_0\right)^n+\right.$$

$$\left.+\frac{1}{\hat{L}_1}\left[Z_0+\sum_{n=1}^{m}\left|r_n^{(1)}\right|n\left(I_0\right)^{n-1}+\frac{2\sqrt{\Phi_1}}{\mu c_1\sqrt{\Phi_1-\phi_0}}\frac{1}{\hat{L}_1}\right]\right\}\equiv$$

$$\equiv\dot{K}_I\hat{\rho}_\mu(((U,\dot{U}),(I,\dot{I})),((\overline{U},\overline{I}),(\dot{\overline{U}},\dot{\overline{I}}))).$$

Therefore, denoting by $K=\max\left\{e^{\mu_0}K_U,\dot{K}_U,e^{\mu_0}K_I,\dot{K}_I\right\}$ we obtain

$$\hat{\rho}_\mu((B(U,I),\dot{B}(U,I)),(B(\overline{U},\overline{I}),\dot{B}(\overline{U},\overline{I})))\le K\,\hat{\rho}_\mu(((U,\dot{U}),(I,\dot{I})),((\overline{U},\overline{I}),(\dot{\overline{U}},\dot{\overline{I}})))$$

For sufficiently large μ we obtain $K<1$. Then B is contractive operator, and in view of Lemma 3.4.5 its fixed point is a periodic solution of (3.5.1).

Theorem 3.5.1 is thus proved.

3.6. NUMERICAL EXAMPLE

We collect all inequalities guaranteeing the conditions of the above-mentioned Theorem 3.5.1:

$$\frac{U_0e^{-\beta}+I_0}{2Z_0}\le I_0;\quad\frac{U_0+I_0e^{-\beta}}{2Z_0}\le I_0;$$

$$\left(1+\frac{e^{\mu_0}-1}{\mu_0}\right)I_0 e^{-\beta}+$$

$$+\left(1+\frac{e^{\mu_0}-1}{2}+\frac{e^{\mu_0}-1}{\mu_0}\right)\frac{1}{\mu\hat{L}_0}\left[\left(Z_0+\frac{H_0}{\mu}\right)(U_0+I_0 e^{-\beta})+2Z_0\sum_{n=1}^{m}\left|r_n^{(0)}\right|(I_0)^n e^{(n-1)\mu_0}+2Z_0 U_{E0}\right]\le U_0;$$

$$\left(1+\frac{e^{\mu_0}-1}{\mu_0}\right)U_0 e^{-\beta}+$$

$$+\left(1+\frac{e^{\mu_0}-1}{2}+\frac{e^{\mu_0}-1}{\mu_0}\right)\frac{1}{\mu\hat{L}_1}\left[\left(Z_0+\frac{H_1}{\mu}\right)(U_0 e^{-\beta}+I_0)+2Z_0\sum_{n=1}^{m}\left|r_n^{(1)}\right|(I_0)^n e^{(n-1)\mu_0}+2Z_0 U_{E1}\right]\le I_0;$$

$$e^{\mu_0}K_U=\left(1+\frac{e^{\mu_0}-1}{2}+\frac{e^{\mu_0}-1}{\mu_0}\right)\frac{e^{2\mu_0}}{\mu^2}\left[\frac{U_{E0}+e^{\mu_0}(U_0+I_0 e^{-\beta})+\sum_{n=1}^{m}\left|r_n^{(0)}\right|(I_0)^n+U_0}{\hat{L}_0^{\,2}}\sum_{n=1}^{m}\left|\tilde{l}_n^{(0)}\right|(I_0)^n+\right.$$

$$\left.+\frac{1}{\hat{L}_0}\left(Z_0+\sum_{n=1}^{m}\left|r_n^{(0)}\right|n.(I_0)^{n-1}+\frac{2\sqrt{\Phi_0}}{\mu\hat{L}_0 c_0\sqrt{\Phi_0-\phi_0}}\right)\right]<1;$$

$$e^{\mu_0}K_I=\left(1+\frac{e^{\mu_0}-1}{2}+\frac{e^{\mu_0}-1}{\mu_0}\right)\frac{e^{2\mu_0}}{\mu^2}\left[\frac{U_{E1}+e^{\mu_0}(U_0 e^{-\beta}+I_0)+\sum_{n=1}^{m}\left|r_n^{(1)}\right|(I_0)^n+U_0}{\hat{L}_1^{\,2}}\sum_{n=1}^{m}\left|\tilde{l}_n^{(1)}\right|(I_0)^n+\right.$$

$$\left.+\frac{1}{\hat{L}_1}\left(Z_0+\sum_{n=1}^{m}\left|r_n^{(1)}\right|n(I_0)^{n-1}+\frac{2\sqrt{\Phi_1}}{\mu\hat{L}_1 c_1\sqrt{\Phi_1-\phi_0}}\right)\right]<1;$$

$$\dot{K}_U=\left(1+\frac{e^{\mu_0}-1}{\mu_0}\right)\frac{1}{\mu}\left\{\frac{1}{\hat{L}_0^{\,2}}\left[U_{E0}+e^{\mu_0}(U_0+I_0 e^{-\beta})+\sum_{n=1}^{m}\left|r_n^{0}\right|(I_0)^n+U_0\right]\sum_{n=1}^{m}\left|\tilde{l}_n^{(0)}\right|(I_0)^n+\right.$$

$$\left.+\frac{1}{\hat{L}_0}\left[Z_0+\sum_{n=1}^{m}\left|r_n^{0}\right|n(I_0)^{n-1}+\frac{2\sqrt{\Phi_0}}{\mu\hat{L}_0 c_0\sqrt{\Phi_0-\phi_0}}\right]\right\}<1;$$

$$\dot{K}_I=\left(1+\frac{e^{\mu_0}-1}{\mu_0}\right)\frac{1}{\mu}\left\{\frac{1}{\hat{L}_1^{\,2}}\left[2U_{E1}+e^{\mu_0}(U_0 e^{-\beta}+I_0)+\sum_{n=1}^{m}\left|r_n^{(1)}\right|(I_0)^n+U_0\right]\sum_{n=1}^{m}\left|\tilde{l}_n^{(1)}\right|(I_0)^n+\right.$$

$$\left.+\frac{1}{\hat{L}_1}\left[Z_0+\sum_{n=1}^{m}\left|r_n^{(1)}\right|n(I_0)^{n-1}+\frac{2\sqrt{\Phi_1}}{\mu\hat{L}_1 c_1\sqrt{\Phi_1-\phi_0}}\right]\right\}<1.$$

Let $\Lambda=1m$, $L=0,2\,\mu H/m$, $C=80\,pF/m$, $Z_0=\sqrt{L/C}=50\,\Omega$,

$$v = 1/\sqrt{LC} = 1/\sqrt{0,2.10^{-6}.80.10^{-12}} = 1/\left(4.10^{-9}\right) = 2,5.10^8 \,;$$

$$T = \Lambda\sqrt{LC} = 4.10^{-9}\ \text{sec}.$$

Let us check the propagation of millimeter waves $\lambda_0 = (1/4)10^{-3}\,m$. We have

$$f_0 = 1/\left(\lambda_0\sqrt{LC}\right) = 1/\left((1/4)10^{-3}.4.10^{-9}\right) = 10^{12}\,Hz \Rightarrow T_0 = 1/f_0 = 10^{-12}\ \text{sec}.$$

If we choose $\mu = 10^{12}$, then $\mu T_0 = \mu_0 = 1$ and

$$T = 4.10^{-9}.10^{12}T_0 = 4000.T_0 \Rightarrow m = 4000\,.$$

We choose resistive elements with the following *V-I* characteristics

$$R_0(i) = R_1(i) = 0,028u - 0,125u^3\,,$$

i.e. $r_1^{(0)} = r_1^{(1)} = 0,028;\ r_2^{(0)} = r_2^{(1)} = 0;\ r_3^{(0)} = r_3^{(1)} = 0,125\,.$

If the inductive element is $L_0(i) = L_1(i) = 3i - (1/12)i^3$, then

$$\widetilde{L}_0(i) = i\left(dL_0(i)/di\right) + L_0(i) = i(3 - (1/4)i^2) + 3i - (1/12)i^3 = 6i - (1/3)i^3$$

If we choose $I_0 = 1$ one obtains $6i - (1/3)i^3 > 6 - (1/3) = 17/3$ and consequently

$$\frac{1}{\widehat{L}_0} = \frac{1}{\widehat{L}_1} = \frac{3}{17}\,;$$

$$\frac{d\widetilde{L}_p}{di} = (i+1)\frac{dL_p(i)}{di} + i\frac{d^2 L_p(i)}{di^2} = (i+1)\left(3 - \frac{i^2}{4}\right) - i\frac{1}{2}i = 3i + 3 - \frac{i^3}{4} - \frac{i^2}{4} \Rightarrow$$

$$\Rightarrow \widetilde{l}_1^{(p)} = 3,\ \widetilde{l}_2^{(p)} = 3,\ \widetilde{l}_3^{(p)} = -(1/4),\ \widetilde{l}_4^{(p)} = -(1/4).$$

If we take $C_0(u) = C_1(u) = c/\sqrt{1 - (u/\Phi)} =$

$c\sqrt{\Phi}/\sqrt{\Phi - u}$, where $h = 2$, $c_0 = c_1 = 50\,pF = 5.10^{-11}\,F$, $\Phi = \Phi_0 = \Phi_1 = 0,4V \Rightarrow$

$$U_0 \le \phi_0 = 0,3 < 0,4\,;\ H_0 = H_1 = \frac{2\sqrt{\Phi_0}}{\mu c_0\sqrt{\Phi_0 - \phi_0}} \approx \frac{2}{25} \Rightarrow \frac{H_0}{\mu} \approx 0\,.$$

The above inequalities become as follows

$$I_0 e^{-(\beta-1)} + \frac{(3e-1)Z_0}{\mu \hat{L}_0}\left[\left(Z_0 + \frac{H_0}{\mu}\right)\mathbf{I}_0 + \sum_{n=1}^{3}\left|r_n^{(0)}\right|\left(\mathbf{I}_0\right)^n e^{n-1} + U_{E_0}\right] \le U_0;$$

$$U_0 e^{-(\beta-1)} + \frac{(3e-1)Z_0}{\mu \hat{L}_1}\left[\left(Z_0 + \frac{H_1}{\mu}\right)\mathbf{I}_0 + \sum_{n=1}^{m}\left|r_n^{(1)}\right|\left(\mathbf{I}_0\right)^n e^{n-1} + U_{E_1}\right] \le I_0.$$

We will not consider $e^{\mu 0}K_U$ and $e^{\mu 0}K_I$ because they are of order $\dfrac{1}{\mu^2}$.

Therefore

$$\dot{K}_U = \frac{3e}{17.10^{12}}\left[\frac{0.1 + 2Z_0 e + \sum\limits_{n=1}^{3}\left|r_n^{(0)}\right| + U_0}{\hat{L}_0}\sum_{n=1}^{m}\left|\tilde{l}_n^{(0)}\right| + Z_0 + \sum_{n=1}^{3}\left|r_n^0\right|n + \frac{2}{25}\cdot\frac{3}{17}\right] < 1;$$

$$\dot{K}_I = \frac{3e}{17.10^{12}}\left[\frac{0.1 + 2Z_0 e + \sum\limits_{n=1}^{3}\left|r_n^{(1)}\right| + U_0}{\hat{L}_1}\sum_{n=1}^{m}\left|\tilde{l}_n^{(1)}\right| + Z_0 + \sum_{n=1}^{3}\left|r_n^{(1)}\right|n + \frac{2}{25}\cdot\frac{3}{17}\right] < 1$$

or for $\beta > 1$ and $U_0 = I_0 = 0.1 = 2U_E$ we have

$$I_0 e^{-(\beta-1)} + \frac{50(3e-1)}{10^{12}}\frac{3}{17}\left[50 + 0.028 + 0.125e^2 + 0.05\right] \le U_0;$$

$$U_0 e^{-(\beta-1)} + \frac{50(3e-1)}{10^{12}}\frac{3}{17}\left[50 + 0.028 + 0.125e^2 + 0.05\right] \le I_0;$$

$$\dot{K}_U = \dot{K}_I = \frac{3e}{17.10^{12}}\left[\frac{3(0.2 + 100e + 0.028 + 0.125)}{17}6.5 + 50 + 0.028 + 3.0.125 + \frac{2}{25}\cdot\frac{3}{17}\right] = 2.54.10^{-10} < 1.$$

The initial approximation is typically chosen to be a set of simple functions. In order to satisfy the conformity condition **(CC)** we smooth the function on a sufficiently small interval, that is, for sufficiently small $\delta > 0$ we obtain $\dot{U}^{(0)}(0) = 0, \dot{I}^{(0)}(0) = 0$, defining

$$\tilde{U}^{(0)}(t) = \begin{cases} \upsilon_l(t), t \in [0,\delta] \\ U_0 \sin \omega_0 t, \, t \in [\delta, T-\delta] \\ \upsilon_r(t), t \in [T-\delta, T] \\ 0, \, t \in [T,\infty) \end{cases}, \tilde{I}^{(0)}(t) = \begin{cases} \iota_l(t), t \in [0,\delta] \\ I_0 \sin \omega_0 t, \, t \in [\delta, T-\delta] \\ \iota_r(t), t \in [T-\delta, T] \\ 0, \, t \in [T,\infty) \end{cases} \left(\omega_0 = \frac{2\pi}{T_0}\right)$$

and $E_0(t) = E_1(t) = U_E \sin \omega_0 t$, $U_E \leq U_0$. We point out that here, by U_0 and I_0, we denote $U_0 = U_0 e^{-\beta}$, $I_0 = I_0 e^{-\beta}$.

Then we have $U^{(n+1)}(t) = B(U^{(n)}, I^{(n)})$, $I^{(n+1)}(t) = B(U^{(n)}, I^{(n)})$ $(n = 0,1,2,...)$ and

$$\rho_\mu^{(k)}((U^{(n+1)}, I^{(n+1)}),(U^{(n)}, I^{(n)})) \leq \frac{K^n}{1-K}\rho_\mu^{(k)}((U^{(1)}, I^{(1)}),(U^{(0)}, I^{(0)})),(n = 0,1,...;k = 0,1,2...) \cdot$$

We find the first approximation:

$$U^{(1)}(t) = B_U^{(k)}(U^{(0)}, I^{(0)})(t) = \int_{T+kT_0}^{t} V(U^{(0)}, I^{(0)})(s)ds - \left(\frac{t-T-kT_0}{T_0} - \frac{1}{2}\right)\int_{T+kT_0}^{T+(k+1)T_0} V(U^{(0)}, I^{(0)})(s)ds -$$

$$-\frac{1}{T_0}\int_{T+kT_0}^{T+(k+1)T_0}\int_{T+kT_0}^{t} V(U^{(0)}, I^{(0)})(s)ds\,dt =$$

$$= \int_{T+kT_0}^{t}\left(\frac{dI(s-T)}{dt} - Z_0\frac{U(s)+I(s-T)-2E_0(s)+2R_0(i_0(s))}{\tilde{L}_0} - \frac{2Z_0}{\tilde{L}_0}\tilde{C}_0^{-1}\left(\frac{1}{2Z_0}\int_{T+kT_0}^{s}i_0(\theta)d\theta\right)\right)ds -$$

$$-\left(\frac{t-T-kT_0}{T_0} - \frac{1}{2}\right)\int_{T+kT_0}^{T+(k+1)T_0}\left(\frac{dI(s-T)}{dt} - Z_0\frac{U(s)+I(s-T)-2E_0(s)+2R_0(i_0(s))}{\tilde{L}_0} -\right.$$

$$\left.-\frac{2Z_0}{\tilde{L}_0}\tilde{C}_0^{-1}\left(\frac{1}{2Z_0}\int_{T+kT_0}^{s}i_0(\theta)d\theta\right)\right)ds -$$

$$-\frac{1}{T_0}\int_{T+kT_0}^{T+(k+1)T_0}\int_{T+kT_0}^{t}\left(\frac{dI(s-T)}{dt} - Z_0\frac{U(s)+I(s-T)-2E_0(s)+2R_0(i_0(s))}{\tilde{L}_0} - \frac{2Z_0}{\tilde{L}_0}\tilde{C}_0^{-1}\left(\frac{1}{2Z_0}\int_{T+kT_0}^{s}i_0(\theta)d\theta\right)\right)ds\,dt=$$

$$= \int_{T+kT_0}^{t}\frac{dI(s-T)}{dt}ds - Z_0\int_{T+kT_0}^{t}\frac{U(s)}{\tilde{L}_0(i_0(s))}ds + 2\int_{T+kT_0}^{t}\frac{I(s-T)}{\tilde{L}_0(i_0(s))}ds + 2\int_{T+kT_0}^{t}\frac{E_0(s)}{\tilde{L}_0(i_0(s))}ds -$$

$$-2\int_{T+kT_0}^{t}\frac{R_0(i_0(s))}{\tilde{L}_0(i_0(s))}ds + 2Z_0\int_{T+kT_0}^{t}\frac{1}{\tilde{L}_0}\tilde{C}_0^{-1}\left(\frac{1}{2Z_0}\int_{T+kT_0}^{s}i_0(\theta)d\theta\right)ds -$$

$$-\left(\frac{t-T-kT_0}{T_0} - \frac{1}{2}\right)\left[\int_{T+kT_0}^{T+(k+1)T_0}\frac{dI(s-T)}{dt}ds - Z_0\int_{T+kT_0}^{T+(k+1)T_0}\frac{U(s)}{\tilde{L}_0(i_0(s))}ds + 2\int_{T+kT_0}^{T+(k+1)T_0}\frac{I(s-T)}{\tilde{L}_0(i_0(s))}ds +\right.$$

$$\left.+2\int_{T+kT_0}^{T+(k+1)T_0}\frac{E_0(s)}{\tilde{L}_0(i_0(s))}ds - 2\int_{T+kT_0}^{T+(k+1)T_0}\frac{R_0(i_0(s))}{\tilde{L}_0(i_0(s))}ds + 2Z_0\int_{T+kT_0}^{T+(k+1)T_0}\frac{1}{\tilde{L}_0}\tilde{C}_0^{-1}\left(\frac{1}{2Z_0}\int_{T+kT_0}^{s}i_0(\theta)d\theta\right)ds\right] -$$

$$-\frac{1}{T_0}\int_{T+kT_0}^{T+(k+1)T_0}\left[\int_{T+kT_0}^{t}\frac{dI(s-T)}{dt}ds - Z_0\int_{T+kT_0}^{t}\frac{U(s)}{\tilde{L}_0(i_0(s))}ds + 2\int_{T+kT_0}^{t}\frac{I(s-T)}{\tilde{L}_0(i_0(s))}ds + 2\int_{T+kT_0}^{t}\frac{E_0(s)}{\tilde{L}_0(i_0(s))}ds -\right.$$

$$\left.-2\int_{T+kT_0}^{t}\frac{+2R_0(i_0(s))}{\tilde{L}_0(i_0(s))}ds + 2Z_0\int_{T+kT_0}^{t}\frac{1}{\tilde{L}_0}\tilde{C}_0^{-1}\left(\frac{1}{2Z_0}\int_{T+kT_0}^{s}i_0(\theta)d\theta\right)ds\right]dt =$$

$$= I_0 \sin \omega_0 t + 2 \int\limits_{T+kT_0}^{t} \frac{I_0 \sin \omega_0 s}{\widetilde{L}_0(i_0(s))} ds + 2 \int\limits_{T+kT_0}^{t} \frac{U_0 \sin \omega_0 s}{\widetilde{L}_0(i_0(s))} ds - 2 \int\limits_{T+kT_0}^{t} \frac{0{,}028 i_0(s) - 0{,}125(i_0(s))^3}{\widetilde{L}_0(i_0(s))} ds +$$

$$-\left(\frac{t-T-kT_0}{T_0} - \frac{1}{2} \right) \left[2 \int\limits_{T+kT_0}^{T+(k+1)T_0} \frac{I_0 \sin \omega_0 s}{\widetilde{L}_0(i_0(s))} ds + 2 \int\limits_{T+kT_0}^{T+(k+1)T_0} \frac{U_0 \sin \omega_0 s}{\widetilde{L}_0(i_0(s))} ds - \right.$$

$$-2 \int\limits_{T+kT_0}^{T+(k+1)T_0} \frac{0{,}028 i_0(s) - 0{,}125(i_0(s))^3}{\widetilde{L}_0(i_0(s))} ds + 2Z_0 \int\limits_{T+kT_0}^{T+(k+1)T_0} \frac{1}{\widetilde{L}_0(i_0(s))} \widetilde{C}_0^{-1}\left(\frac{1}{2Z_0} \int\limits_{T+kT_0}^{s} i_0(\theta)d\theta \right) ds \left. \right] -$$

$$-\frac{1}{T_0} \int\limits_{T+kT_0}^{T+(k+1)T_0} \left[I_0 \sin \omega_0 t + 2 \int\limits_{T+kT_0}^{t} \frac{I_0 \sin \omega_0 s}{\widetilde{L}_0(i_0(s))} ds + 2 \int\limits_{T+kT_0}^{t} \frac{U_0 \sin \omega_0 s}{\widetilde{L}_0(i_0(s))} ds - \right.$$

$$-2 \int\limits_{T+kT_0}^{t} \frac{0{,}028 i_0(s) - 0{,}125(i_0(s))^3}{\widetilde{L}_0(i_0(s))} ds + 2Z_0 \int\limits_{T+kT_0}^{t} \frac{1}{\widetilde{L}_0(i_0(s))} \widetilde{C}_0^{-1}\left(\frac{1}{2Z_0} \int\limits_{T+kT_0}^{s} i_0(\theta)d\theta \right) ds \left. \right] dt \; .$$

We have to estimate every term in order to simplify each of the following approximations:

$$\left| I_0 \sin \omega_0 t \right| \le I_0 \; ; \quad \left| 2 \int\limits_{T+kT_0}^{t} \frac{I_0 \sin \omega_0 s}{\widetilde{L}_0(i_0(s))} ds \right| \le \frac{2}{\hat{L}_0} \frac{e^{\mu_0}-1}{\mu} I_0 \approx \frac{1}{10^{12}} I_0 \; ;$$

$$\left| 2 \int\limits_{T+kT_0}^{t} \frac{|U_{E0} \sin \omega_0 s}{\widetilde{L}_0(i_0(s))} ds \right| \approx \frac{1}{10^{12}} U_0 \; ;$$

$$\left| 2 \int\limits_{T+kT_0}^{t} \frac{0{,}028 i_0(s) - 0{,}125(i_0(s))^3}{\widetilde{L}_0(i_0(s))} ds \right| \le \frac{6}{17}\left(0{,}028 \frac{eU_0}{100} + 0{,}125 \left(\frac{eU_0}{100} \right)^3 \right) \frac{e-1}{\mu} \approx U_0 \frac{1}{10^{12}} \; ;$$

$$\left| Z_0 \int\limits_{T+kT_0}^{T+(k+1)T_0} \frac{1}{\widetilde{L}_0(i_0(s))} \widetilde{C}_0^{-1}\left(\frac{1}{2Z_0} \int\limits_{T+kT_0}^{s} i_0(\theta)d\theta \right) ds \right| \approx \frac{Z_0}{\hat{L}_0} \frac{e-1}{10^{12}} U_0 \; .$$

Consequently

$$U^{(1)}(t) = B_U^{(k)}(U^{(0)}, I^{(0)})(t) = I_0 \sin \omega_0 t - \frac{I_0}{T_0} \int\limits_{T+kT_0}^{T+(k+1)T_0} \sin \omega_0 t\, dt = I_0 \sin \omega_0 t - \frac{I_0}{\omega_0 T_0} \int\limits_{T+kT_0}^{T+(k+1)T_0} \sin \omega_0 t\, d(\omega_0 t) =$$

$$= I_0 \sin \omega_0 t + \frac{I_0}{\omega_0 T_0} \cos \omega_0 t \Big|_{T+kT_0}^{T+(k+1)T_0} = I_0 \sin \omega_0 t + \frac{I_0}{2\pi}\left[\cos \omega_0 (T+(k+1)T_0) - \cos \omega_0 (T+kT_0) \right] =$$

$$= I_0 \sin \omega_0 t + \frac{I_0}{2\pi}\left[\cos(2\pi(m+k+1)) - \cos(2\pi(m+k)) \right] = I_0 \sin \omega_0 t \; .$$

$$I^{(1)}(t) = B_I^{(k)}(U^{(0)}, I^{(0)})(t) = \int\limits_{T+kT_0}^{t} J(U^{(0)}, I^{(0)})(s)ds - \left(\frac{t-T-kT_0}{T_0} - \frac{1}{2}\right)\int\limits_{T+kT_0}^{T+(k+1)T_0} J(U^{(0)}, I^{(0)})(s)ds -$$

$$-\frac{1}{T_0}\int\limits_{T+kT_0}^{T+(k+1)T_0}\int\limits_{T+kT_0}^{t} J(U^{(0)}, I^{(0)})(s)ds\,dt =$$

$$= \int\limits_{T+kT_0}^{t}\left(\frac{dU(s-T)}{dt} - Z_0\frac{I(s)+U(s-T)-2E_1(s-T)+2R_1(\tilde{i}_1(s))}{\tilde{L}_1} - \frac{2Z_0}{\tilde{L}_1}\tilde{C}_1^{-1}\left(\frac{1}{2Z_0}\int\limits_{T+kT_0}^{s} i_1(\theta)d\theta\right)\right)ds -$$

$$-\left(\frac{t-T-kT_0}{T_0} - \frac{1}{2}\right)\int\limits_{T+kT_0}^{T+(k+1)T_0}\left(\frac{dU(s-T)}{dt} - Z_0\frac{U(s-T)+I(s)-2E_1(s)+2R_1(i_1(s))}{\tilde{L}_1} - \frac{2Z_0}{\tilde{L}_1}\tilde{C}_1^{-1}\left(\frac{1}{2Z_0}\int\limits_{T+kT_0}^{s} i_1(\theta)d\theta\right)\right)ds -$$

$$-\frac{1}{T_0}\int\limits_{T+kT_0}^{T+(k+1)T_0}\int\limits_{T+kT_0}^{t}\left(\frac{dU(s-T)}{dt} - Z_0\frac{U(s-T)+I(s)-2E_1(s)+2R_1(i_1(s))}{\tilde{L}_1} - \frac{2Z_0}{\tilde{L}_1}\tilde{C}_1^{-1}\left(\frac{1}{2Z_0}\int\limits_{T+kT_0}^{s} i_1(\theta)d\theta\right)\right)ds\,dt =$$

$$= U(t-T) - Z_0\int\limits_{T+kT_0}^{t}\frac{I(s)}{\tilde{L}_1(i_1(s))}ds + 2\int\limits_{T+kT_0}^{t}\frac{U(s-T)}{\tilde{L}_1(i_1(s))}ds + 2\int\limits_{T+kT_0}^{t}\frac{E_1(s)}{\tilde{L}_1(i_1(s))}ds -$$

$$-2\int\limits_{T+kT_0}^{t}\frac{R_1(i_1(s))}{\tilde{L}_1(i_1(s))}ds + 2Z_0\int\limits_{T+kT_0}^{t}\frac{1}{\tilde{L}_1}\tilde{C}_1^{-1}\left(\frac{1}{2Z_0}\int\limits_{T+kT_0}^{s} i_1(\theta)d\theta\right)ds -$$

$$-\left(\frac{t-T-kT_0}{T_0} - \frac{1}{2}\right)\left[\int\limits_{T+kT_0}^{T+(k+1)T_0}\frac{dU(s-T)}{dt}ds - Z_0\int\limits_{T+kT_0}^{T+(k+1)T_0}\frac{I(s)}{\tilde{L}_1(i_1(s))}ds + 2\int\limits_{T+kT_0}^{T+(k+1)T_0}\frac{U(s-T)}{\tilde{L}_1(i_1(s))}ds + 2\int\limits_{T+kT_0}^{T+(k+1)T_0}\frac{E_1(s)}{\tilde{L}_1(i_1(s))}ds -$$

$$-2\int\limits_{T+kT_0}^{T+(k+1)T_0}\frac{R_1(i_1(s))}{\tilde{L}_1(i_1(s))}ds + 2Z_0\int\limits_{T+kT_0}^{T+(k+1)T_0}\frac{1}{\tilde{L}_1}\tilde{C}_1^{-1}\left(\frac{1}{2Z_0}\int\limits_{T+kT_0}^{s} i_1(\theta)d\theta\right)ds\right] -$$

$$-\frac{1}{T_0}\int\limits_{T+kT_0}^{T+(k+1)T_0}\left[\int\limits_{T+kT_0}^{t}\frac{dU(s-T)}{dt}ds - Z_0\int\limits_{T+kT_0}^{t}\frac{I(s)}{\tilde{L}_1(i_1(s))}ds + 2\int\limits_{T+kT_0}^{t}\frac{U(s-T)}{\tilde{L}_1(i_1(s))}ds + 2\int\limits_{T+kT_0}^{t}\frac{E_1(s)}{\tilde{L}_1(i_1(s))}ds -$$

$$-2\int\limits_{T+kT_0}^{t}\frac{+2R_1(i_1(s))}{\tilde{L}_1(i_1(s))}ds + 2Z_0\int\limits_{T+kT_0}^{t}\frac{1}{\tilde{L}_1}\tilde{C}_1^{-1}\left(\frac{1}{2Z_0}\int\limits_{T+kT_0}^{s} i_1(\theta)d\theta\right)ds\right]dt .$$

Repeating the above reasoning, we obtain $I^{(1)}(t) = U_0\cos\omega_0 t$.

$$\dot{U}^{(1)}(t) = \dot{B}_U^{(k)}(U^{(0)}, I^{(0)})(t) = V(U^{(0)}, I^{(0)})(t) - \frac{1}{T_0}\int\limits_{T+kT_0}^{T+(k+1)T_0} V(U^{(0)}, I^{(0)})(s)ds =$$

$$= \frac{dI(t-T)}{dt} - Z_0\frac{U(t)+I(t-T)-2E_0(t)+2R_0(\tilde{i}_0(t))}{\tilde{L}_0} - \frac{2Z_0}{\tilde{L}_0}\tilde{C}_0^{-1}\left(\frac{1}{2Z_0}\int\limits_{T}^{t} i_0(s)ds\right) -$$

$$-\frac{1}{T_0}\int\limits_{T+kT_0}^{T+(k+1)T_0}\left(\frac{dI(s-T)}{dt} - Z_0\frac{U(s)+I(s-T)-2E_0(s)+2R_0(i_0(s))}{\tilde{L}_0} - \frac{2Z_0}{\tilde{L}_0}\tilde{C}_0^{-1}\left(\frac{1}{2Z_0}\int\limits_{T+kT_0}^{s} i_0(\theta)d\theta\right)\right)ds =$$

$$= \omega_0 I_0\cos\omega_0 t - Z_0\frac{U_0\sin\omega_0 t + I_0\cos\omega_0 t - 2U_0\sin\omega_0 t + 2R_0(i_0(t))}{\tilde{L}_0} - \frac{2Z_0}{\tilde{L}_0}\tilde{C}_0^{-1}\left(\frac{1}{2Z_0}\int\limits_{T}^{t} i_0(s)ds\right) -$$

$$-\frac{1}{T_0}\int_{T+kT_0}^{T+(k+1)T_0}\left(-Z_0\frac{U_0\sin\omega_0 s+I_0\cos\omega_0 s-2U_0\sin\omega_0 s+2R_0\big(i_0(s)\big)}{\widetilde{L}_0}-\frac{2Z_0}{\widetilde{L}_0}\widetilde{C}_0^{-1}\left(\frac{1}{2Z_0}\int_{T+kT_0}^{s}i_0(\theta)d\theta\right)\right)ds\ ;$$

$$\dot{I}^{(1)}(t)=\dot{B}_I^{(k)}(U^{(0)},I^{(0)})(t)=J(U^{(0)},I^{(0)})(t)-\frac{1}{T_0}\int_{T+kT_0}^{T+(k+1)T_0}J(U^{(0)},I^{(0)})(s)ds=$$

$$=\frac{dU(t-T)}{dt}-Z_0\frac{I(t)+U(t-T)-2E_1(t)+2R_1\big(i_1(t)\big)}{\widetilde{L}_1}-\frac{2Z_0}{\widetilde{L}_1}\widetilde{C}_1^{-1}\left(\frac{1}{2Z_0}\int_{T+kT_0}^{t}i_1(s)ds\right)-$$

$$-\frac{1}{T_0}\int_{T+kT_0}^{T+(k+1)T_0}\left(\frac{dU(s-T)}{dt}-Z_0\frac{I(s)+U(s-T)-2E_1(s)+2R_1\big(i_1(s)\big)}{\widetilde{L}_1}-\frac{2Z_0}{\widetilde{L}_1}\widetilde{C}_1^{-1}\left(\frac{1}{2Z_0}\int_{T+kT_0}^{s}i_1(\theta)d\theta\right)\right)ds=$$

$$=\omega_0 U_0\cos\omega_0 t-Z_0\frac{I_0\sin\omega_0 t+U_0\cos\omega_0 t-2U_0\sin\omega_0 t+2R_1\big(i_1(t)\big)}{\widetilde{L}_0}-\frac{2Z_0}{\widetilde{L}_1}\widetilde{C}_1^{-1}\left(\frac{1}{2Z_0}\int_{T}^{t}i_1(s)ds\right)-$$

$$-\frac{1}{T_0}\int_{T+kT_0}^{T+(k+1)T_0}\left(-Z_0\frac{I_0\sin\omega_0 s+U_0\cos\omega_0 s-2U_0\sin\omega_0 s+2R_1\big(i_1(s)\big)}{\widetilde{L}_1}-\frac{2Z_0}{\widetilde{L}_1}\widetilde{C}_1^{-1}\left(\frac{1}{2Z_0}\int_{T+kT_0}^{s}i_1(\theta)d\theta\right)\right)ds.$$

Then

$$\left|U^{(1)}(t)-U^{(0)}(t)\right|\le\left|U_0\sin\omega_0 t-I_0\cos\omega_0 t\right|\le U_0+I_0\ ;\ \left|I^{(1)}(t)-I^{(0)}(t)\right|\le U_0+I_0\ .$$

For the derivatives, we have

$$\left|\dot{U}^{(1)}(t)-\dot{U}^{(0)}(t)\right|\le$$

$$\le\omega_0 U_0+\omega_0 I_0+\frac{Z_0 U_0+I_0+2U_0+0{,}306}{\hat{L}_0}+\frac{2Z_0}{\hat{L}_0}U_0+$$

$$+\frac{1}{T_0}\int_{T+kT_0}^{T+(k+1)T_0}\frac{e^{\mu(t-T-kT_0)}-1}{\mu}dt\left(\frac{Z_0 U_0}{\hat{L}_0}+\frac{2I_0}{\hat{L}_0}+\frac{2U_0}{\hat{L}_0}+\frac{0{,}306}{\hat{L}_0}+\frac{4Z_0}{c_0\hat{L}_0}U_0\right)\le$$

$$\le\omega_0 U_0+\omega_0 I_0+\left(1+\frac{e^{\mu_0}-1}{\mu_0}\right)\frac{1}{\hat{L}_0}\left(Z_0 U_0+2I_0+2U_0+0{,}306+\frac{4Z_0}{c_0}U_0\right)=\dot{P}_U$$

and

$$\left|\dot{I}^{(1)}(t) - \dot{I}^{(0)}(t)\right| \le$$

$$\omega_0 I_0 + \omega_0 U_0 + \left(1 + \frac{e^{\mu_0} - 1}{\mu_0}\right)\frac{1}{\hat{L}_1}\left(Z_0 I_0 + U_0 + 2U_0 + 0,306 + \frac{4Z_0}{c_1}U_0\right) = \dot{P}_I.$$

Finally we obtain

$$P_U \le U_0 + I_0; \ P_I \le U_0 + I_0; \ \dot{P}_U \le 3,22.10^{13}; \ \dot{P}_I \le 3,22.10^{13}.$$

Consequently

$$\rho_\mu^{(k)}((U^{(n+1)}, I^{(n+1)}),(U^{(n)}, I^{(n)})) \le \left(\frac{2,5}{10^{11}}\right)^n \frac{1}{1 - 2,5.10^{-11}} 33.10^{12} \approx \left(\frac{254}{10^{12}}\right)^n 33.10^{12}$$

$$(n = 0,1,\ldots; k = 0,1,2\ldots).$$

3.7. THE EXISTENCE-UNIQUENESS OF A PERIODIC SOLUTION OF A SYSTEM WITH LINEAR AND TIME-VARYING IN SERIES CONNECTED *RLC*-LOADS

The main purpose of the present considerations is to show that our approach is applicable to both nonlinear and linear problems. We formulate the problem for existence-uniqueness of a periodic solution of the system obtained under assumptions $L_k(i) = L_k = \text{const}$, $R_k(i) = R_k i$, $C_k(u) = C_k = \text{const}$ $(k = 0,1)$: to find a solution from $M_U^* \times M_I^*$ of the linear system (we use denotation for the derivatives $\frac{dU(t)}{dt} = \dot{U}(t)$) obtained from (3.5.1):

$$\dot{U}(t) = \dot{I}(t-T) - \frac{Z_0 + R_0}{L_0}U(t) - \frac{Z_0 - R_0}{L_0}I(t-T) -$$

$$-\frac{1}{L_0 C_0}\int_T^t U(s)ds + \frac{1}{L_0 C_0}\int_T^t I(s-T)ds + \frac{2Z_0}{L_0}E_0(t) \equiv V(U,I)(t), \ t \in [T, T + T_0]$$

$$\dot{I}(t) = \dot{U}(t-T) - \frac{Z_0 - R_1}{L_1}U(t-T) - \frac{Z_0 + R_1}{L_1}I(t) + \qquad (3.7.1)$$

$$+\frac{1}{L_1 C_1}\int_T^t U(s-T)ds - \frac{1}{L_1 C_1}\int_T^t I(s)ds + \frac{2Z_0}{L_1}E_1(t) \equiv J(U,I)(t), \ t \in [T, T + T_0],$$

$$U(t) = U_0(t), \ \dot{U}(t) = \dot{U}_0(t), \ t \in [0,T], \ I(t) = I_0(t), \ \dot{I}(t) = \dot{I}_0(t), \ t \in [0,T].$$

Here the initial functions are obtained from $U_0(x), I_0(x)$ after translation along the characteristics.

Define the operator $B = (B_U, B_I)$ on $M_U^* \times M_I^*$ in the same way as in § 3.5.

Lemma 3.2.4 is also valid, that is, its fixed point is a solution of (3.7.1).

Theorem 3.7.1. Let be the following assumptions be valid:

$$U_0(-T) = 0, \ I_0(-T) = 0 \ ,$$

$$E_p(.) \in C_{T_0}^1[T, \infty), \left|E_p(t)\right| \le U_{E_p} \, e^{\mu(t-T-kT_0)} \le U_0 e^{\mu(t-T-kT_0)}, (k = 0,1,\ldots) \ ;$$

$$\left|U_0(t)\right| \le e^{-\beta} U_0 e^{\mu(t-kT_0)}, \left|I_0(t)\right| \le e^{-\beta} I_0 e^{\mu(t-kT_0)} \ ;$$

$$t \in [kT_0, (k+1)T_0], \ (k = 0,1,2,\ldots, m-1) \ .$$

Then there exists a unique T_o-periodic solution of (3.7.1).

Proof: We have to show that $B = (B_U, B_I)$ maps $M_U^* \times M_I^*$ into itself. We get

$$\left|B_U^{(k)}(U,I)(t)\right| \le \left|\int_{T+kT_0}^{t} V(U,I)(s)ds\right| + \left|\left(\frac{t-T-kT_0}{T_0} - \frac{1}{2}\right)\int_{T+kT_0}^{T+(k+1)T_0} V(U,I)(s)ds\right| + \left|\frac{1}{T_0}\int_{T+kT_0}^{T+(k+1)T_0}\int_{T+kT_0}^{t} V(U,I)(s)dsdt\right| \equiv$$

$$\equiv V_1 + V_2 + V_3 \ .$$

But

$$V_1 \le \left|\int_{T+kT_0}^{t} \dot{I}(s-T)ds\right| + \frac{Z_0 + R_0}{L_0} \int_{T+kT_0}^{t} |U(s)|ds + \frac{Z_0 + R_0}{L_0} \int_{T+kT_0}^{t} |I(s-T)|ds +$$

$$+ \frac{1}{L_0 C_0} \int_{T+kT_0}^{t}\int_{T+kT_0}^{s} |U(\tau)|d\tau ds + \frac{1}{L_0 C_0} \int_{T+kT_0}^{t}\int_{T+kT_0}^{s} |I(\tau-T)|d\tau ds + \frac{2Z_0}{L_0} \int_{T+kT_0}^{t} |E_0(s-T)|ds \le$$

$$\le \left|I(t-T) - I(0)\right| + \frac{Z_0 + R_0}{L_0} U_0 \frac{e^{\mu(t-T-kT_0)} - 1}{\mu} + \frac{Z_0 + R_0}{L_0} I_0 e^{-\beta} \frac{e^{\mu(t-T-kT_0)} - 1}{\mu} +$$

$$+ \frac{U_0}{L_0 C_0} \int_{T+kT_0}^{t} \frac{e^{\mu(s-T-kT_0)} - 1}{\mu}ds + \frac{I_0 e^{-\beta}}{L_0 C_0} \int_{T+kT_0}^{t} \frac{e^{\mu(s-T-kT_0)} - 1}{\mu}ds + \frac{2Z_0 U_0 e^{-\beta}}{L_0} \frac{e^{\mu(t-T-kT_0)} - 1}{\mu} \le$$

$$\le e^{\mu(t-T-kT_0)}\left[I_0 e^{-\beta} + \frac{(Z_0 + R_0)(U_0 + I_0 e^{-\beta})}{\mu L_0} + \frac{U_0 + I_0 e^{-\beta}}{\mu^2 L_0 C_0} + \frac{2Z_0 U_{E0}}{\mu L_0}\right] \ ;$$

$$V_2 \le \frac{1}{2}\left|\int_{T+kT_0}^{T+(k+1)T_0} V(U,I)(s)ds\right| \le$$

$$\leq \frac{1}{2}\left[\left|\int\limits_{T+kT_0}^{T+(k+1)T_0} \dot{I}(s-T)ds\right| + \frac{Z_0+R_0}{L_0}\int\limits_{T+kT_0}^{T+(k+1)T_0}\left|U(s)\right|ds + \frac{Z_0+R_0}{L_0}\int\limits_{T+kT_0}^{T+(k+1)T_0}\left|I(s-T)\right|ds + \right.$$

$$\left. +\frac{1}{L_0C_0}\int\limits_{T+kT_0}^{T+(k+1)T_0}\int\limits_{T+kT_0}^{s}\left|U(\tau)\right|d\tau ds + \frac{1}{L_0C_0}\int\limits_{T+kT_0}^{T+(k+1)T_0}\int\limits_{T+kT_0}^{s}\left|I(\tau-T)\right|d\tau ds + \frac{2Z_0}{L_0}\int\limits_{T+kT_0}^{T+(k+1)T_0}\left|E_0(s-T)\right|ds\right] \leq$$

$$\leq \frac{1}{2}\left[\frac{(Z_0+R_0)U_0}{L_0}\frac{e^{\mu T_0}-1}{\mu} + \frac{(Z_0+R_0)I_0e^{-\beta}}{L_0}\frac{e^{\mu T_0}-1}{\mu} + \right.$$

$$\left. +\frac{U_0}{L_0C_0}\int\limits_{T+kT_0}^{T+(k+1)T_0}\frac{e^{\mu(s-T-kT_0)}-1}{\mu}ds + \frac{I_0e^{-\beta}}{L_0C_0}\int\limits_{T+kT_0}^{T+(k+1)T_0}\frac{e^{\mu(s-T-kT_0)}-1}{\mu}ds + \frac{2Z_0U_{E0}}{L_0}\frac{e^{\mu T_0}-1}{\mu}\right] \leq$$

$$\leq \frac{1}{2}\left[\frac{(Z_0+R_0)U_0}{L_0}\frac{e^{\mu T_0}-1}{\mu} + \frac{(Z_0+R_0)I_0e^{-\beta}}{L_0}\frac{e^{\mu T_0}-1}{\mu} + \right.$$

$$\left. +\frac{U_0}{\mu L_0C_0}\frac{e^{\mu T_0}-1}{\mu} + \frac{I_0e^{-\beta}}{\mu L_0C_0}\frac{e^{\mu T_0}-1}{\mu} + \frac{2Z_0U_{E0}}{L_0}\frac{e^{\mu T_0}-1}{\mu}\right] \leq$$

$$\leq \frac{e^{\mu T_0}-1}{2}\left[\frac{(Z_0+R_0)(U_0+I_0e^{-\beta})}{\mu L_0} + \frac{U_0+I_0e^{-\beta}}{\mu^2 L_0C_0} + \frac{2Z_0U_{E0}}{\mu L_0}\right]$$

and

$$V_3 \leq \frac{1}{T_0}\int\limits_{T+kT_0}^{T+(k+1)T_0}\left|\int\limits_{T+kT_0}^{t}V(U,I)(s)ds\right|dt \leq$$

$$\leq e^{\mu(t-T-kT_0)}\frac{e^{\mu_0}-1}{\mu_0}\left[I_0e^{-\beta} + \frac{(Z_0+R_0)(U_0+I_0e^{-\beta})}{\mu L_0} + \frac{U_0+I_0e^{-\beta}}{\mu^2 L_0C_0} + \frac{2Z_0U_{E0}}{\mu L_0}\right].$$

Consequently, the following inequality is satisfied for sufficiently large μ:

$$\left|B_U^{(k)}(U,I)(t)\right| \leq +$$

$$\leq e^{\mu(t-T-kT_0)}\left[I_0e^{-\beta} + \frac{(Z_0+R_0)(U_0+I_0e^{-\beta})}{\mu L_0} + \frac{U_0+I_0e^{-\beta}}{\mu^2 L_0C_0} + \frac{2Z_0U_{E0}}{\mu L_0}\right] +$$

$$+ e^{\mu(t-T-kT_0)}\frac{e^{\mu T_0}-1}{2}\left[\frac{(Z_0+R_0)(U_0+I_0e^{-\beta})}{\mu L_0} + \frac{U_0+I_0e^{-\beta}}{\mu^2 L_0C_0} + \frac{2Z_0U_{E0}}{\mu L_0}\right] +$$

$$+ e^{\mu(t-T-kT_0)} \frac{e^{\mu_0}-1}{\mu_0} \left[I_0 e^{-\beta} + \frac{(Z_0+R_0)(U_0+I_0 e^{-\beta})}{\mu L_0} + \frac{U_0+I_0 e^{-\beta}}{\mu^2 L_0 C_0} + \frac{2Z_0 U_{E0}}{\mu L_0} \right] \le e^{\mu(t-T-kT_0)} \times$$

$$\times \left[\left(1+\frac{e^{\mu_0}-1}{\mu_0}\right) I_0 e^{-\beta} + \left(1+\frac{e^{\mu_0}-1}{2}+\frac{e^{\mu_0}-1}{\mu_0}\right) \left(\frac{(Z_0+R_0)(U_0+I_0 e^{-\beta})}{\mu L_0} + \frac{U_0+I_0 e^{-\beta}}{\mu^2 L_0 C_0} + \frac{2Z_0 U_{E0}}{\mu L_0}\right) \right] \le$$

$$\le U_0 e^{\mu(t-T-kT_0)} .$$

Analogously

$$\left| B_I^{(k)}(U,I)(t) \right| \le \left| \int_{T+kT_0}^{t} J(U,I)(s)ds \right| + \left| \left(\frac{t-T-kT_0}{T_0}-\frac{1}{2}\right) \int_{T+kT_0}^{T+(k+1)T_0} J(U,I)(s)ds \right| + \left| \frac{1}{T_0} \int_{T+kT_0}^{T+(k+1)T_0} \int_{T+kT_0}^{t} J(U,I)(s)dsdt \right| \equiv$$

$$\equiv J_1 + J_2 + J_3 .$$

But

$$J_1 \le \left| \int_{T+kT_0}^{t} \dot{U}(s-T)ds \right| + \frac{Z_0+R_1}{L_1} \int_{T+kT_0}^{t} |U(s-T)|ds + \frac{Z_0+R_1}{L_1} \int_{T+kT_0}^{t} |I(s)|ds +$$

$$+ \frac{1}{L_1 C_1} \int_{T+kT_0}^{t} \int_{T+kT_0}^{s} |U(\tau-T)|d\tau ds + \frac{1}{L_1 C_1} \int_{T+kT_0}^{t} \int_{T+kT_0}^{s} |I(\tau)|d\tau ds + \frac{2Z_0}{L_1} \int_{T+kT_0}^{t} |E_1(s-T)|ds \le$$

$$\le |U(t-T)-U(0)| + \frac{(Z_0+R_1)U_0 e^{-\beta}}{L_1} \frac{e^{\mu(t-T-kT_0)}-1}{\mu} + \frac{(Z_0+R_1)I_0}{L_1} \frac{e^{\mu(t-T-kT_0)}-1}{\mu} +$$

$$+ \frac{U_0 e^{-\beta}}{L_1 C_1} \int_{T+kT_0}^{t} \frac{e^{\mu(s-T-kT_0)}-1}{\mu} ds + \frac{I_0}{L_1 C_1} \int_{T+kT_0}^{t} \frac{e^{\mu(s-T-kT_0)}-1}{\mu} ds + \frac{2Z_0 U_{E_1}}{L_1} \frac{e^{\mu(t-T-kT_0)}-1}{\mu} \le$$

$$\le e^{\mu(t-T-kT_0)} U_0 e^{-\beta} + \frac{(Z_0+R_1)U_0 e^{-\beta}}{L_1} \frac{e^{\mu(t-T-kT_0)}}{\mu} + \frac{(Z_0+R_1)I_0}{L_1} \frac{e^{\mu(t-T-kT_0)}}{\mu} +$$

$$+ \frac{U_0 e^{-\beta}}{\mu L_1 C_1} \frac{e^{\mu(t-T-kT_0)}-1}{\mu} + \frac{I_0}{\mu L_1 C_1} \frac{e^{\mu(t-T-kT_0)}-1}{\mu} + \frac{2Z_0 U_{E_1}}{L_1} \frac{e^{\mu(t-T-kT_0)}-1}{\mu} \le$$

$$\le e^{\mu(t-T-kT_0)} \left[U_0 e^{-\beta} + \frac{(Z_0+R_1)(U_0 e^{-\beta}+I_0)}{\mu L_1} + \frac{U_0 e^{-\beta}+I_0}{\mu^2 L_1 C_1} + \frac{2Z_0 U_{E_1}}{\mu L_1} \right] ;$$

$$J_2 \le \frac{1}{2} \left| \int_{T+kT_0}^{T+(k+1)T_0} J(U,I)(s)ds \right| \le$$

$$\le \frac{1}{2} \left[\left| \int_{T+kT_0}^{T+(k+1)T_0} \dot{U}(s-T)ds \right| + \frac{Z_0+R_1}{L_1} \int_{T+kT_0}^{T+(k+1)T_0} |U(s-T)|ds + \frac{Z_0+R_1}{L_1} \int_{T+kT_0}^{T+(k+1)T_0} |I(s)|ds + \right.$$

$$+\frac{1}{L_1 C_1}\int_{T+kT_0}^{T+(k+1)T_0}\int_{T+kT_0}^{s}|U(\tau-T)|d\tau ds+\frac{1}{L_1 C_1}\int_{T+kT_0}^{T+(k+1)T_0}\int_{T+kT_0}^{s}|I(\tau)|d\tau ds+\frac{2Z_0}{L_1}\int_{T+kT_0}^{T+(k+1)T_0}|E_1(s-T)|ds\Bigg]\le$$

$$\le\frac{1}{2}\Bigg[|U(T+T_0-T)-U(0)|+\frac{(Z_0+R_1)U_0 e^{-\beta}}{L_1}\frac{e^{\mu T_0}-1}{\mu}+\frac{(Z_0+R_1)I_0}{L_1}\frac{e^{\mu T_0}-1}{\mu}+$$

$$+\frac{U_0 e^{-\beta}}{L_1 C_1}\int_{T+kT_0}^{T+(k+1)T_0}\frac{e^{\mu(s-T-kT_0)}-1}{\mu}ds+\frac{I_0}{L_1 C_1}\int_{T+kT_0}^{T+(k+1)T_0}\frac{e^{\mu(s-T-kT_0)}-1}{\mu}ds+\frac{2Z_0 U_{E1}}{L_1}\frac{e^{\mu T_0}-1}{\mu}\Bigg]\le$$

$$\le e^{\mu(t-T-kT_0)}\frac{e^{\mu T_0}-1}{2}\Bigg[\frac{(Z_0+R_1)U_0 e^{-\beta}}{\mu L_1}+\frac{(Z_0+R_1)I_0}{\mu L_1}+\frac{U_0 e^{-\beta}+I_0}{\mu^2 L_1 C_1}+\frac{2Z_0 U_{E1}}{\mu L_1}\Bigg]$$

and

$$J_3\le\frac{1}{T_0}\int_{T+kT_0}^{T+(k+1)T_0}\left|\int_{T+kT_0}^{t}J(U,I)(s)ds\right|dt\le$$

$$\le\Bigg[U_0 e^{-\beta}+\frac{(Z_0+R_1)(U_0 e^{-\beta}+I_0)}{\mu L_1}+\frac{(U_0 e^{-\beta}+I_0)}{\mu^2 L_1 C_1}+\frac{2Z_0 U_{E1}}{\mu L_1}\Bigg]\frac{1}{T_0}\int_{T+kT_0}^{T+(k+1)T_0}e^{\mu(t-T-kT_0)}dt\le$$

$$\le\frac{e^{\mu_0}-1}{\mu_0}\Bigg[U_0 e^{-\beta}+\frac{(Z_0+R_1)(U_0 e^{-\beta}+I_0)}{\mu L_1}+\frac{(U_0 e^{-\beta}+I_0)}{\mu^2 L_1 C_1}+\frac{2Z_0 U_{E1}}{\mu L_1}\Bigg].$$

Therefore for sufficiently large μ:

$$\left|B_I^{(k)}(U,I)(t)\right|\le$$

$$\le e^{\mu(t-T-kT_0)}\Bigg[U_0 e^{-\beta}+\frac{(Z_0+R_1)\big(U_0 e^{-\beta}+I_0\big)}{\mu L_1}+\frac{U_0 e^{-\beta}+I_0}{\mu^2 L_1 C_1}+\frac{2Z_0 U_{E1}}{\mu L_1}\Bigg]+$$

$$+e^{\mu(t-T-kT_0)}\frac{e^{\mu_0}-1}{2}\Bigg[\frac{(Z_0+R_1)\big(U_0 e^{-\beta}+I_0\big)}{\mu L_1}+\frac{U_0 e^{-\beta}+I_0}{\mu^2 L_1 C_1}+\frac{2Z_0 U_{E1}}{\mu L_1}\Bigg]+$$

$$+e^{\mu(t-T-kT_0)}\frac{e^{\mu_0}-1}{\mu_0}\Bigg[U_0 e^{-\beta}+\frac{(Z_0+R_1)(U_0 e^{-\beta}+I_0)}{\mu L_1}+\frac{(U_0 e^{-\beta}+I_0)}{\mu^2 L_1 C_1}+\frac{2Z_0 U_{E1}}{\mu L_1}\Bigg]\le$$

$$\le e^{\mu(t-T-kT_0)}\Bigg[\bigg(1+\frac{e^{\mu_0}-1}{\mu_0}\bigg)U_0 e^{-\beta}+$$

$$+\left(1+\frac{e^{\mu_0}-1}{2}+\frac{e^{\mu_0}-1}{\mu_0}\right)\left(\frac{(Z_0+R_1)(U_0e^{-\beta}+I_0)}{\mu L_1}+\frac{(U_0e^{-\beta}+I_0)}{\mu^2 L_1 C_1}+\frac{2Z_0 U_{E_1}}{\mu L_1}\right)\right]\le I_0 e^{\mu(t-T-kT_0)}.$$

So we have ascertained that $B=(B_U,B_I)$ maps $M_U^* \times M_I^*$ into itself.

We prove that $B=(B_U,B_I)$ is a contractive operator. Indeed

$$\left|B_U^{(k)}(U,I)(t)-B_U^{(k)}(\overline{U},\overline{I})(t)\right|\le\left|\int_{T+kT_0}^{t}\left(V(U,I)(s)-V(\overline{U},\overline{I})(s)\right)ds\right|+$$

$$+\left|\left(\frac{t-T-kT_0}{T_0}-\frac{1}{2}\right)\left[\int_{T+kT_0}^{T+(k+1)T_0}\left(V(U,I)(s)-V(\overline{U},\overline{I})(s)\right)ds\right]\right|+$$

$$+\frac{1}{T_0}\int_{T+kT_0}^{T+(k+1)T_0}\left|\int_{T+kT_0}^{t}\left(V(U,I)(s)-V(\overline{U},\overline{I})(s)\right)ds\right|dt\equiv G_1+G_2+G_3.$$

But

$$G_1\le\left|\int_{T+kT_0}^{t}\left(\dot{I}(s-T)-\dot{\overline{I}}(s-T)\right)ds\right|+\frac{Z_0+R_0}{L_0}\int_{T+kT_0}^{t}\left|U(s)-\overline{U}(s)\right|ds+\frac{Z_0+R_0}{L_0}\int_{T+kT_0}^{t}\left|I(s-T)-\overline{I}(s-T)\right|ds+$$

$$+\frac{C_0}{L_0}\int_{T+kT_0}^{t}\int_{T+kT_0}^{s}\left|U(\tau)-\overline{U}(\tau)\right|d\tau ds+\frac{C_0}{L_0}\int_{T+kT_0}^{t}\int_{T+kT_0}^{s}\left|I(\tau-T)-\overline{I}(\tau-T)\right|d\tau ds\le$$

$$\le\frac{Z_0+R_0}{L_0}\rho^{(k)}(U,\overline{U})\frac{e^{\mu(t-T-kT_0)}-1}{\mu}+\frac{\rho^{(k)}(U,\overline{U})}{L_0 C_0}\int_{T+kT_0}^{t}\frac{e^{\mu(s-T-kT_0)}-1}{\mu}ds\le$$

$$\le\frac{(Z_0+R_0)e^{\mu_0}\rho_\mu^{(k)}(U,\overline{U})}{L_0}\frac{e^{\mu(t-T-kT_0)}-1}{\mu}+\frac{e^{\mu_0}\rho_\mu^{(k)}(U,\overline{U})}{\mu L_0 C_0}\frac{e^{\mu(t-T-kT_0)}-1}{\mu}\le$$

$$\le\frac{(Z_0+R_0)e^{\mu_0}\rho_\mu^{(k)}(\dot{U},\dot{\overline{U}})}{\mu L_0}\frac{e^{\mu(t-T-kT_0)}-1}{\mu}+\frac{e^{\mu_0}\rho_\mu^{(k)}(\dot{U},\dot{\overline{U}})}{\mu^2 L_0 C_0}\frac{e^{\mu(t-T-kT_0)}-1}{\mu}\le$$

$$\le e^{\mu(t-T-kT_0)}\hat{\rho}_\mu((U,I),(\dot{U},\dot{I}),(\overline{U},\overline{I}),(\dot{\overline{U}},\dot{\overline{I}}))e^{\mu_0}\left(\frac{Z_0+R_0}{\mu^2 L_0}+\frac{1}{\mu^3 L_0 C_0}\right);$$

$$G_2\le\frac{1}{2}\left|\int_{T+kT_0}^{T+(k+1)T_0}V(U,I)(s)ds-\int_{T+kT_0}^{T+(k+1)T_0}V(\overline{U},\overline{I})(s)ds\right|\le$$

$$\leq \frac{1}{2}\left[\left|\int_{T+kT_0}^{T+(k+1)T_0}\left(\dot{I}(s-T)-\dot{\bar{I}}(s-T)\right)ds\right|+\frac{Z_0+R_0}{L_0}\int_{T+kT_0}^{T+(k+1)T_0}\left|U(s)-\overline{U}(s)\right|ds+\right.$$

$$+\frac{Z_0+R_0}{L_0}\int_{T+kT_0}^{T+(k+1)T_0}\left|I(s-T)-\bar{I}(s-T)\right|ds+$$

$$\left.+\frac{1}{L_0C_0}\int_{T+kT_0}^{T+(k+1)T_0}\int_{T+kT_0}^{s}\left|U(\tau)-\overline{U}(\tau)\right|d\tau ds+\frac{1}{L_0C_0}\int_{T+kT_0}^{T+(k+1)T_0}\int_{T+kT_0}^{s}\left|I(\tau-T)-\bar{I}(\tau-T)\right|d\tau ds\right]\leq$$

$$\leq\frac{1}{2}\left[\frac{Z_0+R_0}{L_0}\rho^{(k)}(U,\overline{U})\frac{e^{\mu T_0}-1}{\mu}+\frac{\rho^{(k)}(U,\overline{U})}{L_0C_0}\int_{T+kT_0}^{T+(k+1)T_0}\frac{e^{\mu(s-T-kT_0)}-1}{\mu}ds\right]\leq$$

$$\leq\frac{1}{2}\left[\frac{Z_0+R_0}{L_0}e^{\mu_0}\rho_\mu^{(k)}(U,\overline{U})\frac{e^{\mu T_0}-1}{\mu}+\frac{e^{\mu_0}\rho_\mu^{(k)}(U,\overline{U})}{L_0C_0}\int_{T+kT_0}^{T+(k+1)T_0}\frac{e^{\mu(s-T-kT_0)}-1}{\mu}ds\right]\leq$$

$$\leq\frac{1}{2}\left[\frac{Z_0+R_0}{L_0}e^{\mu_0}\rho_\mu^{(k)}(\dot{U},\dot{\overline{U}})\frac{e^{\mu T_0}-1}{\mu^2}+\frac{e^{\mu_0}\rho_\mu^{(k)}(\dot{U},\dot{\overline{U}})}{\mu^2L_0C_0}\frac{e^{\mu T_0}-1}{\mu}\right]\leq$$

$$\leq\hat{\rho}_\mu((U,I),(\dot{U},\dot{I}),(\overline{U},\bar{I}),(\dot{\overline{U}},\dot{\bar{I}}))\frac{e^{\mu_0}-1}{2}e^{\mu_0}\left(\frac{Z_0+R_0}{\mu^2L_0}+\frac{1}{\mu^3L_0C_0}\right)$$

and

$$G_3\leq\frac{1}{T_0}\int_{T+kT_0}^{T+(k+1)T_0}\left|\int_{T+kT_0}^{t}\left(V(U,I)(s)-V(\overline{U},\bar{I})(s)\right)ds\right|dt\leq$$

$$\leq e^{\mu(t-T-kT_0)}\rho_\mu^{(k)}((U,I),(\overline{U},\bar{I}))e^{\mu_0}\left(\frac{Z_0+R_0}{\mu^2L_0}+\frac{1}{\mu^3L_0C_0}\right)\frac{1}{T_0}\int_{T+kT_0}^{T+(k+1)T_0}e^{\mu(t-T-kT_0)}dt\leq$$

$$\leq e^{\mu(t-T-kT_0)}\hat{\rho}_\mu((U,I),(\dot{U},\dot{I}),(\overline{U},\bar{I}),(\dot{\overline{U}},\dot{\bar{I}}))\frac{e^{\mu_0}-1}{\mu_0}e^{\mu_0}\left(\frac{Z_0+R_0}{\mu^2L_0}+\frac{1}{\mu^3L_0C_0}\right)$$

Then

$$\left|B_U^{(k)}(U,I)(t)-B_U^{(k)}(\overline{U},\bar{I})(t)\right|\leq$$

$$\leq e^{\mu(t-T-kT_0)}\rho_\mu^{(k)}((U,I),(\overline{U},\overline{I}))e^{\mu_0}\left(\frac{Z_0+R_0}{\mu^2 L_0}+\frac{1}{\mu^3 L_0 C_0}\right)+$$

$$+e^{\mu(t-T-kT_0)}\rho_\mu^{(k)}((U,I),(\overline{U},\overline{I}))\frac{\left(e^{\mu_0}-1\right)}{2}e^{\mu_0}\left(\frac{Z_0+R_0}{\mu^2 L_0}+\frac{1}{\mu^3 L_0 C_0}\right)+$$

$$+e^{\mu(t-T-kT_0)}\rho_\mu^{(k)}((U,I),(\overline{U},\overline{I}))\frac{e^{\mu_0}-1}{\mu_0}e^{\mu_0}\left(\frac{Z_0+R_0}{\mu^2 L_0}+\frac{1}{\mu^3 L_0 C_0}\right)\leq$$

$$\leq e^{\mu(t-T-kT_0)}\rho_\mu^{(k)}((U,I),(\overline{U},\overline{I}))\left(1+\frac{e^{\mu_0}-1}{2}+\frac{e^{\mu_0}-1}{\mu_0}\right)e^{\mu_0}\left(\frac{Z_0+R_0}{\mu^2 L_0}+\frac{1}{\mu^3 L_0 C_0}\right)\equiv$$

$$\equiv e^{\mu(t-T-kT_0)}K_U\,\hat{\rho}_\mu((U,I),(\dot{U},\dot{I}),(\overline{U},\overline{I}),(\dot{\overline{U}},\dot{\overline{I}}))\leq e^{\mu T_0}K_U\,\hat{\rho}_\mu((U,I),(\dot{U},\dot{I}),(\overline{U},\overline{I}),(\dot{\overline{U}},\dot{\overline{I}})).$$

It follows

$$\hat{\rho}(B_U(U,I),B_U(\overline{U},\overline{I}))\leq e^{\mu T_0}K_U\,\hat{\rho}_\mu((U,I),(\dot{U},\dot{I}),(\overline{U},\overline{I}),(\dot{\overline{U}},\dot{\overline{I}})).$$

Further on we have

$$\left|B_I^{(k)}(U,I)(t)-B_I^{(k)}(\overline{U},\overline{I})(t)\right|\leq\left|\int_{T+kT_0}^{t}J(U,I)(s)ds-\int_{T+kT_0}^{t}J(\overline{U},\overline{I})(s)ds\right|+$$

$$+\left|\left(\frac{t-T-kT_0}{T_0}-\frac{1}{2}\right)\left[\int_{T+kT_0}^{T+(k+1)T_0}J(U,I)(s)ds-\int_{T+kT_0}^{T+(k+1)T_0}J(\overline{U},\overline{I})(s)ds\right]\right|+$$

$$+\frac{1}{T_0}\int_{T+kT_0}^{T+(k+1)T_0}\left|\int_{T+kT_0}^{t}J(U,I)(s)dsdt-\int_{T+kT_0}^{t}J(\overline{U},\overline{I})(s)dsdt\right|\equiv$$

$$\equiv H_1+H_2+H_3.$$

Now

$$H_1\leq\left|\int_{T+kT_0}^{t}\dot{U}(s-T)ds-\int_{T+kT_0}^{t}\dot{\overline{U}}(s-T)ds\right|+\frac{Z_0+R_1}{L_0}\int_{T+kT_0}^{t}\left|U(s-T)-\overline{U}(s-T)\right|ds+$$

$$+\frac{1}{L_1 C_1}\int_{T+kT_0}^{t}\int_{T+kT_0}^{s}\left|U(\tau-T)-\overline{U}(\tau-T)\right|d\tau ds+\frac{1}{L_1 C_1}\int_{T+kT_0}^{t}\int_{T+kT_0}^{s}\left|I(\tau)-\overline{I}(\tau)\right|d\tau ds\leq$$

$$\leq \frac{Z_0 + R_1}{L_1} \rho^{(k)}(I,\bar{I}) \int_{T+kT_0}^{t} e^{\mu(s-T-kT_0)} ds + \frac{\rho^{(k)}(I,\bar{I})}{L_1 C_1} \int_{T+kT_0}^{t} \frac{e^{\mu(s-T-kT_0)} - 1}{\mu} ds \leq$$

$$\leq \frac{Z_0 + R_1}{L_1} e^{\mu_0} \rho_\mu^{(k)}(I,\bar{I}) \int_{T+kT_0}^{t} e^{\mu(s-T-kT_0)} ds + \frac{e^{\mu_0} \rho_\mu^{(k)}(I,\bar{I})}{L_1 C_1} \int_{T+kT_0}^{t} \frac{e^{\mu(s-T-kT_0)} - 1}{\mu} ds \leq$$

$$\leq \frac{(Z_0 + R_1) e^{\mu_0} \rho_\mu^{(k)}(\dot{I},\dot{\bar{I}})}{\mu L_1} \frac{e^{\mu(t-T-kT_0)} - 1}{\mu} + \frac{e^{\mu_0} \rho_\mu^{(k)}(\dot{I},\dot{\bar{I}})}{\mu L_1 C_1} \frac{e^{\mu(t-T-kT_0)} - 1}{\mu} \leq$$

$$\leq e^{\mu(t-T-kT_0)} \hat{\rho}_\mu((U,I),(\dot{U},\dot{I}),(\overline{U},\bar{I}),(\dot{\overline{U}},\dot{\bar{I}})) e^{\mu_0} \left(\frac{Z_0 + R_1}{\mu^2 L_1} + \frac{1}{\mu^2 L_1 C_1} \right);$$

$$\text{H}_2 \leq \frac{1}{2} \left| \int_{T+kT_0}^{T+(k+1)T_0} J(U,I)(s) ds - \int_{T+kT_0}^{T+(k+1)T_0} J(\overline{U},\bar{I})(s) ds \right| \leq$$

$$\leq \frac{1}{2} \left[\left| \int_{T+kT_0}^{T+(k+1)T_0} \dot{U}(s-T) ds - \int_{T+kT_0}^{T+(k+1)T_0} \dot{\overline{U}}(s-T) ds \right| + \frac{Z_0 + R_1}{L_1} \int_{T+kT_0}^{T+(k+1)T_0} \left| U(s-T) - \overline{U}(s-T) \right| ds + \right.$$

$$+ \frac{Z_0 + R_1}{L_1} \int_{T+kT_0}^{T+(k+1)T_0} \left| I(s) - \bar{I}(s) \right| ds +$$

$$+ \frac{1}{L_1 C_1} \int_{T+kT_0}^{T+(k+1)T_0} \int_{T+kT_0}^{s} \left| U(\tau-T) - \overline{U}(\tau-T) \right| d\tau ds + \frac{1}{L_1 C_1} \int_{T+kT_0}^{T+(k+1)T_0} \int_{T+kT_0}^{s} \left| I(\tau) - \bar{I}(\tau) \right| d\tau ds \left. \right] \leq$$

$$\leq \frac{1}{2} \left[\frac{Z_0 + R_1}{L_1} \rho^{(k)}(I,\bar{I}) \frac{e^{\mu T_0} - 1}{\mu} + \frac{\rho^{(k)}(I,\bar{I})}{L_1 C_1} \int_{T+kT_0}^{T+(k+1)T_0} \frac{e^{\mu(s-T-kT_0)} - 1}{\mu} ds \right] \leq$$

$$\leq \frac{1}{2} \left[\frac{Z_0 + R_1}{L_1} e^{\mu_0} \rho_\mu^{(k)}(I,\bar{I}) \frac{e^{\mu T_0} - 1}{\mu} + \frac{e^{\mu_0} \rho_\mu^{(k)}(I,\bar{I})}{L_1 C_1} \int_{T+kT_0}^{T+(k+1)T_0} \frac{e^{\mu(s-T-kT_0)} - 1}{\mu} ds \right] \leq$$

$$\leq \frac{1}{2} \left[\frac{Z_0 + R_1}{L_1} \frac{e^{\mu_0} \rho_\mu^{(k)}(\dot{I},\dot{\bar{I}})}{\mu} \frac{e^{\mu T_0} - 1}{\mu} + \frac{e^{\mu_0} \rho_\mu^{(k)}(\dot{I},\dot{\bar{I}})}{\mu L_1 C_1} \frac{e^{\mu T_0} - 1}{\mu} \right] \leq$$

$$\leq \hat{\rho}_\mu((U,I),(\dot{U},\dot{I}),(\overline{U},\bar{I}),(\dot{\overline{U}},\dot{\bar{I}})) \frac{e^{\mu_0} - 1}{2\mu} e^{\mu_0} \left(\frac{Z_0 + R_1}{\mu L_1} + \frac{1}{\mu L_1 C_1} \right) \leq$$

$$\leq e^{\mu(t-T-kT_0)}\hat{\rho}_\mu((U,I),(\dot{U},\dot{I}),(\overline{U},\overline{I}),(\dot{\overline{U}},\dot{\overline{I}}))\frac{e^{\mu_0}-1}{2}e^{\mu_0}\left(\frac{Z_0+R_1}{\mu^2 L_1}+\frac{1}{\mu^2 L_1 C_1}\right)$$

and

$$H_3\leq\frac{1}{T_0}\int_{T+kT_0}^{T+(k+1)T_0}\left|\int_{T+kT_0}^{t}\left(J(U,I)(s)-J(\overline{U},\overline{I})(s)\right)ds\right|dt\leq$$

$$\leq e^{\mu(t-T-kT_0)}\rho_\mu^{(k)}((U,I),(\overline{U},\overline{I}))e^{\mu_0}\left(\frac{Z_0+R_1}{\mu^2 L_1}+\frac{1}{\mu^2 L_1 C_1}\right)\frac{1}{T_0}\int_{T+kT_0}^{T+(k+1)T_0}e^{\mu(t-T-kT_0)}dt\leq$$

$$\leq e^{\mu(t-T-kT_0)}\hat{\rho}_\mu((U,I),(\dot{U},\dot{I}),(\overline{U},\overline{I}),(\dot{\overline{U}},\dot{\overline{I}}))\frac{e^{\mu_0}-1}{\mu_0}e^{\mu_0}\left(\frac{Z_0+R_1}{\mu^2 L_1}+\frac{1}{\mu^2 L_1 C_1}\right).$$

Then

$$\left|B_I^{(k)}(U,I)(t)-B_I^{(k)}(\overline{U},\overline{I})(t)\right|\leq$$

$$\leq e^{\mu(t-T-kT_0)}\hat{\rho}_\mu((U,I),(\dot{U},\dot{I}),(\overline{U},\overline{I}),(\dot{\overline{U}},\dot{\overline{I}}))e^{\mu_0}\left(\frac{Z_0+R_1}{\mu^2 L_1}+\frac{1}{\mu^2 L_1 C_1}\right)+$$

$$+e^{\mu(t-T-kT_0)}\hat{\rho}_\mu((U,I),(\dot{U},\dot{I}),(\overline{U},\overline{I}),(\dot{\overline{U}},\dot{\overline{I}}))\frac{e^{\mu_0}-1}{2}e^{\mu_0}\left(\frac{Z_0+R_1}{\mu^2 L_1}+\frac{1}{\mu^2 L_1 C_1}\right)+$$

$$+e^{\mu(t-T-kT_0)}\hat{\rho}_\mu((U,I),(\dot{U},\dot{I}),(\overline{U},\overline{I}),(\dot{\overline{U}},\dot{\overline{I}}))\frac{e^{\mu_0}-1}{\mu_0}e^{\mu_0}\left(\frac{Z_0+R_1}{\mu^2 L_1}+\frac{1}{\mu^2 L_1 C_1}\right)\leq$$

$$+e^{\mu(t-T-kT_0)}\hat{\rho}_\mu((U,I),(\dot{U},\dot{I}),(\overline{U},\overline{I}),(\dot{\overline{U}},\dot{\overline{I}}))\left(1+\frac{e^{\mu_0}-1}{2}+\frac{e^{\mu_0}-1}{\mu_0}\right)e^{\mu_0}\left(\frac{Z_0+R_1}{\mu^2 L_1}+\frac{1}{\mu^2 L_1 C_1}\right)\equiv$$

$$\equiv e^{\mu(t-T-kT_0)}K_I\,\hat{\rho}_\mu((U,I),(\dot{U},\dot{I}),(\overline{U},\overline{I}),(\dot{\overline{U}},\dot{\overline{I}}))\leq e^{\mu T_0}K_I\,\hat{\rho}_\mu((U,I),(\dot{U},\dot{I}),(\overline{U},\overline{I}),(\dot{\overline{U}},\dot{\overline{I}}))\cdot$$

It follows

$$\hat{\rho}(B_I(U,I),B_I(\overline{U},\overline{I}))\leq e^{\mu T_0}K_U\,\hat{\rho}_\mu((U,I),(\dot{U},\dot{I}),(\overline{U},\overline{I}),(\dot{\overline{U}},\dot{\overline{I}})).$$

Finally, we obtain Lipschitz estimates for the derivatives.

$$\left|\dot{B}_U^{(k)}(U,I)(t)-\dot{B}_U^{(k)}(\overline{U},\overline{I})(t)\right|\leq$$

$$\leq \left| V(U,I)(s) - V(\overline{U},\overline{I})(s) \right| + \frac{1}{T_0} \left| \int\limits_{T+kT_0}^{T+(k+1)T_0} V(U,I)(s)ds - \int\limits_{T+kT_0}^{T+(k+1)T_0} V(\overline{U},\overline{I})(s)ds \right| \equiv \dot{G}_1 + \dot{G}_2 .$$

But

$$\dot{G}_1 \leq \left| \dot{I}(t-T) - \dot{\overline{I}}(t-T) \right| + \frac{Z_0 + R_0}{L_0} \left| U(t) - \overline{U}(t) \right| + \frac{Z_0 + R_0}{L_0} \left| I(t-T) - \overline{I}(t-T) \right| +$$

$$+ \frac{1}{L_0 C_0} \int\limits_{T+kT_0}^{t} \left| U(s) - \overline{U}(s) \right| ds + \frac{1}{L_0 C_0} \int\limits_{T+kT_0}^{t} \left| I(s-T) - \overline{I}(s-T) \right| ds \leq$$

$$\leq e^{\mu(t-T-kT_0)} \frac{Z_0 + R_0}{L_0} \rho^{(k)}(U,\overline{U}) + \frac{1}{L_0 C_0} \rho^{(k)}(U,\overline{U}) \frac{e^{\mu(t-T-kT_0)} - 1}{\mu} \leq$$

$$\leq e^{\mu(t-T-kT_0)} \frac{Z_0 + R_0}{L_0} e^{\mu_0} \rho_\mu^{(k)}(U,\overline{U}) + \frac{1}{L_0 C_0} e^{\mu_0} \rho_\mu^{(k)}(U,\overline{U}) \frac{e^{\mu(t-T-kT_0)} - 1}{\mu} \leq$$

$$\leq e^{\mu(t-T-kT_0)} \frac{Z_0 + R_0}{L_0} \frac{e^{\mu_0} \rho_\mu^{(k)}(\dot{U},\dot{\overline{U}})}{\mu} + \frac{1}{L_0 C_0} \frac{e^{\mu_0} \rho_\mu^{(k)}(\dot{U},\dot{\overline{U}})}{\mu} \frac{e^{\mu(t-T-kT_0)} - 1}{\mu} \leq$$

$$\leq e^{\mu(t-T-kT_0)} \rho_\mu^{(k)}((\dot{U},\dot{I}),(\dot{\overline{U}},\dot{\overline{I}})) e^{\mu_0} \left(\frac{Z_0 + R_0}{\mu L_0} + \frac{1}{\mu^2 L_0 C_0} \right)$$

and

$$\dot{G}_2 \leq \frac{1}{T_0} \int\limits_{T}^{T+T_0} \left| V(U,I)(s) - V(\overline{U},\overline{I})(s) \right| ds \leq$$

$$\leq \rho_\mu^{(k)}((U,I),(\overline{U},\overline{I})) e^{\mu_0} \left[\frac{Z_0 + R_0}{\mu L_0} + \frac{1}{\mu^2 L_0 C_0} \right] \frac{1}{T_0} \int\limits_{T+kT_0}^{T+(k+1)T_0} e^{\mu(t-T-kT_0)} dt \leq$$

$$\leq e^{\mu(t-T-kT_0)} \rho_\mu^{(k)}((\dot{U},\dot{I}),(\dot{\overline{U}},\dot{\overline{I}})) \frac{e^{\mu_0} - 1}{\mu_0} e^{\mu_0} \left(\frac{Z_0 + R_0}{\mu L_0} + \frac{1}{\mu^2 L_0 C_0} \right).$$

Therefore

$$\left| \dot{B}_U^{(k)}(U,I)(t) - \dot{B}_U^{(k)}(\overline{U},\overline{I})(t) \right| \leq$$

$$\leq e^{\mu(t-T-kT_0)}\hat{\rho}_\mu((U,I),(\dot{U},\dot{I}),(\overline{U},\overline{I}),(\dot{\overline{U}},\dot{\overline{I}}))e^{\mu_0}\left(\frac{Z_0+R_0}{\mu L_0}+\frac{1}{\mu^2 L_0 C_0}\right)+$$

$$+ e^{\mu(t-T-kT_0)}\hat{\rho}_\mu((U,I),(\dot{U},\dot{I}),(\overline{U},\overline{I}),(\dot{\overline{U}},\dot{\overline{I}}))\frac{e^{\mu_0}-1}{\mu_0}e^{\mu_0}\left(\frac{Z_0+R_0}{\mu L_0}+\frac{1}{\mu^2 L_0 C_0}\right)\leq$$

$$\leq e^{\mu(t-T-kT_0)}\hat{\rho}_\mu((U,I),(\dot{U},\dot{I}),(\overline{U},\overline{I}),(\dot{\overline{U}},\dot{\overline{I}}))\left[\left(1+\frac{e^{\mu_0}-1}{\mu_0}\right)e^{\mu_0}\left(\frac{Z_0+R_0}{\mu L_0}+\frac{1}{\mu^2 L_0 C_0}\right)\right]\equiv$$

$$\equiv e^{\mu(t-T-kT_0)}\dot{K}_U\hat{\rho}_\mu((U,I),(\dot{U},\dot{I}),(\overline{U},\overline{I}),(\dot{\overline{U}},\dot{\overline{I}})).$$

It follows

$$\rho_\mu^{(k)}(\dot{B}_U^{(k)}(U,I),\dot{B}_U^{(k)}(\overline{U},\overline{I}))\leq \dot{K}_U\hat{\rho}_\mu((U,I),(\dot{U},\dot{I}),(\overline{U},\overline{I}),(\dot{\overline{U}},\dot{\overline{I}})).$$

Analogously

$$\left|\dot{B}_I^{(k)}(U,I)(t)-\dot{B}_I^{(k)}(\overline{U},\overline{I})(t)\right|\leq\left|\mathrm{I}(U,I)(t)-\mathrm{I}(\overline{U},\overline{I})(t)\right|+$$

$$+\frac{1}{T_0}\int_{T+kT_0}^{T+(k+1)T_0}\left|J(U,I)(s)-J(\overline{U},\overline{I})(s)\right|ds\equiv\dot{\mathrm{H}}_1+\dot{\mathrm{H}}_2.$$

We have

$$\dot{\mathrm{H}}_1\leq e^{\mu(t-T-kT_0)}\left|\dot{U}(t-T)-\dot{\overline{U}}(t-T)\right|+e^{\mu(t-T-kT_0)}\frac{Z_0+R_1}{L_1}\left|U(t-T)-\overline{U}(t-T)\right|+$$

$$+e^{\mu(t-T-kT_0)}\frac{Z_0+R_1}{L_1}\left|I(t)-\overline{I}(t)\right|+$$

$$\frac{1}{L_1 C_1}\int_{T+kT_0}^{t}\left|U(s-T)-\overline{U}(s-T)\right|ds+\frac{1}{L_1 C_1}\int_{T+kT_0}^{t}\left|I(s)-\overline{I}(s)\right|ds\leq$$

$$\leq e^{\mu(t-T-kT_0)}\frac{Z_0+R_1}{L_1}\rho^{(k)}(I,\overline{I})+\frac{\rho^{(k)}(I,\overline{I})}{L_1 C_1}\int_{T+kT_0}^{t}e^{\mu(s-T-kT_0)}ds\leq$$

$$\leq e^{\mu(t-T-kT_0)}\frac{Z_0+R_1}{L_1}e^{\mu_0}\rho_\mu^{(k)}(I,\overline{I})+\frac{e^{\mu_0}\rho_\mu^{(k)}(I,\overline{I})}{L_1 C_1}\int_{T+kT_0}^{t}e^{\mu(s-T-kT_0)}ds\leq$$

$$\leq e^{\mu(t-T-kT_0)}\frac{Z_0+R_1}{L_1}\frac{e^{\mu_0}\rho_\mu^{(k)}(\dot{I},\dot{\bar{I}})}{\mu}+\frac{e^{\mu_0}\rho_\mu^{(k)}(\dot{I},\dot{\bar{I}})}{\mu L_1 C_1}\frac{e^{\mu(t-T-kT_0)}-1}{\mu}\leq$$

$$\leq e^{\mu(t-T-kT_0)}\dot{K}_U\hat{\rho}_\mu((U,I),(\dot{U},\dot{I}),(\overline{U},\overline{I}),(\dot{\overline{U}},\dot{\overline{I}}))e^{\mu_0}\left(\frac{Z_0+R_1}{\mu L_1}+\frac{1}{\mu^2 L_1 C_1}\right)$$

and

$$\dot{H}_2\leq\frac{1}{T_0}\int\limits_{T+kT_0}^{T+(k+1)T_0}\left|J(U,I)(s)-J(\overline{U},\overline{I})(s)\right|ds\leq$$

$$\leq\rho((U,I),(\overline{U},\overline{I}))e^{\mu_0}\left(\frac{Z_0+R_1}{\mu L_1}+\frac{1}{\mu^2 L_1 C_1}\right)\frac{1}{T_0}\int\limits_{T+kT_0}^{T+(k+1)T_0}e^{\mu(t-T-kT_0)}dt\leq$$

$$\leq e^{\mu(t-T-kT_0)}\dot{K}_U\hat{\rho}_\mu((U,I),(\dot{U},\dot{I}),(\overline{U},\overline{I}),(\dot{\overline{U}},\dot{\overline{I}}))\frac{e^{\mu_0}-1}{\mu_0}e^{\mu_0}\left(\frac{Z_0+R_1}{\mu L_1}+\frac{1}{\mu^2 L_1 C_1}\right).$$

Then

$$\left|\dot{B}_I^{(k)}(U,I)(t)-\dot{B}_I^{(k)}(\overline{U},\overline{I})(t)\right|\leq$$

$$e^{\mu(t-T-kT_0)}\dot{K}_U\hat{\rho}_\mu((U,I),(\dot{U},\dot{I}),(\overline{U},\overline{I}),(\dot{\overline{U}},\dot{\overline{I}}))\left(1+\frac{e^{\mu_0}-1}{\mu_0}\right)e^{\mu_0}\left(\frac{Z_0+R_1}{\mu L_1}+\frac{1}{\mu^2 L_1 C_1}\right)\equiv$$

$$\equiv e^{\mu(t-T-kT_0)}\dot{K}_I\hat{\rho}_\mu((U,I),(\dot{U},\dot{I}),(\overline{U},\overline{I}),(\dot{\overline{U}},\dot{\overline{I}})).$$

It is easy to verify that $K=\max\left\{e^{\mu_0}K_U,e^{\mu_0}K_I,\dot{K}_U,\dot{K}_I\right\}<1$ for sufficiently large μ. Therefore

$$\hat{\rho}_\mu(B_U(U,I),B_I(U,I),\dot{B}_U(U,I),\dot{B}_I(U,I),B_U(\overline{U},\overline{I}),B_I(\overline{U},\overline{I}),\dot{B}_U(\overline{U},\overline{I}),\dot{B}_I(\overline{U},\overline{I})\leq$$

$$\leq K\hat{\rho}_\mu((U,I),(\dot{U},\dot{I}),(\overline{U},\overline{I}),(\dot{\overline{U}},\dot{\overline{I}})).$$

This means the operator $B=(B_U,B_I)$, being contractive, has a unique fixed point. Theorem 3.7.1 is thus proved.

Remark 3.7.1. In view of the potential applications, we have to collect all inequalities implying an existence-uniqueness of the periodic solutions:

$$\left(1+\frac{e^{\mu_0}-1}{\mu_0}\right)I_0 e^{-\beta}+\left(1+\frac{e^{\mu_0}-1}{2}+\frac{e^{\mu_0}-1}{\mu_0}\right)\left(\frac{(Z_0+R_0)(U_0+I_0 e^{-\beta})}{\mu L_0}+\frac{U_0+I_0 e^{-\beta}}{\mu^2 L_0 C_0}+\frac{2Z_0 U_{E0}}{\mu L_0}\right)\leq U_0;$$

$$\left(1+\frac{e^{\mu_0}-1}{\mu_0}\right)U_0 e^{-\beta}+\left(1+\frac{e^{\mu_0}-1}{2}+\frac{e^{\mu_0}-1}{\mu_0}\right)\left(\frac{(Z_0+R_1)(U_0 e^{-\beta}+I_0)}{\mu L_1}+\frac{U_0 e^{-\beta}+I_0}{\mu^2 L_1 C_1}+\frac{2Z_0 U_{E1}}{\mu L_1}\right)\leq I_0;$$

$$e^{\mu_0}K_U=e^{\mu_0}\left(1+\frac{e^{\mu_0}-1}{2}+\frac{e^{\mu_0}-1}{\mu_0}\right)e^{\mu_0}\left(\frac{Z_0+R_0}{\mu^2 L_0}+\frac{1}{\mu^2 L_0 C_0}\right)<1;$$

$$e^{\mu_0}K_I=e^{\mu_0}\left(1+\frac{e^{\mu_0}-1}{2}+\frac{e^{\mu_0}-1}{\mu_0}\right)e^{\mu_0}\left(\frac{Z_0+R_1}{\mu^2 L_1}+\frac{1}{\mu^2 L_1 C_1}\right)<1;$$

$$\dot{K}_U=\left[\left(1+\frac{e^{\mu_0}-1}{\mu_0}\right)\left(\frac{Z_0+R_0}{\mu L_0}+\frac{1}{\mu^2 L_0 C_0}\right)\right]<1;$$

$$\dot{K}_I=\left(1+\frac{e^{\mu_0}-1}{\mu_0}\right)\left(\frac{Z_0+R_1}{\mu L_1}+\frac{1}{\mu^2 L_1 C_1}\right)<1.$$

For the transmission line with specific parameters noted in the previous paragraph, and $U_0=I_0$ we obtain

$$e^{-(\beta-1)}+\frac{3e-1}{2}\left(\frac{(Z_0+R_0)(1+e^{-\beta})}{\mu L_0}+\frac{1+e^{-\beta}}{\mu^2 L_0 C_0}+\frac{2Z_0}{\mu L_0}\right)\leq 1;$$

$$e^{-(\beta-1)}+\frac{3e-1}{2}\left(\frac{(Z_0+R_1)(e^{-\beta}+1)}{\mu L_1}+\frac{e^{-\beta}+1}{\mu^2 L_1 C_1}+\frac{2Z_0}{\mu L_1}\right)\leq 1;$$

$$\dot{K}_U=e\left(\frac{Z_0+R_0}{\mu L_0}+\frac{1}{\mu^2 L_0 C_0}\right)<1;\quad \dot{K}_I=e\left(\frac{Z_0+R_1}{\mu L_1}+\frac{1}{\mu^2 L_1 C_1}\right)<1.$$

Remark 3.7.2. Let us consider linear time-varying loads

$$\widetilde{L}_p(i)=i.L_p(t)\ (p=0,1),\ C_p(u)=C_p(t)u,\ R_p(i)=R_p(t)i$$

Then $\dfrac{d\widetilde{L}_p(i)}{di} = L_p(t)$ and we are able to derive the corresponding equations to (3.2.12). Indeed, we obtain

$$\frac{dU(t)}{dt} = \frac{dI(t-T)}{dt} - Z_0 \frac{U(t) + I(t-T) - 2E_0(t-T) + 2R_0(t)i_0(U,I)(t)}{L_0(t).i_0(U,I)(t)} -$$

$$- \frac{2Z_0}{L_0(t).i_0(U,I)(t)} \frac{1}{C_0(t)} \int_T^t i_0(U,I)(s)ds$$

$$\frac{dI(t)}{dt} = \frac{dU(t-T)}{dt} - Z_0 \frac{U(t-T) + I(t) - 2E_1(t-T) - 2R_1(t)i_1(U,I)(t)}{L_1(t).i_1(U,I)(t)} +$$

$$+ \frac{2Z_0}{L_1(t).i_1(U,I)(t)} \frac{1}{C_1(t)} \int_T^t i_1(U,I)(s)ds.$$

In order to formulate a theorem for the existence of a periodic or oscillatory solution, we have to ensure that $i_0(U,I)(t) = \dfrac{U(t) - I(t-T)}{2Z_0} \neq 0, \ i_1(U,I)(t) = \dfrac{U(t-T) - I(t)}{2Z_0} \neq 0$ in some class of functions. We have also to assume that $L_p = L_p(t), \ C_p = C_p(t)$ have a strict positive lower bound, that is, $L_p(t) \geq \hat{L}_p > 0, \ C_p(t) \geq \hat{C}_p > 0$.

3.8. Applications for Non-Uniform Lossless Transmission Lines

As in Chapter II we note that the developed theory is applicable in cases of non-uniform transmission lines. Specifically (cf. J.E. Rowe [96]) for lossless transmission line analog of electromagnetic wave propagation on helical conductors:

1) $(u(x,t), i(x,t)) = (V, I_x)$

$$\frac{\partial I_x}{\partial x} + C_e \frac{\partial V}{\partial t} = 0, \quad \frac{\partial V}{\partial x} + L_e \frac{\partial I_x}{\partial t} = 0$$

and TM-oscillations related to the axial current I_x with

$$C_e = \left(\frac{\beta}{\gamma}\right)^2 \frac{2\pi\varepsilon_0}{I_0(\gamma a)K_0(\gamma a)} [F/m], \ L_e = \frac{\mu_0 \cot^2 \psi}{2\pi} I_1(\gamma a)K_1(\gamma a) \ [H/m], \ v = 1/\sqrt{L_e C_e} \ ;$$

2) a transmission line analog of an electron beam [25]:

$$\frac{\partial V}{\partial z} = \frac{p^2}{\omega^2 \varepsilon_0 \sigma} \frac{\partial I}{\partial t}, \qquad \frac{\partial I}{\partial z} = \frac{\omega \varepsilon_0 \sigma}{p^2} \left(\frac{2\pi}{\lambda_p} \right)^2 \frac{\partial V}{\partial t}$$

with

$$L_e = \frac{1}{I_0} \sqrt{\frac{2V_0}{\eta}} \left(\frac{\omega_p}{\omega} \right)^2 [H/m], \; C_e = \frac{\sigma \varepsilon_0}{p^2} \left(\frac{2\pi}{\lambda_p} \right)^2 [F/m];$$

3) a transmission line analog of plasma column in tube

$$L_e = \frac{\tau}{2\pi b} \frac{1}{\omega_p^2 \varepsilon_0} \frac{I_0(\tau b)}{I_1(\tau b)} [H/m], \; C_e = \frac{2\pi b \gamma \varepsilon_0 K_1(\gamma b)}{\left[1 - \left(\omega / \omega_p \right)^2 \right] K_0(\gamma b)} \; [F/m].$$

For details and denotations cf. [96].

In all the above cases, one can proceed as in § 3.5 and formulate a mixed problem for the lossless transmission line. For 1), with the transformation

$$V(x,t) = \frac{1}{2} U(x,t) + \frac{1}{2} I(x,t), \; I_x(x,t) = \frac{1}{2Z_0} U(x,t) - \frac{1}{2Z_0} I(x,t).$$

The mixed problem can be reduced to an initial value problem on the boundary for a system of neutral equations (in the case of in series connected elements):

$$\frac{dU(t)}{dt} = \frac{dI(t-T)}{dt} + Z_0 \frac{-U(t) - I(t-T) + 2E_0(t-T) - 2R_0\left(\widetilde{i_0}(t)\right)}{\widetilde{L_0}(\widetilde{i_0})} - \frac{2Z_0}{\widetilde{L_0}(\widetilde{i_0})} \, \overline{C_0}^{-1} \left(\frac{1}{2Z_0} \int_T^t \widetilde{i_0}(s) ds \right);$$

$$\frac{dI(t)}{dt} = \frac{dU(t-T)}{dt} - Z_0 \frac{U(t-T) + I(t) - 2E_1(t-T) - 2R_1\left(\widetilde{i_1}(t)\right)}{\widetilde{L_1}(\widetilde{i_1})} + \frac{2Z_0}{\widetilde{L_1}(\widetilde{i_1})} \, \overline{C_1}^{-1} \left(\frac{1}{2Z_0} \int_T^t \widetilde{i_1}(s) ds \right).$$

The existence-uniqueness of a periodic solution follows from Theorem 3.5.1. In the case of non-uniform transmission lines $Z_0 = \sqrt{L_e / C_e}$ is not a constant. It decreases as the frequency γa increases. This can be compensated by choosing sufficiently large $\mu > 0$. Therefore we only need estimates for Z_0. We see that Z_0 always has a strict positive lower bound (cf. J. E. Rowe [96]).

230   Vasil G. Angelov

Indeed, let $\quad C_e = \dfrac{2\pi\varepsilon_0}{I_0(\gamma a)K_0(\gamma a)}[F/m], \quad L_e = \dfrac{\mu_0 \cot^2\psi}{2\pi}I_1(\gamma a)K_1(\gamma a)\,[H/m],$

$$\Lambda = 1m, \quad \sqrt{L_e C_e} = \frac{\cot\psi}{c}\sqrt{\frac{I_1(\gamma a)K_1(\gamma a)}{I_0(\gamma a)K_0(\gamma a)}}, \quad v = \frac{1}{\sqrt{L_e C_e}} = \frac{c}{\cot\psi}\sqrt{\frac{I_0(\gamma a)K_0(\gamma a)}{I_1(\gamma a)K_1(\gamma a)}},$$

$$Z_0 = Z_e = \sqrt{L_e/C_e} = \sqrt{\frac{\mu_0}{\varepsilon_0}}\,\frac{\cot\psi}{2\pi}\sqrt{I_1(\gamma a)K_1(\gamma a)I_0(\gamma a)K_0(\gamma a)},$$

$$T = \Lambda\sqrt{L_e C_e} = \frac{\cot\psi}{c}\sqrt{\frac{I_1(\gamma a)K_1(\gamma a)}{I_0(\gamma a)K_0(\gamma a)}},$$

where μ_0 is magnetic permeability and coincides with $\mu_0 = \mu T_0$.

We would like to point out that the above formulas are derived under the assumption that the solutions are harmonic ones, that is, dependence on time is $u(x,t) = u(x)e^{j\omega t}$. Our consideration is more general because we have proved the existence of general periodic solutions. If we use the dependence of L_e, C_e and $\dfrac{v}{c}$ on the frequency (cf. J. E. Rowe [96]) one can see that for $\cot\psi = 5$ it follows $\dfrac{v}{c} = 0,2$ when the frequency increases. For the numerical example from § 3.6 we have

$$\frac{3e-1}{2\mu\hat{L}_0}\sqrt{\frac{\mu_0}{\varepsilon_0}}\,\frac{\cot\psi}{2\pi}\sqrt{I_1(\gamma a)K_1(\gamma a)I_0(\gamma a)K_0(\gamma a)}\left[U_0 + 2\sum_{n=1}^{m}\left|r_n^{(0)}\right|\left(\frac{U_0}{2Z_e}\right)^n e^{(n-1)}\right] \le U_0,$$

$$\frac{3e-1}{2\mu\hat{L}_1}\sqrt{\frac{\mu_0}{\varepsilon_0}}\,\frac{\cot\psi}{2\pi}\sqrt{I_1(\gamma a)K_1(\gamma a)I_0(\gamma a)K_0(\gamma a)}\left(I_0 + 2\sum_{n=1}^{m}\left|r_n^{(1)}\right|\left(\frac{I_0}{2Z_e}\right)^n e^{(n-1)}\right) \le I_0,$$

$$\frac{3e-1}{2\mu\hat{L}_0}\left[\sqrt{\frac{\mu_0}{\varepsilon_0}}\,\frac{\cot\psi}{2\pi}\sqrt{I_1(\gamma a)K_1(\gamma a)I_0(\gamma a)K_0(\gamma a)} + \left|r_1^{(0)}\right| + \left|r_2^{(0)}\right|\frac{U_0 e}{2Z_e} + \left|r_3^{(0)}\right|\left(\frac{U_0}{2Z_e}\right)^2 e^2\right] \le 1,$$

$$\frac{3e-1}{2\mu\hat{L}_1}\left(\sqrt{\frac{\mu_0}{\varepsilon_0}}\,\frac{\cot\psi}{2\pi}\sqrt{I_1(\gamma a)K_1(\gamma a)I_0(\gamma a)K_0(\gamma a)} + \left|r_1^{(1)}\right| + \left|r_2^{(1)}\right|\frac{I_0 e}{2Z_e} + \left|r_3^{(1)}\right|\left(\frac{I_0}{2Z_e}\right)^2 e^2\right) \le 1$$

and for contractive constants

$$K_U = e\frac{3e-1}{2\mu^2}\left[\frac{1}{\hat{L}_0^2}\left(eU_0 + \sum_{n=1}^{m}\left|r_n^{(0)}\right|\left(\frac{U_0}{2Z_e}\right)^n + U_0\sum_{n=1}^{4}\left|\tilde{l}_n^{(0)}\right|\left(\frac{U_0}{2Z_e}\right)^n\right) + \right.$$

$$+\frac{1}{\hat{L}_0}\left(\sqrt{L_e/C_e}+\sum_{n=1}^{3}\left|r_n^{(0)}\right|n.\left(\frac{U_0}{2\sqrt{L_e/C_e}}\right)^{n-1}+\frac{2\sqrt{\Phi_0}}{\mu\hat{L}_0 c_0\sqrt{\Phi_0-\phi_0}}\right)\Bigg]<1\,;$$

$$K_I=e\frac{3e-1}{2\mu^2}\left[eI_0+\sum_{n=1}^{m}\left|r_n^{(1)}\right|\left(\frac{I_0}{2\sqrt{L_e/C_e}}\right)^{n}+U_0\sum_{n=1}^{m}\left|\widetilde{l}_n^{(1)}\right|\left(\frac{I_0}{2\sqrt{L_e/C_e}}\right)^{n}+\right.$$

$$\left.+\frac{1}{\hat{L}_1}\left[Z_e+\sum_{n=1}^{m}\left|r_n^{(1)}\right|n\left(\frac{I_0}{2Z_e}\right)^{n-1}+\frac{2\sqrt{\Phi_1}}{\mu\hat{L}_1 c_1\sqrt{\Phi_1-\phi_0}}\right]\right\}<1\,;$$

$$\dot{K}_U=\frac{e}{\mu}\left\{\frac{1}{\hat{L}_0^2}\left[eU_0+\sum_{n=1}^{3}\left|r_n^0\right|\left(\frac{U_0}{2Z_e}\right)^{n}+U_0\right]\sum_{n=1}^{4}\left|\widetilde{l}_n^{(0)}\right|\left(\frac{U_0}{2Z_e}\right)^{n}+\right.$$

$$\left.+\frac{1}{\hat{L}_0}\left[Z_e+\sum_{n=1}^{3}\left|r_n^0\right|n\left(\frac{U_0}{2Z_e}\right)^{n-1}+\frac{2\sqrt{\Phi_0}}{\mu\hat{L}_0 c_0\sqrt{\Phi_0-\phi_0}}\right]\right\}<1\,;$$

$$\dot{K}_I=\frac{e}{\mu}\left\{\frac{1}{\hat{L}_1^2}\left[eI_0+\sum_{n=1}^{3}\left|r_n^{(1)}\right|\left(\frac{I_0}{2Z_e}\right)^{n}+U_0\right]\sum_{n=1}^{4}\left|\widetilde{l}_n^{(1)}\right|\left(\frac{I_0}{2Z_e}\right)^{n}+\right.$$

$$\left.+\frac{1}{\hat{L}_1}\left[Z_e+\sum_{n=1}^{3}\left|r_n^{(1)}\right|n\left(\frac{I_0}{2Z_e}\right)^{n-1}+\frac{2\sqrt{\Phi_1}}{\mu c_1\sqrt{\Phi_1-\phi_0}}\right]\right\}<1\,.$$

Our consideration implies a possible propagation of ultraviolet waves, namely $\lambda_0=10^{-7}\,m$. If $\mu=6.10^{14}$, then $\mu T_0=\mu_0=1$ and

$$\mu T=6.10^{14}.\left(1/6\right)10^{-7}=10^{7}\,,e^{-\mu T}\approx 0\,.$$

For resistive elements with *V-I* characteristics

$$R_0(i)=R_1(i)=0{,}028u-0{,}125u^3\,,\text{ i.e. }r_1=0{,}028,r_2=0,\,r_3=0{,}125$$

and inductive element $L_0(i)=L_1(i)=3i-\left(1/12\right)i^3$ (choosing $I_0=1$) one obtains

$$6i-\left(1/3\right)i^3>6-\left(1/3\right)=17/3\text{ and consequently }\frac{1}{\hat{L}_0}=\frac{1}{\hat{L}_1}=\frac{3}{17}\,.$$

One can verify that the above inequalities are satisfied for $U_0=0{,}1$ and $I_0=1$ and

$$C_0(u)=C_1(u)=c/\sqrt{1-\left(u/\Phi\right)}=c\sqrt{\Phi}/\sqrt{\Phi-u}\,,\text{ where}$$

$$h = 2, \; c_0 = c_1 = 50\,pF = 5.10^{-11}F \;\text{ and }\; \Phi = \Phi_0 = \Phi_1 = 0{,}4V \Rightarrow U_0 < 0{,}4.$$

3.9. Oscillatory Regimes for the Nonlinear System

Here we deal with the problem of oscillating solutions of the system:

$$\frac{dU(t)}{dt} = \frac{dI(t-T)}{dt} + \frac{2Z_0}{\widetilde{L}_0(i_0)(t)}E_0(t-T) - \frac{Z_0}{\widetilde{L}_0(i_0)(t)}U(t) - \frac{Z_0}{\widetilde{L}_0(i_0)(t)}I(t-T) -$$

$$- \frac{2Z_0}{\widetilde{L}_0(i_0)(t)}R_0(i_0) - \frac{2Z_0}{\widetilde{L}_0(i_0)(t)}\widetilde{C}_0^{-1}\left(\int_T^t i_0 ds\right); t \in [T,2T],\tag{3.9.1}$$

$$\frac{dI(t)}{dt} = \frac{dU(t-T)}{dt} + \frac{2Z_0}{\widetilde{L}_1(i_1)(t)}E_1(t-T) - \frac{Z_0}{\widetilde{L}_1(i_1)(t)}U(t-T) - \frac{Z_0}{\widetilde{L}_1(i_1)(t)}I(t) +$$

$$+ \frac{2Z_0}{\widetilde{L}_1(i_1)(t)}R_1(i_1) + \frac{2Z_0}{\widetilde{L}_1(i_1)(t)}\widetilde{C}_1^{-1}\left(\frac{1}{2Z_0}\int_T^t i_1 ds\right), t \in [T,2T];$$

$$U(t) = U_0(t), \; \frac{dU(t)}{dt} = \frac{dU_0(t)}{dt}, \; t \in [0,T];$$

$$I(t) = I_0(t), \; \frac{dI(t)}{dt} = \frac{dI_0(t)}{dt}, \; t \in [0,T]$$

where $U_0(t), I_0(t)$ are prescribed T_0-periodic initial functions.

Let us put $t_0 \equiv T$. Now we are able to formulate the main problem: to find an oscillatory solution of problem (3.9.1) with advanced prescribed zeros on an interval $[t_0, \infty)$, where $\upsilon_0(t)$ is a prescribed initial oscillating function on the interval $[0, t_0]$.

Let $S_T = \{\tau_k\}_{k=0}^n, n \in N$ be the set of zeros of the initial function, that is, $U_0(\tau_k) = I_0(\tau_k) = 0$ such that $\tau_0 = 0$, $\tau_n = T \equiv t_0$. Besides $\max\{\tau_{k+1} - \tau_k : k = 0,1,...,n\} \le T_0$.

Let $S = \{t_k\}_{k=0}^\infty$ be a strictly increasing sequence of real numbers satisfying the following conditions (C):

(C1) $\lim\limits_{k\to\infty} t_k = \infty$;

(C2) for every k $\left(t_k - T \ge \tau_0\right)$ there is $s < k$ such that $t_k - T = t_s$ where $t_s \in S_T \cup S$.

Condition (**C2**) implies

$$0 < \inf\{t_{k+1} - t_k : k = 0,1,2,...\} \le \sup\{t_{k+1} - t_k : k = 0,1,2,...\} = T_0 < \infty.$$

Introduce the set $C^1[t_0,\infty)$ consisting of all continuous functions whose derivatives are bounded and continuous on every interval $[t_k,t_{k+1}]$. We note the right and left derivatives at t_k of these functions may not coincide. This is why we introduce below a topology of uniform convergence on every interval $[t_k,t_{k+1}]$ of the derivatives that needs the introduction of uniform spaces (cf. [14]). In fact we introduce a countable family of pseudo-metrics as there are the intervals $[t_k,t_{k+1}]$.

Consider the set

$$M_{SU} = \left\{U(.) \in C^1[t_0,2T]:U(t_k) = 0\,(k = 0,1,2,...)\right\},$$
$$M_{SI} = \left\{I(.) \in C^1[t_0,2T]:I(t_k) = 0\,(k = 0,1,2,...)\right\}$$

and

$$M_{SU}^* = \left\{U(.) \in M_S :|U(t)| \le U_0 e^{\mu(t-t_k)}, t \in [t_k,t_{k=1}]\right\},$$
$$M_{SI}^* = \left\{I(.) \in M_S :|I(t)| \le I_0 e^{\mu(t-t_k)}, t \in [t_k,t_{k=1}]\right\}$$

where U_0,I_0,μ are positive constants. Recall that $\mu T_0 = \mu_0 = \text{const.} > 0$.

Introduce the following family of pseudo-metrics

$$\rho^{(k)}(U,\overline{U}) = \max\left\{|U(t) - \overline{U}(t)| : t \in [t_k,t_{k+1}]\right\},$$
$$\hat\rho^{(k)}(U,\overline{U}) = \max\left\{|U(t) - \overline{U}(t)| : t \in [t_0,t_{k+1}]\right\},$$

$$\rho_\mu^{(k)}(U,\overline{U}) = \max\left\{e^{-\mu(t-t_k)}|U(t) - \overline{U}(t)| : t \in [t_k,t_{k+1}]\right\},$$
$$\hat\rho_\mu^{(k)}(U,\overline{U}) = \max\left\{\rho_\mu^{(0)}(U,\overline{U}), \rho_\mu^{(1)}(U,\overline{U}),...,\rho_\mu^{(k)}(U,\overline{U})\right\},$$

$$\rho_\mu^{(k)}(\dot U,\dot{\overline{U}}) = \max\left\{e^{-\mu(t-t_k)}|\dot U(t) - \dot{\overline{U}}(t)| : t \in [t_k,t_{k+1}]\right\},$$
$$\hat\rho_\mu^{(k)}(\dot U,\dot{\overline{U}}) = \max\left\{\rho_\mu^{(0)}(\dot U,\dot{\overline{U}}), \rho_\mu^{(1)}(\dot U,\dot{\overline{U}}),...,\rho_\mu^{(k)}(\dot U,\dot{\overline{U}})\right\},$$

$$\rho^{(k)}(I,\overline{I}) = \max\left\{|I(t) - \overline{I}(t)| : t \in [t_k,t_{k+1}]\right\},$$
$$\hat\rho^{(k)}(I,\overline{I}) = \max\left\{|I(t) - \overline{I}(t)| : t \in [t_0,t_{k+1}]\right\},$$

$$\rho_\mu^{(k)}(I,\bar{I}) = \max\left\{e^{-\mu(t-t_k)}\left|I(t)-\bar{I}(t)\right| : t \in [t_k,t_{k+1}]\right\},$$

$$\hat{\rho}_\mu^{(k)}(I,\bar{I}) = \max\left\{\rho_\mu^{(0)}(I,\bar{I}),\rho_\mu^{(1)}(I,\bar{I}),\ldots,\rho_\mu^{(k)}(I,\bar{I})\right\},$$

$$\rho_\mu^{(k)}(\dot{I},\dot{\bar{I}}) = \max\left\{e^{-\mu(t-t_k)}\left|\dot{I}(t)-\dot{\bar{I}}(t)\right| : t \in [t_k,t_{k+1}]\right\},$$

$$\hat{\rho}_\mu^{(k)}(\dot{I},\dot{\bar{I}}) = \max\left\{\rho_\mu^{(0)}(\dot{I},\dot{\bar{I}}),\rho_\mu^{(1)}(\dot{I},\dot{\bar{I}}),\ldots,\rho_\mu^{(k)}(\dot{I},\dot{\bar{I}})\right\}.$$

The set $M_{SU}^* \times M_{SI}^*$ turns into a complete uniform space with respect to the countable saturated family of pseudo-metrics

$$\hat{\rho}_\mu^{(k)}((U,\dot{U},I,\dot{I}),(\bar{U},\dot{\bar{U}},\bar{I},\dot{\bar{I}})) = \max\left\{\hat{\rho}^{(k)}(U,I,\bar{U},\bar{I}),\hat{\rho}_\mu^{(k)}(\dot{U},\dot{\bar{U}}),\hat{\rho}_\mu^{(k)}(\dot{I},\dot{\bar{I}}) : k = 0,1,2,\ldots \right\}.$$

Since we look for a continuously differentiable solution we have to require the conformity condition:

$$\widetilde{L}_0 \frac{dU(t_0)}{dt} = \widetilde{L}_0 \frac{dI(0)}{dt} + 2Z_0 E_0(0) - Z_0 U(t_0) - Z_0 I(0) - 2Z_0 R_0\left(\frac{U(t_0)-I(0)}{2Z_0}\right), \tag{CC}$$

$$\widetilde{L}_1 \frac{dI(t_0)}{dt} = \widetilde{L}_1 \frac{dU(0)}{dt} + 2Z_0 E_1(0) - Z_0 U(0) - Z_0 I(t_0) + 2Z_0 R_1\left(\frac{U(0)-I(t_0)}{2Z_0}\right).$$

It is satisfied provided we choose the initial functions such that $U_0(0)=0, I_0(0)=0,$ $\dot{U}_0(0)=0, \dot{I}_0(0)=0$ and $E_0(0)=E_1(0)=0$. The condition **(C2)** $t_k - T = t_s$ implies (because $t_0 - T = 0$) $U_0(t_0)=0, I_0(t_0)=0, \dot{U}_0(t_0)=0, \dot{I}_0(t_0)=0$.

In order to avoid the restrictive condition $\dot{U}_0(t_0)=0, \dot{I}_0(t_0)=0$ we define an operator $B = (B_U(U,I), B_I(U,I))$ using the formulas

$$B_U^{(k)}(U,I)(t) := \int_{t_k}^{t} V(U,I)(s)ds - \frac{t-t_k}{t_{k+1}-t_k}\int_{t_k}^{t_{k+1}} V(U,I)(s)ds, \quad t \in [t_k,t_{k+1}],$$

$$B_I^{(k)}(U,I)(t) := \int_{t_k}^{t} J(U,I)(s)ds - \frac{t-t_k}{t_{k+1}-t_k}\int_{t_k}^{t_{k+1}} J(U,I)(s)ds, \quad t \in [t_k,t_{k+1}]$$

$(k = 0,1,2,\ldots)$, where

$$V(U,I)(t) = \frac{d\bar{I}_0(t)}{dt} + \frac{2Z_0}{\widetilde{L}_0(i_0)(t)}E_0(t) - \frac{Z_0}{\widetilde{L}_0(i_0)(t)}U(t) - \frac{Z_0}{\widetilde{L}_0(i_0)(t)}\bar{I}_0(t) -$$

$$-\frac{2Z_0}{\widetilde{L}_0(i_0)(t)}R_0(i_0)-\frac{2Z_0}{\widetilde{L}_0(i_0)(t)}\widetilde{C}_0^{-1}\left(\int_T^t i_0 ds\right); t\in[T,2T],\qquad(3.9.1)$$

$$J(U,I)(t)=\frac{d\bar{U}_0(t)}{dt}+\frac{2Z_0}{\widetilde{L}_1(i_1)(t)}E_1(t)-\frac{Z_0}{\widetilde{L}_1(i_1)(t)}\bar{U}_0(t)-\frac{Z_0}{\widetilde{L}_1(i_1)(t)}I(t)+$$

$$+\frac{2Z_0}{\widetilde{L}_1(i_1)(t)}R_1(i_1)+\frac{2Z_0}{\widetilde{L}_1(i_1)(t)}\widetilde{C}_1^{-1}\left(\frac{1}{2Z_0}\int_T^t i_1 ds\right)\ t\in[T,2T]$$

$$i_0(U,I)(t)=\frac{U(t)-\bar{I}_0(t)}{2Z_0}\neq 0,\qquad i_1(U,I)(t)=\frac{\bar{U}_0(t)-I(t)}{2Z_0}\neq 0,$$

$$\bar{U}_0(t)=U_0(t-T),\qquad \bar{I}_0(t)=I_0(t-T),\ t\in[T,2T]$$

and

$$\left|U_0(t)\right|\le U_0 e^{-\beta}e^{\mu(t-\tau_n)}, t\in[\tau_n,\tau_{n+1}], \left|I_0(t)\right|\le I_0 e^{-\beta}e^{\mu(t-\tau_n)}, t\in[\tau_n,\tau_{n+1}]$$

which implies

$$\left|\bar{U}_0(t)\right|\le U_0 e^{-\beta}e^{\mu(t-t_k)}, \left|\bar{I}_0(t)\right|\le I_0 e^{-\beta}e^{\mu(t-t_k)}, t\in[t_k,t_{k+1}]\subset\{T,2T\}.$$

One can verify that $M_{SU}^*\times M_{SI}^*$ is closed subset of $C^1[t_0,\infty)\times C^1[t_0,\infty)$ with respect to the above family of pseudo-metrics (cf. K. Zima [126]).

Lemma 3.9.1. Let $E(.)\in C_{ST\cup S}^1[0,\infty), \left|E_p(t)\right|\le U_{E_p}e^{\mu(t-t_k)}, t\in[t_k,t_{k+1}], (p=0,1)$.

Problem (3.9.1) has a solution $(U,I)\in M_{SU}^*\times M_{SI}^*$ iff the operator B has a fixed point in $M_{SU}^*\times M_{SI}^*$, that is,

$$(U,I)=(B_U(U,I),B_I(U,I)).\qquad(3.9.2)$$

Proof: Let $(U,I)\in M_{SU}^*\times M_{SI}^*$ be a solution of (3.9.1). Then, integrating (3.9.1) on every interval $[t_k,t]\subset[t_k,t_{k+1}]$ $(k=0,1,2\ ...\ )$ we obtain

$$U(t)-U(t_k)=\int_{t_k}^t V(U,I)(s)ds\ \Leftrightarrow\ U(t)=\int_{t_k}^t V(U,I)(s)ds,$$

$$I(t) - I(t_k) = \int_{t_k}^{t} J(U,I)(s)ds \quad \Leftrightarrow \quad I(t) = \int_{t_k}^{t} J(U,I)(s)ds.$$

Consequently

$$U(t) = \int_{t_k}^{t} V(U,I)(s)ds \Rightarrow 0 = U(t_{k+1}) = \int_{t_k}^{t_{k+1}} V(U,I)(s)ds \Rightarrow \int_{t_k}^{t_{k+1}} V(U,I)(s)ds = 0;$$

$$I(t) = \int_{t_k}^{t} J(U,I)(s)ds \Rightarrow 0 = I(t_{k+1}) = \int_{t_k}^{t_{k+1}} J(U,I)(s)ds \Rightarrow \int_{t_k}^{t_{k+1}} J(U,I)(s)ds = 0. \tag{3.9.3}$$

Therefore, the functions $U(t), I(t)$ satisfys

$$U(t) = \int_{t_k}^{t} V(U,I)(s)ds - \frac{t-t_k}{t_{k+1}-t_k} \int_{t_k}^{t_{k+1}} V(U,I)(s)ds, \quad t \in [t_k, t_{k+1}],$$

$$I(t) = \int_{t_k}^{t} J(U,I)(s)ds - \frac{t-t_k}{t_{k+1}-t_k} \int_{t_k}^{t_{k+1}} J(U,I)(s)ds, \quad t \in [t_k, t_{k+1}]$$

that is, $(U(t), I(t))$ is a fixed point of B.

Conversely, let $(U,I) \in M_{SU}^* \times M_{SI}^*$ be a solution of $U = B_U(U,I), I = B_I(U,I)$, that is,

$$U(t) = \int_{t_k}^{t} V(U,I)(s)ds - \frac{t-t_k}{t_{k+1}-t_k} \int_{t_k}^{t_{k+1}} V(U,I)(s)ds, t \in [t_k, t_{k+1}],$$

$$I(t) = \int_{t_k}^{t} J(U,I)(s)ds - \frac{t-t_k}{t_{k+1}-t_k} \int_{t_k}^{t_{k+1}} J(U,I)(s)ds, t \in [t_k, t_{k+1}].$$

Then recall that $\mu_0 = \mu T_0$, we can prove $\left| \int_{t_k}^{t_{k+1}} V(U,I)(s)ds \right| = 0, \left| \int_{t_k}^{t_{k+1}} J(U,I)(s)ds \right| = 0$

and after a differentiation we obtain (3.9.1).

Lemma 3.9.1 is thus proved.

Theorem 3.9.1. Let the following conditions be fulfilled:

1) $U_0(.), I_0(.) \in C^1[0,T]$ and

$$\left| U_0(t) \right| \le e^{-\beta} U_0 e^{\mu(t-\tau_n)}, \left| I_0(t) \right| \le e^{-\beta} I_0 e^{\mu(t-\tau_n)}, t \in [\tau_n, \tau_{n+1}];$$

and $U_0(0) = I_0(0) = 0 \; (\Rightarrow U_0(T) = I_0(T) = 0)$;

2) The source function

satisfies $E_p(.) \in C_S^1[0,\infty), \left|E_p(t)\right| \le U_{E_p} e^{\mu(t-t_k)}, t \in [t_k, t_{k+1}], E_p(t_k) = 0, (p = 0,1)$.

3) $\dfrac{U_0 + I_0 e^{-\beta}}{2Z_0} \le \mathrm{I}_0, \dfrac{U_0 e^{-\beta} + I_0}{2Z_0} \le \mathrm{I}_0$ (cf. (3.3.4)).

Then there exists a unique oscillatory solution of the initial value problem (3.9.1), belonging to $M_{SU}^* \times M_{SI}^*$ but only on interval $[T, 2T]$.

Sketch of Proof: We show that B maps $M_{SU}^* \times M_{SI}^*$ into itself.

First we notice that $B(U, I)(t)$ is continuous on $[t_0, \infty)$:

$$B_U^{(0)}(U,I)(t_0) = \int_{t_0}^{t_0} V(U,I)(s)ds - \frac{t_0 - t_0}{t_1 - t_0}\int_{t_0}^{t_1} V(U,I)(s)ds = 0,$$

$$\lim_{t \to t_{k+1}(t<t_{k+1})} B_U^{(k)}(U,I)(t) = \lim_{t \to t_{k+1}(t<t_{k+1})}\left(\int_{t_k}^{t} V(U,I)(s)ds - \frac{t - t_k}{t_{k+1} - t_k}\int_{t_k}^{t_{k+1}} V(U,I)(s)ds\right) = 0,$$

$$\lim_{t \to t_{k+1}(t>t_{k+1})} B_U^{(k+1)}(u)(t) = \lim_{t \to t_{k+1}(t>t_{k+1})}\left(\int_{t_{k+1}}^{t} V(U,I)(s)ds - \frac{t - t_{k+1}}{t_{k+2} - t_{k+1}}\int_{t_{k+1}}^{t_{k+2}} V(U,I)(s)ds\right) = 0,$$

$$B_I^{(0)}(U,I)(t_0) = \int_{t_0}^{t_0} J U,I)(s)ds - \frac{t_0 - t_0}{t_1 - t_0}\int_{t_0}^{t_1} J(U,I)(s)ds = 0,$$

$$\lim_{t \to t_{k+1}(t<t_{k+1})} B_I^{(k)}(U,I)(t) = \lim_{t \to t_{k+1}(t<t_{k+1})}\left(\int_{t_k}^{t} J(U,I)(s)ds - \frac{t - t_k}{t_{k+1} - t_k}\int_{t_k}^{t_{k+1}} J(U,I)(s)ds\right) = 0$$

$$\lim_{t \to t_{k+1}(t>t_{k+1})} B_I^{(k+1)}(u)(t) = \lim_{t \to t_{k+1}(t>t_{k+1})}\left(\int_{t_{k+1}}^{t} J(U,I)(s)ds - \frac{t - t_{k+1}}{t_{k+2} - t_{k+1}}\int_{t_{k+1}}^{t_{k+2}} J(U,I)(s)ds\right) = 0$$

and $\left(B_U(U,I)(t), B_I(U,I)(t)\right)$ is differentiable on every (t_k, t_{k+1}).

Besides it is easy to verify that

$$B_U(U,I)(t_k) = 0, \ B_U(U,I)(t_{k+1}) = 0, B_I(U,I)(t_k) = 0, \ B_I(U,I)(t_{k+1}) = 0.$$

Recall $\left|C_p^{-1}(I)\right| \le H_p|I| \equiv \dfrac{2\Phi_p - \phi_0}{2c_p\sqrt{\Phi_p}\sqrt{\Phi_p - \phi_0}}|I|, (p = 0,1)$ and

$$i_0(U,I)(t) = \frac{U(t) - I(t-T)}{2Z_0}, \quad i_1(U,I)(t) = \frac{U(t-T) - I(t)}{2Z_0}.$$

We note that if $t \in [t_k, t_{k+1}] \subset [T, 2T] \Rightarrow t - T \in [t_q, t_{q+1}] \subset [0, T]$ since $t_k - T = t_q$. In fact $I(t-T) = \bar{I}_0(t)$, where $\bar{I}_0(t) = I_0(t-T)$, $t \in [T, 2T]$. Therefore

$$\left|\bar{U}_0(t)\right| \le e^{-\beta} U_0 e^{\mu(t-t_k)}, t \in [t_k, t_{k+1}] \text{ and } \left|\bar{I}_0(t)\right| \le e^{-\beta} I_0 e^{\mu(t-t_k)}, t \in [t_k, t_{k+1}].$$

We must establish that

$$\left|B_U^{(k)}(U,I)(t)\right| \le U_0 e^{\mu(t-t_k)}, \left|B_I^{(k)}(U,I)(t)\right| \le I_0 e^{\mu(t-t_k)} \; t \in [t_k, t_{k+1}].$$

Indeed,

$$\left|B_U^{(k)}(U,I)(t)\right| \le \left|\int_{t_k}^{t} V(U,I)(s)ds\right| + \left|\frac{t-t_k}{t_{k+1}-t_k} \int_{t_k}^{t_{k+1}} V(U,I)(s)ds\right| \equiv V_1 + V_2.$$

$$V_1 \le \left|\int_{t_k}^{t} \frac{dI(s-T)}{dt}ds\right| + \frac{2Z_0}{\hat{L}_0} \int_{t_k}^{t} |E_0(s-T)|ds + \frac{Z_0}{\hat{L}_0} \int_{t_k}^{t} |U(s)|ds + \frac{Z_0}{\hat{L}_0} \int_{t_k}^{t} |I(s-T)|ds +$$

$$+\frac{2Z_0}{\hat{L}_0} \int_{t_k}^{t} |R_0(i_0)|ds + \frac{2Z_0}{\hat{L}_0} \int_{t_k}^{t} \left|\widetilde{C}_0^{-1}\left(\int_{t_k}^{s} i_0 d\theta\right)\right|ds \le$$

$$\le |I(t-T)| + \frac{2Z_0}{\hat{L}_0} U_{E0} \int_{t_k}^{t} e^{\mu(s-t_k)}ds + \frac{Z_0 U_0}{\hat{L}_0} \int_{t_k}^{t} e^{\mu(s-t_k)}ds + \frac{Z_0 I_0}{\hat{L}_0} e^{-\beta} \int_{t_k}^{t} e^{\mu(s-t_k)}ds +$$

$$+\frac{2Z_0}{\hat{L}_0} \sum_{n=1}^{m} \left|r_n^{(0)}\right| \left|\int_{t_k}^{t} \left|\frac{U(s) - I(s-T)}{2Z_0}\right|^n ds\right| + \frac{2Z_0 H_0}{\hat{L}_0} \int_{t_k}^{t} \left|\int_{t_k}^{s} \frac{U(\theta) - I(\theta-T)}{2Z_0} d\theta\right|ds \le$$

$$\le e^{\mu(t-t_k)} I_0 e^{-\beta} + \frac{2Z_0}{\hat{L}_0} U_{E0} \int_{t_k}^{t} e^{\mu(s-t_k)}ds + \frac{Z_0 U_0}{\hat{L}_0} \int_{t_k}^{t} e^{\mu(s-t_k)}ds + \frac{Z_0 I_0}{\hat{L}_0} e^{-\beta} \int_{t_k}^{t} e^{\mu(s-t_k)}ds +$$

$$+\frac{1}{\hat{L}_0} \sum_{n=1}^{m} \frac{\left|r_n^{(0)}\right|}{(2Z_0)^{n-1}} \int_{t_k}^{t} \left|U_0 e^{\mu(s-t_k)} + e^{-\beta} I_0 e^{\mu(s-t_k)}\right|^n ds + \frac{H_0}{\hat{L}_0} \int_{t_k}^{t} \left|\int_{t_k}^{s} U_0 e^{\mu(\theta-t_k)} + I_0 e^{-\beta} e^{\mu(\theta-t_k)} d\theta\right|ds \le$$

$$\le I_0 e^{\mu(t-t_k)} e^{-\beta} + \frac{2Z_0 U_{E0}}{\hat{L}_0} \int_{t_k}^{t} e^{\mu(s-t_k)}ds + \frac{Z_0 U_0}{\hat{L}_0} \int_{t_k}^{t} e^{\mu(s-t_k)}ds + e^{-\beta} \frac{Z_0 I_0}{\hat{L}_0} \int_{t_k}^{t} e^{\mu(s-t_k)}ds +$$

$$+\frac{1}{\hat{L}_0}\sum_{n=1}^{m}\frac{\left|r_n^{(0)}\right|\left(U_0+e^{-\beta}I_0\right)^n}{\left(2Z_0\right)^{n-1}}\int_{t_k}^{t}e^{n\mu(s-t_k)}ds+\frac{H_0}{\hat{L}_0}\left(U_0+e^{-\beta}I_0\right)\int_{t_k}^{t}\int_{t_k}^{s}e^{\mu(\theta-t_k)}d\theta ds\le$$

$$\le e^{-\beta}I_0e^{\mu(t-t_k)}+\frac{2Z_0U_{E0}}{\hat{L}_0}\frac{e^{\mu(t-t_k)}-1}{\mu}+\frac{Z_0U_0}{\hat{L}_0}\frac{e^{\mu(t-t_k)}-1}{\mu}+e^{-\beta}\frac{Z_0I_0}{\hat{L}_0}\frac{e^{\mu(t-t_k)}-1}{\mu}+$$

$$+\frac{1}{\hat{L}_0}\sum_{n=1}^{m}\frac{\left|r_n^{(0)}\right|\left(U_0+e^{-\beta}I_0\right)^n}{\left(2Z_0\right)^{n-1}}\frac{e^{n\mu(t-t_k)}-1}{n\mu}+\frac{H_0}{\hat{L}_0}\left(U_0+e^{-\beta}I_0\right)\int_{t_k}^{t}\frac{e^{\mu(s-t_k)}-1}{\mu}ds\le$$

$$\le e^{-\beta}I_0e^{\mu(t-t_k)}+e^{\mu(t-t_k)}\frac{2Z_0U_{E0}}{\mu\hat{L}_0}+e^{\mu(t-t_k)}\frac{Z_0U_0}{\mu\hat{L}_0}+e^{\mu(t-t_k)}e^{-\beta}\frac{Z_0I_0}{\mu\hat{L}_0}+$$

$$+\frac{e^{\mu(t-t_k)}-1}{\mu}\frac{1}{\hat{L}_0}\sum_{n=1}^{m}\frac{\left|r_n^{(0)}\right|\left(U_0+e^{-\beta}I_0\right)^n}{\left(2Z_0\right)^{n-1}}e^{(n-1)\mu_0}+\frac{H_0}{\mu\hat{L}_0}\left(U_0+e^{-\beta}I_0\right)\frac{e^{\mu(t-t_k)}-1}{\mu}\le$$

$$\le e^{\mu(t-t_k)}\left[e^{-\beta}I_0+\frac{2Z_0U_{E0}}{\mu\hat{L}_0}+\frac{U_0+e^{-\beta}I_0}{\mu\hat{L}_0}\left(Z_0+\sum_{n=1}^{m}\left|r_n^{(0)}\right|\left(I_0\right)^{n-1}e^{(n-1)\mu_0}+\frac{H_0}{\mu}\right)\right];$$

$$V_2\le\left|\int_{t_k}^{t_{k+1}}\frac{dI(s-T)}{dt}ds\right|+\frac{2Z_0}{\hat{L}_0}\int_{t_k}^{t_{k+1}}\left|E_0(s-T)\right|ds+\frac{Z_0}{\hat{L}_0}\int_{t_k}^{t_{k+1}}\left|U(s)\right|ds+\frac{Z_0}{\hat{L}_0}\int_{t_k}^{t_{k+1}}\left|I(s-T)\right|ds+$$

$$+\frac{2Z_0}{\hat{L}_0}\int_{t_k}^{t_{k+1}}\left|R_0(i_0)\right|ds+\frac{2Z_0}{\hat{L}_0}\int_{t_k}^{t_{k+1}}\left|\widetilde{C}_0^{-1}\left(\int_{t_k}^{s}i_0d\theta\right)\right|ds\le$$

$$\le\frac{2Z_0}{\hat{L}_0}U_{E0}\int_{t_k}^{t_{k+1}}e^{\mu(s-t_k)}ds+\frac{Z_0U_0}{\hat{L}_0}\int_{t_k}^{t_{k+1}}e^{\mu(s-t_k)}ds+\frac{Z_0I_0}{\hat{L}_0}e^{-\beta}\int_{t_k}^{t_{k+1}}e^{\mu(s-t_k)}ds+$$

$$+\frac{2Z_0}{\hat{L}_0}\sum_{n=1}^{m}\left|r_n^{(0)}\right|\int_{t_k}^{t_{k+1}}\left|\frac{U(s)-I(s-T)}{2Z_0}\right|^n ds+\frac{2Z_0H_0}{\hat{L}_0}\int_{t_k}^{t_{k+1}}\left|\int_{t_k}^{s}\frac{U(\theta)-I(\theta-T)}{2Z_0}d\theta\right|ds\le$$

$$\le\frac{2Z_0}{\hat{L}_0}U_{E0}\int_{t_k}^{t_{k+1}}e^{\mu(s-t_k)}ds+\frac{Z_0U_0}{\hat{L}_0}\int_{t_k}^{t_{k+1}}e^{\mu(s-t_k)}ds+\frac{Z_0I_0}{\hat{L}_0}e^{-\beta}\int_{t_k}^{t_{k+1}}e^{\mu(s-t_k)}ds+$$

$$+\frac{1}{\hat{L}_0}\sum_{n=1}^{m}\frac{\left|r_n^{(0)}\right|}{\left(2Z_0\right)^{n-1}}\int_{t_k}^{t_{k+1}}\left|U_0e^{\mu(s-t_k)}+e^{-\beta}I_0e^{\mu(s-t_k)}\right|^n ds+\frac{H_0}{\hat{L}_0}\int_{t_k}^{t_{k+1}}\left|\int_{t_k}^{s}U_0e^{\mu(\theta-t_k)}+I_0e^{-\beta}e^{\mu(\theta-t_k)}d\theta\right|ds\le$$

$$\le\frac{2Z_0U_{E0}}{\hat{L}_0}\frac{e^{\mu(t_{k+1}-t_k)}-1}{\mu}+\frac{Z_0U_0}{\hat{L}_0}\frac{e^{\mu(t_{k+1}-t_k)}-1}{\mu}+e^{-\beta}\frac{Z_0I_0}{\hat{L}_0}\frac{e^{\mu(t_{k+1}-t_k)}-1}{\mu}+$$

$$+\frac{1}{\hat{L}_0}\sum_{n=1}^{m}\frac{\left|r_n^{(0)}\right|\left(U_0+e^{-\beta}I_0\right)^n}{\left(2Z_0\right)^{n-1}}\frac{e^{n\mu(t_{k+1}-t_k)}-1}{n\mu}+\frac{H_0}{\hat{L}_0}\left(U_0+e^{-\beta}I_0\right)\int_{t_k}^{t_{k+1}}\frac{e^{\mu(s-t_k)}-1}{\mu}ds\le$$

$$\le\frac{e^{\mu_0}-1}{\mu}\frac{2Z_0U_{E0}}{\hat{L}_0}+\frac{e^{\mu_0}-1}{\mu}\frac{Z_0U_0}{\hat{L}_0}+\frac{e^{\mu_0}-1}{\mu}e^{-\beta}\frac{Z_0I_0}{\hat{L}_0}+$$

$$+\frac{e^{\mu_0}-1}{\mu}\frac{1}{\hat{L}_0}\sum_{n=1}^{m}\frac{\left|r_n^{(0)}\right|\left(U_0+e^{-\beta}I_0\right)^n}{\left(2Z_0\right)^{n-1}}e^{(n-1)\mu_0}+\frac{H_0}{\mu\hat{L}_0}\left(U_0+e^{-\beta}I_0\right)\frac{e^{\mu_0}-1}{\mu}\le$$

$$\le e^{\mu(t-t_k)}\left(e^{\mu_0}-1\right)\left[\frac{2Z_0U_{E0}}{\mu\hat{L}_0}+\frac{U_0+e^{-\beta}I_0}{\mu\hat{L}_0}\left(Z_0+\sum_{n=1}^{m}\left|r_n^{(0)}\right|I_0^{n-1}e^{(n-1)\mu_0}+\frac{H_0}{\mu}\right)\right]$$

$$\left|B_U^{(k)}(U,I)(t)\right|\le$$

$$\le e^{\mu(t-t_k)}\left[e^{-\beta}I_0+\frac{2Z_0U_{E0}}{\mu\hat{L}_0}+\frac{U_0+e^{-\beta}I_0}{\mu\hat{L}_0}\left(Z_0+\sum_{n=1}^{m}\left|r_n^{(0)}\right|I_0^{n-1}e^{(n-1)\mu_0}+\frac{H_0}{\mu}\right)\right]+$$

$$+e^{\mu(t-t_k)}\left(e^{\mu_0}-1\right)\left[\frac{2Z_0U_{E0}}{\mu\hat{L}_0}+\frac{U_0+e^{-\beta}I_0}{\mu\hat{L}_0}\left(Z_0+\sum_{n=1}^{m}\left|r_n^{(0)}\right|I_0^{n-1}e^{(n-1)\mu_0}+\frac{H_0}{\mu}\right)\right]\le$$

$$\le e^{\mu(t-t_k)}\left[e^{-\beta}I_0+e^{\mu_0}\frac{2Z_0U_{E0}}{\mu\hat{L}_0}+e^{\mu_0}\frac{U_0+e^{-\beta}I_0}{\mu\hat{L}_0}\left(Z_0+\sum_{n=1}^{m}\left|r_n^{(0)}\right|I_0^{n-1}e^{(n-1)\mu_0}+\frac{H_0}{\mu}\right)\right]\le e^{\mu(t-t_k)}U_0$$

Analogously

$$\left|B_I^{(k)}(U,I)(t)\right|\le$$

$$\le e^{\mu(t-t_k)}\left[e^{-\beta}U_0+e^{\mu_0}\frac{2Z_0U_{E1}}{\mu\hat{L}_1}+e^{\mu_0}\frac{U_0e^{-\beta}+I_0}{\mu\hat{L}_1}\left(Z_0+\sum_{n=1}^{m}\left|r_n^{(1)}\right|I_0^{n-1}e^{(n-1)\mu_0}+\frac{H_1}{\mu}\right)\right]\le e^{\mu(t-t_k)}U_0.$$

If we proceed as in § 3.5, implying the above inequalities we obtain the following conditions:

The contraction conditions (just for the derivatives) are:

$$\dot{K}_U=\left(1+\frac{e^{\mu_0}-1}{\mu_0}\right)\frac{1}{\mu}\left\{\frac{1}{\hat{L}_0^{\,2}}\left[U_{E0}+e^{\mu_0}\left(U_0+I_0e^{-\beta}\right)+\sum_{n=1}^{m}\left|r_n^0\right|I_0^n+U_0\right]\sum_{n=1}^{m}\left|\tilde{l}_n^{(0)}\right|I_0^n+\right.$$

$$+ \frac{1}{\hat{L}_0}\left[Z_0 + \sum_{n=1}^{m}\left|r_n^0\right|n\,\mathrm{I}_0^{n-1} + \frac{2\sqrt{\Phi_0}}{\mu\hat{L}_0 c_0\sqrt{\Phi_0 - \phi_0}} \right] \Bigg\} < 1,$$

$$\dot{K}_I = \left(1 + \frac{e^{\mu_0}-1}{\mu_0}\right)\frac{1}{\mu}\Bigg\{\frac{1}{\hat{L}_1^2}\left[U_{E_1} + e^{\mu_0}\left(U_0 e^{-\beta} + I_0\right) + \sum_{n=1}^{m}\left|r_n^{(1)}\right|\left(\mathrm{I}_0\right)^n + U_0\right]\times$$

$$\times \sum_{n=1}^{m}\left|\widetilde{l}_n^{(1)}\right|\left(\mathrm{I}_0\right)^n + \frac{1}{\hat{L}_1}\left[Z_0 + \sum_{n=1}^{m}\left|r_n^{(1)}\right|n\left(\mathrm{I}_0\right)^{n-1} + \frac{2\sqrt{\Phi_1}}{\mu c_1\sqrt{\Phi_1 - \phi_0}} \right]\Bigg\} < 1$$

that is, B is a contractive operator in a uniform space which completes the proof.

Remark 3.9.1. Take the obtained solution on $[T,2T]$ as an initial function, and obtain an oscillatory solution $[2T,3T]$ and so we obtain a global oscillatory solution.

CONCLUSION

Unlike Chapter II, the method outlined in the beginning of § 3.2 leads to system (3.2.4) where the Lipschitz constant is $\left|Z_0\right| > 1$. In that case, the neutral system cannot have a solution (cf. Chapter I, § 1.5). The second method however, gives a good rate of convergence for both nonlinear and linear cases (cf. § 3.6, § 3.7) though only in the case of "small" initial conditions.

Chapter IV

LOSSLESS TRANSMISSION LINES TERMINATED BY PARALLEL CONNECTED *RLC*-LOADS

ABSTRACT

The main purpose of the present chapter is to analyse the processes in a lossless transmission line terminated by parallel connected nonlinear *RLC*-loads at both ends (cf. Fig. 4.1). First we derive the boundary conditions based upon Kirchhoff's law. Then we formulate a mixed problem for the lossless transmission line equations and reduce it to a problem for a neutral system on the boundary. Then we formulate an existence-uniqueness theorem for a periodic solution in both linear and nonlinear cases. In this manner we demonstrate a unique approach to analysis of linear and nonlinear circuits. Finally we formulate conditions for existence-uniqueness of oscillatory solution of the nonlinear neutral system.

INTRODUCTION

The lossless transmission lines terminated by parallel connected nonlinear (and linear) *RLC*-elements have numerous applications. Therefore we investigate the mixed problem for Telegrapher's equations with boundary conditions generated by these loads. So in § 4.1 we derive the boundary conditions as a consequence of Kirchhoff's law, formulate the mixed problem for transmission line equations and propose a manner to reduce the mixed problem to an initial value problem for a neutral equation on the boundary. In § 4.2 we give an operator presentation of the periodic problem for neutral system of equations. In § 4.3 we prove an existence-uniqueness theorem for periodic regimes of a transmission line terminated by linear *RLC*-loads. In § 4.4 we analyze arising nonlinearities in view of the nonlinear characteristics of the loads. In § 4.5 we prove a theorem for existence-uniqueness of periodic solutions of the nonlinear neutral system, that is, when the loads are nonlinear ones. In § 4.6 we propose a different approach of reducing the mixed problem to an initial value problem for nonlinear neutral system. Finally in § 4.7 we prove an existence-uniqueness theorem for oscillatory solution of the neutral system.

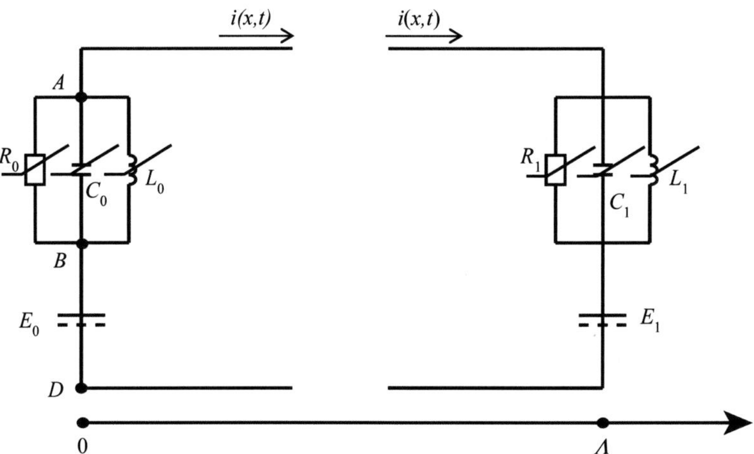

Figure 4.1.

4.1. DERIVATION OF BOUNDARY CONDITIONS, FORMULATION OF THE MIXED PROBLEM AND REDUCTION TO A PERIODIC PROBLEM FOR A NEUTRAL SYSTEM

Here, we consider again ideal (lossless) transmission lines terminated by parallel connected RLC-loads (cf. Fig. 4.1). We rely on, as before, the system of Telegrapher's equations

$$\frac{\partial u(x,t)}{\partial x} + L\frac{\partial i(x,t)}{\partial t} = 0,$$

$$\frac{\partial i(x,t)}{\partial x} + C\frac{\partial u(x,t)}{dt} = 0 \tag{4.1.1}$$

for the unknown voltage and current, where L and C are respectively, the inductance and capacitance of the guiding structure. In order to formulate the mixed problem for (4.1.1) we need initial conditions

$$u(x,0) = u_0(x), \quad i(x,0) = i_0(x), \quad x \in [0, \Lambda] \tag{4.1.2}$$

where $u_0(x)$ and $i_0(x)$ are prescribed voltages and current functions for $x \in [0, \Lambda]$, $t = 0$. The boundary conditions depend on the terminated parallel connected RLC-loads for $x = 0, t \geq 0$ and $x = \Lambda$, $t \geq 0$ (cf. Fig. 4.1).

In view of Fig. 4.1 and Kirchhoff's law we have
$u_{BA} = u(0,t) - E_0(t)$ and $i(0,t) = i_L + i_C + i_R$, where i_L, i_C, i_R are the currents through the coil, condenser and resistor, respectively, for $x = 0$.

We suppose, in general, that *RLC*-loads are nonlinear ones. This means they have nonlinear characteristics, that is, $C_0(u_{BA})$, $L_0(i_L)$, $G_0(u_{BA})$ are nonlinear functions. In view of

$$i_C = \frac{d[C_0(u_{BA})u_{BA}]}{dt}, \quad u_{BA} = \frac{d[L_0(i_L)i_L]}{dt}, \quad i_R = G_0(u_{BA})$$

we obtain

$$i_C = \frac{d[C_0(u_{BA})u_{BA}]}{dt} = \left[u_{BA}\frac{dC_0(u_{BA})}{du_{BA}} + C_0(u_{BA}) \right]\frac{du_{BA}}{dt},$$

$$\widetilde{L}_0(i_L(t)) \equiv i_L(t)L_0(i_L(t)) = \int_T^t u_{BA}(\tau)d\tau$$

and therefore $i_L(t) = \widetilde{L}_0^{-1}\left(\int_T^t u_{BA}(\tau)d\tau \right)$.

In view of $u_{BA} = u(0,t) - E_0(t)$ for the left end of the line, one obtains:

$$-i(0,t) = -\left[(u(0,t) - E_0(t))\frac{dC_0(u(0,t) - E_0(t))}{du} + C_0(u(0,t) - E_0(t)) \right]\left(\frac{du(0,t)}{dt} - \frac{dE_0(t)}{dt} \right) -$$

$$- \widetilde{L}_0^{-1}\left(\int_T^t (u(0,\tau) - E_0(\tau))d\tau \right) - G_0(u(0,t) - E_0(t)) \tag{4.1.3}$$

and for the right end

$$-i(\Lambda,t) = -\left[(u(\Lambda,t) - E_1(t))\frac{dC_1(u(\Lambda,t) - E_1(t))}{du} + C_1(u(\Lambda,t) - E_1(t)) \right]\left(\frac{du(\Lambda,t)}{dt} - \frac{dE_1(t)}{dt} \right) -$$

$$- \widetilde{L}_1^{-1}\left(\int_T^t (u(\Lambda,\tau) - E_1(\tau))d\tau \right) - G_1(u(\Lambda,t) - E_1(t)) \tag{4.1.4}$$

where $C_1(.)$, $L_1(.)$, $G_1(.)$ are, in general, also nonlinear functions.

Recalling denotations from the previous chapters $\widetilde{C}_p(u) = C_p(u)u \ (p = 0,1)$; $\widetilde{L}_p(i) = L_p(i)i \ (p = 0,1)$ we rewrite (4.1.3), (4.1.4) in the form:

$$-i(0,t) = -\frac{d\widetilde{C}_0(u(0,t) - E_0(t))}{du}\left(\frac{du(0,t)}{dt} - \frac{dE_0(t)}{dt} \right) - \widetilde{L}_0^{-1}\left(\int_T^t (u(0,\tau) - E_0(\tau))d\tau \right) - G_0(u(0,t) - E_0(t))$$

$$-i(\Lambda,t) = -\frac{d\widetilde{C}_1(u(\Lambda,t)-E_1(t))}{du}\left(\frac{du(\Lambda,t)}{dt}-\frac{dE_1(t)}{dt}\right)-\widetilde{L}_1^{-1}\left(\int\limits_T^t (u(\Lambda,\tau)-E_1(\tau))d\tau\right)-G_1(u(\Lambda,t)-E_1(t)).$$

In order to reduce the above mixed problem to an initial value problem for a neutral system on the boundary, we first rewrite (4.1.1) in the form:

$$\frac{\partial u(x,t)}{\partial t}+\frac{1}{C}\frac{\partial i(x,t)}{\partial x}=0,$$
$$\frac{\partial i(x,t)}{\partial t}+\frac{1}{L}\frac{\partial u(x,t)}{\partial x}=0 \tag{4.1.5}$$

and, repeating the reasoning from the beginning of Chapter II we obtain

$$\frac{\partial}{\partial t}\big(u(x,t)+Z_0\,i(x,t)\big)+v\frac{\partial}{\partial x}\big(u(x,t)+Z_0\,i(x,t)\big)=0,$$
$$\frac{\partial}{\partial t}\big(u(x,t)-Z_0\,i(x,t)\big)-v\frac{\partial}{\partial x}\big(u(x,t)-Z_0\,i(x,t)\big)=0 \tag{4.1.6}$$

where $Z_0 = \sqrt{L/C}$, $v = 1/\sqrt{LC}$.

Let us put

$$U(x,t) = u(x,t)+Z_0\,i(x,t), \quad I(x,t)=u(x,t)-Z_0\,i(x,t)$$

and then solve the last system with respect to $u(x,t)$, $i(x,t)$:

$$u(x,t) = \frac{1}{2}U(x,t)+\frac{1}{2}I(x,t),$$
$$i(x,t) = \frac{1}{2Z_0}U(x,t)-\frac{1}{2Z_0}I(x,t).$$

The last transformation leads to the system

$$\frac{\partial U(x,t)}{\partial t}+v\frac{\partial U(x,t)}{\partial x}=0,$$
$$\frac{\partial I(x,t)}{\partial t}-v\frac{\partial I(x,t)}{\partial x}=0$$

whose solutions are $\Phi(x-vt), \Psi(x+vt)$ where $\Phi(.), \Psi(.)$ are arbitrary differentiable functions.

Then

$$u(x,t) = \frac{1}{2}\big[\Phi(x-vt) + \Psi(x+vt)\big],$$

$$i(x,t) = \frac{1}{2Z_0}\big[\Phi(x-vt) - \Psi(x+vt)\big].$$

Let us put $x = \Lambda$. Then

$$u(\Lambda,t) = \frac{1}{2}\big[\Phi(\Lambda-vt) + \Psi(\Lambda+vt)\big],$$

$$i(\Lambda,t) = \frac{1}{2Z_0}\big[\Phi(\Lambda-vt) - \Psi(\Lambda+vt)\big].$$

$$(4.1.7)$$

For $x = 0$ we have

$$u(0,t) = \frac{1}{2}\big[\Phi(-vt) + \Psi(vt)\big],$$

$$i(0,t) = \frac{1}{2Z_0}\big[\Phi(-vt) - \Psi(vt)\big].$$

$$(4.1.8)$$

It follows

$$\Phi(\Lambda-vt) = u(\Lambda,t) + Z_0 i(\Lambda,t), \qquad \Psi(\Lambda+vt) = u(\Lambda,t) - Z_0 i(\Lambda,t) \qquad (4.1.9)$$

and

$$\Phi(-vt) = u(0,t) + Z_0 i(0,t), \qquad \Psi(vt) = u(0,t) - Z_0 i(0,t). \qquad (4.1.10)$$

We put

$$\Lambda - vt = -vt' \;\Rightarrow\; \Lambda + vt' = vt \;\Rightarrow\; t = \frac{\Lambda + vt'}{v} \;\; \text{or} \;\; t = t' + T$$

and

$$\Lambda + vt = vt'' \;\Rightarrow\; -\Lambda + vt'' = vt \;\Rightarrow\; t = \frac{-\Lambda + vt''}{v} \;\Leftrightarrow\; t = t'' - T.$$

Replace in (4.1.9) and get

$$\Phi(-vt') = u\big(\Lambda, t' + T\big) + Z_0 i\big(\Lambda, t' + T\big),$$

$$\Psi(vt'') = u(\Lambda, t''-T) - Z_0 i(\Lambda, t''-T).$$

or

$$\Phi(-vt) = u(\Lambda, t+T) + Z_0 i(\Lambda, t+T),$$
$$\Psi(vt) = u(\Lambda, t-T) - Z_0 i(\Lambda, t-T). \tag{4.1.11}$$

Replace $\Phi(-vt)$ and $\Psi(vt)$ from (4.1.11) in (4.1.10) and solve with respect to $u(0,t)$, $i(0,t)$:

$$u(0,t) = \frac{u(\Lambda, t+T) + Z_0 i(\Lambda, t+T) + u(\Lambda, t-T) - Z_0 i(\Lambda, t-T)}{2},$$

$$i(0,t) = \frac{u(\Lambda, t+T)}{2Z_0} + \frac{i(\Lambda, t+T)}{2} - \frac{u(\Lambda, t-T)}{2Z_0} + \frac{i(\Lambda, t-T)}{2}.$$

Now replace $u(0,t), i(0,t)$ into the first boundary condition (4.1.3):

$$-i(0,t) =$$
$$= -\frac{d\widetilde{C}_0(u(0,t) - E_0(t))}{du}\left(\frac{du(0,t)}{dt} - \frac{dE_0(t)}{dt}\right) - \widetilde{L}_0^{-1}\left(\int_T^t (u(0,\tau) - E_0(\tau))d\tau\right) - G_0(u(0,t) - E_0(t))$$

and obtain

$$-\frac{1}{2}\left[\frac{1}{Z_0}u(\Lambda, t+T) + i(\Lambda, t+T) - \frac{1}{Z_0}u(\Lambda, t-T) + i(\Lambda, t-T)\right] =$$

$$= -\frac{d\widetilde{C}_0\left(\dfrac{u(\Lambda, t+T) + Z_0 i(\Lambda, t+T) + u(\Lambda, t-T) - Z_0 i(\Lambda, t-T)}{2} - E_0(t)\right)}{du} \times$$

$$\times\left[\frac{1}{2}\left(\frac{du(\Lambda, t+T)}{dt} + Z_0\frac{di(\Lambda, t+T)}{dt} + \frac{du(\Lambda, t-T)}{dt} - Z_0\frac{di(\Lambda, t-T)}{dt}\right) - \frac{dE_0(t)}{dt}\right] -$$

$$- \widetilde{L}_0^{-1}\left(\int_T^t\left[\frac{u(\Lambda, \tau+T) + Z_0 i(\Lambda, \tau+T) + u(\Lambda, \tau-T) - Z_0 i(\Lambda, \tau-T)}{2} - E_0(\tau)\right]d\tau\right) -$$

$$- G_0\left(\frac{u(\Lambda, t+T) + Z_0 i(\Lambda, t+T) + u(\Lambda, t-T) - Z_0 i(\Lambda, t-T)}{2} - E_0(t)\right)$$

or

$$\frac{d\widetilde{C}_0\left(\dfrac{u(\Lambda,t+T)+Z_0i(\Lambda,t+T)+u(\Lambda,t-T)-Z_0i(\Lambda,t-T)}{2}-E_0(t)\right)}{du}\times$$

$$\times\left[\frac{1}{2}\left(\frac{du(\Lambda,t+T)}{dt}+Z_0\frac{di(\Lambda,t+T)}{dt}+\frac{du(\Lambda,t-T)}{dt}-Z_0\frac{di(\Lambda,t-T)}{dt}\right)-\frac{dE_0(t)}{dt}\right]=$$

$$=\frac{1}{2}\left[\frac{1}{Z_0}u(\Lambda,t+T)+i(\Lambda,t+T)-\frac{1}{Z_0}u(\Lambda,t-T)+i(\Lambda,t-T)\right]-$$

$$-\widetilde{L}_0^{-1}\left(\int_{2T}^{t+T}\left[\frac{u(\Lambda,\theta)+Z_0i(\Lambda,\theta)+u(\Lambda,\theta-2T)-Z_0i(\Lambda,\theta-2T)}{2}-E_0(\theta-T)\right]d\theta\right)-$$

$$-G_0\left(\frac{u(\Lambda,t+T)+Z_0i(\Lambda,t+T)+u(\Lambda,t-T)-Z_0i(\Lambda,t-T)}{2}-E_0(t)\right).$$

Replace in the last equation $t+T$ by t:

$$\frac{d\widetilde{C}_0}{du}\left(\frac{u(\Lambda,t)+Z_0i(\Lambda,t)+u(\Lambda,t-2T)-Z_0i(\Lambda,t-2T)}{2}-E_0(t-T)\right)\times$$

$$\times\left[\frac{1}{2}\left(\frac{du(\Lambda,t)}{dt}+Z_0\frac{di(\Lambda,t)}{dt}+\frac{du(\Lambda,t-2T)}{dt}-Z_0\frac{di(\Lambda,t-2T)}{dt}\right)-\frac{dE_0(t-T)}{dt}\right]=$$

$$=\frac{1}{2}\left[\frac{1}{Z_0}u(\Lambda,t)+i(\Lambda,t)-\frac{1}{Z_0}u(\Lambda,t-2T)+i(\Lambda,t-2T)\right]- \tag{4.1.12}$$

$$-\widetilde{L}_0^{-1}\left(\int_{T}^{t}\left[\frac{u(\Lambda,\theta)+Z_0i(\Lambda,\theta)+u(\Lambda,\theta-2T)-Z_0i(\Lambda,\theta-2T)}{2}-E_0(\theta-T)\right]d\theta\right)-$$

$$-G_0\left(\frac{u(\Lambda,t)+Z_0i(\Lambda,t)+u(\Lambda,t-2T)-Z_0i(\Lambda,t-2T)}{2}-E_0(t-T)\right).$$

Choose $u(\Lambda,t)$, $i(\Lambda,t)$ to be the unknown functions and introduce the denotations $u(t)\equiv u(\Lambda,t)$, $i(t)\equiv i(\Lambda,t)$. So we obtain the system:

$$\frac{d\widetilde{C}_0}{du}\left(\frac{u(t)+Z_0i(t)+u(t-2T)-Z_0i(t-2T)-2E_0(t-T)}{2}\right)\times$$

$$\times\frac{1}{2}\left(\frac{du(t)}{dt}+Z_0\frac{di(t)}{dt}+\frac{du(t-2T)}{dt}-Z_0\frac{di(t-2T)}{dt}-2\frac{dE_0(t-T)}{dt}\right)=$$

$$= \frac{1}{2Z_0} u(t) + \frac{1}{2} i(t) - \frac{1}{2Z_0} u(t-2T) + \frac{1}{2} i(t-2T) - \tag{4.1.13}$$

$$- \widetilde{L}_0^{-1} \left(\int_T^t \left(\frac{u(s) + Z_0 i(s) + u(s-2T) - Z_0 i(s-2T) - 2E_0(s-T)}{2} \right) ds \right) -$$

$$- G_0 \left(\frac{u(t) + Z_0 i(t) + u(t-2T) - Z_0 i(t-2T) - 2E_0(t-T)}{2} \right).$$

Now for the right-hand side, we obtain in an analogous way:

$$- i(\Lambda, t) = - \frac{d\widetilde{C}_1(u(\Lambda, t) - E_1(t))}{du} \left(\frac{du(\Lambda, t)}{dt} - \frac{dE_1(t)}{dt} \right) -$$

$$- \widetilde{L}_1^{-1} \left(\int_T^t \left(u(\Lambda, \tau) - E_1(\tau) \right) d\tau \right) - R_1 \left(u(\Lambda, t) - E_1(t) \right)$$

or

$$\frac{d\widetilde{C}_1(u(t) - E_1(t))}{du} \left(\frac{du(t)}{dt} - \frac{dE_1(t)}{dt} \right) = i(t) - \widetilde{L}_1^{-1} \left(\int_T^t \left(u(\tau) - E_1(\tau) \right) d\tau \right) - G_1 \left(u(t) - E_1(t) \right)$$

4.2. Operator Presentation of the Periodic Problem for a Neutral System with Linear *RLC*-Loads

Here we consider the particular case when the transmission line is terminated by linear *RLC*-loads. Our goal is to show a unified approach to treat linear and nonlinear problems.

When the loads are linear in the last system we obtain

$$\frac{C_0}{2} \frac{du(t)}{dt} + \frac{C_0 Z_0}{2} \frac{di(t)}{dt} + \frac{C_0}{2} \frac{du(t-2T)}{dt} - \frac{C_0 Z_0}{2} \frac{di(t-2T)}{dt} - C_0 \dot{E}_0(t-T) =$$

$$= \frac{1}{2Z_0} u(t) + \frac{1}{2} i(t) - \frac{1}{2Z_0} u(t-2T) + \frac{1}{2} i(t-2T) -$$

$$- \frac{1}{L_0} \int_T^t \left[E_0(\tau-T) - \frac{1}{2} \left(u(\tau) + Z_0 i(\tau) + u(\tau-2T) - Z_0 i(\tau-2T) \right) \right] d\tau +$$

$$+ G_0 E_0(t-T) - \frac{G_0}{2} u(t) - \frac{G_0 Z_0}{2} i(t) - \frac{G_0}{2} u(t-2T) + \frac{G_0 Z_0}{2} i(t-2T) \quad ;$$

$$C_1 \frac{du(t)}{dt} - C_1\dot{E}_1(t) = i(t) - \frac{1}{L_1}\int_T^t (E_1(\tau) - u(\tau))d\tau - G_1 E_1(t) + G_1 u(t).$$

We solve the above system with respect the derivatives of the unknown functions, and formulate the initial value problem for the obtained system: to find a solution of (4.2.1), (4.2.2) for prescribed initial functions (4.2.3). Recalling that with $\frac{du(t)}{dt} \equiv \dot{u}(t)$ we have

$$\frac{du(t)}{dt} = \frac{dE_1(t)}{dt} + \frac{1}{C_1}i(t) + \frac{G_1}{C_1}u(t) - \frac{1}{L_1 C_1}\int_T^t (E_1(\tau) - u(\tau))d\tau - \frac{G_1}{C_1}E_1(t)(t), \ t \in [T, 3T] \quad (4.2.1)$$

$$\frac{di(t)}{dt} = \frac{2}{Z_0}\frac{dE_0(t)}{dt} - \frac{1}{Z_0}\frac{du(t)}{dt} - \frac{1}{Z_0}\frac{du(t-2T)}{dt} + \frac{di(t-2T)}{dt} -$$

$$- \frac{1}{C_0 Z_0^2}u(t) - \frac{1}{C_0 Z_0}i(t) + \frac{1}{C_0 Z_0^2}u(t-2T) - \frac{1}{C_0 Z_0}i(t-2T) + \quad (4.2.2)$$

$$- \frac{2}{L_0 C_0 Z_0}\int_T^t \left(E_0(\tau) - \frac{u(\tau) + Z_0 i(\tau) + u(\tau - 2T) - Z_0 i(\tau - 2T)}{2} \right)d\tau +$$

$$+ \frac{2G_0}{C_0 Z_0}E_0(t-T) - \frac{G_0}{C_0 Z_0}u(t) + \frac{G_0}{C_0}i(t) + \frac{G_0}{C_0 Z_0}u(t-2T) - \frac{G_0}{C_0}i(t-2T), \ t \in [T, 3T],$$

$$\begin{aligned} u(t) &= \upsilon_0(t), \dot{u}(t) = \dot{\upsilon}_0(t) \\ i(t) &= \iota_0(t), \dot{i}(t) = \dot{\iota}_0(t) \end{aligned} \ , t \in [-T, T] \quad (4.2.3)$$

where the initial functions $\upsilon_0(t), \iota_0(t)$ might be obtained by the initial functions $u_0(x)$, $i_0(x)$ after translation along the characteristics as in Chapter II and $E_p(t)(p = 0,1)$ are prescribed functions such that $E_p(t-T) = E_p(t)$.

By $C_{T_0}^1[T, 3T]$ we mean the space of all continuous T_0-periodic functions with continuous derivatives. The main difficulty is defining a suitable operator whose fixed points are the solutions sought.

First we introduce the sets (recalling $T = mT_0$):

$$M_U = \left\{ u(.) \in C_{T_0}^1[T, 3T]: \int_{T+kT_0}^{T+(k+1)T_0} u(t)dt = 0 \ , k = 0,1,2,3,...,2m-1 \right\},$$

$$M_I = \left\{ i(.) \in C_{T_0}^1[T, 3T]: \int_{T+kT_0}^{T+(k+1)T_0} i(t)dt = 0 \ , k = 0,1,2,3,...,2m-1 \right\},$$

$$M_U^* = \left\{ u(.) \in M_U : |u(t)| \le U_0 e^{\mu(t-T-kT_0)} k = 0,1,2,3,...,2m-1 \right\},$$

$$M_I^* = \left\{ i(.) \in M_I : |i(t)| \le I_0 e^{\mu(t-T-kT_0)}, k = 0,1,2,3,...,2m-1 \right\}.$$

The set $M_U^* \times M_I^*$ turns into a complete metric space with respect to the metric:

$$\hat{\rho}_\mu((u,\dot{u},i,\dot{i}),(\bar{u},\dot{\bar{u}},\bar{i},\dot{\bar{i}})) = \max \left\{ \hat{\rho}(u,\bar{u}), \rho_\mu^{(k)}(\dot{u},\dot{\bar{u}}), \hat{\rho}(i,\bar{i}), \rho_\mu^{(k)}(i,\dot{\bar{i}}) : k = 0,1,2,3,...,2m-1 \right\}$$

where

$$\rho^{(k)}(u,\bar{u}) = \max \left\{ |u(t) - \bar{u}(t)| : t \in [T + kT_0, T + (k+1)T_0] \right\},$$

$$\hat{\rho}(u,\bar{u}) = \max \left\{ |u(t) - \bar{u}(t)| : t \in [T,3T] \right\},$$

$$\rho_\mu^{(k)}(\dot{u},\dot{\bar{u}}) = \max \left\{ e^{-\mu(t-T-kT_0)} |\dot{u}(t) - \dot{\bar{u}}(t)| : t \in [T + kT_0, T + (k+1)T_0] \right\},$$

$$\rho^{(k)}(i,\bar{i}) = \max \left\{ |i(t) - \vec{i}(t)| : t \in [T + kT_0, T + (k+1)T_0] \right\},$$

$$\hat{\rho}(i,\bar{i}) = \max \left\{ |i(t) - \bar{i}(t)| : t \in [T,3T] \right\},$$

$$\rho_\mu^{(k)}(i,\dot{\bar{i}}) = \max \left\{ e^{-\mu(t-T-kT_0)} |\dot{i}(t) - \dot{\bar{i}}(t)| : t \in [T + kT_0, T + (k+1)T_0] \right\}.$$

It is easy to see $\rho^{(k)}(u,\bar{u}) \le e^{\mu 0} \rho_\mu^{(k)}(u,\bar{u}) \le e^{\mu 0} \dfrac{\rho_\mu^{(k)}(\dot{u},\dot{\bar{u}})}{\mu}$ and

$$\rho^{(k)}(i,\bar{i}) \le e^{\mu 0} \rho_\mu^{(k)}(i,\bar{i}) \le e^{\mu 0} \dfrac{\rho_\mu^{(k)}(i,\dot{\bar{i}})}{\mu}.$$

Introduce the operator B as a pair $B(u,i)(t) = \left(B_u(u,i)(t), B_i(u,i)(t) \right)$, $t \in [T,3T]$ defined on every interval $[T + kT_0, T + (k+1)T_0]$ (for every $k = 0, 1, 2, ..., 2m-1$) by the expressions (recalling $u(T + kT_0) = u(T) = 0$, $i(T + kT_0) = i(T) = 0$):

$$B_u^{(k)}(u,i)(t) := \int_{T+kT_0}^{t} U(u,i)(s)ds - \left(\frac{t-T-kT_0}{T_0} - \frac{1}{2} \right) \int_{T+kT_0}^{T+(k+1)T_0} U(u,i)(s)ds - \frac{1}{T_0} \int_{T+kT_0}^{T+(k+1)T_0} \int_{T+kT_0}^{t} U(u,i)(s)dsdt$$

$$B_i^{(k)}(U,I)(t) := \int_{T+kT_0}^{t} I(u,i)(s)ds - \left(\frac{t-T-kT_0}{T_0} - \frac{1}{2} \right) \int_{T+kT_0}^{T+(k+1)T_0} I(u,i)(s)ds - \frac{1}{T_0} \int_{T+kT_0}^{T+(k+1)T_0} \int_{T+kT_0}^{t} I(u,i)(s)dsdt$$

where

$$U(u,i)(t) \equiv \dot{E}_1(t) + \frac{1}{C_1}i(t) + \frac{G_1}{C_1}u(t) - \frac{1}{L_1 C_1}\int_T^t \big(E_1(\tau) - u(\tau)\big)d\tau - \frac{G_1}{C_1}E_1(t),$$

$$I(u,i)(t) \equiv \frac{2}{Z_0}\dot{E}_0(t-T) - \frac{1}{Z_0}\frac{du(t)}{dt} - \frac{1}{Z_0}\frac{d\bar{\upsilon}_0(t)}{dt} + \frac{d\bar{\iota}_0(t)}{dt} -$$

$$-\frac{1}{C_0 Z_0^2}u(t) - \frac{1}{C_0 Z_0}i(t) + \frac{1}{C_0 Z_0^2}\bar{\upsilon}_0(t) - \frac{1}{C_0 Z_0}\bar{\iota}_0(t) +$$

$$-\frac{2}{L_0 C_0 Z_0}\int_T^t \left[E_0(\tau - T) - \frac{u(\tau) + Z_0 i(\tau) + \bar{\upsilon}_0(t) - Z_0\bar{\iota}_0(t)}{2}\right]d\tau +$$

$$+\frac{2G_0}{C_0 Z_0}E_0(t-T) - \frac{G_0}{C_0 Z_0}u(t) + \frac{G_0}{C_0}i(t) + \frac{G_0}{C_0 Z_0}\bar{\upsilon}_0(t) - \frac{G_0}{C_0}\bar{\iota}_0(t)$$

and $\bar{\upsilon}_0(t) = \upsilon_0(t - 2T), \bar{\iota}_0(t) = \iota_0(t - 2T), t \in [T, 3T]$, that is, by $\bar{\upsilon}_0(t), \bar{\iota}_0(t); t \in [T, 3T]$ are denoted, translated to the right, initial functions over $[T, 3T]$ (cf. Chapter II). As in the previous considerations we avoid conformity condition **(CC)**.

Recall that $\mu T_0 = \mu_0 = \text{const.}$

From now on, we suppose that the following assumptions are fulfilled:

Assumptions **(IN)**: $\upsilon_0(.), \iota_0(.) \in C_{T_0}^1[-T, T], \ \upsilon_0(-T) = 0, \ \iota_0(-T) = 0$,

$$\int_{-T+pT_0}^{-T+(p+1)T_0}\upsilon_0(t)dt = 0, \quad \int_{-T+pT_0}^{-T+(p+1)T_0}\iota_0(t)dt = 0; \ (p = 0,1,...,2m-1) ;$$

$$\big|\upsilon_0(t)\big| \le U_0 e^{-\beta}e^{\mu(t+T-pT_0)}, \ \big|\iota_0(t)\big| \le I_0 e^{-\beta}e^{\mu(t+T-pT_0)}, t \in [-T+nT_0, \ -T+(n+1)T_0], \beta = \text{const.} > 0$$

$(n = 0,1,...,2m-1)$.

Assumptions **(E)**: $E_p(.) \in C_{T_0}^1[0,\infty), (p = 0,1); \ \int_{nT_0}^{(n+1)T_0}E_p(t)dt = 0 \quad (n = 0,1,2,...),$

$$E_p(0) = 0; E_p(t-T) = E_p(t); \big|E_p(t)\big| \le U_{E_p}e^{\mu(t-T-kT_0)}, t \in [T+kT_0; T+(k+1)T_0];$$

$(k = 0,1,2,...)$.

Remark 4.2.1. It is easy to verify that

$$\big|\bar{\upsilon}_0(t)\big| \le U_0 e^{-\beta}e^{\mu(t-T-kT_0)}, \big|\bar{\iota}_0(t)\big| \le I_0 e^{-\beta}e^{\mu(t-T-kT_0)}, \ t \in [T+kT_0; T+(k+1)T_0].$$

Remark 4.2.2. Obviously

$$\left| u(t-2T) \right| \le U_0 e^{-\beta} e^{\mu(t-T-kT_0)}, \left| i(t-2T) \right| \le I_0 e^{-\beta} e^{\mu(t-T-kT_0)}.$$

Lemma 4.2.1. If $f(.) \in M_U^*$ or M_I^* then $F(t) = \int_T^t f(\tau)d\tau$ is a T_0 periodic function.

The proof is the same as that of Lemma 3.4.1.

Lemma 4.2.2. If the assumptions **(IN)** and **(E)** are satisfied and $(u,i) \in M_U^* \times M_I^*$ then $U(u,i)(t)$, $I(u,i)(t)$ are T_0 periodic functions.

The proof is the same as that of Lemma 3.4.2.

Remark 4.2.3. It is clear that by changing the integration order one obtains

$$\int_{T+kT_0}^{T+(k+1)T_0} \int_{T+kT_0}^{t} U(u,i)(s)ds\,dt = (T+(k+1)T_0) \int_{T+kT_0}^{T+(k+1)T_0} U(u,i)(s)ds - \int_{T+kT_0}^{T+(k+1)T_0} s\,U(u,i)(s)ds =$$

$$= (T+(k+1)T_0) \int_{T+kT_0}^{T+(k+1)T_0} \frac{du(s)}{ds}ds - \int_{T+kT_0}^{T+(k+1)T_0} s\,U(u,i)(s)ds = (T+(k+1)T_0)\big(u(T+(k+1)T_0)-u(T+kT_0)\big) -$$

$$- \int_{T+kT_0}^{T+(k+1)T_0} s\,U(u,i)(s)ds = - \int_{T+kT_0}^{T+(k+1)T_0} s\,U(u,i)(s)ds$$

and analogously

$$\int_{T+kT_0}^{T+(k+1)T_0} \int_{T+kT_0}^{t} I(u,i)(s)ds\,dt = (T+(k+1)T_0) \int_{T+kT_0}^{T+(k+1)T_0} I(u,i)(s)ds - \int_{T+kT_0}^{T+(k+1)T_0} s\,I(u,i)(s)ds = - \int_{T+kT_0}^{T+(k+1)T_0} s\,I(u,i)(s)ds$$

Lemma 4.2.3. For every $\big(u(.),i(.)\big) \in M_U^* \times M_I^*$ it follows

$$\int_{T+kT_0}^{T+(k+1)T_0} \int_{T+kT_0}^{s} U(u,i)(\theta)d\theta\,ds = \int_{T+(k+1)T_0}^{T+(k+2)T_0} \int_{T+(k+1)T_0}^{s} U(u,i)(\theta)d\theta\,ds \quad (k = 0,1,2,...),$$

$$\int_{T+kT_0}^{T+(k+1)T_0} \int_{T+kT_0}^{s} I(u,i)(\theta)d\theta\,ds = \int_{T+(k+1)T_0}^{T+(k+2)T_0} \int_{T+(k+1)T_0}^{s} I(u,i)(\theta)d\theta\,ds \quad (k = 0,1,2,...).$$

The proof is similar to one of Lemma 3.4.3.

Lemma 4.2.4. The initial value problem (4.2.1)-(4.2.3) has a solution $(u,i) \in M_U^* \times M_I^*$ iff the operator B has a fixed point $(u,i) \in M_U^* \times M_I^*$, that is, $u = B_u(u,i)$, $i = B_i(u,i)$.

Proof: Let $(u(.),i(.)) \in M_U^* \times M_I^*$ be a solution of the initial value problem:

$$\frac{du(t)}{dt} = U(u,i)(t), t \in [T,3T],$$ (4.2.4)

$$u(t) = \upsilon_0(t), \frac{du(t)}{dt} = \frac{d\upsilon_0(t)}{dt}, t \in [-T,T]$$

$$\frac{di(t)}{dt} = I(u,i)(t), t \in [T,3T]$$ (4.2.5)

$$i(t) = \iota_0(t), \frac{di(t)}{dt} = \frac{d\iota_0(t)}{dt}, t \in [-T,T].$$

Then after integration of (4.2.4) we have:

$$u(t) - u(T + kT_0) = \int_{T+kT_0}^{t} U(u,i)(t)dt \Rightarrow$$

$$u(T + (k+1)T_0) - u(T + kT_0) = \int_{T+kT_0}^{T+(k+1)T_0} U(u,i)(t)dt \Rightarrow \Rightarrow \int_{T+kT_0}^{T+(k+1)T_0} U(u,i)(t)dt = 0$$

and

$$i(t) - i(T + kT_0) = \int_{T+kT_0}^{t} I(u,i)(t)dt \Rightarrow$$

$$i(T + (k+1)T_0) - i(T + kT_0) = \int_{T+kT_0}^{T+(k+1)T_0} I(u,i)(t)dt \Rightarrow \Rightarrow \int_{T+kT_0}^{T+(k+1)T_0} I(u,i)(t)dt = 0.$$

From (4.2.4), (4.2.5), multiplying by t and integrating from $T + kT_0$ to $T + (k+1)T_0$

we obtain

$$\int_{T+kT_0}^{T+(k+1)T_0} t \frac{du(t)}{dt} dt = \int_{T+kT_0}^{T+(k+1)T_0} t\, U(u,i)(t)dt \text{ and } \int_{T+kT_0}^{T+(k+1)T_0} t \frac{di(t)}{dt} dt = \int_{T+kT_0}^{T+(k+1)T_0} t\, I(u,i)(t)dt$$

or

$$\int_{T+kT_0}^{T+(k+1)T_0} t\, du(t) = \int_{T+kT_0}^{T+(k+1)T_0} t\, U(u,i)(t)dt \text{ and } \int_{T+kT_0}^{T+(k+1)T_0} t\, di(t) = \int_{T+kT_0}^{T+(k+1)T_0} t\, I(u,i)(t)dt,$$

$$0 = (T + (k+1)T_0)u(T + (k+1)T_0) - (T + kT_0)u(T + kT_0) = \int_{T+kT_0}^{T+(k+1)T_0} t\, U(u,i)(t)\, dt$$

and

$$0 = (T + (k+1)T_0)i(T + (k+1)T_0) - (T + kT_0)i(T + kT_0) = \int_{T+kT_0}^{T+(k+1)T_0} t\, I(u,i)(t)\, dt .$$

It follows

$$0 = \int_{T+kT_0}^{T+(k+1)T_0} t\, U(u,i)(t)\, dt \text{ and } 0 = \int_{T+kT_0}^{T+(k+1)T_0} t\, I(u,i)(t)\, dt .$$

In view of Remark 4.2.3 it follows

$$\int_{T+kT_0}^{T+(k+1)T_0} \int_{T+kT_0}^{t} U(u,i)(s)\, ds\, dt = 0 \text{ and } \int_{T+kT_0}^{T+(k+1)T_0} \int_{T+kT_0}^{t} I(u,i)(s)\, ds\, dt = 0 .$$

Therefore

$$u(t) = \int_{T+kT_0}^{t} U(u,i)(s)\, ds , \quad i(t) = \int_{T+kT_0}^{t} I(u,i)(s)\, ds$$

is equivalent to

$$u(t) = \int_{T+kT_0}^{t} U(u,i)(s)\, ds - \left(\frac{t - T - kT_0}{T_0} - \frac{1}{2} \right) \int_{T+kT_0}^{T+(k+1)T_0} U(u,i)(s)\, ds - \frac{1}{T_0} \int_{T+kT_0}^{T+(k+1)T_0} \int_{T+kT_0}^{t} U(u,i)(s)\, ds\, dt,$$

$$i(t) = \int_{T+kT_0}^{t} I(u,i)(s)\, ds - \left(\frac{t - T - kT_0}{T_0} - \frac{1}{2} \right) \int_{T+kT_0}^{T+(k+1)T_0} I(u,i)(s)\, ds - \frac{1}{T_0} \int_{T+kT_0}^{T+(k+1)T_0} \int_{T+kT_0}^{t} I(u,i)(s)\, ds\, dt$$

and then thus the solution (u, i) of (4.2.4), (4.2.5) is a fixed point of B.

Conversely, let $(u(.),i(.)) \in M_U^* \times M_I^*$ be a fixed point B, that is,

$$u = B_u^{(k)}(u,i) , \ i = B_i^{(k)}(u,i) , \ t \in [T + kT_0, T + (k+1)T_0] .$$

Therefore

$$u(T + kT_0) = B_u^{(k)}(u,i)(T + (k+1)T_0), \ i(T + kT_0) = B_i^{(k)}(u,i)(T + (k+1)T_0)$$

or

$$0 = u(T + kT_0) = \int\limits_{T+kT_0}^{T+kT_0} U(u,i)(s)ds - \left(\frac{T + kT_0 - T - kT_0}{T_0} - \frac{1}{2} \right) \int\limits_{T+kT_0}^{T+(k+1)T_0} U(u,i)(s)ds -$$

$$- \frac{1}{T_0} \int\limits_{T+kT_0}^{T+(k+1)T_0} \int\limits_{T+kT_0}^{t} U(u,i)(s)ds\,dt = \frac{1}{2} \int\limits_{T+kT_0}^{T+(k+1)T_0} U(u,i)(s)ds - \frac{1}{T_0} \int\limits_{T+kT_0}^{T+(k+1)T_0} \int\limits_{T+kT_0}^{t} U(u,i)(s)ds\,dt$$

and

$$0 = i(T + kT_0) = \int\limits_{T+kT_0}^{T+kT_0} I(u,i)(s)ds - \left(\frac{T + kT_0 - T - kT_0}{T_0} - \frac{1}{2} \right) \int\limits_{T+kT_0}^{T+(k+1)T_0} I(u,i)(s)ds -$$

$$- \frac{1}{T_0} \int\limits_{T+kT_0}^{T+(k+1)T_0} \int\limits_{T+kT_0}^{t} I(u,i)(s)ds\,dt = \frac{1}{2} \int\limits_{T+kT_0}^{T+(k+1)T_0} I(u,i)(s)ds - \frac{1}{T_0} \int\limits_{T+kT_0}^{T+(k+1)T_0} \int\limits_{T+kT_0}^{t} I(u,i)(s)ds\,dt.$$

We show that $\int\limits_{T+kT_0}^{T+(k+1)T_0} I(u,i)(s)ds = 0$ which implies $\int\limits_{T+kT_0}^{T+(k+1)T_0} \int\limits_{T+kT_0}^{t} I(u,i)(s)ds\,dt = 0.$

Indeed,

$$\left| \int\limits_{T+kT_0}^{T+(k+1)T_0} U(u,i)(t)dt \right| \leq \left| \int\limits_{T+kT_0}^{T+(k+1)T_0} \dot{E}_1(t)dt \right| + \frac{1}{C_1} \int\limits_{T+kT_0}^{T+(k+1)T_0} |i(t)|dt + \frac{G_1}{C_1} \int\limits_{T+kT_0}^{T+(k+1)T_0} |u(t)|dt + \frac{1}{L_1 C_1} \int\limits_{T+kT_0}^{T+(k+1)T_0} \int\limits_{T}^{t} (E_1(\tau) - u(\tau))d\tau\,dt +$$

$$+ \frac{G_1}{C_1} \int\limits_{T+kT_0}^{T+(k+1)T_0} |E_1(t)|dt$$

$$\leq \frac{I_0}{C_1} \int\limits_{T+kT_0}^{T+(k+1)T_0} e^{\mu(t-T-kT_0)}dt + \frac{U_0 G_1}{C_1} \int\limits_{T+kT_0}^{T+(k+1)T_0} e^{\mu(t-T-kT_0)}dt + \frac{E_1 + U_0}{L_1 C_1} \int\limits_{T+kT_0}^{T+(k+1)T_0} \int\limits_{T}^{t} e^{\mu(\tau-T-kT_0)}d\tau\,dt \leq$$

$$\leq \frac{I_0}{C_1} \frac{e^{\mu T_0} - 1}{\mu} + \frac{U_0 G_1}{C_1} \frac{e^{\mu T_0} - 1}{\mu} + \frac{U_{E_1} + U_0}{L_1 C_1} \int\limits_{T+kT_0}^{T+(k+1)T_0} \frac{e^{\mu(t-T-kT_0)} - 1}{\mu}dt \leq$$

$$\leq \frac{I_0}{C_1} \frac{e^{\mu T_0} - 1}{\mu} + \frac{U_0 G_1}{C_1} \frac{e^{\mu T_0} - 1}{\mu} + \frac{U_{E_1} + U_0}{L_1 C_1} \frac{e^{\mu T_0} - 1}{\mu^2} \leq \frac{e^{\mu_0} - 1}{\mu C_1} \left(I_0 + U_0 G_1 + \frac{E_1 + U_0}{\mu L_1} \right) \equiv M_u(\mu)$$

and

$$\left| \int\limits_{T+kT_0}^{T+(k+1)T_0} I(u,i)(t)dt \right| \leq$$

$$\leq \frac{2}{Z_0} \left| \int\limits_{T+kT_0}^{T+(k+1)T_0} \dot{E}_0(t)dt \right| + \frac{1}{Z_0} \left| \int\limits_{T+kT_0}^{T+(k+1)T_0} \frac{du(t)}{dt}dt \right| + \frac{1}{Z_0} \left| \int\limits_{T+kT_0}^{T+(k+1)T_0} \frac{du(t-2T)}{dt}dt \right| + \left| \int\limits_{T+kT_0}^{T+(k+1)T_0} \frac{di(t-2T)}{dt}dt \right| +$$

$$+ \frac{1}{C_0 Z_0^2} \int\limits_{T+kT_0}^{T+(k+1)T_0} |u(t)|dt + \frac{1}{C_0 Z_0} \int\limits_{T+kT_0}^{T+(k+1)T_0} |i(t)|dt + \frac{1}{C_0 Z_0^2} \int\limits_{T+kT_0}^{T+(k+1)T_0} |u(t-2T)|dt + \frac{1}{C_0 Z_0} \int\limits_{T+kT_0}^{T+(k+1)T_0} |i(t-2T)|dt +$$

$$+\frac{2}{L_0C_0Z_0}\int\limits_{T+kT_0}^{T+(k+1)T_0}\int\limits_{T+kT_0}^{t}\left(|E_0(\tau)|+\frac{|u(\tau)|}{2}+\frac{Z_0|i(\tau)|}{2}+\frac{|u(\tau-2T)|}{2}+\frac{Z_0|i(\tau-2T)|}{2}\right)d\tau dt +$$

$$+\frac{2G_0}{C_0Z_0}\int\limits_{T+kT_0}^{T+(k+1)T_0}|E_0(t)|dt+\frac{G_0}{C_0Z_0}\int\limits_{T+kT_0}^{T+(k+1)T_0}|u(t)|dt+\frac{G_0}{C_0}\int\limits_{T+kT_0}^{T+(k+1)T_0}|i(t)|dt +$$

$$+\frac{G_0}{C_0Z_0}\int\limits_{T+kT_0}^{T+(k+1)T_0}|u(t-2T)|dt+\frac{G_0}{C_0}\int\limits_{T+kT_0}^{T+(k+1)T_0}|i(t-2T)|dt \le$$

$$\le\frac{U_0}{C_0Z_0^2}\int\limits_{T+kT_0}^{T+(k+1)T_0}e^{\mu(t-T-kT_0)}dt+\frac{I_0}{C_0Z_0}\int\limits_{T+kT_0}^{T+(k+1)T_0}e^{\mu(t-T-kT_0)}dt+\frac{U_0}{C_0Z_0^2}\int\limits_{T+kT_0}^{T+(k+1)T_0}e^{\mu(t-T-kT_0)}dt +$$

$$+\frac{I_0}{C_0Z_0}\int\limits_{T+kT_0}^{T+(k+1)T_0}e^{\mu(t-T-kT_0)}dt +$$

$$\frac{2}{L_0C_0Z_0}\left(U_{E_0}+\frac{U_0}{2}+\frac{Z_0I_0}{2}+\frac{U_0}{2}+\frac{Z_0I_0}{2}\right)\int\limits_{T+kT_0}^{T+(k+1)T_0}\frac{e^{\mu(t-T-kT_0)}-1}{\mu}dt +$$

$$+\frac{2G_0U_{E_0}}{C_0Z_0}\int\limits_{T+kT_0}^{T+(k+1)T_0}e^{\mu(t-T-kT_0)}dt+\frac{G_0U_0}{C_0Z_0}\int\limits_{T+kT_0}^{T+(k+1)T_0}e^{\mu(t-T-kT_0)}dt+\frac{G_0I_0}{C_0}\int\limits_{T+kT_0}^{T+(k+1)T_0}e^{\mu(t-T-kT_0)}dt +$$

$$+\frac{G_0U_0}{C_0Z_0}\int\limits_{T+kT_0}^{T+(k+1)T_0}e^{\mu(t-T-kT_0)}dt+\frac{G_0I_0}{C_0}\int\limits_{T+kT_0}^{T+(k+1)T_0}e^{\mu(t-T-kT_0)}dt \le$$

$$\le\frac{e^{\mu_0}-1}{\mu C_0}\left(\frac{2U_0}{Z_0^2}+\frac{2(I_0+G_0U_{E_0}+G_0U_0)}{Z_0}+\frac{2(U_{E_0}+U_0+Z_0I_0)}{\mu L_0Z_0}+2G_0I_0\right)\equiv M_i(\mu)\cdot$$

Since $M_u(\mu)\xrightarrow[\mu\to\infty]{}0,\ M_u(\mu)\xrightarrow[\mu\to\infty]{}0$ it follows

$$\int\limits_{t+kT_0}^{t+(k+1)T_0}U(u,i)(t)dt=0 \ \text{ and } \ \int\limits_{t+kT_0}^{t+(k+1)T_0}I(u,i)(t)dt=0.$$

Therefore

$$u=B_u^{(k)}(u,i)\,,\ i=B_i^{(k)}(u,i)\ ,\ t\in[T+kT_0,T+(k+1)T_0]$$

becomes

$$u(t)=\int\limits_{T+kT_0}^{t}U(u,i)(t)dt \quad i(t)=\int\limits_{T+kT_0}^{t}I(u,i)(t)dt$$
,

Differentiating the last equalities we obtain (4.2.4), (4.2.5).
Lemma 4.2.4 is thus proved.

Remark 4.2.2. Let us point out the previous Lemma 4.2.4 is valid for linear as well as nonlinear operators $B = (B_u, B_i)$.

4.3. PERIODIC SOLUTIONS OF THE LINEAR NEUTRAL SYSTEM

Here we formulate the problem for the existence-uniqueness of a periodic solution of the system obtained under the assumption that $L_p(.), G_p(.), C_p(.)$ ($p = 0,1$) are constant functions. For this reason we consider the following problem: to find a solution from $M_U^* \times M_I^*$ of the linear system (4.3.1) - (4.3.3):

$$\frac{du(t)}{dt} = \dot{E}_1(t) + \frac{1}{C_1}\left(i(t) + \frac{1}{L_1}\int_T^t \left(E_1(\tau) - u(\tau)\right)d\tau + G_1 E_1(t) - G_1 u(t)\right) \equiv U(u,i)(t), \ t \in [T,3T] \quad (4.3.1)$$

$$\frac{di(t)}{dt} = \frac{2}{Z_0}\dot{E}_0(t) - \frac{1}{Z_0}\frac{du(t)}{dt} - \frac{1}{Z_0}\frac{du(t-2T)}{dt} + \frac{di(t-2T)}{dt} -$$

$$- \frac{1}{C_0 Z_0^2}u(t) - \frac{1}{C_0 Z_0}i(t) + \frac{1}{C_0 Z_0^2}u(t-2T) - \frac{1}{C_0 Z_0}i(t-2T) + \quad (4.3.2)$$

$$- \frac{2}{L_0 C_0 Z_0}\int_T^t \left[E_0(\tau) - \frac{u(\tau) - Z_0 i(\tau) - u(\tau-2T) + Z_0 i(\tau-2T)}{2}\right]d\tau +$$

$$+ \frac{2G_0}{C_0 Z_0}E_0(t) - \frac{G_0}{C_0 Z_0}u(t) + \frac{G_0}{C_0}i(t) + \frac{G_0}{C_0 Z_0}u(t-2T) - \frac{G_0}{C_0}i(t-2T) \equiv I(u,i)(t), \ t \in [T,3T]$$

$$\begin{aligned}u(t) &= \upsilon_0(t), \frac{du(t)}{dt} = \frac{d\upsilon_0(t)}{dt}, \\ i(t) &= \iota_0(t), \frac{di(t)}{dt} = \frac{d\iota_0(t)}{dt}.\end{aligned} \quad t \in [-T,T]. \quad (4.3.3)$$

In the following we prove an existence-uniqueness result for T_0-periodic solution of (4.3.1) - (4.3.3) showing that the operator $B \equiv (B_u, B_i): M_U^* \times M_I^* \to M_U^* \times M_I^*$: 1) maps $M_U^* \times M_I^*$ into itself; 2) is a contractive operator with respect to the family of pseudo-metrics introduced above. The fixed-point theorems from Chapter I imply the existence-uniqueness of a periodic solution.

Theorem 4.3.1. Let the assumptions **(IN)**, **(E)** be fulfilled and

$$\left(1 + \frac{e^{\mu_0} - 1}{\mu_0}\right)U_{E_1} < U_0; \ \left(1 + \frac{e^{\mu_0} - 1}{\mu_0}\right)\frac{2U_{E_0} + U_0}{Z_0} < I_0; \ \frac{e^{\mu_0} - 1}{\mu_0}\frac{1}{Z_0} < 1.$$

Then there exists a unique T_0-periodic solution of (4.3.1) - (4.3.3).

Proof: We have to show that $B = (B_u, B_i)$ maps $M_U^* \times M_I^*$ into itself.

We can establish that the functions $(B_u(t), B_i(t))$ are continuously differentiable on $[T, \infty)$ as in Chapter III (cf. Lemma 3.4.4.).

In addition, a direct integration yields $\displaystyle\int_{T+kT_0}^{T+(k+1)T_0} B_u^{(k)}(u,i)(t)dt = 0$ and

$$\int_{T+kT_0}^{T+(k+1)T_0} B_i^{(k)}(u,i)(t)dt = 0 \, .$$

Further on we have

$$\left| B_u^{(k)}(u,i)(t) \right| \le \left| \int_{T+kT_0}^{t} U(u,i)(s)ds \right| + \left| \left(\frac{t-T-kT_0}{T_0} - \frac{1}{2} \right) \int_{T+kT_0}^{T+(k+1)T_0} U(u,i)(s)ds \right| + \left| \frac{1}{T_0} \int_{T+kT_0}^{T+(k+1)T_0} \int_{T+kT_0}^{t} U(u,i)(s)dsdt \right| \equiv$$

$$\equiv U_1 + U_2 + U_3 \, .$$

But

$$U_1 \le \left| \int_{T+kT_0}^{t} \dot{E}_1(s)ds \right| + \frac{1}{C_1} \int_{T+kT_0}^{t} |i(s)|ds + \frac{1}{L_1C_1} \int_{T+kT_0}^{t} \int_{T+kT_0}^{s} |E_1(\tau)-u(\tau)|d\tau ds + \frac{G_1}{C_1} \int_{T+kT_0}^{t} |E_1(s)|ds + \frac{G_1}{C_1} \int_{T+kT_0}^{t} |u(s)|ds \le$$

$$\le |E_1(t)| + \frac{I_0}{C_1} \frac{e^{\mu(t-T-kT_0)}-1}{\mu} + \frac{1}{L_1C_1}(U_{E_1}+U_0) \int_{T+kT_0}^{t} \int_{T+kT_0}^{s} e^{\mu(\tau-T-kT_0)}d\tau ds + \frac{G_1 U_{E_1}}{C_1} \int_{T+kT_0}^{t} e^{\mu(s-T-kT_0)}ds + \frac{G_1 U_0}{C_1} \int_{T+kT_0}^{t} e^{\mu(s-T-kT_0)}ds \le$$

$$\le |E_1(t)| + \frac{I_0}{C_1} \frac{e^{\mu(t-T-kT_0)}-1}{\mu} + \frac{U_{E_1}+U_0}{L_1C_1} \frac{e^{\mu(t-T-kT_0)}-1}{\mu^2} + \frac{G_1 U_{E_1}}{C_1} \frac{e^{\mu(t-T-kT_0)}-1}{\mu} + \frac{G_1 U_0}{C_1} \frac{e^{\mu(t-T-kT_0)}-1}{\mu} \le$$

$$\le e^{\mu(t-T-kT_0)}\left(U_{E_1} + \frac{I_0+G_1(U_{E_1}+U_0)}{\mu C_1} + \frac{U_{E_1}+U_0}{\mu^2 C_1 L_1} \right);$$

$$U_2 \le \frac{1}{2} \left| \int_{T+kT_0}^{T+(k+1)T_0} U(u,i)(s)ds \right| \le$$

$$\le \frac{1}{2}\left(\left| \int_{T+kT_0}^{T+(k+1)T_0} \dot{E}_1(s)ds \right| + \frac{1}{C_1} \int_{T+kT_0}^{T+(k+1)T_0} |i(s)|ds + \frac{1}{L_1C_1} \int_{T+kT_0}^{T+(k+1)T_0} \int_{T+kT_0}^{s} |E_1(\tau)-u(\tau)|d\tau ds + \frac{G_1}{C_1} \int_{T+kT_0}^{T+(k+1)T_0} |E_1(s)|ds + \frac{G_1}{C_1} \int_{T+kT_0}^{T+(k+1)T_0} |u(s)|ds \right) \le$$

$$\leq \frac{1}{2}\left(\frac{I_0}{C_1}\frac{e^{\mu T_0}-1}{\mu} + \frac{(U_{E_1}+U_0)}{L_1 C_1}\int\limits_{T+kT_0}^{T+(k+1)T_0}\frac{e^{\mu(s-T-kT_0)}-1}{\mu}ds + \frac{G_1(U_{E_1}+U_0)}{C_1}\int\limits_{T+kT_0}^{T+(k+1)T_0}e^{\mu(s-T-kT_0)}ds \right) \leq$$

$$\leq \frac{1}{2}\left(\frac{e^{\mu T_0}-1}{\mu}\frac{I_0}{C_1} + \frac{(U_{E_1}+U_0)}{\mu L_1 C_1}\frac{e^{\mu T_0}-1}{\mu} + \frac{G_1(U_{E_1}+U_0)}{C_1}\frac{e^{\mu T_0}-1}{\mu} \right) \leq \frac{e^{\mu_0}-1}{2}\left(\frac{I_0+G_1(U_{E_1}+U_0)}{\mu C_1} + \frac{U_{E_1}+U_0}{\mu^2 L_1 C_1} \right)$$

and

$$U_3 = \left| \frac{1}{T_0}\int\limits_{T+kT_0}^{T+(k+1)T_0}\int\limits_{T+kT_0}^{t}U(u,i)(s)\,ds\,dt \right| \leq \frac{1}{T_0}\int\limits_{T+kT_0}^{T+(k+1)T_0}\left| \int\limits_{T+kT_0}^{t}U(u,i)(s)\,ds \right|dt \leq$$

$$\leq \frac{1}{T_0}\left(U_{E_1} + \frac{I_0+G_1(U_{E_1}+U_0)}{\mu C_1} + \frac{U_{E_1}+U_0}{\mu^2 C_1 L_1} \right)\int\limits_{T+kT_0}^{T+(k+1)T_0}e^{\mu(t-T-kT_0)}dt \leq$$

$$\leq \frac{e^{\mu_0}-1}{\mu_0}\left(U_{E_1} + \frac{I_0+G_1(U_{E_1}+U_0)}{\mu C_1} + \frac{U_{E_1}+U_0}{\mu^2 C_1 L_1} \right).$$

Therefore

$$\left| B_u^{(k)}(u,i)(t) \right| \leq e^{\mu(t-T-kT_0)}\left(U_{E_1} + \frac{I_0+G_1(U_{E_1}+U_0)}{\mu C_1} + \frac{U_{E_1}+U_0}{\mu^2 C_1 L_1} \right) + e^{\mu(t-T-kT_0)}\frac{e^{\mu_0}-1}{2}\left(\frac{I_0}{\mu C_1} + \frac{G_1(U_{E_1}+U_0)}{\mu C_1} + \frac{U_{E_1}+U_0}{\mu^2 L_1 C_1} \right) +$$

$$+ e^{\mu(t-T-kT_0)}\frac{e^{\mu_0}-1}{\mu_0}\left(U_{E_1} + \frac{I_0+G_1(U_{E_1}+U_0)}{\mu C_1} + \frac{U_{E_1}+U_0}{\mu^2 C_1 L_1} \right) \leq$$

$$\leq e^{\mu(t-T-kT_0)}\left[\left(1+\frac{e^{\mu_0}-1}{\mu_0}\right)U_{E_1} + \left(1+\frac{e^{\mu_0}-1}{2}+\frac{e^{\mu_0}-1}{\mu_0}\right)\left(\frac{I_0+G_1(U_{E_1}+U_0)}{\mu C_1} + \frac{U_{E_1}+U_0}{\mu^2 C_1 L_1} \right) \right] \leq e^{\mu(t-T-kT_0)}U_0$$

because

$$\left(1+\frac{e^{\mu_0}-1}{\mu_0}\right)U_{E_1} < U_0 .$$

Recalling Remarks 4.1.1 and 4.1.2, namely

$$\left| \bar{v}_0(t) \right| = \left| u(t-2T) \right| \leq U_0 e^{-\beta}e^{\mu(t-T-kT_0)},$$

$$\left| \bar{\iota}_0(t) \right| = \left| i(t-2T) \right| \leq I_0 e^{-\beta}e^{\mu(t-T-kT_0)}$$

$t \in [T+kT_0,\ T+(k+1)T_0]$, we have

$$\left|B_i^{(k)}(u,i)(t)\right| \le \left|\int_{T+kT_0}^{t} I(u,i)(s)ds\right| + \left|\left(\frac{t-T-kT_0}{T_0}-\frac{1}{2}\right)\int_{T+kT_0}^{T+T_0} I(u,i)(s)ds\right| + \left|\frac{1}{T_0}\int_{T+kT_0}^{T+(k+1)T_0}\int_{T+kT_0}^{t} I(u,i)(s)dsdt\right| \equiv$$

$$\equiv I_1 + I_2 + I_3.$$

But

$$I_1 \le \frac{2}{Z_0}\left|\int_{T+kT_0}^{t}\dot{E}_0(s)ds\right| + \frac{1}{Z_0}\left|\int_{T+kT_0}^{t}\dot{u}(s)ds\right| + \frac{1}{Z_0}\left|\int_{T+kT_0}^{t}\dot{\vec{\upsilon}}_0(s)ds\right| + \left|\int_{T+kT_0}^{t}\dot{\vec{\iota}}_0(s)ds\right| +$$

$$+\frac{1}{C_0 Z_0^2}\int_{T+kT_0}^{t}|u(s)|ds + \frac{1}{C_0 Z_0}\int_{T+kT_0}^{t}|i(s)|ds + \frac{1}{C_0 Z_0^2}\int_{T+kT_0}^{t}|\vec{\upsilon}_0(s)|ds + \frac{1}{C_0 Z_0}\int_{T+kT_0}^{t}|\vec{\iota}_0(s)|ds +$$

$$+\frac{2}{L_0 C_0 Z_0}\int_{T+kT_0}^{t}\int_{T+kT_0}^{s}\left|E_0(\tau)-\frac{u(\tau)-Z_0 i(\tau)-\vec{\upsilon}_0(\tau)+Z_0\vec{\iota}_0(\tau)}{2}\right|d\tau ds +$$

$$+\int_{T+kT_0}^{t}\left|\frac{2G_0}{C_0 Z_0}E_0(s)-\frac{G_0}{C_0 Z_0}u(s)+\frac{G_0}{C_0}i(s)+\frac{G_0}{C_0 Z_0}\vec{\upsilon}_0(s)-\frac{G_0}{C_0}\vec{\iota}_0(s)\right|ds \le$$

$$\le \frac{2}{Z_0}|E_0(t)| + \frac{1}{Z_0}|u(t)| + \frac{1}{Z_0}|\vec{\upsilon}_0(t)| + |\vec{\iota}_0(t)| +$$

$$+\frac{U_0}{C_0 Z_0^2}\frac{e^{\mu(t-T-kT_0)}-1}{\mu} + \frac{I_0}{C_0 Z_0}\frac{e^{\mu(t-T-kT_0)}-1}{\mu} + \frac{U_0 e^{-\beta}}{C_0 Z_0^2}\frac{e^{\mu(t-T-kT_0)}-1}{\mu} + \frac{I_0 e^{-\beta}}{C_0 Z_0}\frac{e^{\mu(t-T-kT_0)}-1}{\mu} +$$

$$+\frac{2}{L_0 C_0 Z_0}\left(U_{E_0}+\frac{U_0}{2}+\frac{I_0 Z_0}{2}+\frac{U_0 e^{-\beta}}{2}+\frac{I_0 Z_0 e^{-\beta}}{2}\right)\int_{T+kT_0}^{t}\int_{T+kT_0}^{s}e^{\mu(\tau-T-kT_0)}d\tau ds +$$

$$+\frac{2G_0}{C_0 Z_0}\int_{T+kT_0}^{t}|E_0(s)|ds + \frac{G_0}{C_0 Z_0}\int_{T+kT_0}^{t}|u(s)|ds + \frac{G_0}{C_0}\int_{T+kT_0}^{t}|i(s)|ds + \frac{G_0}{C_0 Z_0}\int_{T+kT_0}^{t}|\vec{\upsilon}_0(s)|ds + \frac{G_0}{C_0}\int_{T+kT_0}^{t}|\vec{\iota}_0(s)|ds \le$$

$$\le e^{\mu(t-T-kT_0)}\left\{\left(\frac{2U_{E_0}+U_0+e^{-\beta}(U_0+Z_0 I_0)}{Z_0}\right)+\right.$$

$$\left.+\frac{1}{\mu C_0 Z_0}\left[\frac{2U_{E_0}}{\mu L_0}+2G_0 U_{E_0}+\left(1+e^{-\beta}\right)\left(U_0+Z_0 I_0\right)\left(\frac{1}{Z_0}+\frac{1}{\mu L_0}+G_0\right)\right]\right\};$$

$$I_2 \le \frac{1}{2}\left(\frac{2}{Z_0}\left|\int_{T+kT_0}^{T+(k+1)T_0}\dot{E}_0(s)ds\right| + \frac{1}{Z_0}\left|\int_{T+kT_0}^{T+(k+1)T_0}\dot{u}(s)ds\right| + \frac{1}{Z_0}\left|\int_{T+kT_0}^{T+(k+1)T_0}\dot{\vec{\upsilon}}_0(s)ds\right| + \left|\int_{T+kT_0}^{T+(k+1)T_0}\dot{\vec{\iota}}_0(s)ds\right| +\right.$$

$$+\frac{1}{C_0 Z_0^2}\int_{T+kT_0}^{T+(k+1)T_0}|u(s)|ds + \frac{1}{C_0 Z_0}\int_{T+kT_0}^{T+(k+1)T_0}|i(s)|ds + \frac{1}{C_0 Z_0^2}\int_{T+kT_0}^{T+(k+1)T_0}|\vec{\upsilon}_0(s)|ds + \frac{1}{C_0 Z_0}\int_{T+kT_0}^{T+(k+1)T_0}|\vec{\iota}_0(s)|ds +$$

$$+\frac{2}{L_0 C_0 Z_0}\int_{T+kT_0}^{T+(k+1)T_0}\int_{T+kT_0}^{s}\left|E_0(\tau)-\frac{u(\tau)+Z_0 i(\tau)+\vec{\upsilon}_0(\tau)-Z_0\vec{\iota}_0(\tau)}{2}\right|d\tau ds +$$

$$+\int_{T+kT_0}^{T+(k+1)T_0}\left|\frac{2G_0}{C_0Z_0}E_0(s)-G_0\frac{u(s)+Z_0i(s)+\vec{\upsilon}_0(s)-Z_0\vec{\iota}_0(s)}{2C_0Z_0}\right|ds\Bigg)\leq$$

$$\leq\frac{e^{\mu T_0}-1}{\mu C_0Z_0}\left[\frac{2U_{E_0}}{\mu L_0}+2U_{E_0}G_0+\left(1+e^{-\beta}\right)\!\left(U_0+Z_0I_0\right)\!\left(\frac{1}{Z_0}+\frac{1}{\mu L_0}+\frac{G_0}{2L_0}\right)\right]$$

and

$$I_3\leq\frac{1}{T_0}\int_{T+kT_0}^{T+(k+1)T_0}\left|\int_{T+kT_0}^{t}I(u,i)(s)ds\right|dt\leq$$

$$\leq\frac{1}{T_0}\int_{T+kT_0}^{T+(k+1)T_0}e^{\mu(t-T-kT_0)}dt\Bigg\{\left(\frac{2U_{E_0}+U_0+e^{-\beta}(U_0+Z_0I_0)}{Z_0}\right)+$$

$$+\frac{1}{\mu C_0Z_0}\left[\frac{2U_{E_0}}{\mu L_0}+2G_0U_{E_0}+\left(1+e^{-\beta}\right)\!\left(U_0+Z_0I_0\right)\!\left(\frac{1}{Z_0}+\frac{1}{\mu L_0}+G_0\right)\right]\Bigg\}\leq$$

$$\leq\frac{e^{\mu_0}-1}{\mu_0}\Bigg\{\left(\frac{2U_{E_0}+U_0+e^{-\beta}(U_0+Z_0I_0)}{Z_0}\right)+$$

$$+\frac{1}{\mu C_0Z_0}\left[\frac{2U_{E_0}}{\mu L_0}+2G_0U_{E_0}+\left(1+e^{-\beta}\right)\!\left(U_0+Z_0I_0\right)\!\left(\frac{1}{Z_0}+\frac{1}{\mu L_0}+G_0\right)\right]\Bigg\}.$$

Then

$$\left|B_i^{(k)}(u,i)(t)\right|\leq I_1+I_2+I_3\leq$$

$$\leq e^{\mu(t-T-kT_0)}\Bigg\{\left(\frac{2U_{E_0}+U_0+e^{-\beta}(U_0+Z_0I_0)}{Z_0}\right)+$$

$$+\frac{1}{\mu C_0Z_0}\left[\frac{2U_{E_0}}{\mu L_0}+2G_0U_{E_0}+\left(1+e^{-\beta}\right)\!\left(U_0+Z_0I_0\right)\!\left(\frac{1}{Z_0}+\frac{1}{\mu L_0}+G_0\right)\right]\Bigg\}+$$

$$+e^{\mu(t-T-kT_0)}\left(e^{\mu_0}-1\right)\frac{1}{\mu C_0Z_0}\left[\frac{2U_{E_0}}{\mu L_0}+2U_{E_0}G_0+\left(1+e^{-\beta}\right)\!\left(U_0+Z_0I_0\right)\!\left(\frac{1}{Z_0}+\frac{1}{\mu L_0}+\frac{G_0}{2L_0}\right)\right]+$$

$$+ e^{\mu(t-T-kT_0)} \frac{e^{\mu_0}-1}{\mu_0} \left\{ \left(\frac{2U_{E_0}+U_0+e^{-\beta}(U_0+Z_0I_0)}{Z_0} \right) + \right.$$

$$\left. + \frac{1}{\mu C_0 Z_0} \left[\frac{2U_{E_0}}{\mu L_0} + 2G_0 U_{E_0} + \left(1+e^{-\beta}\right)\left(U_0+Z_0I_0\right)\left(\frac{1}{Z_0}+\frac{1}{\mu L_0}+G_0\right) \right] \right\} \leq$$

$$\leq e^{\mu(t-T-kT_0)} \left\{ \left(1+\frac{e^{\mu_0}-1}{\mu_0}\right)\left(\frac{2U_{E_0}+U_0+e^{-\beta}(U_0+Z_0I_0)}{Z_0}\right) + \right.$$

$$\left. + \left(e^{\mu_0}+\frac{e^{\mu_0}-1}{\mu_0}\right)\frac{1}{\mu C_0 Z_0}\left[\frac{2U_{E_0}}{\mu L_0}+2G_0U_{E_0}+\left(1+e^{-\beta}\right)\left(U_0+Z_0I_0\right)\left(\frac{1}{Z_0}+\frac{1}{\mu L_0}+G_0\right)\right] \right\} \leq I_0 e^{\mu(t-T-kT_0)}$$

since $\left(1+\dfrac{e^{\mu_0}-1}{\mu_0}\right)\dfrac{2U_{E_0}+U_0}{Z_0} < I_0$.

It remains to show that the operator $B=(B_u,B_i)$ is contractive on $M_U^* \times M_I^*$.

Indeed, recalling 4) from 1.6.1 we have

$$\left| B_u^{(k)}(u,i)(t) - B_u^{(k)}(\bar{u},\bar{i})(t) \right| \leq \left| \int_{T+kT_0}^{t} \left(U(u,i)(s)-U(\bar{u},\bar{i})(s)\right)ds \right| +$$

$$+ \left| \left(\frac{t-T-kT_0}{T_0}-\frac{1}{2}\right) \int_{T+kT_0}^{T+(k+1)T_0} \left(U(u,i)(s)-U(\bar{u},\bar{i})(s)\right)ds \right| +$$

$$+ \left| \frac{1}{T_0} \int_{T+kT_0}^{T+(k+1)T_0} \int_{T+kT_0}^{t} \left(U(u,i)(s)-U(\bar{u},\bar{i})(s)\right)ds\,dt \right| \equiv U_1 + U_2 + U_3.$$

But

$$U_1 \leq \frac{1}{C_1}\int_{T+kT_0}^{t}\left|i(s)-\bar{i}(s)\right|ds + \frac{1}{C_1L_1}\int_{T+kT_0}^{t}\int_{T+kT_0}^{s}\left|u(\tau)-\bar{u}(\tau)\right|d\tau\,ds + \frac{G_1}{C_1}\int_{T+kT_0}^{t}\left|u(s)-\bar{u}(s)\right|ds \leq$$

$$\leq \frac{\rho^{(k)}(i,\bar{i})}{C_1}\frac{e^{\mu(t-T-kT_0)}-1}{\mu} + \frac{\rho^{(k)}(u,\bar{u})}{L_1C_1}\frac{e^{\mu(t-T-kT_0)}-1}{\mu^2} + \frac{G_1\rho^{(k)}(u,\bar{u})}{C_1}\frac{e^{\mu(t-T-kT_0)}-1}{\mu} \leq$$

$$\leq \frac{e^{\mu_0}\rho_\mu^{(k)}(i,\bar{i})}{\mu C_1}\frac{e^{\mu(t-T-kT_0)}-1}{\mu} + \frac{e^{\mu_0}\rho_\mu^{(k)}(\dot{u},\bar{\dot{u}})}{\mu L_1C_1}\frac{e^{\mu(t-T-kT_0)}-1}{\mu^2} + \frac{G_1e^{\mu_0}\rho_\mu^{(k)}(\dot{u},\bar{\dot{u}})}{\mu C_1}\frac{e^{\mu(t-T-kT_0)}-1}{\mu} \leq$$

$$\leq e^{\mu(t-T-kT_0)}\hat{\rho}_\mu((u,\dot{u},i,\dot{i}),(\bar{u},\dot{\bar{u}},\bar{i},\dot{\bar{i}}))\frac{e^{\mu T_0}}{\mu^2 C_1}\left(1+\frac{1}{\mu L_1}+G_1\right);$$

$$U_2 \leq \frac{1}{2}\left|\int_{T+kT_0}^{T+(k+1)T_0}\left(U(u,i)(s)-U(\bar{u},\bar{i})(s)\right)ds\right| \leq$$

$$\leq \frac{1}{2C_1}\int_{T+kT_0}^{T+(k+1)T_0}\left|i(s)-\bar{i}(s)\right|ds+\frac{1}{2L_1C_1}\int_{T+kT_0}^{T+(k+1)T_0}\int_{T+kT_0}^{s}\left|u(\tau)-\bar{u}(\tau)\right|d\tau ds+\frac{G_1}{2C_1}\int_{T+kT_0}^{T+(k+1)T_0}\left|u(s)-\bar{u}(s)\right|ds \leq$$

$$\leq \frac{\rho^{(k)}(i,\bar{i})}{2C_1}\frac{e^{\mu T_0}-1}{\mu}+\frac{\rho^{(k)}(u,\bar{u})}{2\mu L_1 C_1}\frac{e^{\mu T_0}-1}{\mu}+\frac{G_1\rho^{(k)}(u,\bar{u})}{2C_1}\frac{e^{\mu T_0}-1}{\mu} \leq$$

$$\leq \frac{e^{\mu T_0}\rho_\mu^{(k)}(i,\dot{\bar{i}})}{2\mu C_1}\frac{e^{\mu T_0}-1}{\mu}+\frac{e^{\mu T_0}\rho_\mu^{(k)}(\dot{u},\dot{\bar{u}})}{2\mu^2 L_1 C_1}\frac{e^{\mu T_0}-1}{\mu}+\frac{G_1 e^{\mu T_0}\rho_\mu^{(k)}(\dot{u},\dot{\bar{u}})}{2\mu C_1}\frac{e^{\mu T_0}-1}{\mu} \leq$$

$$\leq e^{\mu(t-T-kT_0)}\hat{\rho}_\mu((u,\dot{u},i,\dot{i}),(\bar{u},\dot{\bar{u}},\bar{i},\dot{\bar{i}}))\frac{e^{\mu T_0}-1}{2}\frac{e^{\mu T_0}}{\mu^2 C_1}\left(1+\frac{1}{\mu L_1}+G_1\right)$$

and

$$U_3 \leq \frac{1}{T_0}\int_{T+kT_0}^{T+(k+1)T_0}\left|\int_{T+kT_0}^{t}\left(U(u,i)(s)-U(\bar{u},\bar{i})(s)\right)ds\right|dt \leq$$

$$\leq \hat{\rho}_\mu((u,i),(\bar{u},\bar{i}))\left[\frac{1}{\mu^2 C_1}+\frac{1}{\mu^3 L_1 C_1}+\frac{G_1}{\mu^2 C_1}\right]\frac{1}{T_0}\int_{T+kT_0}^{T+(k+1)T_0}e^{\mu(t-T-kT_0)}dt \leq$$

$$\leq \hat{\rho}_\mu((u,\dot{u},i,\dot{i}),(\bar{u},\dot{\bar{u}},\bar{i},\dot{\bar{i}}))\frac{e^{\mu T_0}-1}{\mu_0}\frac{1}{\mu^2 C_1}\left(1+\frac{1}{\mu L_1}+G_1\right).$$

So we obtain

$$\left|B_u^{(k)}(u,i)(t)-B_u^{(k)}(\bar{u},\bar{i})(t)\right| \leq$$

$$\leq e^{\mu(t-T-kT_0)}\hat{\rho}_\mu((u,i),(\bar{u},\bar{i}))\left(1+\frac{e^{\mu T_0}-1}{2}+\frac{e^{\mu T_0}-1}{\mu_0}\right)\frac{1}{\mu^2 C_1}\left(1+\frac{1}{\mu L_1}+G_1\right) \equiv$$

$$\equiv e^{\mu(t-T-kT_0)}K_u\hat{\rho}_\mu((u,\dot{u},i,\dot{i}),(\bar{u},\dot{\bar{u}},\bar{i},\dot{\bar{i}}))$$

and consequently

$$\rho^{(k)}(B_u^{(k)}(u,i), B_u^{(k)}(\overline{u},\overline{i})) \leq e^{\mu T_0} K_u \hat{\rho}_\mu((u,\dot{u},i,\dot{i}),(\overline{u},\dot{\overline{u}},\overline{i},\dot{\overline{i}})).$$

Therefore $\hat{\rho}(B_u(u,i), B_u(\overline{u},\overline{i})) \leq e^{\mu_0} K_u \hat{\rho}_\mu((u,\dot{u},i,\dot{i}),(\overline{u},\dot{\overline{u}},\overline{i},\dot{\overline{i}}))$.

For the derivative we obtain

$$\left| \dot{B}_u^{(k)}(u,i)(t) - \dot{B}_u^{(k)}(\overline{u},\overline{i})(t) \right| \leq \left| U(u,i)(t) - U(\overline{u},\overline{i})(t) \right| +$$

$$\left| \frac{1}{T_0} \int_{T+kT_0}^{T+(k+1)T_0} \left(U(u,i)(s) - U(\overline{u},\overline{i})(s) \right) ds \right| \equiv U_1 + U_2.$$

But

$$U_1 \leq \frac{1}{C_1} \left| i(t) - \overline{i}(t) \right| + \frac{1}{L_1 C_1} \int_T^t \left| u(\tau) - \overline{u}(\tau) \right| d\tau + \frac{G_1}{C_1} \left| u(t) - \overline{u}(t) \right| \leq$$

$$\leq \frac{e^{\mu(t-T-kT_0)}}{C_1} \rho^{(k)}(i,\overline{i}) + \frac{1}{L_1 C_1} \rho^{(k)}(u,\overline{u}) \int_{T+kT_0}^t e^{\mu(\tau-T-kT_0)} d\tau + e^{\mu(t-T-kT_0)} \frac{G_1}{C_1} \rho^{(k)}(u,\overline{u}) \leq$$

$$\leq \frac{e^{\mu(t-T-kT_0)}}{C_1} \frac{\rho_\mu^{(k)}(i,\overline{i})}{\mu} + \frac{1}{L_1 C_1} \frac{\rho_\mu^{(k)}(u,\overline{u})}{\mu} \frac{e^{\mu(t-T-kT_0)}-1}{\mu} + e^{\mu(t-T-kT_0)} \frac{G_1}{C_1} \frac{\rho_\mu^{(k)}(u,\overline{u})}{\mu} \leq$$

$$\leq e^{\mu(t-T-kT_0)} \hat{\rho}_\mu((u,\dot{u},i,\dot{i}),(\overline{u},\dot{\overline{u}},\overline{i},\dot{\overline{i}})) \frac{1}{\mu C_1} \left(1 + \frac{1}{\mu L_1} + G_1 \right)$$

and

$$U_2 \leq \frac{1}{T_0} \int_{T+kT_0}^{T+(k+1)T_0} \left| U(u,i)(s) - U(\overline{u},\overline{i})(s) \right| ds \leq \hat{\rho}_\mu((u,i),(\overline{u},\overline{i})) \left(\frac{1}{\mu C_1} + \frac{1}{\mu^2 L_1 C_1} + \frac{G_1}{\mu C_1} \right) \frac{1}{T_0} \int_{T+kT_0}^{T+(k+1)T_0} e^{\mu(s-T-kT_0)} ds \leq$$

$$\leq e^{\mu(t-T-kT_0)} \hat{\rho}_\mu((u,\dot{u},i,\dot{i}),(\overline{u},\dot{\overline{u}},\overline{i},\dot{\overline{i}})) \frac{e^{\mu_0}-1}{\mu_0} \frac{1}{\mu C_1} \left(1 + \frac{1}{\mu L_1} + G_1 \right).$$

Then

$$\left| \dot{B}_u^{(k)}(u,i)(t) - \dot{B}_u^{(k)}(\overline{u},\overline{i})(t) \right| \leq$$

$$\le e^{\mu(t-T-kT_0)}\hat{\rho}_\mu((u,i),(\overline{u},\overline{i}))\left(1+\frac{e^{\mu_0}-1}{\mu_0}\right)\frac{1}{\mu C_1}\left(1+\frac{1}{\mu L_1}+G_1\right)\equiv e^{\mu(t-T-kT_0)}\dot{K}_u\,\hat{\rho}_\mu((u,\dot{u},i,\dot{i}),(\overline{u},\dot{\overline{u}},\overline{i},\dot{\overline{i}}))$$

and therefore

$$\rho_\mu^{(k)}\left(\dot{B}_u^{(k)}(u,i),\dot{B}_u^{(k)}(\overline{u},\overline{i})\right)\le \dot{K}_u\,\hat{\rho}_\mu((u,\dot{u},i,\dot{i}),(\overline{u},\dot{\overline{u}},\overline{i},\dot{\overline{i}}))\,.$$

Further on we have

$$\left|B_i^{(k)}(u,i)(t)-B_i^{(k)}(\overline{u},\overline{i})(t)\right|\le\left|\int\limits_{T+kT_0}^{t}\left(I(u,i)(s)-I(\overline{u},\overline{i})(s)\right)ds\right|+$$

$$+\left|\left(\frac{t-T-kT_0}{T_0}-\frac{1}{2}\right)\int\limits_{T+kT_0}^{T+(k+1)T_0}\left(I(u,i)(s)-I(\overline{u},\overline{i})(s)\right)ds\right|+$$

$$+\left|\frac{1}{T_0}\int\limits_{T+kT_0}^{T+(k+1)T_0}\int\limits_{T+kT_0}^{t}\left(I(u,i)(s)-I(\overline{u},\overline{i})(s)\right)dsdt\right|\equiv I_1+I_2+I_3\,.$$

But

$$I_1\le\frac{1}{Z_0}\left|\int\limits_{T+kT_0}^{t}(\dot{u}(s)-\dot{\overline{u}}(s))ds\right|+\frac{1}{Z_0}\left|\int\limits_{T+kT_0}^{t}(\dot{\vec{\upsilon}}_0(s)-\dot{\vec{\upsilon}}_0(s))ds\right|+$$

$$+\left|\int\limits_{T+kT_0}^{t}(\dot{\vec{\imath}}_0(s)-\dot{\vec{\imath}}_0(s))ds\right|+\frac{1}{C_0Z_0^2}\int\limits_{T+kT_0}^{t}\left|u(s)-\overline{u}(s)\right|ds+$$

$$+\frac{1}{C_0Z_0}\int\limits_{T+kT_0}^{t}\left|i(s)-\overline{i}(s)\right|ds+\frac{1}{C_0Z_0^2}\int\limits_{T+kT_0}^{t}\left|\vec{\upsilon}_0(s)-\vec{\upsilon}_0(s)\right|ds+$$

$$+\frac{1}{L_0C_0Z_0}\int\limits_{T+kT_0}^{t}\int\limits_{T+kT_0}^{s}\left|u(\tau)-\overline{u}(\tau)\right|d\tau ds+\frac{1}{L_0C_0}\int\limits_{T+kT_0}^{t}\int\limits_{T+kT_0}^{s}\left|i(\tau)-\overline{i}(\tau)\right|d\tau ds+$$

$$+\frac{1}{L_0C_0Z_0}\int\limits_{T+kT_0}^{t}\int\limits_{T+kT_0}^{s}\left|\vec{\upsilon}_0(\tau)-\vec{\upsilon}_0(\tau)\right|d\tau ds+\frac{1}{L_0C_0}\int\limits_{T+kT_0}^{t}\int\limits_{T+kT_0}^{s}\left|\vec{\imath}_0(\tau)-\vec{\imath}_0(\tau)\right|d\tau ds+$$

$$\frac{1}{C_0Z_0}\int\limits_{T+kT_0}^{t}\left|\vec{\imath}_0(\tau)-\vec{\imath}_0(\tau)\right|ds+$$

$$+\frac{G_0}{C_0Z_0}\int\limits_{T+kT_0}^{t}\left|u(s)-\overline{u}(s)\right|ds+\frac{G_0}{C_0}\int\limits_{T+kT_0}^{t}\left|i(s)-\overline{i}(s)\right|ds+\frac{G_0}{C_0Z_0}\int\limits_{T+kT_0}^{t}\left|\vec{\upsilon}_0(s)-\vec{\upsilon}_0(s)\right|ds+$$

$$+\frac{G_0}{C_0}\int\limits_{T+kT_0}^{t}\left|\vec{\imath}_0(s)-\vec{\imath}_0(s)\right|ds\le$$

$$\leq \frac{1}{Z_0}\left|\int_{T+kT_0}^{t}(\dot{u}(s)-\dot{\bar{u}}(s))ds\right| + \frac{1}{C_0 Z_0^2}\int_{T+kT_0}^{t}|u(s)-u(s)|ds + \frac{1}{C_0 Z_0}\int_{T+kT_0}^{t}|i(s)-\bar{i}(s)|ds +$$

$$+ \frac{1}{L_0 C_0 Z_0}\int_{T+kT_0}^{t}\int_{T+kT_0}^{s}|u(\tau)-\bar{u}(\tau)|d\tau ds + \frac{1}{L_0 C_0}\int_{T+kT_0}^{t}\int_{T+kT_0}^{s}|i(\tau)-\bar{i}(\tau)|d\tau ds +$$

$$+ \frac{G_0}{C_0 Z_0}\int_{T+kT_0}^{t}|u(s)-\bar{u}(s)|ds + \frac{G_0}{C_0}\int_{T+kT_0}^{t}|i(s)-\bar{i}(s)|ds \leq$$

$$\leq \frac{e^{\mu_0}\rho_\mu^{(k)}(\dot{u},\dot{\bar{u}})}{Z_0}\frac{e^{\mu(t-T-kT_0)}-1}{\mu} + \frac{e^{\mu_0}\rho_\mu^{(k)}(u,\bar{u})}{C_0 Z_0^2}\frac{e^{\mu(t-T-kT_0)}-1}{\mu} + \frac{e^{\mu_0}\rho_\mu^{(k)}(i,\bar{i})}{C_0 Z_0}\frac{e^{\mu(t-T-kT_0)}-1}{\mu} +$$

$$+ \frac{e^{\mu_0}\rho_\mu^{(k)}(u,\bar{u})}{L_0 C_0 Z_0}\frac{e^{\mu(t-T-kT_0)}-1}{\mu^2} + \frac{e^{\mu_0}\rho_\mu^{(k)}(i,\bar{i})}{L_0 C_0}\frac{e^{\mu(t-T-kT_0)}-1}{\mu^2} +$$

$$+ \frac{G_0 e^{\mu_0}\rho_\mu^{(k)}(u,\bar{u})}{C_0 Z_0}\frac{e^{\mu(t-T-kT_0)}-1}{\mu} + \frac{G_0 e^{\mu_0}\rho_\mu^{(k)}(i,\bar{i})}{C_0}\frac{e^{\mu(t-T-kT_0)}-1}{\mu} \leq$$

$$\leq \frac{e^{\mu(t-T-kT_0)}}{\mu}e^{\mu_0}\left[\frac{\rho_\mu^{(k)}(\dot{u},\dot{\bar{u}})}{Z_0} + \frac{\rho_\mu^{(k)}(\dot{u},\dot{\bar{u}})}{\mu C_0 Z_0^2} + \frac{\rho_\mu^{(k)}(i,\dot{\bar{i}})}{\mu C_0 Z_0} + \frac{\rho_\mu^{(k)}(\dot{u},\dot{\bar{u}})}{\mu^2 L_0 C_0 Z_0} + \frac{\rho_\mu^{(k)}(i,\dot{\bar{i}})}{\mu^2 L_0 C_0} +\right.$$

$$\left.+ \frac{G_0\rho_\mu^{(k)}(\dot{u},\dot{\bar{u}})}{\mu C_0 Z_0} + \frac{G_0\rho_\mu^{(k)}(i,\dot{\bar{i}})}{\mu C_0}\right] \leq$$

$$\leq \hat{\rho}_\mu((u,\dot{u},i,i),(\bar{u},\dot{\bar{u}},\bar{i},\dot{\bar{i}}))e^{\mu(t-T-kT_0)}e^{\mu_0}\left[\frac{1}{\mu Z_0} + \frac{1+Z_0}{\mu^2 C_0 Z_0}\left(\frac{1}{Z_0} + \frac{1}{\mu L_0} + G_0\right)\right];$$

$$I_2 \leq \frac{1}{2}\int_{T+kT_0}^{T+(k+1)T_0}|I(u,i)(s)-I(\bar{u},\bar{i})(s)|ds \leq$$

$$\leq \frac{1}{2Z_0}\left|\int_{T+kT_0}^{T+(k+1)T_0}(\dot{u}(s)-\dot{\bar{u}}(s))ds\right| + \frac{1}{2Z_0}\left|\int_{T+kT_0}^{T+(k+1)T_0}(\dot{\bar{v}}_0(s)-\dot{v}_0(s))ds\right| +$$

$$+ \frac{1}{2}\left|\int_{T+kT_0}^{T+(k+1)T_0}(\dot{\bar{i}}_0(s)-\dot{i}_0(s))ds\right| + \frac{1}{2C_0 Z_0^2}\int_{T+kT_0}^{T+(k+1)T_0}|u(s)-\bar{u}(s)|ds +$$

$$+ \frac{1}{2C_0 Z_0}\int_{T+kT_0}^{T+(k+1)T_0}|i(s)-\bar{i}(s)|ds + \frac{1}{2C_0 Z_0^2}\int_{T+kT_0}^{T+(k+1)T_0}|\bar{v}_0(s)-v_0(s)|ds +$$

$$\frac{1}{2C_0 Z_0}\int_{T+kT_0}^{T+(k+1)T_0}|\bar{i}_0(s)-i_0(s)|ds +$$

$$+\frac{1}{2L_0C_0Z_0}\int\limits_{T+kT_0}^{T+(k+1)T_0}\int\limits_{T+kT_0}^{s}|u(\tau)-\overline{u}(\tau)|d\tau ds+\frac{1}{2L_0C_0}\int\limits_{T+kT_0}^{T+(k+1)T_0}\int\limits_{T+kT_0}^{s}|i(\tau)-\overline{i}(\tau)|d\tau ds+$$

$$+\frac{1}{2L_0C_0Z_0}\int\limits_{T+kT_0}^{T+(k+1)T_0}\int\limits_{T+kT_0}^{s}|\overline{\upsilon}_0(\tau)-\vec{\upsilon}_0(\tau)|d\tau ds+\frac{1}{2L_0C_0}\int\limits_{T+kT_0}^{T+(k+1)T_0}\int\limits_{T+kT_0}^{s}|\overline{i}_0(\tau)-\vec{i}_0(\tau)|d\tau ds+$$

$$+\frac{G_0}{2C_0Z_0}\int\limits_{T+kT_0}^{T+(k+1)T_0}|u(s)-\overline{u}(s)|ds+\frac{G_0}{2C_0}\int\limits_{T+kT_0}^{T+(k+1)T_0}|i(s)-\overline{i}(s)|ds+\frac{G_0}{2C_0Z_0}\int\limits_{T+kT_0}^{T+(k+1)T_0}|\overline{\upsilon}_0(s)-\vec{\upsilon}_0(s)|ds+$$

$$+\frac{G_0}{2C_0}\int\limits_{T+kT_0}^{T+(k+1)T_0}|\overline{i}_0(s)-\vec{i}_0(s)|ds\le\frac{\rho_\mu^{(k)}(u,\overline{u})}{2C_0Z_0^2}\frac{e^{\mu T_0}-1}{\mu}+\frac{\rho_\mu^{(k)}(i,\overline{i})}{2C_0Z_0}\frac{e^{\mu T_0}-1}{\mu}+$$

$$+\frac{1}{2L_0C_0Z_0}\int\limits_{T+kT_0}^{T+(k+1)T_0}\int\limits_{T+kT_0}^{s}|u(\tau)-\overline{u}(\tau)|d\tau ds+\frac{1}{2L_0C_0}\int\limits_{T+kT_0}^{T+(k+1)T_0}\int\limits_{T+kT_0}^{s}|i(\tau)-\overline{i}(\tau)|d\tau ds+$$

$$+\frac{G_0e^{\mu 0}\rho_\mu^{(k)}(u,\overline{u})}{2C_0Z_0}\frac{e^{\mu T_0}-1}{\mu}+\frac{G_0e^{\mu 0}\rho_\mu^{(k)}(i,\overline{i})}{2C_0}\frac{e^{\mu T_0}-1}{\mu}\le$$

$$\le\frac{e^{\mu T_0}-1}{2\mu}e^{\mu 0}\left[\frac{\rho_\mu^{(k)}(u,\overline{u})}{C_0Z_0^2}+\frac{\rho_\mu^{(k)}(i,\overline{i})}{C_0Z_0}+\frac{\rho_\mu^{(k)}(u,\overline{u})}{\mu L_0C_0Z_0}+\frac{\rho_\mu^{(k)}(i,\overline{i})}{\mu L_0C_0}+\right.$$

$$\left.+\frac{\rho_\mu^{(k)}(i,\overline{i})e^{-2\mu T}}{L_0C_0}+\frac{G_0\rho_\mu^{(k)}(u,\overline{u})}{C_0Z_0}+\frac{G_0\rho_\mu^{(k)}(i,\overline{i})}{C_0}+\right]\le$$

$$\le\frac{e^{\mu T_0}-1}{2\mu}e^{\mu 0}\left[\frac{\rho_\mu^{(k)}(\dot{u},\dot{\overline{u}})}{\mu C_0Z_0^2}+\frac{\rho_\mu^{(k)}(\dot{i},\dot{\overline{i}})}{\mu C_0Z_0}+\frac{\rho_\mu^{(k)}(\dot{u},\dot{\overline{u}})}{\mu^2 L_0C_0Z_0}+\frac{\rho_\mu^{(k)}(\dot{i},\dot{\overline{i}})}{\mu^2 L_0C_0}+\frac{G_0\rho_\mu^{(k)}(\dot{u},\dot{\overline{u}})}{\mu C_0Z_0}+\frac{G_0\rho_\mu^{(k)}(\dot{i},\dot{\overline{i}})}{\mu C_0}\right]\le$$

$$\le\hat{\rho}_\mu((u,\dot{u},i,\dot{i}),(\overline{u},\dot{\overline{u}},\overline{i},\dot{\overline{i}}))e^{\mu(t-T-kT_0)}\frac{e^{\mu 0}-1}{2}e^{\mu 0}\frac{1+Z_0}{\mu^2C_0Z_0}\left(\frac{1}{Z_0}+\frac{1}{\mu L_0}+G_0\right)$$

and

$$I_3\le\frac{1}{T_0}\int\limits_{T+kT_0}^{T+(k+1)T_0}\left|\int\limits_{T+kT_0}^{t}\big(I(u,i)(s)-I(\overline{u},\overline{i})(s)\big)ds\right|dt\le$$

$$\le e^{\mu(t-T-kT_0)}\hat{\rho}_\mu((u,i),(\overline{u},\overline{i}))\frac{1}{T_0}\int\limits_{T+kT_0}^{T+(k+1)T_0}e^{\mu(t-T-kT_0)}dt\times$$

$$\left[\frac{1}{\mu Z_0}+\frac{1+Z_0}{\mu^2C_0}\left(\frac{1}{Z_0^2}+\frac{1}{\mu L_0Z_0}+\frac{G_0}{Z_0}\right)\right]\le$$

$$\leq e^{\mu(t-T-kT_0)}\hat{\rho}_\mu((u,\dot{u},i,\dot{i}),(\overline{u},\dot{\overline{u}},\overline{i},\dot{\overline{i}}))\frac{e^{\mu_0}-1}{\mu_0}e^{\mu_0}\left[\frac{1}{\mu Z_0}+\frac{1+Z_0}{\mu^2 C_0 Z_0}\left(\frac{1}{Z_0}+\frac{1}{\mu L_0}+G_0\right)\right].$$

Therefore

$$\left|B_i^{(k)}(u,i)(t)-B_i^{(k)}(\overline{u},\overline{i})(t)\right|\leq$$

$$\leq \hat{\rho}_\mu((u,\dot{u},i,\dot{i}),(\overline{u},\dot{\overline{u}},\overline{i},\dot{\overline{i}}))e^{\mu(t-T-kT_0)}e^{\mu_0}\left[\frac{1}{\mu Z_0}+\frac{1+Z_0}{\mu^2 C_0 Z_0}\left(\frac{1}{Z_0}+\frac{1}{\mu L_0}+G_0\right)\right]+$$

$$+\hat{\rho}_\mu((u,\dot{u},i,\dot{i}),(\overline{u},\dot{\overline{u}},\overline{i},\dot{\overline{i}}))e^{\mu(t-T-kT_0)}\frac{e^{\mu_0}-1}{2}e^{\mu_0}\frac{1+Z_0}{\mu^2 C_0 Z_0}\left(\frac{1}{Z_0}+\frac{1}{\mu L_0}+G_0\right)+$$

$$\leq e^{\mu(t-T-kT_0)}\hat{\rho}_\mu((u,\dot{u},i,\dot{i}),(\overline{u},\dot{\overline{u}},\overline{i},\dot{\overline{i}}))\frac{e^{\mu_0}-1}{\mu_0}e^{\mu_0}\left[\frac{1}{\mu Z_0}+\frac{1+Z_0}{\mu^2 C_0 Z_0}\left(\frac{1}{Z_0}+\frac{1}{\mu L_0}+G_0\right)\right]\leq$$

$$\leq e^{\mu(t-T)}\hat{\rho}_\mu((u,\dot{u},i,\dot{i}),(\overline{u},\dot{\overline{u}},\overline{i},\dot{\overline{i}}))\left[\left(1+\frac{e^{\mu_0}-1}{\mu_0}\right)\frac{e^{\mu_0}}{\mu Z_0}+\right.$$

$$\left.+\left(1+\frac{e^{\mu_0}-1}{2}+\frac{e^{\mu_0}-1}{\mu_0}\right)e^{\mu_0}\frac{1+Z_0}{\mu^2 Z_0}\left(\frac{1}{Z_0}+\frac{1}{\mu L_0}+G_0\right)\right]\equiv$$

$$\equiv e^{\mu(t-T-kT_0)}K_i\hat{\rho}_\mu((u,\dot{u},i,\dot{i}),(\overline{u},\dot{\overline{u}},\overline{i},\dot{\overline{i}})).$$

It follows

$$\hat{\rho}(B_i(u,i),B_i(\overline{u},\overline{i}))\leq e^{\mu_0}K_i\hat{\rho}_\mu((u,\dot{u},i,\dot{i}),(\overline{u},\dot{\overline{u}},\overline{i},\dot{\overline{i}})).$$

For the derivative we obtain

$$\left|\dot{B}_i^{(k)}(u,i)(t)-\dot{B}_i^{(k)}(\overline{u},\overline{i})(t)\right|\leq\left|I(u,i)(t)-I(\overline{u},\overline{i})(t)\right|+$$

$$\left|\frac{1}{T_0}\int_{T+kT_0}^{T+(k+1)T_0}\left(I(u,i)(s)-I(\overline{u},\overline{i})(s)\right)ds\right|\equiv I_1+I_2.$$

Since

$$I_1 \le \frac{1}{Z_0}\left|\dot{u}(t)-\dot{\overline{u}}(t)\right| + \frac{1}{Z_0}\left|\dot{\vec{\upsilon}}_0(t)-\dot{\vec{\upsilon}}_0(t)\right| + \left|\dot{\vec{i}}(t)-\dot{\overline{i}}(t)\right| +$$

$$+\frac{1}{C_0Z_0^2}\left|u(t)-\overline{u}(t)\right| + \frac{1}{C_0Z_0}\left|i(t)-\overline{i}(t)\right| + \frac{1}{C_0Z_0^2}\left|\vec{\upsilon}_0(t)-\vec{\upsilon}_0(t)\right| +$$

$$+\frac{1}{C_0Z_0}\left|\vec{i}_0(t)-\vec{i}_0(t)\right| + \frac{1}{L_0C_0Z_0}\int\limits_{T+kT_0}^{t}\left(\left|u(\tau)-\overline{u}(\tau)\right| + Z_0\left|i(\tau)-\overline{i}(\tau)\right| +\right.$$

$$\left.+\left|\vec{\upsilon}_0(t)-\vec{\upsilon}_0(t)\right) + Z_0\left|\vec{i}_0(t)-\vec{i}_0(t)\right|\right)d\tau +$$

$$+\frac{G_0}{C_0Z_0}\left|u(t)-\overline{u}(t)\right| + \frac{G_0}{C_0}\left|i(t)-\overline{i}(t)\right| + \frac{G_0}{C_0Z_0}\left|\vec{\upsilon}_0(t)-\vec{\upsilon}_0(t)\right| + \frac{G_0}{C_0}\left|\vec{i}_0(t)-\vec{i}_0(t)\right| \le$$

$$\le e^{\mu(t-T-kT_0)}\frac{\rho_\mu^{(k)}(\dot{u},\dot{\overline{u}})}{Z_0} + e^{\mu(t-T-kT_0)}\frac{\rho_\mu^{(k)}(u,\overline{u})}{C_0Z_0^2} + e^{\mu(t-T-kT_0)}\frac{\rho_\mu^{(k)}(i,\overline{i})}{C_0Z_0} +$$

$$+\frac{\rho_\mu^{(k)}(u,\overline{u})}{L_0C_0Z_0}\frac{e^{\mu(t-T-kT_0)}-1}{\mu} + \frac{Z_0\rho_\mu^{(k)}(i,\overline{i})}{L_0C_0Z_0}\frac{e^{\mu(t-T-kT_0)}-1}{\mu} +$$

$$+e^{\mu(t-T-kT_0)}\frac{G_0\rho_\mu^{(k)}(u,\overline{u})}{C_0Z_0} + e^{\mu(t-T-kT_0)}\frac{G_0\rho_\mu^{(k)}(i,\overline{i})}{C_0} \le$$

$$\le e^{\mu(t-T-kT_0)}\hat{\rho}_\mu((u,\dot{u},i,\dot{i}),(\overline{u},\dot{\overline{u}},\overline{i},\dot{\overline{i}}))\left[\frac{1}{Z_0} + \frac{1+Z_0}{\mu C_0Z_0}\left(\frac{1}{Z_0} + \frac{1}{\mu L_0} + G_0\right)\right]$$

and

$$I_2 \le \frac{1}{T_0}\int\limits_{T+kT_0}^{T+(k+1)T_0}\left|I(u,i)(t) - I(\overline{u},\overline{i})(t)\right|dt \le$$

$$\le \hat{\rho}_\mu((u,i),(\overline{u},\overline{i}))\frac{e^{\mu_0}-1}{T_0}\frac{1+Z_0}{\mu^2C_0}\left(\frac{1}{Z_0^2} + \frac{1}{\mu L_0Z_0} + \frac{G_0}{Z_0}\right) \le$$

$$\le e^{\mu(t-T-kT_0)}\hat{\rho}_\mu((u,\dot{u},i,\dot{i}),(\overline{u},\dot{\overline{u}},\overline{i},\dot{\overline{i}}))\frac{e^{\mu_0}-1}{\mu_0}\frac{1+Z_0}{\mu C_0Z_0}\left(\frac{1}{Z_0} + \frac{1}{\mu L_0} + G_0\right)$$

then

$$\left|\dot{B}_i^{(k)}(u,i)(t) - \dot{B}_i^{(k)}(\overline{u},\overline{i})(t)\right| \le$$

$$\leq e^{\mu(t-T-kT_0)}\hat{\rho}_\mu((u,\dot{u},i,\dot{i}),(\overline{u},\dot{\overline{u}},\overline{i},\dot{\overline{i}}))\left[\frac{1}{Z_0}+\frac{1+Z_0}{\mu C_0 Z_0}\left(\frac{1}{Z_0}+\frac{1}{\mu L_0}+G_0\right)\right]+$$

$$+e^{\mu(t-T-kT_0)}\hat{\rho}_\mu((u,\dot{u},i,\dot{i}),(\overline{u},\dot{\overline{u}},\overline{i},\dot{\overline{i}}))\frac{e^{\mu_0}-1}{\mu_0}\frac{1+Z_0}{\mu C_0 Z_0}\left(\frac{1}{Z_0}+\frac{1}{\mu L_0}+G_0\right)\leq$$

$$\leq e^{\mu(t-T-kT_0)}\hat{\rho}_\mu((u,\dot{u},i,\dot{i}),(\overline{u},\dot{\overline{u}},\overline{i},\dot{\overline{i}}))\left[\frac{1}{Z_0}+\left(1+\frac{e^{\mu_0}-1}{\mu_0}\right)\frac{1+Z_0}{\mu C_0 Z_0}\left(\frac{1}{Z_0}+\frac{1}{\mu L_0}+G_0\right)\right]\equiv$$

$$\equiv e^{\mu(t-T-kT_0)}\dot{K}_i\hat{\rho}_\mu((u,\dot{u},i,\dot{i}),(\overline{u},\dot{\overline{u}},\overline{i},\dot{\overline{i}}))$$

and therefore

$$\rho_\mu^{(k)}(\dot{B}_i^{(k)}(u,i),\dot{B}_i^{(k)}(\overline{u},\overline{i}))\leq\dot{K}_i\hat{\rho}_\mu((u,\dot{u},i,\dot{i}),(\overline{u},\dot{\overline{u}},\overline{i},\dot{\overline{i}})).$$

For sufficiently large $\mu>1$ it follows $K=\max\left\{e^{\mu_0}K_u,\dot{K}_u,e^{\mu_0}K_i,\dot{K}_i\right\}<1$. Then

$$\hat{\rho}_\mu((B_u,\dot{B}_u.B_i,\dot{B}_i).(\overline{B}_u,\dot{\overline{B}}_u,\overline{B}_i,\dot{\overline{B}}_i)\leq K\hat{\rho}_\mu((u,\dot{u},i,\dot{i}),(\overline{u},\dot{\overline{u}},\overline{i},\dot{\overline{i}})).$$

Consequently, the operator $B=(B_u,B_i)$ is contractive on the set $M_U^*\times M_I^*$. Its fixed point is a periodic solution of the initial value problem (4.2.1)-(4.2.3).

Theorem 4.3.1 is thus proved.

4.3.1. Numerical Example

For the applications we have to collect all inequalities guaranteeing an existence-uniqueness of a periodic solution:

$$\left(1+\frac{e^{\mu_0}-1}{\mu_0}\right)U_{E_1}<U_0;\ \left(1+\frac{e^{\mu_0}-1}{\mu_0}\right)\frac{2U_{E_0}+U_0}{Z_0}<I_0;\ \frac{e^{\mu_0}-1}{\mu_0}\frac{1}{Z_0}<1;$$

$$\left(1+\frac{e^{\mu_0}-1}{\mu_0}\right)U_{E_1}+\left(1+\frac{e^{\mu_0}-1}{2}+\frac{e^{\mu_0}-1}{\mu_0}\right)\left(\frac{I_0+G_1(U_{E_1}+U_0)}{\mu C_1}+\frac{U_{E_1}+U_0}{\mu^2 C_1 L_1}\right)\leq U_0;$$

$$\left(1+\frac{e^{\mu_0}-1}{\mu_0}\right)\left(\frac{2U_{E_0}+U_0+e^{-\beta}(U_0+Z_0 I_0)}{Z_0}\right)+$$

$$+\left(e^{\mu_0}+\frac{e^{\mu_0}-1}{\mu_0}\right)\frac{1}{\mu C_0 Z_0}\left[\frac{2U_{E_0}}{\mu L_0}+2G_0 U_{E_0}+\left(1+e^{-\beta}\right)\left(U_0+Z_0 I_0\right)\left(\frac{1}{Z_0}+\frac{1}{\mu L_0}+G_0\right)\right]\le I_0\,;$$

$$K_u=\left(1+\frac{e^{\mu_0}-1}{2}+\frac{e^{\mu_0}-1}{\mu_0}\right)\frac{1}{\mu^2 C_1}\left(1+\frac{1}{\mu L_1}+G_1\right)<1\,;$$

$$\dot{K}_u=\left(1+\frac{e^{\mu_0}-1}{\mu_0}\right)\frac{1}{\mu C_1}\left(1+\frac{1}{\mu L_1}+G_1\right)<1\,;$$

$$K_i=\left(1+\frac{e^{\mu_0}-1}{\mu_0}\right)\frac{1}{\mu^2 Z_0}+\left(1+\frac{e^{\mu_0}-1}{2}+\frac{e^{\mu_0}-1}{\mu_0}\right)\frac{1+Z_0}{\mu^2 Z_0}\left(\frac{1}{Z_0}+\frac{1}{\mu L_0}+G_0\right)<1\,;$$

$$\dot{K}_i=\frac{1}{Z_0}+\left(1+\frac{e^{\mu_0}-1}{\mu_0}\right)\frac{1+Z_0}{\mu C_0 Z_0}\left(\frac{1}{Z_0}+\frac{1}{\mu L_0}+G_0\right)<1.$$

Consider a line with the following specific parameters $\Lambda=1\,m,\ L=0{,}2\,\mu H/m$, $v=1/\sqrt{LC}=2{,}5.10^8\ Z_0=\sqrt{L/C}=50\,\Omega$. Then $T=\Lambda\sqrt{LC}=4.10^{-9}$ sec. Let us check the propagation of millimeter waves $\lambda_0=(1/4)10^{-3}\,m$. We have $f_0=1/\left(\lambda_0\sqrt{LC}\right)=10^{12}\,Hz\Rightarrow\ T_0=1/f_0=10^{-12}$ sec.

Choose $\mu=10^{12}$ and $\beta>0$ sufficiently large. Then $\mu T_0=\mu_0=1,$ and

$$T=4.10^{-9}.10^{12}T_0=4000.T_0\,.$$

Consequently the above inequalities become

$$eU_{E_1}<U_0\,;\ e\frac{2U_{E_0}+U_0}{Z_0}<I_0\,;\frac{e}{Z_0}<1\,;$$

$$eU_{E_1}+\frac{3e-1}{2\mu C_1}\left(I_0+G_1(U_{E_1}+U_0)+\frac{U_{E_1}+U_0}{\mu L_1}\right)\le U_0\,;$$

$$2\left[\frac{U_{E_0}+U_0}{Z_0}\left(e+\frac{3e-1}{2\mu C_0}\left(\frac{1}{\mu L_0}+G_0\right)\right)+\frac{3e-1}{2}\frac{U_0}{Z_0}\frac{1}{\mu C_0 Z_0}\right]\le I_0\,;$$

$$K_u = \frac{3e-1}{2} \frac{1}{\mu^2 C_1} \left(1 + \frac{1}{\mu L_1} + G_1\right) < 1;$$

$$\dot{K}_u = \frac{e}{\mu C_1}\left(1 + \frac{1}{\mu L_1} + G_1\right) < 1 \,;\, K_i = \frac{e}{\mu^2 Z_0} + \frac{3e-1}{2\mu^2}\frac{1+Z_0}{Z_0}\left(\frac{1}{Z_0} + \frac{1}{\mu L_0} + G_0\right) < 1;$$

$$\dot{K}_i = \frac{1}{Z_0} + e\frac{1+Z_0}{\mu C_0 Z_0}\left(\frac{1}{Z_0} + \frac{1}{\mu L_0} + G_0\right) < 1.$$

Obviously the last inequality plays a crucial role because it contains a term not dependent on μ. Indeed, since

$$\mu \hat{C} = 10^{12}.10^{-11}.1,92 = 19,2; \; \mu^2 \hat{C} = 10^{24}.10^{-11}.1,92 = 1,92.10^{13};$$

$$\frac{1}{L_0} = \frac{1}{L_1} \approx \frac{1}{2}; G_0 = G_1 = 0,25$$

we have $\dot{K}_i = \dfrac{1}{50} + e\dfrac{51}{19,2.50}\left(0,02 + 0,25\right) = 0,06 < 1.$

Obviously the method has a better rate of convergence for larger Z_0.

Remark 4.3.2. The above Theorem 4.3.1 can be applied to circuits with time-varying elements, provided the functions $C_p(t) \geq \hat{C}_p > 0$, $L_p(t) \geq \hat{L}_p > 0$ $(p = 0,1)$ possess strictly positive lower bounds.

4.4. ANALYSIS OF THE ARISING NONLINEARITIES

Here we consider the general nonlinear system

$$\left[\left(u(t) - E_1(t)\right)\frac{dC_1\left(u(t) - E_1(t)\right)}{du} + C_1(u(t) - E_1(t))\right]\left(\frac{du(t)}{dt} - \frac{dE_1(t)}{dt}\right) = \quad (4.4.1)$$

$$= -i(t) - \widetilde{L}_1^{-1}\left(\int_T^t [E_1(\tau) - u(\tau)]d\tau\right) - G_1(E_1(t) - u(t)),$$

$$\left[\left(\frac{u(t) + Z_0 i(t) + u(t - 2T) - Z_0 i(t - 2T)}{2} - E_0(t)\right)\frac{dC_0}{du} + \right.$$

$$+ C_0\left(\frac{u(t)+Z_0 i(t)+u(t-2T)-Z_0 i(t-2T)}{2}-E_0(t)\right)\Bigg]\times$$

$$\times\left[\frac{1}{2}\frac{du(t)}{dt}+\frac{Z_0}{2}\frac{di(t)}{dt}+\frac{1}{2}\frac{du(t-2T)}{dt}-\frac{Z_0}{2}\frac{di(t-2T)}{dt}-\frac{dE_0(t)}{dt}\right]=$$

$$=\frac{1}{2Z_0}u(t)+\frac{1}{2}i(t)-\frac{1}{2Z_0}u(t-2T)+\frac{1}{2}i(t-2T)- \tag{4.4.2}$$

$$-\tilde{L}_0^{-1}\left(\frac{1}{2}\int_T^t \big(u(\theta)+Z_0 i(\theta)+u(\theta-2T)-Z_0 i(\theta-2T)-2E_0(\theta)\big)d\theta\right)-$$

$$-G_0\left(\frac{u(t)+Z_0 i(t)+u(t-2T)-Z_0 i(t-2T)}{2}-E_0(t)\right),$$

$$u(t)=\upsilon_0(t),\frac{du(t)}{dt}=\frac{d\upsilon_0(t)}{dt},$$
$$i(t)=\iota_0(t),\frac{di(t)}{dt}=\frac{d\iota_0(t)}{dt} \qquad t\in[-T,T].$$

In what follows we suppose that $C_p(.),G_p(.),L_p(.)\,(p=0,1)$ are nonlinear functions, as in Chapter III. For the capacitance function we take

$$C_p(u)=\frac{c_p}{\big(1-(u/\Phi_p)\big)^{\frac{1}{h}}}=\frac{c_p\sqrt[h]{\Phi_p}}{\sqrt[h]{\Phi_p-u}}\,(p=0,1)$$

where c_p,Φ_p are positive constants and $h\in[2,3]$.

We assume that

$$(U_0):|u|\le\phi_0<\Phi=\min\{\Phi_1,\Phi_2\}.$$

Then

$$\frac{dC_p(u)}{du}=c_p\frac{d\big(1-(u/\Phi_p)\big)^{-\frac{1}{h}}}{du}=c_p\left(-\frac{1}{h}\right)\left(1-\frac{u}{\Phi_p}\right)^{-\frac{1}{h}-1}\left(-\frac{1}{\Phi_p}\right)=\frac{c_p}{h\Phi_p}\left(1-\frac{u}{\Phi_p}\right)^{-\frac{1}{h}-1};$$

$$\frac{d^2C_p(u)}{du^2}=\frac{d}{du}\left(\frac{c_p}{h\Phi_p}\left(1-\frac{u}{\Phi_p}\right)^{-\frac{1}{h}-1}\right)=\frac{c_p}{h\Phi_p}\left(-\frac{1}{h}-1\right)\left(1-\frac{u}{\Phi_p}\right)^{-\frac{1}{h}-2}\left(-\frac{1}{\Phi_p}\right)=$$

$$= \frac{c_p}{\Phi_p^{\,2}}\frac{1+h}{h^2}\left(1-\frac{u}{\Phi_p}\right)^{-\frac{1}{h}-2}.$$

We have to estimate the coefficients before derivatives in (4.4.1) and (4.4.2).

Recalling $\widetilde{C}_p(u)=uC_p(u),\,(p=0,1)$ we obtain

$$\widetilde{C}_p(u)=\frac{d(uC_p(u))}{du}=C_p(u)+u\frac{dC_p(u)}{du}=\frac{c_p\sqrt[h]{\Phi_p}}{\sqrt[h]{\Phi_p}-u}+u\frac{c_p}{h\Phi_p}\frac{\sqrt[h]{\Phi_p}^{\,h+1}}{\sqrt[h]{(\Phi_p-u)^{h+1}}}=$$

$$=c_p\sqrt[h]{\Phi_p}\,\frac{h\Phi_p-(h-1)u}{h(\Phi_p-u)^{1+\frac{1}{h}}}>0 .$$

Then $|u|\le\phi_0<\Phi$ implies $h\Phi_p-hu+u>0\Leftrightarrow u<\dfrac{h\Phi_p}{h-1}.$

For the derivative we obtain

$$\frac{d\widetilde{C}_p(u)}{du}=2\frac{dC_p(u)}{du}+u\frac{d^2C_p(u)}{du^2}=$$

$$=\frac{2c_p}{h\Phi_p}\left(1-\frac{u}{\Phi_p}\right)^{-\frac{1}{h}-1}+u\frac{c_p}{\Phi_p^{\,2}}\frac{1+h}{h^2}\left(1-\frac{u}{\Phi_p}\right)^{-\frac{1}{h}-2}=\frac{c_p\sqrt[h]{\Phi_p}}{h^2}\frac{2h\Phi_p-(h-1)u}{(\Phi_p-u)^{\frac{1+2h}{h}}}\Rightarrow$$

$$\left|\frac{d\widetilde{C}_p(u)}{du}\right|\le\frac{c_p\sqrt[h]{\Phi_p}}{h^2}\frac{2h\Phi_p+(h-1)\phi_0}{(\Phi_p-\phi_0)^{\frac{1+2h}{h}}}\equiv D_p\;(p=0,1).$$

It is easy to see that $-\phi_0\le u\le\phi_0\Rightarrow\dfrac{d\widetilde{C}_p(u)}{du}>0 .$

Therefore

$$\min\left\{\widetilde{C}_p(u):u\in[-\phi_0,\phi_0]\right\}=\widetilde{C}_p(-\phi_0)=c_p\sqrt[h]{\Phi_p}\,\frac{h\Phi_p+(h-1)\phi_0}{h(\Phi_p+\phi_0)^{1+\frac{1}{h}}}=\hat{C}_p>0 .$$

Therefore the arguments of the functions $\widetilde{C}_p(u)$ should satisfy

$$\left| u(t) - E_1(t) \right| \le \phi_0 \tag{4.4.3}$$

and

$$\left| \frac{u(t) + Z_0 i(t) + u(t-2T) - Z_0 i(t-2T)}{2} - E_0(t) \right| \le \phi_0 . \tag{4.4.4}$$

We have to formulate sufficient conditions that imply the above inequalities. Indeed,

$$\left| u(t) - E_1(t) \right| \le U_0 e^{\mu(t-T-kT_0)} + U_{E_1} e^{\mu(t-T-kT_0)} \le e^{\mu T_0}\left(U_0 + U_{E_1} \right) \le \phi_0$$

and

$$\left| \frac{u(t) + Z_0 i(t) + u(t-2T) - Z_0 i(t-2T) - 2E_0(t)}{2} \right| \le$$

$$\le e^{\mu(t-T-kT_0)} \frac{2U_{E_0} + \left(U_0 + Z_0 I_0 \right)\left(1 + e^{-\beta}\right)}{2} \le e^{\mu_0} \phi_0 .$$

Therefore we have to assume
Assumptions **(V)**:

(V1): $e^{\mu_0}\left(U_0 + U_{E_1} \right) \le \phi_0$; **(V0):** $e^{\mu_0}\left(U_0 + Z_0 I_0 + U_{E_0} \right) \le \phi_0 .$

It is clear that Assumptions **(V)** imply (4.4.3) and (4.4.4).
We consider polynomial type *I-V* characteristics for

$$i = G_p(u) = \sum_{n=1}^{m} g_n^{(p)} u^n, (p = 0,1) .$$

We need estimates for $\widetilde{L}_p \equiv i.L_p(i)(p = 0,1)$, where $L_p(i) = \sum_{n=1}^{m-1} l_n^{(p)} i^n$, that is,

$$\widetilde{L}_p(i) = i . L_p(i) = \sum_{n=1}^{m-1} l_n^{(p)} i^{n+1} \quad i \in [-I_0, I_0] .$$

Since $u = \dfrac{d\Psi_p}{dt} = \dfrac{d\left(L_p(i).i\right)}{dt} = \dfrac{d\left(\widetilde{L}_p(i)\right)}{dt}$ it follows

$$\widetilde{L}_p(i) = \int_T^t u(s)\,ds \Rightarrow i = \widetilde{L}_p^{-1}\left(\int_T^t u(s)\,ds\right).$$

We choose I_0 such that for $|i| \le I_0$ it follows $\dfrac{d\widetilde{L}_p(i)}{di} > 0$. Therefore the inverse function

$\widetilde{L}_p^{-1}(.)$ exists and $\widetilde{L}_p^{-1}(.): \left[-\sum_{n=1}^{m-1} l_n^{(p)} \mathrm{I}_0^{n+1}, \sum_{n=1}^{m-1} l_n^{(p)} \mathrm{I}_0^{n+1} \right] \to [-I_0, I_0]$.

The explicit form of the inverse function can be found in the applications.

Further on we use the following estimates:

$$\left| \widetilde{L}_p^{-1}(.) \right| \le \mathrm{I}_0 \le I_0 \quad (p = 0,1) \tag{4.4.5}$$

and

$$\left| \frac{d\widetilde{L}_p^{-1}(l)}{dl} \right| = 1 \Big/ \left| \frac{d\widetilde{L}_p(i)}{di} \right| \le 1 \Big/ \min\left\{ \sum_{n=1}^{m-1} l_n^{(p)} i^{n+1} : i \in [-I_0, I_0] \right\} = \frac{1}{\hat{L}_p} \le \frac{1}{\hat{L}} = \max\left\{ \frac{1}{\hat{L}_p} : p = 0,1 \right\}. \tag{4.4.6}$$

Introduce denotations

$$\widetilde{C}_1(u) \equiv \widetilde{C}_1\big(u(t) - E_1(t)\big);$$

$$\widetilde{C}_0(u,i) \equiv \widetilde{C}_0\left(\frac{u(t) + Z_0 i(t) + u(t-2T) - Z_0 i(t-2T)}{2} - E_0(t) \right).$$

Then we may use the last denotations to rewrite (4.4.1) as

$$\frac{du(t)}{dt} = \frac{dE_1(t)}{dt} + \frac{1}{\widetilde{C}_1(u)}\left[-i(t) - \widetilde{L}_1^{-1}\left(\int_T^t [E_1(\tau) - u(\tau)]\,d\tau \right) - G_1(E_1(t) - u(t)) \right], \quad t \in [T, 3T], \tag{4.4.7}$$

$$\frac{di(t)}{dt} = \frac{2}{Z_0}\frac{dE_0(t)}{dt} - \frac{1}{Z_0}\frac{du(t)}{dt} - \frac{1}{Z_0}\frac{du(t-2T)}{dt} + \frac{di(t-2T)}{dt} +$$

$$+ \frac{2}{Z_0 \widetilde{C}_0(u,i)}\left[\frac{u(t) + Z_0 i(t) - u(t-2T) + Z_0 i(t-2T)}{2Z_0} - \right.$$

$$+ - \widetilde{L}_0^{-1}\left(\int_T^t \left(\frac{u(\theta)}{2} + \frac{Z_0 i(\theta)}{2} + \frac{u(\theta - 2T)}{2} - \frac{Z_0 i(\theta - 2T)}{2} - E_0(\theta) \right) d\theta \right) -$$

$$\left. - G_0\left(\frac{u(t)}{2} + \frac{Z_0 i(t)}{2} + \frac{u(t-2T)}{2} - \frac{Z_0 i(t-2T)}{2} - E_0(t) \right) \right], \quad t \in [T, 3T],$$

$$u(t) = \upsilon_0(t), \quad \frac{du(t)}{dt} = \frac{d\upsilon_0(t)}{dt}, \quad i(t) = \iota_0(t), \quad \frac{di(t)}{dt} = \frac{d\iota_0(t)}{dt} \quad t \in [-T,T].$$

Introduce the operator B as a pair

$$B(u,i)(t) = \big(B_u(u,i)(t),\, B_i(u,i)(t)\big), \quad t \in [T,3T]$$

defined on every interval $[T + kT_0, T + (k+1)T_0]$ (for every $k = 0,1,2,\ldots,2m-1)$) by the expressions (recalling $u(T) = u(T + kT_0) = 0$, $i(T) = i(T + kT_0) = 0$) :

$$B_u^{(k)}(u,i)(t) := \int_{T+kT_0}^{t} U(u,i)(s)\,ds - \left(\frac{t-T-kT_0}{T_0} - \frac{1}{2}\right)\int_{T+kT_0}^{T+(k+1)T_0} U(u,i)(s)\,ds - \frac{1}{T_0}\int_{T+kT_0}^{T+(k+1)T_0}\int_{T+kT_0}^{t} U(u,i)(s)\,ds\,dt,$$

$$B_i^{(k)}(u,i)(t) := \int_{T+kT_0}^{t} I(u,i)(s)\,ds - \left(\frac{t-T-kT_0}{T_0} - \frac{1}{2}\right)\int_{T+kT_0}^{T+(k+1)T_0} I(u,i)(s)\,ds - \frac{1}{T_0}\int_{T+kT_0}^{T+(k+1)T_0}\int_{T+kT_0}^{t} I(u,i)(s)\,ds\,dt$$

where

$$U(u,i)(t) = \frac{dE_1(t)}{dt} + \frac{1}{\widetilde{C}_1(u)}\left[-i(t) - \widetilde{L}_1^{-1}\left(\int_T^t [E_1(\tau) - u(\tau)]d\tau\right) - G_1(E_1(t) - u(t))\right],$$

$$I(u,i) = \frac{2}{Z_0}\frac{dE_0(t)}{dt} - \frac{1}{Z_0}\frac{du(t)}{dt} - \frac{1}{Z_0}\frac{d\vec{\upsilon}_0(t)}{dt} + \frac{d\vec{\iota}_0(t)}{dt} +$$

$$+ \frac{2}{Z_0\overline{C}_0(u,i)}\left[\frac{u(t)}{2Z_0} + \frac{i(t)}{2} - \frac{\vec{\upsilon}_0(t)}{2Z_0} + \frac{\vec{\iota}_0(t)}{2} -\right.$$

$$- \widetilde{L}_0^{-1}\left(\int_T^t \left(\frac{1}{2}u(\theta) + \frac{Z_0}{2}i(\theta) + \frac{1}{2}\vec{\upsilon}_0(\theta) - \frac{Z_0}{2}\vec{\iota}_0(\theta) - E_0(\theta)\right)d\theta\right) -$$

$$\left. - G_0\left(\frac{1}{2}u(t) + \frac{Z_0}{2}i(t) + \frac{1}{2}\vec{\upsilon}_0(t) - \frac{Z_0}{2}\vec{\iota}_0(t) - E_0(t)\right)\right]$$

and $\vec{\upsilon}_0(t)$, $\vec{\iota}_0(t)$ are translated to the right initial functions on $[T,3T]$.

4.5. PERIODIC SOLUTIONS OF THE NONLINEAR NEUTRAL SYSTEM

We proceed as we did in Chapter III and use functional spaces with metrics as in § 4.2.

Lemma 4.5.1. Under Assumptions **(V)** the initial value problem (4.4.7) has a periodic solution $(u(.),i(.)) \in M_u^* \times M_i^*$ iff the operator B has a fixed point $(u,i) \in M_u^* \times M_i^*$, that is, $u = B_u(u,i), i = B_i(u,i)$.

Recall

Assumptions **(IN)**: $\upsilon_0(.), \iota_0(.) \in C_{T_0}^1[-T,T]$,

$$\int_{-T+pT_0}^{-T+(p+1)T_0} \upsilon_0(t)dt = 0, \quad \int_{-T+pT_0}^{-T+(p+1)T_0} \iota_0(t)dt = 0; (p = 0,1,...,2m-1); \upsilon_0(-T) = 0, \ \iota_0(-T) = 0 \ ;$$

$$|\upsilon_0(t)| \le U_0 e^{-\beta} e^{\mu(t+T-pT_0)}, |\iota_0(t)| \le I_0 e^{-\beta} e^{\mu(t+T-pT_0)}, t \in [-T+nT_0, \ -T+(n+1)T_0].$$

Recall that it follows

$$|\bar{\upsilon}_0(t)| \le U_0 e^{-\beta} e^{\mu(t-T-kT_0)}, |\bar{\iota}_0(t)| \le I_0 e^{-\beta} e^{\mu(t-T-kT_0)}, t \in [T+kT_0, \ T+(k+1)T_0].$$

Assumptions (E) and **(V)** are the same ones as in §4.4.

Theorem 4.5.1. Let assumptions **(E)**, **(V)** and **(IN)** be satisfied and

$$\frac{e^{\mu_0} + \mu_0 - 1}{\mu_0 Z_0} < 1, \ \left(1 + \frac{e^{\mu_0} - 1}{\mu_0}\right)U_{E_1} < U_0 ; \ \left(1 + \frac{e^{\mu_0} - 1}{\mu_0}\right)\frac{2U_{E_0}}{Z_0} \le I_0 .$$

Then there exists a unique T_0-periodic solution of (4.4.7).

Proof: We omit the proof that $B_u(t), B_i(t)$ are continuously differentiable functions. This can be accomplished as shown in Chapter III.

In order to show that $B = (B_u, B_i)$ maps $M_u^* \times M_i^*$ into itself we need the following estimates for $t \in [T+kT_0, T+(k+1)T_0]$

$$\left|B_u^{(k)}(u,i)(t)\right| \le \left|\int_{T+kT_0}^{t} U(u,i)(s)ds\right| + \left|\left(\frac{t-T-kT_0}{T_0} - \frac{1}{2}\right)\int_{T+kT_0}^{T+(k+1)T_0} U(u,i)(s)ds\right| +$$

$$+ \left|\frac{1}{T_0}\int_{t+kT_0}^{t+(k+1)T_0}\int_{t+kT_0}^{t} U(u,i)(s)dsdt\right| \equiv U_1 + U_2 + U_3 .$$

In view of (4.4.3) we have

$$U_1 \le \left|\int_{T+kT_0}^{t}\dot{E}_1(s)ds\right| + \frac{1}{\hat{C}_1}\int_{T+kT_0}^{t}|i(s)|ds + \frac{1}{\hat{C}_1}I_0\int_{T+kT_0}^{t}e^{\mu(s-T-kT_0)}ds + \frac{1}{\hat{C}_1}\sum_{n=1}^{m}|g_n^{(1)}|\int_{T+kT_0}^{t}\int_{T+kT_0}^{s}|E_1(\tau)-u(\tau)|^n d\tau ds \le$$

$$\leq U_{E_1}e^{\mu(t-T-kT_0)}+\frac{I_0}{\hat{C}_1}\frac{e^{\mu(t-T-kT_0)}-1}{\mu}+\frac{I_0}{\hat{C}_1}\frac{e^{\mu(t-T-kT_0)}-1}{\mu}+\frac{1}{\hat{C}_1}\sum_{n=1}^{m}|g_n^{(1)}|\left(U_{E_1}+U_0\right)^n\int_{T+kT_0}^{t}\int_{T+kT_0}^{s}e^{n\mu(\tau-T-kT_0)}d\tau ds\leq$$

$$\leq U_{E_1}e^{\mu(t-T-kT_0)}+\frac{e^{\mu(t-T-kT_0)}-1}{\mu}\frac{2I_0}{\mu\hat{C}_1}+\frac{1}{\hat{C}_1}\sum_{n=1}^{m}|g_n^{(1)}|\left(U_{E_1}+U_0\right)^n\int_{T+kT_0}^{t}\frac{e^{n\mu(s-T-kT_0)}-1}{n\mu}ds\leq$$

$$\leq U_{E_1}e^{\mu(t-T-kT_0)}+\frac{e^{\mu(t-T-kT_0)}-1}{\mu}\frac{2I_0}{\hat{C}_1}+\frac{1}{\hat{C}_1}\sum_{n=1}^{m}|g_n^{(1)}|(\phi_0)^n\int_{T+kT_0}^{t}\frac{\left(e^{\mu(s-T-kT_0)}-1\right)ne^{(n-1)\mu(s-T-kT_0)}}{n\mu}ds\leq$$

$$\leq U_{E_1}e^{\mu(t-T-kT_0)}+\frac{e^{\mu(t-T-kT_0)}-1}{\mu}\frac{2I_0}{\mu\hat{C}_1}+\frac{1}{\hat{C}_1}\sum_{n=1}^{m}|g_n^{(1)}|(\phi_0)^n e^{(n-1)\mu T_0}\int_{T+kT_0}^{t}\frac{e^{\mu(s-T-kT_0)}}{\mu}ds\leq$$

$$\leq U_{E_1}e^{\mu(t-T-kT_0)}+\frac{e^{\mu(t-T-kT_0)}-1}{\mu}\frac{2I_0}{\hat{C}_1}+\frac{e^{\mu(t-T-kT_0)}-1}{\mu^2}\frac{1}{\hat{C}_1}\sum_{n=1}^{m}|g_n^{(1)}|(\phi_0)^n e^{(n-1)\mu T_0}\leq$$

$$\leq e^{\mu(t-T-kT_0)}\left[U_{E_1}+\frac{1}{\mu\hat{C}_1}\left(2I_0+\frac{1}{\mu}\sum_{n=1}^{m}|g_n^{(1)}|(\phi_0)^n e^{(n-1)\mu_0}\right)\right];$$

$$U_2\leq\frac{1}{2}\left|\int_{T+kT_0}^{T+(k+1)T_0}\dot{E}_1(s)ds\right|+\frac{1}{2\hat{C}_1}\int_{T+kT_0}^{T+(k+1)T_0}|i(s)|ds+\frac{1}{2\hat{C}_1}I_0\int_{T+kT_0}^{T+(k+1)T_0}e^{\mu(s-T-kT_0)}ds+$$

$$+\frac{1}{2\hat{C}_1}\sum_{n=1}^{m}|g_n^{(1)}|\int_{T+kT_0}^{T+(k+1)T_0}\int_{T+kT_0}^{s}|E_1(\tau)-u(\tau)|^n d\tau ds\leq$$

$$\leq e^{\mu(t-T-kT_0)}\frac{e^{\mu_0}-1}{2}\frac{1}{\mu\hat{C}_1}\left(2I_0+\frac{1}{\mu}\sum_{n=1}^{m}|g_n^{(1)}|(\phi_0)^n e^{(n-1)\mu T_0}\right)$$

and

$$U_3\leq\frac{1}{T_0}\int_{T+kT_0}^{T+(k+1)T_0}e^{\mu(t-T-kT_0)}dt\left[U_{E_1}+\frac{1}{\mu\hat{C}_1}\left(2I_0+\frac{1}{\mu}\sum_{n=1}^{m}|g_n^{(1)}|(\phi_0)^n e^{(n-1)\mu_0}\right)\right]$$

$$\leq e^{\mu(t-T-kT_0))}\frac{e^{\mu_0}-1}{\mu_0}\left[U_{E_1}+\frac{1}{\mu\hat{C}_1}\left(2I_0+\frac{1}{\mu}\sum_{n=1}^{m}|g_n^{(1)}|(\phi_0)^n e^{(n-1)\mu_0}\right)\right].$$

Therefore we get

$$\left|B_u^{(k)}(u,i)(t)\right| \le e^{\mu(t-T-kT_0)}\left[U_{E_1} + \frac{1}{\mu\hat{C}_1}\left(2I_0 + \frac{1}{\mu}\sum_{n=1}^{m}\left|g_n^{(1)}\right|(\phi_0)^n e^{(n-1)\mu_0}\right)\right] +$$

$$+ e^{\mu(t-T-kT_0)}\frac{e^{\mu_0}-1}{2}\frac{1}{\mu\hat{C}_1}\left(2I_0 + \frac{1}{\mu}\sum_{n=1}^{m}\left|g_n^{(1)}\right|(\phi_0)^n e^{(n-1)\mu_0}\right) +$$

$$+ e^{\mu(t-T-kT_0))}\frac{e^{\mu_0}-1}{\mu_0}\left[U_{E_1} + \frac{1}{\mu\hat{C}_1}\left(2I_0 + \frac{1}{\mu}\sum_{n=1}^{m}\left|g_n^{(1)}\right|(\phi_0)^n e^{(n-1)\mu_0}\right)\right] \le$$

$$\le e^{\mu(t-T-kT_0))}\left[\left(1 + \frac{e^{\mu_0}-1}{\mu_0}\right)U_{E_1} + \frac{1}{\mu\hat{C}_1}\left(1 + \frac{e^{\mu_0}-1}{2} + \frac{e^{\mu_0}-1}{\mu_0}\right)\left(2I_0 + \frac{1}{\mu}\sum_{n=1}^{m}\left|g_n^{(1)}\right|(\phi_0)^n e^{(n-1)\mu_0}\right)\right] \le$$

$$\le U_0 e^{\mu(t-T-kT_0))}.$$

For $B_i(u,i)$ we obtain

$$\left|B_i^{(k)}(u,i)(t)\right| \le \left|\int_{T+kT_0}^{t} I(u,i)(s)\,ds\right| + \left|\left(\frac{t-T-kT_0}{T_0} - \frac{1}{2}\right)\int_{T+kT_0}^{T+(k+1)T_0} I(u,i)(s)\,ds\right| +$$

$$+ \left|\frac{1}{T_0}\int_{T+kT_0}^{T+(k+1)T_0}\int_{T+kT_0}^{t} I(u,i)(s)\,ds\,dt\right| \equiv I_1 + I_2 + I_3.$$

In view of Assumption **(V0)** we obtain

$$I_1 \le \frac{2}{Z_0}\left|\int_{T+kT_0}^{t}\dot{E}_0(s)\,ds\right| + \frac{1}{Z_0}\left|\int_{T+kT_0}^{t}\dot{u}(s)\,ds\right| + \frac{1}{Z_0}\left|\int_{T+kT_0}^{t}\dot{u}(s-2T)\,ds\right| + \left|\int_{T+kT_0}^{t}\dot{i}(s-2T)\,ds\right| +$$

$$+ \frac{2}{Z_0\hat{C}_0}\left[\frac{1}{2Z_0}\int_{T+kT_0}^{t}|u(s)|\,ds + \frac{1}{2}\int_{T+kT_0}^{t}|i(s)|\,ds + \frac{1}{2Z_0}\int_{T+kT_0}^{t}|u(s-2T)|\,ds + \frac{1}{2}\int_{T+kT_0}^{t}|i(s-2T)|\,ds + \right.$$

$$\left. + \int_{T+kT_0}^{t} I_0 e^{\mu(s-T-kT_0)}\,ds + \sum_{n=1}^{m}\left|g_n^{(0)}\right|\left(\frac{2U_{E_0} + (U_0 + Z_0 I_0)(1 + e^{-\beta})}{2}\right)^n \int_{T+kT_0}^{t} e^{n\mu(s-T-kT_0)}\,ds\right] \le$$

$$\le \frac{2}{Z_0}|E_0(t)| + \frac{1}{Z_0}|u(t)| + \frac{1}{Z_0}|u(t-2T)| + |i(t-2T)| +$$

$$+ \frac{U_0}{Z_0^2\hat{C}_0}\frac{e^{\mu(t-T-kT_0)}-1}{\mu} + \frac{I_0}{Z_0\hat{C}_0}\frac{e^{\mu(t-T-kT_0)}-1}{\mu} + \frac{U_0 e^{-\beta}}{Z_0^2\hat{C}_0}\frac{e^{\mu(t-T-kT_0)}-1}{\mu} + \frac{I_0 e^{-\beta}}{Z_0\hat{C}_0}\frac{e^{\mu(t-T-kT_0)}-1}{\mu} +$$

$$+ \frac{2}{Z_0 \hat{C}_0} I_0 \frac{e^{\mu(t-T-kT_0)}-1}{\mu}$$

$$+ \frac{2}{Z_0 \hat{C}_0} \frac{e^{\mu(t-T-kT_0)}-1}{\mu} \sum_{n=1}^{m} \left|g_n^{(0)}\right| \left(\frac{2U_{E_0}+(U_0+Z_0 I_0)(1+e^{-\beta})}{2}\right)^n e^{(n-1)\mu(t-T-kT_0)} \le$$

$$\le e^{\mu(t-T-kT_0)} \frac{2U_{E_0}+U_0}{Z_0} + e^{\mu(t-T-kT_0)} \frac{U_0 e^{-\beta}+Z_0 I_0 e^{-\beta}}{Z_0}+$$

$$+ \frac{e^{\mu(t-T-kT_0)}-1}{\mu}\left[\frac{U_0}{Z_0^2 \hat{C}_0}+\frac{I_0}{Z_0 \hat{C}_0}+\frac{U_0 e^{-\beta}}{Z_0^2 \hat{C}_0}+\frac{I_0 e^{-\beta}}{Z_0 \hat{C}_0}+\frac{2}{Z_0 \hat{C}_0}I_0+\frac{2}{Z_0 \hat{C}_0}\sum_{n=1}^{m}\left|g_n^{(0)}\right|\left(\phi_0\right)^n e^{(n-1)\mu T_0}\right] \le$$

$$\le e^{\mu(t-T-kT_0)} \frac{2U_{E_0}+U_0+e^{-\beta}(U_0+Z_0 I_0)}{Z_0}+$$

$$+ \frac{e^{\mu(t-T-kT_0)}}{\mu Z_0 \hat{C}_0}\left[\frac{\left(1+e^{-\beta}\right)(U_0+I_0 Z_0)}{Z_0}+2I_0+2\sum_{n=1}^{m}\left|g_n^{(0)}\right|\left(\phi_0\right)^n e^{(n-1)\mu T_0}\right] \le$$

$$\le e^{\mu(t-T-kT_0)}\left[\frac{2U_{E_0}+e^{-\beta}(U_0+Z_0 I_0)}{Z_0}+\frac{1}{\mu Z_0 \hat{C}_0}\left(\frac{\left(1+e^{-\beta}\right)(U_0+I_0 Z_0)}{Z_0}+2I_0+2\sum_{n=1}^{m}\left|g_n^{(0)}\right|\left(\phi_0\right)^n e^{(n-1)\mu_0}\right)\right],$$

$$I_2 \le \frac{1}{Z_0}\left|\int_{T+kT_0}^{T+(k+1)T_0}\dot{E}_0(s)ds\right|+\frac{1}{2Z_0}\left|\int_{T+kT_0}^{T+(k+1)T_0}\dot{u}(s)ds\right|+\frac{1}{2Z_0}\left|\int_{T+kT_0}^{T+(k+1)T_0}\dot{u}(s-2T)ds\right|+\frac{1}{2}\left|\int_{T+kT_0}^{T+(k+1)T_0}\dot{i}(s-2T)ds\right|+$$

$$+ \frac{2}{Z_0 \hat{C}_0}\left[\frac{1}{2Z_0}\int_{T+kT_0}^{T+(k+1)T_0}\left|u(s)\right|ds+\frac{1}{2}\int_{T+kT_0}^{T+(k+1)T_0}\left|i(s)\right|ds+\frac{1}{2Z_0}\int_{T+kT_0}^{T+(k+1)T_0}\left|u(s-2T)\right|ds+\frac{1}{2}\int_{T+kT_0}^{T+(k+1)T_0}\left|i(s-2T)\right|ds+\right.$$

$$\left.+ \int_{t+kT_0}^{t+(k+1)T_0}I_0 e^{\mu(s-T-kT_0)}ds+\sum_{n=1}^{m}\left|g_n^{(0)}\right|\left(U_{E_0}+U_0+Z_0 I_0\right)^n\int_{t+kT_0}^{t+(k+1)T_0}e^{n\mu(s-T-kT_0)}ds\right] \le$$

$$\le \frac{1}{2Z_0}\left|\int_{T+kT_0}^{T+(k+1)T_0}\dot{u}(s)ds\right|+\frac{1}{2Z_0}\left|\int_{T+kT_0}^{T+(k+1)T_0}\dot{u}(s-2T)ds\right|+\frac{1}{2}\left|\int_{T+kT_0}^{T+(k+1)T_0}\dot{i}(s-2T)ds\right|+$$

$$+ \frac{1}{Z_0 \hat{C}_0}\left[\frac{U_0}{2Z_0}\int_{T+kT_0}^{T+(k+1)T_0}e^{\mu(s-T-kT_0)}ds+\frac{I_0}{2}\int_{T+kT_0}^{T+(k+1)T_0}e^{\mu(s-T-kT_0)}ds+\frac{U_0}{2Z_0}\int_{T+kT_0}^{T+(k+1)T_0}e^{\mu(s-T-kT_0)}ds+\right.$$

$$+ \frac{I_0}{2}\int_{T+kT_0}^{T+(k+1)T_0}e^{\mu(s-T-kT_0)}ds+I_0\int_{T+kT_0}^{T+(k+1)T_0}e^{\mu(s-T-kT_0)}ds+$$

$$\left.+ \sum_{n=1}^{m}\left|g_n^{(0)}\right|\phi_0^n\frac{e^{n\mu_0}-1}{n\mu}\right] \le$$

$$\leq \frac{1}{Z_0 \hat{C}_0}\left[\frac{U_0}{2Z_0}\frac{e^{\mu 0}-1}{\mu}+\frac{I_0}{2}\frac{e^{\mu 0}-1}{\mu}+\frac{U_0 e^{-\beta}}{2Z_0}\frac{e^{\mu 0}-1}{\mu}+\frac{I_0 e^{-\beta}}{2}\frac{e^{\mu 0}-1}{\mu}+I_0\frac{e^{\mu 0}-1}{\mu}+\right.$$

$$\left.+\frac{e^{\mu 0}-1}{\mu}\sum_{n=1}^{m}\left|g_n^{(0)}\right|\phi_0^n e^{(n-1)\mu 0}\right]\leq$$

$$\leq \frac{e^{\mu 0}-1}{2}\frac{1}{\mu Z_0 \hat{C}_0}\left[\frac{\left(1+e^{-\beta}\right)\left(U_0+Z_0 I_0\right)}{Z_0}+2I_0+2\sum_{n=1}^{m}\left|g_n^{(0)}\right|\phi_0^n e^{(n-1)\mu 0}\right]$$

and

$$I_3 \leq \frac{1}{T_0}\int_{T+kT_0}^{T+(k+1)T_0}\left|\int_{T+kT_0}^{t}I(u,i)(s)ds\right|dt \leq$$

$$\leq \frac{e^{\mu 0}-1}{\mu_0}\left[\frac{2U_{E_0}+e^{-\beta}\left(U_0+Z_0 I_0\right)}{Z_0}+\frac{1}{\mu Z_0 \hat{C}_0}\left(\frac{\left(1+e^{-\beta}\right)\left(U_0+I_0 Z_0\right)}{Z_0}+2I_0+2\sum_{n=1}^{m}\left|g_n^{(0)}\right|\phi_0^n e^{(n-1)\mu 0}\right)\right].$$

Therefore we obtain

$$\left|B_i^{(k)}(u,i)(t)\right|\leq I_1+I_2+I_3 \leq$$

$$\leq e^{\mu(t-T-kT_0)}\left[\frac{2U_{E_0}+e^{-\beta}\left(U_0+Z_0 I_0\right)}{Z_0}+\frac{1}{\mu Z_0 \hat{C}_0}\left(\frac{\left(1+e^{-\beta}\right)\left(U_0+I_0 Z_0\right)}{Z_0}+2I_0+2\sum_{n=1}^{m}\left|g_n^{(0)}\right|\phi_0^n e^{(n-1)\mu 0}\right)\right]+$$

$$+e^{\mu(t-T-kT_0)}\frac{e^{\mu 0}-1}{2}\frac{1}{\mu Z_0 \hat{C}_0}\left(\frac{\left(1+e^{-\beta}\right)\left(U_0+Z_0 I_0\right)}{Z_0}+2I_0+2\sum_{n=1}^{m}\left|g_n^{(0)}\right|\phi_0^n e^{(n-1)\mu 0}\right)+$$

$$+e^{\mu(t-T-kT_0)}\frac{e^{\mu 0}-1}{\mu_0}\left[\frac{2U_{E_0}+e^{-\beta}\left(U_0+Z_0 I_0\right)}{Z_0}+\frac{1}{\mu Z_0 \hat{C}_0}\left(\frac{\left(1+e^{-\beta}\right)\left(U_0+I_0 Z_0\right)}{Z_0}+2I_0+2\sum_{n=1}^{m}\left|g_n^{(0)}\right|\phi_0^n e^{(n-1)\mu 0}\right)\right]\leq$$

$$\leq e^{\mu(t-T-kT_0)}\left[\left(1+\frac{e^{\mu 0}-1}{\mu_0}\right)\frac{2U_{E_0}+e^{-\beta}\left(U_0+Z_0 I_0\right)}{Z_0}+\right.$$

$$\left.+\left(1+\frac{e^{\mu 0}-1}{2}+\frac{e^{\mu 0}-1}{\mu_0}\right)\frac{1}{\mu Z_0 \hat{C}_0}\left(\frac{\left(1+e^{-\beta}\right)\left(U_0+I_0 Z_0\right)}{Z_0}+2I_0+2\sum_{n=1}^{m}\left|g_n^{(0)}\right|\phi_0^n e^{(n-1)\mu 0}\right)\right]\leq I_0 e^{\mu(t-T-kT_0)}.$$

It remains to show that the operator B is a contractive one.

Recalling inequalities $\left|\widetilde{L}_p^{-1}(.)\right|\leq I_0 \leq I_0 \;(p=0,1)$ and

$$\left|\frac{d\widetilde{L}_p^{-1}(i)}{di}\right| \le \frac{1}{\hat{L}_p} \le \frac{1}{\hat{L}}, \ \left|\frac{d\widetilde{C}_p(u)}{du}\right| \le D_p \ \ (p=0,1)$$

we obtain

$$\left|B_u^{(k)}(u,i)(t) - B_u^{(k)}(\overline{u},\overline{i})(t)\right| \le \left|\int_{T+kT_0}^{t}\left(U(u,i)(s) - U(\overline{u},\overline{i})(s)\right)ds\right| +$$

$$+\left|\left(\frac{t-T-kT_0}{T_0}-\frac{1}{2}\right)\int_{T+kT_0}^{T+(k+1)T_0}\left(U(u,i)(s) - U(\overline{u},\overline{i})(s)\right)ds\right| +$$

$$+\left|\frac{1}{T_0}\int_{T+kT_0}^{T+(k+1)T_0}\int_{T+kT_0}^{t}\left(U(u,i)(s) - U(\overline{u},\overline{i})(s)\right)dsdt\right| \equiv U_1 + U_2 + U_3.$$

We need the following inequalities

$$\left|G_1\big((E_1(t)-u(t))\big) - G_1\big((E_1(t)-\overline{u}(t))\big)\right| =$$

$$\left|\sum_{n=1}^{m} g_n^{(1)}\big(E_1(t)-u(t)\big)^n - \sum_{n=1}^{m} g_n^{(1)}\big(E_1(t)-\overline{u}(t)\big)^n\right| \le$$

$$\le \sum_{n=1}^{m}\left|g_n^{(1)}\right| n.\mathrm{ess\,sup}\left\{\left|E_1(t)-u(t)\right|^{n-1}\right\}\left|u(\tau)-\overline{u}(\tau)\right| \le$$

$$\le \rho^{(k)}(u,\overline{u})\sum_{n=1}^{m} n\left|g_n^{(1)}\right|\big(U_{E_1}+U_0\big)^{n-1}.\big(e^{\mu(t-T-kT_0)}\big)^{n-1} e^{\mu(t-T-kT_0)} \le$$

$$\le e^{\mu(t-T-kT_0)}\rho_\mu^{(k)}(u,\overline{u})\sum_{n=1}^{m} n\left|g_n^{(1)}\right|\big(U_{E_1}+U_0\big)^{n-1} e^{(n-1)\mu T_0}.$$

Therefore

$$U_1 \le \int_{T+kT_0}^{t}\left|\frac{i(s)}{\widetilde{C}_1(u)(s)} - \frac{\overline{i}(s)}{\widetilde{C}_1(\overline{u})(s)}\right|ds + \int_{T+kT_0}^{t}\left|\frac{\widetilde{L}_1^{-1}\left(\int_{T+kT_0}^{s}[E_1(\tau)-u(\tau)]d\tau\right)}{\widetilde{C}_1(u)(s)} - \frac{\widetilde{L}_1^{-1}\left(\int_{T+kT_0}^{s}[E_1(\tau)-\overline{u}(\tau)]d\tau\right)}{\widetilde{C}_1(\overline{u})(s)}\right|ds +$$

$$+ \int_{T+kT_0}^{t}\left|\frac{G_1(E_1(s)-u(s))}{\widetilde{C}_1(u)(s)} - \frac{G_1(E_1(s)-\overline{u}(s))}{\widetilde{C}_1(\overline{u})(s)}\right|ds \le$$

$$\leq \int_{T+kT_0}^{t}\left|\frac{i(s)}{\widetilde{C}_1(u)(s)}-\frac{\bar{i}(s)}{\widetilde{C}_1(u)(s)}\right|ds + \int_{T+kT_0}^{t}\left|\frac{\bar{i}(s)}{\widetilde{C}_1(u)(s)}-\frac{\bar{i}(s)}{C_1(\bar{u})(s)}\right|ds +$$

$$+ \int_{T+kT_0}^{t}\left|\frac{\widetilde{L}_1^{-1}\left(\int_{T+kT_0}^{s}[E_1(\tau)-u(\tau)]d\tau\right)}{\widetilde{C}_1(u)(s)}-\frac{\widetilde{L}_1^{-1}\left(\int_{T+kT_0}^{s}[E_1(\tau)-u(\tau)]d\tau\right)}{\widetilde{C}_1(\bar{u})(s)}\right|ds +$$

$$+ \int_{T+kT_0}^{t}\left|\frac{\widetilde{L}_1^{-1}\left(\int_{T+kT_0}^{s}[E_1(\tau)-u(\tau)]d\tau\right)}{\widetilde{C}_1(\bar{u})(s)}-\frac{\widetilde{L}_1^{-1}\left(\int_{T+kT_0}^{s}[E_1(\tau)-\bar{u}(\tau)]d\tau\right)}{\widetilde{C}_1(\bar{u})(s)}\right|ds +$$

$$+ \int_{T+kT_0}^{t}\left|\frac{G_1(E_1(s)-u(s))}{\widetilde{C}_1(u)(s)}-\frac{G_1(E_1(s)-u(s))}{\widetilde{C}_1(\bar{u})(s)}\right|ds+$$

$$\int_{T+kT_0}^{t}\left|\frac{G_1(E_1(s)-u(s))}{\widetilde{C}_1(\bar{u})(s)}-\frac{G_1(E_1(s)-\bar{u}(s))}{\widetilde{C}_1(\bar{u})(s)}\right|ds \leq$$

$$\leq \frac{1}{\hat{C}_1}\int_{T+kT_0}^{t}\left|i(s)-\bar{i}(s)\right|ds + \int_{T+kT_0}^{t}\left|\bar{i}(s)\right|\frac{1}{\left(\widetilde{C}_1(u)(s)\right)^2}\left|\frac{d\widetilde{C}_1(u)}{du}\right||u(s)-\bar{u}(s)|ds +$$

$$+ I_0 \int_{T+kT_0}^{t}\frac{1}{\left(\widetilde{C}_1(u)(s)\right)^2}\left|\frac{d\widetilde{C}_1(u)}{du}\right||u(s)-\bar{u}(s)|ds + \frac{1}{\hat{L}_1\hat{C}_1}\int_{T+kT_0}^{t}\left|\int_{T+kT_0}^{s}(u(\tau)-\bar{u}(\tau))d\tau\right|ds +$$

$$+ \sum_{n=1}^{m}\left|g_n^{(1)}\right|\left(U_{E_1}+U_0\right)^n e^{n\mu T_0}\int_{T+kT_0}^{t}\frac{1}{\left(\widetilde{C}_1(u)(s)\right)^2}\left|\frac{d\widetilde{C}_1(u)}{du}\right||u(s)-\bar{u}(s)|ds+$$

$$+ \frac{1}{\hat{C}_1}\sum_{n=1}^{m}n\left|g_n^{(1)}\right|\left(U_{E_1}+U_0\right)^{n-1}e^{(n-1)\mu T_0}\int_{T+kT_0}^{t}|u(s))-\bar{u}(s)|ds \leq$$

$$\leq \frac{1}{\hat{C}_1}\rho^{(k)}(i,\bar{i})\int_{T+kT_0}^{t}e^{\mu(s-T-kT_0)}ds + \frac{I_0 D_1}{\hat{C}_1^2}\rho^{(k)}(u,\bar{u})\int_{T+kT_0}^{t}e^{2\mu(s-T-kT_0)}ds +$$

$$+ \frac{I_0 D_1}{\hat{C}_1^2}\rho^{(k)}(u,\bar{u})\int_{T+kT_0}^{t}e^{\mu(s-T-kT_0)}ds + \frac{\rho^{(k)}(u,\bar{u})}{\mu\hat{L}_1\hat{C}_1}\int_{T+kT_0}^{t}e^{\mu(s-T-kT_0)}ds +$$

$$+ \frac{D_1}{\hat{C}_1^2}\rho^{(k)}(u,\bar{u})\sum_{n=1}^{m}\left|g_n^{(1)}\right|\left(U_{E_1}+U_0\right)^n e^{n\mu T_0}\int_{T+kT_0}^{t}e^{\mu(s-T-kT_0)}ds+$$

$$+\frac{\rho^{(k)}(u,\overline{u})}{\hat{C}_1}\sum_{n=1}^{m} n\left|g_n^{(1)}\right|\left(U_{E_1}+U_0\right)^{n-1} e^{(n-1)\mu T_0}\int_{T+kT_0}^{t} e^{\mu(s-T-kT_0)}ds \le$$

$$\le \frac{1}{\hat{C}_1}\frac{e^{\mu_0}\rho_\mu^{(k)}(\dot{i},\dot{\overline{i}})}{\mu}\frac{e^{\mu(t-T-kT_0)}-1}{\mu}+\frac{I_0 D_1}{\hat{C}_1^2}\frac{e^{\mu_0}\rho_\mu^{(k)}(\dot{u},\overline{u})}{\mu}\frac{e^{2\mu(t-T-kT_0)}-1}{2\mu}+$$

$$+\frac{I_0 D_1}{\hat{C}_1^2}\frac{e^{\mu_0}\rho_\mu^{(k)}(\dot{u},\dot{\overline{u}})}{\mu}\frac{e^{\mu(t-T-kT_0)}-1}{\mu}+\frac{e^{\mu_0}\rho_\mu^{(k)}(\dot{u},\dot{\overline{u}})}{\mu^2 \hat{L}_1 \hat{C}_1}\frac{e^{\mu(t-T-kT_0)}-1}{\mu}+$$

$$+\frac{D_1}{\hat{C}_1^2}\frac{e^{\mu_0}\rho_\mu^{(k)}(\dot{u},\dot{\overline{u}})}{\mu}\sum_{n=1}^{m}\left|g_n^{(1)}\right|\left(U_{E_1}+U_0\right)^n e^{n\mu_0}\frac{e^{\mu(t-T-kT_0)}-1}{\mu}+$$

$$+\frac{e^{\mu_0}\rho_\mu^{(k)}(\dot{u},\dot{\overline{u}})}{\mu \hat{C}_1}\sum_{n=1}^{m} n\left|g_n^{(1)}\right|\left(U_{E_1}+U_0\right)^{n-1} e^{(n-1)\mu T_0}\int_{T+kT_0}^{t} e^{\mu(s-T-kT_0)}ds \le$$

$$\le \hat{\rho}_\mu((u,i),(\overline{u},\overline{i}))\frac{e^{\mu(t-T-kT_0)}-1}{\mu^2 \hat{C}_1}\left[1+\frac{I_0 D_1}{\hat{C}_1}\left(e^{\mu_0}+1\right)+\right.$$

$$\left.+\frac{1}{\mu\hat{L}_1}+D_1\sum_{n=1}^{m}\left|g_n^{(1)}\right|\left(U_{E_1}+U_0\right)^n e^{n\mu T_0}+\frac{1}{\hat{C}_1}\sum_{n=1}^{m} n\left|g_n^{(1)}\right|\left(U_{E_1}+U_0\right)^{n-1} e^{(n-1)\mu T_0}\right] \le$$

$$\le e^{\mu(t-T-kT_0)}\hat{\rho}_\mu((u,\dot{u},i,\dot{i}),(\overline{u},\dot{\overline{u}},\overline{i},\dot{\overline{i}}))\frac{1}{\mu^2 \hat{C}_1}\left[1+\frac{I_0 D_1}{\hat{C}_1}\left(e^{\mu_0}+1\right)+\right.$$

$$\left.+\frac{1}{\mu\hat{L}_1}+D_1\sum_{n=1}^{m}\left|g_n^{(1)}\right|\phi_0^n e^{n\mu T_0}+\frac{1}{\hat{C}_1}\sum_{n=1}^{m} n\left|g_n^{(1)}\right|\phi_0^{n-1} e^{(n-1)\mu T_0}\right];$$

$$U_2 \le \frac{1}{2}\left|\int_{T+kT_0}^{T+(k+1)T_0}\left(U(u,i)(s)-U(\overline{u},\overline{i})(s)\right)ds\right| \le$$

$$\le \hat{\rho}_\mu((u,\dot{u},i,\dot{i}),(\overline{u},\dot{\overline{u}},\overline{i},\dot{\overline{i}}))\frac{e^{\mu_0}-1}{2}\frac{1}{\mu^2 \hat{C}}\left[1+\frac{I_0 D_1}{\hat{C}_1}\left(e^{\mu_0}+1\right)+\right.$$

$$\left.+\frac{1}{\mu\hat{L}_1}+D_1\sum_{n=1}^{m}\left|g_n^{(1)}\right|\phi_0^n e^{n\mu_0}+\frac{1}{\hat{C}_1}\sum_{n=1}^{m} n\left|g_n^{(1)}\right|\phi_0^{n-1} e^{(n-1)\mu_0}\right]$$

and

$$U_3 \le \frac{1}{T_0} \int\limits_{T+kT_0}^{T+(k+1)T_0} \left| \int\limits_{T+kT_0}^{t} \left(U(u,i)(s) - U(\overline{u},\overline{i})(s) \right) ds \right| dt \le$$

$$\le e^{\mu(t-T-kT_0)} \hat{\rho}_\mu((u,\dot{u},i,\dot{i}),(\overline{u},\dot{\overline{u}},\overline{i},\dot{\overline{i}})) \frac{e^{\mu_0}-1}{\mu_0} \frac{1}{\mu^2 \hat{C}_1} \left[1 + \frac{I_0 D_1}{\hat{C}_1}\left(e^{\mu_0}+1\right) + \right.$$

$$\left. + \frac{1}{\mu \hat{L}_1} + D_1 \sum_{n=1}^{m} \left|g_n^{(1)}\right| \phi_0^n e^{n\mu T_0} + \frac{1}{\hat{C}_1} \sum_{n=1}^{m} n\left|g_n^{(1)}\right| \phi_0^{n-1} e^{(n-1)\mu_0} \right].$$

Therefore

$$\left| B_u^{(k)}(u,i)(t) - B_u^{(k)}(\overline{u},\overline{i})(t) \right| \le U_1 + U_2 + U_3 \le$$

$$\le e^{\mu(t-T-kT_0)} \hat{\rho}_\mu((u,\dot{u},i,\dot{i}),(\overline{u},\dot{\overline{u}},\overline{i},\dot{\overline{i}})) \left(1 + \frac{e^{\mu_0}-1}{2} + \frac{e^{\mu_0}-1}{\mu_0} \right) \frac{e^{\mu_0}}{\mu^2 \hat{C}_1} \left[1 + \frac{I_0 D_1}{\hat{C}_1}\left(e^{\mu_0}+1\right) + \right.$$

$$\left. + \frac{1}{\mu \hat{L}_1} + D_1 \sum_{n=1}^{m} \left|g_n^{(1)}\right| (\phi_0)^n e^{n\mu_0} + \frac{1}{\hat{C}_1} \sum_{n=1}^{m} n\left|g_n^{(1)}\right| \phi_0^{n-1} e^{(n-1)\mu_0} \right] \equiv$$

$$\equiv K_u \, e^{\mu(t-T-kT_0)} \hat{\rho}_\mu((u,\dot{u},i,\dot{i}),(\overline{u},\dot{\overline{u}},\overline{i},\dot{\overline{i}})).$$

It follows

$$\hat{\rho}(B_u(u,i),B_u(\overline{u},\overline{i})) \le e^{\mu_0} K_u \, \hat{\rho}_\mu((u,\dot{u},i,\dot{i}),(\overline{u},\dot{\overline{u}},\overline{i},\dot{\overline{i}})).$$

For the derivative we obtain

$$\left| \dot{B}_u^{(k)}(u,i)(t) - \dot{B}_u^{(k)}(\overline{u},\overline{i})(t) \right| \le \left| U(u,i)(t) - U(\overline{u},\overline{i})(t) \right| +$$

$$\left| \frac{1}{T_0} \int\limits_{T+kT_0}^{T+(k+1)T_0} \left(U(u,i)(s) - U(\overline{u},\overline{i})(s) \right) ds \right| \equiv \dot{U}_1 + \dot{U}_2.$$

But

$$\dot{U}_1 \le \left| U(u,i)(t) - U(\overline{u},\overline{i})(t) \right| \le \left| \frac{i(t)}{\widetilde{C}_1(u)(t)} - \frac{\overline{i}(t)}{\widetilde{C}_1(u)(t)} \right| + \left| \frac{\overline{i}(t)}{\widetilde{C}_1(u)(t)} - \frac{\overline{i}(t)}{\widetilde{C}_1(\overline{u})(t)} \right| +$$

$$+\left|\frac{\widetilde{L}_1^{-1}\left(\int\limits_{T+kT_0}^{t}[E_1(\tau)-u(\tau)]d\tau\right)}{\widetilde{C}_1(u)(t)}-\frac{\widetilde{L}_1^{-1}\left(\int\limits_{T+kT_0}^{t}[E_1(\tau)-u(\tau)]d\tau\right)}{\widetilde{C}_1(\overline{u})(t)}\right|+$$

$$+\left|\frac{\widetilde{L}_1^{-1}\left(\int\limits_{T+kT_0}^{t}[E_1(\tau)-u(\tau)]d\tau\right)}{\widetilde{C}_1(\overline{u})(t)}-\frac{\widetilde{L}_1^{-1}\left(\int\limits_{T+kT_0}^{t}[E_1(\tau)-\overline{u}(\tau)]d\tau\right)}{\widetilde{C}_1(\overline{u})(t)}\right|+$$

$$+\left|\frac{G_1(E_1(t)-u(t))}{\widetilde{C}_1(u)(t)}-\frac{G_1(E_1(t)-u(t))}{\widetilde{C}_1(\overline{u})(t)}\right|+\left|\frac{G_1(E_1(t)-u(t))}{\widetilde{C}_1(\overline{u})(t)}-\frac{G_1(E_1(t)-\overline{u}(t))}{\widetilde{C}_1(\overline{u})(t)}\right|\leq$$

$$\leq\frac{1}{\hat{C}_1}\left|i(t)-\overline{i}(t)\right|+\frac{\left|\overline{i}(t)\right|}{\widetilde{C}_1^2(u)(t)}\left|\frac{d\widetilde{C}_1(u)}{du}\right|\left|u(t)-\overline{u}(t)\right|+$$

$$+\frac{I_0}{\widetilde{C}_1^2(u)(t)}\left|\frac{d\widetilde{C}_1(u)}{du}\right|\left|u(t)-\overline{u}(t)\right|+\frac{1}{\hat{L}_1\hat{C}_1}\left|\int\limits_{T+kT_0}^{t}\left(u(\tau)-\overline{u}(\tau)\right)d\tau\right|+$$

$$+\frac{1}{\widetilde{C}_1^2(u)(t)}\left|\frac{d\widetilde{C}_1(u)}{du}\right|\left|u(t)-\overline{u}(t)\right|\sum_{n=1}^{m}\left|g_n^{(1)}\right|\left(U_{E_1}+U_0\right)^n e^{n\mu T_0}+$$

$$+\frac{1}{\hat{C}_1}\sum_{n=1}^{m}n\left|g_n^{(1)}\right|\left(U_{E_1}+U_0\right)^{n-1}e^{(n-1)\mu T_0}\left|u(t)-\overline{u}(t)\right|\leq$$

$$\leq e^{\mu(t-T-kT_0)}\frac{\rho_\mu^{(k)}(i,\overline{i})}{\hat{C}_1}+\frac{I_0 e^{2\mu(t-T-kT_0)}}{\hat{C}_1^2}D_1\rho_\mu^{(k)}(u,\overline{u})+$$

$$+\frac{I_0 e^{\mu(t-T-kT_0)}}{\hat{C}_1^2}D_1\rho_\mu^{(k)}(u,\overline{u})+\frac{\rho_\mu^{(k)}(u,\overline{u})}{\hat{L}_1\hat{C}_1}\frac{e^{\mu(t-T-kT_0)}-1}{\mu}+$$

$$+e^{\mu(t-T-kT_0)}\frac{\rho_\mu^{(k)}(u,\overline{u})}{\hat{C}_1^2}D_1\sum_{n=1}^{m}\left|g_n^{(1)}\right|\left(U_{E_1}+U_0\right)^n e^{n\mu T_0}+$$

$$+e^{\mu(t-T-kT_0)}\frac{\rho_\mu^{(k)}(u,\overline{u})}{\hat{C}_1}\sum_{n=1}^{m}n\left|g_n^{(1)}\right|\left(U_{E_1}+U_0\right)^{n-1}e^{(n-1)\mu T_0}\leq$$

$$\leq e^{\mu(t-T-kT_0)}\frac{\rho_\mu^{(k)}(\dot{i},\dot{\bar{i}})}{\mu\hat{C}_1}+\frac{I_0 e^{2\mu(t-T-kT_0)}}{\hat{C}_1^2}D_1\frac{\rho_\mu^{(k)}(\dot{u},\dot{\bar{u}})}{\mu}+$$

$$+\frac{I_0 e^{\mu(t-T-kT_0)}}{\hat{C}_1^2}D_1\frac{\rho_\mu^{(k)}(\dot{u},\dot{\bar{u}})}{\mu}+\frac{\rho_\mu^{(k)}(\dot{u},\dot{\bar{u}})}{\mu\hat{L}_1\hat{C}_1}\frac{e^{\mu(t-T-kT_0)}-1}{\mu}+$$

$$+e^{\mu(t-T-kT_0)}\frac{\rho_\mu^{(k)}(\dot{u},\dot{\bar{u}})}{\mu\,\hat{C}_1^2}D_1\sum_{n=1}^{m}\left|g_n^{(1)}\right|\left(U_{E_1}+U_0\right)^n e^{n\mu T_0}+$$

$$+e^{\mu(t-T-kT_0)}\frac{\rho_\mu^{(k)}(\dot{u},\dot{\bar{u}})}{\mu\hat{C}_1}\sum_{n=1}^{m}n\left|g_n^{(1)}\right|\left(U_{E_1}+U_0\right)^{n-1}e^{(n-1)\mu T_0}\leq$$

$$\leq e^{\mu(t-T-kT_0)}\hat{\rho}_\mu((u,\dot{u},i,\dot{i}),(\bar{u},\dot{\bar{u}},\bar{i},\dot{\bar{i}}))\frac{1}{\mu\hat{C}_1}\left[1+\frac{I_0 e^{\mu T_0}}{\hat{C}_1}D_1+\frac{I_0}{\hat{C}_1}D_1+\right.$$

$$\left.+\frac{1}{\mu\hat{L}_1}+\frac{1}{\hat{C}_1}D_1\sum_{n=1}^{m}\left|g_n^{(1)}\right|\left(U_{E_1}+U_0\right)^n e^{n\mu T_0}+\sum_{n=1}^{m}n\left|g_n^{(1)}\right|\left(U_{E_1}+U_0\right)^{n-1}e^{(n-1)\mu T_0}\right]\leq$$

$$\leq e^{\mu(t-T-kT_0)}\hat{\rho}_\mu((u,\dot{u},i,\dot{i}),(\bar{u},\dot{\bar{u}},\bar{i},\dot{\bar{i}}))\frac{1}{\mu\hat{C}_1}\left[1+D_1\left(I_0 e^{\mu_0}+I_0+\sum_{n=1}^{m}\left|g_n^{(1)}\right|\phi_0^n e^{n\mu_0}\right)+\frac{1}{\mu\hat{L}_1}+\sum_{n=1}^{m}n\left|g_n^{(1)}\right|\phi_0^{n-1}e^{(n-1)\mu_0}\right]$$

and

$$\dot{U}_2\leq\frac{1}{T_0}\int_{T+kT_0}^{T+(k+1)T_0}\left|U(u,i)(s)-U(\bar{u},\bar{i})(s)\right|ds\leq\frac{1}{T_0}\int_{T+kT_0}^{T+(k+10T_0}e^{\mu(s-T-kT_0)}ds\,\hat{\rho}_\mu((u,\dot{u},i,\dot{i}),(\bar{u},\dot{\bar{u}},\bar{i},\dot{\bar{i}}))\times$$

$$\times\frac{1}{\mu\hat{C}_1}\left[1+D_1\left(I_0 e^{\mu_0}+I_0+\sum_{n=1}^{m}\left|g_n^{(1)}\right|\left(U_{E_1}+U_0\right)^n e^{n\mu_0}\right)+\right.$$

$$\left.\frac{1}{\mu\hat{L}_1}+\sum_{n=1}^{m}n\left|g_n^{(1)}\right|\left(U_{E_1}+U_0\right)^{n-1}e^{(n-1)\mu_0}\right]\leq$$

$$\leq e^{\mu(t-T-kT_0)}\hat{\rho}_\mu((u,\dot{u},i,\dot{i}),(\bar{u},\dot{\bar{u}},\bar{i},\dot{\bar{i}}))\times$$

$$\times\frac{e^{\mu_0}-1}{\mu_0}\frac{1}{\mu\hat{C}_1}\left[1+D_1\left(I_0 e^{\mu_0}+I_0+\sum_{n=1}^{m}\left|g_n^{(1)}\right|\phi_0^n e^{n\mu_0}\right)+\frac{1}{\mu\hat{L}_1}+\sum_{n=1}^{m}n\left|g_n^{(1)}\right|\phi_0^{n-1}e^{(n-1)\mu_0}\right].$$

Then

$$\left|\dot{B}_u^{(k)}(u,i)(t)-\dot{B}_u^{(k)}(\bar{u},\bar{i})(t)\right|\leq e^{\mu(t-T-kT_0)}\hat{\rho}_\mu((u,\dot{u},i,\dot{i}),(\bar{u},\dot{\bar{u}},\bar{i},\dot{\bar{i}}))\times$$

$$\times\left(1+\frac{e^{\mu_0}-1}{\mu_0}\right)\frac{1}{\mu\hat{C}_1}\left[1+D_1\left(I_0e^{\mu_0}+I_0+\sum_{n=1}^{m}\left|g_n^{(1)}\right|\phi_0^n e^{n\mu_0}\right)+\frac{1}{\mu\hat{L}_1}+\sum_{n=1}^{m}n\left|g_n^{(1)}\right|\phi_0^{n-1}e^{(n-1)\mu_0}\right]\equiv$$

$$\equiv e^{\mu(t-T-kT_0)}\dot{K}_u\,\hat{\rho}_\mu((u,\dot{u},i,\dot{i}),(\overline{u},\dot{\overline{u}},\overline{i},\dot{\overline{i}}))\,.$$

It follows

$$\rho_\mu^{(k)}(\dot{B}_u^{(k)}(u,i),\dot{B}_u^{(k)}(\overline{u},\overline{i}))\le\dot{K}_u\,\hat{\rho}_\mu((u,\dot{u},i,\dot{i}),(\overline{u},\dot{\overline{u}},\overline{i},\dot{\overline{i}}))\,.$$

Further on we have

$$\left|B_i^{(k)}(u,i)(t)-B_i^{(k)}(\overline{u},\overline{i})(t)\right|\le\left|\int_{T+kT_0}^{t}\left(I(u,i)(s)-I(\overline{u},\overline{i})(s)\right)ds\right|+$$

$$+\left|\left(\frac{t-T-kT_0}{T_0}-\frac{1}{2}\right)\int_{T+kT_0}^{T+(k+1)T_0}\left(I(u,i)(s)-I(\overline{u},\overline{i})(s)\right)ds\right|+$$

$$+\left|\frac{1}{T_0}\int_{T+kT_0}^{T+(k+1)T_0}\int_{T+kT_0}^{t}\left(I(u,i)(s)-I(\overline{u},\overline{i})(s)\right)dsdt\right|\equiv I_1+I_2+I_3\,.$$

We need the following denotations and estimates

$$P(u,i)(\tau)\equiv E_0(\tau)-\frac{1}{2}u(\tau)-\frac{Z_0}{2}i(\tau)-\frac{1}{2}u(\tau-2T)+\frac{Z_0}{2}i(\tau-2T)\,;$$

$$\left|P(u,i)(\tau)-P(\overline{u},\overline{i})(\tau)\right|\le\frac{1}{2}\left|u(\tau)-\overline{u}(\tau)\right|+\frac{Z_0}{2}\left|i(\tau)-\overline{i}(\tau)\right|+\frac{1}{2}\left|\overline{\upsilon}_0(\tau)-\overline{\upsilon}_0(\tau)\right|+$$

$$+\frac{Z_0}{2}\left|\overline{\iota}_0(\tau)-\overline{\iota}_0(\tau)\right|\le e^{\mu(\tau-T-kT_0)}\frac{1}{2}\rho_\mu^{(k)}(u,\overline{u})+e^{\mu(\tau-T-kT_0)}\frac{Z_0}{2}\rho_\mu^{(k)}(i,\overline{i})\le$$

$$\le e^{\mu(\tau-T-kT_0)}\left(\frac{1}{2}\frac{\rho_\mu^{(k)}(\dot{u},\dot{\overline{u}})}{\mu}+\frac{Z_0}{2}\frac{\rho_\mu^{(k)}(\dot{i},\dot{\overline{i}})}{\mu}\right)\le e^{\mu(\tau-T-kT_0)}\hat{\rho}_\mu((u,\dot{u},i,\dot{i}),(\overline{u},\dot{\overline{u}},\overline{i},\dot{\overline{i}}))\frac{1+Z_0}{2\mu}\,;$$

$$\left|G_0\big(P(u,i)(s)\big)\right|\le\sum_{n=1}^{m}\left|g_n^{(0)}\right|\left|P(u,i)(s)\right|^n\le e^{\mu(\tau-T-kT_0)}\sum_{n=1}^{m}\left|g_n^{(0)}\right|\left(U_{E_0}+U_0+Z_0I_0\right)^n e^{(n-1)\mu T_0}\,;$$

$$\left|G_0\big(P(u,i)(s)\big)-G_0\big(P(\overline{u},\overline{i})(s)\big)\right|\le\left|\sum_{n=1}^{m}g_n^{(0)}\big(P(u,i)(s)\big)^n-\sum_{n=1}^{m}g_n^{(0)}\big(P(\overline{u},\overline{i})(s)\big)^n\right|\le$$

$$\leq \sum_{n=1}^{m} \left| g_n^{(0)} \right| n \ . \mathrm{ess\,sup} \left| P(u,i)(s) \right|^{n-1} \left| P(u,i)(s) - P(\overline{u},\overline{i})(s) \right| \leq$$

$$\leq \sum_{n=1}^{m} \left| g_n^{(0)} \right| n \ . \mathrm{ess\,sup} \left\{ \left| \left| E_0(s) \right| + \frac{1}{2} \left| u(s) \right| + \frac{Z_0}{2} \left| i(s) \right| + \frac{1}{2} \left| \vec{\upsilon}_0(s) \right| + \frac{Z_0}{2} \left| \vec{\imath}_0(s) \right| \right|^{n-1} \right\} \times$$

$$\times \, e^{\mu(s-T-kT_0)} \hat{\rho}_\mu ((u,\dot{u},i,\dot{i}),(\overline{u},\dot{\overline{u}},\overline{i},\dot{\overline{i}})) \frac{1+Z_0}{2\mu} \leq$$

$$\leq e^{\mu(s-T-kT_0)} \hat{\rho}_\mu ((u,\dot{u},i,\dot{i}),(\overline{u},\dot{\overline{u}},\overline{i},\dot{\overline{i}})) \frac{1+Z_0}{2\mu} \sum_{n=1}^{m} \left| g_n^{(0)} \right| n \ . \phi_0^{n-1} e^{(n-1)\mu_0} \ .$$

Then

$$I_1 \leq \frac{1}{Z_0} \left| \int_{T+kT_0}^{t} (\dot{u}(s) - \dot{\overline{u}}(s)) ds \right| + \frac{1}{Z_0} \left| \int_{T+kT_0}^{t} (\dot{\vec{\upsilon}}_0(s) - \dot{\vec{\upsilon}}_0(s)) ds \right| + \frac{1}{Z_0} \left| \int_{T+kT_0}^{t} (\dot{\vec{\imath}}_0(s) - \dot{\vec{\imath}}_0(s)) ds \right| +$$

$$+ \frac{1}{Z_0^2} \int_{T+kT_0}^{t} \left| \frac{u(s)}{\widetilde{C}_0(u)(s)} - \frac{\overline{u}(s)}{\widetilde{C}_0(\overline{u})(s)} \right| ds + \frac{1}{Z_0} \int_{T+kT_0}^{t} \left| \frac{i(s)}{\widetilde{C}_0(u)(s)} - \frac{\overline{i}(s)}{\widetilde{C}_0(\overline{u})(s)} \right| ds +$$

$$+ \frac{1}{Z_0^2} \int_{T+kT_0}^{t} \left| \frac{\vec{\upsilon}_0(s)}{\widetilde{C}_0(u)(s)} - \frac{\vec{\upsilon}_0(s)}{\widetilde{C}_0(\overline{u})(s)} \right| ds + \frac{1}{Z_0} \int_{T+kT_0}^{t} \left| \frac{\vec{\imath}_0(s)}{\widetilde{C}_0(u)(s)} - \frac{\vec{\imath}_0(s)}{\widetilde{C}_0(\overline{u})(s)} \right| ds +$$

$$+ \frac{1}{Z_0 \hat{L}_0} \int_{T+kT_0}^{t} \int_{T+kT_0}^{s} \left(\left| \frac{u(\theta)}{\widetilde{C}_0(u)(\theta)} - \frac{\overline{u}(\theta)}{\widetilde{C}_0(\overline{u})(\theta)} \right| + Z_0 \left| \frac{i(\theta)}{\widetilde{C}_0(u)(\theta)} - \frac{\overline{i}(\theta)}{\widetilde{C}_0(\overline{u})(\theta)} \right| + \right.$$

$$\left. + \left| \frac{\vec{\upsilon}_0(\theta)}{\widetilde{C}_0(u)(\theta)} - \frac{\vec{\upsilon}_0(\theta)}{\widetilde{C}_0(\overline{u})(\theta)} \right| + Z_0 \left| \frac{\vec{\imath}_0(\theta)}{\widetilde{C}_0(u)(\theta)} - \frac{\vec{\imath}_0(\theta)}{\widetilde{C}_0(\overline{u})(\theta)} \right| \right) d\theta ds$$

$$+ \frac{2}{Z_0} \int_{T+kT_0}^{t} \left| \frac{G_0 \left(E_0(s-T) - \dfrac{u(s) + Z_0 i(s) + \vec{\upsilon}_0(s) - Z_0 \vec{\imath}_0(\theta)}{2} \right)}{\widetilde{C}_0(u)(s)} - \right.$$

$$\left. - \frac{G_0 \left(E_0(s-T) - \dfrac{\overline{u}(s) + Z_0 \overline{i}(s) + \vec{\upsilon}_0(s) - Z_0 \vec{\imath}_0(\theta)}{2} \right)}{\widetilde{C}_0(\overline{u})(s)} \right| ds \leq$$

$$\leq \frac{\rho_\mu^{(k)}(\dot{u},\dot{\overline{u}})}{Z_0}\frac{e^{\mu(t-T-kT_0)}-1}{\mu}+$$

$$+\frac{1}{Z_0^2}\int_{T+kT_0}^t\left|\frac{u(s)}{\widetilde{C}_0(u)(s)}-\frac{\overline{u}(s)}{\widetilde{C}_0(u)(s)}\right|ds+\frac{1}{Z_0^2}\int_{T+kT_0}^t\left|\frac{\overline{u}(s)}{\widetilde{C}_0(u)(s)}-\frac{\overline{u}(s)}{\widetilde{C}_0(\overline{u})(s)}\right|ds+$$

$$+\frac{1}{Z_0}\int_{T+kT_0}^t\left|\frac{i(s)}{\widetilde{C}_0(u)(s)}-\frac{\overline{i}(s)}{\widetilde{C}_0(u)(s)}\right|ds+\frac{1}{Z_0}\int_{T+kT_0}^t\left|\frac{\overline{i}(s)}{\widetilde{C}_0(u)(s)}-\frac{\overline{i}(s)}{\widetilde{C}_0(\overline{u})(s)}\right|ds+$$

$$+\frac{1}{Z_0^2}\int_{T+kT_0}^t\left|\frac{\vec{\upsilon}_0(s)}{\widetilde{C}_0(u)(s)}-\frac{\vec{\upsilon}_0(s)}{\widetilde{C}_0(u)(s)}\right|ds+\frac{1}{Z_0^2}\int_{T+kT_0}^t\left|\frac{\vec{\upsilon}_0(s)}{\widetilde{C}_0(u)(s)}-\frac{\vec{\upsilon}_0(s)}{\widetilde{C}_0(\overline{u})(s)}\right|ds+$$

$$+\frac{1}{Z_0}\int_{T+kT_0}^t\left|\frac{\vec{\imath}_0(s)}{\widetilde{C}_0(u)(s)}-\frac{\vec{\imath}_0(s)}{\widetilde{C}_0(u)(s)}\right|ds+\frac{1}{Z_0}\int_{T+kT_0}^t\left|\frac{\vec{\imath}_0(s)}{\widetilde{C}_0(u)(s)}-\frac{\vec{\imath}_0(s)}{\widetilde{C}_0(\overline{u})(s)}\right|ds+$$

$$+\frac{2}{Z_0}\int_{T+kT_0}^t\left|\frac{G_0\left(E_0(s)-\dfrac{u(s)+Z_0i(s)+\vec{\upsilon}_0(s)-Z_0\vec{\imath}_0(s)}{2}\right)}{\widetilde{C}_0(u)(s)}-\right.$$

$$\left.\frac{G_0\left(E_0(s)-\dfrac{\overline{u}(s)+Z_0\overline{i}(s)+\vec{\upsilon}_0(s)-Z_0\vec{\imath}_0(s)}{2}\right)}{\widetilde{C}_0(u)(s)}\right|ds+$$

$$+\frac{2}{Z_0}\int_{T+kT_0}^t\left|\frac{G_0\left(E_0(s)-\dfrac{\overline{u}(s)+Z_0\overline{i}(s)+\vec{\upsilon}_0(s)-Z_0\vec{\imath}_0(s)}{2}\right)}{\widetilde{C}_0(u)(s)}-\right.$$

$$\left.-\frac{G_0\left(E_0(s)-\dfrac{\overline{u}(s)+Z_0\overline{i}(s)+\vec{\upsilon}_0(s)-Z_0\vec{\imath}_0(s)}{2}\right)}{\widetilde{C}_0(\overline{u})(s)}\right|ds\leq$$

$$\leq\frac{e^{\mu_0}\rho_\mu^{(k)}(\dot{u},\dot{\overline{u}})}{Z_0}\frac{e^{\mu(t-T-kT_0)}-1}{\mu}+$$

$$+\frac{\rho_\mu^{(k)}(\dot{u},\dot{\overline{u}})}{\mu Z_0^2\hat{C}_0}\frac{e^{\mu(t-T-kT_0)}-1}{\mu}+\frac{U_0}{Z_0^2}\int_{T+kT_0}^t e^{\mu(s-T-kT_0)}\frac{1}{\widetilde{C}_0^2(u)(s)}\left|\frac{d\widetilde{C}_0(u)}{du}\right||u(s)-\overline{u}(s)|ds+$$

$$+\frac{e^{\mu_0}\rho_\mu^{(k)}(\dot{i},\dot{\overline{i}})}{\mu Z_0\hat{C}_0}\frac{e^{\mu(t-T-kT_0)}-1}{\mu}+\frac{I_0}{Z_0}\int_{T+kT_0}^t e^{\mu(s-T-kT_0)}\frac{1}{\widetilde{C}_0^2(u)(s)}\left|\frac{d\widetilde{C}_0(u)}{du}\right||u(s)-\overline{u}(s)|ds+$$

$$+\frac{U_0}{Z_0^2}\int_{T+kT_0}^{t}e^{\mu(s-T-kT_0)}\frac{1}{\widetilde{C}_0^2(u)(s)}\left|\frac{d\widetilde{C}_0(u)}{du}\right|\left|u(s)-\overline{u}(s)\right|ds+$$

$$+\frac{I_0}{Z_0}\int_{T+kT_0}^{t}e^{\mu(s-T-kT_0)}\frac{1}{\widetilde{C}_0^2(u)(s)}\left|\frac{d\widetilde{C}_0(u)}{du}\right|\left|u(s)-\overline{u}(s)\right|ds+$$

$$+\hat{\rho}_\mu((u,\dot{u},i,\dot{i}),(\overline{u},\dot{\overline{u}},\overline{i},\dot{\overline{i}}))\frac{2}{Z_0\hat{C}_0}\sum_{n=1}^{m}\left|g_n^{(0)}\right|n\cdot\left(U_{E_0}+U_0+Z_0I_0\right)^{n-1}e^{(n-1)\mu T_0}\frac{(1+Z_0)}{2\mu}\int_{T+kT_0}^{t}e^{\mu(s-T-kT_0)}ds+$$

$$+\frac{2}{Z_0}\sum_{n=1}^{m}\left|g_n^{(0)}\right|\left(U_{E_0}+U_0+Z_0I_0\right)^n e^{(n-1)\mu T_0}\int_{T+kT_0}^{t}e^{\mu(s-T-kT_0)}\frac{1}{\widetilde{C}_0^2(u)(s)}\left|\frac{d\widetilde{C}_0(u)}{du}\right|\left|u(s)-\overline{u}(s)\right|ds\le$$

$$\le\frac{e^{\mu_0}\rho_\mu^{(k)}(\dot{u},\dot{\overline{u}})}{Z_0}\frac{e^{\mu(t-T-kT_0)}-1}{\mu}+$$

$$+\frac{e^{\mu_0}\rho_\mu^{(k)}(\dot{u},\dot{\overline{u}})}{\mu Z_0^2\hat{C}_0}\frac{e^{\mu(t-T-kT_0)}-1}{\mu}+\frac{U_0}{Z_0^2\hat{C}_0^2}D_0\frac{e^{\mu_0}\rho_\mu^{(k)}(\dot{u},\dot{\overline{u}})}{\mu}\int_{T+kT_0}^{t}e^{2\mu(s-T-kT_0)}ds+$$

$$+\frac{e^{\mu_0}\rho_\mu^{(k)}(\dot{i},\dot{\overline{i}})}{\mu Z_0\hat{C}_0}\frac{e^{\mu(t-T-kT_0)}-1}{\mu}+\frac{I_0}{Z_0\left(\hat{C}_0\right)^2}D_0\frac{e^{\mu_0}\rho_\mu^{(k)}(\dot{u},\dot{\overline{u}})}{\mu}\int_{T+kT_0}^{t}e^{2\mu(s-T-kT_0)}ds+$$

$$+\frac{U_0}{Z_0^2\hat{C}_0^2}D_0\frac{e^{\mu_0}\rho_\mu^{(k)}(\dot{u},\dot{\overline{u}})}{\mu}\int_{T+kT_0}^{t}e^{2\mu(s-T-kT_0)}ds+$$

$$+\frac{I_0}{Z_0\hat{C}_0^2}D_0\frac{e^{\mu_0}\rho_\mu^{(k)}(\dot{u},\dot{\overline{u}})}{\mu}\int_{T+kT_0}^{t}e^{2\mu(s-T-kT_0)}ds+$$

$$+\hat{\rho}_\mu((u,\dot{u},i,\dot{i}),(\overline{u},\dot{\overline{u}},\overline{i},\dot{\overline{i}}))\frac{e^{\mu(t-T-kT_0)}-1}{\mu}\frac{2e^{\mu_0}}{Z_0\hat{C}_0}\frac{(1+Z_0)}{2\mu}\sum_{n=1}^{m}\left|g_n^{(0)}\right|n\cdot\phi_0^{n-1}e^{(n-1)\mu T_0}+$$

$$+\frac{2e^{\mu_0}}{Z_0\hat{C}_0^2}D_0\sum_{n=1}^{m}\left|g_n^{(0)}\right|\phi_0^n e^{(n-1)\mu T_0}+\frac{\rho_\mu^{(k)}(\dot{u},\dot{\overline{u}})}{\mu}\int_{T+kT_0}^{t}e^{2\mu(s-T-kT_0)}ds\le\frac{e^{\mu_0}\rho_\mu^{(k)}(\dot{u},\dot{\overline{u}})}{Z_0}\frac{e^{\mu(t-T-kT_0)}-1}{\mu}+$$

$$+\frac{e^{\mu_0}\rho_\mu^{(k)}(\dot{u},\dot{\overline{u}})}{\mu Z_0^2\hat{C}_0}\frac{e^{\mu(t-T-kT_0)}-1}{\mu}+\frac{U_0}{Z_0^2\hat{C}_0^2}D_0\frac{e^{\mu_0}\rho_\mu^{(k)}(\dot{u},\dot{\overline{u}})}{\mu}\frac{e^{2\mu(t-T-kT_0)}-1}{2\mu}+$$

$$+\frac{e^{\mu_0}\rho_\mu^{(k)}(\dot{i},\dot{\overline{i}})}{\mu Z_0\hat{C}_0}\frac{e^{\mu(t-T-kT_0)}-1}{\mu}+\frac{I_0}{Z_0\hat{C}_0^2}D_0\frac{e^{\mu_0}\rho_\mu^{(k)}(\dot{u},\dot{\overline{u}})}{\mu}\frac{e^{2\mu(t-T-kT_0)}-1}{2\mu}+$$

$$+\frac{U_0}{Z_0^2\hat{C}_0^2}D_0\frac{\rho_\mu^{(k)}(\dot{u},\dot{\overline{u}})}{\mu}\frac{e^{2\mu(t-T-kT_0)}-1}{2\mu}+$$

$$+ \frac{I_0}{Z_0 \hat{C}_0^2} D_0 \frac{e^{\mu_0} \rho_\mu^{(k)}(\dot{u},\dot{\bar{u}})}{\mu} \frac{e^{2\mu(t-T-kT_0)}-1}{2\mu} +$$

$$+ \hat{\rho}_\mu((u,\dot{u},i,\dot{i}),(\bar{u},\dot{\bar{u}},\bar{i},\dot{\bar{i}})) \frac{e^{\mu(t-T-kT_0)}-1}{\mu} \frac{2e^{\mu_0}}{Z_0 \hat{C}_0} \sum_{n=1}^{m} \left| g_n^{(0)} \right| n . \phi_0^{n-1} e^{(n-1)\mu T_0} \frac{(1+Z_0)}{2\mu} +$$

$$+ \frac{2e^{\mu_0}}{Z_0 \hat{C}_0^2} D_0 \sum_{n=1}^{m} \left| g_n^{(0)} \right| (\phi_0)^n e^{(n-1)\mu T_0} \frac{\rho_\mu^{(k)}(\dot{u},\dot{\bar{u}})}{\mu} \frac{e^{2\mu(t-T-kT_0)}-1}{2\mu} \le$$

$$\le \hat{\rho}_\mu((u,\dot{u},i,\dot{i}),(\bar{u},\dot{\bar{u}},\bar{i},\dot{\bar{i}})) \frac{e^{\mu(t-T-kT_0)}-1}{\mu} e^{\mu_0} \left[\frac{1}{Z_0} + \frac{1}{\mu Z_0^2 \hat{C}_0} + \frac{U_0}{\mu Z_0^2 \hat{C}_0^2} D_0 e^{\mu_0} + \right.$$

$$+ \frac{1}{\mu Z_0 \hat{C}_0} + \frac{I_0}{\mu Z_0 \hat{C}_0^2} D_0 e^{\mu_0} + \frac{U_0}{\mu Z_0^2 \hat{C}_0^2} D_0 e^{\mu_0} + \frac{I_0}{\mu Z_0 \hat{C}_0^2} D_0 e^{\mu_0} +$$

$$\left. + \frac{(1+Z_0)}{\mu Z_0 \hat{C}_0} \sum_{n=1}^{m} \left| g_n^{(0)} \right| n . \phi_0^{n-1} e^{(n-1)\mu T_0} + \frac{2}{\mu Z_0 \hat{C}_0^2} D_0 \sum_{n=1}^{m} \left| g_n^{(0)} \right| \phi_0^n e^{n\mu T_0} \right] \le$$

$$\le e^{\mu(t-T-kT_0)} \hat{\rho}_\mu((u,\dot{u},i,\dot{i}),(\bar{u},\dot{\bar{u}},\bar{i},\dot{\bar{i}})) \frac{e^{\mu_0}}{\mu} \left[\frac{1}{Z_0} + \frac{1+Z_0}{\mu Z_0^2 \hat{C}_0} + \frac{2(U_0 + Z_0 I_0)D_0 e^{\mu_0}}{\mu Z_0^2 \hat{C}_0^2} + \right.$$

$$\left. + \frac{(1+Z_0)}{\mu Z_0 \hat{C}_0} \sum_{n=1}^{m} \left| g_n^{(0)} \right| n . \phi_0^{n-1} e^{(n-1)\mu_0} + \frac{2D_0}{\mu Z_0 \hat{C}_0^2} \sum_{n=1}^{m} \left| g_n^{(0)} \right| \phi_0^n e^{n\mu_0} \right];$$

$$I_2 \le \frac{1}{2} \int_{T+kT_0}^{T+(k+1)T_0} \left| I(u,i)(s) - I(\bar{u},\bar{i})(s) \right| ds \le$$

$$\le \hat{\rho}_\mu((u,\dot{u},i,\dot{i}),(\bar{u},\dot{\bar{u}},\bar{i},\dot{\bar{i}})) \frac{e^{\mu_0}-1}{2} \frac{e^{\mu_0}}{\mu} \left[\frac{1}{Z_0} + \frac{1+Z_0}{\mu Z_0^2 \hat{C}_0} + \frac{2(U_0 + Z_0 I_0)D_0 e^{\mu_0}}{\mu Z_0^2 \hat{C}_0^2} + \right.$$

$$\left. + \frac{(1+Z_0)}{\mu Z_0 \hat{C}_0} \sum_{n=1}^{m} \left| g_n^{(0)} \right| n . \phi_0^{n-1} e^{(n-1)\mu_0} + \frac{2D_0}{\mu Z_0 \hat{C}_0^2} \sum_{n=1}^{m} \left| g_n^{(0)} \right| \phi_0^n e^{n\mu_0} \right];$$

$$I_3 \le \frac{1}{T_0} \int_{T+kT_0}^{T+(k+1)T_0} \left| \int_{T+kT_0}^{t} \left(I(u,i)(s) - I(\bar{u},\bar{i})(s) \right) ds \right| dt \le$$

$$\le e^{\mu(t-T-kT_0)} \hat{\rho}_\mu((u,\dot{u},i,\dot{i}),(\bar{u},\dot{\bar{u}},\bar{i},\dot{\bar{i}})) \frac{e^{\mu_0}-1}{\mu_0} \frac{e^{\mu_0}}{\mu} \left[\frac{1}{Z_0} + \frac{1+Z_0}{\mu Z_0^2 \hat{C}_0} + \frac{2(U_0 + Z_0 I_0)D_0 e^{\mu_0}}{\mu Z_0^2 \hat{C}_0^2} + \right.$$

$$+\frac{(1+Z_0)}{\mu Z_0 \hat{C}_0}\sum_{n=1}^{m}\left|g_n^{(0)}\right|n.\phi_0^{n-1}e^{(n-1)\mu_0}+\frac{2D_0}{\mu Z_0 \hat{C}_0^{\ 2}}\sum_{n=1}^{m}\left|g_n^{(0)}\right|\phi_0^n e^{n\mu_0}\Bigg].$$

Therefore

$$\left|B_i^{(k)}(u,i)(t)-B_i^{(k)}(\overline{u},\overline{i})(t)\right|\le I_1+I_2+I_3\le$$

$$\le e^{\mu(t-T-kT_0)}\hat{\rho}_\mu((u,\dot{u},i,\dot{i}),(\overline{u},\dot{\overline{u}},\overline{i},\dot{\overline{i}}))\frac{e^{\mu_0}}{\mu}\left[\frac{1}{Z_0}+\frac{1+Z_0}{\mu Z_0^2\hat{C}_0}+\frac{2(U_0+Z_0 I_0)D_0 e^{\mu_0}}{\mu Z_0^2\hat{C}_0^{\ 2}}+\right.$$

$$\left.+\frac{(1+Z_0)}{\mu Z_0 \hat{C}_0}\sum_{n=1}^{m}\left|g_n^{(0)}\right|n.\phi_0^{n-1}e^{(n-1)\mu_0}+\frac{2D_0}{\mu Z_0 \hat{C}_0^{\ 2}}\sum_{n=1}^{m}\left|g_n^{(0)}\right|\phi_0^n e^{n\mu_0}\right]+$$

$$+\hat{\rho}_\mu((u,\dot{u},i,\dot{i}),(\overline{u},\dot{\overline{u}},\overline{i},\dot{\overline{i}}))\frac{e^{\mu_0}-1}{2}\frac{e^{\mu_0}}{\mu}\left[\frac{1}{Z_0}+\frac{1+Z_0}{\mu Z_0^2\hat{C}_0}+\frac{2(U_0+Z_0 I_0)e^{\mu_0}}{\mu Z_0^2\hat{C}_0^{\ 2}}+\right.$$

$$\left.+\frac{(1+Z_0)}{\mu Z_0 \hat{C}_0}\sum_{n=1}^{m}\left|g_n^{(0)}\right|n.\phi_0^{n-1}e^{(n-1)\mu_0}+\frac{2D_0}{\mu Z_0 \hat{C}_0^{\ 2}}\sum_{n=1}^{m}\left|g_n^{(0)}\right|\phi_0^n e^{n\mu_0}\right]+$$

$$+e^{\mu(t-T-kT_0)}\hat{\rho}_\mu((u,\dot{u},i,\dot{i}),(\overline{u},\dot{\overline{u}},\overline{i},\dot{\overline{i}}))\frac{e^{\mu_0}-1}{\mu_0}\frac{e^{\mu_0}}{\mu}\left[\frac{1}{Z_0}+\frac{1+Z_0}{\mu Z_0^2\hat{C}_0}+\frac{2(U_0+Z_0 I_0)D_0 e^{\mu_0}}{\mu Z_0^2\hat{C}_0^{\ 2}}+\right.$$

$$\left.+\frac{(1+Z_0)}{\mu Z_0 \hat{C}_0}\sum_{n=1}^{m}\left|g_n^{(0)}\right|n.\phi_0^{n-1}e^{(n-1)\mu_0}+\frac{2D_0}{\mu Z_0 \hat{C}_0^{\ 2}}\sum_{n=1}^{m}\left|g_n^{(0)}\right|\phi_0^n e^{n\mu_0}\right]\le$$

$$+e^{\mu(t-T-kT_0)}\hat{\rho}_\mu((u,\dot{u},i,\dot{i}),(\overline{u},\dot{\overline{u}},\overline{i},\dot{\overline{i}}))\left(1+\frac{e^{\mu_0}-1}{2}+\frac{e^{\mu_0}-1}{\mu_0}\right)\frac{e^{\mu_0}}{\mu}\left[\frac{1}{Z_0}+\frac{1+Z_0}{\mu Z_0^2\hat{C}_0}+\frac{2(U_0+Z_0 I_0)D_0 e^{\mu_0}}{\mu Z_0^2\hat{C}_0^{\ 2}}+\right.$$

$$\left.+\frac{(1+Z_0)}{\mu Z_0 \hat{C}_0}\sum_{n=1}^{m}\left|g_n^{(0)}\right|n.\phi_0^{n-1}e^{(n-1)\mu_0}+\frac{2D_0}{\mu Z_0 \hat{C}_0^{\ 2}}\sum_{n=1}^{m}\left|g_n^{(0)}\right|\phi_0^n e^{n\mu_0}\right]\equiv$$

$$\equiv e^{\mu(t-T-kT_0)}K_i\ \hat{\rho}_\mu((u,\dot{u},i,\dot{i}),(\overline{u},\dot{\overline{u}},\overline{i},\dot{\overline{i}})).$$

It follows

$$\hat{\rho}(B_i((u,i),B_i(\overline{u},\overline{i}))\le e^{\mu_0}K_i\ \hat{\rho}_\mu((u,\dot{u},i,\dot{i}),(\overline{u},\dot{\overline{u}},\overline{i},\dot{\overline{i}})).$$

For the derivative we obtain

$$\left| \dot{B}_i^{(k)}(u,i)(t) - \dot{B}_i^{(k)}(\overline{u},\overline{i})(t) \right| \le \left| I(u,i)(t) - I(\overline{u},\overline{i})(t) \right| +$$

$$\left| \frac{1}{T_0} \int_{T+kT_0}^{T+(k+1)T_0} \left(I(u,i)(s) - I(\overline{u},\overline{i})(s) \right) ds \right| \equiv I_1 + I_2 .$$

But

$$\dot{I}_1 \le \frac{1}{Z_0}\left| \dot{u}(t) - \dot{\overline{u}}(t) \right| + \frac{1}{Z_0}\left| \dot{\upsilon}_0(t) - \dot{\overline{\upsilon}}_0(t) \right| + \frac{1}{Z_0}\left| \dot{\overline{i}}_0(t) - \dot{\overline{i}}_0(t) \right| +$$

$$+ \frac{1}{Z_0^2}\left| \frac{u(t)}{\widetilde{C}_0(u)(t)} - \frac{\overline{u}(t)}{\widetilde{C}_0(\overline{u})(t)} \right| + \frac{1}{Z_0}\left| \frac{i(t)}{\widetilde{C}_0(u)(t)} - \frac{\overline{i}(t)}{\widetilde{C}_0(\overline{u})(t)} \right| +$$

$$+ \frac{1}{Z_0^2}\left| \frac{\overline{\upsilon}_0(t)}{\widetilde{C}_0(u)(t)} - \frac{\overline{\upsilon}_0(t)}{\widetilde{C}_0(\overline{u})(t)} \right| + \frac{1}{Z_0}\left| \frac{\overline{i}_0(t)}{\widetilde{C}_0(u)(t)} - \frac{\overline{i}_0(t)}{\widetilde{C}_0(\overline{u})(t)} \right| +$$

$$+ \frac{1}{Z_0\hat{L}_0} \int_{T+kT_0}^{t} \left(\left| \frac{u(\theta)}{\widetilde{C}_0(u)(\theta)} - \frac{\overline{u}(\theta)}{\widetilde{C}_0(\overline{u})(\theta)} \right| + Z_0 \left| \frac{i(\theta)}{\widetilde{C}_0(u)(\theta)} - \frac{\overline{i}(\theta)}{\widetilde{C}_0(\overline{u})(\theta)} \right| + \right.$$

$$\left. + \left| \frac{\overline{\upsilon}_0(\theta)}{\widetilde{C}_0(u)(\theta)} - \frac{\overline{\upsilon}_0(\theta)}{\widetilde{C}_0(\overline{u})(\theta)} \right| + Z_0 \left| \frac{\overline{i}_0(\theta)}{\widetilde{C}_0(u)(\theta)} - \frac{\overline{i}_0(\theta)}{\widetilde{C}_0(\overline{u})(\theta)} \right| \right) d\theta +$$

$$+ \frac{2}{Z_0} \left| \frac{G_0\left(E_0(t) - \dfrac{u(t)+Z_0 i(t)+\overline{\upsilon}_0(t)-Z_0\overline{i}_0(t)}{2} \right)}{\widetilde{C}_0(u)(t)} - \frac{G_0\left(E_0(t) - \dfrac{\overline{u}(t)+Z_0\overline{i}(t)+\overline{\upsilon}_0(t)-Z_0\overline{i}_0(t)}{2} \right)}{\widetilde{C}_0(\overline{u})(t)} \right| \le$$

$$\le \frac{\rho_\mu^{(k)}(\dot{u},\dot{\overline{u}})}{Z_0} + \frac{\rho_\mu^{(k)}(\dot{u},\dot{\overline{u}})}{\mu Z_0^2 \hat{C}_0} + \frac{U_0}{Z_0^2} e^{\mu(t-T-kT_0)} \frac{1}{\widetilde{C}_0^2(u)(t)} \left| \frac{d\widetilde{C}_0(u)}{du} \right| \left| u(t) - \overline{u}(t) \right| +$$

$$+ \frac{\rho_\mu^{(k)}(\dot{i},\dot{\overline{i}})}{\mu Z_0 \hat{C}_0} + \frac{I_0}{Z_0} e^{\mu(t-T-kT_0)} \frac{1}{\widetilde{C}_0^2(u)(t)} \left| \frac{d\widetilde{C}_0(u)}{du} \right| \left| u(t) - \overline{u}(t) \right| +$$

$$+ \frac{U_0}{Z_0^2} e^{\mu(t-T-kT_0)} \frac{1}{\widetilde{C}_0^2(u)(t)} \left| \frac{d\widetilde{C}_0(u)}{du} \right| \left| u(t) - \overline{u}(t) \right| +$$

$$+ \frac{I_0}{Z_0} e^{\mu(t-T-kT_0)} \frac{1}{\widetilde{C}_0^2(u)(t)} \left| \frac{d\widetilde{C}_0(u)}{du} \right| \left| u(t) - \overline{u}(t) \right| +$$

$$+ \hat{\rho}_\mu((u,\dot{u},i,\dot{i}),(\overline{u},\dot{\overline{u}},\overline{i},\dot{\overline{i}})) \frac{2}{Z_0\hat{C}_0} \frac{(1+Z_0)}{2\mu} \sum_{n=1}^{m} \left| g_n^{(0)} \right| n .\left(U_{E_0} + U_0 + Z_0 I_0 \right)^{n-1} e^{(n-1)\mu T_0} +$$

$$+ \frac{2}{Z_0} \sum_{n=1}^{m} \left| g_n^{(0)} \right| \left(U_{E_0} + U_0 + Z_0 I_0 \right)^{n} e^{(n-1)\mu T_0} \frac{1}{\widetilde{C}_0^2(u)(t)} \left| \frac{d\widetilde{C}_0(u)}{du} \right| \left| u(t) - \overline{u}(t) \right| \le$$

$$\leq \frac{\rho_\mu^{(k)}(\dot{u},\ddot{u})}{Z_0} + \frac{\rho_\mu^{(k)}(\dot{u},\ddot{u})}{\mu Z_0^2 \hat{C}_0} + e^{2\mu(t-T-kT_0)} \frac{U_0}{Z_0^2 \hat{C}_0^2} D_0 \frac{\rho_\mu^{(k)}(\dot{u},\ddot{u})}{\mu} +$$

$$+ \frac{\rho_\mu^{(k)}(\dot{i},\ddot{i})}{\mu Z_0 \hat{C}_0} + e^{2\mu(t-T-kT_0)} \frac{I_0}{Z_0 \hat{C}_0^2} D_0 \frac{\rho_\mu^{(k)}(\dot{u},\ddot{u})}{\mu} +$$

$$+ e^{2\mu(t-T-kT_0)} \frac{U_0}{Z_0^2 \hat{C}_0^2} D_0 \frac{\rho_\mu^{(k)}(\dot{u},\ddot{u})}{\mu} + e^{2\mu(t-T-kT_0)} \frac{I_0}{Z_0 \hat{C}_0^2} D_0 \frac{\rho_\mu^{(k)}(\dot{u},\ddot{u})}{\mu} +$$

$$+ e^{2\mu(t-T-kT_0)} \frac{I_0}{Z_0 \hat{C}_0^2} D_0 \frac{\rho_\mu^{(k)}(\dot{u},\ddot{u})}{\mu} +$$

$$\hat{\rho}_\mu((u,\dot{u},i,\dot{i}),(\overline{u},\dot{\overline{u}},\overline{i},\dot{\overline{i}})) \frac{1+Z_0}{\mu Z_0 \hat{C}_0} \sum_{n=1}^{m} \left|g_n^{(0)}\right| n \cdot \phi_0^{n-1} e^{(n-1)\mu T_0} +$$

$$+ e^{\mu(t-T-kT_0)} \frac{\rho_\mu^{(k)}(\dot{u},\ddot{u})}{\mu} \frac{2}{Z_0 \hat{C}_0^2} \sum_{n=1}^{m} \left|g_n^{(0)}\right| \phi_0^n e^{(n-1)\mu T_0} D_0 \leq$$

$$\leq e^{\mu(t-T-kT_0)} \hat{\rho}_\mu((u,\dot{u},i,\dot{i}),(\overline{u},\dot{\overline{u}},\overline{i},\dot{\overline{i}})) \left[\frac{1}{Z_0} + \frac{1}{\mu Z_0^2 \hat{C}_0} + \frac{1}{\mu Z_0 \hat{C}_0} + 2\frac{U_0+Z_0 I_0}{\mu Z_0^2 \hat{C}_0^2} D_0 e^{\mu_0} + \right.$$

$$\left. + \frac{1+Z_0}{\mu Z_0 \hat{C}_0} \sum_{n=1}^{m} \left|g_n^{(0)}\right| n \cdot \phi_0^{n-1} e^{(n-1)\mu T_0} + \frac{2}{\mu Z_0 \hat{C}_0^2} D_0 \sum_{n=1}^{m} \left|g_n^{(0)}\right| \phi_0^n e^{(n-1)\mu T_0} \right];$$

$$\dot{I}_2 \leq e^{\mu(t-T-kT_0)} \hat{\rho}_\mu((u,\dot{u},i,\dot{i}),(\overline{u},\dot{\overline{u}},\overline{i},\dot{\overline{i}})) \frac{e^{\mu_0}-1}{\mu_0} \left[\frac{1}{Z_0} + \frac{1}{\mu Z_0^2 \hat{C}_0} + \frac{1}{\mu Z_0 \hat{C}_0} + 2\frac{U_0+Z_0 I_0}{\mu Z_0^2 \hat{C}_0^2} D_0 e^{\mu_0} + \right.$$

$$\left. + \frac{1+Z_0}{\mu Z_0 \hat{C}_0} \sum_{n=1}^{m} \left|g_n^{(0)}\right| n \cdot \phi_0^{n-1} e^{(n-1)\mu T_0} + \frac{2D_0}{\mu Z_0 \hat{C}_0^2} \sum_{n=1}^{m} \left|g_n^{(0)}\right| \phi_0^n e^{(n-1)\mu T_0} \right].$$

Therefore

$$\left| \dot{B}_i^{(k)}(u,i)(t) - \dot{B}_i^{(k)}(\overline{u},\overline{i})(t) \right| \leq$$

$$\leq e^{\mu(t-T-kT_0)} \hat{\rho}_\mu((u,\dot{u},i,\dot{i}),(\overline{u},\dot{\overline{u}},\overline{i},\dot{\overline{i}})) \left(1 + \frac{e^{\mu_0}-1}{\mu_0}\right) \left[\frac{1}{Z_0} + \frac{1}{\mu Z_0^2 \hat{C}_0} + \frac{1}{\mu Z_0 \hat{C}_0} + 2\frac{U_0+Z_0 I_0}{\mu Z_0^2 \hat{C}_0^2} D_0 e^{\mu_0} + \right.$$

$$\left. + \frac{1+Z_0}{\mu Z_0 \hat{C}_0} \sum_{n=1}^{m} \left|g_n^{(0)}\right| n \cdot \phi_0^{n-1} e^{(n-1)\mu T_0} + \frac{2}{\mu Z_0 \hat{C}_0^2} D_0 \sum_{n=1}^{m} \left|g_n^{(0)}\right| \phi_0^n e^{(n-1)\mu T_0} \right] \equiv$$

$$\equiv e^{\mu(t-T-kT_0)} \dot{K}_i \, \hat{\rho}_\mu((u,\dot{u},i,\dot{i}),(\overline{u},\dot{\overline{u}},\overline{i},\dot{\overline{i}})).$$

It follows

$$\rho_\mu^{(k)}(B_i^{(k)}(u,i),B_i^{(k)}(\overline{u},\overline{i})) \le \dot{K}_i \hat{\rho}_\mu((u,\dot{u},i,\dot{i}),(\overline{u},\dot{\overline{u}},\overline{i},\dot{\overline{i}}))$$

and therefore

$$\hat{\rho}_\mu((B_u,\dot{B}_u,B_i,\dot{B}_i),(\overline{B}_u,\dot{\overline{B}}_u,\overline{B}_i,\dot{\overline{B}}_i)) \le K\hat{\rho}_\mu((u,\dot{u},i,\dot{i}),(\overline{u},\dot{\overline{u}},\overline{i},\dot{\overline{i}}))$$

where $K = \max\left\{e^{\mu_0}K_u,\dot{K}_u,e^{\mu_0}K_i,\dot{K}_i\right\}$. Since $\dfrac{e^{\mu_0}+\mu_0-1}{\mu_0 Z_0}<1$ for sufficiently large μ we notice that $K = \max\left\{K_u,\dot{K}_u,K_i,\dot{K}_i\right\}<1$, that is, the operator $B=(B_u,B_i)$ is contractive and has a unique fixed point. Therefore (4.4.7) has a unique T_0-periodic solution.

Theorem 4.5.1 is thus proved.

4.5.1. Numerical Example

We collect all inequalities guaranteeing an existence-uniqueness of T_0-periodic solution:

$$e^{\mu_0}\left(U_0+Z_0 I_0+U_{E_0}\right)\le \phi_0 ;\quad \frac{e^{\mu_0}+\mu_0-1}{\mu_0}\frac{U_0}{Z_0}<I_0 ;\quad \frac{e^{\mu_0}+\mu_0-1}{\mu_0 Z_0}<1;$$

$$\left(1+\frac{e^{\mu_0}-1}{\mu_0}\right)U_{E_1}+\frac{1}{\mu\hat{C}_1}\left(1+\frac{e^{\mu_0}-1}{2}+\frac{e^{\mu_0}-1}{\mu_0}\right)\left(2I_0+\frac{1}{\mu}\sum_{n=1}^{m}\left|g_n^{(1)}\right|\phi_0^n e^{(n-1)\mu_0}\right)\le U_0 ;$$

$$\left(1+\frac{e^{\mu_0}-1}{\mu_0}\right)\frac{2U_{E_0}+e^{-\beta}(U_0+Z_0 I_0)}{Z_0}+$$
$$+\left(1+\frac{e^{\mu_0}-1}{2}+\frac{e^{\mu_0}-1}{\mu_0}\right)\frac{1}{\mu Z_0 \hat{C}_0}\left(\frac{\left(1+e^{-\beta}\right)(U_0+I_0 Z_0)}{Z_0}+2I_0+2\sum_{n=1}^{m}\left|g_n^{(0)}\right|\phi_0^n e^{(n-1)\mu_0}\right)\le I_0 ;$$

$$K_u=\left(1+\frac{e^{\mu_0}-1}{2}+\frac{e^{\mu_0}-1}{\mu_0}\right)\frac{e^{\mu_0}}{\mu^2\hat{C}_1}\left[1+\frac{1}{\mu\hat{L}_1}+\frac{1}{\hat{C}_1}\sum_{n=1}^{m}n\left|g_n^{(1)}\right|\phi_0^{n-1}e^{(n-1)\mu_0}+D_1\left(\frac{I_0\left(e^{\mu_0}+1\right)}{\hat{C}_1}+\sum_{n=1}^{m}\left|g_n^{(1)}\right|\phi_0^n e^{n\mu_0}\right)\right]<1;$$

$$\dot{K}_u=\left(1+\frac{e^{\mu_0}-1}{\mu_0}\right)\frac{1}{\mu\hat{C}_1}\left[1+D_1\left(I_0 e^{\mu_0}+I_0+\sum_{n=1}^{m}\left|g_n^{(1)}\right|\phi_0^n e^{n\mu_0}\right)+\frac{1}{\mu\hat{L}_1}+\sum_{n=1}^{m}n\left|g_n^{(1)}\right|\phi_0^{n-1}e^{(n-1)\mu_0}\right]<1 ;$$

$$K_i = \left(1 + \frac{e^{\mu_0}-1}{2} + \frac{e^{\mu_0}-1}{\mu_0}\right)\frac{e^{\mu_0}}{\mu}\left[\frac{1}{Z_0} + \frac{1+Z_0}{\mu Z_0 \hat{C}_0}\left(\frac{1}{Z_0} + \sum_{n=1}^{m}\left|g_n^{(0)}\right|n\,.\phi_0^{n-1}e^{(n-1)\mu_0}\right) + \right.$$
$$\left. + \frac{2D_0}{\mu Z_0 \hat{C}_0{}^2}\left(\frac{(U_0 + Z_0 I_0)e^{\mu_0}}{Z_0} + \sum_{n=1}^{m}\left|g_n^{(0)}\right|\phi_0^n e^{n\mu_0}\right)\right] < 1;$$

$$\dot{K}_i = \left(1 + \frac{e^{\mu_0}-1}{\mu_0}\right)\left[\frac{1}{Z_0} + \frac{1+Z_0}{\mu Z_0 \hat{C}_0}\left(\frac{1}{Z_0} + \sum_{n=1}^{m}\left|g_n^{(0)}\right|n\,.\phi_0^{n-1}e^{(n-1)\mu_0}\right) + \right.$$
$$\left. + \frac{2D_0}{\mu Z_0 \hat{C}_0{}^2}\left(\frac{U_0 + Z_0 I_0}{Z_0}e^{\mu_0} + \sum_{n=1}^{m}\left|g_n^{(0)}\right|\phi_0^n e^{(n-1)\mu_0}\right)\right] < 1$$

Consider a line with the following specific parameters

$$\Lambda = 1\,m,\ L = 0{,}2\,\mu H/m,\ C = 80\,pF/m,\ v = 1/\sqrt{LC} = 1/(4.10^{-9}) = 2{,}5.10^{8},$$
$$Z_0 = \sqrt{L/C} = 50\,\Omega.$$

Then $T = \Lambda\sqrt{LC} = 4.10^{-9}\,s$.

Let us check the propagation of millimeter waves $\lambda_0 = (1/4)10^{-3}\,m$. We have

$$f_0 = 1/(\lambda_0\sqrt{LC}) = 1/(0{,}25.10^{-3}.4.10^{-9}) = 10^{12}\,Hz \Rightarrow T_0 = 1/f_0 = 10^{-12}\,\text{sec}.$$

If we choose $\mu = 10^{12}$, then $\mu T_0 = \mu_0 = 1$ and $T = 4.10^{-9}.10^{12}T_0 = 4000.T_0$.
The data lead to the following simplification of the above inequalities

$$e(U_0 + Z_0 I_0) \le \phi_0;\ e\frac{U_0}{Z_0} < I_0;\ \frac{e}{Z_0} < 1;$$

$$\frac{1}{\mu^2 \hat{C}_1}\left[eU_{E_1} + \frac{3e-1}{2}\left(2I_0 + \sum_{n=1}^{3}\left|g_n^{(1)}\right|\phi_0{}^n e^{n-1}\right)\right] \le U_0;$$

$$e\frac{2\phi_0}{Z_0} + \frac{3e-1}{2}\frac{2}{\mu Z_0 \hat{C}_0}\left(\frac{\phi_0}{Z_0} + I_0 + \sum_{n=1}^{3}\left|g_n^{(0)}\right|\phi_0{}^n e^{n-1}\right) \le I_0;$$

$$K_u = \frac{3e-1}{2}\frac{e}{\mu^2 \hat{C}_1}\left[1 + \frac{1}{\mu\hat{L}_1} + \frac{1}{\hat{C}_1}\sum_{n=1}^{3}n\left|g_n^{(1)}\right|\phi_0{}^{n-1}e^{n-1} + D_1\left(I_0\frac{e+1}{\hat{C}_1} + \sum_{n=1}^{3}\left|g_n^{(1)}\right|\phi_0{}^n e^n\right)\right] < 1;$$

$$\dot{K}_u = \frac{e}{\mu \hat{C}_1}\left[1 + D_1\left(I_0 e + I_0 + \sum_{n=1}^{3}\left|g_n^{(1)}\right|\phi_0^{\,n}e^n\right) + \frac{1}{\mu \hat{L}_1} + \sum_{n=1}^{3} n\left|g_n^{(1)}\right|\phi_0^{\,n-1}e^{n-1}\right] < 1 \; ;$$

$$K_i = \frac{3e-1}{2}\frac{e}{\mu Z_0}\left[1 + \frac{1+Z_0}{\mu \hat{C}_0}\left(\frac{1}{Z_0} + \sum_{n=1}^{3}\left|g_n^{(0)}\right|n\,.\phi_0^{\,n-1}e^{n-1}\right) + \frac{2D_0}{\mu \hat{C}_0^{\,2}}\left(e\frac{U_0+Z_0I_0}{Z_0} + \sum_{n=1}^{3}\left|g_n^{(0)}\right|\phi_0^{\,n}e^n\right)\right] < 1;$$

$$\dot{K}_i = e\left[\frac{1}{Z_0} + \frac{1+Z_0}{\mu Z_0 \hat{C}_0}\left(\frac{1}{Z_0} + \sum_{n=1}^{3}\left|g_n^{(0)}\right|n\,.\phi_0^{\,n-1}e^{n-1}\right) + \frac{2D_0}{\mu Z_0 \hat{C}_0^{\,2}}\left(e\frac{U_0+Z_0I_0}{Z_0} + \sum_{n=1}^{3}\left|g_n^{(0)}\right|\phi_0^{\,n}e^{n-1}\right)\right] < 1$$

Let us consider the case $\widetilde{C}_0(u) = \widetilde{C}_1(u) = \widetilde{C}(u)$, or, $c_0 = c_1$, $\Phi = \Phi_0 = \Phi_1$, that is

$$C(u) = \frac{c_0\sqrt{\Phi}}{\sqrt{\Phi - u}} \;,$$ where *pn*-junction capacity $c_0 = c_1 \in [0,05.10^{-12}\,F;\ 50.10^{-12}\,F]$,

while the *pn*-junction potential $\Phi \in [0,4\ V;\ 0,9\ V]$.

Since $\dfrac{d\widetilde{C}(u)}{du} = \dfrac{c_0\sqrt{\Phi}(4\Phi - u)}{4(\Phi - u)^{\frac{5}{2}}} > 0$ for $-\phi_0 \le u \le \phi_0 = 0,3V$ ($\phi_0 < \Phi$) if we choose

$c_0 = 30\,pF = 3.10^{-11}\,F$ and $\Phi = 0,5V$ then the minimal value is

$$\widetilde{C}(-0,3) = \hat{C} = 3.10^{-11}\frac{\sqrt{0,5}(2.0,5 + 0,3)}{2(0,5 + 0,3)^{\frac{3}{2}}} = 1,92.10^{-11}\,.$$

For $h = 2$ we have $D_0 = D_1 = \dfrac{c_0\sqrt{\Phi_0}}{4}\dfrac{4\Phi_0 + \phi_0}{(\Phi_0 - \phi_0)^{\frac{5}{2}}} = 5,6.\,10^{-10}\,.$

Here, Assumptions **(V)** are

$$U_0 + U_{E_1} \le 2U_0 \le 0,3\;;\ U_0 + I_0 Z_0 + U_{E_0} \le 2U_0 + 50I_0 \le 0,3\,.$$

For the *I-V* characteristics we assume that

$$i = G_p(u) = G(u) = \sum_{n=1}^{3} g_n u^n = 0,25u - 0,14u^3 \quad (p = 0,1)$$

and for *L-I*-characteristics $L_0(i) = 2 - 0,1i \equiv L_1(i) = 2 - 0,1i$. Then

$$\widetilde{L}_p = \widetilde{L}(i) = i.L(i) = 2i - 0,1i^2 \;,i \in [-I_0, I_0]\ (p = 0,1)\,.$$

Recall that $u = \dfrac{d\left(\widetilde{L}(i)\right)}{dt}$ and consequently $\widetilde{L}(i) = \int\limits_T^t u(s)\,ds \Rightarrow i = \widetilde{L}^{-1}\left(\int\limits_T^t u(s)\,ds\right)$.

We choose I_0 such that $|i| \le I_0$ implies $\dfrac{d\widetilde{L}(i)}{di} = 2 - 0{,}2.i \ge 2 - 0{,}2 I_0 = \dfrac{1}{\hat{\widetilde{L}}} > 0$. For instance $|i| \le i_0 \le I_0 = \sqrt{10}$. Therefore the inverse function $\widetilde{L}^{-1}(.)$ exists and

$$\widetilde{L}^{-1}(\lambda) : \left[-2I_0 - 0{,}1 I_0^2 ;\ 2I_0 - 0{,}1\ I_0^2\right] \to [-I_0, I_0]\ .$$

The explicit form of the inverse function can be found solving the following equation with respect to i: $2i - 0{,}1 i^2 = \lambda$. We need, however, just the estimates:

$$\left|\widetilde{L}_p^{-1}(.)\right| \le I_0 \le I_0 \ (p = 0{,}1)$$

and we can choose $I_0 = I_0 = 0{,}01 < \sqrt{10}$;

$$\left|\dfrac{d\widetilde{L}^{-1}(\lambda)}{d\lambda}\right| = \dfrac{1}{\left|\dfrac{d\widetilde{L}(i)}{di}\right|} \le \dfrac{1}{\min\left\{\left|d\widetilde{L}(i)/di\right| : i \in [-I_0; I_0]\right\}} = \dfrac{1}{\min\left\{|2 - 0{,}2.i| : i \in [-0{,}01;\ 0{,}01]\right\}} = \dfrac{1}{\hat{\widetilde{L}}} =$$

$$= \dfrac{1}{|2 - 0{,}2.0{,}01|} \approx \dfrac{1}{2}\ .$$

Then the above inequalities become

$$e\dfrac{U_0}{Z_0} < I_0 ;\ \dfrac{e}{Z_0} < 1;\ e\left(U_0 + U_{E_1}\right) \le \phi_0\ .$$

But $\mu\hat{C} = 10^{12}.10^{-11}.1{,}92 = 19{,}2;\ \mu^2\hat{C} = 10^{24}.10^{-11}.1{,}92 = 1{,}92.10^{13}$;

$\mu\hat{L} = 2.10^{12} \Rightarrow \dfrac{1}{\mu\hat{L}} \approx 0$ and $g_1 = 0{,}25;\ g_2 = 0;\ g_3 = 0{,}14$. Therefore

$$e\dfrac{U_0}{50} < 0{,}01 ;\ \dfrac{e}{50} < 1;\ e\left(U_0 + U_{E_1}\right) \le \phi_0 ;\ \dfrac{1}{2}\left(U_0 + 50.0{,}01\right) \le \phi_0 = 0{,}3\ .$$

If we can choose $U_0 = I_0 = 0{,}1;\ U_{E_1} = 0{,}001$ then

$$e\frac{2\phi_0}{Z_0}+\frac{3e-1}{2}\frac{2}{\mu Z_0\hat{C}_0}\left(\frac{\phi_0}{Z_0}+I_0+\sum_{n=1}^{3}\left|g_n^{(0)}\right|\phi_0^{\,n}e^{(n-1)}\right)\le I_0$$

becomes $0{,}033+0{,}0016<0{,}1$. Fot the contractive constants we have

$$K_u\approx\frac{10^{11}.0{,}24}{10^{13}}=0{,}0024<1\,;\;\dot{K}_u=\frac{3{,}44}{19{,}2}=0{,}179<1\,;\;K_i\approx\frac{3{,}54}{10^{10}}<1\,;$$

$$\dot{K}_i=\frac{e}{50}3{,}141=0{,}17\,.$$

Consequently $K=0{,}179.$

In what follows we calculate the first approximation and then

$$\rho_\mu^{(k)}((u^{(n)},i^{(n)}),(u^{(n+1)},i^{(n+1)}))\le\frac{K^n}{1-K}\rho_\mu^{(k)}((u^{(0)},i^{(0)}),(u^{(1)},i^{(1)}))\,,$$

where $(u^{(n+1)},i^{(n+1)})=(B_u(u^{(n)},i^{(n)}),B_i(u^{(n)},i^{(n)}))\,(n=0,1,2,\ldots).$

Indeed, let us choose

$$u^{(0)}(t)=U_0\sin\omega_0 t,\quad i^{(0)}(t)=-I_0\cos\omega_0 t,\;E_p(t)=e^{-\mu^T}U_{E_p}\sin\omega_0 t\,(p=0,1)$$

Then

$$u^{(1)}=B_u^{(k)}(u^{(0)},i^{(0)})=$$

$$=\int_{T+kT_0}^{t}U(u^{(0)},i^{(0)})(s)ds-\left(\frac{t-T-kT_0}{T_0}-\frac{1}{2}\right)\int_{T+kT_0}^{T+(k+1)T_0}U(u^{(0)},i^{(0)})(s)ds-\frac{1}{T_0}\int_{T+kT_0}^{T+(k+1)T_0}\int_{T+kT_0}^{t}U(u^{(0)},i^{(0)})(s)dsdt.$$

We have to estimate the differences

$$\left|u^{(0)}(t)-u^{(1)}(t)\right|\le U_0+e^{\mu(t-T-kT_0)}U_0\Rightarrow\rho_\mu^{(k)}(u^{(0)},u^{(1)})\le 2U_0\,;$$

$$\left|\dot{u}^{(0)}(t)-\dot{u}^{(1)}(t)\right|\le\omega_0 U_0+\left|U(u^{(0)},i^{(0)})(t)\right|+\frac{1}{T_0}\left|\int_{T+kT_0}^{T+(k+1)T_0}U(u^{(0)},i^{(0)})(s)ds\right|\le$$

$$\le\omega_0 U_0+\omega_0 U_{E_1}+\frac{1}{\hat{C}_1}\left[I_0+\left|\tilde{L}_1^{-1}\left(\int_{T}^{t}[E_1(\tau)-u(\tau)]d\tau\right)\right|+\left|G_1(E_1(t)-u(t))\right|\right]+$$

$$+\frac{2I_0}{\hat{C}_1}\frac{e^{\mu 0}-1}{\mu_0}+\frac{1}{T_0\mu^2\hat{C}_1}\sum_{n=1}^{m}\frac{\left|g_n^{(1)}\right|(\phi_0)^n}{n}\frac{e^{n\mu 0}-1}{n}\le$$

$$\le\omega_0 U_0+\omega_0 U_{E_1}+\frac{2I_0+\sum_{n=1}^{m}\left|g_n^{(1)}\right|\left\|E_1(t)-u(t)\right\|^n}{\hat{C}_1}+\frac{2I_0}{\hat{C}_1}\frac{e^{\mu 0}-1}{\mu_0}+\frac{1}{\mu\hat{C}_1}\frac{e^{\mu 0}-1}{\mu_0}\sum_{n=1}^{m}\frac{\left|g_n^{(1)}\right|(\phi_0)^n}{n}e^{(n-1)\mu 0}\le$$

$$\le\omega_0 U_0+\omega_0 U_{E_1}+\frac{e^{\mu 0}-1}{\mu_0}\left(\frac{2eI_0}{\hat{C}_1}+\frac{1}{\hat{C}_1}\sum_{n=1}^{m}\left|g_n^{(1)}\right|(\phi_0)^n e^{(n-1)\mu 0}\right)\approx$$

$$\approx\frac{2\pi}{10^{-12}}0,3+\frac{1}{\hat{C}_1}\left(2eI_0+0,25U_0 e^{-1}+0,00019\right)=3.10^{12}+10^9.2\approx3.10^{12}$$

Finally we obtain

$$\rho_\mu^{(k)}((u^{(n)},i^{(n)}),(u^{(n+1)},i^{(n+1)}))\le\frac{(0,179)^n}{1-0,179}.3.10^{12}.$$

Obviously, in order to improve the rate of convergence we have to enlarge $\mu\hat{C}_1$ and Z_0.

4.5.2. Applications to Transmission Line Analogs of Electromagnetic Propagation on Helical Conductors and Electron Beams

We can see that all of the problems considered above could be formulated for the examples from Chapter III, namely transmission line analogs terminated by resonator circuits:

$$\frac{\partial I_x}{\partial x}+C_e\frac{\partial V}{\partial t}=0,\quad\frac{\partial V}{\partial x}+L_e\frac{\partial I_x}{\partial t}=0$$

with

$$C_e=\left(\frac{\beta}{\gamma}\right)^2\frac{2\pi\varepsilon_0}{I_0(\gamma a)K_0(\gamma a)}\left[F/m\right],\qquad L_e=\frac{\mu_0\cot^2\psi}{2\pi}I_1(\gamma a)K_1(\gamma a)\left[H/m\right],$$

$$v=1/\sqrt{L_e C_e}.$$

The transmission line analog of an electron beam (cf. [25]) is:

$$\frac{\partial V}{\partial x} = \frac{p^2}{\omega^2 \varepsilon_0 \sigma} \frac{\partial I}{\partial t}, \qquad \frac{\partial I}{\partial x} = \frac{\omega \varepsilon_0 \sigma}{p^2} \left(\frac{2\pi}{\lambda_p} \right)^2 \frac{\partial V}{\partial t}$$

with

$$L_e = \frac{1}{I_0} \sqrt{\frac{2V_0}{\eta}} \left(\frac{\omega_p}{\omega} \right)^2 [H/m], \quad C_e = \frac{\sigma \varepsilon_0}{p^2} \left(\frac{2\pi}{\lambda_p} \right)^2 [F/m]$$

and the transmission line analog of a plasma column in a tube is

$$L_e = \frac{\tau}{2\pi b} \frac{1}{\omega_p^2 \varepsilon_0} \frac{I_0(\tau b)}{I_1(\tau b)} [H/m], \quad C_e = \frac{2\pi b \gamma \varepsilon_0 K_1(\gamma b)}{\left[1 - (\omega/\omega_p)^2 \right] K_0(\gamma b)} [F/m].$$

Here $v = 1/\sqrt{L_e C_e}$ and $Z_0 = \sqrt{L_e/C_e} = \dfrac{1}{2\pi b \varepsilon_0 \omega_p} \sqrt{\dfrac{\tau \left[1 - (\omega/\omega_p)^2 \right] I_0(\tau b) K_0(\gamma b)}{\gamma I_1(\tau b) K_1(\gamma b)}}$

depend on the frequency.

Introducing denotations

$$u_{E_1}(t) = V(t) - E_1(t),$$

$$Y_{E_0}(u,i) = \frac{1}{2} V(t) + \frac{Z_0}{2} I(t) + \frac{1}{2} V(t-2T) - \frac{Z_0}{2} I(t-2T) - E_0(t-T)$$

we reduce the mixed problem to the system

$$\frac{dV(t)}{dt} = \frac{dE_1(t)}{dt} + \frac{-I(t) - \widetilde{L}_1^{-1}\left(\int_T^t u_{E_1}(\tau) d\tau \right) - G_1(u_{E_1}(t))}{d\widetilde{C}_1(u_{E_1}(t))/du},$$

$$\frac{dI(t)}{dt} = -\frac{1}{Z_0} \frac{dV(t)}{dt} - \frac{1}{Z_0} \frac{dV(t-2T)}{dt} + \frac{dI(t-2T)}{dt} + \frac{2}{Z_0} \frac{dE_0(t)}{dt}$$

$$= \frac{\dfrac{1}{Z_0} V(t) + I(t) - \dfrac{1}{Z_0} V(t-2T) + I(t-2T) - 2\widetilde{L}_0^{-1}\left(\int_T^t Y_{E_0}(V,I)(\theta) d\theta \right) - 2G_0\left(Y_{E_0}(V,I) \right)}{Z_0 d\widetilde{C}_0\left(Y_{E_0}(V,I) \right)/du},$$

$$V(t) = \upsilon_0(t), \frac{dV(t)}{dt} = \frac{d\upsilon_0(t)}{dt}, \quad I(t) = \iota_0(t), \frac{dI(t)}{dt} = \frac{d\iota_0(t)}{dt}, t \in [-T, T].$$

It is known that by decreasing Z_0 as frequency γa increases one can compensate by choosing sufficiently large $\mu > 0$, though not completely. If the above inequalities are satisfied then a periodic solution does exist.

Obviously, the most restrictive condition is

$$2\pi b \varepsilon_0 \omega_p \, e \sqrt{\frac{\gamma I_1(\tau b) K_1(\gamma b)}{\tau [1 - (\omega / \omega_p)^2] I_0(\tau b) K_0(\gamma b)}} < 1 .$$

4.6. A Different Approach to Obtaining a Nonlinear Neutral System

Our goal here is to show that each problem requires specifics. Indeed, proceeding from Telegrapher's system

$$\frac{\partial u(x,t)}{\partial t} + \frac{1}{C} \frac{\partial i(x,t)}{\partial x} = 0,$$

$$\frac{\partial i(x,t)}{\partial t} + \frac{1}{L} \frac{\partial u(x,t)}{\partial x} = 0$$

and introducing new variables (U, I) the initial conditions can be obtained from

$$U(x,t) = u(x,t) + Z_0 \, i(x,t),$$

$$I(x,t) = u(x,t) - Z_0 \, i(x,t)$$

after setting $t = 0$:

$$U(x,0) = u(x,0) + Z_0 \, i(x,0) = u_0(x) + Z_0 \, i_0(x) \equiv U_0(x),$$

$$I(x,0) = u(x,0) - Z_0 \, i(x,0) = u_0(x) - Z_0 \, i_0(x) \equiv I_0(x).$$

Further on we have

$$u(0,t) = \frac{1}{2} U(0,t) + \frac{1}{2} I(0,t), \qquad u(\Lambda,t) = \frac{1}{2} U(\Lambda,t) + \frac{1}{2} I(\Lambda,t),$$

$$\text{and}$$

$$i(0,t) = \frac{1}{2Z_0} U(0,t) - \frac{1}{2Z_0} I(0,t) \qquad i(\Lambda,t) = \frac{1}{2Z_0} U(\Lambda,t) - \frac{1}{2Z_0} I(\Lambda,t)$$

$$-\left(\frac{1}{2Z_0}U(0,t)-\frac{1}{2Z_0}I(0,t)\right)=-C_0\left(\dot{E}_0(t)-\frac{1}{2}\frac{dU(0,t)}{dt}-\frac{1}{2}\frac{dI(0,t)}{dt}\right)-$$

$$-L_0\left(\int_T^t[E_0(\tau)-\frac{1}{2}U(0,\tau)-\frac{1}{2}I(0,\tau)]d\tau\right)-G_0\left(E_0(t)-\frac{1}{2}U(0,t)-\frac{1}{2}I(0,t)\right)$$

$$\frac{1}{2Z_0}U(\Lambda,t)-\frac{1}{2Z_0}I(\Lambda,t)=-C_1\left(\dot{E}_1(t)-\frac{1}{2}\frac{dU(\Lambda,t)}{dt}-\frac{1}{2}\frac{dI(\Lambda,t)}{dt}\right)-$$

$$-L_1\left(\int_T^t[E_1(\tau)-\frac{1}{2}U(\Lambda,\tau)-\frac{1}{2}I(\Lambda,\tau)]d\tau\right)-G_1\left(E_1(t)-\frac{1}{2}U(\Lambda,t)-\frac{1}{2}I(\Lambda,t)\right).$$

Integration along the characteristics yields

$$U(\Lambda,t+T)=U(0,t) \text{ and } I(\Lambda,t)=I(0,t+T).$$

Taking into account these relations we obtain

$$-\left(\frac{1}{2Z_0}U(\Lambda,t+T)-\frac{1}{2Z_0}I(\Lambda,t-T)\right)=-C_0\left(\dot{E}_0(t)-\frac{1}{2}\frac{dU(\Lambda,t+T)}{dt}-\frac{1}{2}\frac{dI(\Lambda,t-T)}{dt}\right)-$$

$$-L_0\left(\int_T^t[E_0(\tau)-\frac{1}{2}U(\Lambda,\tau+T)-\frac{1}{2}I(\Lambda,\tau-T)]d\tau\right)-$$

$$G_0(E_0(t)-\frac{1}{2}U(\Lambda,t+T)-\frac{1}{2}I(\Lambda,t-T)).$$

In the integral

$$J=\int_T^t[E_0(\tau)-\frac{1}{2}U(\Lambda,\tau+T)-\frac{1}{2}I(\Lambda,\tau-T)]d\tau$$

we change the variable setting $\tau+T=\theta$. Then

$$J=\int_{2T}^{t+T}[E_0(\theta-T)-\frac{1}{2}U(\Lambda,\theta)-\frac{1}{2}I(\Lambda,\theta-2T)]d\theta$$

Replacing $t+T$ by t we get

$$-\left(\frac{1}{2Z_0}U(\Lambda,t)-\frac{1}{2Z_0}I(\Lambda,t-2T)\right)=$$

$$-C_0\left(\dot{E}_0(t-T)-\frac{1}{2}\frac{dU(\Lambda,t)}{dt}-\frac{1}{2}\frac{dI(\Lambda,t-2T)}{dt}\right)-$$

$$-L_0\left(\int_{2T}^{t}[E_0(\theta-T)-\frac{1}{2}U(\Lambda,\theta)-\frac{1}{2}I(\Lambda,\tau-2T)]d\tau\right)$$

$$-G_0(E_0(t-T)-\frac{1}{2}U(\Lambda,t)-\frac{1}{2}I(\Lambda,t-2T))\,.$$

We choose unknown functions to be $U(\Lambda,t)$, $I(\Lambda,t)$ and we set $U(t)=U(\Lambda,t)$, $I(t)=I(\Lambda,t)$. So we get the system

$$-\left(\frac{1}{2Z_0}U(t)-\frac{1}{2Z_0}I(t-2T)\right)=$$

$$-C_0\left(\dot{E}_0(t-T)-\frac{1}{2}\frac{dU(t)}{dt}-\frac{1}{2}\frac{dI(t-2T)}{dt}\right)-$$

$$-L_0\left(\int_{2T}^{t}[E_0(\theta-T)-\frac{1}{2}U(\theta)-\frac{1}{2}I(\tau-2T)]d\tau\right)-$$

$$G_0(E_0(t-T)-\frac{1}{2}U(t)-\frac{1}{2}I(t-2T))$$

and

$$\frac{1}{2Z_0}U(t)-\frac{1}{2Z_0}I(t)=-C_1\left(\dot{E}_1(t)-\frac{1}{2}\frac{dU(t)}{dt}-\frac{1}{2}\frac{dI(t)}{dt}\right)-$$

$$-L_1\left(\int_{T}^{t}[E_1(\tau)-\frac{1}{2}U(\tau)-\frac{1}{2}I(\tau)]d\tau\right)-R_1\left(E_1(t)-\frac{1}{2}U(t)-\frac{1}{2}I(t)\right)$$

or

$$C_0\left(\dot{E}_0(t-T)-\frac{1}{2}\frac{dU(t)}{dt}-\frac{1}{2}\frac{dI(t-2T)}{dt}\right)=\left(\frac{1}{2Z_0}U(t)-\frac{1}{2Z_0}I(t-2T)\right)-$$

$$-L_0\left(\int_{2T}^{t}[E_0(\theta-T)-\frac{1}{2}U(\theta)-\frac{1}{2}I(\tau-2T)]d\tau\right)-$$

$$-G_0(E_0(t-T)-\frac{1}{2}U(t)-\frac{1}{2}I(t-2T))\equiv A(U,I)(t),$$

$$C_1\left(\dot{E}_1(t)-\frac{1}{2}\frac{dU(t)}{dt}-\frac{1}{2}\frac{dI(t)}{dt}\right)=-\frac{1}{2Z_0}U(t)+\frac{1}{2Z_0}I(t)-$$

$$-L_1\left(\int_T^t[E_1(\tau)-\frac{1}{2}U(\tau)-\frac{1}{2}I(\tau)]d\tau\right)-G_1\left(E_1(t)-\frac{1}{2}U(t)-\frac{1}{2}I(t)\right)\equiv B(U,I)(t).$$

But we have already established that $C_k(.)=C_k$ have inverse functions. Then

$$\dot{E}_0(t-T)-\frac{1}{2}\frac{dU(t)}{dt}-\frac{1}{2}\frac{dI(t-2T)}{dt}=C_0^{-1}(A(U,I)(t)),$$

$$\dot{E}_1(t)-\frac{1}{2}\frac{dU(t)}{dt}-\frac{1}{2}\frac{dI(t)}{dt}=C_1^{-1}(B(U,I)(t))$$

or

$$\frac{dU(t)}{dt}=2\dot{E}_0(t-T)-\frac{dI(t-2T)}{dt}-2C_0^{-1}(A(U,I)(t)),$$

$$\frac{dI(t)}{dt}=2\dot{E}_1(t)-\frac{dU(t)}{dt}-2C_1^{-1}(B(U,I)(t)).$$

It is clear that the constant before the neutral part in the second equation is $K_u=\left|-1\right|$.

But in Chapter III we obtained the same type of system. Here we can avoid the same difficulty by using the same method (cf. § 3.4).

4.7. OSCILLATORY REGIMES FOR THE NONLINEAR NEUTRAL SYSTEM

Here we deal with the problem of existence-uniqueness of oscillatory solution of (4.4.7), namely

$$\frac{du(t)}{dt}=\frac{dE_1(t)}{dt}+\frac{1}{\widetilde{C}_1(E_1(t)-u(t))}\left[-i(t)-\widetilde{L}_1^{-1}\left(\int_{t_0}^t[E_1(\tau)-u(\tau)]d\tau\right)-G_1(E_1(t)-u(t))\right]\equiv U(u,i),\ t\in[t_0,\infty),$$

$$\frac{di(t)}{dt}=\frac{2}{Z_0}\frac{dE_0(t)}{dt}-\frac{1}{Z_0}\frac{du(t)}{dt}-\frac{1}{Z_0}\frac{du(t-2T)}{dt}+\frac{di(t-2T)}{dt}+$$

$$+ \frac{2}{Z_0 \widetilde{C}_0 \left(\dfrac{u(t)+Z_0 i(t)+u(t-2T)-Z_0 i(t-2T)}{2} - E_0(t) \right)} \left[\frac{u(t)+Z_0 i(t)-u(t-2T)+Z_0 i(t-2T)}{2Z_0} - \right.$$

$$\left. + -\widetilde{L}_0^{-1} \left(\int_{t_0}^{t} \left(\frac{u(s)+Z_0 i(s)+u(s-2T)-Z_0 i(s-2T)}{2} - E_0(s) \right) ds \right) - \right.$$

$$\left. - G_0 \left(\frac{u(t)+Z_0 i(t)+u(t-2T)-Z_0 i(t-2T)}{2} - E_0(t) \right) \right] \equiv I(u,i),\ t \in [t_0,\infty), \tag{4.7.1}$$

$$u(t) = \upsilon_0(t),\ \frac{du(t)}{dt} = \frac{d\upsilon_0(t)}{dt},\ i(t) = \iota_0(t),\ \frac{di(t)}{dt} = \frac{d\iota_0(t)}{dt},\ t \in [-t_0, t_0].$$

Let us put $t_0 \equiv T$. Now we are able to formulate the main problem: to find an oscillatory solution of problem (2.4.1) with advanced prescribed zeros on an interval $[t_0,\infty)$, where $\upsilon_0(t)$ is a prescribed initial oscillatory function on the interval $[-t_0, t_0]$.

Let $S_T = \{\tau_k\}_{k=0}^{n}$, $n \in N$ be the set of zeros of the initial function, that is, $U_0(\tau_k) = I_0(\tau_k) = 0$ such that $\tau_0 = -T$, $\tau_n = T \equiv t_0$. Besides $\max\{\tau_{k+1} - \tau_k : k = 0,1,...,n\} \leq T_0$.

Let $S = \{t_k\}_{k=0}^{\infty}$ be a strictly increasing sequence of real numbers, satisfying the following conditions (C):

(C1) $\lim\limits_{k \to \infty} t_k = \infty$;

(C2) for every k $\left(t_k - T \geq \tau_0 \right)$ there is $s < k$ such that $t_k - T = t_s$ where $t_s \in S_T \cup S$.

It is easy to verify that (C2) implies

$$0 < \inf\{t_{k+1} - t_k : k = 0,1,2,...\} \leq \sup\{t_{k+1} - t_k : k = 0,1,2,...\} = T_0 < \infty.$$

Introduce the set $C^1[t_0,\infty)$ consisting of all continuous functions whose derivatives are bounded and continuous on every interval $[t_k, t_{k+1}]$. We note the right and left derivatives at t_k of these functions may not coincide. That is why we introduce below a topology of uniform convergence on every interval $[t_k, t_{k+1}]$ of the derivatives which need the introduction of uniform spaces (cf. [14]).

Consider the set

$$M_{SU} = \left\{ u(.) \in C^1[t_0,\infty) : u(t_k) = 0 \right\}, \quad M_{SI} = \left\{ i(.) \in C^1[t_0,\infty) : i(t_k) = 0 \right\}$$

and

$$M_{SU}^* = \left\{ u(.) \in M_{SU} : |u(t)| \le U_0 e^{\mu(t-t_k)}, t \in [t_k, t_{k+1}] \right\},$$

$$M_{SI}^* = \left\{ i(.) \in M_{SI} : |i(t)| \le I_0 e^{\mu(t-t_k)}, t \in [t_k, t_{k+1}] \right\}$$

where U_0, I_0, μ are positive constants. Recall that $\mu T_0 = \mu_0 = \text{const.} > 0$.

Introduce the following family of pseudo-metrics

$$\rho^{(k)}(u,\overline{u}) = \max\left\{ |u(t) - \overline{u}(t)| : t \in [t_k, t_{k+1}] \right\},$$

$$\hat{\rho}^{(k)}(u,\overline{u}) = \max\left\{ |u(t) - \overline{u}(t)| : t \in [t_0, t_{k+1}] \right\},$$

$$\rho_\mu^{(k)}(u,\overline{u}) = \max\left\{ e^{-\mu(t-t_k)}|u(t) - \overline{u}(t)| : t \in [t_k, t_{k+1}] \right\},$$

$$\hat{\rho}_\mu^{(k)}(u,\overline{u}) = \max\left\{ \rho_\mu^{(0)}(u,\overline{u}), \rho_\mu^{(1)}(u,\overline{u}),..., \rho_\mu^{(k)}(u,\overline{u}) \right\},$$

$$\rho_\mu^{(k)}(\dot{u},\dot{\overline{u}}) = \max\left\{ e^{-\mu(t-t_k)}|\dot{u}(t) - \dot{\overline{u}}(t)| : t \in [t_k, t_{k+1}] \right\},$$

$$\hat{\rho}_\mu^{(k)}(\dot{u},\dot{\overline{u}}) = \max\left\{ \rho_\mu^{(0)}(\dot{u},\dot{\overline{u}}), \rho_\mu^{(1)}(\dot{u},\dot{\overline{u}}),..., \rho_\mu^{(k)}(\dot{u},\dot{\overline{u}}) \right\},$$

$$\rho^{(k)}(i,\overline{i}) = \max\left\{ |i(t) - \overline{i}(t)| : t \in [t_k, t_{k+1}] \right\},$$

$$\hat{\rho}^{(k)}(i,\overline{i}) = \max\left\{ |i(t) - \overline{i}(t)| : t \in [t_0, t_{k+1}] \right\},$$

$$\rho_\mu^{(k)}(i,\overline{i}) = \max\left\{ e^{-\mu(t-t_k)}|i(t) - \overline{i}(t)| : t \in [t_k, t_{k+1}] \right\},$$

$$\hat{\rho}_\mu^{(k)}(i,\overline{i}) = \max\left\{ \rho_\mu^{(0)}(i,\overline{i}), \rho_\mu^{(1)}(i,\overline{i}),..., \rho_\mu^{(k)}(i,\overline{i}) \right\},$$

$$\rho_\mu^{(k)}(\dot{i},\dot{\overline{i}}) = \max\left\{ e^{-\mu(t-t_k)}|\dot{i}(t) - \dot{\overline{i}}(t)| : t \in [t_k, t_{k+1}] \right\},$$

$$\hat{\rho}_\mu^{(k)}(\dot{i},\dot{\overline{i}}) = \max\left\{ \rho_\mu^{(0)}(\dot{i},\dot{\overline{i}}), \rho_\mu^{(1)}(\dot{i},\dot{\overline{i}}),..., \rho_\mu^{(k)}(\dot{i},\dot{\overline{i}}) \right\}.$$

The set $M_{SU}^* \times M_{SI}^*$ turns into a complete uniform space with respect to the countable saturated family of pseudo-metrics

$$\hat{\rho}_\mu^{(k)}((u,\dot{u},i,\dot{i}),(\overline{u},\dot{\overline{u}},\overline{i},\dot{\overline{i}})) = \max\left\{ \hat{\rho}^{(k)}(u,i,\overline{u},\overline{i}), \hat{\rho}_\mu^{(k)}(\dot{u},\dot{\overline{u}}), \hat{\rho}_\mu^{(k)}(\dot{i},\dot{\overline{i}}) : k = 0,1,2,... \right\}$$

In order to avoid the conformity condition (**CC**) we define an operator

$B = (B_u(u,i), B_i(u,i))$ by the formulas

$$B_u^{(k)}(u,i)(t) := \int_{t_k}^{t} U(u,i)(s)ds - \frac{t-t_k}{t_{k+1}-t_k} \int_{t_k}^{t_{k+1}} U(u,i)(s)ds, \quad t \in [t_k, t_{k+1}]$$

$$B_i^{(k)}(u,i)(t) := \int_{t_k}^{t} I(u,i)(s)ds - \frac{t-t_k}{t_{k+1}-t_k} \int_{t_k}^{t_{k+1}} I(u,i)(s)ds, \quad t \in [t_k, t_{k+1}]$$

$(k = 0,1,2,\dots)$, where

$$U(u,i) = \frac{dE_1(t)}{dt} + \frac{1}{\widetilde{C}_1(E_1(t)-u(t))}\left[-i(t) - \widetilde{L}_1^{-1}\left(\int_{t_0}^{t} [E_1(\tau)-u(\tau)]d\tau \right) - G_1(E_1(t)-u(t)) \right];$$

$$I(u,i) = \frac{2}{Z_0}\frac{dE_0(t)}{dt} - \frac{1}{Z_0}\frac{du(t)}{dt} - \frac{1}{Z_0}\frac{d\bar{u}(t)}{dt} + \frac{d\bar{i}(t)}{dt} +$$

$$+ \frac{2}{Z_0\widetilde{C}_0\left(\dfrac{u(t)+Z_0i(t)+\bar{u}(t)-Z_0\bar{i}(t)}{2} - E_0(t) \right)}\left[\frac{u(t)+Z_0i(t)-\bar{u}(t)+Z_0\bar{i}(t)}{2Z_0} - \right.$$

$$\left. - \widetilde{L}_0^{-1}\left(\int_{t_0}^{t}\left(\frac{u(s)+Z_0i(s)+\bar{u}(s)-Z_0\bar{i}(s)}{2} - E_0(s) \right)ds \right) - G_0\left(\frac{u(t)+Z_0i(t)+\bar{u}(t)-Z_0\bar{i}(t)}{2} - E_0(t) \right) \right]$$

and $\bar{u}(t) = \begin{cases} \upsilon_0(t-2T), & t \in [T,3T] \\ u(t-2T), & t \in [3T,\infty) \end{cases}$. In other words on $[T,3T]$ we have the translated to the right initial function, while for $[3T,\infty)$ we have a translated to the right function $u(t)$. So we can consider the operator to the right from T.

One can verify that $M_{SU}^* \times M_{SI}^*$ is a closed subset of $C^1[t_0,\infty) \times C^1[t_0,\infty)$ with respect to the above family of pseudo-metrics (cf. K. Zima [126]).

Lemma 4.7.1. Let the following conditions be valid:

(IN) $|\upsilon_0(t)| \le U_0 e^{\mu(t-\tau_k)}, |\iota_0(t)| \le I_0 e^{\mu(t-\tau_k)}, t \in [\tau_k, \tau_{k+1}], (k = 0,1,2,\dots,2m-1), \upsilon_0(-T) = \iota_0(-T) = 0$;

(E) $E_p(.) \in C_{ST \cup S}^1[0,\infty), |E_p(t)| \le U_0 e^{\mu(t-t_k)}, t \in [t_k, t_{k+1}], E_p(t-T) = E_p(t), (p = 0,1)$.

Then problem (4.7.1) has a solution $(u,i) \in M_{SU}^* \times M_{SI}^*$ iff the operator B has a fixed point in $M_{SU}^* \times M_{SI}^*$, that is,

$$(u,i) = (B_u(u,i), B_i(u,i)). \tag{4.7.2}$$

Proof: Let $(u,i) \in M_{SU}^* \times M_{SI}^*$ be a solution of (4.7.1). Then, integrating (4.7.1) on every interval $[t_k,t] \subset [t_k,t_{k+1}]$ $(k = 0,1,2 \ldots)$ we obtain

$$u(t) - u(t_k) = \int_{t_k}^t U(u,i)(s)ds \quad \Leftrightarrow \quad u(t) = \int_{t_k}^t U(u,i)(s)ds,$$

$$i(t) - i(t_k) = \int_{t_k}^t I(u,i)(s)ds \quad \Leftrightarrow \quad i(t) = \int_{t_k}^t I(u,i)(s)ds.$$

Consequently

$$u(t) = \int_{t_k}^t U(u,i)(s)ds \Rightarrow 0 = u(t_{k+1}) = \int_{t_k}^{t_{k+1}} U(u,i)(s)ds \Rightarrow \int_{t_k}^{t_{k+1}} U(u,i)(s)ds = 0,$$

$$i(t) = \int_{t_k}^t I(u,i)(s)ds \Rightarrow 0 = i(t_{k+1}) = \int_{t_k}^{t_{k+1}} I(u,i)(s)ds \Rightarrow \int_{t_k}^{t_{k+1}} I(u,i)(s)ds = 0.$$

$$(4.7.3)$$

Then

$$u(t) = \int_{t_k}^t U(u,i)(s)ds, \quad i(t) = \int_{t_k}^t I(u,i)(s)ds$$

is equivalent to

$$u(t) = \int_{t_k}^t U(u,i)(s)ds - \frac{t-t_k}{t_{k+1}-t_k} \int_{t_k}^{t_{k+1}} U(u,i)(s)ds, t \in [t_k,t_{k+1}]$$

$$i(t) = \int_{t_k}^t I(u,i)(s)ds - \frac{t-t_k}{t_{k+1}-t_k} \int_{t_k}^{t_{k+1}} I(u,i)(s)ds, t \in [t_k,t_{k+1}]$$

and then the solution (u,i) of (4.7.1) is a fixed point of B.

Conversely, let $(u(.),i(.)) \in M_{SU}^* \times M_{SI}^*$ be a fixed point B, that is,

$$u = B_u^{(k)}(u,i), \quad i = B_i^{(k)}(u,i), \quad t \in [t_k,t_{k+1}].$$

Then as in the proof of Theorem 4.5.1 one can prove that

$$\left| \int_{t_k}^{t_{k+1}} U(u,i)(t)dt \right| \leq \frac{e^{\mu_0}-1}{2} \frac{1}{\mu^2 \hat{C}_1} \left(2I_0 + \sum_{n=1}^m \left|g_n^{(1)}\right| \left(U_{E_1} + U_0\right)^n e^{(n-1)\mu_0} \right) \Rightarrow \left| \int_{t_k}^{t_{k+1}} U(u,i)(t)dt \right| = 0$$

and analogously $\left| \int\limits_{t_k}^{t_{k+1}} I(u,i)(t)dt \right| = 0$.

Therefore $u(t), i(t)$ satisfy

$$u(t) = \int\limits_{t_k}^{t} U(u,i)(s)ds, \ t \in [t_k, t_{k+1}] \Leftrightarrow u = B_u(u,i),$$

$$i(t) = \int\limits_{t_k}^{t} I(u,i)(s)ds, \ t \in [t_k, t_{k+1}] \Leftrightarrow i = B_i(u,i)$$

that is, $(u(t), i(t))$ is a solution of (4.7.1)

Lemma 4.7.1 is thus proved.

Theorem 4.7.1. Let conditions **(E)** and **(IN)** (from Lemma 4.7.1) be fulfilled. Then there exists a unique oscillatory solution of (4.7.1), belonging to $M_{SU}^* \times M_{SI}^*$.

Proof: We show that B maps $M_{SU}^* \times M_{SI}^*$ into itself.

First we notice that $B(U,I)(t)$ is continuous on $[t_0, \infty)$:

$$B_u^{(0)}(u,i)(t_0) = \int\limits_{t_0}^{t_0} U(u,i)(s)ds - \left(\frac{t_0 - t_0}{t_1 - t_0}\right) \int\limits_{t_0}^{t_1} U(u,i)(s)ds = 0,$$

$$\lim_{t \to t_{k+1}(t < t_{k+1})} B_u^{(k)}(u,i)(t) = \lim_{t \to t_{k+1}(t < t_{k+1})} \left(\int\limits_{t_k}^{t} U(u,i)(s)ds - \frac{t - t_k}{t_{k+1} - t_k} \int\limits_{t_k}^{t_{k+1}} U(u,i)(s)ds \right) = 0,$$

$$\lim_{t \to t_{k+1}(t > t_{k+1})} B_U^{(k+1)}(u)(t) = \lim_{t \to t_{k+1}(t > t_{k+1})} \left(\int\limits_{t_{k+1}}^{t} U(u,i)(s)ds - \frac{t - t_{k+1}}{t_{k+2} - t_{k+1}} \int\limits_{t_{k+1}}^{t_{k+2}} U(u,i)(s)ds \right) = 0,$$

$$B_i^{(0)}(u,i)(t_0) = \int\limits_{t_0}^{t_0} I(u,i)(s)ds - \frac{t_0 - t_0}{t_1 - t_0} \int\limits_{t_0}^{t_1} I(u,i)(s)ds = 0,$$

$$\lim_{t \to t_{k+1}(t < t_{k+1})} B_i^{(k)}(u,i)(t) = \lim_{t \to t_{k+1}(t < t_{k+1})} \left(\int\limits_{t_k}^{t} I(u,i)(s)ds - \frac{t - t_k}{t_{k+1} - t_k} \int\limits_{t_k}^{t_{k+1}} I(u,i)(s)ds \right) = 0,$$

$$\lim_{t \to t_{k+1}(t > t_{k+1})} B_i^{(k+1)}(u,i)(t) = \lim_{t \to t_{k+1}(t > t_{k+1})} \left(\int\limits_{t_{k+1}}^{t} I(u,i)(s)ds - \frac{t - t_{k+1}}{t_{k+2} - t_{k+1}} \int\limits_{t_{k+1}}^{t_{k+2}} I(u,i)(s)ds \right) = 0$$

and $\left(B_u(u,i)(t), B_i(u,i)(t)\right)$ is differentiable on every (t_k, t_{k+1}) .

Besides, it is easy to verify that

$$B_u(u,i)(t_k) = 0, \;\; B_u(u,i)(t_{k+1}) = 0, \; B_i(u,i)(t_k) = 0, \;\; B_i(u,i)(t_{k+1}) = 0 \,.$$

As above, we establish that

$$\left| B_u^{(k)}(u,i)(t) \right| \le U_0 e^{\mu(t-t_k)}, \; t \in [t_k, t_{k+1}]$$

$$\left| B_i^{(k)}(u,i)(t) \right| \le I_0 e^{\mu(t-t_k)}, t \in [t_k, t_{k+1}]$$

that is, B maps is $M_{SU}^* \times M_{SI}^*$ into itself. Indeed,

$$\left| B_u^{(k)}(u,i)(t) \right| \le \left| \int_{t_k}^{t} U(u,i)(s)\,ds \right| + \left| \frac{t-t_k}{t_{k+1}-t_k} \int_{t_k}^{t_{k+1}} U(u,i)(s)\,ds \right| \equiv U_1 + U_2 \,.$$

In view of (4.4.3) we have

$$U_1 \le \left| \int_{t_k}^{t} \dot{E}_1(s)\,ds \right| + \frac{1}{\hat{C}_1} \int_{t_k}^{t} |i(s)|\,ds + \frac{1}{\hat{C}_1} I_0 \int_{t_k}^{t} e^{\mu(s-t_k)}\,ds + \frac{1}{\hat{C}_1} \sum_{n=1}^{m} |g_n^{(1)}| \left| \int_{t_k}^{t}\int_{t_k}^{s} |E_1(\tau) - u(\tau)|^n \, d\tau\,ds \right| \le$$

$$\le U_{E_1} e^{\mu(t-t_k)} + \frac{I_0}{\hat{C}_1} \frac{e^{\mu(t-t_k)}-1}{\mu} + \frac{I_0}{\hat{C}_1} \frac{e^{\mu(t-t_k)}-1}{\mu} + \frac{1}{\hat{C}_1} \sum_{n=1}^{m} |g_n^{(1)}| \left(U_{E_1} + U_0 \right)^n \int_{t_k}^{t}\int_{t_k}^{s} e^{n\mu(\tau-t_k)} \, d\tau\,ds \le$$

$$\le U_{E_1} e^{\mu(t-t_k)} + \frac{e^{\mu(t-t_k)}-1}{\mu} \frac{2I_0}{\mu\hat{C}_1} + \frac{1}{\hat{C}_1} \sum_{n=1}^{m} |g_n^{(1)}| \left(U_{E_1} + U_0 \right)^n \int_{T+kT_0}^{t} \frac{e^{n\mu(s-t_k)}-1}{n\mu} \, ds \le$$

$$\le U_{E_1} e^{\mu(t-t_k)} + \frac{e^{\mu(t-t_k)}-1}{\mu} \frac{2I_0}{\mu\hat{C}_1} + \frac{1}{\hat{C}_1} \sum_{n=1}^{m} |g_n^{(1)}| \phi_0^n \int_{T+kT_0}^{t} \frac{\left(e^{\mu(s-t_k)}-1 \right) n e^{(n-1)\mu(s-t_k)}}{n\mu} \, ds \le$$

$$\le U_{E_1} e^{\mu(t-t_k)} + \frac{e^{\mu(t-t_k)}-1}{\mu} \frac{2I_0}{\mu\hat{C}_1} + \frac{1}{\hat{C}_1} \sum_{n=1}^{m} |g_n^{(1)}| \phi_0^n e^{(n-1)\mu T_0} \int_{T+kT_0}^{t} \frac{e^{\mu(s-t_k)}}{\mu} \, ds \le$$

$$\le U_{E_1} e^{\mu(t-t_k)} + \frac{e^{\mu(t-t_k)}-1}{\mu} \frac{2I_0}{\mu\hat{C}_1} + \frac{e^{\mu(t-t_k)}-1}{\mu^2} \frac{1}{\hat{C}_1} \sum_{n=1}^{m} |g_n^{(1)}| \phi_0^n e^{(n-1)\mu T_0} \le$$

$$\le e^{\mu(t-t_k)} \left[U_{E_1} + \frac{1}{\mu^2 \hat{C}_1} \left(2I_0 + \sum_{n=1}^{m} |g_n^{(1)}| \phi_0^n e^{(n-1)\mu_0} \right) \right];$$

$$U_2 \leq \left| \int_{t_k}^{t_{k+1}} \dot{E}_1(s)\,ds \right| + \frac{1}{\hat{C}_1} \int_{t_k}^{t_{k+1}} |i(s)|\,ds + \frac{1}{\hat{C}_1} I_0 \int_{t_k}^{t_{k+1}} e^{\mu(s-T-kT_0)}\,ds +$$

$$\frac{1}{\hat{C}_1} \sum_{n=1}^{m} \left| g_n^{(1)} \right| \int_{t_k}^{t_{k+1}} \int_{t_k}^{s} |E_1(\tau) - u(\tau)|^n \, d\tau\, ds \leq$$

$$\leq e^{\mu(t-t_k)} \frac{e^{\mu_0} - 1}{\mu^2 \hat{C}_1} \left(2I_0 + \sum_{n=1}^{m} \left| g_n^{(1)} \right| \phi_0^n e^{(n-1)\mu_0} \right)$$

Therefore we get

$$\left| B_u^{(k)}(u,i)(t) \right| \leq e^{\mu(t-t_k)} \left[U_{E_1} + \frac{1}{\mu^2 \hat{C}_1} \left(2I_0 + \sum_{n=1}^{m} \left| g_n^{(1)} \right| \phi_0^n e^{(n-1)\mu_0} \right) \right] +$$

$$+ e^{\mu(t-t_k)} \frac{e^{\mu_0} - 1}{\mu^2 \hat{C}_1} \left(2I_0 + \sum_{n=1}^{m} \left| g_n^{(1)} \right| \phi_0^n e^{(n-1)\mu_0} \right) \leq$$

$$\leq e^{\mu(t-t_k)} \left[U_{E_1} + \frac{e^{\mu_0}}{\mu^2 \hat{C}_1} \left(2I_0 + \sum_{n=1}^{m} \left| g_n^{(1)} \right| \phi_0^n e^{(n-1)\mu_0} \right) \right] \leq U_0 e^{\mu(t-t_k)}.$$

Prior to estimating $B_i(u,i)$ we have

$$\left| \frac{1}{2} u(t) + \frac{Z_0}{2} i(t) + \frac{1}{2} u(t - 2T) - \frac{Z_0}{2} i(t - 2T) - E_0(t) \right| \leq$$

$$\leq e^{\mu_0} \left(U_{E_0} + U_0 + Z_0 I_0 \right) \leq \phi_0.$$

Then

$$\left| B_i^{(k)}(u,i)(t) \right| \leq \left| \int_{t_k}^{t} I(u,i)(s)\,ds \right| + \left| \frac{t - t_k}{t_{k+1} - t_k} \int_{t_k}^{t_{k+1}} I(u,i)(s)\,ds \right| \equiv I_1 + I_2.$$

But

$$I_1 \leq \frac{2}{Z_0} \left| \int_{t_k}^{t} \dot{E}_0(s)\,ds \right| + \frac{1}{Z_0} \left| \int_{t_k}^{t} \dot{u}(s)\,ds \right| + \frac{1}{Z_0} \left| \int_{t_k}^{t} \dot{u}(s - 2T)\,ds \right| + \left| \int_{t_k}^{t} i(s - 2T)\,ds \right| +$$

$$+ \frac{1}{\hat{C}_0 Z_0^{\ 2}} \int_{t_k}^{t} |u(s)| ds + \frac{1}{Z_0 \hat{C}_0} \int_{t_k}^{t} |i(s)| ds + \frac{1}{\hat{C}_0 Z_0^{\ 2}} \int_{t_k}^{t} |u(s-2T)| ds + \frac{1}{Z_0 \hat{C}_0} \int_{t_k}^{t} |i(s-2T)| ds +$$

$$+ \frac{2}{Z_0 \hat{C}_0} \int_{t_k}^{t} I_0 e^{\mu(s-t_k)} ds + \frac{2}{Z_0 \hat{C}_0} \sum_{n=1}^{m} \left| g_n^{(0)} \right| \phi_0^n \int_{T+kT_0}^{t} e^{n\mu(s-t_k)} ds \le$$

$$\le \frac{2}{Z_0} \left| E_0(t) \right| + \frac{1}{Z_0} |u(t)| + \frac{1}{Z_0} |u(t-2T)| + |i(t-2T)| +$$

$$+ \frac{U_0}{Z_0^2 \hat{C}_0} \frac{e^{\mu(t-t_k)}-1}{\mu} + \frac{I_0}{Z_0 \hat{C}_0} \frac{e^{\mu(t-t_k)}-1}{\mu} + \frac{U_0}{Z_0^2 \hat{C}_0} \frac{e^{\mu(t-t_k)}-1}{\mu} + \frac{I_0}{Z_0 \hat{C}_0} \frac{e^{\mu(t-t_k)}-1}{\mu} +$$

$$+ \frac{2}{Z_0 \hat{C}_0} I_0 \frac{e^{\mu(t-t_k)}-1}{\mu} + \frac{2}{Z_0 \hat{C}_0} \frac{e^{\mu(t-t_k)}-1}{\mu} \sum_{n=1}^{m} \left| g_n^{(0)} \right| \phi_0^n e^{(n-1)\mu(t-t_k)} \le$$

$$\le e^{\mu(t-t_k)} \frac{2U_{E0}}{Z_0} + e^{\mu(t-t_{k0})} \frac{U_0}{Z_0} + e^{\mu(t-t_k)} \frac{U_0}{Z_0} + e^{\mu(t-t_k)} I_0 +$$

$$+ \frac{e^{\mu(t-t_k)}-1}{\mu} \left[\frac{U_0}{Z_0^2 \hat{C}_0} + \frac{I_0}{Z_0 \hat{C}_0} + \frac{U_0}{Z_0^2 \hat{C}_0} + \frac{I_0}{Z_0 \hat{C}_0} + \frac{2}{Z_0 \hat{C}_0} I_0 + \frac{2}{Z_0 \hat{C}_0} \sum_{n=1}^{m} \left| g_n^{(0)} \right| \phi_0^n e^{(n-1)\mu_0} \right] \le$$

$$\le e^{\mu(t-t_k)} \frac{2U_{E0} + 2(U_0 + Z_0 I_0)}{Z_0} + \frac{e^{\mu(t-t_k)}}{\mu Z_0 \hat{C}_0} \left[\frac{2(U_0 + I_0 Z_0)}{Z_0} + 2I_0 + 2\sum_{n=1}^{m} \left| g_n^{(0)} \right| \phi_0^n e^{(n-1)\mu_0} \right] \le$$

$$\le e^{\mu(t-t_k)} \left[\frac{2\phi_0}{Z_0} + \frac{2}{\mu Z_0 \hat{C}_0} \left(\frac{\phi_0}{Z_0} + I_0 + \sum_{n=1}^{m} \left| g_n^{(0)} \right| \phi_0^n e^{(n-1)\mu_0} \right) \right];$$

$$I_2 \le \frac{2}{Z_0} \left| \int_{t_k}^{t_{k+1}} \dot{E}_0(s) ds \right| + \frac{1}{Z_0} \left| \int_{t_k}^{t_{k+1}} \dot{u}(s) ds \right| + \frac{1}{Z_0} \left| \int_{t_k}^{t_{k+1}} \dot{u}(s-2T) ds \right| + \left| \int_{t_k}^{t_{k+1}} \dot{i}(s-2T) ds \right| +$$

$$+ \frac{1}{\hat{C}_0 Z_0^{\ 2}} \int_{t_k}^{t_{k+1}} |u(s)| ds + \frac{1}{Z_0 \hat{C}_0} \int_{t_k}^{t_{k+1}} |i(s)| ds + \frac{1}{\hat{C}_0 Z_0^{\ 2}} \int_{t_k}^{t_{k+1}} |u(s-2T)| ds + \frac{1}{Z_0 \hat{C}_0} \int_{t_k}^{t_{k+1}} |i(s-2T)| ds +$$

$$+ \frac{2}{Z_0 \hat{C}_0} \int_{t_k}^{t_{k+1}} I_0 e^{\mu(s-T-kT_0)} ds + \frac{2}{Z_0 \hat{C}_0} \sum_{n=1}^{m} \left| g_n^{(0)} \right| \phi_0^n \int_{T+kT_0}^{t_{k+1}} e^{n\mu(s-t_k)} ds \le$$

$$\le \frac{U_0}{\hat{C}_0 Z_0^{\ 2}} \frac{e^{\mu(t_{k+1}-t_k)}-1}{\mu} + \frac{I_0}{Z_0 \hat{C}_0} \frac{e^{\mu(t_{k+1}-t_k)}-1}{\mu} + \frac{U_0}{\hat{C}_0 Z_0^{\ 2}} \frac{e^{\mu(t_{k+1}-t_k)}-1}{\mu} + \frac{I_0}{Z_0 \hat{C}_0} \frac{e^{\mu(t_{k+1}-t_k)}-1}{\mu} +$$

$$+ \frac{2I_0}{Z_0 \hat{C}_0} \frac{e^{\mu(t_{k+1}-t_k)}-1}{\mu} + \frac{2}{Z_0 \hat{C}_0} \sum_{n=1}^{m} \left| g_n^{(0)} \right| \phi_0^n \frac{e^{n\mu(t_{k+1}-t_k)}-1}{n\mu} \le$$

$$\leq \frac{e^{\mu(t_{k+1}-t_k)}-1}{\mu}\frac{2}{Z_0\hat{C}_0}\left[\frac{1}{Z_0}\frac{2U_{E_0}+2(U_0+Z_0 I_0)}{2}+I_0+\sum_{n=1}^{m}\left|g_n^{(0)}\right|\phi_0^n e^{(n-1)\mu_0}\right]\leq$$

$$\leq \frac{2\left(e^{\mu_0}-1\right)}{\mu Z_0\hat{C}_0}\left[\frac{\phi_0}{Z_0}+I_0+\sum_{n=1}^{m}\left|g_n^{(0)}\right|\phi_0^n e^{(n-1)\mu_0}\right].$$

Therefore we obtain

$$\left|B_i^{(k)}(u,i)(t)\right|\leq I_1+I_2+I_3\leq$$

$$e^{\mu(t-t_k)}\left[\frac{2\phi_0}{Z_0}+\frac{2}{\mu Z_0\hat{C}_0}\left(\frac{\phi_0}{Z_0}+I_0+\sum_{n=1}^{m}\left|g_n^{(0)}\right|\phi_0^n e^{(n-1)\mu_0}\right)\right]+$$

$$+\frac{2\left(e^{\mu_0}-1\right)}{\mu Z_0\hat{C}_0}\left(\frac{\phi_0}{Z_0}+I_0+\sum_{n=1}^{m}\left|g_n^{(0)}\right|\phi_0^n e^{(n-1)\mu_0}\right)\leq$$

$$\leq e^{\mu(t-t_k)}\left[\frac{2\phi_0}{Z_0}+\frac{2e^{\mu_0}}{\mu Z_0\hat{C}_0}\left(\frac{\phi_0}{Z_0}+I_0+\sum_{n=1}^{m}\left|g_n^{(0)}\right|\phi_0^n e^{(n-1)\mu_0}\right)\right]\leq I_0 e^{\mu(t-T-kT_0)}.$$

Consequently, the inequalities guaranteeing that operator B maps the set $M_{SU}^* \times M_{SI}^*$ into itself are

$$U_{E_1}+\frac{e^{\mu_0}}{\mu^2\hat{C}_1}\left(2I_0+\sum_{n=1}^{m}\left|g_n^{(1)}\right|\phi_0^n e^{(n-1)\mu_0}\right)\leq U_0;$$

$$\frac{2\phi_0}{Z_0}+\frac{2e^{\mu_0}}{\mu Z_0\hat{C}_0}\left(\frac{\phi_0}{Z_0}+I_0+\sum_{n=1}^{m}\left|g_n^{(0)}\right|\phi_0^n e^{(n-1)\mu_0}\right)\leq I_0.$$

Obviously for U_{E_1} small and for Z_0 and μ large we obtain the required result.

The contractiveness of the operator B can be accomplished in the analogous way:

$$\hat{\rho}_\mu^{(k)}(B_u(u,i),\dot{B}_u(u,i),B_u(\overline{u},\overline{i}),\dot{B}_u(\overline{u},\overline{i}))\leq K\hat{\rho}_\mu^{(k)}((u,\dot{u},i,\dot{i}),(\overline{u},\dot{\overline{u}},\overline{i},\dot{\overline{i}}))$$

$(k=0,1,2,\ldots)$.

By using the fixed-point theorem for contractive mappings in uniform spaces one can conclude that B has a unique fixed point that is an oscillating solution of the problem stated.

CONCLUSION

Here we find that the first method of reduction from Chapter II considered in § 4.2 leads to a system with good rate of convergence (cf. § 4.5 with $K = 0{,}179$), while the second method shown in § 4.6 leads to neutral equations with Lipschitz constant equals 1. This means that such a system cannot be solved. Obviously, every particular circuit (in the present chapter of parallel connected elements) needs a specific method for solution of the neutral system obtained.

Chapter V

Lossy Transmission Lines Terminated by a Nonlinear Resistive Element

Abstract

This Chapter is devoted to same problem as in Chapter II, but here we take into account the lossy transmission lines. This means that we proceed from the hyperbolic system (5.1.1) and formulate a mixed problem based on the same initial and boundary conditions. This method of reducing the mixed problem to a periodic problem on the boundary is more complicated than the lossless case. We consider again two different cases: for R-load with polynomial, and exponential nonlinearities in the V-I characteristics. We also investigate oscillatory regimes. Finally, we introduce an extension of the Heaviside condition. This implies an analogous handling of transmission lines with time-varying specific parameters.

Introduction

It is known that transmission lines model various guiding structures such as power lines, wires, cables, printed circuit board traces, buses for carrying digital data in electronic circuits, VLSI interconnections, and microwave circuits and so on. In the previous chapters we have addressed our investigations on ideal or lossless transmission lines. But since in the work of every technical device there is lossiness the problem arising is: can the lossiness be neglected or not? The main goal of the present chapter is to investigate transmission lines taking lossiness into account. We consider the same configuration shown in Chapter II but for lossy transmission lines. This means that we proceed from a more complicated system of partial differential equations, which implies more difficulties in solving the same problems.

In Chapter V we extend our technique of reducing the mixed problem to an initial value one, and solve the obtained neutral equation by using the fixed-point method the obtained neutral equation. In § 5.1 first we formulate the mixed problem for a lossy transmission line system, and then in § 5.2 propose a reduction of the mixed problem to a periodic initial value problem on the boundary with an unknown function for the voltage of the line. A theorem for existence-uniqueness of T_0-periodic solution of the obtained neutral equation is proved. Here, the V-I characteristic of the R-load is of a polynomial type. The proof does not depend on the degree of the polynomial. In § 5.3 we give conditions for the existence-uniqueness of an oscillatory solution of the neutral equation with polynomial nonlinearity. In § 5.4 we propose another manner of reducing the mixed problem to an oscillatory problem for a neutral equation. In § 5.5 we formulate conditions for existence-uniqueness of an oscillatory regime for a neutral equation with exponential nonlinearity. In § 5.6 we propose a new approach of reducing the mixed problem for lossy transmission lines to an oscillatory problem for a neutral equation with exponential nonlinearity. In § 5.7 we consider a transmission line with time-varying specific parameters. We introduce a generalization of the Heaviside condition that allows us to reduce the mixed problem with time varying parameters to the previous case.

5.1. Formulation of the Mixed Problem for Lossy Transmission Lines Terminated by a Nonlinear R-Element

In this section we formulate the same mixed problem considered in Chapter II but for lossy transmission line system of equations.

We proceed from the circuit shown on Fig. 5.1 where E is the source, R_0 and C_0 – linear loads, while the resistive load (at the right end) has a nonlinear V-I characteristic $i = f(u)$. We consider two cases: 1) nonlinear characteristics of a polynomial type and 2) nonlinear characteristics of an exponential type.

Recall (cf. Chapter I) that a lossy transmission line can be described by using the following hyperbolic system

$$C\frac{\partial u(x,t)}{\partial t} + \frac{\partial i(x,t)}{\partial x} + Gu(x,t) = 0,$$
$$L\frac{\partial i(x,t)}{\partial t} + \frac{\partial u(x,t)}{\partial x} + Ri(x,t) = 0 \tag{5.1.1}$$

$$(x,t) \in \Pi = \left\{(x,t) \in \Pi^2 : (x,t) \in [0,\Lambda] \times [0,\infty)\right\}$$

where $u(x,t)$ and $i(x,t)$ are the unknown voltage and current, while L, C, R and G are distributed specific parameters of the line and $\Lambda > 0$ is its length. For the above system (5.1.1) can be formulated the following mixed problem: to find $u(x,t)$ and $i(x,t)$ in Π such that

$$u(x,0) = u_0(x), \ i(x,0) = i_0(x), \ x \in [0,\Lambda], \tag{5.1.2}$$

$$E(t) - u(0,t) - R_0 i(0,t) = 0, \ t \geq 0, \tag{5.1.3}$$

$$C_0 \frac{du(\Lambda,t)}{dt} = i(\Lambda,t) - f(u(\Lambda,t)), \ t \geq 0 \tag{5.1.4}$$

where $i_0(x)$, $u_0(x)$ are prescribed initial functions − the current and voltage at the initial instant, and (5.1.3), (5.1.4) are boundary conditions corresponding to the loads from Fig. 5.1.

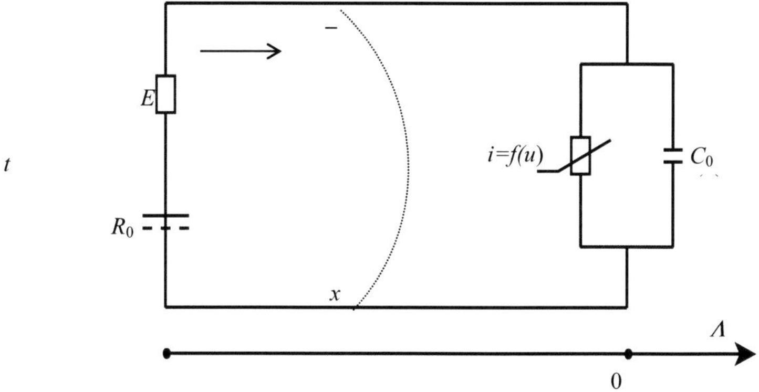

Figure 5.1.

Most often, the third-order polynomials are used for the approximation in the oscillator circuits where $f(u)$ is a polynomial with a partially negative differential resistance (cf. R. K. Brayton [28], [29], L. A. Bessonov [24], L. O. Chua, C. A. Desoer & E. S. Kuh [37], L. V. Danilov [45], L. Jiang, Y. Nashio & A. Ushida [68], P. N. Matkhanov [84], [59], P. Vizmuller [111]) of tunnel diodes (or L. Esaki diodes), p-n-p-n diodes, Gunn-diodes and others).

We present a method for reducing the mixed problem for lossy transmission lines to an initial value problem on the boundary for a neutral functional differential equation. The results are valid for transmission lines without dispersion, i.e. the Heaviside condition is fulfilled $R/L = G/C$.

First we present (5.1.1) in the form

$$\frac{\partial u(x,t)}{\partial t} + \frac{1}{C}\frac{\partial i(x,t)}{\partial x} + \frac{G}{C}u(x,t) = 0,$$

$$\frac{\partial i(x,t)}{\partial t} + \frac{1}{L}\frac{\partial u(x,t)}{\partial x} + \frac{R}{L}i(x,t) = 0$$

and then in a matrix form:

$$\frac{\partial U}{\partial t} + A_1\frac{\partial U}{\partial x} + A_2 U = 0 \tag{5.1.5}$$

where

$$U = \begin{bmatrix} u \\ i \end{bmatrix}, \quad \frac{\partial U}{\partial t} = \begin{bmatrix} \dfrac{\partial u}{\partial t} \\[4pt] \dfrac{\partial i}{\partial t} \end{bmatrix}, \quad \frac{\partial U}{\partial x} = \begin{bmatrix} \dfrac{\partial u}{\partial x} \\[4pt] \dfrac{\partial i}{\partial x} \end{bmatrix}, \ A_1 = \begin{bmatrix} 0 & 1/C \\ 1/L & 0 \end{bmatrix}, \quad A_2 = \begin{bmatrix} G/C & 0 \\ 0 & R/L \end{bmatrix}. \quad (5.1.6)$$

In order to transform the matrix $A_1 = \begin{bmatrix} 0 & 1/C \\ 1/L & 0 \end{bmatrix}$ into a diagonal form we have to solve

the characteristic equation: $\begin{vmatrix} -\lambda & 1/C \\ 1/L & -\lambda \end{vmatrix} = 0$. The roots are $\lambda_1 = 1/\sqrt{LC}, \ \lambda_2 = -1/\sqrt{LC}$.

For the eigenvectors we obtain the following systems:

$$\begin{vmatrix} -\dfrac{1}{\sqrt{LC}}\xi_1 + \dfrac{1}{L}\xi_2 = 0 \\[8pt] \dfrac{1}{C}\xi_1 - \dfrac{1}{\sqrt{LC}}\xi_2 = 0 \end{vmatrix} \quad \text{and} \quad \begin{vmatrix} \dfrac{1}{\sqrt{LC}}\xi_1 + \dfrac{1}{L}\xi_2 = 0 \\[8pt] \dfrac{1}{C}\xi_1 + \dfrac{1}{\sqrt{LC}}\xi_2 = 0 \end{vmatrix} .$$

Hence $\left(\xi_1^{(1)}, \xi_2^{(1)}\right) = \left(\sqrt{C}, \sqrt{L}\right),$ $\quad \left(\xi_1^{(2)}, \xi_2^{(2)}\right) = \left(-\sqrt{C}, \sqrt{L}\right).$

Denote by H the matrix formed by eigenvectors $H = \begin{bmatrix} \sqrt{C} & \sqrt{L} \\ -\sqrt{C} & \sqrt{L} \end{bmatrix}$. Its inverse one is

$$H^{-1} = \begin{bmatrix} \dfrac{1}{2\sqrt{C}} & -\dfrac{1}{2\sqrt{C}} \\[8pt] \dfrac{1}{2\sqrt{L}} & \dfrac{1}{2\sqrt{L}} \end{bmatrix}. \ \text{Then } A^{\text{can}} = HAH^{-1} \text{ where } A^{\text{can}} = \begin{bmatrix} \dfrac{1}{\sqrt{LC}} & 0 \\[8pt] 0 & -\dfrac{1}{\sqrt{LC}} \end{bmatrix}.$$

Introduce new variables $Z = HU$, (or $U = H^{-1}Z$)

$$Z = \begin{bmatrix} V(x,t) \\ I(x,t) \end{bmatrix}, \quad H = \begin{bmatrix} \sqrt{C} & \sqrt{L} \\ -\sqrt{C} & \sqrt{L} \end{bmatrix}, \quad U = \begin{bmatrix} u(x,t) \\ i(x,t) \end{bmatrix}.$$

Then

$$\begin{vmatrix} V(x,t) = \sqrt{C}\,u(x,t) + \sqrt{L}\,i(x,t) \\[4pt] I(x,t) = -\sqrt{C}\,u(x,t) + \sqrt{L}\,i(x,t) \end{vmatrix} \qquad (5.1.7)$$

or

$$\left|\begin{array}{l} u(x,t) = \dfrac{1}{2\sqrt{C}} V(x,t) - \dfrac{1}{2\sqrt{C}} I(x,t) \\[4mm] i(x,t) = \dfrac{1}{2\sqrt{L}} V(x,t) + \dfrac{1}{2\sqrt{L}} I(x,t). \end{array}\right. \tag{5.1.8}$$

Substituting $U = H^{1}Z$ in (5.1.6) we obtain $\dfrac{\partial\left(H^{-1}Z\right)}{\partial t} + A\dfrac{\partial\left(H^{-1}Z\right)}{\partial x} + B\left(H^{-1}Z\right) = 0$.

Recalling reasoning's our reasoning from Chapter I and in view of Heaviside condition $\dfrac{R}{L} = \dfrac{G}{C}$ we have

$$\frac{\partial Z}{\partial t} + A^{can}\frac{\partial Z}{\partial x} + \left(HBH^{-1}\right)Z = 0 \tag{5.1.9}$$

where $HBH^{-1} = \begin{bmatrix} R/L & 0 \\ 0 & R/L \end{bmatrix}$.

Then the explicit form of (5.1.9) becomes:

$$\begin{bmatrix} \dfrac{\partial V}{\partial t} \\[3mm] \dfrac{\partial I}{\partial t} \end{bmatrix} + \begin{bmatrix} \dfrac{1}{\sqrt{LC}} & 0 \\[3mm] 0 & -\dfrac{1}{\sqrt{LC}} \end{bmatrix} \begin{bmatrix} \dfrac{\partial V}{\partial x} \\[3mm] \dfrac{\partial I}{\partial x} \end{bmatrix} + \begin{bmatrix} \dfrac{R}{L} & 0 \\[3mm] 0 & \dfrac{R}{L} \end{bmatrix} \begin{bmatrix} V \\[3mm] I \end{bmatrix} = \begin{bmatrix} 0 \\[3mm] 0 \end{bmatrix}. \tag{5.1.10}$$

The new initial conditions we obtain from (5.1.7):

$$V(x,0) = \sqrt{C}\, u(x,0) + \sqrt{L}\, i(x,0) = \sqrt{C}\, u_0(x) + \sqrt{L}\, i_0(x) \equiv V_0(x), \tag{5.1.7-1}$$

$$I(x,0) = -\sqrt{C}\, u(x,0) + \sqrt{L}\, i(x,0) = -\sqrt{C}\, u_0(x) + \sqrt{L}\, i_0(x) \equiv I_0(x),\, x \in [0, \Lambda]. \tag{5.1.7-2}$$

Using (5.1.8) we reach the new boundary conditions

$$\begin{aligned} & E(t) - \frac{V(0,t) - I(0,t)}{2\sqrt{C}} - R_0 \frac{V(0,t) + I(0,t)}{2\sqrt{L}} = 0, \\[4mm] & C_0 \frac{1}{2\sqrt{C}}\left[\frac{dV(\Lambda,t)}{dt} - \frac{dI(\Lambda,t)}{dt}\right] = \frac{V(\Lambda,t) + I(\Lambda,t)}{2\sqrt{L}} - f\left(\frac{V(\Lambda,t) - I(\Lambda,t)}{2\sqrt{C}}\right), t \geq 0. \end{aligned} \tag{5.1.11}$$

Now we have reduced the mixed problem (5.1.1) - (5.1.4) to (5.1.10), (5.1.7-1), (5.1.7-2), (5.1.11).

System (5.1.10) can be further simplified one more time by the substitution:

$$W(x,t) = e^{\frac{R}{L}t}V(x,t), \quad J(x,t) = e^{\frac{R}{L}t}I(x,t),$$

or

$$V(x,t) = e^{-\frac{R}{L}t}W(x,t), \quad I(x,t) = e^{-\frac{R}{L}t}J(x,t). \qquad (5.1.12)$$

Replacing in (5.1.8) we obtain the following transformation:

$$u(x,t) = \frac{1}{2\sqrt{C}}e^{-\frac{R}{L}t}W(x,t) - \frac{1}{2\sqrt{C}}e^{-\frac{R}{L}t}J(x,t),$$

$$i(x,t) = \frac{1}{2\sqrt{L}}e^{-\frac{R}{L}t}W(x,t) + \frac{1}{2\sqrt{L}}e^{-\frac{R}{L}t}J(x,t) \qquad (5.1.13)$$

and

$$W(x,t) = e^{\frac{R}{L}t}\left(\sqrt{C}\,u(x,t) + \sqrt{L}\,i(x,t)\right),$$

$$J(x,t) = e^{\frac{R}{L}t}\left(\sqrt{L}\,i(x,t) - \sqrt{C}\,u(x,t)\right).$$

Rewrite (5.1.10) in the form:

$$\frac{\partial V(x,t)}{\partial t} + \frac{1}{\sqrt{LC}}\frac{\partial V(x,t)}{\partial x} + \frac{R}{L}V(x,t) = 0,$$

$$\frac{\partial I(x,t)}{\partial t} - \frac{1}{\sqrt{LC}}\frac{\partial I(x,t)}{\partial x} + \frac{R}{L}I(x,t) = 0 \qquad (5.1.14)$$

and putting $V(x,t)$ and $I(x,t)$ from (5.1.12) into (5.1.14), we are given

$$\frac{\partial W(x,t)}{\partial t} + \frac{1}{\sqrt{LC}}\frac{\partial W(x,t)}{\partial x} = 0,$$

$$\frac{\partial J(x,t)}{\partial t} - \frac{1}{\sqrt{LC}}\frac{\partial J(x,t)}{\partial x} = 0. \qquad (5.1.15)$$

Since the initial conditions remain the same ones in view of

$$W(x,0) = e^{\frac{R}{L}\cdot 0}V(x,0) = V_0(x) = \sqrt{C}\,u_0(x) + \sqrt{L}\,i_0(x),$$

$$J(x,0) = e^{\frac{R_0}{L}} I(x,0) = I_0(x) = -\sqrt{C}\, u_0(x) + \sqrt{L}\, i_0(x),\, x \in [0, \Lambda]$$

we have to transform the boundary conditions

$$E(t) - \frac{V(0,t) - I(0,t)}{2\sqrt{C}} - R_0 \frac{V(0,t) + I(0,t)}{2\sqrt{L}} = 0,$$

$$C_0 \frac{1}{2\sqrt{C}} \left[\frac{dV(\Lambda,t)}{dt} - \frac{dI(\Lambda,t)}{dt} \right] = \frac{V(\Lambda,t) + I(\Lambda,t)}{2\sqrt{L}}\, f\left(\frac{V(\Lambda,t) - I(\Lambda,t)}{2\sqrt{C}} \right), t \geq 0.$$

Indeed, in view of (5.1.12) we obtain

$$E(t) - \frac{e^{-\frac{R}{L}t} W(0,t) - e^{-\frac{R}{L}t} J(0,t)}{2\sqrt{C}} - R_0 \frac{e^{-\frac{R}{L}t} W(0,t) + e^{-\frac{R}{L}t} J(0,t)}{2\sqrt{L}} = 0,$$

$$C_0 \frac{1}{2\sqrt{C}} \left[\frac{d}{dt}\left(e^{-\frac{R}{L}t} W(\Lambda,t) \right) - \frac{d}{dt}\left(e^{-\frac{R}{L}t} J(\Lambda,t) \right) \right] =$$

$$= \frac{e^{-\frac{R}{L}t} W(\Lambda,t) + e^{-\frac{R}{L}t} J(\Lambda,t)}{2\sqrt{L}} - f\left(\frac{e^{-\frac{R}{L}t} W(\Lambda,t) - e^{-\frac{R}{L}t} J(\Lambda,t)}{2\sqrt{C}} \right), t \geq 0. \tag{5.1.16}$$

5.2. Reducing the Mixed Problem to an Initial Value Problem for a Neutral Equation

The solution of (5.1.15) is a pair of arbitrary differentiable functions

$$W(x,t) = \Phi_W(x - vt),\ J(x,t) = \Phi_J(x + vt).$$

Substitute $W(x,t)$ and $J(x,t)$ in (5.1.13) we obtain

$$u(x,t) = \frac{e^{-\frac{R}{L}t}}{2\sqrt{C}} \left[\Phi_W(x - vt) - \Phi_J(x + vt) \right],$$

$$i(x,t) = \frac{e^{-\frac{R}{L}t}}{2\sqrt{L}} \left[\Phi_W(x - vt) + \Phi_J(x + vt) \right]. \tag{5.2.1}$$

Hence

$$\Phi_W(x-vt)=e^{\frac{R}{L}t}\left(\sqrt{C}\,u(x,t)+\sqrt{L}\,i(x,t)\right)$$
$$\Phi_J(x+vt)=e^{\frac{R}{L}t}\left(\sqrt{L}\,i(x,t)-\sqrt{C}\,u(x,t)\right)$$

(5.2.2)

For $x=\Lambda$ we obtain

$$\Phi_W(\Lambda-vt)=e^{\frac{R}{L}t}\left[\sqrt{C}\,u(\Lambda,t)+\sqrt{L}\,i(\Lambda,t)\right],$$
$$\Phi_J(\Lambda+vt)=e^{\frac{R}{L}t}\left[\sqrt{L}\,i(\Lambda,t)-\sqrt{C}\,u(\Lambda,t)\right]$$

(5.2.3)

Let us put

$$\Lambda-vt=-vt'\quad\Rightarrow\quad t=t'+\Lambda/v\equiv t'+T\quad(T=\Lambda/v)$$

and then substituting t by $t'+T$ in the first equation of (5.2.3) we get

$$\Phi_W(-vt')=e^{\frac{R}{L}(t'+T)}\left[\sqrt{C}\,u(\Lambda,t'+T)+\sqrt{L}\,i(\Lambda,t'+T)\right].$$

For the second equation of (5.2.3) we put $\Lambda+vt=vt''\Rightarrow t=t''-\Lambda/v\equiv t''-T$ and $(T=\Lambda/v)$ then we have

$$\Phi_J(vt'')=e^{\frac{R}{L}(t''-T)}[\sqrt{L}i(\Lambda,t''-T)-\sqrt{C}u(\Lambda,t''-T)].$$

We consider the above equations with the same argument t:

$$\Phi_W(-vt)=e^{\frac{R}{L}(t+T)}\left[\sqrt{C}\,u(\Lambda,t+T)+\sqrt{L}\,i(\Lambda,t+T)\right]$$

(5.2.4)

and

$$\Phi_J(vt)=e^{\frac{R}{L}(t-T)}\left[\sqrt{L}\,i(\Lambda,t-T)-\sqrt{C}\,u(\Lambda,t-T)\right].$$

(5.2.5)

From (5.2.1) for $x=0$ we have

$$u(0,t) = e^{-\frac{R}{L}t}\,\frac{\Phi_W(-vt) - \Phi_J(vt)}{2\sqrt{C}},$$

$$i(0,t) = e^{-\frac{R}{L}t}\,\frac{\Phi_W(-vt) + \Phi_J(vt)}{2\sqrt{L}}.$$

$$(5.2.6)$$

Substituting $\Phi_W(-vt)$ and $\Phi_J(vt)$ from (5.2.4) and (5.2.5) into (5.2.6) we obtain:

$$u(0,t) =$$

$$= \frac{e^{-\frac{R}{L}t}}{2\sqrt{C}}\left[e^{\frac{R}{L}(t+T)}\sqrt{C}\,u(\Lambda,t+T) + e^{\frac{R}{L}(t+T)}\sqrt{L}\,i(\Lambda,t+T) - e^{\frac{R}{L}(t-T)}\sqrt{L}\,i(\Lambda,t-T) + e^{\frac{R}{L}(t-T)}\sqrt{C}\,u(\Lambda,t-T) \right],$$

$$i(0,t) = \frac{e^{-\frac{R}{L}t}}{2\sqrt{L}}\left[e^{\frac{R}{L}(t+T)}\sqrt{C}\,u(\Lambda,t+T) + e^{\frac{R}{L}(t+T)}\sqrt{L}\,i(\Lambda,t+T) + e^{\frac{R}{L}(t-T)}\sqrt{L}\,i(\Lambda,t-T) - e^{\frac{R}{L}(t-T)}\sqrt{C}\,u(\Lambda,t-T) \right].$$

Recalling that $Z_0 = \sqrt{\dfrac{L}{C}}$ and $\dfrac{RT}{L} = \dfrac{R\Lambda\sqrt{LC}}{L} = \dfrac{R\Lambda}{Z_0}$ we substitute the above

expressions into the first boundary condition (5.1.3):

$$E(t) - u(0,t) - R_0 i(0,t) = 0$$

that is,

$$E(t) -$$

$$- \frac{e^{-\frac{R}{L}t}}{2\sqrt{C}}\left[e^{\frac{R}{L}(t+T)}\sqrt{C}\,u(\Lambda,t+T) + e^{\frac{R}{L}(t+T)}\sqrt{L}\,i(\Lambda,t+T) - e^{\frac{R}{L}(t-T)}\sqrt{L}\,i(\Lambda,t-T) + e^{\frac{R}{L}(t-T)}\sqrt{C}\,u(\Lambda,t-T) \right] -$$

$$- \frac{R_0 e^{-\frac{R}{L}t}}{2\sqrt{L}}\left[e^{\frac{R}{L}(t+T)}\sqrt{C}\,u(\Lambda,t+T) + e^{\frac{R}{L}(t+T)}\sqrt{L}\,i(\Lambda,t+T) + e^{\frac{R}{L}(t-T)}\sqrt{L}\,i(\Lambda,t-T) - e^{\frac{R}{L}(t-T)}\sqrt{C}\,u(\Lambda,t-T) \right] = 0.$$

Let us put $t' = t + T$. In view of $t - T = t' - 2T$ (and again replace t' by t) we obtain:

$$E(t-T) -$$

$$- \frac{e^{-\frac{R}{L}(t-T)}}{2\sqrt{C}}\left[e^{\frac{R}{L}t}\sqrt{C}\,u(\Lambda,t) + e^{\frac{R}{L}t}\sqrt{L}\,i(\Lambda,t) - e^{\frac{R}{L}(t-2T)}\sqrt{L}\,i(\Lambda,t-2T) + e^{\frac{R}{L}(t-2T)}\sqrt{C}\,u(\Lambda,t-2T) \right] -$$

$$- \frac{R_0 e^{-\frac{R}{L}(t-T)}}{2\sqrt{L}}\left[e^{\frac{R}{L}t}\sqrt{C}\,u(\Lambda,t) + e^{\frac{R}{L}t}\sqrt{L}\,i(\Lambda,t) + e^{\frac{R}{L}(t-2T)}\sqrt{L}\,i(\Lambda,t-2T) - e^{\frac{R}{L}(t-2T)}\sqrt{C}\,u(\Lambda,t-2T) \right] = 0$$

or

$$\frac{2E(t-T)}{Z_0+R_0} - e^{\frac{R}{L}T}\frac{u(\Lambda,t)}{Z_0} - e^{\frac{R}{L}T}i(\Lambda,t) + e^{-\frac{R}{L}T}\frac{Z_0-R_0}{Z_0+R_0}i(\Lambda,t-2T) + e^{-\frac{R}{L}T}\frac{R_0-Z_0}{Z_0(Z_0+R_0)}u(\Lambda,t-2T) = 0 \qquad (5.2.7)$$

The second boundary condition is $C_0\dfrac{du(\Lambda,t)}{dt} = i(\Lambda,t) - f\big(u(\Lambda,t)\big)$ or

$$i(\Lambda,t) = C_0\frac{du(\Lambda,t)}{dt} + f\big(u(\Lambda,t)\big) \qquad (5.2.8)$$

and

$$i(\Lambda,t-2T) = C_0\frac{du(\Lambda,t-2T)}{dt} + f\big(u(\Lambda,t-2T)\big). \qquad (5.2.9)$$

Substituting $i(\Lambda,t)$ and $i(\Lambda,t-2T)$ from (5.2.8) and (5.2.9) into (5.2.7) and in view of $E(t-T) = E(t)$ we obtain:

$$\frac{du(\Lambda,t)}{dt} = \frac{2E(t)e^{-\frac{R}{L}T}}{C_0(Z_0+R_0)} - \frac{u(\Lambda,t)}{C_0Z_0} - \frac{1}{C_0}f\big(u(\Lambda,t)\big) + e^{-2\frac{R}{L}T}\frac{R_0-Z_0}{C_0Z_0(Z_0+R_0)}u(\Lambda,t-2T) +$$

$$+ e^{-2\frac{R}{L}T}\frac{Z_0-R_0}{C_0(Z_0+R_0)}f\big(u(\Lambda,t-2T)\big) + e^{-2\frac{R}{L}T}\frac{Z_0-R_0}{Z_0+R_0}\frac{du(\Lambda,t-2T)}{dt}.$$

Denoting by $\delta = \dfrac{Z_0-R_0}{Z_0+R_0}$ and choosing the unknown function to be $u(t) = u(\Lambda,t)$ we obtain the following neutral equation on the boundary $x = \Lambda$:

$$\frac{du(t)}{dt} = \frac{2E(t-T)e^{-\frac{R}{L}T}}{C_0(Z_0+R_0)} - \frac{u(t)}{C_0Z_0} - \frac{1}{C_0}f\big(u(t)\big) - e^{-2\frac{R}{L}T}\frac{\delta}{C_0Z_0}u(t-2T) +$$

$$+ e^{-2\frac{R}{L}T}\frac{\delta}{C_0}f\big(u(t-2T)\big) + e^{-2\frac{R}{L}T}\delta\frac{du(t-2T)}{dt}. \qquad (5.2.10)$$

We shall consider the above equation for $t \geq T$ and $x = \Lambda$.

5.3. EXPONENTIALLY VANISHING OSCILLATORY REGIMES FOR NEUTRAL EQUATIONS WITH POLYNOMIAL NONLINEARITIES

Here we choose a suitable function space endowed with a saturated family of pseudo-metrics. We propose such operators whose fixed points are oscillatory solutions of the neutral equation. Then using the fixed-point theorem in uniform spaces we prove the existence-uniqueness of an oscillatory solution. In the present paragraph we consider the case where the *V-I* characteristic of the nonlinear resistive load is a polynomial of an arbitrary order, that is,

$$f(u) = \sum_{n=1}^{p} r_n u^n .$$

Further on in (5.2.11) we substitute $f(u) = \sum_{n=1}^{p} r_n u^n$ in (5.2.10) and look for oscillatory solutions for the following equation:

$$\frac{du(t)}{dt} = \frac{2E(t)e^{-\frac{R}{L}T}}{C_0(Z_0+R_0)} - \frac{u(t)}{C_0 Z_0} - \frac{1}{C_0}\sum_{n=1}^{p} r_n (u(t))^n - e^{-2\frac{R}{L}T}\frac{\delta}{C_0 Z_0}u(t-2T) +$$

$$+ e^{-2\frac{R}{L}T}\frac{\delta}{C_0}\sum_{n=1}^{p} r_n(u(t-2T))^n + e^{-2\frac{R}{L}T}\delta\frac{du(t-2T)}{dt} , t\in[t_0,\infty)$$

$$u(t) = \upsilon_0(t),\ t\in[-T,T],\ \dot{u}(t) = \dot{\upsilon}_0(t), t\in[-T,T]. \tag{5.3.1}$$

The initial function is defined as in Chapter II.

Now we are able to formulate the main problem: to finding a solution of (5.3.1) with advanced prescribed zeros on an interval $[t_0,\infty), T = t_0$, where $\upsilon_0(t)$ is a prescribed oscillating function on the interval $[-T,T]$.

Let $S_T = \{\tau_k\}_{k=0}^n, n\in N$ be the set of zeros of the initial function, that is, $\upsilon_0(\tau_k)=0$ such that $\tau_0 = -T$, $\tau_n = T \equiv t_0$. Besides $\max\{\tau_{k+1}-\tau_k : k=0,1,...,n\} \leq T_0$.

Let $S = \{t_k\}_{k=0}^\infty$ be a strictly increasing sequence of real numbers satisfying the following conditions **(C)**:

(C1) $\lim\limits_{k\to\infty} t_k = \infty$;

(C2) for every k there is $q < k$ such that $t_k - T = t_q$ where $t_q \in S_T \cup S$.

It follows

(C3) $0 < \inf\{t_{k+1}-t_k : k=0,1,2,...\} \leq \sup\{t_{k+1}-t_k : k=0,1,2,...\} = T_0 < \infty$ and $t_k - T = t_q \Rightarrow t_k - 2T = t_p$ for some p.

Remark 5.3.1. Condition **(C3)** implies that operator

$$K : u(t) \to \vec{u}(t) = \begin{cases} \upsilon_0(t-2T), t \in [T,3T] \\ u(t-2T), \ t \in [3T,\infty) \end{cases}$$

maps the set of all continuous oscillating functions on $[t_0,\infty)$, $(t_0 \equiv T)$ into itself.

Introduce the set $C^1[t_0,\infty)$, $(t_0 \equiv T)$ consisting of all continuous and bounded functions differentiable piece-wise with bounded derivatives on every interval $[t_k,t_{k+1}]$. Let us note that the functions from $C^1[t_0,\infty)$ may be not differentiable at t_k. This yields the necessity to introducing a topology of uniform convergence of the function on $[t_0,t_{k+1}]$ and their derivatives on every interval $[t_k,t_{k+1}]$. Thus it is to consider uniform spaces (cf. Chapter I).

Introduce the set

$$M_U = \left\{ u(.) \in C^1[t_0,\infty) : u(t_k) = 0 \wedge |u(t)| \le U_0 e^{-\frac{R}{L}t}, t \in [t_0,\infty) \right\}.$$

It is easy to see that if $u(.) \in M_U \Rightarrow |u(t)| \le U_0 e^{\mu(t-t_k)}, t \in [t_k,t_{k+1}], (k=0,1,2,...)$ where $U_0, \mu, \mu T_0 = \mu_0 = \text{const.}$ are positive constants.

Assumption **(IN)** $|\upsilon_0(t)| \le U_0 e^{-\frac{R}{L}t}, t \in [-t_0,t_0], \ |\upsilon_0(t)| \le U_0 e^{\mu(t-\tau_k)}, t \in [\tau_k,\tau_{k+1}]$

$(k = 0,1,2,...,2m-1)$.

Assumption **(E)** $|E(t)| \le U_E e^{-\frac{R}{L}t}, t \in [-T;\infty); |E(t)| \le U_E e^{\mu(t-\tau_k)}, t \in [\tau_k,\tau_{k+1}]$

and $|E(t)| \le U_E e^{\mu(t-t_k)}, t \in [t_k,t_{k+1}](k=0,1,...,);U_E \le U_0$.

Introduce the following family of pseudo-metrics

$$\rho^{(k)}(u,\bar{u}) = \max\left\{|u(t)-\bar{u}(t)| : t \in [t_k,t_{k+1}]\right\},$$

$$\hat{\rho}^{(k)}(u,\bar{u}) = \max\left\{|u(t)-\bar{u}(t)| : t \in [t_0,t_{k+1}]\right\},$$

$$\rho_\mu^{(k)}(u,\bar{u}) = \max\left\{e^{-\mu(t-t_k)}|u(t)-\bar{u}(t)| : t \in [t_k,t_{k+1}]\right\},$$

$$\hat{\rho}_\mu^{(k)}(u,\bar{u}) = \max\left\{\rho_\mu^{(0)}(u,\bar{u}), \rho_\mu^{(1)}(u,\bar{u}),..., \rho_\mu^{(k)}(u,\bar{u})\right\},$$

$$\rho_\mu^{(k)}(\dot{u},\dot{\bar{u}}) = \max\left\{e^{-\mu(t-t_k)}|\dot{u}(t)-\dot{\bar{u}}(t)| : t \in [t_k,t_{k+1}]\right\},$$

$$\hat{\rho}_{\mu}^{(k)}(\dot{u},\dot{\overline{u}}) = \max\left\{\rho_{\mu}^{(0)}(\dot{u},\dot{\overline{u}}), \rho_{\mu}^{(1)}(\dot{u},\dot{\overline{u}}),...,\rho_{\mu}^{(k)}(\dot{u},\dot{\overline{u}})\right\}.$$

Remark 5.3.1. The following inequalities imply the equivalence of the both families of pseudo-metrics:

$$\rho_{\mu}^{(k)}(u,\overline{u}) \le \rho^{(k)}(u,\overline{u}) \le e^{\mu T_0}\rho_{\mu}^{(k)}(u,\overline{u}), \ (k = 0,1,2,...)$$

and

$$\hat{\rho}^{(k)}(u,\overline{u}) = \max\left\{\rho^{(0)}(u,\overline{u}), \rho^{(1)}(u,\overline{u}),...,\rho^{(k)}(u,\overline{u})\right\} \le$$
$$\le e^{\mu T_0}\max\left\{\rho_{\mu}^{(0)}(u,\overline{u}),\rho_{\mu}^{(1)}(u,\overline{u}),...,\rho_{\mu}^{(k)}(u,\overline{u})\right\} = e^{\mu T_0}\hat{\rho}_{\mu}^{(k)}(u,\overline{u}).$$

The set M_U turns out into a complete uniform space with respect to the saturated family of pseudo-metrics

$$\hat{\rho}_{\mu}^{(k)}((u,\dot{u}),(\overline{u},\dot{\overline{u}})) = \max\left\{\hat{\rho}^{(k)}(u,\overline{u}), \hat{\rho}_{\mu}^{(k)}(\dot{u},\dot{\overline{u}})\right\}, (k = 0,1,2,...).$$

Define an operator B by the formulas

$$B(u)(t) := B_k(u)(t), \ \ t \in [t_k,t_{k+1}], \ (k = 0,1,2,...)$$

where

$$B_k(u)(t) \equiv \int_{t_k}^{t} U(u)(s)ds - \frac{t-t_k}{t_{k+1}-t_k}\int_{t_k}^{t_{k+1}} U(u)(s)ds, \ \ t \in [t_k,t_{k+1}], \ (k = 0,1,2,...)$$

and $U(u)$ is the right-hand side of (5.3.1):

$$U(u)(t) = \frac{2E(t)e^{-\frac{R}{L}T}}{C_0(Z_0 + R_0)} - \frac{u(t)}{C_0 Z_0} - \frac{1}{C_0}\sum_{n=1}^{P} r_n\big(u(t)\big)^n - e^{-2\frac{R}{L}T}\frac{\delta}{C_0 Z_0}\overline{u}(t) +$$
$$+ e^{-2\frac{R}{L}T}\frac{\delta}{C_0}\sum_{n=1}^{P} r_n\big(\overline{u}(t)\big)^n + e^{-2\frac{R}{L}T}\delta\frac{d\overline{u}(t)}{dt}, t \in [t_0,\infty)$$

Let us recall that $K : M_U \to M_U$ assigns to every $u(.) \in M_U$ a function

$$\overline{u}(t) = \begin{cases} \upsilon_0(t-2T), t \in [T,3T] \\ u(t-2T), \ t \in [3T,\infty) \end{cases}.$$ In fact $u(t-2T)$ is replaced by $\overline{u}(t)$. In this way we overcome a difficulty caused by the conformity condition **(CC)**.

Lemma 5.3.1. Problem (5.3.1) has a solution $u(.) \in M_U$ iff the operator B has a fixed point in M_U, that is,

$$u = B(u) . \tag{5.3.2}$$

Proof: Let $u(.) \in M_U$ be a solution of (5.3.1). Then integrating (5.3.1) on every interval $[t_k, t] \subset [t_k, t_{k+1}]$ $(k = 0,1,2 \dots)$ we obtain

$$u(t) - u(t_k) = \int_{t_k}^{t} U(u)(s)ds \Leftrightarrow u(t) = \int_{t_k}^{t} U(u)(s)ds$$

and then

$$u(t) = \int_{t_k}^{t} U(u)(s)ds \Rightarrow 0 = u(t_{k+1}) = \int_{t_k}^{t} U(u)(s)ds \Rightarrow \int_{t_k}^{t_{k+1}} U(u)(s)ds = 0 . \tag{5.3.3}$$

Therefore $u(t)$ satisfies

$$u(t) = \int_{t_k}^{t} U(u)(s)ds - \frac{t - t_k}{t_{k+1} - t_k} \int_{t_k}^{t_{k+1}} U(u)(s)ds, \ t \in [t_k, t_{k+1}] \Leftrightarrow u = B(u),$$

that is, $u(t)$ is a fixed point of B.

Conversely, let u be a fixed point of B, that is, $u(.) \in M_U$ be a solution of (5.3.2) or

$$u(t) = \int_{t_k}^{t} U(u)(s)ds - \frac{t - t_k}{t_{k+1} - t_k} \int_{t_k}^{t_{k+1}} U(u)(s)ds, \ \ t \in [t_k, t_{k+1}], \ (k = 0,1,2,\dots) .$$

We notice that for every $k \in \{0,1,2,\dots\} \Rightarrow t_k - 2T = t_q$ and $t_{k+1} - 2T = t_{q+1}$, $t_k - T = t_m$ and $t_{k+1} - T = t_{m+1}$. This means that the delay generates a map which to every $[t_k, t_{k+1}]$ assigns $[t_q, t_{q+1}]$ and respectively $[t_m, t_{m+1}]$. The last ones are to the left with respect to $[t_k, t_{k+1}]$.

Then keeping in mind that $\mu_0 = \mu T_0 = \text{const.}$ and $t_0 = T$ we obtain

$$\left| \int_{t_k}^{t_{k+1}} U(u)(s)ds \right| \le$$

$$\leq \frac{2}{(Z_0+R_0)C_0}\int_{t_k}^{t_{k+1}}|E(t)|dt+\frac{1}{Z_0C_0}\int_{t_k}^{t_{k+1}}|u(t)|dt+\frac{|\delta|}{Z_0C_0}\int_{t_k}^{t_{k+1}}|u(t-2T)|dt+$$

$$+\frac{1}{C_0}\sum_{n=1}^{p}|r_n|\int_{t_k}^{t_{k+1}}|u(t)|^n dt+\frac{|\delta|}{C_0}\sum_{n=1}^{p}|r_n|\int_{t_k}^{t_{k+1}}|u(t-2T)|^n dt+|\delta|\left|\int_{t_k}^{t_{k+1}}\frac{du(t-2T)}{dt}dt\right|\leq$$

$$\leq \frac{2U_E}{(Z_0+R_0)C_0}\sup\{e^{\mu(t-t_k)}:t\in[t_k,t_{k+1}]\}\int_{t_k}^{t_{k+1}}e^{\mu(t-t_k)}dt+\frac{1}{Z_0C_0}U_0\int_{t_k}^{t_{k+1}}e^{\mu(t-t_k)}dt+$$

$$+\frac{|\delta|}{Z_0C_0}\sup\{e^{\mu(t-t_q)}:t\in[t_q,t_{q+1}]\}\int_{t_k}^{t_{k+1}}e^{\mu(t-t_k)}dt+$$

$$+\frac{1}{C_0}\sum_{n=1}^{p}|r_n|U_0^{\,n}\int_{t_k}^{t_{k+1}}e^{n\mu(t-t_k)}dt+\frac{|\delta|}{C_0}\sum_{n=1}^{p}|r_n|U_0^{\,n}\sup\{e^{n\mu(t-t_q)}:t\in[t_q,t_{q+1}]\}\int_{t_k}^{t_{k+1}}e^{\mu(t-t_k)}dt+$$

$$+|\delta|\,|u(t_{k+1}-2T)-u(t_k-2T)|\leq$$

$$\leq \frac{2U_E e^{\mu 0}}{(Z_0+R_0)C_0}\frac{e^{\mu(t_{k+1}-t_k)}-1}{\mu}+\frac{U_0}{Z_0C_0}\frac{e^{\mu(t_{k+1}-t_k)}-1}{\mu}+\frac{|\delta|}{Z_0C_0}U_0 e^{\mu 0}\frac{e^{\mu(t_{k+1}-t_k)}-1}{\mu}+$$

$$+\frac{1}{C_0}\sum_{n=1}^{p}|r_n|U_0^{\,n}\frac{e^{n\mu(t_{k+1}-t_k)}-1}{n\mu}+\frac{|\delta|}{C_0}\frac{e^{\mu(t_{k+1}-t_k)}-1}{\mu}\sum_{n=1}^{p}|r_n|U_0^{\,n}e^{n\mu 0}\leq$$

$$\leq \frac{2U_E e^{\mu 0}}{(Z_0+R_0)C_0}\frac{e^{\mu(t_{k+1}-t_k)}-1}{\mu}+\frac{U_0}{Z_0C_0}\frac{e^{\mu(t_{k+1}-t_k)}-1}{\mu}+\frac{|\delta|}{Z_0C_0}U_0 e^{\mu 0}\frac{e^{\mu(t_{k+1}-t_k)}-1}{\mu}+$$

$$+\frac{1}{C_0}\frac{e^{\mu(t_{k+1}-t_k)}-1}{\mu}\sum_{n=1}^{p}|r_n|U_0^{\,n}\frac{ne^{(n-1)\mu(t_{k+1}-t_k)}}{n}+\frac{|\delta|}{C_0}\frac{e^{\mu(t_{k+1}-t_k)}-1}{\mu}\sum_{n=1}^{p}|r_n|U_0^{\,n}e^{n\mu 0}\leq$$

$$\leq \frac{e^{\mu 0}-1}{\mu C_0}\left[\frac{2U_E e^{\mu 0}}{Z_0+R_0}+\frac{U_0}{Z_0}+\frac{|\delta|U_0 e^{\mu 0}}{Z_0}+\sum_{n=1}^{p}|r_n|U_0^{\,n}e^{(n-1)\mu 0}+|\delta|\sum_{n=1}^{p}|r_n|U_0^{\,n}e^{n\mu 0}\right]\equiv M_0(\mu)\cdot$$

Since $\displaystyle\lim_{\mu\to\infty}M_0(\mu)=0$ as in Chapter II we conclude that $\displaystyle\int_{t_k}^{t_{k+1}}U(u)(s)ds=0$. Therefore

$$u(t)=\int_{t_k}^{t}U(u)(s)ds-\frac{t-t_k}{t_{k+1}-t_k}\int_{t_k}^{t_{k+1}}U(u)(s)ds\Leftrightarrow u(t)=\int_{t_k}^{t}U(u)(s)ds\,.$$

Differentiating the last integral equation we obtain (5.3.1).
Lemma 5.3.1 is thus proved.

Remark 5.3.2. Since $u(t-2T)$ coincides with $\upsilon_0(t)$ on $[t_0,t_0+2T]$ and all zeros of $u(t-2T)$ belong to $S_T \cup S$ then the zeros of $u(t-2T)$ belong to $S_T \cup S$.

Theorem 5.3.1. Let the following conditions be fulfilled:

1) The initial function $\upsilon_0(.) \in C^1[-T,T]$ satisfies

$$\left|\upsilon_0(t)\right| \leq U_0 e^{-\frac{R}{L}t}, t \in [-T,T], \ \left|\dot{\upsilon}_0(t)\right| \leq U_0 e^{\mu(t-\tau_k)}, t \in [\tau_k,\tau_{k+1}], (k=0,1,2,...,2m),$$

$\upsilon_0(t_0)=0$;

2) $E(.) \in C_S^1[t_0,\infty), \left|E(t)\right| \leq U_0 e^{-\frac{R}{L}t} \ t \in [t_0,\infty)$;

3) $\dfrac{4L}{C_0 R}\sinh\left(\dfrac{RT_0}{L}\right)\left[\dfrac{2}{Z_0+R_0}+\left(1+|\delta|e^{-2\frac{R}{L}T}\right)\left(\dfrac{1}{Z_0}+\sum_{n=1}^{p}|r_n|U_0^{n-1}\right)\right]+|\delta|e^{-2\frac{R}{L}T} \leq 1$.

Then there exists a unique oscillatory solution of the initial value problem (5.3.1), belonging to M_U.

Proof: We show that B maps M_U into itself, that is, $u \in M_U$ implies $B(u) \in M_U$.

$B(u)(t)$ is continuous on $[-t_0,\infty)$ since

$$\lim_{t\to t_{k+1}-0} B_k(u)(t) = \int_{t_k}^{t_{k+1}} U(u)(s)ds - \frac{t_{k+1}-t_k}{t_{k+1}-t_k}\int_{t_k}^{t_{k+1}} U(u)(s)ds = 0,$$

$$\lim_{t\to t_{k+1}+0} B_{k+1}(u)(t) = \int_{t_{k+1}}^{t_{k+1}} U(u)(s)ds - \frac{t_{k+1}-t_{k+1}}{t_{k+2}-t_{k+1}}\int_{t_{k+1}}^{t_{k+2}} U(u)(s)ds = 0.$$

$B(u)(t)$ is differentiable on every interval $[t_k,t_{k+1}]$ in view of the continuity of $U(u)(t)$. We notice that in general $\lim\limits_{t\to t_{k+1}-0} \dot{B}_k(u)(t) \neq \lim\limits_{t\to t_{k+1}+0} \dot{B}_{k+1}(u)(t)$.

We have also $B(u)(t_k)=0$ and $B(u)(t_{k+1})=0$.

We have to show $\left|(Bu)(t)\right| \leq U_0 e^{-\frac{R}{L}t}, t \in [t_0,\infty)$.

Indeed, for every $u(.) \in M_U$ the function $B(u)(t)$ is continuous and differentiable on every interval (t_k, t_{k+1}). We notice that $\left| \dfrac{t - t_k}{t_{k+1} - t_k} \right| \le 1$, $t \in [t_k, t_{k+1}]$. We obtain for $t \in [t_k, t_{k+1}]$:

$$\left| B_k(u)(t) \right| \le \left| \int_{t_k}^{t} U(u)(s)\,ds \right| + \left| \int_{t_k}^{t_{k+1}} U(u)(s)\,ds \right| \equiv P_1 + P_2.$$

In view of $e^{\frac{R}{L}T_0} - 1 \le e^{\frac{R}{L}T_0} - e^{-\frac{R}{L}T_0}$ we have

$$P_1 \le \frac{2}{(Z_0 + R_0)C_0} \int_{t_k}^{t} |E(s)|\,ds + \frac{1}{Z_0 C_0} \int_{t_k}^{t} |u(s)|\,ds + \frac{\delta}{Z_0 C_0} e^{-2\frac{R}{L}T} \int_{t_k}^{t} |u(s - 2T)|\,ds +$$

$$+ \frac{1}{C_0} \sum_{n=1}^{p} |r_n| \int_{t_k}^{t} |u(s)|^n\,ds + \frac{|\delta|}{C_0} e^{-2\frac{R}{L}T} \sum_{n=1}^{p} |r_n| \int_{t_k}^{t} |u(s - 2T)|^n\,ds + e^{-2\frac{R}{L}T} |\delta| \left| \int_{t_k}^{t} \frac{du(s - 2T)}{dt}\,ds \right| \le$$

$$\le \frac{2U_0}{(Z_0 + R_0)C_0} \int_{t_k}^{t} e^{-\frac{R}{L}s}\,ds + \frac{U_0}{Z_0 C_0} \int_{t_k}^{t} e^{-\frac{R}{L}s}\,ds + e^{-2\frac{R}{L}T} \frac{|\delta| U_0}{Z_0 C_0} \int_{t_k}^{t} e^{-\frac{R}{L}s}\,ds +$$

$$+ \frac{1}{C_0} \sum_{n=1}^{p} |r_n| U_0^n \int_{t_k}^{t} e^{-n\frac{R}{L}s}\,ds + \frac{|\delta|}{C_0} e^{-2\frac{R}{L}T} \sum_{n=1}^{p} |r_n| U_0^n \int_{t_k}^{t} e^{-n\frac{R}{L}s}\,ds + |\delta| e^{-2\frac{R}{L}T} |u(t - 2T)| \le$$

$$\le \frac{2U_0}{(Z_0 + R_0)C_0} \frac{L}{R} \left(e^{-\frac{R}{L}t_k} - e^{-\frac{R}{L}t} \right) + \frac{U_0}{Z_0 C_0} \frac{L}{R} \left(e^{-\frac{R}{L}t_k} - e^{-\frac{R}{L}t} \right) + \frac{|\delta| U_0}{Z_0 C_0} e^{-2\frac{R}{L}T} \frac{L}{R} \left(e^{-\frac{R}{L}t_k} - e^{-\frac{R}{L}t} \right) +$$

$$+ \frac{1}{C_0} \sum_{n=1}^{p} |r_n| U_0^n \frac{L}{nR} \left(e^{-\frac{R}{L}t_k} - e^{-\frac{R}{L}t} \right) \left(e^{-(n-1)\frac{R}{L}t_k} + e^{-(n-2)\frac{R}{L}t_k} e^{-\frac{R}{L}t} + \ldots + e^{-(n-1)\frac{R}{L}t} \right) +$$

$$+ \frac{|\delta|}{C_0} e^{-2\frac{R}{L}T} \sum_{n=1}^{p} |r_n| U_0^n \frac{L}{nR} \left(e^{-\frac{R}{L}t_k} - e^{-\frac{R}{L}t} \right) \left(e^{-(n-1)\frac{R}{L}t_k} + e^{-(n-2)\frac{R}{L}t_k} e^{-\frac{R}{L}t} + \ldots + e^{-(n-1)\frac{R}{L}t} \right) + |\delta| e^{-2\frac{R}{L}T} U_0 e^{-\frac{R}{L}t} \le$$

$$\le \frac{L}{R} \left(e^{-\frac{R}{L}t_k} - e^{-\frac{R}{L}t} \right) \left[\frac{2U_0}{(Z_0 + R_0)C_0} + \frac{U_0}{Z_0 C_0} + \frac{|\delta| U_0}{Z_0 C_0} e^{-2\frac{R}{L}T} + \right.$$

$$\left. + \frac{1}{C_0} \sum_{n=1}^{p} |r_n| U_0^n \frac{1}{n} n + \frac{|\delta|}{C_0} e^{-2\frac{R}{L}T} \sum_{n=1}^{p} |r_n| U_0^n \frac{1}{n} n + |\delta| e^{-2\frac{R}{L}T} U_0 e^{-\frac{R}{L}t} \right] \le$$

$$\le \frac{L}{R} e^{-\frac{R}{L}t} \left(e^{\frac{R}{L}(t - t_k)} - 1 \right) \left[\frac{2U_0}{(Z_0 + R_0)C_0} + \frac{1 + e^{-2\frac{R}{L}T} |\delta|}{Z_0 C_0} U_0 + \frac{1 + |\delta| e^{-2\frac{R}{L}T}}{C_0} \sum_{n=1}^{p} |r_n| U_0^n \right] + |\delta| e^{-2\frac{R}{L}T} U_0 e^{-\frac{R}{L}t} \le$$

$$\leq U_0 e^{-\frac{R}{L}t}\left\{\frac{L}{C_0 R}\left(e^{\frac{R}{L}T_0}-1\right)\left[\frac{2}{Z_0+R_0}+\left(1+|\delta|e^{-2\frac{R}{L}T}\right)\left(\frac{1}{Z_0}+\sum_{n=1}^{p}|r_n|U_0^{n-1}\right)\right]+e^{-2\frac{R}{L}T}|\delta|\right\}\leq$$

$$\leq U_0 e^{-\frac{R}{L}t}\left\{\frac{L}{C_0 R}\left(e^{\frac{R}{L}T_0}-e^{-\frac{R}{L}T_0}\right)\left[\frac{2}{Z_0+R_0}+\left(1+|\delta|e^{-2\frac{R}{L}T}\right)\left(\frac{1}{Z_0}+\sum_{n=1}^{p}|r_n|U_0^{n-1}\right)\right]+e^{-2\frac{R}{L}T}|\delta|\right\}$$

and

$$P_2 \leq \frac{2}{(Z_0+R_0)C_0}\int_{t_k}^{t_{k+1}}|E(s)|ds + \frac{1}{Z_0 C_0}\int_{t_k}^{t_{k+1}}|u(s)|ds + \frac{|\delta|}{Z_0 C_0}e^{-2\frac{R}{L}T}\int_{t_k}^{t}|u(s-2T)|ds +$$

$$+\frac{1}{C_0}\sum_{n=1}^{p}|r_n|\int_{t_k}^{t_{k+1}}|u(s)|^n ds + \frac{|\delta|}{C_0}e^{-2\frac{R}{L}T}\sum_{n=1}^{p}|r_n|\int_{t_k}^{t_{k+1}}|u(s-2T)|^n ds + |\delta|e^{-2\frac{R}{L}T}\left|\int_{t_k}^{t_{k+1}}\frac{du(s-2T)}{dt}ds\right|\leq$$

$$\leq \frac{2U_0}{(Z_0+R_0)C_0}\int_{t_k}^{t_{k+1}}e^{-\frac{R}{L}s}ds + \frac{U_0}{Z_0 C_0}\int_{t_k}^{t_{k+1}}e^{-\frac{R}{L}s}ds + \frac{|\delta|U_0}{Z_0 C_0}e^{-2\frac{R}{L}T}\int_{t_k}^{t_{k+1}}e^{-\frac{R}{L}s}ds +$$

$$+\frac{1}{C_0}\sum_{n=1}^{p}|r_n|U_0^n\int_{t_k}^{t_{k+1}}e^{-n\frac{R}{L}s}ds + \frac{|\delta|}{C_0}e^{-2\frac{R}{L}T}\sum_{n=1}^{p}|r_n|U_0^n\int_{t_k}^{t_{k+1}}e^{-n\frac{R}{L}s}ds \leq$$

$$\leq \frac{2U_0}{(Z_0+R_0)C_0}\frac{L}{R}\left(e^{-\frac{R}{L}t_k}-e^{-\frac{R}{L}t_{k+1}}\right)+\frac{U_0}{Z_0 C_0}\frac{L}{R}\left(e^{-\frac{R}{L}t_k}-e^{-\frac{R}{L}t_{k+1}}\right)+\frac{|\delta|U_0}{Z_0 C_0}e^{-2\frac{R}{L}T}\frac{L}{R}\left(e^{-\frac{R}{L}t_k}-e^{-\frac{R}{L}t_{k+1}}\right)+$$

$$+\frac{1}{C_0}\sum_{n=1}^{p}|r_n|U_0^n\frac{L}{nR}\left(e^{-\frac{R}{L}t_k}-e^{-\frac{R}{L}t_{k+1}}\right)\left(e^{-(n-1)\frac{R}{L}t_k}+e^{-(n-2)\frac{R}{L}t_k}e^{-\frac{R}{L}t_{k+1}}+...+e^{-(n-1)\frac{R}{L}t_{k+1}}\right)+$$

$$+\frac{|\delta|}{C_0}e^{-2\frac{R}{L}T}\sum_{n=1}^{p}|r_n|U_0^n\frac{L}{nR}\left(e^{-\frac{R}{L}t_k}-e^{-\frac{R}{L}t_{k+1}}\right)\left(e^{-(n-1)\frac{R}{L}t_k}+e^{-(n-2)\frac{R}{L}t_k}e^{-\frac{R}{L}t_{k+1}}+...+e^{-(n-1)\frac{R}{L}t_{k+1}}\right)\leq$$

$$\leq \frac{L}{R}\left(e^{-\frac{R}{L}t_k}-e^{-\frac{R}{L}t_{k+1}}\right)\left[\frac{2U_0}{(Z_0+R_0)C_0}+\frac{U_0}{Z_0 C_0}+\frac{|\delta|U_0}{Z_0 C_0}e^{-2\frac{R}{L}T}+\right.$$

$$+\frac{1}{C_0}\sum_{n=1}^{p}|r_n|U_0^n\frac{1}{n}n+\frac{|\delta|}{C_0}e^{-2\frac{R}{L}T}\sum_{n=1}^{p}|r_n|U_0^n\frac{1}{n}n\right]\leq$$

$$\leq \frac{L}{R}\left(e^{-\frac{R}{L}t_k}-e^{-\frac{R}{L}t}+e^{-\frac{R}{L}t}-e^{-\frac{R}{L}t_{k+1}}\right)\left[\frac{2U_0}{(Z_0+R_0)C_0}+\frac{U_0}{Z_0 C_0}+\frac{|\delta|U_0}{Z_0 C_0}e^{-2\frac{R}{L}T}+\right.$$

$$+\frac{1}{C_0}\sum_{n=1}^{p}|r_n|U_0^n+\frac{|\delta|}{C_0}e^{-2\frac{R}{L}T}\sum_{n=1}^{p}|r_n|U_0^n\right]\leq$$

$$\leq \frac{L}{R}\left(e^{-\frac{R}{L}t_k}-e^{-\frac{R}{L}t}\right)\left[\frac{2U_0}{(Z_0+R_0)C_0}+\frac{U_0}{Z_0C_0}+\frac{|\delta|e^{-2\frac{R}{L}T}U_0}{Z_0C_0}+\frac{1}{C_0}\sum_{n=1}^{P}|r_n||U_0^n|+\frac{|\delta|e^{-2\frac{R}{L}T}}{C_0}\sum_{n=1}^{P}|r_n||U_0^n|\right]+$$

$$+\frac{L}{R}\left(e^{-\frac{R}{L}t}-e^{-\frac{R}{L}t_{k+1}}\right)\left[\frac{2U_0}{(Z_0+R_0)C_0}+\frac{U_0}{Z_0C_0}+\frac{|\delta|e^{-2\frac{R}{L}T}U_0}{Z_0C_0}+\frac{1}{C_0}\sum_{n=1}^{P}|r_n||U_0^n|+\frac{|\delta|e^{-2\frac{R}{L}T}}{C_0}\sum_{n=1}^{P}|r_n||U_0^n|\right]\leq$$

$$\leq e^{-\frac{R}{L}t}\frac{L}{R}\left(e^{\frac{R}{L}(t-t_k)}-1\right)\left[\frac{2U_0}{(Z_0+R_0)C_0}+\frac{U_0}{Z_0C_0}+\frac{|\delta|U_0}{Z_0C_0}e^{-2\frac{R}{L}T}+\frac{1}{C_0}\sum_{n=1}^{P}|r_n||U_0^n|+\frac{|\delta|}{C_0}e^{-2\frac{R}{L}T}\sum_{n=1}^{P}|r_n||U_0^n|\right]+$$

$$+e^{-\frac{R}{L}t}\frac{L}{R}\left(1-e^{-\frac{R}{L}(t_{k+1}-t)}\right)\left[\frac{2U_0}{(Z_0+R_0)C_0}+\frac{U_0}{Z_0C_0}+\frac{|\delta|U_0}{Z_0C_0}e^{-2\frac{R}{L}T}+\frac{1}{C_0}\sum_{n=1}^{P}|r_n||U_0^n|+\frac{|\delta|}{C_0}e^{-2\frac{R}{L}T}\sum_{n=1}^{P}|r_n||U_0^n|\right]\leq$$

$$\leq e^{-\frac{R}{L}t}\frac{L}{R}\left(e^{\frac{R}{L}T_0}-1\right)\left[\frac{2U_0}{(Z_0+R_0)C_0}+\frac{U_0}{Z_0C_0}+\frac{|\delta|U_0}{Z_0C_0}e^{-2\frac{R}{L}T}+\frac{1}{C_0}\sum_{n=1}^{P}|r_n||U_0^n|+\frac{|\delta|}{C_0}e^{-2\frac{R}{L}T}\sum_{n=1}^{P}|r_n||U_0^n|\right]+$$

$$+e^{-\frac{R}{L}t}\frac{L}{R}\left(1-e^{-\frac{R}{L}T_0}\right)\left[\frac{2U_0}{(Z_0+R_0)C_0}+\frac{U_0}{Z_0C_0}+\frac{|\delta|U_0e^{-2\frac{R}{L}T}}{Z_0C_0}+\frac{1}{C_0}\sum_{n=1}^{P}|r_n||U_0^n|+\frac{|\delta|e^{-2\frac{R}{L}T}}{C_0}\sum_{n=1}^{P}|r_n||U_0^n|\right]\leq$$

$$\leq U_0e^{-\frac{R}{L}t}\frac{L}{C_0R}\left(e^{\frac{R}{L}T_0}-e^{-\frac{R}{L}T_0}\right)\left[\frac{2}{Z_0+R_0}+\left(1+|\delta|e^{-2\frac{R}{L}T}\right)\left(\frac{1}{Z_0}+\sum_{n=1}^{P}|r_n||U_0^{n-1}|\right)\right].$$

Then we obtain

$$|B_k(u)(t)|\leq U_0e^{-\frac{R}{L}t}\left\{\frac{L}{C_0R}\left(e^{\frac{R}{L}T_0}-e^{-\frac{R}{L}T_0}\right)\left[\frac{2}{Z_0+R_0}+\left(1+|\delta|e^{-2\frac{R}{L}T}\right)\left(\frac{1}{Z_0}+\sum_{n=1}^{P}|r_n||U_0^{n-1}|\right)\right]+|\delta|e^{-2\frac{R}{L}T}\right\}+$$

$$+U_0e^{-\frac{R}{L}t}\frac{L}{C_0R}\left(e^{\frac{R}{L}T_0}-e^{-\frac{R}{L}T_0}\right)\left[\frac{2}{Z_0+R_0}+\left(1+|\delta|e^{-2\frac{R}{L}T}\right)\left(\frac{1}{Z_0}+\sum_{n=1}^{P}|r_n||U_0^{n-1}|\right)\right]\leq$$

$$\leq U_0e^{-\frac{R}{L}t}\left\{\frac{4L}{C_0R}\sinh\left(\frac{RT_0}{L}\right)\left[\frac{2}{Z_0+R_0}+\left(1+|\delta|e^{-2\frac{R}{L}T}\right)\left(\frac{1}{Z_0}+\sum_{n=1}^{P}|r_n||U_0^{n-1}|\right)\right]+|\delta|e^{-2\frac{R}{L}T}\right\}\leq U_0e^{-\frac{R}{L}t}.$$

Consequently the operator B maps M_U into itself.

In what follows we show that B is a contractive operator.

Indeed,

$$\left|B_k(u)(t) - B_k(\overline{u})(t)\right| \le \left|\int_{t_k}^{t}\left(U(u)(s) - U(\overline{u})(s)\right)ds\right| +$$

$$\left|\frac{t - t_k}{t_{k+1} - t_k}\right|\left\|\int_{t_k}^{t_{k+1}}\left(U(u)(s) - U(\overline{u})(s)\right)ds\right\| \equiv$$

$$\equiv P_1 + P_2, \ t \in [t_k, t_{k+1}].$$

Let us note that if $t_k - 2T = t_q$, $t_{k+1} - 2T = t_{q+1}$ $(q < k; q + n = k)$ then $t \in [t_k, t_{k+1}] \Rightarrow t - 2T \in [t_q, t_{q+1}]$.

Then

$$P_1 \le \frac{1}{Z_0 C_0}\int_{t_k}^{t}\left|u(s) - \overline{u}(s)\right|ds + \frac{|\delta|}{Z_0 C_0}e^{-2\frac{R}{L}T}\int_{t_k}^{t}\left|u(s - 2T) - \overline{u}(s - 2T)\right|ds +$$

$$+\frac{1}{C_0}\sum_{n=1}^{p}|r_n|\left|\int_{t_k}^{t}\left|(u(s))^n - (\overline{u}(s))^n\right|ds +$$

$$+\frac{|\delta|}{C_0}e^{-2\frac{R}{L}T}\sum_{n=1}^{p}|r_n|\int_{t_k}^{t}\left|(u(s-2T))^n - (\overline{u}(s-2T))^n\right|ds + |\delta|e^{-2\frac{R}{L}T}\left|\int_{t_k}^{t}\left(\frac{du(s-2T)}{dt} - \frac{d\overline{u}(s-2T)}{dt}\right)ds\right| \le$$

$$\le \frac{1}{Z_0 C_0}\rho^{(k)}(u,\overline{u})\int_{t_k}^{t}e^{\mu(s-t_k)}ds + \frac{|\delta|}{Z_0 C_0}e^{-2\frac{R}{L}T}\rho^{(q)}(u,\overline{u})\sup\left\{u(\theta)\left|e^{\mu(\theta - t_q)} : \theta \in [t_q, t_{q+1}]\right\}\int_{t_k}^{t}e^{\mu(s-t_k)}ds +\right.$$

$$+\frac{1}{C_0}\sum_{n=1}^{p}n|r_n|\sup\left\{|u(s)|^{n-1}e^{(n-1)\mu(s-t_k)} : s \in [t_k, t_{k+1}]\right\}\rho^{(k)}(u,\overline{u})\int_{t_k}^{t}e^{\mu(s-t_k)}ds +$$

$$+\frac{|\delta|}{C_0}e^{-2\frac{R}{L}T}\sum_{n=1}^{p}n|r_n|\sup\left\{|u(\theta)|^{n-1}e^{(n-1)\mu(\theta - t_k)} : \theta \in [t_q, t_{q+1}]\right\}\rho^{(q)}(u,\overline{u})\int_{t_k}^{t}e^{\mu(s-t_k)}ds +$$

$$+|\delta|e^{-2\frac{R}{L}T}\rho_{\mu}^{(q)}(\dot{u},\dot{\overline{u}})\int_{t_k}^{t}e^{\mu(s-t_k)}ds \le$$

$$\le \frac{e^{\mu_0}\rho_{\mu}^{(k)}(u,\overline{u})}{Z_0 C_0}\frac{e^{\mu(t-t_k)} - 1}{\mu} + \frac{|\delta|}{Z_0 C_0}e^{-2\frac{R}{L}T}e^{\mu_0}\rho_{\mu}^{(q)}(u,\overline{u})\frac{e^{\mu(t-t_k)} - 1}{\mu} +$$

$$+\frac{e^{\mu_0}\rho_{\mu}^{(k)}(u,\overline{u})}{C_0}\frac{e^{\mu(t-t_k)} - 1}{\mu}\sum_{n=1}^{p}|r_n|nU_0^{n-1}e^{(n-1)\mu_0} +$$

$$+\frac{|\delta|e^{\mu_0}\rho_{\mu}^{(q)}(u,\overline{u})}{C_0}e^{-2\frac{R}{L}T}\frac{e^{\mu(t-t_k)} - 1}{\mu}\sum_{n=1}^{p}|r_n|nU_0^{n-1}e^{(n-1)\mu_0} + |\delta|e^{-2\frac{R}{L}T}\rho_{\mu}^{(q)}(\dot{u},\dot{\overline{u}})\frac{e^{\mu(t-t_k)} - 1}{\mu} \le$$

$$\leq \frac{e^{\mu(t-t_k)}-1}{\mu}\left[\frac{e^{\mu_0}\rho_\mu^{(k)}(u,\overline{u})}{Z_0 C_0}+\frac{|\delta|}{Z_0 C_0}e^{-2\frac{R}{L}T}e^{\mu_0}\rho_\mu^{(q)}(u,\overline{u})+\frac{e^{\mu_0}\rho_\mu^{(k)}(u,\overline{u})}{C_0}\sum_{n=1}^{p}|r_n|nU_0^{n-1}e^{(n-1)\mu_0}+\right.$$

$$+\left.\frac{|\delta|e^{\mu_0}\rho_\mu^{(q)}(u,\overline{u})}{C_0}e^{-2\frac{R}{L}T}\sum_{n=1}^{p}|r_n|nU_0^{n-1}e^{(n-1)\mu_0}+|\delta|e^{-2\frac{R}{L}T}\rho_\mu^{(q)}(\dot{u},\dot{\overline{u}})\right]\leq$$

$$\leq \frac{e^{\mu(t-t_k)}-1}{\mu}\left[\frac{e^{\mu_0}\rho_\mu^{(k)}(\dot{u},\dot{\overline{u}})}{\mu Z_0 C_0}+\frac{|\delta|}{\mu Z_0 C_0}e^{-2\frac{R}{L}T}e^{\mu_0}\rho_\mu^{(q)}(\dot{u},\dot{\overline{u}})+\frac{e^{\mu_0}\rho_\mu^{(k)}(\dot{u},\dot{\overline{u}})}{\mu C_0}\sum_{n=1}^{p}|r_n|nU_0^{n-1}e^{(n-1)\mu_0}+\right.$$

$$+\left.\frac{|\delta|e^{\mu_0}\rho_\mu^{(q)}(\dot{u},\dot{\overline{u}})}{\mu C_0}e^{-2\frac{R}{L}T}\sum_{n=1}^{p}|r_n|nU_0^{n-1}e^{(n-1)\mu_0}+|\delta|e^{-2\frac{R}{L}T}\rho_\mu^{(q)}(\dot{u},\dot{\overline{u}})\right]\leq$$

$$\leq e^{\mu(t-t_k)}\hat{\rho}_\mu^{(k)}(\dot{u},\dot{\overline{u}})\left[\frac{e^{\mu_0}}{\mu^2 C_0}\left(1+|\delta|e^{-2\frac{R}{L}T}\right)\left(\frac{1}{Z_0}+\sum_{n=1}^{p}|r_n|nU_0^{n-1}e^{(n-1)\mu_0}\right)+\frac{|\delta|}{\mu}e^{-2\frac{R}{L}T}\right].$$

Note that we have used $\rho_\mu^{(q)}(\dot{u},\dot{\overline{u}})\leq \hat{\rho}_\mu^{(k)}(\dot{u},\dot{\overline{u}})$ because $q<k$.

Further on we have

$$P_2 \leq \frac{1}{Z_0 C_0}\int_{t_k}^{t_{k+1}}|u(s)-\overline{u}(s)|ds+\frac{|\delta|}{Z_0 C_0}e^{-2\frac{R}{L}T}\int_{t_k}^{t_{k+1}}|u(s-2T)-\overline{u}(s-2T)|ds+$$

$$+\frac{1}{C_0}\sum_{n=1}^{p}|r_n|\int_{t_k}^{t_{k+1}}|(u(s))^n-(\overline{u}(s))^n|ds+$$

$$+\frac{|\delta|}{C_0}e^{-2\frac{R}{L}T}\sum_{n=1}^{p}|r_n|\int_{t_k}^{t_{k+1}}|(u(s-2T))^n-(\overline{u}(s-2T))^n|ds+|\delta|e^{-2\frac{R}{L}T}\left|\int_{t_k}^{t_{k+1}}\left(\frac{du(s-2T)}{dt}-\frac{d\overline{u}(s-2T)}{dt}\right)ds\right|\leq$$

$$\leq \frac{e^{\mu_0}-1}{\mu}\left[\frac{e^{\mu_0}\rho_\mu^{(k)}(\dot{u},\dot{\overline{u}})}{\mu Z_0 C_0}+\frac{|\delta|}{\mu Z_0 C_0}e^{-2\frac{R}{L}T}e^{\mu_0}\rho_\mu^{(q)}(\dot{u},\dot{\overline{u}})+\frac{e^{\mu_0}\rho_\mu^{(k)}(\dot{u},\dot{\overline{u}})}{\mu C_0}\sum_{n=1}^{p}|r_n|nU_0^{n-1}e^{(n-1)\mu_0}+\right.$$

$$+\left.\frac{|\delta|e^{\mu_0}\rho_\mu^{(q)}(\dot{u},\dot{\overline{u}})}{\mu C_0}e^{-2\frac{R}{L}T}\sum_{n=1}^{p}|r_n|nU_0^{n-1}e^{(n-1)\mu_0}\right]\leq$$

$$\leq e^{\mu(t-t_k)}\hat{\rho}_\mu^{(k)}(\dot{u},\dot{\overline{u}})\left(e^{\mu_0}-1\right)\frac{e^{\mu_0}}{\mu^2 C_0}\left(1+|\delta|e^{-2\frac{R}{L}T}\right)\left(\frac{1}{Z_0}+\sum_{n=1}^{p}|r_n|nU_0^{n-1}e^{(n-1)\mu_0}\right)$$

Consequently

$$|B_k(u)(t)-B_k(\overline{u})(t)|\leq$$

$$\leq e^{\mu(t-t_k)}\hat{\rho}_\mu^{(k)}(\dot{u},\dot{\bar{u}})\left[\frac{e^{\mu_0}}{\mu^2 C_0}\left(1+|\delta|e^{-2\frac{R}{L}T}\right)\left(\frac{1}{Z_0}+\sum_{n=1}^{p}|r_n|nU_0^{n-1}e^{(n-1)\mu_0}\right)+\frac{|\delta|}{\mu}e^{-2\frac{R}{L}T}\right]+$$

$$+e^{\mu(t-t_k)}\hat{\rho}_\mu^{(k)}(\dot{u},\dot{\bar{u}})\left(e^{\mu_0}-1\right)\frac{e^{\mu_0}}{\mu^2 C_0}\left(1+|\delta|e^{-2\frac{R}{L}T}\right)\left(\frac{1}{Z_0}+\sum_{n=1}^{p}|r_n|nU_0^{n-1}e^{(n-1)\mu_0}\right)\leq$$

$$\leq e^{\mu(t-t_k)}\hat{\rho}_\mu^{(k)}\left((u,\dot{u}),(\bar{u},\dot{\bar{u}})\right)\frac{e^{2\mu_0}}{\mu^2 C_0}\left(1+|\delta|e^{-\frac{2RT}{L}}\right)\left(\frac{1}{Z_0}+\sum_{n=1}^{p}|r_n|nU_0^{n-1}e^{(n-1)\mu_0}\right)\equiv$$

$$\equiv e^{\mu(t-t_k)}K_u\hat{\rho}_\mu^{(k)}\left((u,\dot{u}),(\bar{u},\dot{\bar{u}})\right).$$

It follows

$$\hat{\rho}^{(k)}(B(u),B(\bar{u}))\leq e^{\mu_0}K_u\hat{\rho}_\mu^{(k)}((u,\dot{u}),(\bar{u},\dot{\bar{u}})).$$

For the derivative we obtain

$$\left|\dot{B}_k(u)(t)-\dot{B}_k(\bar{u})(t)\right|\leq$$

$$\left|U(u)(t)-U(\bar{u})(t)\right|+\frac{1}{t_{k+1}-t_k}\left|\int_{t_k}^{t_{k+1}}U(u)(s)ds-\int_{t_k}^{t_{k+1}}U(\bar{u})(s)ds\right|\equiv \dot{P}_1+\dot{P}_2.$$

But

$$\dot{P}_1\leq\frac{1}{C_0 Z_0}|u(t)-\bar{u}(t)|+\frac{|\delta|}{C_0 Z_0}e^{-2\frac{R}{L}T}|u(t-2T)-\bar{u}(t-2T)|+\frac{1}{C_0}\sum_{n=1}^{p}|r_n|\left|u^n(t)-\bar{u}^n(t)\right|+$$

$$+\frac{|\delta|}{C_0}e^{-2\frac{R}{L}T}\sum_{n=1}^{p}|r_n|\left|u^n(t-2T)-\bar{u}^n(t-2T)\right|+|\delta|e^{-2\frac{R}{L}T}\left|\dot{u}(t-2T)-\dot{\bar{u}}(t-2T)\right|\leq$$

$$\leq e^{\mu(t-t_k)}\frac{\rho_\mu^{(k)}(u,\bar{u})}{C_0 Z_0}+e^{\mu(t-t_q)}\frac{|\delta|}{C_0 Z_0}e^{-2\frac{R}{L}T}\rho_\mu^{(q)}(u,\bar{u})+$$

$$+e^{\mu(t-t_k)}\frac{1}{C_0}\sum_{n=1}^{p}|r_n|n\,\mathrm{ess\,sup}\left\{|u^{n-1}(t)|e^{(n-1)\mu(t-t_k)}:t\in[t_k,t_{k+1}]\right\}\rho_\mu^{(k)}(u,\bar{u})+$$

$$+e^{\mu(t-t_q)}\frac{|\delta|}{C_0}e^{-2\frac{R}{L}T}\sum_{n=1}^{p}|r_n|n\,\mathrm{ess\,sup}\left\{|u^{n-1}(s)|e^{(n-1)\mu(s-t_q)}:s\in[t_q,t_{q+1}]\right\}\rho_\mu^{(q)}(u,\bar{u})+$$

$$+e^{\mu(t-t_q)}|\delta|e^{-2\frac{R}{L}T}\rho_\mu^{(q)}(\dot{u},\dot{\bar{u}})\leq$$

$$\leq e^{\mu(t-t_k)}\frac{\rho_\mu^{(k)}(\dot u,\dot{\bar u})}{\mu C_0 Z_0}+e^{\mu_0}\frac{|\delta|\rho_\mu^{(q)}(\dot u,\dot{\bar u})}{\mu C_0 Z_0}e^{-2\frac{R}{L}T}+e^{\mu(t-t_k)}\frac{\rho_\mu^{(k)}(\dot u,\dot{\bar u})}{\mu C_0}\sum_{n=1}^{p}|r_n|n\,U_0^{n-1}e^{(n-1)\mu_0}+$$

$$+e^{\mu_0}\frac{|\delta|\rho_\mu^{(q)}(\dot u,\dot{\bar u})}{\mu C_0}e^{-2\frac{R}{L}T}\sum_{n=1}^{p}|r_n|n\,U_0^{n-1}e^{(n-1)\mu_0}+e^{\mu_0}|\delta|e^{-2\frac{R}{L}T}\rho_\mu^{(q)}(\dot u,\dot{\bar u})\leq$$

$$\leq e^{\mu(t-t_k)}\hat\rho_\mu^{(k)}(\dot u,\dot{\bar u})\left[\frac{1}{\mu C_0}\left(\frac{1}{Z_0}+\sum_{n=1}^{p}|r_n|n\,U_0^{n-1}e^{(n-1)\mu_0}+e^{\mu_0}\frac{|\delta|e^{-2\frac{R}{L}T}}{Z_0}+e^{\mu_0}|\delta|e^{-2\frac{R}{L}T}\sum_{n=1}^{p}|r_n|n\,U_0^{n-1}e^{(n-1)\mu_0}\right)+\right.$$

$$\left.+e^{\mu_0}|\delta|e^{-2\frac{R}{L}T}\right]\leq$$

$$\leq e^{\mu(t-t_k)}\hat\rho_\mu^{(k)}(\dot u,\dot{\bar u})\left[\frac{e^{\mu_0}}{\mu C_0}\left(1+|\delta|e^{\mu_0}e^{-2\frac{R}{L}T}\right)\left(\frac{1}{Z_0}+\sum_{n=1}^{p}|r_n|n\,U_0^{n-1}e^{(n-1)\mu_0}\right)+|\delta|e^{-2\frac{R}{L}T}e^{\mu_0}\right]$$

and

$$\dot B_2=\frac{1}{t_{k+1}-t_k}\left|\int_{t_k}^{t_{k+1}}U(u)(s)ds-\int_{t_k}^{t_{k+1}}U(\bar u)(s)ds\right|\leq$$

$$\leq\frac{1}{t_{k+1}-t_k}\frac{1}{Z_0 C_0}\int_{t_k}^{t_{k+1}}|u(s)-\bar u(s)|ds+\frac{1}{t_{k+1}-t_k}\frac{|\delta|}{Z_0 C_0}e^{-2\frac{R}{L}T}\int_{t_k}^{t_{k+1}}|u(s-2T)-\bar u(s-2T)|ds+$$

$$+\frac{1}{t_{k+1}-t_k}\frac{1}{C_0}\sum_{n=1}^{p}|r_n|\int_{t_k}^{t_{k+1}}\left|(u(s))^n-(\bar u(s))^n\right|ds+$$

$$+\frac{1}{t_{k+1}-t_k}\frac{|\delta|}{C_0}e^{-2\frac{R}{L}T}\sum_{n=1}^{p}|r_n|\int_{t_k}^{t_{k+1}}\left|(u(s-2T))^n-(\bar u(s-2T))^n\right|ds+\frac{e^{-2\frac{R}{L}T}}{t_{k+1}-t_k}|\delta|\left|\int_{t_k}^{t_{k+1}}(\dot u(s-2T)-\dot{\bar u}(s-2T))ds\right|\leq$$

$$\leq\hat\rho_\mu^{(k)}(\dot u,\dot{\bar u})\frac{e^{\mu(t_{k+1}-t_k)}-1}{\mu(t_{k+1}-t_k)}\frac{e^{\mu_0}}{\mu C_0}\left(1+|\delta|e^{\mu_0}e^{-2\frac{R}{L}T}\right)\left(\frac{1}{Z_0}+\sum_{n=1}^{p}|r_n|n\,U_0^{n-1}e^{(n-1)\mu_0}\right)\leq$$

$$\leq\hat\rho_\mu^{(k)}(\dot u,\dot{\bar u})\frac{e^{\mu_0}-1}{\mu_0}\frac{e^{\mu_0}}{\mu C_0}\left(1+|\delta|e^{\mu_0}e^{-2\frac{R}{L}T}\right)\left(\frac{1}{Z_0}+\sum_{n=1}^{p}|r_n|n\,U_0^{n-1}e^{(n-1)\mu_0}\right).$$

In the last inequality we have applied Remark 5.3.3, that is, $\xi(h)=\dfrac{e^h-1}{h},h\geq 0$ is increasing one.

Then

$$\left| \dot{B}_k(u)(t) - \dot{B}_k(\overline{u})(t) \right| \le$$

$$\le e^{\mu(t-t_k)} \hat{\rho}_\mu^{(k)}(\dot{u},\dot{\overline{u}}) \left[\frac{e^{\mu_0}}{\mu C_0} \left(1 + |\delta| e^{\mu_0} e^{-2\frac{R}{L}T} \right) \left(\frac{1}{Z_0} + \sum_{n=1}^{p} |r_n| n \, U_0^{n-1} e^{(n-1)\mu_0} \right) + |\delta| e^{-2\frac{R}{L}T} e^{\mu_0} \right] +$$

$$+ e^{\mu(t-t_k)} \hat{\rho}_\mu^{(k)}(\dot{u},\dot{\overline{u}}) \frac{e^{\mu_0}-1}{\mu_0} \frac{e^{\mu_0}}{\mu C_0} \left(1 + |\delta| e^{\mu_0} e^{-2\frac{R}{L}T} \right) \left(\frac{1}{Z_0} + \sum_{n=1}^{p} |r_n| n \, U_0^{n-1} e^{(n-1)\mu_0} \right) \le$$

$$\le e^{\mu(t-t_k)} \hat{\rho}_\mu^{(k)}(\dot{u},\dot{\overline{u}}) \left(1 + \frac{e^{\mu_0}-1}{\mu_0} \right) e^{\mu_0} \left[\frac{1}{\mu C_0} \left(1 + |\delta| e^{\mu_0} e^{-2\frac{R}{L}T} \right) \left(\frac{1}{Z_0} + \sum_{n=1}^{p} |r_n| n \, U_0^{n-1} e^{(n-1)\mu_0} \right) + |\delta| e^{-2\frac{R}{L}T} \right] \equiv$$

$$\equiv e^{\mu(t-t_k)} \dot{K}_u \, \hat{\rho}_\mu^{(k)}(\dot{u},\dot{\overline{u}}).$$

It follows

$$\hat{\rho}_\mu^{(k)}(\dot{B}_k(u),\dot{B}_k(\overline{u})) \le \dot{K}_u \hat{\rho}_\mu^{(k)}((u,\dot{u}),(\overline{u},\dot{\overline{u}})).$$

Then

$$\hat{\rho}_\mu^{(k)}(B_k(u),\dot{B}_k(u),B_k(\overline{u}),\dot{B}_k(\overline{u})) \le \max\left\{ e^{\mu_0} K_u, \dot{K}_u \right\} \hat{\rho}_\mu^{(k)}((u,\dot{u}),(\overline{u},\dot{\overline{u}})).$$

We have to verify that M_U is j-bounded. Indeed, since j is an identity mapping then

$$\hat{\rho}_\mu^{j^n (k)}((u,\dot{u}),(\overline{u},\dot{\overline{u}})) \le \hat{\rho}_\mu^{(k)}((u,\dot{u}),(\overline{u},\dot{\overline{u}})) < \infty \quad (n = 0,1,2,...).$$

Therefore in view of the fixed-point theorem for contractive mappings in uniform spaces (cf. Chapter I) the operator B has a unique fixed point, and it is an oscillating solution of (5.3.1).

Theorem 5.3.1 is thus proved.

5.3.1. Numerical Example

Let us collect all inequalities guaranteeing the existence-uniqueness of an oscillatory solution:

$$\frac{4L}{C_0 R} \sinh\left(\frac{RT_0}{L} \right) \left[\frac{2}{Z_0 + R_0} + \left(1 + |\delta| e^{-2\frac{R}{L}T} \right) \left(\frac{1}{Z_0} + \sum_{n=1}^{p} |r_n| U_0^{n-1} \right) \right] + |\delta| e^{-2\frac{R}{L}T} \le 1 ;$$

$$e^{\mu_0}K_u = e^{\mu_0}\frac{e^{2\mu_0}}{\mu^2 C_0}\left(1+|\delta|e^{-2\frac{R}{L}T}\right)\left(\frac{1}{Z_0}+\sum_{n=1}^{P}|r_n|nU_0^{n-1}e^{(n-1)\mu_0}\right)<1\;;$$

$$\dot{K}_u = \left(1+\frac{e^{\mu_0}-1}{\mu_0}\right)e^{\mu_0}\left[\frac{1}{\mu C_0}\left(1+|\delta|e^{\mu_0}e^{-2\frac{R}{L}T}\right)\left(\frac{1}{Z_0}+\sum_{n=1}^{P}|r_n|n\,U_0^{n-1}e^{(n-1)\mu_0}\right)+|\delta|e^{-2\frac{R}{L}T}\right]<1\;.$$

Consider a transmission line with length $\Lambda = 1000m,$ and cross-section area $S = 2mm^2$. The specific resistance for the copper is $\rho_c = 0{,}0175$. Then the resistance per-unit length is $R = \dfrac{\rho_c \Lambda}{S} = \dfrac{0{,}0175.1000}{2} \approx 8{,}75\Omega$.

If $L=0{,}2\,\mu H/m,\; C=80pF/m,$ then

$$Z_0 = \sqrt{L/C} = 50\ \Omega, \quad \frac{R}{L} = \frac{8{,}75}{0{,}2.10^{-6}} \approx 4{,}375.10^7\;;\frac{L}{R} = \frac{0{,}2.10^{-6}}{8{,}75} \approx 2{,}3.10^{-8}\;;$$

$$v = 1/\sqrt{LC} = 1/(4.10^{-9}) = 2{,}5.10^8,$$

$$R_0 = 30\Omega,\;\; \delta = \frac{50-30}{50+30} = 0{,}25;\; C_0 = 10.10^{-12}F\;;\quad E = 1{,}5V;$$

$$T = \Lambda/v = 16.10^{-9} = 1{,}6.10^{-8}\;\text{sec}.,$$

$$\frac{RT}{L} = \frac{R\Lambda}{Z_0} = \frac{8{,}75.1000}{50} = 175 \Rightarrow e^{-2\frac{RT}{L}} = e^{-300} \approx 0.$$

Let be the resistive element have the following *V-I* characteristic

$$f(u) = 0{,}028u - 0{,}125u^2 + 0{,}14u^3\,,\text{i.e. } r_1 = 0{,}028, r_2 = 0{,}125,\; r_3 = 0{,}14.$$

Let us check the propagation of centimeter waves $\lambda_0 = (1/4)10^{-2}m$. Then

$$f_0 = 1/(\lambda_0\sqrt{LC}) = 10^{11}\,Hz \Rightarrow T_0 = 1/f_0 = 10^{-11}\;\text{sec}., l_0 = 0{,}8.10^{-11}\;\text{sec}.$$

If we choose $\mu = 10^{11}$, then $\mu T_0 = \mu_0 = 1$ and $T = 16.10^{-9}.10^{11}T_0 = 1600.T_0$.

We have also $\mu T = 10^{11}.16.10^{-9} = 16.10^2 = 1600,$

$$\mu C_0 = 10^{11}.10^{-11} = 1;\; \mu^2 C_0 = 10^{22}.10^{-11} = 10^{11},$$

$$\frac{e^{\frac{R}{L}T_0} - e^{-\frac{R}{L}T_0}}{2} = \frac{RT_0/L}{1!} + \frac{(RT_0/L)^3}{3!} + ... \approx \frac{RT_0}{L} = \frac{8{,}75.10^{-11}}{0{,}2.10^{-6}} \approx 4{,}38.10^{-4}$$

Then the above inequalities for $U_0 = 0,01$ become

$$\frac{4}{10^{-11}}\frac{0,2.10^{-6}}{8,75}\frac{8,75.10^{-4}}{2}\left[\frac{2}{80}+\left(\frac{1}{50}+0,028+0,125U_0+0,14U_0^2\right)\right] \approx 0,3 \leq 1;$$

$$e^{\mu_0}K_u = \frac{e^3}{10^{22}.10^{-11}}\left(\frac{1}{Z_0}+\sum_{n=1}^{3}|r_n|nU_0^{n-1}e^{n-1}\right) < 1;$$

$$\dot{K}_u = \frac{e^2}{\mu C_0}\left(\frac{1}{Z_0}+\sum_{n=1}^{3}|r_n|n\ U_0^{n-1}e^{n-1}\right) = e^2\left(0,02+0,028+2.0,125U_0e+3.0,14U_0^2e^2\right) =$$

$$= e^2\left(0,048+0,0068+0,0003\right) \approx 0,4 < 1$$

.

Then $K = \max\left\{eK_u, \dot{K}_u\right\} = \dot{K}_u = 0,4$.

The successive approximations may be obtained as in Chapter II.

5.4. A Different Manner to Reducing the Mixed Problem to an Oscillatory Problem for a Neutral Equation

Here we consider the equation already obtained in § 5.1.
We proceed from the boundary condition assuming

$$E(t-T) = E(t)\left(\Rightarrow E(t_k) = 0,\ E(t_k-T) = 0\right):$$

$$E(t)-u(0,t)-R_0i(0,t) = 0,\ t \geq T,$$

$$C_0\frac{du(\Lambda,t)}{dt} = i(\Lambda,t)-f(\Lambda,t),\ t \geq T.$$

Using transformation (5.1.13) we obtain for $x = 0$

$$\left|\begin{array}{l} u(0,t) = \dfrac{1}{2\sqrt{C}}e^{-\frac{R}{L}t}W(0,t)-\dfrac{1}{2\sqrt{C}}e^{-\frac{R}{L}t}J(0,t), \\[4mm] i(0,t) = \dfrac{1}{2\sqrt{L}}e^{-\frac{R}{L}t}W(0,t)+\dfrac{1}{2\sqrt{L}}e^{-\frac{R}{L}t}J(0,t) \end{array}\right.$$

and for $x = \Lambda$

$$u(\Lambda,t) = \frac{1}{2\sqrt{C}}e^{-\frac{R}{L}t}W(\Lambda,t) - \frac{1}{2\sqrt{C}}e^{-\frac{R}{L}t}J(\Lambda,t),$$

$$i(\Lambda,t) = \frac{1}{2\sqrt{L}}e^{-\frac{R}{L}t}W(\Lambda,t) + \frac{1}{2\sqrt{L}}e^{-\frac{R}{L}t}J(\Lambda,t).$$

Substituting $u(0,t)$, $u(0,t)$, $u(\Lambda,t)$, $i(\Lambda,t)$ into the above boundary conditions, we obtain

$$E(t) - \frac{e^{-\frac{R}{L}t}W(0,t) - e^{-\frac{R}{L}t}J(0,t)}{2\sqrt{C}} - R_0\frac{e^{-\frac{R}{L}t}W(0,t) + e^{-\frac{R}{L}t}J(0,t)}{2\sqrt{L}} = 0,$$

$$C_0\frac{1}{2\sqrt{C}}\left[\frac{d}{dt}\left(e^{-\frac{R}{L}t}W(\Lambda,t)\right) - \frac{d}{dt}\left(e^{-\frac{R}{L}t}J(\Lambda,t)\right)\right] =$$

$$= \frac{e^{-\frac{R}{L}t}W(\Lambda,t) + e^{-\frac{R}{L}t}J(\Lambda,t)}{2\sqrt{L}} - f\left(\frac{e^{-\frac{R}{L}t}W(\Lambda,t) - e^{-\frac{R}{L}t}J(\Lambda,t)}{2\sqrt{C}}\right), t \geq 0.$$

For the polynomial with a type *V-I* characteristic $i = f(u) = \sum_{n=1}^{p} r_n u^n$ the above equation yields the type:

$$E(t) - \frac{W(0,t)}{2\sqrt{C}}e^{-\frac{R}{L}t} + \frac{J(0,t)}{2\sqrt{C}}e^{-\frac{R}{L}t} - \frac{R_0}{2\sqrt{L}}e^{-\frac{R}{L}t}W(0,t) - \frac{R_0}{2\sqrt{L}}e^{-\frac{R}{L}t}J(0,t) = 0,$$

$$C_0\frac{d}{dt}\left(\frac{1}{2\sqrt{C}}e^{-\frac{R}{L}t}W(\Lambda,t) - \frac{1}{2\sqrt{C}}e^{-\frac{R}{L}t}J(\Lambda,t)\right) =$$

$$\frac{1}{2\sqrt{L}}e^{-\frac{R}{L}t}W(\Lambda,t) + \frac{1}{2\sqrt{L}}e^{-\frac{R}{L}t}J(\Lambda,t) - -\sum_{n=1}^{p}r_n\left(\frac{W(\Lambda,t) - J(\Lambda,t)}{2\sqrt{C}}e^{-\frac{R}{L}t}\right)^n.$$

Let the unknown functions be $W(t) = W(\Lambda,t)$, $J(t) = J(0,t)$.

The integration on the characteristics implies

$$W(0,t) = W(\Lambda,t+T), J(\Lambda,t) = J(0,t+T).$$

Then

$$E(t)-\frac{1}{2\sqrt{C}}e^{-\frac{R}{L}(t+T)}W(\Lambda,t+T)+\frac{1}{2\sqrt{C}}e^{-\frac{R}{L}t}J(0,t)-$$

$$-\frac{R_0}{2\sqrt{L}}e^{-\frac{R}{L}(t+T)}W(\Lambda,t+T)-\frac{R_0}{2\sqrt{L}}e^{-\frac{R}{L}t}J(0,t)=0,$$

$$\frac{C_0}{2\sqrt{C}}\frac{d}{dt}\left(e^{-\frac{R}{L}t}W(\Lambda,t)-e^{-\frac{R}{L}(t+T)}J(0,t+T)\right)=$$

$$\frac{1}{2\sqrt{L}}e^{-\frac{R}{L}t}W(\Lambda,t)+\frac{1}{2\sqrt{L}}e^{-\frac{R}{L}(t+T)}J(0,t+T)-$$

$$-\sum_{n=1}^{p}r_n\left(\frac{1}{2\sqrt{C}}e^{-\frac{R}{L}t}W(\Lambda,t)-\frac{1}{2\sqrt{C}}e^{-\frac{R}{L}(t+T)}J(0,t+T)\right)^n$$

or

$$E(t)-\frac{1}{2\sqrt{C}}e^{-\frac{R}{L}(t+T)}W(t+T)+\frac{1}{2\sqrt{C}}e^{-\frac{R}{L}t}J(t)-$$

$$\frac{R_0}{2\sqrt{L}}e^{-\frac{R}{L}(t+T)}W(t+T)-\frac{R_0}{2\sqrt{L}}e^{-\frac{R}{L}t}J(t)=0,$$

$$\frac{C_0}{2\sqrt{C}}\frac{d}{dt}\left(e^{-\frac{R}{L}t}W(t)-e^{-\frac{R}{L}(t+T)}J(t+T)\right)=\frac{1}{2\sqrt{L}}e^{-\frac{R}{L}t}W(t)+\frac{1}{2\sqrt{L}}e^{-\frac{R}{L}(t+T)}J(t+T)-$$

$$-\sum_{n=1}^{p}r_n\left(\frac{1}{2\sqrt{C}}e^{-\frac{R}{L}t}W(t)-\frac{1}{2\sqrt{C}}e^{-\frac{R}{L}(t+T)}J(t+T)\right)^n.$$

Let us put $t+T\equiv t$. Then

$$2\sqrt{C}E(t-T)-e^{-\frac{R}{L}t}W(t)+e^{-\frac{R}{L}(t-T)}J(t-T)-\frac{R_0}{Z_0}e^{-\frac{R}{L}t}W(t)-\frac{R_0}{Z_0}e^{-\frac{R}{L}(t-T)}J(t-T)=0,$$

$$\frac{d}{dt}\left(e^{-\frac{R}{L}(t-T)}W(t-T)-e^{-\frac{R}{L}t}J(t)\right)=\frac{1}{C_0Z_0}e^{-\frac{R}{L}(t-T)}W(t-T)+\frac{1}{C_0Z_0}e^{-\frac{R}{L}t}J(t)-$$

$$-\frac{2\sqrt{C}}{C_0}\sum_{n=1}^{p}r_n\left(\frac{1}{2\sqrt{C}}e^{-\frac{R}{L}(t-T)}W(t-T)-\frac{1}{2\sqrt{C}}e^{-\frac{R}{L}t}J(t)\right)^n.$$

The system obtained is not autonomous one and obviously $e^{-\frac{R}{L}t}$ is not a periodic function, so that the right-hand side of the second equation is not a periodic function either. We notice, however, the following

Principal Remark. Let us set $\breve{W}(t) = e^{-\frac{R}{L}t}W(t)$, $\breve{J}(t) = e^{-\frac{R}{L}t}J(t)$. If $W(t), J(t)$ are periodic functions then $\breve{W}(t), \breve{J}(t)$ are oscillating ones and satisfy the inequalities $\left|\breve{W}(t)\right| \le e^{-\frac{R}{L}t}W_0$, $\left|\breve{J}(t)\right| \le e^{-\frac{R}{L}t}J_0$. Therefore we can formulate the problem for the existence of oscillatory solutions satisfying the previous inequalities.

Further on we again use the denotations (W, J) instead of $(\breve{W}, \breve{J})$.

First we rewrite the above system as:

$$2\sqrt{C}E(t) - W(t) + J(t-T) - \frac{R_0}{Z_0}W(t) - \frac{R_0}{Z_0}J(t-T) = 0,$$

$$\frac{d}{dt}\big(W(t-T) - J(t)\big) = \frac{1}{C_0 Z_0}W(t-T) + \frac{1}{C_0 Z_0}J(t) - \frac{2\sqrt{C}}{C_0}\sum_{n=1}^{p}r_n\left(\frac{1}{2\sqrt{C}}W(t-T) - \frac{1}{2\sqrt{C}}J(t)\right)^n$$

or

$$W(t) = \frac{2Z_0\sqrt{C}E(t)}{Z_0 + R_0} + \frac{Z_0 - R_0}{Z_0 + R_0}J(t-T),$$

$$\frac{dJ(t)}{dt} = \frac{dW(t-T)}{dt} - \frac{1}{C_0 Z_0}W(t-T) - \frac{1}{C_0 Z_0}J(t) + \frac{2\sqrt{C}}{C_0}\sum_{n=1}^{p}r_n\left(\frac{1}{2\sqrt{C}}W(t-T) - \frac{1}{2\sqrt{C}}J(t)\right)^n$$

Let transform the first equation, changing the argument $\tau - T \equiv t \Rightarrow t - T = \tau - 2T$

$$W(\tau - T) = \frac{2Z_0\sqrt{C}E(\tau)}{Z_0 + R_0} + \frac{Z_0 - R_0}{Z_0 + R_0}J(\tau - 2T) \Leftrightarrow W(t-T) = \frac{2Z_0\sqrt{C}E(t-2T)}{Z_0 + R_0} + \frac{Z_0 - R_0}{Z_0 + R_0}J(t-2T)$$

and substitutinge in the second equation:

$$\frac{dJ(t)}{dt} = \frac{2Z_0\sqrt{C}}{(Z_0 + R_0)}\frac{dE(t)}{dt} + \frac{Z_0 - R_0}{Z_0 + R_0}\frac{dJ(t-2T)}{dt} -$$

$$- \frac{1}{C_0 Z_0}\left(\frac{2Z_0\sqrt{C}E(t)}{Z_0 + R_0} + \frac{Z_0 - R_0}{Z_0 + R_0}J(t-2T)\right) - \frac{1}{C_0 Z_0}J(t) +$$

$$+ \frac{1}{C_0}\sum_{n=1}^{p}r_n\left(\frac{1}{2\sqrt{C}}\right)^{n-1}\left(\frac{2Z_0\sqrt{C}E(t)}{Z_0 + R_0} + \frac{Z_0 - R_0}{Z_0 + R_0}J(t-2T) - J(t)\right)^n$$

or

$$\frac{dJ(t)}{dt} = \frac{Z_0 - R_0}{Z_0 + R_0} \frac{dJ(t-2T)}{dt} + \frac{2Z_0\sqrt{C}}{(Z_0 + R_0)} \frac{dE(t)}{dt} - \frac{2\sqrt{C}}{C_0(Z_0 + R_0)} E(t) -$$

$$- \frac{Z_0 - R_0}{C_0 Z_0 (Z_0 + R_0)} J(t-2T) - \frac{1}{C_0 Z_0} J(t) +$$

$$+ \frac{1}{C_0} \sum_{n=1}^{p} r_n \left(\frac{1}{2\sqrt{C}}\right)^{n-1} \left(\frac{2Z_0\sqrt{C}}{Z_0 + R_0} E(t) + \frac{Z_0 - R_0}{Z_0 + R_0} J(t-2T) - J(t)\right)^n \qquad (3.4.1)$$

The initial conditions can be found from

$$\left| \begin{array}{l} W(x,t) = e^{\frac{R}{L}t}\sqrt{C}u(x,t) + e^{\frac{R}{L}t}\sqrt{L}i(x,t), \\ J(x,t) = -e^{\frac{R}{L}t}\sqrt{C}u(x,t) + e^{\frac{R}{L}t}\sqrt{L}i(x,t) \end{array} \right.$$

that is,

$$\tilde{J}_0(x) = J(x,0) = u(x,0) - Z_0\, i(x,0) = u_0(x) - Z_0\, i_0(x)\ ,\ x \in [0,\Lambda].$$

As above, one can shift the initial functions along the characteristics on the interval $[0,T]$ and obtain $J(t) = \tilde{J}_0(t),\ t \in [-T,T]$.

Now we are able to formulate the main problem: to find a solution of (5.4.1) with advanced prescribed zeros on an interval $[t_0,\infty),\ T = t_0$, where $\tilde{W}_0(t),\ \tilde{J}_0(t)$ are prescribed oscillating functions on the interval $[-T,T]$.

As above $S_T = \{\tau_k\}_{k=0}^{n}, n \in N$ is the set of zeros of for the initial function, that is, $\tilde{W}_0(\tau_k) = 0,\ \tilde{J}_0(\tau_k) = 0$ such that $\tau_0 = -T,\ \tau_n = T \equiv t_0$. Besides $\max\{\tau_{k+1} - \tau_k : k = 0,1,...,n\} \le T_0$.

Let $S = \{t_k\}_{k=0}^{\infty}$ be a strictly increasing sequence of real numbers satisfying the following conditions **(C)**:

(C1) $\lim\limits_{k \to \infty} t_k = \infty$;

(C2 for every k there is $s < k$ such that $t_k - T = t_s$ where $t_s \in S_T \cup S$.

It follows

(C3) $0 \le \inf\{t_{k+1} - t_k : k = 0,1,2,...\} \le \sup\{t_{k+1} - t_k : k = 0,1,2,...\} = T_0 < \infty$.

Introduce the set $C^1[t_0,\infty)$ consisting of all continuous and bounded functions differentiable with bounded derivatives on every interval $[t_k, t_{k+1}]$. Let us note that the functions from $C^1[t_0,\infty)$ can be not differentiable at t_k. That is why, we introduce below a topology of uniform convergence on every interval $[t_k, t_{k+1}]$ which that needs the introduction of uniform spaces (cf. Chapter I).

$$M_J = \left\{ J(.) \in C^1[t_0,\infty) : J(t_k) = 0 \wedge \left|J(t)\right| \le I_0 e^{-\frac{R}{L}t}, t \in [t_0,\infty) \right\},$$

where $\beta > 0$ chosen is a sufficiently large positive constant chosen sufficiently large.

Since we look are looking for continuously differentiable solutions, we have to take into account the conformity condition (CC):

$$\frac{dJ(T)}{dt} = \frac{Z_0 - R_0}{Z_0 + R_0}\frac{dJ(-T)}{dt} + \frac{2Z_0\sqrt{C}}{(Z_0 + R_0)}\frac{dE(T)}{dt} - \frac{2\sqrt{C}}{C_0(Z_0 + R_0)}E(T) -$$

$$- \frac{Z_0 - R_0}{C_0 Z_0(Z_0 + R_0)}J(-T) - \frac{1}{C_0 Z_0}J(T) +$$

$$+ \frac{1}{C_0}\sum_{n=1}^{p} r_n \left(\frac{1}{2\sqrt{C}}\right)^{n-1}\left(\frac{2Z_0\sqrt{C}}{Z_0 + R_0}E(T) + \frac{Z_0 - R_0}{Z_0 + R_0}J(-T) - J(T)\right)^{n} .$$

The last condition is satisfied because the points $-T = -t_0 = \tau_0, T = t_0$ are zeros of the initial function and $E(t)$. Then **(CC)** becomes $\dfrac{dJ(T)}{dt} = \dfrac{Z_0 - R_0}{Z_0 + R_0}\dfrac{dJ(-T)}{dt}$. If we choose the initial function such that $\dfrac{dJ(-T)}{dt} = 0$ then one should be assumed $\dfrac{dJ(T)}{dt} = 0$. The last condition is too restrictive. We show how to overcome this difficulty below.

Remark 5.4.1. It follows that the functions from the above set satisfy the inequalities

$$\left|J(t)\right| \le I_0 e^{\mu(t-t_k)}, t \in [t_k, t_{k+1}]$$

where U_0, μ are positive constants and $k = 0,1,2,...$.

The set $M_W \times M_J$ turns out into a complete uniform space (cf. Chapter I) with respect to a family of pseudo-metrics.

Introduce the following family of pseudo-metrics

$$\rho^{(k)}(J,\bar{J}) = \max\left\{\left|J(t) - \bar{J}(t)\right| : t \in [t_k, t_{k+1}]\right\},$$

$$\hat{\rho}^{(k)}(J,\bar{J}) = \max\left\{\left|J(t) - \bar{J}(t)\right| : t \in [t_0, t_{k+1}]\right\},$$

$$\rho_\mu^{(k)}(J,\bar{J}) = \max\left\{e^{-\mu(t-t_k)}\left|J(t) - \bar{J}(t)\right| : t \in [t_k, t_{k+1}]\right\},$$

$$\hat{\rho}_\mu^{(k)}(J,\bar{J}) = \max\left\{\rho_\mu^{(0)}(J,\bar{J}), \rho_\mu^{(1)}(J,\bar{J}),...,\rho_\mu^{(k)}(J,\bar{J})\right\},$$

$$\rho_\mu^{(k)}(\dot{J},\dot{\bar{J}}) = \max\left\{e^{-\mu(t-t_k)}\left|\dot{J}(t) - \dot{\bar{J}}(t)\right| : t \in [t_k, t_{k+1}]\right\},$$

$$\hat{\rho}_\mu^{(k)}(\dot{J},\dot{\bar{J}}) = \max\left\{\rho_\mu^{(0)}(\dot{J},\dot{\bar{J}}), \rho_\mu^{(1)}(\dot{J},\dot{\bar{J}}),...,\rho_\mu^{(k)}(\dot{J},\dot{\bar{J}})\right\}.$$

Remark 5.4.2. The following inequalities imply the equivalence of the both families of pseudo-metrics

$$\rho_\mu^{(k)}(J,\bar{J}) \le \rho^{(k)}(J,\bar{J}) \le e^{\mu_0}\rho_\mu^{(k)}(J,\bar{J}), \ (k = 0,1,2,...).$$

It is easy to verify that

$$\hat{\rho}^{(k)}(J,\bar{J}) = \max\left\{\rho^{(0)}(J,\bar{J}), \rho^{(1)}(J,\bar{J}),...,\rho^{(k)}(J,\bar{J})\right\} \le$$
$$\le e^{\mu_0}\max\left\{\rho_\mu^{(0)}(J,\bar{J}), \rho_\mu^{(1)}(J,\bar{J}),...,\rho_\mu^{(k)}(J,\bar{J})\right\} = e^{\mu_0}\hat{\rho}_\mu^{(k)}(J,\bar{J}).$$

The set M_J turns out into a complete uniform space with respect to the saturated family of pseudo-metrics

$$\hat{\rho}_\mu^{(k)}((J,\dot{J}),(\bar{J},\dot{\bar{J}})) = \max\left\{\hat{\rho}^{(k)}(J,\bar{J}), \hat{\rho}_\mu^{(k)}(\dot{J},\dot{\bar{J}})\right\} (k = 0,1,2,...).$$

One can verify that M_J is closed subset of $C^1[t_0,\infty)$ with respect to the above family of pseudo-metrics.

In order to overcome the difficulty caused by **(CC)** we define the operator $B_J(J)$ by the formulas

$$B_J(J)(t) = B_J^{(k)}(J)(t) := \int_{t_k}^{t} I(J)(s)ds - \frac{t-t_k}{t_{k+1}-t_k}\int_{t_k}^{t_{k+1}} I(J)(s)ds, \ t \in [t_k, t_{k+1}], \ (k = 0,1,2,...)$$

where $\delta = \dfrac{Z_0 - R_0}{Z_0 + R_0},$

$$I(J)(t) \equiv \delta \frac{d\bar{J}(t)}{dt} + \frac{2Z_0\sqrt{C}}{(Z_0+R_0)}\frac{dE(t)}{dt} - \frac{2\sqrt{C}\,E(t)}{C_0(Z_0+R_0)} -$$

$$-\frac{\delta}{C_0 Z_0}\bar{J}(t) - \frac{1}{C_0 Z_0}J(t) + \frac{1}{C_0}\sum_{n=1}^{p} r_n \left(\frac{1}{2\sqrt{C}}\right)^{n-1}\left(\frac{2Z_0\sqrt{C}}{Z_0+R_0}E(t) + \delta\bar{J}(t) - J(t)\right)^n$$

(5.4.2)

Here the following operator $K : M_J \to M_J$ is introduced: to every function $J(.) \in M_J$

assigns a function $\bar{J}(t) = \begin{cases} \tilde{J}_0(t-2T), t \in [T,3T] \\ J(t-2T), \ t \in [3T,\infty) \end{cases}$. This means that for $t \in [T,3T]$ $\bar{J}(t)$

coincides with an initial function translated to the right initial function, while for $t \in [3T,\infty)$

$\bar{J}(t)$ coincides with a translated to the right $J(t)$. In fact $J(t-2T)$ is replaced by $\bar{J}(t)$.

Remark 5.4.3. It is easy to verify that the functions from M_J are not necessary

differentiable at the points $S = \{t_k\}_{k=0}^{\infty}$ which implies that we have to consider a space with a

countable family of pseudo-metrics, but not with one metric. For this reason we involve the

fixed-point theory in uniform spaces (cf. Chapter I).

Lemma 5.4.1. If $\left|E(t)\right| \leq U_E e^{\mu(t-t_k)}, t \in [t_k, t_{k+1}](k=0,1,...)$ problem (5.4.1) has a

solution $J \in M_J$ iff the operator B has a fixed point in M_J, that is,

$$J = B_J(J).$$

(5.4.3)

Proof: Let $J \in M_J$ be a solution of (5.4.1). We consider only the second component.

Integrating the second equation of (5.4.1) on every interval $[t_k, t] \subset [t_k, t_{k+1}]$

$(k=0,1,2\ ...\ )$ we obtain

$$J(t) - J(t_k) = \int_{t_k}^{t} I(J)(s)ds \Leftrightarrow J(t) = \int_{t_k}^{t} I(J)(s)ds$$

and then

$$J(t) = \int_{t_k}^{t} I(J)(s)ds \Rightarrow 0 = J(t_{k+1}) = \int_{t_k}^{t} I(J)(s)ds \Rightarrow \int_{t_k}^{t_{k+1}} I(J)(s)ds = 0.$$

(5.4.4)

Therefore J satisfies

$$J(t) = \int_{t_k}^{t} I(J)(s)ds - \frac{t-t_k}{t_{k+1}-t_k}\int_{t_k}^{t_{k+1}} I(J)(s)ds$$

for $t \in [t_k, t_{k+1}]$ iff $J = B_J(J)$, that is, J is a fixed point of B.

Conversely, let J be a fixed point of B or

$$J(t) = \int_{t_k}^{t} I(J)(s)\,ds - \frac{t - t_k}{t_{k+1} - t_k} \int_{t_k}^{t_{k+1}} I(J)(s)\,ds .$$

Then in view of $\mu_0 = \mu T_0 = \text{const.}$ and $t_k - 2T = t_q, t_{k+1} - 2T = t_{q+1}$ we obtain

$$\left| \int_{t_k}^{t_{k+1}} I(J)(s)\,ds \right| \le$$

$$\le |\delta| \left| \int_{t_k}^{t_{k+1}} \frac{dJ(t-2T)}{dt}\,dt \right| + \frac{2Z_0\sqrt{C}}{(Z_0 + R_0)} \left| \int_{t_k}^{t_{k+1}} \frac{dE(t)}{dt}\,dt \right| + \frac{2\sqrt{C}}{C_0(Z_0 + R_0)} \int_{t_k}^{t_{k+1}} |E(t)|\,dt +$$

$$+ \frac{|\delta|}{C_0 Z_0} \int_{t_k}^{t_{k+1}} |J(t-2T)|\,dt + \frac{1}{C_0 Z_0} \int_{t_k}^{t_{k+1}} |J(t)|\,dt +$$

$$+ \frac{1}{C_0} \sum_{n=1}^{p} r_n \left(\frac{1}{2\sqrt{C}} \right)^{n-1} \int_{t_k}^{t_{k+1}} \left(\frac{2Z_0\sqrt{C}}{Z_0 + R_0} |E(t)| + |\delta||J(t-2T)| + |J(t)| \right)^n dt \le$$

$$\le \frac{2\sqrt{C}}{C_0(Z_0 + R_0)} U_E \int_{t_k}^{t_{k+1}} e^{\mu(t-t_k)}\,dt + \frac{|\delta|}{C_0 Z_0} I_0 \int_{t_k}^{t_{k+1}} e^{\mu(t-2T-t_q)}\,dt + \frac{1}{C_0 Z_0} I_0 \int_{t_k}^{t_{k+1}} \left| e^{\mu(t-t_k)} \right| dt +$$

$$+ \frac{1}{C_0} \sum_{n=1}^{p} r_n \left(\frac{1}{2\sqrt{C}} \right)^{n-1} \int_{t_k}^{t_{k+1}} \left(\frac{2Z_0\sqrt{C}}{Z_0 + R_0} U_E e^{\mu(t-t_k)} + |\delta| I_0 e^{\mu(t-2T-t_q)} + I_0 e^{\mu(t-t_k)} \right)^n dt \le$$

$$\le \frac{2\sqrt{C}}{C_0(Z_0 + R_0)} U_E \int_{t_k}^{t_{k+1}} e^{\mu(t-t_k)}\,dt + \frac{|\delta|}{C_0 Z_0} I_0 \int_{t_k}^{t_{k+1}} e^{\mu(t-t_k)}\,dt + \frac{1}{C_0 Z_0} I_0 \int_{t_k}^{t_{k+1}} e^{\mu(t-t_k)}\,dt +$$

$$+ \frac{1}{C_0} \sum_{n=1}^{p} r_n \left(\frac{1}{2\sqrt{C}} \right)^{n-1} \left(\frac{2Z_0\sqrt{C}}{Z_0 + R_0} U_E + |\delta| I_0 + I_0 \right)^n \int_{t_k}^{t_{k+1}} e^{n\mu(t-t_k)}\,dt \le$$

$$\le \frac{2\sqrt{C}}{C_0(Z_0 + R_0)} U_E \frac{e^{\mu(t_{k+1}-t_k)} - 1}{\mu} + \frac{|\delta|}{C_0 Z_0} I_0 \frac{e^{\mu(t_{k+1}-t_k)} - 1}{\mu} + \frac{1}{C_0 Z_0} I_0 \frac{e^{\mu(t_{k+1}-t_k)} - 1}{\mu} +$$

$$+ \frac{1}{C_0} \sum_{n=1}^{p} r_n \left(\frac{1}{2\sqrt{C}} \right)^{n-1} \left(\frac{2Z_0\sqrt{C}}{Z_0 + R_0} U_E + |\delta| I_0 + I_0 \right)^n \frac{e^{\mu(t_{k+1}-t_k)} - 1}{n\mu} n e^{(n-1)\mu_0} \le$$

$$\leq \frac{e^{\mu_0}-1}{\mu C_0}\left[\frac{2\sqrt{C}U_E}{Z_0+R_0}+\frac{1+|\delta|}{Z_0}I_0+\sum_{n=1}^{P}r_n\left(\frac{1}{2\sqrt{C}}\right)^{n-1}\left(\frac{2Z_0\sqrt{C}}{Z_0+R_0}U_E+|\delta|I_0+I_0\right)^n e^{(n-1)\mu_0}\right]\equiv M_0(\mu)$$

As in Chapter II we conclude that $\displaystyle\int_{t_k}^{t_{k+1}}I(J)(s)ds=0$. Therefore

$$J(t)=\int_{t_k}^{t}I(J)(s)ds-\frac{t-t_k}{t_{k+1}-t_k}\int_{t_k}^{t_{k+1}}I(J)(s)ds \Leftrightarrow J(t)=\int_{t_k}^{t}I(J)(s)ds\,.$$

Differentiating the last integral equation we obtain (5.4.1).

Lemma 5.4.1 is thus proved.

The following theorem guarantees an existence-uniqueness of for the solution belonging to M_J.

Theorem 5.4.1. Let the following conditions be fulfilled:

(IN) The initial function $\widetilde{J}_0(t)\in C^1[-T,T]$ satisfy $\left|\widetilde{J}_0(t)\right|\leq I_0 e^{-\frac{R}{L}t}$

(E) $E(.)\in C^1[T,\infty),|E(t)|\leq U_E e^{-\frac{R}{L}t},t\in[T,\infty);E(t-T)=E(t),E(t_k-T)=E(t_q)(q<k)$;

(R) $\dfrac{eR_0}{Z_0}<1$;

$$|\delta|I_0+\frac{2Z_0\sqrt{C}}{Z_0+R_0}U_E+$$

$$+4\sinh\left(\frac{RT_0}{L}\right)\frac{L}{RC_0}\left[\frac{2\sqrt{C}}{Z_0+R_0}U_E+\frac{(1+|\delta|)I_0}{Z_0}+\sum_{n=1}^{P}r_n\left(\frac{1}{2\sqrt{C}}\right)^{n-1}\left(\frac{2Z_0\sqrt{C}}{Z_0+R_0}U_E+|\delta|I_0+I_0\right)^n\right]\leq I_0\,.$$

Then there exists an unique oscillatory solution of (5.4.1), belonging to M_J.

Proof: We omit the proof that $B_J(J)(t)$ are continuous and differentiable on every interval $t\in[t_k,t_{k+1}],(k=0,1,2,...)$.

We use $\left|\dfrac{t-t_k}{t_{k+1}-t_k}\right|\leq1$ for $t\in[t_k,t_{k+1}],(k=0,1,2,...)$ and have

$$J_1\leq\int_{t_k}^{t}|I(J)(s)|ds\leq$$

$$
\leq |\delta| \left| \int_{t_k}^{t} \frac{dJ(s-2T)}{dt} ds \right| + \frac{2Z_0\sqrt{C}}{(Z_0+R_0)} \left| \int_{t_k}^{t} \frac{dE(s)}{ds} ds \right| + \frac{2\sqrt{C}}{C_0(Z_0+R_0)} \int_{t_k}^{t} |E(s)| ds +
$$

$$
+ \frac{|\delta|}{C_0 Z_0} \int_{t_k}^{t} |J(s-2T)| ds + \frac{1}{C_0 Z_0} \int_{t_k}^{t} |J(s)| ds +
$$

$$
+ \frac{1}{C_0} \sum_{n=1}^{p} r_n \left(\frac{1}{2\sqrt{C}} \right)^{n-1} \int_{t_k}^{t} \left(\frac{2Z_0\sqrt{C}}{Z_0+R_0} |E(s)| + |\delta| |J(s-2T)| + |J(s)| \right)^n ds \leq
$$

$$
\leq |\delta| |J(t-2T)| + \frac{2Z_0\sqrt{C}}{(Z_0+R_0)} |E(t)| + \frac{2\sqrt{C}}{C_0(Z_0+R_0)} U_E \int_{t_k}^{t} e^{-\frac{R}{L}s} ds +
$$

$$
+ \frac{|\delta| I_0}{C_0 Z_0} \int_{t_k}^{t} e^{-\frac{R}{L}s} ds + \frac{I_0}{C_0 Z_0} \int_{t_k}^{t} e^{-\frac{R}{L}s} ds + \frac{1}{C_0} \sum_{n=1}^{p} r_n \left(\frac{1}{2\sqrt{C}} \right)^{n-1} \left(\frac{2Z_0\sqrt{C}}{Z_0+R_0} U_E + |\delta| I_0 + I_0 \right)^n \int_{t_k}^{t} e^{-n\frac{R}{L}s} dt \leq
$$

$$
\leq |\delta| I_0 e^{-\frac{R}{L}t} + \frac{2Z_0\sqrt{C}}{(Z_0+R_0)} U_E e^{-\frac{R}{L}t} + \frac{2\sqrt{C}}{C_0(Z_0+R_0)} U_E \frac{L}{R} \left(e^{-\frac{R}{L}t_k} - e^{-\frac{R}{L}t} \right) +
$$

$$
+ \frac{|\delta| I_0}{C_0 Z_0} \frac{L}{R} \left(e^{-\frac{R}{L}t_k} - e^{-\frac{R}{L}t} \right) + \frac{I_0}{C_0 Z_0} \frac{L}{R} \left(e^{-\frac{R}{L}t_k} - e^{-\frac{R}{L}t} \right) +
$$

$$
+ \frac{1}{C_0} \sum_{n=1}^{p} r_n \left(\frac{1}{2\sqrt{C}} \right)^{n-1} \left(\frac{2Z_0\sqrt{C}}{Z_0+R_0} U_E + |\delta| I_0 + I_0 \right)^n \frac{L}{nR} \left(e^{-n\frac{R}{L}t_k} - e^{-n\frac{R}{L}t} \right) \leq
$$

$$
\leq |\delta| I_0 e^{-\frac{R}{L}t} + \frac{2Z_0\sqrt{C}}{(Z_0+R_0)} U_E e^{-\frac{R}{L}t} + \frac{2\sqrt{C}}{C_0(Z_0+R_0)} U_E \frac{L}{R} \left(e^{-\frac{R}{L}t_k} - e^{-\frac{R}{L}t} \right) +
$$

$$
+ \frac{|\delta| I_0}{C_0 Z_0} \frac{L}{R} \left(e^{-\frac{R}{L}t_k} - e^{-\frac{R}{L}t} \right) + \frac{I_0}{C_0 Z_0} \frac{L}{R} \left(e^{-\frac{R}{L}t_k} - e^{-\frac{R}{L}t} \right) +
$$

$$
+ \frac{1}{C_0} \sum_{n=1}^{p} r_n \left(\frac{1}{2\sqrt{C}} \right)^{n-1} \left(\frac{2Z_0\sqrt{C}}{Z_0+R_0} U_E + |\delta| I_0 + I_0 \right)^n \frac{L}{nR} \left(e^{-\frac{R}{L}t_k} - e^{-\frac{R}{L}t} \right) n \leq
$$

$$
\leq |\delta| I_0 e^{-\frac{R}{L}t} + \frac{2Z_0\sqrt{C}}{(Z_0+R_0)} U_E e^{-\frac{R}{L}t} + \frac{L}{RC_0} e^{-\frac{R}{L}t} \left(e^{\frac{R}{L}(t-t_k)} - 1 \right) \left[\frac{2\sqrt{C}}{Z_0+R_0} U_E + \frac{(1+|\delta|) I_0}{Z_0} + \right.
$$

$$
\left. + \sum_{n=1}^{p} r_n \left(\frac{1}{2\sqrt{C}} \right)^{n-1} \left(\frac{2Z_0\sqrt{C}}{Z_0+R_0} U_E + |\delta| I_0 + I_0 \right)^n \right] \leq
$$

$$
\leq e^{-\frac{R}{L}t} \left\{ |\delta| I_0 + \frac{2Z_0\sqrt{C}}{(Z_0+R_0)} U_E + \frac{L}{RC_0} \left(e^{\frac{R}{L}T_0} - e^{-\frac{R}{L}T_0} \right) \left[\frac{2\sqrt{C}}{Z_0+R_0} U_E + \frac{(1+|\delta|) I_0}{Z_0} + \right. \right.
$$

$$+ \sum_{n=1}^{p} r_n \left(\frac{1}{2\sqrt{C}} \right)^{n-1} \left(\frac{2Z_0\sqrt{C}}{Z_0 + R_0} U_E + |\delta| I_0 + I_0 \right)^n \bigg]\bigg\}.$$

Using the inequality

$$e^{-\frac{R}{L}t_k} - e^{-\frac{R}{L}t_{k+1}} = e^{-\frac{R}{L}t_k} - e^{-\frac{R}{L}t} + e^{-\frac{R}{L}t} - e^{-\frac{R}{L}t_{k+1}} = e^{-\frac{R}{L}t}\left(e^{-\frac{R}{L}(t_k-t)} - 1 \right) + e^{-\frac{R}{L}t}\left(1 - e^{-\frac{R}{L}(t_{k+1}-t)} \right) \leq$$

$$\leq e^{-\frac{R}{L}t}\left(e^{\frac{R}{L}(t-t_k)} - 1 + 1 - e^{-\frac{R}{L}(t_{k+1}-t_k)} \right) \leq e^{-\frac{R}{L}t}\left(e^{\frac{R}{L}T_0} - e^{-\frac{R}{L}T_0} \right)$$

we obtain

$$J_2 \leq \int_{t_k}^{t_{k+1}} |I(J)(s)| \, ds \leq$$

$$\leq |\delta| \left| \int_{t_k}^{t_{k+1}} \frac{dJ(t-2T)}{dt} \, dt \right| + \frac{2Z_0\sqrt{C}}{(Z_0+R_0)} \left| \int_{t_k}^{t_{k+1}} \frac{dE(t)}{dt} \, dt \right| + \frac{2\sqrt{C}}{C_0(Z_0+R_0)} \int_{t_k}^{t_{k+1}} |E(t)| \, dt +$$

$$+ \frac{|\delta|}{C_0 Z_0} \int_{t_k}^{t_{k+1}} |J(t-2T)| \, dt + \frac{1}{C_0 Z_0} \int_{t_k}^{t_{k+1}} |J(t)| \, dt +$$

$$+ \frac{1}{C_0} \sum_{n=1}^{p} r_n \left(\frac{1}{2\sqrt{C}} \right)^{n-1} \int_{t_k}^{t_{k+1}} \left(\frac{2Z_0\sqrt{C}}{Z_0+R_0} |E(t)| + |\delta| |J(t-2T)| + |J(t)| \right)^n dt \leq$$

$$\leq \frac{2\sqrt{C}}{C_0(Z_0+R_0)} U_E \frac{L}{R} \left(e^{-\frac{R}{L}t_k} - e^{-\frac{R}{L}t_{k+1}} \right) +$$

$$+ \frac{|\delta| I_0}{C_0 Z_0} \frac{L}{R} \left(e^{-\frac{R}{L}t_k} - e^{-\frac{R}{L}t_{k+1}} \right) + \frac{I_0}{C_0 Z_0} \frac{L}{R} \left(e^{-\frac{R}{L}t_k} - e^{-\frac{R}{L}t_{k+1}} \right) +$$

$$+ \frac{1}{C_0} \sum_{n=1}^{p} r_n \left(\frac{1}{2\sqrt{C}} \right)^{n-1} \left(\frac{2Z_0\sqrt{C}}{Z_0+R_0} U_E + |\delta| I_0 + I_0 \right)^n \frac{L}{nR} \left(e^{-n\frac{R}{L}t_k} - e^{-n\frac{R}{L}t_{k+1}} \right) \leq$$

$$\leq \frac{L}{R} \left(e^{-\frac{R}{L}t_k} - e^{-\frac{R}{L}t_{k+1}} \right) \bigg[\frac{2\sqrt{C}}{C_0(Z_0+R_0)} U_E + \frac{|\delta| I_0}{C_0 Z_0} + \frac{I_0}{C_0 Z_0} +$$

$$+ \frac{1}{C_0} \sum_{n=1}^{p} r_n \left(\frac{1}{2\sqrt{C}} \right)^{n-1} \left(\frac{2Z_0\sqrt{C}}{Z_0+R_0} U_E + |\delta| I_0 + I_0 \right)^n \bigg] \leq$$

$$\le e^{-\frac{R}{L}t}\left(e^{\frac{R}{L}T_0}-e^{-\frac{R}{L}T_0}\right)\frac{L}{RC_0}\left[\frac{2\sqrt{C}}{Z_0+R_0}U_E+\frac{(1+|\delta|)I_0}{Z_0}+\sum_{n=1}^{p}r_n\left(\frac{1}{2\sqrt{C}}\right)^{n-1}\left(\frac{2Z_0\sqrt{C}}{Z_0+R_0}U_E+|\delta|I_0+I_0\right)^n\right].$$

Consequently

$$\left|B_J^{(k)}(W,J)(t)\right|\le$$

$$\le e^{-\frac{R}{L}t}\left\{|\delta|I_0+\frac{2Z_0\sqrt{C}}{(Z_0+R_0)}U_E+\right.$$

$$+\left(e^{\frac{R}{L}T_0}-e^{-\frac{R}{L}T_0}\right)\frac{L}{RC_0}\left[\frac{2\sqrt{C}}{Z_0+R_0}U_E+\frac{(1+|\delta|)I_0}{Z_0}++\sum_{n=1}^{p}r_n\left(\frac{1}{2\sqrt{C}}\right)^{n-1}\left(\frac{2Z_0\sqrt{C}}{Z_0+R_0}U_E+|\delta|I_0+I_0\right)^n\right]\right\}+$$

$$+e^{-\frac{R}{L}t}\left(e^{\frac{R}{L}T_0}-e^{-\frac{R}{L}T_0}\right)\frac{L}{RC_0}\left[\frac{2\sqrt{C}}{Z_0+R_0}U_E+\frac{(1+|\delta|)I_0}{Z_0}+\sum_{n=1}^{p}r_n\left(\frac{1}{2\sqrt{C}}\right)^{n-1}\left(\frac{2Z_0\sqrt{C}}{Z_0+R_0}U_E+|\delta|I_0+I_0\right)^n\right]\le$$

$$\le e^{-\frac{R}{L}t}\left\{|\delta|I_0+\frac{2Z_0\sqrt{C}}{Z_0+R_0}U_E+\right.$$

$$\left.+4\sinh\left(\frac{RT_0}{L}\right)\frac{L}{RC_0}\left[\frac{2\sqrt{C}}{Z_0+R_0}U_E+\frac{(1+|\delta|)I_0}{Z_0}++\sum_{n=1}^{p}r_n\left(\frac{1}{2\sqrt{C}}\right)^{n-1}\left(\frac{2Z_0\sqrt{C}}{Z_0+R_0}U_E+|\delta|I_0+I_0\right)^n\right]\right\}\le$$

$$\le I_0 e^{-\frac{R}{L}t}.$$

Therefore the operator B maps M_J into itself.

We show that B is a contractive one. Indeed, recalling $t_{k+1}-T=t_{q+1}$, $t_k-T=t_q$ $(q<k)$ we have

$$\left|B_J^{(k)}(J)(t)-B_J^{(k)}(\bar{J})(t)\right|\le\left|\int_{t_k}^{t}\left(I(J)(s)-I(\bar{J})(s)\right)ds\right|+$$

$$\left|\frac{t-t_k}{t_{k+1}-t_k}\int_{t_k}^{t_{k+1}}\left(I(J)(s)-I(\bar{J})(s)\right)ds\right|\equiv I_1+I_2.$$

Then

$$I_1\le\int_{t_k}^{t}\left|I(J)(s)-I(\bar{J})(s)\right|ds\le$$

$$\leq |\delta| \left| \int_{t_k}^{t} \left(\frac{dJ(s-2T)}{dt} - \frac{d\bar{J}(s-2T)}{dt} \right) ds \right| +$$

$$+ \frac{|\delta|}{C_0 Z_0} \int_{t_k}^{t} |J(s-2T) - \bar{J}(s-2T)| ds + \frac{1}{C_0 Z_0} \int_{t_k}^{t} |J(s) - \bar{J}(s)| ds +$$

$$+ \frac{1}{C_0} \sum_{n=1}^{p} r_n \left(\frac{1}{2\sqrt{C}} \right)^{n-1} \left(\frac{2Z_0\sqrt{C}}{Z_0 + R_0} U_E + |\delta| I_0 + I_0 \right)^{n-1} e^{(n-1)\mu_0} \left(|\delta| \int_{t_k}^{t} |J(s-2T) - \bar{J}(s-2T)| ds + \int_{t_k}^{t} |J(s) - \bar{J}(s)| ds \right) \leq$$

$$\leq |\delta| \rho_\mu^{(q)}(\dot{J}, \dot{\bar{J}}) \int_{t_k}^{t} e^{\mu(s-t_k)} ds + \frac{|\delta|}{C_0 Z_0} \rho_\mu^{(q)}(J, \bar{J}) \int_{t_k}^{t} e^{\mu(s-t_k)} ds + \frac{1}{C_0 Z_0} \rho_\mu^{(k)}(J, \bar{J}) \int_{t_k}^{t} e^{\mu(s-t_k)} ds +$$

$$+ \frac{1}{C_0} \sum_{n=1}^{p} r_n \left(\frac{1}{2\sqrt{C}} \right)^{n-1} \left(\frac{2Z_0\sqrt{C}}{Z_0 + R_0} U_E + |\delta| I_0 + I_0 \right)^{n-1} e^{(n-1)\mu_0} \left(|\delta| \rho_\mu^{(q)}(J, \bar{J}) \int_{t_k}^{t} e^{\mu(s-t_k)} ds + \rho_\mu^{(k)}(J, \bar{J}) \int_{t_k}^{t} e^{\mu(s-t_k)} ds \right) \leq$$

$$\leq |\delta| \rho_\mu^{(q)}(\dot{J}, \dot{\bar{J}}) \int_{t_k}^{t} e^{\mu(s-t_k)} ds + \frac{|\delta|}{\mu C_0 Z_0} \rho_\mu^{(q)}(\dot{J}, \dot{\bar{J}}) \int_{t_k}^{t} e^{\mu(s-t_k)} ds + \frac{1}{\mu C_0 Z_0} \rho_\mu^{(k)}(\dot{J}, \dot{\bar{J}}) \int_{t_k}^{t} e^{\mu(s-t_k)} ds +$$

$$+ \frac{1}{C_0} \sum_{n=1}^{p} r_n \left(\frac{1}{2\sqrt{C}} \right)^{n-1} \left(\frac{2Z_0\sqrt{C}}{Z_0 + R_0} U_E + |\delta| I_0 + I_0 \right)^{n-1} e^{(n-1)\mu_0} \left(\frac{|\delta| \rho_\mu^{(q)}(\dot{J}, \dot{\bar{J}})}{\mu} \int_{t_k}^{t} e^{\mu(s-t_k)} ds + \frac{\rho_\mu^{(k)}(\dot{J}, \dot{\bar{J}})}{\mu} \int_{t_k}^{t} e^{\mu(s-t_k)} ds \right) \leq$$

$$\leq \int_{t_k}^{t} e^{\mu(s-t_k)} ds \left[|\delta| \rho_\mu^{(q)}(\dot{J}, \dot{\bar{J}}) + \frac{|\delta|}{\mu C_0 Z_0} \rho_\mu^{(q)}(\dot{J}, \dot{\bar{J}}) + \frac{1}{\mu C_0 Z_0} \rho_\mu^{(k)}(\dot{J}, \dot{\bar{J}}) + \right.$$

$$\left. + \frac{1}{C_0} \left(\frac{|\delta| \rho_\mu^{(q)}(\dot{J}, \dot{\bar{J}})}{\mu} + \frac{\rho_\mu^{(k)}(\dot{J}, \dot{\bar{J}})}{\mu} \right) \sum_{n=1}^{p} r_n \left(\frac{1}{2\sqrt{C}} \right)^{n-1} \left(\frac{2Z_0\sqrt{C}}{Z_0 + R_0} U_E + |\delta| I_0 + I_0 \right)^{n-1} e^{(n-1)\mu_0} \right] \leq$$

$$\leq e^{\mu(t-t_k)} \hat{\rho}_\mu^{(k)}((J, \dot{J}), (\bar{J}, \dot{\bar{J}})) \frac{1}{\mu} \left[|\delta| + \frac{|\delta| + 1}{\mu C_0 Z_0} + \frac{|\delta| + 1}{\mu C_0} \sum_{n=1}^{p} r_n \left(\frac{e^{\mu_0}}{2\sqrt{C}} \right)^{n-1} \left(\frac{2Z_0\sqrt{C}}{Z_0 + R_0} U_E + |\delta| I_0 + I_0 \right)^{n-1} \right]$$

$$I_2 \leq |\delta| \left| \int_{t_k}^{t_{k+1}} \left(\frac{dJ(s-2T)}{dt} - \frac{d\bar{J}(s-2T)}{dt} \right) ds \right| +$$

$$+ \frac{|\delta|}{C_0 Z_0} \int_{t_k}^{t_{k+1}} |J(s-2T) - \bar{J}(s-2T)| ds + \frac{1}{C_0 Z_0} \int_{t_k}^{t_{k+1}} |J(s) - \bar{J}(s)| ds +$$

$$+ \frac{1}{C_0} \sum_{n=1}^{p} r_n \left(\frac{1}{2\sqrt{C}} \right)^{n-1} \left(\frac{2Z_0\sqrt{C}}{Z_0 + R_0} U_E + |\delta| I_0 + I_0 \right)^{n-1} e^{(n-1)\mu_0} \left(|\delta| \int_{t_k}^{t_{k+1}} |J(s-2T) - \bar{J}(s-2T)| ds + \int_{t_k}^{t_{k+1}} |J(s) - \bar{J}(s)| ds \right) \leq$$

$$\leq \frac{|\delta|}{C_0 Z_0} \int_{t_k}^{t_{k+1}} |J(s-2T) - \bar{J}(s-2T)| ds + \frac{1}{C_0 Z_0} \int_{t_k}^{t_{k+1}} |J(s) - \bar{J}(s)| ds +$$

$$+ \frac{1}{C_0} \sum_{n=1}^{p} r_n \left(\frac{1}{2\sqrt{C}} \right)^{n-1} \left(\frac{2Z_0\sqrt{C}}{Z_0 + R_0} U_E + |\delta| I_0 + I_0 \right)^{n-1} e^{(n-1)\mu_0} \left(|\delta| \int_{t_k}^{t_{k+1}} |J(s-2T) - \bar{J}(s-2T)| ds + \int_{t_k}^{t_{k+1}} |J(s) - \bar{J}(s)| ds \right) \leq$$

$$\leq \int_{t_k}^{t_{k+1}} e^{\mu(s-t_k)} ds \left[\frac{|\delta|}{\mu C_0 Z_0} \rho_\mu^{(q)}(\dot{J}, \dot{\bar{J}}) + \frac{1}{\mu C_0 Z_0} \rho_\mu^{(k)}(\dot{J}, \dot{\bar{J}}) + \right.$$

$$\left. + \frac{1}{C_0} \left(\frac{|\delta| \rho_\mu^{(q)}(\dot{J}, \dot{\bar{J}})}{\mu} + \frac{\rho_\mu^{(k)}(\dot{J}, \dot{\bar{J}})}{\mu} \right) \sum_{n=1}^{p} r_n \left(\frac{1}{2\sqrt{C}} \right)^{n-1} \left(\frac{2Z_0\sqrt{C}}{Z_0 + R_0} U_E + |\delta| I_0 + I_0 \right)^{n-1} e^{(n-1)\mu_0} \right] \leq$$

$$\leq e^{\mu(t-t_k)} \hat{\rho}_\mu^{(k)}((J, \dot{J}), (\bar{J}, \dot{\bar{J}})) \frac{e^{\mu_0} - 1}{\mu} \left[\frac{|\delta| + 1}{\mu C_0 Z_0} + \frac{|\delta| + 1}{\mu C_0} \sum_{n=1}^{p} r_n \left(\frac{e^{\mu_0}}{2\sqrt{C}} \right)^{n-1} \left(\frac{2Z_0\sqrt{C}}{Z_0 + R_0} U_E + |\delta| I_0 + I_0 \right)^{n-1} \right]$$

Consequently

$$\left| B_J^{(k)}(J)(t) - B_J^{(k)}(\bar{J})(t) \right| \le$$

$$\le e^{\mu(t-t_k)} \hat{\rho}_\mu^{(k)}((J,\dot{J}),(\bar{J},\dot{\bar{J}})) \frac{1}{\mu} \left[|\delta| + \frac{|\delta|+1}{\mu C_0} \left(\frac{1}{Z_0} + \sum_{n=1}^{p} r_n \left(\frac{e^{\mu_0}}{2\sqrt{C}} \right)^{n-1} \left(\frac{2Z_0\sqrt{C}}{Z_0+R_0} U_E + |\delta| I_0 + I_0 \right)^{n-1} \right) \right] +$$

$$+ e^{\mu(t-t_k)} \hat{\rho}_\mu^{(k)}((J,\dot{J}),(\bar{J},\dot{\bar{J}})) \frac{e^{\mu_0}-1}{\mu} \frac{|\delta|+1}{\mu C_0} \left(\frac{1}{Z_0} + \sum_{n=1}^{p} r_n \left(\frac{e^{\mu_0}}{2\sqrt{C}} \right)^{n-1} \left(\frac{2Z_0\sqrt{C}}{Z_0+R_0} U_E + |\delta| I_0 + I_0 \right)^{n-1} \right) \le$$

$$\le e^{\mu(t-t_k)} \hat{\rho}_\mu^{(k)}((J,\dot{J}),(\bar{J},\dot{\bar{J}})) \frac{1}{\mu} \left[|\delta| + e^{\mu_0} \frac{|\delta|+1}{\mu C_0} \left(\frac{1}{Z_0} + \sum_{n=1}^{p} r_n \left(\frac{e^{\mu_0}}{2\sqrt{C}} \right)^{n-1} \left(\frac{2Z_0\sqrt{C}}{Z_0+R_0} U_E + |\delta| I_0 + I_0 \right)^{n-1} \right) \right] \equiv$$

$$\equiv e^{\mu_0} K_J \hat{\rho}_\mu^{(k)}((J,\dot{J}),(\bar{J},\dot{\bar{J}})),$$

that is,

$$\hat{\rho}^{(k)}(B_J(J), B_J(\bar{J})) \le e^{\mu_0} K_J \hat{\rho}_\mu^{(k)}((J,\dot{J}),(\bar{J},\dot{\bar{J}})).$$

For the derivative we obtain

$$\left| \dot{B}_J^{(k)}(J)(t) - \dot{B}_J^{(k)}(\bar{J})(t) \right| \le \left| I(J)(t) - I(\bar{J})(t) \right| + \frac{1}{t_{k+1}-t_k} \left| \int_{t_k}^{t_{k+1}} \left(I(J)(s) - I(\bar{J})(s) \right) ds \right| \equiv \dot{I}_1 + \dot{I}_2.$$

But

$$\dot{I}_1 \le |\delta| \left| \frac{dJ(t-2T)}{dt} - \frac{d\bar{J}(t-2T)}{dt} \right| + \frac{|\delta|}{C_0 Z_0} \left| J(t-2T) - \bar{J}(t-2T) \right| + \frac{1}{C_0 Z_0} \left| J(t) - \bar{J}(t) \right| +$$

$$+ \frac{1}{C_0} \sum_{n=1}^{p} r_n \left(\frac{1}{2\sqrt{C}} \right)^{n-1} \left(\frac{2Z_0\sqrt{C}}{Z_0+R_0} U_E + |\delta| I_0 + I_0 \right)^{n-1} e^{(n-1)\mu_0} \left(|\delta| \left| J(t-2T) - \bar{J}(t-2T) \right| + \left| J(t) - \bar{J}(t) \right| \right) \le$$

$$\le e^{\mu(t-t_k)} \left| |\delta| \rho_\mu^{(q)}(\dot{J},\dot{\bar{J}}) + \frac{|\delta|}{C_0 Z_0} \rho_\mu^{(q)}(J,\bar{J}) + \frac{1}{C_0 Z_0} \rho_\mu^{(k)}(J,\bar{J}) + \right.$$

$$+ \frac{1}{C_0} \sum_{n=1}^{p} r_n \left(\frac{1}{2\sqrt{C}} \right)^{n-1} \left(\frac{2Z_0\sqrt{C}}{Z_0+R_0} U_E + |\delta| I_0 + I_0 \right)^{n-1} e^{(n-1)\mu_0} \left(|\delta| \rho_\mu^{(q)}(J,\bar{J}) + \rho_\mu^{(k)}(J,\bar{J}) \right) \le$$

$$\leq e^{\mu(t-t_k)}|\delta|\rho_\mu^{(q)}(\dot J,\dot{\bar J})+\frac{|\delta|}{\mu C_0 Z_0}\rho_\mu^{(q)}(\dot J,\dot{\bar J})+\frac{1}{\mu C_0 Z_0}\rho_\mu^{(k)}(\dot J,\dot{\bar J})+$$

$$+\frac{1}{C_0}\frac{|\delta|\rho_\mu^{(q)}(\dot J,\dot{\bar J})+\rho_\mu^{(k)}(\dot J,\dot{\bar J})}{\mu}\sum_{n=1}^{p}r_n\left(\frac{1}{2\sqrt C}\right)^{n-1}\left(\frac{2Z_0\sqrt C}{Z_0+R_0}U_E+|\delta|I_0+I_0\right)^{n-1}e^{(n-1)\mu_0}\leq$$

$$\leq e^{\mu(t-t_k)}\hat\rho_\mu^{(k)}((J,\dot J),(\bar J,\dot{\bar J}))\left[|\delta|+\frac{|\delta|+1}{\mu C_0}\left(\frac{1}{Z_0}+\sum_{n=1}^{p}r_n\left(\frac{1}{2\sqrt C}\right)^{n-1}\left(\frac{2Z_0\sqrt C}{Z_0+R_0}U_E+|\delta|I_0+I_0\right)^{n-1}e^{(n-1)\mu_0}\right)\right].$$

Prior to estimating the second term we should remember Remark 5.3.3:

$$\dot I_2\leq\frac{1}{t_{k+1}-t_k}\int_{t_k}^{t_{k+1}}e^{\mu(s-t_k)}ds\left[\frac{|\delta|}{\mu C_0 Z_0}\rho_\mu^{(q)}(\dot J,\dot{\bar J})+\frac{1}{\mu C_0 Z_0}\rho_\mu^{(k)}(\dot J,\dot{\bar J})+\right.$$

$$\left.+\frac{1}{C_0}\left(\frac{|\delta|\rho_\mu^{(q)}(\dot J,\dot{\bar J})}{\mu}+\frac{\rho_\mu^{(k)}(\dot J,\dot{\bar J})}{\mu}\right)\sum_{n=1}^{p}r_n\left(\frac{1}{2\sqrt C}\right)^{n-1}\left(\frac{2Z_0\sqrt C}{Z_0+R_0}U_E+|\delta|I_0+I_0\right)^{n-1}e^{(n-1)\mu_0}\right]\leq$$

$$\leq\hat\rho_\mu^{(k)}((J,\dot J),(\bar J,\dot{\bar J}))\frac{e^{\mu_0}-1}{\mu_0}\frac{|\delta|+1}{\mu C_0}\left(\frac{1}{Z_0}+\sum_{n=1}^{p}r_n\left(\frac{1}{2\sqrt C}\right)^{n-1}\left(\frac{2Z_0\sqrt C}{Z_0+R_0}U_E+|\delta|I_0+I_0\right)^{n-1}e^{(n-1)\mu_0}\right);$$

Then

$$\left|\dot B_J^{(k)}(J)(t)-\dot B_J^{(k)}(\bar J)(t)\right|\leq$$

$$\leq e^{\mu(t-t_k)}\hat\rho_\mu^{(k)}((J,\dot J),(\bar J,\dot{\bar J}))\left[|\delta|+\frac{|\delta|+1}{\mu C_0}\left(\frac{1}{Z_0}+\sum_{n=1}^{p}r_n\left(\frac{1}{2\sqrt C}\right)^{n-1}\left(\frac{2Z_0\sqrt C}{Z_0+R_0}U_E+|\delta|I_0+I_0\right)^{n-1}e^{(n-1)\mu_0}\right)\right]+$$

$$+e^{\mu(t-t_k)}\hat\rho_\mu^{(k)}((J,\dot J),(\bar J,\dot{\bar J}))\frac{e^{\mu_0}-1}{\mu_0}\frac{|\delta|+1}{\mu C_0}\left(\frac{1}{Z_0}+\sum_{n=1}^{p}r_n\left(\frac{1}{2\sqrt C}\right)^{n-1}\left(\frac{2Z_0\sqrt C}{Z_0+R_0}U_E+|\delta|I_0+I_0\right)^{n-1}e^{(n-1)\mu_0}\right)\leq$$

$$\leq e^{\mu(t-t_k)}\hat\rho_\mu^{(k)}((J,\dot J),(\bar J,\dot{\bar J}))\left[|\delta|+\left(1+\frac{e^{\mu_0}-1}{\mu_0}\right)\frac{|\delta|+1}{\mu C_0}\left(\frac{1}{Z_0}+\sum_{n=1}^{p}r_n\left(\frac{1}{2\sqrt C}\right)^{n-1}\left(\frac{2Z_0\sqrt C}{Z_0+R_0}U_E+|\delta|I_0+I_0\right)^{n-1}e^{(n-1)\mu_0}\right)\right]\equiv$$

$$\equiv e^{\mu(t-t_k)}\dot K_J\,\hat\rho_\mu^{(k)}((J,\dot J),(\bar J,\dot{\bar J})).$$

The above inequalities imply

$$\rho_\mu^{(k)}(\dot B_J^{(k)}(J),\dot B_J^{(k)}(\bar J))\leq\dot K_J\,\hat\rho_\mu^{(k)}((J,\dot J),(\bar J,\dot{\bar J}))$$

or finally

$$\hat{\rho}_{\mu}^{(k)}(B_J(J),\dot{B}_J(J),B_J(\overline{J}),\dot{B}_J(\overline{J})) \le K\,\hat{\rho}_{\mu}^{(k)}((J,\dot{J}),(\overline{J},\dot{\overline{J}}))$$

where $K = \max\{e^{\mu_0}K_J(\mu),\dot{K}_J(\mu)\} < 1$.

Theorem 5.4.1 is thus proved.

5.4.1. Numerical Example

We collect all inequalities implying an existence-uniqueness of a periodic solution:

$$|\delta|I_0 + \frac{2Z_0\sqrt{C}}{Z_0+R_0}U_E +$$

$$+ 4\sinh\left(\frac{RT_0}{L}\right)\frac{L}{RC_0}\left[\frac{2\sqrt{C}}{Z_0+R_0}U_E + \frac{(1+|\delta|)I_0}{Z_0} + 2\sqrt{C}\sum_{n=1}^{p}\left|r_n\right|\left(\frac{Z_0}{Z_0+R_0}U_E + \frac{|\delta|+1}{2\sqrt{C}}I_0\right)^{n}\right] \le I_0$$

$$e^{\mu_0}K_J = e^{\mu_0}\left[\frac{|\delta|}{\mu} + e^{\mu_0}\frac{|\delta|+1}{\mu^2 C_0}\left(\frac{1}{Z_0} + \sum_{n=1}^{p}r_n\left(\frac{e^{\mu_0}}{2\sqrt{C}}\right)^{n}\left(\frac{2Z_0\sqrt{C}}{Z_0+R_0}U_E + |\delta|I_0+I_0\right)^{n-1}\right)\right] < 1;$$

$$\dot{K}_J = |\delta| + \left(1+\frac{e^{\mu_0}-1}{\mu_0}\right)\frac{|\delta|+1}{\mu C_0}\left(\frac{1}{Z_0} + \sum_{n=1}^{p}r_n\left(\frac{Z_0}{Z_0+R_0}U_E + \frac{|\delta|+1}{2\sqrt{C}}I_0\right)^{n-1}e^{(n-1)\mu_0}\right) < 1$$

For the specific parameters $\Lambda = 10\,m$, $S = 4\,mm^2$, $\rho_c = 0{,}0175$,

$$R = \frac{\rho_c\Lambda}{S} = \frac{0{,}0175.100}{4} \approx 0{,}438\,\Omega, \quad L = 0{,}45\,\mu H/m, \quad C = 80\,pF/m,$$

$$v = \frac{1}{\sqrt{LC}} = \frac{1}{\sqrt{0{,}45.10^{-6}.80.10^{-12}}} = \frac{1}{6.10^{-9}} = 1{,}66.10^{8},$$

$$Z_0 = \sqrt{\frac{L}{C}} = \sqrt{\frac{0{,}45.10^{-6}}{80.10^{-12}}} = 75\,\Omega, \quad R_0 = 70\,\Omega, \quad \delta = \frac{15}{135}, \quad C_0 = 10.10^{-12}\,F, \quad E = 1V,$$

$$T = \frac{\Lambda}{v} = \frac{10}{1{,}66.10^{8}} = 6.10^{-8}\,\text{sec .},$$

$$f(u) = -0{,}12u + 0{,}8u^3 \Rightarrow r_1 = -0{,}12; r_2 = 0; r_3 = 0{,}8; |\delta| = \frac{75-70}{75+70} = \frac{5}{145} = \frac{1}{29}.$$

If $\lambda_0 = \dfrac{1}{4} 10^{-3} m$ then

$$f_0 = \frac{1}{\lambda_0 \sqrt{LC}} = \frac{1}{(1/4)10^{-3}.4.10^{-9}} = 10^{12}\, Hz \Rightarrow T_0 = \frac{1}{f_0} = 10^{-12}\, sec.$$

If we choose $\mu = 10^{12}$, then

$$\mu T_0 = 1\,,\ \mu C_0 = 10^{12}.10^{-11} = 10\,;\ \mu^2 C_0 = 10^{24}.10^{-11} = 10^{13}\,,$$

$$\frac{R}{L} = \frac{4,375}{0,2.10^{-6}} \approx 2,2.10^{7}\,,\ \frac{RT_0}{L} = 4,375.10^{6}.10^{-12} = 4,38.10^{-6}\,,\ \frac{RT}{L} = 4,375.10^{6}.6.10^{-8} = 0,2625\,\cdot$$

$$e^{0,2625} \approx 1,3\,,\ \sinh\left(\frac{RT_0}{L}\right) \approx \frac{RT_0}{L}\,;\ \frac{U_E}{Z_0 + R_0} \approx I_0\,.$$

If we take $I_0 = 10^{-6}, U_E = 0,1$ then the above inequalities become

$$|\delta| I_0 + 150\sqrt{80.10^{-12}}\,I_0 +$$

$$+ 4\frac{T_0}{C_0}\left[2\sqrt{80.10^{-12}}\,I_0 + \frac{(1+|\delta|)I_0}{75} + \sum_{n=1}^{3}|r_n|\left(\frac{1}{2\sqrt{C}}\right)^{n-1}\left(150\sqrt{80.10^{-12}}\,I_0 + |\delta| I_0 + I_0\right)^n\right] \leq I_0 \Leftrightarrow$$

$$\Leftrightarrow \frac{1}{29} + 150\sqrt{80.10^{-12}}\,\frac{4.10^{-12}}{10.10^{-12}}\left[2\sqrt{80.10^{-12}} + \frac{1}{75}\frac{30}{29} + \sum_{n=1}^{3}|r_n|\left(\frac{I_0}{2\sqrt{80.10^{-12}}}\right)^{n-1}\left(150\sqrt{80.10^{-12}} + \frac{30}{29}\right)^n\right] \leq 1$$

$$\dot{K}_J = 0,03 + 0,38 \approx 0,41 < 1\,.$$

5.5. OSCILLATORY REGIMES FOR NEUTRAL EQUATIONS WITH EXPONENTIAL NONLINEARITY

Here we consider

$$\frac{du(t)}{dt} = \frac{2E(t)}{(Z_0 + R_0)C_0} - \frac{1}{Z_0 C_0}u(t) - \frac{1}{C_0}f(u(t)) - e^{-2\frac{R}{L}T}\frac{\delta}{Z_0 C_0}u(t - 2T) + \frac{\delta}{C_0}e^{-2\frac{R}{L}T}f(u(t-2T)) +$$

$$+ \delta e^{-2\frac{R}{L}T}\frac{du(t-2T)}{dt} \tag{5.5.1}$$

with an exponential *V-I* characteristic of to the nonlinear resistive element:

$$i = f(u) = i_{diff} + i_{tunn} = \hat{I}_0(e^{\frac{q}{k\tau}u} - 1) + \frac{I_{max}}{U_{max}}u\,e^{1 - \frac{u}{U_{max}}}\,,$$

and using denotations $\alpha = \dfrac{q}{k\tau}$, $U_m = U_{max}$, $I_m = I_{max}$ (5.5.1), which becomes

$$i = f(u) = \hat{I}_0(e^{\alpha u} - 1) + \frac{I_m}{U_m} u e^{1 - \frac{u}{U_m}}$$

where $\hat{I}_0, I_m, \alpha, U_m$ are positive constants.

Next, we deal with the problem of an existence-uniqueness of an oscillatory solution of the neutral equation with exponential nonlinearity:

$$\frac{du(t)}{dt} = \frac{2E(t)}{(Z_0 + R_0)C_0} - \frac{1}{Z_0 C_0} u(t) - \frac{\hat{I}_0}{C_0}\left(e^{\alpha u(t)} - 1\right) - \frac{I_m}{C_0 U_m} u e^{1 - \frac{u(t)}{U_m}} - e^{-2\frac{R}{L}T} \frac{\delta}{Z_0 C_0} u(t - 2T) +$$

$$+ \frac{\delta}{C_0} e^{-2\frac{R}{L}T} \hat{I}_0\left(e^{\alpha u(t - 2T)} - 1\right) + \frac{\delta}{C_0} e^{-2\frac{R}{L}T} \frac{I_m}{U_m} u(t - 2T) e^{1 - \frac{u(t - 2T)}{U_m}} + \delta e^{-2\frac{R}{L}T} \frac{du(t - 2T)}{dt} \qquad (5.5.2)$$

$$u(t) = \upsilon_0(t), \ \dot{u}(t) = \dot{\upsilon}_0(t) \ \text{for } t \in [-t_0, t_0].$$

Now we are able to formulate the main problem: to find a solution of (5.5.2) with advanced prescribed zeros on an interval $[t_0, \infty)$, $T = t_0$, where $\upsilon_0(t)$ is a prescribed oscillating function on the interval $[-T, T]$.

Let $S_T = \{\tau_k\}_{k=0}^n, n \in N$ be the set of zeros of the initial function, that is, $\upsilon_0(\tau_k) = 0$ such that $\tau_0 = -T$, $\tau_n = T \equiv t_0$. In addition $\max\{\tau_{k+1} - \tau_k : k = 0,1,\ldots,n\} \le T_0$.

Let $S = \{t_k\}_{k=0}^\infty$ be a strictly increasing sequence of real numbers satisfying the following conditions **(C)**:

(C1) $\lim\limits_{k \to \infty} t_k = \infty$;

(C2) for every k there is $s < k$ such that $t_k - T = t_s$ where $t_s \in S_T \cup S$.

It follows

(C3) $0 < \inf\{t_{k+1} - t_k : k = 0,1,2,\ldots\} \le \sup\{t_{k+1} - t_k : k = 0,1,2,\ldots\} = T_0 < \infty$.

The above introduces the sets $C^1[t_0, \infty)$ consisting of all continuous and bounded functions differentiable with bounded derivatives on every interval $[t_k, t_{k+1}]$ and

$$M_S = \left\{u(.) \in C^1[t_0, \infty) : u(t_k) = 0; k = 0,1,2,\ldots \right\},$$

$$M_{SU} = \left\{ u(.) \in M_S : |u(t)| \leq U_0 e^{-\frac{R}{L}t}, t \in [t_k, t_{k+1}] \right\},$$

where U_0, μ are positive constants.

Recall that

(U) $|u(t)| < U_0 e^{\mu T_0} \leq U_m$.

(G) $g(u) = u e^{1-\frac{u}{U_m}}$,

(G1) $g'(u) = e^{1-\frac{u}{U_m}}\left(1 - \frac{u}{U_m}\right) \Rightarrow |g'(u)| \leq \left(1 + \frac{U_0 e^{\mu 0}}{U_m}\right) e^{1+\frac{U_0 e^{\mu 0}}{U_m}} \leq \left(1 + \frac{U_m}{U_m}\right) e^{1+\frac{U_m}{U_m}} \leq 2e^2$

and also $|v_0(t)| \leq U_0 e^{-\frac{R}{L}t}, t \in [-t_0, t_0]$.

Introduce the following family of pseudo-metrics

$$\rho^{(k)}(u, \overline{u}) = \max\left\{|u(t) - \overline{u}(t)| : t \in [t_k, t_{k+1}]\right\},$$
$$\hat{\rho}^{(k)}(u, \overline{u}) = \max\left\{|u(t) - \overline{u}(t)| : t \in [t_0, t_{k+1}]\right\},$$

$$\rho_\mu^{(k)}(u, \overline{u}) = \max\left\{e^{-\mu(t-t_k)}|u(t) - \overline{u}(t)| : t \in [t_k, t_{k+1}]\right\},$$
$$\hat{\rho}_\mu^{(k)}(u, \overline{u}) = \max\left\{\rho_\mu^{(0)}(u, \overline{u}), \rho_\mu^{(1)}(u, \overline{u}), ..., \rho_\mu^{(k)}(u, \overline{u})\right\},$$

$$\rho_\mu^{(k)}(\dot{u}, \dot{\overline{u}}) = \max\left\{e^{-\mu(t-t_k)}|\dot{u}(t) - \dot{\overline{u}}(t)| : t \in [t_k, t_{k+1}]\right\},$$
$$\hat{\rho}_\mu^{(k)}(\dot{u}, \dot{\overline{u}}) = \max\left\{\rho_\mu^{(0)}(\dot{u}, \dot{\overline{u}}), \rho_\mu^{(1)}(\dot{u}, \dot{\overline{u}}), ..., \rho_\mu^{(k)}(\dot{u}, \dot{\overline{u}})\right\}.$$

Remark 5.5.1. The following inequalities imply the equivalence of the both families of pseudo-metrics

$$\rho_\mu^{(k)}(u, \overline{u}) \leq \rho^{(k)}(u, \overline{u}) \leq e^{\mu T_0} \rho_\mu^{(k)}(u, \overline{u}), \ (k = 0, 1, 2, ...).$$

It is easy to verify that

$$\hat{\rho}^{(k)}(u, \overline{u}) = \max\left\{\rho^{(0)}(u, \overline{u}), \rho^{(1)}(u, \overline{u}), ..., \rho^{(k)}(u, \overline{u})\right\} \leq$$
$$\leq e^{\mu T_0} \max\left\{\rho_\mu^{(0)}(u, \overline{u}), \rho_\mu^{(1)}(u, \overline{u}), ..., \rho_\mu^{(k)}(u, \overline{u})\right\} = e^{\mu T_0} \hat{\rho}_\mu^{(k)}(u, \overline{u}).$$

The set M_{SU} turns out into a complete uniform space with respect to the saturated family of pseudo-metrics

$$\hat{\rho}_{\mu}^{(k)}((u,\dot{u}),(\overline{u},\dot{\overline{u}})) = \max\{\hat{\rho}^{(k)}(u,\overline{u}),\hat{\rho}_{\mu}^{(k)}(\dot{u},\dot{\overline{u}})\}, (k = 0,1,2,...) .$$

Define an operator B by the formulas

$$B(u)(t):= \int_{t_k}^{t} U(u)(s)ds - \frac{t-t_k}{t_{k+1}-t_k} \int_{t_k}^{t_{k+1}} U(u)(s)ds, \quad t \in [t_k,t_{k+1}], (k = 0,1,2,...) ,$$

where

$$U(u)(t) = \frac{2E(t)}{(Z_0 + R_0)C_0} - \frac{1}{Z_0 C_0}u(t) - \frac{\hat{I}_0}{C_0}\left(e^{\alpha u(t)} - 1\right) - \frac{I_m}{C_0 U_m}ue^{1-\frac{u(t)}{U_m}} - e^{-2\frac{R}{L}T}\frac{\delta}{Z_0 C_0}\ddot{u}(t)+$$

$$+\frac{\delta}{C_0}e^{-2\frac{R}{L}T}\hat{I}_0\left(e^{\alpha\ddot{u}(t)} - 1\right)+\frac{\delta}{C_0}e^{-2\frac{R}{L}T}\frac{I_m}{U_m}\ddot{u}(t)e^{1-\frac{\ddot{u}(t)}{U_m}} + \delta e^{-2\frac{R}{L}T}\frac{d\ddot{u}(t)}{dt} \qquad (5.5.3)$$

where $\ddot{u}(t)$ is defined in § 5.3.

Since $t_k - T = t_s$ we have $t_k - 2T = t_m$.

Obviously $u(t - 2T)$ coincides with $\upsilon_0(t)$ on $[t_0,t_0 + 2T]$ but all zeros of $u(t-2T)$ belong to $S_T \cup S$. In other words $u(t - 2T)\in M_{SU}$.

Lemma 5.5.1. Problem (5.5.2) has a solution $u(.)\in M_{SU}$ iff the operator B has a fixed point in M_{SU} , that is,

$$u = B(u) .$$

Proof: Let $u(.)\in M_{SU}$ be a solution of (5.5.2). Then integrating (5.5.2) on every interval $[t_k,t]\subset [t_k,t_{k+1}]$ $(k = 0,1,2 ...)$ we obtain

$$u(t)-u(t_k) = \int_{t_k}^{t}U(u)(s)ds \Leftrightarrow u(t) = \int_{t_k}^{t}U(u)(s)ds \text{ and then}$$

$$u(t)=\int_{t_k}^{t}U(u)(s)ds \Rightarrow 0 = u(t_{k+1}) = \int_{t_k}^{t}U(u)(s)ds \Rightarrow \int_{t_k}^{t_{k+1}}U(u)(s)ds = 0 \qquad (5.5.4)$$

Therefore $u(t)$ satisfies

$$u(t) = \int_{t_k}^{t} U(u)(s)ds - \frac{t-t_k}{t_{k+1}-t_k}\int_{t_k}^{t_{k+1}} U(u)(s)ds,\ t\in[t_k,t_{k+1}]\ \Leftrightarrow\ u(t)=B(u)(t),$$

that is, $u(t)$ is a fixed point of B.

Conversely, let $u(.)\in M_{SU}$ be a solution of $u(t)=B(u)(t)$ or

$$u(t) = \int_{t_k}^{t} U(u)(s)ds - \frac{t-t_k}{t_{k+1}-t_k}\int_{t_k}^{t_{k+1}} U(u)(s)ds.$$

Then having in mind that $\mu_0 = \mu T_0$, $|\delta| = \left|\dfrac{Z_0 - R_0}{Z_0 + R_0}\right| < 1$, $\left|e^u - 1\right| < e^{|u|} - 1 = |u|\xi(|u|)$,

$\xi(h) = \dfrac{e^h - 1}{h}$ and $t_k - 2T = t_q$, $t_{k+1} - 2T = t_{q+1}$, $t_k - T = t_p, t_{k+1} - T = t_{p+1}$ we obtain

$$\left|\int_{t_k}^{t_{k+1}} U(u)(s)ds\right| \le \frac{2}{C_0(Z_0+R_0)}\int_{t_k}^{t_{k+1}}|E(t-T)|dt + \frac{1}{C_0 Z_0}\int_{t_k}^{t_{k+1}}|u(t)|dt +$$

$$+\frac{1}{C_0}\left(\widehat{I}_0(e^{\alpha U_m}-1)+\frac{I_m}{U_m}e^2\right)\int_{t_k}^{t_{k+1}}e^{\mu(s-t_k)}ds + \frac{|\delta|}{Z_0 C_0}e^{-2\frac{R}{L}T}\int_{t_k}^{t_{k+1}}|u(t-2T)|dt +$$

$$+\frac{|\delta|}{C_0}e^{-2\frac{R}{L}T}\left(\widehat{I}_0(e^{\alpha U_m}-1)+\frac{I_m}{U_m}e^2\right)\int_{t_k}^{t_{k+1}}e^{\mu(s-t_k)}ds + |\delta|e^{-2\frac{R}{L}T}\left|\int_{t_k}^{t_{k+1}}\dot{u}(s-2T)ds\right| \le$$

$$\le \frac{2U_E e^{\mu_0}}{C_0(Z_0+R_0)}\frac{e^{\mu(t_{k+1}-t_k)}-1}{\mu} + \frac{U_0}{C_0 Z_0}\frac{e^{\mu(t_{k+1}-t_k)}-1}{\mu} +$$

$$+\frac{1}{C_0}\left(\widehat{I}_0(e^{\alpha U_m}-1)+\frac{I_m}{U_m}e^2\right)\frac{e^{\mu(t_{k+1}-t_k)}-1}{\mu} + \frac{|\delta|U_0}{Z_0 C_0}e^{-2\frac{R}{L}T}\frac{e^{\mu(t_{k+1}-t_k)}-1}{\mu} +$$

$$+\frac{|\delta|}{C_0}e^{-2\frac{R}{L}T}\left(\widehat{I}_0(e^{\alpha U_m}-1)+\frac{I_m}{U_m}e^2\right)\frac{e^{\mu(t_{k+1}-t_k)}-1}{\mu} + |\delta|e^{-2\frac{R}{L}T}\left|u(t_{k+1}-2T)-u(t_k-2T)\right| \le$$

$$\le \frac{e^{\mu_0}-1}{\mu C_0}\left[\frac{2U_E}{Z_0+R_0} + \left(1+|\delta|e^{-2\frac{R}{L}T}\right)\left(\frac{U_0}{Z_0}+\widehat{I}_0(e^{\alpha U_m}-1)+\frac{I_m e^2}{U_m}\right)\right] \equiv M_0(\mu).$$

Obviously $\lim_{\mu \to \infty} M_0(\mu) = 0$ and consequently $\int_{t_k}^{t_{k+1}} U(u)(s)ds = 0$. Then

$$u(t) = \int_{t_k}^{t} U(u)(s)ds - \frac{t - t_k}{t_{k+1} - t_k} \int_{t_k}^{t_{k+1}} U(u)(s)ds \Leftrightarrow u(t) = \int_{t_k}^{t} U(u)(s)ds.$$

Differentiating the last integral equation we obtain (5.5.2).

Lemma 5.5.1 is thus proved.

Theorem 5.5.1. Let the following conditions be fulfilled:

1) The initial function $\upsilon_0(.) \in C^1[-T,T]$ satisfies

$$|\upsilon_0(t)| \leq U_0 e^{-\frac{R}{L}t}, t \in [-t_0, t_0], \upsilon_0(t_0) = 0;$$

2) $E(.) \in C_S^1[0,\infty), |E(t)| \leq U_E e^{-\frac{R}{L}t}, t \in [0,\infty)$.

Then there exists a unique oscillatory solution of the initial value problem (5.5.2), belonging to M_{SU}.

Proof: We show that B maps M_{SU} into itself, that is, $u \in M_{SU}$ implies $B(u) \in M_{SU}$.

Indeed, $B(u)(t)$ is continuous on $[t_0, \infty)$ and differentiable on every $[t_k, t_{k+1}]$. We also have $B(u)(t_k) = 0$ and $B(u)(t_{k+1}) = 0$.

We show that $|B_k(u)(t)| \leq U_0 e^{-\frac{R}{L}t}, t \in [t_0, \infty)$.

Note that for every $u(.) \in M_{SU}$ the function $B(u)(t)$ is continuous and differentiable

on every (t_k, t_{k+1}). We notice that $\left|\dfrac{t - t_k}{t_{k+1} - t_k}\right| \leq 1, t \in [t_k, t_{k+1}]$. We show that for

$t \in [t_k, t_{k+1}]$:

$$|B_k(u)(t)| \leq U_0 e^{-\frac{R}{L}t}.$$

Indeed, for every $u(.) \in M_{SU}$ in view of the inequalities from 1.6.1

$$|e^u - 1| < e^{|u|} - 1 = |u|\left(1 + \frac{|u|}{2!} + \frac{|u|^2}{3!} + ...\right) \equiv |u|\xi(|u|)$$

and

(U) $\left|u(t)\right| < U_0 e^{\mu T_0} \le U_m$

we obtain for every $t \in [t_k, t_{k+1}]$

$$\left|B_k(u)(t)\right| \le \left|\int_{t_k}^{t} U(u)(s)\,ds\right| + \left|\frac{t - t_k}{t_{k+1} - t_k}\right| \left|\int_{t_k}^{t_{k+1}} U(u)(s)\,ds\right| \equiv B_1 + B_2.$$

But

$$B_1 \le \frac{2}{C_0(Z_0 + R_0)} \int_{t_k}^{t} |E(s)|\,ds + \frac{1}{C_0 Z_0} \int_{t_k}^{t} |u(s)|\,ds + \frac{\widehat{I}_0}{C_0} \int_{t_k}^{t} \left|e^{\alpha u(s)} - 1\right|\,ds + \frac{I_m}{C_0 U_m} \int_{t_k}^{t} |u(s)| e^{1 - \frac{u(s)}{U_m}}\,ds +$$

$$+ \frac{|\delta|}{C_0 Z_0} e^{-2\frac{R}{L}T} \int_{t_k}^{t} |u(s - 2T)|\,ds + \frac{|\delta|\widehat{I}_0}{C_0} e^{-2\frac{R}{L}T} \int_{t_k}^{t} \left|e^{\alpha u(s-2T)} - 1\right|\,ds + \frac{|\delta|I_m}{C_0 U_m} e^{-2\frac{R}{L}T} \int_{t_k}^{t} |u(s-2T)| e^{1 - \frac{u(s-2T)}{U_m}}\,ds +$$

$$+ |\delta| e^{-2\frac{R}{L}T} \left|\int_{t_k}^{t} \dot{u}(s - 2T)\,ds\right| \le$$

$$\le \frac{2}{C_0(Z_0 + R_0)} \int_{t_k}^{t} |E(s)|\,ds + \frac{1}{C_0 Z_0} \int_{t_k}^{t} |u(s)|\,ds + \frac{\alpha \widehat{I}_0}{C_0} \int_{t_k}^{t} |u(s)|\xi(\alpha|u(s)|)\,ds + \frac{I_m}{C_0 U_m} \int_{t_k}^{t} |u(s)| e^{1 - \frac{u(s)}{U_m}}\,ds +$$

$$+ \frac{|\delta|}{C_0 Z_0} e^{-2\frac{R}{L}T} \int_{t_k}^{t} |u(s - 2T)|\,ds + \frac{\alpha|\delta|\widehat{I}_0}{C_0} e^{-2\frac{R}{L}T} \int_{t_k}^{t} |u(s-2T)|\xi(\alpha|u(s-2T)|)\,ds +$$

$$+ \frac{|\delta|I_m}{C_0 U_m} e^{-2\frac{R}{L}T} \int_{t_k}^{t} |u(s-2T)| e^{1 - \frac{u(s-2T)}{U_m}}\,ds + |\delta| e^{-2\frac{R}{L}T} |u(t - 2T)| \le$$

$$\le \frac{2U_E}{C_0(Z_0 + R_0)} \int_{t_k}^{t} e^{-\frac{R}{L}s}\,ds + \left(\frac{1}{C_0 Z_0} + \frac{\alpha \widehat{I}_0}{C_0}\xi(\alpha U_m) + \frac{I_m}{C_0 U_m} e^2\right) \int_{t_k}^{t} |u(s)|\,ds +$$

$$+ |\delta| e^{-2\frac{R}{L}T} \left(\frac{1}{C_0 Z_0} + \frac{\alpha \widehat{I}_0}{C_0}\xi(\alpha U_m) + \frac{I_m}{C_0 U_m} e^2\right) \int_{t_k}^{t} |u(s - 2T)|\,ds + |\delta| e^{-2\frac{R}{L}T} |u(t - 2T)| \le$$

$$\le \frac{2U_E}{C_0(Z_0 + R_0)} \int_{t_k}^{t} e^{-\frac{R}{L}s}\,ds + \left(\frac{1}{C_0 Z_0} + \frac{\alpha \widehat{I}_0}{C_0}\xi(\alpha U_m) + \frac{I_m}{C_0 U_m} e^2\right) U_0 \int_{t_k}^{t} e^{-\frac{R}{L}s}\,ds +$$

$$+ |\delta| e^{-2\frac{R}{L}T} \left(\frac{1}{C_0 Z_0} + \frac{\alpha \widehat{I}_0}{C_0}\xi(\alpha U_m) + \frac{I_m}{C_0 U_m} e^2\right) U_0 \int_{t_k}^{t} e^{-\frac{R}{L}(s-2T)}\,ds + |\delta| e^{-2\frac{R}{L}T} U_0 e^{-\frac{R}{L}(t-2T)} \le$$

$$\le \int_{t_k}^{t} e^{-\frac{R}{L}s}\,ds \left[\frac{2U_E}{C_0(Z_0 + R_0)} + U_0 \frac{(1 + |\delta|)}{C_0}\left(\frac{1}{Z_0} + \alpha \widehat{I}_0 \xi(\alpha U_m) + \frac{I_m}{U_m} e^2\right) + |\delta| U_0\right] \le$$

$$\leq \frac{L}{R}\left(e^{-\frac{R}{L}t_k} - e^{-\frac{R}{L}t}\right)\left[\frac{2U_E}{C_0(Z_0+R_0)} + \frac{U_0\left(1+|\delta|\right)}{C_0}\left(\frac{1}{Z_0} + \alpha\widehat{I}_0\xi(\alpha U_m) + \frac{I_m}{U_m}e^2\right) + |\delta|U_0\right] \leq$$

$$\leq e^{-\frac{R}{L}t}\frac{L}{R}\left(e^{\frac{R}{L}(t-t_k)} - 1\right)\left[\frac{2U_E}{C_0(Z_0+R_0)} + \frac{U_0\left(1+|\delta|\right)}{C_0}\left(\frac{1}{Z_0} + \alpha\widehat{I}_0\xi(\alpha U_m) + \frac{I_m}{U_m}e^2\right) + |\delta|U_0\right] \leq$$

$$\leq e^{-\frac{R}{L}t}\frac{L}{R}\left(e^{\frac{R}{L}T_0} - 1\right)\left[\frac{2U_E}{C_0(Z_0+R_0)} + \frac{U_0\left(1+|\delta|\right)}{C_0}\left(\frac{1}{Z_0} + \alpha\widehat{I}_0\xi(\alpha U_m) + \frac{I_m}{U_m}e^2\right) + |\delta|U_0\right] \leq$$

$$\leq e^{-\frac{R}{L}t}\frac{L}{R}\left(e^{\frac{R}{L}T_0} - e^{-\frac{R}{L}T_0}\right)\left[\frac{2U_E}{C_0(Z_0+R_0)} + \frac{U_0\left(1+|\delta|\right)}{C_0}\left(\frac{1}{Z_0} + \alpha\widehat{I}_0\xi(\alpha U_m) + \frac{I_m}{U_m}e^2\right) + |\delta|U_0\right].$$

Further on we use

$$e^{-\frac{R}{L}t_k} - e^{-\frac{R}{L}t_{k+1}} \leq e^{-\frac{R}{L}t}\left(e^{\frac{R}{L}T_0} - e^{-\frac{R}{L}T_0}\right).$$

Then

$$B_2 \leq \frac{2}{C_0(Z_0+R_0)}\int_{t_k}^{t_{k+1}}|E(s)|ds + \frac{1}{C_0Z_0}\int_{t_k}^{t_{k+1}}|u(s)|ds + \frac{\widehat{I}_0}{C_0}\int_{t_k}^{t_{k+1}}\left|e^{\alpha u(s)} - 1\right|ds + \frac{I_m}{C_0U_m}\int_{t_k}^{t_{k+1}}|u(s)|e^{1-\frac{u(s)}{U_m}}ds +$$

$$+ \frac{|\delta|}{C_0Z_0}e^{-2\frac{R}{L}T}\int_{t_k}^{t_{k+1}}|u(s-2T)|ds + \frac{|\delta|\widehat{I}_0}{C_0}e^{-2\frac{R}{L}T}\int_{t_k}^{t_{k+1}}\left|e^{\alpha u(s-2T)} - 1\right|ds + \frac{|\delta|I_m}{C_0U_m}e^{-2\frac{R}{L}T}\int_{t_k}^{t_{k+1}}|u(s-2T)|e^{1-\frac{u(s-2T)}{U_m}}ds +$$

$$+ |\delta|e^{-2\frac{R}{L}T}\left|\int_{t_k}^{t_{k+1}}\dot{u}(s-2T)ds\right| \leq$$

$$\leq \frac{2}{C_0(Z_0+R_0)}\int_{t_k}^{t_{k+1}}|E(s)|ds + \frac{1}{C_0Z_0}\int_{t_k}^{t_{k+1}}|u(s)|ds + \frac{\alpha\widehat{I}_0}{C_0}\int_{t_k}^{t_{k+1}}|u(s)|\xi(\alpha|u(s)|)ds + \frac{I_m}{C_0U_m}\int_{t_k}^{t_{k+1}}|u(s)|e^{1-\frac{u(s)}{U_m}}ds +$$

$$+ \frac{|\delta|}{C_0Z_0}e^{-2\frac{R}{L}T}\int_{t_k}^{t_{k+1}}|u(s-2T)|ds + \frac{\alpha|\delta|\widehat{I}_0}{C_0}e^{-2\frac{R}{L}T}\int_{t_k}^{t_{k+1}}|u(s-2T)|\xi(\alpha|u(s-2T)|)ds +$$

$$+ \frac{|\delta|I_m}{C_0U_m}e^{-2\frac{R}{L}T}\int_{t_k}^{t_{k+1}}|u(s-2T)|e^{1-\frac{u(s-2T)}{U_m}}ds \leq$$

$$\leq \frac{2U_E}{C_0(Z_0+R_0)}\int_{t_k}^{t_{k+1}}e^{-\frac{R}{L}s}ds + \left(\frac{1}{C_0Z_0} + \frac{\alpha\widehat{I}_0}{C_0}\xi(\alpha U_m) + \frac{I_m}{C_0U_m}e^2\right)U_0\int_{t_k}^{t_{k+1}}e^{-\frac{R}{L}s}ds +$$

$$+|\delta|e^{-2\frac{R}{L}T}\left(\frac{1}{C_0Z_0}+\frac{\alpha\hat{I}_0}{C_0}\xi(\alpha U_m)+\frac{I_m}{C_0U_m}e^2\right)U_0\int_{t_k}^{t_{k+1}}e^{-\frac{R}{L}(s-2T)}ds\le$$

$$\le\int_{t_k}^{t_{k+1}}e^{-\frac{R}{L}s}ds\left[\frac{2U_E}{C_0(Z_0+R_0)}+U_0\frac{(1+|\delta|)}{C_0}\left(\frac{1}{Z_0}+\alpha\hat{I}_0\xi(\alpha U_m)+\frac{I_m}{U_m}e^2\right)\right]\le$$

$$\le\frac{L}{C_0R}\left(e^{-\frac{R}{L}t_k}-e^{-\frac{R}{L}t_{k+1}}\right)\left[\frac{2U_E}{Z_0+R_0}+U_0(1+|\delta|)\left(\frac{1}{Z_0}+\alpha\hat{I}_0\xi(\alpha U_m)+\frac{I_m}{U_m}e^2\right)\right]\le$$

$$\le e^{-\frac{R}{L}t}\left(e^{\frac{R}{L}T_0}-e^{-\frac{R}{L}T_0}\right)\frac{L}{C_0R}\left[\frac{2U_E}{Z_0+R_0}+U_0(1+|\delta|)\left(\frac{1}{Z_0}+\alpha\hat{I}_0\xi(\alpha U_m)+\frac{I_m}{U_m}e^2\right)\right]$$

Therefore

$$|B_k(u)(t)|\le$$

$$\le e^{-\frac{R}{L}t}\left(e^{\frac{R}{L}T_0}-e^{\frac{R}{L}T_0}\right)\frac{L}{R}\left[\frac{2U_E}{C_0(Z_0+R_0)}+\frac{U_0(1+|\delta|)}{C_0}\left(\frac{1}{Z_0}+\alpha\hat{I}_0\xi(\alpha U_m)+\frac{I_m}{U_m}e^2\right)+|\delta|U_0\right]+$$

$$+e^{-\frac{R}{L}t}\left(e^{\frac{R}{L}T_0}-e^{-\frac{R}{L}T_0}\right)\frac{L}{R}\left[\frac{2U_E}{C_0(Z_0+R_0)}+\frac{U_0(1+|\delta|)}{C_0}\left(\frac{1}{Z_0}+\alpha\hat{I}_0\xi(\alpha U_m)+\frac{I_m}{U_m}e^2\right)\right]\le$$

$$\le e^{-\frac{R}{L}t}2\sinh\left(\frac{RT_0}{L}\right)\frac{L}{R}\left[\frac{4U_E}{C_0(Z_0+R_0)}+\frac{2U_0(1+|\delta|)}{C_0}\left(\frac{1}{Z_0}+\alpha\hat{I}_0\xi(\alpha U_m)+\frac{I_m}{U_m}e^2\right)+|\delta|U_0\right]\le e^{-\frac{R}{L}t}U_0$$

Consequently the operator B maps M_{SU} into itself.

We show that B is a contractive operator:

$$|B_k(u)(t)-B_k(\overline{u})(t)|\le\left|\int_{t_k}^{t}(U(u)(s)-U(\overline{u})(s))ds\right|+$$

$$\left|\frac{t-t_k}{t_{k+1}-t_k}\right|\left|\int_{t_k}^{t_{k+1}}(U(u)(s)-U(\overline{u})(s))ds\right|\equiv B_1+B_2.$$

Let $t\in[t_k,t_{k+1}]$. Then recalling that $0<U_0e^{\mu T_0}=U_0e^{\mu 0}\le U_m$ and

$$g(u)=ue^{1-\frac{u}{U_m}},\ g'(u)=e^{1-\frac{u}{U_m}}\left(1-\frac{u}{U_m}\right)\Rightarrow|g'(u)|\le\left(1+\frac{U_0e^{\mu T_0}}{U_m}\right)e^{1+\frac{U_0e^{\mu T_0}}{U_m}}\le 2e^2$$

we obtain

$$
B_1 \le \frac{1}{C_0 Z_0} \int\limits_{t_k}^{t} |u(s) - \overline{u}(s)| ds + \frac{\widehat{I}_0}{C_0} \int\limits_{t_k}^{t} \left| e^{\alpha u(s)} - e^{\alpha \overline{u}(s)} \right| ds +
$$

$$
+ \frac{I_m}{C_0 U_m} \int\limits_{t_k}^{t} \left| u(s) e^{1 - \frac{u(s)}{U_m}} - \overline{u}(s) e^{1 - \frac{\overline{u}(s)}{U_m}} \right| ds +
$$

$$
+ \frac{|\delta|}{Z_0 C_0} e^{-2\frac{RT}{L}} \int\limits_{t_k}^{t} |u(s - 2T) - \overline{u}(s - 2T)| ds + \frac{|\delta|\widehat{I}_0}{C_0} e^{-2\frac{RT}{L}} \int\limits_{t_k}^{t} \left| e^{\alpha u(s - 2T)} - e^{\alpha \overline{u}(s - 2T)} \right| ds +
$$

$$
+ \frac{|\delta| I_m}{C_0 U_m} e^{-2\frac{RT}{L}} \int\limits_{t_k}^{t} \left| u(s - 2T) e^{1 - \frac{u(s - 2T)}{U_m}} - \overline{u}(s - 2T) e^{1 - \frac{\overline{u}(s - 2T)}{U_m}} \right| ds + |\delta| e^{-2\frac{RT}{L}} \int\limits_{t_k}^{t} |\dot{u}(s - 2T) - \dot{\overline{u}}(s - 2T)| ds \le
$$

$$
\le \frac{\rho^{(k)}(u,\overline{u})}{C_0 Z_0} \frac{e^{\mu(t - t_k)} - 1}{\mu} + \frac{\alpha e^{\alpha U_m} \widehat{I}_0}{C_0} \int\limits_{t_k}^{t} |u(s) - \overline{u}(s)| ds + \frac{2e^2 I_m}{C_0 U_m} \int\limits_{t_k}^{t} |u(s) - \overline{u}(s)| ds +
$$

$$
+ \rho^{(q)}(u,\overline{u}) \frac{|\delta|}{Z_0 C_0} e^{-2\frac{RT}{L}} \frac{e^{\mu(t - t_k)} - 1}{\mu} + \frac{\alpha e^{\alpha U_m} |\delta| \widehat{I}_0}{C_0} e^{-2\frac{RT}{L}} \rho^{(q)}(u,\overline{u}) \frac{e^{\mu(t - t_k)} - 1}{\mu} +
$$

$$
+ \frac{2e^2 |\delta| I_m}{C_0 U_m} \rho^{(q)}(u,\overline{u}) e^{-2\frac{RT}{L}} \frac{e^{\mu(t - t_k)} - 1}{\mu} + |\delta| e^{-2\frac{RT}{L}} e^{\mu 0} \rho_\mu^{(q)}(\dot{u}, \dot{\overline{u}}) \le
$$

$$
\le \frac{\rho^{(k)}(u,\overline{u})}{C_0 Z_0} \frac{e^{\mu(t - t_k)} - 1}{\mu} + \frac{\alpha e^{\alpha U_m} \widehat{I}_0}{C_0} \rho^{(k)}(u,\overline{u}) \frac{e^{\mu(t - t_k)} - 1}{\mu} + \frac{2e^2 I_m}{C_0 U_m} \rho^{(k)}(u,\overline{u}) \frac{e^{\mu(t - t_k)} - 1}{\mu} +
$$

$$
+ \rho^{(q)}(u,\overline{u}) \frac{|\delta|}{Z_0 C_0} e^{-2\frac{RT}{L}} \frac{e^{\mu(t - t_k)} - 1}{\mu} + \frac{\alpha e^{\alpha U_m} |\delta| \widehat{I}_0}{C_0} e^{-2\frac{RT}{L}} \rho^{(q)}(u,\overline{u}) \frac{e^{\mu(t - t_k)} - 1}{\mu} +
$$

$$
+ \frac{2e^2 |\delta| I_m}{C_0 U_m} \rho^{(q)}(u,\overline{u}) e^{-2\frac{RT}{L}} \frac{e^{\mu(t - t_k)} - 1}{\mu} + |\delta| e^{-2\frac{RT}{L}} e^{\mu 0} \rho_\mu^{(q)}(\dot{u}, \dot{\overline{u}}) \le
$$

$$
\le \frac{e^{\mu 0} \widehat{\rho}_\mu^{(k)}(\dot{u}, \dot{\overline{u}})}{\mu} \frac{e^{\mu(t - t_k)}}{\mu C_0} \left(\frac{1}{Z_0} + \alpha e^{\alpha U_m} I_0 + \frac{2e^2 I_m}{U_m} + \frac{|\delta| e^{-2\frac{RT}{L}}}{Z_0} + \alpha e^{\alpha U_m} \widehat{I}_0 |\delta| e^{-2\frac{RT}{L}} + \frac{2e^2 I_m |\delta| e^{-2\frac{RT}{L}}}{U_m} \right) +
$$

$$
+ e^{\mu 0} |\delta| e^{-2\frac{RT}{L}} \widehat{\rho}_\mu^{(k)}(\dot{u}, \dot{\overline{u}}) \le
$$

$$
\le e^{\mu(t - t_k)} \widehat{\rho}_\mu^{(k)}(\dot{u}, \dot{\overline{u}}) e^{\mu 0} \left[\frac{1}{\mu^2 C_0} \left(1 + e^{-2\frac{RT}{L}} |\delta| \right) \left(\frac{1}{Z_0} + \alpha e^{\alpha U_m} \widehat{I}_0 + \frac{2e^2 I_m}{U_m} \right) + |\delta| e^{-2\frac{RT}{L}} \right]
$$

and

$$B_2 \le \frac{1}{C_0 Z_0} \int_{t_k}^{t_{k+1}} |u(s) - \overline{u}(s)| ds + \frac{\widehat{I}_0}{C_0} \int_{t_k}^{t_{k+1}} \left| e^{\alpha u(s)} - e^{\alpha \overline{u}(s)} \right| ds + \frac{I_m}{C_0 U_m} \int_{t_k}^{t_{k+1}} \left| u(s) e^{1 - \frac{u(s)}{U_m}} - \overline{u}(s) e^{1 - \frac{\overline{u}(s)}{U_m}} \right| ds +$$

$$+ \frac{|\delta|}{Z_0 C_0} e^{-2\frac{RT}{L}} \int_{t_k}^{t_{k+1}} |u(s - 2T) - \overline{u}(s - 2T)| ds + \frac{|\delta|\widehat{I}_0}{C_0} e^{-2\frac{RT}{L}} \int_{t_k}^{t_{k+1}} \left| e^{\alpha u(s - 2T)} - e^{\alpha \overline{u}(s - 2T)} \right| ds +$$

$$+ \frac{|\delta|I_m}{C_0 U_m} e^{-2\frac{RT}{L}} \int_{t_k}^{t_{k+1}} \left| u(s - 2T) e^{1 - \frac{u(s - 2T)}{U_m}} - \overline{u}(s - 2T) e^{1 - \frac{\overline{u}(s - 2T)}{U_m}} \right| ds + |\delta| e^{-2\frac{RT}{L}} \int_{t_k}^{t_{k+1}} |\dot{u}(s - 2T) - \dot{\overline{u}}(s - 2T)| ds \le$$

$$\le \frac{\rho^{(k)}(u,\overline{u})}{C_0 Z_0} \frac{e^{\mu(t_{k+1} - t_k)} - 1}{\mu} + \frac{\alpha e^{\alpha U_m}\widehat{I}_0}{C_0} \rho^{(k)}(u,\overline{u}) \frac{e^{\mu(t_{k+1} - t_k)} - 1}{\mu} + \frac{2e^2 I_m}{C_0 U_m} \rho^{(k)}(u,\overline{u}) \frac{e^{\mu(t_{k+1} - t_k)} - 1}{\mu} +$$

$$+ \rho^{(q)}(u,\overline{u}) \frac{|\delta|}{Z_0 C_0} e^{-2\frac{RT}{L}} \frac{e^{\mu(t_{k+1} - t_k)} - 1}{\mu} + \frac{\alpha e^{\alpha U_m}|\delta|\widehat{I}_0}{C_0} \rho^{(q)}(u,\overline{u}) e^{-2\frac{RT}{L}} \frac{e^{\mu(t_{k+1} - t_k)} - 1}{\mu} +$$

$$+ \frac{2e^2 |\delta| I_m}{C_0 U_m} \rho^{(q)}(u,\overline{u}) e^{-2\frac{RT}{L}} \frac{e^{\mu(t_{k+1} - t_k)} - 1}{\mu} \le$$

$$\le \frac{e^{\mu_0} - 1}{\mu C_0} \frac{e^{\mu_0} \hat{\rho}_\mu^{(k)}(\dot{u}, \dot{\overline{u}})}{\mu} \left(\frac{1}{Z_0} + \alpha e^{\alpha U_m}\widehat{I}_0 + \frac{2e^2 I_m}{U_m} + \frac{|\delta| e^{-2\frac{RT}{L}}}{Z_0} + \alpha e^{\alpha U_m}|\delta|\widehat{I}_0 e^{-2\frac{RT}{L}} + \frac{2e^2 |\delta| I_m e^{-2\frac{RT}{L}}}{U_m} \right) \le$$

$$\le e^{\mu(t - t_k)} \hat{\rho}_\mu^{(k)}(\dot{u}, \dot{\overline{u}}) e^{\mu_0} \left[\frac{e^{\mu_0} - 1}{\mu^2 C_0} \left(1 + e^{-2\frac{RT}{L}}|\delta| \right) \left(\frac{1}{Z_0} + \alpha e^{\alpha U_m}\widehat{I}_0 + \frac{2e^2 I_m}{U_m} \right) \right].$$

Consequently

$$\left| B_k(u)(t) - B_k(\overline{u})(t) \right| \le$$

$$\le e^{\mu(t - t_k)} \hat{\rho}_\mu^{(k)}(\dot{u}, \dot{\overline{u}}) e^{\mu_0} \left[\frac{1}{\mu^2 C_0} \left(1 + e^{-2\frac{RT}{L}}|\delta| \right) \left(\frac{1}{Z_0} + \alpha e^{\alpha U_m}\widehat{I}_0 + \frac{2e^2 I_m}{U_m} \right) + |\delta| e^{-2\frac{RT}{L}} \right] +$$

$$+ e^{\mu(t - t_k)} \hat{\rho}_\mu^{(k)}(\dot{u}, \dot{\overline{u}}) e^{\mu_0} \left[\frac{e^{\mu_0} - 1}{\mu^2 C_0} \left(1 + e^{-2\frac{RT}{L}}|\delta| \right) \left(\frac{1}{Z_0} + \alpha e^{\alpha U_m}\widehat{I}_0 + \frac{2e^2 I_m}{U_m} \right) \right] \le$$

$$\le e^{\mu(t - t_k)} \hat{\rho}_\mu^{(k)}(\dot{u}, \dot{\overline{u}}) e^{\mu_0} \left[\frac{e^{\mu_0}}{\mu^2 C_0} \left(1 + e^{-2\frac{RT}{L}}|\delta| \right) \left(\frac{1}{Z_0} + \alpha e^{\alpha U_m}\widehat{I}_0 + \frac{2e^2 I_m}{U_m} \right) + |\delta| e^{-2\frac{RT}{L}} \right] \le$$

$$\le e^{\mu_0} \hat{\rho}_\mu^{(k)}((u,\dot{u}),(\overline{u},\dot{\overline{u}})) e^{\mu_0} \left[\frac{e^{\mu_0}}{\mu^2 C_0} \left(1 + e^{-2\frac{RT}{L}}|\delta| \right) \left(\frac{1}{Z_0} + \alpha e^{\alpha U_m}\widehat{I}_0 + \frac{2e^2 I_m}{U_m} \right) + |\delta| e^{-2\frac{RT}{L}} \right] \equiv$$

$$\equiv e^{\mu_0} K_u \hat{\rho}_\mu^{(k)}((u,\dot{u}),(\overline{u},\dot{\overline{u}})).$$

It follows

$$\hat{\rho}^{(k)}(B(u), B(\overline{u})) \le e^{\mu_0} K_u \hat{\rho}^{(k)}_\mu((u, \dot{u}), (\overline{u}, \dot{\overline{u}})) .$$

For the derivatives we obtain

$$\left| \dot{B}_k(u)(t) - \dot{B}_k(\overline{u})(t) \right| \le$$

$$\left| U(u)(t) - U(\overline{u})(t) \right| + \frac{1}{t_{k+1} - t_k} \left| \int_{t_k}^{t_{k+1}} U(u)(s)ds - \int_{t_k}^{t_{k+1}} U(\overline{u})(s)ds \right| \equiv \dot{P}_1 + \dot{P}_2 .$$

But

$$\dot{P}_1 \le \frac{1}{C_0 Z_0} |u(t) - \overline{u}(t)| + \frac{\widehat{I}_0}{C_0} \left| e^{\alpha u(t)} - e^{\alpha \overline{u}(t)} \right| + \frac{I_m}{C_0 U_m} \left| u(t) e^{1 - \frac{u(t)}{U_m}} - \overline{u}(t) e^{1 - \frac{\overline{u}(t)}{U_m}} \right| +$$

$$+ \frac{|\delta|}{C_0 Z_0} e^{-2\frac{RT}{L}} |u(t - 2T) - \overline{u}(t - 2T)| + \frac{|\delta| \widehat{I}_0}{C_0} e^{-2\frac{RT}{L}} \left| e^{\alpha u(t - 2T)} - e^{\alpha \overline{u}(t - 2T)} \right| +$$

$$+ \frac{|\delta| I_m}{C_0 U_m} e^{-2\frac{RT}{L}} \left| u(t - 2T) e^{1 - \frac{u(t - 2T)}{U_m}} - \overline{u}(t - 2T) e^{1 - \frac{\overline{u}(t - 2T)}{U_m}} \right| + |\delta| e^{-2\frac{RT}{L}} |\dot{u}(t - 2T) - \dot{\overline{u}}(t - 2T)| \le$$

$$\le \frac{e^{\mu(t - t_k)} \rho^{(k)}_\mu(u, \overline{u})}{C_0 Z_0} + \frac{\alpha \widehat{I}_0 e^{\alpha U_m}}{C_0} e^{\mu(t - t_k)} \rho^{(k)}_\mu(u, \overline{u}) + \frac{2e^2 I_m}{C_0 U_m} e^{\mu(t - t_k)} \rho^{(k)}_\mu(u, \overline{u}) +$$

$$+ e^{\mu(t - t_k)} \frac{|\delta|}{C_0 Z_0} e^{-2\frac{RT}{L}} e^{\mu_0} \rho^{(q)}_\mu(u, \overline{u}) + e^{\mu(t - t_k)} \frac{\alpha |\delta| \widehat{I}_0}{C_0} e^{-2\frac{RT}{L}} e^{\mu_0} \rho^{(q)}_\mu(u, \overline{u}) +$$

$$+ e^{\mu(t - t_k)} \frac{2e^2 |\delta| I_m}{C_0 U_m} e^{-2\frac{RT}{L}} e^{\mu_0} \rho^{(q)}_\mu(u, \overline{u}) + e^{\mu(t - t_k)} |\delta| e^{\mu_0} \rho^{(q)}_\mu(\dot{u}, \dot{\overline{u}}) e^{-2\frac{RT}{L}} \le$$

$$\le e^{\mu(t - t_k)} \frac{\hat{\rho}^{(k)}_\mu(\dot{u}, \dot{\overline{u}})}{\mu C_0} \left(\frac{1}{Z_0} + \alpha \widehat{I}_0 e^{\alpha U_m} + \frac{2e^2 I_m}{U_m} + e^{\mu_0} \frac{|\delta| e^{-2\frac{RT}{L}}}{Z_0} + e^{\mu_0} \alpha \widehat{I}_0 |\delta| e^{-2\frac{RT}{L}} + e^{\mu_0} \frac{2e^2 I_m |\delta| e^{-2\frac{RT}{L}}}{U_m} \right) +$$

$$+ e^{\mu(t - t_k)} |\delta| e^{\mu_0} \hat{\rho}^{(k)}_\mu(\dot{u}, \dot{\overline{u}}) e^{-2\frac{RT}{L}} \le$$

$$\le e^{\mu(t - t_k)} \hat{\rho}^{(k)}_\mu(\dot{u}, \dot{\overline{u}}) \left[\frac{1}{\mu C_0} \left(1 + |\delta| e^{\mu_0} e^{-2\frac{RT}{L}} \right) \left(\frac{1}{Z_0} + \alpha \widehat{I}_0 e^{\alpha U_m} + \frac{2e^2 I_m}{U_m} \right) + |\delta| e^{-2\frac{RT}{L}} e^{\mu_0} \right]$$

and

$$\dot{P}_2 \leq \frac{1}{t_{k+1}-t_k}\left(\frac{1}{C_0 Z_0}\int_{t_k}^{t_{k+1}}|u(s)-\overline{u}(s)|ds + \frac{\widehat{I}_0}{C_0}\int_{t_k}^{t_{k+1}}\left|e^{\alpha u(s)}-e^{\alpha \overline{u}(s)}\right|ds + \frac{I_m}{C_0 U_m}\int_{t_k}^{t_{k+1}}\left|u(s)e^{1-\frac{u(s)}{U_m}}-\overline{u}(s)e^{1-\frac{\overline{u}(s)}{U_m}}\right|ds + \right.$$

$$+\frac{|\delta|}{Z_0 C_0}e^{-2\frac{RT}{L}}\int_{t_k}^{t_{k+1}}|u(s-2T)-\overline{u}(s-2T)|ds + \frac{|\delta|\widehat{I}_0}{C_0}e^{-2\frac{RT}{L}}\int_{t_k}^{t_{k+1}}\left|e^{\alpha u(s-2T)}-e^{\alpha \overline{u}(s-2T)}\right|ds +$$

$$+\frac{|\delta|I_m}{C_0 U_m}e^{-2\frac{RT}{L}}\int_{t_k}^{t_{k+1}}\left|u(s-2T)e^{1-\frac{u(s-2T)}{U_m}}-\overline{u}(s-2T)e^{1-\frac{\overline{u}(s-2T)}{U_m}}\right|ds + |\delta|e^{-2\frac{RT}{L}}\int_{t_k}^{t_{k+1}}|\dot{u}(s-2T)-\dot{\overline{u}}(s-2T)|ds\right) \leq$$

$$\leq \frac{\hat{\rho}_\mu^{(k)}(\dot{u},\dot{\overline{u}})}{\mu}\frac{e^{\mu(t-t_k)}-1}{\mu(t_{k+1}-t_k)}\left(\frac{1}{C_0 Z_0}+\frac{\alpha e^{\alpha U_m}\widehat{I}_0}{C_0}+\frac{2e^2 I_m}{C_0 U_m}+\frac{|\delta|e^{\mu_0}e^{-2\frac{RT}{L}}}{Z_0 C_0}+\frac{\alpha e^{\alpha U_m}\widehat{I}_0|\delta|e^{\mu_0}e^{-2\frac{RT}{L}}}{C_0}+\frac{2e^2 I_m|\delta|e^{\mu_0}e^{-2\frac{RT}{L}}}{C_0 U_m}\right) \leq$$

$$\leq \hat{\rho}_\mu^{(k)}(\dot{u},\dot{\overline{u}})\frac{e^{\mu_0}-1}{\mu_0}\left[\frac{1}{\mu^2 C_0}\left(1+|\delta|e^{\mu_0}e^{-2\frac{RT}{L}}\right)\left(\frac{1}{Z_0}+\alpha\widehat{I}_0 e^{\alpha U_m}+\frac{2e^2 I_m}{U_m}\right)\right].$$

Therefore

$$\left|\dot{B}_k(u)(t)-\dot{B}_k(\overline{u})(t)\right| \leq$$

$$\leq e^{\mu(t-t_k)}\hat{\rho}_\mu^{(k)}(\dot{u},\dot{\overline{u}})\left[\frac{1+|\delta|e^{\mu_0}e^{-2\frac{RT}{L}}}{\mu C_0}\left(\frac{1}{Z_0}+\alpha\widehat{I}_0 e^{\alpha U_m}+\frac{2e^2 I_m}{U_m}\right)+|\delta|e^{-2\frac{RT}{L}}e^{\mu_0}\right]+$$

$$+\hat{\rho}_\mu^{(k)}(\dot{u},\dot{\overline{u}})\frac{e^{\mu_0}-1}{\mu_0}\left[\frac{1}{\mu^2 C_0}\left(1+|\delta|e^{\mu_0}e^{-2\frac{RT}{L}}\right)\left(\frac{1}{Z_0}+\alpha\widehat{I}_0 e^{\alpha U_m}+\frac{2e^2 I_m}{U_m}\right)\right] \leq$$

$$\leq e^{\mu(t-t_k)}\hat{\rho}_\mu^{(k)}((u,\dot{u}),(\overline{u},\dot{\overline{u}}))\left[\left(1+\frac{e^{\mu_0}-1}{\mu_0}\right)\frac{1+|\delta|e^{\mu_0}e^{-2\frac{RT}{L}}}{\mu C_0}\left(\frac{1}{Z_0}+\alpha\widehat{I}_0 e^{\alpha U_m}+\frac{2e^2 I_m}{U_m}\right)+|\delta|e^{-2\frac{RT}{L}}e^{\mu_0}\right] \equiv$$

$$\equiv e^{\mu(t-t_k)}\dot{K}_u\,\hat{\rho}_\mu^{(k)}((u,\dot{u}),(\overline{u},\dot{\overline{u}})).$$

It follows

$$\hat{\rho}_\mu^{(k)}(\dot{B}u,\dot{B}\overline{u}) \leq \dot{K}_u\hat{\rho}_\mu^{(k)}((u,\dot{u}),(\overline{u},\dot{\overline{u}})).$$

Therefore

$$\hat{\rho}_{\mu}^{(k)}(B(u),\dot{B}(u),B(\overline{u}),\dot{B}(\overline{u})) \leq \max\left\{e^{\mu_0}K_u,\dot{K}_u\right\}\hat{\rho}_{\mu}^{(k)}((u,\dot{u}),(\overline{u},\dot{\overline{u}})).$$

If $K = \max\left\{e^{\mu_0}K_u,\dot{K}_u\right\} < 1$ for sufficiently large $\mu > 0$ then operator B is contractive one.

We have to verify that M_{SU} is j-bounded. Indeed, since j is an identity mapping then

$$\hat{\rho}_{\mu}^{j^n(k)}((u,\dot{u}),(\overline{u},\dot{\overline{u}})) \leq \hat{\rho}_{\mu}^{(k)}((u,\dot{u}),(\overline{u},\dot{\overline{u}})) < \infty \quad (n = 0,1,2,\ldots).$$

Therefore in view of the fixed-point theorem for contractive mappings in uniform spaces (cf. Chapter I), the operator B has a unique fixed point and it is an oscillating solution of (5.5.2).

Theorem 5.5.1 is thus proved.

5.5.1. Numerical Example

We collect all inequalities implying the above theorem, assuming $U_E \leq U_0$

$$\frac{4L}{C_0 R}\sinh\left(\frac{RT_0}{L}\right)\left[\frac{2}{Z_0+R_0}+\left(1+|\delta|\right)\left(\frac{1}{Z_0}+\alpha\hat{I}_0\,\xi(\alpha U_m)+\frac{I_m e^2}{U_m}\right)\right]+|\delta| \leq 1;$$

$$e^{\mu_0}K_u = e^{2\mu_0}\left[\frac{e^{\mu_0}}{\mu^2 C_0}\left(1+e^{-2\frac{RT}{L}}|\delta|\right)\left(\frac{1}{Z_0}+\alpha e^{\alpha U_m}\hat{I}_0+\frac{2e^2 I_m}{U_m}\right)+|\delta|e^{-2\frac{RT}{L}}\right] < 1;$$

$$\dot{K}_u = \left(1+\frac{e^{\mu_0}-1}{\mu_0}\right)\frac{1+|\delta|e^{\mu_0}e^{-2\frac{RT}{L}}}{\mu C_0}\left(\frac{1}{Z_0}+\alpha\hat{I}_0 e^{\alpha U_m}+\frac{2e^2 I_m}{U_m}\right)+|\delta|e^{-2\frac{RT}{L}}e^{\mu_0} < 1$$

Consider a transmission line with a length $\Lambda = 1000m$, a cross-section area $S = 10mm^2$, a specific resistance for the copper of $\rho_c = 0,0175$, and a the resistance per-unit length is of $R = \dfrac{\rho_c \Lambda}{S} = \dfrac{0,0175.1000}{10} = 1,75\Omega$. Let

$L = 0,2\,\mu H/m$, $C = 80\,pF/m$, then $C_0 = 10.10^{-12}F$,

$v = 1/\sqrt{LC} = 1/\sqrt{0,2.10^{-6}.80.10^{-12}} = 1/(4.10^{-9}) = 2,5.10^8$, $R_0 = 25\Omega$, $\delta = 1/3$,

$E_0 = 1,5V$, $Z_0 = \sqrt{L/C} = 50\ \Omega$, $T = \Lambda/v = 1000.4.10^{-9} = 4.10^{-6}$ sec.,

$\dfrac{L}{R} = \dfrac{0,2.10^{-6}}{1,75} \approx 1,14.10^{-7}$.

Let us check the propagation of waves with length $\lambda_0 = (1/4)10^{-3}\,m$. Then

$$f_0 = \frac{1}{\lambda_0\sqrt{LC}} = \frac{1}{(1/4)10^{-3}.4.10^{-9}} = 10^{12}\,Hz \Rightarrow T_0 = \frac{1}{f_0} = 10^{-12}\,sec., \; l_0 = (1/4)10^{-12}\,sec$$

If we choose $\mu = 10^{12}$, then $\mu T_0 = \mu_0 = 1$ and $T = 4.10^{-6}.10^{12}T_0 = 4\,000\,000.T_0$. We have also

$$\mu T = 10^{12}.4.10^{-6} = 4.10^6 = 4000000,\; \mu C_0 = 10^{12}.10^{-11} = 10;\; \mu^2 C_0 = 10^{24}.10^{-11} = 10^{13},$$

$$e^{-\mu T} \approx 0;\; \hat{I}_0 = 10^{-8}\,A,\; \alpha = \frac{q}{k\tau} = \frac{1}{2.26.10^{-3}} = 19{,}23\;;\; \alpha I_0 = 19{,}23.10^{-8}\;;$$

$$e^{\frac{R}{L}T_0} - e^{-\frac{R}{L}T_0} \approx \frac{RT_0}{L} = 1{,}14.10^{-7}.10^{-12} = 1{,}14.10^{-19}\;;$$

$$\frac{I_m}{U_m} = \frac{1{,}9.10^{-3}\,A}{0{,}15V} = 12{,}67.10^{-3},\; U_m = 0{,}15;\; \frac{1}{U_m} = 6{,}66;\; \alpha U_m = 2{,}8845;\; e^{\alpha U_m} \le e^3\;;$$

$$\frac{I_0}{C_0} = 10^3 \; \text{and} \; \xi(u) = \frac{e^u - 1}{u} \Rightarrow \xi(\alpha U_m) = \frac{e^{\alpha U_m} - 1}{\alpha U_m}\;.$$

Then for $U_0 = 0{,}01$ the above inequalities become:

$$2{,}6.10^{-15}\left[0{,}027 + 1{,}33\left(0{,}02 + 12{,}67.10^{-3}e^2\right)\right] + \frac{1}{3} \le 1;$$

$$e^{\mu_0}K_u = \frac{e^3}{10^{22}.10^{-11}}\left(\frac{1}{50} + 19{,}23.10^{-8}e^{2,8845} + 2e^2 12{,}67.10^{-3}\right) < 1;$$

$$\dot{K}_u = \frac{e}{10^{12}.10^{-11}}\left(\frac{1}{50} + 19{,}23.10^{-8}e^{2,8845} + 2e^2.12{,}67.10^{-3}\right) \approx 0{,}077 < 1$$

that is, $K = 0{,}077$.

For the successive approximations we have

$$u^{(0)}(t) = \begin{cases} \dfrac{\sin(\omega_0 t)}{\omega_0}, & t \in [-T,T] \\ 0, & t \in [T, T+T_0] \end{cases},\; \omega_0 = \frac{2\pi}{T_0} = \frac{10^{12}\pi}{2}\,.$$

Obviously $\upsilon_0(T) = \upsilon_0(-T) = 0$.

Then if we choose interval $[t_0 = T, t_1 = T + T_0]$ we have

$$u^{(1)}(t) = \frac{2}{C_0(Z_0 + R_0)} \int_T^t U_0 \sin(\omega_0(s-T))ds - \frac{1}{Z_0 C_0} \int_T^t u^{(0)}(s)ds - \frac{\widehat{I}_0}{C_0} \int_T^t (e^{\alpha u^{(0)}(s)} - 1)ds +$$

$$+ \frac{I_m}{U_m C_0} \int_T^t u^{(0)}(s) e^{1 - \frac{u^{(0)}(s)}{U_m}} ds - \frac{\delta}{C_0 Z_0} \int_T^t u^{(0)}(s - 2T)ds +$$

$$+ \frac{\delta \widehat{I}_0}{C_0} e^{-2\frac{RT}{L}} \int_T^t (e^{\alpha u^{(0)}(s-2T)} - 1)ds + \frac{\delta I_m}{C_0 U_m} e^{-2\frac{RT}{L}} \int_T^t u^{(0)}(s-2T) e^{1 - \frac{u^{(0)}(s-2T)}{U_m}} ds + \delta e^{-2\frac{RT}{L}} \int_T^t \dot{u}^{(0)}(s-2T)ds -$$

$$- \left(\frac{t-T}{T_0} - \frac{1}{2} \right) \left[\frac{2}{C_0(Z_0 + R_0)} \int_T^{T+T_0} U_0 \sin(\omega_0(s-T))ds - \frac{1}{Z_0 C_0} \int_T^{T+T_0} u^{(0)}(s)ds - \frac{\widehat{I}_0}{C_0} \int_T^{T+T_0} (e^{\alpha u^{(0)}(s)} - 1)ds - \right.$$

$$- \frac{I_m}{C_0 U_m} \int_T^{T+T_0} u^{(0)}(s) e^{1 - \frac{u^{(0)}(s)}{U_m}} ds - \frac{\delta}{C_0 Z_0} e^{-2\frac{RT}{L}} \int_T^{T+T_0} u^{(0)}(s - 2T)ds +$$

$$+ \frac{\delta . \widehat{I}_0}{C_0} e^{-2\frac{RT}{L}} \int_T^{T+T_0} (e^{\alpha u^{(0)}(s-2T)} - 1)ds + \frac{\delta . I_m}{C_0 U_m} e^{-2\frac{RT}{L}} \int_T^{T+T_0} u^{(0)}(s-2T) e^{1 - \frac{u^{(0)}(s-2T)}{U_m}} ds +$$

$$+ \delta e^{-\frac{2RT}{L}} \int_T^{T+T_0} \dot{u}^{(0)}(s-2T)ds = \frac{2E_0(1 - \cos\omega_0 t)}{\omega_0 C_0(Z_0 + R_0)} - \frac{\delta}{\omega_0 C_0 Z_0} \int_T^t \sin(\omega_0 \tau)d\tau +$$

$$+ \frac{\delta . \widehat{I}_0}{C_0} \int_T^t (e^{\alpha \frac{\sin(\omega_0 \tau)}{\omega_0}} - 1)d\tau + \frac{\delta . I_m e}{C_0 U_m} \int_T^t \frac{\sin(\omega_0 \tau)}{\omega_0} e^{-\frac{\sin(\omega_0 \tau)}{\omega_0 U_m}} d\tau + e^{-2\frac{RT}{L}} \frac{\delta \sin(\omega_0 t)}{\omega_0} -$$

$$- \left(\frac{t-T}{T_0} - \frac{1}{2} \right) \left[-\frac{\delta}{\Omega_0 C_0 Z_0} \int_T^{T+T_0} \frac{\sin(\omega_0 \tau)}{\omega_0} d\tau + \frac{\delta . \widehat{I}_0}{C_0} \int_T^{T+T_0} (e^{\alpha \frac{\sin(\omega_0 \tau)}{\omega_0}} - 1)d\tau + \right.$$

$$\left. + \frac{\delta . I_m e}{C_0 U_m} \int_T^{T+T_0} \frac{\sin(\omega_0 \tau)}{\omega_0} e^{-\frac{\sin(\omega_0 \tau)}{\omega_0 U_m}} d\tau \right] = \frac{2U_0\left(1 - \cos\omega_0(t-T)\right)}{\omega_0 C_0(Z_0 + R_0)} - \frac{\delta(1 - \cos(\omega_0 t))}{\omega_0^2 C_0 Z_0} +$$

$$+ \frac{\delta . \widehat{I}_0}{C_0} \sum_{n=1}^{\infty} \frac{\alpha^n}{n! \omega_0^n} \int_T^t (\sin(\omega_0 \tau))^n d\tau + \frac{\delta . I_m e}{C_0 U_m} \sum_{n=1}^{\infty} \frac{(-1)^{n-1}}{(n-1)!(U_m)^{n-1}\omega_0^n} \int_T^t (\sin(\omega_0 \tau))^n d\tau + e^{-2\frac{RT}{L}} \frac{\delta \sin(\omega_0 t)}{\omega_0} -$$

$$- \left(\frac{t-T}{T_0} - \frac{1}{2} \right) \left[\frac{\delta . \widehat{I}_0}{C_0} \sum_{n=1}^{\infty} \frac{\alpha^n}{n! \omega_0^n} \int_T^t (\sin(\omega_0 \tau))^n d\tau + \frac{\delta . I_m e}{C_0 U_m} \sum_{n=1}^{\infty} \frac{(-1)^{n-1}}{(n-1)!(U_m)^{n-1}\omega_0^n} \int_T^t (\sin(\omega_0 \tau))^n d\tau \right] =$$

$$= \frac{2U_0\left(1 - \cos\omega_0(t-T)\right)}{\omega_0 C_0(Z_0 + R_0)} - \frac{\delta(1 - \cos(\omega_0 t))}{\omega_0 C_0 Z_0} +$$

$$\frac{\delta}{C_0} \sum_{n=1}^{\infty} \left(\widehat{I}_0 \frac{\alpha^n}{n!} + \frac{I_m e . (-1)^{n-1}}{(n-1)!(U_m)^n} \right) \int_T^t (\sin(\omega_0 \tau))^n d\tau +$$

$$+ \delta \sin(\omega_0 t) - \frac{\delta}{C_0}\left(\frac{t-T}{T_0} - \frac{1}{2}\right)\sum_{n=1}^{\infty}\left(\hat{I}_0 \frac{\alpha^n}{n!} + \frac{I_m e.(-1)^{n-1}}{(n-1)!(U_m)^n}\right)\int_T^{T+T_0}\left(\sin(\omega_0\tau)\right)^n d\tau$$

and

$$\dot{u}^{(1)}(t) = \frac{2U_0 \sin\omega_0(t-T)}{C_0(Z_0+R_0)} - \frac{\delta \sin(\omega_0 t)}{\omega_0 C_0 Z_0} + \frac{\delta}{C_0}\sum_{n=1}^{\infty}\left(\hat{I}_0\frac{\alpha^n}{n!} + \frac{I_m e.(-1)^{n-1}}{(n-1)!(U_m)^n}\right)\left(\sin(\omega_0 t)\right)^n +$$

$$+ \delta\cos(\omega_0 t) - \frac{\delta}{C_0}\frac{1}{T_0}\sum_{n=1}^{\infty}\left(\hat{I}_0\frac{\alpha^n}{n!} + \frac{I_m e.(-1)^{n-1}}{(n-1)!(U_m)^n}\right)\int_T^{T+T_0}\left(\sin(\omega_0\tau)\right)^n d\tau \quad .$$

It is known that if $u^*(t)$ is the solution then $\rho_\mu^{(k)}(u^{(n)}, u^*) \le \frac{0018^n}{1-0,018}\hat{\rho}_\mu(u^{(0)}, u^{(1)})$.

It is easy to verify that

$$\left|u^{(1)}(t) - u^{(0)}(t)\right| \le e^{\mu(t-T)}\left[\frac{4U_0}{\omega_0 C_0(Z_0+R_0)} + \frac{2|\delta|}{\omega_0^2 C_0 Z_0} + \right.$$

$$\left. + \frac{3|\delta|.2\pi}{2C_0\omega_0}\sum_{n=1}^{\infty}\left(\hat{I}_0\frac{\alpha^n}{n!} + \frac{I_m e.(-1)^{n-1}}{(n-1)!(U_m)^n}\right) + \frac{|\delta|+1}{\omega_0}\right] \le$$

$$\le e^{\mu 0}\frac{1}{C_0\omega_0}\left[\frac{4U_0}{Z_0+R_0} + 2C_0^{-1} + |\delta|\left(\frac{2}{\omega_0 Z_0} + 3\pi\hat{I}_0(e^\alpha - 1) + \frac{3\pi I_m}{U_m}e^{1-\frac{1}{U_m}}\right)\right] \le$$

$$\le \frac{e}{10^{-11}\frac{10^{12}\pi}{2}}\left[\frac{6}{50+25} + 2.10^{11} + \frac{1}{3}\left(\frac{2}{\frac{10^{12}\pi}{2}50} + 3\pi10^{-8}(e^{20}-1) + 3\pi.12,67.10^{-3}e^{1-6,66}\right)\right] \approx$$

$$\le 0,17.[0,08 + 2.10^{11} + 4.10^8\pi10^{-8}] \approx 2.10^{11} = P.$$

It follows $\rho(u^{(1)}, u^{(0)}) \le P$.

For the derivative we obtain

$$\left|\dot{u}^{(1)}(t) - \dot{u}^{(0)}(t)\right| \le e^{\mu(t-T)}\left(\frac{2U_0}{C_0(Z_0+R_0)} + \frac{|\delta|}{\omega_0 C_0 Z_0} + \right.$$

$$+\frac{2|\delta|}{C_0}\sum_{n=1}^{\infty}\left(\hat{I}_0\frac{\alpha^n}{n!}+\frac{I_m e}{U_m}\frac{(-1)^{n-1}}{(n-1)!(U_m)^{n-1}}\right)+|\delta|+1\right)\le$$

$$\le e^{\mu_0}\frac{1}{C_0}\left[\frac{2U_0}{Z_0+R_0}+\frac{|\delta|}{\omega_0 Z_0}+2|\delta|\left(\hat{I}_0(e^\alpha-1)+\frac{I_m}{U_m}e^{1-\frac{1}{U_m}}\right)+|\delta|+1\right]\approx$$

$$\le\frac{e}{3.10^{-11}}\left[\frac{6}{5}+\frac{2}{75.10^{12}\pi}+2.10^{-8}(e^{19,23}-1)+2.12,67.10^{-3}e^{1-6,66}+4\right]\le$$

$$\le 10^{11}\left(5,2+2.10^{-8}e^{20}\right)\approx 10^{11}\left(5,2+2.5.10^8 10^{-8}\right)\approx 1,52.10^{12}=\dot{P}$$

or

$$\rho(\dot{u}^{(1)},\dot{u}^{(0)})\le\dot{P}\Rightarrow\rho_\mu(u^{(1)},u^{(0)})\le\max\{P,\dot{P}\}=\max\{2.10^{11}\ ;\ 1,52.10^{12}\}=1,52.10^{12}$$

This implies $\left|u^{(n)}(t)-u^*(t)\right|\le\dfrac{(0,077)^n}{1-0,077}.1,52.10^{12}$.

5.6. ANOTHER APPROACH TO REDUCING THE MIXED PROBLEM TO AN OSCILLATORY ONE FOR A NEUTRAL EQUATION WITH AN EXPONENTIAL NONLINEARITY

Here we consider the problem from § 5.5.
We proceed from the boundary condition

$$E(t)-u(0,t)-R_0 i(0,t)=0, t\ge T\ ,$$

$$C_0\frac{du(\Lambda,t)}{dt}=i(\Lambda,t)-f(u(\Lambda,t))\ ,$$

where $f(u)=\hat{I}_0(e^{\alpha u}-1)+\dfrac{I_m}{U_m}ue^{1-\frac{u}{U_m}}$.

Taking into account (5.2.8), namely

$$\left| \begin{array}{l} u(x,t) = \dfrac{1}{2\sqrt{C}} e^{-\frac{R}{L}t} W(x,t) - \dfrac{1}{2\sqrt{C}} e^{-\frac{R}{L}t} J(x,t) \\[3mm] i(x,t) = \dfrac{1}{2\sqrt{L}} e^{-\frac{R}{L}t} W(x,t) + \dfrac{1}{2\sqrt{L}} e^{-\frac{R}{L}t} J(x,t) \end{array} \right.$$

$$(5.6.1)$$

we have for $x = 0$

$$\left| \begin{array}{l} u(0,t) = \dfrac{1}{2\sqrt{C}} e^{-\frac{R}{L}t} W(0,t) - \dfrac{1}{2\sqrt{C}} e^{-\frac{R}{L}t} J(0,t), \\[3mm] i(0,t) = \dfrac{1}{2\sqrt{L}} e^{-\frac{R}{L}t} W(0,t) + \dfrac{1}{2\sqrt{L}} e^{-\frac{R}{L}t} J(0,t) \end{array} \right.$$

and for $x = \Lambda$

$$\left| \begin{array}{l} u(\Lambda,t) = \dfrac{1}{2\sqrt{C}} e^{-\frac{R}{L}t} W(\Lambda,t) - \dfrac{1}{2\sqrt{C}} e^{-\frac{R}{L}t} J(\Lambda,t), \\[3mm] i(\Lambda,t) = \dfrac{1}{2\sqrt{L}} e^{-\frac{R}{L}t} W(\Lambda,t) + \dfrac{1}{2\sqrt{L}} e^{-\frac{R}{L}t} J(\Lambda,t). \end{array} \right.$$

Taking into account that

$$W(0,t) = W(\Lambda,t+T), \quad J(\Lambda,t) = J(0,t+T)$$

we reach the system

$$E(t) - \frac{1}{2\sqrt{C}} e^{-\frac{R}{L}(t+T)} W(\Lambda,t+T) + \frac{1}{2\sqrt{C}} e^{-\frac{R}{L}t} J(0,t)$$

$$- \frac{R_0}{2\sqrt{L}} e^{-\frac{R}{L}(t+T)} W(\Lambda,t+T) - \frac{R_0}{2\sqrt{L}} e^{-\frac{R}{L}t} J(0,t) = 0,$$

$$\frac{C_0}{2\sqrt{C}} \frac{d}{dt}\left(e^{-\frac{R}{L}t} W(\Lambda,t) - e^{-\frac{R}{L}(t+T)} J(0,t+T) \right) = \frac{1}{2\sqrt{L}} e^{-\frac{R}{L}t} W(\Lambda,t) + \frac{1}{2\sqrt{L}} e^{-\frac{R}{L}(t+T)} J(0,t+T) -$$

$$- f\left(\frac{1}{2\sqrt{C}} e^{-\frac{R}{L}t} W(\Lambda,t) - \frac{1}{2\sqrt{C}} e^{-\frac{R}{L}(t+T)} J(0,t+T) \right).$$

Let us put $t + T \equiv t$ and assume the unknown functions are

$$W(t) = W(\Lambda, t), \quad J(t) = J(0, t).$$

Then

$$E(t-T) - \frac{1}{2\sqrt{C}} e^{-\frac{R}{L}t} W(t) + \frac{1}{2\sqrt{C}} e^{-\frac{R}{L}(t-T)} J(t-T) - \frac{R_0}{2\sqrt{L}} e^{-\frac{R}{L}t} W(t) - \frac{R_0}{2\sqrt{L}} e^{-\frac{R}{L}(t-T)} J(t-T) = 0,$$

$$\frac{C_0}{2\sqrt{C}} \frac{d}{dt}\left(e^{-\frac{R}{L}(t-T)} W(t-T) - e^{-\frac{R}{L}t} J(t) \right) = \frac{1}{2\sqrt{L}} e^{-\frac{R}{L}(t-T)} W(t-T) + \frac{1}{2\sqrt{L}} e^{-\frac{R}{L}t} J(t) -$$

$$- f\left(\frac{1}{2\sqrt{C}} e^{-\frac{R}{L}(t-T)} W(t-T) - \frac{1}{2\sqrt{C}} e^{-\frac{R}{L}t} J(t) \right).$$

The system obtained is not autonomous one, and obviously $e^{-\frac{R}{L}(t-T)}$ is not a periodic function, so that the right-hand side of the second equation is also not a periodic function. Let us set $\breve{W}(t) = e^{-\frac{R}{L}t} W(t), \breve{J}(t) = e^{-\frac{R}{L}t} J(t)$. If $W(t), J(t)$ are periodic functions then $\breve{W}(t), \breve{J}(t)$ are oscillating ones functions and satisfy the inequalities $\left|\breve{W}(t)\right| \le e^{-\frac{R}{L}t} W_0$, $\left|\breve{J}(t)\right| \le e^{-\frac{R}{L}t} J_0$. Therefore we can formulate the problem for the existence of oscillatory solutions satisfying the last inequalities.

Further on we again use the denotations (W, J) instead of $(\breve{W}, \breve{J})$.

First we rewrite the above system as:

$$E(t-T) - \frac{W(t)}{2\sqrt{C}} + \frac{J(t-T)}{2\sqrt{C}} - \frac{R_0}{2\sqrt{L}} W(t) - \frac{R_0}{2\sqrt{L}} J(t-T) = 0,$$

$$\frac{C_0}{2\sqrt{C}} \frac{d}{dt}\left(W(t-T) - J(t) \right) = \frac{W(t-T)}{2\sqrt{L}} + \frac{J(t)}{2\sqrt{L}} - f\left(\frac{W(t-T) - J(t)}{2\sqrt{C}} \right)$$

or in view of the explicit type of f

$$f(u) = \hat{I}_0(e^{\alpha u} - 1) + \frac{I_m}{U_m} u e^{1 - \frac{u}{U_m}}$$

we obtain, assuming $E(t-T) = E(t)$:

$$W(t) = \frac{2\sqrt{L}}{Z_0 + R_0} E(t) + \frac{Z_0 - R_0}{Z_0 + R_0} J(t - T),$$

$$\frac{dJ(t)}{dt} = \frac{dW(t-T)}{dt} - \frac{W(t-T)}{C_0 Z_0} - \frac{J(t)}{C_0 Z_0} + \frac{2\sqrt{C}}{C_0} \hat{I}_0 (e^{\alpha \frac{W(t-T)-J(t)}{2\sqrt{C}}} - 1) + \frac{I_m}{U_m C_0} (W(t-T) - J(t)) e^{1 - \frac{W(t-T)-J(t)}{2U_m \sqrt{C}}}$$

We can exclude $W(t)$ proceeding by in the following mannerreasoning: recalling that

$$\delta = \frac{Z_0 - R_0}{Z_0 + R_0} \text{ we substitute}$$

$$W(t - T) = \frac{2\sqrt{L}}{Z_0 + R_0} E(t) + \frac{Z_0 - R_0}{Z_0 + R_0} J(t - 2T)$$

and

$$\frac{dW(t-T)}{dt} = \frac{2\sqrt{L}}{Z_0 + R_0} \frac{dE(t)}{dt} + \frac{Z_0 - R_0}{Z_0 + R_0} \frac{dJ(t - 2T)}{dt}$$

into the second above equation. But first we introduce a denotation:

$$W(t - T) - J(t) = \frac{2\sqrt{L}}{Z_0 + R_0} E(t) + \frac{Z_0 - R_0}{Z_0 + R_0} J(t - 2T) - J(t) \equiv H(J)(t).$$

Then we obtain

$$\frac{dJ(t)}{dt} = \frac{2\sqrt{L}}{Z_0 + R_0} \frac{dE(t)}{dt} + \delta \frac{dJ(t - 2T)}{dt} - \frac{1}{C_0 Z_0} H(J)(t) + \frac{2\sqrt{C} \hat{I}_0}{C_0} (e^{\alpha \frac{H(J)(t)}{2\sqrt{C}}} - 1) + \frac{I_m}{U_m C_0} H(J)(t) e^{1 - \frac{H(J)(t)}{2U_m \sqrt{C}}}$$

The initial conditions can be found from (5.6.1), that is,

$$\widetilde{W}_0(x) = W(x,0) = u(x,0) + Z_0\, i(x,0) = u_0(x) + Z_0\, i_0(x),$$

$$\widetilde{J}_0(x) = J(x,0) = u(x,0) - Z_0\, i(x,0) = u_0(x) - Z_0\, i_0(x)\, , \ x \in [0, \Lambda].$$

As above, one can translate the initial functions along the characteristics on the interval $[0, T]$ and obtain $J(t) = \widetilde{J}_0(t),\ t \in [0, T]$.

Now we are able to formulate the main problem: to find an oscillatory solution with advanced prescribed zeros on an interval $[t_0, \infty)$, $T = t_0$ of the system

$$\frac{dJ(t)}{dt} = \frac{2\sqrt{L}}{Z_0+R_0}\frac{dE(t)}{dt} - \frac{1}{C_0 Z_0}H(J)(t) + \frac{2\sqrt{C}\,\hat{I}_0}{C_0}(e^{\alpha\frac{H(J)(t)}{2\sqrt{C}}} - 1) + \frac{I_m}{U_m C_0}H(J)(t)e^{1-\frac{H(J)(t)}{2U_m\sqrt{C}}} +$$

$$+\delta\frac{dJ(t-2T)}{dt}, t\in[t_0,\infty) \tag{5.6.2}$$

$$J(t) = \tilde{J}_0(t), \frac{dJ(t)}{dt} = \frac{d\tilde{J}_0(t)}{dt}, \; t\in[-t_0,t_0]$$

where $\tilde{J}_0(t)$, $t \in [-t_0,t_0]$ is prescribed oscillating function on the interval $[-t_0,t_0]$.

Let $S_T = \{\tau_k\}_{k=0}^n, n \in N$ be the set of zeros of the initial function, that is, $\tilde{W}_0(\tau_k) = 0$, $\tilde{J}_0(\tau_k) = 0$ such that $\tau_0 = 0$, $\tau_n = T \equiv t_0$. Besides $\max\{\tau_{k+1} - \tau_k : k = 0,1,...,n\} \le T_0$.

Let $S = \{t_k\}_{k=0}^\infty$ be a strictly increasing sequence of real numbers satisfying the following conditions **(C)**:

(C1) $\lim\limits_{k\to\infty} t_k = \infty$;

(C2) for every k there is $s < k$ such that $t_k - T = t_s$ where $t_s \in S_T \cup S$.

It follows
(C3) $0 \le \inf\{t_{k+1} - t_k : k = 0,1,2,...\} \le \sup\{t_{k+1} - t_k : k = 0,1,2,...\} = T_0 < \infty$.

We assume also

(E) $E(t-T) = E(t), E(t_k) = 0, E(t_k - T) = 0,$.

Introduce the sets $C^1[t_0,\infty)$ consisting of all continuous and bounded functions differentiable with bounded derivatives on every interval $[t_k,t_{k+1}]$ (Note the functions from $C^1[t_0,\infty)$ could be not differentiable at t_k),

$$M_J = \left\{ J(.) \in C^1[t_0,\infty) : J(t_k) = 0 \wedge |J(t)| \le I_0 e^{-\frac{R}{L}t}, t\in[t_0,\infty) \right\}$$

where $I_0 > 0$ is a positive constant.

Remark 5.6.1. It follows that the functions from the above set satisfy the inequalities

$$|J(t)| \le I_0 e^{\mu(t-t_k)}, t\in[t_k,t_{k+1}]$$

where I_0, μ are positive constants and $k = 0,1,2,\dots$.

Introduce the following family of pseudo-metrics

$$\rho^{(k)}(J,\bar{J}) = \max\left\{\left|J(t)-\bar{J}(t)\right| : t \in [t_k,t_{k+1}]\right\},$$

$$\hat{\rho}^{(k)}(J,\bar{J}) = \max\left\{\left|J(t)-\bar{J}(t)\right| : t \in [t_0,t_{k+1}]\right\},$$

$$\rho_\mu^{(k)}(J,\bar{J}) = \max\left\{e^{-\mu(t-t_k)}\left|J(t)-\bar{J}(t)\right| : t \in [t_k,t_{k+1}]\right\},$$

$$\hat{\rho}_\mu^{(k)}(J,\bar{J}) = \max\left\{\rho_\mu^{(0)}(J,\bar{J}), \rho_\mu^{(1)}(J,\bar{J}),\dots, \rho_\mu^{(k)}(J,\bar{J})\right\},$$

$$\rho_\mu^{(k)}(\dot{J},\dot{\bar{J}}) = \max\left\{e^{-\mu(t-t_k)}e^\beta\left|\dot{J}(t)-\dot{\bar{J}}(t)\right| : t \in [t_k,t_{k+1}]\right\},$$

$$\hat{\rho}_\mu^{(k)}(\dot{J},\dot{\bar{J}}) = \max\left\{\rho_\mu^{(0)}(\dot{J},\dot{\bar{J}}), \rho_\mu^{(1)}(\dot{J},\dot{\bar{J}}),\dots, \rho_\mu^{(k)}(\dot{J},\dot{\bar{J}})\right\}.$$

Remark 5.6.2. The following inequalities imply the equivalence of the both families of pseudo-metrics

$$\rho_\mu^{(k)}(J,\bar{J}) \le \rho^{(k)}(J,\bar{J}) \le e^{\mu T_0}\rho_\mu^{(k)}(J,\bar{J}), \ (k = 0,1,2,\dots).$$

It is easy to verify that

$$\hat{\rho}^{(k)}(J,\bar{J}) = \max\left\{\rho^{(0)}(J,\bar{J}), \rho^{(1)}(J,\bar{J}),\dots, \rho^{(k)}(J,\bar{J})\right\} \le$$
$$\le e^{\mu T_0}\max\left\{\rho_\mu^{(0)}(J,\bar{J}), \rho_\mu^{(1)}(J,\bar{J}),\dots, \rho_\mu^{(k)}(J,\bar{J})\right\} = e^{\mu T_0}\hat{\rho}_\mu^{(k)}(J,\bar{J}).$$

The set M_J turns out into a complete uniform space (cf. [14]) with respect to the family of pseudo-metrics

$$\hat{\rho}_\mu^{(k)}(J,\dot{J}) = \max\left\{\hat{\rho}^{(k)}(J,\bar{J}), \hat{\rho}_\mu^{(k)}(\dot{J},\dot{\bar{J}})\right\} (k = 0,1,2,\dots).$$

One can verify that M_J is closed subset of $C^1[t_0,\infty)$ with respect to the above family of pseudo-metrics.

Since we look for continuously differentiable solutions, we have to take into account the conformity condition

(CC):
$$\frac{dJ(T)}{dt} = \frac{2\sqrt{L}}{Z_0+R_0}\frac{dE(T)}{dt} + \frac{Z_0-R_0}{Z_0+R_0}\frac{dJ(-T)}{dt} - \frac{1}{C_0Z_0}H(J)(T)+$$
$$+\frac{2\sqrt{C}\,\hat{I}_0}{C_0}(e^{\alpha\frac{H(J)(T)}{2\sqrt{C}}}-1)+\frac{I_m}{U_mC_0}H(J)(T)e^{1-\frac{H(J)(T)}{2U_m\sqrt{C}}}, \; t\in[t_0,\infty).$$

If we choose $\{0,T\}\in S_T \cup S$ then the above condition becomes $\dfrac{dJ(t_0)}{dt} = \dfrac{dJ(-t_0)}{dt}$.

In order to avoid the conformity condition **(CC)** we define operator $B = B_J(J)$ by the formulas:

$$B_J(J)(t):=\tilde{J}_0(t), t\in[-T,T]$$

$$B_J(J)(t)= B_J^{(k)}(J)(t):= \int_{t_k}^{t} I(J)(s)ds - \frac{t-t_k}{t_{k+1}-t_k}\int_{t_k}^{t_{k+1}} I(J)(s)ds$$

$$t\in[t_k,t_{k+1}], \; (k=0,1,2,\dots)$$

where

$$I(J)(t)\equiv \frac{2\sqrt{L}}{Z_0+R_0}\frac{dE(t)}{dt} + \delta\frac{d\bar{J}(t)}{dt} - \frac{1}{C_0Z_0}\bar{H}(J)(t)+$$
$$+\frac{2\sqrt{C}\,\hat{I}_0}{C_0}(e^{\alpha\frac{\bar{H}(J)(t)}{2\sqrt{C}}}-1)+\frac{I_m}{U_mC_0}\bar{H}(J)(t)e^{1-\frac{\bar{H}(J)(t)}{2U_m\sqrt{C}}}, \tag{5.6.3}$$

$$\bar{H}(J)(t) = \frac{2\sqrt{L}}{Z_0+R_0}E(t)+\delta\,\bar{J}(t)-J(t), \; \bar{J}(t)=\begin{cases}\tilde{J}_0(t-2T), t\in[T,3T]\\ J(t-2T), \; t\in[3T,\infty)\end{cases} \quad \text{(cf. § 5.4).}$$

Remark 5.6.3. It is easy to verify that the functions from M_J are not necessary differentiable at the points $S = \{t_k\}_{k=0}^{\infty}$, which implies that we have to consider a space with a countable family of pseudo-metrics, but not with one a single metric. Therefore we have to involve the fixed-point theory in uniform spaces [14].

Preliminary inequalities:

$$\left|\bar{H}(J)(t)\right| \le \frac{2\sqrt{L}}{Z_0+R_0}\left|E(t)\right|+\left|\delta\right|\left|\bar{J}(t)\right|+\left|J(t)\right| \le e^{-\frac{R}{L}t}\left(\frac{2\sqrt{L}\,U_E}{Z_0+R_0}+(\left|\delta\right|+1)I_0\right),$$

$$\frac{\alpha\left|W(t-T)-J(t)\right|}{2\sqrt{C}} = \frac{\alpha\left|\vec{H}(J)(t)\right|}{2\sqrt{C}} \le e^{-\frac{R}{L}t}\,\frac{\alpha}{2\sqrt{C}}\left(\frac{2\sqrt{L}\,U_E}{Z_0+R_0}+(|\delta|+1)I_0\right) \le \alpha U_m \, ,$$

$$\left|e^{\alpha\frac{W(t-T)-J(t)}{2\sqrt{C}}}-1\right| \le e^{\alpha\frac{|W(t-T)-J(t)|}{2\sqrt{C}}}-1 \le e^{-\frac{R}{L}t}\,\frac{\alpha}{2\sqrt{C}}\left(\frac{2\sqrt{L}\,U_E}{Z_0+R_0}+(|\delta|+1)I_0\right)\xi(\alpha U_m),$$

$$\left(\xi(h)=\frac{e^h-1}{h}\right).$$

Lemma 5.6.1. Problem (5.6.2) has a solution $J \in M_J$ iff the operator B has a fixed point in M_J, that is, $J = B_J(J)$.

Proof: Let $J \in M_J$ be a solution of (5.6.2). We consider only the second component. Then integrating the second equation of (5.6.2) on every interval $[t_k,t] \subset [t_k,t_{k+1}]$ $(k=0,1,2 \ldots)$ we obtain

$$J(t)-J(t_k)=\int_{t_k}^{t} I(J)(s)ds \Leftrightarrow J(t)-\int_{t_k}^{t} I(J)(s)ds$$

and then

$$J(t)=\int_{t_k}^{t} I(J)(s)ds \Rightarrow 0 = J(t_{k+1})=\int_{t_k}^{t} I(J)(s)ds \Rightarrow \int_{t_k}^{t_{k+1}} I(J)(s)ds = 0 \,. \; (5.6.4)$$

Therefore J satisfies

$$J(t) = \int_{t_k}^{t} I(J)(s)ds - \frac{t-t_k}{t_{k+1}-t_k}\int_{t_k}^{t_{k+1}} I(J)(s)ds,\; t \in [t_k,t_{k+1}] \Leftrightarrow$$

$$J = B_J(J)$$

that is, J is a fixed point of B.

Conversely, let J be a fixed point of B or

$$J(t) = \int_{t_k}^{t} I(J)(s)ds - \frac{t-t_k}{t_{k+1}-t_k}\int_{t_k}^{t_{k+1}} I(J)(s)ds \,.$$

Bearing in mind that $\mu_0 = \mu T_0 = \text{const.}$ we obtain

$$\left| \int_{t_k}^{t_{k+1}} I(J)(s)\,ds \right| \le$$

$$\le \frac{2\sqrt{L}}{Z_0+R_0} \left| \int_{t_k}^{t_{k+1}} \frac{dE(s)}{ds}\,ds \right| + |\delta| \left| \int_{t_k}^{t_{k+1}} \frac{d\vec{J}(s)}{ds}\,ds \right| + \frac{1}{C_0 Z_0} \left| \int_{t_k}^{t_{k+1}} \vec{H}(J)(s)\,ds \right| +$$

$$+ \frac{2\sqrt{C}\,\hat{I}_0}{C_0} \int_{t_k}^{t_{k+1}} \left| e^{\alpha \frac{\vec{H}(J)(s)}{2\sqrt{C}}} - 1 \right| ds + \frac{I_m}{U_m C_0} \int_{t_k}^{t_{k+1}} \left| \vec{H}(J)(s) e^{1 - \frac{\vec{H}(J)(s)}{2U_m\sqrt{C}}} \right| ds \le$$

$$\le \frac{1}{C_0 Z_0} \left(\frac{2\sqrt{L}\,U_E}{Z_0+R_0} + (|\delta|+1)I_0 \right) \int_{t_k}^{t_{k+1}} e^{\mu(s-t_k)}\,ds + \frac{2\sqrt{C}\,\hat{I}_0}{C_0} \frac{\alpha}{2\sqrt{C}} \left(\frac{2\sqrt{L}\,U_E}{Z_0+R_0} + (|\delta|+1)I_0 \right) \xi(\alpha U_m) \int_{t_k}^{t_{k+1}} e^{\mu(s-t_k)}\,ds +$$

$$+ \frac{I_m e^2}{U_m C_0} \left(\frac{2\sqrt{L}\,U_E}{Z_0+R_0} + (|\delta|+1)I_0 \right) \int_{t_k}^{t_{k+1}} e^{\mu(s-t_k)}\,ds \le$$

$$\le \frac{e^{\mu_0}-1}{\mu C_0} \left(\frac{2\sqrt{L}\,U_E}{Z_0+R_0} + (|\delta|+1)I_0 \right) \left(\frac{1}{Z_0} + \alpha \hat{I}_0 \xi(\alpha U_m) + \frac{I_m e^2}{U_m} \right) \equiv M_0(\mu).$$

As in Chapter II we conclude that $\int_{t_k}^{t_{k+1}} I(J)(s)\,ds = 0$. Therefore

$$J(t) = \int_{t_k}^{t} I(J)(s)\,ds - \frac{t-t_k}{t_{k+1}-t_k} \int_{t_k}^{t_{k+1}} I(J)(s)\,ds \Leftrightarrow J(t) = \int_{t_k}^{t} I(J)(s)\,ds .$$

Differentiating the last integral equation we obtain (5.6.2).

Lemma 5.6.1 is thus proved.

Theorem 5.6.1. Let the following conditions be fulfilled:

1) The initial function $\tilde{J}_0(t) \in C^1[-T,T]$ satisfies $\left| \tilde{J}_0(t) \right| \le I_0 e^{-\frac{R}{L}t}$;

2) $E(t) \in C^1[-T,\infty), \left| E(t) \right| \le U_E e^{-\frac{R}{L}t}, t \in [-T,\infty)$;

3) $\dfrac{R_0}{Z_0} < 1; \quad \dfrac{\alpha}{2\sqrt{C}} \left(\dfrac{2\sqrt{L}\,U_E}{Z_0+R_0} + (|\delta|+1)I_0 \right) \le \alpha U_m$;

$$\frac{2\sqrt{L}\,U_E}{Z_0+R_0}+|\delta|I_0+\frac{4L}{C_0R}\sinh\!\left(\frac{RT_0}{L}\right)\!\left(\frac{2\sqrt{L}\,U_E}{Z_0+R_0}+(|\delta|+1)I_0\right)\!\left(\frac{1}{Z_0}+\alpha\hat{I}_0\,\xi(\alpha U_m)+\frac{I_m e^2}{U_m}\right)\le I_0\,.$$

Then there exists a periodic solution of the initial value problem (5.6.2), belonging to M_J.

Proof: Recall the definition of the operator B by the formulas

$$B_J(J)(t)=B_J^{(k)}(J)(t):=\int_{t_k}^{t}I(J)(s)ds-\frac{t-t_k}{t_{k+1}-t_k}\int_{t_k}^{t_{k+1}}I(J)(s)ds\,,$$

$t\in[t_k,t_{k+1}],\ (k=0,1,2,\dots)$

where $I(J)$ is (5.9.3).

We show that $\left|B_J^{(k)}(J)(t)\right|\le I_0 e^{-\frac{R}{L}t}$. Indeed,

$$\left|B_J^{(k)}(J)(t)\right|\le\left|\int_{t_k}^{t}I(J)(s)ds\right|+\left|\int_{t_k}^{t_{k+1}}I(J)(s)ds\right|\equiv J_1+J_2\,.$$

But

$$J_1\le\frac{2\sqrt{L}}{Z_0+R_0}\left|\int_{t_k}^{t}\frac{dE(s)}{ds}ds\right|+|\delta|\left|\int_{t_k}^{t}\frac{d\bar{J}(s)}{ds}ds\right|+\frac{1}{C_0Z_0}\left|\int_{t_k}^{t}\bar{H}(J)(s)ds\right|+$$

$$+\frac{2\sqrt{C}\,\hat{I}_0}{C_0}\left|\int_{t_k}^{t}\left(e^{\alpha\frac{\bar{H}(J)(s)}{2\sqrt{C}}}-1\right)ds\right|+\frac{I_m}{U_m C_0}\left|\int_{t_k}^{t}\bar{H}(J)(s)e^{1-\frac{\bar{H}(J)(s)}{2U_m\sqrt{C}}}ds\right|\le$$

$$\le\frac{2\sqrt{L}}{Z_0+R_0}|E(t)|+|\delta||\bar{J}(t)|+\frac{1}{C_0Z_0}\left(\frac{2\sqrt{L}\,U_E}{Z_0+R_0}+(|\delta|+1)I_0\right)\int_{t_k}^{t}e^{-\frac{R}{L}s}ds+$$

$$+\frac{2\sqrt{C}\,\hat{I}_0}{C_0}\frac{\alpha}{2\sqrt{C}}\left(\frac{2\sqrt{L}\,U_E}{Z_0+R_0}+(|\delta|+1)I_0\right)\xi(\alpha U_m)\int_{t_k}^{t}e^{-\frac{R}{L}s}ds+\frac{I_m e^2}{U_m C_0}\left(\frac{2\sqrt{L}\,U_E}{Z_0+R_0}+(|\delta|+1)I_0\right)\int_{t_k}^{t}e^{-\frac{R}{L}s}ds\le$$

$$\le\frac{2\sqrt{L}\,U_E}{Z_0+R_0}e^{-\frac{R}{L}t}+|\delta|I_0 e^{-\frac{R}{L}t}+\frac{1}{C_0Z_0}\left(\frac{2\sqrt{L}\,U_E}{Z_0+R_0}+(|\delta|+1)I_0\right)\frac{L}{R}\left(e^{-\frac{R}{L}t_k}-e^{-\frac{R}{L}t}\right)+$$

$$+\frac{\alpha\hat{I}_0}{C_0}\left(\frac{2\sqrt{L}\,U_E}{Z_0+R_0}+(|\delta|+1)I_0\right)\xi(\alpha U_m)\left(e^{-\frac{R}{L}t_k}-e^{-\frac{R}{L}t}\right)+\frac{I_m e^2}{U_m C_0}\left(\frac{2\sqrt{L}\,U_E}{Z_0+R_0}+(|\delta|+1)I_0\right)\left(e^{-\frac{R}{L}t_k}-e^{-\frac{R}{L}t}\right)\le$$

$$\leq \frac{2\sqrt{L}\,U_E}{Z_0+R_0}e^{-\frac{R}{L}t}+|\delta|I_0 e^{-\frac{R}{L}t}+e^{-\frac{R}{L}t}\left(\frac{2\sqrt{L}\,U_E}{Z_0+R_0}+(|\delta|+1)I_0\right)\frac{L}{C_0 R}\left(e^{\frac{R}{L}(t-t_k)}-1\right)\left(\frac{1}{Z_0}+\alpha\,\widehat{I}_0\,\xi(\alpha U_m)+\frac{I_m e^2}{U_m}\right)\leq$$

$$\leq e^{-\frac{R}{L}t}\left[\frac{2\sqrt{L}\,U_E}{Z_0+R_0}+|\delta|I_0+\left(\frac{2\sqrt{L}\,U_E}{Z_0+R_0}+(|\delta|+1)I_0\right)\frac{L}{C_0 R}\left(e^{\frac{R}{L}T_0}-e^{-\frac{R}{L}T_0}\right)\left(\frac{1}{Z_0}+\alpha\,\widehat{I}_0\,\xi(\alpha U_m)+\frac{I_m e^2}{U_m}\right)\right].$$

Further on we need the following inequalities

$$e^{-\frac{R}{L}t_k}-e^{-\frac{R}{L}t_{k+1}}=e^{-\frac{R}{L}t_k}-e^{-\frac{R}{L}t}+e^{-\frac{R}{L}t}-e^{-\frac{R}{L}t_{k+1}}=e^{-\frac{R}{L}t}\left(e^{-\frac{R}{L}(t_k-t)}-1\right)+e^{-\frac{R}{L}t}\left(1-e^{-\frac{R}{L}(t_{k+1}-t)}\right)\leq$$

$$\leq e^{-\frac{R}{L}t}\left(e^{\frac{R}{L}T_0}-1+1-e^{-\frac{R}{L}T_0}\right)=e^{-\frac{R}{L}t}\left(e^{\frac{R}{L}T_0}-e^{-\frac{R}{L}T_0}\right).$$

Then

$$J_2\leq\frac{2\sqrt{L}}{Z_0+R_0}\left|\int_{t_k}^{t_{k+1}}\frac{dE(s)}{ds}ds\right|+|\delta|\left|\int_{t_k}^{t_{k+1}}\frac{d\vec{J}(s)}{ds}ds\right|+\frac{1}{C_0 Z_0}\left|\int_{t_k}^{t_{k+1}}\bar{H}(J)(s)ds\right|+$$

$$+\frac{2\sqrt{C}\,\widehat{I}_0}{C_0}\int_{t_k}^{t_{k+1}}\left|e^{\alpha\frac{\bar{H}(J)(s)}{2\sqrt{C}}}-1\right|ds+\frac{I_m}{U_m C_0}\int_{t_k}^{t_{k+1}}\left|\bar{H}(J)(s)e^{1-\frac{\bar{H}(J)(s)}{2U_m\sqrt{C}}}\right|ds\leq$$

$$\leq\frac{1}{C_0 Z_0}\left(\frac{2\sqrt{L}\,U_E}{Z_0+R_0}+(|\delta|+1)I_0\right)\int_{t_k}^{t_{k+1}}e^{-\frac{R}{L}s}ds+\frac{2\sqrt{C}\,\widehat{I}_0}{C_0}\frac{\alpha}{2\sqrt{C}}\left(\frac{2\sqrt{L}\,U_E}{Z_0+R_0}+(|\delta|+1)I_0\right)\xi(\alpha U_m)\int_{t_k}^{t_{k+1}}e^{-\frac{R}{L}s}ds+$$

$$+\frac{I_m e^2}{U_m C_0}\left(\frac{2\sqrt{L}\,U_E}{Z_0+R_0}+(|\delta|+1)I_0\right)\int_{t_k}^{t_{k+1}}e^{-\frac{R}{L}s}ds\leq$$

$$\leq\frac{1}{C_0 Z_0}\left(\frac{2\sqrt{L}\,U_E}{Z_0+R_0}+(|\delta|+1)I_0\right)\frac{L}{R}\left(e^{-\frac{R}{L}t_k}-e^{-\frac{R}{L}t_{k+1}}\right)+\frac{2\sqrt{C}\,\widehat{I}_0}{C_0}\frac{\alpha}{2\sqrt{C}}\left(\frac{2\sqrt{L}\,U_E}{Z_0+R_0}+(|\delta|+1)I_0\right)\xi(\alpha U_m)\frac{L}{R}\left(e^{-\frac{R}{L}t_k}-e^{-\frac{R}{L}t_{k+1}}\right)+$$

$$\leq e^{-\frac{R}{L}t}\left(e^{\frac{R}{L}T_0}-e^{-\frac{R}{L}T_0}\right)\frac{L}{R C_0}\left(\frac{2\sqrt{L}\,U_E}{Z_0+R_0}+(|\delta|+1)I_0\right)\left(\frac{1}{Z_0}+\alpha\,\widehat{I}_0\,\xi(\alpha U_m)+\frac{I_m e^2}{U_m}\right).$$

Thus

$$\left|B_J^{(k)}(J)(t)\right|\leq$$

$$\le e^{-\frac{R}{L}t}\left[\frac{2\sqrt{L}\,U_E}{Z_0+R_0}+|\delta|I_0+\frac{2L}{C_0R}\sinh\!\left(\frac{RT_0}{L}\right)\!\left(\frac{2\sqrt{L}\,U_E}{Z_0+R_0}+(|\delta|+1)I_0\right)\!\left(\frac{1}{Z_0}+\alpha\,\hat{I}_0\,\xi(\alpha U_m)+\frac{I_m e^2}{U_m}\right)\right]+$$

$$+e^{-\frac{R}{L}t}\frac{2L}{RC_0}\sinh\!\left(\frac{RT_0}{L}\right)\!\left(\frac{2\sqrt{L}\,U_E}{Z_0+R_0}+(|\delta|+1)I_0\right)\!\left(\frac{1}{Z_0}+\alpha\,\hat{I}_0\xi(\alpha U_m)+\frac{I_m e^2}{U_m}\right)\le$$

$$\le e^{-\frac{R}{L}t}\left[\frac{2\sqrt{L}\,U_E}{Z_0+R_0}+|\delta|I_0+\frac{4L}{C_0R}\sinh\!\left(\frac{RT_0}{L}\right)\!\left(\frac{2\sqrt{L}\,U_E}{Z_0+R_0}+(|\delta|+1)I_0\right)\!\left(\frac{1}{Z_0}+\alpha\,\hat{I}_0\,\xi(\alpha U_m)+\frac{I_m e^2}{U_m}\right)\right]\le I_0 e^{-\frac{R}{L}t}$$

Therefore the operator B maps M_J into itself.

We show that B is a contractive operator. Since $t_{k+1}-T=t_{q+1}$, $t_k-T=t_q$ $(q<k)$

and $\left|\vec{H}(J)(s)-\vec{H}(\tilde{J})(s)\right|\le|\delta|\left|\vec{\tilde{J}}(t)(s)-\vec{\tilde{\tilde{J}}}(t)(s)\right|+\left|J(s)-\tilde{J}(s)\right|$ we have

$$\left|B_J^{(k)}(J)(t)-B_J^{(k)}(\tilde{J})(t)\right|\le\int_{t_k}^{t}\left|I(J)(s)-I(\tilde{J})(s)\right|ds+\left|\int_{t_k}^{t_{k+1}}\!\!\left(I(J)(s)-I(\tilde{J})(s)\right)ds\right|\equiv I_1+I_2.$$

But

$$I_1=\int_{t_k}^{t}\left|I(J)(s)-I(\tilde{J})(s)\right|ds\le|\delta|\int_{t_k}^{t}\left|\frac{d\vec{J}(s)}{ds}-\frac{d\vec{\tilde{J}}(s)}{ds}\right|ds+\frac{1}{C_0Z_0}\int_{t_k}^{t}\left|\vec{H}(J)(s)-\vec{H}(\tilde{J})(s)\right|ds+$$

$$+\frac{2\sqrt{C}\,\hat{I}_0}{C_0}\int_{t_k}^{t}\left|e^{\alpha\frac{\vec{H}(J)(s)}{2\sqrt{C}}}-e^{\alpha\frac{\vec{H}(\tilde{J})(s)}{2\sqrt{C}}}\right|ds+\frac{I_m}{U_mC_0}\int_{t_k}^{t}\left|\vec{H}(J)(s)e^{1-\frac{\vec{H}(J)(s)}{2U_m\sqrt{C}}}-\vec{H}(\tilde{J})(s)e^{1-\frac{\vec{H}(\tilde{J})(s)}{2U_m\sqrt{C}}}\right|ds\le$$

$$\le|\delta|\rho_\mu^{(q)}(\dot{J},\dot{\tilde{J}})\int_{t_k}^{t}e^{\mu(s-t_k)}ds+\frac{1}{C_0Z_0}\int_{t_k}^{t}\left(|\delta|\left|\vec{\tilde{J}}(s)-\vec{\tilde{\tilde{J}}}(s)\right|+\left|J(s)-\tilde{J}(s)\right|\right)ds+$$

$$+\frac{2\sqrt{C}\,\hat{I}_0}{C_0}\frac{\alpha}{2\sqrt{C}}e^{\alpha U_m}\int_{t_k}^{t}\left(|\delta|\left|\vec{\tilde{J}}(s)-\vec{\tilde{\tilde{J}}}(s)\right|+\left|J(s)-\tilde{J}(s)\right|\right)ds+$$

$$+\frac{2e^2I_m}{U_mC_0}\int_{t_k}^{t}\left(|\delta|\left|\vec{\tilde{J}}(s)-\vec{\tilde{\tilde{J}}}(s)\right|+\left|J(s)-\tilde{J}(s)\right|\right)ds\le$$

$$\le\int_{t_k}^{t}e^{\mu(s-t_k)}ds|\delta|\rho_\mu^{(q)}(\dot{J},\dot{\tilde{J}})+\int_{t_k}^{t}e^{\mu(s-t_k)}ds\frac{|\delta|\rho_\mu^{(q)}(\dot{J},\dot{\tilde{J}})+\rho_\mu^{(k)}(\dot{J},\dot{\tilde{J}})}{\mu}\left(\frac{1}{C_0Z_0}+\frac{\alpha\,\hat{I}_0}{C_0}e^{\alpha U_m}+\frac{2e^2I_m}{U_mC_0}\right)\le$$

$$\le\hat{\rho}_\mu^{(k)}((J,\dot{J}),(\tilde{J},\dot{\tilde{J}}))\frac{e^{\mu(t-t_k)}-1}{\mu}\left[|\delta|+\frac{|\delta|+1}{\mu C_0}\left(\frac{1}{Z_0}+\alpha\,\hat{I}_0e^{\alpha U_m}+\frac{2e^2I_m}{U_m}\right)\right];$$

$$I_2 = \int_{t_k}^{t_{k+1}} \left| I(J)(s) - I(\widetilde{J})(s) \right| ds \leq \left| \delta \right| \int_{t_k}^{t_{k+1}} \left| \frac{d\vec{J}(s)}{ds} - \frac{d\vec{\widetilde{J}}(s)}{ds} \right| ds + \frac{1}{C_0 Z_0} \int_{t_k}^{t_{k+1}} \left| \vec{H}(J)(s) - \vec{H}(\widetilde{J})(s) \right| ds +$$

$$+ \frac{2\sqrt{C}\,\widehat{I}_0}{C_0} \int_{t_k}^{t_{k+1}} \left| e^{\alpha \frac{\vec{H}(J)(s)}{2\sqrt{C}}} - e^{\alpha \frac{\vec{H}(\widetilde{J})(s)}{2\sqrt{C}}} \right| ds + \frac{I_m}{U_m C_0} \int_{t_k}^{t_{k+1}} \left| \vec{H}(J)(s) e^{1 - \frac{\vec{H}(J)(s)}{2U_m\sqrt{C}}} - \vec{H}(\widetilde{J})(s) e^{1 - \frac{\vec{H}(\widetilde{J})(s)}{2U_m\sqrt{C}}} \right| ds \leq$$

$$\leq \int_{t_k}^{t_{k+1}} e^{\mu(s-t_k)} ds \, \frac{\left| \delta \right| \rho_\mu^{(q)}(\dot{J},\dot{\widetilde{J}}) + \rho_\mu^{(k)}(\dot{J},\dot{\widetilde{J}})}{\mu} \left(\frac{1}{C_0 Z_0} + \frac{\alpha \widehat{I}_0}{C_0} e^{\alpha U_m} + \frac{2e^2 I_m}{U_m C_0} \right) \leq$$

$$\leq \widehat{\rho}_\mu^{(k)}((J,\dot{J}),(\widetilde{J},\dot{\widetilde{J}})) \frac{e^{\mu_0} - 1}{\mu} \left[\frac{\left| \delta \right| + 1}{\mu C_0} \left(\frac{1}{Z_0} + \alpha \widehat{I}_0 e^{\alpha U_m} + \frac{2e^2 I_m}{U_m} \right) \right].$$

Therefore

$$\left| B_J^{(k)}(J)(t) - B_J^{(k)}(\widetilde{J})(t) \right| \leq$$

$$\leq e^{\mu(t-t_k)} \widehat{\rho}_\mu^{(k)}((J,\dot{J}),(\widetilde{J},\dot{\widetilde{J}})) \left[\frac{\left| \delta \right|}{\mu} + \frac{\left| \delta \right| + 1}{\mu^2 C_0} \left(\frac{1}{Z_0} + \alpha \widehat{I}_0 e^{\alpha U_m} + \frac{2e^2 I_m}{U_m} \right) \right] +$$

$$+ e^{\mu(t-t_k)} \widehat{\rho}_\mu^{(k)}((J,\dot{J}),(\widetilde{J},\dot{\widetilde{J}}))(e^{\mu_0} - 1) \left[\frac{\left| \delta \right| + 1}{\mu^2 C_0} \left(\frac{1}{Z_0} + \alpha \widehat{I}_0 e^{\alpha U_m} + \frac{2e^2 I_m}{U_m} \right) \right] \leq$$

$$\leq e^{\mu(t-t_k)} \widehat{\rho}_\mu^{(k)}((J,\dot{J}),(\widetilde{J},\dot{\widetilde{J}})) \left[\frac{\left| \delta \right|}{\mu} + e^{\mu_0} \frac{\left| \delta \right| + 1}{\mu^2 C_0} \left(\frac{1}{Z_0} + \alpha \widehat{I}_0 e^{\alpha U_m} + \frac{2e^2 I_m}{U_m} \right) \right] \equiv$$

$$\equiv e^{\mu(t-t_k)} K_J \widehat{\rho}_\mu^{(k)}((J,\dot{J}),(\widetilde{J},\dot{\widetilde{J}})) \leq e^{\mu_0} K_J \widehat{\rho}_\mu^{(k)}((J,\dot{J}),(\widetilde{J},\dot{\widetilde{J}}))$$

or

$$\widehat{\rho}^{(k)}(B_J(J), B_J(\widetilde{J})) \leq e^{\mu_0} K_J \widehat{\rho}_\mu^{(k)}((J,\dot{J}),(\widetilde{J},\dot{\widetilde{J}})).$$

For the derivative we obtain

$$\left| \dot{B}_J^{(k)}(J)(t) - \dot{B}_J^{(k)}(\widetilde{J})(t) \right| \leq \left| I(J)(t) - I(\widetilde{J})(t) \right| + \frac{1}{t_{k+1} - t_k} \left| \int_{t_k}^{t_{k+1}} \left(I(J)(s) - I(\widetilde{J})(s) \right) ds \right| \equiv \dot{i}_1 + \dot{i}_2.$$

But

$$\dot{i}_1 \leq |\delta|\left|\frac{d\vec{J}(t)}{ds} - \frac{d\tilde{\vec{J}}(t)}{ds}\right| + \frac{1}{C_0 Z_0}\left|\vec{H}(J)(t) - \vec{H}(\tilde{J})(t)\right| +$$

$$+ \frac{2\sqrt{C}\,\hat{I}_0}{C_0}\left|e^{\alpha\frac{\vec{H}(J)(t)}{2\sqrt{C}}} - e^{\alpha\frac{\vec{H}(\tilde{J})(t)}{2\sqrt{C}}}\right| + \frac{I_m}{U_m C_0}\left|\vec{H}(J)(t)e^{1-\frac{\vec{H}(J)(t)}{2U_m\sqrt{C}}} - \vec{H}(\tilde{J})(t)e^{1-\frac{\vec{H}(\tilde{J})(t)}{2U_m\sqrt{C}}}\right| \leq$$

$$\leq |\delta|\rho_\mu^{(q)}(\dot{J},\dot{\tilde{J}}) + \frac{1}{C_0 Z_0}\left(|\delta|\left\|\vec{J}(t) - \tilde{\vec{J}}(t)\right\| + \left|J(t) - \tilde{J}(t)\right|\right) +$$

$$+ \frac{2\sqrt{C}\,\hat{I}_0}{C_0}\frac{\alpha}{2\sqrt{C}}e^{\alpha U_m}\left(|\delta|\left\|\vec{J}(t) - \tilde{\vec{J}}(t)\right\| + \left|J(t) - \tilde{J}(t)\right|\right) + \frac{2e^2 I_m}{U_m C_0}\left(|\delta|\left\|\vec{J}(t) - \tilde{\vec{J}}(t)\right\| + \left|J(t) - \tilde{J}(t)\right|\right) \leq$$

$$\leq |\delta|\rho_\mu^{(q)}(\dot{J},\dot{\tilde{J}}) + \frac{|\delta|\rho_\mu^{(q)}(J,\tilde{J}) + \rho_\mu^{(k)}(\dot{J},\dot{\tilde{J}})}{\mu}\left(\frac{1}{C_0 Z_0} + \frac{\alpha\,\hat{I}_0}{C_0}e^{\alpha U_m} + \frac{2e^2 I_m}{U_m C_0}\right) \leq$$

$$\leq e^{\mu(t-t_k)}\hat{\rho}_\mu^{(k)}((J,\dot{J}),(\tilde{J},\dot{\tilde{J}}))\left[|\delta| + \frac{|\delta|+1}{\mu C_0}\left(\frac{1}{Z_0} + \alpha\,\hat{I}_0 e^{\alpha U_m} + \frac{2e^2 I_m}{U_m}\right)\right]$$

and

$$\dot{i}_2 \leq \frac{1}{t_{k+1} - t_k}\int_{t_k}^{t_{k+1}}\left|I(J)(s) - I(\tilde{J})(s)\right|ds \leq$$

$$\leq \frac{1}{t_{k+1} - t_k}\left(|\delta|\left|\int_{t_k}^{t_{k+1}}\left(\frac{d\vec{J}(s)}{ds} - \frac{d\tilde{\vec{J}}(s)}{ds}\right)ds\right| + \frac{1}{C_0 Z_0}\int_{t_k}^{t_{k+1}}\left|\vec{H}(J)(s) - \vec{H}(\tilde{J})(s)\right|ds + \right.$$

$$\left. + \frac{2\sqrt{C}\,\hat{I}_0}{C_0}\int_{t_k}^{t_{k+1}}\left|e^{\alpha\frac{\vec{H}(J)(s)}{2\sqrt{C}}} - e^{\alpha\frac{\vec{H}(\tilde{J})(s)}{2\sqrt{C}}}\right|ds + \frac{I_m}{U_m C_0}\int_{t_k}^{t_{k+1}}\left|\vec{H}(J)(s)e^{1-\frac{\vec{H}(J)(s)}{2U_m\sqrt{C}}} - \vec{H}(\tilde{J})(s)e^{1-\frac{\vec{H}(\tilde{J})(s)}{2U_m\sqrt{C}}}\right|ds\right) \leq$$

$$\leq \frac{1}{t_{k+1} - t_k}\int_{t_k}^{t_{k+1}}e^{\mu(s-t_k)}ds\,\frac{|\delta|\rho_\mu^{(q)}(\dot{J},\tilde{J}) + \rho_\mu^{(k)}(\dot{J},\dot{\tilde{J}})}{\mu}\left(\frac{1}{C_0 Z_0} + \frac{\alpha\,\hat{I}_0}{C_0}e^{\alpha U_m} + \frac{2e^2 I_m}{U_m C_0}\right) \leq$$

$$\leq \hat{\rho}_\mu^{(k)}((J,\dot{J}),(\tilde{J},\dot{\tilde{J}}))\frac{e^{\mu_0} - 1}{\mu_0}\frac{|\delta|+1}{\mu C_0}\left(\frac{1}{Z_0} + \alpha\,\hat{I}_0 e^{\alpha U_m} + \frac{2e^2 I_m}{U_m}\right).$$

Therefore

$$\left|\dot{B}_J^{(k)}(J)(t) - \dot{B}_J^{(k)}(\tilde{J})(t)\right| \leq$$

$$\le e^{\mu(t-t_k)}\hat{\rho}_\mu^{(k)}((J,\dot{J}),(\widetilde{J},\dot{\widetilde{J}}))\left[|\delta|+\frac{|\delta|+1}{\mu C_0}\left(\frac{1}{Z_0}+\alpha\,\widehat{I}_0 e^{\alpha U_m}+\frac{2e^2 I_m}{U_m}\right)\right]+$$

$$+\hat{\rho}_\mu^{(k)}((J,\dot{J}),(\widetilde{J},\dot{\widetilde{J}}))\frac{e^{\mu_0}-1}{\mu_0}\frac{|\delta|+1}{\mu C_0}\left(\frac{1}{Z_0}+\alpha\,\widehat{I}_0 e^{\alpha U_m}+\frac{2e^2 I_m}{U_m}\right)\le$$

$$\le e^{\mu(t-t_k)}\hat{\rho}_\mu^{(k)}((J,\dot{J}),(\widetilde{J},\dot{\widetilde{J}}))\left[|\delta|+\left(1+\frac{e^{\mu_0}-1}{\mu_0}\right)\frac{|\delta|+1}{\mu C_0}\left(\frac{1}{Z_0}+\alpha\,\widehat{I}_0 e^{\alpha U_m}+\frac{2e^2 I_m}{U_m}\right)\right]\equiv$$

$$\equiv e^{\mu(t-t_k)}\dot{K}_J\hat{\rho}_\mu^{(k)}((J,\dot{J}),(\widetilde{J},\dot{\widetilde{J}}))$$

or

$$\rho_\mu^{(k)}(\dot{B}_J^{(k)}(J),\dot{B}_J^{(k)}(\widetilde{J}))\le\dot{K}_J\hat{\rho}_\mu^{(k)}((J,\dot{J}),(\widetilde{J},\dot{\widetilde{J}}))) .$$

Finally we obtain

$$\rho_\mu^{(k)}((B_J^{(k)}(J),\dot{B}_J^{(k)}(J)),(B_J^{(k)}(\widetilde{J}),\dot{B}_J^{(k)}(\widetilde{J}))\le K\hat{\rho}_\mu^{(k)}((J,\dot{J}),(\widetilde{J},\dot{\widetilde{J}}))$$

where $K=\max\left\{e^{\mu_0}K_J,\dot{K}_J\right\}<1$ for sufficiently large $\mu>0$.

Theorem 5.6.1 is thus proved.

5.6.1. Numerical Example

Finally, we collect all inequalities guaranteeing conditions of the existence-uniqueness theorem:

$$\frac{Z_0 U_E}{Z_0+R_0}+\frac{|\delta|+1}{2\sqrt{C}}I_0\le U_m;$$

$$\frac{2\sqrt{L}\,U_E}{Z_0+R_0}+|\delta|I_0+\frac{4L}{C_0 R}\sinh\left(\frac{RT_0}{L}\right)\left(\frac{2\sqrt{L}\,U_E}{Z_0+R_0}+(|\delta|+1)I_0\right)\left(\frac{1}{Z_0}+\alpha\widehat{I}_0\,\xi(\alpha U_m)+\frac{I_m e^2}{U_m}\right)\le I_0;$$

$$e^{\mu_0}K_J=e^{\mu_0}\left[\frac{|\delta|}{\mu}+e^{\mu_0}\frac{|\delta|+1}{\mu^2 C_0}\left(\frac{1}{Z_0}+\alpha\,\widehat{I}_0 e^{\alpha U_m}+\frac{2e^2 I_m}{U_m}\right)\right]<1;$$

$$\dot{K}_J = |\delta| + \left(1 + \frac{e^{\mu_0}-1}{\mu_0}\right)\frac{|\delta|+1}{\mu C_0}\left(\frac{1}{Z_0} + \alpha\,\widehat{I}_0 e^{\alpha U_m} + \frac{2e^2 I_m}{U_m}\right) < 1.$$

Consider a line with specific parameters as in the previous numerical examples:

$$\Lambda = 100\,m,\ \ S = 4\,mm^2,\ \ \rho_c = 0,0175,\ \ R = \frac{\rho_c \Lambda}{S} = \frac{0,0175.100}{4} \approx 0,64\Omega,$$

$$L = 0,45\,\mu H/m,\ C = 80\,pF/m,\ Z_0 = 75\ \Omega.\,,\ C_0 = 10.10^{-12}\,F,$$

$$v = 1/\sqrt{LC} = (1/6).10^9,\ R_0 = 25\Omega,\ \ \delta = \frac{|Z_0 - R_0|}{Z_0 + R_0} = \frac{50}{100} = 0,5\,;\frac{L}{R} \approx 1,14.10^{-7}.$$

Let us choose $\lambda_0 = 6.10^{-3}\,m$; $f_0 = \dfrac{1}{\lambda_0\sqrt{LC}} = 10^{12}\,Hz \Rightarrow T_0 = \dfrac{1}{f_0} = 10^{-12}\,\text{sec}.$

Choose $\mu = 10^{12}$. Then

$$\mu T_0 = \mu_0 = 1,\ \mu C_0 = 10^{12}.10^{-11} = 10;\ \mu^2 C_0 = 10^{24}.10^{-11} = 10^{13};$$

$$\widehat{I}_0 = 10^{-8}\,A,\ \alpha = 19,23\,;\ \alpha\widehat{I}_0 = 19,23.10^{-8}\,;$$

$$e^{\frac{R}{L}T_0} - e^{-\frac{R}{L}T_0} \approx \frac{RT_0}{L} = \frac{1}{1,14.10^{-7}}10^{-12} \approx 10^{-5},\ \sinh\!\left(\frac{RT_0}{L}\right) \approx \frac{RT_0}{L},\ \frac{I_m}{U_m} = 12,67.10^{-3},$$

$$U_m = 0,15;\ \frac{1}{U_m} = 6,66;\ \ \alpha U_m = 2,8845;\ e^{\alpha U_m} \le e^3;\ \frac{I_0}{C_0} = 10^3\ \text{and}$$

$$\xi(u) = \frac{e^u - 1}{u} \Rightarrow \xi(\alpha U_m) = \frac{e^{\alpha U_m}-1}{\alpha U_m}\ .$$

For $U_0 = I_0 = 10^{-6}$ the above inequalities become

$$\frac{75}{100}U_E + \frac{1,5}{2\sqrt{80.10^{-12}}}I_0 \le 0,15;$$

$$\frac{2\sqrt{0,45.10^{-6}}}{100}U_E + 0,5I_0 + \frac{4.10^{-12}}{10.10^{-12}}\left(\frac{2\sqrt{0,45.10^{-6}}}{100}U_E + 1,5I_0\right)\left(\frac{1}{75} + 19,23.10^{-8}\frac{e^3-1}{3} + 12,67.10^{-3}e^2\right) \le I_0 \approx$$

$$\approx 1,375.10^{-5}\,U_E + 0,568 I_0 \le I_0\,;$$

$$eK_J = e\left[\frac{0,5}{10^{12}} + e\frac{1,5}{10^{24}.10.10^{-12}}\left(\frac{1}{75} + 19,23.10^{-8}\,e^3 + 2e^2.12,67.10^{-3}\right)\right] < 1;$$

$$\dot{K}_J = 0,5 + 0,4\left(0,013 + 19,23.10^{-8}\ e^3 + 2e^2.12,67.10^{-3}\right) \approx 0,58 < 1;$$

$$K = \dot{K}_J = 0,58 < 1.$$

5.7. Transmission Lines with Time-Varying Specific Parameters

We proceed again from (5.1.1) in the form

$$\frac{\partial u(x,t)}{\partial t} + \frac{1}{C}\frac{\partial i(x,t)}{\partial x} + \frac{G}{C}u(x,t) = 0,$$

$$\frac{\partial i(x,t)}{\partial t} + \frac{1}{L}\frac{\partial u(x,t)}{\partial x} + \frac{R}{L}i(x,t) = 0$$

only in this case, the parameters depend on the time
$C = C(t), L = L(t), G = G(t), R = R(t)$.

Clearly, we have to assume that $C(t), L(t)$ have strictly positive lower bounds. The Heaviside condition is also satisfied $R/L = G/C$.

The above system in a matrix form is:

$$\frac{\partial U}{\partial t} + A_1\frac{\partial U}{\partial x} + A_2 U = 0 \tag{5.7.1}$$

where

$$U = \begin{bmatrix} u \\ i \end{bmatrix}, \quad \frac{\partial U}{\partial t} = \begin{bmatrix} \dfrac{\partial u}{\partial t} \\ \dfrac{\partial i}{\partial t} \end{bmatrix}, \quad \frac{\partial U}{\partial x} = \begin{bmatrix} \dfrac{\partial u}{\partial x} \\ \dfrac{\partial i}{\partial x} \end{bmatrix}, \quad A_1(t) = \begin{bmatrix} 0 & 1/C \\ 1/L & 0 \end{bmatrix}, \quad A_2(t) = \begin{bmatrix} G/C & 0 \\ 0 & R/L \end{bmatrix}.$$

Since $A_1(t) = \begin{bmatrix} 0 & 1/C \\ 1/L & 0 \end{bmatrix}$ has eigenvalues $\lambda_1 = 1/\sqrt{LC}$, $\lambda_2 = -1/\sqrt{LC}$ from the systems:

$$\left| \begin{aligned} -\frac{1}{\sqrt{LC}}\xi_1 + \frac{1}{L}\xi_2 &= 0 \\ \frac{1}{C}\xi_1 - \frac{1}{\sqrt{LC}}\xi_2 &= 0 \end{aligned} \right. \quad \text{and} \quad \left| \begin{aligned} \frac{1}{\sqrt{LC}}\xi_1 + \frac{1}{L}\xi_2 &= 0 \\ \frac{1}{C}\xi_1 + \frac{1}{\sqrt{LC}}\xi_2 &= 0 \end{aligned} \right. .$$

we obtain eigenvectors $\left(\xi_1^{(1)},\xi_2^{(1)}\right)=\left(\sqrt{C},\sqrt{L}\right),\qquad \left(\xi_1^{(2)},\xi_2^{(2)}\right)=\left(-\sqrt{C},\sqrt{L}\right).$

Here, the matrix H formed by the eigenvectors depends on time $H(t)=\begin{bmatrix}\sqrt{C} & \sqrt{L}\\ -\sqrt{C} & \sqrt{L}\end{bmatrix}$

and its inverse one $H^{-1}(t)=\begin{bmatrix}\dfrac{1}{2\sqrt{C}} & -\dfrac{1}{2\sqrt{C}}\\[2mm] \dfrac{1}{2\sqrt{L}} & \dfrac{1}{2\sqrt{L}}\end{bmatrix}.$

Then $A^{can}=HA_1H^{-1}$ where $A^{can}(t)=\begin{bmatrix}1/\sqrt{LC} & 0\\ 0 & -1/\sqrt{LC}\end{bmatrix}.$

Introduce new variables $Z=HU,$ (or $U=H^1Z$)

$$Z=\begin{bmatrix}V(x,t)\\ I(x,t)\end{bmatrix},\quad H(t)=\begin{bmatrix}\sqrt{C} & \sqrt{L}\\ -\sqrt{C} & \sqrt{L}\end{bmatrix},\quad U=\begin{bmatrix}u(x,t)\\ i(x,t)\end{bmatrix}.$$

Then

$$\left|\begin{aligned}V(x,t)&=\sqrt{C}\,u(x,t)+\sqrt{L}\,i(x,t)\\ I(x,t)&=-\sqrt{C}\,u(x,t)+\sqrt{L}\,i(x,t)\end{aligned}\right.$$

or

$$\left|\begin{aligned}u(x,t)&=\frac{1}{2\sqrt{C}}V(x,t)-\frac{1}{2\sqrt{C}}I(x,t)\\[2mm] i(x,t)&=\frac{1}{2\sqrt{L}}V(x,t)+\frac{1}{2\sqrt{L}}I(x,t).\end{aligned}\right.$$

Substituting $U=H^{-1}(t)Z$ in (5.7.1) we obtain

$$\frac{\partial\left(H^{-1}(t)Z\right)}{\partial t}+A_1\frac{\partial\left(H^{-1}(t)Z\right)}{\partial x}+A_2\left(H^{-1}(t)Z\right)=0.$$

In contrast to transformations from Chapter I, we have

$$H^{-1}(t)\frac{\partial Z}{\partial t} + \frac{\partial H^{-1}(t)}{\partial t}Z + A_1 H^{-1}(t)\frac{\partial Z}{\partial x} + A_2 H^{-1}(t)Z = 0.$$

Multiply the last equality by $H(t)$ from the left

$$\frac{\partial Z}{\partial t} + H(t)\frac{\partial H^{-1}(t)}{\partial t}Z + H(t)A_1 H^{-1}(t)\frac{\partial Z}{\partial x} + \left(H(t)A_2 H^{-1}(t)\right)Z = 0$$

where $H(t)A_2(t)H^{-1}(t) = \dfrac{1}{2}\begin{bmatrix} \dfrac{G(t)}{C(t)}+\dfrac{R(t)}{L(t)} & -\dfrac{G(t)}{C(t)}+\dfrac{R(t)}{L(t)} \\[3mm] -\dfrac{G(t)}{C(t)}+\dfrac{R(t)}{L(t)} & \dfrac{G(t)}{C(t)}+\dfrac{R(t)}{L(t)} \end{bmatrix}$ and

$$H(t)\frac{\partial H^{-1}(t)}{\partial t} = \frac{1}{4}\begin{bmatrix} -\dfrac{\dot C(t)}{C(t)}-\dfrac{\dot L(t)}{L(t)} & \dfrac{\dot C(t)}{C(t)}-\dfrac{\dot L(t)}{L(t)} \\[3mm] \dfrac{\dot C(t)}{C(t)}-\dfrac{\dot L(t)}{L(t)} & -\dfrac{\dot C(t)}{C(t)}-\dfrac{\dot L(t)}{L(t)} \end{bmatrix}.$$

The dot means a differentiation with respect the time. We see that

$$H(t)A_2 H^{-1}(t) + H(t)\frac{\partial H^{-1}(t)}{\partial t} =$$

$$= \frac{1}{2}\begin{bmatrix} \dfrac{G(t)}{C(t)}+\dfrac{R(t)}{L(t)} & -\dfrac{G(t)}{C(t)}+\dfrac{R(t)}{L(t)} \\[3mm] -\dfrac{G(t)}{C(t)}+\dfrac{R(t)}{L(t)} & \dfrac{G(t)}{C(t)}+\dfrac{R(t)}{L(t)} \end{bmatrix} + \frac{1}{2}\begin{bmatrix} -\dfrac{\dot C(t)}{2C(t)}-\dfrac{\dot L(t)}{2L(t)} & \dfrac{\dot C(t)}{2C(t)}-\dfrac{\dot L(t)}{2L(t)} \\[3mm] \dfrac{\dot C(t)}{2C(t)}-\dfrac{\dot L(t)}{2L(t)} & -\dfrac{\dot C(t)}{2C(t)}-\dfrac{\dot L(t)}{2L(t)} \end{bmatrix} =$$

$$= \frac{1}{2}\begin{bmatrix} \dfrac{G(t)}{C(t)}+\dfrac{R(t)}{L(t)}-\dfrac{\dot C(t)}{2C(t)}-\dfrac{\dot L(t)}{2L(t)} & -\dfrac{G(t)}{C(t)}+\dfrac{R(t)}{L(t)}+\dfrac{\dot C(t)}{2C(t)}-\dfrac{\dot L(t)}{2L(t)} \\[3mm] -\dfrac{G(t)}{C(t)}+\dfrac{R(t)}{L(t)}+\dfrac{\dot C(t)}{2C(t)}-\dfrac{\dot L(t)}{2L(t)} & \dfrac{G(t)}{C(t)}+\dfrac{R(t)}{L(t)}-\dfrac{\dot C(t)}{2C(t)}-\dfrac{\dot L(t)}{2L(t)} \end{bmatrix}.$$

So the system

$$\frac{\partial Z(x,t)}{\partial t} + A^{can}(t)\frac{\partial Z(x,t)}{\partial x} + \left[H(t)A_2 H^{-1}(t) + H(t)\frac{\partial H^{-1}(t)}{\partial t}\right]Z(x,t) = 0$$

becomes

$$\begin{bmatrix} \dfrac{\partial V}{\partial t} \\[2ex] \dfrac{\partial I}{\partial t} \end{bmatrix} + \begin{bmatrix} \dfrac{1}{\sqrt{LC}} & 0 \\[2ex] 0 & -\dfrac{1}{\sqrt{LC}} \end{bmatrix} \begin{bmatrix} \dfrac{\partial V}{\partial x} \\[2ex] \dfrac{\partial I}{\partial x} \end{bmatrix} + \frac{1}{2} \begin{bmatrix} \dfrac{G}{C}+\dfrac{R}{L}-\dfrac{\dot C}{2C}-\dfrac{\dot L}{2L} & -\dfrac{G}{C}+\dfrac{R}{L}+\dfrac{\dot C}{2C}-\dfrac{\dot L}{2L} \\[2ex] -\dfrac{G}{C}+\dfrac{R}{L}+\dfrac{\dot C}{2C}-\dfrac{\dot L}{2L} & \dfrac{G}{C}+\dfrac{R}{L}-\dfrac{\dot C}{2C}-\dfrac{\dot L}{2L} \end{bmatrix} \begin{bmatrix} V \\ I \end{bmatrix} = \begin{bmatrix} 0 \\ 0 \end{bmatrix}$$

or, in an explicit form

$$\frac{\partial V(x,t)}{\partial t} + \frac{1}{\sqrt{L(t)C(t)}}\frac{\partial V(x,t)}{\partial x} + \left(\frac{G(t)}{C(t)} + \frac{R(t)}{L(t)} - \frac{\dot C(t)}{2C(t)} - \frac{\dot L(t)}{2L(t)} \right) V(x,t) +$$

$$+ \left(-\frac{G}{C(t)} + \frac{R}{L(t)} + \frac{\dot C}{2C(t)} - \frac{\dot L}{2L(t)} \right) I(x,t) = 0$$

$$\frac{\partial I(x,t)}{\partial t} - \frac{1}{\sqrt{L(t)C(t)}}\frac{\partial I(x,t)}{\partial x} + \left(-\frac{G(t)}{C(t)} + \frac{R(t)}{L(t)} + \frac{\dot C(t)}{2C(t)} - \frac{\dot L(t)}{2L(t)} \right) V(x,t) +$$

$$+ \left(\frac{G(t)}{C(t)} + \frac{R(t)}{L(t)} - \frac{\dot C(t)}{2C(t)} - \frac{\dot L(t)}{2L(t)} \right) I(x,t) = 0.$$

Let us suppose that the following condition be fulfilled:

$$-\frac{G(t)}{C(t)} + \frac{R(t)}{L(t)} + \frac{\dot C(t)}{2C(t)} - \frac{\dot L(t)}{2L(t)} = 0.$$

Then the previous system can be rewritten in the form

$$\frac{\partial V(x,t)}{\partial t} + \frac{1}{\sqrt{L(t)C(t)}}\frac{\partial V(x,t)}{\partial x} + \frac{2G(t)-\dot C(t)}{C(t)} V(x,t) = 0,$$

$$\frac{\partial I(x,t)}{\partial t} - \frac{1}{\sqrt{L(t)C(t)}}\frac{\partial I(x,t)}{\partial x} + \frac{2G(t)-\dot C(t)}{C(t)} I(x,t) = 0.$$

Put $h(t) = \dfrac{2G(t)-\dot C(t)}{C(t)}$ and rewrite the above system in the form:

$$\frac{\partial V}{\partial t} + \frac{1}{\sqrt{LC}}\frac{\partial V}{\partial x} + hV = 0 , \quad \frac{\partial I}{\partial t} - \frac{1}{\sqrt{LC}}\frac{\partial I}{\partial x} + hI = 0.$$

This system might be simplified one more time by the substitution:

$$W(x,t) = e^{ht}V(x,t), \quad J(x,t) = e^{ht}I(x,t),$$

or

$$V(x,t) = e^{-ht}W(x,t) , \quad I(x,t) = e^{-ht}J(x,t).$$

Consequently we obtain the following transformation:

$$u(x,t) = \frac{1}{2\sqrt{C}} e^{-ht}W(x,t) - \frac{1}{2\sqrt{C}} e^{-ht}J(x,t)$$

$$i(x,t) = \frac{1}{2\sqrt{L}} e^{-ht}W(x,t) + \frac{1}{2\sqrt{L}} e^{-ht}J(x,t)$$

and

$$W(x,t) = e^{ht}\left(\sqrt{C}\, u(x,t) + \sqrt{L}\, i(x,t)\right),$$
$$J(x,t) = e^{ht}\left(\sqrt{L}\, i(x,t) - \sqrt{C}\, u(x,t)\right).$$

Substitutinge $V(x,t)$ and $I(x,t)$ into the above system we have

$$\frac{\partial\left(e^{-h(t)t}W(x,t)\right)}{\partial t} + \frac{1}{\sqrt{LC}}\frac{\partial\left(e^{-h(t)t}W(x,t)\right)}{\partial x} + e^{-h(t)t}h(t)W(x,t) = 0 ,$$

$$\frac{\partial\left(e^{-h(t)t}J(x,t)\right)}{\partial t} - \frac{1}{\sqrt{LC}}\frac{\partial\left(e^{-h(t)t}J(x,t)\right)}{\partial x} + e^{-h(t)t}h(t)J(x,t) = 0$$

which leads to

$$\frac{\partial W}{\partial t} + \frac{1}{\sqrt{LC}}\frac{\partial W}{\partial x} - t\dot{h}(t)W = 0 , \quad \frac{\partial J}{\partial t} - \frac{1}{\sqrt{LC}}\frac{\partial J}{\partial x} - t\dot{h}(t)J = 0 .$$

The previous system may take the form

$$\frac{\partial W}{\partial t} + \frac{1}{\sqrt{LC}}\frac{\partial W}{\partial x} = 0 , \quad \frac{\partial J}{\partial t} - \frac{1}{\sqrt{LC}}\frac{\partial J}{\partial x} = 0$$

under condition $\dot{h}(t) = 0 \Leftrightarrow h(t) = \text{const.} = \gamma$ or $\dfrac{2G(t) - \dot{C}(t)}{C(t)} = \text{const.} = \gamma$.

Since the initial conditions remain the same ones in view of

$$W(x,0) = e^{\frac{R_0}{L}} V(x,0) = V_0(x) = \sqrt{C}\, u_0(x) + \sqrt{L}\, i_0(x),$$

$$J(x,0) = e^{\frac{R_0}{L}} I(x,0) = I_0(x) = -\sqrt{C}\, u_0(x) + \sqrt{L}\, i_0(x), x \in [0,\Lambda]$$

we have to transform the boundary conditions

$$E(t) - \frac{e^{-\prime\prime} W(0,t) - e^{-\prime\prime} J(0,t)}{2\sqrt{C}} - R_0 \frac{e^{-\prime\prime} W(0,t) + e^{-\prime\prime} J(0,t)}{2\sqrt{L}} = 0,$$

$$C_0 \frac{1}{2\sqrt{C}} \left[\frac{d}{dt}\left(e^{-\prime\prime} W(\Lambda,t)\right) - \frac{d}{dt}\left(e^{-\prime\prime} J(\Lambda,t)\right) \right] =$$

$$= \frac{e^{-\prime\prime} W(\Lambda,t) + e^{-\prime\prime} J(\Lambda,t)}{2\sqrt{L}} - f\left(\frac{e^{-\prime\prime} W(\Lambda,t) - e^{-\prime\prime} J(\Lambda,t)}{2\sqrt{C}} \right), t \geq 0.$$

So we have obtained the same problem, and thus it may be treated by the same way. In fact we derived conditions

$$-\frac{G(t)}{C(t)} + \frac{R(t)}{L(t)} + \frac{\dot{C}(t)}{2C(t)} - \frac{\dot{L}(t)}{2L(t)} = 0, \frac{G(t)}{C(t)} - \frac{\dot{C}(t)}{2C(t)} = \frac{1}{2}\gamma$$

which yield a Heaviside condition, provided $\dfrac{\dot{C}(t)}{C(t)} = \dfrac{\dot{L}(t)}{L(t)}$.

Remark 5.7.1. We notice that the above system yields $\dot{L} + \gamma L = 2R$. But this ordinary differential equation has a solution $L(t) = e^{-\prime\prime}\left(L_T + 2\int_T^t R(s)e^{\gamma s}\, ds \right)$.

CONCLUSION

The main goal of the present chapter is to expose two manners of reducing the mixed problem for lossy transmission line equations to periodic and oscillatory problems on the boundary. Both methods can be compared by simply substituting one and the same parameter values provided in the obtained inequalities.

The generalized Heaviside condition (introduced here) allows us to extend previous methods to transmission lines with time-varying specific parameters.

DISTORTIONLESS LOSSY TRANSMISSION LINES TERMINATED BY PARALLEL CONNECTED *RLC*-LOADS

ABSTRACT

In this chapter we investigate lossy transmission lines terminated by parallel connected *RLC*-loads at both ends. Although we proceed as in Chapter V, here we reduce the mixed problem for a hyperbolic system to a neutral system whose right-hand sides contain integrals of unknown functions. This requires us to introduce a special class of functions where the solution must be found. Using suitable fixed-point theorems we prove an existence-uniqueness result for the periodic and oscillatory solution.

INTRODUCTION

In the previous Chapters III and IV we have investigated lossless transmission lines terminated by series connected and parallel connected *RLC*-loads. Here we take into account the losses for transmission lines terminated by parallel connected nonlinear *RLC*-loads at both ends (Fig. 6.1) which means $R \neq 0, G \neq 0$ in (6.1.2). From the mathematical point of view, this leads to a more complicated transformation of the mixed problem, and hence to a more complicated neutral system. Our considerations are valid under the Heaviside condition, that is, $R/L = G/C$. This implies that the line is without distortion.

In this chapter we consider lossy transmission lines terminated by parallel connected nonlinear *RLC*-loads at both ends. This means $R \neq 0, G \neq 0$ which generates additional difficulties. So in § 6.1 we derive boundary conditions corresponding to the circuit configuration of the loads. Then we reduce the mixed problem for the lossy transmission line system to a periodic problem on the boundary. In § 6.2 we make an analysis of the arising nonlinearities. In § 6.3 we propose an operator presentation of the periodic problem. In § 6.4 we prove an existence-uniqueness theorem for a periodic solution for the nonlinear neutral system. In § 6.5 we consider a numerical example. In § 6.6 the existence-uniqueness of an oscillatory solution for a neutral system is obtained. In § 6.7 we give another method for reducing the mixed problem to an oscillatory one on the boundary.

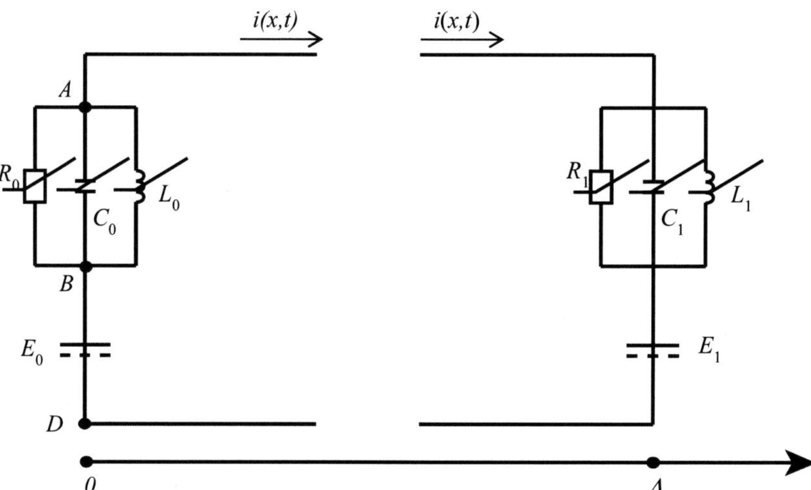

Figure 6. 1.

Our analysis is based on the reducing the mixed problem for the first-order hyperbolic partial differential systems describing the distortionless lossy transmission lines to an equivalent initial value problem for a neutral functional differential system. We formulate and solve the problem for the existence of a periodic solution of this system. The main difficulties are the following: 1) the availability of polynomial nonlinearities arising most often in the applications (cf. [37]-[39], [51], [55], [81], [89], [96], [119]) and 2), unlike the problems considered in Chapters II and V the right-hand side of the neutral system contain integrals of the unknown functions. The first difficulty is by using the fixed-point method to overcome this, while the second one is by choosing a suitable function space.

6.1. DERIVATION OF THE BOUNDARY CONDITIONS AND REDUCING THE MIXED PROBLEM TO AN INITIAL VALUE PROBLEM

We proceed from the lossy transmission line system of equations:

$$C\frac{\partial u(x,t)}{\partial t} + \frac{\partial i(x,t)}{\partial x} + Gu(x,t) = 0$$

$$L\frac{\partial i(x,t)}{\partial t} + \frac{\partial u(x,t)}{\partial x} + Ri(x,t) = 0 \tag{6.1.1}$$

$$(x,t) \in \Pi = \left\{ (x,t) \in R^2 : (x,t) \in [0,\Lambda] \times [0,\infty) \right\} \tag{6.1.2}$$

$$u(x,0) = u_0(x),\ i(x,0) = i_0(x),\ x \in [0,\Lambda]$$

where L, C, R and G are prescribed specific parameters of the line and $\Lambda > 0$ is its length. Here $u(x,t), i(x,t)$ are unknown functions – voltage and current respectively. The initial conditions for the foregoing system (6.1.1) are prescribed functions $u_0(x)$, $i_0(x)$.

The boundary conditions can be derived from the loads and sources at the ends of the line (cf. Fig.6.1):

$$u(0,t) = u_{AD}, \ E_0(t) = u_{BD}, \ i(0,t) = I_L + I_C + I_R, \ i_C = \frac{d\left[C_0\left(u_{BA}\right)u_{BA}\right]}{dt},$$

$$u_{BA} = \frac{d\left[L_0\left(i_L\right).i_L\right]}{dt}, \ i_R = G_0(u_{BA}).$$

Recall denotations

$$\widetilde{C}_p(u) = uC_p(u), \ \frac{d\widetilde{C}_p(u)}{du} = \frac{d\left(C_p(u)u\right)}{du} = u\frac{dC_p(u)}{du} + C_p(u), \ (p = 0,1),$$

$$\frac{d\widetilde{C}_p(u)}{dt} = \frac{d\left(C_p(u)u\right)}{du}\frac{du}{dt} = \left(u\frac{dC_p(u)}{du} + C_p(u)\right)\frac{du}{dt}, \ (p = 0,1),$$

$$\widetilde{L}_p(u) = iL_p(i), \ \frac{d\widetilde{L}_p(i)}{di} = \frac{d\left(iL_p(i)\right)}{di}, \ (p = 0,1).$$

Then

$$\widetilde{L}_0(i_L(t)) \equiv L_0\left(i_L(t)\right)i_L(t) = \int_T^t u_{BA}(\tau)d\tau.$$

Therefore $i_L(t) = \widetilde{L}_0^{-1}\left(\int_T^t u_{BA}(\tau)d\tau\right)$. In view of $u_{BA} = u(0,t) - E_0(t)$ for the left end of the line, one obtains:

$$-i(0,t) = -\frac{d\widetilde{C}_0(u(0,t) - E_0(t))}{du}\left(\frac{du(0,t)}{dt} - \frac{dE_0(t)}{dt}\right) - \widetilde{L}_0^{-1}\left(\int_T^t (u(0,\tau) - E_0(\tau))d\tau\right) - G_0(u(0,t) - E_0(t)) \qquad (6.1.3)$$

where $C_0(.)$, $L_0(.)$, $G_0(.)$ are prescribed nonlinear functions (cf. below).

Similarly we obtain for the right end the equality

$$-i(\Lambda,t) = -\frac{d\widetilde{C}_1(u(\Lambda,t) - E_1(t))}{du}\left(\frac{du(\Lambda,t)}{dt} - \frac{dE_1(t)}{dt}\right) - \widetilde{L}_1^{-1}\left(\int_T^t (u(\Lambda,\tau) - E_1(\tau))d\tau\right) - G_1(u(\Lambda,t) - E_1(t)) \qquad (6.1.4)$$

where $C_1(.), L_1(.), G_1(.)$ are also prescribed nonlinear functions.

Thus the system (6.1.1)-(6.1.4) forms a mixed problem for the lossy transmission line.

To reduce the mixed problem to an initial value one, we follow the derivation from the Chapter I.

Rewrite (6.1.1) in the form:

$$\frac{\partial u(x,t)}{\partial t} + \frac{1}{C}\frac{\partial i(x,t)}{\partial x} + \frac{G}{C}u(x,t) = 0,$$

$$\frac{\partial i(x,t)}{\partial t} + \frac{1}{L}\frac{\partial u(x,t)}{\partial x} + \frac{R}{L}i(x,t) = 0$$

and then in a matrix form

$$\frac{\partial U}{\partial t} + A\frac{\partial U}{\partial x} + BU = 0 \qquad\qquad (6.1.5)$$

where $U = \begin{bmatrix} u \\ i \end{bmatrix}$, $\quad \dfrac{\partial U}{\partial t} = \begin{bmatrix} \dfrac{\partial u}{\partial t} \\ \dfrac{\partial i}{\partial t} \end{bmatrix}$, $\quad \dfrac{\partial U}{\partial x} = \begin{bmatrix} \dfrac{\partial u}{\partial x} \\ \dfrac{\partial i}{\partial x} \end{bmatrix}$ $\quad A = \begin{bmatrix} 0 & 1/C \\ 1/L & 0 \end{bmatrix}$, $\quad B = \begin{bmatrix} G/C & 0 \\ 0 & R/L \end{bmatrix}$.

To transform the matrix $A = \begin{bmatrix} 0 & 1/C \\ 1/L & 0 \end{bmatrix}$ in a diagonal form we solve the characteristic

equation $\begin{vmatrix} -\lambda & 1/C \\ 1/L & -\lambda \end{vmatrix} = 0$. The roots are $\lambda_1 = \dfrac{1}{\sqrt{LC}}$ and $\lambda_2 = -\dfrac{1}{\sqrt{LC}}$. Denote by H the

matrix formed by eigenvectors $H = \begin{bmatrix} \sqrt{C} & \sqrt{L} \\ -\sqrt{C} & \sqrt{L} \end{bmatrix}$. Its inverse is

$$H^{-1} = \begin{bmatrix} \dfrac{1}{2\sqrt{C}} & -\dfrac{1}{2\sqrt{C}} \\ \dfrac{1}{2\sqrt{L}} & \dfrac{1}{2\sqrt{L}} \end{bmatrix}.$$

If we denote this by $A^{can} = \begin{bmatrix} \dfrac{1}{\sqrt{LC}} & 0 \\ 0 & -\dfrac{1}{\sqrt{LC}} \end{bmatrix}$, then $A^{can} = HAH^{-1}$.

Introduce new variables $Z = HU$, (or $U = H^{-1}Z$)

where

$$Z = \begin{bmatrix} V(x,t) \\ I(x,t) \end{bmatrix}, \ U = \begin{bmatrix} u(x,t) \\ i(x,t) \end{bmatrix}. \ \text{Then}$$

$$\begin{vmatrix} V(x,t) = \sqrt{C}\, u(x,t) + \sqrt{L}\, i(x,t) \\ I(x,t) = -\sqrt{C}\, u(x,t) + \sqrt{L}\, i(x,t) \end{vmatrix} \tag{6.1.6}$$

and

$$\begin{vmatrix} u(x,t) = \dfrac{1}{2\sqrt{C}} V(x,t) - \dfrac{1}{2\sqrt{C}} I(x,t) \\ i(x,t) = \dfrac{1}{2\sqrt{L}} V(x,t) + \dfrac{1}{2\sqrt{L}} I(x,t). \end{vmatrix} \tag{6.1.7}$$

Substitute $U = H^{-1}Z$ in (6.1.5) and we obtain

$$\frac{\partial\left(H^{-1}Z\right)}{\partial t} + A\frac{\partial\left(H^{-1}Z\right)}{\partial x} + B\left(H^{-1}Z\right) = 0 .$$

Since H^{1} is a constant matrix we have:

$$H^{-1}\frac{\partial Z}{\partial t} + \left(AH^{-1}\right)\frac{\partial Z}{\partial x} + \left(BH^{-1}\right)Z = 0 .$$

After multiplication from the left by H we obtain

$$\frac{\partial Z}{\partial t} + H\left(AH^{-1}\right)\frac{\partial Z}{\partial x} + H\left(BH^{-1}\right)Z = 0,$$

i.e.

$$\frac{\partial Z}{\partial t} + A^{\text{can}}\frac{\partial Z}{\partial x} + H\left(BH^{-1}\right)Z = 0 . \tag{6.1.8}$$

But $HBH^{-1} = \begin{bmatrix} \dfrac{1}{2}\left(\dfrac{G}{C}+\dfrac{R}{L}\right) & \dfrac{1}{2}\left(-\dfrac{G}{C}+\dfrac{R}{L}\right) \\ \dfrac{1}{2}\left(-\dfrac{G}{C}+\dfrac{R}{L}\right) & \dfrac{1}{2}\left(\dfrac{G}{C}+\dfrac{R}{L}\right) \end{bmatrix}$ and then

$$\begin{bmatrix} \dfrac{\partial V}{\partial t} \\ \dfrac{\partial I}{\partial t} \end{bmatrix} + \begin{bmatrix} \dfrac{1}{\sqrt{LC}} & 0 \\ 0 & -\dfrac{1}{\sqrt{LC}} \end{bmatrix} \begin{bmatrix} \dfrac{\partial V}{\partial x} \\ \dfrac{\partial I}{\partial x} \end{bmatrix} + \begin{bmatrix} \dfrac{1}{2}\left(\dfrac{R}{L}+\dfrac{G}{C}\right) & \dfrac{1}{2}\left(\dfrac{R}{L}-\dfrac{G}{C}\right) \\ \dfrac{1}{2}\left(\dfrac{R}{L}-\dfrac{G}{C}\right) & \dfrac{1}{2}\left(\dfrac{R}{L}+\dfrac{G}{C}\right) \end{bmatrix} \begin{bmatrix} V \\ I \end{bmatrix} = \begin{bmatrix} 0 \\ 0 \end{bmatrix}.$$

We consider distortionless lossy transmission lines, which means the following Heaviside condition is fulfilled:

$$\frac{R}{L} = \frac{G}{C}. \qquad\qquad\qquad\text{(H)}$$

Then HBH^{-1} might may be simplified, and (6.1.8) becomes:

$$\begin{bmatrix} \dfrac{\partial V}{\partial t} \\ \dfrac{\partial I}{\partial t} \end{bmatrix} + \begin{bmatrix} \dfrac{1}{\sqrt{LC}} & 0 \\ 0 & -\dfrac{1}{\sqrt{LC}} \end{bmatrix} \begin{bmatrix} \dfrac{\partial V}{\partial x} \\ \dfrac{\partial I}{\partial x} \end{bmatrix} + \begin{bmatrix} \dfrac{R}{L} & 0 \\ 0 & \dfrac{R}{L} \end{bmatrix} \begin{bmatrix} V \\ I \end{bmatrix} = \begin{bmatrix} 0 \\ 0 \end{bmatrix}. \qquad (6.1.9)$$

The new initial conditions obtained from (6.1.7) and (6.1.2) are:

$$V(x,0) = \sqrt{C}\, u(x,0) + \sqrt{L}\, i(x,0) = \sqrt{C}\, u_0(x) + \sqrt{L}\, i_0(x) \equiv V_0(x),\, x \in [0,\Lambda], \quad (6.1.10)$$

$$I(x,0) = -\sqrt{C}\, u(x,0) + \sqrt{L}\, i(x,0) = -\sqrt{C}\, u_0(x) + \sqrt{L}\, i_0(x) \equiv I_0(x),\, x \in [0,\Lambda]. \quad (6.1.11)$$

System (6.1.9) can be written in the form:

$$\frac{\partial V(x,t)}{\partial t} + \frac{1}{\sqrt{LC}}\frac{\partial V(x,t)}{\partial x} + \frac{R}{L}V(x,t) = 0,$$

$$\frac{\partial I(x,t)}{\partial t} - \frac{1}{\sqrt{LC}}\frac{\partial I(x,t)}{\partial x} + \frac{R}{L}I(x,t) = 0. \qquad (6.1.12)$$

The next substitution simplifies (6.1.12) one more time:

$$W(x,t) = e^{\frac{R}{L}t} V(x,t), \quad J(x,t) = e^{\frac{R}{L}t} I(x,t),$$

or

$$V(x,t) = e^{-\frac{R}{L}t} W(x,t), \quad I(x,t) = e^{-\frac{R}{L}t} J(x,t). \tag{6.1.13}$$

Then replacing (6.1.13) in (6.1.12) we obtain the following homogenous system

$$\begin{aligned}
&\frac{\partial W(x,t)}{\partial t} + \frac{1}{\sqrt{LC}}\frac{\partial W(x,t)}{\partial x} = 0, \\[2mm]
&\frac{\partial J(x,t)}{\partial t} - \frac{1}{\sqrt{LC}}\frac{\partial J(x,t)}{\partial x} = 0.
\end{aligned} \tag{6.1.14}$$

The transformation $\big(u(.),i(.)\big) \Rightarrow \big(W(.),J(.)\big)$ is given by (6.1.13) and (6.1.6) while the inverse one by the formulas

$$\left|\begin{aligned}
&u(x,t) = \frac{e^{-\frac{R}{L}t}}{2\sqrt{C}} W(x,t) - \frac{e^{-\frac{R}{L}t}}{2\sqrt{C}} J(x,t), \\[3mm]
&i(x,t) = \frac{e^{-\frac{R}{L}t}}{2\sqrt{L}} W(x,t) + \frac{e^{-\frac{R}{L}t}}{2\sqrt{L}} J(x,t).
\end{aligned}\right.$$

The initial conditions remain unaltered because

$$W(0,t) = e^{\frac{R}{L}0} V(0,t) = V_0(x), \quad J(0,t) = e^{\frac{R}{L}0} I(0,t) = I_0(x).$$

System (6.1.14) corresponds to a lossless transmission line, and then from there one we can reduce the mixed problem for (6.1.14) to an initial value problem for a system of functional differential equations of neutral type on the right boundary $x = \Lambda$. The system obtained is a nonlinear one in view of the nonlinear characteristics of the RLC-loads.

It is known that the solution of (6.1.14) is a pair of functions

$$W(x,t) = \Phi_W(x - vt) \text{ and } J(x,t) = \Phi_J(x + vt)$$

where $\Phi_W(.)$ and $\Phi_J(.)$ are arbitrary smooth functions while $v = 1/\sqrt{LC}$ is the speed of propagation of waves. In view of (6.1.7) and (6.1.3) we obtain

$$u(x,t) = \frac{e^{-\frac{R}{L}t}}{2\sqrt{C}}\left[\Phi_W(x-vt)-\Phi_J(x+vt)\right],$$

$$i(x,t) = \frac{e^{-\frac{R}{L}t}}{2\sqrt{L}}\left[\Phi_W(x-vt)+\Phi_J(x+vt)\right].$$

(6.1.15)

Hence

$$\Phi_W(x-vt) = e^{\frac{R}{L}t}\left(\sqrt{C}\,u(x,t)+\sqrt{L}\,i(x,t)\right),$$

$$\Phi_J(x+vt) = e^{\frac{R}{L}t}\left(\sqrt{L}\,i(x,t)-\sqrt{C}\,u(x,t)\right).$$

(6.1.16)

For $x = \Lambda$ we obtain

$$\Phi_W(\Lambda-vt) = e^{\frac{R}{L}t}\left[\sqrt{C}\,u(\Lambda,t)+\sqrt{L}\,i(\Lambda,t)\right],$$

$$\Phi_J(\Lambda+vt) = e^{\frac{R}{L}t}\left[\sqrt{L}\,i(\Lambda,t)-\sqrt{C}\,u(\Lambda,t)\right].$$

(6.1.17)

Denote by $T = \Lambda/v = \Lambda\sqrt{LC}$ and put $\Lambda - vt = -vt' \Rightarrow t = t'+\Lambda/v \equiv t'+T$. Substituting t in the first equation of (6.1.16) we get

$$\Phi_W(-vt') = e^{\frac{R}{L}(t'+T)}\left[\sqrt{C}\,u(\Lambda,t'+T)+\sqrt{L}\,i(\Lambda,t'+T)\right].$$

Now let us put $\Lambda + vt = vt'' \Rightarrow t = t''-T$. Then, substituting in the second equation of (6.1.17), we obtain

$$\Phi_J(vt'') = e^{\frac{R}{L}(t''-T)}\left[\sqrt{L}\,i(\Lambda,t''-T)-\sqrt{C}\,u(\Lambda,t''-T)\right].$$

Therefore

$$\Phi_W(-vt) = e^{\frac{R}{L}(t+T)}\left[\sqrt{C}\,u(\Lambda,t+T)+\sqrt{L}\,i(\Lambda,t+T)\right],$$

(6.1.18)

$$\Phi_J(vt) = e^{\frac{R}{L}(t-T)}\left[\sqrt{L}\,i(\Lambda,t-T)-\sqrt{C}\,u(\Lambda,t-T)\right].$$

(6.1.19)

From (6.1.15), by from $x = 0$ we have

$$u(0,t) = \frac{e^{-\frac{R}{L}t}}{2\sqrt{C}}\left[\Phi_W(-vt) - \Phi_J(vt)\right],$$

$$i(0,t) = \frac{e^{-\frac{R}{L}t}}{2\sqrt{L}}\left[\Phi_W(-vt) + \Phi_J(vt)\right]. \tag{6.1.20}$$

Substituting $\Phi_W(-vt)$ and $\Phi_J(vt)$ from (6.1.18) and (6.1.19) into (6.1.20) we obtain (recall that $Z_0 = \sqrt{L/C}$):

$$u(0,t) = \frac{e^{-\frac{R}{L}t}}{2\sqrt{C}}\left[e^{\frac{R}{L}(t+T)}\left(\sqrt{C}\,u(\Lambda,t+T) + \sqrt{L}\,i(\Lambda,t+T)\right) - e^{\frac{R}{L}(t-T)}\left(\sqrt{L}\,i(\Lambda,t-T) - \sqrt{C}\,u(\Lambda,t-T)\right)\right] \tag{6.1.21}$$

$$i(0,t) = \frac{e^{-\frac{R}{L}t}}{2\sqrt{L}}\left[e^{\frac{R}{L}(t+T)}\left(\sqrt{C}\,u(\Lambda,t+T) + \sqrt{L}\,i(\Lambda,t+T)\right) + e^{\frac{R}{L}(t-T)}\left(\sqrt{L}\,i(\Lambda,t-T) - \sqrt{C}\,u(\Lambda,t-T)\right)\right] \tag{6.1.22}$$

We put $t + T \equiv t$. Then t should be replaced by $t - T$ and $t - T$ by $t - 2T$ in (6.1.21) and (6.1.22). Therefore we obtain

$$u(0,t-T) = e^{-\frac{R}{L}(t-T)}\left(e^{\frac{R}{L}t}\,\frac{u(\Lambda,t) + Z_0\,i(\Lambda,t)}{2} - e^{\frac{R}{L}(t-2T)}\,\frac{Z_0\,i(\Lambda,t-2T) - u(\Lambda,t-2T)}{2}\right), \tag{6.1.23}$$

$$i(0,t-T) = e^{-\frac{R}{L}(t-T)}\left(e^{\frac{R}{L}t}\,\frac{u(\Lambda,t) + Z_0\,i(\Lambda,t)}{2Z_0} + e^{\frac{R}{L}(t-2T)}\,\frac{Z_0 i(\Lambda,t-2T) - u(\Lambda,t-2T)}{2Z_0}\right). \tag{6.1.24}$$

Then replacing $t + T \equiv t$ into the first boundary condition (6.1.3) we obtain

$$i(0,t-T) = \frac{d\widetilde{C}_0\big(u(0,t-T) - E_0(t-T)\big)}{du}\left(\frac{du(0,t-T)}{dt} - \frac{dE_0(t-T)}{dt}\right) +$$

$$+ \widetilde{L}_0^{-1}\left(\int_0^{t-T}\big(u(0,\tau) - E_0(\tau)\big)d\tau\right) + G_0\big(u(0,t-T) - E_0(t-T)\big). \tag{6.1.25}$$

Introducing denotation

$$\widetilde{u}(t) \equiv u(0,t) - E_0(t), \quad \widetilde{u}(t-T) = u(0,t-T) - E_0(t-T)$$

we simplify the above equation

$$i(0, t-T) = \frac{d\widetilde{C}_0(\widetilde{u}(t-T))}{du}\frac{d\widetilde{u}(t-T)}{dt} + \widetilde{L}_0^{-1}\left(\int_0^{t-T}\widetilde{u}(\tau)d\tau\right) + G_0(\widetilde{u}(t-T)). \quad (6.1.26)$$

Take $i(0, t-T)$ from (6.1.24) and replace in (6.1.26):

$$\frac{e^{\frac{RT}{L}}}{2Z_0}u(\Lambda,t) + \frac{e^{\frac{RT}{L}}}{2}i(\Lambda,t) + \frac{e^{-\frac{RT}{L}}}{2}i(\Lambda,t-2T) - \frac{e^{-\frac{RT}{L}}}{2Z_0}u(\Lambda,t-2T) =$$

$$\quad (6.1.27)$$

$$= \frac{d\widetilde{C}_0(\widetilde{u}(t-T))}{du}\frac{d\widetilde{u}(t-T)}{dt} + \widetilde{L}_0^{-1}\left(\int_0^{t-T}\widetilde{u}(\tau)d\tau\right) + G_0(\widetilde{u}(t-T)).$$

From (6.1.27) we obtain

$$\frac{d\widetilde{u}(t-T)}{dt} = -\frac{1}{dC_0(\widetilde{u}(t-T))/du}L_0^{-1}\left(\int_0^{t-T}\widetilde{u}(\tau)d\tau\right) - \frac{1}{d\widetilde{C}_0(\widetilde{u}(t-T))/du}G_0(\widetilde{u}(t-T)) +$$

$$+ \frac{1}{d\widetilde{C}_0(\widetilde{u}(t-T))/du}\left(\frac{e^{\frac{RT}{L}}}{2Z_0}u(\Lambda,t) + \frac{e^{\frac{RT}{L}}}{2}i(\Lambda,t) + \frac{e^{-\frac{RT}{L}}}{2}i(\Lambda,t-2T) - \frac{e^{-\frac{RT}{L}}}{2Z_0}u(\Lambda,t-2T)\right).$$

Since we have to solve the last equation with respect to $\dfrac{di(\Lambda,t)}{dt}$ we calculate

$$\frac{d\widetilde{u}(t-T)}{dt} = \frac{e^{\frac{RT}{L}}}{2}\frac{du(\Lambda,t)}{dt} + \frac{e^{\frac{RT}{L}}}{2}Z_0\frac{di(\Lambda,t)}{dt} - \frac{e^{-\frac{RT}{L}}}{2}Z_0\frac{di(\Lambda,t-2T)}{dt} + \frac{e^{\frac{RT}{L}}}{2}\frac{du(\Lambda,t-2T)}{dt} - \frac{dE_0(t-T)}{dt}.$$

Replace the last expression in (6.1.27) and solve with respect to $\dfrac{di(\Lambda,t)}{dt}$:

$$\frac{di(\Lambda,t)}{dt} = -\frac{1}{Z_0}\frac{du(\Lambda,t)}{dt} + e^{-2\frac{RT}{L}}\frac{di(\Lambda,t-2T)}{dt} - e^{-2\frac{RT}{L}}\frac{1}{Z_0}\frac{du(\Lambda,t-2T)}{dt} + \frac{2}{Z_0}e^{-\frac{RT}{L}}\dot{E}_0(t-T) -$$

$$- \frac{2}{Z_0}e^{-\frac{RT}{L}}\frac{1}{d\widetilde{C}_0(\widetilde{u}(t-T))/du}L_0^{-1}\left(\int_0^{t-T}\widetilde{u}(\tau)d\tau\right) - \frac{2}{Z_0}e^{-\frac{RT}{L}}\frac{G_0(\widetilde{u}(t-T))}{d\widetilde{C}_0(\widetilde{u}(t-T))/du} \quad (6.1.28)$$

We notice that $\displaystyle\int_0^{t-T}\widetilde{u}(\tau)d\tau = \int_T^{t}\widetilde{u}(\theta-T)d\theta \Rightarrow L_0^{-1}\left(\int_0^{t-T}\widetilde{u}(\tau)d\tau\right) = L_0^{-1}\left(\int_T^{t}\widetilde{u}(\theta-T)d\theta\right).$

We consider the system (6.1.28), (6.1.4). In order to solve this system with respect to $\dfrac{du(\Lambda,t)}{dt}$ we assume that $\dfrac{d\widetilde{C}_1(u(\Lambda,t)-E_1(t))}{du} \neq 0$.

Then finally the neutral system becomes:

$$\frac{di(\Lambda,t)}{dt} = -\frac{1}{Z_0}\frac{du(\Lambda,t)}{dt} + e^{-2\frac{RT}{L}}\frac{di(\Lambda,t-2T)}{dt} - e^{-2\frac{RT}{L}}\frac{1}{Z_0}\frac{du(\Lambda,t-2T)}{dt} + \frac{2}{Z_0}e^{-\frac{RT}{L}}\dot{E}_0(t-T) -$$

$$-\frac{2}{Z_0}e^{-\frac{RT}{L}}\frac{1}{d\widetilde{C}_0(\widetilde{u}(t-T))/du}L_0^{-1}\left(\int_T^t \widetilde{u}(\theta-T)d\theta\right) - \frac{2}{Z_0}e^{-\frac{RT}{L}}\frac{G_0(\widetilde{u}(t-T))}{d\widetilde{C}_0(\widetilde{u}(t-T))/du} \qquad (6.1.29\ )$$

$$\frac{du(\Lambda,t)}{dt} = \frac{i(\Lambda,t)}{d\overline{C}_1(u(\Lambda,t)-E_1(t))/du} + \frac{dE_1(t)}{dt} - \frac{\overline{L}_1^{-1}\left(\int_T^t (u(\Lambda,\tau)-E_1(\tau))d\tau\right)}{d\overline{C}_1(u(\Lambda,t)-E_1(t))/du} - \frac{G_1(u(\Lambda,t)-E_1(t))}{d\overline{C}_1(u(\Lambda,t)-E_1(t))/du}.$$

6.2. ANALYSIS OF THE ARISING NONLINEARITIES

To obtain (6.1.29) we have to formulate conditions implying

$$\frac{d\widetilde{C}_0(\widetilde{u}(t-T))}{du} = \frac{\widetilde{C}_0(u(0,t-T)-E_0(t-T))}{du} \neq 0 \ \text{ and } \ \frac{\widetilde{C}_1(u(\Lambda,t-T)-E_1(t-T))}{du} \neq 0 .$$

The capacity functions are

$$C_p(u) = \frac{c_p}{\sqrt[h]{1-\dfrac{u}{\Phi_p}}} = \frac{c_p\sqrt[h]{\Phi_p}}{\sqrt[h]{\Phi_p-u}}, (p=0,1),$$

where $c_p, \Phi_p, h \in [2,3]$ are positive constants.

We make the following

Assumption (U_0) : $|u| \leq \phi_0 < \Phi = \min\{\Phi_0, \Phi_1\}$.

Since $\dfrac{dC_p(u)}{du} = -\dfrac{1}{h}c_p\sqrt[h]{\Phi_p}(\Phi_p-u)^{-\frac{1}{h}-1} = \dfrac{c_p\sqrt[h]{\Phi_p}}{h\sqrt[h]{(\Phi_p-u)^{1+h}}} > 0, (p=0,1)$ therefore the

minimal value of $C_p(u)$ is $C_p(-\phi_0) = \dfrac{c_p\sqrt[h]{\Phi_p}}{\sqrt[h]{\Phi_p+\phi_0}} = \breve{C}_p > 0$ and

$$\left|\frac{dC_p(u)}{du}\right| \le \frac{c_p \sqrt[h]{\Phi_0}}{h(\Phi_p - \phi_0)^{\frac{1+h}{h}}}.$$

For $\dfrac{d\widetilde{C}_p(u)}{du} = \dfrac{d(uC_p(u))}{du}$ we have

$$\frac{d\widetilde{C}_p(u)}{du} = u\frac{dC_p(u)}{du} + C_p(u) = \frac{c_p}{h}\frac{\sqrt[h]{\Phi_p}\,u}{\left(\sqrt[h]{\Phi_p}-u\right)^{h+1}} + \frac{c_p\sqrt[h]{\Phi_p}}{\sqrt[h]{\Phi_p}-u} = c_p\sqrt[h]{\Phi_p}\,\frac{\Phi_p - \dfrac{h-1}{h}u}{(\Phi_p - u)^{\frac{h+1}{h}}};$$

$$\frac{d^2\widetilde{C}_p(u)}{du^2} = c_p\sqrt[h]{\Phi_p}\,\frac{d}{du}\left(\frac{\Phi_p - \dfrac{h-1}{h}u}{(\Phi_p - u)^{\frac{h+1}{h}}}\right) =$$

$$= c_p\sqrt[h]{\Phi_p}\,\frac{-\dfrac{h-1}{h}(\Phi_p - u)^{\frac{h+1}{h}} + \left(\Phi_p - \dfrac{h-1}{h}u\right)\dfrac{h+1}{h}(\Phi_p - u)^{\frac{h+1}{h}-1}}{(\Phi_p - u)^{\frac{2(h+1)}{h}}} = \frac{2c_p\sqrt[h]{\Phi_p}}{h}\,\frac{\Phi_p - \dfrac{h-1}{2h}u}{(\Phi_p - u)^{\frac{2h+1}{h}}}.$$

Since $\dfrac{h-1}{2h}|u| \le |u| \le \phi_0 \Rightarrow \dfrac{d^2\widetilde{C}_p(u)}{du^2} > 0$. Therefore $\dfrac{d\widetilde{C}_p(u)}{du}$ is an increasing

function and the lower bound attains to $u = -\phi_0$, that is,

$$\frac{d\widetilde{C}_p(-\phi_0)}{du} = c_p\sqrt[h]{\Phi_p}\,\frac{\Phi_p + \dfrac{h-1}{h}\phi_0}{(\Phi_p + \phi_0)^{\frac{h+1}{h}}} \equiv \dot{\widetilde{C}}_p > 0.$$

We also need the estimate

$$\left|\frac{d^2\widetilde{C}_p(u)}{du^2}\right| \le \frac{2c_p\sqrt[h]{\Phi_p}}{h}\,\frac{\Phi_p + \dfrac{h-1}{2h}\phi_0}{(\Phi_p - \phi_0)^{\frac{2h+1}{h}}} = H_p.$$

The minimal value of $\widetilde{C}_p(u)$ is:

$$\min\left\{\widetilde{C}_p(u) : u \in [-\phi_0, \phi_0]\right\} = \widetilde{C}_p(-\phi_0) = \frac{c_p\sqrt[h]{\Phi_p}}{h^2}\,\frac{2h + (h^2-1)\phi_0}{\sqrt[h]{(\Phi_p + \phi_0)^{h+1}}} \equiv \hat{C}_p > 0.$$

We need to find also the upper bound for $\dfrac{d\widetilde{C}_p(u)}{du}$, that is,

$$\left|\frac{d\widetilde{C}_p(u)}{du}\right| = \frac{c_p \sqrt[h]{\Phi_p}}{h^2}\frac{2h-(h^2-1)u}{\sqrt[h]{(\Phi_p-u)^{h+1}}} \leq \frac{c_p \sqrt[h]{\Phi_p}}{h^2}\frac{2h-(h^2-1)\phi_0}{\sqrt[h]{(\Phi_p-\phi_0)^{h+1}}} \tag{6.2.1}$$

because $\dfrac{d\widetilde{C}_p(u)}{du}$ is increasing for $-\phi_0 \leq u \leq \phi_0 \Rightarrow u < \Phi_1$.

For the *I-L* characteristics we obtain $\widetilde{L}_p \equiv i.L_p(i)\,(p=0,1)$, where $L_p(i) = \sum\limits_{n=1}^{m-1} l_n^{(p)} i^n$,

that is, $\widetilde{L}_p(i) = i\,.L_p(i) = \sum\limits_{n=1}^{m-1} l_n^{(p)} i^{n+1}\quad i \in [-i_0, i_0]$.

Since $u = \dfrac{d\widetilde{L}_p(i)}{dt} = \dfrac{d\big(L_p(i).i\big)}{dt} = i\dfrac{dL_p(i)}{di} + L_p(i)$ it follows

$$\widetilde{L}_p(i) = \int\limits_{T}^{t} u(s)\,ds \Rightarrow i(t) = \widetilde{L}_p^{-1}\left(\int\limits_{T}^{t} u(s)\,ds\right).$$

We choose i_0 such that $\dfrac{d\widetilde{L}_p(i)}{di} > 0$ for $|i| \leq i_0$. Therefore the inverse function $\widetilde{L}_p^{-1}(\lambda)$

exists and $\widetilde{L}_p^{-1}(\lambda):\left[-\sum\limits_{n=1}^{m-1} l_n^{(p)} i_0^{n+1}, \sum\limits_{n=1}^{m-1} l_n^{(p)} i_0^{n+1}\right] \to [-i_0, i_0]$.

The explicit form of the inverse function is found in the applications.
Further on we use the following estimates:

$$\left|\widetilde{L}_p^{-1}(.)\right| \leq i_0 \leq I_0 \ (p=0,1) \tag{6.2.2}$$

and

$$\left|\frac{\widetilde{L}_p^{-1}(\lambda)}{d\lambda}\right| = \frac{1}{\left|\dfrac{d\widetilde{L}_p(i)}{di}\right|} \leq \frac{1}{\min\left\{\sum\limits_{n=1}^{m-1} l_n^{(p)} i^{n+1} \,:\, i \in [-i_0, i_0]\right\}} = \frac{1}{\hat{L}_p} \leq \frac{1}{\hat{L}} = \max\left\{\frac{1}{\hat{L}_p} : p=0,1\right\} \tag{6.2.3}$$

For the conductivity functions, we shall assume that they are of a polynomial type

$$G_p(u) = \sum_{n=1}^{m} g_n^{(p)} u^n \quad (p = 0,1).$$

Let change the variable in the integral $\widetilde{L}_0^{-1}\left(\int\limits_0^{t-T} \widetilde{u}(\tau)d\tau \right)$ putting $\theta = \tau + T$. Then

$$\widetilde{L}_0^{-1}\left(\int\limits_0^{t-T} \widetilde{u}(\tau)d\tau \right) = \widetilde{L}_0^{-1}\left(\int\limits_T^{t} \widetilde{u}(\theta - T)d\theta \right).$$

Remark 6.2.1. We use $u(t) = u(\Lambda,t)$, $i(t) = i(\Lambda,t)$ and assume that they are unknown functions. Then $\widetilde{u}(t-T)$ becomes

$$\widetilde{u}(t-T) = e^{\frac{RT}{L}}\frac{u(t) + Z_0\, i(t)}{2} - e^{-\frac{RT}{L}}\frac{Z_0\, i(t-2T) - u(t-2T)}{2} - E_0(t-T).$$

Remark 6.2.2. We need below the following estimate $\left| \widetilde{L}_p^{-1}(\lambda) \right| \le I_0|\lambda|$ $(p = 0,1)$. The previous one is satisfied for $|\lambda| \le 1$. Since in our case $\lambda = \int\limits_T^{t} \widetilde{u}(\theta - T)d\theta$ we have to assure the inequality $\left| \int\limits_T^{t} \widetilde{u}(\theta - T)d\theta \right| \le 1$ for sufficiently large μ.

6.3. AN OPERATOR PRESENTATION OF THE PERIODIC PROBLEM

In § 6.2 we have derived a neutral system of functional differential equations.

We suppose that $(u(\Lambda,t), i(\Lambda,t))$ are the unknown functions and introduce denotation $(u(t), i(t)) \equiv (u(\Lambda,t), i(\Lambda,t))$. Then (6.1.29) can be rewritten in the form:

$$\frac{du(t)}{dt} = \frac{dE_1(t)}{dt} + \frac{i(t)}{\dfrac{d\widetilde{C}_1(u(t) - E_1(t))}{du}} - \frac{\widetilde{L}_1^{-1}\left(\int\limits_T^{t} (u(\tau) - E_1(\tau))d\tau \right)}{\dfrac{d\widetilde{C}_1(u(t) - E_1(t))}{du}} - \frac{G_1(u(t) - E_1(t))}{\dfrac{d\widetilde{C}_1(u(t) - E_1(t))}{du}},$$

$$\frac{di(t)}{dt} = \frac{2}{Z_0} e^{-\frac{RT}{L}} \dot{E}_0(t-T) - \frac{1}{Z_0}\frac{du(t)}{dt} + e^{-2\frac{RT}{L}}\frac{di(t-2T)}{dt} - e^{-2\frac{RT}{L}}\frac{1}{Z_0}\frac{du(t-2T)}{dt} -$$

$$-\frac{2}{Z_0}e^{-\frac{RT}{L}}\frac{1}{\dfrac{d\widetilde{C}_0(\widetilde{u}(t-T))}{du}}\widetilde{L}_0^{-1}\left(\int_0^{t-T}\widetilde{u}(\tau)d\tau\right)-\frac{2}{Z_0}e^{-\frac{RT}{L}}\frac{1}{\dfrac{d\widetilde{C}_0(\widetilde{u}(t-T))}{du}}G_0(\widetilde{u}(t-T)),$$

$$\text{(6.3.1)}$$

$$t\in[T,3T]$$

$$u(t)=\upsilon_0(t),\quad \frac{du(t)}{dt}=\frac{d\upsilon_0(t)}{dt},\quad i(t)=\iota_0(t),\quad \frac{di(t)}{dt}=\frac{d\iota_0(t)}{dt},\ t\in\left[-T,T\right].$$

(recalling that

$$\widetilde{u}(t-T)=e^{\frac{RT}{L}}\frac{u(t)+Z_0\,i(t)}{2}-e^{-\frac{RT}{L}}\frac{Z_0\,i(t-2T)-u(t-2T)}{2}-E_0(t-T)).$$

We are able to look for a periodic solution of (6.1.29) on the interval $[T,3T]$.

The initial functions $\upsilon_0(t)$, $\iota_0(t)$ are obtained from the initial conditions (6.1.2) as done in Chapter II.

Now we formulate the main problem: to find a T_0-periodic solution of (6.3.1) on the interval $[T,3T]$ coinciding with prescribed initial periodic functions on $[-T,T]$.

By $C_{T_0}^1[T,3T]$ we mean the space of all continuous T_0-periodic functions with continuous derivatives. First we introduce the sets (recall that $T=mT_0$)

$$M_U=\left\{u(.)\in C_{T_0}^1[T,3T]:\ \int_{T+kT_0}^{T+(k+1)T_0}u(t)dt=0\ (k=0,1,2,3,...,2m-1)\right\},$$

$$M_I=\left\{i(.)\in C_{T_0}^1[T,3T]):\ \int_{T+kT_0}^{T+(k+1)T_0}i(t)dt=0\ (k=0,1,2,3,...,2m-1)\right\},$$

$$M_U^*=\left\{u(.)\in M_U:|u(t)|\le U_0e^{\mu(t-T-kT_0)}(k=0,1,2,3,...,2m-1)\right\},$$

$$M_I^*=\left\{i(.)\in M_I:|i(t)|\le I_0e^{\mu(t-T-kT_0)}(k=0,1,2,3,...,2m-1)\right\}.$$

The set $M_U^*\times M_I^*$ turns out into a complete metric space (cf. Chapter I, § 1.2) with respect to the metric:

$$\hat{\rho}_\mu((u,\dot{u},i,\dot{i}),(\bar{u},\bar{i},\dot{\bar{u}},\dot{\bar{i}}))=\max\left\{\hat{\rho}(u,\bar{u}),\rho^{(k)}(\dot{u},\dot{\bar{u}}),\hat{\rho}(i,\bar{i}),\rho^{(k)}(\dot{i},\dot{\bar{i}}):k=0,1,2,...,2m-1\right\},$$

where

$$\rho^{(k)}(u,\overline{u}) = \max\left\{|u(t) - \overline{u}(t)| : t \in [T + kT_0, T + (k+1)T_0]\right\},$$

$$\rho_\mu^{(k)}(u,\overline{u}) = \max\left\{e^{-\mu(t-T-kT_0)}|u(t) - \overline{u}(t)| : t \in [T + kT_0, T + (k+1)T_0]\right\},$$

$$\hat{\rho}(u,\overline{u}) = \max\left\{|u(t) - \overline{u}(t)| : t \in [T, 3T]\right\},$$

$$\rho_\mu^{(k)}(\dot{u},\dot{\overline{u}}) = \max\left\{e^{-\mu(t-T-kT_0)}|\dot{u}(t) - \dot{\overline{u}}(t)| : t \in [T + kT_0, T + (k+1)T_0]\right\},$$

$$\rho^{(k)}(i,\overline{i}) = \max\left\{|i(t) - \overline{i}(t)| : t \in [T + kT_0, T + (k+1)T_0]\right\},$$

$$\rho_\mu^{(k)}(i,\overline{i}) = \max\left\{e^{-\mu(t-T-kT_0)}|i(t) - \overline{i}(t)| : t \in [T + kT_0, T + (k+1)T_0]\right\},$$

$$\hat{\rho}(i,\overline{i}) = \max\left\{|i(t) - \overline{i}(t)| : t \in [T, 3T]\right\},$$

$$\rho_\mu^{(k)}(\dot{i},\dot{\overline{i}}) = \max\left\{e^{-\mu(t-T-kT_0)}|\dot{i}(t) - \dot{\overline{i}}(t)| : t \in [T + kT_0, T + (k+1)T_0]\right\}.$$

From now on we suppose that the following assumptions are fulfilled:

Assumptions **(IN)**: $\upsilon_0(.), \iota_0(.) \in C_{T_0}^1[-T, T]$, and

$$\int_{-T+nT_0}^{-T+(n+1)T_0} \upsilon_0(t)dt = 0, \quad \int_{-T+nT_0}^{-T+(n+1)T_0} \iota_0(t)dt = 0; \ (n = 0,1,...,2m-1); \upsilon_0(-T) = 0, \ \iota_0(-T) = 0 \ ;$$

$$|\upsilon_0(t)| \le U_0 e^{-\beta} e^{\mu(t+T-nT_0)}, \ |\iota_0(t)| \le I_0 e^{-\beta} e^{\mu(t+T-nT_0)}, t \in [-T + nT_0, \ -T + (n+1)T_0].$$

Assumptions **(E)**: $E_p(.) \in C_{T_0}^1[T, \infty), (p = 0,1); \ \int_{T+kT_0}^{T+(k+1)T_0} E_p(t)dt = 0 \ \ (k = 0,1,2,...),$

$$E_p(0) = 0; \ |E_p(t)| \le U_{E_p} e^{\mu(t-T-kT_0)} \ (k = 0,1,2,3,...,2m-1); U_{E_p} < U_0.$$

Remark 6.3.1. It follows

$$|u(t - 2T)| = |\overline{\upsilon}_0(t)| \le U_0 e^{-\beta} e^{\mu(t-T-kT_0)}, |i(t - 2T)| = |\overline{\iota}_0(t)| \le I_0 e^{-\beta} e^{\mu(t-T-kT_0)} \ (k = 0,1,2,3,...,2m-1)$$

where

$$\overline{\upsilon}_0(t) = \upsilon_0(t - 2T), \ \overline{\iota}_0(t) = \iota_0(t - 2T), \ t \in [T, 3T].$$

The arguments of the capacity functions $C_0(u), \widetilde{C}_0(u)$ and $C_1(u), \widetilde{C}_1(u)$ are respectively $\widetilde{u}(t) = u(t) - E_1(t)$ and $\widetilde{u}(t - T)$ so we have to assume that the following inequalities should be satisfied:

$$\left| u(t) - E_p(t) \right| \le \left(U_0 + U_{E_p} \right) e^{\mu(t-T-kT_0)} \le e^{\mu T_0} \left(U_0 + U_{E_p} \right) \le \phi_0 \ (p = 0,1) \qquad (6.3.2)$$

and

$$\left| \tilde{u}(t-T) \right| = \left| e^{\frac{RT}{L}} \frac{u(t) + Z_0\, i(t)}{2} - e^{-\frac{RT}{L}} \frac{Z_0\, i(t-2T) - u(t-2T)}{2} - E_0(t) \right| \le$$

$$\le \frac{\left(e^{\frac{RT}{L}} + e^{-\frac{RT}{L}} \right) e^{-\beta}\left(U_0 + Z_0\, I_0 \right) + 2U_{E_0}}{2} e^{\mu(t-T-kT_0)} = \left(\cosh\left(\frac{RT}{L}\right) e^{-\beta}\left(U_0 + Z_0\, I_0 \right) + U_{E_0} \right) e^{\mu(t-T-kT_0)} \le$$

$$\le U_0 e^{\mu T_0} \le \phi_0.$$

Besides, in view of Remark 6.2.2 we have

$$\int_{T+kT_0}^{t} \left| \tilde{u}(s-T) \right| ds \le \left(\cosh\left(\frac{RT}{L}\right) e^{-\beta}\left(U_0 + Z_0\, I_0 \right) + U_{E_0} \right) \int_{T+kT_0}^{t} e^{\mu(s-T-kT_0)} ds \le$$

$$\le \frac{e^{\mu(t-T-kT_0)} - 1}{\mu} \left(\cosh\left(\frac{RT}{L}\right) e^{-\beta}\left(U_0 + Z_0\, I_0 \right) + U_{E_0} \right) \le \frac{e^{\mu 0}}{\mu} \left(\cosh\left(\frac{RT}{L}\right) e^{-\beta}\left(U_0 + Z_0\, I_0 \right) + U_{E_0} \right) \le \frac{\phi_0}{\mu}.$$

Then $\displaystyle \int_{T+kT_0}^{t} \left| \tilde{u}(s-T) \right| ds \le 1$ for sufficiently large μ.

Assumption (V_1): $U_0 + U_{E_1} \le \phi_0$;

Assumption (V_0): $\cosh\left(\dfrac{RT}{L}\right) e^{-\beta}\left(U_0 + Z_0\, I_0 \right) + U_{E_0} \le U_0 e^{\mu 0} \le \phi_0$.

Introduce the operator B as a pair of functions

$$B(u,i)(t) = \left(B_u(u,i)(t),\ B_i(u,i)(t) \right),\ t \in [T, 3T]$$

defined on every interval $[T + kT_0, T + (k+1)T_0]$ for every $k = 0, 1, 2,..., 2m\text{-}1$ by the expressions:

$$B_u^{(k)}(u,i)(t) := \int_{T+kT_0}^{t} U(u,i)(s)ds - \left(\frac{t-T-kT_0}{T_0} - \frac{1}{2} \right) \int_{T+kT_0}^{T+(k+1)T_0} U(u,i)(s)ds - \frac{1}{T_0} \int_{T+kT_0}^{T+(k+1)T_0} \int_{T+kT_0}^{t} U(u,i)(s)dsdt$$

$$B_i^{(k)}(U,I)(t) := \int_{T+kT_0}^{t} I(u,i)(s)ds - \left(\frac{t-T-kT_0}{T_0} - \frac{1}{2} \right) \int_{T+kT_0}^{T+(k+1)T_0} I(u,i)(s)ds - \frac{1}{T_0} \int_{T+kT_0}^{T+(k+1)T_0} \int_{T+kT_0}^{t} I(u,i)(s)dsdt$$

Recalling $u(T) = 0$, $i(T) = 0 \Rightarrow u(T + kT_0) = u(T) = 0$, $i(T + kT_0) = i(T) = 0$

where

$$U(u,i)(t) \equiv \frac{dE_1(t)}{dt} + \frac{i(t)}{\dfrac{d\widetilde{C}_1(u(t) - E_1(t))}{du}} - \frac{\widetilde{L}_1^{-1}\left(\displaystyle\int_{T+kT_0}^{t}(u(\tau) - E_1(\tau))d\tau\right)}{\dfrac{d\widetilde{C}_1(u(t) - E_1(t))}{du}} - \frac{G_1(u(t) - E_1(t))}{\dfrac{d\widetilde{C}_1(u(t) - E_1(t))}{du}}$$

and

$$I(u,i)(t) \equiv \frac{2}{Z_0} e^{-\frac{RT}{L}} \dot{E}_0(t) - \frac{1}{Z_0}\frac{du(t)}{dt} + e^{-2\frac{RT}{L}}\frac{d\bar{i}_0(t)}{dt} - e^{-2\frac{RT}{L}}\frac{1}{Z_0}\frac{d\bar{v}_0(t)}{dt} +$$

$$-\frac{2}{Z_0} e^{-\frac{RT}{L}} \frac{1}{\dfrac{d\widetilde{C}_0(\widetilde{u}(t-T))}{du}} \widetilde{L}_0^{-1}\left(\displaystyle\int_{T+kT_0}^{t}\widetilde{u}(\tau - T)d\tau\right) - \frac{2}{Z_0} e^{-\frac{RT}{L}} \frac{1}{\dfrac{d\widetilde{C}_0(\widetilde{u}(t-T))}{du}} G_0(\widetilde{u}(t-T))$$

and

$$\widetilde{u}(t-T) = e^{\frac{RT}{L}}\frac{u(t) + Z_0 i(t)}{2} - e^{-\frac{RT}{L}}\frac{Z_0 \bar{i}_0(t) - \bar{v}_0(t)}{2} - E_0(t).$$

Lemma 6.3.1. If $f(.) \in M_U$ or M_I then $F(t) = \displaystyle\int_T^t f(\tau)d\tau$ is T_0-periodic function.

The proof is the same as that of Lemma 4.2.1.

Lemma 6.3.2. If the assumptions **(IN)** and **(E)** are satisfied and $(u,i) \in M_U^* \times M_I^*$ then $U(u,i)(t)$, $I(u,i)(t)$ are T_0 periodic functions.

The proof is the same as that of Lemma 4.2.2.

Remark 6.3.2. It is easy to see that by changing the integration order, one obtains

$$\int_{T+kT_0}^{T+(k+1)T_0}\int_{T+kT_0}^{t} U(u,i)(s)\,ds\,dt = (T + (k+1)T_0)\int_{T+kT_0}^{T+(k+1)T_0} U(u,i)(s)\,ds - \int_{T+kT_0}^{T+(k+1)T_0} s\,U(u,i)(s)\,ds$$

and

$$\int_{T+kT_0}^{T+(k+1)T_0}\int_{T+kT_0}^{t} I(u,i)(s)\,ds\,dt = (T + (k+1)T_0)\int_{T+kT_0}^{T+(k+1)T_0} I(u,i)(s)\,ds - \int_{T+kT_0}^{T+(k+1)T_0} s\,I(u,i)(s)\,ds .$$

Lemma 6.3.3. For every $\big(u(.),i(.)\big) \in M_U^* \times M_I^*$ it follows

$$\int_{T+kT_0}^{T+(k+1)T_0}\int_{T+kT_0}^{s} U(u,i)(\theta)\,d\theta\,ds = \int_{T+(k+1)T_0}^{T+(k+2)T_0}\int_{T+(k+1)T_0}^{s} U(u,i)(\theta)\,d\theta\,ds \quad (k = 0,1,2,...)$$

$$\int_{T+kT_0}^{T+(k+1)T_0}\int_{T+kT_0}^{s}I(u,i)(\theta)d\theta ds = \int_{T+(k+1)T_0}^{T+(k+2)T_0}\int_{T+(k+1)T_0}^{s}I(u,i)(\theta)d\theta ds \ \ (k=0,1,2,...).$$

The proof is the same as the one of lemma 3.4.3.

Lemma 6.3.4. The initial value problem (6.2.4) has a solution $(u,i)\in M_U^*\times M_I^*$ iff the operator B has a fixed point $(u,i)\in M_U^*\times M_I^*$, that is, $u=B_u(u,i),\ i=B_i(u,i)$.

Proof: Let $(u(.),i(.))\in M_U^*\times M_I^*$ be a solution of the initial value problem:

$$\frac{du(t)}{dt}=U(u,i)(t), t\in[T,3T],\ u(T)=0\,,\tag{6.3.4}$$

$$\frac{di(t)}{dt}=I(u,i)(t), t\in[T,3T], i(T)=0\,.\tag{6.3.5}$$

Then after integration of (6.3.4), (6.3.5) we have:

$$u(t)=\int_{T+kT_0}^{t}U(u,i)(t)dt \Rightarrow$$

$$0=u(T+(k+1)T_0)=\int_{T+kT_0}^{T+(k+1)T_0}U(u,i)(t)dt \Rightarrow \int_{T+kT_0}^{T+(k+1)T_0}U(u,i)(t)dt = 0$$

and

$$i(t)=\int_{T+kT_0}^{t}I(u,i)(t)dt \Rightarrow$$

$$0=i(T+(k+1)T_0)=\int_{T+kT_0}^{T+(k+1)T_0}I(u,i)(t)dt \Rightarrow \int_{T+kT_0}^{T+(k+1)T_0}I(u,i)(t)dt = 0.$$

From (6.3.4) and (6.3.5), multiplying by t and integrating from $T+kT_0$ to $T+(k+1)T_0$ we obtain

$$\int_{T+kT_0}^{T+(k+1)T_0}t\frac{du(t)}{dt}dt = \int_{T+kT_0}^{T+(k+1)T_0}t\,U(u,i)(t)dt \ \text{ and } \ \int_{T+kT_0}^{T+(k+1)T_0}t\frac{di(t)}{dt}dt = \int_{T+kT_0}^{T+(k+1)T_0}t\,I(u,i)(t)dt$$

or

$$\int_{T+kT_0}^{T+(k+1)T_0}t\,du(t) = \int_{T+kT_0}^{T+(k+1)T_0}t\,U(u,i)(t)dt \ \text{ and } \ \int_{T+kT_0}^{T+(k+1)T_0}t\,di(t) = \int_{T+kT_0}^{T+(k+1)T_0}t\,I(u,i)(t)dt$$

$$(T+(k+1)T_0)u(T+(k+1)T_0)-(T+kT_0)u(T+kT_0) = \int_{T+kT_0}^{T+(k+1)T_0}t\,U(u,i)(t)dt$$

and

$$(T + (k+1)T_0)i(T + (k+1)T_0) - (T + kT_0)i(T + kT_0) = \int_{T+kT_0}^{T+(k+1)T_0} t\, I(u,i)(t)\,dt .$$

It follows

$$0 = \int_{T+kT_0}^{T+(k+1)T_0} t\, U(u,i)(t)\,dt \text{ and } 0 = \int_{T+kT_0}^{T+(k+1)T_0} t\, I(u,i)(t)\,dt .$$

In view of Remark 6.3.2 it follows

$$\int_{T+kT_0}^{T+(k+1)T_0} \int_{T+kT_0}^{t} U(u,i)(s)\,ds\,dt = 0 \text{ and } \int_{T+kT_0}^{T+(k+1)T_0} \int_{T+kT_0}^{t} I(u,i)(s)\,ds\,dt = 0 .$$

Therefore

$$u(t) = \int_{T+kT_0}^{t} U(u,i)(s)\,ds, \; i(t) = \int_{T+kT_0}^{t} I(u,i)(s)\,ds$$

is equivalent to

$$u(t) = \int_{T+kT_0}^{t} U(u,i)(s)\,ds - \left(\frac{t-T-kT_0}{T_0} - \frac{1}{2}\right)\int_{T+kT_0}^{T+(k+1)T_0} U(u,i)(s)\,ds - \frac{1}{T_0}\int_{T+kT_0}^{T+(k+1)T_0}\int_{T+kT_0}^{t} U(u,i)(s)\,ds\,dt,$$

$$i(t) = \int_{T+kT_0}^{t} I(u,i)(s)\,ds - \left(\frac{t-T-kT_0}{T_0} - \frac{1}{2}\right)\int_{T+kT_0}^{T+(k+1)T_0} I(u,i)(s)\,ds - \frac{1}{T_0}\int_{T+kT_0}^{T+(k+1)T_0}\int_{T+kT_0}^{t} I(u,i)(s)\,ds\,dt$$

and then the solution (u,i) of (6.3.4), (6.3.5) is a fixed point of B.

Conversely, let $(u(.),i(.)) \in M_U^* \times M_I^*$ be a fixed point B, that is,

$$u = B_u^{(k)}(u,i) \,, \; i = B_i^{(k)}(u,i) \,, \; t \in [T + kT_0, T + (k+1)T_0] .$$

Therefore

$$u(T + kT_0) = B_u^{(k)}(u,i)(T + kT_0), \; i(T + kT_0) = B_i^{(k)}(u,i)(T + kT_0)$$

or

$$0 = u(T + kT_0) = \int\limits_{T+kT_0}^{T+kT_0} U(u,i)(s)ds - \left(\frac{T + kT_0 - T - kT_0}{T_0} - \frac{1}{2}\right)\int\limits_{T+kT_0}^{T+(k+1)T_0} U(u,i)(s)ds -$$

$$-\frac{1}{T_0}\int\limits_{T+kT_0}^{T+(k+1)T_0}\int\limits_{T+kT_0}^{t} U(u,i)(s)dsdt = \frac{1}{2}\int\limits_{T+kT_0}^{T+(k+1)T_0} U(u,i)(s)ds - \frac{1}{T_0}\int\limits_{T+kT_0}^{T+(k+1)T_0}\int\limits_{T+kT_0}^{t} U(u,i)(s)dsdt$$

$$= \frac{1}{2}\int\limits_{T+kT_0}^{T+(k+1)T_0} U(u,i)(s)ds - \frac{1}{T_0}\int\limits_{T+kT_0}^{T+(k+1)T_0}\int\limits_{T+kT_0}^{t} U(u,i)(s)dsdt$$

and

$$0 = i(T + kT_0) = \int\limits_{T+kT_0}^{T+kT_0} I(u,i)(s)ds - \left(\frac{T + kT_0 - T - kT_0}{T_0} - \frac{1}{2}\right)\int\limits_{T+kT_0}^{T+(k+1)T_0} I(u,i)(s)ds -$$

$$-\frac{1}{T_0}\int\limits_{T+kT_0}^{T+(k+1)T_0}\int\limits_{T+kT_0}^{t} I(u,i)(s)dsdt = \frac{1}{2}\int\limits_{T+kT_0}^{T+(k+1)T_0} I(u,i)(s)ds - \frac{1}{T_0}\int\limits_{T+kT_0}^{T+(k+1)T_0}\int\limits_{T+kT_0}^{t} I(u,i)(s)dsdt.$$

We show that $\int\limits_{T+kT_0}^{T+(k+1)T_0} I(u,i)(s)ds = 0$ which implies $\int\limits_{T+kT_0}^{T+(k+1)T_0}\int\limits_{T+kT_0}^{t} I(u,i)(s)dsdt = 0$.

Indeed,

$$\left|\int\limits_{T+kT_0}^{T+(k+1)T_0} U(u,i)(t)dt\right| \leq \left|\int\limits_{T+kT_0}^{T+(k+1)T_0} \dot{E}_1(t)dt\right| + \frac{1}{\dot{C}_1}\int\limits_{T+kT_0}^{T+(k+1)T_0} |i(t)|dt +$$

$$+\frac{1}{\dot{C}_1}\int\limits_{T+kT_0}^{T+(k+1)T_0} \left|\widetilde{L}_1^{-1}\left(\int\limits_{T+kT_0}^{s} (u(\tau) - E_1(\tau))d\tau\right)\right|ds + \frac{1}{\dot{C}_1}\int\limits_{T+kT_0}^{T+(k+1)T_0} |G_1(u(t) - E_1(t))|dt \leq$$

$$\leq \frac{I_0}{\dot{C}_1}\int\limits_{T+kT_0}^{T+(k+1)T_0} e^{\mu(t-T-kT_0)}dt + \frac{I_0}{\dot{C}_1}\int\limits_{T+kT_0}^{T+(k+1)T_0} e^{\mu(t-T-kT_0)}dt + \frac{1}{\dot{C}_1}\sum\limits_{n=1}^{m}\left|g_n^{(1)}\right|\int\limits_{T+kT_0}^{T+(k+1)T_0}\int\limits_{T+kT_0}^{t} e^{n\mu(\tau-T-kT_0)}|u(\tau) - E_1(\tau)|^n d\tau dt \leq$$

$$\leq \frac{I_0}{\dot{C}_1}\frac{e^{\mu T_0} - 1}{\mu} + \frac{I_0}{\dot{C}_1}\frac{e^{\mu T_0} - 1}{\mu} + \frac{1}{\dot{C}_1}\sum\limits_{n=1}^{m}\left|g_n^{(1)}\right|\int\limits_{T+kT_0}^{T+(k+1)T_0}\frac{e^{n\mu(t-T-kT_0)} - 1}{n\mu}(\phi_0)^n dt \leq$$

$$\leq \frac{I_0}{\dot{C}_1}\frac{e^{\mu T_0} - 1}{\mu} + \frac{I_0}{\dot{C}_1}\frac{e^{\mu T_0} - 1}{\mu} + \frac{1}{\dot{C}_1}\sum\limits_{n=1}^{m}\left|g_n^{(1)}\right|(\phi_0)^n \frac{e^{(n-1)\mu_0}}{\mu}\int\limits_{T+kT_0}^{T+(k+1)T_0} e^{\mu(t-T-kT_0)}dt \leq$$

$$\leq \frac{2I_0}{\dot{C}_1}\frac{e^{\mu T_0} - 1}{\mu} + \frac{1}{\mu\dot{C}_1}\frac{e^{\mu T_0} - 1}{\mu}\sum\limits_{n=1}^{m}\left|g_n^{(1)}\right|\phi_0^n e^{(n-1)\mu_0} \leq \frac{2I_0}{\dot{C}_1}\frac{e^{\mu T_0} - 1}{\mu} + \frac{1}{\mu\dot{C}_1}\frac{e^{\mu T_0} - 1}{\mu}\sum\limits_{n=1}^{m}\left|g_n^{(1)}\right|\phi_0^n e^{(n-1)\mu_0} \leq$$

424   Vasil G. Angelov

$$\leq \frac{e^{\mu_0}-1}{\mu \dot{\bar{C}}_1}\left(2I_0 + \frac{1}{\mu}\sum_{n=1}^{m}\left|g_n^{(1)}\right|\phi_0^n e^{(n-1)\mu_0}\right) \equiv M_u(\mu)$$

and

$$\left|\int_{T+kT_0}^{T+(k+1)T_0} I(u,i)(t)dt\right| \leq$$

$$\leq \frac{1}{Z_0}\left|\int_{T+kT_0}^{T+(k+1)T_0}\frac{du(t)}{dt}dt\right| + e^{-2\frac{RT}{L}}\left|\int_{T+kT_0}^{T+(k+1)T_0}\frac{di(t-2T)}{dt}dt\right| + e^{-2\frac{RT}{L}}\frac{1}{Z_0}\left|\int_{T+kT_0}^{T+(k+1)T_0}\frac{du(t-2T)}{dt}dt\right| +$$

$$+ \frac{2}{Z_0}e^{-\frac{RT}{L}}\left|\int_{T+kT_0}^{T+(k+1)T_0}\dot{E}_0(t)dt\right| + \frac{2}{Z_0\bar{C}_0}e^{-\frac{RT}{L}}\int_{T+kT_0}^{T+(k+1)T_0}\left|\tilde{L}_0^{-1}\left(\int_T^t \tilde{u}(\tau)d\tau\right)\right| + \frac{2}{Z_0\bar{C}_0}e^{-\frac{RT}{L}}\int_{T+kT_0}^{T+(k+1)T_0}\left|G_0(\tilde{u}(t-T))\right|dt \leq$$

$$\leq \frac{2I_0}{Z_0\dot{\bar{C}}_0}e^{-\frac{RT}{L}}\int_{T+kT_0}^{T+(k+1)T_0}e^{\mu(t-T-kT_0)}dt + \frac{2}{Z_0\dot{\bar{C}}_0}e^{-\frac{RT}{L}}\sum_{n=1}^{p}\left|g_n^{(0)}\right|\int_{T+kT_0}^{T+(k+1)T_0}\left|\tilde{u}(t-T)\right|^n dt \leq$$

$$\leq \frac{2I_0}{Z_0\dot{\bar{C}}_0}e^{-\frac{RT}{L}}\frac{e^{\mu T_0}-1}{\mu} + \frac{2}{Z_0\dot{\bar{C}}_0}e^{-\frac{RT}{L}}\sum_{n=1}^{p}\left|g_n^{(0)}\right|\phi_0^n\int_{T+kT_0}^{T+(k+1)T_0}e^{\mu(t-T-kT_0)}dt \leq$$

$$\leq 2e^{-\frac{RT}{L}}\frac{e^{\mu_0}-1}{\mu Z_0\dot{\bar{C}}_0}\left(I_0 + \sum_{n=1}^{p}\left|g_n^{(0)}\right|\phi_0^n\right) \equiv M_i(\mu)$$

Since $M_u(\mu) \xrightarrow[\mu\to\infty]{} 0$, $M_i(\mu) \xrightarrow[\mu\to\infty]{} 0$ it follows

$$\int_{T+kT_0}^{T+(k+1)T_0}U(u,i)(t)dt = 0 \text{ and } \int_{T+kT_0}^{T+(k+1)T_0}I(u,i)(t)dt = 0.$$

Therefore

$$u = B_u^{(k)}(u,i)\,,\ i = B_i^{(k)}(u,i)\,,\ t \in [T+kT_0, T+(k+1)T_0]$$

become

$$u(t) = \int_{T+kT_0}^{t}U(u,i)(t)dt\,,\ i(t) = \int_{T+kT_0}^{t}I(u,i)(t)dt.$$

Differentiating the last equalities we obtain (6.3.4), (6.3.5).
Lemma 6.3.4 is thus proved.

Remark 6.3.3. Let us point out the previous Lemma 6.3.4 is valid for linear as well as nonlinear operators $B = (B_u, B_i)$.

6.4. EXISTENCE-UNIQUENESS OF A PERIODIC SOLUTION FOR THE NONLINEAR NEUTRAL SYSTEM

Here we prove the following

Theorem 6.4.1. Let Assumptions **(E)** and **(IN)** be fulfilled and $U_0 + U_{E_1} \le \phi_0$;

$$\left(1 + \frac{e^{\mu_0} - 1}{\mu_0}\right)\frac{U_0}{Z_0} < I_0 \; ; \; \left(1 + \frac{e^{\mu_0} - 1}{\mu_0}\right)U_{E_1} < U_0 \; ;$$

$$\cosh\left(\frac{RT}{L}\right)e^{-\beta}\left(U_0 + Z_0\,I_0\right) + U_{E_0} \le U_0 e^{\mu_0} \le \phi_0 .$$

Then there exists a unique T_0 -periodic solution of (6.2.4).

Proof: Define the operator $B \equiv (B_u, B_i) : M_U^* \times M_I^* \to M_U^* \times M_I^*$ by the formulas:

$$B_u^{(k)}(u,i)(t) := \int_{T+kT_0}^{t} U(u,i)(s)ds - \left(\frac{t - T - kT_0}{T_0} - \frac{1}{2}\right)\int_{T+kT_0}^{T+(k+1)T_0} U(u,i)(s)ds - \frac{1}{T_0}\int_{T+kT_0}^{T+(k+1)T_0}\int_{T+kT_0}^{t} U(u,i)(s)dsdt$$

$$t \in [T + kT_0, T + (k+1)T_0] \; (k = 0,1,2,\dots) ; \; u(T + kT_0) = u(T) = 0 ,$$

$$B_i^{(k)}(U,I)(t) := \int_{T+kT_0}^{t} I(u,i)(s)ds - \left(\frac{t - T - kT_0}{T_0} - \frac{1}{2}\right)\int_{T+kT_0}^{T+(k+1)T_0} I(u,i)(s)ds - \frac{1}{T_0}\int_{T+kT_0}^{T+(k+1)T_0}\int_{T+kT_0}^{t} I(u,i)(s)dsdt$$

$$t \in [T + kT_0, T + (k+1)T_0] \; (k = 0,1,2,\dots) ; \; i(T + kT_0) = i(T) = 0$$

where in the right-hand sides of the system, the unknown functions $U(u,i), I(u,i)$ with retarded arguments are replaced by the translated initial functions (cf. Chapter II).

Let us check the continuity of $B_u(t)$ at the points $T + kT_0 \; (k = 1,2,\dots)$. Indeed,

$$B_u^{(k)}(u,i)(T + (k+1)T_0) := \int_{T+kT_0}^{T+(k+1)T_0} U(u,i)(s)ds - \frac{1}{2}\int_{T+kT_0}^{T+(k+1)T_0} U(u,i)(s)ds - \frac{1}{T_0}\int_{T+kT_0}^{T+(k+1)T_0}\int_{T+kT_0}^{t} U(u,i)(s)dsdt =$$

$$= \frac{1}{2}\int_{T+kT_0}^{T+(k+1)T_0} U(u,i)(s)ds - \frac{1}{T_0}\int_{T+kT_0}^{T+(k+1)T_0}\int_{T+kT_0}^{t} U(u,i)(s)dsdt \; ;$$

$$B_u^{(k+1)}(u,i)(T+(k+1)T_0):=\int_{T+(k+1)T_0}^{T+(k+1)T_0}U(u,i)(s)ds-\left(\frac{T+(k+1)T_0-T-(k+1)T_0}{T_0}-\frac{1}{2}\right)\int_{T+(k+1)T_0}^{T+(k+2)T_0}U(u,i)(s)ds-$$

$$-\frac{1}{T_0}\int_{T+(k+1)T_0}^{T+(k+2)T_0}\int_{T+(k+1)T_0}^{t}U(u,i)(s)ds\,dt=$$

$$=\frac{1}{2}\int_{T+(k+1)T_0}^{T+(k+2)T_0}U(u,i)(s)ds-\frac{1}{T_0}\int_{T+(k+1)T_0}^{T+(k+2)T_0}\int_{T+(k+1)T_0}^{t}U(u,i)(s)ds\,dt\ .$$

But $U(u,i)(s)$ is T_0-periodic and therefore

$$\int_{T+kT_0}^{T+(k+1)T_0}U(u,i)(s)ds=\int_{T+(k+1)T_0}^{T+(k+2)T_0}U(u,i)(s)ds\ .$$

Lemma 6.3.3 implies the conclusion.

The equalities $\displaystyle\int_{T+kT_0}^{T+(k+1)T_0}B_u^{(k)}(u,i)(t)dt=0,\ \int_{T+kT_0}^{T+(k+1)T_0}B_i^{(k)}(u,i)(t)dt=0$ can be verified by a direct integration.

In order to establish that the operator B maps $M_U^*\times M_I^*$ into itself, it remains to show the inequalities

$$|u(t)|\le U_0 e^{\mu(t-T-kT_0)}\wedge|i(t)|\le I_0 e^{\mu(t-T-kT_0)}\Rightarrow|B_u|\le U_0 e^{\mu(t-T-kT_0)}\wedge|B_i|\le I_0 e^{\mu(t-T-kT_0)}$$

Recalling that $\left|\dfrac{t-T-kT_0}{T_0}-\dfrac{1}{2}\right|\le\dfrac{1}{2},\ t\in[T+kT_0,T+(k+1)T_0]$, we have

$$\left|B_u^{(k)}(u,i)(t)\right|\le\left|\int_{T+kT_0}^{t}U(u,i)(s)ds\right|+$$

$$+\left|\int_{T+kT_0}^{T+(k+1)T_0}U(u,i)(s)ds\right|+\frac{1}{T_0}\int_{T+kT_0}^{T+(k+1)T_0}\left|\int_{T+kT_0}^{t}U(u,i)(s)ds\right|dt\equiv U_1+U_2+U_3\ .$$

Since

$$U_1\le\left|\int_{T+kT_0}^{t}\dot E_1(s)ds\right|+\frac{1}{\tilde C_1}\int_{T+kT_0}^{t}|i(s)|ds+\frac{1}{\tilde C_1}\int_{T+kT_0}^{t}\left|\tilde L_1^{-1}\left(\int_{T+kT_0}^{t}(u(\tau)-E_1(\tau))d\tau\right)\right|ds+\frac{1}{\tilde C_1}\int_{T+kT_0}^{t}|G_1(u(s)-E_1(s))|ds\le$$

$$\leq \left|E_1(t)\right| + \frac{I_0}{\dot{\overline{C}}_1} \int\limits_{T+kT_0}^{t} e^{\mu(s-T-kT_0)}ds + \frac{I_0}{\dot{\overline{C}}_1} \int\limits_{T+kT_0}^{t} e^{\mu(s-T-kT_0)}ds + \frac{I_0}{\dot{\overline{C}}_1}\sum_{n=1}^{m}\left|g_n^{(1)}\right| \int\limits_{T+kT_0}^{t}\left|u(s)-E_1(s)\right|^n ds \leq$$

$$\leq U_{E_1}e^{\mu(t-T-kT_0)} + \frac{I_0}{\dot{\overline{C}}_1}\frac{e^{\mu(t-T-kT_0)}-1}{\mu}\int\limits_{T+kT_0}^{t}ds + \frac{I_0}{\dot{\overline{C}}_1}\frac{e^{\mu(t-T-kT_0)}-1}{\mu} + \frac{1}{\dot{\overline{C}}_1}\sum_{n=1}^{m}\left|g_n^{(1)}\right|\left\|U_0+U_{E_1}\right\|^n \int\limits_{T+kT_0}^{t} e^{n\mu(s-T-kT_0)}ds \leq$$

$$\leq U_{E_1}e^{\mu(t-T-kT_0)} + \frac{I_0}{\dot{\overline{C}}_1}\frac{e^{\mu(t-T-kT_0)}-1}{\mu} + \frac{I_0}{\dot{\overline{C}}_1}\frac{e^{\mu(t-T-kT_0)}-1}{\mu} + \frac{1}{\dot{\overline{C}}_1}\sum_{n=1}^{m}\left|g_n^{(1)}\right|\left(U_0+U_{E_1}\right)^n \frac{e^{n\mu(t-T-kT_0)}-1}{n\mu} \leq$$

$$\leq U_{E_1}e^{\mu(t-T-kT_0)} + \frac{I_0}{\dot{\overline{C}}_1}\frac{e^{\mu(t-T-kT_0)}-1}{\mu} + \frac{I_0}{\dot{\overline{C}}_1}\frac{e^{\mu(t-T-kT_0)}-1}{\mu} +$$

$$+\frac{1}{\dot{\overline{C}}_1}\frac{e^{\mu(t-T-kT_0)}-1}{\mu}\sum_{n=1}^{m}\left|g_n^{1}\right|\left(U_0+U_{E_1}\right)^n e^{(n-1)\mu(t-T-kT_0)} \leq$$

$$\leq U_{E_1}e^{\mu(t-T-kT_0)} + \frac{e^{\mu(t-T-kT_0)}-1}{\mu\dot{\overline{C}}_1}\left(2I_0 + \sum_{n=1}^{m}\left|g_n^{(1)}\right|\left(U_0+U_{E_1}\right)^n e^{(n-1)\mu T_0}\right) \leq$$

$$\leq e^{\mu(t-T-kT_0)}\left[U_{E_1} + \frac{1}{\mu\dot{\overline{C}}_1}\left(2I_0 + \sum_{n=1}^{m}\left|g_n^{(1)}\right|\phi_0^n e^{(n-1)\mu_0}\right)\right];$$

$$U_2 \leq \frac{1}{2}\left|\int\limits_{T+kT_0}^{T+(k+1)T_0} U(u,i)(s)ds\right| \leq$$

$$\leq \frac{1}{2}\left|\int\limits_{T+kT_0}^{T+(k+1)T_0}\dot{E}_1(s)ds\right| + \frac{1}{2\dot{\overline{C}}_1}\int\limits_{T+kT_0}^{T+(k+1)T_0}\left|i(s)\right|ds + \frac{1}{2\dot{\overline{C}}_1}\int\limits_{T+kT_0}^{T+(k+1)T_0}\left|\widetilde{L}_1^{-1}\left(\int\limits_{T+kT_0}^{t}\left(u(\tau)-E_1(\tau)\right)d\tau\right)\right|ds +$$

$$+\frac{1}{2\dot{\overline{C}}_1}\int\limits_{T+kT_0}^{T+(k+1)T_0}\left|G_1(u(s)-E_1(s))\right|ds \leq$$

$$\leq \frac{I_0}{2\dot{\overline{C}}_1}\frac{e^{\mu T_0}-1}{\mu} + \frac{1}{2\dot{\overline{C}}_1}\frac{e^{\mu T_0}-1}{\mu} + \frac{1}{2\dot{\overline{C}}_1}\sum_{n=1}^{m}\left|g_n^{(1)}\right|\left(U_0+U_{E_1}\right)^n \frac{e^{n\mu T_0}-1}{n\mu} \leq$$

$$\leq \frac{e^{\mu_0}-1}{2\mu\dot{\overline{C}}_1}\left(2I_0 + \sum_{n=1}^{m}\left|g_n^{(1)}\right|\phi_0^n e^{(n-1)\mu_0}\right)$$

and

$$U_3 \le \frac{1}{T_0} \int\limits_{T+kT_0}^{T+(k+1)T_0} \left| \int\limits_{T+kT_0}^{t} U(u,i)(s)\,ds \right| dt \le$$

$$\le e^{\mu(t-T-kT_0)} \left[U_{E_1} + \frac{1}{\mu \dot{\overline{C}}_1}\left(2I_0 + \sum_{n=1}^{m} \left| g_n^{(1)} \right|\left(U_0 + U_{E_1} \right)^n e^{(n-1)\mu T_0} \right) \right] \frac{1}{T_0} \int\limits_{T+kT_0}^{T+(k+1)T_0} e^{\mu(t-T-kT_0)}\,dt \le$$

$$\le e^{\mu(t-T-kT_0)} \frac{e^{\mu_0}-1}{\mu_0} \left[U_{E_1} + \frac{1}{\mu \dot{\overline{C}}_1}\left(2I_0 + \sum_{n=1}^{m} \left| g_n^{(1)} \right| \phi_0^n e^{(n-1)\mu_0} \right) \right]$$

then

$$\left| B_u^{(k)}(u,i)(t) \right| \le$$

$$\le e^{\mu(t-T-kT_0)} \left[U_{E_1} + \frac{1}{\mu \dot{\overline{C}}_1}\left(2I_0 + \sum_{n=1}^{m} \left| g_n^{(1)} \right|\left(U_0 + U_{E_1} \right)^n e^{(n-1)\mu_0} \right) \right] +$$

$$+ e^{\mu(t-T-kT_0)} \frac{e^{\mu_0}-1}{2\mu \dot{\overline{C}}_1}\left(2I_0 + \sum_{n=1}^{m} \left| g_n^{(1)} \right|\left(U_0 + U_{E_1} \right)^n e^{(n-1)\mu_0} \right) +$$

$$+ e^{\mu(t-T-kT_0)} \frac{e^{\mu_0}-1}{\mu_0} \left[U_{E_1} + \frac{1}{\mu \dot{\overline{C}}_1}\left(2I_0 + \sum_{n=1}^{m} \left| g_n^{(1)} \right|\left(U_0 + U_{E_1} \right)^n e^{(n-1)\mu_0} \right) \right] \le$$

$$\le e^{\mu(t-T-kT_0)} \left[\left(1 + \frac{e^{\mu_0}-1}{\mu_0} \right) U_{E_1} + \left(1 + \frac{e^{\mu_0}-1}{2} + \frac{e^{\mu_0}-1}{\mu_0} \right) \frac{1}{\mu \dot{\overline{C}}_1}\left(2I_0 + \sum_{n=1}^{m} \left| g_n^{(1)} \right| \phi_0^n e^{(n-1)\mu_0} \right) \right] \le$$

$$\le e^{\mu(t-T-kT_0)} U_0$$

because $\left(1 + \dfrac{e^{\mu_0}-1}{\mu_0} \right) U_{E_1} < U_0$ and sufficiently large $\mu > 1$.

Further on, in view of the condition of the present theorem we have

$$U_0 + U_{E_1} \le \phi_0 \; ; \; \left| \widetilde{u}(t-T) \right| \le \cosh\left(\frac{RT}{L} \right) e^{-\beta}\left(U_0 + Z_0\, I_0 \right) + U_{E_0} \le U_0 e^{\mu_0} \le \phi_0$$

and then

$$\left| B_i^{(k)}(u,i)(t) \right| \le \left| \int_{T+kT_0}^{t} I(u,i)(s)ds \right| +$$

$$+\frac{1}{2}\left| \int_{T+kT_0}^{T+(k+1)T_0} I(u,i)(s)ds \right| + \frac{1}{T_0} \int_{T+kT_0}^{T+(k+1)T_0} \left| \int_{T+kT_0}^{t} I(u,i)(s)ds \right| dt \equiv I_1 + I_2 + I_3 .$$

Since

$$I_1 \le \frac{2e^{\frac{RT}{L}}}{Z_0}\left| \int_{T+kT_0}^{t} \dot{E}_0(s)ds \right| + \frac{1}{Z_0}\left| \int_{T+kT_0}^{t} \ddot{u}(s)ds \right| + e^{-\frac{2RT}{L}}\left| \int_{T+kT_0}^{t} \dot{i}(s-2T)ds \right| + \frac{e^{-\frac{2RT}{L}}}{Z_0}\left| \int_{T+kT_0}^{t} \ddot{u}(s-2T)ds \right| +$$

$$+\frac{2e^{-\frac{RT}{L}}}{Z_0\dot{\overline{C}}_0} \int_{T+kT_0}^{t} \left| \widetilde{L}_0^{-1}\left(\int_{T}^{s} \widetilde{u}(\theta-T)d\theta \right) \right| ds + \frac{2e^{-\frac{RT}{L}}}{Z_0\dot{\overline{C}}_0} \int_{T+kT_0}^{t} \left| G_0(\widetilde{u}(s-T)) \right| ds \le$$

$$\le \frac{2e^{-\frac{RT}{L}}}{Z_0}U_{E_0}e^{\mu(t-T-kT_0)} + \frac{U_0 e^{\mu(t-T-kT_0)}}{Z_0} + e^{-\frac{2RT}{L}}I_0 e^{\mu(t-T-kT_0)} + \frac{e^{-\frac{2RT}{L}}U_0 e^{\mu(t-T-kT_0)}}{Z_0} +$$

$$+\frac{2e^{-\frac{RT}{L}}I_0}{Z_0\dot{\overline{C}}_0}\frac{e^{\mu(t-T-kT_0)}-1}{\mu} + \frac{2e^{-\frac{RT}{L}}}{Z_0\dot{\overline{C}}_0}\sum_{n=1}^{m}\left| g_n^{(0)} \right| \phi_0^{n}\frac{e^{\mu(t-T-kT_0)}-1}{\mu} \le$$

$$\le e^{\mu(t-T-kT_0)}\frac{1}{Z_0}\left[2e^{-\frac{RT}{L}}U_{E_0} + U_0 + e^{-\frac{2RT}{L}}Z_0 I_0 + e^{-\frac{2RT}{L}}U_0 + \frac{2e^{-\frac{RT}{L}}}{\mu\dot{\overline{C}}_0}\left(I_0 + \sum_{n=1}^{m}\left| g_n^{(0)} \right| \phi_0^{n} \right) \right];$$

$$I_2 \le \frac{e^{-\frac{RT}{L}}}{Z_0}\left| \int_{T+kT_0}^{T+(k+1)T_0} \dot{E}_0(s-T)ds \right| + \frac{1}{2Z_0}\left| \int_{T+kT_0}^{T+(k+1)T_0} \ddot{u}(s)ds \right| + \frac{1}{2}e^{-\frac{2RT}{L}}\left| \int_{T+kT_0}^{T+(k+1)T_0} \dot{i}(s-2T)ds \right| + \frac{e^{-\frac{2RT}{L}}}{2Z_0}\left| \int_{T+kT_0}^{T+(k+1)T_0} \ddot{u}(s-2T)ds \right| +$$

$$+\frac{e^{-\frac{RT}{L}}}{Z_0\dot{\overline{C}}_0} \int_{T+kT_0}^{T+(k+1)T_0} \left| \widetilde{L}_0^{-1}\left(\int_{T}^{s} \widetilde{u}(\theta-T)d\theta \right) \right| ds + \frac{e^{-\frac{RT}{L}}}{Z_0\dot{\overline{C}}_0} \int_{T+kT_0}^{T+(k+1)T_0} \left| G_0(\widetilde{u}(s-T)) \right| ds \le$$

$$\le \frac{e^{-\frac{RT}{L}}}{Z_0\dot{\overline{C}}_0}I_0 \int_{T+kT_0}^{T+(k+1)T_0} e^{\mu(s-T-kT_0)}ds + \frac{e^{-\frac{RT}{L}}}{Z_0\dot{\overline{C}}_0}\sum_{n=1}^{m}\left| g_n^{(0)} \right| \phi_0^{n} \int_{T+kT_0}^{T+(k+1)T_0} e^{\mu(s-T-kT_0)}ds \le$$

$$\frac{e^{\mu T_0}-1}{2\mu}\frac{e^{-\frac{RT}{L}}}{Z_0\dot{\overline{C}}_0}\left(I_0 + \sum_{n=1}^{m}\left| g_n^{(0)} \right| \phi_0^{n} \right)$$

and

$$I_3 \le \frac{1}{T_0}\int\limits_{T+kT_0}^{T+(k+1)T_0}\left|\int\limits_{T+kT_0}^{t} I(u,i)(s)\,ds\right|dt \le \frac{1}{T_0}\int\limits_{T+kT_0}^{T+(k+1)T_0} e^{\mu(t-T-kT_0)}\,dt\,\frac{1}{Z_0}\times$$

$$\times\left[2e^{-\frac{RT}{L}}U_{E_0}+U_0+e^{-\frac{2RT}{L}}Z_0I_0+e^{-\frac{2RT}{L}}U_0+\frac{2e^{-\frac{RT}{L}}}{\mu\dot{\bar{C}}_0}\left(I_0+\sum_{n=1}^{m}\left|g_n^{(0)}\right|\phi_0^{\,n}\right)\right]\le$$

$$\le\frac{e^{\mu_0}-1}{\mu_0}\frac{1}{Z_0}\left[2e^{-\frac{RT}{L}}U_{E_0}+U_0+e^{-\frac{2RT}{L}}Z_0I_0+e^{-\frac{2RT}{L}}U_0+\frac{2e^{-\frac{RT}{L}}}{\mu\dot{\bar{C}}_0}\left(I_0+\sum_{n=1}^{m}\left|g_n^{(0)}\right|\phi_0^{\,n}\right)\right]$$

we get

$$\left|B_i^{(k)}(u,i)(t)\right|\le$$

$$\le e^{\mu(t-T-kT_0)}\left[\frac{2e^{-\frac{RT}{L}}U_{E_0}+U_0+e^{-\frac{2RT}{L}}Z_0I_0+e^{-\frac{2RT}{L}}U_0}{Z_0}+\frac{2e^{-\frac{RT}{L}}}{\mu\dot{\bar{C}}_0Z_0}\left(I_0+\sum_{n=1}^{m}\left|g_n^{(0)}\right|\phi_0^{\,n}\right)\right]+$$

$$+e^{\mu(t-T-kT_0)}\frac{e^{\mu_0}-1}{2}\frac{e^{-\frac{RT}{L}}}{\mu Z_0\dot{\bar{C}}_0}\left(I_0+\sum_{n=1}^{m}\left|g_n^{(0)}\right|\phi_0^{\,n}\right)+$$

$$+\frac{e^{\mu_0}-1}{\mu_0}\left[\frac{2e^{-\frac{RT}{L}}U_{E_0}+U_0+e^{-\frac{2RT}{L}}Z_0I_0+e^{-\frac{2RT}{L}}U_0}{Z_0}+\frac{2e^{-\frac{RT}{L}}}{\mu\dot{\bar{C}}_0Z_0}\left(I_0+\sum_{n=1}^{m}\left|g_n^{(0)}\right|\phi_0^{\,n}\right)\right]\le$$

$$\le\left[\left(1+\frac{e^{\mu_0}-1}{\mu_0}\right)\frac{1}{Z_0}\left(2e^{-\frac{RT}{L}}U_{E_0}+U_0+e^{-\frac{2RT}{L}}Z_0I_0+e^{-\frac{2RT}{L}}U_0\right)+\right.$$

$$\left.+\left(1+\frac{e^{\mu_0}-1}{2}+\frac{e^{\mu_0}-1}{\mu_0}\right)\frac{e^{-\frac{RT}{L}}}{\mu\dot{\bar{C}}Z_0}\left(I_0+\sum_{n=1}^{m}\left|g_n^{(0)}\right|\phi_0^{\,n}\right)\right]e^{\mu(t-T-kT_0)}\le I_0e^{\mu(t-T-kT_0)}.$$

Therefore the operator B maps the set $M_U^* \times M_I^*$ into itself.

In what follows we show that B is contractive operator.

In view of $\left|\dfrac{d\widetilde{C}_p(u)}{du}\right| \le \dfrac{c_p\sqrt[h]{\Phi_p}}{h^2}\,\dfrac{2h-(h^2-1)\phi_0}{\sqrt[h]{(\Phi_p-\phi_0)^{h+1}}}$, $\left|\widetilde{L}_p^{-1}(.)\right| \le i_0 \le I_0$ $(p=0,1)$ for the

first operator component we obtain

$$\left|B_u^{(k)}(u,i)(t)-B_u^{(k)}(\overline{u},\overline{i})(t)\right| \le$$

$$\le \int_{T+kT_0}^{t}\left|U(u,i)(s)-U(\overline{u},\overline{i})(s)\right|ds +$$

$$+\left(\dfrac{t-T-kT_0}{T_0}-\dfrac{1}{2}\right)\left|\int_{T+kT_0}^{T+(k+1)T_0}\left(U(u,i)(s)-U(\overline{u},\overline{i})(s)\right)ds\right| +$$

$$+\dfrac{1}{T_0}\int_{T+kT_0}^{T+(k+1)T_0}\left|\int_{T+kT_0}^{t}\left(U(u,i)(s)-U(\overline{u},\overline{i})(s)\right)ds\right|dt \equiv U_1+U_2+U_3.$$

But

$$U_1 \le \int_{T+kT_0}^{t}\dfrac{\left|i(s)-\overline{i}(s)\right|}{d\widetilde{C}_1(u(s)-E_1(s))/du}ds + \int_{T+kT_0}^{t}\left|\dfrac{\overline{i}(s)}{d\widetilde{C}_1(u(s)-E_1(s))/du}-\dfrac{\overline{i}(s)}{d\widetilde{C}_1(u(s)-E_1(s))/du}\right|ds +$$

$$+\int_{T+kT_0}^{t}\left|\dfrac{\widetilde{L}_1^{-1}\left(\displaystyle\int_{T+kT_0}^{s}(u(\tau)-E_1(\tau))d\tau\right)}{d\widetilde{C}_1(u(s)-E_1(s))/du}-\dfrac{\widetilde{L}_1^{-1}\left(\displaystyle\int_{T+kT_0}^{s}(\overline{u}(\tau)-E_1(\tau))d\tau\right)}{d\widetilde{C}_1(u(s)-E_1(s))/du}\right|ds +$$

$$+\int_{T+kT_0}^{t}\left|\widetilde{L}_1^{-1}\left(\int_{T+kT_0}^{s}(\overline{u}(\tau)-E_1(\tau))d\tau\right)\right|\left|\dfrac{1}{d\widetilde{C}_1(u(s)-E_1(s))/du}-\dfrac{1}{d\widetilde{C}_1(\overline{u}(s)-E_1(s))/du}\right|ds +$$

$$+\int_{T+kT_0}^{t}\left|\dfrac{G_1(u(s)-E_1(s))}{d\widetilde{C}_1(u(s)-E_1(s))/du}-\dfrac{G_1(\overline{u}(s)-E_1(s))}{d\widetilde{C}_1(\overline{u}(s)-E_1(s))/du}\right|ds +$$

$$+\int_{T+kT_0}^{t}\left|\dfrac{G_1(\overline{u}(s)-E_1(s))}{d\widetilde{C}_1(u(s)-E_1(s))/du}-\dfrac{G_1(\overline{u}(s)-E_1(s))}{d\widetilde{C}_1(\overline{u}(s)-E_1(s))/du}\right| \le$$

$$\le \dfrac{\rho^{(k)}(i,\overline{i})}{\overset{\cdot}{\overline{C}}_1}\dfrac{e^{\mu(t-T-kT_0)}-1}{\mu}+\int_{T+kT_0}^{t}\dfrac{\left|\overline{i}(s)\right|}{\left(d\widetilde{C}_1(u(s)-E_1(s))/du\right)^2}\left|\dfrac{d^2\widetilde{C}_1}{du^2}\right|\left|u(s)-\overline{u}(s)\right|ds +$$

$$+ \frac{1}{\hat{L}_1 \dot{\overline{C}}_1} \int\limits_{T+kT_0}^{t} \left| \int\limits_{T+kT_0}^{s} (u(\tau) - \overline{u}(\tau)) d\tau \right| ds +$$

$$+ \int\limits_{T+kT_0}^{t} \frac{I_0}{\left(d\widetilde{C}_1 (u(s) - E_1(s))/du \right)^2} \left| \frac{d^2 \widetilde{C}_1}{du^2} \right| |u(s) - \overline{u}(s)| ds +$$

$$+ \frac{1}{\dot{\overline{C}}_1} \sum_{n=1}^{m} n |g_n^1| \|U_0 + U_{E_1}\|^{n-1} \int\limits_{T+kT_0}^{t} e^{(n-1)\mu(s-T-kT_0)} |u(s) - \overline{u}(s)| ds +$$

$$+ \sup \left\{ |G_1(u(s) - E_1(s))| : s \in [T+kT_0, T+(k+1)T_0] \right\} \int\limits_{T+kT_0}^{t} \frac{1}{\left(d\widetilde{C}_1(u(s) - E_1(s))/du \right)^2} \left| \frac{d^2 \widetilde{C}_1}{du^2} \right| |u(s) - \overline{u}(s)| ds \le$$

$$\le \frac{\rho^{(k)}(i,\overline{i})}{\dot{\overline{C}}_1} \frac{e^{\mu(t-T-kT_0)}-1}{\mu} + \frac{I_0}{\dot{\overline{C}}_1^2} H_1 \rho^{(k)}(u,\overline{u}) \int\limits_{T+kT_0}^{t} e^{\mu(s-T-kT_0)} ds +$$

$$+ \frac{\rho^{(k)}(u,\overline{u})}{\hat{L}_1 \dot{\overline{C}}_1} \int\limits_{T+kT_0}^{t} \frac{e^{\mu(s-T-kT_0)}-1}{\mu} ds + \frac{I_0}{\dot{\overline{C}}_1^2} H_1 \rho^{(k)}(u,\overline{u}) \int\limits_{T+kT_0}^{t} e^{\mu(s-T-kT_0)} ds +$$

$$+ \frac{1}{\hat{C}_1} \sum_{n=1}^{m} n |g_n^{(1)}| \|U_0 + U_{E_1}\|^{n-1} \rho^{(k)}(u,\overline{u}) \int\limits_{T+kT_0}^{t} e^{n\mu(s-T-kT_0)} ds +$$

$$+ \sum_{n=1}^{m} |g_n^0| (U_0 + U_{E_1})^n e^{n\mu T_0} \frac{1}{\dot{\overline{C}}_1^2} H_1 \rho^{(k)}(u,\overline{u}) \int\limits_{T+kT_0}^{t} e^{\mu(s-T-kT_0)} ds \le$$

$$\le \frac{\rho^{(k)}(i,\overline{i})}{\hat{C}_1} \frac{e^{\mu(t-T-kT_0)}-1}{\mu} + \rho^{(k)}(u,\overline{u}) \frac{I_0}{\dot{\overline{C}}_1^2} H_1 \frac{e^{\mu(t-T-kT_0)}-1}{\mu} +$$

$$+ \frac{\rho^{(k)}(u,\overline{u})}{\mu \hat{L}_1 \hat{C}_1} \frac{e^{\mu(t-T-kT_0)}-1}{\mu} + \frac{I_0}{\dot{\overline{C}}_1^2} H_1 \rho^{(k)}(u,\overline{u}) \frac{e^{\mu(t-T-kT_0)}-1}{\mu} +$$

$$+ \frac{1}{\hat{C}_1} \sum_{n=1}^{m} n |g_n^{(1)}| (U_0 + U_{E_1})^{n-1} \rho^{(k)}(u,\overline{u}) \frac{e^{n\mu(t-T-kT_0)}-1}{n\mu} +$$

$$+ \sum_{n=1}^{m} |g_n^{(1)}| (U_0 + U_{E_1})^n e^{n\mu T_0} \frac{1}{\dot{\overline{C}}_1^2} H_1 \rho^{(k)}(u,\overline{u}) \frac{e^{\mu(t-T-kT_0)}-1}{\mu} \le$$

$$\le \frac{e^{\mu(t-T-kT_0)}-1}{\mu} e^{\mu_0} \left(\frac{\rho_\mu^{(k)}(i,\overline{i})}{\hat{C}_1} + \rho_\mu^{(k)}(u,\overline{u}) \frac{I_0}{\dot{\overline{C}}_1^2} H_1 + \right.$$

$$+\frac{\rho_\mu^{(k)}(u,\overline{u})}{\mu\hat{L}_1\hat{C}_1}+\frac{I_0}{\dot{C}_1^{\,2}}H_1\rho_\mu^{(k)}(u,\overline{u})+\frac{1}{\hat{C}_1}\sum_{n=1}^m n\big|g_n^1\big|\big(U_0+U_{E_1}\big)^{n-1}\rho_\mu^{(k)}(u,\overline{u})e^{(n-1)\mu T_0}+$$

$$+\sum_{n=1}^m \big|g_n^{(1)}\big|\big(U_0+U_{E_1}\big)^n e^{n\mu T_0}\frac{1}{\dot{C}_1^{\,2}}H_1\rho_\mu^{(k)}(u,\overline{u})\Bigg)\le$$

$$\le\frac{e^{\mu(t-T-kT_0)}-1}{\mu}e^{\mu_0}\Bigg(\frac{\rho_\mu^{(k)}(\dot{i},\dot{\overline{i}})}{\mu\hat{C}_1}+\frac{\rho_\mu^{(k)}(\dot{u},\dot{\overline{u}})}{\mu}\frac{I_0}{\dot{C}_1^{\,2}}H_1+\frac{\rho_\mu^{(k)}(\dot{u},\dot{\overline{u}})}{\mu^2\hat{L}_1\hat{C}_1}+H_1\frac{\rho_\mu^{(k)}(\dot{u},\dot{\overline{u}})}{\mu}+$$

$$+\frac{1}{\hat{C}_1}\sum_{n=1}^m n\big|g_n^{(1)}\big|\big|U_0+U_{E_1}\big|^{n-1}\frac{\rho_\mu^{(k)}(\dot{u},\dot{\overline{u}})}{\mu}e^{(n-1)\mu T_0}+\sum_{n=1}^m \big|g_n^{(1)}\big|\big(U_0+U_{E_1}\big)^n e^{n\mu T_0}\frac{1}{\dot{C}_1^{\,2}}H_1\frac{\rho_\mu^{(k)}(\dot{u},\dot{\overline{u}})}{\mu}\Bigg)\le$$

$$\le e^{\mu(t-T-kT_0)}\hat{\rho}_\mu\big((u,\dot{u},i,\dot{i}),(\overline{u},\overline{i},\dot{\overline{u}},\dot{\overline{i}})\big)\frac{e^{\mu_0}}{\mu^2\dot{\tilde{C}}_1}\Bigg[1+\Bigg(2I_0+\sum_{n=1}^m\big|g_n^{(1)}\big|\phi_0^n e^{n\mu T_0}\Bigg)H_1+\frac{1}{\hat{L}_1}+2\sum_{n=1}^m n\big|g_n^{(1)}\big|\phi_0^{n-1}e^{(n-1)\mu T_0}\Bigg];$$

$$U_2\le\frac{1}{2}\int_{T+kT_0}^{T+(k+1)T_0}\frac{|i(s)-\overline{i}(s)|}{d\tilde{C}_1(u(s)-E_1(s))/du}\,ds+\frac{1}{2}\int_{T+kT_0}^{T+(k+1)T_0}\left|\frac{\overline{i}(s)}{d\tilde{C}_1(u(s)-E_1(s))/du}-\frac{\overline{i}(s)}{d\tilde{C}_1(\overline{u}(s)-E_1(s))/du}\right|ds+$$

$$+\frac{1}{2}\int_{T+kT_0}^{T+(k+1)T_0}\left|\frac{\tilde{L}_1^{-1}\left(\displaystyle\int_{T+kT_0}^s\big(u(\tau)-E_1(\tau)\big)d\tau\right)}{d\tilde{C}_1(u(s)-E_1(s))/du}-\frac{\tilde{L}_1^{-1}\left(\displaystyle\int_{T+kT_0}^s\big(\overline{u}(\tau)-E_1(\tau)\big)d\tau\right)}{d\tilde{C}_1(\overline{u}(s)-E_1(s))/du}\right|ds+$$

$$+\frac{1}{2}\int_{T+kT_0}^{T+(k+1)T_0}\left|\tilde{L}_1^{-1}\left(\int_{T+kT_0}^s\big(\overline{u}(\tau)-E_1(\tau)\big)d\tau\right)\right|\left|\frac{1}{d\tilde{C}_1(u(s)-E_1(s))/du}-\frac{1}{d\tilde{C}_1(\overline{u}(s)-E_1(s))/du}\right|ds+$$

$$+\frac{1}{2}\int_{T+kT_0}^{T+(k+1)T_0}\left|\frac{G_1(u(s)-E_1(s))}{d\tilde{C}_1(u(s)-E_1(s))/du}-\frac{G_1(\overline{u}(s)-E_1(s))}{d\tilde{C}_1(\overline{u}(s)-E_1(s))/du}\right|ds+$$

$$+\frac{1}{2}\int_{T+kT_0}^{T+(k+1)T_0}\left|\frac{G_1(\overline{u}(s)-E_1(s))}{d\tilde{C}_1(u(s)-E_1(s))/du}-\frac{G_1(\overline{u}(s)-E_1(s))}{d\tilde{C}_1(\overline{u}(s)-E_1(s))/du}\right|\le$$

$$\le\frac{1}{2}\frac{\rho^{(k)}(i,\overline{i})}{\dot{C}_1}\frac{e^{\mu T_0}-1}{\mu}+\frac{1}{2}\frac{I_0}{\dot{C}_1^{\,2}}H_1\rho^{(k)}(u,\overline{u})\int_{T+kT_0}^{T+(k+1)T_0}e^{\mu(s-T-kT_0)}ds+$$

$$+\frac{1}{2}\frac{\rho^{(k)}(u,\overline{u})}{\hat{L}_1\dot{C}_1}\int_{T+kT_0}^{T+(k+1)T_0}\frac{e^{\mu(s-T-kT_0)}-1}{\mu}\,ds+\frac{1}{2}\frac{I_0 H_1}{\dot{C}_1^{\,2}}\rho^{(k)}(u,\overline{u})\int_{T+kT_0}^{T+(k+1)T_0}e^{\mu(s-T-kT_0)}ds+$$

$$+\frac{1}{2\dot{C}_1}\sum_{n=1}^m n\big|g_n^{(1)}\big|\big(U_0+U_{E_1}\big)^{n-1}\rho^{(k)}(u,\overline{u})\int_{T+kT_0}^{T+(k+1)T_0}e^{n\mu(s-T-kT_0)}ds+$$

$$+\frac{1}{2}\frac{H_1}{\dot{\bar{C}}_1^2}\rho_\mu^{(k)}(u,\bar{u})\int_{T+kT_0}^{T+(k+1)T_0}e^{\mu(s-T-kT_0)}ds\sum_{n=1}^{m}\left|g_n^{(1)}\right|\left(U_0+U_{E_1}\right)^n e^{n\mu T_0}\le$$

$$\le\frac{e^{\mu T_0}-1}{2\mu}e^{\mu_0}\left(\frac{\rho_\mu^{(k)}(\dot{i},\dot{\bar{i}})}{\mu\dot{\bar{C}}_1}+\frac{\rho_\mu^{(k)}(\dot{u},\dot{\bar{u}})}{\mu}\frac{I_0 H_1}{\dot{\bar{C}}_1^2}+\frac{\rho_\mu^{(k)}(\dot{u},\dot{\bar{u}})}{\mu^2\hat{L}_1\dot{\bar{C}}_1}+\frac{\rho_\mu^{(k)}(\dot{u},\dot{\bar{u}})}{\mu}\frac{I_0 H_1}{\dot{\bar{C}}_1^2}+\right.$$

$$\left.+\frac{1}{\dot{\bar{C}}_1}\frac{\rho_\mu^{(k)}(\dot{u},\dot{\bar{u}})}{\mu}\sum_{n=1}^{m}n\left|g_n^{(1)}\right|\left(U_0+U_{E_1}\right)^{n-1}e^{(n-1)\mu T_0}+\frac{H_1}{\dot{\bar{C}}_1^2}\frac{\rho_\mu^{(k)}(\dot{u},\dot{\bar{u}})}{\mu}\sum_{n=1}^{m}\left|g_n^{(1)}\right|\left(U_0+U_{E_1}\right)^n e^{n\mu T_0}\right)\le$$

$$\le e^{\mu(t-T-kT_0)}\hat{\rho}_\mu((u,\dot{u},i,\dot{i}),(\bar{u},\dot{\bar{i}},\dot{\bar{u}},\dot{\bar{i}}))\frac{e^{\mu T_0}-1}{2\mu^2\dot{\bar{C}}_1}e^{\mu_0}\left[1+\left(2I_0+\sum_{n=1}^{m}\left|g_n^{(1)}\right|\phi_0^n e^{n\mu_0}\right)H_1+\right.$$

$$\left.+\frac{1}{\hat{L}_1}+\frac{1}{\dot{\bar{C}}_1}\sum_{n=1}^{m}n\left|g_n^{(1)}\right|\phi_0^{n-1}e^{(n-1)\mu T_0}\right]$$

and

$$U_3\le\frac{1}{T_0}\int_{T+kT_0}^{T+(k+1)T_0}\left|\int_{T+kT_0}^{t}U(u,i)(s)ds-\int_{T+kT_0}^{t}U(\bar{u},\bar{i})(s)ds\right|dt\le$$

$$\le\hat{\rho}_\mu((u,\dot{u},i,\dot{i}),(\bar{u},\dot{\bar{i}},\dot{\bar{u}},\dot{\bar{i}}))\frac{e^{\mu_0}-1}{\mu_0}\frac{1}{\mu^2\hat{C}_1}e^{\mu_0}\left[1+\left(2I_0+\sum_{n=1}^{m}\left|g_n^{(1)}\right|\phi_0^n e^{n\mu_0}\right)H_1+\right.$$

$$\left.+\frac{1}{\hat{L}_1}+2\sum_{n=1}^{m}n\left|g_n^{(1)}\right|\phi_0^{n-1}e^{(n-1)\mu T_0}\right]$$

then

$$\left|B_u^{(k)}(u,i)(t)-B_u^{(k)}(\bar{u},\bar{i})(t)\right|\le$$

$$\le e^{\mu(t-T-kT_0)}\hat{\rho}_\mu((u,\dot{u},i,\dot{i}),(\bar{u},\dot{\bar{i}},\dot{\bar{u}},\dot{\bar{i}}))\frac{e^{\mu_0}}{\mu^2\dot{\bar{C}}_1}\left[1+\left(2I_0+\sum_{n=1}^{m}\left|g_n^{(1)}\right|\left(U_0+U_{E_1}\right)^n e^{n\mu T_0}\right)H_1+\right.$$

$$\left.+\frac{1}{\hat{L}_1}+2\sum_{n=1}^{m}n\left|g_n^{(1)}\right|\left(U_0+U_{E_1}\right)^{n-1}e^{(n-1)\mu_0}\right]+$$

$$+e^{\mu(t-T-kT_0)}\hat{\rho}_\mu((u,\dot{u},i,\dot{i}),(\bar{u},\dot{\bar{i}},\dot{\bar{u}},\dot{\bar{i}}))\frac{e^{\mu_0}-1}{2}\frac{e^{\mu_0}}{\mu^2\dot{\bar{C}}_1}\left[1+\left(2I_0+\sum_{n=1}^{m}\left|g_n^{(1)}\right|\left(U_0+U_{E_1}\right)^n e^{n\mu_0}\right)H_1+\right.$$

$$+\frac{1}{\hat{L}_1}+\frac{1}{\dot{\overline{C}}_1}\sum_{n=1}^{m}n\left|g_n^{(1)}\right|\left(U_0+U_{E_1}\right)^{n-1}e^{(n-1)\mu_0}\Bigg]+$$

$$+e^{\mu(t-T-kT_0)}\hat{\rho}_\mu((u,\dot{u},i,\dot{i}),(\overline{u},\overline{i},\dot{\overline{u}},\dot{\overline{i}}))\frac{e^{\mu_0}-1}{\mu_0}\frac{e^{\mu_0}}{\mu^2\dot{\overline{C}}_1}\left[1+\left(2I_0+\sum_{n=1}^{m}\left|g_n^{(1)}\right|\left(U_0+U_{E_1}\right)^n e^{n\mu_0}\right)H_1+\right.$$

$$+\frac{1}{\hat{L}_1}+2\sum_{n=1}^{m}n\left|g_n^{(1)}\right|\left(U_0+U_{E_1}\right)^{n-1}e^{(n-1)\mu_0}\Bigg]\leq$$

$$\leq e^{\mu(t-T-kT_0)}\hat{\rho}_\mu((u,\dot{u},i,\dot{i}),(\overline{u},\overline{i},\dot{\overline{u}},\dot{\overline{i}}))\left(1+\frac{e^{\mu_0}-1}{2}+\frac{e^{\mu_0}-1}{\mu_0}\right)\frac{e^{\mu_0}}{\mu^2\hat{C}_1}\left[1+\left(2I_0+\sum_{n=1}^{m}\left|g_n^{(1)}\right|\phi_0^n e^{n\mu_0}\right)H_1+\right.$$

$$+\frac{1}{\hat{L}_1}+2\sum_{n=1}^{m}n\left|g_n^{(1)}\right|\left(\phi_0\right)^{n-1}e^{(n-1)\mu_0}\Bigg]\equiv e^{\mu(t-T-kT_0)}K_u\hat{\rho}_\mu((u,\dot{u},i,\dot{i}),(\overline{u},\overline{i},\dot{\overline{u}},\dot{\overline{i}}))\,.$$

Consequently

$$\hat{\rho}(B(u,i),B(\overline{u},\overline{i}))\leq e^{\mu_0}K_u\hat{\rho}_\mu((u,\dot{u},i,\dot{i}),(\overline{u},\overline{i},\dot{\overline{u}},\dot{\overline{i}}))\,.$$

We recall the following inequalities

$$\left|\widetilde{u}(t-T)\right|\leq U_0 e^{\mu T_0}\leq\phi_0\,;$$

$$\left|\widetilde{u}(t-T)-\overline{\widetilde{u}}(t-T)\right|\leq$$

$$\leq\frac{1}{2}e^{\frac{RT}{L}}\left|u(t)-\overline{u}(t)\right|+\frac{1}{2}e^{\frac{RT}{L}}Z_0\left|i(t)-\overline{i}(t)\right|\leq e^{\mu(t-T-kT_0)}\hat{\rho}_\mu((u,\dot{u},i,\dot{i}),(\overline{u},\overline{i},\dot{\overline{u}},\dot{\overline{i}}))e^{\frac{RT}{L}}\frac{(1+Z_0)}{2\mu}\,.$$

Further on we have

$$\left|B_i^{(k)}(u,i)(t)-B_i^{(k)}(\overline{u},\overline{i})(t)\right|\leq$$

$$\leq\int_{T+kT_0}^{t}\left|I(u,i)(s)-I(\overline{u},\overline{i})(s)\right|ds+$$

$$\left(\frac{t-T-kT_0}{T_0}-\frac{1}{2}\right)\left|\int_{T+kT_0}^{T+(k+1)T_0}I(u,i)(s)ds-\int_{T+kT_0}^{T+(k+1)T_0}I(\overline{u},\overline{i})(s)ds\right|+$$

$$+\frac{1}{T_0}\int_{T+kT_0}^{T+(k+1)T_0}\left|\int_{T+kT_0}^{t}I(u,i)(s)ds-\int_{T+kT_0}^{t}I(\overline{u},\overline{i})(s)ds\right|dt\equiv I_1+I_2+I_3\,.$$

We get

$$
I_1 \le \frac{1}{Z_0}\int_{T+kT_0}^{t}|\dot{u}(s)-\dot{\bar{u}}(s)|ds + e^{-\frac{2RT}{L}}\left|\int_{T+kT_0}^{t}\big(\dot{i}(s-2T)-\dot{\bar{i}}(s-2T)\big)ds\right| + \frac{e^{-\frac{2RT}{L}}}{Z_0}\left|\int_{T+kT_0}^{t}\big(\dot{u}(s-2T)-\dot{\bar{u}}(s-2T)\big)ds\right| +
$$

$$
+\frac{2e^{-\frac{RT}{L}}}{Z_0}\int_{T+kT_0}^{t}\left|\frac{\widetilde{L}_0^{-1}\left(\int_{T+kT_0}^{s}\widetilde{u}(\theta-T)d\theta\right)}{d\widetilde{C}_0(\widetilde{u}(s-T))/du} - \frac{\widetilde{L}_0^{-1}\left(\int_{T+kT_0}^{s}\overline{\widetilde{u}}(\theta-T)d\theta\right)}{d\widetilde{C}_0(\overline{\widetilde{u}}(s-T))/du}\right|ds +
$$

$$
+\frac{2e^{-\frac{RT}{L}}}{Z_0}\int_{T+kT_0}^{t}\left|\frac{G_0(\widetilde{u}(s-T))}{d\widetilde{C}_0(\widetilde{u}(s-T))/du} - \frac{G_0(\overline{\widetilde{u}}(s-T))}{d\widetilde{C}_0(\overline{\widetilde{u}}(s-T))/du}\right|ds \le
$$

$$
\le \frac{\rho_\mu^{(k)}(\dot{u},\dot{\bar{u}})}{Z_0}\frac{e^{\mu(t-T-kT_0)}-1}{\mu} + \rho_\mu^{(k)}(\dot{i},\dot{\bar{i}})e^{-\frac{2RT}{L}}\frac{e^{\mu(t-T-kT_0)}-1}{\mu} + \frac{\rho_\mu^{(k)}(\dot{u},\dot{\bar{u}})}{Z_0}e^{-\frac{2RT}{L}}\frac{e^{\mu(t-T-kT_0)}-1}{\mu} +
$$

$$
+\frac{2e^{-\frac{RT}{L}}}{Z_0\dot{\bar{C}}_0}\int_{T+kT_0}^{t}\left|L_0^{-1}\left(\int_{T+kT_0}^{s}\widetilde{u}(\theta-T)d\theta\right)-\widetilde{L}_0^{-1}\left(\int_{T+kT_0}^{s}\overline{\widetilde{u}}(\theta-T)d\theta\right)\right|ds +
$$

$$
+\frac{2e^{-\frac{RT}{L}}I_0}{Z_0}\int_{T+kT_0}^{t}\left|\frac{1}{d\widetilde{C}_0(\widetilde{u}(s-T))/du}-\frac{1}{d\widetilde{C}_0(\overline{\widetilde{u}}(s-T))/du}\right|ds +
$$

$$
+\frac{2e^{-\frac{RT}{L}}}{Z_0\dot{\bar{C}}_0}\int_{T+kT_0}^{t}\left|G_0(\widetilde{u}(s-T))-G_0(\overline{\widetilde{u}}(s-T))\right|ds +
$$

$$
+\frac{2e^{-\frac{RT}{L}}}{Z_0}\max\left|G_0(\overline{\widetilde{u}}(s-T))\right|\int_{T+kT_0}^{t}\left|\frac{1}{d\widetilde{C}_0(\widetilde{u}(s-T))/du}-\frac{1}{d\widetilde{C}_0(\overline{\widetilde{u}}(s-T))/du}\right|ds \le
$$

$$
\le \frac{e^{\mu(t-T-kT_0)}-1}{\mu}\left(\frac{\rho_\mu^{(k)}(\dot{u},\dot{\bar{u}})}{Z_0}+\rho_\mu^{(k)}(\dot{i},\dot{\bar{i}})e^{-\frac{2RT}{L}}+\frac{\rho_\mu^{(k)}(\dot{u},\dot{\bar{u}})}{Z_0}e^{-\frac{2RT}{L}}\right)+
$$

$$
+\frac{2e^{-\frac{RT}{L}}}{Z_0\dot{\bar{C}}_0\hat{L}_0}\int_{T+kT_0}^{t}\int_{T+kT_0}^{s}\left|\widetilde{u}(\theta-T)-\overline{\widetilde{u}}(\theta-T)\right|d\theta ds +
$$

$$
+\frac{2e^{-\frac{RT}{L}}I_0}{Z_0}\int_{T+kT_0}^{t}\sup\left\{\frac{1}{\left(d\widetilde{C}_0(\widetilde{u}(s-T))/du\right)^2}\left|\frac{d^2\widetilde{C}_0(\widetilde{u}(s-T))}{du^2}\right|\right\}\left|\widetilde{u}(s-T)-\overline{\widetilde{u}}(s-T)\right|ds +
$$

$$
+\frac{2e^{-\frac{RT}{L}}}{Z_0\dot{\bar{C}}_0}\sum_{n=1}^{m}n\left|g_n^{(0)}\right|\sup\left\{\left|\widetilde{u}(s-T)\right|^{n-1}\right\}\int_{T+kT_0}^{t}\left|\widetilde{u}(s-T)-\overline{\widetilde{u}}(s-T)\right|ds +
$$

$$+\frac{2e^{-\frac{RT}{L}}}{Z_0}\max\left|G_0(\overline{\overline{u}}(s-T))\right|\int\limits_{T+kT_0}^{t}\sup\left\{\frac{1}{\left(d\widetilde{C}_0(\widetilde{u}(s-T))/du\right)^2}\left|\frac{d^2\widetilde{C}_0(\widetilde{u}(s-T))}{du^2}\right|\right\}\left|\widetilde{u}(s-T)-\overline{\overline{u}}(s-T)\right|ds\le$$

$$\le\frac{e^{\mu(t-T-kT_0)}-1}{\mu}\left(\frac{1}{Z_0}+e^{-\frac{2RT}{L}}+\frac{1}{Z_0}e^{-\frac{2RT}{L}}\right)\hat{\rho}_\mu((u,\dot{u},i,\dot{i}),(\overline{u},\overline{i},\dot{\overline{u}},\dot{\overline{i}}))+$$

$$+\frac{2e^{-\frac{RT}{L}}}{Z_0\dot{\overline{C}}_0\hat{L}_0}e^{\frac{RT}{L}}\frac{(1+Z_0)}{\mu}\frac{e^{\mu(t-T-kT_0)}-1}{\mu^2}\hat{\rho}_\mu((u,\dot{u},i,\dot{i}),(\overline{u},\overline{i},\dot{\overline{u}},\dot{\overline{i}}))+$$

$$+\frac{2e^{-\frac{RT}{L}}I_0}{Z_0\dot{\overline{C}}_0^2}H_0e^{\frac{RT}{L}}\frac{(1+Z_0)}{\mu}\frac{e^{\mu(t-T-kT_0)}-1}{\mu}\hat{\rho}_\mu((u,\dot{u},i,\dot{i}),(\overline{u},\overline{i},\dot{\overline{u}},\dot{\overline{i}}))+$$

$$+\frac{2e^{-\frac{RT}{L}}}{Z_0\dot{\overline{C}}_0}e^{\frac{RT}{L}}\frac{(1+Z_0)}{\mu}\frac{e^{\mu(t-T-kT_0)}-1}{\mu}\hat{\rho}_\mu((u,\dot{u},i,\dot{i}),(\overline{u},\overline{i},\dot{\overline{u}},\dot{\overline{i}}))\sum_{n=1}^{m}n\left|g_n^{(0)}\right|\phi_0^{n-1}+$$

$$+\frac{2e^{-\frac{RT}{L}}}{Z_0\dot{\overline{C}}_0^2}H_0e^{\frac{RT}{L}}\frac{(1+Z_0)}{\mu}\frac{e^{\mu(t-T-kT_0)}-1}{\mu}\hat{\rho}_\mu((u,\dot{u},i,\dot{i}),(\overline{u},\overline{i},\dot{\overline{u}},\dot{\overline{i}}))\sum_{n=1}^{m}\left|g_n^{(0)}\right|\phi_0^{n}\le$$

$$\le e^{\mu(t-T-kT_0)}\hat{\rho}_\mu((u,\dot{u},i,\dot{i}),(\overline{u},\overline{i},\dot{\overline{u}},\dot{\overline{i}}))\left[\frac{1}{\mu}\left(\frac{1}{Z_0}+e^{-\frac{2RT}{L}}+\frac{1}{Z_0}e^{-\frac{2RT}{L}}\right)+\right.$$

$$\left.+2\frac{1+Z_0}{\mu^2Z_0\dot{\overline{C}}_0}\left(\frac{1}{\mu\hat{L}_0}+\frac{I_0H_0}{\dot{\overline{C}}_0}+\sum_{n=1}^{m}n\left|g_n^{(0)}\right|\phi_0^{n-1}+\frac{H_0}{\dot{\overline{C}}_0}\sum_{n=1}^{m}\left|g_n^{(0)}\right|\phi_0^{n}\right)\right]$$

and

$$I_2\le\frac{1}{2Z_0}\left|\int\limits_{T+kT_0}^{T+(k+1)T_0}\left(\dot{u}(s)-\dot{\overline{u}}(s)\right)ds\right|+\frac{1}{2}e^{-\frac{2RT}{L}}\left|\int\limits_{T+kT_0}^{T+(k+1)T_0}\left(\dot{i}(s-2T)-\dot{\overline{i}}(s-2T)\right)ds\right|+$$

$$+\frac{e^{-\frac{2RT}{L}}}{2Z_0}\left|\int\limits_{T+kT_0}^{T+(k+1)T_0}\left(\dot{u}(s-2T)-\dot{\overline{u}}(s-2T)\right)ds\right|+$$

$$+\frac{e^{-\frac{RT}{L}}}{Z_0}\int\limits_{T+kT_0}^{T+(k+1)T_0}\left|\frac{\widetilde{L}_0^{-1}\left(\int\limits_{T+kT_0}^{s}\widetilde{u}(\theta-T))d\theta\right)}{d\widetilde{C}_0(\widetilde{u}(s-T))/du}-\frac{\widetilde{L}_0^{-1}\left(\int\limits_{T+kT_0}^{s}\overline{\widetilde{u}}(\theta-T)d\theta\right)}{d\widetilde{C}_0(\overline{\widetilde{u}}(s-T))/du}\right|ds+$$

$$+\frac{e^{-\frac{RT}{L}}}{Z_0}\int_{T+kT_0}^{T+(k+1)T_0}\left|\frac{G_0(\widetilde{u}(s-T))}{d\widetilde{C}_0(\widetilde{u}(s-T))/du}-\frac{G_0(\overline{\overline{u}}(s-T))}{d\widetilde{C}_0(\overline{\overline{u}}(s-T))/du}\right|ds\le$$

$$\le\frac{e^{-\frac{RT}{L}}}{Z_0\dot{\overline{C}}_0}\int_{T+kT_0}^{T+(k+1)T_0}\left|\widetilde{L}_0^{-1}\left(\int_{T+kT_0}^{s}\widetilde{u}(\theta-T)d\theta\right)-\widetilde{L}_0^{-1}\left(\int_{T+kT_0}^{s}\overline{\overline{u}}(\theta-T)d\theta\right)\right|ds+$$

$$+\frac{e^{-\frac{RT}{L}}I_0}{Z_0}\int_{T+kT_0}^{T+(k+1)T_0}\left|\frac{1}{d\widetilde{C}_0(\widetilde{u}(s-T))/du}-\frac{1}{d\widetilde{C}_0(\overline{\overline{u}}(s-T))/du}\right|ds+$$

$$+\frac{e^{-\frac{RT}{L}}}{Z_0\dot{\overline{C}}_0}\int_{T+kT_0}^{T+(k+1)T_0}\left|G_0(\widetilde{u}(s-T))-G_0(\overline{\overline{u}}(s-T))\right|ds+$$

$$+\frac{e^{-\frac{RT}{L}}}{Z_0}\max\left|G_0(\overline{\overline{u}}(s-T))\right|\int_{T+kT_0}^{T+(k+1)T_0}\left|\frac{1}{d\widetilde{C}_0(\widetilde{u}(s-T))/du}-\frac{1}{d\widetilde{C}_0(\overline{\overline{u}}(s-T))/du}\right|ds\le$$

$$\le\frac{e^{-\frac{RT}{L}}}{Z_0\dot{\overline{C}}_0\hat{L}_0}e^{\frac{RT}{L}}\frac{(1+Z_0)}{\mu}\int_{T+kT_0}^{T+(k+1)T_0}\int_{T+kT_0}^{s}e^{\mu(\theta-T-kT_0)}d\theta ds\hat{\rho}_\mu((u,\dot{u},i,\dot{i}),(\overline{u},\overline{i},\dot{\overline{u}},\dot{\overline{i}}))+$$

$$+\frac{e^{-\frac{RT}{L}}}{Z_0\dot{\overline{C}}_0^2}H_0\sum_{n=1}^{m}\left|g_n^{(0)}\right|(\phi_0)^n\int_{T+kT_0}^{T+(k+1)T_0}\left|\widetilde{u}(s-T)-\overline{\overline{u}}(s-T)\right|ds\le$$

$$\le e^{\mu(t-T-kT_0)}\frac{e^{\mu_0}-1}{2}2\left[\frac{1+Z_0}{\mu^2Z_0\dot{\overline{C}}_0}\left(\frac{1}{\mu\hat{L}_0}+\frac{I_0H_0}{\dot{\overline{C}}_0}+\sum_{n=1}^{m}n\left|g_n^{(0)}\right|\phi_0^{n-1}+\frac{H_0}{\dot{\overline{C}}_0}\sum_{n=1}^{m}\left|g_n^{(0)}\right|\phi_0^n\right)\right].$$

For the third summand we obtain

$$I_3\le\frac{1}{T_0}\int_{T+kT_0}^{T+(k+1)T_0}\int_{T+kT_0}^{t}\left|I(u,i)(s)-I(\overline{u},\overline{i})(s)\right|dsdt\le$$

$$\le e^{\mu(t-T-kT_0)}\frac{e^{\mu_0}-1}{\mu_0}\hat{\rho}_\mu((u,\dot{u},i,\dot{i}),(\overline{u},\overline{i},\dot{\overline{u}},\dot{\overline{i}}))\left[\frac{1}{\mu}\left(\frac{1}{Z_0}+e^{-\frac{2RT}{L}}+\frac{1}{Z_0}e^{-\frac{2RT}{L}}\right)+\right.$$

$$\left.+2\frac{1+Z_0}{\mu^2Z_0\dot{\overline{C}}_0}\left(\frac{1}{\mu\hat{L}_0}+\frac{I_0H_0}{\dot{\overline{C}}_0}+\sum_{n=1}^{m}n\left|g_n^{(0)}\right|\phi_0^{n-1}+\frac{H_0}{\dot{\overline{C}}_0}\sum_{n=1}^{m}\left|g_n^{(0)}\right|\phi_0^n\right)\right].$$

Therefore

$$\left| B_i^{(k)}(u,i)(t) - B_i^{(k)}(\overline{u},\overline{i})(t) \right| \le I_1 + I_2 + I_3 \le$$

$$\le e^{\mu(t-T-kT_0)} \hat{\rho}_\mu((u,\dot{u},i,\dot{i}),(\overline{u},\overline{i},\dot{\overline{u}},\dot{\overline{i}})) \left[\frac{1}{\mu}\left(\frac{1}{Z_0} + e^{-\frac{2RT}{L}} + \frac{1}{Z_0} e^{-\frac{2RT}{L}} \right) + \right.$$

$$\left. + 2\frac{1+Z_0}{\mu^2 Z_0 \dot{\overline{C}}_0}\left(\frac{1}{\mu \hat{L}_0} + \frac{I_0 H_0}{\dot{\overline{C}}_0} + \sum_{n=1}^{m} n \left| g_n^{(0)} \right| \phi_0^{n-1} + \frac{H_0}{\dot{\overline{C}}_0} \sum_{n=1}^{m} \left| g_n^{(0)} \right| \phi_0^n \right) \right] +$$

$$+ e^{\mu(t-T-kT_0)} \frac{e^{\mu_0}-1}{2} 2\left[\frac{1+Z_0}{\mu^2 Z_0 \dot{\overline{C}}_0}\left(\frac{1}{\mu \hat{L}_0} + \frac{I_0 H_0}{\dot{\overline{C}}_0} + \sum_{n=1}^{m} n \left| g_n^{(0)} \right| \phi_0^{n-1} + \frac{H_0}{\dot{\overline{C}}_0} \sum_{n=1}^{m} \left| g_n^{(0)} \right| \phi_0^n \right) \right] +$$

$$+ e^{\mu(t-T-kT_0)} \frac{e^{\mu_0}-1}{\mu_0} \hat{\rho}_\mu((u,\dot{u},i,\dot{i}),(\overline{u},\overline{i},\dot{\overline{u}},\dot{\overline{i}})) \left[\frac{1}{\mu}\left(\frac{1}{Z_0} + e^{-\frac{2RT}{L}} + \frac{1}{Z_0} e^{-\frac{2RT}{L}} \right) + \right.$$

$$\left. + 2\frac{1+Z_0}{\mu^2 Z_0 \dot{\overline{C}}_0}\left(\frac{1}{\mu \hat{L}_0} + \frac{I_0 H_0}{\dot{\overline{C}}_0} + \sum_{n=1}^{m} n \left| g_n^{(0)} \right| \phi_0^{n-1} + \frac{H_0}{\dot{\overline{C}}_0} \sum_{n=1}^{m} \left| g_n^{(0)} \right| \phi_0^n \right) \right] \le$$

$$\le e^{\mu_0} \hat{\rho}_\mu((u,\dot{u},i,\dot{i}),(\overline{u},\overline{i},\dot{\overline{u}},\dot{\overline{i}})) \left[\left(1 + \frac{e^{\mu_0}-1}{\mu_0} \right) \frac{1}{\mu}\left(\frac{1}{Z_0} + e^{-\frac{2RT}{L}} + \frac{1}{Z_0} e^{-\frac{2RT}{L}} \right) + \right.$$

$$\left. + \left(1 + \frac{e^{\mu_0}-1}{2} + \frac{e^{\mu_0}-1}{\mu_0} \right) 2\frac{1+Z_0}{\mu^2 Z_0 \dot{\overline{C}}_0}\left(\frac{1}{\mu \hat{L}_0} + \frac{I_0 H_0}{\dot{\overline{C}}_0} + \sum_{n=1}^{m} n \left| g_n^{(0)} \right| \phi_0^{n-1} + \frac{H_0}{\dot{\overline{C}}_0} \sum_{n=1}^{m} \left| g_n^{(0)} \right| \phi_0^n \right) \right] \equiv$$

$$\equiv e^{\mu_0} K_i \hat{\rho}_\mu((u,\dot{u},i,\dot{i}),(\overline{u},\overline{i},\dot{\overline{u}},\dot{\overline{i}})).$$

It follows

$$\hat{\rho}(B_i(u,i), B_i(\overline{u},\overline{i})) \le e^{\mu_0} K_i \, \hat{\rho}_\mu((u,\dot{u},i,\dot{i}),(\overline{u},\overline{i},\dot{\overline{u}},\dot{\overline{i}})).$$

Finally we have to obtain an estimate for $t \in [T + kT_0, T + (k+1)T_0]$ of the derivatives.

$$\left| \dot{B}_u^{(k)}(u,i)(t) - \dot{B}_u^{(k)}(\overline{u},\overline{i})(t) \right| \le$$

$$\le \left| U(u,i)(t) - U(\overline{u},\overline{i})(t) \right| + \frac{1}{T_0}\left| \int_{T+kT_0}^{T+(k+1)T_0} \left(U(u,i)(s) - U(\overline{u},\overline{i})(s) \right) ds \right| \equiv \dot{U}_1 + \dot{U}_2.$$

We get

$$\dot{U}_1 \leq \frac{\left|i(t)-\bar{i}(t)\right|}{\left|d\widetilde{C}_1(u(t)-E_1(t))/du\right|} + \left|\frac{\bar{i}(t)}{d\widetilde{C}_1(u(t)-E_1(t))/du} - \frac{\bar{i}(t)}{d\widetilde{C}_1(\bar{u}(t)-E_1(t))/du}\right| +$$

$$+ \left|\frac{\widetilde{L}_1^{-1}\left(\int\limits_{T+kT_0}^{t}(u(\tau)-E_1(\tau))d\tau\right)}{d\widetilde{C}_1(u(s)-E_1(s))/du} - \frac{\widetilde{L}_1^{-1}\left(\int\limits_{T+kT_0}^{t}(\bar{u}(\tau)-E_1(\tau))d\tau\right)}{d\widetilde{C}_1(u(s)-E_1(s))/du}\right| +$$

$$+ \left|\widetilde{L}_1^{-1}\left(\int\limits_{T+kT_0}^{t}(\bar{u}(\tau)-E_1(\tau))d\tau\right)\right|\left|\frac{1}{d\widetilde{C}_1(u(t)-E_1(t))/du} - \frac{1}{d\widetilde{C}_1(\bar{u}(t)-E_1(t))/du}\right| +$$

$$+ \left|\frac{G_1(u(t)-E_1(t))}{d\widetilde{C}_1(\bar{u}(t)-E_1(t))/du} - \frac{G_1(\bar{u}(t)-E_1(t))}{d\widetilde{C}_1(\bar{u}(t)-E_1(t))/du}\right| + \left|\frac{G_1(\bar{u}(t)-E_1(t))}{d\widetilde{C}_1(\bar{u}(t)-E_1(t))/du} - \frac{G_1(\bar{u}(t)-E_1(t))}{d\widetilde{C}_1(\bar{u}(t)-E_1(t))/du}\right| \leq$$

$$\leq e^{\mu(t-T-kT_0)}\frac{\rho_\mu^{(k)}(i,\bar{i})}{\overset{\cdot}{\widetilde{C}}_1} + \frac{\left|\bar{i}(s)\right|}{\left(d\widetilde{C}_1(\bar{u}(t)-E_1(t))/du\right)^2}\left|\frac{d^2\widetilde{C}_1}{du^2}\right|\left|u(s)-\bar{u}(s)\right| +$$

$$+ \frac{1}{\hat{L}_1\overset{\cdot}{\widetilde{C}}_1}\left|\int\limits_{T+kT_0}^{t}(u(\tau)-\bar{u}(\tau))d\tau\right| + \frac{I_0}{\left(d\widetilde{C}_1(\bar{u}(t)-E_1(t))/du\right)^2}\left|\frac{d^2\widetilde{C}_1}{du^2}\right|\left|u(t)-\bar{u}(t)\right| +$$

$$+ \frac{1}{\overset{\cdot}{\widetilde{C}}_1}\sum_{n=1}^{m}n\left|g_n^1\right|\left(U_0+U_{E_1}\right)^{n-1}e^{(n-1)\mu(t-T-kT_0)}\left|u(t)-\bar{u}(t)\right| +$$

$$+ \sup\left\{\left|G_1(u(t)-E_1(t))\right| : t\in[T+kT_0,T+(k+1)T_0]\right\}\frac{1}{\left(d\widetilde{C}_1(\bar{u}(t)-E_1(t))/du\right)^2}\left|\frac{d^2\widetilde{C}_1}{du^2}\right|\left|u(t)-\bar{u}(t)\right| \leq$$

$$\leq e^{\mu(t-T-kT_0)}\frac{\rho_\mu^{(k)}(i,\bar{i})}{\overset{\cdot}{\widetilde{C}}_1} + e^{\mu(t-T-kT_0)}\frac{I_0 H_1}{\overset{\cdot}{\widetilde{C}}_1^2}\rho_\mu^{(k)}(u,\bar{u}) +$$

$$+ e^{\mu(t-T-kT_0)}\frac{\rho_\mu^{(k)}(u,\bar{u})}{\mu\hat{L}_1\overset{\cdot}{\widetilde{C}}_1} + e^{\mu(t-T-kT_0)}\frac{I_0 H_1}{\overset{\cdot}{\widetilde{C}}_1^2}\rho_\mu^{(k)}(u,\bar{u}) +$$

$$+ e^{\mu(t-T-kT_0)}\rho_\mu^{(k)}(u,\bar{u})\frac{1}{\hat{C}_1}\sum_{n=1}^{m}n\left|g_n^{(1)}\right|\left(U_0+U_{E_1}\right)^{n-1}e^{(n-1)\mu T_0} +$$

$$+ e^{\mu(t-T-kT_0)}\rho_\mu^{(k)}(u,\bar{u})\frac{H_1}{\overset{\cdot}{\widetilde{C}}_1^2}\sum_{n=1}^{m}\left|g_n^{(1)}\right|\left(U_0+U_{E_1}\right)^{n}e^{n\mu T_0} \leq$$

$$\leq e^{\mu(t-T-kT_0)}\frac{\rho_\mu^{(k)}(i,\bar{i})}{\mu\overset{\cdot}{\widetilde{C}}_1} + e^{\mu(t-T-kT_0)}\frac{\rho_\mu^{(k)}(\dot{u},\overset{\cdot}{\bar{u}})}{\mu}\frac{I_0 H_1}{\overset{\cdot}{\widetilde{C}}_1^2} +$$

$$+ e^{\mu(t-T-kT_0)} \frac{\rho_\mu^{(k)}(\dot{u},\dot{\overline{u}})}{\mu^2 \hat{L}_1 \dot{\overline{C}}_1} + e^{\mu(t-T-kT_0)} \frac{I_0 H_1}{\dot{\overline{C}}_1^2} \frac{\rho_\mu^{(k)}(\dot{u},\dot{\overline{u}})}{\mu} +$$

$$+ e^{\mu(t-T-kT_0)} \frac{1}{\dot{\overline{C}}_1} \frac{\rho_\mu^{(k)}(\dot{u},\dot{\overline{u}})}{\mu} \sum_{n=1}^{m} n \left| g_n^{(1)} \right| \left(U_0 + U_{E_1} \right)^{n-1} e^{(n-1)\mu T_0} +$$

$$+ e^{\mu(t-T-kT_0)} \frac{H_1}{\dot{\overline{C}}_1^2} \frac{\rho_\mu^{(k)}(\dot{u},\dot{\overline{u}})}{\mu} \sum_{n=1}^{m} \left| g_n^{(1)} \right| \left(U_0 + U_{E_1} \right)^{n} e^{n\mu T_0} \le$$

$$\le e^{\mu(t-T-kT_0)} \hat{\rho}_\mu((u,\dot{u},i,\dot{i}),(\overline{u},\overline{i},\dot{\overline{u}},\dot{\overline{i}})) \frac{e^{\mu_0}}{\mu \dot{\overline{C}}_1} \left[1 + \frac{H_1}{\dot{\overline{C}}_1} \left(2I_0 + \sum_{n=1}^{m} \left| g_n^{(1)} \right| \phi_0^n e^{n\mu_0} \right) + \frac{1}{\hat{L}_1} + \sum_{n=1}^{m} n \left| g_n^{(1)} \right| \phi_0^{n-1} e^{(n-1)\mu_0} \right]$$

and

$$\dot{U}_2 \le \frac{1}{T_0} \left| \int_{T+kT_0}^{T+(k+1)T_0} \left(U(u,i)(s) - U(\overline{u},\overline{i})(s) \right) ds \right| \le$$

$$\le e^{\mu(t-T-kT_0)} \hat{\rho}_\mu((u,\dot{u},i,\dot{i}),(\overline{u},\overline{i},\dot{\overline{u}},\dot{\overline{i}})) \frac{e^{\mu_0}-1}{2\mu_0} \frac{e^{\mu_0}}{\mu \dot{\overline{C}}_1} \left(1 + \left(2I_0 + \sum_{n=1}^{m} \left| g_n^{(1)} \right| \phi_0^n e^{n\mu_0} \right) H_1 + \right.$$

$$\left. + \frac{1}{\hat{L}_1} + \sum_{n=1}^{m} n \left| g_n^{(1)} \right| \phi_0^{n-1} e^{(n-1)\mu_0} \right)$$

then

$$\left| \dot{B}_u^{(k)}(u,i)(t) - \dot{B}_u^{(k)}(\overline{u},\overline{i})(t) \right| \le$$

$$\le e^{\mu(t-T-kT_0)} \hat{\rho}_\mu((u,\dot{u},i,\dot{i}),(\overline{u},\overline{i},\dot{\overline{u}},\dot{\overline{i}})) \frac{e^{\mu_0}}{\mu \dot{\overline{C}}_1} \left[1 + \frac{H_1}{\dot{\overline{C}}_1} \left(2I_0 + \sum_{n=1}^{m} \left| g_n^{(1)} \right| \phi_0^n e^{n\mu_0} \right) + \frac{1}{\hat{L}_1} + \sum_{n=1}^{m} n \left| g_n^{(1)} \right| \phi_0^{n-1} e^{(n-1)\mu_0} \right] +$$

$$+ e^{\mu(t-T-kT_0)} \hat{\rho}_\mu((u,\dot{u},i,\dot{i}),(\overline{u},\overline{i},\dot{\overline{u}},\dot{\overline{i}})) \frac{e^{\mu_0}-1}{2\mu_0} \frac{e^{\mu_0}}{\mu \dot{\overline{C}}_1} \left(1 + \left(2I_0 + \sum_{n=1}^{m} \left| g_n^{(1)} \right| \phi_0^n e^{n\mu_0} \right) H_1 + \frac{1}{\hat{L}_1} + \sum_{n=1}^{m} n \left| g_n^{(1)} \right| \phi_0^{n-1} e^{(n-1)\mu_0} \right) \le$$

$$\le e^{\mu(t-T-kT_0)} \hat{\rho}_\mu((u,\dot{u},i,\dot{i}),(\overline{u},\overline{i},\dot{\overline{u}},\dot{\overline{i}})) \left(1 + \frac{e^{\mu_0}-1}{2\mu_0} \right) \frac{e^{\mu_0}}{\mu \dot{\overline{C}}_1} \left[1 + \left(2I_0 + \sum_{n=1}^{m} \left| g_n^{(1)} \right| \phi_0^n e^{n\mu_0} \right) H_1 + \frac{1}{\hat{L}_1} + \sum_{n=1}^{m} n \left| g_n^{(1)} \right| \phi_0^{n-1} e^{(n-1)\mu_0} \right] \equiv$$

$$\equiv e^{\mu(t-T-kT_0)} \dot{K}_u \hat{\rho}_\mu((u,\dot{u},i,\dot{i}),(\overline{u},\overline{i},\dot{\overline{u}},\dot{\overline{i}})) .$$

It follows

$$\rho_\mu^{(k)}(\dot{B}_u^{(k)}(u,i),\dot{B}_u^{(k)}(\overline{u},\overline{i})) \le \dot{K}_u \hat{\rho}_\mu((u,\dot{u},i,\dot{i}),(\overline{u},\overline{i},\dot{\overline{u}},\dot{\overline{i}})).$$

Finally we have

$$\left| \dot{B}_i^{(k)}(u,i)(t) - \dot{B}_i^{(k)}(\overline{u},\overline{i})(t) \right| \le$$

$$\le \left| I(u,i)(t) - I(\overline{u},\overline{i})(t) \right| + \frac{1}{T_0} \left| \int_{T+kT_0}^{T+(k+1)T_0} \left(I(u,i)(s) - I(\overline{u},\overline{i})(s) \right) ds \right| \equiv \dot{I}_1 + \dot{I}_2.$$

Since

$$\dot{I}_1 \le \frac{1}{Z_0} \left| \dot{u}(t) - \dot{\overline{u}}(t) \right| + e^{-\frac{2RT}{L}} \left| \dot{i}(t-2T) - \dot{\overline{i}}(t-2T) \right| + \frac{e^{-\frac{2RT}{L}}}{Z_0} \left| \dot{u}(t-2T) - \dot{\overline{u}}(t-2T) \right| +$$

$$+ \frac{2e^{-\frac{RT}{L}}}{Z_0} \left| \frac{\widetilde{L}_0^{-1}\left(\int_{T+kT_0}^{t} \widetilde{u}(\theta-T)d\theta \right)}{d\widetilde{C}_0(\widetilde{u}(t-T))/du} - \frac{\widetilde{L}_0^{-1}\left(\int_{T+kT_0}^{t} \overline{\widetilde{u}}(\theta-T)d\theta \right)}{d\widetilde{C}_0(\overline{\widetilde{u}}(t-T))/du} \right| +$$

$$+ \frac{2e^{-\frac{RT}{L}}}{Z_0} \left| \frac{G_0(\widetilde{u}(t-T))}{d\widetilde{C}_0(\widetilde{u}(t-T))/du} - \frac{G_0(\overline{\widetilde{u}}(t-T))}{d\widetilde{C}_0(\overline{\widetilde{u}}(t-T))/du} \right| \le$$

$$\le e^{\mu(t-T-kT_0)} \frac{\rho_\mu^{(k)}(\dot{u},\dot{\overline{u}})}{Z_0} + \frac{2e^{-\frac{RT}{L}}}{Z_0 \dot{\overline{C}}_0} \left| \widetilde{L}_0^{-1}\left(\int_{T+kT_0}^{t} \widetilde{u}(\theta-T)d\theta \right) - \widetilde{L}_0^{-1}\left(\int_{T+kT_0}^{t} \overline{\widetilde{u}}(\theta-T)d\theta \right) \right| +$$

$$+ \frac{2e^{-\frac{RT}{L}} I_0}{Z_0} \left| \frac{1}{d\widetilde{C}_0(\widetilde{u}(t-T))/du} - \frac{1}{d\widetilde{C}_0(\overline{\widetilde{u}}(t-T))/du} \right| +$$

$$\frac{2e^{-\frac{RT}{L}}}{Z_0 \dot{\overline{C}}_0} \left| G_0(\widetilde{u}(t-T)) - G_0(\overline{\widetilde{u}}(t-T)) \right| +$$

$$+ \frac{2e^{-\frac{RT}{L}}}{Z_0} \max\left| G_0(\widetilde{u}(t-T)) \right| \left| \frac{1}{d\widetilde{C}_0(\widetilde{u}(t-T))/du} - \frac{1}{d\widetilde{C}_0(\overline{\widetilde{u}}(t-T))/du} \right| \le$$

$$\leq e^{\mu(t-T-kT_0)}\frac{\rho_\mu^{(k)}(\dot{u},\dot{\bar{u}})}{Z_0}+\frac{2e^{-\frac{RT}{L}}}{Z_0\dot{\bar{C}}_0\hat{L}_0}\int_{T+kT_0}^{t}\left|\widetilde{u}(\theta-T)-\overline{\widetilde{u}}(\theta-T)\right|d\theta+$$

$$+\frac{2e^{-\frac{RT}{L}}I_0}{Z_0}\sup\left\{\frac{1}{\left(d\widetilde{C}_0(\widetilde{u}(t-T))/du\right)^2}\left|\frac{d^2\widetilde{C}_0(\widetilde{u}(s-T))}{du^2}\right|\right\}\left|\widetilde{u}(t-T)-\overline{\widetilde{u}}(t-T)\right|+$$

$$+\frac{2e^{-\frac{RT}{L}}}{Z_0\dot{\bar{C}}_0}\sum_{n=1}^{m}n\left|g_n^{(0)}\right|\sup\left\{\left|\widetilde{u}(s-T)\right|^{n-1}\right\}\left|\widetilde{u}(t-T)-\overline{\widetilde{u}}(t-T)\right|+$$

$$+\frac{2e^{-\frac{RT}{L}}}{Z_0}\max\left|G_0(\overline{\widetilde{u}}(s-T))\right|\sup\left\{\frac{1}{\left(d\widetilde{C}_0(\widetilde{u}(t-T))/du\right)^2}\left|\frac{d^2\widetilde{C}_0(\widetilde{u}(s-T))}{du^2}\right|\right\}\left|\widetilde{u}(t-T)-\overline{\widetilde{u}}(t-T)\right|\leq$$

$$\leq e^{\mu(t-T-kT_0)}\frac{1}{Z_0}\hat{\rho}_\mu((u,\dot{u},i,\dot{i}),(\overline{u},\overline{i},\dot{\overline{u}},\dot{\overline{i}}))+$$

$$+\frac{2e^{-\frac{RT}{L}}}{Z_0\dot{\bar{C}}_0\hat{L}_0}e^{\frac{RT}{L}}\frac{(1+Z_0)}{\mu}e^{\mu(t-T-kT_0)}\hat{\rho}_\mu((u,\dot{u},i,\dot{i}),(\overline{u},\overline{i},\dot{\overline{u}},\dot{\overline{i}}))+$$

$$+e^{\mu(t-T-kT_0)}\hat{\rho}_\mu((u,\dot{u},i,\dot{i}),(\overline{u},\overline{i},\dot{\overline{u}},\dot{\overline{i}}))\frac{2e^{-\frac{RT}{L}}I_0H_0}{Z_0\dot{\bar{C}}_0^2}e^{\frac{RT}{L}}\frac{(1+Z_0)}{\mu}+$$

$$+e^{\mu(t-T-kT_0)}\hat{\rho}_\mu((u,\dot{u},i,\dot{i}),(\overline{u},\overline{i},\dot{\overline{u}},\dot{\overline{i}}))\frac{2e^{-\frac{RT}{L}}}{Z_0\dot{\bar{C}}_0}e^{\frac{RT}{L}}\frac{(1+Z_0)}{\mu}\sum_{n=1}^{m}n\left|g_n^{(0)}\right|\phi_0^{n-1}+$$

$$+e^{\frac{RT}{L}}\frac{(1+Z_0)}{\mu}e^{\mu(t-T-kT_0)}\hat{\rho}_\mu((u,\dot{u},i,\dot{i}),(\overline{u},\overline{i},\dot{\overline{u}},\dot{\overline{i}}))\frac{2e^{-\frac{RT}{L}}H_0}{Z_0\dot{\bar{C}}_0^2}\sum_{n=1}^{m}\left|g_n^{(0)}\right|\phi_0^n\leq$$

$$\leq e^{\mu(t-T-kT_0)}\hat{\rho}_\mu((u,\dot{u},i,\dot{i}),(\overline{u},\overline{i},\dot{\overline{u}},\dot{\overline{i}}))\left[\frac{1}{Z_0}+\frac{2(1+Z_0)}{\mu Z_0\dot{\bar{C}}_0}\left(\frac{1}{\hat{L}_0}+\frac{I_0H_0}{\dot{\bar{C}}_0}+\sum_{n=1}^{m}n\left|g_n^{(0)}\right|\phi_0^{n-1}+\frac{H_0}{\dot{\bar{C}}_0}\sum_{n=1}^{m}\left|g_n^{(0)}\right|\phi_0^n\right)\right]$$

and

$$\dot{i}_2\leq\frac{1}{T_0}\left[\frac{1}{2Z_0}\left|\int_{T+kT_0}^{T+(k+1)T_0}\left(\dot{u}(s)-\dot{\overline{u}}(s)\right)ds\right|+\frac{1}{2}e^{-\frac{2RT}{L}}\left|\int_{T+kT_0}^{T+(k+1)T_0}\left(\dot{i}(s-2T)-\dot{\overline{i}}(s-2T)\right)ds\right|+\right.$$

$$+\frac{e^{-\frac{2RT}{L}}}{2Z_0}\left|\int_{T+kT_0}^{T+(k+1)T_0}\left(\dot{u}(s-2T)-\dot{\overline{u}}(s-2T)\right)ds\right|+$$

$$
+\frac{e^{-\frac{RT}{L}}}{Z_0}\int\limits_{T+kT_0}^{T+(k+1)T_0}\left|\frac{\widetilde{L}_0^{-1}\left(\int\limits_{T+kT_0}^{s}\widetilde{u}(\theta-T)d\theta\right)}{d\widetilde{C}_0(\widetilde{u}(t-T))/du}-\frac{\widetilde{L}_0^{-1}\left(\int\limits_{T+kT_0}^{s}\overline{\widetilde{u}}(\theta-T)d\theta\right)}{d\widetilde{C}_0(\overline{\widetilde{u}}(t-T))/du}\right|ds+
$$

$$
+\frac{e^{-\frac{RT}{L}}}{Z_0}\int\limits_{T+kT_0}^{T+(k+1)T_0}\left|\frac{G_0(\widetilde{u}(s-T))}{d\widetilde{C}_0(\widetilde{u}(t-T))/du}-\frac{G_0(\overline{\widetilde{u}}(s-T))}{d\widetilde{C}_0(\overline{\widetilde{u}}(t-T))/du}\right|ds\right]\le
$$

$$
\le\frac{1}{T_0}\left[\frac{e^{-\frac{RT}{L}}}{Z_0\dot{\overline{C}}_0}\int\limits_{T+kT_0}^{T+(k+1)T_0}\left|\widetilde{L}_0^{-1}\left(\int\limits_{T+kT_0}^{s}\widetilde{u}(\theta-T)(\theta)d\theta\right)-\widetilde{L}_0^{-1}\left(\int\limits_{T+kT_0}^{s}\overline{\widetilde{u}}(\theta-T)(\theta)d\theta\right)\right|ds+\right.
$$

$$
+\frac{e^{-\frac{RT}{L}}I_0}{Z_0}\int\limits_{T+kT_0}^{T+(k+1)T_0}\left|\frac{1}{d\widetilde{C}_0(\widetilde{u}(t-T))/du}-\frac{1}{d\widetilde{C}_0(\overline{\widetilde{u}}(t-T))/du}\right|ds+
$$

$$
+\frac{e^{-\frac{RT}{L}}}{Z_0\dot{\overline{C}}_0}\int\limits_{T+kT_0}^{T+(k+1)T_0}\left|G_0(\widetilde{u}(s-T))-G_0(\overline{\widetilde{u}}(s-T))\right|ds+
$$

$$
\left.+\frac{e^{-\frac{RT}{L}}}{Z_0}\max\left|G_0(\overline{\widetilde{u}}(s-T)(s))\right|\int\limits_{T+kT_0}^{T+(k+1)T_0}\left|\frac{1}{d\widetilde{C}_0(\widetilde{u}(t-T))/du}-\frac{1}{d\widetilde{C}_0(\overline{\widetilde{u}}(t-T))/du}\right|ds\right]\le
$$

$$
\le\frac{e^{-\frac{RT}{L}}}{Z_0\dot{\overline{C}}_0\hat{L}_0}e^{\frac{RT}{L}}\frac{(1+Z_0)}{\mu T_0}\hat{\rho}_\mu((u,\dot{u},i,\dot{i}),(\overline{u},\overline{i},\dot{\overline{u}},\dot{\overline{i}}))\int\limits_{T+kT_0}^{T+(k+1)T_0}\frac{e^{\mu(s-T-kT_0)}}{\mu}ds+
$$

$$
+\frac{e^{-\frac{RT}{L}}I_0}{Z_0\dot{\overline{C}}_0^2}H_0e^{\frac{RT}{L}}\frac{(1+Z_0)}{\mu T_0}\hat{\rho}_\mu((u,\dot{u},i,\dot{i}),(\overline{u},\overline{i},\dot{\overline{u}},\dot{\overline{i}}))\int\limits_{T+kT_0}^{T+(k+1)T_0}e^{\mu(s-T-kT_0)}ds+
$$

$$
+\frac{e^{-\frac{RT}{L}}}{Z_0\dot{\overline{C}}_0}e^{\frac{RT}{L}}\frac{(1+Z_0)}{\mu T_0}\hat{\rho}_\mu((u,\dot{u},i,\dot{i}),(\overline{u},\overline{i},\dot{\overline{u}},\dot{\overline{i}}))\int\limits_{T+kT_0}^{T+(k+1)T_0}e^{\mu(s-T-kT_0)}ds\sum_{n=1}^{m}n\left|g_n^{(0)}\right|\phi_0^{n-1}+
$$

$$
+\frac{e^{-\frac{RT}{L}}}{Z_0\dot{\overline{C}}_0^2}H_0e^{\frac{RT}{L}}\frac{(1+Z_0)}{\mu T_0}\hat{\rho}_\mu((u,\dot{u},i,\dot{i}),(\overline{u},\overline{i},\dot{\overline{u}},\dot{\overline{i}}))\int\limits_{T+kT_0}^{T+(k+1)T_0}e^{\mu(s-T-kT_0)}ds\sum_{n=1}^{m}\left|g_n^{(0)}\right|\phi_0^{n}\le
$$

$$
\le\hat{\rho}_\mu((u,\dot{u},i,\dot{i}),(\overline{u},\overline{i},\dot{\overline{u}},\dot{\overline{i}}))\frac{e^{\mu_0}-1}{\mu_0}\frac{(1+Z_0)}{\mu^2Z_0\dot{\overline{C}}_0\hat{L}_0}+
$$

$$+ \hat{\rho}_\mu((u,\dot{u},i,\dot{i}),(\overline{u},\overline{i},\dot{\overline{u}},\dot{\overline{i}}))\frac{e^{\mu_0}-1}{\mu_0}I_0 H_0 \frac{(1+Z_0)}{\mu Z_0 \dot{\overline{C}}_0^2}+$$

$$+ \hat{\rho}_\mu((u,\dot{u},i,\dot{i}),(\overline{u},\overline{i},\dot{\overline{u}},\dot{\overline{i}}))\frac{e^{\mu_0}-1}{\mu_0}\frac{(1+Z_0)}{\mu Z_0 \dot{\overline{C}}_0}\sum_{n=1}^{m} n\left|g_n^{(0)}\right|\phi_0^{n-1}+$$

$$+ \hat{\rho}_\mu((u,\dot{u},i,\dot{i}),(\overline{u},\overline{i},\dot{\overline{u}},\dot{\overline{i}}))\frac{e^{\mu_0}-1}{\mu_0}H_0 \frac{(1+Z_0)}{\mu Z_0 \dot{\overline{C}}_0^2}\sum_{n=1}^{m}\left|g_n^{(0)}\right|\phi_0^{n} \le$$

$$\le \hat{\rho}_\mu((u,\dot{u},i,\dot{i}),(\overline{u},\overline{i},\dot{\overline{u}},\dot{\overline{i}}))\frac{e^{\mu_0}-1}{\mu_0}\frac{1+Z_0}{\mu Z_0 \dot{\overline{C}}_0}\left[\frac{1}{\mu \hat{L}_0}+\frac{I_0 H_0}{\dot{\overline{C}}_0}+\sum_{n=1}^{m}n\left|g_n^{(0)}\right|\phi_0^{n-1}+\frac{H_0}{\dot{\overline{C}}_0}\sum_{n=1}^{m}\left|g_n^{(0)}\right|\phi_0^{n}\right].$$

Therefore

$$\left|\dot{B}_i(u,i)(t)-\dot{B}_i(\overline{u},\overline{i})(t)\right| \le$$

$$\le e^{\mu(t-T-kT_0)}\hat{\rho}_\mu((u,\dot{u},i,\dot{i}),(\overline{u},\overline{i},\dot{\overline{u}},\dot{\overline{i}}))\left[\frac{1}{Z_0}+\frac{2(1+Z_0)}{\mu Z_0 \dot{\overline{C}}_0}\left(\frac{1}{\mu \hat{L}_0}+\frac{I_0 H_0}{\dot{\overline{C}}_0}+\sum_{n=1}^{m}n\left|g_n^{(0)}\right|\phi_0^{n-1}+\frac{H_0}{\dot{\overline{C}}_0}\sum_{n=1}^{m}\left|g_n^{(0)}\right|\phi_0^{n}\right)\right]+$$

$$+ \hat{\rho}_\mu((u,\dot{u},i,\dot{i}),(\overline{u},\overline{i},\dot{\overline{u}},\dot{\overline{i}}))\frac{e^{\mu_0}-1}{\mu_0}\frac{1+Z_0}{\mu Z_0 \dot{\overline{C}}_0}\left[\frac{1}{\mu \hat{L}_0}+\frac{I_0 H_0}{\dot{\overline{C}}_0}+\sum_{n=1}^{m}n\left|g_n^{(0)}\right|\phi_0^{n-1}+\frac{H_0}{\dot{\overline{C}}_0}\sum_{n=1}^{m}\left|g_n^{(0)}\right|\phi_0^{n}\right] \le$$

$$\le e^{\mu(t-T-kT_0)}\hat{\rho}_\mu((u,\dot{u},i,\dot{i}),(\overline{u},\overline{i},\dot{\overline{u}},\dot{\overline{i}}))\times$$

$$\times\left[\frac{1}{Z_0}+\left(2+\frac{e^{\mu_0}-1}{\mu_0}\right)\frac{1+Z_0}{\mu Z_0 \dot{\overline{C}}_0}\left(\frac{1}{\mu \hat{L}_0}+\sum_{n=1}^{m}n\left|g_n^{(0)}\right|\phi_0^{n-1}+\frac{H_0}{\dot{\overline{C}}_0}\left(I_0+\sum_{n=1}^{m}\left|g_n^{(0)}\right|\phi_0^{n}\right)\right)\right]\equiv$$

$$\equiv e^{\mu(t-T-kT_0)}\dot{K}_i \hat{\rho}_\mu((u,\dot{u},i,\dot{i}),(\overline{u},\overline{i},\dot{\overline{u}},\dot{\overline{i}})).$$

It follows $\rho_\mu^{(k)}(\dot{B}_i^{(k)}(u,i),\dot{B}_i^{(k)}(\overline{u},\overline{i})) \le \dot{K}_i \hat{\rho}_\mu((u,\dot{u},i,\dot{i}),(\overline{u},\overline{i},\dot{\overline{u}},\dot{\overline{i}}))$.

Denote by $K = \max\left\{e^{\mu_0}K_u, \dot{K}_u, e^{\mu_0}K_i, \dot{K}_i\right\}$.

Then

$$\hat{\rho}_\mu\left(B_u(u,i),\dot{B}_u(u,i),B_i(u,i),\dot{B}_i(u,i),B_u(\overline{u},\overline{i}),\dot{B}_u(\overline{u},\overline{i}),B_i(\overline{u},\overline{i}),\dot{B}_i(\overline{u},\overline{i})\right)\le K\hat{\rho}_\mu((u,\dot{u},i,\dot{i}),(\overline{u},\overline{i},\dot{\overline{u}},\dot{\overline{i}})).$$

Therefore B is contractive operator and has a unique fixed point in $M_u \times M_i$ (cf. [14]).
With accordance to Lemma 6.3.7 its fixed point is a periodic solution of (6.3.1).
   Thus Theorem 6.4.1 is thus proved.

6.5. NUMERICAL EXAMPLE

We collect all inequalities used in the proof of Theorem 6.4.1:

$$\left(1+\frac{e^{\mu_0}-1}{\mu_0}\right)\frac{U_0}{Z_0}<I_0,\ U_0+U_{E_1}\leq\phi_0;\ \left(1+\frac{e^{\mu_0}-1}{\mu_0}\right)U_{E_1}<U_0;$$

$$\cosh\left(\frac{RT}{L}\right)e^{-\beta}\left(U_0+Z_0I_0\right)+U_{E_0}\leq U_0e^{\mu_0}\leq\phi_0;$$

$$\left(1+\frac{e^{\mu_0}-1}{\mu_0}\right)U_{E_1}+\left(1+\frac{e^{\mu_0}-1}{2}+\frac{e^{\mu_0}-1}{\mu_0}\right)\frac{1}{\mu\dot{\bar{C}}_1}\left(2I_0+\sum_{n=1}^{m}\left|g_n^{(1)}\right|\phi_0^n e^{(n-1)\mu_0}\right)\leq U_0;$$

$$\left(1+\frac{e^{\mu_0}-1}{\mu_0}\right)\frac{1}{Z_0}\left(2e^{-\frac{RT}{L}}U_{E_0}+U_0+e^{-\frac{2RT}{L}}Z_0I_0+e^{-\frac{2RT}{L}}U_0\right)+$$

$$+\left(1+\frac{e^{\mu_0}-1}{2}+\frac{e^{\mu_0}-1}{\mu_0}\right)\frac{e^{-\frac{RT}{L}}}{\mu\dot{\bar{C}}Z_0}\left(I_0+\sum_{n=1}^{m}\left|g_n^{(0)}\right|\phi_0^n\right)\leq I_0;$$

$$e^{\mu_0}K_u=e^{\mu_0}\left(1+\frac{e^{\mu_0}-1}{2}+\frac{e^{\mu_0}-1}{\mu_0}\right)\frac{e^{\mu_0}}{\mu^2\hat{C}_1}\left[1+\left(2I_0+\sum_{n=1}^{m}\left|g_n^{(1)}\right|\phi_0^n e^{n\mu_0}\right)H_1+\right.$$

$$\left.+\frac{1}{\hat{L}_1}+2\sum_{n=1}^{m}n\left|g_n^{(1)}\right|\phi_0^{n-1}e^{(n-1)\mu_0}\right]<1;$$

$$e^{\mu_0}K_i=e^{\mu_0}\left[\left(1+\frac{e^{\mu_0}-1}{\mu_0}\right)\frac{1}{\mu}\left(\frac{1}{Z_0}+e^{-\frac{2RT}{L}}+\frac{1}{Z_0}e^{-\frac{2RT}{L}}\right)+\right.$$

$$\left.+\left(1+\frac{e^{\mu_0}-1}{2}+\frac{e^{\mu_0}-1}{\mu_0}\right)2\frac{1+Z_0}{\mu^2 Z_0\dot{\bar{C}}_0}\left(\frac{1}{\mu\hat{L}_0}+\frac{I_0H_0}{\dot{\bar{C}}_0}+\sum_{n=1}^{m}n\left|g_n^{(0)}\right|\phi_0^{n-1}+\frac{H_0}{\dot{\bar{C}}_0}\sum_{n=1}^{m}\left|g_n^{(0)}\right|\phi_0^n\right)\right]<1;$$

$$\dot{K}_u=\left(1+\frac{e^{\mu_0}-1}{2\mu_0}\right)\frac{e^{\mu_0}}{\mu\dot{\bar{C}}_1}\left[1+\left(2I_0+\sum_{n=1}^{m}\left|g_n^{(1)}\right|\phi_0^n e^{n\mu_0}\right)H_1+\frac{1}{\hat{L}_1}+\sum_{n=1}^{m}n\left|g_n^{(1)}\right|\phi_0^{n-1}e^{(n-1)\mu_0}\right]<1;$$

$$\dot{K}_i=\frac{1}{Z_0}+\left(2+\frac{e^{\mu_0}-1}{\mu_0}\right)\frac{1+Z_0}{\mu Z_0\dot{\bar{C}}_0}\left(\frac{1}{\mu\hat{L}_0}+\sum_{n=1}^{m}n\left|g_n^{(0)}\right|\phi_0^{n-1}+\frac{H_0}{\dot{\bar{C}}_0}\left(I_0+\sum_{n=1}^{m}\left|g_n^{(0)}\right|\phi_0^n\right)\right)<1.$$

For a transmission line with length $\Lambda = 100\, m$, cross-section area $S = 4\, mm^2$, specific resistance for the copper $\rho_c = 0,0175$, the resistance per-unit length is

$$R = \frac{\rho_c \Lambda}{S} = \frac{0,0175.100}{4} \approx 0,44\, \Omega. \text{ If } L = 0,45\, \mu H/m,\ C = 80\, pF/m,$$

$$v = \frac{1}{\sqrt{LC}} = \frac{1}{\sqrt{0,45.10^{-6}.80.10^{-12}}} = \frac{1}{6.10^{-9}} = 1,66.10^8.$$

Then $\quad \dfrac{R}{L} = \dfrac{0,44}{0,45.10^{-6}} \approx 10^6,\ Z_0 = 75\ \Omega, \quad T = \Lambda\sqrt{LC} = 100.6.10^{-9} = 6.10^{-7}\, s;$

$$\frac{R\Lambda}{Z_0} = \frac{RT}{L} = 10^6.6.10^{-7} = 0,6; \quad e^{-\frac{RT}{L}} = e^{-0,6}.$$

We choose a resistive element with the following *V-I* characteristic

$$G(u) = G_0(u) = G_1(u) = 0,02u - 0,12u^2 + 0,14u^3,$$

i.e. $g_1 = 0,02,\, g_2 = -0,12;\, g_3 = 0,1.$

Let us check the propagation of waves with $\lambda_0 = (1/6)10^{-4}\, m$. Then

$$f_0 = \frac{1}{\lambda_0 \sqrt{LC}} = \frac{1}{(1/6)10^{-4}.6.10^{-9}} = 10^{13}\, Hz \Rightarrow T_0 = \frac{1}{f_0} = 10^{-13}\, \text{sec.};$$

$$\frac{RT_0}{L} = 10^6.10^{-13} = 10^{-7}.$$

If we choose $\mu = 10^{13}$, then $\mu T_0 = \mu_0 = 1$ and $T = 6.10^{-7}.10^{13}T_0 = 10^6.T_0$.

We also have $\mu T = 10^{13}.6.10^{-7} = 6.10^6 \Rightarrow e^{-\mu T} \approx 0$.

Let us consider the case $\widetilde{C}_0(u) = \widetilde{C}_1(u) = \widetilde{C}(u)$, that is, $c_0 = c_1,\ \Phi = \Phi_0 = \Phi_1$ where

$$\frac{2c_0 \sqrt[h]{\Phi_0}}{h} \frac{\Phi_0 + \dfrac{h-1}{2h}\phi_0}{(\Phi_0 - \phi_0)^{\frac{2h+1}{h}}} = H_0. \text{ The } pn\text{-junction capacity } c_0 \in [0,05.10^{-12}\, F;\ 50.10^{-12}\, F],$$

while the pn-junction potentials $\Phi_0, \Phi_1 \in [0,4V;\ 0,9V]$. For $h = 2$ we get

$$H_0 = c_0 \sqrt{\Phi_0}\, \frac{\Phi_0 + 0,25\phi_0}{(\Phi_0 - \phi_0)^2 \sqrt{\Phi_0 - \phi_0}} \quad \text{for} \quad -\phi_0 \le u \le \phi_0\ (\phi_0 < \Phi) \quad \text{and if we choose}$$

$c_0 = 50\, pF = 5.10^{-11}\, F$ and $\Phi = 0,4$ and $\phi_0 = 0,3$, then

$$H_0 = H_1 = c_0 \sqrt{\Phi_0} \, \frac{\Phi_0 + 0,25\phi_0}{\left(\Phi_0 - \phi_0\right)^2 \sqrt{\Phi_0 - \phi_0}} = 10^{-11} 1,5.10^3 = 1,5.10^{-8}$$

The minimal values are

$$\min\left\{ \widetilde{C}_0(u) : u \in [-\phi_0, \phi_0] \right\} = \widetilde{C}_0(-0,1) = 5.10^{-11} \equiv \hat{C}_0 > 0;$$

$$\min \frac{d\widetilde{C}_0(.)}{du} = \frac{d\widetilde{C}_0(-\phi_0)}{du} = c_0 \sqrt[h]{\Phi_0} \, \frac{\Phi_0 + \frac{h-1}{h}\phi_0}{\left(\Phi_0 + \phi_0\right)^{\frac{h+1}{h}}} = 5.10^{-11}\sqrt{0,4} \, \frac{0,4 + 0,5.0,3}{(0,7)^{\frac{3}{2}}} \approx 3.10^{-11} \equiv \dot{\widetilde{C}}_0 > 0 \cdot$$

Then $\mu \dot{\widetilde{C}}_0 = 10^{13}.3.10^{-11} \approx 3.10^2; \quad \mu^2 \dot{\widetilde{C}}_0 = 10^{26}.10^{-11}.3 \approx 3.10^{15}$.

We have to estimate

$$\widetilde{L}_p = \widetilde{L}(i) = i.L(i) = i(2 - 0,1i) = 2i - 10^{-1} i^2 \ , i \in [-i_0, i_0] \ (p = 0,1).$$

Recall that $u = \dfrac{d\left(\widetilde{L}(i)\right)}{dt}$ and consequently $\widetilde{L}(i) = \displaystyle\int_T^t u(s)ds \Rightarrow i = \widetilde{L}^{-1}\left(\int_T^t u(s)ds \right)$.

We choose i_0 such that for $|i| \le i_0$ it follows $\dfrac{d\overline{L}(i)}{di} = 2 - 0,2.i > 0$, that is, $|i| \le i_0 < \sqrt{10}$. Therefore the inverse function $\widetilde{L}^{-1}(.)$ exists and

$$\widetilde{L}^{-1}(\lambda) : \left[-2i_0 - 0,1.i_0^2 \ ; \ 2i_0 - 0,1.i_0^2\right] \to [-i_0, i_0] \ .$$

The explicit form of the inverse function can be found solving the following equation with respect to i: $2i - 0,1i^2 = \lambda$. We need, however, just the estimates:

$$\left| \widetilde{L}^{-1}(.) \right| \le i_0 \le I_0 \ (k = 0,1).$$

We can choose $i_0 = I_0 = 0,01 < \sqrt{10}$;

$$\left| \frac{d\widetilde{L}^{-1}(\lambda)}{d\lambda} \right| = \frac{1}{\left| \dfrac{d\widetilde{L}(i)}{di} \right|} \le \frac{1}{\min\left\{ \left| \dfrac{d\widetilde{L}(i)}{di} \right| : i \in [-i_0; i_0] \right\}} = \frac{1}{\min\left\{ |2 - 0,2.i| : i \in [-0,01; 0,01] \right\}} =$$

$$= \frac{1}{|2 - 0,2.0,01|} = \frac{1}{1,998} \approx \frac{1}{2} = \frac{1}{\hat{L}} \quad \Rightarrow \quad \frac{1}{\mu\hat{L}} \approx \frac{1}{2.10^{13}} \approx 0.$$

Then for $U_0 = 10^{-4}$, $I_0 = 10^{-4}$ and $U_{E_1} = 10^{-5}$ the above inequalities are satisfied.

Most often the initial approximation is chosen to be the solution of the linearized equations. In some cases, however, even for linear systems the solution is not easily found. That is why, we chose simple trigonometric functions as an initial approximation, namely:

$$u^{(0)}(t) = U_0 \sin \omega_0 t, t \in [-T,T], \ i^{(0)}(t) = I_0 \sin \omega_0 t, t \in [-T,T] \ (\omega_0 = 2\pi / T_0),$$

$$E_p(t) = U_{E_p} \sin \omega_0 t \ (p = 0,1).$$

We calculate the first approximation:

$$u^{(1)}(t) = B_u^{(k)}(u^{(0)},i^{(0)}) = \int_{T+kT_0}^{t} U(u^{(0)},i^{(0)})(s)ds - \left(\frac{t-T-kT_0}{T_0} - \frac{1}{2}\right) \int_{T+kT_0}^{T+(k+1)T_0} U(u^{(0)},i^{(0)})(s)ds -$$

$$- \frac{1}{T_0} \int_{T+kT_0}^{T+(k+1)T_0} \int_{T+kT_0}^{t} U(u^{(0)},i^{(0)})(s)ds\,dt =$$

$$= \int_{T+kT_0}^{t} \frac{dE_1(s)}{ds}ds + \int_{T+kT_0}^{t} \frac{i^{(0)}(s)}{d\widetilde{C}_1(u(s)-E_1(s))/du}ds - \int_{T+kT_0}^{t} \frac{\widetilde{L}_1^{-1}\left(\int_{T+kT_0}^{s}\left(u^{(0)}(\tau)-E_1(\tau)\right)d\tau\right)}{d\widetilde{C}_1(u(s)-E_1(s))/du}ds -$$

$$- \int_{T+kT_0}^{t} \frac{G_1(u^{(0)}(s)-E_1(s))}{d\widetilde{C}_1(u(s)-E_1(s))/du}ds -$$

$$- \left(\frac{t-T-kT_0}{T_0} - \frac{1}{2}\right)\left[\int_{T+kT_0}^{T+(k+1)T_0} \frac{dE_1(s)}{ds}ds + \int_{T+kT_0}^{T+(k+1)T_0} \frac{i^{(0)}(s)}{d\widetilde{C}_1(u(s)-E_1(s))/du}ds - \right.$$

$$- \int_{T+kT_0}^{T+(k+1)T_0} \frac{\widetilde{L}_1^{-1}\left(\int_{T}^{s}\left(u^{(0)}(\tau)-E_1(\tau)\right)d\tau\right)}{d\widetilde{C}_1(u(s)-E_1(s))/du}dt - \int_{T+kT_0}^{T+(k+1)T_0} \frac{G_1(u^{(0)}(s)-E_1(s))}{d\widetilde{C}_1(u(s)-E_1(s))/du}ds \left.\right] -$$

$$- \frac{1}{T_0}\int_{T+kT_0}^{T+(k+1)T_0}\left(\int_{T+kT_0}^{t} \frac{dE_1(s)}{ds}ds + \int_{T+kT_0}^{t} \frac{i^{(0)}(s)}{d\widetilde{C}_1(u(s)-E_1(s))/du}ds - \right.$$

$$- \int_{T+kT_0}^{t} \frac{\widetilde{L}_1^{-1}\left(\int_{T+kT_0}^{s}\left(u^{(0)}(\theta)-E_1(\theta)\right)d\theta\right)}{d\widetilde{C}_1(u(s)-E_1(s))/du}ds - \int_{T+kT_0}^{t} \frac{G_1(u^{(0)}(s)-E_1(s))}{d\widetilde{C}_1(u(s)-E_1(s))/du}ds \left.\right)dt \,.$$

We could estimate all terms in the above approximation as for instance

$$\left| U_{E_1} \sin \omega_0 t \right| \le U_{E_1} \, ;$$

$$\left| \int_{T+kT_0}^{t} \frac{i^{(0)}(s)}{d\widetilde{C}_1(u(s)-E_1(s))/du} ds \right| \le I_0 \frac{e^{\mu_0}-1}{\mu \dot{\widetilde{C}}_1} = I_0 \frac{e-1}{10^{13}.3.10^{-11}} \, ;$$

$$\left| \int_{T+kT_0}^{t} \frac{\widetilde{L}_1^{-1}\left(\int_{T+kT_0}^{s}\left(u^{(0)}(\tau)-E_1(\tau)\right)d\tau \right)}{\widetilde{C}_1(u(s)-E_1(s))} ds \right| \le \frac{I_0\left(e^{\mu_0}-1\right)}{\mu \hat{C}_1} \approx \frac{I_0(e-1)}{10^{13}.3.10^{-11}} \, ;$$

$$\left| \int_{T+kT_0}^{t} \frac{G_1(u^{(0)}(s)-E_1(s))}{\widetilde{C}_1(u(s)-E_1(s))} ds \right| \le \frac{1}{\hat{C}_1}\sum_{n=1}^{3}\left|g_n^{(1)}\right| \left| \int_{T+kT_0}^{t}\left|U_{E_1}\sin\omega_0 s\right| ds \right|^n \le$$

$$\le \frac{1}{\hat{C}_1}\frac{e^{\mu(t-T-kT_0)}-1}{\mu}\sum_{n=1}^{3}\left|g_n^{(1)}\right|\left(U_{E_1}\right)^n \le \frac{e^{\mu T_0}-1}{\mu \hat{C}_1}\sum_{n=1}^{3}\left|g_n^{(1)}\right|\left(U_{E_1}\right)^n \approx \frac{e-1}{4,5.10^2}U_{E_1}\sum_{n=1}^{3}\left|g_n^{(1)}\right|\left(U_{E_1}\right)^{n-1}$$

and so on. Depending on the accuracy some members might be neglected which simplifies the first approximation.

In analogous way the small terms in the next approximations might be neglected after obvious estimates

$$i^{(1)}(t) = B_i^{(k)}(u^{(0)},i^{(0)}) = \int_{T+kT_0}^{t} I(u^{(0)},i^{(0)})(s)ds - \left(\frac{t-T-kT_0}{T_0} - \frac{1}{2} \right)\int_{T+kT_0}^{T+(k+1)T_0} I(u^{(0)},i^{(0)})(s)ds -$$

$$- \frac{1}{T_0}\int_{T+kT_0}^{T+(k+1)T_0}\int_{T+kT_0}^{t} I(u^{(0)},i^{(0)})(s)ds\,dt \, .$$

In view of the estimates in the proof of the last theorem we obtain

$$\left| u^{(1)}(t)-u^{(0)}(t)\right| \le 2U_0 e^{\mu_0} \, , \quad \left| i^{(1)}(t)-i^{(0)}(t)\right| \le 2I_0 e^{\mu_0} \, .$$

For the derivatives we have

$$\left| \frac{du^{(1)}(t)}{dt} - \frac{du^{(0)}(t)}{dt} \right| \le$$

$$\leq \left|\frac{dE_1(t)}{dt}\right| + \left|\frac{i^{(0)}(t)}{\widetilde{C}_1(u(s)-E_1(s))}\right| + \left|\frac{\widetilde{L}_1^{-1}\left(\int\limits_T^t \left(u^{(0)}(\tau)-E_1(\tau)\right)d\tau\right)}{\widetilde{C}_1(u(s)-E_1(s))}\right| + \left|\frac{G_1(u^{(0)}(t)-E_1(t))}{\widetilde{C}_1(u(s)-E_1(s))}\right| +$$

$$+\frac{1}{T_0}\left|\int\limits_{T+kT_0}^{T+(k+1)T_0}\frac{dE_1(s)}{ds}ds\right| + \frac{1}{T_0}\left|\int\limits_{T+kT_0}^{T+(k+1)T_0}\frac{i^{(0)}(s)}{d\widetilde{C}_1(u(s)-E_1(s))/du}ds\right| +$$

$$+\frac{1}{T_0}\int\limits_{T+kT_0}^{T+(k+1)T_0}\left|\frac{\widetilde{L}_1^{-1}\left(\int\limits_T^s\left(u^{(0)}(\tau)-E_1(\tau)\right)d\tau\right)}{d\widetilde{C}_1(u(s)-E_1(s))/du}\right|dt + \frac{1}{T_0}\int\limits_{T+kT_0}^{T+(k+1)T_0}\left|\frac{G_1(u^{(0)}(s)-E_1(s))}{d\widetilde{C}_1(u(s)-E_1(s))/du}\right|ds + \left|\ddot{u}^{(0)}(t)\right| \leq$$

$$\leq U_{E_1}\omega_0 + \frac{I_0 e^{\mu_0}}{\dot{\widetilde{C}}_1} + \frac{I_0}{\dot{\widetilde{C}}_1} + \frac{1}{\dot{\widetilde{C}}_1}\sum_{n=1}^3\left|g_n^{(0)}\right|\left|u^{(0)}(t)-E_1(t)\right|^n + \frac{I_0}{\dot{\widetilde{C}}_1 T_0}\left|\int\limits_{T+kT_0}^{T+(k+1)T_0}e^{\mu(s-T-kT_0)}ds\right| +$$

$$+\frac{I_0}{\dot{\widetilde{C}}_1 T_0}\int\limits_{T+kT_0}^{T+(k+1)T_0}e^{\mu(t-T-kT_0)}dt + \frac{1}{\dot{\widetilde{C}}_1 T_0}\sum_{n=1}^3\left|g_n^{(0)}\right|\int\limits_{T+kT_0}^{T+(k+1)T_0}\left|u^{(0)}(s)-E_1(s)\right|^n ds + U_0\omega_0 \leq$$

$$\leq \left(U_{E_1}+U_0\right)\omega_0 + \frac{I_0\left(e^{\mu_0}+1\right)}{\dot{\widetilde{C}}_1} + \frac{1}{\dot{\widetilde{C}}_1}\sum_{n=1}^3\left|g_n^{(0)}\right|\phi_0^n + \frac{I_0}{\dot{\widetilde{C}}_1 T_0}\frac{e^{\mu T_0}-1}{\mu} +$$

$$+\frac{I_0}{\dot{\widetilde{C}}_1 T_0}\frac{e^{\mu T_0}-1}{\mu} + \frac{1}{\dot{\widetilde{C}}_1 T_0}\sum_{n=1}^3\left|g_n^{(0)}\right|\phi_0^n\int\limits_{T+kT_0}^{T+(k+1)T_0}e^{\mu(s-T-kT_0)}ds \leq$$

$$\leq \left(U_{E_1}+U_0\right)\omega_0 + \frac{1}{\dot{\widetilde{C}}_1}\left(e^{\mu_0}I_0 + \left(1+\frac{e^{\mu_0}-1}{\mu_0}\right)\left(2I_0 + \sum_{n=1}^3\left|g_n^{(0)}\right|\phi_0^n\right)\right) \leq$$

$$\leq 10^{-1}\frac{2\pi}{10^{-13}} + \frac{e}{3.10^{-11}}\left(3.10^{-4}+0,006+0,0009+0,00027\right) \approx 2\pi.10^{12}$$

or $\rho_\mu^{(k)}\left(\ddot{u}^{(1)},\ddot{u}^{(0)}\right)\leq 2\pi.10^{12}$.

The order of the current derivatives is the same one $\rho_\mu^{(k)}\left(\ddot{i}^{(1)},\ddot{i}^{(0)}\right)\leq 2\pi.10^{12}$.

Finally we conclude that

$$\rho_\mu^{(k)}\left((u^{(1)},i^{(1)}),(u^{(0)},i^{(0)})\right)\leq 2\pi.10^{12}.$$

Therefore

$$\rho_\mu^{(k)}((u^{(n+1)}, i^{(n+1)}), (u^{(n)}, i^{(n)})) \le \frac{(0,326)^n}{1 - 0,326} 2\pi.10^{12}, (n = 0,1,2,\dots).$$

6.6. OSCILLATORY REGIMES FOR THE NEUTRAL SYSTEM

Here we deal with the problem of oscillating vanishing solutions of the same system (6.2.1):

$$\frac{du(t)}{dt} = \frac{dE_1(t)}{dt} + \frac{i(t)}{\dfrac{d\widetilde{C}_1(u(t) - E_1(t))}{du}} - \frac{\widetilde{L}_1^{-1}\left(\displaystyle\int_T^t (u(\tau) - E_1(\tau))d\tau\right)}{\dfrac{d\widetilde{C}_1(u(t) - E_1(t))}{du}} - \frac{G_1(u(t) - E_1(t))}{\dfrac{d\widetilde{C}_1(u(t) - E_1(t))}{du}}, t \in [T, \infty),$$

$$(6.6.1)$$

$$\frac{di(t)}{dt} = \frac{2}{Z_0} e^{-\frac{RT}{L}} \dot{E}_0(t - T) - \frac{1}{Z_0} \frac{du(t)}{dt} + e^{-2\frac{RT}{L}} \frac{di(t - 2T)}{dt} - e^{-2\frac{RT}{L}} \frac{1}{Z_0} \frac{du(t - 2T)}{dt} -$$

$$- \frac{2}{Z_0} e^{-\frac{RT}{L}} \frac{1}{\dfrac{d\widetilde{C}_0(\widetilde{u}(t - T))}{du}} \widetilde{L}_0^{-1}\left(\displaystyle\int_T^t \widetilde{u}(\tau - T)d\tau\right) - \frac{2}{Z_0} e^{-\frac{RT}{L}} \frac{1}{\dfrac{d\widetilde{C}_0(\widetilde{u}(t - T))}{du}} G_0(\widetilde{u}(t - T)), t \in [T, \infty)$$

$$u(t) = \upsilon_0(t), \quad \frac{du(t)}{dt} = \frac{d\upsilon_0(t)}{dt}, \quad i(t) = \iota_0(t), \quad \frac{di(t)}{dt} = \frac{d\iota_0(t)}{dt}, \quad t \in [-T, T].$$

Let us put $t_0 \equiv T$. Now we are able to formulate the main problem: to find an oscillatory solution of problem (6.7.1) with advanced prescribed zeros on an interval $[t_0, \infty)$, where $\upsilon_0(t)$ is a prescribed initial oscillating function on the interval $[-t_0, t_0]$.

Let $S_T = \{\tau_k\}_{k=0}^n, n \in N$ be the set of zeros of the initial function, that is, $U_0(\tau_k) = I_0(\tau_k) = 0$ such that $\tau_0 = -T$, $\tau_n = T \equiv t_0$. Besides $\max\{\tau_{k+1} - \tau_k : k = 0,1,\dots,n\} \le T_0$.

Let $S = \{t_k\}_{k=0}^\infty$ be a strictly increasing sequence of real numbers satisfying the following conditions (C):

(C1) $\displaystyle\lim_{k \to \infty} t_k = \infty$;

(C2) for every k $\left(t_k - T \ge \tau_0\right)$ there is $s < k$ such that $t_k - T = t_s$ where $t_s \in S_T \cup S$.

Condition (C2) implies

$$0 < \inf\{t_{k+1} - t_k : k = 0,1,2,\dots\} \le \sup\{t_{k+1} - t_k : k = 0,1,2,\dots\} = T_0 < \infty .$$

Introduce the set $C^1[t_0, \infty)$ consisting of all continuous functions whose derivatives are bounded and continuous on every interval $[t_k, t_{k+1}]$. We note the right and left derivatives at t_k of these functions might not coincide. That is why, we introduce below a topology of uniform convergence on every interval $[t_k, t_{k+1}]$ of the derivatives which needs introducing of uniform spaces (cf. [14]).).

Consider the sets

$$M_{SU} = \{u(.) \in C^1[t_0, \infty) : u(t_k) = 0 \ (k = 0,1,2,\dots)\},$$
$$M_{SI} = \{i(.) \in C^1[t_0, \infty) : i(t_k) = 0 \ (k = 0,1,2,\dots)\}$$

and

$$M_{SU}^* = \left\{ u(.) \in M_{SU} : |u(t)| \le U_0 e^{-\frac{R}{L}t}, t \in [t_0, \infty) \right\},$$

$$M_{SI}^* = \left\{ i(.) \in M_{SI} : |i(t)| \le I_0 e^{-\frac{R}{L}t}, t \in [t_0, \infty) \right\}$$

where U_0, I_0, μ are positive constants. Recall that $\mu T_0 = \mu_0 = \text{const.} > 0$.

Introduce the following family of pseudo-metrics

$$\rho^{(k)}(u, \overline{u}) = \max\{|u(t) - \overline{u}(t)| : t \in [t_k, t_{k+1}]\},$$
$$\hat{\rho}^{(k)}(u, \overline{u}) = \max\{|u(t) - \overline{u}(t)| : t \in [t_0, t_{k+1}]\},$$

$$\rho_\mu^{(k)}(u, \overline{u}) = \max\{e^{-\mu(t-t_k)}|u(t) - \overline{u}(t)| : t \in [t_k, t_{k+1}]\},$$
$$\hat{\rho}_\mu^{(k)}(u, \overline{u}) = \max\{\rho_\mu^{(0)}(u, \overline{u}), \rho_\mu^{(1)}(u, \overline{u}), \dots, \rho_\mu^{(k)}(u, \overline{u})\},$$

$$\rho_\mu^{(k)}(\dot{u}, \dot{\overline{u}}) = \max\{e^{-\mu(t-t_k)}|\dot{u}(t) - \dot{\overline{u}}(t)| : t \in [t_k, t_{k+1}]\},$$
$$\hat{\rho}_\mu^{(k)}(\dot{u}, \dot{\overline{u}}) = \max\{\rho_\mu^{(0)}(\dot{u}, \dot{\overline{u}}), \rho_\mu^{(1)}(\dot{u}, \dot{\overline{u}}), \dots, \rho_\mu^{(k)}(\dot{u}, \dot{\overline{u}})\},$$

$$\rho^{(k)}(i, \overline{i}) = \max\{|i(t) - \overline{i}(t)| : t \in [t_k, t_{k+1}]\},$$
$$\hat{\rho}^{(k)}(i, \overline{i}) = \max\{|i(t) - \overline{i}(t)| : t \in [t_0, t_{k+1}]\},$$

$$\rho_\mu^{(k)}(i,\bar{i}) = \max\left\{e^{-\mu(t-t_k)}\left|i(t)-\bar{i}(t)\right| : t \in [t_k, t_{k+1}]\right\},$$

$$\hat{\rho}_\mu^{(k)}(i,\bar{i}) = \max\left\{\rho_\mu^{(0)}(i,\bar{i}), \rho_\mu^{(1)}(i,\bar{i}), \dots, \rho_\mu^{(k)}(i,\bar{i})\right\},$$

$$\rho_\mu^{(k)}(\dot{i},\dot{\bar{i}}) = \max\left\{e^{-\mu(t-t_k)}\left|\dot{i}(t)-\dot{\bar{i}}(t)\right| : t \in [t_k, t_{k+1}]\right\},$$

$$\hat{\rho}_\mu^{(k)}(\dot{i},\dot{\bar{i}}) = \max\left\{\rho_\mu^{(0)}(\dot{i},\dot{\bar{i}}), \rho_\mu^{(1)}(\dot{i},\dot{\bar{i}}), \dots, \rho_\mu^{(k)}(\dot{i},\dot{\bar{i}})\right\}.$$

The set $M_{SU}^* \times M_{SI}^*$ turns out into a complete uniform space with respect to the countable saturated family of pseudo-metrics

$$\hat{\rho}_\mu^{(k)}((u,\dot{u},i,\dot{i}),(\bar{u},\dot{\bar{u}},\bar{i},\dot{\bar{i}})) = \max\left\{\hat{\rho}^{(k)}(u,i,\bar{u},\bar{i}), \hat{\rho}_\mu^{(k)}(\dot{u},\dot{\bar{u}}), \hat{\rho}_\mu^{(k)}(\dot{i},\dot{\bar{i}}) : k = 0,1,2,\dots \right\}.$$

Define an operator $B = (B_u(u,i), B_i(u,i))$ by the formulas

$$B_u^{(k)}(u,i)(t) := \int_{t_k}^{t} U(u,i)(s)\,ds - \frac{t-t_k}{t_{k+1}-t_k}\int_{t_k}^{t_{k+1}} U(u,i)(s)\,ds, \quad t \in [t_k, t_{k+1}],$$

$$B_i^{(k)}(u,i)(t) := \int_{t_k}^{t} I(u,i)(s)\,ds - \frac{t-t_k}{t_{k+1}-t_k}\int_{t_k}^{t_{k+1}} I(u,i)(s)\,ds, \quad t \in [t_k, t_{k+1}]$$

$(k = 0,1,2,\dots)$., where

$$U(u,i) = \frac{dE_1(t)}{dt} + \frac{i(t)}{\dfrac{d\tilde{C}_1(u(t)-E_1(t))}{du}} - \frac{\tilde{L}_1^{-1}\left(\displaystyle\int_T^t (u(\tau)-E_1(\tau))\,d\tau\right)}{\dfrac{d\tilde{C}_1(u(t)-E_1(t))}{du}} - \frac{G_1(u(t)-E_1(t))}{\dfrac{d\tilde{C}_1(u(t)-E_1(t))}{du}}, \quad t \in [T,\infty),$$

$$I(u,i) = \frac{2}{Z_0}e^{-\frac{RT}{L}}\dot{E}_0(t) - \frac{1}{Z_0}\frac{du(t)}{dt} + e^{-2\frac{RT}{L}}\frac{d\bar{i}(t)}{dt} - e^{-2\frac{RT}{L}}\frac{1}{Z_0}\frac{d\bar{u}(t)}{dt} -$$

$$- \frac{2}{Z_0}e^{-\frac{RT}{L}}\frac{1}{\dfrac{d\tilde{C}_0(\tilde{u}(t-T))}{du}}\tilde{L}_0^{-1}\left(\int_T^t \tilde{u}(\tau-T)\,d\tau\right) - \frac{2}{Z_0}e^{-\frac{RT}{L}}\frac{1}{\dfrac{d\tilde{C}_0(\tilde{u}(t-T))}{du}}G_0(\tilde{u}(t-T)), \quad t \in [T,\infty).$$

In order to avoid conformity condition **(CC)**, that is, the difficulty with the derivative continuity at the initial point we introduce functions

$$\bar{u}(t) = \begin{cases} \upsilon_0(t-2T), & t \in [T,3T] \\ u(t-2T), & t \in [3T,\infty) \end{cases}, \quad \bar{i}(t) = \begin{cases} \iota_0(t-2T), & t \in [T,3T] \\ i(t-2T), & t \in [3T,\infty) \end{cases}.$$

In other words on $[T,3T]$ we have the translated to the right initial functions, while for $[3T,\infty)$ we have the translated to the right functions $u(t)$ and $i(t)$. So we can consider the operator function to the right from T, where

$$\tilde{u}(t-T) = e^{\frac{RT}{L}}\frac{u(t)+Z_0\,i(t)}{2} - e^{-\frac{RT}{L}}\frac{Z_0\,\vec{i}(t)-\vec{u}(t)}{2} - E_0(t).$$

Further on we assume:

(IN) $\left|\upsilon_0(t)\right| \leq U_0 e^{-\frac{R}{L}(t+2T)}, \left|\iota_0(t)\right| \leq I_0 e^{-\frac{R}{L}(t+2T)}, t \in [-T,T] \equiv [-t_0,t_0].$

It follows

$$\left|\vec{\upsilon}_0(t)\right| = \left|\upsilon_0(t-2T)\right| \leq U_0 e^{-\frac{R}{L}t}, \left|\vec{\iota}_0(t)\right| = \left|\iota_0(t-2T)\right| \leq I_0 e^{-\frac{R}{L}t}, t \in [T,3T].$$

One can verify that $M_{SU}^* \times M_{SI}^*$ is closed subset of $C^1[t_0,\infty) \times C^1[t_0,\infty)$ with respect to the above family of pseudo-metrics (cf. K. Zima [126]).

Lemma 6.6.1. Let $E(.) \in C^1_{S_T \cup S}[0,\infty), \left|E(t)\right| \leq U_0 e^{\mu(t-t_k)}, t \in [t_k, t_{k+1}]$. Problem (6.6.1) has a solution $(u,i) \in M_{SU}^* \times M_{SI}^*$ iff the operator B has a fixed point in $M_{SU}^* \times M_{SI}^*$, that is,

$$(u,i) = (B_u(u,i), B_i(u,i)). \tag{6.6.2}$$

Proof: Let $(u,i) \in M_{SU}^* \times M_{SI}^*$ be a solution of (6.6.1). Then, integrating (6.6.1) on every interval $[t_k,t] \subset [t_k,t_{k+1}]$ $(k = 0,1,2 ...)$ we obtain

$$u(t) - u(t_k) = \int_{t_k}^t U(u,i)(s)ds \iff u(t) = \int_{t_k}^t U(u,i)(s)ds,$$

$$i(t) - i(t_k) = \int_{t_k}^t I(u,i)(s)ds \iff i(t) = \int_{t_k}^t I(u,i)(s)ds.$$

Consequently

$$u(t)=\int_{t_k}^{t}U(u,i)(s)ds \;\Rightarrow\; 0=u(t_{k+1})=\int_{t_k}^{t_{k+1}}U(u,i)(s)ds \;\Rightarrow\; \int_{t_k}^{t_{k+1}}U(u,i)(s)ds=0,$$

$$i(t)=\int_{t_k}^{t}I(u,i)(s)ds \;\Rightarrow\; 0=i(t_{k+1})=\int_{t_k}^{t_{k+1}}I(u,i)(s)ds \;\Rightarrow\; \int_{t_k}^{t_{k+1}}I(u,i)(s)ds=0. \qquad (6.6.3)$$

Then

$$u(t)=\int_{t_k}^{t}U(u,i)(s)ds\ , \quad i(t)=\int_{t_k}^{t}I(u,i)(s)ds$$

is equivalent to

$$u(t)=\int_{t_k}^{t}U(u,i)(s)ds-\frac{t-t_k}{t_{k+1}-t_k}\int_{t_k}^{t_{k+1}}U(u,i)(s)ds\,,t\in[t_k,t_{k+1}]$$

$$i(t)=\int_{t_k}^{t}I(u,i)(s)ds-\frac{t-t_k}{t_{k+1}-t_k}\int_{t_k}^{t_{k+1}}I(u,i)(s)ds\,,t\in[t_k,t_{k+1}]$$

and then the solution (u, i) of (6.6.1) is a fixed point of B.

Conversely, let $(u(.),i(.))\in M_{SU}^{*}\times M_{SI}^{*}$ be a fixed point of B, that is,

$$u=B_{u}^{(k)}(u,i)\,,\; i=B_{i}^{(k)}(u,i)\ ,\; t\in[t_k,t_{k+1}].$$

Then as in the proof of Theorem 4.5.1 one can prove that

$$\left|\int_{t_k}^{t_{k+1}}U(u,i)(t)dt\right|\le\frac{e^{\mu_0}-1}{2}\frac{1}{\mu^2\hat{C}_1}\left(2I_0+\sum_{n=1}^{m}\left|g_n^{(1)}\right|\left(U_{E_1}+U_0\right)^n e^{(n-1)\mu_0}\right)\Rightarrow\left|\int_{t_k}^{t_{k+1}}U(u,i)(t)dt\right|=0$$

and analogously $\left|\int_{t_k}^{t_{k+1}}I(u,i)(t)dt\right|=0$.

Therefore $u(t),i(t)$ satisfy

$$u(t)=\int_{t_k}^{t}U(u,i)(s)ds,\; t\in[t_k,t_{k+1}]\Leftrightarrow u=B_u(u,i),$$

$$i(t)=\int_{t_k}^{t}I(u,i)(s)ds,\; t\in[t_k,t_{k+1}]\Leftrightarrow i=B_i(u,i)$$

that is, $(u(t),i(t))$ is a solution of (6.6.1).

Lemma 6.6.1 is thus proved.

Theorem 6.6.1. Let the following conditions be fulfilled:

1) The initial functions $\upsilon_0(.), \iota_0(.) \in C^1[-T,T]$ satisfy

$$\left|\upsilon_0(t)\right| \le U_0 e^{-\frac{R}{L}(t+2T)}, \left|\iota_0(t)\right| \le I_0 e^{-\frac{R}{L}(t+2T)} \text{ for } t \in [-T,T] \equiv [-t_0,t_0], \text{ and}$$

$$\upsilon_0(-T) = \iota_0(-T) = 0 \ (\Rightarrow \upsilon_0(T) = \iota_0(T) = 0);$$

2) The source function

$$E(.) \in C_S^1[0,\infty), E(t-T) = E(t); \left|E(t)\right| \le E_0 e^{-\frac{R}{L}t}, t \in [t_0,\infty), E(t_k) = 0;$$

3) $\sinh\left(\dfrac{RT_0}{L}\right) e^{\frac{RT}{L}} \dfrac{2L}{R}\left(U_0 + Z_0 I_0 + U_{E_0}\right) \le 1; \dfrac{2e^{-\frac{RT}{L}} U_{E_0} + U_0}{Z_0} < I_0.$

Then there exists a unique oscillatory solution of the initial value problem (6.6.1), belonging to $M_{SU}^* \times M_{SI}^*$ on the interval $[T,3T]$.

Proof: We show that B maps $M_{SU}^* \times M_{SI}^*$ into itself.

First we notice that $B(u,i)(t)$ is continuous on $[t_0,\infty)$:

$$B_u^{(0)}(u,i)(t_0) = \int_{t_0}^{t_0} U(u,i)(s)ds - \frac{t_0 - t_0}{t_1 - t_0} \int_{t_0}^{t_1} U(u,i)(s)ds = 0,$$

$$\lim_{t \to t_{k+1}(t<t_{k+1})} B_u^{(k)}(u,i)(t) = \lim_{t \to t_{k+1}(t<t_{k+1})}\left(\int_{t_k}^{t} U(u,i)(s)ds - \frac{t-t_k}{t_{k+1}-t_k} \int_{t_k}^{t_{k+1}} U(u,i)(s)ds\right) = 0,$$

$$\lim_{t \to t_{k+1}(t>t_{k+1})} B_U^{(k+1)}(u)(t) =$$

$$= \lim_{t \to t_{k+1}(t>t_{k+1})}\left(\int_{t_{k+1}}^{t} U(u,i)(s)ds - \frac{t-t_{k+1}}{t_{k+2}-t_{k+1}} \int_{t_{k+1}}^{t_{k+2}} U(u,i)(s)ds\right) = 0,$$

$$B_i^{(0)}(u,i)(t_0) = \begin{cases} \displaystyle\int_{t_0}^{t_0} I(u,i)(s)ds - \frac{t_0 - t_0}{t_1 - t_0} \int_{t_0}^{t_1} I(u,i)(s)ds = 0 \\ \iota_0(t_0) = 0 \end{cases},$$

$$\lim_{t \to t_{k+1}(t<t_{k+1})} B_i^{(k)}(u,i)(t) = \lim_{t \to t_{k+1}(t<t_{k+1})}\left(\int_{t_k}^{t} I(u,i)(s)ds - \frac{t-t_k}{t_{k+1}-t_k} \int_{t_k}^{t_{k+1}} I(u,i)(s)ds\right) = 0,$$

$$\lim_{t \to t_{k+1}(t>t_{k+1})} B_i^{(k+1)}(u,i)(t) = \lim_{t \to t_{k+1}(t>t_{k+1})} \left(\int_{t_{k+1}}^{t} I(u,i)(s)ds - \frac{t-t_{k+1}}{t_{k+2}-t_{k+1}} \int_{t_{k+1}}^{t_{k+2}} I(u,i)(s)ds \right) = 0$$

and $\big(B_u(u,i)(t), B_i(u,i)(t)\big)$ is differentiable on every (t_k, t_{k+1}).

Besides it is easy to verify that

$$B_u(u,i)(t_k) = 0, \ B_u(u,i)(t_{k+1}) = 0, B_i(u,i)(t_k) = 0, \ B_i(u,i)(t_{k+1}) = 0 .$$

We have to establish that

$$\left| B_u^{(k)}(u,i)(t) \right| \le U_0 e^{\mu(t-t_k)}, \left| B_i^{(k)}(u,i)(t) \right| \le I_0 e^{\mu(t-t_k)} \ t \in [t_k, t_{k+1}] .$$

Indeed

$$\left| B_u^{(k)}(u,i)(t) \right| \le \left| \int_{t_k}^{t} U(u,i)(s)ds \right| + \int_{t_k}^{t_{k+1}} U(u,i)(s)ds \equiv U_1 + U_2 .$$

We get

$$U_1 \le \left| \int_{t_k}^{t} \dot{E}_1(s)ds \right| + \frac{1}{\widetilde{C}_1} \int_{t_k}^{t} |i(s)| ds + \frac{1}{\widetilde{C}_1} \left| \int_{t_k}^{t} \widetilde{L}_1^{-1} \left(\int_{t_k}^{t} (u(\tau) - E_1(\tau))d\tau \right) ds \right| + \frac{1}{\widetilde{C}_1} \int_{t_k}^{t} |G_1(u(s) - E_1(s))| ds \le$$

$$\le \left| E_1(t) \right| + \frac{I_0}{\widetilde{C}_1} \int_{t_k}^{t} e^{-\frac{R}{L}s} ds + \frac{I_0}{\widetilde{C}_1} \int_{t_k}^{t} e^{-\frac{R}{L}s} ds + \frac{I_0}{\widetilde{C}_1} \sum_{n=1}^{m} |g_n^{(1)}| \left| \int_{t_k}^{t} |u(s) - E_1(s)|^n ds \le$$

$$\le U_{E_1} e^{-\frac{R}{L}t} + \frac{I_0}{\widetilde{C}_1} \frac{L}{R} \left(e^{-\frac{R}{L}t_k} - e^{-\frac{R}{L}t} \right) + \frac{I_0}{\widetilde{C}_1} \frac{L}{R} \left(e^{-\frac{R}{L}t_k} - e^{-\frac{R}{L}t} \right) + \frac{I_0}{\widetilde{C}_1} \sum_{n=1}^{m} |g_n^{(1)}| (U_0 + U_{E_1})^n \int_{T+kT_0}^{t} e^{-n\frac{R}{L}t} ds \le$$

$$\le U_{E_1} e^{-\frac{R}{L}t} + \frac{I_0}{\widetilde{C}_1} \frac{L}{R} \left(e^{-\frac{R}{L}t_k} - e^{-\frac{R}{L}t} \right) + \frac{I_0}{\widetilde{C}_1} \frac{L}{R} \left(e^{-\frac{R}{L}t_k} - e^{-\frac{R}{L}t} \right) +$$

$$+ \frac{I_0}{\widetilde{C}_1} \sum_{n=1}^{m} |g_n^{(1)}| (U_0 + U_{E_1})^n \frac{L}{nR} \left(e^{-n\frac{R}{L}t_k} - e^{-n\frac{R}{L}t} \right) \le$$

$$\leq U_{E_1} e^{-\frac{R}{L}t} + \frac{I_0}{\dot{C}_1}\frac{L}{R}\left(e^{-\frac{R}{L}t_k} - e^{-\frac{R}{L}t}\right) + \frac{I_0}{\dot{C}_1}\frac{L}{R}\left(e^{-\frac{R}{L}t_k} - e^{-\frac{R}{L}t}\right) +$$

$$+ \frac{I_0}{\dot{C}_1}\left(e^{-\frac{R}{L}t_k} - e^{-\frac{R}{L}t}\right)\sum_{n=1}^{m}\left|g_n^1\right|\left(U_0 + U_{E_1}\right)^n \frac{L}{nR} n \leq$$

$$\leq U_{E_1} e^{-\frac{R}{L}t} + e^{-\frac{R}{L}t}\frac{L}{R}\left(e^{\frac{R}{L}(t-t_k)} - 1\right)\left[\frac{2I_0}{\dot{C}_1} + \frac{I_0}{\dot{C}_1}\sum_{n=1}^{m}\left|g_n^{(1)}\right|\phi_0^n\right] \leq$$

$$\leq e^{-\frac{R}{L}t}\left[U_{E_1} + \frac{L}{R}\frac{I_0}{\dot{C}_1}\left(e^{\frac{R}{L}T_0} - e^{-\frac{R}{L}T_0}\right)\left(2 + \sum_{n=1}^{m}\left|g_n^{(1)}\right|\phi_0^n\right)\right].$$

We use the inequality

$$e^{-\frac{R}{L}t_k} - e^{-\frac{R}{L}t_{k+1}} \leq e^{-\frac{R}{L}t}\left(e^{\frac{R}{L}T_0} - e^{-\frac{R}{L}T_0}\right) \leq 2e^{-\frac{R}{L}t}\sinh\left(\frac{RT_0}{L}\right)$$

and obtain

$$U_2 \leq \left|\int_{t_k}^{t_{k+1}}\dot{E}_1(s)ds\right| + \frac{1}{\dot{C}_1}\int_{t_k}^{t_{k+1}}\left|i(s)\right|ds + \frac{1}{\dot{C}_1}\int_{t_k}^{t_{k+1}}\left|\widetilde{L}_1^{-1}\left(\int_{t_k}^{t_{k+1}}\left(u(\tau) - E_1(\tau)\right)d\tau\right)\right|ds + \frac{1}{\dot{C}_1}\int_{t_k}^{t_{k+1}}\left|G_1(u(s) - E_1(s))\right|ds \leq$$

$$\leq \left|E_1(t)\right| + \frac{I_0}{\dot{C}_1}\int_{t_k}^{t_{k+1}}e^{-\frac{R}{L}s}ds + \frac{I_0}{\dot{C}_1}\int_{t_k}^{t_{k+1}}e^{-\frac{R}{L}s}ds + \frac{I_0}{\dot{C}_1}\sum_{n=1}^{m}\left|g_n^{(1)}\right|\int_{t_k}^{t_{k+1}}\left|u(s) - E_1(s)\right|^n ds \leq$$

$$\leq U_{E_1}e^{-\frac{R}{L}t} + \frac{I_0}{\dot{C}_1}\frac{L}{R}\left(e^{-\frac{R}{L}t_k} - e^{-\frac{R}{L}t_{k+1}}\right) + \frac{I_0}{\dot{C}_1}\frac{L}{R}\left(e^{-\frac{R}{L}t_k} - e^{-\frac{R}{L}t_{k+1}}\right) + \frac{I_0}{\dot{C}_1}\sum_{n=1}^{m}\left|g_n^{(1)}\right|\left(U_0 + U_{E_1}\right)^n\int_{t_k}^{t_{k+1}}e^{-n\frac{R}{L}t}ds \leq$$

$$\leq U_{E_1}e^{-\frac{R}{L}t} + \frac{I_0}{\dot{C}_1}\frac{L}{R}\left(e^{-\frac{R}{L}t_k} - e^{-\frac{R}{L}t_{k+1}}\right) + \frac{I_0}{\dot{C}_1}\frac{L}{R}\left(e^{-\frac{R}{L}t_k} - e^{-\frac{R}{L}t_{k+1}}\right) +$$

$$+ \frac{I_0}{\dot{C}_1}\sum_{n=1}^{m}\left|g_n^{(1)}\right|\left(U_0 + U_{E_1}\right)^n\frac{L}{nR}\left(e^{-n\frac{R}{L}t_k} - e^{-n\frac{R}{L}t_{k+1}}\right) \leq$$

$$\leq U_{E_1} e^{-\frac{R}{L}t} + \frac{I_0}{\dot{C}_1} \frac{L}{R}\left(e^{-\frac{R}{L}t_k} - e^{-\frac{R}{L}t_{k+1}}\right) + \frac{I_0}{\dot{C}_1}\frac{L}{R}\left(e^{-\frac{R}{L}t_k} - e^{-\frac{R}{L}t_{k+1}}\right) +$$

$$+ \frac{I_0}{\dot{C}_1}\left(e^{-\frac{R}{L}t_k} - e^{-\frac{R}{L}t_{k+1}}\right)\sum_{n=1}^{m}\left|g_n^{(1)}\right|\phi_0^n \frac{L}{nR} n \leq$$

$$\leq U_{E_1} e^{-\frac{R}{L}t} + \frac{L}{R}\frac{I_0}{\dot{C}_1}\left(e^{-\frac{R}{L}t_k} - e^{-\frac{R}{L}t_{k+1}}\right)\left(2 + \sum_{n=1}^{m}\left|g_n^{(1)}\right|\phi_0^n\right)$$

$$\leq e^{-\frac{R}{L}t}\left[U_{E_1} + \left(e^{\frac{R}{L}T_0} - e^{-\frac{R}{L}T_0}\right)\frac{L}{R}\frac{I_0}{\dot{C}_1}\left(2 + \sum_{n=1}^{m}\left|g_n^{(1)}\right|\phi_0^n\right)\right].$$

Therefore

$$\left|B_u^{(k)}(u,i)(t)\right| \leq e^{-\frac{R}{L}t}\left[U_{E_1} + \frac{L}{R}\frac{I_0}{\dot{C}_1}\left(e^{\frac{R}{L}T_0} - e^{-\frac{R}{L}T_0}\right)\left(2 + \sum_{n=1}^{m}\left|g_n^{(1)}\right|\phi_0^n\right)\right] +$$

$$+ e^{-\frac{R}{L}t}\left[U_{E_1} + \frac{L}{R}\frac{I_0}{\dot{C}_1}\left(e^{\frac{R}{L}T_0} - e^{-\frac{R}{L}T_0}\right)\left(2 + \sum_{n=1}^{m}\left|g_n^{(1)}\right|\phi_0^n\right)\right] \leq$$

$$\leq e^{-\frac{R}{L}t} 2\left[U_{E_1} + \frac{2L}{R}\frac{I_0}{\dot{C}_1}\sinh\left(\frac{RT_0}{L}\right)\left(2 + \sum_{n=1}^{m}\left|g_n^{(1)}\right|\phi_0^n\right)\right] \leq U_0 e^{-\frac{R}{L}t}.$$

Remark 6.6.1. We need the following more fine inequalities

$$\left|\tilde{u}(t-T)\right| = \left|e^{\frac{RT}{L}}\frac{u(t) + Z_0 i(t)}{2} - e^{\frac{RT}{L}}\frac{Z_0 i(t-2T) - u(t-2T)}{2} - E_0(t-T)\right| \leq$$

$$\leq e^{\frac{RT}{L}}\frac{U_0 e^{-\frac{R}{L}t} + Z_0 I_0 e^{-\frac{R}{L}t}}{2} + e^{-\frac{RT}{L}}\frac{Z_0 I_0 e^{-\frac{R}{L}t} + U_0 e^{-\frac{R}{L}t}}{2} + U_{E_0} e^{-\frac{R}{L}t} \leq$$

$$\leq \cosh\left(\frac{RT}{L}\right)(U_0 + Z_0 I_0) + U_{E_0} \leq \phi_0.$$

Remark 6.6.2. We see that since $\left|\tilde{L}_0^{-1}(.)\right| \leq I_0$ then $\left|\tilde{L}_0^{-1}(\lambda.)\right| \leq I_0|\lambda|$ for $|\lambda| \leq 1$. But the argument $\left|\tilde{u}(t-T)\right|$ of $\tilde{L}_0^{-1}(.)$ satisfies this condition because $\left|\tilde{u}(t-T)\right| \leq \phi_0 < 1$. Then

$$\left| B_i^{(k)}(u,i)(t) \right| \le \left| \int_{t_k}^{t} I(u,i)(s)ds \right| + \left| \int_{t_k}^{t_{k+1}} I(u,i)(s)ds \right| \equiv I_1 + I_2 .$$

Since

$$I_1 \le \frac{2e^{\frac{RT}{L}}}{Z_0}\left| \int_{t_k}^{t} \dot{E}_0(s)ds \right| + \frac{1}{Z_0}\left| \int_{t_k}^{t} \dot{u}(s)ds \right| + e^{-\frac{2RT}{L}}\left| \int_{t_k}^{t} \dot{i}(s-2T)ds \right| + \frac{e^{\frac{2RT}{L}}}{Z_0}\left| \int_{t_k}^{t} \dot{u}(s-2T)ds \right| +$$

$$+ \frac{2e^{-\frac{RT}{L}}}{Z_0\dot{\bar{C}}_0}\int_{t_k}^{t}\left| \widetilde{L}_0^{-1}\left(\int_{t_k}^{s} \widetilde{u}(\theta-T)d\theta \right) \right| ds + \frac{2e^{-\frac{RT}{L}}}{Z_0\dot{\bar{C}}_0}\int_{t_k}^{t} \left| G_0(\widetilde{u}(s-T)) \right| ds \le$$

$$\le \frac{2e^{\frac{RT}{L}}}{Z_0}\left| E_0(t) \right| + \frac{1}{Z_0}\left| u(t) \right| + e^{-\frac{2RT}{L}}\left| i(t-2T) \right| + \frac{e^{\frac{2RT}{L}}}{Z_0}\left| u(t-2T) \right| +$$

$$+ \frac{2e^{-\frac{RT}{L}}}{Z_0\dot{\bar{C}}_0} I_0 \int_{t_k}^{t}\int_{t_k}^{s} \left| (\widetilde{u}(\theta-T)) \right| d\theta ds + \frac{2e^{-\frac{RT}{L}}}{Z_0\dot{\bar{C}}_0}\sum_{n=1}^{m}\left| g_n^{(0)} \right|\left| \int_{t_k}^{t} \left| \widetilde{u}(s-T) \right| \right|^{n} ds \le$$

$$\le \frac{2e^{-\frac{RT}{L}}}{Z_0}U_{E_0}e^{-\frac{R}{L}t} + \frac{U_0}{Z_0}e^{-\frac{R}{L}t} + e^{-\frac{2RT}{L}}I_0 e^{-\frac{R}{L}t} + \frac{e^{-\frac{2RT}{L}}U_0}{Z_0}e^{-\frac{R}{L}t} +$$

$$+ \frac{2e^{-\frac{RT}{L}}I_0}{Z_0\dot{\bar{C}}_0}\left(\cosh\left(\frac{RT}{L}\right)(U_0 + Z_0 I_0) + U_{E_0} \right)\int_{t_k}^{t}\int_{t_k}^{s} e^{-\frac{R}{L}\theta}d\theta ds +$$

$$+ \frac{2e^{-\frac{RT}{L}}}{Z_0\dot{\bar{C}}_0}\sum_{n=1}^{m}\left| g_n^{(0)} \right|\left(\cosh\left(\frac{RT}{L}\right)(U_0 + Z_0 I_0) + U_{E_0} \right)^{n}\int_{t_k}^{t} e^{-n\frac{R}{L}s}ds \le$$

$$\le e^{-\frac{R}{L}t}\left(\frac{2e^{-\frac{RT}{L}}U_{E_0}}{Z_0} + \frac{U_0}{Z_0} + e^{-\frac{2RT}{L}}\frac{Z_0 I_0}{Z_0} + e^{-\frac{2RT}{L}}\frac{U_0}{Z_0} \right) +$$

$$+ e^{-\frac{R}{L}t}\frac{2e^{-\frac{RT}{L}}}{Z_0\dot{\bar{C}}_0}I_0\left(\cosh\left(\frac{RT}{L}\right)(U_0 + Z_0 I_0) + U_{E_0} \right)\left(\frac{L}{R}\right)^2\left(e^{\frac{RT_0}{L}} - e^{-\frac{RT_0}{L}} \right)^2 +$$

$$+ \frac{2e^{-\frac{RT}{L}}}{Z_0\dot{\bar{C}}_0}\sum_{n=1}^{m}\left| g_n^{(0)} \right|\left(\cosh\left(\frac{RT}{L}\right)(U_0 + Z_0 I_0) + U_{E_0} \right)^{n}\frac{L}{nR}\left(e^{-n\frac{R}{L}t_k} - e^{-n\frac{R}{L}t} \right) \le$$

$$\leq e^{-\frac{R}{L}t}\left(\frac{2e^{-\frac{RT}{L}}U_{E_0}}{Z_0}+\frac{U_0}{Z_0}+e^{-\frac{2RT}{L}}\frac{Z_0I_0}{Z_0}+e^{-\frac{2RT}{L}}\frac{U_0}{Z_0}\right)+$$

$$+e^{-\frac{R}{L}t}\frac{2e^{-\frac{RT}{L}}I_0\phi_0}{Z_0\dot{C}_0}\left(\frac{L}{R}\right)^2\left(e^{\frac{RT_0}{L}}-e^{-\frac{RT_0}{L}}\right)^2+\frac{2e^{-\frac{RT}{L}}}{Z_0\dot{C}_0}\sum_{n=1}^{m}\left|g_n^{(0)}\right|\phi_0^n\frac{L}{nR}\left(e^{-\frac{R}{L}t_k}-e^{-\frac{R}{L}t}\right)n\leq$$

$$\leq e^{-\frac{R}{L}t}\left(\frac{2e^{-\frac{RT}{L}}U_{E_0}}{Z_0}+\frac{U_0}{Z_0}+e^{-\frac{2RT}{L}}\frac{Z_0I_0}{Z_0}+e^{-\frac{2RT}{L}}\frac{U_0}{Z_0}+\frac{2e^{-\frac{RT}{L}}I_0\phi_0}{Z_0\dot{C}_0}\left(\frac{L}{R}\right)^2\left(e^{\frac{RT_0}{L}}-e^{-\frac{RT_0}{L}}\right)^2\right)+$$

$$+e^{-\frac{R}{L}t}\frac{2e^{-\frac{RT}{L}}}{Z_0\dot{C}_0}\sum_{n=1}^{m}\left|g_n^{(0)}\right|\phi_0^n\frac{L}{R}\left(e^{-\frac{R}{L}t_k}-e^{\frac{R}{L}t}\right)\leq$$

$$\leq e^{-\frac{R}{L}t}\left[\frac{2e^{\frac{RT}{L}}}{Z_0}U_{E_0}+\frac{U_0}{Z_0}+e^{-\frac{2RT}{L}}\frac{U_0+Z_0I_0}{Z_0}+\frac{4e^{-\frac{RT}{L}}}{Z_0\dot{C}_0}I_0\phi_0\left(\frac{L}{R}\right)^2\sinh^2\left(\frac{RT_0}{L}\right)+\right.$$

$$\left.+\frac{4e^{-\frac{RT}{L}}}{Z_0\dot{C}_0}\frac{L}{R}\sinh\left(\frac{RT_0}{L}\right)\sum_{n=1}^{m}\left|g_n^{(0)}\right|\phi_0^n\right]$$

and

$$I_2\leq\frac{e^{\frac{RT}{L}}}{Z_0}\left|\int_{t_k}^{t_{k+1}}\dot{E}_0(s)ds\right|+\frac{1}{2Z_0}\left|\int_{t_k}^{t_{k+1}}\ddot{u}(s)ds\right|+\frac{1}{2}e^{-\frac{2RT}{L}}\left|\int_{t_k}^{t_{k+1}}\dot{i}(s-2T)ds\right|+\frac{e^{\frac{2RT}{L}}}{2Z_0}\left|\int_{t_k}^{t_{k+1}}\dot{u}(s-2T)ds\right|+$$

$$+\frac{e^{-\frac{RT}{L}}}{Z_0\dot{C}_0}\int_{t_k}^{t_{k+1}}\left|\widetilde{L}_0^{-1}\left(\int_{t_k}^{s}\widetilde{u}(\theta-T)d\theta\right)\right|ds+\frac{e^{-\frac{RT}{L}}}{Z_0\dot{C}_0}\int_{t_k}^{t_{k+1}}\left|G_0(\widetilde{u}(s-T))\right|ds\leq$$

$$\leq\frac{e^{-\frac{RT}{L}}}{Z_0\dot{C}_0}\int_{t_k}^{t_{k+1}}\left|\overline{L}_0^{-1}\left(\int_{t_k}^{s}\widetilde{u}(\theta-T)d\theta\right)\right|ds+\frac{e^{-\frac{RT}{L}}}{Z_0\dot{C}_0}\int_{t_k}^{t_{k+1}}\left|G_0(\widetilde{u}(s-T))\right|ds\leq$$

$$\leq\frac{2e^{-\frac{RT}{L}}}{Z_0\dot{C}_0}I_0\int_{t_k}^{t_{k+1}}\int_{t_k}^{s}\left|(\widetilde{u}(\theta-T))\right|d\theta ds+\frac{2e^{-\frac{RT}{L}}}{Z_0\dot{C}_0}\sum_{n=1}^{m}\left|g_n^{(0)}\right|\int_{t_k}^{t_{k+1}}\left|\widetilde{u}(s-T)\right|^n ds\leq$$

$$\leq \frac{2e^{-\frac{RT}{L}}}{Z_0 \dot{\overline{C}}_0} I_0 \left(\left(e^{\frac{RT}{L}} + e^{-\frac{RT}{L}} \right) \frac{U_0 + Z_0 I_0}{2} + U_{E0} \right)^{t_{k+1}} \int\limits_{t_k}^{s} \int\limits_{t_k}^{s} e^{-\frac{R}{L}\theta} d\theta ds +$$

$$+ \frac{2e^{-\frac{RT}{L}}}{Z_0 \dot{\overline{C}}_0} \sum_{n=1}^{m} \left| g_n^{(0)} \right| \left(\left(e^{\frac{RT}{L}} + e^{-\frac{RT}{L}} \right) \frac{U_0 + Z_0 I_0}{2} + U_{E0} \right)^{n} \int\limits_{t_k}^{t_{k+1}} e^{-n\frac{R}{L}s} ds \leq$$

$$\leq e^{-\frac{R}{L}t} \frac{4e^{-\frac{RT}{L}}}{Z_0 \dot{\overline{C}}_0} \frac{L}{R} \sinh\left(\frac{RT_0}{L} \right) \left[2 I_0 \phi_0 \frac{L}{R} \sinh\left(\frac{RT_0}{L} \right) + \sum_{n=1}^{m} \left| g_n^{(0)} \right| \phi_0^n \right]$$

then

$$\left| B_i^{(k)}(u,i)(t) \right| \leq$$

$$\leq e^{-\frac{R}{L}t} \left[\frac{2e^{\frac{RT}{L}}}{Z_0} U_{E0} + \frac{U_0}{Z_0} + e^{-\frac{2RT}{L}} \frac{U_0 + Z_0 I_0}{Z_0} + \frac{8e^{-\frac{RT}{L}}}{Z_0 \dot{\overline{C}}_0} I_0 \phi_0 \left(\frac{L}{R} \right)^2 \left(\sinh\left(\frac{RT_0}{L} \right) \right)^2 + \right.$$

$$\left. + \frac{4e^{-\frac{RT}{L}}}{Z_0 \dot{\overline{C}}_0} \frac{L}{R} \sinh\left(\frac{RT_0}{L} \right) \sum_{n=1}^{m} \left| g_n^{(0)} \right| \phi_0^n \right] +$$

$$+ e^{-\frac{R}{L}t} \frac{4e^{-\frac{RT}{L}}}{Z_0 \dot{\overline{C}}_0} \frac{L}{R} \sinh\left(\frac{RT_0}{L} \right) \left[2 I_0 \phi_0 \frac{L}{R} \sinh\left(\frac{RT_0}{L} \right) + \sum_{n=1}^{m} \left| g_n^{(0)} \right| \phi_0^n \right] \leq$$

$$\leq e^{-\frac{R}{L}t} \left[\frac{2e^{-\frac{RT}{L}} U_{E0} + U_0 + e^{-\frac{2RT}{L}} (U_0 + Z_0 I_0)}{Z_0} + \right.$$

$$\left. + \frac{4e^{-\frac{RT}{L}}}{Z_0 \dot{\overline{C}}_0} \frac{L}{R} \sinh\left(\frac{RT_0}{L} \right) \left(2 I_0 \phi_0 \frac{L}{R} \sinh\left(\frac{RT_0}{L} \right) + \sum_{n=1}^{m} \left| g_n^{(0)} \right| \phi_0^n \right) \right] \leq I_0 e^{-\frac{R}{L}t}.$$

Therefore the operator B maps the set $M_{SU}^* \times M_{SI}^*$ into itself.

The inequalities implying that B is an contractive operator in a uniform space are analogous to the one from the previous paragraph.

Theorem 6.6.1 is thus proved.

Remark 6.6.3. We have to check the new inequalities for the data from Numerical example:

$$2U_{E_1} + 4\frac{L}{R}\frac{I_0}{\dot{\overline{C}}_1}\sinh\left(\frac{RT_0}{L}\right)\left(2 + \sum_{n=1}^{m}\left|g_n^{(1)}\right|\phi_0^n\right) \le U_0 \, ;$$

$$\frac{2e^{-\frac{RT}{L}}U_{E_0} + U_0 + e^{-\frac{2RT}{L}}\left(U_0 + Z_0 I_0\right)}{Z_0} + \frac{4e^{-\frac{RT}{L}}}{Z_0\dot{\overline{C}}_0}\frac{L}{R}\sinh\left(\frac{RT_0}{L}\right)\left(2I_0\phi_0\frac{L}{R}\sinh\left(\frac{RT_0}{L}\right) + \sum_{n=1}^{m}\left|g_n^{(0)}\right|\phi_0^n\right) \le I_0 .$$

In view of $2\sinh\left(\dfrac{RT_0}{L}\right) \approx \dfrac{R}{L}T_0 \approx 10^{-7} ; \dot{\overline{C}}_1 = 3.10^{-11} ; \dfrac{L}{R} = 10^{-6}$ we obtain

$$2U_{E_1} + 2.10^{-6}\frac{1}{3.10^{-11}}I_0 10^{-7}\left(2 + 0{,}02(0{,}3) + 0{,}12(0{,}3)^2 + 0{,}1(0{,}3)^3\right) \le U_0 \, ;$$

$$\frac{1}{75}U_{E_0} + \frac{U_0}{75}\left(1 + e^{-0,.6}\right) + e^{-1,2}I_0 +$$

$$+ \frac{1}{75.3.10^{-11}}10^{-6}.10^{-7}\left(0{,}3.0^{-6}.10^{-7}I_0 + 0{,}02(0{,}3) + 0{,}12(0{,}3)^2 + 0{,}1(0{,}3)^3\right) \le I_0$$

.

6.7. ANOTHER MANNER TO REDUCING THE MIXED PROBLEM TO AN OSCILLATORY ONE

We would like to propose another manner to reduce the above mixed problem to an initial value problem on the boundary. In fact we establish that the obtained system may have only oscillating solutions vanishing at infinity.

We proceed from boundary conditions (6.1.3), (6.1.4).

$$-i(0,t) = -\left[\left(u(0,t) - E_0(t)\right)\frac{dC_0(u(0,t) - E_0(t))}{du} + C_0(u(0,t) - E_0(t))\right]\left(\frac{du(0,t)}{dt} - \frac{dE_0(t)}{dt}\right) -$$

$$-\tilde{L}_0^{-1}\left(\int_T^t \left(u(0,\tau) - E_0(\tau)\right)d\tau\right) - G_0(u(0,t) - E_0(t)) \, ,$$

$$-i(\Lambda,t) = -\left[\left(u(\Lambda,t) - E_1(t)\right)\frac{dC_1(u(\Lambda,t) - E_1(t))}{du} + C_1(u(\Lambda,t) - E_1(t))\right]\left(\frac{du(\Lambda,t)}{dt} - \frac{dE_1(t)}{dt}\right) -$$

$$-\tilde{L}_1^{-1}\left(\int_T^t \left(u(\Lambda,\tau) - E_1(\tau)\right)d\tau\right) - G_1(u(\Lambda,t) - E_1(t)) \, .$$

Using transformation

$$u(x,t) = \frac{e^{-\frac{R}{L}t}}{2\sqrt{C}}W(x,t) - \frac{e^{-\frac{R}{L}t}}{2\sqrt{C}}J(x,t)$$

$$i(x,t) = \frac{e^{-\frac{R}{L}t}}{2\sqrt{L}}W(x,t) + \frac{e^{-\frac{R}{L}t}}{2\sqrt{L}}J(x,t)$$

we have

$$u(0,t) = \frac{e^{-\frac{R}{L}t}}{2\sqrt{C}}W(0,t) - \frac{e^{-\frac{R}{L}t}}{2\sqrt{C}}J(0,t) \quad u(\Lambda,t) = \frac{e^{-\frac{R}{L}t}}{2\sqrt{C}}W(\Lambda,t) - \frac{e^{-\frac{R}{L}t}}{2\sqrt{C}}J(\Lambda,t)$$

$$i(0,t) = \frac{e^{-\frac{R}{L}t}}{2\sqrt{L}}W(0,t) + \frac{e^{-\frac{R}{L}t}}{2\sqrt{L}}J(0,t) \quad i(\Lambda,t) = \frac{e^{-\frac{R}{L}t}}{2\sqrt{L}}W(\Lambda,t) + \frac{e^{-\frac{R}{L}t}}{2\sqrt{L}}J(\Lambda,t)$$

Then substituting the last expressions into the boundary conditions we obtain

$$-\frac{e^{-\frac{R}{L}t}W(0,t)}{2\sqrt{L}} - \frac{e^{-\frac{R}{L}t}J(0,t)}{2\sqrt{L}} =$$

$$= -\frac{d\widetilde{C}_0}{du}\left(\frac{e^{-\frac{R}{L}t}W(0,t) - e^{-\frac{R}{L}t}J(0,t)}{2\sqrt{C}} - E_0(t)\right)\left(\frac{d}{dt}\left(\frac{e^{-\frac{R}{L}t}}{2\sqrt{C}}W(0,t) - \frac{e^{-\frac{R}{L}t}}{2\sqrt{C}}J(0,t)\right) - \frac{dE_0(t)}{dt}\right) -$$

$$-\widetilde{L}_0^{-1}\left(\int_T^t\left(\frac{e^{-\frac{R}{L}\tau}W(0,\tau) - e^{-\frac{R}{L}\tau}J(0,\tau)}{2\sqrt{C}} - E_0(\tau)\right)d\tau\right) - G_0\left(\frac{e^{-\frac{R}{L}t}W(0,t) - e^{-\frac{R}{L}t}J(0,t)}{2\sqrt{C}} - E_0(t)\right)$$

and

$$-\frac{e^{-\frac{R}{L}t}}{2\sqrt{L}}W(\Lambda,t) - \frac{e^{-\frac{R}{L}t}}{2\sqrt{L}}J(\Lambda,t) =$$

$$= -\frac{d\widetilde{C}_1\left(\frac{e^{\frac{R}{L}t}W(\Lambda,t) - e^{\frac{R}{L}t}J(\Lambda,t)}{2\sqrt{C}} - E_1(t)\right)}{du}\left(\frac{d}{dt}\left(\frac{e^{-\frac{R}{L}t}(W(\Lambda,t) - J(\Lambda,t))}{2\sqrt{C}}\right) - \frac{dE_1(t)}{dt}\right) -$$

$$-\widetilde{L}_1^{-1}\left(\int_T^t\left(\frac{e^{-\frac{R}{L}\tau}(W(\Lambda,\tau) - J(\Lambda,\tau))}{2\sqrt{C}} - E_1(\tau)\right)d\tau\right) - G_1\left(\frac{e^{-\frac{R}{L}t}(W(\Lambda,t) - J(\Lambda,t))}{2\sqrt{C}} - E_1(t)\right).$$

We know that the integration along the characteristics yields

$$W(0,t) = W(\Lambda, t+T),\ J(\Lambda, t) = J(0, t+T) \Leftrightarrow J(\Lambda, t-T) = J(0,t).$$

Then the above system becomes

$$-\frac{e^{-\frac{R}{L}(t+T)}W(\Lambda, t+T)}{2\sqrt{L}} - \frac{e^{-\frac{R}{L}t}J(0,t)}{2\sqrt{L}} =$$

$$= -\frac{d\widetilde{C}_0}{du}\left(\frac{e^{-\frac{R}{L}(t+T)}W(\Lambda, t+T) - e^{-\frac{R}{L}t}J(0,t)}{2\sqrt{C}} - E_0(t) \right)\left(\frac{d}{dt}\left(\frac{e^{-\frac{R}{L}(t+T)}W(\Lambda, t+T) - e^{-\frac{R}{L}t}J(0,t)}{2\sqrt{C}} \right) - \frac{dE_0(t)}{dt} \right) -$$

$$- \widetilde{L}_0^{-1}\left(\int_T^t \left(\frac{e^{-\frac{R}{L}(\tau+T)}W(\Lambda, \tau+T) - e^{-\frac{R}{L}\tau}J(0,\tau)}{2\sqrt{C}} - E_0(\tau) \right)d\tau \right) - G_0\left(\frac{e^{-\frac{R}{L}(t+T)}W(\Lambda, t+T) - e^{-\frac{R}{L}t}J(0,t)}{2\sqrt{C}} - E_0(t) \right);$$

$$-\frac{e^{-\frac{R}{L}t}}{2\sqrt{L}}W(\Lambda, t) - \frac{e^{-\frac{R}{L}(t+T)}}{2\sqrt{L}}J(0, t+T) =$$

$$= -\frac{d\widetilde{C}_1}{du}\left(\frac{e^{-\frac{R}{L}t}W(\Lambda, t) - e^{-\frac{R}{L}(t+T)}J(0, t+T)}{2\sqrt{C}} - E_1(t) \right)\left(\frac{d}{dt}\left(\frac{e^{-\frac{R}{L}t}W(\Lambda, t) - e^{-\frac{R}{L}(t+T)}J(0, t+T)}{2\sqrt{C}} \right) - \frac{dE_1(t)}{dt} \right) -$$

$$- \widetilde{L}_1^{-1}\left(\int_T^t \left(\frac{e^{-\frac{R}{L}\tau}W(\Lambda, \tau) - e^{-\frac{R}{L}(\tau+T)}J(0, \tau+T)}{2\sqrt{C}} - E_1(\tau) \right)d\tau \right) - G_1\left(\frac{e^{-\frac{R}{L}t}W(\Lambda, t) - e^{-\frac{R}{L}(t+T)}J(0, t+T)}{2\sqrt{C}} - E_1(t) \right).$$

Let us put $t + T = t$. Then

$$-\frac{e^{-\frac{R}{L}t}W(\Lambda, t)}{2\sqrt{L}} - \frac{e^{-\frac{R}{L}(t-T)}J(0, t-T)}{2\sqrt{L}} =$$

$$= -\frac{d\widetilde{C}_0}{du}\left(\frac{e^{-\frac{R}{L}t}W(\Lambda, t) - e^{-\frac{R}{L}(t-T)}J(0, t-T)}{2\sqrt{C}} - E_0(t-T) \right)\left(\frac{d}{dt}\left(\frac{e^{-\frac{R}{L}t}W(\Lambda, t) - e^{-\frac{R}{L}(t-T)}J(0, t-T)}{2\sqrt{C}} \right) - \frac{dE_0(t-T)}{dt} \right) -$$

$$- \widetilde{L}_0^{-1}\left(\int_0^{t-T} \left(\frac{e^{-\frac{R}{L}\tau}W(\Lambda, \tau) - e^{-\frac{R}{L}(\tau-T)}J(0, \tau-T)}{2\sqrt{C}} - E_0(\tau-T) \right)d\tau \right) -$$

$$- G_0\left(\frac{e^{-\frac{R}{L}t}W(\Lambda, t) - e^{-\frac{R}{L}(t-T)}J(0, t-T)}{2\sqrt{C}} - E_0(t-T) \right);$$

$$-\frac{e^{-\frac{R}{L}(t-T)}}{2\sqrt{L}}W(\Lambda,t-T)-\frac{e^{-\frac{R}{L}t}}{2\sqrt{L}}J(0,t)=$$

$$=-\frac{d\widetilde{C}_1}{du}\left(\frac{e^{-\frac{R}{L}(t-T)}W(\Lambda,t-T)-e^{-\frac{R}{L}t}J(0,t)}{2\sqrt{C}}-E_1(t-T)\right)\left(\frac{d}{dt}\left(\frac{e^{-\frac{R}{L}(t-T)}W(\Lambda,t-T)-e^{-\frac{R}{L}t}J(0,t)}{2\sqrt{C}}\right)-\frac{dE_1(t-T)}{dt}\right)-$$

$$-\widetilde{L}_1^{-1}\left(\int_0^{t-T}\left(\frac{e^{-\frac{R}{L}(\tau-T)}W(\Lambda,\tau-T)-e^{-\frac{R}{L}\tau}J(0,\tau)}{2\sqrt{C}}-E_1(\tau-T)\right)d\tau\right)-$$

$$-G_1\left(\frac{e^{-\frac{R}{L}(t-T)}W(\Lambda,t-T)-e^{-\frac{R}{L}t}J(0,t)}{2\sqrt{C}}-E_1(t-T)\right).$$

We assume that the unknown functions are $W(\Lambda,t)\equiv W(t)$ and $J(0,t)\equiv J(t)$. Then the last system becomes

$$-\frac{e^{-\frac{R}{L}t}W(t)+e^{-\frac{R}{L}(t-T)}J(t-T)}{Z_0}\frac{1}{d\widetilde{C}_0(.)/du}=$$

$$=-\frac{d}{dt}\left(e^{-\frac{R}{L}t}W(t)-e^{-\frac{R}{L}(t-T)}J(t-T)\right)+\frac{dE_0(t-T)}{dt}-$$

$$-\frac{2\sqrt{C}}{d\widetilde{C}_0(.)/du}\widetilde{L}_0^{-1}\left(\int_T^t\left(\frac{e^{-\frac{R}{L}(s-T)}W(s-T)-e^{-\frac{R}{L}(s-2T)}J(s-2T)}{2\sqrt{C}}-E_0(s-2T)\right)d\tau\right)-$$

$$-\frac{2\sqrt{C}}{d\widetilde{C}_0(.)/du}G_0\left(\frac{e^{-\frac{R}{L}t}W(t)-e^{-\frac{R}{L}(t-T)}J(t-T)}{2\sqrt{C}}-E_0(t-T)\right);$$

$$-\frac{e^{-\frac{R}{L}(t-T)}W(t-T)+e^{-\frac{R}{L}t}J(t)}{2\sqrt{L}}=$$

$$=-\frac{d\widetilde{C}_1}{du}\left(\frac{e^{-\frac{R}{L}(t-T)}W(t-T)-e^{-\frac{R}{L}t}J(t)}{2\sqrt{C}}-E_1(t-T)\right)\left(\frac{d}{dt}\left(\frac{e^{-\frac{R}{L}(t-T)}W(t-T)-e^{-\frac{R}{L}t}J(t)}{2\sqrt{C}}\right)-\frac{dE_1(t-T)}{dt}\right)-$$

$$-\tilde{L}_1^{-1}\left(\int_0^{t-T}\left(\frac{e^{-\frac{R}{L}(\tau-T)}W(\tau-T)-e^{-\frac{R}{L}\tau}J(\tau)}{2\sqrt{C}}-E_1(\tau-T)\right)d\tau\right)-$$

$$-G_1\left(\frac{e^{-\frac{R}{L}(t-T)}W(t-T)-e^{-\frac{R}{L}t}J(t)}{2\sqrt{C}}-E_1(t-T)\right).$$

We notice that one can set $\widehat{W}(t)=W(t)e^{-\frac{R}{L}t}$, $\hat{J}(t)=J(t)e^{-\frac{R}{L}t}$ as in the previous chapter.

Consequently the last system can be rewritten as:

$$-\frac{\widehat{W}(t)}{2\sqrt{L}}-\frac{\hat{J}(t-T)}{2\sqrt{L}}=\frac{d\tilde{C}_0\left(\frac{\widehat{W}(t)-\hat{J}(t-T)}{2\sqrt{C}}-E_0(t)\right)}{du}\left(\frac{d}{dt}\left(\frac{\widehat{W}(t)-\hat{J}(t-T)}{2\sqrt{C}}\right)-\frac{dE_0(t)}{dt}\right)-$$

$$-\tilde{L}_0^{-1}\left(\int_0^{t-T}\left(\frac{\widehat{W}(\tau)-\hat{J}(\tau-T)}{2\sqrt{C}}-E_0(\tau)\right)d\tau\right)-G_0\left(\frac{\widehat{W}(t)-\hat{J}(t-T)}{2\sqrt{C}}-E_0(t)\right);$$

$$-\frac{\widehat{W}(t-T)+\hat{J}(t)}{2\sqrt{L}}=-=\frac{d\tilde{C}_1\left(\frac{\widehat{W}(t-T)-\hat{J}(t)}{2\sqrt{C}}-E_1(t)\right)}{du}\left(\frac{d}{dt}\left(\frac{\widehat{W}(t-T)-\hat{J}(t)}{2\sqrt{C}}\right)-\frac{dE_1(t)}{dt}\right)-$$

$$-\tilde{L}_1^{-1}\left(\int_0^{t-T}\left(\frac{\widehat{W}(\tau-T)-\hat{J}(\tau)}{2\sqrt{C}}-E_1(\tau)\right)d\tau\right)-G_1\left(\frac{\widehat{W}(t-T)-\hat{J}(t)}{2\sqrt{C}}-E_1(t)\right).$$

Remark 6.7.1. Further on we use denotations $W(t), J(t)$ instead of $\widehat{W}(t), \hat{J}(t)$ but we look for solution satisfying the estimate $|W(t)|\le U_0 e^{-\frac{R}{L}t}, |J(t)|\le I_0 e^{-\frac{R}{L}t}$.

Solving the first equation with respect to $\frac{dW(t)}{dt}$ while the second one – to $\frac{dJ(t)}{dt}$ and change the variables in the integrals we reach the system (recalling $E_p(t-T)=E_p(t)$)

$$\frac{dW(t)}{dt}=-\frac{1}{Z_0}\frac{W(t)+J(t-T)}{d\tilde{C}_0\left(\frac{W(t)-J(t-T)}{2\sqrt{C}}-E_0(t)\right)/du}+\frac{dJ(t-T)}{dt}+2\sqrt{C}\frac{dE_0(t)}{dt}+$$

$$+2\sqrt{C}\frac{\tilde{L}_0^{-1}\left(\int_T^t\left(\frac{W(\tau)-J(\tau-T)}{2\sqrt{C}}-E_0(\tau)\right)d\tau\right)}{d\tilde{C}_0\left(\frac{W(t)-J(t-T)}{2\sqrt{C}}-E_0(t)\right)/du}+2\sqrt{C}\frac{G_0\left(\frac{W(t)-J(t-T)}{2\sqrt{C}}-E_0(t)\right)}{d\tilde{C}_0\left(\frac{W(t)-J(t-T)}{2\sqrt{C}}-E_0(t)\right)/du};$$

$$(6.7.1)$$

$$\frac{dJ(t)}{dt} = \frac{dW(t-T)}{dt} - 2\sqrt{C}\,\frac{dE_1(t)}{dt} + \frac{1}{Z_0}\,\frac{W(t-T)+J(t)}{d\widetilde{C}_1\left(\dfrac{\widehat{W}(t-T)-\widehat{J}(t)}{2\sqrt{C}} - E_1(t)\right)/du} -$$

$$-2\sqrt{C}\,\frac{\widetilde{L}_1^{-1}\left(\displaystyle\int_T^t\left(\dfrac{W(\tau-T)-J(\tau)}{2\sqrt{C}} - E_1(\tau)\right)d\tau\right)}{d\widetilde{C}_1\left(\dfrac{W(t-T)-J(t)}{2\sqrt{C}} - E_1(t)\right)/du} - 2\sqrt{C}\,\frac{G_1\left(\dfrac{W(t-T)-J(t)}{2\sqrt{C}} - E_1(t)\right)}{d\widetilde{C}_1\left(\dfrac{W(t-T)-J(t)}{2\sqrt{C}} - E_1(t)\right)/du}\,.$$

As above one can shift the initial functions along the characteristics on the interval $[0,T]$ and obtain $W(t) = \widetilde{W}_0(t),\ J(t) = \widetilde{J}_0(t),\ t \in [0,T]$.

Now we are able to formulate the main problem: to find a solution of (6.7.1) with advanced prescribed zeros on an interval $[t_0,\infty)$, $T = t_0$, where $\widetilde{W}_0(t),\ \widetilde{J}_0(t),\ t \in [0,T]$ are prescribed oscillating functions on the interval $[0,T]$.

Let $S_T = \{\tau_k\}_{k=0}^n, n \in N$ be the set of zeros of the initial function, that is, $\widetilde{W}_0(\tau_k) = 0,\ \widetilde{J}_0(\tau_k) = 0$ such that $\tau_0 = 0,\ \tau_n = T \equiv t_0$.

Besides $\max\{\tau_{k+1} - \tau_k : k = 0,1,...,n\} \le T_0$.

Let $S = \{t_k\}_{k=0}^\infty$ be a strictly increasing sequence of real numbers satisfying the following conditions **(C)**:

(C1) $\displaystyle\lim_{k\to\infty} t_k = \infty$;

(C2) for every k there is $s < k$ such that $t_k - T = t_s$ where $t_s \in S_T \cup S$;

(E) $E(t_k) = 0,\ E(t_k - T) = 0$.

It follows

(C3) $0 \le \inf\{t_{k+1} - t_k : k = 0,1,2,...\} \le \sup\{t_{k+1} - t_k : k = 0,1,2,...\} = T_0 < \infty$.

Introduce the set $C^1[t_0,\infty)$ consisting of all continuous and bounded functions differentiable with bounded derivatives on every interval $[t_k,t_{k+1}]$. Let us note that the functions from $C^1[t_0,\infty)$ could be not differentiable at t_k. That is why, we introduce below a topology of uniform convergence on every interval $[t_k,t_{k+1}]$ which needs introducing of uniform spaces (cf. Chapter I).

$$M_W = \left\{ W(.) \in C^1[t_0,\infty) : W(t_k) = 0 \wedge \left| W(t) \right| \le U_0 e^{-\beta} e^{-\frac{R}{L}t}, t \in [t_0,\infty) \right\},$$

$$M_J = \left\{ J(.) \in C^1[t_0,\infty) : J(t_k) = 0 \wedge \left| J(t) \right| \le I_0 e^{-\beta} e^{-\frac{R}{L}t}, t \in [t_0,\infty) \right\}.$$

Remark 6.7.2. It follows that the functions from the above set satisfy the inequalities

$$\left| W(t) \right| \le U_0 e^{\mu(t-t_k)}, t \in [t_k,t_{k+1}] \,, \left| J(t) \right| \le I_0 e^{\mu(t-t_k)}, t \in [t_k,t_{k+1}]$$

where U_0, μ are positive constants and $k = 0,1,2,\dots$.

The set $M_W \times M_J$ turns out into a complete uniform space (cf. Chapter I) with respect to the family of pseudo-metrics

Introduce the following family of pseudo-metrics

$$\rho^{(k)}(W,\overline{W}) = \max\left\{ \left| W(t) - \overline{W}(t) \right| : t \in [t_k,t_{k+1}] \right\},$$

$$\rho^{(k)}(J,\overline{J}) = \max\left\{ \left| J(t) - \overline{J}(t) \right| : t \in [t_k,t_{k+1}] \right\},$$

$$\hat{\rho}^{(k)}(W,\overline{W}) = \max\left\{ \left| W(t) - \overline{W}(t) \right| : t \in [t_0,t_{k+1}] \right\},$$

$$\hat{\rho}^{(k)}(J,\overline{J}) = \max\left\{ \left| J(t) - \overline{J}(t) \right| : t \in [t_0,t_{k+1}] \right\},$$

$$\rho_\mu^{(k)}(W,\overline{W}) = \max\left\{ e^{-\mu(t-t_k)} \left| W(t) - \overline{W}(t) \right| : t \in [t_k,t_{k+1}] \right\},$$

$$\rho_\mu^{(k)}(J,\overline{J}) = \max\left\{ e^{-\mu(t-t_k)} \left| J(t) - \overline{J}(t) \right| : t \in [t_k,t_{k+1}] \right\},$$

$$\hat{\rho}_\mu^{(k)}(W,\overline{W}) = \max\left\{ \rho_\mu^{(0)}(W,\overline{W}), \rho_\mu^{(1)}(W,\overline{W}), \dots, \rho_\mu^{(k)}(W,\overline{W}) \right\},$$

$$\hat{\rho}_\mu^{(k)}(J,\overline{J}) = \max\left\{ \rho_\mu^{(0)}(J,\overline{J}), \rho_\mu^{(1)}(J,\overline{J}), \dots, \rho_\mu^{(k)}(J,\overline{J}) \right\},$$

$$\rho_\mu^{(k)}(\dot{W},\dot{\overline{W}}) = \max\left\{ e^{-\mu(t-t_k)} \left| \dot{W}(t) - \dot{\overline{W}}(t) \right| : t \in [t_k,t_{k+1}] \right\},$$

$$\rho_\mu^{(k)}(\dot{J},\dot{\overline{J}}) = \max\left\{ e^{-\mu(t-t_k)} \left| \dot{J}(t) - \dot{\overline{J}}(t) \right| : t \in [t_k,t_{k+1}] \right\},$$

$$\hat{\rho}_\mu^{(k)}(\dot{W},\dot{\overline{W}}) = \max\left\{ \rho_\mu^{(0)}(\dot{W},\dot{\overline{W}}), \rho_\mu^{(1)}(\dot{W},\dot{\overline{W}}), \dots, \rho_\mu^{(k)}(\dot{W},\dot{\overline{W}}) \right\},$$

$$\hat{\rho}_\mu^{(k)}(\dot{J},\dot{\overline{J}}) = \max\left\{ \rho_\mu^{(0)}(\dot{J},\dot{\overline{J}}), \rho_\mu^{(1)}(\dot{J},\dot{\overline{J}}), \dots, \rho_\mu^{(k)}(\dot{J},\dot{\overline{J}}) \right\}.$$

Remark 6.7.3. The following inequalities imply the equivalence of the both families of pseudo-metrics

$$\rho_\mu^{(k)}(W,\overline{W}) \le \rho^{(k)}(W,\overline{W}) \le e^{\mu_0}\rho_\mu^{(k)}(W,\overline{W}), \ (k=0,1,2,...),$$

$$\rho_\mu^{(k)}(J,\overline{J}) \le \rho^{(k)}(J,\overline{J}) \le e^{\mu_0}\rho_\mu^{(k)}(J,\overline{J}), \ (k=0,1,2,...).$$

It is easy to verify that

$$\hat{\rho}^{(k)}(W,\overline{W}) = \max\left\{\rho^{(0)}(W,\overline{W}),\rho^{(1)}(W,\overline{W}),...,\rho^{(k)}(W,\overline{W})\right\} \le$$
$$\le e^{\mu_0}\max\left\{\rho_\mu^{(0)}(W,\overline{W}),\rho_\mu^{(1)}(W,\overline{W}),...,\rho_\mu^{(k)}(W,\overline{W})\right\} = e^{\mu_0}\hat{\rho}_\mu^{(k)}(W,\overline{W});$$

$$\hat{\rho}^{(k)}(J,\overline{J}) = \max\left\{\rho^{(0)}(J,\overline{J}),\rho^{(1)}(J,\overline{J}),...,\rho^{(k)}(J,\overline{J})\right\} \le$$
$$\le e^{\mu_0}\max\left\{\rho_\mu^{(0)}(J,\overline{J}),\rho_\mu^{(1)}(J,\overline{J}),...,\rho_\mu^{(k)}(J,\overline{J})\right\} = e^{\mu_0}\hat{\rho}_\mu^{(k)}(J,\overline{J}).$$

The set M_U turns out into a complete uniform space with respect to the saturated family of pseudo-metrics

$$\hat{\rho}_\mu^{(k)}((W,\dot{W},J,\dot{J}),(\overline{W},\dot{\overline{W}},\overline{J},\dot{\overline{J}})) = \max\left\{\hat{\rho}^{(k)}(W,\overline{W}),\hat{\rho}^{(k)}(J,\overline{J}),\hat{\rho}_\mu^{(k)}(\dot{W},\dot{\overline{W}}),\hat{\rho}_\mu^{(k)}(\dot{J},\dot{\overline{J}})\right\},$$
$$(k=0,1,2,...).$$

One can verify that $M_W \times M_J$ is closed subset of $C^1[t_0,\infty) \times C^1[t_0,\infty)$ with respect to the above family of pseudo-metrics.

Since we look for continuously differentiable solutions we have to take into account the conformity condition **(CC)**. If we choose $\{0,T\} \in S_T \cup S$ then **(CC)** becomes

$$\frac{dW(T)}{dt} = \frac{dJ(0)}{dt}, \ \frac{dJ(T)}{dt} = \frac{dW(0)}{dt}.$$

In order to avoid the just mentioned difficulty we define the operator $B = \left(B_W(W,J), B_J(W,J)\right)$ by the formulas

$$B_W(W,J)(t) = B_W^{(k)}(W,J)(t) := \int_{t_k}^{t} U(W,J)(s)\,ds - \frac{t-t_k}{t_{k+1}-t_k}\int_{t_k}^{t_{k+1}} U(W,J)(s)\,ds,$$

$$t \in [t_k,t_{k+1}], \ (k=0,1,2,...);$$

472   Vasil G. Angelov

$$B_J(W,J)(t) = B_J^{(k)}(W,J)(t) := \int_{t_k}^{t} I(W,J)(s)\,ds - \frac{t-t_k}{t_{k+1}-t_k}\int_{t_k}^{t_{k+1}} I(W,J)(s)\,ds\,,$$

$t \in [t_k, t_{k+1}], \ (k=0,1,2,\dots)$

where

$$U(W,J)(t) \equiv \frac{d\bar{J}(t)}{dt} + 2\sqrt{C}\,\frac{dE_0(t)}{dt} - \frac{1}{Z_0}\frac{W(t)+\bar{J}(t)}{d\widetilde{C}_0\left(\dfrac{W(t)-\bar{J}(t)}{2\sqrt{C}} - E_0(t)\right)/du} +$$

$$+2\sqrt{C}\,\frac{\widetilde{L}_0^{-1}\left(\int_{t_k}^{t}\left(\dfrac{W(\tau)-\bar{J}(\tau)}{2\sqrt{C}}-E_0(\tau)\right)d\tau\right)}{d\widetilde{C}_0\left(\dfrac{W(t)-\bar{J}(t)}{2\sqrt{C}}-E_0(t)\right)/du} + 2\sqrt{C}\,\frac{G_0\left(\dfrac{W(t)-\bar{J}(t)}{2\sqrt{C}}-E_0(t)\right)}{d\widetilde{C}_0\left(\dfrac{W(t)-\bar{J}(t)}{2\sqrt{C}}-E_0(t)\right)/du}\,;$$

$$I(W,J)(t) \equiv \frac{d\bar{W}(t)}{dt} - 2\sqrt{C}\,\frac{dE_1(t)}{dt} + \frac{1}{Z_0}\frac{\bar{W}(t)+J(t)}{d\widetilde{C}_1\left(\dfrac{\bar{W}(t)-J(t)}{2\sqrt{C}} - E_1(t)\right)/du} -$$

$$-2\sqrt{C}\,\frac{\widetilde{L}_1^{-1}\left(\int_{t_k}^{t}\left(\dfrac{\bar{W}(\tau)-J(\tau)}{2\sqrt{C}}-E_1(\tau)\right)d\tau\right)}{d\widetilde{C}_1\left(\dfrac{\bar{W}(t)-J(t)}{2\sqrt{C}}-E_1(t)\right)/du} - 2\sqrt{C}\,\frac{G_1\left(\dfrac{\bar{W}(t)-J(t)}{2\sqrt{C}}-E_1(t)\right)}{d\widetilde{C}_1\left(\dfrac{\bar{W}(t)-J(t)}{2\sqrt{C}}-E_1(t)\right)/du}$$

where $\displaystyle \bar{W}(t) = \begin{cases} \widetilde{W}_0(t-T), t\in[T,2T] \\ W(t-T),\ t\in[2T,\infty) \end{cases}, \ \bar{J}(t) = \begin{cases} \widetilde{J}_0(t-T), t\in[T,2T] \\ J(t-T),\ t\in[2T,\infty) \end{cases}$. The last operators

avoid the difficulties caused by conformity condition.

Remark 6.7.4. It is easy to verify that the functions from M_W and M_J are not necessary differentiable at the points of $S = \{t_k\}_{k=0}^{\infty}$.

Lemma 6.7.1. Problem (6.7.1) has a solution $(W,J) \in M_W \times M_J$ iff the operator B has a fixed point in $M_W \times M_J$, that is,

$$(W,J) = (B_W(W,J), B_J(W,J))\,. \tag{6.7.2}$$

Proof: Let $(W,J) \in M_W \times M_J$ be a solution of (6.7.1). We consider only the first component. Then integrating the first equation of (6.7.1) on every interval $[t_k,t] \subset [t_k, t_{k+1}]$ $(k=0,1,2\dots)$ we obtain

$$W(t) - W(t_k) = \int_{t_k}^{t} U(W,J)(s)\,ds \iff J(t) = \int_{t_k}^{t} U(W,J)(s)\,ds$$

and then

$$J(t) = \int_{t_k}^{t} U(W,J)(s)\,ds \Rightarrow 0 = J(t_{k+1}) = \int_{t_k}^{t} U(W,J)(s)\,ds \Rightarrow \int_{t_k}^{t_{k+1}} U(W,J)(s)\,ds = 0 \qquad (6.7.3)$$

Therefore the pair (W,J) satisfies

$$W = B_W(W,J), J = B_J(W,J) \iff (W,J) = (B_W(W,J), B_J(W,J))$$

that is, (W,J) is a fixed point of B.

Conversely, let (W,J) be a fixed point of B or

$$W(t) = \int_{t_k}^{t} U(W,J)(s)\,ds - \frac{t-t_k}{t_{k+1}-t_k} \int_{t_k}^{t_{k+1}} U(W,J)(s)\,ds),$$

$$J(t) = \int_{t_k}^{t} I(W,J)(s)\,ds - \frac{t-t_k}{t_{k+1}-t_k} \int_{t_k}^{t_{k+1}} I(W,J)(s)\,ds .$$

Then in view of $\mu_0 = \mu T_0 = \text{const.}$ we obtain

$$\left| \int_{t_k}^{t_{k+1}} U(W,J)(s)\,ds \right| \le$$

$$\le \frac{1}{Z_0 \dot{\bar{C}}_0} \int_{t_k}^{t_{k+1}} |W(t)|\,dt + \frac{1}{Z_0 \dot{\bar{C}}_0} \int_{t_k}^{t_{k+1}} |J(t-T)|\,dt + \left| \int_{t_k}^{t_{k+1}} \frac{dJ(t-T)}{dt}\,dt \right| + 2\sqrt{C} \left| \int_{t_k}^{t_{k+1}} \frac{dE_0(t)}{dt}\,dt \right| +$$

$$+ \frac{2\sqrt{C}}{\dot{\bar{C}}_0} \int_{t_k}^{t_{k+1}} \left| \tilde{L}_0^{-1} \left(\int_{T}^{t} \left(\frac{W(\tau) - J(\tau-T)}{2\sqrt{C}} - E_0(\tau) \right) d\tau \right) \right| dt + \frac{2\sqrt{C}}{\dot{\bar{C}}_0} \int_{t_k}^{t_{k+1}} \left| G_0 \left(\frac{W(t) - J(t-T)}{2\sqrt{C}} - E_0(t) \right) \right| dt \le$$

$$\le \frac{U_0}{Z_0 \dot{\bar{C}}_0} \frac{e^{\mu_0} - 1}{\mu} + \frac{I_0}{Z_0 \dot{\bar{C}}_0} \frac{e^{\mu_0} - 1}{\mu} + \frac{2\sqrt{C}}{\dot{\bar{C}}_0} I_0 \int_{t_k}^{t_{k+1}} \int_{t_k}^{t} \left(\frac{|W(\tau)| + |J(\tau-T)|}{2\sqrt{C}} + |E_0(\tau)| \right) d\tau dt +$$

$$+ \frac{2\sqrt{C}}{\dot{\bar{C}}_0} \sum_{n=1}^{m} \left| g_n^{(0)} \right| \int_{t_k}^{t_{k+1}} \left| \frac{W(t) - J(t-T)}{2\sqrt{C}} - E_0(t) \right|^n dt \le$$

$$\leq \frac{U_0 + I_0 e^{\mu 0}}{Z_0 \dot{C}_0} \frac{e^{\mu 0} - 1}{\mu} + \frac{2\sqrt{C}}{\dot{C}_0} I_0 \left(\frac{U_0 + I_0 e^{\mu 0}}{2\sqrt{C}} + U_{E0} \right) \int\limits_{t_k}^{t_{k+1}} \int\limits_{t_k}^{t} e^{\mu(\tau - t_k)} d\tau dt +$$

$$+ \frac{2\sqrt{C}}{\dot{C}_0} \sum_{n=1}^{m} \left| g_n^{(0)} \right| \left(\frac{U_0 + I_0 e^{\mu 0}}{2\sqrt{C}} + U_{E0} \right)^n \int\limits_{t_k}^{t_{k+1}} e^{\mu n(t - t_k)} dt \leq$$

$$\leq \frac{e^{\mu T_0} - 1}{\mu} \frac{U_0 + I_0 e^{\mu 0}}{Z_0 \dot{C}_0} + \frac{e^{\mu 0} - 1}{\mu^2} \frac{2\sqrt{C}}{\dot{C}_0} I_0 \left(\frac{U_0 + I_0 e^{\mu 0}}{2\sqrt{C}} + U_{E0} \right) +$$

$$+ \frac{2\sqrt{C}}{\dot{C}_0} \sum_{n=1}^{m} \left| g_n^{(0)} \right| \left(\frac{U_0 + I_0 e^{\mu 0}}{2\sqrt{C}} + U_{E0} \right)^n \frac{e^{n \mu T_0} - 1}{n\mu} \leq$$

$$\leq \frac{e^{\mu 0} - 1}{\mu} \frac{U_0 + I_0 e^{\mu 0}}{Z_0 \dot{C}_0} + \frac{e^{\mu 0} - 1}{\mu^2} \frac{2\sqrt{C}}{\dot{C}_0} I_0 \left(\frac{U_0 + I_0 e^{\mu 0}}{2\sqrt{C}} + U_{E0} \right) +$$

$$+ \frac{2\sqrt{C}}{\dot{C}_0} \sum_{n=1}^{m} \left| g_n^{(0)} \right| \left(\frac{U_0 + I_0 e^{\mu 0}}{2\sqrt{C}} + U_{E0} \right)^n \frac{\left(e^{\mu 0} - 1 \right) n}{n\mu} \leq$$

$$\leq \frac{e^{\mu 0} - 1}{\mu} \left[\frac{U_0 + I_0 e^{\mu 0}}{Z_0 \dot{C}_0} + \frac{2\sqrt{C}}{\mu \dot{C}_0} I_0 \left(\frac{U_0 + I_0 e^{\mu 0}}{2\sqrt{C}} + U_{E0} \right) + \frac{2\sqrt{C}}{\dot{C}_0} \sum_{n=1}^{m} \left| g_n^{(0)} \right| \left(\frac{U_0 + I_0 e^{\mu 0}}{2\sqrt{C}} + U_{E0} \right)^n \right] \equiv M_0(\mu).$$

As in Chapter II we conclude that $\int\limits_{t_k}^{t_{k+1}} U(W, J)(s) ds = 0$. Therefore

$$W(t) = \int\limits_{t_k}^{t} U(W, J)(s) ds - \frac{t - t_k}{t_{k+1} - t_k} \int\limits_{t_k}^{t_{k+1}} U(W, J)(s) ds \Leftrightarrow W(t) = \int\limits_{t_k}^{t} U(W, J)(s) ds$$

and analogously

$$J(t) = \int\limits_{t_k}^{t} I(W, J)(s) ds - \frac{t - t_k}{t_{k+1} - t_k} \int\limits_{t_k}^{t_{k+1}} I(W, J)(s) ds \Leftrightarrow J(t) = \int\limits_{t_k}^{t} I(W, J)(s) ds .$$

Differentiating the last integral equation we obtain (6.7.1).

Lemma 6.7.1 is thus proved.

Theorem 6.7.1. Let the following conditions be fulfilled:

(IN) The initial functions satisfy the inequalities

$$\widetilde{W}_0(t), \widetilde{J}_0(t) \in C^1[0,T], \left| \widetilde{W}_0(t) \right| \leq e^{-\beta} U_0 e^{-\frac{R}{L} t}; \left| \widetilde{J}_0(t) \right| \leq e^{-\beta} I_0 e^{-\frac{R}{L} t};$$

(E) $E_p(.) \in C^1[0,\infty), \left| E_p(t) \right| \le U_{E_p} e^{-\frac{R}{L}t}, E_p(t-T) = E_p(t), t \in [0,\infty), (p = 0,1)$;

(R) $\dfrac{U_0 + I_0}{2\sqrt{C}} + U_{E_1} \le \phi_0 ; \dfrac{2L}{R} e^{\frac{RT}{L}} \sinh\!\left(\dfrac{RT_0}{L} \right)\!\left(U_0 + Z_0 I_0 + U_{E_0} \right) \le \phi_0 \le 1$.

Then there exists an unique oscillatory solution of (6.7.1), belonging to $M = M_W \times M_J$ on the interval $[T, 2T]$.

Proof: We omit the proof that $B_W(W,J)(t)$ and $B_J(W,J)(t)$ are continuous and differentiable on every $[t_k, t_{k+1}]$, $(k = 0,1,2,\dots)$. Consider for $t \in [t_k, t_{k+1}]$, $(k = 0,1,2,\dots)$

$$\left| B_W^{(k)}(W,J)(t) \right| \le \left| \int_{t_k}^{t} U(W,J)(s)\,ds \right| + \left| \int_{t_k}^{t_{k+1}} U(W,J)(s)\,ds \right| \equiv P_1 + P_2 .$$

We have

$$P_1 \le \left| \int_{t_k}^{t} \frac{dJ(s-T)}{dt}\,ds \right| + 2\sqrt{C}\left| \int_{t_k}^{t} \frac{dE_0(t)}{dt}\,dt \right| + \frac{1}{Z_0 \dot{\overline{C}}_0} \int_{t_k}^{t} \left| W(t) \right| dt + \frac{1}{Z_0 \dot{\overline{C}}_0} \int_{t_k}^{t} \left| J(t-T) \right| dt +$$

$$+ \frac{2\sqrt{C}}{\dot{\overline{C}}_0} \int_{t_k}^{t} \left| \widetilde{L}_0^{-1}\!\left(\int_{T}^{s}\!\left(\frac{W(\tau) - J(\tau-T)}{2\sqrt{C}} - E_0(\tau) \right) d\tau \right) \right| ds + \frac{2\sqrt{C}}{\dot{\overline{C}}_0} \int_{t_k}^{t} \left| G_0\!\left(\frac{W(s) - J(s-T)}{2\sqrt{C}} - E_0(s) \right) \right| ds \le$$

$$\le \left| J(t-T) \right| + 2\sqrt{C}\left| E_0(t) \right| + \frac{U_0}{Z_0 \dot{\overline{C}}_0} \frac{L}{R}\!\left(e^{-\frac{R}{L}t_k} - e^{-\frac{R}{L}t} \right) + \frac{I_0 e^{-\beta}}{Z_0 \dot{\overline{C}}_0} \frac{L}{R}\!\left(e^{-\frac{R}{L}t_k} - e^{-\frac{R}{L}t} \right) +$$

$$+ \frac{2\sqrt{C}\,I_0}{\dot{\overline{C}}_0} \int_{t_k}^{t}\!\int_{t_k}^{s}\!\left(\frac{\left| W(\tau) \right| + \left| J(\tau-T) \right|}{2\sqrt{C}} + \left| E_0(\tau) \right| \right) d\tau\,ds +$$

$$+ \frac{2\sqrt{C}}{\dot{\overline{C}}_0} \sum_{n=1}^{m} \left| g_n^{(0)} \right| \left| \int_{t_k}^{t}\!\left(\frac{\left| W(s) \right| + \left| J(s-T) \right|}{2\sqrt{C}} + \left| E_0(s) \right| \right) \right|^n ds \le$$

$$\le \left| \vec{J}(t) \right| + 2\sqrt{C}\,U_{E_0} e^{-\frac{R}{L}t} + \frac{U_0 + I_0 e^{-\beta}}{Z_0 \dot{\overline{C}}_0} \frac{L}{R}\!\left(e^{-\frac{R}{L}t_k} - e^{-\frac{R}{L}t} \right) + \frac{2\sqrt{C}\,I_0}{\dot{\overline{C}}_0}\!\left(\frac{U_0 + I_0 e^{-\beta}}{2\sqrt{C}} + U_{E_0} \right) \int_{t_k}^{t}\!\int_{t_k}^{s} e^{-\frac{R}{L}\tau}\,d\tau\,ds +$$

$$+ \frac{2\sqrt{C}}{\dot{\overline{C}}_0} \sum_{n=1}^{m} \left| g_n^{(0)} \right| \left(\frac{U_0 + I_0 e^{-\beta}}{2\sqrt{C}} + U_{E_0} \right)^n \int_{t_k}^{t} e^{-n\frac{R}{L}s}\,ds \le$$

$$\leq I_0 e^{-\beta} e^{-\frac{R}{L}t} + 2\sqrt{C}\,U_{E_0}\, e^{-\frac{R}{L}t} + \frac{L}{R}\left(e^{-\frac{R}{L}t_k} - e^{-\frac{R}{L}t}\right)\frac{U_0 + I_0 e^{-\beta}}{Z_0 \dot{\bar{C}}_0} +$$

$$+\left(\frac{L}{R}\right)^2\left(e^{-\frac{R}{L}t_k} - e^{-\frac{R}{L}t}\right)^2 \frac{2\sqrt{C}\,I_0}{\dot{\bar{C}}_0}\left(\frac{U_0 + I_0 e^{-\beta}}{2\sqrt{C}} + U_{E_0}\right) +$$

$$+\frac{2\sqrt{C}}{\dot{\bar{C}}_0}\sum_{n=1}^{m}\left|g_n^{(0)}\right|\left(\frac{U_0 + I_0 e^{-\beta}}{2\sqrt{C}} + U_{E_0}\right)^n \frac{L}{nR}\left(e^{-n\frac{R}{L}t_k} - e^{-n\frac{R}{L}t}\right) \leq$$

$$\leq I_0 e^{-\beta} e^{-\frac{R}{L}t} + 2\sqrt{C}\,U_{E_0}\, e^{-\frac{R}{L}t} + \frac{L}{R}\left(e^{-\frac{R}{L}t_k} - e^{-\frac{R}{L}t}\right)\frac{U_0 + I_0 e^{-\beta}}{Z_0 \dot{\bar{C}}_0} +$$

$$+\left(\frac{L}{R}\right)^2\left(e^{-\frac{R}{L}t_k} - e^{-\frac{R}{L}t}\right)^2 \frac{2\sqrt{C}\,I_0}{\dot{\bar{C}}_0}\left(\frac{U_0 + I_0 e^{-\beta}}{2\sqrt{C}} + U_{E_0}\right) +$$

$$+\frac{2\sqrt{C}}{\dot{\bar{C}}_0}\sum_{n=1}^{m}\left|g_n^{(0)}\right|\left(\frac{U_0 + I_0 e^{-\beta}}{2\sqrt{C}} + U_{E_0}\right)^n \frac{L}{R}\left(e^{-\frac{R}{L}t_k} - e^{-\frac{R}{L}t}\right) \leq$$

$$\leq e^{-\frac{R}{L}t}\left[I_0 e^{-\beta} + 2\sqrt{C}\,U_{E_0} + \frac{2L}{\dot{\bar{C}}_0 R}\sinh\left(\frac{RT_0}{L}\right)\left(\frac{U_0 + I_0 e^{-\beta}}{Z_0} + \frac{4L}{R}\sinh\left(\frac{RT_0}{L}\right)2\sqrt{C}\,I_0\phi_0 + 2\sqrt{C}\sum_{n=1}^{m}\left|g_n^{(0)}\right|\phi_0^n\right)\right].$$

Having in mind that $e^{-\frac{R}{L}t_k} - e^{-\frac{R}{L}t_{k+1}} \leq e^{-\frac{R}{L}t}\left(e^{\frac{R}{L}T_0} - e^{-\frac{R}{L}T_0}\right)$ we obtain

$$P_2 \leq \left|\int_{t_k}^{t_{k+1}}\frac{dJ(t-T)}{dt}dt\right| + 2\sqrt{C}\left|\int_{t_k}^{t_{k+1}}\frac{dE_0(t)}{dt}dt\right| + \frac{1}{Z_0\dot{\bar{C}}_0}\int_{t_k}^{t_{k+1}}\left|W(t)\right|dt + \frac{1}{Z_0\dot{\bar{C}}_0}\int_{t_k}^{t_{k+1}}\left|J(t-T)\right|dt +$$

$$+\frac{2\sqrt{C}}{\dot{\bar{C}}_0}\int_{t_k}^{t_{k+1}}\left|\widetilde{L}_0^{-1}\left(\int_{T}^{s}\left(\frac{W(\tau)-J(\tau-T)}{2\sqrt{C}} - E_0(\tau)\right)d\tau\right)\right|ds + \frac{2\sqrt{C}}{\dot{\bar{C}}_0}\int_{t_k}^{t_{k+1}}\left|G_0\left(\frac{W(s)-J(s-T)}{2\sqrt{C}} - E_0(s)\right)\right|ds \leq$$

$$\leq \frac{U_0}{Z_0\dot{\bar{C}}_0}\frac{L}{R}\left(e^{-\frac{R}{L}t_k} - e^{-\frac{R}{L}t_{k+1}}\right) + \frac{I_0 e^{-\frac{R}{L}T}}{Z_0\dot{\bar{C}}_0}\frac{L}{R}\left(e^{-\frac{R}{L}t_k} - e^{-\frac{R}{L}t_{k+1}}\right) +$$

$$+\frac{2\sqrt{C}\,I_0}{\dot{\bar{C}}_0}\int_{t_k}^{t_{k+1}}\int_{t_k}^{s}\left(\frac{\left|W(\tau)\right| + \left|J(\tau-T)\right|}{2\sqrt{C}} + \left|E_0(\tau)\right|\right)d\tau ds +$$

$$+\frac{2\sqrt{C}}{\dot{\bar{C}}_0}\sum_{n=1}^{m}\left|g_n^{(0)}\right|\left|\int_{t_k}^{t_{k+1}}\left(\frac{\left|W(s)\right| + \left|J(s-T)\right|}{2\sqrt{C}} + \left|E_0(s)\right|\right)\right|^n ds \leq$$

$$\leq \frac{U_0 + I_0}{Z_0 \dot{C}_0} \frac{L}{R}\left(e^{-\frac{R}{L}t_k} - e^{-\frac{R}{L}t_{k+1}}\right) + \frac{2\sqrt{C}}{\dot{C}_0} I_0 \left(\frac{U_0 + I_0 e^{-\beta}}{2\sqrt{C}} + U_{E0}\right) \int_{t_k}^{t_{k+1}} \int_{t_k}^{s} e^{-\frac{R}{L}\tau} \, d\tau \, ds +$$

$$+ \frac{2\sqrt{C}}{\dot{C}_0} \sum_{n=1}^{m} \left|g_n^{(0)}\right| \left(\frac{U_0 + I_0 e^{-\beta}}{2\sqrt{C}} + U_{E0}\right)^n \int_{t_k}^{t_{k+1}} e^{-n\frac{R}{L}s} \, ds \leq$$

$$\leq \frac{L}{R}\left(e^{-\frac{R}{L}t_k} - e^{-\frac{R}{L}t_{k+1}}\right)\frac{U_0 + I_0 e^{-\beta}}{Z_0 \dot{C}_0} + \left(\frac{L}{R}\right)^2 \left(e^{-\frac{R}{L}t_k} - e^{-\frac{R}{L}t_{k+1}}\right)^2 \frac{2\sqrt{C}}{\dot{C}_0} I_0 \left(\frac{U_0 + I_0 e^{-\beta}}{2\sqrt{C}} + U_{E0}\right) +$$

$$+ \frac{2\sqrt{C}}{\dot{C}_0} \sum_{n=1}^{m} \left|g_n^{(0)}\right| \left(\frac{U_0 + I_0 e^{-\beta}}{2\sqrt{C}} + U_{E0}\right)^n \frac{L}{nR}\left(e^{-n\frac{R}{L}t_k} - e^{-n\frac{R}{L}t_{k+1}}\right) \leq$$

$$\leq \frac{L}{R}\left(e^{-\frac{R}{L}t_k} - e^{-\frac{R}{L}t_{k+1}}\right)\frac{U_0 + I_0 e^{-\beta}}{Z_0 \dot{C}_0} + \left(\frac{L}{R}\right)^2 \left(e^{-\frac{R}{L}t_k} - e^{-\frac{R}{L}t_{k+1}}\right)^2 \frac{2\sqrt{C}}{\dot{C}_0} I_0 \left(\frac{U_0 + I_0 e^{-\beta}}{2\sqrt{C}} + U_{E0}\right) +$$

$$+ \frac{2\sqrt{C}}{\dot{C}_0} \sum_{n=1}^{m} \left|g_n^{(0)}\right| \left(\frac{U_0 + I_0 e^{-\beta}}{2\sqrt{C}} + U_{E0}\right)^n \frac{L}{R}\left(e^{-\frac{R}{L}t_k} - e^{-\frac{R}{L}t_{k+1}}\right) \leq$$

$$\leq e^{-\frac{R}{L}t} \frac{2L}{\dot{C}_0 R} \sinh\left(\frac{RT_0}{L}\right)\left[\frac{U_0 + I_0 e^{-\beta}}{Z_0} + \frac{4L\sqrt{C}}{R} I_0 \phi_0 \sinh\left(\frac{RT_0}{L}\right) + 2\sqrt{C}\sum_{n=1}^{m}\left|g_n^{(0)}\right|\phi_0^n\right].$$

Then

$$\left|B_W^{(k)}(W, J)(t)\right| \leq e^{-\frac{R}{L}t}\left[I_0 e^{-\beta} + 2\sqrt{C} U_{E0} + \frac{2L}{\dot{C}_0 R}\sinh\left(\frac{RT_0}{L}\right) \times \right.$$

$$\times \left(\frac{U_0 + I_0 e^{-\beta}}{Z_0} + 4\frac{L}{R}\sinh\left(\frac{RT_0}{L}\right)\sqrt{C}\, I_0 \phi_0 + 2\sqrt{C}\sum_{n=1}^{m}\left|g_n^{(0)}\right|\phi_0^n\right)\Bigg] +$$

$$+ e^{-\frac{R}{L}t}\frac{2L}{\dot{C}_0 R}\sinh\left(\frac{RT_0}{L}\right)\left[\frac{U_0 + I_0 e^{-\beta}}{Z_0} + 4\frac{L}{R}\sinh\left(\frac{RT_0}{L}\right)\sqrt{C}\, I_0(\phi_0) + 2\sqrt{C}\sum_{n=1}^{m}\left|g_n^{(0)}\right|\phi_0^n\right] \leq$$

$$\leq e^{-\frac{R}{L}t}\left[I_0 e^{-\beta} + 2\sqrt{C} U_{E0} + \frac{4L}{\dot{C}_0 R}\sinh\left(\frac{RT_0}{L}\right)\left(\frac{U_0 + I_0 e^{-\beta}}{Z_0} + 4\frac{L}{R}\sinh\left(\frac{RT_0}{L}\right)\sqrt{C}\, I_0 \phi_0 + 2\sqrt{C}\sum_{n=1}^{m}\left|g_n^{(0)}\right|\phi_0^n\right)\right] \leq$$

$$\leq U_0 e^{-\frac{R}{L}t}.$$

We have also

$$\left|B_J^{(k)}(W, J)(t)\right| \leq \left|\int_{t_k}^{t} I(W, J)(s)\,ds\right| + \left|\int_{t_k}^{t_{k+1}} I(W, J)(s)\,ds\right| \leq J_1 + J_2.$$

478 Vasil G. Angelov

Since

$$J_1 \leq \left| \int_{t_k}^{t} \frac{dW(s-T)}{dt} ds \right| + 2\sqrt{C} \left| \int_{t_k}^{t} \frac{dE_1(s)}{dt} ds \right| + \frac{1}{Z_0 \dot{\bar{C}}_1} \int_{t_k}^{t} |W(s-T)| ds + \frac{1}{Z_0 \dot{\bar{C}}_1} \int_{t_k}^{t} |J(s)| ds +$$

$$+ \frac{2\sqrt{C}}{\dot{\bar{C}}_1} \int_{t_k}^{t} \left| \widetilde{L}_1^{-1} \left(\int_{t_k}^{s} \left(\frac{W(\tau-T)-J(\tau)}{2\sqrt{C}} - E_1(\tau) \right) d\tau \right) \right| ds + \frac{2\sqrt{C}}{\dot{\bar{C}}_1} \int_{t_k}^{t} \left| G_1 \left(\frac{W(s-T)-J(s)}{2\sqrt{C}} - E_1(s) \right) \right| ds \leq$$

$$\leq |W(t-T)| + 2\sqrt{C} |E_1(t)| + \frac{U_0}{Z_0 \dot{\bar{C}}_1} \int_{t_k}^{t} e^{-\frac{R}{L}s} ds + \frac{I_0}{Z_0 \dot{\bar{C}}_1} \int_{t_k}^{t} e^{-\frac{R}{L}s} ds +$$

$$+ \frac{2\sqrt{C} I_0}{\dot{\bar{C}}_1} \int_{t_k}^{t} \int_{t_k}^{s} \left| \frac{W(\tau-T)-J(\tau)}{2\sqrt{C}} - E_1(\tau) \right| d\tau ds + \frac{2\sqrt{C}}{\dot{\bar{C}}_1} \sum |g_n^{(1)}| \left| \int_{t_k}^{t} \left| \frac{W(s-T)-J(s)}{2\sqrt{C}} - E_1(s) \right| ds \right|^n \leq$$

$$\leq U_0 e^{-\beta} e^{-\frac{R}{L}t} + 2\sqrt{C} U_{E_1} e^{-\frac{R}{L}t} + \frac{U_0 e^{-\beta}}{Z_0 \dot{\bar{C}}_1} \frac{L}{R} \left(e^{-\frac{R}{L}t_k} - e^{-\frac{R}{L}t} \right) + \frac{I_0}{Z_0 \dot{\bar{C}}_1} \frac{L}{R} \left(e^{-\frac{R}{L}t_k} - e^{-\frac{R}{L}t} \right) +$$

$$+ \frac{2\sqrt{C} I_0}{\dot{\bar{C}}_1} \left(\frac{U_0 e^{-\beta} + I_0}{2\sqrt{C}} + U_{E_1} \right) \int_{t_k}^{t} \int_{t_k}^{s} e^{-\frac{R}{L}\tau} d\tau ds + \frac{2\sqrt{C}}{\dot{\bar{C}}_1} \sum_{n=1}^{m} |g_n^{(1)}| \left(\frac{U_0 e^{-\beta} + I_0}{2\sqrt{C}} + U_{E_1} \right)^n \int_{t_k}^{t} e^{-\frac{R}{L}s} ds \leq$$

$$\leq e^{-\frac{R}{L}t} \left[U_0 e^{-\beta} + 2\sqrt{C} U_{E_1} + \frac{L}{\dot{\bar{C}}_1 R} \left(e^{\frac{R}{L}T_0} - e^{-\frac{R}{L}T_0} \right) \times \right.$$

$$\left. \times \left(\frac{U_0 e^{-\beta} + I_0}{Z_0} + 2\sqrt{C} I_0 \frac{L}{R} \left(e^{\frac{R}{L}T_0} - e^{-\frac{R}{L}T_0} \right) \phi_0 + 2\sqrt{C} \sum_{n=1}^{m} |g_n^{(1)}| \phi_0^n \right) \right]$$

and

$$J_2 \leq \left| \int_{t_k}^{t_{k+1}} \frac{dW(s-T)}{dt} ds \right| + 2\sqrt{C} \left| \int_{t_k}^{t_{k+1}} \frac{dE_1(s)}{dt} ds \right| + \frac{1}{Z_0 \dot{\bar{C}}_1} \int_{t_k}^{t_{k+1}} |W(s-T)| ds + \frac{1}{Z_0 \dot{\bar{C}}_1} \int_{t_k}^{t_{k+1}} |J(s)| ds +$$

$$+ \frac{2\sqrt{C}}{\dot{\bar{C}}_1} \int_{t_k}^{t_{k+1}} \left| \widetilde{L}_1^{-1} \left(\int_{t_k}^{s} \left(\frac{W(\tau-T)-J(\tau)}{2\sqrt{C}} - E_1(\tau) \right) d\tau \right) \right| ds + \frac{2\sqrt{C}}{\dot{\bar{C}}_1} \int_{t_k}^{t_{k+1}} \left| G_1 \left(\frac{W(s-T)-J(s)}{2\sqrt{C}} - E_1(s) \right) \right| ds \leq$$

$$\leq \frac{U_0 e^{\mu_0}}{Z_0 \dot{\bar{C}}_1} \frac{L}{R} \left(e^{-\frac{R}{L}t_k} - e^{-\frac{R}{L}t_{k+1}} \right) + \frac{I_0}{Z_0 \dot{\bar{C}}_1} \frac{L}{R} \left(e^{-\frac{R}{L}t_k} - e^{-\frac{R}{L}t_{k+1}} \right) +$$

$$+ \frac{2\sqrt{C} I_0}{\dot{\bar{C}}_1} \left(\frac{U_0 e^{-\beta} + I_0}{2\sqrt{C}} + U_{E_1} \right) \left(\frac{L}{R} \right)^2 \left(e^{-\frac{R}{L}t_k} - e^{-\frac{R}{L}t_{k+1}} \right)^2 +$$

$$+ \frac{2\sqrt{C}}{\dot{\bar{C}}_1} \sum_{n=1}^{m} |g_n^{(1)}| \left(\frac{U_0 e^{-\beta} + I_0}{2\sqrt{C}} + U_{E_1} \right)^n \frac{L}{R} \left(e^{-\frac{R}{L}t_k} - e^{-\frac{R}{L}t_{k+1}} \right) \leq$$

$$\le e^{-\frac{R}{L}t}\frac{L}{\dot{\widetilde{C}}_1 R}\left(e^{\frac{RT_0}{L}}-e^{-\frac{RT_0}{L}}\right)\left(\frac{U_0 e^{-\beta}+I_0}{Z_0}+2\sqrt{C}I_0\phi_0\frac{L}{R}\left(e^{\frac{RT_0}{L}}-e^{-\frac{RT_0}{L}}\right)+2\sqrt{C}\sum_{n=1}^{m}\left|g_n^{(1)}\right|\phi_0^n\right)$$

then

$$\left|B_J^{(k)}(W,J)(t)\right|\le e^{-\frac{R}{L}t}\left[U_0 e^{-\beta}+2\sqrt{C}U_{E_1}+\frac{2L}{\dot{\widetilde{C}}_1 R}\left(e^{\frac{R}{L}T_0}-e^{-\frac{R}{L}T_0}\right)\times\right.$$

$$\left.\times\left(\frac{U_0 e^{-\beta}+I_0}{Z_0}+2\sqrt{C}I_0\frac{L}{R}\left(e^{\frac{R}{L}T_0}-e^{-\frac{R}{L}T_0}\right)\phi_0+2\sqrt{C}\sum_{n=1}^{m}\left|g_n^{(1)}\right|\phi_0^n\right)\right]\le I_0 e^{-\frac{R}{L}t}.$$

Therefore the operator B maps $M = M_W \times M_J$ into itself.

We show that B is a contractive one. We need the following inequalities

$$\sup\widetilde{L}_0^{-1}(.)\equiv\sup\left\{\widetilde{L}_0^{-1}\left(\int_{t_k}^{t}\left(\frac{W(\tau)-J(\tau-T)}{2\sqrt{C}}-E_0(\tau-T)\right)d\tau\right):t\in[t_k,t_{k+1}]\right\}\le I_0$$

$$\sup G_0(.)\equiv\sup\left\{G_0\left(\frac{W(t)-J(t-T)}{2\sqrt{C}}-E_0(t-T)\right):t\in[t_k,t_{k+1}]\right\}\le$$

$$\le\sum_{n=1}^{m}\left|g_n^{(0)}\right|\left(\frac{U_0+I_0 e^{-\beta}}{2\sqrt{C}}+U_{E0}\right)^n e^{n\mu_0}\le\sum_{n=1}^{m}\left|g_n^{(0)}\right|\phi_0^n e^{n\mu_0}.$$

We have

$$\left|B_W^{(k)}(W,J)(t)-B_W^{(k)}(\overline{W},\overline{J})(t)\right|\le\left|\int_{t_k}^{t}\left(U(W,J)(s)-U(\overline{W},\overline{J})(s)\right)ds\right|+$$

$$+\left|\int_{t_k}^{t_{k+1}}\left(U(W,J)(s)-U(\overline{W},\overline{J})(s)\right)ds\right|\equiv P_1+P_2$$

and then

$$P_1\le\left|\int_{t_k}^{t}\left(\frac{dJ(s-T)}{dt}-\frac{d\overline{J}(s-T)}{dt}\right)ds\right|+\frac{1}{Z_0\dot{\overline{C}}_0}\int_{t_k}^{t}\left|W(s)-\overline{W}(s)\right|ds+\frac{1}{Z_0\overline{C}_0}\int_{t_k}^{t}\left|J(s-T)-\overline{J}(s-T)\right|ds+$$

$$+\frac{U_0+I_0 e^{-\beta}}{Z_0}\int_{t_k}^{t}\frac{1}{\left(d\widetilde{C}_0\left(\frac{W(t)-J(t-T)}{2\sqrt{C}}-E_0(t).\right)/du\right)^2}\left|\frac{d^2\widetilde{C}_0}{du^2}\right|\left(\left|W(s)-\overline{W}(s)\right|+\left|J(s-T)-\overline{J}(s-T)\right|\right)ds+$$

$$+\frac{2\sqrt{C}}{\dot{\overline{C}}_0}\int_{t_k}^{t}\left|\widetilde{L}_0^{-1}\left(\int_{T}^{s}\left(\frac{W(\tau)-J(\tau-T)}{2\sqrt{C}}-E_0(\tau)\right)d\tau\right)-\widetilde{L}_0^{-1}\left(\int_{T}^{s}\left(\frac{\overline{W}(\tau)-\overline{J}(\tau-T)}{2\sqrt{C}}-E_0(\tau)\right)d\tau\right)\right|ds+$$

$$+ \frac{2\sqrt{C}}{\dot{\overline{C}}_0} \int_{t_k}^{t} \left| G_0 \left(\frac{W(s) - J(s-T)}{2\sqrt{C}} - E_0(s) \right) - G_0 \left(\frac{\overline{W}(s) - \overline{J}(s-T)}{2\sqrt{C}} - E_0(s) \right) \right| ds +$$

$$+ \frac{\sup G_0(.)}{Z_0} \int_{t_k}^{t} \frac{1}{\left(d\widetilde{C}_0 \left(\frac{W(t) - J(t-T)}{2\sqrt{C}} - E_0(t) \right) / du \right)^2} \left| \frac{d^2 \widetilde{C}_0}{du^2} \right| \left(|W(s) - \overline{W}(s)| + |J(s-T) - \overline{J}(s-T)| \right) ds \leq$$

$$\leq \frac{\rho^{(k)}(W, \overline{W})}{Z_0 \dot{\overline{C}}_0} \int_{t_k}^{t} e^{\mu(s-t_k)} ds + \frac{U_0 + I_0 e^{-\beta}}{Z_0 \dot{\overline{C}}_0^2} H_0 \rho^{(k)}(W, \overline{W}) \int_{t_k}^{t} e^{\mu(s-t_k)} ds +$$

$$+ \frac{2\sqrt{C}}{\dot{\overline{C}}_0} \frac{1}{\hat{L}_0} \frac{1}{2\sqrt{C}} \int_{t_k}^{t} \int_{t_k}^{s} |W(\tau) - \overline{W}(\tau)| d\tau ds + \frac{\sup \widetilde{L}_0^{-1}(.)}{Z_0 \dot{\overline{C}}_0^2} H_0 \rho^{(k)}(W, \overline{W}) \int_{t_k}^{t} e^{\mu(s-t_k)} ds +$$

$$+ \frac{2\sqrt{C}}{\dot{\overline{C}}_0} \sum_{n=1}^{m-1} n |g_n^{(0)}| (\phi_0)^{n-1} \frac{\rho^{(k)}(W, \overline{W})}{2\sqrt{C}} \int_{t_k}^{t} e^{\mu(s-t_k)} ds + \frac{\sup G_0(.)}{Z_0 \dot{\overline{C}}_0^2} H_0 \rho^{(k)}(W, \overline{W}) \int_{t_k}^{t} e^{\mu(s-t_k)} ds \leq$$

$$\leq \int_{t_k}^{t} e^{\mu(s-t_k)} ds \, e^{\mu_0} \rho_\mu^{(k)}(W, \overline{W}) \left[\frac{1}{Z_0 \dot{\overline{C}}_0} + \frac{U_0 + I_0 e^{-\beta}}{Z_0 \dot{\overline{C}}_0^2} H_0 + \frac{1}{\mu \hat{L}_0 \dot{\overline{C}}_0} + \right.$$

$$\left. + \frac{\sup \widetilde{L}_0^{-1}(.)}{Z_0 \dot{\overline{C}}_0^2} H_0 + \frac{1}{\dot{\overline{C}}_0} \sum_{n=1}^{m-1} n |g_n^{(0)}| (\phi_0)^{n-1} + \frac{\sup G_0(.)}{Z_0 \dot{\overline{C}}_0^2} H_0 \right] \leq$$

$$\leq \frac{e^{\mu(s-t_k)} - 1}{\mu} e^{\mu_0} \frac{\rho_\mu^{(k)}(\dot{W}, \dot{\overline{W}})}{\mu} \left[\frac{1}{Z_0 \dot{\overline{C}}_0} + \frac{U_0 + I_0 e^{-\beta}}{Z_0 \dot{\overline{C}}_0^2} H_0 + \frac{1}{\mu \hat{L}_0 \dot{\overline{C}}_0} + \right.$$

$$\left. + \frac{I_0}{Z_0 \dot{\overline{C}}_0^2} H_0 + \frac{1}{\mu \dot{\overline{C}}_0} \sum_{n=1}^{m-1} n |g_n^{(0)}| (\phi_0)^{n-1} + \frac{H_0}{Z_0 \dot{\overline{C}}_0^2} \sum_{n=1}^{m} |g_n^{(0)}| \phi_0^{\, n} \right] \leq$$

$$\leq e^{\mu(s-t_k)} \hat{\rho}_\mu^{(k)}((W, \dot{W}, J, \dot{J}), (\overline{W}, \dot{\overline{W}}, \overline{J}, \dot{\overline{J}})) \frac{e^{\mu_0}}{\mu^2 \dot{\overline{C}}_0} \left[\frac{1}{Z_0} + \frac{1}{\mu \hat{L}_0} + \sum_{n=1}^{m-1} n |g_n^{(0)}| \phi_0^{n-1} + \right.$$

$$\left. + \frac{H_0}{Z_0 \dot{\overline{C}}_0} \left(U_0 + I_0 e^{-\beta} + I_0 + \sum_{n=1}^{m} |g_n^{(0)}| \phi_0^{\, n} \right) \right];$$

$$P_2 \leq \left| \int_{t_k}^{t_{k+1}} \left(\frac{dJ(s-T)}{dt} - \frac{dJ(s-T)}{dt} \right) ds \right| + \frac{1}{Z_0 \dot{\overline{C}}_0} \int_{t_k}^{t_{k+1}} |W(s) - \overline{W}(s)| ds + \frac{1}{Z_0 \dot{\overline{C}}_0} \int_{t_k}^{t_{k+1}} |J(s-T) - \overline{J}(s-T)| ds +$$

$$+ \frac{U_0 + I_0 e^{-\beta}}{Z_0} \int_{t_k}^{t_{k+1}} \frac{1}{\left(d\widetilde{C}_0 \left(\frac{W(t) - J(t-T)}{2\sqrt{C}} - E_0(t) \right) / du \right)^2} \left| \frac{d^2 \widetilde{C}_0}{du^2} \right| \left(|W(s) - \overline{W}(s)| + |J(s-T) - \overline{J}(s-T)| \right) ds +$$

$$+\frac{2\sqrt{C}}{\dot{\overline{C}}_0}\int_{t_k}^{t_{k+1}}\left|\widetilde{L}_0^{-1}\left(\int_T^s\left(\frac{W(\tau)-J(\tau-T)}{2\sqrt{C}}-E_0(\tau)\right)d\tau\right)-\widetilde{L}_0^{-1}\left(\int_T^s\left(\frac{\overline{W}(\tau)-\overline{J}(\tau-T)}{2\sqrt{C}}-E_0(\tau)\right)d\tau\right)\right|ds+$$

$$+\frac{2\sqrt{C}}{\dot{\overline{C}}_0}\int_{t_k}^{t_{k+1}}\left|G_0\left(\frac{W(s)-J(s-T)}{2\sqrt{C}}-E_0(s)\right)-G_0\left(\frac{\overline{W}(s)-\overline{J}(s-T)}{2\sqrt{C}}-E_0(s)\right)\right|ds\le$$

$$\le e^{\mu(s-t_k)}\,\hat{\rho}_\mu^{(k)}((W,\dot{W},J,\dot{J}),(\overline{W},\dot{\overline{W}},\overline{J},\dot{\overline{J}}))(e^{\mu_0}-1)\frac{e^{\mu_0}}{\mu^2\dot{\overline{C}}_0}\left[\frac{1}{Z_0}+\frac{1}{\mu\hat{L}_0}+\sum_{n=1}^{m-1}n\left|g_n^{(0)}\right|\phi_0^{n-1}+\right.$$

$$\left.+\frac{H_0}{Z_0\dot{\overline{C}}_0}\left(U_0+I_0e^{-\beta}+I_0+\sum_{n=1}^m\left|g_n^{(0)}\right|\phi_0^{\,n}\right)\right].$$

Then

$$\left|B_W^{(k)}(W,J)(t)-B_W^{(k)}(\overline{W},\overline{J})(t)\right|\le$$

$$\le e^{\mu_0}\hat{\rho}_\mu^{(k)}((W,\dot{W},J,\dot{J}),(\overline{W},\dot{\overline{W}},\overline{J},\dot{\overline{J}}))\times$$

$$\times e^{\mu(s-t_k)}\frac{e^{2\mu_0}}{\mu^2\dot{\overline{C}}_0}\left[\frac{1}{Z_0}+\frac{1}{\mu\hat{L}_0}+\sum_{n=1}^{m-1}n\left|g_n^{(0)}\right|\phi_0^{n-1}+\frac{H_0}{Z_0\dot{\overline{C}}_0}\left(U_0+I_0e^{-\beta}+I_0+\sum_{n=1}^m\left|g_n^{(0)}\right|\phi_0^{\,n}\right)\right]\le$$

$$\le e^{\mu_0}K_W\hat{\rho}_\mu^{(k)}((W,\dot{W},J,\dot{J}),(\overline{W},\dot{\overline{W}},\overline{J},\dot{\overline{J}}))\,.$$

It follows

$$\hat{\rho}^{(k)}(B_W(W,J),B_W(\overline{W},\overline{J}))\le e^{\mu_0}K_W\hat{\rho}_\mu^{(k)}((W,\dot{W},J,\dot{J}),(\overline{W},\dot{\overline{W}},\overline{J},\dot{\overline{J}}))\,.$$

Further on we have

$$\left|B_J^{(k)}(W,J)(t)-B_J^{(k)}(\overline{W},\overline{J})(t)\right|\le\left|\int_{t_k}^t\left(I(W,J)(s)-I(\overline{W},\overline{J})(s)\right)ds\right|+$$

$$+\left|\frac{t-t_k}{t_{k+1}-t_k}\int_{t_k}^{t_{k+1}}\left(I(W,J)(s)-I(\overline{W},\overline{J})(s)\right)ds\right|\equiv I_1+I_2$$

But

$$I_1\le\left|\int_{t_k}^t\left(\frac{dW(s-T)}{dt}-\frac{d\overline{W}(s-T)}{dt}\right)ds\right|+\frac{1}{Z_0\dot{\overline{C}}_1}\int_{t_k}^t\left|W(s-T)-\overline{W}(s-T)\right|ds+\frac{1}{Z_0\dot{\overline{C}}_1}\int_{t_k}^t\left|J(s)-\overline{J}(s)\right|ds+$$

$$+\frac{U_0 e^{-\beta}+I_0}{Z_0}\int\limits_{t_k}^{t}\frac{1}{\left(d\widetilde{C}_1\left(\frac{W(t-T)-J(t)}{2\sqrt{C}}-E_1(t)\right)/du\right)^2}\left|\frac{d^2\widetilde{C}_1}{du^2}\right|\left(\left|W(s-T)-\overline{W}(s-T)\right|+\left|J(s)-\overline{J}(s)\right|\right)ds+$$

$$+\frac{2\sqrt{C}}{\overset{\cdot}{\overline{C}}_1}\int\limits_{t_k}^{t}\left|\widetilde{L}_1^{-1}\left(\int\limits_{T}^{s}\left(\frac{W(\tau-T)-J(\tau)}{2\sqrt{C}}-E_1(\tau)\right)d\tau\right)-\widetilde{L}_1^{-1}\left(\int\limits_{T}^{s}\left(\frac{\overline{W}(\tau-T)-\overline{J}(\tau)}{2\sqrt{C}}-E_1(\tau)\right)d\tau\right)\right|ds+$$

$$+\frac{2\sqrt{C}}{\overset{\cdot}{\overline{C}}_1}\int\limits_{t_k}^{t}\left|G_1\left(\frac{W(s-T)-J(s)}{2\sqrt{C}}-E_1(s)\right)-G_1\left(\frac{W(s-T)-J(s)}{2\sqrt{C}}-E_1(s)\right)\right|ds+$$

$$+\frac{\sup G_1(.)}{Z_0}\int\limits_{t_k}^{t}\frac{1}{\left(d\widetilde{C}_1\left(\frac{W(t-T)-J(t)}{2\sqrt{C}}-E_1(t)\right)/du\right)^2}\left|\frac{d^2\widetilde{C}_1}{du^2}\right|\left(\left|W(s-T)-\overline{W}(s-T)\right|+\left|J(s)-\overline{J}(s)\right|\right)ds\leq$$

$$\leq\frac{\rho^{(k)}(J,\overline{J})}{Z_0\overset{\cdot}{\overline{C}}_1}\int\limits_{t_k}^{t}e^{\mu(s-t_k)}ds+\frac{U_0 e^{-\beta}+I_0}{Z_0}\frac{\sup G_1(.)}{Z_0\overset{\cdot}{\overline{C}}_1^{2}}H_1\rho^{(k)}(J,\overline{J})\int\limits_{t_k}^{t}e^{\mu(s-t_k)}ds\leq$$

$$+\frac{2\sqrt{C}}{\overset{\cdot}{\overline{C}}_1}\frac{1}{\hat{L}_1}\frac{1}{2\sqrt{C}}\int\limits_{t_k}^{t}\int\limits_{t_k}^{s}\left|J(\tau)-\overline{J}(\tau)\right|d\tau ds+\frac{\sup\widetilde{L}_1^{-1}(.)}{Z_0\overset{\cdot}{\overline{C}}_1^{2}}H_1\rho^{(k)}(J,\overline{J})\int\limits_{t_k}^{t}e^{\mu(s-t_k)}ds+$$

$$+\frac{2\sqrt{C}}{\overset{\cdot}{\overline{C}}_1}\sum_{n=1}^{m-1}n\left|g_n^{(1)}\right|(\phi_0)^{n-1}\frac{1}{2\sqrt{C}}\rho^{(k)}(J,\overline{J})\int\limits_{t_k}^{t}e^{\mu(s-t_k)}ds+\frac{\sup G_1(.)}{Z_0\overset{\cdot}{\overline{C}}_1^{2}}H_1\rho^{(k)}(J,\overline{J})\int\limits_{t_k}^{t}e^{\mu(s-t_k)}ds\leq$$

$$\leq e^{\mu(t-t_k)}\rho_{\mu}^{(k)}(\dot{J},\dot{\overline{J}})\frac{e^{2\mu_0}}{\mu^2\overset{\cdot}{\overline{C}}_1}\left[\frac{1}{Z_0}+\frac{1}{\mu\hat{L}_1}+\frac{1}{\overset{\cdot}{\overline{C}}_1}\sum_{n=1}^{m-1}n\left|g_n^{(1)}\right|\phi_0^{n-1}+\frac{H_1}{Z_0\overset{\cdot}{\overline{C}}_1}\left(U_0 e^{-\beta}+I_0+I_0+\sum_{n=1}^{m}\left|g_n^{(1)}\right|\phi_0^{n}\right)\right]\leq$$

$$\leq e^{\mu(t-t_k)}\hat{\rho}_{\mu}^{(k)}((W,\dot{W},J,\dot{J}),(\overline{W},\dot{\overline{W}},\overline{J},\dot{\overline{J}}))\times$$

$$\times\frac{e^{2\mu_0}}{\mu^2\overset{\cdot}{\overline{C}}_1}\left[\frac{1}{Z_0}+\frac{1}{\mu\hat{L}_1}+\frac{1}{\overset{\cdot}{\overline{C}}_1}\sum_{n=1}^{m-1}n\left|g_n^{(1)}\right|\phi_0^{n-1}+\frac{H_1}{Z_0\overset{\cdot}{\overline{C}}_1}\left(U_0 e^{-\beta}+I_0+I_0+\sum_{n=1}^{m}\left|g_n^{(1)}\right|\phi_0^{n}\right)\right]$$

and

$$P_2\leq\left|\int\limits_{t_k}^{t_{k+1}}\left(\frac{dW(s-T)}{dt}-\frac{d\overline{W}(s-T)}{dt}\right)ds\right|+\frac{1}{Z_0\overset{\cdot}{\overline{C}}_1}\int\limits_{t_k}^{t_{k+1}}\left|W(s-T)-\overline{W}(s-T)\right|ds+\frac{1}{Z_0\overset{\cdot}{\overline{C}}_1}\int\limits_{t_k}^{t_{k+1}}\left|J(s)-\overline{J}(s)\right|ds+$$

$$+\frac{2\sqrt{C}}{\overset{\cdot}{\overline{C}}_1}\int\limits_{t_k}^{t_{k+1}}\left|\widetilde{L}_1^{-1}\left(\int\limits_{T}^{s}\left(\frac{W(\tau-T)-J(\tau)}{2\sqrt{C}}-E_1(\tau)\right)d\tau\right)-\widetilde{L}_1^{-1}\left(\int\limits_{T}^{s}\left(\frac{\overline{W}(\tau-T)-\overline{J}(\tau)}{2\sqrt{C}}-E_1(\tau)\right)d\tau\right)\right|ds+$$

$$+\frac{2\sqrt{C}}{\overset{\cdot}{\overline{C}}_1}\int\limits_{t_k}^{t_{k+1}}\left|G_1\left(\frac{W(s-T)-J(s)}{2\sqrt{C}}-E_1(s)\right)-G_1\left(\frac{\overline{W}(s-T)-\overline{J}(s)}{2\sqrt{C}}-E_1(s)\right)\right|ds\leq$$

$$\leq e^{\mu(t-t_k)}\hat{\rho}_\mu^{(k)}((W,\dot{W},J,\dot{J}),(\overline{W},\dot{\overline{W}},\overline{J},\dot{\overline{J}}))\times$$

$$\times\left(e^{\mu_0}-1\right)\frac{e^{2\mu_0}}{\mu^2\dot{\overline{C}}_1}\left[\frac{1}{Z_0}+\frac{1}{\mu\hat{L}_1}+\frac{1}{\dot{\overline{C}}_1}\sum_{n=1}^{m-1}n\left|g_n^{(1)}\right|\phi_0^{n-1}+\frac{H_1}{Z_0\dot{\overline{C}}_1}\left(U_0e^{-\beta}+I_0+I_0+\sum_{n=1}^{m}\left|g_n^{(1)}\right|\phi_0^{\ n}\right)\right].$$

Consequently

$$\left|B_J^{(k)}(W,J)(t)-B_J^{(k)}(\overline{W},\overline{J})(t)\right|\leq$$

$$\leq e^{\mu(t-t_k)}\hat{\rho}_\mu^{(k)}((W,\dot{W},J,\dot{J}),(\overline{W},\dot{\overline{W}},\overline{J},\dot{\overline{J}}))\times$$

$$\times\frac{e^{3\mu_0}}{\mu^2\dot{\overline{C}}_1}\left[\frac{1}{Z_0}+\frac{1}{\mu\hat{L}_1}+\frac{1}{\dot{\overline{C}}_1}\sum_{n=1}^{m-1}n\left|g_n^{(1)}\right|\phi_0^{n-1}+\frac{H_1}{Z_0\dot{\overline{C}}_1}\left(U_0e^{-\beta}+I_0+I_0+\sum_{n=1}^{m}\left|g_n^{(1)}\right|\phi_0^{\ n}\right)\right]\equiv$$

$$\equiv e^{\mu_0}K_J\hat{\rho}_\mu^{(k)}((W,\dot{W},J,\dot{J}),(\overline{W},\dot{\overline{W}},\overline{J},\dot{\overline{J}}))$$

that is,

$$\hat{\rho}^{(k)}(B_J(W,J),B_J(\overline{W},\overline{J}))\leq e^{\mu_0}K_J\hat{\rho}_\mu^{(k)}((W,\dot{W},J,\dot{J}),(\overline{W},\dot{\overline{W}},\overline{J},\dot{\overline{J}})).$$

Remark 6.7.4. We notice the function $\xi(h)=\dfrac{e^h-1}{h}$ is well defined at $h=0$ because

$\lim\limits_{h\to 0}\dfrac{e^h-1}{h}=1$ and $\xi(h)$ is monotone increasing. Indeed, $\xi'(h)=\dfrac{he^h-e^h+1}{h^2}$. For $h\geq 1$

obviously $\xi'(h)>0$. For $0\leq h<1$ we have

$$he^h-e^h+1>0\Leftrightarrow\left(1-h\right)e^h<1\Leftrightarrow g(h)<g(0),$$

where $g(h)=\left(1-h\right)e^h$.

Therefore $\lim\limits_{(t_{k+1}-t_k)\to 0}\dfrac{e^{\mu(t_{k+1}-t_k)}-1}{\mu\left(t_{k+1}-t_k\right)}=1, k\to\infty$ and then $\dfrac{e^{\mu(t_{k+1}-t_k)}-1}{\mu\left(t_{k+1}-t_k\right)}\leq\dfrac{e^{\mu T_0}-1}{\mu T_0}$.

For the derivative we obtain

$$\left|\dot{B}_W^{(k)}(W,J)(t)-\dot{B}_W^{(k)}(\overline{W},\overline{J})(t)\right|\leq\dot{P}_1+\dot{P}_2.$$

Since

$$\dot{P_1} \le \left| \frac{dJ(s-T)}{dt} - \frac{d\overline{J}(s-T)}{dt} \right| + \frac{1}{Z_0 \dot{\overline{C}}_0} \left| W(t) - \overline{W}(t) \right| + \frac{1}{Z_0 \dot{\overline{C}}_0} \left| J(t-T) - \overline{J}(t-T) \right| +$$

$$+ \frac{U_0 + I_0 e^{\mu_0}}{Z_0} \frac{1}{\left(d\widetilde{C}_0 \left(\frac{W(t)-J(t-T)}{2\sqrt{C}} - E_0(t) \right)/du \right)^2} \left| \frac{d^2\widetilde{C}_0}{du^2} \right| \left(\left| W(t) - \overline{W}(t) \right| + \left| J(t-T) - \overline{J}(t-T) \right| \right) +$$

$$+ \frac{2\sqrt{C}}{\dot{\overline{C}}_0} \left| \widetilde{L}_0^{-1} \left(\int_{t_k}^{t} \left(\frac{W(\tau)-J(\tau-T)}{2\sqrt{C}} - E_0(\tau) \right) d\tau \right) - \widetilde{L}_0^{-1} \left(\int_{t_k}^{t} \left(\frac{\overline{W}(\tau)-\overline{J}(\tau-T)}{2\sqrt{C}} - E_0(\tau) \right) d\tau \right) \right| +$$

$$+ \frac{2\sqrt{C}}{\dot{\overline{C}}_0} \left| G_0 \left(\frac{W(t)-J(t-T)}{2\sqrt{C}} - E_0(t) \right) - G_0 \left(\frac{W(t)-J(t-T)}{2\sqrt{C}} - E_0(t) \right) \right| +$$

$$+ \frac{\sup G_0(.)}{Z_0} \frac{1}{\left(d\widetilde{C}_0 \left(\frac{W(t)-J(t-T)}{2\sqrt{C}} - E_0(t-T) \right)/du \right)^2} \left| \frac{d^2\widetilde{C}_0}{du^2} \right| \left(\left| W(s) - \overline{W}(s) \right| + \left| J(s-T) - \overline{J}(s-T) \right| \right) \le$$

$$\le \frac{\rho_\mu^{(k)}(W,\overline{W}) e^{\mu_0}}{Z_0 \dot{\overline{C}}_0} + \frac{U_0 + I_0 e^{-\beta}}{Z_0 \dot{\overline{C}}_0^2} H_0 e^{\mu_0} \rho_\mu^{(k)}(W,\overline{W}) +$$

$$+ \frac{1}{\dot{\overline{C}}_0 \hat{L}_0} e^{\mu_0} \rho_\mu^{(k)}(W,\overline{W}) \int_{t_k}^{t} e^{\mu(s-t_k)} d\tau + \frac{\sup \widetilde{L}_0^{-1}(.)}{Z_0 \dot{\overline{C}}_0^2} H_0 e^{\mu_0} \rho_\mu^{(k)}(W,\overline{W}) +$$

$$+ \frac{e^{\mu_0} \rho_\mu^{(k)}(W,\overline{W})}{\dot{\overline{C}}_0} \sum_{n=1}^{m-1} n \left| g_n^{(0)} \right| \phi_0^{n-1} + \frac{\sup G_0(.)}{Z_0 \dot{\overline{C}}_0^2} H_0 e^{\mu_0} \rho_\mu^{(k)}(W,\overline{W}) \le$$

$$\le e^{\mu_0} \frac{\rho_\mu^{(k)}(\dot{W},\dot{\overline{W}})}{\mu \dot{\overline{C}}_0} \left(\frac{1}{Z_0} + \frac{U_0 + I_0 e^{-\beta}}{Z_0 \dot{\overline{C}}_0} H_0 + \frac{1}{\mu \hat{L}_0} + \frac{I_0 H_0}{Z_0 \dot{\overline{C}}_0} + \sum_{n=1}^{m-1} n \left| g_n^{(0)} \right| \phi_0^{n-1} + \frac{H_0}{Z_0 \dot{\overline{C}}_0} \sum_{n=1}^{m} \left| g_n^{(0)} \right| \phi_0^{n} \right) \le$$

$$\le \hat{\rho}_\mu^{(k)} \left((W,\dot{W},J,\dot{J}), (\overline{W},\dot{\overline{W}},\overline{J},\dot{\overline{J}}) \right) \times$$

$$\times \frac{e^{\mu_0}}{\mu \dot{\overline{C}}_0} \left[\frac{1}{Z_0} + \frac{1}{\mu \hat{L}_0} + \sum_{n=1}^{m-1} n \left| g_n^{(0)} \right| \phi_0^{n-1} + \frac{H_0}{Z_0 \dot{\overline{C}}_0} \left(U_0 + I_0 e^{-\beta} + I_0 + \sum_{n=1}^{m} \left| g_n^{(0)} \right| \phi_0^{n} \right) \right]$$

and

$$\dot{P_2} \le e^{\mu(t-t_k)} \hat{\rho}_\mu^{(k)} \left((W,\dot{W},J,\dot{J}), (\overline{W},\dot{\overline{W}},\overline{J},\dot{\overline{J}}) \right) \times$$

$$\times \frac{e^{\mu_0}-1}{\mu_0} \frac{e^{\mu_0}}{\mu \dot{\overline{C}}_0} \left[\frac{1}{Z_0} + \frac{1}{\mu \hat{L}_0} + \sum_{n=1}^{m-1} n \left| g_n^{(0)} \right| \phi_0^{n-1} + \frac{H_0}{Z_0 \dot{\overline{C}}_0} \left(U_0 + I_0 e^{-\beta} + I_0 + \sum_{n=1}^{m} \left| g_n^{(0)} \right| \phi_0^{n} \right) \right].$$

Then

$$\left| \dot{B}_W^{(k)}(W,J)(t) - \dot{B}_W^{(k)}(\overline{W},\overline{J})(t) \right| \le$$

$$\le e^{\mu(t-t_k)} \hat{\rho}_\mu^{(k)}((W,\dot{W},J,\dot{J}),(\overline{W},\dot{\overline{W}},\overline{J},\dot{\overline{J}})) \times$$

$$\times \left(1 + \frac{e^{\mu_0}-1}{\mu_0} \right) \frac{e^{\mu_0}}{\mu \dot{\overline{C}}_0} \left[\frac{1}{Z_0} + \frac{1}{\mu \hat{L}_0} + \sum_{n=1}^{m-1} n \left| g_n^{(0)} \right| \phi_0^{\,n-1} + \frac{H_0}{Z_0 \dot{\overline{C}}_0} \left(U_0 + I_0 e^{-\beta} + I_0 + \sum_{n=1}^{m} \left| g_n^{(0)} \right| \phi_0^{\,n} \right) \right] \equiv$$

$$\equiv e^{\mu(t-t_k)} \dot{K}_W \hat{\rho}_\mu^{(k)}((W,\dot{W},J,\dot{J}),(\overline{W},\dot{\overline{W}},\overline{J},\dot{\overline{J}})).$$

Therefore

$$\rho_\mu^{(k)}(\dot{B}_W^{(k)}(W,J), \dot{B}_W^{(k)}(\overline{W},\overline{J})) \le \dot{K}_W \hat{\rho}_\mu^{(k)}((W,\dot{W},J,\dot{J}),(\overline{W},\dot{\overline{W}},\overline{J},\dot{\overline{J}})).$$

For the second component we have

$$\left| \dot{B}_J^{(k)}(W,J)(t) - \dot{B}_J^{(k)}(\overline{W},\overline{J})(t) \right| \le \left| I(W,J)(t) - I(\overline{W},\overline{J})(t) \right| + \frac{1}{t_{k+1}-t_k} \left| \int_{t_k}^{t_{k+1}} \left(I(W,J)(s) - I(\overline{W},\overline{J})(s) \right) ds \right| \equiv$$

$$\equiv \dot{I}_1 + \dot{I}_2.$$

Since

$$\dot{I}_1 \le \left| \frac{dW(t-T)}{dt} - \frac{d\overline{W}(t-T)}{dt} \right| + \frac{1}{Z_0 \dot{\overline{C}}_1} \left| W(t-T) - \overline{W}(t-T) \right| + \frac{1}{Z_0 \dot{\overline{C}}_1} \left| J(t) - \overline{J}(t) \right| +$$

$$+ \frac{U_0 e^{-\beta} + I_0}{Z_0} \frac{1}{\left(d\tilde{C}_1 \left(\frac{W(t-T)-J(t)}{2\sqrt{C}} - E_1(t) \right) / du \right)^2} \left| \frac{d^2 \tilde{C}_1}{du^2} \right| \left(\left| W(t-T) - \overline{W}(t-T) \right| + \left| J(t) - \overline{J}(t) \right| \right) +$$

$$+ \frac{2\sqrt{C}}{\dot{\overline{C}}_1} \tilde{L}_1^{-1} \left(\int_{t_k}^{t} \left(\frac{W(\tau-T)-J(\tau)}{2\sqrt{C}} - E_1(\tau) \right) d\tau \right) - \tilde{L}_1^{-1} \left(\int_{t_k}^{t} \left(\frac{\overline{W}(\tau-T)-\overline{J}(\tau)}{2\sqrt{C}} - E_1(\tau) \right) d\tau \right) +$$

$$+ \frac{2\sqrt{C}}{\dot{\overline{C}}_1} \left| G_1 \left(\frac{W(t-T)-J(t)}{2\sqrt{C}} - E_1(t) \right) - G_1 \left(\frac{W(t-T)-J(t)}{2\sqrt{C}} - E_1(t) \right) \right| +$$

$$+ \frac{\sup G_1(.)}{Z_0} \frac{1}{\left(d\tilde{C}_1 \left(\frac{W(t-T)-J(t)}{2\sqrt{C}} - E_1(t) \right) / du \right)^2} \left| \frac{d^2 \tilde{C}_1}{du^2} \right| \left(\left| W(t-T) - \overline{W}(t-T) \right| + \left| J(t) - \overline{J}(t) \right| \right) \le$$

$$\le e^{\mu_0} \frac{\rho_\mu^{(k)}(J,\overline{J})}{Z_0 \dot{\overline{C}}_1} + \frac{U_0 e^{-\beta} + I_0}{Z_0 \dot{\overline{C}}_1^{\,2}} H_1 e^{\mu_0} \rho_\mu^{(k)}(J,\overline{J}) +$$

$$+\frac{2\sqrt{C}}{\dot{\overline{C}}_1}\frac{1}{\hat{L}_1}\frac{1}{2\sqrt{C}}e^{\mu_0}\rho_\mu^{(k)}(J,\overline{J})\int_{t_k}^{t}e^{\mu(s-t_k)}d\tau+\frac{\sup\widetilde{L}_1^{-1}(.)}{Z_0\dot{\overline{C}}_1^2}H_1e^{\mu_0}\rho_\mu^{(k)}(J,\overline{J})+$$

$$+\frac{2\sqrt{C}}{\dot{\overline{C}}_1}\sum_{n=1}^{m-1}n\left|g_n^{(1)}\right|\phi_0^{n-1}\frac{1}{2\sqrt{C}}e^{\mu_0}\rho_\mu^{(k)}(J,\overline{J})+\frac{\sup G_1(.)}{Z_0\dot{\overline{C}}_1^2}H_1e^{\mu_0}\rho_\mu^{(k)}(J,\overline{J})\le$$

$$\le\rho_\mu^{(k)}(\dot{J},\dot{\overline{J}})\frac{e^{\mu_0}}{\mu\dot{\overline{C}}_1}\left(\frac{1}{Z_0}+\frac{U_0e^{-\beta}+I_0}{Z_0\dot{\overline{C}}_1}H_1+\frac{1}{\mu\hat{L}_1}+\frac{I_0H_1}{Z_0\dot{\overline{C}}_1}+\sum_{n=1}^{m}n\left|g_n^{(1)}\right|\phi_0^{n-1}+\frac{H_1}{Z_0\dot{\overline{C}}_1}\sum_{n=1}^{m}\left|g_n^{(1)}\right|\phi_0^n\right)$$

$$\le e^{\mu(t-t_k)}\hat{\rho}_\mu^{(k)}((W,\dot{W},J,\dot{J}),(\overline{W},\dot{\overline{W}},\overline{J},\dot{\overline{J}}))\times$$

$$\times\frac{e^{\mu_0}}{\mu\dot{\overline{C}}_1}\left[\frac{1}{Z_0}+\frac{1}{\mu\hat{L}_1}+\sum_{n=1}^{m}n\left|g_n^{(1)}\right|\phi_0^{n-1}+\frac{H_1}{Z_0\dot{\overline{C}}_1}\left(U_0e^{-\beta}+I_0+I_0+\sum_{n=1}^{m}\left|g_n^{(1)}\right|\phi_0^n\right)\right]$$

and

$$\dot{I}_2\le e^{\mu(t-t_k)}\hat{\rho}_\mu^{(k)}((W,\dot{W},J,\dot{J}),(\overline{W},\dot{\overline{W}},\overline{J},\dot{\overline{J}}))\times$$

$$\times\frac{e^{\mu_0}-1}{\mu_0}\frac{e^{\mu_0}}{\mu\dot{\overline{C}}_1}\left[\frac{1}{Z_0}+\frac{1}{\mu\hat{L}_1}+\sum_{n=1}^{m}n\left|g_n^{(1)}\right|\phi_0^{n-1}+\frac{H_1}{Z_0\dot{\overline{C}}_1}\left(U_0e^{-\beta}+I_0+I_0+\sum_{n=1}^{m}\left|g_n^{(1)}\right|\phi_0^n\right)\right]$$

Then

$$\left|\dot{B}_J^{(k)}(W,J)(t)-\dot{B}_J^{(k)}(\overline{W},\overline{J})(t)\right|\le$$

$$\le e^{\mu(t-t_k)}\hat{\rho}_\mu^{(k)}((W,\dot{W},J,\dot{J}),(\overline{W},\dot{\overline{W}},\overline{J},\dot{\overline{J}}))\times$$

$$\times\left(1+\frac{e^{\mu_0}-1}{\mu_0}\right)\frac{e^{\mu_0}}{\mu\dot{\overline{C}}_1}\left[\frac{1}{Z_0}+\frac{1}{\mu\hat{L}_1}+\sum_{n=1}^{m}n\left|g_n^{(1)}\right|\phi_0^{n-1}+\frac{H_1}{Z_0\dot{\overline{C}}_1}\left(U_0e^{-\beta}+2I_0+\sum_{n=1}^{m}\left|g_n^{(1)}\right|\phi_0^n\right)\right]\equiv$$

$$\equiv e^{\mu(t-t_k)}\dot{K}_J\,\hat{\rho}_\mu^{(k)}((W,\dot{W},J,\dot{J}),(\overline{W},\dot{\overline{W}},\overline{J},\dot{\overline{J}}))$$

or

$$\rho_\mu^{(k)}(\dot{B}_J^{(k)}(W,J),\dot{B}_J^{(k)}(\overline{W},\overline{J}))\le\dot{K}_J\,\hat{\rho}_\mu^{(k)}((W,\dot{W},J,\dot{J}),(\overline{W},\dot{\overline{W}},\overline{J},\dot{\overline{J}})).$$

Finally we conclude

$$\hat{\rho}_{\mu}^{(k)}(B_W(W,J),\dot{B}_W(W,J),B_J(W,J),\dot{B}_J(W,J),B_W(\overline{W},\overline{J}),\dot{B}_W(\overline{W},\overline{J}),B_J(\overline{W},\overline{J}),\dot{B}_J(\overline{W},\overline{J})) \le$$

$$\le K\,\hat{\rho}_{\mu}^{(k)}((W,\dot{W},J,\dot{J}),(\overline{W},\dot{\overline{W}},\overline{J},\dot{\overline{J}}))$$

where $K = \max\left\{e^{\mu 0}K_W(\mu),e^{\mu 0}K_J(\mu),\dot{K}_W(\mu),\dot{K}_J(\mu)\right\} < 1$.

Theorem 6.7.1 is thus proved.

6.7.1. Numerical Example

We collect all inequalities guaranteeing existence-uniqueness of oscillatory vanishing solution.

$$\frac{U_0+I_0}{2\sqrt{C}}+U_{E_1} \le \phi_0\,;\;\frac{2L}{R}e^{\frac{RT}{L}}\sinh\left(\frac{RT_0}{L}\right)\left(U_0+Z_0I_0+U_{E_0}\right)\le \phi_0 \le 1\,;$$

$$I_0e^{-\beta}+2\sqrt{C}U_{E_0}+\frac{4L}{\dot{C}_0 R}\sinh\left(\frac{RT_0}{L}\right)\left(\frac{U_0+I_0e^{-\beta}}{Z_0}+4\frac{L}{R}\sinh\left(\frac{RT_0}{L}\right)\sqrt{C}\;I_0\phi_0+2\sqrt{C}\sum_{n=1}^{m}\left|g_n^{(0)}\right|\phi_0^{\;n}\right)\le U_0\,;$$

$$U_0e^{-\beta}+2\sqrt{C}U_{E_1}+\frac{4L}{\dot{C}_1 R}\sinh\left(\frac{RT_0}{L}\right)\left(\frac{U_0e^{-\beta}+I_0}{Z_0}+4\sqrt{C}\;I_0\sinh\left(\frac{RT_0}{L}\right)\phi_0+2\sqrt{C}\sum_{n=1}^{m}\left|g_n^{(1)}\right|\phi_0^{\;n}\right)\le I_0\,;$$

$$e^{\mu 0}K_W = e^{\mu 0}\frac{e^{2\mu 0}}{\mu^2\dot{C}_0}\left[\frac{1}{Z_0}+\frac{1}{\mu\hat{L}_0}+\sum_{n=1}^{m-1}n\left|g_n^{(0)}\right|\phi_0^{\;n-1}+\frac{H_0}{Z_0\dot{C}_0}\left(U_0+I_0e^{-\beta}+I_0+\sum_{n=1}^{m}\left|g_n^{(0)}\right|\phi_0^{\;n}\right)\right]<1\,;$$

$$e^{\mu 0}K_J = e^{\mu 0}\frac{e^{2\mu 0}}{\mu^2\dot{C}_1}\left[\frac{1}{Z_0}+\frac{1}{\mu\hat{L}_1}+\frac{1}{\dot{C}_1}\sum_{n=1}^{m-1}n\left|g_n^{(1)}\right|\phi_0^{\;n-1}+\frac{H_1}{Z_0\dot{C}_1}\left(U_0e^{-\beta}+I_0+I_0+\sum_{n=1}^{m}\left|g_n^{(1)}\right|\phi_0^{\;n}\right)\right]<1\,;$$

$$\dot{K}_W = \left[\left(1+\frac{e^{\mu 0}-1}{\mu_0}\right)\frac{e^{\mu 0}}{\mu\dot{C}_0}\left(\frac{1}{Z_0}+\frac{1}{\mu\hat{L}_0}+\sum_{n=1}^{m}n\left|g_n^{(0)}\right|\phi_0^{\;n-1}+\frac{H_0}{Z_0\dot{C}_0}\left(U_0+I_0e^{-\beta}+I_0+\sum_{n=1}^{m}\left|g_n^{(0)}\right|(\phi_0)^n\right)\right)\right]<1;$$

$$\dot{K}_J = \left(1+\frac{e^{\mu 0}-1}{\mu_0}\right)\frac{e^{\mu 0}}{\mu\dot{C}_1}\left[\frac{1}{Z_0}+\frac{1}{\mu\hat{L}_1}+\sum_{n=1}^{m}n\left|g_n^{(1)}\right|\phi_0^{\;n-1}+\frac{H_1}{Z_0\dot{C}_1}\left(U_0e^{-\beta}+2I_0+\sum_{n=1}^{m}\left|g_n^{(1)}\right|\phi_0^{\;n}\right)\right]<1\,.$$

We check the above inequalities for the line with specific parameters as in the last numerical example:

$$\Lambda = 100\,m,\quad R=\frac{\rho_c\Lambda}{S}=\frac{0{,}0175\cdot 100}{4}\approx 0{,}44\,\Omega;\;L=0{,}45\,\mu H/m,\;C=80\,pF/m,$$

$$v=\frac{1}{\sqrt{LC}}=\frac{1}{6.10^{-9}}=1{,}66.10^{8};\;\frac{R}{L}=\frac{0{,}44}{0{,}45.10^{-6}}\approx 10^{6},\;Z_0=75\,\Omega,$$

$$T = \Lambda\sqrt{LC} = 100.6.10^{-9} = 6.10^{-7}s; \quad \frac{R\Lambda}{Z_0} = \frac{RT}{L} = 10^6.6.10^{-7} = 0{,}6; \quad e^{-\frac{RT}{L}} = e^{-0{,}6}.$$

$$G(u) = G_0(u) = G_1(u) = 0{,}02u - 0{,}12u^2 + 0{,}14u^3,$$

i.e. $g_1 = 0{,}02, g_2 = -0{,}12; g_3 = 0{,}1.$

For $\lambda_0 = (1/6)10^{-4}\,m$ we have $f_0 = \dfrac{1}{\lambda_0\sqrt{LC}} = 10^{13}\,Hz \Rightarrow T_0 = \dfrac{1}{f_0} = 10^{-13}$ sec.;

$$\frac{RT_0}{L} = 10^6.10^{-13} = 10^{-7}.e^{\frac{R}{L}T_0} - e^{-\frac{R}{L}T_0} \approx \frac{R}{L}T_0 = 10^{-7}. \text{ Choose } \mu = 10^{13}, \text{ then}$$

$\mu T_0 = \mu_0 = 1$ and $T = 6.10^{-7}.10^{13}T_0 = 10^6.T_0 = m.T_0 \Rightarrow m = 10^6.$

We assume $\widetilde{C}_0(u) = \widetilde{C}_1(u) = \widetilde{C}(u)$, that is, $c_0 = c_1$, $\Phi = \Phi_0 = \Phi_1$ where the *pn*-junction capacity $c_0 \in [0{,}05.10^{-12}F; \ 50.10^{-12}F]$, while the *pn*-junction potentials $\Phi_0, \Phi_1 \in [0{,}4\,V; \ 0{,}9\,V]$.

For $h = 2$ $\quad H_0 = \dfrac{2c_0\sqrt[h]{\Phi_0}}{h}\dfrac{\Phi_0 + \dfrac{h-1}{2h}\phi_0}{(\Phi_0 - \phi_0)^{\frac{2h+1}{h}}}$ becomes $\quad H_0 = c_0\sqrt{\Phi_0}\dfrac{\Phi_0 + 0{,}25\phi_0}{(\Phi_0 - \phi_0)^{5/2}}$,

$-\phi_0 \leq u \leq \phi_0$ ($\phi_0 < \Phi$) and if we choose $c_0 = 50pF = 5.10^{-11}F$ and $\Phi = 0{,}4$ and

$$H_0 = c_0\sqrt{\Phi_0}\frac{\Phi_0 + 0{,}25\phi_0}{(\Phi_0 - \phi_0)^2\sqrt{\Phi_0 - \phi_0}} = 10^{-11}1{,}5.10^3 = 1{,}5.10^{-8}$$

$\phi_0 = 0{,}3$ then   .

Besides

$$\min \frac{d\overline{C}_0(.)}{du} = \frac{d\overline{C}_0(-\phi_0)}{du} = c_0\sqrt[h]{\Phi_0}\frac{\Phi_0 + (h-1)/h\phi_0}{(\Phi_0 + \phi_0)^{\frac{h+1}{h}}} = 5.10^{-11}\sqrt{0{,}4}\frac{0{,}4 + 0{,}5.0{,}3}{(0{,}7)^{\frac{3}{2}}} \approx 3.10^{-11} \equiv \dot{\overline{C}}_0 > 0.$$

Then $\mu\dot{\overline{C}}_0 = 10^{13}.3.10^{-11} \approx 3.10^2$; $\quad \mu^2\dot{\overline{C}}_0 = 10^{26}.10^{-11}.3 \approx 3.10^{15}.$

We have to estimate

$$\widetilde{L}_p = \widetilde{L}(i) = i.L(i) = i(2 - 0{,}1i) = 2i - 10^{-1}i^2 \ , i \in [-i_0, i_0] \ (p = 0{,}1).$$

We choose i_0 such that for $|i| \leq i_0$ it follows $\dfrac{d\widetilde{L}(i)}{di} = 2 - 0{,}2.i > 0$, that is, $|i| \leq i_0 < \sqrt{10}$. Therefore the inverse function $\widetilde{L}^{-1}(.)$ exists and

$$\widetilde{L}^{-1}(\lambda):\left[-2i_0-0,1.i_0^2\,;\ 2i_0-0,1.i_0^2\right]\to\left[-i_0,i_0\right]\,.$$

The explicit form of the inverse function can be found solving the following equation with respect to i: $2i-0,1i^2=\lambda$. We need, however, just the estimates:

$$\left|\widetilde{L}^{-1}(.)\right|\le i_0\le I_0\ (k=0,1)\,.$$

We can choose $i_0=I_0=0,01<\sqrt{10}$

$$\left|\frac{d\widetilde{L}^{-1}(\lambda)}{d\lambda}\right|=\frac{1}{\left|\frac{d\widetilde{L}(i)}{di}\right|}\le\frac{1}{\min\left\{\left|\frac{d\widetilde{L}(i)}{di}\right|:i\in[-i_0;i_0]\right\}}=\frac{1}{\min\left\{\left|2-0,2.i\right|:i\in[-0,01;\,0,01]\right\}}=\frac{1}{\hat{L}}\approx\frac{1}{2}\,;$$

$$\frac{1}{\mu\hat{L}}\approx\frac{1}{2.10^{13}}\approx0\,.$$ Then for $U_0=10^{-4}$, $I_0=10^{-4}$ and $U_{E_p}=10^{-5}\,(p=0,1)$, $\beta=4$

we can verify that above inequalities are satisfied.

6.8. LOSSY TRANSMISSION LINES WITH TIME-VARYING SPECIFIC PARAMETERS

We proceed again from the lossy transmission line system

$$\frac{\partial u(x,t)}{\partial t}+\frac{1}{C(t)}\frac{\partial i(x,t)}{\partial x}+\frac{G(t)}{C(t)}u(x,t)=0,$$

$$\frac{\partial i(x,t)}{\partial t}+\frac{1}{L(t)}\frac{\partial u(x,t)}{\partial x}+\frac{R}{L(t)}i(x,t)=0$$

where the parameters depend on the time $C=C(t),L=L(t),G=G(t),R=R(t)$.

Obviously we have to assume that $C(t),L(t)$ have strictly positive lower bounds.

Here we use the generalized Heaviside condition.

The above system in a matrix form is:

$$\frac{\partial U}{\partial t}+A_1\frac{\partial U}{\partial x}+A_2U=0 \qquad (5.6.1)$$

where

$$U = \begin{bmatrix} u \\ i \end{bmatrix}, \quad \frac{\partial U}{\partial t} = \begin{bmatrix} \dfrac{\partial u}{\partial t} \\ \dfrac{\partial i}{\partial t} \end{bmatrix}, \quad \frac{\partial U}{\partial x} = \begin{bmatrix} \dfrac{\partial u}{\partial x} \\ \dfrac{\partial i}{\partial x} \end{bmatrix}, \quad A_1(t) = \begin{bmatrix} 0 & 1/C \\ 1/L & 0 \end{bmatrix}, \quad A_2(t) = \begin{bmatrix} G/C & 0 \\ 0 & R/L \end{bmatrix}.$$

The eigenvalues of $A_1(t) = \begin{bmatrix} 0 & 1/C \\ 1/L & 0 \end{bmatrix}$ are $\lambda_1 = 1/\sqrt{LC}$, $\lambda_2 = -1/\sqrt{LC}$ whose eigenvectors are $\left(\xi_1^{(1)},\xi_2^{(1)}\right) = \left(\sqrt{C},\sqrt{L}\right)$, $\left(\xi_1^{(2)},\xi_2^{(2)}\right) = \left(-\sqrt{C},\sqrt{L}\right)$. The matrix H formed by the eigenvectors depends on time $H(t) = \begin{bmatrix} \sqrt{C(t)} & \sqrt{L(t)} \\ -\sqrt{C(t)} & \sqrt{L(t)} \end{bmatrix}$ and its inverse one too

$$H^{-1}(t) = \begin{bmatrix} \dfrac{1}{2\sqrt{C(t)}} & -\dfrac{1}{2\sqrt{C(t)}} \\ \dfrac{1}{2\sqrt{L(t)}} & \dfrac{1}{2\sqrt{L(t)}} \end{bmatrix}.$$

Then $A^{\mathrm{can}} = HA_1H^{-1}$, where $A^{\mathrm{can}}(t) = \begin{bmatrix} 1/\sqrt{L(t)C(t)} & 0 \\ 0 & -1/\sqrt{L(t)C(t)} \end{bmatrix}.$

Introduce new variables $Z = HU$, (or $U = H^1Z$)

$$Z = \begin{bmatrix} V(x,t) \\ I(x,t) \end{bmatrix}, \qquad U = \begin{bmatrix} u(x,t) \\ i(x,t) \end{bmatrix}.$$

where

$$\left|\begin{aligned} V(x,t) &= \sqrt{C(t)}\, u(x,t) + \sqrt{L(t)}\, i(x,t) \\ I(x,t) &= -\sqrt{C(t)}\, u(x,t) + \sqrt{L(t)}\, i(x,t) \end{aligned}\right.$$

and

$$\left|\begin{aligned} u(x,t) &= \frac{1}{2\sqrt{C(t)}}V(x,t) - \frac{1}{2\sqrt{C(t)}}I(x,t) \\ i(x,t) &= \frac{1}{2\sqrt{L(t)}}V(x,t) + \frac{1}{2\sqrt{L(t)}}I(x,t). \end{aligned}\right.$$

Substituting $U = H^{-1}(t)Z$ in (5.6.1) we obtain

$$\frac{\partial\left(H^{-1}(t)Z\right)}{\partial t} + A_1\frac{\partial\left(H^{-1}(t)Z\right)}{\partial x} + A_2\left(H^{-1}(t)Z\right) = 0$$

and multiply the last equality by $H(t)$ from the left

$$\frac{\partial Z}{\partial t} + H(t)\frac{\partial H^{-1}(t)}{\partial t}Z + H(t)A_1H^{-1}(t)\frac{\partial Z}{\partial x} + \left(H(t)A_2H^{-1}(t)\right)Z = 0$$

where

$$HA_2H^{-1} = \frac{1}{2}\begin{bmatrix} \dfrac{G(t)}{C(t)}+\dfrac{R(t)}{L(t)} & -\dfrac{G(t)}{C(t)}+\dfrac{R(t)}{L(t)} \\[2mm] -\dfrac{G(t)}{C(t)}+\dfrac{R(t)}{L(t)} & \dfrac{G(t)}{C(t)}+\dfrac{R(t)}{L(t)} \end{bmatrix}, \; H(t)\frac{\partial H^{-1}(t)}{\partial t} = \frac{1}{4}\begin{bmatrix} -\dfrac{\dot C(t)}{C(t)}-\dfrac{\dot L(t)}{L(t)} & \dfrac{\dot C(t)}{C(t)}-\dfrac{\dot L(t)}{L(t)} \\[2mm] \dfrac{\dot C(t)}{C(t)}-\dfrac{\dot L(t)}{L(t)} & -\dfrac{\dot C(t)}{C(t)}-\dfrac{\dot L(t)}{L(t)} \end{bmatrix}.$$

We see that

$$H(t)A_2H^{-1}(t) + H(t)\frac{\partial H^{-1}(t)}{\partial t} =$$

$$\frac{1}{2}\begin{bmatrix} \dfrac{G(t)}{C(t)}+\dfrac{R(t)}{L(t)}-\dfrac{\dot C(t)}{2C(t)}-\dfrac{\dot L(t)}{2L(t)} & -\dfrac{G(t)}{C(t)}+\dfrac{R(t)}{L(t)}+\dfrac{\dot C(t)}{2C(t)}-\dfrac{\dot L(t)}{2L(t)} \\[2mm] -\dfrac{G(t)}{C(t)}+\dfrac{R(t)}{L(t)}+\dfrac{\dot C(t)}{2C(t)}-\dfrac{\dot L(t)}{2L(t)} & \dfrac{G(t)}{C(t)}+\dfrac{R(t)}{L(t)}-\dfrac{\dot C(t)}{2C(t)}-\dfrac{\dot L(t)}{2L(t)} \end{bmatrix}.$$

So the system

$$\frac{\partial Z}{\partial t} + A^{\mathrm{can}}(t)\frac{\partial Z}{\partial x} + \left[H(t)A_2H^{-1}(t) + H(t)\frac{\partial H^{-1}(t)}{\partial t}\right]Z = 0$$

becomes in an explicit form

$$\frac{\partial V}{\partial t} + \frac{1}{\sqrt{LC}}\frac{\partial V}{\partial x} + \left(\frac{G}{C}+\frac{R}{L}-\frac{\dot C}{2C}-\frac{\dot L}{2L}\right)V + \left(-\frac{G}{C}+\frac{R}{L}+\frac{\dot C}{2C}-\frac{\dot L}{2L}\right)I = 0$$

$$\frac{\partial I}{\partial t} - \frac{1}{\sqrt{LC}}\frac{\partial I}{\partial x} + \left(-\frac{G}{C}+\frac{R}{L}+\frac{\dot C}{2C}-\frac{\dot L}{2L}\right)V + \left(\frac{G}{C}+\frac{R}{L}-\frac{\dot C}{2C}-\frac{\dot L}{2L}\right)I = 0.$$

We use a generalized Heaviside condition (introduced in Chapter V):

$$-\frac{G(t)}{C(t)}+\frac{R(t)}{L(t)}+\frac{\dot{C}(t)}{2C(t)}-\frac{\dot{L}(t)}{2L(t)}=0.$$

Then the last system can be rewritten in the form

$$\frac{\partial V(x,t)}{\partial t}+\frac{1}{\sqrt{L(t)C(t)}}\frac{\partial V(x,t)}{\partial x}+\frac{2G(t)-\dot{C}(t)}{C(t)}V(x,t)=0,$$

$$\frac{\partial I(x,t)}{\partial t}-\frac{1}{\sqrt{L(t)C(t)}}\frac{\partial I(x,t)}{\partial x}+\frac{2G(t)-\dot{C}(t)}{C(t)}I(x,t)=0.$$

For the new initial conditions we obtain:

$$V(x,0)=\sqrt{C}\,u(x,0)+\sqrt{L}\,i(x,0)=\sqrt{C}\,u_0(x)+\sqrt{L}\,i_0(x)\equiv V_0(x),$$

$$I(x,0)=-\sqrt{C}\,u(x,0)+\sqrt{L}\,i(x,0)=-\sqrt{C}\,u_0(x)+\sqrt{L}\,i_0(x)\equiv I_0(x),x\in\left[0,\Lambda\right].$$

Put $h(t)=\dfrac{2G(t)-\dot{C}(t)}{C(t)}$ and rewrite the above system in the form:

$$\frac{\partial V(x,t)}{\partial t}+\frac{1}{\sqrt{L(t)C(t)}}\frac{\partial V(x,t)}{\partial x}+h(t)V(x,t)=0,$$

$$\frac{\partial I(x,t)}{\partial t}-\frac{1}{\sqrt{L(t)C(t)}}\frac{\partial I(x,t)}{\partial x}+h(t)I(x,t)=0.$$

The last system might be simplified one more time by the substitution:

$$W(x,t)=e^{h(t)t}V(x,t),\ J(x,t)=e^{h(t)t}I(x,t),$$

or

$$V(x,t)=e^{-h(t)t}W(x,t),\ I(x,t)=e^{-h(t)t}J(x,t).$$

Consequently we obtain the following transformation:

$$u(x,t)=\frac{1}{2\sqrt{C(t)}}e^{-h(t)t}W(x,t)-\frac{1}{2\sqrt{C(t)}}e^{-h(t)t}J(x,t),$$

$$i(x,t) = \frac{1}{2\sqrt{L(t)}} e^{-h(t)t} W(x,t) + \frac{1}{2\sqrt{L(t)}} e^{-h(t)t} J(x,t)$$

and

$$W(x,t) = e^{h(t)t}\left(\sqrt{C(t)}\, u(x,t) + \sqrt{L(t)}\, i(x,t)\right),$$
$$J(x,t) = e^{h(t)t}\left(\sqrt{L(t)}\, i(x,t) - \sqrt{C(t)}\, u(x,t)\right).$$

Substitute $V(x,t)$ and $I(x,t)$ into the last system we have

$$\frac{\partial\left(e^{-h(t)t}W(x,t)\right)}{\partial t} + \frac{1}{\sqrt{LC}}\frac{\partial\left(e^{-h(t)t}W(x,t)\right)}{\partial x} + e^{-h(t)t}h(t)W(x,t) = 0\ ,$$
$$\frac{\partial\left(e^{-h(t)t}J(x,t)\right)}{\partial t} - \frac{1}{\sqrt{LC}}\frac{\partial\left(e^{-h(t)t}J(x,t)\right)}{\partial x} + e^{-h(t)t}h(t)J(x,t) = 0$$

which leads to

$$\frac{\partial W}{\partial t} + \frac{1}{\sqrt{LC}}\frac{\partial W}{\partial x} - t\dot{h}(t)W = 0\ ,\quad \frac{\partial J}{\partial t} - \frac{1}{\sqrt{LC}}\frac{\partial J}{\partial x} - t\dot{h}(t)J = 0.$$

The last system may take the form

$$\frac{\partial W}{\partial t} + \frac{1}{\sqrt{LC}}\frac{\partial W}{\partial x} = 0\ ,\quad \frac{\partial J}{\partial t} - \frac{1}{\sqrt{LC}}\frac{\partial J}{\partial x} = 0\ .$$

If we assume $\dot{h}(t) = 0 \Leftrightarrow h(t) = \text{const.} = \gamma$ or $\dfrac{2G - \dot{C}}{C} = \text{const.} = \gamma$.

We are able to obtain an explicit form of the relation between capacity function and conductivity one. Indeed, the linear differential equation $\dot{C}(t) + \gamma C(t) = 2G(t)$ has a solution

$$C(t) = e^{-\gamma t}\left(C_T + 2\int_T^t G(s)e^{\gamma s}\,ds \right), C_T = C(T)\ .$$

Since the initial conditions remain the same ones in view of

$$W(x,0) = e^{\frac{R}{L}.0} V(x,0) = V_0(x) = \sqrt{C}\, u_0(x) + \sqrt{L}\, i_0(x),$$

$$J(x,0) = e^{\frac{R}{L}.0} I(x,0) = I_0(x) = -\sqrt{C}\, u_0(x) + \sqrt{L}\, i_0(x),\ x \in [0,\Lambda]$$

we have to transform the boundary conditions using

$$\frac{d\widetilde{C}_p(u(0,t)-E_p(t))}{dt} = C_p(t)\frac{d(u(0,t)-E_p(t))}{dt};\ \ i(0,t) = \frac{1}{L_p(t)}\int_T^t u(0,\tau)d\tau .$$

Then they can be written in the form:

$$-i(0,t) = -C_0(t)\frac{d(u(0,t)-E_0(t))}{dt} - \frac{1}{L_0(t)}\int_T^t u(0,\tau)d\tau - G_0(u(0,t)-E_0(t))$$

$$-i(\Lambda,t) = -C_1(t)\frac{d(u(\Lambda,t)-E_1(t))}{dt} - \frac{1}{L_1(t)}\int_T^t u(\Lambda,\tau)d\tau - G_1(u(\Lambda,t)-E_1(t))$$
.

Since

$$i(0,t) = \frac{e^{-h(t)t}W(0,t)}{2\sqrt{L(t)}} + \frac{e^{-h(t)t}J(0,t)}{2\sqrt{L(t)}},$$

$$u(0,t) = \frac{e^{-h(t)t}W(0,t)}{2\sqrt{C(t)}} - \frac{e^{-h(t)t}J(0,t)}{2\sqrt{C(t)}}$$

and

$$i(\Lambda,t) = \frac{e^{-h(t)t}W(\Lambda,t)}{2\sqrt{L(t)}} + \frac{e^{-h(t)t}J(\Lambda,t)}{2\sqrt{L(t)}},$$

$$u(\Lambda,t) = \frac{e^{-h(t)t}W(\Lambda,t)}{2\sqrt{C(t)}} - \frac{e^{-h(t)t}J(\Lambda,t)}{2\sqrt{C(t)}}$$

we obtain the system

$$-\frac{e^{-h}W(0,t)}{2\sqrt{L(t)}} - \frac{e^{-h}J(0,t)}{2\sqrt{L(t)}} = -C_0(t)\frac{d}{dt}\left(\frac{e^{-h(t)t}W(0,t)}{2\sqrt{C(t)}} - \frac{e^{-h(t)t}J(0,t)}{2\sqrt{C(t)}}\right) + C_0(t)\frac{dE_0(t)}{dt} -$$

$$-\frac{1}{L_0(t)}\int_T^t\left(\frac{e^{-h}W(0,s)}{2\sqrt{L(s)}} - \frac{e^{-h}J(0,s)}{2\sqrt{L(s)}}\right)ds - G_0(t)\left(\frac{e^{-h}W(0,t)}{2\sqrt{L(t)}} - \frac{e^{-h}J(0,t)}{2\sqrt{L(t)}} - E_0(t)\right),$$

$$-\frac{e^{-\pi}W(\Lambda,t)}{2\sqrt{L(t)}}-\frac{e^{-\pi}J(\Lambda,t)}{2\sqrt{L(t)}}=-C_1(t)\left(\frac{e^{-\pi}W(\Lambda,t)}{2\sqrt{C(t)}}-\frac{e^{-\pi}J(\Lambda,t)}{2\sqrt{C(t)}}\right)+C_1(t)\frac{dE_1(t)}{dt}-$$

$$-\frac{1}{L_0(t)}\int_T^t\left(\frac{e^{-\pi s}W(\Lambda,s)}{2\sqrt{L(s)}}-\frac{e^{-\pi s}J(\Lambda,s)}{2\sqrt{L(s)}}\right)ds-G_1(t)\left(\frac{e^{-\pi}W(\Lambda,t)}{2\sqrt{L(t)}}-\frac{e^{-\pi}J(\Lambda,t)}{2\sqrt{L(t)}}-E_1(t)\right).$$

But in view of substitutuion $e^{-\pi}W(\Lambda,t)\equiv W(t), e^{-\pi}J(0,t)\equiv J(t)$ and in view of

$$W(\Lambda,t+T)=W(0,t),\; J(\Lambda,t)=J(0,t+T)$$

we have

$$-\frac{W(\Lambda,t+T)}{2\sqrt{L(t)}}-\frac{J(0,t)}{2\sqrt{L(t)}}=-C_0(t)\frac{d}{dt}\left(\frac{W(\Lambda,t+T)}{2\sqrt{C(t)}}-\frac{J(0,t)}{2\sqrt{C(t)}}\right)+C_0(t)\frac{dE_0(t)}{dt}-$$

$$-\frac{1}{L_0(t)}\int_T^t\left(\frac{W(\Lambda,s+T)}{2\sqrt{L(s)}}-\frac{J(0,s)}{2\sqrt{L(s)}}\right)ds-G_0(t)\left(\frac{W(\Lambda,t+T)}{2\sqrt{L(t)}}-\frac{J(0,t)}{2\sqrt{L(t)}}-E_0(t)\right),$$

$$-\frac{W(\Lambda,t)}{2\sqrt{L(t)}}-\frac{J(0,t+T)}{2\sqrt{L(t)}}=-C_1(t)\left(\frac{W(\Lambda,t)}{2\sqrt{C(t)}}-\frac{J(0,t+T)}{2\sqrt{C(t)}}\right)+C_1(t)\frac{dE_1(t)}{dt}-$$

$$-\frac{1}{L_0(t)}\int_T^t\left(\frac{W(\Lambda,s)}{2\sqrt{L(s)}}-\frac{J(0,s+T))}{2\sqrt{L(s)}}\right)ds-G_1(t)\left(\frac{W(\Lambda,t)}{2\sqrt{L(t)}}-\frac{J(0,t+T)}{2\sqrt{L(t)}}-E_1(t)\right).$$

Let us put $t+T\equiv t$ and finally obtain the system

$$-\frac{W(\Lambda,t)}{2\sqrt{L(t-T)}}-\frac{J(0,t-T)}{2\sqrt{L(t-T)}}=-C_0(t-T)\frac{d}{dt}\left(\frac{W(\Lambda,t)}{2\sqrt{C(t-T)}}-\frac{J(0,t-T)}{2\sqrt{C(t-T)}}\right)+C_0(t-T)\frac{dE_0(t-T)}{dt}-$$

$$-\frac{1}{L_0(t)}\int_0^{t-T}\left(\frac{W(\Lambda,s+T)}{2\sqrt{L(s)}}-\frac{J(0,s)}{2\sqrt{L(s)}}\right)ds-G_0(t)\left(\frac{W(\Lambda,t)}{2\sqrt{L(t-T)}}-\frac{J(0,t-T)}{2\sqrt{L(t-T)}}-E_0(t-T)\right),$$

$$-\frac{W(\Lambda,t-T)}{2\sqrt{L(t-T)}}-\frac{J(0,t)}{2\sqrt{L(t-T)}}=-C_1(t-T)\frac{d}{dt}\left(\frac{W(\Lambda,t-T)}{2\sqrt{C(t-T)}}-\frac{J(0,t)}{2\sqrt{C(t-T)}}\right)+C_1(t-T)\frac{dE_1(t-T)}{dt}-$$

$$-\frac{1}{L_0(t)}\int_0^{t-T}\left(\frac{W(\Lambda,s)}{2\sqrt{L(s)}}-\frac{J(0,s+T))}{2\sqrt{L(s)}}\right)ds-G_1(t)\left(\frac{W(\Lambda,t-T)}{2\sqrt{L(t-T)}}-\frac{J(0,t)}{2\sqrt{L(t-T)}}-E_1(t-T)\right).$$

Let us assume that the unknown functions are $W(\Lambda,t)\equiv W(t), J(0,t)\equiv J(t)$. Changing the variable in the integrals and taking into account the periodicity of the source and parametric functions we obtain:

$$-\frac{W(t)}{2\sqrt{L(t)}}-\frac{J(t-T)}{2\sqrt{L(t)}}=-C_0(t)\frac{d}{dt}\left(\frac{W(t)-J(t-T)}{2\sqrt{C(t)}}\right)+C_0(t)\frac{dE_0(t)}{dt}-$$

$$-\frac{1}{L_0(t)}\int_T^t\left(\frac{W(s)}{2\sqrt{L(s)}}-\frac{J(s-T)}{2\sqrt{L(s)}}\right)ds-G_0(t)\left(\frac{W(t)}{2\sqrt{L(t)}}-\frac{J(t-T)}{2\sqrt{L(t)}}-E_0(t)\right),$$

$$-\frac{W(t-T)}{2\sqrt{L(t)}}-\frac{J(t)}{2\sqrt{L(t)}}=-C_1(t)\frac{d}{dt}\left(\frac{W(t-T)-J(t)}{2\sqrt{C(t)}}\right)+C_1(t)\frac{dE_1(t)}{dt}-$$

$$-\frac{1}{L_0(t)}\int_T^t\left(\frac{W(s-T)}{2\sqrt{L(s)}}-\frac{J(s))}{2\sqrt{L(s)}}\right)ds-G_1(t)\left(\frac{W(t-T)}{2\sqrt{L(t)}}-\frac{J(t)}{2\sqrt{L(t)}}-E_1(t)\right).$$

Solve with respect to the derivatives we obtain

$$\frac{dW(t)}{dt}=\frac{dJ(t-T)}{dt}+\frac{W(t)}{C_0(t)\sqrt{Z_0(t)}}+\frac{J(t-T)}{C_0(t)\sqrt{Z_0(t)}}-\frac{W(t)-J(t-T)}{2}\frac{\dot{C}(t)}{\sqrt{C(t)}}+$$

$$+2\sqrt{C(t)}\frac{dE_0(t)}{dt}-\frac{2\sqrt{C(t)}}{C_0(t)L_0(t)}\int_T^t\left(\frac{W(s)}{2\sqrt{L(s)}}-\frac{J(s-T)}{2\sqrt{L(s)}}\right)ds-\frac{G_0(t)}{C_0(t)}\left(\frac{W(t)}{\sqrt{Z_0(t)}}-\frac{J(t-T)}{\sqrt{Z_0(t)}}-E_0(t)\right),$$

$$\frac{dJ(t)}{dt}=\frac{dW(t-T)}{dt}+\frac{J(t)}{C_1(t)\sqrt{Z_0(t)}}+\frac{W(t-T)}{C_1(t)\sqrt{Z_0(t)}}-\frac{J(t)-W(t-T)}{2}\frac{\dot{C}(t)}{\sqrt{C(t)}}+$$

$$+2\sqrt{C(t)}\frac{dE_1(t)}{dt}-\frac{2\sqrt{C(t)}}{C_1(t)L_1(t)}\int_T^t\left(\frac{W(s-T)}{2\sqrt{L(s)}}-\frac{J(s)}{2\sqrt{L(s)}}\right)ds-\frac{G_1(t)}{C_1(t)}\left(\frac{W(t-T)}{\sqrt{Z_0(t)}}-\frac{J(t)}{\sqrt{Z_0(t)}}-E_1(t)\right).$$

CONCLUSION

We notice that the choice of applying the first or the second method depends on the specific circuit and the values for the specific parameters. We formulate an existence-uniqueness theorem for a periodic and oscillatory solution whose conditions are easy verifiable.

We point out that the procedure for treating the treatment of time-varying specific parameters might may also be applied to the present circuit (cf. § 5.7).

DISTORTIONLESS LOSSY TRANSMISSION LINES TERMINATED BY IN SERIES CONNECTED *RLC*-LOADS

ABSTRACT

In this chapter we consider lossy transmission lines terminated by in series connected *RLC*-loads at both ends. First we derive boundary conditions using Kirchhoff's law, and then we formulate a mixed problem for the hyperbolic system corresponding to lossy transmission lines. It is shown that there is only one way to reduce the mixed problem to an initial value problem on the boundary. It is clear that we have to look for oscillatory solutions vanishing at infinity.

INTRODUCTION

In Chapter III we formulated the same problem but for lossless lines. If we take the losses into account, new mathematical difficulties arise. In contrast to parallel connected loads (cf. Chapter VI), here the multiplier cannot be eliminated. This implies that in the case of parallel connected loads, solutions may be both periodic and oscillating, whereas here they are necessarily oscillating, vanishing at infinity. The attenuation is due to losses. In [16] we have reduced the mixed problem for the lossy transmission line equations and formulated an existence-uniqueness result for the obtained neutral system on the boundary. Here we formulate the problem globally on the whole interval and find conditions for the existence-uniqueness of an oscillatory solution vanishing exponentially at infinity. As in the previous chapter we assume the Heaviside condition is satisfied. This implies that the line is without distortion. By applying the fixed-point method we find a solution and show the rate of convergence of the successive approximations in a suitable pseudo-metrics. As in the previous chapters the main difficulty is to define an operator whose fixed point is a solution of the system.

In this chapter we consider distortionless lossy transmission lines terminated by in series connected *RLC*-loads. In § 7.1 we derive boundary conditions corresponding to the terminated nonlinear loads. Then we formulate the mixed problem for the transmission line system. In § 7.2 we reduce the mixed problem for the transmission line system to an initial value problem for a nonlinear neutral system. In § 7.3 we propose a different manner of reducing the mixed problem to an oscillatory problem on the boundary. In § 7.4 an existence-uniqueness theorem for an oscillatory solution for the neutral system is proved. As in the previous chapters we use the fixed-point method in suitable metric space. In § 7.5 we show how to apply the method to the specific problems. Finally in § 7.6 we apply the same method to transmission lines terminated by linear and time-varying loads.

7.1. DERIVATION OF THE BOUNDARY CONDITIONS, FORMULATION OF THE MIXED PROBLEM AND ANALYSIS OF THE NONLINEARITIES

We proceed again from the system of lossy transmission line equations:

$$C\frac{\partial u(x,t)}{\partial t} + \frac{\partial i(x,t)}{\partial x} + Gu(x,t) = 0,$$

$$L\frac{\partial i(x,t)}{\partial t} + \frac{\partial u(x,t)}{\partial x} + Ri(x,t) = 0$$

$$(x,t) \in \Pi = \left\{ (x,t) \in R^2 : (x,t) \in [0,\Lambda] \times [0,\infty) \right\}$$

where L, C, R and G are prescribed specific parameters of the line and $\Lambda > 0$ is its length.

Here $i(x,t), u(x,t)$ are unknown functions – current and voltage respectively. Using Kirchhoff's law we derive boundary conditions proceeding from the transmission line and elements with which it is loaded shown on Fig. 7.1.

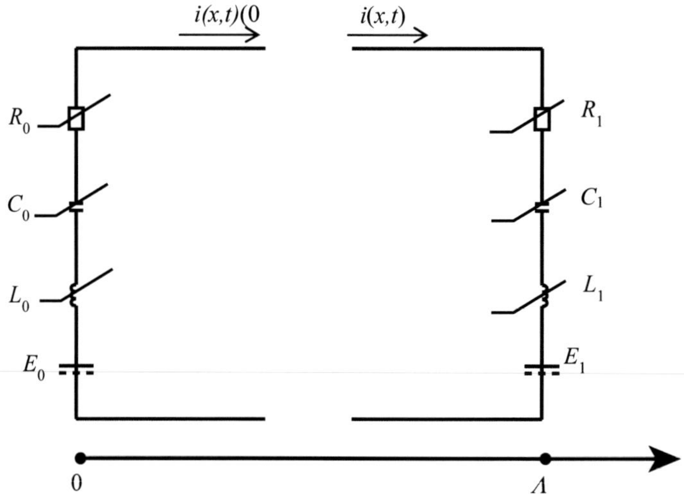

Figure 7.1.

For the voltages for $x = 0$ we obtain:

$$-u(0,t) = u_{R_0} + u_{C_0} + u_{\Psi_0} - E_0(t),$$
(7.1.1)

where $E_0(t)$ is the source voltage.

Recall that $u_{C_p}(T) \equiv u(T) = 0$, $\widetilde{C}_p(u) = C_p(u).u$, $\widetilde{L}_p(i) = L_p(i).i$ $(p = 0,1)$ and $T = \Lambda / v = \Lambda\sqrt{LC}$. To find the voltages of the condenser $C_p (p = 0,1)$ we proceed from the relation:

$$i = \frac{dq}{dt} = \frac{d\widetilde{C}_p(u)}{dt} \Rightarrow \int_T^t i(x,\tau)d\tau = \widetilde{C}_p(u).$$

We suppose that the function $\widetilde{C}_p(u)$ possesses an inverse one. Therefore

$$u_{C_p} = u(0,t) = \widetilde{C}_p^{-1}\left(\int_T^t i(0,\tau)d\tau\right).$$

For the voltages of the inductors u_{Ψ_p} $(p = 0,1)$ we have

$$u_{\Psi_p} = \frac{d\Psi_p}{dt} = \frac{d\widetilde{L}_p(i)}{dt} \equiv \frac{d\left(L_p(i).i\right)}{dt} = \left(i\frac{dL_p(i)}{di} + L_p(i)\right)\frac{di}{dt} \ (p = 0,1),$$

where $L_p(i) = \sum_{n=1}^{m} l_n^{(p)} i^n$.

For the *I-V* characteristics of the resistive elements we assume a polynomial dependence:

$$u_{R_p} = R_p(i) = \sum_{n=1}^{m} r_n^{(p)} i^n, (p = 0,1).$$

Then the first boundary condition becomes:

$$u(0,t) = E_0(t) - R_0(i(0,t)) - \widetilde{C}_0^{-1}\left(\int_T^t i(0,\tau)d\tau\right) - \left[i(0,t)\frac{dL_0(i(0,t))}{di} + L_0(i(0,t))\right]\frac{di(0,t)}{dt}.$$

Analogously for $x = \Lambda$ (cf. Fig.7.1) one obtains

$$u(\Lambda,t) = E_1(t) + R_1(i(\Lambda,t)) + \tilde{C}_1^{-1}\left(\int_T^t i(\Lambda,\tau)d\tau\right) + \frac{d\tilde{L}_1(i(\Lambda,t))}{di}\frac{di(\Lambda,t)}{dt}$$

or

$$u(\Lambda,t) = E_1(t) + R_1(i(\Lambda,t)) + \tilde{C}_1^{-1}\left(\int_T^t i(\Lambda,\tau)d\tau\right) + \left[i(\Lambda,t)\frac{dL_1(i(\Lambda,t))}{di} + L_1(i(\Lambda,t))\right]\frac{di(\Lambda,t)}{dt}.$$

Now we are able to formulate the mixed problem for the lossless transmission line system: to find a solution $(u(x,t),i(x,t))$ of the following system

$$C\frac{\partial u(x,t)}{\partial t} + \frac{\partial i(x,t)}{\partial x} + Gu(x,t) = 0,$$

$$L\frac{\partial i(x,t)}{\partial t} + \frac{\partial u(x,t)}{\partial x} + Ri(x,t) = 0 \tag{7.1.2}$$

for $(x,t) \in \Pi = \{(x,t) \in R^2 : 0 \leq x \leq \Lambda,\ t \geq 0\}$, satisfying the initial conditions

$$u(x,0) = u_0(x),\ i(x,0) = i_0(x)\ \text{ for }\ x \in [0,\Lambda] \tag{7.1.3}$$

where $u_0(x),\ i_0(x)$ are prescribed functions, and the boundary conditions for $x = 0$

$$u(0,t) = E_0(t) - R_0(i(0,t)) - \tilde{C}_0^{-1}\left(\int_T^t i(0,\tau)d\tau\right) - \left[i(0,t)\frac{dL_0(i(0,t))}{di} + L_0(i(0,t))\right]\frac{di(0,t)}{dt} \tag{7.1.4}$$

and for $x = \Lambda$

$$u(\Lambda,t) = E_1(t) + R_1(i(\Lambda,t)) + \tilde{C}_1^{-1}\left(\int_T^t i(\Lambda,\tau)d\tau\right) + \left[i(\Lambda,t)\frac{dL_1(i(\Lambda,t))}{di} + L_1(i(\Lambda,t))\right]\frac{di(\Lambda,t)}{dt}. \tag{7.1.5}$$

We begin with capacitive nonlinearities, where

$$C_p(u) = \frac{c_p\sqrt[h]{\Phi_p}}{\sqrt[h]{\Phi_p - u}},\ (p = 0,1);\, c_p, \Phi_p > 0, h \in [2,3]\ \text{are constants.}$$

First we notice $\lim\limits_{u\to\Phi_p} \dfrac{c_p\sqrt[h]{\Phi_p}}{\sqrt[h]{\Phi_p-u}}=\infty$ and $\lim\limits_{u\to-\infty} \dfrac{c_p\sqrt[h]{\Phi_p}\,u}{\sqrt[h]{\Phi_p-u}}=-\infty$. Consequently we have to

consider functions $C_p(u)$ for $|u|\le\phi_0<\Phi=\min\{\Phi_0,\Phi_1\}\le\dfrac{h}{h-1}\Phi$.

Besides $\dfrac{d\widetilde{C}_p(u)}{du}=\dfrac{c_p\sqrt[h]{\Phi_p}}{\sqrt[h]{\Phi_p-u}}\left(1+\dfrac{u}{h\left(\Phi_p-u\right)}\right)>0,\ u\in[-\phi_0,\phi_0]$, the inverse function

$\widetilde{C}_p^{-1}(I)$ does exist and

$$\widetilde{C}_p(u)=uC_p(u)=\frac{c_p\sqrt[h]{\Phi_p}\,u}{\sqrt[h]{\Phi_p-u}}:[-\phi_0,\phi_0]\to\left[\frac{-c_p\sqrt[h]{\Phi_p}\,\phi_0}{\sqrt[h]{\Phi_p+\phi_0}},\ \frac{c_p\sqrt[h]{\Phi_p}\,\phi_0}{\sqrt[h]{\Phi_p-\phi_0}}\right]$$

that is,

$$\widetilde{C}_p^{-1}(I):\left[\frac{-c_p\sqrt[h]{\Phi_p}\,\phi_0}{\sqrt[h]{\Phi_p+\phi_0}},\ \frac{c_p\sqrt[h]{\Phi_p}\,\phi_0}{\sqrt[h]{\Phi_p-\phi_0}}\right]\to[-\phi_0,\phi_0].$$

Then obviously

$$\left|\widetilde{C}_p^{-1}(I)\right|\le\phi_0. \tag{7.1.6}$$

The explicit form of the inverse function for $h=2$ is

$$\frac{c_p\sqrt{\Phi_p}\,u}{\sqrt{\Phi_p-u}}=I\ \Rightarrow\ c_p^2\Phi_p u^2+I^2 u-\Phi_p I^2=0 ;$$

$$\widetilde{C}_p^{-1}(I)=\frac{1}{2c_p^2\Phi_p}\left(\sqrt{I^4+4c_p^2\Phi_p^2 I^2}-I^2\right)=|I|\frac{\sqrt{I^2+4c_p^2\Phi_p^2}-|I|}{2c_p^2\Phi_p}\ (p=0,1);$$

$$\frac{d\widetilde{C}_p^{-1}(I)}{dI}=\frac{1}{2c_p^2\Phi_p}\left(\frac{4I^3+8c_p^2\Phi_p^2 I}{2\sqrt{I^4+4c_p^2\Phi_p^2 I^2}}-2I\right)=I\frac{I^2+2c_p^2\Phi_p^2-\sqrt{I^4+4c_p^2\Phi_p^2 I^2}}{c_p^2\Phi_p\sqrt{I^4+4c_p^2\Phi_p^2 I^2}} ;$$

$$\left|\widetilde{C}_p^{-1}(I)\right| \le |I|\,\frac{\sqrt{I^2+4c_p^2\Phi_p^2}-|I|}{2c_p^2\Phi_p} \le |I|\,\frac{\sqrt{\left(\dfrac{c_p\sqrt{\Phi_p}\,\phi_0}{\sqrt{\Phi_p-\phi_0}}\right)^2+4c_p^2\Phi_p^2}}{2c_p^2\Phi_p} =$$

$$\frac{2\Phi_p-\phi_0}{2c_p\sqrt{\Phi_p}\sqrt{\Phi_p-\phi_0}}\,|I| \equiv H_p|I|.$$

For for $h=3$ from

$$\frac{u\,c_p\sqrt[3]{\Phi_p}}{\sqrt[3]{\Phi_p-u}} = I \Leftrightarrow u^3 + \frac{I^3}{c_p^3\Phi_p}u - \frac{I^3}{c_p^3} = 0$$

we obtain

$$\widetilde{C}_p^{-1}(I) == \sqrt[3]{\frac{I^3}{2c_p^3}+\sqrt{\frac{I^6}{4c_p^6}+\frac{I^9}{27c_p^9\Phi_p^3}}} + \sqrt[3]{\frac{I^3}{2c_p^3}-\sqrt{\frac{I^6}{4c_p^6}+\frac{I^9}{27c_p^9\Phi_p^3}}}\ .$$

We need also the following estimates:

$$|I| \le \frac{c_p\sqrt{\Phi_p}\,\phi_0}{\sqrt{\Phi_p+\phi_0}} = \min\left\{\left|\frac{-c_p\sqrt{\Phi_p}\,\phi_0}{\sqrt{\Phi_p+\phi_0}}\right|,\left|\frac{c_p\sqrt{\Phi_p}\,\phi_0}{\sqrt{\Phi_p-\phi_0}}\right|\right\};$$

$$\left|\frac{d\widetilde{C}_p^{-1}(I)}{dI}\right| \le \frac{|I|}{c_p^2\Phi_p}\left(\frac{|I|^2+2c_p^2\Phi_p^2}{\sqrt{I^4+4c_p^2\Phi_p^2 I^2}}+1\right) \le \frac{|I|}{c_p^2\Phi_p}\left(\frac{|I|^2+2c_p^2\Phi_p^2}{|I|\sqrt{I^2+4c_p^2\Phi_p^2}}+1\right) \le$$

$$\le \frac{1}{c_p^2\Phi_p}\left(\sqrt{I^2+4c_p^2\Phi_p^2}+|I|\right) \le \frac{1}{c_p^2\Phi_p}\left(\sqrt{\left(\frac{c_p\sqrt{\Phi_p}\,\phi_0}{\sqrt{\Phi_p-\phi_0}}\right)^2+4c_p^2\Phi_p^2}+\frac{c_p\sqrt{\Phi_p}\,\phi_0}{\sqrt{\Phi_p-\phi_0}}\right) = \frac{2\sqrt{\Phi_p}}{c_p\sqrt{\Phi_p-\phi_0}}$$

$$(7.1.7)$$

The voltages of the inductors L_p are

$$u_{\Psi_p} = \frac{d\widetilde{L}_p}{dt} = \frac{d\left(L_p(i).i\right)}{dt} = \left[i\,\frac{dL_p(i)}{di}+L_p(i)\right]\frac{di}{dt}$$

and in view of $L_p(i) = \sum\limits_{n=1}^{m} l_n^{(p)} i^n$ we obtain $\widetilde{L}_p(i) = i \cdot L_p(i) = i \cdot \sum\limits_{n=1}^{m} l_n^{(p)} i^n = \sum\limits_{n=1}^{m} l_n^{(p)} i^{n+1}$ and

$$\frac{d\widetilde{L}_p(i)}{di} = i\frac{dL_p(i)}{di} + L_p(i) = i\sum_{n=1}^{m} n l_n^{(p)} i^{n-1} + \sum_{n=1}^{m} l_n^{(p)} i^n = \sum_{n=1}^{m}(n+1)l_n^{(p)} i^n, \quad (p=0,1).$$

7.2. REDUCING THE MIXED PROBLEM TO AN INITIAL VALUE PROBLEM FOR A NONLINEAR NEUTRAL SYSTEM

First we present (7.1.2) in the form:

$$\begin{aligned} \frac{\partial u(x,t)}{\partial t} + \frac{1}{C}\frac{\partial i(x,t)}{\partial x} + \frac{G}{C}u(x,t) = 0, \\ \frac{\partial i(x,t)}{\partial t} + \frac{1}{L}\frac{\partial u(x,t)}{\partial x} + \frac{R}{L}i(x,t) = 0 \end{aligned} \tag{7.2.1}$$

or

$$\frac{\partial U}{\partial t} + A_1\frac{\partial U}{\partial x} + A_2 U = 0 \tag{7.2.2}$$

where, $U = \begin{bmatrix} u \\ i \end{bmatrix}$, $\dfrac{\partial U}{\partial t} = \begin{bmatrix} \dfrac{\partial u}{\partial t} \\ \dfrac{\partial i}{\partial t} \end{bmatrix}$, $\dfrac{\partial U}{\partial x} = \begin{bmatrix} \dfrac{\partial u}{\partial x} \\ \dfrac{\partial i}{\partial x} \end{bmatrix}$, $A_1 = \begin{bmatrix} 0 & 1/C \\ 1/L & 0 \end{bmatrix}$, $A_2 = \begin{bmatrix} G/C & 0 \\ 0 & R/L \end{bmatrix}$.

Since $|A_1| \neq 0$ to transform $A_1 = \begin{bmatrix} 0 & 1/C \\ 1/L & 0 \end{bmatrix}$ into a diagonal form we solve the

characteristic equation $\begin{vmatrix} -\lambda & 1/C \\ 1/L & -\lambda \end{vmatrix} = 0$. Its roots are $\lambda_1 = 1/\sqrt{LC}$ and $\lambda_2 = -1/\sqrt{LC}$.

For the eigenvectors matrix and its inverse we obtain respectively

$$H = \begin{bmatrix} \sqrt{C} & \sqrt{L} \\ -\sqrt{C} & \sqrt{L} \end{bmatrix}, \quad H^{-1} = \begin{bmatrix} 1/(2\sqrt{C}) & -1/(2\sqrt{C}) \\ 1/(2\sqrt{L}) & 1/(2\sqrt{L}) \end{bmatrix}.$$

Denote by $A^{can} = \begin{bmatrix} \dfrac{1}{\sqrt{LC}} & 0 \\ 0 & -\dfrac{1}{\sqrt{LC}} \end{bmatrix}$. Then $A^{can} = HAH^{-1}$. Introduce new variables

$Z = HU$, (or $U = H^{-1}Z$), where $Z = \begin{bmatrix} V(x,t) \\ I(x,t) \end{bmatrix}$, $U = \begin{bmatrix} u(x,t) \\ i(x,t) \end{bmatrix}$.

Then

$$\left| \begin{aligned} V(x,t) &= \sqrt{C}\, u(x,t) + \sqrt{L}\, i(x,t) \\ I(x,t) &= -\sqrt{C}\, u(x,t) + \sqrt{L}\, i(x,t) \end{aligned} \right. \tag{7.2.3}$$

and

$$\left| \begin{aligned} u(x,t) &= \frac{1}{2\sqrt{C}} V(x,t) - \frac{1}{2\sqrt{C}} I(x,t) \\ i(x,t) &= \frac{1}{2\sqrt{L}} V(x,t) + \frac{1}{2\sqrt{L}} I(x,t). \end{aligned} \right. \tag{7.2.4}$$

Replacing $U = H^{-1}Z$ in (7.2.2) and multiply from the left by H we obtain

$$\frac{\partial Z}{\partial t} + H\left(AH^{-1}\right)\frac{\partial Z}{\partial x} + H\left(BH^{-1}\right)Z = 0,$$

i.e.

$$\frac{\partial Z}{\partial t} + A^{can}\frac{\partial Z}{\partial x} + H\left(BH^{-1}\right)Z = 0. \tag{7.2.5}$$

But $HBH^{-1} = \begin{bmatrix} \dfrac{1}{2}\left(\dfrac{G}{C}+\dfrac{R}{L}\right) & \dfrac{1}{2}\left(-\dfrac{G}{C}+\dfrac{R}{L}\right) \\ \dfrac{1}{2}\left(-\dfrac{G}{C}+\dfrac{R}{L}\right) & \dfrac{1}{2}\left(\dfrac{G}{C}+\dfrac{R}{L}\right) \end{bmatrix}$ and then

$$
\begin{bmatrix} \dfrac{\partial V}{\partial t} \\[2mm] \dfrac{\partial I}{\partial t} \end{bmatrix} + \begin{bmatrix} \dfrac{1}{\sqrt{LC}} & 0 \\[2mm] 0 & -\dfrac{1}{\sqrt{LC}} \end{bmatrix} \begin{bmatrix} \dfrac{\partial V}{\partial x} \\[2mm] \dfrac{\partial I}{\partial x} \end{bmatrix} + \begin{bmatrix} \dfrac{1}{2}\left(\dfrac{R}{L}+\dfrac{G}{C}\right) & \dfrac{1}{2}\left(\dfrac{R}{L}-\dfrac{G}{C}\right) \\[2mm] \dfrac{1}{2}\left(\dfrac{R}{L}-\dfrac{G}{C}\right) & \dfrac{1}{2}\left(\dfrac{R}{L}+\dfrac{G}{C}\right) \end{bmatrix} \begin{bmatrix} V \\ I \end{bmatrix} = \begin{bmatrix} 0 \\ 0 \end{bmatrix}.
$$

We consider distortionless lossy transmission line that means the following Heaviside condition is fulfilled:

$$
R/L = G/C. \tag{H}
$$

Then HBH^{-1} can be simplified and the last system becomes:

$$
\begin{bmatrix} \dfrac{\partial V}{\partial t} \\[2mm] \dfrac{\partial I}{\partial t} \end{bmatrix} + \begin{bmatrix} \dfrac{1}{\sqrt{LC}} & 0 \\[2mm] 0 & -\dfrac{1}{\sqrt{LC}} \end{bmatrix} \begin{bmatrix} \dfrac{\partial V}{\partial x} \\[2mm] \dfrac{\partial I}{\partial x} \end{bmatrix} + \begin{bmatrix} \dfrac{R}{L} & 0 \\[2mm] 0 & \dfrac{R}{L} \end{bmatrix} \begin{bmatrix} V \\ I \end{bmatrix} = \begin{bmatrix} 0 \\ 0 \end{bmatrix} \tag{7.2.6}
$$

or

$$
\begin{aligned}
\frac{\partial V}{\partial t} + \frac{1}{\sqrt{LC}}\frac{\partial V}{\partial x} + \frac{R}{L}V = 0, \\[3mm]
\frac{\partial I}{\partial t} - \frac{1}{\sqrt{LC}}\frac{\partial I}{\partial x} + \frac{R}{L}I = 0.
\end{aligned} \tag{7.2.7}
$$

The new initial conditions we obtain from (7.2.3) and (7.1.3):

$$
\begin{aligned}
V(x,0) = \sqrt{C}\, u(x,0) + \sqrt{L}\, i(x,0) = \sqrt{C}\, u_0(x) + \sqrt{L}\, i_0(x) \equiv V_0(x), \\[2mm]
I(x,0) = -\sqrt{C}\, u(x,0) + \sqrt{L}\, i(x,0) = -\sqrt{C}\, u_0(x) + \sqrt{L}\, i_0(x) \equiv I_0(x)
\end{aligned} ,\ x \in [0,\Lambda] \tag{7.2.8}
$$

In the same way one can transform the boundary conditions. Further on we set

$$
W(x,t) = e^{\frac{R}{L}t} V(x,t),\ J(x,t) = e^{\frac{R}{L}t} I(x,t), \tag{7.2.9}
$$

$$
V(x,t) = e^{-\frac{R}{L}t} W(x,t),\ V(x,t) = e^{-\frac{R}{L}t} W(x,t).
$$

Replacing (7.2.9) in system (7.2.7) we obtain:

$$\frac{\partial W}{\partial t} + \frac{1}{\sqrt{LC}}\frac{\partial W}{\partial x} = 0,$$

$$\frac{\partial J}{\partial t} - \frac{1}{\sqrt{LC}}\frac{\partial J}{\partial x} = 0. \tag{7.2.10}$$

The initial conditions remain unaltered because

$$W(0,t) = V(0,t) = V_0(x), \quad J(0,t) = I(0,t) = I_0(x).$$

System (7.2.10) corresponds to a lossless transmission line and then one can reduce the mixed problem for (7.2.10) to an initial value problem for a system of functional differential equations of neutral type on the right end of the boundary $x = \Lambda$. The system obtained is a nonlinear one in view of the nonlinear characteristics of the RLC-loads.

It is known that the solution of (7.2.10) is a pair of functions

$$W(x,t) = \Phi_W(x - vt) \text{ and } J(x,t) = \Phi_J(x + vt)$$

where Φ_W and Φ_J are arbitrary smooth functions and $v = 1/\sqrt{LC}$.

Since

$$\left| \begin{aligned} u(x,t) &= \frac{1}{2\sqrt{C}}V(x,t) - \frac{1}{2\sqrt{C}}I(x,t) \\ i(x,t) &= \frac{1}{2\sqrt{L}}V(x,t) + \frac{1}{2\sqrt{L}}I(x,t) \end{aligned} \right.$$

we obtain

$$\left| \begin{aligned} u(x,t) &= \frac{e^{-\frac{R}{L}t}}{2\sqrt{C}}W(x,t) - \frac{e^{-\frac{R}{L}t}}{2\sqrt{C}}J(x,t) \\ i(x,t) &= \frac{e^{-\frac{R}{L}t}}{2\sqrt{L}}W(x,t) + \frac{e^{-\frac{R}{L}t}}{2\sqrt{L}}J(x,t) \end{aligned} \right.$$

and consequently

$$\left| \begin{array}{l} u(x,t) = \dfrac{e^{-\frac{R}{L}t}}{2\sqrt{C}}\left[\Phi_W\left(x-vt\right)-\Phi_J\left(x+vt\right)\right], \\[3mm] i(x,t) = \dfrac{e^{-\frac{R}{L}t}}{2\sqrt{L}}\left[\Phi_W\left(x-vt\right)+\Phi_J\left(x+vt\right)\right] \end{array} \right. \tag{7.2.11}$$

and hence

$$\Phi_W(x-vt) = e^{\frac{R}{L}t}\left(\sqrt{C}\,u(x,t)+\sqrt{L}\,i(x,t)\right), \ \Phi_J(x+vt) = e^{\frac{R}{L}t}\left(\sqrt{L}\,i(x,t)-\sqrt{C}\,u(x,t)\right) \tag{7.2.12}$$

For $x = \Lambda$ we obtain

$$\begin{array}{l} \Phi_W(\Lambda-vt) = e^{-\frac{R}{L}t}\left[\sqrt{C}\,u(\Lambda,t)+\sqrt{L}\,i(\Lambda,t)\right], \\[3mm] \Phi_J(\Lambda+vt) = e^{-\frac{R}{L}t}\left[\sqrt{L}\,i(\Lambda,t)-\sqrt{C}\,u(\Lambda,t)\right] \end{array} \tag{7.2.13}$$

Recall that $T = \dfrac{\Lambda}{v} = \Lambda\sqrt{LC}$ and put

$$\Lambda - vt = -vt' \Rightarrow t = t'+\Lambda/v \equiv t'+T.$$

Substituting t in the first equation of (7.2.12) we get

$$\Phi_W(-vt') = e^{\frac{R}{L}(t'+T)}\left[\sqrt{C}\,u(\Lambda,t'+T)+\sqrt{L}\,i(\Lambda,t'+T)\right].$$

Now let us put $T = \dfrac{\Lambda}{v}$, $\Lambda + vt = vt'' \Rightarrow t = t''-T$. Then substitute in the second equation of (7.2.12) we obtain

$$\Phi_J(vt'') = e^{\frac{R}{L}(t''-T)}\left[\sqrt{L}\,i(\Lambda,t''-T)-\sqrt{C}\,u(\Lambda,t''-T)\right].$$

Therefore

$$\Phi_W(-vt) = e^{\frac{R}{L}(t+T)}\left[\sqrt{C}\,u(\Lambda,t+T)+\sqrt{L}\,i(\Lambda,t+T)\right], \tag{7.2.14}$$

$$\Phi_J(vt) = e^{\frac{R}{L}(t-T)}\left[\sqrt{L}\,i(\Lambda,t-T)-\sqrt{C}\,u(\Lambda,t-T)\right]. \tag{7.2.15}$$

From (7.2.11) by $x = 0$ we have

$$u(0,t) = \frac{e^{-\frac{R}{L}t}}{2\sqrt{C}}\left[\Phi_W(-vt) - \Phi_J(vt)\right],$$

$$i(0,t) = \frac{e^{-\frac{R}{L}t}}{2\sqrt{L}}\left[\Phi_W(-vt) + \Phi_J(vt)\right]. \tag{7.2.16}$$

Substituting $\Phi_W(-vt)$ and $\Phi_J(vt)$ from (7.2.14) and (7.2.15) into (7.2.16) we get

$$u(0,t) = \frac{1}{2\sqrt{C}}\left[e^{\frac{RT}{L}}\left(\sqrt{C}\,u(\Lambda,t+T) + \sqrt{L}\,i(\Lambda,t+T)\right) - e^{\frac{RT}{L}}\left(\sqrt{L}\,i(\Lambda,t-T) - \sqrt{C}\,u(\Lambda,t-T)\right)\right] \tag{7.2.17}$$

$$i(0,t) = \frac{1}{2\sqrt{L}}\left[e^{\frac{RT}{L}}\left(\sqrt{C}\,u(\Lambda,t+T) + \sqrt{L}\,i(\Lambda,t+T)\right) + e^{\frac{RT}{L}}\left(\sqrt{L}\,i(\Lambda,t-T) - \sqrt{C}\,u(\Lambda,t-T)\right)\right. \tag{7.2.18}$$

We put $t + T \equiv t$. Then t becomes $t - T$ and $t - T$ becomes $t - 2T$ and therefore

$$u(0,t-T) = \frac{1}{2\sqrt{C}}\left[e^{\frac{RT}{L}}\left(\sqrt{C}\,u(\Lambda,t) + \sqrt{L}\,i(\Lambda,t)\right) - \right.$$

$$e^{-\frac{RT}{L}}\left(\sqrt{L}\,i(\Lambda,t-2T) - \sqrt{C}\,u(\Lambda,t-2T)\right)\Big]; \tag{7.2.19}$$

$$i(0,t-T) = \frac{1}{2\sqrt{L}}\left[e^{\frac{RT}{L}}\left(\sqrt{C}\,u(\Lambda,t) + \sqrt{L}\,i(\Lambda,t)\right) + \right.$$

$$e^{-\frac{RT}{L}}\left(\sqrt{L}\,i(\Lambda,t-2T) - \sqrt{C}\,u(\Lambda,t-2T)\right)\Big]. \tag{7.2.20}$$

The boundary condition (7.1.10) after $t + T \equiv t$ becomes

$$u(0,t-T) = E_0(t-T) - R_0(i(0,t-T)) - \tilde{C}_0^{-1}\left(\int_0^{t-T} i(0,\tau)d\tau\right) -$$

$$-\left[i(0,t-T)\frac{dL_0(i(0,t-T))}{di} + L_0(i(0,t-T))\right]\frac{di(0,t-T)}{dt}$$

or

$$u(0,t-T) = E_0(t-T) - R_0(i(0,t-T)) - \widetilde{C}_0^{-1}\left(\int_T^t i(0,\theta-T)d\theta\right) - \frac{d\widetilde{L}_0(i(0,t-T))}{di}\frac{di(0,t-T)}{dt}.$$

We add the second boundary condition (7.1.11)

$$u(\Lambda,t) = E_1(t) + R_1(i(\Lambda,t)) + \widetilde{C}_1^{-1}\left(\int_T^t i(\Lambda,\tau)d\tau\right) + \left[i(\Lambda,t)\frac{dL_1(i(\Lambda,t))}{di} + L_1(i(\Lambda,t))\right]\frac{di(\Lambda,t)}{dt}.$$

Recall that $Z_0 = \sqrt{L/C}$ and setting $u(t) \equiv u(\Lambda,t)$, $i(t) \equiv i(\Lambda,t)$ we obtain the following system of neutral type with respect to the unknown functions $u(t)$ and $i(t)$:

$$\frac{e^{\frac{RT}{L}}\left(u(t)+Z_0\sqrt{L}\,i(t)\right) - e^{-\frac{RT}{L}}\left(Z_0\,i(t-2T)-u(t-2T)\right)}{2} =$$

$$= E_0(t-T) - R_0(i(0,t-T)) - \widetilde{C}_0^{-1}\left(\int_T^t i(0,\theta-T)d\theta\right) -$$

$$- \frac{i(0,t-T)dL_0/di + L_0(i(0,t-T))}{2}\left[\frac{e^{\frac{RT}{L}}}{Z_0}\frac{du(t)}{dt} + e^{\frac{RT}{L}}\frac{di(t)}{dt} + \right.$$

$$\left. + e^{-\frac{RT}{L}}\frac{di(t-2T)}{dt} - e^{-\frac{RT}{L}}\frac{1}{Z_0}\frac{du(t-2T)}{dt}\right],$$

$$u(t) = E_1(t) + R_1(i(t)) + \widetilde{C}_1^{-1}\left(\int_T^t i(\tau)d\tau\right) + \left[i(t)\frac{dL_1(i(t))}{di} + L_1(i(t))\right]\frac{di(t)}{dt}.$$

We solve the last system with respect to the derivatives $\dfrac{du(t)}{dt}, \dfrac{di(t)}{dt}$ at the present instant:

$$\frac{du(t)}{dt} = 2Z_0 e^{-\frac{RT}{L}}\frac{E_0(t) - R_0(i(0,t-T)) - \widetilde{C}_0^{-1}\left(\int_T^t i(0,\theta-T)d\theta\right)}{i(0,t-T)\dfrac{dL_0(i(0,t-T))}{di} + L_0(i(0,t-T))} -$$

$$- Z_0 e^{-\frac{RT}{L}}\frac{e^{\frac{RT}{L}}\left(u(t)+Z_0\sqrt{L}\,i(t)\right) - e^{\frac{RT}{L}}\left(Z_0\,i(t-2T)-u(t-2T)\right)}{i(0,t-T)\dfrac{dL_0(i(0,t-T))}{di} + L_0(i(0,t-T))} -$$

$$-Z_0 \frac{di(t)}{dt} - e^{-2\frac{RT}{L}} Z_0 \frac{di(t-2T)}{dt} + e^{-2\frac{RT}{L}} \frac{du(t-2T)}{dt},$$

$$\frac{di(t)}{dt} = \frac{u(t) - E_1(t) - R_1(i(t)) - \widetilde{C}_1^{-1}\left(\int_T^t i(\tau)d\tau\right)}{i(t)\dfrac{dL_1(i(t))}{di} + L_1(i(t))}.$$

We notice that the Lipschitz coefficient before $\dfrac{di(t)}{dt}$ is Z_0. In the applications $Z_0 > 1$.

This shows that we have to abandon the above system and proceed in § 7.3 as in the previous chapter.

7.3. ANOTHER MANNER TO REDUCING THE MIXED PROBLEM TO AN OSCILLATORY ONE

We proceed from the already obtained in Chapter VI relations

$$\left|\begin{array}{l} u(x,t) = \dfrac{e^{-\frac{R}{L}t}}{2\sqrt{C}} W(x,t) - \dfrac{e^{-\frac{R}{L}t}}{2\sqrt{C}} J(x,t) \\[3mm] i(x,t) = \dfrac{e^{-\frac{R}{L}t}}{2\sqrt{L}} W(x,t) + \dfrac{e^{-\frac{R}{L}t}}{2\sqrt{L}} J(x,t). \end{array}\right.$$

For $x = 0$ we have

$$\left|\begin{array}{l} u(0,t) = \dfrac{e^{-\frac{R}{L}t}}{2\sqrt{C}} W(0,t) - \dfrac{e^{-\frac{R}{L}t}}{2\sqrt{C}} J(0,t) \\[3mm] i(0,t) = \dfrac{e^{-\frac{R}{L}t}}{2\sqrt{L}} W(0,t) + \dfrac{e^{-\frac{R}{L}t}}{2\sqrt{L}} J(0,t) \end{array}\right.$$

$$(7.3.1)$$

and for $x = \Lambda$

$$\left|\begin{array}{l} u(\Lambda,t) = \dfrac{e^{-\frac{R}{L}t}}{2\sqrt{C}}W(\Lambda,t) - \dfrac{e^{-\frac{R}{L}t}}{2\sqrt{C}}J(\Lambda,t) \\[4ex] i(\Lambda,t) = \dfrac{e^{-\frac{R}{L}t}}{2\sqrt{L}}W(\Lambda,t) + \dfrac{e^{-\frac{R}{L}t}}{2\sqrt{L}}J(\Lambda,t). \end{array}\right. \tag{7.3.2}$$

Replacing (7.3.1) into the first boundary condition

$$u(0,t) = E_0(t) - R_0(i(0,t)) - \tilde{C}_0^{-1}\left(\int_T^t i(0,\tau)d\tau\right) - \frac{d\tilde{L}_0(i(0,t))}{di}\frac{di(0,t)}{dt}$$

we obtain

$$\frac{e^{-\frac{R}{L}t}}{2\sqrt{C}}W(0,t) - \frac{e^{-\frac{R}{L}t}}{2\sqrt{C}}J(0,t) = E_0(t) - R_0\left(\frac{e^{-\frac{R}{L}t}\left(W(0,t)+J(0,t)\right)}{2\sqrt{L}}\right) -$$

$$-\tilde{C}_0^{-1}\left(\int_T^t\left(\frac{e^{-\frac{R}{L}\tau}}{2\sqrt{L}}W(0,\tau) + \frac{e^{-\frac{R}{L}\tau}}{2\sqrt{L}}J(0,\tau)\right)d\tau\right) - \tag{7.3.3}$$

$$-\frac{d\tilde{L}_0\left(\dfrac{e^{-\frac{R}{L}t}W(0,t)+e^{-\frac{R}{L}t}J(0,t)}{2\sqrt{L}}\right)}{di}\frac{d}{dt}\left(\frac{e^{-\frac{R}{L}t}W(0,t)+e^{-\frac{R}{L}t}J(0,t)}{2\sqrt{L}}\right).$$

Repeating the procedure with (7.2.2) for the second boundary condition

$$u(\Lambda,t) = E_1(t) + R_1(i(\Lambda,t)) + \tilde{C}_1^{-1}\left(\int_T^t i(\Lambda,\tau)d\tau\right) + \frac{d\tilde{L}_1(i(\Lambda,t))}{di}\frac{di(\Lambda,t)}{dt}$$

we obtain a second equation

$$\frac{e^{-\frac{R}{L}t}}{2\sqrt{C}}W(\Lambda,t) - \frac{e^{-\frac{R}{L}t}}{2\sqrt{C}}J(\Lambda,t) = E_1(t) + R_1\left(\frac{e^{-\frac{R}{L}t}\left(W(\Lambda,t)+J(\Lambda,t)\right)}{2\sqrt{L}}\right) +$$

$$+\widetilde{C}_1^{-1}\left(\int_T^t\left(\frac{e^{-\frac{R}{L}t}}{2\sqrt{L}}W(\Lambda,t)+\frac{e^{-\frac{R}{L}t}}{2\sqrt{L}}J(\Lambda,t)\right)d\tau\right)+$$

$$+\frac{d\widetilde{L}_1\left(\dfrac{e^{-\frac{R}{L}t}W(\Lambda,t)+e^{-\frac{R}{L}t}J(\Lambda,t)}{2\sqrt{L}}\right)}{di}\frac{d}{dt}\left(\frac{e^{-\frac{R}{L}t}W(\Lambda,t)+e^{-\frac{R}{L}t}J(\Lambda,t)}{2\sqrt{L}}\right). \tag{7.3.4}$$

An integration along the characteristics yields

$$W(\Lambda,t+T)=W(0,t),\ J(\Lambda,t)=J(0,t+T)$$

and then (7.3.3) and (7.3.4) become

$$\frac{e^{-\frac{R}{L}(t+T)}}{2\sqrt{C}}W(\Lambda,t+T)-\frac{e^{-\frac{R}{L}t}}{2\sqrt{C}}J(0,t)=E_0(t)-R_0\left(\left(e^{-\frac{R}{L}t}W(\Lambda,t)+e^{-\frac{R}{L}(t+T)}J(0,t+T)\right)/\left(2\sqrt{L}\right)\right)-$$

$$-\widetilde{C}_0^{-1}\left(\frac{1}{2\sqrt{L}}\int_0^{t-T}\left(e^{-\frac{R}{L}(\tau+T)}W(\Lambda,\tau+T)+e^{-\frac{R}{L}\tau}J(0,\tau)\right)d\tau\right)-$$

$$-\left[\frac{e^{-\frac{R}{L}(t+T)}W(\Lambda,t+T)+e^{-\frac{R}{L}t}J(0,t)}{2\sqrt{L}}\frac{d}{di}L_0\left(\left(e^{-\frac{R}{L}(t+T)}W(\Lambda,t+T)+e^{-\frac{R}{L}t}J(0,t)\right)/\left(2\sqrt{L}\right)\right)+\right.$$

$$\left.+L_0\left(\left(e^{-\frac{R}{L}(t+T)}W(\Lambda,t+T)+e^{-\frac{R}{L}t}J(0,t)\right)/2\sqrt{L}\right)\right]\times\frac{d}{dt}\left(\left(e^{-\frac{R}{L}(t+T)}W(\Lambda,t+T)+e^{-\frac{R}{L}t}J(0,t)\right)/\left(2\sqrt{L}\right)\right)$$

and

$$\frac{e^{-\frac{R}{L}t}}{2\sqrt{C}}W(\Lambda,t)-\frac{e^{-\frac{R}{L}(t+T)}}{2\sqrt{C}}J(0,t+T)=E_1(t)+R_1\left(\left(e^{-\frac{R}{L}t}W(\Lambda,t)+e^{-\frac{R}{L}(t+T)}J(0,t+T)\right)/\left(2\sqrt{L}\right)\right)+$$

$$+\widetilde{C}_1^{-1}\left(\frac{1}{2\sqrt{L}}\int_0^{t-T}\left(e^{-\frac{R}{L}\tau}W(\Lambda,\tau)+e^{-\frac{R}{L}(\tau+T)}J(0,\tau+T)\right)d\tau\right)+$$

$$+\left[\frac{e^{-\frac{R}{L}t}W(\Lambda,t)+e^{-\frac{R}{L}(t+T)}J(0,t+T)}{2\sqrt{L}}\frac{d}{di}L_1\left(\left(e^{-\frac{R}{L}t}W(\Lambda,t)+e^{-\frac{R}{L}(t+T)}J(0,t+T)\right)/\left(2\sqrt{L}\right)\right)+\right.$$

$$\left.+L_1\left(\left(e^{-\frac{R}{L}t}W(\Lambda,t)+e^{-\frac{R}{L}(t+T)}J(0,t+T)\right)/2\sqrt{L}\right)\right]\times\frac{d}{dt}\left(\left(e^{-\frac{R}{L}t}W(\Lambda,t)+e^{-\frac{R}{L}(t+T)}J(0,t+T)\right)/\left(2\sqrt{L}\right)\right).$$

Then we put $t+T\equiv t$ and assuming that the unknown functions are

$$W(\Lambda,t) = W(t),\ J(0,t) = J(t)$$

we obtain

$$\frac{e^{-\frac{R}{L}t}}{2\sqrt{C}}W(t) - \frac{e^{-\frac{R}{L}(t-T)}}{2\sqrt{C}}J(t-T) = E_0(t-T) - R_0\left(\left(e^{-\frac{R}{L}t}W(t) + e^{-\frac{R}{L}(t-T)}J(t-T)\right)\big/\left(2\sqrt{L}\right)\right) -$$

$$-\widetilde{C}_0^{-1}\left(\frac{1}{2\sqrt{L}}\int_0^{t-T}\left(e^{-\frac{R}{L}(\tau+T)}W(\tau+T) + e^{-\frac{R}{L}\tau}J(\tau)\right)d\tau\right) -$$

$$-\left[\frac{e^{-\frac{R}{L}t}W(t) + e^{-\frac{R}{L}(t-T)}J(t-T)}{2\sqrt{L}}\frac{d}{di}L_0\left(\left(e^{-\frac{R}{L}t}W(t) + e^{-\frac{R}{L}(t-T)}J(t-T)\right)\big/\left(2\sqrt{L}\right)\right) +\right.$$

$$\left.+ L_0\left(\left(e^{-\frac{R}{L}t}W(t) + e^{-\frac{R}{L}(t-T)}J(t-T)\right)\big/2\sqrt{L}\right)\right] \times \frac{d}{dt}\left(\left(e^{-\frac{R}{L}t}W(t) + e^{-\frac{R}{L}(t-T)}J(t-T)\right)\big/\left(2\sqrt{L}\right)\right)$$

and

$$\frac{e^{-\frac{R}{L}(t-T)}}{2\sqrt{C}}W(t-T) - \frac{e^{-\frac{R}{L}t}}{2\sqrt{C}}J(t) = E_1(t-T) + R_1\left(\left(e^{-\frac{R}{L}(t-T)}W(t-T) + e^{-\frac{R}{L}t}J(t)\right)\big/\left(2\sqrt{L}\right)\right) +$$

$$+ \widetilde{C}_1^{-1}\left(\frac{1}{2\sqrt{L}}\int_0^{t-T}\left(e^{-\frac{R}{L}\tau}W(\tau) + e^{-\frac{R}{L}(\tau+T)}J(\tau+T)\right)d\tau\right) +$$

$$+\left[\frac{e^{-\frac{R}{L}(t-T)}W(t-T) + e^{-\frac{R}{L}t}J(t)}{2\sqrt{L}}\cdot\frac{d}{di}L_1\left(\left(e^{-\frac{R}{L}(t-T)}W(t-T) + e^{-\frac{R}{L}t}J(t)\right)\big/\left(2\sqrt{L}\right)\right) +\right.$$

$$\left.+ L_1\left(\frac{e^{-\frac{R}{L}(t-T)}W(t-T) + e^{-\frac{R}{L}t}J(t)}{2\sqrt{L}}\right)\right]\cdot\frac{d}{dt}\left(\left(e^{-\frac{R}{L}(t-T)}W(t-T) + e^{-\frac{R}{L}t}J(t)\right)\big/\left(2\sqrt{L}\right)\right).$$

Changing the variable in the integrals $\theta = \tau + T$ we finally obtain

$$\frac{e^{-\frac{R}{L}t}}{2\sqrt{C}}W(t) - \frac{e^{-\frac{R}{L}(t-T)}}{2\sqrt{C}}J(t-T) = E_0(t-T) - R_0\left(\left(e^{-\frac{R}{L}t}W(t) + e^{-\frac{R}{L}(t-T)}J(t-T)\right)\big/\left(2\sqrt{L}\right)\right) -$$

$$-\widetilde{C}_0^{-1}\left(\frac{1}{2\sqrt{L}}\int_T^{t}\left(e^{-\frac{R}{L}\theta}W(\theta) + e^{-\frac{R}{L}(\theta-T)}J(\theta-T)\right)d\theta\right) -$$

$$
-\left[\frac{e^{-\frac{R}{L}t}W(t)+e^{-\frac{R}{L}(t-T)}J(t-T)}{2\sqrt{L}}\frac{d}{di}L_0\left(\left(e^{-\frac{R}{L}t}W(t)+e^{-\frac{R}{L}(t-T)}J(t-T)\right)/\left(2\sqrt{L}\right)\right)\right.+
$$

$$
\left.+L_0\left(\left(e^{-\frac{R}{L}t}W(t)+e^{-\frac{R}{L}(t-T)}J(t-T)\right)/2\sqrt{L}\right)\right].
$$

$$
\frac{d}{dt}\left(\left(e^{-\frac{R}{L}t}W(t)+e^{-\frac{R}{L}(t-T)}J(t-T)\right)/\left(2\sqrt{L}\right)\right),
$$

$$
\frac{e^{-\frac{R}{L}(t-T)}}{2\sqrt{C}}W(t-T)-\frac{e^{-\frac{R}{L}t}}{2\sqrt{C}}J(t)=E_1(t-T)+R_1\left(\left(e^{-\frac{R}{L}(t-T)}W(t-T)+e^{-\frac{R}{L}t}J(t)\right)/\left(2\sqrt{L}\right)\right)+
$$

$$
+\tilde{C}_1^{-1}\left(\frac{1}{2\sqrt{L}}\int_T^t\left(e^{-\frac{R}{L}(\theta-T)}W(\theta-T)+e^{-\frac{R}{L}\theta}J(\theta)\right)d\theta\right)+
$$

$$
+\left[\frac{e^{-\frac{R}{L}(t-T)}W(t-T)+e^{-\frac{R}{L}t}J(t)}{2\sqrt{L}}\frac{d}{di}L_1\left(\left(e^{-\frac{R}{L}(t-T)}W(t-T)+e^{-\frac{R}{L}t}J(t)\right)/\left(2\sqrt{L}\right)\right)\right.+
$$

$$
\left.+L_1\left(\left(e^{-\frac{R}{L}(t-T)}W(t-T)+e^{-\frac{R}{L}t}J(t)\right)/2\sqrt{L}\right)\right]\times
$$

$$
\frac{d}{dt}\left(\left(e^{-\frac{R}{L}(t-T)}W(t-T)+e^{-\frac{R}{L}t}J(t)\right)/\left(2\sqrt{L}\right)\right).
$$

Principal Remark 7.3.1. We have to find periodic functions $W(t), J(t)$ of the above system. It follows that $e^{-\frac{R}{L}t}W(t), e^{-\frac{R}{L}t}J(t)$ are oscillatory solutions. That is why we set $\breve{W}(t)\equiv e^{-\frac{R}{L}t}W(t), \breve{J}(t)\equiv e^{-\frac{R}{L}t}J(t)$, where $\breve{W}(t), \breve{J}(t)$ are the new unknown functions. Further on we again denote them by $W(t), J(t)$. Obviously $W(t), J(t)$ should satisfy the inequalities

$$
\left|\breve{W}(t)\right|\le W_0 e^{-\frac{R}{L}t},\left|\breve{J}(t)\right|\le J_0 e^{-\frac{R}{L}t}\quad\Leftrightarrow\quad\left|W(t)\right|\le W_0 e^{-\frac{R}{L}t},\left|J(t)\right|\le J_0 e^{-\frac{R}{L}t}.
$$

Then the last system can be rewritten as

$$\frac{W(t)}{2\sqrt{C}} - \frac{J(t-T)}{2\sqrt{C}} = E_0(t-T) - R_0\left(\frac{W(t)+J(t-T)}{2\sqrt{L}}\right) - \widetilde{C}_0^{-1}\left(\frac{1}{2\sqrt{L}}\int_T^t (W(\theta)+J(\theta-T))d\theta\right) -$$

$$-\left[\frac{W(t)+J(t-T)}{2\sqrt{L}}\frac{dL_0\big((W(t)+J(t-T))/(2\sqrt{L})\big)}{di} + L_0\left(\frac{W(t)+J(t-T)}{2\sqrt{L}}\right)\right]\cdot\frac{d}{dt}\left(\frac{W(t)+J(t-T)}{2\sqrt{L}}\right), \qquad (7.3.5)$$

$$\frac{W(t-T)-J(t)}{2\sqrt{C}} = E_1(t-T) + R_1\left(\frac{W(t-T)+J(t)}{2\sqrt{C}}\right) + \widetilde{C}_1^{-1}\left(\frac{1}{2\sqrt{L}}\int_T^t (W(\theta-T)+J(\theta))d\theta\right) +$$

$$+\left[\frac{W(t-T)+J(t)}{2\sqrt{C}}\frac{dL_1\big((W(t-T)+J(t))/(2\sqrt{L})\big)}{di} + L_1\left(\frac{W(t-T)+J(t)}{2\sqrt{L}}\right)\right]\cdot\frac{d}{dt}\left(\frac{W(t-T)+J(t)}{2\sqrt{L}}\right).$$

Introduce the denotations

$$\widetilde{L}_0(W,J)(t) = \frac{W(t)+J(t-T)}{2\sqrt{L}}\frac{dL_0\big((W(t)+J(t-T))/2\sqrt{L}\big)}{di} + L_0\left(\frac{W(t)+J(t-T)}{2\sqrt{L}}\right) =$$

$$= \sum_{n=0}^{m}(n+1)l_n^{(0)}\left(\frac{W(t)+J(t-T)}{2\sqrt{L}}\right)^n;$$

$$\widetilde{L}_1(W,J)(t) = \frac{W(t-T)+J(t)}{2\sqrt{L}}\frac{dL_1\big((W(t-T)+J(t))/2\sqrt{L}\big)}{di} + L_1\left(\frac{W(t-T)+J(t)}{2\sqrt{L}}\right) =$$

$$= \sum_{n=0}^{m}(n+1)l_n^{(1)}\left(\frac{W(t-T)+J(t)}{2\sqrt{L}}\right)^n.$$

Then equations (7.3.5) could be rewritten as

$$\frac{W(t)}{2\sqrt{C}} - \frac{J(t-T)}{2\sqrt{C}} = E_0(t-T) - R_0\left(\frac{W(t)+J(t-T)}{2\sqrt{L}}\right) -$$

$$\widetilde{C}_0^{-1}\left(\frac{1}{2\sqrt{L}}\int_T^t (W(\theta)+J(\theta-T))d\theta\right) -$$

$$- \widetilde{L}_0(W,J)(t)\frac{1}{2\sqrt{L}}\left(\frac{dW(t)}{dt} + \frac{dJ(t-T)}{dt}\right), \qquad (7.3.6)$$

$$\frac{W(t-T)-J(t)}{2\sqrt{C}} = E_1(t-T) + R_1\left(\frac{W(t-T)+J(t)}{2\sqrt{C}}\right) + \widetilde{C}_1^{-1}\left(\frac{1}{2\sqrt{L}}\int_T^t (W(\theta-T)+J(\theta))d\theta\right) +$$

$$+ \widetilde{L}_1(W,J)(t).\frac{1}{2\sqrt{L}}\left(\frac{dW(t-T)}{dt} + \frac{dJ(t)}{dt}\right).$$

In order to solve the above equations with respect to the derivatives $\dfrac{dW(t)}{dt}$ and $\dfrac{dJ(t)}{dt}$ we have to divide the above equations to $\widetilde{L}_0(W,J)(t)$ and $\widetilde{L}_1(W,J)(t)$ respectively. Therefore we have to formulate conditions implying strict positive lower bounds for $\widetilde{L}_0(W,J)(t)$ and $\widetilde{L}_1(W,J)(t)$ respectively.

Further on we assume that:

$$\left|\widetilde{W}_0(t)\right| \le e^{-\frac{R}{L}(t+T)}e^{-\beta}W_0;\ \left|\widetilde{J}_0(t)\right| \le e^{-\frac{R}{L}(t+T)}e^{-\beta}J_0.$$

It follows

$$\left|W(t-T)\right| = \left|\widetilde{W}_0(t-T)\right| \le e^{-\frac{R}{L}t}W_0 e^{-\beta};\ \left|J(t-T)\right| = \left|\widetilde{J}_0(t-T)\right| \le e^{-\frac{R}{L}t}J_0 e^{-\beta}.$$

Let us assume that we can find an interval $\left|\dfrac{W(t)+J(t-T)}{2\sqrt{L}}\right| \le i_0$ such that the

inequalities to imply

$$\widetilde{L}_0(t-T) = \sum_{n=0}^{m}(n+1)l_n^{(0)}\left(\frac{W(t)+J(t-T)}{2\sqrt{L}}\right)^n \ge \hat{L}_0 > 0 \Rightarrow \frac{1}{\widetilde{L}_0(W,J)(t)} \le \frac{1}{\hat{L}_0}$$

and

$$\widetilde{L}_1(t-T) = \sum_{n=0}^{m}(n+1)l_n^{(1)}\left(\frac{W(t-T)+J(t)}{2\sqrt{L}}\right)^n \ge \hat{L}_1 > 0 \Rightarrow \frac{1}{\widetilde{L}_1(W,J)(t)} \le \frac{1}{\hat{L}_1}.$$

This can be done if the polynomial has suitable properties (cf. Numerical example).

Solve (7.3.6) with respect to $\dfrac{dW(t)}{dt}$ and $\dfrac{dJ(t)}{dt}$:

$$\frac{dW(t)}{dt} = -\frac{dJ(t-T)}{dt} + \frac{2\sqrt{L}}{\widetilde{L}_0(W,J)(t)}E_0(t) - \frac{2\sqrt{L}}{\widetilde{L}_0(W,J)(t)}R_0\left(\frac{W(t)+J(t-T)}{2\sqrt{L}}\right) -$$

$$-\frac{2\sqrt{L}}{\widetilde{L}_0(W,J)(t)}\frac{W(t)}{2\sqrt{C}} + \frac{2\sqrt{L}}{\widetilde{L}_0(W,J)(t)}\frac{J(t-T)}{2\sqrt{C}} - \frac{2\sqrt{L}}{\widetilde{L}_0(W,J)(t)}\widetilde{C}_0^{-1}\left(\int_T^t\left(\frac{W(\theta)}{2\sqrt{L}}+\frac{J(\theta-T)}{2\sqrt{L}}\right)d\theta\right);$$

$$\frac{dJ(t)}{dt} = -\frac{dW(t-T)}{dt} - \frac{2\sqrt{L}}{\widetilde{L}_1(W,J)(t)}E_1(t) - \frac{2\sqrt{L}}{\widetilde{L}_1(W,J)(t)}R_1\left(\frac{W(t-T)}{2\sqrt{L}}+\frac{J(t)}{2\sqrt{L}}\right) +$$

$$+\frac{2\sqrt{L}}{\widetilde{L}_1(W,J)(t)}\frac{W(t-T)}{2\sqrt{C}}-\frac{2\sqrt{L}}{\widetilde{L}_1(W,J)(t)}\frac{J(t)}{2\sqrt{C}}-\frac{2\sqrt{L}}{\widetilde{L}_1(W,J)(t)}\widetilde{C}_1^{-1}\left(\int_T^t\left(\frac{W(\theta-T)}{2\sqrt{L}}+\frac{J(\theta)}{2\sqrt{L}}\right)d\theta\right)$$

or in view of $Z_0=\sqrt{\dfrac{L}{C}}$ we have

$$\frac{dW(t)}{dt}=-\frac{dJ(t-T)}{dt}+\frac{2\sqrt{L}}{\widetilde{L}_0(W,J)(t)}E_0(t)-\frac{2\sqrt{L}}{\widetilde{L}_0(W,J)(t)}R_0\left(\frac{W(t)+J(t-T)}{2\sqrt{L}}\right)-$$

$$-\frac{Z_0W(t)}{\widetilde{L}_0(W,J)(t)}+\frac{Z_0J(t-T)}{\widetilde{L}_0(W,J)(t)}-\frac{2\sqrt{L}}{\widetilde{L}_0(W,J)(t)}\widetilde{C}_0^{-1}\left(\int_T^t\left(\frac{W(\theta)+J(\theta-T)}{2\sqrt{L}}\right)d\theta\right)\equiv U(W,J)(t),\quad(7.3.7)$$

$$\frac{dJ(t)}{dt}=-\frac{dW(t-T)}{dt}-\frac{2\sqrt{L}}{\widetilde{L}_1(W,J)(t)}E_1(t)-\frac{2\sqrt{L}}{\widetilde{L}_1(W,J)(t)}R_1\left(\frac{W(t-T)+J(t)}{2\sqrt{L}}\right)+$$

$$+\frac{Z_0W(t-T)}{\widetilde{L}_1(W,J)(t)}-\frac{Z_0J(t)}{\widetilde{L}_1(W,J)(t)}-\frac{2\sqrt{L}}{\widetilde{L}_1(W,J)(t)}\widetilde{C}_1^{-1}\left(\int_T^t\left(\frac{W(\theta-T)+J(\theta)}{2\sqrt{L}}\right)d\theta\right)\equiv I(W,J)(t),$$

$$W(t)=\widetilde{W}_0(t),\ \dot{W}(t)=\dot{\widetilde{W}}_0(t),\ \ J(t)=\widetilde{J}_0(t),\dot{J}(t)=\dot{\widetilde{J}}_0(t)\ ,\ t\in[0,T].$$

The initial functions are obtained from the initial conditions of the mixed problem as in Chapter II.

We formulate the conditions for the initial functions

(IN) $\left|\widetilde{W}_0(t)\right|\le e^{-\frac{R}{L}(t+T)}W_0;\ \ \left|\widetilde{J}_0(t)\right|\le e^{-\frac{R}{L}(t+T)}J_0\,,t\in[0,T]\,.$

7.4. THE EXISTENCE-UNIQUENESS OF AN OSCILLATORY SOLUTION FOR THE NONLINEAR NEUTRAL SYSTEM

Here we formulate an oscillatory problem for the neutral system obtained. The sources $E_0(t),E_1(t)$ are continuously differentiable oscillatory functions. We introduce an operator presentation of the oscillatory problem and by a fixed point theorem in uniform spaces [14] we prove an existence-uniqueness theorem.

The main problem is: to find a solution of the system (7.3.7) with advanced prescribed zeros on an interval $[t_0,\infty)$, $T\equiv t_0$, where $\widetilde{W}_0(t)$, $\widetilde{J}_0(t)$ are prescribed initial oscillating functions on the interval $[0,t_0]$.

Let $S_T=\{\tau_k\}_{k=0}^n,n\in N$ $(\tau_0=0,\ \tau_n=T\equiv t_0)$ be the set of zeros of the initial function, that is, $\widetilde{W}_0(\tau_k)=0,\ \widetilde{J}_0(\tau_k)=0$. Besides $\max\{\tau_{k+1}-\tau_k:k=0,1,...,n\}\le T_0$.

Let $S = \{t_k\}_{k=0}^{\infty}$ be a strictly increasing sequence of real numbers satisfying the following conditions **(C)**:

(C1) $\lim_{k \to \infty} t_k = \infty$;

(C2) for every k there is $s < k$ such that $t_k - T = t_s$ where $t_s \in S_T \cup S$.

It follows

(C3) $0 \le \inf\{t_{k+1} - t_k : k = 0,1,2,...\} \le \sup\{t_{k+1} - t_k : k = 0,1,2,...\} = T_0 < \infty$.

We assume also

(E) $E_p(t_k - T) = E_p(t_s) = 0 \ (p = 0,1)$.

Introduce the set $C^1[t_0, \infty)$ consisting of all continuous and bounded functions differentiable with bounded derivatives on every interval $[t_k, t_{k+1}]$.

Remark 7.4.1. Let us note that the left and right derivative at t_k of any $W(.), J(.) \in C^1[t_0, \infty)$ may not coincide. That is why we introduce a suitable topology for continuous functions with piece wise continuous derivatives. This is one reason to introduce uniform spaces (cf. Chapter I, also [14]).

Introduce the sets

$$M_W = \left\{ W(.) \in C^1[t_0, \infty) : W(t_k) = 0 \wedge |W(t)| \le W_0 e^{-\frac{R}{L}t}, t \in [t_0, \infty) \right\},$$

$$M_J = \left\{ J(.) \in C^1[t_0, \infty) : J(t_k) = 0 \wedge |J(t)| \le J_0 e^{-\frac{R}{L}t}, t \in [t_0, \infty) \right\}.$$

Remark 7.4.2. Let us comment the conformity condition **(CC)**. It could be obtained replacing $t = t_0$ in (7.3.7) and in view of

$$W(0) = W(t_0) = J(0) = J(t_0) = E_p(0) = E_p(t_0) = 0$$

we have to assume

$$\tilde{C}_p^{-1}(0) = 0, \ \dot{W}(t_0) = -\dot{J}(0), \ \dot{J}(t_0) = -\dot{W}(0)$$

(where $\dot{W}(t) = dW(t)/dt$). We notice that **(CC)** becomes a relation between the initial functions $\dot{\tilde{W}}_0(t_0) = -\dot{\tilde{J}}_0(0)$, $\dot{\tilde{J}}_0(t_0) = -\dot{\tilde{W}}_0(0)$. If the last condition is not satisfied then the jump of the derivative at $t = t_0$ propagates to the right and it falls at some zero point because of $t_k - T = t_s$. We do not go beyond our function space because the derivative of our functions might have jumps at $t = t_k$.

Remark 7.4.3. It follows that the functions from M_W and M_J satisfy the inequalities

$$\left|W(t)\right| \le W_0 e^{\mu(t-t_k)}, t \in [t_k, t_{k+1}] \ , \ \left|J(t)\right| \le J_0 e^{\mu(t-t_k)}, t \in [t_k, t_{k+1}], (k = 0,1,2,\ldots)$$

where $W_0, J_0, \mu, \ \mu T_0 = \mu_0 = const.$ are positive constants and $W_0 e^{\mu_0} \le \phi_0 < \infty$.

Introduce the following families of pseudo-metrics

$$\rho^{(k)}(W,\overline{W}) = \max\left\{\left|W(t) - \overline{W}(t)\right| : t \in [t_k, t_{k+1}]\right\},$$
$$\rho^{(k)}(J,\overline{J}) = \max\left\{\left|J(t) - \overline{J}(t)\right| : t \in [t_k, t_{k+1}]\right\},$$

$$\hat{\rho}^{(k)}(W,\overline{W}) = \max\left\{\left|W(t) - \overline{W}(t)\right| : t \in [t_0, t_{k+1}]\right\},$$
$$\hat{\rho}^{(k)}(J,\overline{J}) = \max\left\{\left|J(t) - \overline{J}(t)\right| : t \in [t_0, t_{k+1}]\right\},$$

$$\rho_\mu^{(k)}(W,\overline{W}) = \max\left\{e^{-\mu(t-t_k)}\left|W(t) - \overline{W}(t)\right| : t \in [t_k, t_{k+1}]\right\},$$
$$\rho_\mu^{(k)}(J,\overline{J}) = \max\left\{e^{-\mu(t-t_k)}\left|J(t) - \overline{J}(t)\right| : t \in [t_k, t_{k+1}]\right\},$$

$$\hat{\rho}_\mu^{(k)}(W,\overline{W}) = \max\left\{\rho_\mu^{(0)}(W,\overline{W}), \rho_\mu^{(1)}(W,\overline{W}),\ldots,\rho_\mu^{(k)}(W,\overline{W})\right\},$$
$$\hat{\rho}_\mu^{(k)}(J,\overline{J}) = \max\left\{\rho_\mu^{(0)}(J,\overline{J}), \rho_\mu^{(1)}(J,\overline{J}),\ldots,\rho_\mu^{(k)}(J,\overline{J})\right\},$$

$$\rho_\mu^{(k)}(\dot{W},\dot{\overline{W}}) = \max\left\{e^{-\mu(t-t_k)}\left|\dot{W}(t) - \dot{\overline{W}}(t)\right| : t \in [t_k, t_{k+1}]\right\},$$
$$\rho_\mu^{(k)}(\dot{J},\dot{\overline{J}}) = \max\left\{e^{-\mu(t-t_k)}\left|\dot{J}(t) - \dot{\overline{J}}(t)\right| : t \in [t_k, t_{k+1}]\right\},$$

$$\hat{\rho}_\mu^{(k)}(\dot{W},\dot{\overline{W}}) = \max\left\{\rho_\mu^{(0)}(\dot{W},\dot{\overline{W}}), \rho_\mu^{(1)}(\dot{W},\dot{\overline{W}}),\ldots,\rho_\mu^{(k)}(\dot{W},\dot{\overline{W}})\right\},$$
$$\hat{\rho}_\mu^{(k)}(\dot{J},\dot{\overline{J}}) = \max\left\{\rho_\mu^{(0)}(\dot{J},\dot{\overline{J}}), \rho_\mu^{(1)}(\dot{J},\dot{\overline{J}}),\ldots,\rho_\mu^{(k)}(\dot{J},\dot{\overline{J}})\right\}.$$

The following inequalities imply the equivalence of the both families of pseudo-metrics

$$\rho_\mu^{(k)}(W,\overline{W}) \le \rho^{(k)}(W,\overline{W}) \le e^{\mu T_0}\rho_\mu^{(k)}(W,\overline{W}), \ (k=0,1,2,\ldots),$$

$$\rho_\mu^{(k)}(J,\overline{J}) \le \rho^{(k)}(J,\overline{J}) \le e^{\mu T_0}\rho_\mu^{(k)}(J,\overline{J}), \ (k=0,1,2,\ldots).$$

It is easy to verify that

$$\hat{\rho}^{(k)}(W,\overline{W}) = \max\left\{\rho^{(0)}(W,\overline{W}),\rho^{(1)}(W,\overline{W}),\ldots,\rho^{(k)}(W,\overline{W})\right\} \le$$
$$\le e^{\mu T_0}\max\left\{\rho_\mu^{(0)}(W,\overline{W}),\rho_\mu^{(1)}(W,\overline{W}),\ldots,\rho_\mu^{(k)}(W,\overline{W})\right\} = e^{\mu T_0}\hat{\rho}_\mu^{(k)}(W,\overline{W})$$

$$\hat{\rho}^{(k)}(J,\overline{J}) = \max\left\{\rho^{(0)}(J,\overline{J}),\rho^{(1)}(J,\overline{J}),\ldots,\rho^{(k)}(J,\overline{J})\right\} \le$$
$$\le e^{\mu T_0}\max\left\{\rho_\mu^{(0)}(J,\overline{J}),\rho_\mu^{(1)}(J,\overline{J}),\ldots,\rho_\mu^{(k)}(J,\overline{J})\right\} = e^{\mu T_0}\hat{\rho}_\mu^{(k)}(J,\overline{J}).$$

$$(7.4.1)$$

The set $M_W \times M_J$ turns out into a complete uniform space with respect to the family of pseudo-metrics

$$\hat{\rho}_\mu^{(k)}((W,\dot{W},J,\dot{J}),(\overline{W},\dot{\overline{W}},\overline{J},\dot{\overline{J}})) = \max\left\{\hat{\rho}^{(k)}(W,\overline{W}),\hat{\rho}^{(k)}(J,\overline{J}),\hat{\rho}_\mu^{(k)}(\dot{W},\dot{\overline{W}}),\hat{\rho}_\mu^{(k)}(\dot{J},\dot{\overline{J}})\right\}$$
$$(k=0,1,2,\ldots).$$

We define the operator

$$B = \left(B_W(W,J),B_J(W,J)\right)$$

by the formulas

$$B_W(W,J)(t) = B_W^{(k)}(W,J)(t) := \int_{t_k}^{t}U(W,J)(s)ds - \frac{t-t_k}{t_{k+1}-t_k}\int_{t_k}^{t_{k+1}}U(W,J)(s)ds,$$

$$t \in [t_k,t_{k+1}], \ (k=0,1,2,\ldots)$$

$$B_J(W,J)(t) = B_J^{(k)}(W,J)(t) := \int_{t_k}^{t}I(W,J)(s)ds - \frac{t-t_k}{t_{k+1}-t_k}\int_{t_k}^{t_{k+1}}I(W,J)(s)ds,$$

$$t \in [t_k,t_{k+1}], \ (k=0,1,2,\ldots).$$

Further on the following assumptions will be hold:

Assumtion **(IN)**:

$$\left|\widetilde{W}_0(t)\right| \le W_0 e^{-\beta}e^{-\frac{R}{L}(t+T)}; \left|\widetilde{J}_0(t)\right| \le J_0 e^{-\beta}e^{-\frac{R}{L}(t+T)}, t \in [0,T],(k=01,2,\ldots,n-1);$$

Assumtion **(E)**: $\left|E_p(t)\right| \le U_{E_p} e^{-\frac{R}{L}t}, t \in [0,\infty), E_p(t-T) = E_p(t), (p = 0,1); U_{E_p} \le W_0;$

Assumption (Π): $\dfrac{W_0 + J_0}{2\sqrt{L}} \le i_0;\ W_0 e^{\mu 0} \le \phi_0.$

Lemma 7.4.1. Problem (7.3.8) has a solution $(W,J) \in M_W \times M_J$ iff the operator B has a fixed point in $M_W \times M_J$, that is,

$$(W,J) = (B_W(W,J), B_J(W,J)). \tag{7.4.2}$$

Proof: Let $(W,J) \in M_W \times M_J$ be a solution of the system (7.3.7). Then integrating every equation of (7.3.7) on every interval $[t_k, t] \subset [t_k, t_{k+1}]$ $(k = 0,1,2 \dots)$ we obtain

$$W(t) - W(t_k) = \int_{t_k}^{t} U(W,J)(s)ds \Leftrightarrow W(t) = \int_{t_k}^{t} U(W,J)(s)ds,$$

$$J(t) - J(t_k) = \int_{t_k}^{t} I(W,J)(s)ds \Leftrightarrow J(t) = \int_{t_k}^{t} I(W,J)(s)ds$$

and then

$$W(t) = \int_{t_k}^{t} U(W,J)(s)ds \Rightarrow 0 = W(t_{k+1}) = \int_{t_k}^{t} U(W,J)(s)ds \Rightarrow \int_{t_k}^{t_{k+1}} U(W,J)(s)ds = 0,$$

$$J(t) = \int_{t_k}^{t} I(W,J)(s)ds \Rightarrow 0 = J(t_{k+1}) = \int_{t_k}^{t} I(W,J)(s)ds \Rightarrow \int_{t_k}^{t_{k+1}} I(W,J)(s)ds = 0. \tag{7.4.3}$$

Therefore the pair (W,J) satisfies

$$W(t) = \int_{t_k}^{t} U(W,J)(s)ds - \frac{t-t_k}{t_{k+1}-t_k} \int_{t_k}^{t_{k+1}} U(W,J)(s)ds,\ t \in [t_k, t_{k+1}]$$

$$J(t) = \int_{t_k}^{t} I(W,J)(s)ds - \frac{t-t_k}{t_{k+1}-t_k} \int_{t_k}^{t_{k+1}} I(W,J)(s)ds,\ t \in [t_k, t_{k+1}]$$

$\Leftrightarrow (W,J) = (B_W(W,J), B_J(W,J))$, that is, (W,J) is a fixed point of B.

Conversely, let (W,J) be a fixed point of B or

$$W(t) = \int_{t_k}^{t} U(W,J)(s)\,ds - \frac{t-t_k}{t_{k+1}-t_k} \int_{t_k}^{t_{k+1}} U(W,J)(s)\,ds,$$

$$J(t) = \int_{t_k}^{t} I(W,J)(s)\,ds - \frac{t-t_k}{t_{k+1}-t_k} \int_{t_k}^{t_{k+1}} I(W,J)(s)\,ds.$$

Then having in mind that $\mu_0 = \mu T_0 = \text{const.}$ we obtain

$$\left| \int_{t_k}^{t_{k+1}} U(W,J)(s)\,ds \right| \le$$

$$\le \left| \int_{t_k}^{t_{k+1}} \frac{dJ(t-T)}{dt}\,dt \right| + \frac{2\sqrt{L}}{\hat{L}_0} \int_{t_k}^{t_{k+1}} |E_0(t)|\,dt + \frac{2\sqrt{L}}{\hat{L}_0} \int_{t_k}^{t_{k+1}} \left| R_0\left(\frac{W(t)+J(t-T)}{2\sqrt{L}} \right) \right| dt +$$

$$+ \frac{Z_0}{\hat{L}_0} \int_{t_k}^{t_{k+1}} |W(t)|\,dt + \frac{Z_0}{\hat{L}_0} \int_{t_k}^{t_{k+1}} |J(t-T)|\,dt + \frac{2\sqrt{L}}{\hat{L}_0} \int_{t_k}^{t_{k+1}} \left| \overline{C}_0^{-1}\left(\int_{t_k}^{s} \left(\frac{W(\theta)+J(\theta-T)}{2\sqrt{L}} \right) d\theta \right) \right| ds \le$$

$$\le \frac{2\sqrt{L}}{\hat{L}_0} W_0 \int_{t_k}^{t_{k+1}} e^{\mu(t-t_k)}\,dt + \frac{2\sqrt{L}}{\hat{L}_0} \sum_{n=1}^{m} \left| r_n^{(0)} \right| \int_{t_k}^{t_{k+1}} \left| \frac{W(t)+J(t-T)}{2\sqrt{L}} \right|^n dt +$$

$$+ \frac{Z_0 W_0}{\hat{L}_0} \int_{t_k}^{t_{k+1}} e^{\mu(t-t_k)}\,dt + \frac{Z_0 J_0}{\hat{L}_0} \int_{t_k}^{t_{k+1}} e^{\mu(t-T-t_k)}\,dt + \frac{2\sqrt{L}}{\hat{L}_0} W_0 \int_{t_k}^{t_{k+1}} e^{\mu(s-t_k)}\,ds \le$$

$$\le \frac{2\sqrt{L}}{\hat{L}_0} W_0 \frac{e^{\mu(t_{k+1}-t_k)}-1}{\mu} + \frac{2\sqrt{L}}{\hat{L}_0} \sum_{n=1}^{m} \left| r_n^{(0)} \right| \int_{t_k}^{t_{k+1}} \left| \frac{W_0+J_0}{2\sqrt{L}} \right|^n e^{n\mu(t-t_k)}\,dt +$$

$$+ \frac{Z_0 W_0}{\hat{L}_0} \frac{e^{\mu(t_{k+1}-t_k)}-1}{\mu} + \frac{Z_0 J_0}{\hat{L}_0} \frac{e^{\mu(t_{k+1}-t_k)}-1}{\mu} + \frac{2\sqrt{L}}{\hat{L}_0} W_0 \frac{e^{\mu(t_{k+1}-t_k)}-1}{\mu} \le$$

$$\le \frac{2\sqrt{L}}{\hat{L}_0} W_0 \frac{e^{\mu(t_{k+1}-t_k)}-1}{\mu} + \frac{2\sqrt{L}}{\hat{L}_0} \sum_{n=1}^{m} \left| r_n^{(0)} \right| \left(\frac{W_0+J_0}{2\sqrt{L}} \right)^n \frac{e^{n\mu(t_{k+1}-t_k)}-1}{n\mu} +$$

$$+ \frac{Z_0 W_0}{\hat{L}_0} \frac{e^{\mu(t_{k+1}-t_k)}-1}{\mu} + \frac{Z_0 J_0}{\hat{L}_0} \frac{e^{\mu(t_{k+1}-t_k)}-1}{\mu} + \frac{2\sqrt{L}}{\hat{L}_0} W_0 \frac{e^{\mu(t_{k+1}-t_k)}-1}{\mu} \le$$

$$\le \frac{e^{\mu(t_{k+1}-t_k)}-1}{\mu} \left[\frac{4\sqrt{L}\,W_0}{\hat{L}_0} + \frac{2\sqrt{L}}{\hat{L}_0} \sum_{n=1}^{m} \left| r_n^{(0)} \right| \left(\frac{W_0+J_0}{2\sqrt{L}} \right)^n e^{(n-1)\mu(t_{k+1}-t_k)} + \frac{2Z_0 W_0}{\hat{L}_0} + \frac{Z_0 J_0}{\hat{L}_0} + \frac{2\sqrt{L}\,W_0}{\hat{L}_0} \right] \le$$

$$\leq \frac{e^{\mu_0}-1}{\mu\hat{L}_0}\left(6\sqrt{L}\,W_0 + 2\sqrt{L}\sum_{n=1}^{m}\left|r_n^{(0)}\right|\left(i_0\right)^n e^{(n-1)\mu_0} + 4\sqrt{L}Z_0 i_0\right) \equiv M_0(\mu);$$

$$\left|\int_{t_k}^{t_{k+1}} I(W,J)(s)\,ds\right| \leq$$

$$\leq \int_{t_k}^{t_{k+1}}\left|\frac{dW(t-T)}{dt}\right|dt + \frac{2\sqrt{L}}{\hat{L}_1}\int_{t_k}^{t_{k+1}}\left|E_1(t)\right|dt + \frac{2\sqrt{L}}{\hat{L}_1}\int_{t_k}^{t_{k+1}}\left|R_1\left(\frac{W(t-T)+J(t)}{2\sqrt{L}}\right)\right|dt +$$

$$+ \frac{Z_0}{\hat{L}_1}\int_{t_k}^{t_{k+1}}\left|W(t-T)\right|dt + \frac{Z_0}{\hat{L}_1}\int_{t_k}^{t_{k+1}}\left|J(t)\right|dt + \frac{2\sqrt{L}}{\hat{L}_1}\int_{t_k}^{t_{k+1}}\left|\overline{C}_1^{-1}\left(\int_{t_k}^{s}\left(\frac{W(\theta-T)+J(\theta)}{2\sqrt{L}}\right)d\theta\right)\right|ds \leq$$

$$\leq \frac{2\sqrt{L}\,W_0}{\hat{L}_1}\frac{e^{\mu(t_{k+1}-t_k)}-1}{\mu} + \frac{2\sqrt{L}}{\hat{L}_1}\sum_{n=1}^{m}\left|r_n^{(1)}\right|\left(\frac{W_0+J_0}{2\sqrt{L}}\right)^n \frac{e^{n\mu(t_{k+1}-t_k)}-1}{n\mu} +$$

$$+ \frac{Z_0 W_0}{\hat{L}_1}\frac{e^{\mu(t_{k+1}-t_k)}-1}{\mu} + \frac{Z_0 J_0}{\hat{L}_1}\frac{e^{\mu(t_{k+1}-t_k)}-1}{\mu} + \frac{2\sqrt{L}}{\hat{L}_1}W_0\frac{e^{\mu(t_{k+1}-t_k)}-1}{\mu} \leq$$

$$\leq \frac{e^{\mu_0}-1}{\mu\hat{L}_1}\left(4\sqrt{L}\,W_0 + 2\sqrt{L}\sum_{n=1}^{m}\left|r_n^{(1)}\right|\left(i_0\right)^n e^{(n-1)\mu T_0} + 2\sqrt{L}Z_0 i_0\right) \equiv M_1(\mu).$$

As in Chapter II, we conclude that

$$\int_{t_k}^{t_{k+1}} U(W,J)(s)\,ds = 0 \quad \text{and} \quad \int_{t_k}^{t_{k+1}} I(W,J)(s)\,ds = 0.$$

Therefore

$$W(t) = \int_{t_k}^{t} U(W,J)(s)\,ds - \frac{t-t_k}{t_{k+1}-t_k}\int_{t_k}^{t_{k+1}} U(W,J)(s)\,ds \Leftrightarrow W(t) = \int_{t_k}^{t} U(W,J)(s)\,ds,$$

$$J(t) = \int_{t_k}^{t} I(W,J)(s)\,ds - \frac{t-t_k}{t_{k+1}-t_k}\int_{t_k}^{t_{k+1}} I(W,J)(s)\,ds \Leftrightarrow J(t) = \int_{t_k}^{t} I(W,J)(s)\,ds.$$

Differentiating the last integral equations we obtain (7.3.7).
Lemma 7.4.1 is thus proved.
Preliminary assertions:

1) $\left| \dfrac{t - t_k}{t_{k+1} - t_k} \right| \leq 1,\ t \in [t_k, t_{k+1}]$;

2) $\dfrac{e^{n(t_{k+1} - t_k)} - 1}{n} = \dfrac{(e^{t_{k+1} - t_k} - 1)(e^{(n-1)(t_{k+1} - t_k)} + e^{(n-2)(t_{k+1} - t_k)} + \ldots + 1)}{n} \leq \dfrac{(e^{(t_{k+1} - t_k)} - 1)n e^{(n-1)(t_{k+1} - t_k)}}{n} =$

$= (e^{(t_{k+1} - t_k)} - 1) e^{(n-1)(t_{k+1} - t_k)}$;

3) $\displaystyle\int_{t_k}^{t} e^{-\frac{R}{L} s}\, ds = -\dfrac{L}{R}\left(e^{-\frac{R}{L} t} - e^{-\frac{R}{L} t_k} \right) = \dfrac{L}{R}\left(e^{-\frac{R}{L} t_k} - e^{-\frac{R}{L} t} \right) = e^{-\frac{R}{L} t}\dfrac{L}{R}\left(e^{\frac{R}{L}(t - t_k)} - 1 \right) \leq$

$\leq e^{-\frac{R}{L} t}\dfrac{L}{R}\left(e^{\frac{R}{L} T_0} - 1 \right) \leq e^{-\frac{R}{L} t}\dfrac{L}{R}\left(e^{\frac{R T_0}{L}} - e^{-\frac{R T_0}{L}} \right)$;

$\displaystyle\int_{t_k}^{t} e^{-n\frac{R}{L} s}\, ds = -\dfrac{L}{nR}\left(e^{-n\frac{R}{L} t} - e^{-n\frac{R}{L} t_k} \right) = \dfrac{L}{nR}\left(e^{-n\frac{R}{L} t_k} - e^{-n\frac{R}{L} t} \right) =$

$= e^{-\frac{R}{L} t}\dfrac{L}{nR}\left(e^{\frac{R}{L}(t - t_k)} - 1 \right)\left(e^{-(n-1)\frac{R}{L} t_k} + \ldots + e^{-(n-1)\frac{R}{L} t} \right) \leq e^{-\frac{R}{L} t}\dfrac{L}{nR}\left(e^{\frac{R}{L}(t - t_k)} - 1 \right)n \leq e^{-\frac{R}{L} t}\dfrac{L}{R}\left(e^{\frac{R T_0}{L}} - e^{-\frac{R T_0}{L}} \right)$;

4) $\displaystyle\int_{t_k}^{t_{k+1}} e^{-\frac{R}{L} s}\, ds = -\dfrac{L}{R}\left(e^{-\frac{R}{L} t_{k+1}} - e^{-\frac{R}{L} t_k} \right) = \dfrac{L}{R}\left(e^{-\frac{R}{L} t_k} - e^{-\frac{R}{L} t_{k+1}} \right) = \dfrac{L}{R}\left(e^{-\frac{R}{L} t_k} - e^{-\frac{R}{L} t} + e^{-\frac{R}{L} t} - e^{-\frac{R}{L} t_{k+1}} \right) =$

$= e^{-\frac{R}{L} t}\dfrac{L}{R}\left(e^{\frac{R}{L}(t - t_k)} - 1 + 1 - e^{-\frac{R}{L}(t_{k+1} - t)} \right) \leq e^{-\frac{R}{L} t}\dfrac{L}{R}\left(e^{\frac{R}{L}(t_{k+1} - t_k)} - e^{-\frac{R}{L}(t_{k+1} - t_k)} \right) \leq$

$\leq e^{-\frac{R}{L} t}\dfrac{L}{R}\left(e^{\frac{R}{L} T_0} - e^{-\frac{R}{L} T_0} \right) = e^{-\frac{R}{L} t}\dfrac{2L}{R}\sinh\left(\dfrac{R T_0}{L} \right)$;

5) $\displaystyle\int_{t_k}^{t_{k+1}} e^{-n\frac{R}{L} s}\, ds = \dfrac{L}{nR}\left(e^{-n\frac{R}{L} t_k} - e^{-n\frac{R}{L} t_{k+1}} \right) =$

$= \dfrac{L}{nR}\left(e^{-\frac{R}{L} t_k} - e^{-\frac{R}{L} t_{k+1}} \right)\left(e^{-(n-1)\frac{R}{L} t_k} + e^{-(n-2)\frac{R}{L} t_k}\, e^{-\frac{R}{L} t_{k+1}} + \ldots + e^{-(n-1)\frac{R}{L} t_{k+1}} \right) \leq$

$\leq \dfrac{L}{nR}\left(e^{-\frac{R}{L} t_k} - e^{-\frac{R}{L} t_{k+1}} \right)n = \dfrac{L}{R}\left(e^{-\frac{R}{L} t_k} - e^{-\frac{R}{L} t_{k+1}} \right) \leq e^{-\frac{R}{L} t}\dfrac{2L}{R}\sinh\left(\dfrac{R}{L} T_0 \right)$;

6) the function $\xi(h) = \dfrac{e^h - 1}{h},\ h \geq 0$ is increasing and $\lim\limits_{h \to 0} \xi(h) = 1$;

7) $\displaystyle \sup\left|\frac{d\widetilde{L}_0(i)}{di}\right| \le \sum_{n=1}^{m}(n+1)\left|l_n^{(0)}\right|\left(\frac{W_0+J_0}{2\sqrt{L}}\right)^n e^{-n\frac{R}{L}t} \le \sum_{n=1}^{m}(n+1)\left|l_n^{(0)}\right|i_0^n e^{-n\frac{R}{L}t}$;

8) $\displaystyle \sup\left|\frac{d\widetilde{L}_1(i)}{di}\right| \le \sum_{n=1}^{m}(n+1)\left|l_n^{(1)}\right|\left(\frac{W_0+J_0}{2\sqrt{L}}\right)^n e^{-n\frac{R}{L}t} \le \sum_{n=1}^{m}(n+1)\left|l_n^{(1)}\right|i_0^n e^{-n\frac{R}{L}t}$;

9) $\displaystyle \rho^{(k)}(W,\overline{W}) \le e^{\mu 0}\rho_\mu^{(k)}(W,\overline{W}) \le \frac{\rho_\mu^{(k)}(\dot{W},\dot{\overline{W}})}{\mu}; \ \rho^{(k)}(J,\overline{J}) \le e^{\mu 0}\rho_\mu^{(k)}(J,\overline{J}) \le \frac{\rho_\mu^{(k)}(\dot{J},\dot{\overline{J}})}{\mu}.$

Theorem 7.4.1. Let conditions **(IN)**, **(E)**, $\left(\Pi\right)$ be fulfilled and the following inequalities be satisfied

$$J_0 e^{-\beta} + \frac{4L}{R}\sinh\left(\frac{RT_0}{L}\right)\left[\frac{2\sqrt{L}}{\hat{L}_0}U_{E0} + \frac{W_0+J_0}{\hat{L}_0}\left(\sum_{n=1}^{m}\left|r_n^{(0)}\right|\left(\frac{W_0+J_0}{2\sqrt{L}}\right)^{n-1} + Z_0 + \frac{2H_0 L}{R}\sinh\left(\frac{RT_0}{L}\right)\right)\right] \le W_0;$$

$$W_0 e^{-\beta} + \frac{4L}{R}\sinh\left(\frac{RT_0}{L}\right)\left[\frac{2\sqrt{L}}{\hat{L}_1}U_{E1} + \frac{W_0+J_0}{\hat{L}_1}\left(\sum_{n=1}^{m}\left|r_n^{(1)}\right|\left(\frac{W_0+J_0}{2\sqrt{L}}\right)^{n-1} + Z_0 + \frac{2H_1 L}{R}\sinh\left(\frac{RT_0}{L}\right)\right)\right] \le J_0.$$

Then there exists a unique oscillatory solution of (7.3.8), belonging to $M = M_W \times M_J$.

Proof*:* First we notice that the function $B_W(W,J)(t)$ is continuous on $[t_0,\infty)$. Indeed

$$B_W^{(0)}(W,J)(t_0) = \int_{t_0}^{t_0} U(W,J)(s)ds - \frac{t_0-t_0}{t_1-t_0}\int_{t_0}^{t_1} U(W,J)(s)ds) = 0,$$

$$B_J^{(0)}(W,J)(t_0) = \int_{t_0}^{t_0} I(W,J)(s)ds - \frac{t_0-t_0}{t_1-t_0}\int_{t_0}^{t_1} I(W,J)(s)ds = 0,$$

$$B_W^{(k)}(W,J)(t_{k+1}) = \int_{t_k}^{t_{k+1}} U(W,J)(s)ds - \frac{t_{k+1}-t_k}{t_{k+1}-t_k}\int_{t_k}^{t_{k+1}} U(W,J)(s)ds = 0,$$

$$B_W^{(k+1)}(W,J)(t_{k+1}):= \int_{t_{k+1}}^{t_{k+1}} U(W,J)(s)ds - \frac{t_{k+1}-t_{k+1}}{t_{k+2}-t_{k+1}}\int_{t_{k+1}}^{t_{k+2}} U(W,J)(s)ds = 0,$$

$$B_J^{(k)}(W,J)(t_{k+1}) = \int_{t_k}^{t_{k+1}} I(W,J)(s)ds - \frac{t_{k+1}-t_k}{t_{k+1}-t_k}\int_{t_k}^{t_{k+1}} I(W,J)(s)ds = 0,$$

$$B_J^{(k+1)}(W,J)(t_{k+1}) := \int\limits_{t_{k+1}}^{t_{k+1}} I(W,J)(s)\,ds - \frac{t_{k+1}-t_{k+1}}{t_{k+2}-t_{k+1}} \int\limits_{t_{k+1}}^{t_{k+2}} I(W,J)(s)\,ds = 0.$$

To complete the proof that $\big(B_W(W,J), B_J(W,J)\big) \in M_W \times M_J$ we show $\left| B_W^{(k)}(W,J)(t) \right| \le W_0 e^{-\frac{R}{L}t} e^{\mu(t-t_k)}$. Indeed,

$$\left| B_W^{(k)}(W,J)(t) \right| \le \left| \int\limits_{t_k}^{t} U(W,J)(s)\,ds \right| + \left| \int\limits_{t_k}^{t_{k+1}} U(W,J)(s)\,ds \right| \equiv W_1 + W_2.$$

Recall that $\left| \widetilde{C}_p^{-1}(I) \right| \le H_p |I|$ $\;(p=0,1)$ where $H_p = \dfrac{2\Phi_p - \phi_0}{2c_p\sqrt{\Phi_p}\sqrt{\Phi_p - \phi_0}}$.

Then we have

$$W_1 \le \left| \int\limits_{t_k}^{t} \frac{dJ(s-T)}{ds}\,ds \right| + \frac{2\sqrt{L}}{\hat{L}_0} \int\limits_{t_k}^{t} |E_0(s)|\,ds + \frac{2\sqrt{L}}{\hat{L}_0} \int\limits_{t_k}^{t} \left| R_0\!\left(\frac{W(s)+J(s-T)}{2\sqrt{L}} \right) \right|\,ds +$$

$$+ \frac{Z_0}{\hat{L}_0} \int\limits_{t_k}^{t} |W(s)|\,ds + \frac{Z_0}{\hat{L}_0} \int\limits_{t_k}^{t} |J(s-T)|\,ds + \frac{2\sqrt{L}}{\hat{L}_0} \int\limits_{t_k}^{t} \left| \widetilde{C}_0^{-1}\!\left(\int\limits_{t_k}^{s} \left(\frac{W(\theta)+J(\theta-T)}{2\sqrt{L}} \right) d\theta \right) \right| ds \le$$

$$\le |J(t-T)| + \frac{2\sqrt{L}}{\hat{L}_0} U_{E0} \int\limits_{t_k}^{t} e^{-\frac{R}{L}s}\,ds + \frac{2\sqrt{L}}{\hat{L}_0} \sum\limits_{n=1}^{m} \left| r_n^{(0)} \right| \left| \int\limits_{t_k}^{t} \left(\frac{|W(s)|+|J(s-T)|}{2\sqrt{L}} \right)^n \right| ds +$$

$$+ \frac{Z_0 W_0}{\hat{L}_0} \int\limits_{t_k}^{t} e^{-\frac{R}{L}s}\,ds + \frac{Z_0 J_0}{\hat{L}_0} e^{-\beta} \int\limits_{t_k}^{t} e^{-\frac{R}{L}s}\,ds + \frac{2\sqrt{L}H_0}{\hat{L}_0} \int\limits_{t_k}^{t} \left(\int\limits_{t_k}^{s} \frac{|W(\theta)|+|J(\theta-T)|}{2\sqrt{L}}\,d\theta \right) ds \le$$

$$\le J_0 e^{-\frac{R}{L}t} e^{-\beta} + \frac{2\sqrt{L}}{\hat{L}_0} U_{E0} \int\limits_{t_k}^{t} e^{-\frac{R}{L}s}\,ds + \frac{2\sqrt{L}}{\hat{L}_0} \sum\limits_{n=1}^{m} \left| r_n^{(0)} \right| \left(\frac{W_0+J_0}{2\sqrt{L}} \right)^n \int\limits_{t_k}^{t} e^{-n\frac{R}{L}s}\,ds +$$

$$+ \frac{Z_0 W_0}{\hat{L}_0} \int\limits_{t_k}^{t} e^{-\frac{R}{L}s}\,ds + \frac{Z_0 J_0}{\hat{L}_0} \int\limits_{t_k}^{t} e^{-\frac{R}{L}s}\,ds + \frac{2\sqrt{L}H_0}{\hat{L}_0} \frac{W_0+J_0}{2\sqrt{L}} \int\limits_{t_k}^{t} \left(\int\limits_{t_k}^{s} e^{-\frac{R}{L}\theta}\,d\theta \right) ds \le$$

$$\le J_0 e^{-\frac{R}{L}t} e^{-\beta} + \frac{2\sqrt{L}}{\hat{L}_0} U_{E0} \int\limits_{t_k}^{t} e^{-\frac{R}{L}s}\,ds + \frac{2\sqrt{L}}{\hat{L}_0} \sum\limits_{n=1}^{m} \left| r_n^{(0)} \right| \left(\frac{W_0+J_0}{2\sqrt{L}} \right)^n \int\limits_{t_k}^{t} e^{-n\frac{R}{L}s}\,ds +$$

$$+ \frac{Z_0 W_0}{\hat{L}_0} \int\limits_{t_k}^{t} e^{-\frac{R}{L}s}\,ds + \frac{Z_0 J_0}{\hat{L}_0} \int\limits_{t_k}^{t} e^{-\frac{R}{L}s}\,ds + \frac{2\sqrt{L}H_0}{\hat{L}_0} \frac{W_0+J_0}{2\sqrt{L}} \frac{L}{R} \int\limits_{t_k}^{t} \left(e^{-\frac{R}{L}t_k} - e^{-\frac{R}{L}s} \right) ds \le$$

$$\leq J_0 e^{-\frac{R}{L}t} e^{-\beta} + \frac{2\sqrt{L}}{\hat{L}_0} U_{E_0} \int_{t_k}^{t} e^{-\frac{R}{L}s} ds + \frac{2\sqrt{L}}{\hat{L}_0} \sum_{n=1}^{m} \left| r_n^{(0)} \right| \left(\frac{W_0 + J_0}{2\sqrt{L}} \right)^n \int_{t_k}^{t} e^{-n\frac{R}{L}s} ds +$$

$$+ \frac{Z_0 W_0}{\hat{L}_0} \int_{t_k}^{t} e^{-\frac{R}{L}s} ds + \frac{Z_0 J_0}{\hat{L}_0} \int_{t_k}^{t} e^{-\frac{R}{L}s} ds + \frac{2\sqrt{L} H_0}{\hat{L}_0} \frac{W_0 + J_0}{2\sqrt{L}} \frac{L}{R} \int_{t_k}^{t} e^{-\frac{R}{L}s} \left(e^{\frac{RT_0}{L}} - 1 \right) ds \leq$$

$$\leq J_0 e^{-\frac{R}{L}t} e^{-\beta} + e^{-\frac{R}{L}t} \left(e^{\frac{RT_0}{L}} - e^{-\frac{RT_0}{L}} \right) \left[\frac{2\sqrt{L}}{\hat{L}_0} U_{E_0} + \frac{W_0 + J_0}{\hat{L}_0} \left(\sum_{n=1}^{m} \left| r_n^{(0)} \right| \left(\frac{W_0 + J_0}{2\sqrt{L}} \right)^{n-1} + Z_0 + \frac{H_0 L}{R} \left(e^{\frac{RT_0}{L}} - e^{-\frac{RT_0}{L}} \right) \right) \right] \leq$$

$$\leq e^{-\frac{R}{L}t} \left\{ J_0 e^{-\beta} + 2\frac{L}{R} \sinh\left(\frac{RT_0}{L} \right) \left[\frac{2\sqrt{L}}{\hat{L}_0} U_{E_0} + \frac{W_0 + J_0}{\hat{L}_0} \left(\sum_{n=1}^{m} \left| r_n^{(0)} \right| \left(\frac{W_0 + J_0}{2\sqrt{L}} \right)^{n-1} + Z_0 + \frac{2H_0 L}{R} \sinh\left(\frac{RT_0}{L} \right) \right) \right] \right\}$$

and

$$W_2 \leq \left| \int_{t_k}^{t_{k+1}} \frac{dJ(s-T)}{ds} ds \right| + \frac{2\sqrt{L}}{\hat{L}_0} \int_{t_k}^{t_{k+1}} \left| E_0(s) \right| ds + \frac{2\sqrt{L}}{\hat{L}_0} \int_{t_k}^{t_{k+1}} \left| R_0 \left(\frac{W(s) + J(s-T)}{2\sqrt{L}} \right) \right| ds +$$

$$+ \frac{Z_0}{\hat{L}_0} \int_{t_k}^{t_{k+1}} \left| W(s) \right| ds + \frac{Z_0}{\hat{L}_0} \int_{t_k}^{t_{k+1}} \left| J(s-T) \right| ds + \frac{2\sqrt{L}}{\hat{L}_0} \int_{t_k}^{t_{k+1}} \left| \widetilde{C}_0^{-1} \left(\int_{t_k}^{s} \left(\frac{W(\theta) + J(\theta - T)}{2\sqrt{L}} \right) d\theta \right) \right| ds \leq$$

$$\leq \int_{t_k}^{t_{k+1}} e^{-\frac{R}{L}s} ds \left[\frac{2\sqrt{L}}{\hat{L}_0} U_{E_0} + \frac{W_0 + J_0}{\hat{L}_0} \left(\sum_{n=1}^{m} \left| r_n^{(0)} \right| \left(\frac{W_0 + J_0}{2\sqrt{L}} \right)^{n-1} + Z_0 + \frac{H_0 L}{R} \left(e^{\frac{RT_0}{L}} - e^{-\frac{RT_0}{L}} \right) \right) \right] \leq$$

$$\leq e^{-\frac{R}{L}t} \frac{L}{R} \left(e^{\frac{RT_0}{L}} - e^{-\frac{RT_0}{L}} \right) \left[\frac{2\sqrt{L}}{\hat{L}_0} U_{E_0} + \frac{W_0 + J_0}{\hat{L}_0} \left(\sum_{n=1}^{m} \left| r_n^{(0)} \right| \left(\frac{W_0 + J_0}{2\sqrt{L}} \right)^{n-1} + Z_0 + \frac{H_0 L}{R} \left(e^{\frac{RT_0}{L}} - e^{-\frac{RT_0}{L}} \right) \right) \right] \leq$$

$$\leq e^{-\frac{R}{L}t} \left\{ 2\frac{L}{R} \sinh\left(\frac{RT_0}{L} \right) \left[\frac{2\sqrt{L}}{\hat{L}_0} U_{E_0} + \frac{W_0 + J_0}{\hat{L}_0} \left(\sum_{n=1}^{m} \left| r_n^{(0)} \right| \left(\frac{W_0 + J_0}{2\sqrt{L}} \right)^{n-1} + Z_0 + \frac{2H_0 L}{R} \sinh\left(\frac{RT_0}{L} \right) \right) \right] \right\}.$$

Then

$$\left| B_W^{(k)}(W, J)(t) \right| \leq$$

$$\leq e^{-\frac{R}{L}t} \left\{ J_0 e^{-\beta} + 2\frac{L}{R} \sinh\left(\frac{RT_0}{L} \right) \left[\frac{2\sqrt{L}}{\hat{L}_0} U_{E_0} + \frac{W_0 + J_0}{\hat{L}_0} \left(\sum_{n=1}^{m} \left| r_n^{(0)} \right| \left(\frac{W_0 + J_0}{2\sqrt{L}} \right)^{n-1} + Z_0 + \frac{2H_0 L}{R} \sinh\left(\frac{RT_0}{L} \right) \right) \right] \right\} +$$

$$+ e^{-\frac{R}{L}t} \left\{ 2\frac{L}{R} \sinh\left(\frac{RT_0}{L} \right) \left[\frac{2\sqrt{L}}{\hat{L}_0} U_{E_0} + \frac{W_0 + J_0}{\hat{L}_0} \left(\sum_{n=1}^{m} \left| r_n^{(0)} \right| \left(\frac{W_0 + J_0}{2\sqrt{L}} \right)^{n-1} + Z_0 + \frac{2H_0 L}{R} \sinh\left(\frac{RT_0}{L} \right) \right) \right] \right\} \leq$$

$$\leq e^{-\frac{R}{L}t}\left\{J_0 e^{-\beta}+\frac{4L}{R}\sinh\left(\frac{RT_0}{L}\right)\left[\frac{2\sqrt{L}}{\hat{L}_0}U_{E_0}+\frac{W_0+J_0}{\hat{L}_0}\left(\sum_{n=1}^{m}\left|r_n^{(0)}\right|\left(\frac{W_0+J_0}{2\sqrt{L}}\right)^{n-1}+Z_0+\frac{2H_0 L}{R}\sinh\left(\frac{RT_0}{L}\right)\right)\right]\right\}\leq$$

$$\leq W_0 e^{-\frac{R}{L}t}.$$

For the second component we have

$$\left|B_J^{(k)}(W,J)(t)\right|\leq\left|\int_{t_k}^{t}I(W,J)(s)ds\right|+\left|\int_{t_k}^{t_{k+1}}I(W,J)(s)ds\right|\equiv J_1+J_2.$$

But

$$J_1\leq\left|\int_{t_k}^{t}\frac{dW(s-T)}{ds}ds\right|+\frac{2\sqrt{L}}{\hat{L}_1}\int_{t_k}^{t}\left|E_1(s-T)\right|ds+\frac{2\sqrt{L}}{\hat{L}_1}\sum_{n=1}^{m}\left|r_n^{(1)}\right|\left|\int_{t_k}^{t}\left(\frac{|W(s-T)|+|J(s)|}{2\sqrt{L}}\right)^n ds\right|+$$

$$+\frac{Z_0}{\hat{L}_1}\int_{t_k}^{t}\left|W(s-T)\right|ds+\frac{Z_0}{\hat{L}_1}\int_{t_k}^{t}\left|J(s)\right|ds+\frac{2\sqrt{L}}{\hat{L}_1}\left|\int_{t_k}^{t}\left|\tilde{C}_1^{-1}\left(\int_{T}^{s}\left(\frac{W(\theta-T)+J(\theta)}{2\sqrt{L}}\right)d\theta\right)\right|ds\right|\leq$$

$$\leq\left|W(t-T)\right|+\frac{2\sqrt{L}}{\hat{L}_1}U_{E_1}\int_{t_k}^{t}e^{-\frac{R}{L}s}ds+\frac{2\sqrt{L}}{\hat{L}_1}\sum_{n=1}^{m}\left|r_n^{(1)}\right|\left|\int_{t_k}^{t}\left(\frac{|W(s-T)|+|J(s)|}{2\sqrt{L}}\right)^n ds\right|+$$

$$+\frac{Z_0 W_0 e^{-\beta}}{\hat{L}_1}\int_{t_k}^{t}e^{-\frac{R}{L}s}ds+\frac{Z_0 J_0}{\hat{L}_1}\int_{t_k}^{t}e^{-\frac{R}{L}s}ds+\frac{2\sqrt{L}}{\hat{L}_1}\int_{t_k}^{t}\left|\tilde{C}_1^{-1}\left(\int_{T}^{s}\frac{|W(\theta-T)|+|J(\theta)|}{2\sqrt{L}}d\theta\right)\right|ds\leq$$

$$\leq W_0 e^{-\frac{R}{L}t}e^{-\beta}+\frac{2\sqrt{L}U_{E_1}}{\hat{L}_1}\int_{t_k}^{t}e^{-\frac{R}{L}s}ds+\frac{2\sqrt{L}}{\hat{L}_1}\sum_{n=1}^{m}\left|r_n^{(1)}\right|\left(\frac{W_0+J_0}{2\sqrt{L}}\right)^n\int_{t_k}^{t}e^{-n\frac{R}{L}s}ds+$$

$$+\frac{Z_0 W_0}{\hat{L}_1}\int_{t_k}^{t}e^{-\frac{R}{L}s}ds+\frac{Z_0 J_0}{\hat{L}_1}\int_{t_k}^{t}e^{-\frac{R}{L}s}ds+\frac{2\sqrt{L}H_1}{\hat{L}_1}\frac{W_0+J_0}{2\sqrt{L}}\int_{t_k}^{t}\left(\int_{t_k}^{s}e^{-\frac{R}{L}\theta}d\theta\right)ds\leq$$

$$\leq e^{-\frac{R}{L}t}W_0 e^{-\beta}+\frac{2\sqrt{L}\,U_{E_1}}{\hat{L}_1}\int_{t_k}^{t}e^{-\frac{R}{L}s}ds+\frac{2\sqrt{L}}{\hat{L}_1}\sum_{n=1}^{m}\left|r_n^{(1)}\right|\left(\frac{W_0+J_0}{2\sqrt{L}}\right)^n\int_{t_k}^{t}e^{-n\frac{R}{L}s}ds+$$

$$+\frac{Z_0 W_0}{\hat{L}_1}\int_{t_k}^{t}e^{-\frac{R}{L}s}ds+\frac{Z_0 J_0}{\hat{L}_1}\int_{t_k}^{t}e^{-\frac{R}{L}s}ds+\frac{2\sqrt{L}H_1}{\hat{L}_1}\frac{W_0+J_0}{2\sqrt{L}}\frac{L}{R}\left(e^{\frac{RT_0}{L}}-e^{-\frac{RT_0}{L}}\right)\int_{t_k}^{t}e^{-\frac{R}{L}s}ds\leq$$

$$\leq e^{-\frac{R}{L}t}\left\{W_0 e^{-\beta}+2\frac{L}{R}\sinh\left(\frac{RT_0}{L}\right)\left[\frac{2\sqrt{L}}{\hat{L}_1}U_{E_1}+\frac{W_0+J_0}{\hat{L}_1}\left(\sum_{n=1}^{m}\left|r_n^{(1)}\right|\left(\frac{W_0+J_0}{2\sqrt{L}}\right)^{n-1}+Z_0+\frac{2H_1 L}{R}\sinh\left(\frac{RT_0}{L}\right)\right)\right]\right\}$$

and

$$J_2 \leq \left| \int_{t_k}^{t_{k+1}} \frac{dW(s-T)}{ds} ds \right| + \frac{2\sqrt{L}}{\hat{L}_1} \int_{t_k}^{t_{k+1}} |E_1(s)| ds + \frac{2\sqrt{L}}{\hat{L}_1} \int_{t_k}^{t_{k+1}} \left| R_1\left(\frac{W(s-T)+J(s)}{2\sqrt{L}} \right) \right| ds +$$

$$+ \frac{Z_0}{\hat{L}_1} \int_{t_k}^{t_{k+1}} |W(s-T)| ds + \frac{Z_0}{\hat{L}_1} \int_{t_k}^{t_{k+1}} |J(s)| ds + \frac{2\sqrt{L}}{\hat{L}_1} \int_{t_k}^{t_{k+1}} \left| \widetilde{C}_1^{-1}\left(\int_{t_k}^{s} \left(\frac{W(\theta-T)+J(\theta)}{2\sqrt{L}} \right) d\theta \right) \right| ds \leq$$

$$\leq \frac{2\sqrt{L}}{\hat{L}_1} U_{E_1} \int_{t_k}^{t_{k+1}} e^{-\frac{R}{L}s} ds + \frac{2\sqrt{L}}{\hat{L}_1} \sum_{n=1}^{m} |r_n^{(1)}| \int_{t_k}^{t_{k+1}} \left(\frac{|W(s-T)|+|J(s)|}{2\sqrt{L}} \right)^n ds +$$

$$+ \frac{Z_0 W_0}{\hat{L}_1} \int_{t_k}^{t_{k+1}} e^{-\frac{R}{L}s} ds + \frac{Z_0 J_0}{\hat{L}_1} \int_{t_k}^{t_{k+1}} e^{-\frac{R}{L}s} ds + \frac{2\sqrt{L}\,H_1}{\hat{L}_1} \int_{t_k}^{t_{k+1}} \left(\int_{t_k}^{s} \left| \frac{W(\theta-T)+J(\theta)}{2\sqrt{L}} \right| d\theta \right) ds \leq$$

$$\leq e^{-\frac{R}{L}t} \left\{ \frac{2L}{R} \sinh\left(\frac{RT_0}{L} \right) \left[\frac{2\sqrt{L}}{\hat{L}_1} U_{E_1} + \frac{W_0+J_0}{\hat{L}_1} \left(\sum_{n=1}^{m} |r_n^{(0)}| \left(\frac{W_0+J_0}{2\sqrt{L}} \right)^{n-1} + Z_0 + \frac{2H_0 L}{R} \sinh\left(\frac{RT_0}{L} \right) \right) \right] \right\}.$$

Therefore

$$\left| B_J^{(k)}(W,J)(t) \right| \leq$$

$$\leq e^{-\frac{R}{L}t} \left\{ W_0 e^{-\beta} + 2\frac{L}{R} \sinh\left(\frac{RT_0}{L} \right) \left[\frac{2\sqrt{L}}{\hat{L}_1} U_{E_1} + \frac{W_0+J_0}{\hat{L}_1} \left(\sum_{n=1}^{m} |r_n^{(1)}| \left(\frac{W_0+J_0}{2\sqrt{L}} \right)^{n-1} + Z_0 + \frac{2H_1 L}{R} \sinh\left(\frac{RT_0}{L} \right) \right) \right] \right\} +$$

$$+ e^{-\frac{R}{L}t} \left\{ \frac{2L}{R} \sinh\left(\frac{RT_0}{L} \right) \left[\frac{2\sqrt{L}}{\hat{L}_1} U_{E_1} + \frac{W_0+J_0}{\hat{L}_1} \left(\sum_{n=1}^{m} |r_n^{(0)}| \left(\frac{W_0+J_0}{2\sqrt{L}} \right)^{n-1} + Z_0 + \frac{2H_0 L}{R} \sinh\left(\frac{RT_0}{L} \right) \right) \right] \right\} \leq$$

$$\leq e^{-\frac{R}{L}t} \left\{ W_0 e^{-\beta} + \frac{4L}{R} \sinh\left(\frac{RT_0}{L} \right) \left[\frac{2\sqrt{L}}{\hat{L}_1} U_{E_1} + \frac{W_0+J_0}{\hat{L}_1} \left(\sum_{n=1}^{m} |r_n^{(1)}| \left(\frac{W_0+J_0}{2\sqrt{L}} \right)^{n-1} + Z_0 + \frac{2H_1 L}{R} \sinh\left(\frac{RT_0}{L} \right) \right) \right] \right\} \leq$$

$$\leq J_0 e^{-\frac{R}{L}t}.$$

Consequently the operator B maps $M_W \times M_J$ into itself.

In what follows we show that B is contractive operator.

First we notice $t_k - T = t_q$, $t_{k+1} - T = t_{q+1}$ for some q.

The maximum $\hat{\rho}_\mu^{(k)}(W,\overline{W}) = \max\{\rho_\mu^{(0)}(W,\overline{W}), \rho_\mu^{(1)}(W,\overline{W}),..., \rho_\mu^{(k)}(W,\overline{W})\}$ attains to some $s \in \{0,1,2,...,k\}$.

Then in view of Preliminary assertions 7) and 8) we obtain

$$\left| B_W^{(k)}(W,J)(t) - B_W^{(k)}(\overline{W},\overline{J})(t) \right| \le \int\limits_{t_k}^{t} \left| U(W,J)(s) - U(\overline{W},\overline{J})(s) \right| ds +$$

$$+ \frac{t-t_k}{t_{k+1}-t_k} \left| \int\limits_{t_k}^{t_{k+1}} \left(U(W,J)(s) - U(\overline{W},\overline{J})(s) \right) ds \right| \equiv U_1 + U_2 .$$

But

$$U_1 \le \left| \int\limits_{t_k}^{t} \left(\frac{dJ(s-T)}{ds} - \frac{d\overline{J}(s-T)}{ds} \right) ds \right| + 2\sqrt{L} \left| \int\limits_{t_k}^{t} \left(\frac{E_0(s-T)}{\widetilde{L}_0(W,J)(s)} - \frac{E_0(s-T)}{\widetilde{L}_0(\overline{W},\overline{J})(s)} \right) ds \right| +$$

$$+ 2\sqrt{L} \int\limits_{t_k}^{t} \left| \frac{1}{\widetilde{L}_0(W,J)(t)} R_0\left(\frac{W(s)+J(s-T)}{2\sqrt{L}} \right) - \frac{1}{\widetilde{L}_0(\overline{W},\overline{J})(t)} R_0\left(\frac{\overline{W}(s)+\overline{J}(s-T)}{2\sqrt{L}} \right) \right| ds +$$

$$+ Z_0 \int\limits_{t_k}^{t} \left| \frac{W(s)}{\widetilde{L}_0(W,J)(t)} - \frac{\overline{W}(s)}{\widetilde{L}_0(\overline{W},\overline{J})(t)} \right| ds + Z_0 \int\limits_{t_k}^{t} \left| \frac{J(s-T)}{\widetilde{L}_0(W,J)(t)} - \frac{\overline{J}(s-T)}{\widetilde{L}_0(\overline{W},\overline{J})(t)} \right| ds +$$

$$+ 2\sqrt{L} \int\limits_{t_k}^{t} \left| \frac{1}{\widetilde{L}_0(W,J)(t)} \widetilde{C}_0^{-1}\left(\int\limits_{t_k}^{s} \frac{W(\theta)+J(\theta-T)}{2\sqrt{L}} d\theta \right) - \frac{1}{\widetilde{L}_0(\overline{W},\overline{J})(t)} \widetilde{C}_0^{-1}\left(\int\limits_{t_k}^{s} \frac{\overline{W}(\theta)+\overline{J}(\theta-T)}{2\sqrt{L}} d\theta \right) \right| ds \le$$

$$\le U_{E0} \left| \int\limits_{t_k}^{t} e^{\mu(s-t_k)} \frac{1}{\left(\widetilde{L}_0(W,J) \right)^2} \left| \frac{d\widetilde{L}_0}{di} \right| \left| W(s)+J(s-T)-\overline{W}(s)-\overline{J}(s-T) \right| ds \right| +$$

$$+ 2\sqrt{L} \int\limits_{t_k}^{t} \left| \frac{1}{\widetilde{L}_0(W,J)(s)} R_0\left(\frac{W(s)+J(s-T)}{2\sqrt{L}} \right) - \frac{1}{\widetilde{L}_0(W,J)(s)} R_0\left(\frac{\overline{W}(s)+\overline{J}(s-T)}{2\sqrt{L}} \right) \right| ds +$$

$$+ 2\sqrt{L} \int\limits_{t_k}^{t} \left| \frac{1}{\widetilde{L}_0(W,J)(s)} R_0\left(\frac{\overline{W}(s)+\overline{J}(s-T)}{2\sqrt{L}} \right) - \frac{1}{\widetilde{L}_0(\overline{W},\overline{J})(s)} R_0\left(\frac{\overline{W}(s)+\overline{J}(s-T)}{2\sqrt{L}} \right) \right| ds +$$

$$+ Z_0 \int\limits_{t_k}^{t} \left| \frac{W(s)}{\widetilde{L}_0(W,J)(t)} - \frac{\overline{W}(s)}{\widetilde{L}_0(W,J)(t)} \right| ds + Z_0 \int\limits_{t_k}^{t} \left| \frac{\overline{W}(s)}{\widetilde{L}_0(W,J)(t)} - \frac{\overline{W}(s)}{\widetilde{L}_0(\overline{W},\overline{J})(t)} \right| ds +$$

$$+ Z_0 \int\limits_{t_k}^{t} \left| \frac{J(s-T)}{\widetilde{L}_0(W,J)(s)} - \frac{\overline{J}(s-T)}{\widetilde{L}_0(W,J)(ss)} \right| ds + Z_0 \int\limits_{t_k}^{t} \left| \frac{\overline{J}(s-T)}{\widetilde{L}_0(W,J)(s)} - \frac{\overline{J}(s-T)}{\widetilde{L}_0(\overline{W},\overline{J})(s)} \right| ds +$$

$$+ 2\sqrt{L} \int\limits_{t_k}^{t} \left| \frac{1}{\widetilde{L}_0(W,J)(s)} \widetilde{C}_0^{-1}\left(\int\limits_{t_k}^{s} \frac{W(\theta)+J(\theta-T)}{2\sqrt{L}} d\theta \right) - \frac{1}{\widetilde{L}_0(\overline{W},\overline{J})(s)} \widetilde{C}_0^{-1}\left(\int\limits_{t_k}^{s} \frac{W(\theta)+J(\theta-T)}{2\sqrt{L}} d\theta \right) \right| ds +$$

$$+ 2\sqrt{L} \int\limits_{t_k}^{t} \left| \frac{1}{\widetilde{L}_0(\overline{W},\overline{J})(s)} \widetilde{C}_0^{-1}\left(\int\limits_{t_k}^{s} \frac{W(\theta)+J(\theta-T)}{2\sqrt{L}} d\theta \right) - \frac{1}{\widetilde{L}_0(\overline{W},\overline{J})(s)} \widetilde{C}_0^{-1}\left(\int\limits_{t_k}^{s} \frac{\overline{W}(\theta)+\overline{J}(\theta-T)}{2\sqrt{L}} d\theta \right) \right| ds \le$$

$$\le \rho_\mu^{(k)}(W,\overline{W}) \frac{W_0}{\left(\hat{L}_0 \right)^2} \sup\left| \frac{d\widetilde{L}_0}{di} \right| \left| \int\limits_{t_k}^{t} e^{\mu(s-t_k)} ds \right| +$$

$$+ \frac{2\sqrt{L}}{\hat{L}_0} \sum_{n=1}^{m} \left| r_n^{(0)} \right| n \int_{t_k}^{t} \left| \frac{W_0 e^{\mu(s-t_k)} + J_0 e^{\mu(s-t_k)}}{2\sqrt{L}} \right|^{n-1} \left| \frac{W(s) + J(s-T)}{2\sqrt{L}} - \frac{\overline{W}(s) + \overline{J}(s-T)}{2\sqrt{L}} \right| ds +$$

$$+ 2\sqrt{L} \sum_{n=1}^{m} \left| r_n^{(0)} \right| \frac{1}{\hat{L}_0^2} \left| \frac{d\widetilde{L}_0}{di} \right| \int_{t_k}^{t} \left| \frac{W_0 e^{\mu(s-t_k)} + J_0 e^{\mu(s-t_k)}}{2\sqrt{L}} \right|^{n} \left| \frac{W(s) + J(s-T)}{2\sqrt{L}} - \frac{\overline{W}(s) + \overline{J}(s-T)}{2\sqrt{L}} \right| ds +$$

$$+ \frac{Z_0}{\hat{L}_0} \int_{t_k}^{t} \left| W(s) - \overline{W}(s) \right| ds + \frac{Z_0 W_0}{\hat{L}_0^2} \left| \frac{d\widetilde{L}_0}{di} \right| \int_{t_k}^{t} \left| \frac{W(s) + J(s-T)}{2\sqrt{L}} - \frac{\overline{W}(s) + \overline{J}(s-T)}{2\sqrt{L}} \right| ds +$$

$$+ \frac{Z_0}{\hat{L}_0} \int_{t_k}^{t} \left| J(s-T) - \overline{J}(s-T) \right| ds + \frac{Z_0 J_0}{\hat{L}_0^2} \left| \frac{d\widetilde{L}_0}{di} \right| \int_{t_k}^{t} \left| \frac{W(s) + J(s-T)}{2\sqrt{L}} - \frac{\overline{W}(s) + \overline{J}(s-T)}{2\sqrt{L}} \right| ds +$$

$$+ \frac{2\sqrt{L}}{\hat{L}_0^2} H_0 \int_{t_k}^{t} \left| \int_{t_k}^{s} \frac{W(\theta) + J(\theta-T)}{2\sqrt{L}} d\theta \right| \left| \frac{d\widetilde{L}_0}{di} \right| \left| \frac{W(s) + J(s-T)}{2\sqrt{L}} - \frac{\overline{W}(s) + \overline{J}(s-T)}{2\sqrt{L}} \right| ds +$$

$$+ \frac{2\sqrt{L}}{\hat{L}_0} \frac{2\sqrt{\Phi_0}}{c_0 \sqrt{\Phi_0 - \phi_0}} \int_{t_k}^{t} \left| \int_{t_k}^{s} \frac{W(\theta) + J(\theta-T)}{2\sqrt{L}} d\theta - \int_{t_k}^{s} \frac{\overline{W}(\theta) + \overline{J}(\theta-T)}{2\sqrt{L}} d\theta \right| ds \leq$$

$$\leq \frac{\rho_\mu^{(k)}(\dot{W}, \dot{\overline{W}})}{\mu} \frac{e^{\mu(t-t_k)} - 1}{\mu} \frac{W_0}{\hat{L}_0^2} \sup \left| \frac{d\widetilde{L}_0}{di} \right| +$$

$$\rho_\mu^{(k)}(W, \overline{W}) \frac{e^{\mu(t-t_k)} - 1}{\mu} \frac{1}{\hat{L}_0} \sum_{n=1}^{m} \left| r_n^{(0)} \right| n \left(\frac{W_0 + J_0}{2\sqrt{L}} \right)^{n-1} +$$

$$+ \rho_\mu^{(k)}(W, \overline{W}) \frac{e^{\mu(t-t_k)} - 1}{\mu} \frac{1}{\hat{L}_0^2} \sum_{n=1}^{m} \left| r_n^{(0)} \right| \left(\frac{W_0 + J_0}{2\sqrt{L}} \right)^{n} \sup \left| \frac{d\widetilde{L}_0}{di} \right| + \frac{Z_0}{\hat{L}_0} \rho_\mu^{(k)}(W, \overline{W}) \frac{e^{\mu(t-t_k)} - 1}{\mu} +$$

$$+ \rho_\mu^{(k)}(W, \overline{W}) \frac{e^{\mu(t-t_k)} - 1}{2\mu\sqrt{L}} \frac{Z_0 W_0}{\hat{L}_0^2} \sup \left| \frac{d\widetilde{L}_0}{di} \right| + \rho_\mu^{(k)}(W, \overline{W}) \frac{e^{\mu(t-t_k)} - 1}{2\mu\sqrt{L}} \frac{Z_0 J_0}{\hat{L}_0^2} \sup \left| \frac{d\widetilde{L}_0}{di} \right| +$$

$$+ \frac{\rho_\mu^{(k)}(W, \overline{W})}{2\sqrt{L}} \int_{t_k}^{t} \frac{e^{\mu(s-t_k)} - 1}{\mu} ds \frac{H_0 (W_0 + J_0)}{2\hat{L}_0^2} \sup \left| \frac{d\widetilde{L}_0}{di} \right| +$$

$$\frac{\rho_\mu^{(k)}(W, \overline{W})}{\hat{L}_0} \int_{t_k}^{t} \frac{e^{\mu(s-t_k)} - 1}{\mu} ds \frac{2\sqrt{\Phi_0}}{c_0 \sqrt{\Phi_0 - \phi_0}} \leq$$

$$\leq \frac{\rho_\mu^{(k)}(\dot{W}, \dot{\overline{W}})}{\mu} \frac{e^{\mu(t-t_k)} - 1}{\mu} \frac{W_0}{\hat{L}_0^2} \sup \left| \frac{d\widetilde{L}_0}{di} \right| + \frac{e^{\mu(t-t_k)} - 1}{\mu} \frac{\rho_\mu^{(k)}(\dot{W}, \dot{\overline{W}})}{\mu} \frac{1}{\hat{L}_0} \sum_{n=1}^{m} \left| r_n^{(0)} \right| n \left(\frac{W_0 + J_0}{2\sqrt{L}} \right)^{n-1} +$$

$$+ \frac{e^{\mu(t-t_k)} - 1}{\mu} \frac{\rho_\mu^{(k)}(\dot{W}, \dot{\overline{W}})}{\mu} \frac{1}{\hat{L}_0^2} \sum_{n=1}^{m} \left| r_n^{(0)} \right| \left(\frac{W_0 + J_0}{2\sqrt{L}} \right)^{n} \sup \left| \frac{d\widetilde{L}_0}{di} \right| + \frac{Z_0}{\hat{L}_0} \frac{e^{\mu(t-t_k)} - 1}{\mu} \frac{\rho_\mu^{(k)}(\dot{W}, \dot{\overline{W}})}{\mu} +$$

$$+ \frac{e^{\mu(t-t_k)} - 1}{\mu} \frac{\rho_\mu^{(k)}(\dot{W}, \dot{\overline{W}})}{\mu} \frac{1}{2\sqrt{L}} \frac{Z_0 W_0}{\hat{L}_0^2} \sup \left| \frac{d\widetilde{L}_0}{di} \right| +$$

$$\frac{e^{\mu(t-t_k)}-1}{\mu}\frac{\rho_\mu^{(k)}(\dot{W},\dot{\overline{W}})}{\mu}\frac{1}{2\sqrt{L}}\frac{Z_0 J_0}{\hat{L}_0^2}\sup\left|\frac{d\widetilde{L}_0}{di}\right|+$$

$$+\frac{\rho_\mu^{(k)}(\dot{W},\dot{\overline{W}})}{\mu}\int_{t_k}^{t}\frac{e^{\mu(s-t_k)}-1}{\mu}ds\frac{H_0(W_0+J_0)}{\hat{L}_0^2 2\sqrt{L}}\sup\left|\frac{d\widetilde{L}_0}{di}\right|+\frac{\rho_\mu^{(k)}(\dot{W},\dot{\overline{W}})}{\mu\hat{L}_0}\int_{t_k}^{t}\frac{e^{\mu(s-t_k)}-1}{\mu}ds\frac{2\sqrt{\Phi_0}}{c_0\sqrt{\Phi_0-\phi_0}}\le$$

$$\le\hat{\rho}_\mu^{(k)}((W,\dot{W},J,\dot{J}),(\overline{W},\dot{\overline{W}},\overline{J},\dot{\overline{J}}))\frac{e^{\mu(t-t_k)}-1}{\mu}\left\{\frac{1}{\mu\hat{L}_0}\left[\frac{1}{\hat{L}_0}\left(W_0+\sum_{n=1}^{m}\left|r_n^{(0)}\right|i_0^n+Z_0 i_0+\frac{H_0 i_0}{\mu}\right)\sup\left|\frac{d\widetilde{L}_0}{di}\right|+\right.\right.$$

$$+\left.\left.\sum_{n=1}^{m}\left|r_n^{(0)}\right|n(i_0)^{n-1}+Z_0+\frac{2\sqrt{\Phi_0}}{\mu c_0\sqrt{\Phi_0-\phi_0}}\right]\right\}\le$$

$$\le\hat{\rho}_\mu^{(k)}((W,\dot{W},J,\dot{J}),(\overline{W},\dot{\overline{W}},\overline{J},\dot{\overline{J}}))\frac{e^{\mu_0}-1}{\mu}\left\{\frac{1}{\mu\hat{L}_0}\left[\frac{1}{\hat{L}_0}\left(W_0+\sum_{n=1}^{m}\left|r_n^{(0)}\right|i_0^n+\right.\right.\right.$$

$$+\left.\left.\left.Z_0 i_0+\frac{H_0}{\mu}i_0\right)\sum_{n=1}^{m}(n+1)\left|l_n^{(0)}\right|i_0^n+\sum_{n=1}^{m}\left|r_n^{(0)}\right|n\,i_0^{n-1}+Z_0+\frac{2\sqrt{\Phi_0}}{\mu c_0\sqrt{\Phi_0-\phi_0}}\right]\right\}$$

and

$$U_2\le W_0\left|\int_{t_k}^{t_{k+1}}e^{\mu(s-t_k)}\frac{1}{\left(\widetilde{L}_0(W,J)\right)^2}\left|\frac{d\widetilde{L}_0}{di}\right|\left|W(s)+J(s-T)-\overline{W}(s)-\overline{J}(s-T)\right|ds\right|+$$

$$+2\sqrt{L}\int_{t_k}^{t_{k+1}}\left|\frac{1}{\widetilde{L}_0(W,J)(s)}R_0\left(\frac{W(s)+J(s-T)}{2\sqrt{L}}\right)-\frac{1}{\widetilde{L}_0(W,J)(s)}R_0\left(\frac{\overline{W}(s)+\overline{J}(s-T)}{2\sqrt{L}}\right)\right|ds+$$

$$+2\sqrt{L}\int_{t_k}^{t_{k+1}}\left|\frac{1}{\widetilde{L}_0(W,J)(s)}R_0\left(\frac{\overline{W}(s)+\overline{J}(s-T)}{2\sqrt{L}}\right)-\frac{1}{\widetilde{L}_0(\overline{W},\overline{J})(s)}R_0\left(\frac{\overline{W}(s)+\overline{J}(s-T)}{2\sqrt{L}}\right)\right|ds+$$

$$+Z_0\int_{t_k}^{t_{k+1}}\left|\frac{W(s)}{\widetilde{L}_0(W,J)(t)}-\frac{\overline{W}(s)}{\widetilde{L}_0(W,J)(t)}\right|ds+Z_0\int_{t_k}^{t_{k+1}}\left|\frac{\overline{W}(s)}{\widetilde{L}_0(W,J)(t)}-\frac{\overline{W}(s)}{\widetilde{L}_0(\overline{W},\overline{J})(t)}\right|ds+$$

$$+Z_0\int_{t_k}^{t_{k+1}}\left|\frac{J(s-T)}{\widetilde{L}_0(W,J)(s)}-\frac{\overline{J}(s-T)}{\widetilde{L}_0(W,J)(s)}\right|ds+Z_0\int_{t_k}^{t_{k+1}}\left|\frac{\overline{J}(s-T)}{\widetilde{L}_0(W,J)(s)}-\frac{\overline{J}(s-T)}{\widetilde{L}_0(\overline{W},\overline{J})(s)}\right|ds+$$

$$+2\sqrt{L}\int_{t_k}^{t_{k+1}}\left|\frac{1}{\widetilde{L}_0(W,J)(s)}\widetilde{C}_0^{-1}\left(\int_{t_k}^{s}\frac{W(\theta)+J(\theta-T)}{2\sqrt{L}}d\theta\right)-\frac{1}{\widetilde{L}_0(\overline{W},\overline{J})(s)}\widetilde{C}_0^{-1}\left(\int_{t_k}^{s}\frac{W(\theta)+J(\theta-T)}{2\sqrt{L}}d\theta\right)\right|ds+$$

$$+2\sqrt{L}\int_{t_k}^{t_{k+1}}\left|\frac{1}{\widetilde{L}_0(\overline{W},\overline{J})(s)}\widetilde{C}_0^{-1}\left(\int_{t_k}^{s}\frac{W(\theta)+J(\theta-T)}{2\sqrt{L}}d\theta\right)-\frac{1}{\widetilde{L}_0(\overline{W},\overline{J})(s)}\widetilde{C}_0^{-1}\left(\int_{t_k}^{s}\frac{\overline{W}(\theta)+\overline{J}(\theta-T)}{2\sqrt{L}}d\theta\right)\right|ds\le$$

$$\le e^{\mu(t-t_k)}\hat{\rho}_\mu^{(k)}((W,\dot{W},J,\dot{J}),(\overline{W},\dot{\overline{W}},\overline{J},\dot{\overline{J}}))\frac{e^{\mu_0}-1}{\mu}\left\{\frac{1}{\mu\hat{L}_0}\left[\frac{1}{\hat{L}_0}\left(W_0+\sum_{n=1}^{m}\left|r_n^{(0)}\right|i_0^n+\right.\right.\right.$$

$$+ Z_0 i_0 + \frac{H_0}{\mu} i_0 \Bigg) \sum_{n=1}^{m} (n+1) \left| l_n^{(0)} \right| i_0^n + \sum_{n=1}^{m} \left| r_n^{(0)} \right| n\, i_0^{n-1} + Z_0 + \frac{2\sqrt{\Phi_0}}{\mu c_0 \sqrt{\Phi_0 - \phi_0}} \Bigg] \Bigg\} \le$$

$$\le \hat{\rho}_\mu^{(k)} ((W,\dot{W},J,\dot{J}),(\overline{W},\dot{\overline{W}},\overline{J},\dot{\overline{J}})) \frac{e^{\mu_0}-1}{\mu} \Bigg\{ \frac{1}{\mu \hat{L}_0} \Bigg[\frac{1}{\hat{L}_0} \Bigg(W_0 + \sum_{n=1}^{m} \left| r_n^{(0)} \right| i_0^n +$$

$$+ Z_0 i_0 + \frac{H_0}{\mu} i_0 \Bigg) \sum_{n=1}^{m} (n+1) \left| l_n^{(0)} \right| i_0^n + \sum_{n=1}^{m} \left| r_n^{(0)} \right| n (i_0)^{n-1} + Z_0 + \frac{2\sqrt{\Phi_0}}{\mu c_0 \sqrt{\Phi_0 - \phi_0}} \Bigg] \Bigg\} .$$

Therefore

$$\left| B_W^{(k)}(W,J)(t) - B_W^{(k)}(\overline{W},\overline{J})(t) \right| \le$$

$$\le e^{\mu(t-t_k)} \hat{\rho}_\mu^{(k)} ((W,\dot{W},J,\dot{J}),(\overline{W},\dot{\overline{W}},\overline{J},\dot{\overline{J}})) e^{\mu(t-t_k)} \Bigg\{ \frac{1}{\mu^2 \hat{L}_0} \Bigg[\frac{1}{\hat{L}_0} \Bigg(W_0 + \sum_{n=1}^{m} \left| r_n^{(0)} \right| i_0^n +$$

$$+ Z_0 i_0 + \frac{H_0}{\mu} i_0 \Bigg) \sum_{n=1}^{m} (n+1) \left| l_n^{(0)} \right| i_0^n + \sum_{n=1}^{m} \left| r_n^{(0)} \right| n\, i_0^{n-1} + Z_0 + \frac{2\sqrt{\Phi_0}}{\mu c_0 \sqrt{\Phi_0 - \phi_0}} \Bigg] \Bigg\} +$$

$$+ e^{\mu(t-t_k)} \hat{\rho}_\mu^{(k)} ((W,\dot{W},J,\dot{J}),(\overline{W},\dot{\overline{W}},\overline{J},\dot{\overline{J}})) (e^{\mu_0} - 1) \Bigg\{ \frac{1}{\mu^2 \hat{L}_0} \Bigg[\frac{1}{\hat{L}_0} \Bigg(W_0 + \sum_{n=1}^{m} \left| r_n^{(0)} \right| i_0^n +$$

$$+ Z_0 i_0 + \frac{H_0}{\mu} i_0 \Bigg) \sum_{n=1}^{m} (n+1) \left| l_n^{(0)} \right| (i_0)^n + \sum_{n=1}^{m} \left| r_n^{(0)} \right| n\, i_0^{n-1} + Z_0 + \frac{2\sqrt{\Phi_0}}{\mu c_0 \sqrt{\Phi_0 - \phi_0}} \Bigg] \Bigg\} \le$$

$$\le e^{\mu(t-t_k)} \hat{\rho}_\mu^{(k)} ((W,\dot{W},J,\dot{J}),(\overline{W},\dot{\overline{W}},\overline{J},\dot{\overline{J}})) \frac{e^{\mu_0}}{\mu^2 \hat{L}_0} \Bigg[\frac{1}{\hat{L}_0} \Bigg(W_0 + \sum_{n=1}^{m} \left| r_n^{(0)} \right| i_0^n + Z_0 i_0 + \frac{H_0}{\mu} i_0 \Bigg) \times$$

$$\times \sum_{n=1}^{m} (n+1) \left| l_n^{(0)} \right| i_0^n + \sum_{n=1}^{m} \left| r_n^{(0)} \right| n\, i_0^{n-1} + Z_0 + \frac{2\sqrt{\Phi_0}}{\mu c_0 \sqrt{\Phi_0 - \phi_0}} \Bigg] \equiv$$

$$\equiv e^{\mu(t-t_k)} K_W\, \hat{\rho}_\mu^{(k)} ((W,\dot{W},J,\dot{J}),(\overline{W},\dot{\overline{W}},\overline{J},\dot{\overline{J}})) .$$

It follows

$$\left| B_W^{(k)}(W,\dot{W},J,\dot{J})(t) - B_W^{(k)}(\overline{W},\dot{\overline{W}},\overline{J},\dot{\overline{J}})(t) \right| \le e^{\mu_0} K_W \hat{\rho}_\mu^{(k)} ((W,\dot{W},J,\dot{J}),(\overline{W},\dot{\overline{W}},\overline{J},\dot{\overline{J}})),$$

$$t \in [t_k, t_{k+1}] \quad (k = 0,1,2,...) \text{ and consequently}$$

$$\rho^{(k)}(B_W^{(k)}(W,\dot{W},J,\dot{J}),B_W^{(k)}(\overline{W},\dot{\overline{W}},\overline{J},\dot{\overline{J}})) \le e^{\mu_0} K_W \, \hat{\rho}_\mu^{(k)}((W,\dot{W},J,\dot{J}),(\overline{W},\dot{\overline{W}},\overline{J},\dot{\overline{J}})),$$

$$\hat{\rho}^{(k)}(B_W(W,\dot{W},J,\dot{J}),B_W(\overline{W},\dot{\overline{W}},\overline{J},\dot{\overline{J}})) \le e^{\mu_0} K_W \, \hat{\rho}_\mu^{(k)}((W,\dot{W},J,\dot{J}),(\overline{W},\dot{\overline{W}},\overline{J},\dot{\overline{J}})).$$

Further on we have

$$\left| B_J^{(k)}(W,J)(t) - B_J^{(k)}(\overline{W},\overline{J})(t) \right| \le \int_{t_k}^{t} \left| I(W,J)(s) - I(\overline{W},\overline{J})(s) \right| ds +$$

$$+ \left| \int_{t_k}^{t_{k+1}} I(W,J)(s)ds - \int_{t_k}^{t_{k+1}} I(\overline{W},\overline{J})(s)ds \right| \equiv I_1 + I_2.$$

But

$$I_1 \le \left| \int_{t_k}^{t} \left(\dot{W}(s-T) - \dot{\overline{W}}(s-T) \right) ds \right| + 2\sqrt{L} \left| \int_{t_k}^{t} \left(\frac{E_1(s-T)}{\widetilde{L}_1(W,J)(s)} - \frac{E_1(s-T)}{\widetilde{L}_1(\overline{W},\overline{J})(s)} \right) ds \right| +$$

$$+ 2\sqrt{L} \int_{t_k}^{t} \left| \frac{R_1\left(\dfrac{W(s-T)+J(s)}{2\sqrt{L}} \right)}{\widetilde{L}_1(W,J)(s)} - \frac{R_1\left(\dfrac{\overline{W}(s-T)+\overline{J}(s)}{2\sqrt{L}} \right)}{\widetilde{L}_1(\overline{W},\overline{J})(s)} \right| ds +$$

$$+ Z_0 \int_{t_k}^{t} \left| \frac{W(s-T)}{\widetilde{L}_1(W,J)(s)} - \frac{\overline{W}(s-T)}{\widetilde{L}_1(\overline{W},\overline{J})(s)} \right| ds + Z_0 \int_{t_k}^{t} \left| \frac{J(s)}{\widetilde{L}_1(W,J)(s)} - \frac{\overline{J}(s)}{\widetilde{L}_1(\overline{W},\overline{J})(s)} \right| ds +$$

$$+ 2\sqrt{L} \int_{t_k}^{t} \left| \frac{\widetilde{C}_1^{-1}\left(\dfrac{1}{2\sqrt{L}} \int_T^t (W(\theta-T)+J(\theta))d\theta \right)}{\widetilde{L}_1(W,J)(s)} - \frac{\widetilde{C}_1^{-1}\left(\dfrac{1}{2\sqrt{L}} \int_T^t (\overline{W}(\theta-T)+\overline{J}(\theta))d\theta \right)}{\widetilde{L}_1(\overline{W},\overline{J})(s)} \right| ds \le$$

$$\le W_0 \left| \int_{t_k}^{t} e^{-\frac{R}{L}s} \frac{1}{\left(\widetilde{L}_1(W,J)\right)^2} \left| \frac{d\widetilde{L}_1}{di} \right| \left| W(s-T)+J(s)-\overline{W}(s-T)-\overline{J}(s) \right| ds \right| +$$

$$+ 2\sqrt{L} \int_{t_k}^{t} \left| \frac{1}{\widetilde{L}_1(W,J)(s)} R_1\left(\frac{W(s-T)+J(s)}{2\sqrt{L}} \right) - \frac{1}{\widetilde{L}_1(W,J)(s)} R_1\left(\frac{\overline{W}(s-T)+\overline{J}(s)}{2\sqrt{L}} \right) \right| ds +$$

$$+ 2\sqrt{L} \int_{t_k}^{t} \left| \frac{1}{\widetilde{L}_1(W,J)(s)} R_1\left(\frac{\overline{W}(s-T)+\overline{J}(s)}{2\sqrt{L}} \right) - \frac{1}{\widetilde{L}_1(\overline{W},\overline{J})(s)} R_1\left(\frac{\overline{W}(s-T)+\overline{J}(s)}{2\sqrt{L}} \right) \right| ds +$$

$$+ Z_0 \int_{t_k}^{t} \left| \frac{W(s-T)}{\widetilde{L}_1(W,J)(t)} - \frac{\overline{W}(s-T)}{\widetilde{L}_1(W,J)(t)} \right| ds + Z_0 \int_{t_k}^{t} \left| \frac{\overline{W}(s-T)}{\widetilde{L}_1(W,J)(t)} - \frac{\overline{W}(s-T)}{\widetilde{L}_1(\overline{W},\overline{J})(t)} \right| ds +$$

$$+ Z_0 \int_{t_k}^{t} \left| \frac{J(s)}{\widetilde{L}_1(W,J)(s)} - \frac{\overline{J}(s)}{\widetilde{L}_1(W,J)(s)} \right| ds + Z_0 \int_{t_k}^{t} \left| \frac{\overline{J}(s)}{\widetilde{L}_1(W,J)(s)} - \frac{\overline{J}(s)}{\widetilde{L}_1(\overline{W},\overline{J})(s)} \right| ds +$$

$$+ 2\sqrt{L} \int_{t_k}^{t} \left| \frac{1}{\widetilde{L}_1(W,J)(s)} \widetilde{C}_1^{-1} \left(\int_{t_k}^{s} \frac{W(\theta-T)+J(\theta)}{2\sqrt{L}} d\theta \right) - \frac{1}{\widetilde{L}_1(\overline{W},\overline{J})(s)} \widetilde{C}_1^{-1} \left(\int_{t_k}^{s} \frac{W(\theta-T)+J(\theta)}{2\sqrt{L}} d\theta \right) \right| ds +$$

$$+ 2\sqrt{L} \int_{t_k}^{t} \left| \frac{1}{\widetilde{L}_1(\overline{W},\overline{J})(s)} \widetilde{C}_1^{-1} \left(\int_{t_k}^{s} \frac{W(\theta-T)+J(\theta)}{2\sqrt{L}} d\theta \right) - \frac{1}{\widetilde{L}_1(\overline{W},\overline{J})(s)} \widetilde{C}_1^{-1} \left(\int_{t_k}^{s} \frac{\overline{W}(\theta-T)+\overline{J}(\theta)}{2\sqrt{L}} d\theta \right) \right| ds \leq$$

$$\leq \rho_\mu^{(k)}(J,\overline{J}) \frac{U_{E_1}}{\hat{L}_0^2} \sup \left| \frac{d\widetilde{L}_1}{di} \right| \int_{t_k}^{t} e^{\mu(s-t_k)} ds +$$

$$+ \frac{2\sqrt{L}}{\hat{L}_1} \sum_{n=1}^{m} \left| r_n^{(1)} \right| n \int_{t_k}^{t} \left| \frac{W_0 e^{\mu(s-t_k)} + J_0 e^{\mu(s-t_k)}}{2\sqrt{L}} \right|^{n-1} \left| \frac{W(s-T)+J(s)}{2\sqrt{L}} - \frac{\overline{W}(s-T)+\overline{J}(s)}{2\sqrt{L}} \right| ds +$$

$$+ 2\sqrt{L} \sum_{n=1}^{m} \left| r_n^{(1)} \right| \frac{1}{\hat{L}_0^2} \left| \frac{d\widetilde{L}_1}{di} \right| \int_{t_k}^{t} \left| \frac{W_0 e^{\mu(s-t_k)} + J_0 e^{\mu(s-t_k)}}{2\sqrt{L}} \right|^{n} \left| \frac{W(s-T)+J(s)}{2\sqrt{L}} - \frac{\overline{W}(s-T)+\overline{J}(s)}{2\sqrt{L}} \right| ds +$$

$$+ \frac{Z_0}{\hat{L}_1} \int_{t_k}^{t} \left| W(s-T) - \overline{W}(s-T) \right| ds + \frac{Z_0 W_0}{\hat{L}_0^2} \left| \frac{d\widetilde{L}_1}{di} \right| \int_{t_k}^{t} \left| \frac{W(s-T)+J(s)}{2\sqrt{L}} - \frac{\overline{W}(s-T)+\overline{J}(s)}{2\sqrt{L}} \right| ds +$$

$$+ \frac{Z_0}{\hat{L}_1} \int_{t_k}^{t} \left| J(s) - \overline{J}(s) \right| ds + \frac{Z_0 J_0}{\hat{L}_0^2} \left| \frac{d\widetilde{L}_1}{di} \right| \int_{t_k}^{t} \left| \frac{W(s-T)+J(s)}{2\sqrt{L}} - \frac{\overline{W}(s-T)+\overline{J}(s)}{2\sqrt{L}} \right| ds +$$

$$+ \frac{2\sqrt{L}}{\hat{L}_0^2} H_1 \int_{t_k}^{t} \left| \int_{t_k}^{s} \frac{W(\theta-T)+J(\theta)}{2\sqrt{L}} d\theta \right| \left\| \frac{d\widetilde{L}_1}{di} \right\| \left| \frac{W(s-T)+J(s)}{2\sqrt{L}} - \frac{\overline{W}(s-T)+\overline{J}(s)}{2\sqrt{L}} \right| ds +$$

$$+ \frac{2\sqrt{L}}{\hat{L}_1} \frac{2\sqrt{\Phi_1}}{c_1\sqrt{\Phi_1-\phi_0}} \int_{t_k}^{t} \left| \int_{t_k}^{s} \frac{W(\theta-T)+J(\theta)}{2\sqrt{L}} d\theta - \int_{t_k}^{s} \frac{\overline{W}(\theta-T)+\overline{J}(\theta)}{2\sqrt{L}} d\theta \right| ds \leq$$

$$\leq \frac{\rho_\mu^{(k)}(\dot{J},\dot{\overline{J}})}{\mu} \frac{e^{\mu(t-t_k)}-1}{\mu} \frac{U_{E_1}}{\hat{L}_0^2} \sup \left| \frac{d\widetilde{L}_1}{di} \right| +$$

$$\rho_\mu^{(k)}(J,\overline{J}) \frac{e^{\mu(t-t_k)}-1}{\mu} \frac{1}{\hat{L}_1} \sum_{n=1}^{m} \left| r_n^{(1)} \right| n \left(\frac{W_0 + J_0}{2\sqrt{L}} \right)^{n-1} +$$

$$+ \rho_\mu^{(k)}(J,\overline{J}) \frac{e^{\mu(t-t_k)}-1}{\mu} \frac{1}{\hat{L}_0^2} \sum_{n=1}^{m} \left| r_n^{(1)} \right| \left(\frac{W_0 + J_0}{2\sqrt{L}} \right)^{n} \sup \left| \frac{d\widetilde{L}_1}{di} \right| +$$

$$+ \frac{Z_0}{\hat{L}_1} \rho_\mu^{(k)}(J,\overline{J}) \frac{e^{\mu(t-t_k)}-1}{\mu} + \rho_\mu^{(k)}(J,\overline{J}) \frac{e^{\mu(t-t_k)}-1}{2\mu\sqrt{L}} \frac{Z_0 W_0}{\hat{L}_0^2} \sup \left| \frac{d\widetilde{L}_1}{di} \right| +$$

$$+ \rho_\mu^{(k)}(J,\overline{J}) \frac{e^{\mu(t-t_k)}-1}{2\mu\sqrt{L}} \frac{Z_0 J_0}{\hat{L}_0^2} \sup \left| \frac{d\widetilde{L}_1}{di} \right| +$$

$$
+\frac{\rho_\mu^{(k)}(J,\bar{J})}{2\sqrt{L}}\int_{t_k}^{t}\frac{e^{\mu(s-t_k)}-1}{\mu}ds\,\frac{H_1(W_0+J_0)}{2\hat{L}_0^2}\sup\left|\frac{d\widetilde{L}_1}{di}\right|+\frac{\rho_\mu^{(k)}(J,\bar{J})}{\hat{L}_1}\int_{t_k}^{t}\frac{e^{\mu(s-t_k)}-1}{\mu}ds\,\frac{2\sqrt{\Phi_1}}{c_1\sqrt{\Phi_1-\phi_0}}\le
$$

$$
\le\frac{\rho_\mu^{(k)}(\dot{J},\dot{\bar{J}})}{\mu}\frac{e^{\mu(t-t_k)}-1}{\mu}\frac{U_{E_1}}{\hat{L}_0^2}\sup\left|\frac{d\widetilde{L}_1}{di}\right|+
$$

$$
\frac{e^{\mu(t-t_k)}-1}{\mu}\frac{\rho_\mu^{(k)}(\dot{J},\dot{\bar{J}})}{\mu}\frac{1}{\hat{L}_1}\sum_{n=1}^{m}\left|r_n^{(1)}\right|n\left(\frac{W_0+J_0}{2\sqrt{L}}\right)^{n-1}+
$$

$$
+\frac{e^{\mu(t-t_k)}-1}{\mu}\frac{\rho_\mu^{(k)}(\dot{J},\dot{\bar{J}})}{\mu}\frac{1}{\hat{L}_0^2}\sum_{n=1}^{m}\left|r_n^{(1)}\right|\left(\frac{W_0+J_0}{2\sqrt{L}}\right)^{n}\sup\left|\frac{d\widetilde{L}_1}{di}\right|+\frac{Z_0}{\hat{L}_1}\frac{e^{\mu(t-t_k)}-1}{\mu}\frac{\rho_\mu^{(k)}(\dot{J},\dot{\bar{J}})}{\mu}+
$$

$$
+\frac{e^{\mu(t-t_k)}-1}{\mu}\frac{\rho_\mu^{(k)}(\dot{J},\dot{\bar{J}})}{\mu}\frac{1}{2\sqrt{L}}\frac{Z_0 W_0}{\hat{L}_0^2}\sup\left|\frac{d\widetilde{L}_1}{di}\right|+\frac{e^{\mu(t-t_k)}-1}{\mu}\frac{\rho_\mu^{(k)}(\dot{J},\dot{\bar{J}})}{\mu}\frac{1}{2\sqrt{L}}\frac{Z_0 J_0}{\hat{L}_0^2}\sup\left|\frac{d\widetilde{L}_1}{di}\right|+
$$

$$
+\frac{\rho_\mu^{(k)}(\dot{J},\dot{\bar{J}})}{\mu}\int_{t_k}^{t}\frac{e^{\mu(s-t_k)}-1}{\mu}ds\,\frac{H_1(W_0+J_0)}{\hat{L}_0^2\,2\sqrt{L}}\sup\left|\frac{d\widetilde{L}_1}{di}\right|+\frac{\rho_\mu^{(k)}(\dot{J},\dot{\bar{J}})}{\mu\hat{L}_1}\int_{t_k}^{t}\frac{e^{\mu(s-t_k)}-1}{\mu}ds\,\frac{2\sqrt{\Phi_1}}{c_1\sqrt{\Phi_1-\phi_0}}\le
$$

$$
\le\hat{\rho}_\mu^{(k)}((W,\dot{W},J,\dot{J}),(\overline{W},\dot{\overline{W}},\overline{J},\dot{\overline{J}}))\frac{e^{\mu(t-t_k)}-1}{\mu}\left[\frac{W_0}{\mu\hat{L}_0^2}\sup\left|\frac{d\widetilde{L}_1}{di}\right|+\right.
$$

$$
+\frac{1}{\mu\hat{L}_1}\sum_{n=1}^{m}\left|r_n^{(1)}\right|n(i_0)^{n-1}+\frac{1}{\mu\hat{L}_0^2}\sum_{n=1}^{m}\left|r_n^{(1)}\right|(i_0)^{n}\sup\left|\frac{d\widetilde{L}_1}{di}\right|+
$$

$$
\left.+\frac{Z_0}{\mu\hat{L}_1}+\frac{Z_0 i_0}{\mu\hat{L}_0^2}\sup\left|\frac{d\widetilde{L}_1}{di}\right|+\frac{H_1 i_0}{\mu^2\hat{L}_0^2}\sup\left|\frac{d\widetilde{L}_1}{di}\right|+\frac{1}{\mu^2\hat{L}_1}\frac{2\sqrt{\Phi_1}}{c_1\sqrt{\Phi_1-\phi_0}}\right]\le
$$

$$
\le\hat{\rho}_\mu^{(k)}((W,\dot{W},J,\dot{J}),(\overline{W},\dot{\overline{W}},\overline{J},\dot{\overline{J}}))\frac{e^{\mu(t-t_k)}-1}{\mu}\left\{\frac{1}{\mu\hat{L}_1}\left[\frac{1}{\hat{L}_1}\left(W_0+\sum_{n=1}^{m}\left|r_n^{(1)}\right|i_0^n+\right.\right.\right.
$$

$$
\left.\left.\left.+Z_0 i_0+\frac{H_1 i_0}{\mu}\right)\sup\left|\frac{d\widetilde{L}_1}{di}\right|+\sum_{n=1}^{m}\left|r_n^{(1)}\right|n(i_0)^{n-1}+Z_0+\frac{2\sqrt{\Phi_1}}{\mu c_1\sqrt{\Phi_1-\phi_0}}\right]\right\}\le
$$

$$
\le e^{\mu(t-t_k)}\hat{\rho}_\mu^{(k)}((W,\dot{W},J,\dot{J}),(\overline{W},\dot{\overline{W}},\overline{J},\dot{\overline{J}}))\frac{1}{\mu^2\hat{L}_1}\left[\frac{1}{\hat{L}_1}\left(W_0+\sum_{n=1}^{m}\left|r_n^{(1)}\right|i_0^n+\right.\right.
$$

$$
\left.\left.+Z_0 i_0+\frac{H_1}{\mu}i_0\right)\sum_{n=1}^{m}(n+1)\left|l_n^{(1)}\right|i_0^n+\sum_{n=1}^{m}\left|r_n^{(1)}\right|n\,i_0^{n-1}+Z_0+\frac{2\sqrt{\Phi_1}}{\mu c_1\sqrt{\Phi_1-\phi_0}}\right]
$$

and

$$I_2 \leq 2\sqrt{L}_0 \left| \int_{t_k}^{t_{k+1}} e^{\mu(s-T-t_k)} \frac{1}{\left(\widetilde{L}_1(W,J)\right)^2} \left| \frac{d\widetilde{L}_1}{di} \right| \left| W(s-T)+J(s)-\overline{W}(s-T)-\overline{J}(s) \right| ds \right| +$$

$$+ 2\sqrt{L} \int_{t_k}^{t_{k+1}} \left| \frac{1}{\widetilde{L}_1(W,J)(s)} R_1\!\left(\frac{W(s-T)+J(s)}{2\sqrt{L}} \right) - \frac{1}{\widetilde{L}_1(W,J)(s)} R_1\!\left(\frac{\overline{W}(s-T)+\overline{J}(s)}{2\sqrt{L}} \right) \right| ds +$$

$$+ 2\sqrt{L} \int_{t_k}^{t_{k+1}} \left| \frac{1}{\widetilde{L}_1(W,J)(s)} R_1\!\left(\frac{\overline{W}(s-T)+\overline{J}(s)}{2\sqrt{L}} \right) - \frac{1}{\widetilde{L}_1(\overline{W},\overline{J})(s)} R_1\!\left(\frac{\overline{W}(s-T)+\overline{J}(s)}{2\sqrt{L}} \right) \right| ds +$$

$$+ Z_0 \int_{t_k}^{t_{k+1}} \left| \frac{W(s-T)}{\widetilde{L}_1(W,J)(t)} - \frac{\overline{W}(s-T)}{\widetilde{L}_1(W,J)(t)} \right| ds + Z_0 \int_{t_k}^{t_{k+1}} \left| \frac{\overline{W}(s-T)}{\widetilde{L}_1(W,J)(t)} - \frac{\overline{W}(s-T)}{\widetilde{L}_1(\overline{W},\overline{J})(t)} \right| ds +$$

$$+ Z_0 \int_{t_k}^{t_{k+1}} \left| \frac{J(s)}{\widetilde{L}_1(W,J)(s)} - \frac{\overline{J}(s)}{\widetilde{L}_1(W,J)(s)} \right| ds + Z_0 \int_{t_k}^{t_{k+1}} \left| \frac{\overline{J}(s)}{\widetilde{L}_1(W,J)(s)} - \frac{\overline{J}(s)}{\widetilde{L}_1(\overline{W},\overline{J})(s)} \right| ds +$$

$$+ 2\sqrt{L} \int_{t_k}^{t_{k+1}} \left| \frac{1}{\widetilde{L}_1(W,J)(s)} \widetilde{C}_1^{-1}\!\left(\int_{t_k}^{s} \frac{W(\theta-T)+J(\theta)}{2\sqrt{L}} d\theta \right) - \frac{1}{\widetilde{L}_1(\overline{W},\overline{J})(s)} \widetilde{C}_1^{-1}\!\left(\int_{t_k}^{s} \frac{W(\theta-T)+J(\theta)}{2\sqrt{L}} d\theta \right) \right| ds +$$

$$+ 2\sqrt{L} \int_{t_k}^{t_{k+1}} \left| \frac{1}{\widetilde{L}_1(\overline{W},\overline{J})(s)} \widetilde{C}_1^{-1}\!\left(\int_{t_k}^{s} \frac{W(\theta-T)+J(\theta)}{2\sqrt{L}} d\theta \right) - \frac{1}{\widetilde{L}_1(\overline{W},\overline{J})(s)} \widetilde{C}_1^{-1}\!\left(\int_{t_k}^{s} \frac{\overline{W}(\theta-T)+\overline{J}(\theta)}{2\sqrt{L}} d\theta \right) \right| ds \leq$$

$$\leq e^{\mu(t-t_k)} \hat{\rho}_\mu^{(k)}((W,\dot{W},J,\dot{J}),(\overline{W},\dot{\overline{W}},\overline{J},\dot{\overline{J}})) \frac{e^{\mu_0}-1}{\mu} \left\{ \frac{1+e^{-\mu T}}{\mu \hat{L}_1} \left[\frac{1}{\hat{L}_1} \left(W_0 + \sum_{n=1}^{m} \left|r_n^{(1)}\right| (i_0)^n + Z_0 i_0 + \frac{H_1}{\mu} i_0 \right) \times \right. \right.$$

$$\left. \left. \times \sum_{n=1}^{m} (n+1)\left|l_n^{(1)}\right| i_0^n + \sum_{n=1}^{m} \left|r_n^{(1)}\right| n\, i_0^{n-1} + Z_0 + \frac{2\sqrt{\Phi_1}}{\mu c_1 \sqrt{\Phi_1-\phi_0}} \right] \right\} \leq$$

$$\leq e^{\mu(t-t_k)} \hat{\rho}_\mu^{(k)}((W,\dot{W},J,\dot{J}),(\overline{W},\dot{\overline{W}},\overline{J},\dot{\overline{J}})) \frac{e^{\mu_0}-1}{\mu^2 \hat{L}_1} \left[\frac{1}{\hat{L}_1} \left(W_0 + \sum_{n=1}^{m} \left|r_n^{(1)}\right| i_0^n + Z_0 i_0 + \frac{H_1}{\mu} i_0 \right) \sum_{n=1}^{m} (n+1)\left|l_n^{(1)}\right| i_0^n + \right.$$

$$\left. + \sum_{n=1}^{m} \left|r_n^{(1)}\right| n\, i_0^{n-1} + Z_0 + \frac{2\sqrt{\Phi_1}}{\mu c_1 \sqrt{\Phi_1-\phi_0}} \right].$$

Therefore

$$\left| B_J^{(k)}(W,J)(t) - B_J^{(k)}(\overline{W},\overline{J})(t) \right| \leq$$

$$\leq e^{\mu(t-t_k)} \hat{\rho}_\mu^{(k)}((W,\dot{W},J,\dot{J}),(\overline{W},\dot{\overline{W}},\overline{J},\dot{\overline{J}})) \frac{e^{\mu_0}}{\mu^2 \hat{L}_1} \left[\frac{1}{\hat{L}_1} \left(W_0 + \sum_{n=1}^{m} \left|r_n^{(1)}\right| i_0^n + Z_0 i_0 + \frac{H_1}{\mu} i_0 \right) \sum_{n=1}^{m} (n+1)\left|l_n^{(1)}\right| i_0^n + \right.$$

$$\left. + \sum_{n=1}^{m} \left|r_n^{(1)}\right| n\, i_0^{n-1} + Z_0 + \frac{2\sqrt{\Phi_1}}{\mu c_1 \sqrt{\Phi_1-\phi_0}} \right] \equiv e^{\mu(t-t_k)} K_J\, \hat{\rho}_\mu^{(k)}((W,\dot{W},J,\dot{J}),(\overline{W},\dot{\overline{W}},\overline{J},\dot{\overline{J}})).$$

It follows

$$\left| B_J^{(k)}(W,\dot{W},J,\dot{J})(t) - B_J^{(k)}(\overline{W},\dot{\overline{W}},\overline{J},\dot{\overline{J}})(t) \right| \le e^{\mu_0} K_J \hat{\rho}_\mu^{(k)}((W,\dot{W},J,\dot{J}),(\overline{W},\dot{\overline{W}},\overline{J},\dot{\overline{J}})),$$

$t \in [t_k, t_{k+1}]$ $(k = 0,1,2,...,m-1)$ and consequently

$$\rho^{(k)}(B_J^{(k)}(W,\dot{W},J,\dot{J}), B_J^{(k)}(\overline{W},\dot{\overline{W}},\overline{J},\dot{\overline{J}})) \le e^{\mu_0} K_J \,\hat{\rho}_\mu^{(k)}((W,\dot{W},J,\dot{J}),(\overline{W},\dot{\overline{W}},\overline{J},\dot{\overline{J}})),$$

$$\hat{\rho}^{(k)}(B_J(W,\dot{W},J,\dot{J}), B_J(\overline{W},\dot{\overline{W}},\overline{J},\dot{\overline{J}})) \le e^{\mu_0} K_J \,\hat{\rho}_\mu^{(k)}((W,\dot{W},J,\dot{J}),(\overline{W},\dot{\overline{W}},\overline{J},\dot{\overline{J}})).$$

Finally we have to obtain an estimate of the derivatives for $t \in [t_k, t_{k+1}]$:

$$\left| \dot{B}_W^{(k)}(W,J)(t) - \dot{B}_W^{(k)}(\overline{W},\overline{J})(t) \right| \le \left| U(W,J)(t) - U(\overline{W},\overline{J})(t) \right| +$$

$$+ \frac{1}{t_{k+1} - t_k} \left| \int_{t_k}^{t_{k+1}} U(W,J)(s)ds - \int_{t_k}^{t_{k+1}} U(\overline{W},\overline{J})(s)ds \right| \equiv \dot{U}_1 + \dot{U}_2 .$$

But

$$\dot{U}_1 \le \left| \frac{dJ(t-T)}{dt} - \frac{d\overline{J}(t-T)}{dt} \right| + 2\sqrt{L} \left| \frac{R_0\left(\frac{W(t)+J(t-T)}{2\sqrt{L}} \right)}{\widetilde{L}_0(W,J)(t)} - \frac{R_0\left(\frac{\overline{W}(t)+\overline{J}(t-T)}{2\sqrt{L}} \right)}{\widetilde{L}_0(\overline{W},\overline{J})(t)} \right| +$$

$$+ Z_0 \left| \frac{W(t)}{\widetilde{L}_0(W,J)(t)} - \frac{\overline{W}(t)}{\widetilde{L}_0(\overline{W},\overline{J})(t)} \right| + \frac{Z_0}{\hat{L}_0} \left| \frac{J(t-T)}{\widetilde{L}_0(W,J)(t)} - \frac{\overline{J}(t-T)}{\widetilde{L}_0(\overline{W},\overline{J})(t)} \right| +$$

$$+ 2\sqrt{L} \left| \frac{1}{\widetilde{L}_0(W,J)(t)} \widetilde{C}_0^{-1}\left(\int_{t_k}^{t} \left(\frac{W(\theta)+J(\theta-T)}{2\sqrt{L}} \right) d\theta \right) - \frac{1}{\widetilde{L}_0(\overline{W},\overline{J})(t)} \widetilde{C}_0^{-1}\left(\int_{t_k}^{t} \left(\frac{\overline{W}(\theta)+\overline{J}(\theta-T)}{2\sqrt{L}} \right) d\theta \right) \right| \le$$

$$\le \rho_\mu^{(k)}(W,\overline{W}) \frac{W_0}{\hat{L}_0^2} \sup\left| \frac{d\widetilde{L}_0}{di} \right| +$$

$$\frac{2\sqrt{L}}{\hat{L}_0} \sum_{n=1}^{m} \left| r_n^{(0)} \right| n \left| \frac{W_0+J_0}{2\sqrt{L}} \right|^{n-1} \left| \frac{W(t)+J(t-T)}{2\sqrt{L}} - \frac{\overline{W}(t)+\overline{J}(t-T)}{2\sqrt{L}} \right| +$$

$$+ 2\sqrt{L} \sum_{n=1}^{m} \left| r_n^{(1)} \right| \frac{1}{\hat{L}_0^2} \left| \frac{d\widetilde{L}_0}{di} \right| \left| \frac{W_0+J_0}{2\sqrt{L}} \right|^{n-1} \left| \frac{W(t)+J(t-T)}{2\sqrt{L}} - \frac{\overline{W}(t)+\overline{J}(t-T)}{2\sqrt{L}} \right| +$$

$$+ \frac{Z_0}{\hat{L}_0} \left| W(t) - \overline{W}(t) \right| + \frac{Z_0 W_0}{\hat{L}_0^2} \left| \frac{d\widetilde{L}_0}{di} \right| \left| \frac{W(t)+J(t-T)}{2\sqrt{L}} - \frac{\overline{W}(t)+\overline{J}(t-T)}{2\sqrt{L}} \right| +$$

$$+ \frac{Z_0}{\hat{L}_0}\left|J(t-T)-\bar{J}(t-T)\right| + \frac{Z_0 J_0}{\hat{L}_0^2}\left|\frac{d\widetilde{L}_0}{di}\right|\left\|\frac{W(t)+J(t-T)}{2\sqrt{L}} - \frac{\overline{W}(t)+\bar{J}(t-T)}{2\sqrt{L}}\right\| +$$

$$+ \frac{2\sqrt{L}}{\hat{L}_0^2}H_0\left|\int_{t_k}^{t}\frac{W(\theta)+J(\theta-T)}{2\sqrt{L}}d\theta\right|\left\|\frac{d\widetilde{L}_0}{di}\right\|\left\|\frac{W(t)+J(t-T)}{2\sqrt{L}} - \frac{\overline{W}(t)+\bar{J}(t-T)}{2\sqrt{L}}\right\| +$$

$$+ \frac{2\sqrt{L}}{\hat{L}_1}\frac{2\sqrt{\Phi_0}}{c_0\sqrt{\Phi_0-\phi_0}}\left|\int_{t_k}^{t}\left(\frac{W(\theta)+J(\theta-T)}{2\sqrt{L}} - \frac{\overline{W}(\theta)+\bar{J}(\theta-T)}{2\sqrt{L}}\right)d\theta\right| \le$$

$$\le \rho_\mu^{(k)}(W,\overline{W})\frac{W_0}{\hat{L}_0^2}\sup\left|\frac{d\widetilde{L}_0}{di}\right| + \rho_\mu^{(k)}(W,\overline{W})\frac{1}{\hat{L}_0}\sum_{n=1}^{m}\left|r_n^{(0)}\right|n\left(\frac{W_0+J_0}{2\sqrt{L}}\right)^{n-1} +$$

$$+ \rho_\mu^{(k)}(W,\overline{W})\frac{1}{\hat{L}_0^2}\sum_{n=1}^{m}\left|r_n^{(0)}\right|\left(\frac{W_0+J_0}{2\sqrt{L}}\right)^{n}\sup\left|\frac{d\widetilde{L}_0}{di}\right| +$$

$$+ \frac{Z_0}{\hat{L}_0}\rho_\mu^{(k)}(W,\overline{W}) + \rho_\mu^{(k)}(W,\overline{W})\frac{1}{2\sqrt{L}}\frac{Z_0 W_0}{\hat{L}_0^2}\sup\left|\frac{d\widetilde{L}_0}{di}\right| +$$

$$+ \rho_\mu^{(k)}(W,\overline{W})\frac{1}{2\sqrt{L}}\frac{Z_0 J_0}{\hat{L}_0^2}\sup\left|\frac{d\widetilde{L}_0}{di}\right| + \frac{\rho_\mu^{(k)}(W,\overline{W})}{2\sqrt{L}}\frac{e^{\mu(t-t_k)}-1}{\mu}\frac{2\sqrt{L}}{\hat{L}_0^2}H_0\frac{W_0+J_0}{2\sqrt{L}}\sup\left|\frac{d\widetilde{L}_0}{di}\right| +$$

$$+ \frac{\rho_\mu^{(k)}(W,\overline{W})}{\hat{L}_0}\frac{e^{\mu(t-t_k)}-1}{\mu}\frac{2\sqrt{\Phi_0}}{c_0\sqrt{\Phi_0-\phi_0}} \le$$

$$\le \frac{\rho_\mu^{(k)}(\dot{W},\dot{\overline{W}})}{\mu}\frac{W_0}{\hat{L}_0^2}\sup\left|\frac{d\widetilde{L}_0}{di}\right| + \frac{\rho_\mu^{(k)}(\dot{W},\dot{\overline{W}})}{\mu}\frac{1}{\hat{L}_0}\sum_{n=1}^{m}\left|r_n^{(0)}\right|n\left(\frac{W_0+J_0}{2\sqrt{L}}\right)^{n-1} +$$

$$+ \frac{\rho_\mu^{(k)}(\dot{W},\dot{\overline{W}})}{\mu}\frac{1}{\hat{L}_0^2}\sum_{n=1}^{m}\left|r_n^{(0)}\right|\left(\frac{W_0+J_0}{2\sqrt{L}}\right)^{n}\sup\left|\frac{d\widetilde{L}_0}{di}\right| +$$

$$+ \frac{Z_0}{\hat{L}_0}\frac{\rho_\mu^{(k)}(\dot{W},\dot{\overline{W}})}{\mu} + \frac{\rho_\mu^{(k)}(\dot{W},\dot{\overline{W}})}{\mu}\frac{1}{2\sqrt{L}}\frac{Z_0 W_0}{\hat{L}_0^2}\sup\left|\frac{d\widetilde{L}_0}{di}\right| +$$

$$\frac{\rho_\mu^{(k)}(\dot{W},\dot{\overline{W}})}{\mu}\frac{1}{2\sqrt{L}}\frac{Z_0 W_0}{\hat{L}_0^2}\sup\left|\frac{d\widetilde{L}_0}{di}\right| +$$

$$+ \frac{\rho_\mu^{(k)}(\dot{W},\dot{\overline{W}})}{\mu}\frac{1}{2\sqrt{L}}\frac{Z_0 J_0}{\hat{L}_0^2}\sup\left|\frac{d\widetilde{L}_0}{di}\right| + \frac{\rho_\mu^{(k)}(\dot{W},\dot{\overline{W}})}{2\mu\sqrt{L}}\frac{e^{\mu(t-t_k)}-1}{\mu}\frac{2\sqrt{L}}{\hat{L}_0^2}H_0\frac{W_0+J_0}{2\sqrt{L}}\sup\left|\frac{d\widetilde{L}_0}{di}\right| +$$

$$+ \frac{\rho_\mu^{(k)}(\dot{W},\dot{\overline{W}})}{\mu\hat{L}_0}\frac{e^{\mu(t-t_k)}-1}{\mu}\frac{2\sqrt{\Phi_0}}{c_0\sqrt{\Phi_0-\phi_0}} \le$$

$$\leq e^{\mu(t-t_k)}\hat{\rho}_{\mu}^{(k)}((W,\dot{W},J,\dot{J}),(\overline{W},\dot{\overline{W}},\overline{J},\dot{\overline{J}}))\frac{1}{\mu\hat{L}_0}\left[\frac{1}{\hat{L}_0}\left(W_0+\sum_{n=1}^{m}\left|r_n^{(0)}\right|i_0^n+Z_0i_0+\frac{i_0H_0}{\mu}\right)\sum_{n=1}^{m}(n+1)\left|l_n^{(0)}\right|i_0^n+\right.$$

$$\left.+\sum_{n=1}^{m}\left|r_n^{(0)}\right|n\,i_0^{n-1}+Z_0+\frac{2\sqrt{\Phi_0}}{\mu c_0\sqrt{\Phi_0-\phi_0}}\right]$$

and

$$\dot{U}_2\leq e^{\mu(t-t_k)}\hat{\rho}_{\mu}^{(k)}((W,\dot{W},J,\dot{J}),(\overline{W},\dot{\overline{W}},\overline{J},\dot{\overline{J}}))\frac{e^{\mu(t-t_k)}-1}{\mu(t_{k+1}-t_k)}\left\{\frac{1}{\mu\hat{L}_0}\left[\frac{1}{\hat{L}_0}\left(W_0+\sum_{n=1}^{m}\left|r_n^{(0)}\right|i_0^n+i_0Z_0+i_0\frac{H_0}{\mu}\right)\times\right.\right.$$

$$\left.\left.\times\sum_{n=1}^{m}(n+1)\left|l_n^{(0)}\right|i_0^n+\sum_{n=1}^{m}\left|r_n^{(0)}\right|n\,i_0^{n-1}+Z_0+\frac{2\sqrt{\Phi_0}}{\mu c_0\sqrt{\Phi_0-\phi_0}}\right]\right\}.$$

The last term is bounded because

$$\frac{e^{\mu(t-t_k)}-1}{\mu(t_{k+1}-t_k)}\to 1 \text{ as } (t_{k+1}-t_k)\to 0 \text{ and } \frac{e^{\mu(t-t_k)}-1}{\mu(t_{k+1}-t_k)}\leq\frac{e^{\mu T_0}-1}{\mu T_0}.$$

Consequently

$$\dot{U}_2\leq e^{\mu(t-t_k)}\hat{\rho}_{\mu}^{(k)}((W,\dot{W},J,\dot{J}),(\overline{W},\dot{\overline{W}},\overline{J},\dot{\overline{J}}))\frac{e^{\mu_0}-1}{\mu_0}\left\{\frac{1}{\mu\hat{L}_0}\left[\frac{1}{\hat{L}_0}\left(W_0+\sum_{n=1}^{m}\left|r_n^{(0)}\right|i_0^n+i_0Z_0+i_0\frac{H_0}{\mu}\right)\times\right.\right.$$

$$\left.\left.\times\sum_{n=1}^{m}(n+1)\left|l_n^{(0)}\right|i_0^n+\sum_{n=1}^{m}\left|r_n^{(0)}\right|n\,i_0^{n-1}+Z_0+\frac{2\sqrt{\Phi_0}}{\mu c_0\sqrt{\Phi_0-\phi_0}}\right]\right\}.$$

Therefore

$$\left|\dot{B}_W^{(k)}(W,J)(t)-\dot{B}_W^{(k)}(\overline{W},\overline{J})(t)\right|\leq$$

$$\leq e^{\mu(t-t_k)}\hat{\rho}_{\mu}^{(k)}((W,\dot{W},J,\dot{J}),(\overline{W},\dot{\overline{W}},\overline{J},\dot{\overline{J}}))\left(1+\frac{e^{\mu_0}-1}{\mu_0}\right)\frac{1}{\mu\hat{L}_0}\left[\frac{1}{\hat{L}_0}\left(W_0+\sum_{n=1}^{m}\left|r_n^{(0)}\right|i_0^n+Z_0i_0+\frac{H_0}{\mu}i_0\right)\times\right.$$

$$\left.\times\sum_{n=1}^{m}(n+1)\left|l_n^{(0)}\right|i_0^n+\sum_{n=1}^{m}\left|r_n^{(0)}\right|n\,i_0^{n-1}+Z_0+\frac{2\sqrt{\Phi_0}}{\mu c_0\sqrt{\Phi_0-\phi_0}}\right]\equiv$$

$$\equiv e^{\mu(t-t_k)}\dot{K}_W\hat{\rho}_{\mu}^{(k)}((W,\dot{W},J,\dot{J}),(\overline{W},\dot{\overline{W}},\overline{J},\dot{\overline{J}})),$$

that is,

$$\rho_{\mu}^{(k)}(\dot{B}_W^{(k)}(W,J),\dot{B}_W^{(k)}(\overline{W},\overline{J})) \le \dot{K}_W \hat{\rho}_{\mu}^{(k)}((W,\dot{W},J,\dot{J}),(\overline{W},\dot{\overline{W}},\overline{J},\dot{\overline{J}})).$$

It follows

$$\hat{\rho}_{\mu}^{(k)}(\dot{B}_W(W,J),\dot{B}_W(\overline{W},\overline{J})) \le \dot{K}_W \hat{\rho}_{\mu}^{(k)}((W,\dot{W},J,\dot{J}),(\overline{W},\dot{\overline{W}},\overline{J},\dot{\overline{J}})).$$

For the derivative of the second component of B we obtain

$$\left| \dot{B}_J^{(k)}(W,J)(t) - \dot{B}_J^{(k)}(\overline{W},\overline{J})(t) \right| \le \left| I(W,J)(t) - I(\overline{W},\overline{J})(t) \right| +$$

$$+ \frac{1}{t_{k+1} - t_k} \left| \int_{t_k}^{t_{k+1}} I(W,J)(s)ds - \int_{t_k}^{t_{k+1}} I(\overline{W},\overline{J})(s)ds \right| \equiv \dot{I}_1 + \dot{I}_2.$$

Since

$$\dot{I}_1 \le \rho_{\mu}^{(k)}(J,\overline{J})\frac{W_0}{\hat{L}_1^2}\sup\left|\frac{d\widetilde{L}_1}{di}\right| + \rho_{\mu}^{(k)}(J,\overline{J})\frac{1}{\hat{L}_1^2}\sum_{n=1}^{m}\left|r_n^{(1)}\right|\left(\frac{W_0+J_0}{2\sqrt{L}}\right)^n\sup\left|\frac{d\widetilde{L}_1}{di}\right| +$$

$$+ \frac{Z_0}{\hat{L}_1}\rho_{\mu}^{(k)}(J,\overline{J}) + \rho_{\mu}^{(k)}(J,\overline{J})\frac{1}{2\sqrt{L}}\frac{Z_0 W_0}{\hat{L}_1^2}\sup\left|\frac{d\widetilde{L}_1}{di}\right| +$$

$$+ \rho_{\mu}^{(k)}(J,\overline{J})\frac{1}{2\sqrt{L}}\frac{Z_0 J_0}{\hat{L}_1^2}\sup\left|\frac{d\widetilde{L}_1}{di}\right| + \frac{\rho_{\mu}^{(k)}(J,\overline{J})}{2\sqrt{L}}\frac{e^{\mu(t-t_k)}-1}{\mu}\frac{H_1}{\hat{L}_1^2}(W_0+J_0)\sup\left|\frac{d\widetilde{L}_1}{di}\right| +$$

$$+ \frac{\rho_{\mu}^{(k)}(J,\overline{J})}{\hat{L}_1}\frac{e^{\mu(t-t_k)}-1}{\mu}ds\frac{2\sqrt{\Phi_1}}{c_1\sqrt{\Phi_1-\phi_0}} \le$$

$$\le e^{\mu(t-t_k)}\hat{\rho}_{\mu}^{(k)}((W,\dot{W},J,\dot{J}),(\overline{W},\dot{\overline{W}},\overline{J},\dot{\overline{J}}))\left\{\frac{1}{\mu\hat{L}_1}\left[\frac{1}{\hat{L}_1}\left(W_0+\sum_{n=1}^{m}\left|r_n^{(1)}\right|\left(\frac{W_0+J_0}{2\sqrt{L}}\right)^n +\right.\right.\right.$$

$$\left.\left.\left. + Z_0\frac{W_0+J_0}{2\sqrt{L}}+\frac{W_0+J_0}{2\sqrt{L}}\frac{H_1}{\mu}\right)\sum_{n=1}^{m}(n+1)\left|l_n^{(1)}\right|\left(\frac{W_0+J_0}{2\sqrt{L}}\right)^n + \sum_{n=1}^{m}\left|r_n^{(1)}\right|n\left(\frac{W_0+J_0}{2\sqrt{L}}\right)^{n-1} + Z_0 + \frac{2\sqrt{\Phi_1}}{\mu c_1\sqrt{\Phi_1-\phi_0}}\right]\right\} \le$$

$$\le e^{\mu(t-t_k)}\hat{\rho}_{\mu}^{(k)}((W,\dot{W},J,\dot{J}),(\overline{W},\dot{\overline{W}},\overline{J},\dot{\overline{J}}))\left\{\frac{1}{\mu\hat{L}_1}\left[\frac{1}{\hat{L}_1}\left(W_0+\sum_{n=1}^{m}\left|r_n^{(1)}\right|i_0^n + Z_0 i_0 + i_0\frac{H_1}{\mu}\right)\times\right.\right.$$

$$\left.\left. \times \sum_{n=1}^{m}(n+1)\left|l_n^{(1)}\right|i_0^n + \sum_{n=1}^{m}\left|r_n^{(1)}\right|ni_0^{n-1} + Z_0 + \frac{2\sqrt{\Phi_1}}{\mu c_1\sqrt{\Phi_1-\phi_0}}\right]\right\}$$

and

$$\dot{I}_2 \le e^{\mu(t-t_k)} \hat{\rho}_\mu^{(k)}((W,\dot{W},J,\dot{J}),(\overline{W},\dot{\overline{W}},\overline{J},\dot{\overline{J}})) \frac{e^{\mu_0}-1}{\mu_0}\left\{\frac{1}{\mu\hat{L}_1}\left[\frac{1}{\hat{L}_1}\left(W_0 + \sum_{n=1}^{m}\left|r_n^{(1)}\right|i_0^n + \right.\right.\right.$$

$$\left. + Z_0 i_0 + i_0\frac{H_1}{\mu}\right)\sum_{n=1}^{m}(n+1)\left|l_n^{(1)}\right|i_0^n + \sum_{n=1}^{m}\left|r_n^{(1)}\right|n i_0^{n-1} + Z_0\frac{2\sqrt{\Phi_1}}{\mu c_1\sqrt{\Phi_1-\phi_0}}\bigg]\bigg\}$$

then

$$\left|\dot{B}_J^{(k)}(W,J)(t) - \dot{B}_J^{(k)}(\overline{W},\overline{J})(t)\right| \le$$

$$\le e^{\mu(t-t_k)}\hat{\rho}_\mu^{(k)}((W,\dot{W},J,\dot{J}),(\overline{W},\dot{\overline{W}},\overline{J},\dot{\overline{J}}))e^{\mu_0}\left\{\left(1+\frac{e^{\mu_0}-1}{\mu_0}\right)\frac{1}{\mu\hat{L}_1}\left[\frac{1}{\hat{L}_1}\left(W_0 + \sum_{n=1}^{m}\left|r_n^{(1)}\right|i_0^n + Z_0 i_0 + \frac{H_1}{\mu}i_0\right)\times\right.\right.$$

$$\left.\times\sum_{n=1}^{m}(n+1)\left|l_n^{(1)}\right|i_0^n + \sum_{n=1}^{m}\left|r_n^{(1)}\right|n i_0^{n-1} + Z_0 + \frac{2\sqrt{\Phi_1}}{\mu c_1\sqrt{\Phi_1-\phi_0}}\bigg]\bigg\} \equiv$$

$$\equiv e^{\mu(t-t_k)}\dot{K}_J\hat{\rho}_\mu^{(k)}((W,\dot{W},J,\dot{J}),(\overline{W},\dot{\overline{W}},\overline{J},\dot{\overline{J}})).$$

Then it follows

$$\rho_\mu^{(k)}(\dot{B}_J^{(k)}(W,J),\dot{B}_J^{(k)}(\overline{W},\overline{J})) \le \dot{K}_J\,\hat{\rho}_\mu^{(k)}((W,\dot{W},J,\dot{J}),(\overline{W},\dot{\overline{W}},\overline{J},\dot{\overline{J}}))$$

and

$$\hat{\rho}_\mu^{(k)}(\dot{B}_J(W,J),\dot{B}_J(\overline{W},\overline{J})) \le \dot{K}_J\hat{\rho}_\mu^{(k)}((W,\dot{W},J,\dot{J}),(\overline{W},\dot{\overline{W}},\overline{J},\dot{\overline{J}})).$$

Finally we obtain

$$\hat{\rho}_\mu^{(k)}(B(W,J),\dot{B}(W,J),B(\overline{W},\overline{J}),\dot{B}(\overline{W},\overline{J})) \le K\,\hat{\rho}_\mu^{(k)}((W,\dot{W},J,\dot{J}),(\overline{W},\dot{\overline{W}},\overline{J},\dot{\overline{J}}))$$

where $K = \max\left\{e^{\mu_0}K_W, e^{\mu_0}K_J, \dot{K}_W, \dot{K}_J\right\}$ might be chosen smaller than 1. Since the uniform space is j-bounded in view of the inequalities

$$\hat{\rho}_\mu^{j^n(k)}((W,\dot{W},J,\dot{J}),(\overline{W},\dot{\overline{W}},\overline{J},\dot{\overline{J}})) \le \hat{\rho}_\mu^{(k)}((W,\dot{W},J,\dot{J}),(\overline{W},\dot{\overline{W}},\overline{J},\dot{\overline{J}})) < \infty \quad (n=0,1,2,...)$$

the fixed point theorem for contractive mappings in uniform spaces (cf. Chapter I) the operator B has a unique fixed point and it is an oscillating solution of (7.3.7).

Thus Theorem 7.4.1 is thus proved.

7.5. NUMERICAL EXAMPLE

Finally we demonstrate of how to apply the above theorem to engineering problems. We collect all inequalities implying an the existence-uniqueness of an oscillatory solution:

$$\frac{W_0 + J_0}{2\sqrt{L}} \le i_0 \; ; \; H_p = \frac{2\Phi_p - \phi_0}{2c_p \sqrt{\Phi_p}\sqrt{\Phi_p - \phi_0}} \; (p = 0,1) \; ;$$

$$J_0 e^{-\beta} + \frac{4L}{R}\sinh\left(\frac{RT_0}{L}\right)\left[\frac{2\sqrt{L}}{\hat{L}_0}U_{E_0} + \frac{W_0 + J_0}{\hat{L}_0}\left(\sum_{n=1}^{m}\left|r_n^{(0)}\right|\left(\frac{W_0 + J_0}{2\sqrt{L}}\right)^{n-1} + Z_0 + \frac{2H_0 L}{R}\sinh\left(\frac{RT_0}{L}\right)\right)\right] \le W_0 \; ;$$

$$W_0 e^{-\beta} + \frac{4L}{R}\sinh\left(\frac{RT_0}{L}\right)\left[\frac{2\sqrt{L}}{\hat{L}_1}U_{E_1} + \frac{W_0 + J_0}{\hat{L}_1}\left(\sum_{n=1}^{m}\left|r_n^{(1)}\right|\left(\frac{W_0 + J_0}{2\sqrt{L}}\right)^{n-1} + Z_0 + \frac{2H_1 L}{R}\sinh\left(\frac{RT_0}{L}\right)\right)\right] \le J_0 \; ;$$

$$K_W = \frac{e^{2\mu_0}}{\mu^2 \hat{L}_0}\left[\frac{1}{\hat{L}_0}\left(W_0 + \sum_{n=1}^{m}\left|r_n^{(0)}\right|(i_0)^n + Z_0 i_0 + \frac{H_0}{\mu}i_0\right)\sum_{n=1}^{m}(n+1)\left|l_n^{(0)}\right|i_0^n + \right.$$
$$\left. + \sum_{n=1}^{m}\left|r_n^{(0)}\right|n\, i_0^{n-1} + Z_0 + \frac{2\sqrt{\Phi_0}}{\mu c_0 \sqrt{\Phi_0 - \phi_0}}\right] < 1 \; ;$$

$$e^{\mu_0} K_J = \frac{e^{2\mu_0}}{\mu^2 \hat{L}_1}\left[\frac{1}{\hat{L}_1}\left(W_0 + \sum_{n=1}^{m}\left|r_n^{(1)}\right|i_0^n + Z_0 i_0 + \frac{H_1}{\mu}i_0\right)\sum_{n=1}^{m}(n+1)\left|l_n^{(1)}\right|i_0^n + \right.$$
$$\left. + \sum_{n=1}^{m}\left|r_n^{(1)}\right|n i_0^{n-1} + Z_0 + \frac{2\sqrt{\Phi_1}}{\mu c_1 \sqrt{\Phi_1 - \phi_0}}\right]\right\} < 1 \; ;$$

$$\dot{K}_W = \left(1 + \frac{e^{\mu_0} - 1}{\mu_0}\right)\frac{1}{\mu \hat{L}_0}\left[\frac{1}{\hat{L}_0}\left(W_0 + \sum_{n=1}^{m}\left|r_n^{(0)}\right|i_0^n + Z_0 i_0 + \frac{H_0}{\mu}i_0\right)\sum_{n=1}^{m}(n+1)\left|l_n^{(0)}\right|i_0^n + \right.$$
$$\left. + \sum_{n=1}^{m}\left|r_n^{(0)}\right|n i_0^{n-1} + Z_0 + \frac{2\sqrt{\Phi_0}}{\mu c_0 \sqrt{\Phi_0 - \phi_0}}\right] < 1 \; ;$$

$$\dot{K}_J = \left(1 + \frac{e^{\mu_0} - 1}{\mu_0}\right)\frac{1}{\mu \hat{L}_1}\left[\frac{1}{\hat{L}_1}\left(W_0 + \sum_{n=1}^{m}\left|r_n^{(1)}\right|(i_0)^n + Z_0 i_0 + \frac{H_1}{\mu}i_0\right)\sum_{n=1}^{m}(n+1)\left|l_n^{(1)}\right|i_0^n + \right.$$
$$\left. + \sum_{n=1}^{m}\left|r_n^{(1)}\right|n i_0^{n-1} + Z_0 + \frac{2\sqrt{\Phi_1}}{\mu c_1 \sqrt{\Phi_1 - \phi_0}}\right] < 1.$$

For a transmission line with length $\Lambda = 1000\,m$; $L = 0,45\,\mu H/m$; $C = 80\,pF/m$; cross-section area $S = 6\,mm^2$, copper specific resistance $\rho_c = 0,0175$; one obtains

$$R = (\rho_c \Lambda)/S \approx 3\,\Omega; \quad v = 1/\sqrt{LC} = 1/(6.10^{-9}) = 1,66.10^8; \quad Z_0 = 75\,\Omega \ . \text{ Then}$$

$$T = \Lambda\sqrt{LC} = 6.10^{-6}\ .$$

For waves with $\lambda_0 = (1/6)10^{-3}\,m$ we have

$$f_0 = 1/(\lambda_0 \sqrt{LC}) = 10^{12}\,Hz \Rightarrow T_0 = 1/f_0 = 10^{-12}\ .$$

Let us choose $\mu = 10^{12}$, then $\mu T_0 = \mu_0 = 1$, and $T = 6.10^{-6}.10^{12}\,T_0 = 6.10^6\,T_0$.

Consequently $e^{-\mu T} = e^{-6.10^6} \approx 0$;

$$R/L = 3/(0,45.10^{-6}) \approx 6,6.10^6; \quad L/R = 1/(6,6.10^6);$$

$$R\Lambda/Z_0 = RT/L = 6,6.10^6.6.10^{-6} \approx 40;$$

$$(RT_0/L) \approx 6,6.10^{-6} \Rightarrow e^{(RT_0)/L} - 1 \approx (RT_0)/L \approx 6,6.10^{-6};$$

$$\sinh((RT_0)/L) \approx (RT_0)/L = 6,6.10^{-6},$$

$$e^{-\left(\mu - \frac{R}{L}\right)T} = e^{-\left(10^{12} - 6,6.10^6\right)6.10^{-6}} = e^{-6.10^6\left(1 - 6,6.10^{-6}\right)} \approx e^{-6.10^6}\ .$$

We choose resistive elements with *V-I* characteristics

$$R_0(i) = R_1(i) = 0,028u - 0,125u^3 \ , \text{ i.e. } r_1 = 0,028, r_2 = 0,\ r_3 = 0,125$$

and inductive element with $L_0(i) = L_1(i) = 3i - (1/12)i^3$.

Then $\widetilde{L}_0(i) = i(dL_0(i)/di) + L_0(i) = 6i - (1/3)i^3$ and

$$\sum_{n=1}^{m}(n+1)\left|I_n^{(0)}\right| = 2.3 + 4.\frac{1}{12} = \frac{19}{3}\ .$$

For $i_0 = 1$ one obtains $\widetilde{L}_0(i) = 6i - (1/3)i^3 > 6 - (1/3) = 17/3 = \hat{L}$ and consequently

$1/\hat{L} = 1/\hat{L}_0 = 1/\hat{L}_1 = 3/17$; $\mu\hat{L} = 5,66.10^{12}$; $(2\sqrt{L})/\hat{L}_0 \approx 2,4.10^{-4}$.

Let us take, where

$$h = 2\,, c = c_0 = c_1 = 50\,pF = 5.10^{-11}\,F\,, \Phi_0 = \Phi_1 = 0,9\,V\,, \phi_0 = 0,1\,V\,,$$

$$W_0 e \leq \phi_0 \Leftrightarrow W_0 \leq e^{-1}\phi_0 \approx 0,037; \quad H_0 = H_1 = (2\Phi_0 - \phi_0)/(2c_0\sqrt{\Phi_0}\sqrt{\Phi_0 - \phi_0}) = 2.10^{10};$$

$$(2\sqrt{\Phi_0})/(\mu c_0\sqrt{\Phi_0 - \phi_0}) = 0,0424\ .$$

We can take $W_0 = J_0 = 0{,}5.10^{-3}$. It is easy to verify the above inequalities or the chosen values of specific parameters. $e(W_0 + J_0)/(2\sqrt{L}) \le 1$,

$$W_0 + J_0 \le \frac{2}{e}\sqrt{0{,}45.10^{-6}} = 0{,}493 < i_0 \text{ and so on.}$$

$$\dot{K}_W = \frac{3e}{17.10^{12}}\left[\frac{19}{17}\left(W_0 + 0{,}028 + 0{,}125 + 75 + \frac{2.10^{10}}{10^{12}}\right) + 0{,}028 + 3.0{,}125 + 75 + 0{,}0424\right];$$

$$\dot{K}_J = \frac{3e}{17.10^{12}}\left[\frac{19}{17}\left(W_0 + 0{,}028 + 0{,}125 + 75 + \frac{2.10^{10}}{10^{12}}\right) + 0{,}028 + 3.0{,}125 + 75 + 0{,}0424\right]$$

or $e\dfrac{48\sqrt{0{,}45}}{17}10^{-12}76 \le 1$; $K_W = K_J = \dfrac{208{,}59}{10^{24}}$; $\dot{K}_W = \dot{K}_J = 8.10^{-11}$.

Let the initial approximation be

$$W^{(0)}(t) = \begin{cases} W_0 \sin \omega_0 t, & t \in [0,T] \\ 0, & t \in [T,\infty) \end{cases}, \quad J^{(0)}(t) = \begin{cases} J_0 \sin \omega_0 t, & t \in [0,T] \\ 0, & t \in [T,T+T_0] \end{cases} \quad \left(\omega_0 = \frac{2\pi}{T_0}\right),$$

$$E_0(t) = E_1(t) = E_0 \sin \omega_0 t \text{ and } t_{k+1} - t_k = T_0.$$

Then we have $W^{(n+1)}(t) = B_u(W^{(n)}, J^{(n)})$, $J^{(n+1)}(t) = B_i(W^{(n)}, J^{(n)})$ $(n = 0,1,2,...)$ and

$$W^{(1)}(t) = B_u(W^{(0)}, J^{(0)})(t) = \int_{t_k}^{t} U(W^{(0)}, J^{(0)})(s)ds - \frac{t-t_k}{t_{k+1}-t_k}\int_{t_k}^{t_{k+1}} U(W^{(0)}, J^{(0)})(s)ds =$$

$$= -J^{(0)}(t-T) + 2\sqrt{L}\int_{t_k}^{t}\frac{E_0(s)}{\widetilde{L}_0(s-T)}ds - 2\sqrt{L}\int_{t_k}^{t}\frac{1}{\widetilde{L}_0(s-T)}R_0\left(\frac{W(s)+J(s-T)}{2\sqrt{L}}\right)ds -$$

$$- Z_0\int_{t_k}^{t}\frac{W(s)}{\widetilde{L}_0(s-T)}ds + Z_0\int_{t_k}^{t}\frac{J(s-T)}{\widetilde{L}_0(s-T)}ds - 2\sqrt{L}\int_{t_k}^{t}\frac{1}{\widetilde{L}_0(s-T)}\widetilde{C}_0^{-1}\left(\int_{t_k}^{s}\left(\frac{W(\theta)+J(\theta-T)}{2\sqrt{L}}\right)d\theta\right)ds +$$

$$+ \frac{t-t_k}{t_{k+1}-t_k}\left[\int_{t_k}^{t_{k+1}}\frac{dJ(s-T)}{ds}ds - 2\sqrt{L}\int_{t_k}^{t_{k+1}}\frac{E_0(s)}{\widetilde{L}_0(s-T)}ds +\right.$$

$$+ 2\sqrt{L}\int_{t_k}^{t_{k+1}}\frac{1}{\widetilde{L}_0(s-T)}R_0\left(\frac{W(s)+J(s-T)}{2\sqrt{L}}\right)ds + Z_0\int_{t_k}^{t_{k+1}}\frac{W(s)}{\widetilde{L}_0(s-T)}ds - Z_0\int_{t_k}^{t_{k+1}}\frac{J(s-T)}{\widetilde{L}_0(s-T)}ds +$$

$$+2\sqrt{L}\int_{t_k}^{t_{k+1}}\frac{1}{\widetilde{L}_0(s-T)}\widetilde{C}_0^{-1}\left(\int_{t_k}^{s}\left(\frac{W(\theta)+J(\theta-T)}{2\sqrt{L}}\right)d\theta\right)\Bigg]\,;$$

$$J^{(1)}(t)=B_i(W^{(0)},J^{(0)})(t)=\int_{t_k}^{t}I(W^{(0)},J^{(0)})(s)ds-\frac{t-t_k}{t_{k+1}-t_k}\int_{t_k}^{t_{k+1}}I(W^{(0)},J^{(0)})(s)ds=$$

$$=-W^{(0)}(t-T)-2\sqrt{L}\int_{t_k}^{t}\frac{E_1(s)}{\widetilde{L}_1(s-T)}ds-2\sqrt{L}\int_{t_k}^{t}\frac{1}{\widetilde{L}_1(s-T)}R_1\left(\frac{W(s-T)+J(s)}{2\sqrt{L}}\right)ds+$$

$$+Z_0\int_{t_k}^{t}\frac{W^{(0)}(s-T)}{\widetilde{L}_1(s-T)}ds-Z_0\int_{t_k}^{t}\frac{J^{(0)}(s)}{\widetilde{L}_1(s-T)}ds-2\sqrt{L}\int_{t_k}^{t}\frac{1}{\widetilde{L}_1(s-T)}\widetilde{C}_1^{-1}\left(\int_{t_k}^{s}\frac{W(\theta-T)+J(\theta)}{2\sqrt{L}}d\theta\right)ds-$$

$$-\frac{t-t_k}{t_{k+1}-t_k}\Bigg[-\int_{t_k}^{t_{k+1}}\dot{W}^{(0)}(s-T)ds-Z_0\int_{t_k}^{t_{k+1}}\frac{W^{(0)}(s-T)}{\widetilde{L}_1(s-T)}ds+Z_0\int_{t_k}^{t_{k+1}}\frac{J^{(0)}(s)}{\widetilde{L}_1(s-T)}ds+$$

$$+2\sqrt{L}\int_{t_k}^{t_{k+1}}\frac{E_1(s)}{\widetilde{L}_1(s-T)}ds-2\sqrt{L}\int_{t_k}^{t_{k+1}}\frac{1}{\widetilde{L}_1(s-T)}R_1\left(\frac{W(s-T)+J(s)}{2\sqrt{L}}\right)ds-$$

$$-2\sqrt{L}\int_{t_k}^{t_{k+1}}\frac{1}{\widetilde{L}_1(s-T)}\widetilde{C}_1^{-1}\left(\int_{T}^{s}\frac{W(\theta-T)+J(\theta)}{2\sqrt{L}}d\theta\right)ds\Bigg].$$

It follows

$$\left|2\sqrt{L}\int_{t_k}^{t}\frac{R_0\big((W(s)+J(s-T))/(2\sqrt{L})\big)}{\widetilde{L}_0(s-T)}ds\right|\le 2\sqrt{L}\int_{t_k}^{t}\frac{\left|r_1^{(0)}\right|+\left|r_3^{(0)}\right|}{\hat{L}_0}e^{\mu(s-t_k)}ds=10^{-16}\sqrt{0,45}\approx 0\,;$$

$$\left|2\sqrt{L}\int_{t_k}^{t}\frac{1}{\widetilde{L}_1(s-T)}R_1\left(\frac{W(s-T)+J(s)}{2\sqrt{L}}\right)ds\right|\le\frac{2\sqrt{L}}{\hat{L}_1}\left(\left|r_1^{(1)}\right|+\left|r_3^{(1)}\right|\right)\int_{t_k}^{t}e^{\mu(s-t_k)}ds\le\frac{1,6\sqrt{0,45.10^{-6}}}{17.10^{12}}\approx 0\,;$$

$$2\sqrt{L}\left|\int_{t_k}^{t}\frac{1}{\widetilde{L}_0(s-T)}\widetilde{C}_0^{-1}\left(\frac{J_0}{2\sqrt{L}}\int_{t_k}^{s}\sin\omega_0\theta d\theta\right)dsdt\right|\le W_0\frac{2\sqrt{L}}{\hat{L}_0}\int_{t_k}^{t}e^{\mu(s-t_k)}ds\le 10^{-17}\frac{3.(e-1)}{17}\approx 0\,;$$

$$2\sqrt{L}\left|\int_{t_k}^{t}\frac{1}{\widetilde{L}_1(s-T)}\widetilde{C}_1^{-1}\left(\int_{t_k}^{s}\frac{W(\theta-T)+J(\theta)}{2\sqrt{L}}d\theta\right)dsdt\right|\approx\frac{3.10^{-6}\sqrt{0,45}}{17}\frac{e-1}{10^{12}}\approx 0\,;$$

$$\left|\int_{t_k}^{t}\frac{Z_0J_0\sin\omega_0 s}{\widetilde{L}_0(s-T)}ds\right|\le\frac{Z_0J_0}{\hat{L}_0}\int_{t_k}^{t}e^{\mu(s-t_k)}ds\le Z_0J_0\frac{e^{\mu T_0}-1}{\mu\hat{L}_0}=75.1,5.10^{-3}\frac{(e-1)}{34.10^{12}}\approx 0\,;$$

$$\left| \int_{t_k}^{t} \frac{Z_0 W_0 \sin \omega_0 s}{\widetilde{L}_1(s-T)} \, ds \right| \leq \frac{W_0 J_0}{\hat{L}_1} \int_{t_k}^{t} e^{\mu(s-t_k)} ds \leq W_0 J_0 \frac{e^{\mu T_0}-1}{\mu \hat{L}_1} = 0{,}75.10^{-6}.\frac{(e-1)}{34.10^{12}} \approx 0 \cdot$$

Then the first approximations become

$$W^{(1)}(t) = B_W^{(k)}(W^{(0)}, J^{(0)})(t) = J_0 \sin \omega_0 t + 2U_{E_0}\sqrt{L}\left[\int_{t_k}^{t} \frac{\sin \omega_0 s}{\widetilde{L}_0(s-T)} ds - \frac{t-t_k}{T_0}\int_{t_k}^{t_k+T_0} \frac{\sin \omega_0 s}{\widetilde{L}_0(s-T)} ds\right]$$

$$J^{(1)}(t) = B_i(W^{(0)}, J^{(0)})(t) = -W_0 \sin \omega_0 t - 2U_{E_1}\sqrt{L}\left[\int_{t_k}^{t} \frac{\sin \omega_0 s}{\widetilde{L}_1(s-T)} ds - \frac{t-t_k}{T_0}\int_{t_k}^{t_k+T_0} \frac{\sin \omega_0 s}{\widetilde{L}_1(s-T)} ds\right] ;$$

$$\dot{W}^{(1)}(t) = \omega_0 J_0 \cos \omega_0 t + \frac{2U_{E_0}\sqrt{L}\sin \omega_0 t}{\widetilde{L}_0(t-T)} - \frac{2U_{E_0}\sqrt{L}}{T_0}\int_{t_k}^{t_k+T_0} \frac{\sin \omega_0 s}{\widetilde{L}_0(s-T)} ds ;$$

$$\dot{J}^{(1)}(t) = -\omega_0 W_0 \cos \omega_0 t + \frac{2U_{E_1}\sqrt{L}\sin \omega_0 t}{\widetilde{L}_1(t-T)} - \frac{2U_{E_1}\sqrt{L}}{T_0}\int_{t_k}^{t_k+T_0} \frac{\sin \omega_0 s}{\widetilde{L}_1(s-T)} ds .$$

Then we obtain

$$\left| W^{(1)}(t) - W^{(0)}(t) \right| \leq$$

$$\leq \left| J_0 \sin \omega_0 t + 2U_{E_0}\sqrt{L}\int_{t_k}^{t} \frac{\sin \omega_0 s}{\widetilde{L}_0(s-T)} ds - \left(\frac{t-t_k}{T_0}\right) 2U_{E_0}\sqrt{L}\int_{t_k}^{t_k+T_0} \frac{\sin \omega_0 s}{\widetilde{L}_0(s-T)} ds - W_0 \sin \omega_0 t \right| \leq$$

$$\leq J_0 + W_0 + \frac{2U_{E_0}\sqrt{L}}{\hat{L}_0}\frac{e^{\mu(t-t_k)}-1}{\mu} + \frac{2U_{E_0}\sqrt{L}}{\hat{L}_0}\frac{e^{\mu T_0}-1}{\mu} \leq J_0 + W_0 + \frac{4U_{E_0}\sqrt{L}}{\hat{L}_0}\frac{e^{\mu T_0}-1}{\mu} \approx J_0 + W_0$$

and

$$\left| J^{(1)}(t) - J^{(0)}(t) \right| \leq$$

$$\leq \left| -W_0 \sin \omega_0 t - 2U_{E_1}\sqrt{L}\int_{t_k}^{t} \frac{\sin \omega_0 s}{\widetilde{L}_1(s-T)} ds - \left(\frac{t-t_k}{T_0}\right) 2U_{E_1}\sqrt{L}\int_{t_k}^{t_k+T_0} \frac{\sin \omega_0 s}{\widetilde{L}_1(s-T)} ds - J_0 \sin \omega_0 t \right| \leq$$

$$\leq J_0 + W_0 + \frac{2U_{E_1}\sqrt{L}}{\hat{L}_1}\frac{e^{\mu(t-t_k)}-1}{\mu} + \frac{2U_{E_1}\sqrt{L}}{\hat{L}_1}\frac{e^{\mu T_0}-1}{\mu} \leq J_0 + W_0 + \frac{4U_{E_1}\sqrt{L}}{\hat{L}_1}\frac{e^{\mu T_0}-1}{\mu} \approx J_0 + W_0 \cdot$$

For the derivative we have

$$\left| \dot{W}^{(1)}(t) - \dot{W}^{(0)}(t) \right| \le$$

$$\le \left| \omega_0 J_0 \cos \omega_0 t + \frac{2U_{E_0}\sqrt{L}\sin\omega_0 t}{\widetilde{L}_0(t-T)} - \frac{2U_{E_0}\sqrt{L}}{T_0} \int_{t_k}^{t_k+T_0} \frac{\sin\omega_0 s}{\widetilde{L}_0(s-T)}ds - \omega_0 W_0 \cos\omega_0 s \right| \le$$

$$\le \left[\omega_0(W_0 + J_0) + \frac{2U_{E_0}\sqrt{L}}{\hat{L}_0} + \frac{2U_{E_0}\sqrt{L}}{\hat{L}_0}\frac{e^{\mu T_0}-1}{\mu T_0} \right] \approx \pi.10^9 + 0{,}65.$$

In the same way we can obtain estimates for the second component of the operator B

$$\left| \dot{J}^{(1)}(t) - \dot{J}^{(0)}(t) \right| \le$$

$$\le \left| -\omega_0 W_0 \cos \omega_0 t + \frac{2U_{E_1}\sqrt{L}\sin\omega_0 t}{\widetilde{L}_1(t-T)} - \frac{2U_{E_1}\sqrt{L}}{T_0} \int_{t_k}^{t_k+T_0} \frac{\sin\omega_0 s}{\widetilde{L}_1(s-T)}ds - \omega_0 J_0 \cos\omega_0 t \right| \le$$

$$\le \left[\omega_0(W_0 + J_0) + \frac{2U_{E_1}\sqrt{L}}{\hat{L}_1} + \frac{2U_{E_1}\sqrt{L}}{T_0\hat{L}_1}\frac{e^{\mu T_0}-1}{\mu} \right] \approx \pi.10^9 + 0{,}65.$$

Consequently $\rho_\mu^{(k)}((W,J),(\overline{W},\overline{J})) \le \pi.10^9 + 0{,}65$ and then

$$\hat{\rho}_\mu^{(k)}((W^{(n+1)},J^{(n+1)}),(W^{(n)},J^{(n)})) \le \frac{\left(8.10^{-11}\right)^n}{1-8.10^{-11}}\left(\pi.10^9 + 0{,}65\right), (n=0,1,\dots).$$

7.6. Applications to Transmission Lines Terminated by Linear and Time-Varying Loads

We proceed again from the lossy transmission line system

$$\frac{\partial u(x,t)}{\partial t} + \frac{1}{C(t)}\frac{\partial i(x,t)}{\partial x} + \frac{G(t)}{C(t)}u(x,t) = 0,$$

$$\frac{\partial i(x,t)}{\partial t} + \frac{1}{L(t)}\frac{\partial u(x,t)}{\partial x} + \frac{R}{L(t)}i(x,t) = 0$$

only in this case, the parameters depend on the time

$$C = C(t), L = L(t), G = G(t), R = R(t).$$

Obviously we have to assume that $C(t), L(t)$ have strictly positive lower bounds.
Here we introduce a more general than Heaviside condition $R/L = G/C$.
The above system in a matrix form is:

$$\frac{\partial U}{\partial t} + A_1 \frac{\partial U}{\partial x} + A_2 U = 0 \tag{5.6.1}$$

where

$$U = \begin{bmatrix} u \\ i \end{bmatrix}, \quad \frac{\partial U}{\partial t} = \begin{bmatrix} \dfrac{\partial u}{\partial t} \\ \dfrac{\partial i}{\partial t} \end{bmatrix}, \quad \frac{\partial U}{\partial x} = \begin{bmatrix} \dfrac{\partial u}{\partial x} \\ \dfrac{\partial i}{\partial x} \end{bmatrix}, \quad A_1(t) = \begin{bmatrix} 0 & 1/C \\ 1/L & 0 \end{bmatrix}, \quad A_2(t) = \begin{bmatrix} G/C & 0 \\ 0 & R/L \end{bmatrix}.$$

The eigenvalues of $A_1(t) = \begin{bmatrix} 0 & 1/C \\ 1/L & 0 \end{bmatrix}$ are $\lambda_1 = 1/\sqrt{LC}, \ \lambda_2 = -1/\sqrt{LC}$ whose eigenvectors are: $\left(\xi_1^{(1)}, \xi_2^{(1)}\right) = \left(\sqrt{C}, \sqrt{L}\right), \quad \left(\xi_1^{(2)}, \xi_2^{(2)}\right) = \left(-\sqrt{C}, \sqrt{L}\right).$

Here the matrix H formed by the eigenvectors depends on time $H(t) = \begin{bmatrix} \sqrt{C(t)} & \sqrt{L(t)} \\ -\sqrt{C(t)} & \sqrt{L(t)} \end{bmatrix}$

and its inverse one too $H^{-1}(t) = \begin{bmatrix} \dfrac{1}{2\sqrt{C(t)}} & -\dfrac{1}{2\sqrt{C(t)}} \\ \dfrac{1}{2\sqrt{L(t)}} & \dfrac{1}{2\sqrt{L(t)}} \end{bmatrix}.$

Then $A^{\text{can}} = HA_1 H^{-1}$, where $A^{\text{can}}(t) = \begin{bmatrix} 1/\sqrt{L(t)C(t)} & 0 \\ 0 & -1/\sqrt{L(t)C(t)} \end{bmatrix}.$

Introduce new variables $Z = HU$, (or $U = H^{-1}Z$)

$$Z = \begin{bmatrix} V(x,t) \\ I(x,t) \end{bmatrix}, \quad U = \begin{bmatrix} u(x,t) \\ i(x,t) \end{bmatrix}.$$

Then

$$\left| \begin{aligned} V(x,t) &= \sqrt{C(t)}\, u(x,t) + \sqrt{L(t)}\, i(x,t) \\ I(x,t) &= -\sqrt{C(t)}\, u(x,t) + \sqrt{L(t)}\, i(x,t) \end{aligned} \right.$$

or

$$\left| \begin{array}{l} u(x,t) = \dfrac{1}{2\sqrt{C(t)}}V(x,t) - \dfrac{1}{2\sqrt{C(t)}}I(x,t) \\[2mm] i(x,t) = \dfrac{1}{2\sqrt{L(t)}}V(x,t) + \dfrac{1}{2\sqrt{L(t)}}I(x,t). \end{array}\right.$$

Substituting $U = H^{-1}(t)Z$ in (5.6.1) we obtain

$$\frac{\partial\left(H^{-1}(t)Z\right)}{\partial t} + A_1 \frac{\partial\left(H^{-1}(t)Z\right)}{\partial x} + A_2\left(H^{-1}(t)Z\right) = 0.$$

In contrast to transformations from Chapter I we have

$$H^{-1}(t)\frac{\partial Z}{\partial t} + \frac{\partial H^{-1}(t)}{\partial t}Z + A_1 H^{-1}(t)\frac{\partial Z}{\partial x} + A_2 H^{-1}(t)Z = 0.$$

Multiply the last equality by $H(t)$ from the left

$$\frac{\partial Z}{\partial t} + H(t)\frac{\partial H^{-1}(t)}{\partial t}Z + H(t)A_1 H^{-1}(t)\frac{\partial Z}{\partial x} + \left(H(t)A_2 H^{-1}(t)\right)Z = 0$$

where

$$HA_2H^{-1} = \frac{1}{2}\begin{bmatrix} \dfrac{G(t)}{C(t)}+\dfrac{R(t)}{L(t)} & -\dfrac{G(t)}{C(t)}+\dfrac{R(t)}{L(t)} \\[3mm] -\dfrac{G(t)}{C(t)}+\dfrac{R(t)}{L(t)} & \dfrac{G(t)}{C(t)}+\dfrac{R(t)}{L(t)} \end{bmatrix}, \; H(t)\frac{\partial H^{-1}(t)}{\partial t} = \frac{1}{4}\begin{bmatrix} -\dfrac{\dot{C}(t)}{C(t)}-\dfrac{\dot{L}(t)}{L(t)} & \dfrac{\dot{C}(t)}{C(t)}-\dfrac{\dot{L}(t)}{L(t)} \\[3mm] \dfrac{\dot{C}(t)}{C(t)}-\dfrac{\dot{L}(t)}{L(t)} & -\dfrac{\dot{C}(t)}{C(t)}-\dfrac{\dot{L}(t)}{L(t)} \end{bmatrix}.$$

The dot means a differentiation with respect the time. We see that

$$H(t)A_2 H^{-1}(t) + H(t)\frac{\partial H^{-1}(t)}{\partial t} =$$

$$= \frac{1}{2}\begin{bmatrix} \dfrac{G(t)}{C(t)}+\dfrac{R(t)}{L(t)}-\dfrac{\dot{C}(t)}{2C(t)}-\dfrac{\dot{L}(t)}{2L(t)} & -\dfrac{G(t)}{C(t)}+\dfrac{R(t)}{L(t)}+\dfrac{\dot{C}(t)}{2C(t)}-\dfrac{\dot{L}(t)}{2L(t)} \\[3mm] -\dfrac{G(t)}{C(t)}+\dfrac{R(t)}{L(t)}+\dfrac{\dot{C}(t)}{2C(t)}-\dfrac{\dot{L}(t)}{2L(t)} & \dfrac{G(t)}{C(t)}+\dfrac{R(t)}{L(t)}-\dfrac{\dot{C}(t)}{2C(t)}-\dfrac{\dot{L}(t)}{2L(t)} \end{bmatrix}.$$

So the matrix equation

$$\frac{\partial Z}{\partial t} + A^{\mathrm{can}}(t)\frac{\partial Z}{\partial x} + \left[H(t)A_2 H^{-1}(t) + H(t)\frac{\partial H^{-1}(t)}{\partial t} \right] Z = 0$$

becomes

$$\begin{bmatrix} \dfrac{\partial V}{\partial t} \\[2mm] \dfrac{\partial I}{\partial t} \end{bmatrix} + \begin{bmatrix} \dfrac{1}{\sqrt{LC}} & 0 \\[2mm] 0 & -\dfrac{1}{\sqrt{LC}} \end{bmatrix} \begin{bmatrix} \dfrac{\partial V}{\partial x} \\[2mm] \dfrac{\partial I}{\partial x} \end{bmatrix} + \frac{1}{2} \begin{bmatrix} \dfrac{G}{C}+\dfrac{R}{L}-\dfrac{\dot{C}}{2C}-\dfrac{\dot{L}}{2L} & -\dfrac{G}{C}+\dfrac{R}{L}+\dfrac{\dot{C}}{2C}-\dfrac{\dot{L}}{2L} \\[2mm] -\dfrac{G}{C}+\dfrac{R}{L}+\dfrac{\dot{C}}{2C}-\dfrac{\dot{L}}{2L} & \dfrac{G}{C}+\dfrac{R}{L}-\dfrac{\dot{C}}{2C}-\dfrac{\dot{L}}{2L} \end{bmatrix} \begin{bmatrix} V \\ I \end{bmatrix} = \begin{bmatrix} 0 \\ 0 \end{bmatrix}$$

or in explicit form

$$\frac{\partial V(x,t)}{\partial t} + \frac{1}{\sqrt{LC}}\frac{\partial V(x,t)}{\partial x} + \left(\frac{G}{C}+\frac{R}{L}-\frac{\dot{C}}{2C}-\frac{\dot{L}}{2L}\right)V(x,t) + \left(-\frac{G}{C}+\frac{R}{L}+\frac{\dot{C}}{2C}-\frac{\dot{L}}{2L}\right)I(x,t) = 0,$$

$$\frac{\partial I(x,t)}{\partial t} - \frac{1}{\sqrt{LC}}\frac{\partial I(x,t)}{\partial x} + \left(-\frac{G}{C}+\frac{R}{L}+\frac{\dot{C}}{2C}-\frac{\dot{L}}{2L}\right)V(x,t) + \left(\frac{G}{C}+\frac{R}{L}-\frac{\dot{C}}{2C}-\frac{\dot{L}}{2L}\right)I(x,t) = 0.$$

We introduce a generalized Heaviside condition:

$$-\frac{G(t)}{C(t)} + \frac{R(t)}{L(t)} + \frac{\dot{C}(t)}{2C(t)} - \frac{\dot{L}(t)}{2L(t)} = 0.$$

Then the previous system can be rewritten in the form

$$\frac{\partial V(x,t)}{\partial t} + \frac{1}{\sqrt{L(t)C(t)}}\frac{\partial V(x,t)}{\partial x} + \frac{2G(t)-\dot{C}(t)}{C(t)}V(x,t) = 0,$$

$$\frac{\partial I(x,t)}{\partial t} - \frac{1}{\sqrt{L(t)C(t)}}\frac{\partial I(x,t)}{\partial x} + \frac{2G(t)-\dot{C}(t)}{C(t)}I(x,t) = 0.$$

For the new initial conditions we obtain:

$$V(x,0) = \sqrt{C}\, u(x,0) + \sqrt{L}\, i(x,0) = \sqrt{C}\, u_0(x) + \sqrt{L}\, i_0(x) \equiv V_0(x),$$

$$I(x,0) = -\sqrt{C}\, u(x,0) + \sqrt{L}\, i(x,0) = -\sqrt{C}\, u_0(x) + \sqrt{L}\, i_0(x) \equiv I_0(x),\ x \in [0,\Lambda].$$

Put $h(t) = \dfrac{2G(t)-\dot{C}(t)}{C(t)}$ and rewrite the above system in the form:

$$\frac{\partial V(x,t)}{\partial t} + \frac{1}{\sqrt{L(t)C(t)}} \frac{\partial V(x,t)}{\partial x} + h(t)V(x,t) = 0,$$

$$\frac{\partial I(x,t)}{\partial t} - \frac{1}{\sqrt{L(t)C(t)}} \frac{\partial I(x,t)}{\partial x} + h(t)I(x,t) = 0.$$

The last system might be simplified one more time by the substitution:

$$W(x,t) = e^{h(t)t}V(x,t), \; J(x,t) = e^{h(t)t}I(x,t),$$

or

$$V(x,t) = e^{-h(t)t}W(x,t), \; I(x,t) = e^{-h(t)t}J(x,t).$$

Consequently we obtain the following transformation:

$$u(x,t) = \frac{1}{2\sqrt{C(t)}} e^{-h(t)t}W(x,t) - \frac{1}{2\sqrt{C(t)}} e^{-h(t)t}J(x,t),$$

$$i(x,t) = \frac{1}{2\sqrt{L(t)}} e^{-h(t)t}W(x,t) + \frac{1}{2\sqrt{L(t)}} e^{-h(t)t}J(x,t)$$

and

$$W(x,t) = e^{h(t)t}\left(\sqrt{C(t)}\, u(x,t) + \sqrt{L(t)}\, i(x,t)\right),$$
$$J(x,t) = e^{h(t)t}\left(\sqrt{L(t)}\, i(x,t) - \sqrt{C(t)}\, u(x,t)\right).$$

Substitute $V(x,t)$ and $I(x,t)$ into the last system we have

$$\frac{\partial\left(e^{-h(t)t}W(x,t)\right)}{\partial t} + \frac{1}{\sqrt{LC}} \frac{\partial\left(e^{-h(t)t}W(x,t)\right)}{\partial x} + e^{-h(t)t}h(t)W(x,t) = 0,$$

$$\frac{\partial\left(e^{-h(t)t}J(x,t)\right)}{\partial t} - \frac{1}{\sqrt{LC}} \frac{\partial\left(e^{-h(t)t}J(x,t)\right)}{\partial x} + e^{-h(t)t}h(t)J(x,t) = 0$$

which leads to

$$\frac{\partial W}{\partial t} + \frac{1}{\sqrt{LC}} \frac{\partial W}{\partial x} - t\dot{h}(t)W = 0, \; \frac{\partial J}{\partial t} - \frac{1}{\sqrt{LC}} \frac{\partial J}{\partial x} - t\dot{h}(t)J = 0.$$

The last system takes the form

$$\frac{\partial W(x,t)}{\partial t} + \frac{1}{\sqrt{L(t)C(t)}}\frac{\partial W(x,t)}{\partial x} = 0 ,$$

$$\frac{\partial J(x,t)}{\partial t} - \frac{1}{\sqrt{L(t)C(t)}}\frac{\partial J(x,t)}{\partial x} = 0$$

under condition $\dot{h}(t) = 0 \Leftrightarrow h(t) = \text{const.} = \gamma$ or $\dfrac{2G - \dot{C}}{C} = \text{const.} = \gamma$.

We are able to obtain an explicit form of the relation between capacity function and conductivity one. Indeed, the linear differential equation $\dot{C}(t) + \gamma C(t) = 2G(t)$ has a

solution $C(t) = e^{-\gamma t}\left(C_T + 2\int\limits_{T}^{t} G(s)e^{\gamma s}ds \right), C_T = C(T)$.

Since the initial conditions remain the same ones in view of

$$W(x,0) = e^{\frac{R_0}{L}.0}V(x,0) = V_0(x) = \sqrt{C}\, u_0(x) + \sqrt{L}\, i_0(x) ,$$

$$J(x,0) = e^{\frac{R_0}{L}0}I(x,0) = I_0(x) = -\sqrt{C}\, u_0(x) + \sqrt{L}\, i_0(x)\, , x \in [0,\Lambda]$$

we have to transform the boundary conditions:

$$u(0,t) = E_0(t) - R_0(t)i(0,t) - \frac{1}{C_0(t)}\int\limits_{T}^{t} i(0,\tau)d\tau - 2L_0(t)i(0,t)\frac{di(0,t)}{dt} ,$$

$$u(\Lambda,t) = E_1(t) + R_1(t)i(\Lambda,t) + \frac{1}{C_1(t)}\int\limits_{T}^{t} i(\Lambda,\tau)d\tau + 2L_1(t)i(\Lambda,t)\frac{di(\Lambda,t)}{dt} .$$

Since

$$i(0,t) = \frac{1}{2\sqrt{L(t)}}e^{-h(t)t}W(0,t) + \frac{1}{2\sqrt{L(t)}}e^{-h(t)t}J(0,t) ,$$

$$u(0,t) = \frac{1}{2\sqrt{C(t)}}e^{-h(t)t}W(0,t) - \frac{1}{2\sqrt{C(t)}}e^{-h(t)t}J(0,t)$$

and

$$u(\Lambda,t) = \frac{1}{2\sqrt{C(t)}}e^{-h(t)t}W(\Lambda,t) - \frac{1}{2\sqrt{C(t)}}e^{-h(t)t}J(\Lambda,t) .$$

$$i(\Lambda,t) = \frac{1}{2\sqrt{L(t)}}e^{-h(t)t}W(\Lambda,t) + \frac{1}{2\sqrt{L(t)}}e^{-h(t)t}J(\Lambda,t)$$

we obtain a system

$$\frac{e^{-\gamma t}W(0,t)}{\sqrt{C(t)}} - \frac{e^{-\gamma t}J(0,t)}{\sqrt{C(t)}} = 2E_0(t) - R_0(t)\left(\frac{e^{-\gamma t}W(0,t)}{\sqrt{L(t)}} + \frac{e^{-\gamma t}J(0,t)}{\sqrt{L(t)}}\right) -$$

$$-\frac{1}{C_0(t)}\int_T^t\left(\frac{e^{-\gamma s}W(0,s)}{\sqrt{L(s)}} + \frac{e^{-\gamma s}J(0,s)}{\sqrt{L(s)}}\right)ds - \qquad ;$$

$$-L_0(t)\left(\frac{e^{-\gamma t}W(0,t)}{\sqrt{L(t)}} + \frac{e^{-\gamma t}J(0,t)}{\sqrt{L(t)}}\right)\frac{d}{dt}\left(\frac{e^{-\gamma t}W(0,t)}{\sqrt{L(t)}} + \frac{e^{-\gamma t}J(0,t)}{\sqrt{L(t)}}\right)$$

$$\frac{e^{-\gamma t}W(\Lambda,t)}{2\sqrt{C(t)}} - \frac{e^{-\gamma t}J(\Lambda,t)}{2\sqrt{C(t)}} = E_1(t) + R_1(t)\left(\frac{e^{-\gamma t}W(\Lambda,t)}{2\sqrt{L(t)}} + \frac{e^{-\gamma t}J(\Lambda,t)}{2\sqrt{L(t)}}\right) +$$

$$+\frac{1}{C_1(t)}\int_T^t\left(\frac{e^{-\gamma s}W(\Lambda,s)}{2\sqrt{L(s)}} + \frac{e^{-\gamma s}J(\Lambda,s)}{2\sqrt{L(s)}}\right)ds +$$

$$+L_1(t)\left(\frac{e^{-\gamma t}W(\Lambda,t)}{2\sqrt{L(t)}} + \frac{e^{-\gamma t}J(\Lambda,t)}{2\sqrt{L(t)}}\right)\frac{d}{dt}\left(\frac{e^{-\gamma t}W(\Lambda,t)}{\sqrt{L(t)}} + \frac{e^{-\gamma t}J(\Lambda,t)}{\sqrt{L(t)}}\right)$$

and the above system is of neutral type with initial functions

$$W(t) = \tilde{W}_0(t),\ \dot{W}(t) = \dot{\tilde{W}}_0(t),\ \ J(t) = \tilde{J}_0(t), \dot{J}(t) = \dot{\tilde{J}}_0(t)\ ,\ t \in [0,T]_.$$

But we have

$$W(\Lambda,t+T) = W(0,t),\ J(\Lambda,t) = J(0,t+T).$$

We can rewrite the above equations as follows

$$\frac{e^{-\gamma(t+T)}W(\Lambda,t+T)}{\sqrt{C(t)}} - \frac{e^{-\gamma t}J(0,t)}{\sqrt{C(t)}} = 2E_0(t) - R_0(t)\left(\frac{e^{-\gamma(t+T)}W(\Lambda,t+T)}{\sqrt{L(t)}} + \frac{e^{-\gamma t}J(0,t)}{\sqrt{L(t)}}\right) -$$

$$-\frac{1}{C_0(t)}\int_T^t\left(\frac{e^{-\gamma(s+T)}W(\Lambda,s+T)}{\sqrt{L(s)}} + \frac{e^{-\gamma s}J(0,s)}{\sqrt{L(s)}}\right)ds -$$

$$-L_0(t)\left(\frac{e^{-\gamma(t+T)}W(\Lambda,t+T)}{\sqrt{L(t)}} + \frac{e^{-\gamma t}J(0,t)}{\sqrt{L(t)}}\right)\frac{d}{dt}\left(\frac{e^{-\gamma(t+T)}W(\Lambda,t+T)}{\sqrt{L(t)}} + \frac{e^{-\gamma t}J(0,t)}{\sqrt{L(t)}}\right)$$

$$\frac{e^{-\varkappa}W(\Lambda,t)}{2\sqrt{C(t)}} - \frac{e^{-\gamma(t+T)}J(0,t+T)}{2\sqrt{C(t)}} = E_1(t) + R_1(t)\left(\frac{e^{-\varkappa}W(\Lambda,t)}{2\sqrt{L(t)}} + \frac{e^{-\gamma(t+T)}J(0,t+T)}{2\sqrt{L(t)}}\right) +$$

$$+\frac{1}{C_1(t)}\int_T^t\left(\frac{e^{-\varkappa}W(\Lambda,s)}{2\sqrt{L(s)}} + \frac{e^{-\gamma(s+T)}J(0,s+T)}{2\sqrt{L(s)}}\right)ds +$$

$$+L_1(t)\left(\frac{e^{-\varkappa}W(\Lambda,t)}{2\sqrt{L(t)}} + \frac{e^{-\gamma(s+T)}J(0,s+T)}{2\sqrt{L(t)}}\right)\frac{d}{dt}\left(\frac{e^{-\varkappa}W(\Lambda,t)}{\sqrt{L(t)}} + \frac{e^{-\gamma(t+T)}J(0,t+T)}{\sqrt{L(t)}}\right).$$

Let us put $e^{-\varkappa}W(\Lambda,t) \equiv W(t), e^{-\varkappa}J(0,t) \equiv J(t)$ and then

$$\frac{W(t+T)}{\sqrt{C(t)}} - \frac{J(t)}{\sqrt{C(t)}} = 2E_0(t) - R_0(t)\left(\frac{W(t+T)}{\sqrt{L(t)}} + \frac{J(t)}{\sqrt{L(t)}}\right) -$$

$$-\frac{1}{C_0(t)}\int_T^t\left(\frac{W(s+T)}{\sqrt{L(s)}} + \frac{J(s)}{\sqrt{L(s)}}\right)ds - L_0(t)\left(\frac{W(t+T)}{\sqrt{L(t)}} + \frac{J(t)}{\sqrt{L(t)}}\right)\frac{d}{dt}\left(\frac{W(t+T)}{\sqrt{L(t)}} + \frac{J(t)}{\sqrt{L(t)}}\right)$$

$$\frac{W(t)}{\sqrt{C(t)}} - \frac{J(t+T)}{\sqrt{C(t)}} = 2E_1(t) + R_1(t)\left(\frac{W(t)}{\sqrt{L(t)}} + \frac{J(t+T)}{\sqrt{L(t)}}\right) +$$

$$+\frac{1}{C_1(t)}\int_T^t\left(\frac{W(s)}{\sqrt{L(s)}} + \frac{J(s+T)}{\sqrt{L(s)}}\right)ds + L_1(t)\left(\frac{W(t)}{\sqrt{L(t)}} + \frac{J(s+T)}{\sqrt{L(t)}}\right)\frac{d}{dt}\left(\frac{W(t)}{\sqrt{L(t)}} + \frac{J(t+T)}{\sqrt{L(t)}}\right).$$

Now we put $t+T \equiv t$:

$$\frac{W(t)}{\sqrt{C(t-T)}} - \frac{J(t-T)}{\sqrt{C(t-T)}} = 2E_0(t-T) - R_0(t-T)\left(\frac{W(t)}{\sqrt{L(t-T)}} + \frac{J(t-T)}{\sqrt{L(t-T)}}\right) -$$

$$-\frac{1}{C_0(t-T)}\int_T^t\left(\frac{W(\tau)}{\sqrt{L(\tau-T)}} + \frac{J(\tau-T)}{\sqrt{L(\tau-T)}}\right)d\tau - L_0(t-T)\left(\frac{W(t)}{\sqrt{L(t-T)}} + \frac{J(t-T)}{\sqrt{L(t-T)}}\right)\frac{d}{dt}\left(\frac{W(t)}{\sqrt{L(t-T)}} + \frac{J(t-T)}{\sqrt{L(t-T)}}\right);$$

$$\frac{W(t-T)}{\sqrt{C(t-T)}} - \frac{J(t)}{\sqrt{C(t-T)}} = 2E_1(t-T) + R_1(t-T)\left(\frac{W(t-T)}{\sqrt{L(t-T)}} + \frac{J(t)}{\sqrt{L(t-T)}}\right) +$$

$$+\frac{1}{C_1(t-T)}\int_T^t\left(\frac{W(\tau-T)}{\sqrt{L(\tau-T)}} + \frac{J(\tau)}{\sqrt{L(\tau-T)}}\right)d\tau + L_1(t-T)\left(\frac{W(t-T)}{\sqrt{L(t-T)}} + \frac{J(s)}{\sqrt{L(t-T)}}\right)\frac{d}{dt}\left(\frac{W(t-T)}{\sqrt{L(t-T)}} + \frac{J(t)}{\sqrt{L(t-T)}}\right).$$

An natural assumption is T_0-periodicity of $C_p(t), L_p(t), E_p(t)(p = 0,1)$. In view of $mT_0 = T$ we obtain

$$C_p(t) = C_p(t-T), L_p(t) = (t-T), E_p(t) = (t-T)(p = 0,1)$$

which leads to the system

$$\frac{W(t)-J(t-T)}{\sqrt{C(t)}} = 2E_0(t) - R_0(t)\frac{W(t)+J(t-T)}{\sqrt{L(t)}} -$$

$$-\frac{1}{C_0(t)}\int_T^t \frac{W(\tau)+J(\tau-T)}{\sqrt{L(\tau)}}d\tau - L_0(t)\frac{W(t)+J(t-T)}{\sqrt{L(t)}}\frac{d}{dt}\frac{W(t)+J(t-T)}{\sqrt{L(t)}};$$

$$\frac{W(t-T)-J(t)}{\sqrt{C(t)}} = 2E_1(t) + R_1(t)\frac{W(t-T)+J(t)}{\sqrt{L(t)}} +$$

$$+\frac{1}{C_1(t)}\int_T^t \frac{W(\tau-T)+J(\tau)}{\sqrt{L(\tau)}}d\tau + L_1(t)\frac{W(t-T)+J(s)}{\sqrt{L(t)}}\frac{d}{dt}\frac{W(t-T)+J(t)}{\sqrt{L(t)}}$$

or

$$\frac{dW(t)}{dt} = -\frac{dJ(t-T)}{dt} + \frac{\dot{L}(t)(W(t)+J(t-T))}{2L(t)} +$$

$$+\frac{L(t)}{L_0(t)}\left(\frac{2E_0(t)-Z_0(t)(W(t)-J(t-T))}{W(t)+J(t-T)} - \frac{1}{C_0(t)(W(t)+J(t-T))}\int_T^t \frac{W(\tau)+J(\tau-T)}{\sqrt{L(\tau)}}d\tau - R_0(t)\right);$$

$$\frac{dJ(t)}{dt} = -\frac{dW(t-T)}{dt} + \frac{\dot{L}(t)(J(t)+W(t-T))}{2L(t)} +$$

$$+\frac{L(t)}{L_1(t)}\left(\frac{2E_1(t)-Z_0(t)(J(t)-W(t-T))}{J(t)+W(t-T)} - \frac{1}{C_1(t)(J(t)+W(t-T))}\int_T^t \frac{J(\tau)+W(\tau-T)}{\sqrt{L(\tau)}}d\tau - R_1(t)\right).$$

CONCLUSION

Here we have seen that the first method of reducing the mixed problem for lossy transmission line system is not applicable to the present case because the Lipschitz constant Z_0 before the neutral part is larger than 1. In contrast to the previous chapters we notice that in the case of series connected loads just the second method of reducing is applicable.

We point out that the procedure for treating of time-varying specific parameters might be applied to the present circuit (cf. § 5.7). This is due to the newly introduced generalized Heaviside condition.

LOSSLESS AND LOSSY TRANSMISSION LINES TERMINATED BY IN SERIES CONNECTED *RL*-LOADS PARALLEL TO THE *C*-LOAD

ABSTRACT

Here we consider lossless and lossy transmission lines terminated by a circuit consisting of in series connected *RL*-loads parallel to the *C*-load. As in the previous chapters first we derive boundary conditions and formulate the mixed problems for both lossless and lossy cases. Then we reduce each mixed problem to an initial value problem on the boundary. Here, however, the following difficulty arises. We are not able to exclude some transitional current functions and obtain a system of 4 equations for 4 unknown functions. The system obtained is once again a neutral one. Further on we prove existence-uniqueness theorems for a periodic solution in the lossless case, and an oscillatory ones solution for the lossy case.

INTRODUCTION

The main purpose of the present chapter is to consider transmission lines terminated by a circuit consisting of in series connected *RL*-loads that are parallel to *C*-load. Such configurations arise not only in radio frequencies devices but in various geophysical studies as well (cf. [79]).

Here we consider both cases lossless and lossy transmission lines cases. The loads are in general nonlinear ones, but the theory developed may be applied to the linear loads as well. In § 8.1 we derive boundary conditions and formulate the mixed problem for lossless transmission lines. It should be emphasized that in this case the *RL* current function cannot be excluded, and so we have to consider four equations instead of two ones as in the previous chapters. In § 8.2 we reduce the mixed problem to a periodic initial value problem on the boundary. In § 8.3 we give a different approach for reducing the mixed problem to a periodic one on the boundary. In § 8.4 we analyze the arising nonlinearities. In § 8.5 we give an operator presentation of the periodic problem. In § 8.6 we demonstrate, using numerical examples, how to apply our method to specific problems. In § 8.7 we consider lossy transmission lines terminated by the same configuration of nonlinear *RLC*-loads. In § 8.8 we reduce the mixed problem to an initial value problem on the boundary. Finally we give an operator presentation of the oscillatory problem in § 8.8 and provide a numerical examples in § 8.9.

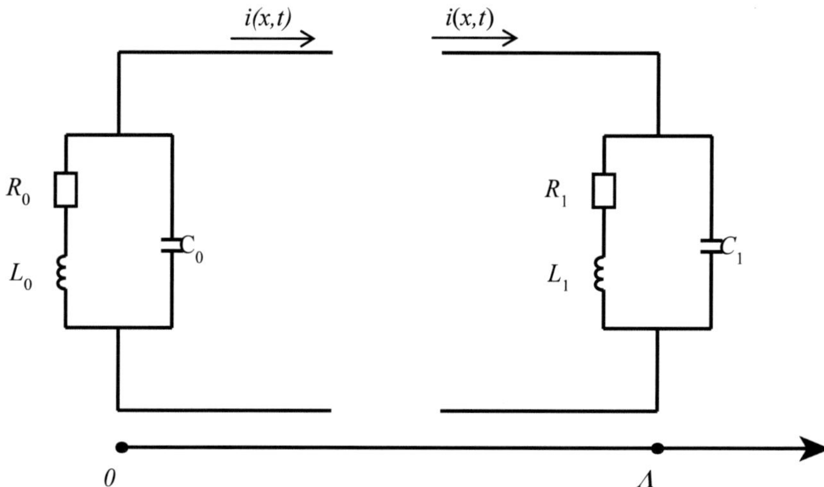

Figure 8.1.

8.1. DERIVATION OF THE BOUNDARY CONDITIONS FOR LOSSLESS TRANSMISSION LINE EQUATIONS AND FORMULATION OF THE MIXED PROBLEM

We proceed from the transmission line terminated by a configuration of nonlinear loads as shown on Fig.8.1. In accordance of Kirchhoff's voltage-law (cf. Fig. 8.1) we have to sum the voltages of the elements R_0 and L_0 after that to sum the current of $R_0 L_0$ with the current of C_0. In the real cases parallel to $R_0 L_0$ and C_0 is connected an input voltage $-g_m \breve{U}_{in}(t)$, where g_m is the amplification coefficient. Since we have to sum the currents one can replace $g_m \breve{U}_{in}(t)$ by an equivalent current source $\breve{I}_{in}(t)$. We assume that the second end is terminated by the same configuration.

Recall that Λ is the length of the transmission line and $T = \Lambda/(1/\sqrt{LC}) = \Lambda\sqrt{LC}$, where L is per unit-length inductance and C – per unit-length capacitance.

Assume that R_p, L_p and C_p $(p = 0,1)$ are nonlinear elements, that is, $R_p = R_p(i)$, $L_p = L_p(i)$ and $C_p = C_p(u)$ are prescribed nonlinear functions.

Using already introduced denotation we have

$$u_{R_p} = R_p(i),\ u_{\psi_p} = \frac{d\Psi_p}{dt} = \frac{d\tilde{L}_p(i)}{dt} \equiv \frac{d(L_p(i).i)}{dt} = \left[i\frac{dL_p(i)}{di} + L_p(i)\right]\frac{di}{dt}$$

and then

$$u_{R_p L_p} = R_p(i_{R_p L_p}) + \left[i_{R_p L_p} \frac{dL_p(i_{R_p L_p})}{di_{R_p L_p}} + L_p(i_{R_p L_p}) \right] \frac{di_{R_p L_p}}{dt} \equiv \frac{d\widetilde{L}_0(i_{R_p L_p})}{di_{R_p L_p}} \frac{di_{R_p L_p}}{dt} \quad (p = 0,1) \cdot$$

Usually input current $I_{input}(t)$ is connected parallel to RL-elements.

Then Kirchhoff's current-law yields

$$-i(0,t) = I_{input}(t) + i_{R_0 L_0} + i_{C_0}. \tag{8.1.1}$$

Since $u_{R_0 L_0}(t) = u(0,t)$ then $i_{R_0 L_0}$ can be found as a solution of differential equation

$$\left[i_{R_0 L_0} \frac{dL_0(i_{R_0 L_0})}{di_{R_0 L_0}} + L_0(i_{R_0 L_0}) \right] \frac{di_{R_0 L_0}(t)}{dt} = u(0,t) - R_0(i_{R_0 L_0}) \cdot$$

Since $i_{C_0} = \dfrac{dq_{C_0}}{dt} = \dfrac{d(C_0(u).u)}{dt} \equiv \dfrac{d\widetilde{C}_0(u)}{dt}$ from (8.1.1) we obtain

$$-i(0,t) = I_{input}(t) + i_{R_0 L_0}(t) + \frac{d\widetilde{C}_0(u(0,t))}{dt}$$

or

$$-i(0,t) = I_{input}(t) + i_{R_0 L_0}(t) + \frac{du(0,t)}{dt} \left[\frac{dC_0(u(0,t))}{du} u(0,t) + C_0(u(0,t)) \right].$$

For the right end we have

$$\left[i_{R_1 L_1}(t) \frac{dL_1(i_{R_1 L_1}(t))}{di_{R_1 L_1}} + L_1(i_{R_1 L_1}(t)) \right] \frac{di_{R_1 L_1}(t)}{dt} = u(\Lambda,t) - R_1(i_{R_1 L_1}(t)),$$

$$\frac{du(\Lambda,t)}{dt} \left[\frac{dC_1(u(\Lambda,t))}{du} u(\Lambda,t) + C_1(u(\Lambda,t)) \right] = -i(\Lambda,t) - I_{output}(t) - i_{R_1 L_1}(t) \cdot$$

Remark 8.1.1. Let us point out that in the previous chapters the currents $i_{R_p L_p}, (p = 0,1)$ can be expressed in an explicit form. In this case this is not possible and therefore we add the above differential equation. In conclusion we wish to emphasize that boundary conditions yield a system of four equations instead of two ones as in the previous chapters.

Now we are able to formulate the mixed problem for the hyperbolic transmission line equations: to find a solution $(u(x,t), i(x,t))$ of the first order partial differential system of hyperbolic type

$$\frac{\partial u(x,t)}{\partial x} + L\frac{\partial i(x,t)}{\partial t} = 0,$$
$$\frac{\partial i(x,t)}{\partial x} + C\frac{\partial u(x,t)}{\partial t} = 0 \tag{8.1.2}$$

for $(x,t) \in \Pi = \{(x,t) \in R^2 : 0 \le x \le \Lambda,\ t \ge 0\}$, satisfying the initial conditions

$$u(x,0) = u_0(x),\ i(x,0) = i_0(x) \text{ for } x \in [0,\Lambda] \tag{8.1.3}$$

and the boundary conditions for $x = 0$

$$\left[i_{R_0 L_0}(t)\frac{dL_0(i_{R_0 L_0}(t))}{di_{R_0 L_0}} + L_0(i_{R_0 L_0}(t)) \right]\frac{di_{R_0 L_0}(t)}{dt} = u(0,t) - R_0(i_{R_0 L_0}(t)),$$
$$\left[\frac{dC_0(u(0,t))}{du}u(0,t) + C_0(u(0,t)) \right]\frac{du(0,t)}{dt} = -i(0,t) - \breve{I}_{in}(t) - i_{R_0 L_0}(t) \tag{8.1.4}$$

and for $x = \Lambda$:

$$\left[i_{R_1 L_1}(t)\frac{dL_1(i_{R_1 L_1}(t))}{di_{R_1 L_1}} + L_1(i_{R_1 L_1}(t)) \right]\frac{di_{R_1 L_1}(t)}{dt} = u(\Lambda,t) - R_1(i_{R_1 L_1}(t)),$$
$$\left[\frac{dC_1(u(\Lambda,t))}{du}u(\Lambda,t) + C_1(u(\Lambda,t)) \right]\frac{du(\Lambda,t)}{dt} = -i(\Lambda,t) - i_{R_1 L_1}(t). \tag{8.1.5}$$

8.2. REDUCING THE MIXED PROBLEM TO AN INITIAL VALUE PROBLEM ON THE BOUNDARY

Here we briefly recall the transformation from Chapters II, III and IV. First rewrite the system

$$\frac{\partial u(x,t)}{\partial x} + L\frac{\partial i(x,t)}{\partial t} = 0,$$
$$\frac{\partial i(x,t)}{\partial x} + C\frac{\partial u(x,t)}{dt} = 0 \tag{8.2.1}$$

in the form

$$\frac{\partial u(x,t)}{\partial t} + \frac{1}{C}\frac{\partial i(x,t)}{\partial x} = 0,$$
$$\sqrt{\frac{L}{C}}\frac{\partial i(x,t)}{\partial t} + \sqrt{\frac{1}{LC}}\frac{\partial u(x,t)}{\partial x} = 0. \tag{8.2.2}$$

Adding the above equations we get:

$$\frac{\partial u(x,t)}{\partial t} + \sqrt{\frac{L}{C}}\frac{\partial i(x,t)}{\partial t} + \frac{1}{C}\frac{\partial i(x,t)}{\partial x} + \frac{1}{\sqrt{LC}}\frac{\partial u(x,t)}{\partial x} = 0 \Rightarrow$$
$$\Rightarrow \frac{\partial}{\partial t}\left(u(x,t) + \sqrt{\frac{L}{C}}\, i(x,t)\right) + \frac{1}{\sqrt{LC}}\frac{\partial}{\partial x}\left(u(x,t) + \sqrt{\frac{L}{C}}\, i(x,t)\right) = 0.$$

We subtract the equations of (8.2.2):

$$\frac{\partial u(x,t)}{\partial t} - \sqrt{\frac{L}{C}}\frac{\partial i(x,t)}{\partial t} + \frac{1}{C}\frac{\partial i(x,t)}{\partial x} - \frac{1}{\sqrt{LC}}\frac{\partial u(x,t)}{\partial x} = 0 \Rightarrow$$
$$\Rightarrow \frac{\partial}{\partial t}\left(u(x,t) - \sqrt{\frac{L}{C}}\, i(x,t)\right) - \frac{1}{\sqrt{LC}}\frac{\partial}{\partial x}\left(u(x,t) - \sqrt{\frac{L}{C}}\, i(x,t)\right) = 0$$

and in view of the usually accepted denotations $Z_0 = \sqrt{L/C}, \ \ v = 1/\sqrt{LC}$ we get:

$$\frac{\partial}{\partial t}\big(u(x,t) + Z_0\, i(x,t)\big) + v\frac{\partial}{\partial x}\big(u(x,t) + Z_0\, i(x,t)\big) = 0,$$
$$\frac{\partial}{\partial t}\big(u(x,t) - Z_0\, i(x,t)\big) - v\frac{\partial}{\partial x}\big(u(x,t) - Z_0\, i(x,t)\big) = 0. \tag{8.2.3}$$

Let us put

$$U(x,t) = u(x,t) + Z_0\, i(x,t), \qquad I(x,t) = u(x,t) - Z_0\, i(x,t)\,,$$

that is,

$$\begin{bmatrix} U(x,t) \\ I(x,t) \end{bmatrix} = \begin{bmatrix} 1 & Z_0 \\ 1 & -Z_0 \end{bmatrix}\begin{bmatrix} u(x,t) \\ i(x,t) \end{bmatrix} \text{ and } \begin{bmatrix} u(x,t) \\ i(x,t) \end{bmatrix} = \begin{bmatrix} 1/2 & 1/2 \\ 1/(2Z_0) & -1/(2Z_0) \end{bmatrix}\begin{bmatrix} U(x,t) \\ I(x,t) \end{bmatrix}$$

or

$$u(x,t) = \frac{1}{2}U(x,t) + \frac{1}{2}I(x,t),$$

$$i(x,t) = \frac{1}{2Z_0}U(x,t) - \frac{1}{2Z_0}I(x,t).$$

Then (8.2.3) becomes

$$\frac{\partial U(x,t)}{\partial t} + v\frac{\partial U(x,t)}{\partial x} = 0,$$

$$\frac{\partial I(x,t)}{\partial t} - v\frac{\partial I(x,t)}{\partial x} = 0.$$

The characteristics of the system are the families of straight lines

$$x - vt = \text{const}, \quad x + vt = \text{const}$$

and the solution is:

$$U(x,t) = \Phi(x - vt), \quad I(x,t) = \Psi(x + vt),$$

where Φ и Ψ are arbitrarily chosen smooth functions. Then

$$u(x,t) + Z_0 i(x,t) = \Phi(x - vt), \quad u(x,t) - Z_0\, i(x,t) = \Psi(x + vt).$$

Solving the previous system with respect to (u, i) we obtain:

$$u(x,t) = \frac{1}{2}\left[\Phi(x - vt) + \Psi(x + vt)\right], \; i(x,t) = \frac{1}{2Z_0}\left[\Phi(x - vt) - \Psi(x + vt)\right]. \quad (8.2.4)$$

Let us put $x = \Lambda$. Then

$$\left|\begin{aligned}
u(\Lambda,t) &= \frac{1}{2}\left[\Phi(\Lambda - vt) + \Psi(\Lambda + vt)\right], \\[2mm]
i(\Lambda,t) &= \frac{1}{2Z_0}\left[\Phi(\Lambda - vt) - \Psi(\Lambda + vt)\right].
\end{aligned}\right. \qquad (8.2.5)$$

For $x = 0$ we have

$$\left| \begin{array}{l} u(0,t) = \dfrac{1}{2}\big[\Phi(-vt) + \Psi(vt)\big], \\[2mm] i(0,t) = \dfrac{1}{2Z_0}\big[\Phi(-vt) - \Psi(vt)\big]. \end{array} \right. \qquad (8.2.6)$$

It follows

$$\Phi(\Lambda - vt) = u(\Lambda,t) + Z_0 i(\Lambda,t),$$
$$\Psi(\Lambda + vt) = u(\Lambda,t) - Z_0 i(\Lambda,t) \qquad (8.2.7)$$

and

$$\Phi(-vt) = u(0,t) + Z_0 i(0,t),$$
$$\Psi(vt) = u(0,t) - Z_0 i(0,t). \qquad (8.2.8)$$

We have

$$\Lambda - vt = -vt' \Rightarrow \Lambda + vt' = vt \Rightarrow t = \frac{\Lambda + vt'}{v} \quad \text{or} \quad t = t' + T$$

and

$$\Lambda + vt = vt'' \Rightarrow -\Lambda + vt'' = vt \Rightarrow t = \frac{-\Lambda + vt''}{v} \quad \Leftrightarrow t = t'' - T \,.$$

Replacing $\Lambda - vt$ and $\Lambda + vt$ in (8.2.7) we get

$$\Phi(-vt') = u\big(\Lambda, t'+T\big) + Z_0 i\big(\Lambda, t'+T\big), \quad \Psi(vt'') = u\big(\Lambda, t''-T\big) - Z_0 i\big(\Lambda, t''-T\big).$$

or

$$\Phi(-vt) = u\big(\Lambda, t+T\big) + Z_0 i\big(\Lambda, t+T\big),$$
$$\Psi(vt) = u\big(\Lambda, t-T\big) - Z_0 i\big(\Lambda, t-T\big). \qquad (8.2.9).$$

Replacing $\Phi(-vt)$ and $\Psi(vt)$ from (8.2.9) in (8.2.6) and obtain:

$$\left| u(0,t) = \frac{u(\Lambda,t+T)}{2} + \frac{Z_0 i(\Lambda,t+T)}{2} + \frac{u(\Lambda,t-T)}{2} - \frac{Z_0 i(\Lambda,t-T)}{2} \right. ,$$

$$\left| i(0,t) = \frac{u(\Lambda,t+T)}{2Z_0} + \frac{i(\Lambda,t+T)}{2} - \frac{u(\Lambda,t-T)}{2Z_0} + \frac{i(\Lambda,t-T)}{2} \right. .$$

Replacing $u(0,t)$ and $i(0,t)$ in (8.1.4) we have:

$$\left[i_{R_0 L_0}(t) \frac{dL_0(i_{R_0 L_0}(t))}{di_{R_0 L_0}} + L_0(i_{R_0 L_0}(t)) \right] \frac{di_{R_0 L_0}(t)}{dt} =$$

$$= \frac{u(\Lambda,t+T)}{2} + \frac{Z_0 i(\Lambda,t+T)}{2} + \frac{u(\Lambda,t-T)}{2} - \frac{Z_0 i(\Lambda,t-T)}{2} - R_0(i_{R_0 L_0}(t)),$$

$$\left[\frac{dC_0(u(0,t))}{du} u(0,t) + C_0(u(0,t)) \right] \frac{d}{dt} \left(\frac{u(\Lambda,t+T)}{2} + \frac{Z_0 i(\Lambda,t+T)}{2} + \frac{u(\Lambda,t-T)}{2} - \frac{Z_0 i(\Lambda,t-T)}{2} \right) =$$

$$= -\frac{u(\Lambda,t+T)}{2Z_0} - \frac{i(\Lambda,t+T)}{2} + \frac{u(\Lambda,t-T)}{2Z_0} - \frac{i(\Lambda,t-T)}{2} - \breve{I}_{in}(t) - i_{R_0 L_0}(t)$$

$$\left[i_{R_1 L_1}(t) \frac{dL_1(i_{R_1 L_1}(t))}{di_{R_1 L_1}} + L_1(i_{R_1 L_1}(t)) \right] \frac{di_{R_1 L_1}(t)}{dt} = u(\Lambda,t) - R_1(i_{R_1 L_1}(t)),$$

$$\left[\frac{dC_1(u(\Lambda,t))}{du} u(\Lambda,t) + C_1(u(\Lambda,t)) \right] \frac{du(\Lambda,t)}{dt} = -i(\Lambda,t) - i_{R_1 L_1}(t).$$

Let us put $t + T = t$ in the first two equations

$$\left[i_{R_0 L_0}(t-T) \frac{dL_0(i_{R_0 L_0}(t-T))}{di_{R_0 L_0}} + L_0(i_{R_0 L_0}(t-T)) \right] \frac{di_{R_0 L_0}(t-T)}{dt} =$$

$$= \frac{u(\Lambda,t)}{2} + \frac{Z_0 i(\Lambda,t)}{2} + \frac{u(\Lambda,t-2T)}{2} - \frac{Z_0 i(\Lambda,t-2T)}{2} - R_0(i_{R_0 L_0}(t-T)),$$

$$\left[\frac{dC_0(u(0,t-T))}{du} u(0,t-T) + C_0(u(0,t-T)) \right] \frac{d}{dt} \left(\frac{u(\Lambda,t)}{2} + \frac{Z_0 i(\Lambda,t)}{2} + \frac{u(\Lambda,t-2T)}{2} - \frac{Z_0 i(\Lambda,t-2T)}{2} \right) =$$

$$= -\frac{u(\Lambda,t)}{2Z_0} - \frac{i(\Lambda,t)}{2} + \frac{u(\Lambda,t-2T)}{2Z_0} - \frac{i(\Lambda,t-2T)}{2} - \breve{I}_{in}(t-T) - i_{R_0 L_0}(t-T)$$

$$\left[i_{R_1L_1}(t)\frac{dL_1(i_{R_1L_1}(t))}{di_{R_1L_1}} + L_1(i_{R_1L_1}(t)) \right]\frac{di_{R_1L_1}(t)}{dt} = u(\Lambda,t) - R_1(i_{R_1L_1}(t)),$$

$$\left[\frac{dC_1(u(\Lambda,t))}{du}u(\Lambda,t) + C_1(u(\Lambda,t)) \right]\frac{du(\Lambda,t)}{dt} = -i(\Lambda,t) - i_{R_1L_1}(t).$$

With respect to the unknown functions

$i_{R_0L_0}(t),\ u(t) \equiv u(\Lambda,t),\ i(t) \equiv i(\Lambda,t),\ i_{R_1L_1}(t)$ the above system becomes

$$\frac{d\widetilde{L}_0(i_{R_0L_0}(t-T))}{di_{R_0L_0}}\frac{di_{R_0L_0}(t-T)}{dt} =$$

$$= \frac{u(t)}{2} + \frac{Z_0 i(t)}{2} + \frac{u(t-2T)}{2} - \frac{Z_0 i(t-2T)}{2} - R_0(i_{R_0L_0}(t-T)),$$

$$\frac{d}{dt}\big(u(t) + Z_0 i(t) + u(t-2T) - Z_0 i(t-2T)\big) =$$

$$= \frac{1}{\dfrac{d\widetilde{C}_0(u(0,t-T))}{du}}\left(-\frac{u(t)}{Z_0} - i(t) + \frac{u(t-2T)}{Z_0} - i(t-2T) - 2\breve{I}_{in}(t-T) - 2i_{R_0L_0}(t-T) \right)$$

$$\frac{d\widetilde{L}_1(i_{R_1L_1}(t))}{di_{R_1L_1}}\frac{di_{R_1L_1}(t)}{dt} = u(t) - R_1(i_{R_1L_1}(t)),$$

$$\frac{du(t)}{dt} = \frac{1}{\dfrac{d\widetilde{C}_1(u(t))}{du}}\big(-i(t) - i_{R_1L_1}(t)\big).$$

It is not easy to formulate a periodic problem for such a system. That is why we propose a different approach in the following section.

8.3. A Different Approach to Reducing the Mixed Problem to a Periodic Problem on the Boundary

Replace in (8.3.1) and (8.3.2) $u(0,t)$ and $i(0,t)$ from

$$u(x,t) = \frac{1}{2}U(x,t) + \frac{1}{2}I(x,t),$$

$$i(x,t) = \frac{1}{2Z_0}U(x,t) - \frac{1}{2Z_0}I(x,t)$$

for $x = 0$:

$$\left[i_{R_0L_0}(t)\frac{dL_0(i_{R_0L_0}(t))}{di_{R_0L_0}} + L_0(i_{R_0L_0}(t)) \right]\frac{di_{R_0L_0}(t)}{dt} = \frac{1}{2}U(0,t) + \frac{1}{2}I(0,t) - R_0(i_{R_0L_0}(t)),$$

$$\left[\frac{dC_0(u(0,t))}{du}u(0,t) + C_0(u(0,t)) \right]\frac{1}{2}\left(\frac{dU(0,t)}{dt} + \frac{dI(0,t)}{dt} \right) = -i\frac{1}{2Z_0}U(0,t) + \frac{1}{2Z_0}I(0,t) - \breve{I}_{in}(t) - i_{R_0L_0}(t)$$

and for $x = \Lambda$:

$$\left[i_{R_1L_1}(t)\frac{dL_1(i_{R_1L_1}(t))}{di_{R_1L_1}} + L_1(i_{R_1L_1}(t)) \right]\frac{di_{R_1L_1}(t)}{dt} = \frac{1}{2}U(\Lambda,t) + \frac{1}{2}I(\Lambda,t) - R_1(i_{R_1L_1}(t)),$$

$$\left[\frac{dC_1(u(\Lambda,t))}{du}u(\Lambda,t) + C_1(u(\Lambda,t)) \right]\frac{1}{2}\left(\frac{dU(\Lambda,t)}{dt} + \frac{dI(\Lambda,t)}{dt} \right) = -\frac{1}{2Z_0}U(\Lambda,t) + \frac{1}{2Z_0}I(\Lambda,t) - i_{R_1L_1}(t).$$

But $U(0,t) = U(\Lambda,t+T)$, $I(0,t+T) = I(\Lambda,t)$.

We assume that the unknown functions are

$$U(0,t) \equiv U(t),\ I(\Lambda,t) \equiv I(t)$$

and then in view of

$$u(0,t) = \frac{U(0,t) + I(0,t)}{2} = \frac{U(t) + I(t-T)}{2};\ u(\Lambda,t) = \frac{U(\Lambda,t) + I(\Lambda,t)}{2} = \frac{U(t-T) + I(t)}{2}$$

we obtain

$$\frac{d\widetilde{L}_0(i_{R_0L_0}(t))}{di_{R_0L_0}}\frac{di_{R_0L_0}(t)}{dt} = \frac{1}{2}U(t) + \frac{1}{2}I(t-T) - R_0(i_{R_0L_0}(t)),$$

$$\frac{d\widetilde{C}_0\left(\dfrac{U(t) + I(t-T)}{2} \right)}{du}\frac{1}{2}\left(\frac{dU(t)}{dt} + \frac{dI(t-T)}{dt} \right) = -i\frac{1}{2Z_0}U(t) + \frac{1}{2Z_0}I(t-T) - \breve{I}_{in}(t) - i_{R_0L_0}(t),$$

$$\frac{d\widetilde{L}_1(i_{R_1L_1}(t))}{di_{R_1L_1}}\frac{di_{R_1L_1}(t)}{dt} = \frac{1}{2}U(t-T) + \frac{1}{2}I(t) - R_1(i_{R_1L_1}(t)),$$

$$\frac{d\widetilde{C}_1\left(\dfrac{U(t-T) + I(t)}{2} \right)}{du}\frac{1}{2}\left(\frac{dU(t-T)}{dt} + \frac{dI(t)}{dt} \right) = -\frac{1}{2Z_0}U(t-T) + \frac{1}{2Z_0}I(t) - i_{R_1L_1}(t).$$

The last system could be solved with respect to the derivatives:

$$\frac{di_{R_0 L_0}(t)}{dt} = \frac{U(t) + I(t-T) - 2R_0(i_{R_0 L_0}(t))}{2d\widetilde{L}_0(i_{R_0 L_0}(t))/di_{R_0 L_0}},$$

$$\frac{dU(t)}{dt} = -\frac{dI(t-T)}{dt} + \frac{-\dfrac{U(t)}{Z_0} + \dfrac{I(t-T)}{Z_0} - 2\breve{I}_{in}(t) - 2i_{R_0 L_0}(t)}{d\widetilde{C}_0\!\left(\dfrac{U(t)+I(t-T)}{2}\right)/du},$$

$$\frac{di_{R_1 L_1}(t)}{dt} = \frac{U(t-T) + I(t) - 2R_1(i_{R_1 L_1}(t))}{2d\widetilde{L}_1(i_{R_1 L_1}(t))/di_{R_1 L_1}},$$

$$\frac{dI(t)}{dt} = -\frac{dU(t-T)}{dt} + \frac{-U(t-T) + I(t) - 2Z_0 i_{R_1 L_1}(t)}{Z_0 d\widetilde{C}_1\!\left(\dfrac{U(t-T)+I(t)}{2}\right)/du}. \tag{8.3.1.}$$

So we have obtained a neutral system of differential equations with retarded arguments.

8.4. ESTIMATES OF THE ARISING NONLINEARITIES AND INTRODUCING METRICS

Recall denotations $\widetilde{C}_p(u) = C_p(u).u\ ,(p = 0,1)$ where

$$C_p(u) = \frac{c_p}{\sqrt[h]{\left(1 - u/\Phi_p\right)}} = \frac{c_p\sqrt[h]{\Phi_p}}{\sqrt[h]{\Phi_p - u}}\ ,\ c_p > 0, \Phi_p > 0,\ h \in [2,3]$$

are constants and $|u| \le \phi_0 < \min\{\Phi_0, \Phi_1\}$. We have to find an interval where $\widetilde{C}_p(u) = C_p(u).u$ has a strictly positive lower bound. We have

$$\frac{d\widetilde{C}_p(u)}{du} = C_p(u) + u\frac{dC_p(u)}{du} =$$

$$= \frac{c_p\sqrt[h]{\Phi_p}}{\sqrt[h]{\Phi_p - u}} + u\frac{c_p\sqrt[h]{\Phi_p}}{h\sqrt[h]{\left(\Phi_p - u\right)^{h+1}}} = c_p\sqrt[h]{\Phi_p}\,\frac{\Phi_p - \left((h-1)/h\right)u}{\left(\Phi_p - u\right)^{\frac{1}{h}+1}}, u \in [-\phi_0, \phi_0];$$

$$\frac{d^2\widetilde{C}_p(u)}{du^2} = 2\frac{dC_p(u)}{du} + u\frac{d^2C_p(u)}{du^2} = \frac{2c_p\sqrt[h]{\Phi_p}}{h\left(\Phi_p - u\right)^{\frac{1}{h}+1}} - u\frac{\left(1+(1/h)\right)2c_p\sqrt[h]{\Phi_p}}{h\left(\Phi_p - u\right)^{\frac{1}{h}+2}} =$$

$$= \frac{2c_p\sqrt[h]{\Phi_p}}{\left(\Phi_p - u\right)^{\frac{1}{h}+2}}\,\frac{\Phi_p - \left(2+\left(1/h\right)\right)u}{h}.$$

We choose also $\phi_0 < \dfrac{h}{1+2h}\Phi_p \Rightarrow \Phi_p - \left(2+(1/h)\right)|u| > 0 \Leftrightarrow |u| < \dfrac{h}{1+2h}\Phi_p < \Phi_p$.

Therefore the minimal value of $\dfrac{d\widetilde{C}_p(u)}{du}$ is

$$\min\left\{\frac{d\widetilde{C}_p(u)}{du}:|u|\le\phi_0\right\} = \frac{d\widetilde{C}_p(-\phi_0)}{du} = c_p\sqrt[h]{\Phi_p}\,\frac{\Phi_p+((h-1)/h)\phi_0}{\left(\Phi_p+\phi_0\right)^{\frac{1}{h}+1}} = \hat{C}_p > 0, (p = 0,1)$$

$$\left|\frac{d\widetilde{C}_p(u)}{du}\right| \le \frac{2c_p\sqrt[h]{\Phi_p}}{\left(\Phi_p-\phi_0\right)^{\frac{1}{h}+2}}\,\frac{\Phi_p+((2h+1)/h)\phi_0}{h} \equiv M_p;$$

$$\left|\frac{d^2\widetilde{C}_p(u)}{du^2}\right| \le \frac{2c_p\sqrt[h]{\Phi_p}}{h\left(\Phi_p-\phi_0\right)^{\frac{1}{h}+1}} + |u|\frac{(1+h)2c_p\sqrt[h]{\Phi_p}}{h^2\left(\Phi_p-\phi_0\right)^{\frac{1}{h}+2}} = \frac{2c_p\sqrt[h]{\Phi_p}\left(h(\Phi_p-\phi_0)+|u|(1+h)\right)}{h^2\sqrt[h]{\left(\Phi_p-\phi_0\right)^{1+2h}}}.$$

It follows

$$\left|\frac{d^2\widetilde{C}_p(u)}{du^2}\right| \le \frac{2c_0\sqrt[h]{\Phi_p}\left[h(\Phi_p-\phi_0)+|u|(1+h)\right]}{h^2\sqrt[h]{\left(\Phi_p-\phi_0\right)^{1+2h}}} \le \frac{2c_p\sqrt[h]{\Phi_p}\left[h(\Phi_p-\phi_0)+\phi_0(1+h)\right]}{h^2\sqrt[h]{\left(\Phi_p-\phi_0\right)^{1+2h}}} =$$

$$= \frac{2c_p\sqrt[h]{\Phi_p}\left(h\Phi_p+\phi_0\right)}{h^2\sqrt[h]{\left(\Phi_p-\phi_0\right)^{1+2h}}} = H_p \ (p = 0,1);$$

$$\left|\frac{d\widetilde{C}_p}{du}\right| > 0 \Rightarrow \min\left\{\widetilde{C}_p(u):u\in[-\phi_0,\phi_0]\right\} = \widetilde{C}_p(-\phi_0) \ge \frac{2c_p\sqrt[h]{\Phi_p}}{\left(\Phi_p+\phi_0\right)^{\frac{1}{h}+2}}\,\frac{\Phi_p-\left(2+(1/h)\right)\phi_0}{h} = \hat{C}_p > 0.$$

For the *I-V* characteristics we assume $R_0(i) = \sum\limits_{n=1}^{m} r_n^{(0)}i^n, (k = 0,1)$ and $L_p(i) = \sum\limits_{n=1}^{m} l_n^{(p)}i^n$.

Then $\widetilde{L}_p(i) = i \cdot L_p(i) = i.\sum\limits_{n=1}^{m} l_n^{(p)}i^n$. For $\widetilde{L}_p(i)$ we get

$$\frac{d\widetilde{L}_p(i)}{di} = i\frac{dL_p(i)}{di} + L_p(i) = i\sum_{n=1}^{m} nl_n^{(p)}i^{n-1} + \sum_{n=1}^{m} l_n^{(p)}i^n = \sum_{n=1}^{m}(n+1)l_n^{(p)}i^n;$$

$$\frac{d^2\widetilde{L}_p(i)}{di^2} = \sum_{n=1}^{m}(n+1)nl_n^{(p)}i^{n-1}.$$

Assumptions **(L)**: $|i(t)| \le i_0 \Rightarrow \dfrac{d\widetilde{L}_p(i(t))}{di} = \sum_{n=1}^{m}(n+1)l_n^{(p)}\big(i(t)\big)^n \ge \hat{L}_p > 0 \ (p = 0,1).$

(IN): $U_0(.), \ I_0(.) \in C_{T_0}^1[0,T],$

$\left|\widetilde{U}_0(t)\right| \le U_0 e^{\mu(t+T-kT_0)}, \left|\widetilde{I}_0(t)\right| \le U_0 e^{\mu(t+T-kT_0)} (k = 0,1,2,\ldots,m-1).$

Let us put $\bar{U}_0(t) = \widetilde{U}_0(t-T), \ \bar{I}_0(t) = \widetilde{I}_0(t-T), \ t \in [T,2T].$ Then

$\left|\bar{U}_0(t)\right| \le U_0 e^{\mu(t-T-kT_0)}, \left|\bar{I}_0(t)\right| \le U_0 e^{\mu(t-T-kT_0)} (k = 0,1,2,\ldots,m-1), t \in [T+kT_0, T+(k+1)T_0]$

Assumptions **(U)**: $e^{\mu T_0} \dfrac{U_0 + I_0}{2} \le \phi_0.$

It follows

$|u(0,t)| \le \dfrac{|U(t)| + |I(t-T)|}{2} \equiv \dfrac{|U(t)| + |\bar{I}_0(t)|}{2} \le e^{\mu(t-T-kT_0)} \dfrac{U_0 + I_0}{2} \le e^{\mu T_0} \dfrac{U_0 + I_0}{2} \le \phi_0;$

$|u(\Lambda,t)| \le \dfrac{|U(t-T)| + |I(t)|}{2} \equiv \dfrac{|U_0(t)| + |I(t)|}{2} \le e^{\mu(t-T-kT_0)} \dfrac{U_0 + I_0}{2} \le e^{\mu T_0} \dfrac{U_0 + I_0}{2} \le \phi_0.$

For the resistors we use *V-I* characteristics with polynomial type nonlinearities:

$$R_0(i_{R_0 L_0}) = \sum_{n=1}^{m} r_n^{(0)} (i_{R_0 L_0})^n, \ R_1(i_{R_1 L_1}) = \sum_{n=1}^{m} r_n^{(1)} (i_{R_1 L_1})^n.$$

We introduce the sets for the unknown functions $i_{R_0 L_0}(t), \ U(t), \ i_{R_1 L_1}(t), \ I(t)$

$M_0 = \left\{ i_{R_0 L_0}(t) \in C_{T_0}^1[T,2T] : \left|i_{R_0 L_0}(t)\right| \le I_{R_0} e^{\mu(t-T-kT_0)}, t \in [T+kT_0, T+(k+1)T_0] \right\} (k = 0,1,2,\ldots,m-1),$

$M_U = \left\{ u \in C_{T_0}^1[T,2T] : |U(t)| \le U_0 e^{\mu(t-T-kT_0)}, t \in [T+kT_0, T+(k+1)T_0] \right\} (k = 0,1,2,\ldots,m-1),$

$M_1 = \left\{ i_{R_1 L_1}(t) \in C_{T_0}^1[T,2T] : \left|i_{R_1 L_1}(t)\right| \le I_{R_1} e^{\mu(t-T-kT_0)}, t \in [T+kT_0, T+(k+1)T_0] \right\} (k = 0,1,2,\ldots,m-1),$

$M_I = \left\{ u \in C_{T_0}^1[T,2T] : |I(t)| \le I_0 e^{\mu(t-T-kT_0)}, t \in [T+kT_0, T+(k+1)T_0] \right\} (k = 0,1,2,\ldots,m-1),$

where $C_{T_0}^1[T,2T]$ is the set of all continuously differentiable T_0-periodic functions and $I_{R_0}, U_0, I_{R_1}, I_0, T_0, \mu$ are positive constants (chosen below) and $\mu T_0 = \mu_0 = \text{const.}$

Introduce the metrics

$$\rho^{(k)}(i_{R_0 L_0}, \bar{i}_{R_0 L_0}) = \max\left\{\left|i_{R_0 L_0}(t) - \bar{i}_{R_0 L_0}(t)\right| : t \in [T + kT_0, T + (k+1)T_0]\right\},$$

$$\hat{\rho}(i_{R_0 L_0}, \bar{i}_{R_0 L_0}) = \max\left\{\left|i_{R_0 L_0}(t) - \bar{i}_{R_0 L_0}(t)\right| : t \in [T, 2T]\right\},$$

$$\rho_\mu^{(k)}(\dot{i}_{R_0 L_0}, \dot{\bar{i}}_{R_0 L_0}) = \max\left\{e^{-\mu(t-T-kT_0)}\left|\dot{i}_{R_0 L_0}(t) - \dot{\bar{i}}_{R_0 L_0}(t)\right| : t \in [T + kT_0, T + (k+1)T_0]\right\},$$

$$\rho^{(k)}(U, \overline{U}) = \max\left\{\left|U(t) - \overline{U}(t)\right| : t \in [T + kT_0, T + (k+1)T_0]\right\},$$

$$\hat{\rho}(U, \overline{U}) = \max\left\{\left|U(t) - \overline{U}(t)\right| : t \in [T, 2T]\right\},$$

$$\rho_\mu^{(k)}(\dot{U}, \dot{\overline{U}}) = \max\left\{e^{-\mu(t-T-kT_0)}\left|\dot{U}(t) - \dot{\overline{U}}(t)\right| : t \in [T + kT_0, T + (k+1)T_0]\right\},$$

$$\rho^{(k)}(i_{R_1 L_1}, \bar{i}_{R_1 L_1}) = \max\left\{\left|i_{R_1 L_1}(t) - \bar{i}_{R_1 L_1}(t)\right| : t \in [T + kT_0, T + (k+1)T_0]\right\},$$

$$\hat{\rho}(i_{R_1 L_1}, \bar{i}_{R_1 L_1}) = \max\left\{\left|i_{R_1 L_1}(t) - \bar{i}_{R_1 L_1}(t)\right| : t \in [T, 2T]\right\},$$

$$\rho_\mu^{(k)}(\dot{i}_{R_1 L_1}, \dot{\bar{i}}_{R_1 L_1}) = \max\left\{e^{-\mu(t-T-kT_0)}\left|\dot{i}_{R_1 L_1}(t) - \dot{\bar{i}}_{R_1 L_1}(t)\right| : t \in [T + kT_0, T + (k+1)T_0]\right\},$$

$$\rho^{(k)}(I, \bar{I}) = \max\left\{\left|I(t) - \bar{I}(t)\right| : t \in [T + kT_0, T + (k+1)T_0]\right\},$$

$$\hat{\rho}(I, \bar{I}) = \max\left\{\left|I(t) - \bar{I}(t)\right| : t \in [T, 2T]\right\},$$

$$\rho_\mu^{(k)}(\dot{I}, \dot{\bar{I}}) = \max\left\{e^{-\mu(t-T-kT_0)}\left|\dot{I}(t) - \dot{\bar{I}}(t)\right| : t \in [T + kT_0, T + (k+1)T_0]\right\}.$$

The set $M_0 \times M_U \times M_1 \times M_I$ turns out into a complete metric space (cf. Chapter I, § 1.2 or for details [14]) with respect to the metric:

$$\hat{\rho}_\mu\left((i_{R_0 L_0}, U, i_{R_1 L_1}, I), (\bar{i}_{R_0 L_0}, \overline{U}, \bar{i}_{R_1 L_1}, \bar{I})\right) =$$

$$= \max\left\{\hat{\rho}(i_{R_0L_0},\bar{i}_{R_0L_0}),\rho_\mu^{(k)}(i_{R_0L_0},\dot{\bar{i}}_{R_0L_0}),\hat{\rho}(U,\bar{U}),\rho_\mu^{(k)}(\dot{U},\dot{\bar{U}}),\hat{\rho}(i_{R_1L_1},\bar{i}_{R_1L_1}),\rho_\mu^{(k)}(\dot{i}_{R_1L_1},\dot{\bar{i}}_{R_1L_1}),\hat{\rho}(I,\bar{I}),\rho_\mu^{(k)}(\dot{I},\dot{\bar{I}}):\right.$$

$$k = 0,1,2,\ldots,m-1\}.$$

8.5. Operator Presentation of the Periodic Problem

Now we formulate the main problem: to find a T_0-periodic solution $\left(i_{R_0L_0}(t),U(t),i_{R_1L_1}(t),I(t)\right)$ of the system (8.3.1) on the interval $[T,2T]$ coinciding with prescribed T_0-periodic initial functions $i_{R_0L_0}^{(0)}(t),\widetilde{U}_0(t),i_{R_1L_1}^{(0)}(t),\widetilde{I}_0(t)$ on the interval respectively:

$$U(t)=\widetilde{U}_0(t),\ \frac{dU(t)}{dt}=\frac{d\widetilde{U}_0(t)}{dt},t\in[0,T]\ ;\ I(t)=\widetilde{I}_0(t),\ \frac{dI(t)}{dt}=\frac{d\widetilde{I}_0(t)}{dt},t\in[0,T]$$

$$i_{R_1L_1}(T)=0,\ i_{R_0L_0}(T)=0,\ \widetilde{U}_0(T)=0,\ \widetilde{I}_0(T)=0.$$

Remark 8.5.1. As in Chapter II one can shift the initial function of the mixed problem from the interval $[0,\Lambda]$ along the characteristic to the interval $[0,T]$.

Then the obtained function might be continued periodically on $[0,\,T]$. Therefore the initial functions are correctly defined.

The main difficulty is to define a suitable operator whose fixed points are solutions sought. We define it in the following way: the quadruplet functions

$$B = (B_0(t),B_U(t),B_1(t),B_I(t))$$

are defined on every interval $[T+kT_0,T+(k+1)T_0]$ (for every k = 0, 1, 2, …) by the expressions

$$B_0^{(k)}(i_{R_0L_0},U,i_{R_1L_1},I)(t):=\int_{T+kT_0}^{t}I_{R_0}(i_{R_0L_0},U,i_{R_1L_1},I)(s)ds-\frac{t-T-kT_0}{T_0}\int_{T+kT_0}^{T+(k+1)T_0}I_{R_0}(i_{R_0L_0},U,i_{R_1L_1},I)(s)ds,$$

$$B_U^{(k)}(i_{R_0L_0},U,i_{R_1L_1},I)(t):=\int_{T+kT_0}^{t}V(i_{R_0L_0},U,i_{R_1L_1},I)(s)ds-\frac{t-T-kT_0}{T_0}\int_{T+kT_0}^{T+(k+1)T_0}V(i_{R_0L_0},U,i_{R_1L_1},I)(s)ds,$$

$$B_1^{(k)}(i_{R_0L_0},U,i_{R_1L_1},I)(t):=\int_{T+kT_0}^{t}I_{R_1}(i_{R_0L_0},U,i_{R_1L_1},I)(s)ds-\frac{t-T-kT_0}{T_0}\int_{T+kT_0}^{T+(k+1)T_0}I_{R_1}(i_{R_0L_0},U,i_{R_1L_1},I)(s)ds$$

$$B_I^{(k)}(i_{R_0L_0},U,i_{R_1L_1},I)(t):=\int_{T+kT_0}^{t}J(i_{R_0L_0},U,i_{R_1L_1},I)(s)ds-\frac{t-T-kT_0}{T_0}\int_{T+kT_0}^{T+(k+1)T_0}J(i_{R_0L_0},U,i_{R_1L_1},I)(s)ds,$$

where

$$I_{R_0}(i_{R_0L_0},U,I)(t) = \frac{U(t) + \vec{I}_0(t) - 2R_0(i_{R_0L_0}(t))}{2\left[i_{R_0L_0}(t)\dfrac{dL_0(i_{R_0L_0}(t))}{di_{R_0L_0}} + L_0(i_{R_0L_0}(t)) \right]} ;$$

$$V(i_{R_0L_0},U,I)(t) = -\frac{d\vec{I}_0(t)}{dt} + \frac{1}{Z_0}\frac{-U(t) + \vec{I}_0(t) - 2Z_0\breve{I}_{in}(t) - 2Z_0i_{R_0L_0}(t)}{d\widetilde{C}_0\left(\dfrac{U(t)+\vec{I}_0(t)}{2}\right)/du} ;$$

$$I_{R_1}(U,i_{R_1L_1},I)(t) = \frac{\vec{U}_0(t) + I(t) - 2R_1(i_{R_1L_1}(t))}{2\left[i_{R_1L_1}(t)\dfrac{dL_1(i_{R_1L_1}(t))}{di_{R_1L_1}} + L_1(i_{R_1L_1}(t)) \right]} ;$$

$$J(U,i_{R_1L_1},I) = -\frac{d\vec{U}_0(t)}{dt} + \frac{1}{Z_0}\frac{-\vec{U}_0(t) + I(t) - 2Z_0i_{R_1L_1}(t)}{d\widetilde{C}_1\left(\dfrac{\vec{U}_0(t)+I(t)}{2}\right)/du}$$

and $\vec{U}_0(t),\vec{I}_0(t)$ are translations to the right of the initial function $U_0(t),I_0(t)$ over $[T,2T]$.

Lemma 8.5.1. If
$\breve{I}_{in}(.)\in C^1_{T_0}[0,\infty); \left|\breve{I}_{in}(t)\right| \le e^{\mu(t-T-kT_0)}I_{R_0}, t\in[T+kT_0,T+(k+1)T_0]$ and **(IN)** are
satisfied and $(i_{R_0L_0},U,i_{R_1L_1},I)\in M_0\times M_U\times M_1\times M_I$ then

$$B_0(i_{R_0L_0},U,i_{R_1L_1},I)(t),\ \ B_U(i_{R_0L_0},U,i_{R_1L_1},I)(t),\ \ B_1(i_{R_0L_0},U,i_{R_1L_1},I)(t),\ \ B_I(i_{R_0L_0},U,i_{R_1L_1},I)(t)$$

are T_0-periodic ones.

Proof: It is easy to verify that $I_{R_0}(t)$, $V(t)$, $I_{R_1}(t)$, $J(T)$ are T_0-periodic functions. Since $I_{R_0}(t)$ is T_0-periodic function we have

$$\int_t^{t+T_0} I_{R_0}(s))ds = \int_{T+kT_0}^{T+kT_0+T_0} I_{R_0}(s))ds .$$

Then for $t+T_0\in[T+(k+1)T_0,T+(k+2)T_0]$ implies
$t\in[T+(k+1)T_0,T+(k+1)T_0]$ and it is easy to verify that

$$B_0^{(k)}(i_{R_0L_0},U,i_{R_1L_1})(t-T_0) = B_0^{(k+1)}(i_{R_0L_0},U,i_{R_1L_1})(t)$$

and analogously for the other components. Lemma 8.5.1 is thus proved.

Lemma 8.5.2. If $(i_{R_0L_0},U,i_{R_1L_1},I) \in M_0 \times M_U \times M_1 \times M_I$ then

$$\left(B_0(i_{R_0L_0},U,i_{R_1L_1},I)(t), B_U(i_{R_0L_0},U,i_{R_1L_1},I)(t), B_1(i_{R_0L_0},U,i_{R_1L_1},I)(t), B_I(i_{R_0L_0},U,i_{R_1L_1},I)(t)\right) \in \left(C_{T_0}^1[T,2T]\right)^4.$$

Proof: We first prove that

$$B_0(i_{R_0L_0},U,i_{R_1L_1},I)(t):= \int_{T+kT_0}^{t} I_{R_0}(s)ds - \frac{t-T-kT_0}{T_0}\int_{T+kT_0}^{T+(k+1)T_0} I_{R_0}(s)ds, \quad t \in [T+kT_0, T+(k+1)T_0]\,,$$

$$B_U(i_{R_0L_0},U,i_{R_1L_1},I)(t):= \int_{T+kT_0}^{t} V(s)ds - \frac{t-T-kT_0}{T_0}\int_{T+kT_0}^{T+(k+1)T_0} V(s)ds, \quad t \in [T+kT_0, T+(k+1)T_0]\,,$$

$$B_1(i_{R_0L_0},U,i_{R_1L_1},I)(t):= \int_{T+kT_0}^{t} I_{R_1}(s)ds - \frac{t-T-kT_0}{T_0}\int_{T+kT_0}^{T+(k+1)T_0} I_{R_1}(s)ds, \quad t \in [T+kT_0, T+(k+1)T_0],$$

$$B_I(i_{R_0L_0},U,i_{R_1L_1},I)(t):= \int_{T+kT_0}^{t} J(s)ds - \frac{t-T-kT_0}{T_0}\int_{T+kT_0}^{T+(k+1)T_0} J(s)ds, \quad t \in [T+kT_0, T+(k+1)T_0]$$

($k = 0,1,2,...,m-1$) are continuously differentiable function.

Indeed, the continuity follows from

$$B_0^{(k)}(i_{R_0L_0},U,i_{R_1L_1},I)(T+(k+1)T_0) = \int_{T+kT_0}^{T+(k+1)T_0} I_{R_0}(s) - \frac{T+(k+1)T_0-T-kT_0}{T_0}\int_{T+kT_0}^{T+(k+1)T_0} I_{R_0}(s)ds = 0,$$

$$B_0^{(k+1)}(i_{R_0L_0},U,i_{R_1L_1},I)(T+(k+1)T_0):= \int_{T+(k+1)T_0}^{T+(k+1)T_0} I_{R_0}(s)ds - \frac{T+(k+1)T_0-T-(k+1)T_0}{T_0}\int_{T+(k+1)T_0}^{T+(k+2)T_0} I_{R_0}(s)ds = 0,$$

$$B_U^{(0)}(i_{R_0L_0},U,i_{R_1L_1},I)(T) = \int_{T}^{T} V(i_{R_0L_0},U,i_{R_1L_1})(s)ds - \frac{T-T}{T_0}\int_{T}^{T+T_0} V(i_{R_0L_0},U,i_{R_1L_1})(s)ds = 0,$$

$$B_U^{(0)}(i_{R_0L_0},U,i_{R_1L_1},I)(T) = U_0(T) = 0,$$

$$B_U^{(k)}(i_{R_0L_0},U,i_{R_1L_1},I)(T+(k+1)T_0) = \int_{T+kT_0}^{T+(k+1)T_0} V(i_{R_0L_0},U,i_{R_1L_1},I)ds - \frac{T+(k+1)T_0-T-kT_0}{T_0}\int_{T+kT_0}^{T+(k+1)T_0} V(i_{R_0L_0},U,i_{R_1L_1},I)(s)ds = 0,$$

$$B_U^{(k+1)}(i_{R_0L_0},U,i_{R_1L_1})(T+(k+1)T_0) = \int_{T+(k+1)T_0}^{T+(k+1)T_0} V(i_{R_0L_0},U,i_{R_1L_1})(s)ds - \frac{T+(k+1)T_0-T-(k+1)T_0}{T_0}\int_{T+(k+1)T_0}^{T+(k+2)T_0} V(i_{R_0L_0},U,i_{R_1L_1})(s)ds = 0,$$

$$B_1^{(k)}(i_{R_0L_0},U,i_{R_1L_1})(T+(k+1)T_0) = \int\limits_{T+kT_0}^{T+(k+1)T_0} I_{R_1}(s)ds - \frac{T+(k+1)T_0-T-kT_0}{T_0}\int\limits_{T+kT_0}^{T+(k+1)T_0} I_{R_1}(s)ds = 0,$$

$$B_1^{(k+1)}(i_{R_0L_0},u,i_{R_1L_1})(T+(k+1)T_0) = \int\limits_{T+(k+1)T_0}^{T+(k+1)T_0} I_{R_1}(s)ds - \frac{T+(k+1)T_0-T-(k+1)T_0}{T_0}\int\limits_{T+(k+1)T_0}^{T+(k+2)T_0} I_{R_1}(s)ds = 0,$$

$$B_I^{(0)}(i_{R_0L_0},U,i_{R_1L_1},I)(T) = \int\limits_{T}^{T} J(i_{R_0L_0},U,i_{R_1L_1})(s)ds - \frac{T-T}{T_0}\int\limits_{T}^{T+T_0} J(i_{R_0L_0},U,i_{R_1L_1})(s)ds = 0,$$

$$B_I^{(0)}(i_{R_0L_0},U,i_{R_1L_1},I)(T) = I_0(T) = 0,$$

$$B_I^{(k)}(i_{R_0L_0},U,i_{R_1L_1},I)(T+(k+1)T_0) = \int\limits_{T+kT_0}^{T+(k+1)T_0} J(i_{R_0L_0},U,i_{R_1L_1},I)ds - \frac{T+(k+1)T_0-T-kT_0}{T_0}\int\limits_{T+kT_0}^{T+(k+1)T_0} J(i_{R_0L_0},U,i_{R_1L_1},I)(s)ds = 0,$$

$$B_I^{(k+1)}(i_{R_0L_0},U,i_{R_1L_1})(T+(k+1)T_0) = \int\limits_{T+(k+1)T_0}^{T+(k+1)T_0} J(i_{R_0L_0},U,i_{R_1L_1})(s)ds - \frac{T+(k+1)T_0-T-(k+1)T_0}{T_0}\int\limits_{T+(k+1)T_0}^{T+(k+2)T_0} J(i_{R_0L_0},U,i_{R_1L_1})(s)ds = 0.$$

The differentiability follows from

$$\frac{dB_0^{(k)}(i_{R_0L_0},U)(T+(k+1)T_0)}{dt} = I_{R_0}(i_{R_0L_0},U,I)(T+(k+1)T_0) - \frac{1}{T_0}\int\limits_{T+kT_0}^{T+(k+1)T_0} I_{R_0}(i_{R_0L_0},U,I)(s)ds =$$

$$= I_{R_0L_0}(i_{R_0L_0},U,I)(T+(k+1)T_0) - \frac{1}{T_0}\int\limits_{T+(k+1)T_0}^{T+(k+2)T_0} I_{R_0L_0}(i_{R_0L_0},U,I)(s)ds = \frac{dB_0^{(k+1)}(i_{R_0L_0},U,I)(T+(k+1)T_0)}{dt}.$$

Lemma 8.5.2 is thus proved.

Remark 8.5.2. We recall the inequality

$$\frac{e^{n\mu T_0}-1}{n} = \frac{(e^{\mu T_0}-1)(e^{(n-1)\mu T_0}+e^{(n-2)\mu T_0}+...+1)}{n} \le \frac{(e^{\mu T_0}-1)ne^{(n-1)\mu T_0}}{n} = (e^{\mu T_0}-1)e^{(n-1)\mu T_0}.$$

Lemma 8.5.3. The periodic problem (8.3.1) has a solution $(i_{R_0L_0},U,i_{R_1L_1},I) \in M_0 \times M_U \times M_1 \times M_I$ iff the operator B has a fixed point $(i_{R_0L_0},U,i_{R_1L_1},I) \in M_0 \times M_U \times M_1 \times M_I$, that is,

$$i_{R_0L_0} = B_0(i_{R_0L_0},U,i_{R_1L_1},I), \quad U = B_U(i_{R_0L_0},U,i_{R_1L_1},I),$$

$$i_{R_1L_1} = B_1(i_{R_0L_0},U,i_{R_1L_1},I), \quad I = B_I(i_{R_0L_0},U,i_{R_1L_1},I).$$

Proof: Let $(i_{R_0L_0},U,i_{R_1L_1},I) \in M_0 \times M_U \times M_1 \times M_I$ be a T_0-periodic solution of (8.3.1). Then after integration of the first equation we have (recall $i_{R_0L_0}(T) = 0$):

$$i_{R_0L_0}(t) = \int\limits_{T+kT_0}^{t} I_{R_0L_0}(t)dt \Rightarrow i_{R_0L_0}(T+(k+1)T_0) = \int\limits_{T+kT_0}^{T+(k+1)T_0} I_{R_0}(t)dt \Rightarrow \int\limits_{T+kT_0}^{T+(k+1)T_0} I_{R_0}(t)dt = 0.$$

Therefore $B_0^{(k)}(i_{R_0L_0}, U, i_{R_1L_1}, I)(t) = \int\limits_{T+kT_0}^{t} I_{R_0L_0}(s)ds$.

Analogously, we obtain

$$B_u^{(k)}(i_{R_0L_0}, U, i_{R_1L_1}, I)(t) = \int\limits_{T+kT_0}^{t} U(s)ds, \quad B_1^{(k)}(i_{R_0L_0}, U, i_{R_1L_1}, I)(t) = \int\limits_{T+kT_0}^{t} I_{R_1}(s)ds$$

that is, $(i_{R_0L_0}, U, i_{R_1L_1}, I) \in M_0 \times M_U \times M_1 \times M_I$ is a fixed point of B.

Conversely, let B have a fixed point $(i_{R_0L_0}, U, i_{R_1L_1}, I) \in M_0 \times M_U \times M_1 \times M_I$
Then

$$\left| \int\limits_{T+kT_0}^{T+(k+1)T_0} I_{R_0}(i_{R_0L_0}, U, I)(s)ds \right| \leq \frac{1}{2} \int\limits_{T+kT_0}^{T+(k+1)T_0} \left| \frac{U(t) + \vec{I}_0(t) - 2R_0(i_{R_0L_0}(t))}{i_{R_0L_0}(t)\dfrac{dL_0(i_{R_0L_0}(t))}{di_{R_0L_0}} + L_0(i_{R_0L_0}(t))} \right| dt \leq$$

$$\leq \frac{1}{2\hat{L}_0} \int\limits_{T+kT_0}^{T+(k+1)T_0} \left(U_0 e^{\mu(t-T-kT_0)} + I_0 e^{\mu(t-T-kT_0)} + 2\sum_{n=1}^{m} \left| r_n^{(0)} \right| \left| i_{R_0L_0}(t) \right|^n \right) dt \leq$$

$$\leq \frac{U_0}{\hat{L}_0} \frac{e^{\mu T_0}-1}{\mu} + \frac{I_0}{\hat{L}_0} \frac{e^{\mu T_0}-1}{\mu} + \frac{2}{\hat{L}_0} \sum_{n=1}^{m} \left| r_n^{(0)} \right| \left| I_{R_0}^n \right| \int\limits_{T+kT_0}^{T+(k+1)T_0} e^{n\mu(t-T-kT_0)} dt \leq$$

$$\leq \frac{U_0 + I_0}{\hat{L}_0} \frac{e^{\mu T_0}-1}{\mu} + \frac{2}{\hat{L}_0} \sum_{n=1}^{m} \left| r_n^{(0)} \right| \left| I_{R_0}^n \right| \frac{e^{n\mu T_0}-1}{n\mu} \leq \frac{e^{\mu T_0}-1}{\mu} \frac{U_0 + I_0}{\hat{L}_0} + \frac{e^{\mu T_0}-1}{\mu} \frac{2}{\hat{L}_0} \sum_{n=1}^{m} \left| r_n^{(0)} \right| \left| I_{R_0}^n e^{(n-1)\mu T_0} \right| \leq$$

$$\leq \frac{e^{\mu_0}-1}{\mu \hat{L}_0} \left(U_0 + I_0 + 2\sum_{n=1}^{m} \left| r_n^{(0)} \right| \left| I_{R_0}^n e^{(n-1)\mu_0} \right| \right) \equiv M_{I_{R_0}}(\mu);$$

$$\left| \int\limits_{T+kT_0}^{T+(k+1)T_0} V(i_{R_0L_0}, U, I)(s)ds \right| \leq$$

$$\leq \left| \int\limits_{T+kT_0}^{T+(k+1)T_0} \left(-\frac{d\vec{I}_0(t)}{dt} + \frac{1}{Z_0} \frac{-U(t) + \vec{I}_0(t) - 2Z_0\breve{I}_{in}(t) - 2Z_0 i_{R_0L_0}(t)}{d\tilde{C}_0(u(0,t))/du} \right) dt \right| \leq$$

$$\leq \left| \int_{T+kT_0}^{T+(k+1)T_0} \frac{d\vec{I}_0(t)}{dt}\,dt \right| + \frac{1}{Z_0} \int_{T+kT_0}^{T+(k+1)T_0} \left| \frac{-U(t) + \vec{I}_0(t) - 2Z_0 \breve{I}_{in}(t) - 2Z_0 i_{R_0 L_0}(t)}{d\widetilde{C}_0(u(0,t))/du} \right| dt \leq$$

$$\leq \frac{1}{Z_0 \hat{C}_0} \int_{T+kT_0}^{T+(k+1)T_0} \left(U_0 e^{\mu(t-T-kT_0)} + I_0 e^{\mu(t-T-kT_0)} + 2Z_0 I_{R_0} e^{\mu(t-T-kT_0)} + 2Z_0 I_{R_0} e^{\mu(t-T-kT_0)} \right) dt \leq$$

$$\leq \frac{1}{Z_0 \hat{C}_0} \left(U_0 + I_0 + 4Z_0 I_{R_0} \right) \int_{T+kT_0}^{T+(k+1)T_0} e^{\mu(t-T-kT_0)}\,dt \leq \frac{e^{\mu_0} - 1}{\mu} \frac{U_0 + I_0 + 4Z_0 I_{R_0}}{Z_0 \hat{C}_0} \equiv M_V(\mu);$$

$$\left| \int_{T+kT_0}^{T+(k+1)T_0} I_{R_1}(U, i_{R_1 L_1}, I)(s)\,ds \right| \leq \int_{T+kT_0}^{T+(k+1)T_0} \left| \frac{\vec{U}_0(t) + I(t) - 2R_1(i_{R_1 L_1}(t))}{2d\widetilde{L}_1(i_{R_1 L_1}(t))/di_{R_1 L_1}} \right| dt \leq$$

$$\leq \frac{U_0}{2\hat{L}_1} \int_{T+kT_0}^{T+(k+1)T_0} e^{\mu(t-T-kT_0)}\,dt + \frac{I_0}{2\hat{L}_1} \int_{T+kT_0}^{T+(k+1)T_0} e^{\mu(t-T-kT_0)}\,dt + \frac{1}{\hat{L}_1} \int_{T+kT_0}^{T+(k+1)T_0} \left| R_1(i_{R_1 L_1}(t)) \right| dt \leq$$

$$\leq \frac{U_0 + I_0}{\hat{L}_1} \frac{e^{\mu_0} - 1}{\mu} + \frac{1}{\hat{L}_1} \sum_{n=1}^{m} \left| r_n^{(1)} \right| \int_{T+kT_0}^{T+(k+1)T_0} \left| i_{R_1 L_1}(t) \right|^n dt \leq$$

$$\leq \frac{U_0 + I_0}{\hat{L}_1} \frac{e^{\mu_0} - 1}{\mu} + \frac{1}{\hat{L}_1} \sum_{n=1}^{m} \left| r_n^{(1)} \right| I_{R_1}^n \int_{T+kT_0}^{T+(k+1)T_0} e^{n\mu(t-T-kT_0)}\,dt \leq \frac{U_0 + I_0}{\hat{L}_1} \frac{e^{\mu T_0} - 1}{\mu} + \frac{1}{\hat{L}_1} \sum_{n=1}^{m} \left| r_n^{(1)} \right| I_{R_1}^n \frac{e^{n\mu T_0} - 1}{n\mu} \leq$$

$$\leq \frac{e^{\mu_0} - 1}{\mu \hat{L}_1} \left(U_0 + I_0 + \sum_{n=1}^{m} \left| r_n^{(1)} \right| I_{R_1}^n e^{(n-1)\mu_0} \right) \equiv M_{I_{R_1}}(\mu);$$

$$\left| \int_{T+kT_0}^{T+(k+1)T_0} J(U, i_{R_1 L_1}, I)(s)\,ds \right| \leq \int_{T+kT_0}^{T+(k+1)T_0} \left| -\frac{d\vec{U}_0(t)}{dt} + \frac{1}{Z_0} \frac{-\vec{U}_0(t) + I(t) - 2Z_0 i_{R_1 L_1}(t)}{d\widetilde{C}_1(u(\Lambda,t))/du} \right| dt \leq$$

$$\leq \left| \int_{T+kT_0}^{T+(k+1)T_0} \frac{d\vec{U}_0(t)}{dt}\,dt \right| + \frac{1}{Z_0} \int_{T+kT_0}^{T+(k+1)T_0} \left| \frac{-\vec{U}_0(t) + I(t) - 2Z_0 i_{R_1 L_1}(t)}{d\widetilde{C}_1(u(\Lambda,t))/du} \right| dt \leq$$

$$\leq \frac{1}{Z_0 \hat{C}_1} \int_{T+kT_0}^{T+(k+1)T_0} \left(\left| \vec{U}_0(t) \right| + \left| I(t) \right| + 2Z_0 \left| i_{R_1 L_1}(t) \right| \right) dt \leq$$

$$\leq \frac{1}{Z_0 \hat{C}_1} \left(U_0 + I_0 + 2Z_0 I_{R_1} \right) \int_{T+kT_0}^{T+(k+1)T_0} e^{\mu(t-T-kT_0)}\,dt \leq \frac{U_0 + I_0 + 2Z_0 I_{R_1}}{Z_0 \hat{C}_1} \frac{e^{\mu_0} - 1}{\mu}.$$

If we assume $\left| \int_{T+kT_0}^{T+(k+1)T_0} I_{R_0}(s)ds \right| \neq 0$ in view of $\lim\limits_{\mu \to \infty} M_{I_{R_0}}(\mu) = 0$ one obtains a

contradiction. It follows $\int_{T+kT_0}^{T+(k+1)T_0} I_{R_0}(s)ds = 0$.

Analogously $\int_{T+kT_0}^{T+(k+1)T_0} V(t)dt = 0, \quad \int_{T+kT_0}^{T+(k+1)T_0} I_{R_1}(s)ds = 0, \quad \int_{T+kT_0}^{T+(k+1)T_0} J(t)dt = 0$.

Consequently

$$i_{R_0L_0}(t) = \int_{T+kT_0}^{t} I_{R_0}(i_{R_0L_0},U,i_{R_1L_1},I)(s)ds, \quad U(t) = \int_{T+kT_0}^{t} V(i_{R_0L_0},U,i_{R_1L_1},I)(s)ds,$$

$$i_{R_1L_1}(t) = \int_{T+kT_0}^{t} I_{R_1}(i_{R_0L_0},U,i_{R_1L_1},I)(s)ds, \quad I(t) = \int_{T+kT_0}^{t} J(i_{R_0L_0},U,i_{R_1L_1},I)(s)ds.$$

Differentiating the last equalities we conclude that (8.3.1) has T_0-periodic solution.

Lemma 8.5.3 is thus proved.

Theorem 8.5.1. Let the following conditions be fulfilled:

(U) $e^{\mu T_0}\dfrac{U_0 + I_0}{2} \leq \phi_0;$

(L) $|i(t)| \leq i_0 \Rightarrow \overline{L}_p(i)(t) = \sum\limits_{n=1}^{m}(n+1)l_n^{(p)}(i(t))^n \geq \hat{L}_p > 0 \quad (p=0,1);$

(E) $E_p(.) \in C_{T_0}^1[0,\infty), p=0,1; |E_p(t)| \leq U_0 e^{\mu(t-T-kT_0)}(k=0,1,2,...), t \in [T+kT_0, T+(k+1)T_0];$

$\breve{I}_{in}(.) \in C_{T_0}^1[0,\infty); |\breve{I}_{in}(t)| \leq e^{\mu(t-T-kT_0)}I_{R_0};$

(IN) $\widetilde{U}_0(.), \widetilde{I}_0(t) \in C_{T_0}^1[0,T],$

$|\widetilde{U}_0(t)| \leq e^{-\beta}U_0 e^{\mu(t+T-kT_0)}; |\widetilde{I}_0(t)| \leq e^{-\beta}I_0 e^{\mu(t+T-kT_0)}, t \in [-T+kT_0, -T+(k+1)T_0],$

$(k=0,1,2,...,m-1), \beta = \text{const.} > 0.$

Then there exists a unique T_0-periodic solution of (8.3.1).

Proof: We show B maps $M_0 \times M_u \times M_1$ into itself. Indeed,

$$\left| B_0^{(k)}(i_{R_0L_0},U,i_{R_1L_1},I)(t) \right| \leq \int_{T+kT_0}^{t}\left| I_{R_0L_0}(s) \right|ds + \left| \int_{T+kT_0}^{T+(k+1)T_0} I_{R_0L_0}(s)ds \right| \equiv J_1 + J_2.$$

We have

$$J_1 \leq \int\limits_{T+kT_0}^{t} \left| \frac{U(t) + \vec{I}_0(t) - 2R_0(i_{R_0L_0}(t))}{2dL_0(i_{R_0L_0}(t))/di_{R_0L_0}} \right| ds \leq \frac{1}{2\hat{L}_0} \int\limits_{T+kT_0}^{t} \left(|U(s)| + |\vec{I}_0(t)| + 2|R_0(i_{R_0L_0}(s))| \right) ds \leq$$

$$\leq \frac{1}{2\hat{L}_0} \int\limits_{T+kT_0}^{t} U_0 e^{\mu(s-T-kT_0)} ds + \frac{1}{2\hat{L}_0} \int\limits_{T+kT_0}^{t} I_0 e^{\mu(s-T-kT_0)} e^{-\beta} ds + \frac{1}{\hat{L}_0} \sum_{n=1}^{m} |r_n^{(0)}| \int\limits_{T+kT_0}^{t} |i_{R_0L_0}(s)|^n ds \leq$$

$$\leq \frac{U_0}{2\hat{L}_0} \frac{e^{\mu(t-T-kT_0)} - 1}{\mu} + \frac{I_0 e^{-\beta}}{2\hat{L}_0} \frac{e^{\mu(t-T-kT_0)} - 1}{\mu} + \frac{1}{\hat{L}_0} \sum_{n=1}^{m} |r_n^{(0)}| I_{R_0}^n \int\limits_{T+kT_0}^{t} e^{n\mu(s-T-kT_0)} ds$$

$$\leq \frac{e^{-\beta} U_0}{2\hat{L}_0} \frac{e^{\mu(t-T-kT_0)} - 1}{\mu} + \frac{I_0 e^{-\beta}}{2\hat{L}_0} \frac{e^{\mu(t-T-kT_0)} - 1}{\mu} + \frac{1}{\hat{L}_0} \sum_{n=1}^{m} |r_n^{(0)}| I_{R_0}^n \frac{e^{n\mu(t-T-kT_0)} - 1}{n\mu} \leq$$

$$\leq \frac{U_0}{2\hat{L}_0} \frac{e^{\mu(t-T-kT_0)} - 1}{\mu} + \frac{I_0 e^{-\beta}}{2\hat{L}_0} \frac{e^{\mu(t-T-kT_0)} - 1}{\mu} + \frac{1}{\hat{L}_0} \frac{e^{\mu(t-T-kT_0)} - 1}{\mu} \sum_{n=1}^{m} |r_n^{(0)}| I_{R_0}^n e^{(n-1)\mu T_0} \leq$$

$$\leq \frac{e^{\mu(t-T-kT_0)} - 1}{\mu\hat{L}_0} \left(\frac{U_0 + e^{-\beta} I_0}{2} + \sum_{n=1}^{m} |r_n^{(0)}| I_{R_0}^n e^{(n-1)\mu_0} \right) \leq e^{\mu(t-T-kT_0)} \frac{1}{\mu\hat{L}_0} \left(\frac{U_0 + e^{-\beta} I_0}{2} + \sum_{n=1}^{m} |r_n^{(0)}| I_{R_0}^n e^{(n-1)\mu_0} \right)$$

and

$$J_2 \leq \frac{1}{2\hat{L}_0} \int\limits_{T+kT_0}^{T+(k+1)T_0} \left(|U(s)| + |\vec{I}_0(t)| + 2|R_0(i_{R_0L_0}(s))| \right) ds \leq$$

$$\leq \frac{1}{2\hat{L}_0} \int\limits_{T+kT_0}^{T+(k+1)T_0} U_0 e^{\mu(s-T-kT_0)} ds + \frac{1}{2\hat{L}_0} \int\limits_{T+kT_0}^{T+(k+1)T_0} e^{-\beta} I_0 e^{\mu(s-T-kT_0)} ds + \frac{1}{\hat{L}_0} \sum_{n=1}^{m} |r_n^{(0)}| \int\limits_{T+kT_0}^{T+(k+1)T_0} |i_{R_0L_0}(s)|^n ds \leq$$

$$\leq \frac{U_0}{2\hat{L}_0} \frac{e^{\mu T_0} - 1}{\mu} + \frac{e^{-\beta} I_0}{2\hat{L}_0} \frac{e^{\mu T_0} - 1}{\mu} + \frac{1}{\hat{L}_0} \sum_{n=1}^{m} |r_n^{(0)}| I_{R_0}^n \int\limits_{T+kT_0}^{T+(k+1)T_0} e^{n\mu(s-T-kT_0)} ds$$

$$\leq \frac{U_0}{2\hat{L}_0} \frac{e^{\mu_0} - 1}{\mu} + \frac{e^{-\beta} I_0}{2\hat{L}_0} \frac{e^{\mu_0} - 1}{\mu} + \frac{1}{\hat{L}_0} \frac{e^{\mu_0} - 1}{\mu} \sum_{n=1}^{m} |r_n^{(0)}| I_{R_0}^n e^{(n-1)\mu_0} \leq$$

$$\leq \frac{e^{\mu_0} - 1}{\mu\hat{L}_0} \left(\frac{U_0 + e^{-\beta} I_0}{2} + \sum_{n=1}^{m} |r_n^{(0)}| I_{R_0}^n e^{(n-1)\mu_0} \right) \leq e^{\mu(t-T-kT_0)} \frac{e^{\mu_0} - 1}{\mu\hat{L}_0} \left(\frac{U_0 + e^{-\beta} I_0}{2} + \sum_{n=1}^{m} |r_n^{(0)}| I_{R_0}^n e^{(n-1)\mu_0} \right).$$

Then

$$\left| B_0^{(k)}(i_{R_0 L_0}, U, i_{R_1 L_1}, I)(t) \right| \le$$

$$\le e^{\mu(t-T-kT_0)} \frac{1}{\mu \hat{L}_0} \left(\frac{U_0 + e^{-\beta} I_0}{2} + \sum_{n=1}^{m} \left| r_n^{(0)} \right| I_{R_0}^n e^{(n-1)\mu T_0} \right) + e^{\mu(t-T-kT_0)} \frac{e^{\mu_0} - 1}{\mu \hat{L}_0} \left(\frac{U_0 + e^{-\beta} I_0}{2} + \sum_{n=1}^{m} \left| r_n^{(0)} \right| I_{R_0}^n e^{(n-1)\mu_0} \right) \le$$

$$\le e^{\mu(t-T-kT_0)} \frac{e^{\mu_0}}{\mu \hat{L}_0} \left(\frac{U_0 + e^{-\beta} I_0}{2} + \sum_{n=1}^{m} \left| r_n^{(0)} \right| I_{R_0}^n e^{(n-1)\mu_0} \right) \le e^{\mu(t-T-kT_0)} I_{R_0} .$$

Further on we have

$$\left| B_U^{(k)}(i_{R_0 L_0}, U, i_{R_1 L_1}, I)(t) \right| \le \left| \int_{T+kT_0}^{t} V(i_{R_0 L_0}, U, i_{R_1 L_1}, I)(s)ds \right| + \left| \int_{T+kT_0}^{T+(k+1)T_0} V(i_{R_0 L_0}, U, i_{R_1 L_1}, I)(s)ds \right| \equiv W_1 + W_2.$$

But

$$W_1 \le \left| \int_{T+kT_0}^{t} \left(-\frac{d\vec{I}_0(s)}{ds} + \frac{1}{Z_0} \frac{-U(s) + \vec{I}_0(s) - 2Z_0 \breve{I}_{in}(s) - 2Z_0 i_{R_0 L_0}(s)}{d\widetilde{C}_0(u(0,s))/du} \right) ds \right| \le \left| \int_{T+kT_0}^{t} \frac{d\vec{I}_0(s)}{ds} ds \right| +$$

$$+ \frac{1}{\hat{C}_0 Z_0} \int_{T+kT_0}^{t} |U(s)| ds + \frac{1}{\hat{C}_0 Z_0} \int_{T+kT_0}^{t} \left| \vec{I}_0(s) \right| ds + \frac{2Z_0}{\hat{C}_0 Z_0} \int_{T+kT_0}^{t} \left| \breve{I}_{in}(s) \right| ds + \frac{2Z_0}{\hat{C}_0 Z_0} \int_{T+kT_0}^{t} \left| i_{R_0 L_0}(s) \right| ds \le$$

$$\le \left| \vec{I}_0(t) \right| + \frac{1}{\hat{C}_0 Z_0} \int_{T+kT_0}^{t} U_0 e^{\mu(s-T-kT_0)} ds + \frac{1}{\hat{C}_0 Z_0} \int_{T+kT_0}^{t} I_0 e^{\mu(s-T-kT_0)} e^{-\beta} ds +$$

$$+ \frac{2Z_0}{\hat{C}_0 Z_0} I_{R_0} \int_{T+kT_0}^{t} e^{\mu(s-T-kT_0)} ds + \frac{2Z_0}{\hat{C}_0 Z_0} I_{R_0} \int_{T+kT_0}^{t} e^{\mu(s-T-kT_0)} ds \le$$

$$\le e^{-\beta} I_0 e^{\mu(t-T-kT_0)} + \frac{U_0}{\hat{C}_0 Z_0} \frac{e^{\mu(t-T-kT_0)} - 1}{\mu} + \frac{e^{-\beta} I_0}{\hat{C}_0 Z_0} \frac{e^{\mu(t-T-kT_0)} - 1}{\mu} + \frac{4Z_0}{\hat{C}_0 Z_0} I_{R_0} \frac{e^{\mu(t-T-kT_0)} - 1}{\mu} \le$$

$$\le e^{\mu(t-T-kT_0)} \left(I_0 e^{-\beta} + \frac{U_0 + I_0 e^{-\beta} + 4Z_0 I_{R_0}}{\mu \hat{C}_0 Z_0} \right)$$

and

$$W_2 \le \left| \int_{T+kT_0}^{T+(k+1)T_0} \frac{d\vec{I}_0(s)}{ds} ds \right| +$$

$$+ \frac{1}{\hat{C}_0 Z_0} \int\limits_{T+kT_0}^{T+(k+1)T_0} |U(s)| ds + \frac{1}{\hat{C}_0 Z_0} \int\limits_{T+kT_0}^{T+(k+1)T_0} |\breve{I}_0(s)|\, ds + \frac{2Z_0}{\hat{C}_0 Z_0} \int\limits_{T+kT_0}^{T+(k+1)T_0} |\breve{I}_{in}(s)| ds + \frac{2Z_0}{\hat{C}_0 Z_0} \int\limits_{T+kT_0}^{T+(k+1)T_0} |i_{R_0 L_0}(s)| ds \le$$

$$\le \frac{1}{\hat{C}_0 Z_0} \int\limits_{T+kT_0}^{T+(k+1)T_0} U_0 e^{\mu(s-T-kT_0)} ds + \frac{1}{\hat{C}_0 Z_0} \int\limits_{T+kT_0}^{T+(k+1)T_0} I_0\, e^{-\beta} e^{\mu(s-T-kT_0)}\, ds +$$

$$+ \frac{2Z_0}{\hat{C}_0 Z_0} I_{R_0} \int\limits_{T+kT_0}^{T+(k+1)T_0} e^{\mu(s-T-kT_0)} ds + \frac{2Z_0}{\hat{C}_0 Z_0} I_{R_0} \int\limits_{T+kT_0}^{T+(k+1)T_0} e^{\mu(s-T-kT_0)} ds \le$$

$$\le \frac{U_0}{\hat{C}_0 Z_0} \frac{e^{\mu T_0} - 1}{\mu} + \frac{I_0 e^{-\beta}}{\hat{C}_0 Z_0} \frac{e^{\mu T_0} - 1}{\mu} + \frac{4Z_0}{\hat{C}_0 Z_0} I_{R_0} \frac{e^{\mu T_0} - 1}{\mu} \le$$

$$e^{\mu(t-T-kT_0)} \left(e^{\mu_0} - 1 \right) \frac{U_0 + I_0 e^{-\beta} + 4Z_0 I_{R_0}}{\mu \hat{C}_0 Z_0} \ .$$

Then

$$\left| B_U^{(k)}(i_{R_0 L_0}, U, I)(t) \right| \le$$

$$\le e^{\mu(t-T-kT_0)} \left[I_0 e^{-\beta} + \frac{U_0 + I_0 e^{-\beta} + 4Z_0 I_{R_0}}{\mu \hat{C}_0 Z_0} \right] + e^{\mu(t-T-kT_0)} \frac{\left(e^{\mu_0} - 1 \right)\left(U_0 e^{-\beta} + I_0 e^{-\beta} + 4Z_0 I_{R_0} \right)}{\mu \hat{C}_0 Z_0} \le$$

$$\le e^{\mu(t-T-kT_0)} \left[I_0 e^{-\beta} + \frac{e^{\mu_0} \left(U_0 + I_0 e^{-\beta} + 4Z_0 I_{R_0} \right)}{\mu \hat{C}_0 Z_0} \right] \le U_0 e^{\mu(t-T-kT_0)} \ .$$

We have

$$\left| B_1^{(k)}(i_{R_0 L_0}, U, i_{R_1 L_1}, I)(t) \right| \le \int\limits_{T+kT_0}^{t} |I_{R_1}(s)| ds + \left| \int\limits_{T+kT_0}^{T+(k+1)T_0} I_{R_1}(s) ds \right| \equiv I_1 + I_2 .$$

But

$$I_1 \le \frac{1}{2\hat{L}_1} \int\limits_{T+kT_0}^{t} \left(|\bar{U}_0(s)| + |I(s)| + |R_1(i_{R_1 L_1}(s))| \right) ds \le$$

$$\le \frac{1}{2\hat{L}_1} \left(\int\limits_{T+kT_0}^{t} U_0 e^{-\beta} e^{\mu(s-T-kT_0)} ds + \int\limits_{T+kT_0}^{t} I_0 e^{\mu(s-T-kT_0)} ds + 2\sum_{n=1}^{m} |r_n^{(1)}| \int\limits_{T+kT_0}^{t} |i_{R_1 L_1}(s)|^n ds \right) \le$$

$$\leq \frac{1}{2\hat{L}_1}\left(U_0 e^{-\beta}\frac{e^{\mu(t-T-kT_0)}-1}{\mu} + I_0\frac{e^{\mu(t-T-kT_0)}-1}{\mu} + 2\sum_{n=1}^{m}\left|r_n^{(1)}\right|\left|I_{R_1}^n\right|\int_{T+kT_0}^{t}e^{n\mu(s-T-kT_0)}ds\right)\leq$$

$$\leq \frac{1}{2\hat{L}_1}\left(U_0 e^{-\beta}\frac{e^{\mu(t-T-kT_0)}-1}{\mu} + I_0\frac{e^{\mu(t-T-kT_0)}-1}{\mu} + 2\sum_{n=1}^{m}\left|r_n^{(1)}\right|\left|I_{R_1}^n\right|\frac{e^{n\mu(t-T-kT_0)}-1}{n\mu}\right)\leq$$

$$\leq \frac{1}{2\hat{L}_1}\left(U_0 e^{-\beta}\frac{e^{\mu(t-T-kT_0)}-1}{\mu} + I_0\frac{e^{\mu(t-T-kT_0)}-1}{\mu} + \frac{e^{\mu(t-T-kT_0)}-1}{\mu}2\sum_{n=1}^{m}\left|r_n^{(1)}\right|\left|I_{R_1}^n\right|e^{(n-1)\mu_0}\right)\leq$$

$$\leq e^{\mu(t-T-kT_0)}\frac{1}{\mu\hat{L}_1}\left(\frac{e^{-\beta}U_0 + I_0}{2} + \sum_{n=1}^{m}\left|r_n^{(1)}\right|\left|I_{R_1}^n\right|e^{(n-1)\mu_0}\right)$$

and

$$I_2 \leq \frac{1}{2\hat{L}_1}\int_{T+kT_0}^{T+(k+1)T_0}\left(\left|\vec{U}_0(s)\right| + \left|I(s)\right| + \left|R_1(i_{R_1L_1}(s))\right|\right)ds \leq$$

$$\leq \frac{1}{2\hat{L}_1}\left(\int_{T+kT_0}^{T+(k+1)T_0}U_0 e^{-\beta}e^{\mu(s-T-kT_0)}ds + \int_{T+kT_0}^{T+(k+1)T_0}I_0 e^{\mu(s-T-kT_0)}ds + 2\sum_{n=1}^{m}\left|r_n^{(1)}\right|\int_{T+kT_0}^{T+(k+1)T_0}\left|i_{R_1L_1}(s)\right|^n ds\right)\leq$$

$$\leq \frac{1}{2\hat{L}_1}\left(U_0 e^{-\beta}\frac{e^{\mu T_0}-1}{\mu} + I_0\frac{e^{\mu T_0}-1}{\mu} + 2\sum_{n=1}^{m}\left|r_n^{(1)}\right|\left|I_{R_1}^n\right|\int_{T+kT_0}^{T+(k+1)T_0}e^{n\mu(s-T-kT_0)}ds\right)\leq$$

$$\leq \frac{1}{2\hat{L}_1}\left(U_0 e^{-\beta}\frac{e^{\mu T_0}-1}{\mu} + I_0\frac{e^{\mu T_0}-1}{\mu} + 2\sum_{n=1}^{m}\left|r_n^{(1)}\right|\left|I_{R_1}^n\right|\frac{e^{n\mu T_0}-1}{n\mu}\right)\leq$$

$$\leq \frac{1}{2\hat{L}_1}\left(U_0 e^{-\beta}\frac{e^{\mu T_0}-1}{\mu} + I_0\frac{e^{\mu T_0}-1}{\mu} + \frac{e^{\mu T_0}-1}{\mu}2\sum_{n=1}^{m}\left|r_n^{(1)}\right|\left|I_{R_1}^n\right|e^{(n-1)\mu_0}\right)\leq$$

$$\leq e^{\mu(t-T-kT_0)}\frac{e^{\mu_0}-1}{\mu\hat{L}_1}\left(\frac{e^{-\beta}U_0 + I_0}{2} + \sum_{n=1}^{m}\left|r_n^{(1)}\right|\left|I_{R_1}^n\right|e^{(n-1)\mu_0}\right).$$

Then

$$\left|B_1^{(k)}(i_{R_0L_0},u,i_{R_1L_1})(t)\right|\leq e^{\mu(t-T-kT_0)}\frac{e^{\mu_0}}{\mu\hat{L}_1}\left(\frac{e^{-\beta}U_0+I_0}{2} + \sum_{n=1}^{m}\left|r_n^{(1)}\right|\left|I_{R_1}^n\right|e^{(n-1)\mu T_0}\right)\leq I_{R_1}e^{\mu(t-T-kT_0)}$$

Finally

$$\left| B_I^{(k)}(i_{R_0L_0}, U, i_{R_1L_1}, I)(t) \right| \leq \left| \int\limits_{T+kT_0}^{t} J(s)ds \right| + \left| \int\limits_{T+kT_0}^{T+(k+1)T_0} J(s)ds \right| \equiv V_1 + V_2.$$

But

$$V_1 \leq \left| \int\limits_{T+kT_0}^{t} \frac{d\bar{U}_0(s)}{ds} ds \right| + \frac{1}{Z_0} \int\limits_{T+kT_0}^{t} \left| \frac{-\bar{U}_0(s) + I(s) - 2Z_0 i_{R_1L_1}(s)}{d\widetilde{C}_1(u(\Lambda,s))/du} \right| ds \leq$$

$$\leq \left| \bar{U}_0(t) \right| + \frac{1}{Z_0\hat{C}_1} \int\limits_{T+kT_0}^{t} \left(\left| \bar{U}_0(s) \right| + \left| I(s) \right| + 2Z_0 \left| i_{R_1L_1}(s) \right| \right) ds \leq$$

$$\leq e^{\mu(t-T-kT_0)} U_0 e^{-\beta} + \frac{1}{Z_0\hat{C}_1} \int\limits_{T+kT_0}^{t} \left(U_0 e^{-\beta} e^{\mu(s-T-kT_0)} + I_0 e^{\mu(s-T-kT_0)} + 2Z_0 I_{R_1} e^{\mu(s-T-kT_0)} \right) ds \leq$$

$$\leq e^{\mu(t-T-kT_0)} U_0 e^{-\beta} + \frac{U_0 e^{-\beta} + I_0 + 2Z_0 I_{R_1}}{Z_0\hat{C}_1} \frac{e^{\mu(t-T-kT_0)}-1}{\mu} \leq$$

$$\leq e^{\mu(t-T-kT_0)} \left(U_0 e^{-\beta} + \frac{U_0 e^{-\beta} + I_0 + 2Z_0 I_{R_1}}{\mu Z_0\hat{C}_1} \right)$$

and

$$V_2 \leq \left| \int\limits_{T+kT_0}^{T+(k+1)T_0} \frac{d\bar{U}_0(s)}{ds} ds \right| + \frac{1}{Z_0} \int\limits_{T+kT_0}^{T+(k+1)T_0} \left| \frac{-\bar{U}_0(s) + I(s) - 2Z_0 i_{R_1L_1}(s)}{d\widetilde{C}_1(u(\Lambda,s))/du} \right| ds \leq$$

$$\leq \frac{1}{Z_0\hat{C}_1} \int\limits_{T+kT_0}^{T+(k+1)T_0} \left(\left| \bar{U}_0(s) \right| + \left| I(s) \right| + 2Z_0 \left| i_{R_1L_1}(s) \right| \right) ds \leq$$

$$\leq \frac{1}{Z_0\hat{C}_1} \int\limits_{T+kT_0}^{T+(k+1)T_0} \left(U_0 e^{-\beta} e^{\mu(s-T-kT_0)} + I_0 e^{\mu(s-T-kT_0)} + 2Z_0 I_{R_1} e^{\mu(s-T-kT_0)} \right) ds \leq$$

$$\leq \frac{U_0 e^{-\beta} + I_0 + 2Z_0 I_{R_1}}{Z_0 \hat{C}_1} \frac{e^{\mu T_0} - 1}{\mu} \leq e^{\mu(t-T-kT_0)} \frac{\left(e^{\mu T_0} - 1\right)\left(U_0 e^{-\beta} + I_0 + 2Z_0 I_{R_1}\right)}{\mu Z_0 \hat{C}_1}.$$

Then

$$\left|B_1^{(k)}(i_{R_0 L_0}, U, i_{R_1 L_1}, I)(t)\right| \leq e^{\mu(t-T-kT_0)}\left(U_0 e^{-\beta} + \frac{U_0 e^{-\beta} + I_0 + 2Z_0 I_{R_1}}{\mu Z_0 \hat{C}_1}\right) +$$

$$+ e^{\mu(t-T-kT_0)} \frac{\left(e^{\mu_0} - 1\right)\left(U_0 e^{-\beta} + I_0 + 2Z_0 I_{R_1}\right)}{\mu Z_0 \hat{C}_1} \leq$$

$$\leq e^{\mu(t-T-kT_0)}\left(U_0 e^{-\beta} + e^{\mu_0} \frac{U_0 e^{-\beta} + I_0 + 2Z_0 I_{R_1}}{\mu Z_0 \hat{C}_1}\right) \leq e^{\mu(t-T-kT_0)} I_0.$$

It remains to obtain the Lipschitz estimates for the right-hand sides of the equations. First we notice

$$\frac{\partial I_{R_0}(i_{R_0 L_0}, U, I)}{\partial i_{R_0 L_0}} = \frac{\partial}{\partial i_{R_0 L_0}}\left(\frac{U(t) + I(t-T) - 2R_0(i_{R_0 L_0}(t))}{d\widetilde{L}_0(i_{R_0 L_0})/di_{R_0 L_0}}\right) =$$

$$= \frac{-2\dfrac{dR_0(i_{R_0 L_0})}{di_{R_0 L_0}}\dfrac{d\widetilde{L}_0(i_{R_0 L_0})}{di_{R_0 L_0}} - \left(U(t) + I(t-T) - 2R_0(i_{R_0 L_0})\right)\dfrac{d^2\widetilde{L}_0(i_{R_0 L_0})}{di_{R_0 L_0}^2}}{\left[d\widetilde{L}_0(i_{R_0 L_0})/di_{R_0 L_0}\right]^2};$$

$$\frac{\partial I_{R_0}(i_{R_0 L_0}, U, I)}{\partial U} = \frac{\partial}{\partial U}\left(\frac{U(t) + I(t-T) - 2R_0(i_{R_0 L_0})}{d\widetilde{L}_0(i_{R_0 L_0})/di_{R_0 L_0}}\right) = \frac{1}{d\widetilde{L}_0(i_{R_0 L_0})/di_{R_0 L_0}};$$

$$\frac{\partial I_{R_0}(i_{R_0 L_0}, U, I)}{\partial I} = \frac{\partial}{\partial I}\left(\frac{U(t) + I(t-T) - 2R_0(i_{R_0 L_0})}{d\widetilde{L}_0(i_{R_0 L_0})/di_{R_0 L_0}}\right) = \frac{1}{d\widetilde{L}_0(i_{R_0 L_0})/di_{R_0 L_0}};$$

$$\frac{\partial V(i_{R_0 L_0}, U, I)}{\partial i_{R_0 L_0}} = -\frac{2}{d\widetilde{C}_0(u)/du};$$

$$\frac{\partial V(i_{R_0L_0},U,I)}{\partial U}=$$

$$=\frac{1}{Z_0}\frac{-d\widetilde{C}_0\left(\frac{U(t)+I(t-T)}{2}\right)/du-\frac{\left(-U(t)+I(t-T)-2Z_0\breve{I}_{in}(t)-2Z_0 i_{R_0L_0}(t)\right)}{2}d^2\widetilde{C}_0\left(\frac{U(t)+I(t-T)}{2}\right)/du^2}{\left(d\widetilde{C}_0\left(\frac{U(t)+I(t-T)}{2}\right)/du\right)^2};$$

$$\frac{\partial V(i_{R_0L_0},U,I)}{\partial I(t-T)}=$$

$$=\frac{1}{Z_0}\frac{d\widetilde{C}_0\left(\frac{U(t)+I(t-T)}{2}\right)/du-\frac{-U(t)+I(t-T)-2Z_0\breve{I}_{in}(t)-2Z_0 i_{R_0L_0}(t)}{2}d^2\widetilde{C}_0\left(\frac{U(t)+I(t-T)}{2}\right)/du^2}{\left(d\widetilde{C}_0\left(\frac{U(t)+I(t-T)}{2}\right)/du\right)^2};$$

$$\frac{\partial V(i_{R_0L_0},U,I)}{\partial \dot{I}(t-T)}=-1;$$

$$\frac{\partial I_{R_1}(U,i_{R_1L_1},I)}{\partial U}=\frac{\partial}{\partial U}\left(\frac{U(t-T)+I(t)-2R_1(i_{R_1L_1}(t))}{2d\widetilde{L}_1(i_{R_1L_1})/di_{R_1L_1}}\right)=\frac{1}{2d\widetilde{L}_1(i_{R_1L_1})/di_{R_1L_1}};$$

$$\frac{\partial I_{R_1}(U,i_{R_1L_1},I)}{\partial i_{R_1L_1}}=\frac{\partial}{\partial i_{R_1L_1}}\left(\frac{U(t-T)+I(t)-2R_1(i_{R_1L_1}(t))}{2d\widetilde{L}_1(i_{R_1L_1})/di_{R_1L_1}}\right)=$$

$$=\frac{1}{2}\frac{-2\frac{dR_1(i_{R_1L_1})}{di_{R_1L_1}}\frac{d\widetilde{L}_1(i_{R_1L_1})}{di_{R_1L_1}}-\left(U(t-T)+I(t)-2R_1(i_{R_1L_1})\right)\frac{d^2\widetilde{L}_1(i_{R_1L_1})}{di_{R_1L_1}^2}}{\left[d\widetilde{L}_1(i_{R_1L_1})/di_{R_1L_1}\right]^2};$$

$$\frac{\partial I_{R_1}(U,i_{R_1L_1},I)}{\partial I}=\frac{\partial}{\partial I}\left(\frac{U(t-T)+I(t)-2R_1(i_{R_1L_1}(t))}{2d\widetilde{L}_1(i_{R_1L_1})/di_{R_1L_1}}\right)=\frac{1}{2d\widetilde{L}_1(i_{R_1L_1})/di_{R_1L_1}};$$

$$\frac{\partial J(U,i_{R_1L_1},I)}{\partial U(t-T)}=$$

$$=\frac{1}{Z_0}\frac{-d\widetilde{C}_1\left(\frac{U(t-T)+I(t)}{2}\right)/du-\frac{-U(t-T)+I(t)-2Z_0 i_{R_1L_1}(t)}{2}d^2\widetilde{C}_1\left(\frac{U(t-T)+I(t)}{2}\right)/du^2}{\left(d\widetilde{C}_1\left(\frac{U(t-T)+I(t)}{2}\right)/du\right)^2};$$

$$\frac{\partial J(U,i_{R_1L_1},I)}{\partial i_{R_1L_1}}=-\frac{2}{d\widetilde{C}_1(u)/du};$$

$$\frac{\partial J(U, i_{R_1 L_1}, I)}{\partial I} =$$

$$= \frac{1}{Z_0} \frac{d\widetilde{C}_1\left(\dfrac{U(t-T)+I(t)}{2}\right)/du - \dfrac{-U(t-T)+I(t)-2Z_0 i_{R_1 L_1}(t)}{2} d^2\widetilde{C}_1\left(\dfrac{U(t-T)+I(t)}{2}\right)/du^2}{\left(d\widetilde{C}_1\left(\dfrac{U(t-T)+I(t)}{2}\right)/du\right)^2};$$

$$\frac{\partial J(U, i_{R_1 L_1}, I)}{\partial \dot{U}(t-T)} = -1.$$

Therefore

$$\left| I_{R_0}(i_{R_0 L_0}, U, I) - I_{R_0}(\bar{i}_{R_0 L_0}, \overline{U}, \overline{I}) \right| \le$$

$$\le \left| \frac{-2\dfrac{dR_0(i_{R_0 L_0})}{di_{R_0 L_0}}\dfrac{d\widetilde{L}_0(i_{R_0 L_0})}{di_{R_0 L_0}} - \left(U(t)+I(t-T)-2R_0(i_{R_0 L_0})\right)\dfrac{d^2\widetilde{L}_0(i_{R_0 L_0})}{di^2_{R_0 L_0}}}{\left[d\widetilde{L}_0(i_{R_0 L_0})/di_{R_0 L_0}\right]^2} \right| \left| i_{R_0 L_0} - \bar{i}_{R_0 L_0} \right| +$$

$$+ \frac{1}{\left| d\widetilde{L}_0(i_{R_0 L_0})/di_{R_0 L_0} \right|}\left| U - \overline{U} \right| + \frac{1}{\left| d\widetilde{L}_0(i_{R_0 L_0})/di_{R_0 L_0} \right|}\left| I(t-T) - \overline{I}(t-T) \right| \le$$

$$\le \frac{1}{\hat{L}_0^2}\left(\left| \frac{dR_0(i_{R_0 L_0})}{di_{R_0 L_0}}\frac{d\widetilde{L}_0(i_{R_0 L_0})}{di_{R_0 L_0}} \right| + \left(|U(t)| + |I(t-T)| + |R_0(i_{R_0 L_0})|\right)\left| \frac{d^2\widetilde{L}_0(i_{R_0 L_0})}{di^2_{R_0 L_0}} \right|\right)\left| i_{R_0 L_0} - \bar{i}_{R_0 L_0} \right| +$$

$$+ \frac{1}{\hat{L}_0}\left| U(t) - \overline{U}(t) \right| + \frac{1}{\hat{L}_0}\left| I(t-T) - \overline{I}(t-T) \right| \le$$

$$\le \frac{1}{\hat{L}_0^2}\left(\sum_{n=1}^{m} n\left|r_n^{(0)}\right|\left|i_{R_0 L_0}\right|^{n-1} \cdot \sum_{n=1}^{m}(n+1)\left|l_n^{(p)}\right|\left|i_{R_0 L_0}\right|^n + \left(|U(t)| + |I(t-T)| + \sum_{n=1}^{m}\left|r_n^{(0)}\right|\left|i_{R_0 L_0}\right|^n\right)\right) \times$$

$$\times \sum_{n=1}^{m}(n+1)n\left|l_n^{(p)}\right|\left|i_{R_0 L_0}\right|^{n-1}\left| i_{R_0 L_0} - \bar{i}_{R_0 L_0} \right| + \frac{1}{\hat{L}_0}\left| U(t) - \overline{U}(t) \right| + \frac{1}{\hat{L}_0}\left| I(t-T) - \overline{I}(t-T) \right|;$$

$$\left| V(i_{R_0 L_0}, U, I) - V(\bar{i}_{R_0 L_0}, \overline{U}, \overline{I}) \right| \le \left| \frac{2}{d\overline{C}_0(u)/du} \right|\left| i_{R_0 L_0}(t) - \bar{i}_{R_0 L_0}(t) \right| +$$

$$+ \frac{1}{Z_0}\left| -\frac{d\widetilde{C}_0(.)}{du} + \frac{U(t)-I(t-T)+2Z_0\breve{I}_{in}(t)+2Z_0 i_{R_0 L_0}(t)}{2\left(d\widetilde{C}_0(.)/du\right)^2}\frac{d^2\widetilde{C}_0(.)}{du^2} \right|\left| U - \overline{U} \right| +$$

$$
+\frac{1}{Z_0}\left|\frac{1}{\dfrac{d\widetilde{C}_0(.)}{du}}+\frac{U(t)-I(t-T)+2Z_0\breve{I}_{in}(t)+2Z_0 i_{R_0 L_0}(t)}{2\left(\dfrac{d\widetilde{C}_0(.)}{du}\right)^2}\frac{d^2\widetilde{C}_0(.)}{du^2}\right|\left|I(t-T)-\bar{I}(t-T)\right|+
$$

$$
+\left|\dot{I}(t-T)-\dot{\bar{I}}(t-T)\right|
$$

$$
\leq \frac{2}{\hat{C}_0}\left|i_{R_0 L_0}(t)-\bar{i}_{R_0 L_0}(t)\right|+
$$

$$
+\frac{1}{Z_0}\left|\frac{1}{\hat{C}_0}+\frac{\left|U(t)\right|+\left|I(t-T)\right|+2Z_0\left|\breve{I}_{in}(t)\right|+2Z_0\left|i_{R_0 L_0}(t)\right|}{2\hat{C}_0^{\,2}}H_0\right|\left|U-\bar{U}\right|+
$$

$$
+\frac{1}{Z_0}\left|\frac{1}{\hat{C}_0}+\frac{\left|U(t)\right|+\left|I(t-T)\right|+2Z_0\left|\breve{I}_{in}(t)\right|+2Z_0\left|i_{R_0 L_0}(t)\right|}{2\hat{C}_0^{\,2}}H_0\right|\left|I(t-T)-\bar{I}(t-T)\right|+
$$

$$
\left|\dot{I}(t-T)-\dot{\bar{I}}(t-T)\right|;
$$

$$
\left|I_{R_1}(U,i_{R_1 L_1},I)-I_{R_1}(\bar{U},\bar{i}_{R_1 L_1},\bar{I})\right|\leq \frac{1}{2\left|d\widetilde{L}_1(i_{R_1 L_1})/di_{R_1 L_1}\right|}\left|U(t-T)-\bar{U}(t-T)\right|+
$$

$$
+\frac{1}{2}\left|\frac{-2\dfrac{dR_1(i_{R_1 L_1})}{di_{R_1 L_1}}\dfrac{d\widetilde{L}_1(i_{R_1 L_1})}{di_{R_1 L_1}}-\left(U(t-T)+I(t)-2R_1(i_{R_1 L_1})\right)\dfrac{d^2\widetilde{L}_1(i_{R_1 L_1})}{di_{R_1 L_1}^2}}{\left(d\widetilde{L}_1(i_{R_1 L_1})/di_{R_1 L_1}\right)^2}\right|\left|i_{R_1 L_1}-\bar{i}_{R_1 L_1}\right|+
$$

$$
+\frac{1}{2\left|d\widetilde{L}_1(i_{R_1 L_1})/di_{R_1 L_1}\right|}\left|I(t)-\bar{I}(t)\right|\leq \frac{1}{2\hat{L}_1}\left|U(t-T)-\bar{U}(t-T)\right|+
$$

$$
+\frac{2\left|\dfrac{dR_1(i_{R_1 L_1})}{di_{R_1 L_1}}\right|\left|\dfrac{d\widetilde{L}_1(i_{R_1 L_1})}{di_{R_1 L_1}}\right|+\left(\left|U(t-T)\right|+\left|I(t)\right|+2\left|R_1(i_{R_1 L_1})\right|\right)\left|\dfrac{d^2\widetilde{L}_1(i_{R_1 L_1})}{di_{R_1 L_1}^2}\right|}{2\hat{L}_1^{\,2}}\left|i_{R_1 L_1}-\bar{i}_{R_1 L_1}\right|+\frac{1}{2\hat{L}_1}\left|I(t)-\bar{I}(t)\right|;
$$

$$
\left|J(U,i_{R_1 L_1},I)-J(\bar{U},\bar{i}_{R_1 L_1},\bar{I})\right|\leq
$$

$$
+\frac{1}{Z_0}\left|\frac{-\dfrac{d\widetilde{C}_1((U(t-T)+I(t))/2)}{du}-\left(-U(t-T)+I(t)-2Z_0 i_{R_1 L_1}(t)\right)\dfrac{1}{2}\dfrac{d^2\widetilde{C}_1((U(t-T)+I(t))/2)}{du^2}}{\left(d\widetilde{C}_1((U(t-T)+I(t))/2)/du\right)^2}\right|\times
$$

$$
\times\left|U(t-T)-\bar{U}(t-T)\right|+\left|-\frac{2}{d\widetilde{C}_1(u)/du}\right|\left|i_{R_1 L_1}-i_{R_1 L_1}\right|+
$$

$$+\frac{|I-\bar{I}|}{Z_0}\left|\frac{\dfrac{d\widetilde{C}_1\big((U(t-T)+I(t))/2\big)}{du}-\big(-U(t-T)+I(t)-2Z_0 i_{R_1L_1}(t)\big)\dfrac{1}{2}\dfrac{d^2\widetilde{C}_1\big((U(t-T)+I(t))/2\big)}{du^2}}{\big(d\widetilde{C}_1\big((U(t-T)+I(t))/2\big)/du\big)^2}\right|+$$

$$+\left|\dot{U}(t-T)-\dot{\overline{U}}(t-T)\right|\le$$

$$\frac{2}{\hat{C}_1}\left|i_{R_1L_1}(t)-\bar{i}_{R_1L_1}(t)\right|+$$

$$+\frac{1}{Z_0}\left[\frac{1}{\hat{C}_1}+\frac{|U(t-T)|+|I(t)|+2Z_0\left|i_{R_1L_1}(t)\right|}{2\hat{C}_1^{\,2}}H_1\right]\left|I-\bar{I}\right|+$$

$$+\frac{1}{Z_0}\left[\frac{1}{\hat{C}_1}+\frac{|U(t-T)|+|I(t)|+2Z_0\left|i_{R_1L_1}(t)\right|}{2\hat{C}_1^{\,2}}H_1\right]\left|U(t-T)-\overline{U}(t-T)\right|+$$

$$\left|\dot{U}(t-T)-\dot{\overline{U}}(t-T)\right|.$$

Then

$$\left|B_0^{(k)}(i_{R_0L_0},U,I)(t)-B_0^{(k)}(\bar{i}_{R_0L_0},\overline{U},\bar{I})(t)\right|\le\int_{T+kT_0}^{t}\left|I_{R_0}(i_{R_0L_0},U,I)(s)-I_{R_0}(\bar{i}_{R_0L_0},\overline{U},\bar{I})(s)\right|ds+$$

$$+\left|\int_{T+kT_0}^{T+(k+1)T_0}\big(I_{R_0}(i_{R_0L_0},U,I)(s)-I_{R_0}(\bar{i}_{R_0L_0},\overline{U},\bar{I})(s)\big)ds\right|\equiv I_1+I_2.$$

But

$$I_1\le\frac{1}{\hat{L}_0^2}\int_{T+kT_0}^{t}\left[\sum_{n=1}^{m}n\left|r_n^{(0)}\right|\left|i_{R_0L_0}(s)\right|^{n-1}\cdot\sum_{n=1}^{m}(n+1)\left|l_n^{(p)}\right|\left|i_{R_0L_0}(s)\right|^{n}+\right.$$

$$+\left.\left(|U(s)|+|I(s-T)|+\sum_{n=1}^{m}\left|r_n^{(0)}\right|\left|i_{R_0L_0}(s)\right|^{n}\right)\sum_{n=1}^{m}(n+1)n\left|l_n^{(p)}\right|\left|i_{R_0L_0}(s)\right|^{n-1}\right]\left|i_{R_0L_0}(s)-\bar{i}_{R_0L_0}(s)\right|ds+$$

$$+\frac{1}{\hat{L}_0}\int_{T+kT_0}^{t}\left|U(s)-\overline{U}(s)\right|ds\le$$

$$\le\frac{1}{\hat{L}_0^2}\int_{T+kT_0}^{t}\left[\sum_{n=1}^{m}n\left|r_n^{(0)}\right|\left|I_{R_0}^{n-1}\right|e^{(n-1)\mu(s-T-kT_0)}\cdot\sum_{n=1}^{m}(n+1)\left|l_n^{(0)}\right|\left|I_{R_0}^{n}\right|e^{n\mu(s-T-kT_0)}+\right.$$

$$+\left.\left(U_0 e^{\mu(s-T-kT_0)}+I_0 e^{-\beta}e^{\mu(s-T-kT_0)}+\sum_{n=1}^{m}\left|r_n^{(0)}\right|\left|I_{R_0}^{n}\right|e^{n\mu(s-T-kT_0)}\right)\sum_{n=1}^{m}(n+1)n\left|l_n^{(0)}\right|\left|I_{R_0}^{n-1}\right|e^{(n-1)\mu(s-T-kT_0)}\right]\times$$

$$\times \left| i_{R_0 L_0}(s) - \bar{i}_{R_0 L_0}(s) \right| ds + \frac{\rho_\mu^{(k)}(U,\overline{U})}{\hat{L}_0} \int_{T+kT_0}^{t} e^{\mu(s-T-kT_0)} ds \le$$

$$\frac{1}{\hat{L}_0^2} \left[\sum_{n=1}^{m} n \left| r_n^{(0)} \right| I_{R_0}^{n-1} e^{(n-1)\mu_0} \cdot \sum_{n=1}^{m} (n+1) \left| l_n^{(0)} \right| I_{R_0}^{n} e^{n\mu_0} + \right.$$

$$+ \left. \left(U_0 e^{\mu_0} + I_0 e^{-\beta} e^{\mu_0} + \sum_{n=1}^{m} \left| r_n^{(0)} \right| I_{R_0}^{n} e^{n\mu_0} \right) \sum_{n=1}^{m} (n+1) n \left| l_n^{(0)} \right| I_{R_0}^{n-1} e^{(n-1)\mu_0} \right] \int_{T+kT_0}^{t} \left| i_{R_0 l_0}(s) - \bar{i}_{R_0 l_0}(s) \right| ds +$$

$$+ \frac{\rho_\mu^{(k)}(U,\overline{U})}{\hat{L}_0} \int_{T+kT_0}^{t} e^{\mu(s-T-kT_0)} ds \le$$

$$\frac{\rho_\mu^{(k)}(i_{R_0 L_0}, \bar{i}_{R_0 L_0})}{\hat{L}_0^2} \frac{e^{\mu(t-T-kT_0)}-1}{\mu} \left[\sum_{n=1}^{m} n \left| r_n^{(0)} \right| I_{R_0}^{n-1} e^{(n-1)\mu_0} \cdot \sum_{n=1}^{m} (n+1) \left| l_n^{(0)} \right| I_{R_0}^{n} e^{n\mu_0} + \right.$$

$$+ \left. \left(U_0 e^{\mu_0} + I_0 e^{-\beta} e^{\mu_0} + \sum_{n=1}^{m} \left| r_n^{(0)} \right| I_{R_0}^{n} e^{n\mu_0} \right) \sum_{n=1}^{m} (n+1) n \left| l_n^{(0)} \right| I_{R_0}^{n-1} e^{(n-1)\mu_0} \right] +$$

$$\frac{\rho_\mu^{(k)}(\dot{U},\dot{\overline{U}})}{\mu \hat{L}_0} \frac{e^{\mu(t-T-kT_0)}-1}{\mu} \le$$

$$\le e^{\mu(t-T-kT_0)} \frac{1}{\mu^2} \left\{ \frac{\rho_\mu^{(k)}(\dot{i}_{R_0 L_0}, \dot{\bar{i}}_{R_0 L_0})}{\hat{L}_0^2} \left[\sum_{n=1}^{m} n \left| r_n^{(0)} \right| I_{R_0}^{n-1} e^{(n-1)\mu_0} \cdot \sum_{n=1}^{m} (n+1) \left| l_n^{(0)} \right| I_{R_0}^{n} e^{n\mu_0} + \right. \right.$$

$$+ \left. \left(U_0 e^{\mu_0} + I_0 e^{-\beta} e^{\mu_0} + \sum_{n=1}^{m} \left| r_n^{(0)} \right| I_{R_0}^{n} e^{n\mu_0} \right) \sum_{n=1}^{m} (n+1) n \left| l_n^{(0)} \right| I_{R_0}^{n-1} e^{(n-1)\mu_0} \right] + \frac{\rho_\mu^{(k)}(\dot{U},\dot{\overline{U}})}{\hat{L}_0} \right\}$$

and

$$I_2 \le \frac{1}{\hat{L}_0^2} \int_{T+kT_0}^{T+(k+1)T_0} \left[\sum_{n=1}^{m} n \left| r_n^{(0)} \right| \left\| i_{R_0 L_0}(s) \right\|^{n-1} \cdot \sum_{n=1}^{m} (n+1) \left| l_n^{(0)} \right| \left\| i_{R_0 L_0}(s) \right\|^{n} + \right.$$

$$+ \left. \left(|U(s)| + |I(s-T)| + \sum_{n=1}^{m} \left| r_n^{(0)} \right| \left\| i_{R_0 L_0}(s) \right\|^{n} \right) \sum_{n=1}^{m} (n+1) n \left| l_n^{(0)} \right| \left\| i_{R_0 L_0}(s) \right\|^{n-1} \right] \left| i_{R_0 L_0}(s) - \bar{i}_{R_0 L_0}(s) \right| ds$$

$$+ \frac{1}{\hat{L}_0} \int_{T+kT_0}^{T+(k+1)T_0} \left| U(s) - \overline{U}(s) \right| ds \le$$

$$\frac{\rho_\mu^{(k)}(i_{R_0 L_0}, \bar{i}_{R_0 L_0})}{\hat{L}_0^2} \frac{e^{\mu T_0}-1}{\mu} \left[\sum_{n=1}^{m} n \left| r_n^{(0)} \right| I_{R_0}^{n-1} e^{(n-1)\mu_0} \cdot \sum_{n=1}^{m} (n+1) \left| l_n^{(0)} \right| I_{R_0}^{n} e^{n\mu_0} + \right.$$

$$+\left(U_0 e^{\mu 0}+I_0 e^{-\beta}e^{\mu 0}+\sum_{n=1}^{m}\left|r_n^{(0)}\right|\left|I_{R_0}^n\right|e^{n\mu 0}\right)\sum_{n=1}^{m}(n+1)n\left|l_n^{(0)}\right|\left|I_{R_0}^{n-1}\right|e^{(n-1)\mu 0}\right]+$$

$$\frac{\rho_\mu^{(k)}(\dot U,\dot{\overline{U}})}{\mu\hat L_0}\frac{e^{\mu 0}-1}{\mu}\le$$

$$\le e^{\mu(t-T-kT_0)}\frac{e^{\mu 0}-1}{\mu^2}\left\{\frac{\rho_\mu^{(k)}(\dot i_{R_0 L_0},\dot{\overline{i}}_{R_0 L_0})}{\hat L_0^2}\left[\sum_{n=1}^{m}n\left|r_n^{(0)}\right|\left|I_{R_0}^{n-1}\right|e^{(n-1)\mu 0}\cdot\sum_{n=1}^{m}(n+1)\left|l_n^{(0)}\right|\left|I_{R_0}^n\right|e^{n\mu 0}+\right.\right.$$

$$+\left(U_0 e^{\mu 0}+I_0 e^{-\beta}e^{\mu 0}+\sum_{n=1}^{m}\left|r_n^{(0)}\right|\left|I_{R_0}^n\right|e^{n\mu 0}\right)\sum_{n=1}^{m}(n+1)n\left|l_n^{(0)}\right|\left|I_{R_0}^{n-1}\right|e^{(n-1)\mu 0}\right]+\frac{\rho_\mu^{(k)}(\dot U,\dot{\overline{U}})}{\hat L_0}\right\}.$$

Thus

$$\left|B_0^{(k)}(i_{R_0 L_0},U,I)(t)-B_0^{(k)}(\overline{i}_{R_0 L_0},\overline{U},\overline{I})(t)\right|\le$$

$$\le e^{\mu(t-T-kT_0)}\frac{e^{\mu 0}}{\mu^2}\left\{\frac{\rho_\mu^{(k)}(\dot i_{R_0 L_0},\dot{\overline{i}}_{R_0 L_0})}{\hat L_0^2}\left[\sum_{n=1}^{m}n\left|r_n^{(0)}\right|\left|I_{R_0}^{n-1}\right|e^{(n-1)\mu 0}\cdot\sum_{n=1}^{m}(n+1)\left|l_n^{(0)}\right|\left|I_{R_0}^n\right|e^{n\mu 0}+\right.\right.$$

$$+\left(U_0 e^{\mu 0}+I_0 e^{-\beta}e^{\mu 0}+\sum_{n=1}^{m}\left|r_n^{(0)}\right|\left|I_{R_0}^n\right|e^{n\mu 0}\right)\sum_{n=1}^{m}(n+1)n\left|l_n^{(0)}\right|\left|I_{R_0}^{n-1}\right|e^{(n-1)\mu 0}\right]+\frac{\rho_\mu^{(k)}(\dot U,\dot{\overline{U}})}{\hat L_0}\right\}\le$$

$$\le e^{\mu(t-T-kT_0)}\frac{e^{\mu 0}}{\mu^2}\left\{\frac{1}{\hat L_0^2}\left[\left(\sum_{n=1}^{m}n\left|r_n^{(0)}\right|\left|I_{R_0}^{n-1}\right|e^{(n-1)\mu 0}\right)\left(\sum_{n=1}^{m}(n+1)\left|l_n^{(0)}\right|\left|I_{R_0}^n\right|e^{n\mu 0}\right)+\right.\right.$$

$$+\left(U_0 e^{\mu 0}+I_0 e^{-\beta}e^{\mu 0}+\sum_{n=1}^{m}\left|r_n^{(0)}\right|\left|I_{R_0}^n\right|e^{n\mu 0}\right)\sum_{n=1}^{m}(n+1)n\left|l_n^{(0)}\right|\left|I_{R_0}^{n-1}\right|e^{(n-1)\mu 0}\right]+\frac{1}{\hat L_0}\right\}\hat\rho_\mu\left((i_{R_0 L_0},U,i_{R_1 L_1},I),(\overline{i}_{R_0 L_0},\overline{U},\overline{i}_{R_1 L_1},\overline{I})\right)\equiv$$

$$\equiv e^{\mu(t-T-kT_0)}K_0\hat\rho_\mu\left((i_{R_0 L_0},U,i_{R_1 L_1},I),(\overline{i}_{R_0 L_0},\overline{U},\overline{i}_{R_1 L_1},\overline{I})\right)\le e^{\mu T_0}K_0\hat\rho_\mu\left((i_{R_0 L_0},U,i_{R_1 L_1},I),(\overline{i}_{R_0 L_0},\overline{U},\overline{i}_{R_1 L_1},\overline{I})\right)$$

It follows

$$\rho^{(k)}(B_0^{(k)}(i_{R_0 L_0},U,I),B_0^{(k)}(\overline{i}_{R_0 L_0},\overline{U},\overline{I}))\le e^{\mu 0}K_0\hat\rho_\mu\left((i_{R_0 L_0},U,i_{R_1 L_1},I),(\overline{i}_{R_0 L_0},\overline{U},\overline{i}_{R_1 L_1},\overline{I})\right)$$

or

$$\hat\rho(B_0(i_{R_0 L_0},U,I),B_0(\overline{i}_{R_0 L_0},\overline{U},\overline{I}))\le e^{\mu 0}K_0\hat\rho_\mu\left((i_{R_0 L_0},U,i_{R_1 L_1},I),(\overline{i}_{R_0 L_0},\overline{U},\overline{i}_{R_1 L_1},\overline{I})\right)$$

Further on we get

$$\left|B_U^{(k)}(i_{R_0 L_0},U,I)(t)-B_U^{(k)}(\overline{i}_{R_0 L_0},\overline{U},\overline{I})(t)\right|\le\int_{T+kT_0}^{t}\left|V(i_{R_0 L_0},U,I)(s)-V(\overline{i}_{R_0 L_0},\overline{U},\overline{I})(s)\right|ds+$$

$$+ \int_{T+kT_0}^{T+(k+1)T_0} \left| V(i_{R_0L_0}, U, I)(s) - V(\bar{i}_{R_0L_0}, \overline{U}, \overline{I})(s) \right| ds \equiv W_1 + W_2.$$

But

$$W_1 \leq \frac{2}{\hat{C}_0} \int_{T+kT_0}^{t} \left| i_{R_0L_0}(s) - \bar{i}_{R_0L_0}(s) \right| ds +$$

$$\frac{1}{Z_0}\left[\frac{1}{\hat{C}_0} + \frac{U_0 + I_0 + 4Z_0 I_{R_0}}{2\hat{C}_0^2} e^{\mu_0} H_0 \right] \int_{T+kT_0}^{t} \left| U(s) - \overline{U}(s) \right| ds +$$

$$+ \frac{1}{Z_0}\left[\frac{1}{\hat{C}_0} + \frac{U_0 + I_0 + 4Z_0 I_{R_0}}{2\hat{C}_0^2} e^{\mu_0} H_0 \right] \int_{T+kT_0}^{t} \left| I(s-T) - \overline{I}(s-T) \right| ds +$$

$$+ \int_{T+kT_0}^{t} \left| \dot{I}(s-T) - \dot{\overline{I}}(s-T) \right| ds \leq \frac{2\rho_\mu^{(k)}(i_{R_0L_0}, \bar{i}_{R_0L_0})}{\hat{C}_0} \frac{e^{\mu(t-T-kT_0)} - 1}{\mu} +$$

$$+ \frac{1}{Z_0}\left[\frac{1}{\hat{C}_0} + \frac{e^{\mu_0}\left(U_0 + I_0 + 4Z_0 I_{R_0}\right) H_0}{2\hat{C}_0^2} \right] \rho_\mu^{(k)}(U, \overline{U}) \frac{e^{\mu(t-T-kT_0)} - 1}{\mu} \leq$$

$$\leq \frac{e^{\mu(t-T-kT_0)} - 1}{\mu} \left\{ \frac{2\rho_\mu^{(k)}(\dot{i}_{R_0L_0}, \dot{\bar{i}}_{R_0L_0})}{\mu\hat{C}_0} + \right.$$

$$\left. \frac{1}{Z_0}\left[\frac{1}{\hat{C}_0} + \frac{e^{\mu_0}\left(U_0 + I_0 + 4Z_0 I_{R_0}\right)}{2\hat{C}_0^2} H_0 \right] \frac{\rho_\mu^{(k)}(\dot{U}, \dot{\overline{U}})}{\mu} \right\} \leq$$

$$\leq e^{\mu(t-T-kT_0)} \hat{\rho}_\mu\left((i_{R_0L_0}, U, i_{R_1L_1}, I), (\bar{i}_{R_0L_0}, \overline{U}, \bar{i}_{R_1L_1}, \overline{I})\right) \frac{1}{\mu^2\hat{C}_0}\left(2 + \frac{1}{Z_0} + \frac{e^{\mu_0}\left(U_0 + I_0 + 4Z_0 I_{R_0}\right) H_0}{Z_0\hat{C}_0} \right)$$

and

$$W_2 \leq e^{\mu(t-T-kT_0)} \hat{\rho}_\mu\left((i_{R_0L_0}, U, i_{R_1L_1}, I), (\bar{i}_{R_0L_0}, \overline{U}, \bar{i}_{R_1L_1}, \overline{I})\right) \times$$

$$\times \frac{e^{\mu_0} - 1}{\mu^2\hat{C}_0}\left(2 + \frac{1}{Z_0} + \frac{e^{\mu_0}\left(U_0 + I_0 + 4Z_0 I_{R_0}\right) H_0}{Z_0\hat{C}_0} \right).$$

Then

$$\left| B_U^{(k)}(i_{R_0L_0},U,I)(t) - B_U^{(k)}(\bar{i}_{R_0L_0},\overline{U},\overline{I})(t) \right| \le$$

$$e^{\mu(t-T-kT_0)}\hat{\boldsymbol{\rho}}_\mu\big((i_{R_0L_0},U,i_{R_1L_1},I),(\bar{i}_{R_0L_0},\overline{U},\bar{i}_{R_1L_1},\overline{I})\big)\times$$

$$\times \frac{1}{\mu^2\hat{C}_0}\left(2+\frac{1}{Z_0}+\frac{e^{\mu_0}\big(U_0+I_0+4Z_0I_{R_0}\big)H_0}{Z_0\hat{C}_0}\right)+$$

$$+ e^{\mu(t-T-kT_0)}\hat{\boldsymbol{\rho}}_\mu\big((i_{R_0L_0},U,i_{R_1L_1},I),(\bar{i}_{R_0L_0},\overline{U},\bar{i}_{R_1L_1},\overline{I})\big)\frac{e^{\mu_0}-1}{\mu^2\hat{C}_0}\left(2+\frac{1}{Z_0}+\frac{e^{\mu_0}\big(U_0+I_0+4Z_0I_{R_0}\big)H_0}{Z_0\hat{C}_0}\right)\le$$

$$\le e^{\mu(t-T-kT_0)}\hat{\boldsymbol{\rho}}_\mu\big((i_{R_0L_0},U,i_{R_1L_1},I),(\bar{i}_{R_0L_0},\overline{U},\bar{i}_{R_1L_1},\overline{I})\big)\frac{e^{\mu_0}}{\mu^2\hat{C}_0}\left(2+\frac{1}{Z_0}+\frac{e^{\mu_0}\big(U_0+I_0+4Z_0I_{R_0}\big)H_0}{Z_0\hat{C}_0}\right)\le$$

$$\le e^{\mu_0}K_U\hat{\boldsymbol{\rho}}_\mu\big((i_{R_0L_0},U,i_{R_1L_1},I),(\bar{i}_{R_0L_0},\overline{U},\bar{i}_{R_1L_1},\overline{I})\big).$$

It follows

$$\hat{\rho}(B_U(i_{R_0L_0},U,I),B_U(\bar{i}_{R_0L_0},\overline{U},\overline{I})) \le e^{\mu_0}K_U\hat{\boldsymbol{\rho}}_\mu\big((i_{R_0L_0},U,i_{R_1L_1},I),(\bar{i}_{R_0L_0},\overline{U},\bar{i}_{R_1L_1},\overline{I})\big).$$

For the third component we obtain:

$$\left| B_1^{(k)}(U,i_{R_1L_1},I)(t) - B_1^{(k)}(\overline{U},\bar{i}_{R_1L_1},\overline{I})(t) \right| \le \int_{T+kT_0}^{t}\left| I_{R_1}(U,i_{R_1L_1},I)(s) - I_{R_1}(\overline{U},\bar{i}_{R_1L_1},\overline{I})(s)\right|ds +$$

$$+ \int_{T+kT_0}^{T+(k+1)T_0}\left| I_{R_1}(U,i_{R_1L_1},I)(s) - I_{R_1}(\overline{U},\bar{i}_{R_1L_1},\overline{I})(s)\right|ds \equiv J_1 + J_2 .$$

But

$$J_1 = \int_{T+kT_0}^{t}\left| I_{R_1}(U,i_{R_1L_1},I)(s) - I_{R_1}(\overline{U},\bar{i}_{R_1L_1},\overline{I})(s)\right|ds \le \frac{1}{2\hat{L}_1}\int_{T+kT_0}^{t}\left| U(s-T)-\overline{U}(s-T)\right|ds +$$

$$+ \int_{T+kT_0}^{t}\frac{2\left|\dfrac{dR_1(i_{R_1L_1})}{di_{R_1L_1}}\right|\left|\dfrac{d\widetilde{L}_1(i_{R_1L_1})}{di_{R_1L_1}}\right| + \big(\left|U(s-T)\right|+\left|I(s)\right|+2\left|R_1(i_{R_1L_1})\right|\big)\left|\dfrac{d^2\widetilde{L}_1(i_{R_1L_1})}{di_{R_1L_1}^2}\right|}{2\hat{L}_1^{2}}\left| i_{R_1L_1}(s)-\bar{i}_{R_1L_1}(s)\right|ds +$$

$$+ \frac{1}{2\hat{L}_1}\int_{T+kT_0}^{t}\left| I(s)-\overline{I}(s)\right|ds \le$$

$$\le \frac{1}{\hat{L}_1^{2}}\left[\left(\sum_{n=1}^{m}n\left|r_n^{(1)}\right|I_{R_1}^{n-1}e^{(n-1)\mu_0}\right)\cdot\left(\sum_{n=1}^{m}(n+1)\left|l_n^{(1)}\right|I_{R_1}^{n}e^{n\mu_0}\right)+\right.$$

$$+ \left.\left(U_0e^{\mu_0}+I_0e^{\mu_0}+\sum_{n=1}^{m}\left|r_n^{(1)}\right|I_{R_1}^{n}e^{n\mu_0}\right)\sum_{n=1}^{m}(n+1)n\left|l_n^{(1)}\right|I_{R_1}^{n-1}e^{(n-1)\mu_0}\right]\rho_\mu^{(k)}(i_{R_1L_1},\bar{i}_{R_1L_1})\frac{e^{\mu(t-T-kT_0)}-1}{\mu}+$$

$$+\frac{1}{2\hat{L}_1}\rho_\mu^{(k)}(I,\bar{I})\frac{e^{\mu(t-T-kT_0)}-1}{\mu}\leq$$

$$\leq e^{\mu(t-T-kT_0)}\frac{\rho_\mu^{(k)}(\dot{i}_{R_1L_1},\dot{\bar{i}}_{R_1L_1})}{\mu^2}\frac{1}{\hat{L}_1^2}\left[\left(\sum_{n=1}^m n\left|r_n^{(1)}\right|I_{R_1}^{n-1}e^{(n-1)\mu_0}\right)\left(\sum_{n=1}^m(n+1)\left|l_n^{(1)}\right|I_{R_1}^n e^{n\mu_0}\right)+\right.$$

$$\left.+\left(U_0 e^{\mu_0}+I_0 e^{\mu_0}+\sum_{n=1}^m\left|r_n^{(1)}\right|I_{R_1}^n e^{n\mu_0}\right)\sum_{n=1}^m(n+1)n\left|l_n^{(1)}\right|I_{R_1}^{n-1}e^{(n-1)\mu_0}\right]+e^{\mu(t-T-kT_0)}\frac{1}{2\hat{L}_1}\frac{\rho_\mu^{(k)}(\dot{I},\dot{\bar{I}})}{\mu^2}\leq$$

$$\leq e^{\mu(t-T-kT_0)}\hat{\rho}_\mu\left((i_{R_0L_0},U,i_{R_1L_1},I),(\bar{i}_{R_0L_0},\bar{U},\bar{i}_{R_1L_1},\bar{I})\right)\frac{1}{\mu^2\hat{L}_1}\left\{\frac{1}{2}+\frac{1}{\hat{L}_1}\left[\left(\sum_{n=1}^m n\left|r_n^{(1)}\right|I_{R_1}^{n-1}e^{(n-1)\mu_0}\right)\left(\sum_{n=1}^m(n+1)\left|l_n^{(1)}\right|I_{R_1}^n e^{n\mu_0}\right)+\right.\right.$$

$$\left.\left.+\left(U_0 e^{\mu_0}+I_0 e^{\mu_0}+\sum_{n=1}^m\left|r_n^{(1)}\right|I_{R_1}^n e^{n\mu_0}\right)\sum_{n=1}^m(n+1)n\left|l_n^{(1)}\right|I_{R_1}^{n-1}e^{(n-1)\mu_0}\right]\right\}$$

and

$$J_2=\int_{T+kT_0}^{T+(k+1)T_0}\left|I_{R_1}(U,i_{R_1L_1},I)(s)-I_{R_1}(\bar{U},\bar{i}_{R_1L_1},\bar{I})(s)\right|ds\leq$$

$$\leq e^{\mu(t-T-kT_0)}\hat{\rho}_\mu\left((i_{R_0L_0},U,i_{R_1L_1},I),(\bar{i}_{R_0L_0},\bar{U},\bar{i}_{R_1L_1},\bar{I})\right)\frac{e^{\mu_0}-1}{\mu^2\hat{L}_1}\left\{\frac{1}{2}+\frac{1}{\hat{L}_1}\left[\left(\sum_{n=1}^m n\left|r_n^{(1)}\right|I_{R_1}^{n-1}e^{(n-1)\mu_0}\right)\left(\sum_{n=1}^m(n+1)\left|l_n^{(1)}\right|I_{R_1}^n e^{n\mu_0}\right)+\right.\right.$$

$$\left.\left.+\left(U_0 e^{\mu_0}+I_0 e^{\mu_0}+\sum_{n=1}^m\left|r_n^{(1)}\right|I_{R_1}^n e^{n\mu_0}\right)\sum_{n=1}^m(n+1)n\left|l_n^{(1)}\right|I_{R_1}^{n-1}e^{(n-1)\mu_0}\right]\right\}.$$

Thus

$$\left|B_1^{(k)}(U,i_{R_1L_1},I)(t)-B_1^{(k)}(\bar{U},\bar{i}_{R_1L_1},I)(t)\right|\leq$$

$$\leq e^{\mu(t-T-kT_0)}\hat{\rho}_\mu\left((i_{R_0L_0},U,i_{R_1L_1},I),(\bar{i}_{R_0L_0},\bar{U},\bar{i}_{R_1L_1},\bar{I})\right)\frac{e^{\mu_0}}{\mu^2\hat{L}_1}\left\{\frac{1}{2}+\frac{1}{\hat{L}_1}\left[\left(\sum_{n=1}^m n\left|r_n^{(1)}\right|I_{R_1}^{n-1}e^{(n-1)\mu_0}\right)\left(\sum_{n=1}^m(n+1)\left|l_n^{(1)}\right|I_{R_1}^n e^{n\mu_0}\right)+\right.\right.$$

$$\left.\left.+\left(U_0 e^{\mu_0}+I_0 e^{\mu_0}+\sum_{n=1}^m\left|r_n^{(1)}\right|I_{R_1}^n e^{n\mu_0}\right)\sum_{n=1}^m(n+1)n\left|l_n^{(1)}\right|I_{R_1}^{n-1}e^{(n-1)\mu_0}\right]\right\}\equiv$$

$$\equiv e^{\mu(t-T-kT_0)}K_1\hat{\rho}_\mu\left((i_{R_0L_0},U,i_{R_1L_1},I),(\bar{i}_{R_0L_0},\bar{U},\bar{i}_{R_1L_1},\bar{I})\right)$$

or

$$\hat{\rho}(B_1(i_{R_0L_0},U,i_{R_1L_1},i_{R_1L_1}),B_1(\bar{i}_{R_0L_0},\bar{U},\bar{i}_{R_1L_1},\bar{I}))\leq e^{\mu_0}K_1\hat{\rho}_\mu\left((i_{R_0L_0},U,i_{R_1L_1},I),(\bar{i}_{R_0L_0},\bar{U},\bar{i}_{R_1L_1},\bar{I})\right)$$

For the fourth component we have

$$\left| B_I^{(k)}(i_{R_0L_0},U,i_{R_1L_1},I)(t) - B_I^{(k)}(\bar{i}_{R_0L_0},\overline{U},\bar{i}_{R_1L_1},\bar{I})(t) \right| \le \int\limits_{T+kT_0}^{t} \left| J(i_{R_0L_0},U,i_{R_1L_1},I)(s) - J(\bar{i}_{R_0L_0},\overline{U},\bar{i}_{R_1L_1},\bar{I})(s) \right| ds +$$

$$+ \left| \int\limits_{T+kT_0}^{T+(k+1)T_0} \left(J(i_{R_0L_0},U,i_{R_1L_1},I)(s) - J(\bar{i}_{R_0L_0},\overline{U},\bar{i}_{R_1L_1},\bar{I})(s) \right) ds \right| \equiv K_1 + K_2.$$

But

$$K_1 \le \frac{2}{\hat{C}_1} \int\limits_{T+kT_0}^{t} \left| i_{R_1L_1}(s) - \bar{i}_{R_1L_1}(s) \right| ds +$$

$$+ \frac{1}{Z_0} \int\limits_{T+kT_0}^{t} \left[\frac{1}{\hat{C}_1} + \frac{|U(s-T)| + |I(s)| + 2Z_0|i_{R_1L_1}(s)|}{2\hat{C}_1^2} H_1 \right] |I(s) - \bar{I}(s)| ds +$$

$$+ \frac{1}{Z_0} \int\limits_{T+kT_0}^{t} \left[\frac{1}{\hat{C}_1} + \frac{|U(s-T)| + |I(s)| + 2Z_0|i_{R_1L_1}(s)|}{2\hat{C}_1^2} H_1 \right] |U(s-T) - \overline{U}(s-T)| ds +$$

$$+ \int\limits_{T+kT_0}^{t} \left| \dot{U}(s-T) - \dot{\overline{U}}(s-T) \right| ds \le$$

$$\le \frac{2}{\hat{C}_1} \rho_\mu^{(k)}(i_{R_1L_1},\bar{i}_{R_1L_1}) \frac{e^{\mu(t-T-kT_0)} - 1}{\mu} +$$

$$+ \frac{1}{Z_0} \left(\frac{1}{\hat{C}_1} + \frac{U_0 e^{\mu 0} + I_0 e^{\mu 0} + 2Z_0 I_{R_1} e^{\mu 0}}{2\hat{C}_1^2} H_1 \right) \rho_\mu^{(k)}(I,\bar{I}) \frac{e^{\mu(t-T-kT_0)} - 1}{\mu} +$$

$$+ \frac{1}{Z_0} \left(\frac{1}{\hat{C}_1} + \frac{U_0 e^{\mu 0} + I_0 e^{\mu 0} + 2Z_0 I_{R_1} e^{\mu 0}}{2\hat{C}_1^2} H_1 \right) \rho_\mu^{(k)}(\vec{U}_0,\vec{U}_0) \frac{e^{\mu(t-T-kT_0)} - 1}{\mu} \le$$

$$\le e^{\mu(t-T-kT_0)} \frac{1}{\mu^2} \left[\frac{2}{\hat{C}_1} \rho_\mu^{(k)}(i_{R_1L_1},\bar{i}_{R_1L_1}) + \frac{1}{Z_0} \left(\frac{1}{\hat{C}_1} + \frac{U_0 e^{\mu 0} + I_0 e^{\mu 0} + 2Z_0 I_{R_1} e^{\mu 0}}{2\hat{C}_1^2} H_1 \right) \frac{\rho_\mu^{(k)}(\dot{I},\dot{\bar{I}})}{\mu} \right]$$

$$\le e^{\mu(t-T-kT_0)} \hat{\rho}_\mu \left((i_{R_0L_0},U,i_{R_1L_1},I),(\bar{i}_{R_0L_0},\overline{U},\bar{i}_{R_1L_1},\bar{I}) \right) \times$$

$$\frac{1}{\mu^2 \hat{C}_1} \left(2 + \frac{1}{Z_0} + \frac{U_0 e^{\mu 0} + I_0 e^{\mu 0} + 2Z_0 I_{R_1} e^{\mu 0}}{2\hat{C}_1 Z_0} H_1 \right)$$

594   Vasil G. Angelov

and

$$K_2 \le e^{\mu(t-T-kT_0)}\hat{\rho}_\mu\big((i_{R_0L_0},U,i_{R_1L_1},I),(\bar{i}_{R_0L_0},\overline{U},\bar{i}_{R_1L_1},\overline{I})\big)\times\frac{e^{\mu_0}-1}{\mu^2\hat{C}_1}\left(2+\frac{1}{Z_0}+\frac{U_0e^{\mu_0}+I_0e^{\mu_0}+2Z_0I_{R_1}e^{\mu_0}}{2Z_0\hat{C}_1}H_1\right).$$

Then

$$\left|B_I^{(k)}(i_{R_0L_0},U,i_{R_1L_1},I)(t)-B_I^{(k)}(\bar{i}_{R_0L_0},\overline{U},\bar{i}_{R_1L_1},\overline{I})(t)\right|\le$$

$$\le e^{\mu_0)}\hat{\rho}_\mu\big((i_{R_0L_0},U,i_{R_1L_1},I),(\bar{i}_{R_0L_0},\overline{U},\bar{i}_{R_1L_1},\overline{I})\big)\frac{e^{\mu_0}}{\mu^2\hat{C}_1}\left(2+\frac{1}{Z_0}+\frac{U_0e^{\mu_0}+I_0e^{\mu_0}+2Z_0I_{R_1}e^{\mu_0}}{2Z_0\hat{C}_1}H_1\right)\equiv$$

$$\equiv e^{\mu_0}K_I\hat{\rho}_\mu\big((i_{R_0L_0},U,i_{R_1L_1},I),(\bar{i}_{R_0L_0},\overline{U},\bar{i}_{R_1L_1},\overline{I})\big).$$

It follows

$$\hat{\rho}(B_I(i_{R_0L_0},U,i_{R_1L_1},i_{R_1L_1}),B_I(\bar{i}_{R_0L_0},\overline{U},\bar{i}_{R_1L_1},\overline{I}))\le e^{\mu_0}K_I\hat{\rho}_\mu\big((i_{R_0L_0},U,i_{R_1L_1},I),(\bar{i}_{R_0L_0},\overline{U},\bar{i}_{R_1L_1},\overline{I})\big).$$

For the derivatives we obtain

$$\left|\dot{B}_0^{(k)}(i_{R_0L_0},U,I)(t)-\dot{B}_0^{(k)}(\bar{i}_{R_0L_0},\overline{U},\overline{I})(t)\right|\le\left|I_{R_0}(i_{R_0L_0},U,I)(t)-I_{R_0}(\bar{i}_{R_0L_0},\overline{U},\overline{I})(t)\right|+$$

$$+\frac{1}{T_0}\left|\int_{T+kT_0}^{T+(k+1)T_0}\big(I_{R_0}(i_{R_0L_0},U,I)(s)-I_{R_0}(\bar{i}_{R_0L_0},\overline{U},\overline{I})(s)\big)ds\right|\equiv\dot{A}_1+\dot{A}_2.$$

But

$$\dot{A}_1\le\frac{1}{\hat{L}_0^2}\left[\sum_{n=1}^m n\left|r_n^{(0)}\right|\left\|i_{R_0L_0}(t)\right\|^{n-1}.\sum_{n=1}^m(n+1)\left|l_n^{(0)}\right|\left\|i_{R_0L_0}(t)\right\|^n+\right.$$

$$+\left(|U(t)|+|I(t-T)|+\sum_{n=1}^m\left|r_n^{(0)}\right|\left\|i_{R_0L_0}(t)\right\|^n\right)\sum_{n=1}^m(n+1)n\left|l_n^{(0)}\right|\left\|i_{R_0L_0}(t)\right\|^{n-1}\right]\left|i_{R_0L_0}(t)-\bar{i}_{R_0L_0}(t)\right|+$$

$$+\frac{1}{\hat{L}_0}\left|U(t)-\overline{U}(t)\right|\le$$

$$\le e^{\mu(t-T-kT_0)}\frac{1}{\hat{L}_0^2}\left[\left(\sum_{n=1}^m n\left|r_n^{(0)}\right|\left|I_{R_0}^{n-1}\right|e^{(n-1)\mu(t-T-kT_0)}\right).\sum_{n=1}^m(n+1)\left|l_n^{(0)}\right|\left|I_{R_0}^n\right|e^{n\mu(t-T-kT_0)}+\right.$$

$$+\left(U_0 e^{\mu(t-T-kT_0)}+I_0 e^{\mu(t-T-kT_0)}+\sum_{n=1}^{m}\left|r_n^{(0)}\right|I_{R_0}^n e^{n\mu(t-T-kT_0)}\right)\sum_{n=1}^{m}(n+1)n\left|l_n^{(0)}\right|I_{R_0}^{n-1} e^{(n-1)\mu(t-T-kT_0)}\right]\rho_\mu^{(k)}(i_{R_0 L_0},\bar{i}_{R_0 L_0})+$$

$$+e^{\mu(t-T-kT_0)}\frac{\rho_\mu^{(k)}(U,\overline{U})}{\hat{L}_0}\le$$

$$e^{\mu(t-T-kT_0)}\frac{\rho_\mu^{(k)}(\dot{i}_{R_0 L_0},\dot{\bar{i}}_{R_0 L_0})}{\mu\hat{L}_0^2}\left[\left(\sum_{n=1}^{m}n\left|r_n^{(0)}\right|I_{R_0}^{n-1} e^{(n-1)\mu_0}\right)\cdot\left(\sum_{n=1}^{m}(n+1)\left|l_n^{(0)}\right|I_{R_0}^n e^{n\mu_0}\right)+\right.$$

$$\left.+\left(U_0 e^{\mu_0}+I_0 e^{\mu_0}+\sum_{n=1}^{m}\left|r_n^{(0)}\right|I_{R_0}^n e^{n\mu_0}\right)\sum_{n=1}^{m}(n+1)n\left|l_n^{(0)}\right|I_{R_0}^{n-1} e^{(n-1)\mu_0}\right]+e^{\mu(t-T-kT_0)}\frac{\rho_\mu^{(k)}(\dot{U},\dot{\overline{U}})}{\mu\hat{L}_0}\le$$

$$\le e^{\mu(t-T-kT_0)}\frac{1}{\mu}\left\{\frac{\rho_\mu^{(k)}(\dot{i}_{R_0 L_0},\dot{\bar{i}}_{R_0 L_0})}{\hat{L}_0^2}\left[\left(\sum_{n=1}^{m}n\left|r_n^{(0)}\right|I_{R_0}^{n-1} e^{(n-1)\mu_0}\right)\cdot\left(\sum_{n=1}^{m}(n+1)\left|l_n^{(0)}\right|I_{R_0}^n e^{n\mu_0}\right)+\right.\right.$$

$$\left.\left.+\left(U_0 e^{\mu_0}+I_0 e^{\mu_0}+\sum_{n=1}^{m}\left|r_n^{(0)}\right|I_{R_0}^n e^{n\mu_0}\right)\sum_{n=1}^{m}(n+1)n\left|l_n^{(0)}\right|I_{R_0}^{n-1} e^{(n-1)\mu_0}\right]+\frac{\rho_\mu^{(k)}(\dot{U},\dot{\overline{U}})}{\hat{L}_0}\right\}\le$$

$$\le e^{\mu(t-T-kT_0)}\hat{\rho}_\mu\left((i_{R_0 L_0},U,i_{R_1 L_1},I),(\bar{i}_{R_0 L_0},\overline{U},\bar{i}_{R_1 L_1},\bar{I})\right)\frac{1}{\mu}\left\{\frac{1}{\hat{L}_0^2}\left[\left(\sum_{n=1}^{m}n\left|r_n^{(0)}\right|I_{R_0}^{n-1} e^{(n-1)\mu_0}\right)\cdot\left(\sum_{n=1}^{m}(n+1)\left|l_n^{(0)}\right|I_{R_0}^n e^{n\mu_0}\right)+\right.\right.$$

$$\left.\left.+\left(U_0 e^{\mu_0}+I_0 e^{\mu_0}+\sum_{n=1}^{m}\left|r_n^{(0)}\right|I_{R_0}^n e^{n\mu_0}\right)\sum_{n=1}^{m}(n+1)n\left|l_n^{(0)}\right|I_{R_0}^{n-1} e^{(n-1)\mu_0}\right]+\frac{1}{\hat{L}_0}\right\}$$

and

$$\dot{A}_2\le e^{\mu(t-T-kT_0)}\hat{\rho}_\mu\left((i_{R_0 L_0},U,i_{R_1 L_1},I),(\bar{i}_{R_0 L_0},\overline{U},\bar{i}_{R_1 L_1},\bar{I})\right)\times$$

$$\times\frac{e^{\mu_0}-1}{\mu_0}\frac{1}{\mu}\left\{\frac{1}{\hat{L}_0^2}\left[\left(\sum_{n=1}^{m}n\left|r_n^{(0)}\right|I_{R_0}^{n-1} e^{(n-1)\mu_0}\right)\cdot\left(\sum_{n=1}^{m}(n+1)\left|l_n^{(0)}\right|I_{R_0}^n e^{n\mu_0}\right)+\right.\right.$$

$$\left.\left.+\left(U_0 e^{\mu_0}+I_0 e^{\mu_0}+\sum_{n=1}^{m}\left|r_n^{(0)}\right|I_{R_0}^n e^{n\mu_0}\right)\sum_{n=1}^{m}(n+1)n\left|l_n^{(0)}\right|I_{R_0}^{n-1} e^{(n-1)\mu_0}\right]+\frac{1}{\hat{L}_0}\right\}.$$

Consequently

$$\left|\dot{B}_0^{(k)}(i_{R_0 L_0},U,I)(t)-\dot{B}_0^{(k)}(\bar{i}_{R_0 L_0},\overline{U},\bar{I})(t)\right|\le$$

$$\le e^{\mu(t-T-kT_0)}\hat{\rho}_\mu\left((i_{R_0 L_0},U,i_{R_1 L_1},I),(\bar{i}_{R_0 L_0},\overline{U},\bar{i}_{R_1 L_1},\bar{I})\right)\times$$

$$\times\left(1+\frac{e^{\mu_0}-1}{\mu_0}\right)\frac{1}{\mu\hat{L}_0}\left\{\frac{1}{\hat{L}_0}\left[\left(\sum_{n=1}^{m}n\left|r_n^{(0)}\right|I_{R_0}^{n-1} e^{(n-1)\mu_0}\right)\cdot\left(\sum_{n=1}^{m}(n+1)\left|l_n^{(0)}\right|I_{R_0}^n e^{n\mu_0}\right)+\right.\right.$$

$$+\left(U_0 e^{\mu 0}+I_0 e^{\mu 0}+\sum_{n=1}^{m}\left|r_n^{(0)}\right|I_{R_0}^n e^{n\mu 0}\right)\sum_{n=1}^{m}(n+1)n\left|l_n^{(0)}\right|I_{R_0}^{n-1}e^{(n-1)\mu 0}\right]+1\right\}\equiv$$

$$\equiv e^{\mu(t-T-kT_0)}\dot{K}_0\hat{\rho}_\mu\big((i_{R_0L_0},U,i_{R_1L_1},I),(\bar{i}_{R_0L_0},\overline{U},\bar{i}_{R_1L_1},\bar{I})\big).$$

It follows

$$\rho_\mu^{(k)}(\dot{B}_0^{(k)}(i_{R_0L_0},U,I),\dot{B}_0^{(k)}(\bar{i}_{R_0L_0},\overline{U},\bar{I}))\le \dot{K}_0\hat{\rho}_\mu\big((i_{R_0L_0},U,i_{R_1L_1},I),(\bar{i}_{R_0L_0},\overline{U},\bar{i}_{R_1L_1},\bar{I})\big).$$

Further on we have

$$\left|\dot{B}_U^{(k)}(i_{R_0L_0},U,I)(t)-\dot{B}_U^{(k)}(\bar{i}_{R_0L_0},\overline{U},\bar{I})(t)\right|\le \left|V(i_{R_0L_0},U,I)(t)-V(\bar{i}_{R_0L_0},\overline{U},\bar{I})(t)\right|+$$

$$+\frac{1}{T_0}\int_{T+kT_0}^{T+(k+1)T_0}\left|V(i_{R_0L_0},U,I)(s)-V(\bar{i}_{R_0L_0},\overline{U},\bar{I})(s)\right|ds\equiv \dot{W}_1+\dot{W}_2.$$

But

$$W_1\le \frac{2}{\hat{C}_0}\left|i_{R_0L_0}(t)-\bar{i}_{R_0L_0}(t)\right|+$$

$$\frac{1}{Z_0}\left(\frac{1}{\hat{C}_0}+\frac{U_0 e^{\mu 0}+I_0 e^{\mu 0}+4Z_0 I_{R_0}e^{\mu 0}}{2\hat{C}_0^2}H_0\right)\left|U(t)-\overline{U}(t)\right|+$$

$$+\frac{1}{Z_0}\left(\frac{1}{\hat{C}_0}+\frac{U_0 e^{\mu 0}+I_0 e^{\mu 0}+4Z_0 I_{R_0}e^{\mu 0}}{2\hat{C}_0^2}H_0\right)\left|I(t-T)-\bar{I}(t-T)\right|+$$

$$\left|\dot{I}(t-T)-\dot{\bar{I}}(t-T)\right|\le$$

$$\le e^{\mu(t-T-kT_0)}\frac{2\rho_\mu^{(k)}(i_{R_0L_0},\bar{i}_{R_0L_0})}{\hat{C}_0}+$$

$$\frac{1}{Z_0}\left(\frac{1}{\hat{C}_0}+\frac{U_0 e^{\mu 0}+I_0 e^{\mu 0}+4Z_0 I_{R_0}e^{\mu 0}}{2\hat{C}_0^2}H_0\right)\rho_\mu^{(k)}(U,\overline{U})\le$$

$$\le e^{\mu(t-T-kT_0)}\left[\frac{2\rho_\mu^{(k)}(\dot{i}_{R_0L_0},\dot{\bar{i}}_{R_0L_0})}{\mu\hat{C}_0}+\frac{1}{Z_0}\left(\frac{1}{\hat{C}_0}+\frac{U_0 e^{\mu 0}+I_0 e^{\mu 0}+4Z_0 I_{R_0}e^{\mu 0}}{2\hat{C}_0^2}H_0\right)\frac{\rho_\mu^{(k)}(\dot{U},\dot{\overline{U}})}{\mu}\right]\le$$

$$\leq e^{\mu(t-T-kT_0)}\hat{\boldsymbol{\rho}}_\mu\big((i_{R_0L_0},U,i_{R_1L_1},I),(\bar{i}_{R_0L_0},\overline{U},\bar{i}_{R_1L_1},\bar{I})\big)\frac{1}{\mu\hat{C}_0}\left(2+\frac{1}{Z_0}+\frac{U_0e^{\mu_0}+I_0e^{\mu_0}+4Z_0I_{R_0}e^{\mu_0}}{2Z_0\hat{C}_0}H_0\right)$$

and

$$\dot{W}_2\leq\frac{1}{T_0}\int_{T+kT_0}^{T+(k+1)T_0}\big|V(i_{R_0L_0},U,I)(s)-V(\bar{i}_{R_0L_0},\overline{U},\bar{I})(s)\big|ds\leq$$

$$\leq e^{\mu(t-T-kT_0)}\hat{\boldsymbol{\rho}}_\mu\big((i_{R_0L_0},U,i_{R_1L_1},I),(\bar{i}_{R_0L_0},\overline{U},\bar{i}_{R_1L_1},\bar{I})\big)\times$$

$$\times\frac{e^{\mu_0}-1}{\mu_0}\frac{1}{\mu\hat{C}_0}\left(2+\frac{1}{Z_0}+\frac{U_0e^{\mu_0}+I_0e^{\mu_0}+4Z_0I_{R_0}e^{\mu_0}}{2Z_0\hat{C}_0}H_0\right).$$

Therefore

$$\left|\dot{B}_U^{(k)}(i_{R_0L_0},U,I)(t)-\dot{B}_U^{(k)}(\bar{i}_{R_0L_0},\overline{U},\bar{I})(t)\right|\leq$$

$$\leq e^{\mu(t-T-kT_0)}\hat{\boldsymbol{\rho}}_\mu\big((i_{R_0L_0},U,i_{R_1L_1},I),(\bar{i}_{R_0L_0},\overline{U},\bar{i}_{R_1L_1},\bar{I})\big)\times$$

$$\times\left(1+\frac{e^{\mu_0}-1}{\mu_0}\right)\frac{1}{\mu\hat{C}_0}\left(2+\frac{1}{Z_0}+\frac{U_0e^{\mu_0}+I_0e^{\mu_0}+4Z_0I_{R_0}e^{\mu_0}}{2Z_0\hat{C}_0}H_0\right)\equiv$$

$$\equiv e^{\mu(t-T-kT_0)}\dot{K}_U\hat{\boldsymbol{\rho}}_\mu\big((i_{R_0L_0},U,i_{R_1L_1},I),(\bar{i}_{R_0L_0},\overline{U},\bar{i}_{R_1L_1},\bar{I})\big).$$

Consequently

$$\rho_\mu^{(k)}(\dot{B}_U^{(k)}(i_{R_0L_0},U,I),\dot{B}_U^{(k)}(\bar{i}_{R_0L_0},\overline{U},\bar{I}))\leq\dot{K}_U\hat{\boldsymbol{\rho}}_\mu\big((i_{R_0L_0},U,i_{R_1L_1},I),(\bar{i}_{R_0L_0},\overline{U},\bar{i}_{R_1L_1},\bar{I})\big).$$

Further on we have

$$\left|\dot{B}_1^{(k)}(U,i_{R_1L_1},I)(t)-\dot{B}_1^{(k)}(\overline{U},\bar{i}_{R_1L_1},\bar{I})(t)\right|\leq\left|I_{R_1}(U,i_{R_1L_1},I)(t)-I_{R_1}(\overline{U},\bar{i}_{R_1L_1},\bar{I})(t)\right|+$$

$$+\frac{1}{T_0}\int_{T+kT_0}^{T+(k+1)T_0}\left|I_{R_1}(U,i_{R_1L_1},I)(s)-I_{R_1}(\overline{U},\bar{i}_{R_1L_1},\bar{I})(s)\right|ds\equiv\dot{J}_1+\dot{J}_2.$$

But

$$\dot{J}_1\leq\frac{1}{2\hat{L}_1}\left|U(t-T)-\overline{U}(t-T)\right|+$$

$$+\frac{2\left|\dfrac{dR_1(i_{R_1L_1})}{di_{R_1L_1}}\right|\left|\dfrac{d\widetilde{L}_1(i_{R_1L_1})}{di_{R_1L_1}}\right|+\left(|U(t-T)|+|I(t)|+2|R_1(i_{R_1L_1})|\right)\left|\dfrac{d^2\widetilde{L}_1(i_{R_1L_1})}{di_{R_1L_1}^2}\right|}{2\hat{L}_1^2}\left|i_{R_1L_1}(t)-\bar{i}_{R_1L_1}(t)\right|+$$

$$+\frac{1}{2\hat{L}_1}\left|I(t)-\hat{I}(t)\right|\leq\frac{1}{\hat{L}_1^2}\Bigg[\left(\sum_{n=1}^{m}n\left|r_n^{(1)}\right|I_{R_1}^{n-1}e^{(n-1)\mu_0}\right)\left(\sum_{n=1}^{m}(n+1)\left|l_n^{(1)}\right|I_{R_1}^{n}e^{n\mu_0}\right)+$$

$$+\left(U_0e^{\mu_0}+I_0e^{\mu_0}+\sum_{n=1}^{m}\left|r_n^{(1)}\right|I_{R_1}^{n}e^{n\mu_0}\right)\sum_{n=1}^{m}(n+1)n\left|l_n^{(1)}\right|I_{R_1}^{n-1}e^{(n-1)\mu_0}\Bigg]e^{\mu(t-T-kT_0)}\rho_\mu^{(k)}(i_{R_1L_1},\bar{i}_{R_1L_1})+$$

$$+e^{\mu(t-T-kT_0)}\frac{1}{2\hat{L}_1}\rho_\mu^{(k)}(I,\bar{I})\leq$$

$$\leq e^{\mu(t-T-kT_0)}\hat{\rho}_\mu\big((i_{R_0L_0},U,i_{R_1L_1},I),(\bar{i}_{R_0L_0},\overline{U},\bar{i}_{R_1L_1},\bar{I})\big)\frac{1}{\mu}\Bigg\{\frac{1}{\hat{L}_1^2}\Bigg[\left(\sum_{n=1}^{m}n\left|r_n^{(1)}\right|I_{R_1}^{n-1}e^{(n-1)\mu_0}\right)\left(\sum_{n=1}^{m}(n+1)\left|l_n^{(1)}\right|I_{R_1}^{n}e^{n\mu_0}\right)+$$

$$+\left(U_0e^{\mu_0}+I_0e^{\mu_0}+\sum_{n=1}^{m}\left|r_n^{(1)}\right|I_{R_1}^{n}e^{n\mu_0}\right)\sum_{n=1}^{m}(n+1)n\left|l_n^{(1)}\right|I_{R_1}^{n-1}e^{(n-1)\mu_0}\Bigg]+\frac{1}{2\hat{L}_1}\Bigg\}$$

and

$$\dot{B}_2\leq e^{\mu(t-T-kT_0)}\hat{\rho}_\mu\big((i_{R_0L_0},U,i_{R_1L_1},I),(\bar{i}_{R_0L_0},\overline{U},\bar{i}_{R_1L_1},\bar{I})\big)\times$$

$$\times\frac{e^{\mu_0}-1}{\mu_0}\frac{1}{\mu}\Bigg\{\frac{1}{\hat{L}_1^2}\Bigg[\left(\sum_{n=1}^{m}n\left|r_n^{(1)}\right|I_{R_1}^{n-1}e^{(n-1)\mu_0}\right)\cdot\left(\sum_{n=1}^{m}(n+1)\left|l_n^{(1)}\right|I_{R_1}^{n}e^{n\mu_0}\right)+$$

$$+\left(U_0e^{\mu_0}+I_0e^{\mu_0}+\sum_{n=1}^{m}\left|r_n^{(1)}\right|I_{R_1}^{n}e^{n\mu_0}\right)\sum_{n=1}^{m}(n+1)n\left|l_n^{(1)}\right|I_{R_1}^{n-1}e^{(n-1)\mu_0}\Bigg]+\frac{1}{2\hat{L}_1}\Bigg\}.$$

Thus

$$\left|\dot{B}_1^{(k)}(U,i_{R_1L_1},I)(t)-\dot{B}_1^{(k)}(\overline{U},\bar{i}_{R_1L_1},\bar{I})(t)\right|\leq$$

$$\leq e^{\mu(t-T-kT_0)}\left(1+\frac{e^{\mu_0}-1}{\mu_0}\right)\frac{1}{\mu}\Bigg\{\frac{1}{\hat{L}_1^2}\Bigg[\left(\sum_{n=1}^{m}n\left|r_n^{(1)}\right|I_{R_1}^{n-1}e^{(n-1)\mu_0}\right)\cdot\left(\sum_{n=1}^{m}(n+1)\left|l_n^{(1)}\right|I_{R_1}^{n}e^{n\mu_0}\right)+$$

$$+\left(U_0e^{\mu_0}+I_0e^{\mu_0}+\sum_{n=1}^{m}\left|r_n^{(1)}\right|I_{R_1}^{n}e^{n\mu_0}\right)\sum_{n=1}^{m}(n+1)n\left|l_n^{(1)}\right|I_{R_1}^{n-1}e^{(n-1)\mu_0}\Bigg]+\frac{1}{2\hat{L}_1}\Bigg\}\equiv$$

$$\equiv e^{\mu(t-T-kT_0)}\dot{K}_1\hat{\rho}_\mu\big((i_{R_0L_0},U,i_{R_1L_1},I),(\bar{i}_{R_0L_0},\overline{U},\bar{i}_{R_1L_1},\bar{I})\big).$$

Then

$$\rho_\mu^{(k)}\left(\dot B_1^{(k)}(U,i_{R_1L_1},I),\dot B_1^{(k)}(\overline U,\overline i_{R_1L_1},\overline I)\right)\le \dot K_1\hat\rho_\mu\left((i_{R_0L_0},U,i_{R_1L_1},I),(\overline i_{R_0L_0},\overline U,\overline i_{R_1L_1},\overline I)\right)\ (k=0,1,2,\ldots,m-1).$$

For the last component of the derivative we obtain

$$\left|\dot B_I^{(k)}(i_{R_0L_0},U,i_{R_1L_1},I)(t)-\dot B_I^{(k)}(\overline i_{R_0L_0},\overline U,\overline i_{R_1L_1},\overline I)(t)\right|\le\left|J(i_{R_0L_0},U,i_{R_1L_1},I)(t)-J(\overline i_{R_0L_0},\overline U,\overline i_{R_1L_1},\overline I)(t)\right|+$$

$$+\frac{1}{T_0}\left|\int_{T+kT_0}^{T+(k+1)T_0}\left(J(i_{R_0L_0},U,i_{R_1L_1},I)(s)-J(\overline i_{R_0L_0},\overline U,\overline i_{R_1L_1},\overline I)(s)\right)ds\right|\equiv\dot K_1+\dot K_2.$$

But

$$\dot K_1\le\frac{2}{\hat C_1}\left|i_{R_1L_1}(t)-\overline i_{R_1L_1}(t)\right|+$$

$$\frac{1}{Z_0}\left(\frac{1}{\hat C_1}+\frac{\left|U(t-T)\right|+\left|I(t)\right|+2Z_0\left|i_{R_1L_1}(s)\right|}{2\hat C_1^2}H_1\right)\left|I(t)-\overline I(t)\right|+$$

$$+\frac{1}{Z_0}\left(\frac{1}{\hat C_1}+\frac{\left|U(t-T)\right|+\left|I(t)\right|+2Z_0\left|i_{R_1L_1}(s)\right|}{2\hat C_1^2}H_1\right)\left|U(t-T)-\overline U(t-T)\right|+$$

$$\left|\dot U(t-T)-\dot{\overline U}(t-T)\right|\le$$

$$\le e^{\mu(t-T-kT_0)}\frac{2}{\hat C_1}\rho_\mu^{(k)}(i_{R_1L_1},\overline i_{R_1L_1})+$$

$$+e^{\mu(t-T-kT_0)}\frac{1}{Z_0}\left(\frac{1}{\hat C_1}+\frac{U_0e^{\mu_0}+I_0e^{\mu_0}+2Z_0I_{R_1}e^{\mu_0}}{2\hat C_1^2}H_1\right)\rho_\mu^{(k)}(I,\overline I)\le$$

$$\le e^{\mu(t-T-kT_0)}\frac{1}{\mu\hat C_1}\left[2\rho_\mu^{(k)}(\dot i_{R_1L_1},\dot{\overline i}_{R_1L_1})+\left(\frac{1}{Z_0}+\frac{U_0e^{\mu_0}+I_0e^{\mu_0}+2Z_0I_{R_1}e^{\mu_0}}{2\hat C_1Z_0}H_1\right)\rho_\mu^{(k)}(\dot I,\dot{\overline I})\right]\le$$

$$\le e^{\mu(t-T-kT_0)}\hat\rho_\mu\left((i_{R_0L_0},U,i_{R_1L_1},I),(\overline i_{R_0L_0},\overline U,\overline i_{R_1L_1},\overline I)\right)\times$$

$$\times\frac{1}{\mu\hat C_1}\left[2+\frac{1}{Z_0}+\frac{U_0e^{\mu_0}+I_0e^{\mu_0}+2Z_0I_{R_1}e^{\mu_0}}{2\hat C_1Z_0}H_1\right]$$

and

$$\dot K_2\le e^{\mu(t-T-kT_0)}\hat\rho_\mu\left((i_{R_0L_0},U,i_{R_1L_1},I),(\overline i_{R_0L_0},\overline U,\overline i_{R_1L_1},\overline I)\right)\times$$

$$\times \frac{e^{\mu_0}-1}{\mu_0}\frac{1}{\mu\hat{C}_1}\left[2+\frac{1}{Z_0}+\frac{U_0 e^{\mu_0}+I_0 e^{\mu_0}+2Z_0 I_{R_1}e^{\mu_0}}{2\hat{C}_1 Z_0}H_1\right].$$

Then

$$\left|\dot{B}_I^{(k)}(i_{R_0 L_0},U,i_{R_1 L_1},I)(t)-\dot{B}_I^{(k)}(\bar{i}_{R_0 L_0},\bar{U},\bar{i}_{R_1 L_1},\bar{I})(t)\right|\le$$

$$\le e^{\mu(t-T-kT_0)}\hat{\rho}_\mu\left((i_{R_0 L_0},U,i_{R_1 L_1},I),(\bar{i}_{R_0 L_0},\bar{U},\bar{i}_{R_1 L_1},\bar{I})\right)\times$$

$$\times\left(1+\frac{e^{\mu_0}-1}{\mu_0}\right)\frac{1}{\mu\hat{C}_1}\left(2+\frac{1}{Z_0}+\frac{U_0 e^{\mu_0}+I_0 e^{\mu_0}+2Z_0 I_{R_1}e^{\mu_0}}{2\hat{C}_1 Z_0}H_1\right)\equiv$$

$$\equiv e^{\mu(t-T-kT_0)}\dot{K}_I\hat{\rho}_\mu\left((i_{R_0 L_0},U,i_{R_1 L_1},I),(\bar{i}_{R_0 L_0},\bar{U},\bar{i}_{R_1 L_1},\bar{I})\right).$$

It follows

$$\rho_\mu^{(k)}(\dot{B}_I^{(k)}(i_{R_0 L_0},U,i_{R_1 L_1},I),\dot{B}_I^{(k)}(\bar{i}_{R_0 L_0},\bar{U},\bar{i}_{R_1 L_1},\bar{I}))\le \dot{K}_I\hat{\rho}_\mu\left((i_{R_0 L_0},U,i_{R_1 L_1},I),(\bar{i}_{R_0 L_0},\bar{U},\bar{i}_{R_1 L_1},\bar{I})\right).$$

Finally we have

$$\hat{\rho}_\mu\left(B_0(i_{R_0 L_0},U,i_{R_1 L_1},I),B_U(i_{R_0 L_0},U,i_{R_1 L_1},I),B_1(i_{R_0 L_0},U,i_{R_1 L_1},I),B_I(i_{R_0 L_0},U,i_{R_1 L_1},I),\right.$$
$$\left.B_0(\bar{i}_{R_0 L_0},\bar{U},\bar{i}_{R_1 L_1},\bar{I}),B_U(\bar{i}_{R_0 L_0},\bar{U},i_{R_1 L_1},\bar{I}),B_1(\bar{i}_{R_0 L_0},\bar{U},\bar{i}_{R_1 L_1},\bar{I}),B_I(\bar{i}_{R_0 L_0},\bar{U},\bar{i}_{R_1 L_1},\bar{I})\right)\le$$
$$\le K\hat{\rho}_\mu\left((i_{R_0 L_0},U,i_{R_1 L_1},I),(\bar{i}_{R_0 L_0},\bar{U},\bar{i}_{R_1 L_1},\bar{I})\right),$$

where

$$K=\max\left\{e^{\mu_0}K_0,e^{\mu_0}K_U,e^{\mu_0}K_1,e^{\mu_0}K_I,\dot{K}_0,\dot{K}_U,\dot{K}_1,\dot{K}_I\right\}<1.$$

Then B has a unique fixed point which is a periodic solution of (8.3.1). Theorem 8.5.1 is thus proved.

8.6. Numerical Example

Here we collect all inequalities guaranteeing an existence-uniqueness result:

$$e^{\mu_0}\frac{U_0+I_0}{2}\le\phi_0;\quad \frac{e^{\mu_0}}{\mu\hat{L}_0}\left(\frac{U_0+I_0}{2}+\sum_{n=1}^m\left|r_n^{(0)}\right|I_{R_0}^n e^{(n-1)\mu_0}\right)\le I_{R_0};$$

$$I_0 e^{-\beta} + e^{\mu_0} \frac{U_0 + I_0 + 4Z_0 I_{R_0}}{\mu \hat{C}_0 Z_0} \le U_0 \; ; \quad \frac{e^{\mu_0}}{\mu \hat{L}_1}\left(\frac{U_0 + I_0}{2} + \sum_{n=1}^{m} \left| r_n^{(0)} \right| I_{R_1}^n \, e^{(n-1)\mu T_0} \right) \le I_{R_1} \; ;$$

$$U_0 e^{-\beta} + e^{\mu_0} \frac{U_0 + I_0 + 2Z_0 I_{R_1}}{\mu Z_0 \hat{C}_1} \le I_0 \; ;$$

$$e^{\mu_0} K_0 = e^{\mu_0} \frac{e^{\mu_0}}{\mu^2 \hat{L}_0} \left\{ \frac{1}{2} + \frac{1}{\hat{L}_0}\left[\left(\sum_{n=1}^{m} n \left| r_n^{(0)} \right| I_{R_0}^{n-1} e^{(n-1)\mu_0} \right)\!\!\left(\sum_{n=1}^{m} (n+1) \left| l_n^{(0)} \right| I_{R_0}^n e^{n\mu_0} \right) + \right.\right.$$

$$\left.\left. + \left(U_0 e^{\mu_0} + I_0 e^{\mu_0} + \sum_{n=1}^{m} \left| r_n^{(0)} \right| I_{R_0}^n e^{n\mu_0} \right)\!\!\left(\sum_{n=1}^{m} (n+1)n \left| l_n^{(0)} \right| I_{R_0}^{n-1} e^{(n-1)\mu_0} \right) \right] \right\} < 1;$$

$$e^{\mu_0} K_U = e^{\mu_0} \frac{e^{\mu_0}}{\mu^2 \hat{C}_0} \left(2 + \frac{1}{Z_0} + \frac{U_0 + I_0 + 2Z_0 I_{R_0}}{2} e^{\mu_0} \frac{H_0}{\hat{C}_0} \right) < 1;$$

$$e^{\mu_0} K_1 = e^{\mu_0} \frac{e^{\mu_0}}{\mu^2 \hat{L}_1} \left\{ \frac{1}{2} + \frac{1}{\hat{L}_1}\left[\left(\sum_{n=1}^{m} n \left| r_n^{(1)} \right| I_{R_1}^{n-1} e^{(n-1)\mu_0} \right)\!\!\left(\sum_{n-1}^{m} (n+1) \left| l_n^{(1)} \right| I_{R_1}^n e^{n\mu_0} \right) + \right.\right.$$

$$\left.\left. + \left(U_0 e^{\mu_0} + I_0 e^{\mu_0} + \sum_{n=1}^{m} \left| r_n^{(1)} \right| I_{R_1}^n e^{n\mu_0} \right)\!\!\left(\sum_{n=1}^{m} (n+1)n \left| l_n^{(1)} \right| I_{R_1}^{n-1} e^{(n-1)\mu_0} \right) \right] \right\} < 1 ;$$

$$e^{\mu_0} K_I = e^{\mu_0} \frac{e^{\mu_0}}{\mu^2 \hat{C}_1} \left(2 + \frac{1}{Z_0} + \frac{U_0 + I_0 + 2Z_0 I_{R_1}}{2} e^{\mu_0} \frac{H_1}{\hat{C}_1} \right) < 1;$$

$$\dot{K}_0 = \left(1 + \frac{e^{\mu_0} - 1}{\mu_0} \right) \frac{1}{\mu} \left\{ \frac{1}{\hat{L}_0^2}\left[\left(\sum_{n=1}^{m} n \left| r_n^{(0)} \right| I_{R_0}^{n-1} e^{(n-1)\mu_0} \right)\!\!\left(\sum_{n=1}^{m} (n+1) \left| l_n^{(0)} \right| I_{R_0}^n e^{n\mu_0} \right) + \right.\right.$$

$$\left.\left. + \left(U_0 e^{\mu_0} + I_0 e^{\mu_0} + \sum_{n=1}^{m} \left| r_n^{(0)} \right| I_{R_0}^n e^{n\mu_0} \right)\!\!\left(\sum_{n=1}^{m} (n+1)n \left| l_n^{(0)} \right| I_{R_0}^{n-1} e^{(n-1)\mu_0} \right) \right] + \frac{1}{\hat{L}_0} \right\} < 1;$$

$$\dot{K}_U = \left(1 + \frac{e^{\mu_0} - 1}{\mu_0} \right) \frac{1}{\mu \hat{C}_0} \left[2 + \frac{1}{Z_0} + \frac{U_0 + I_0 + 2Z_0 I_{R_0}}{2Z_0} e^{\mu_0} \frac{H_0}{\hat{C}_0} \right] < 1 ;$$

$$\dot{K}_1 = \left(1 + \frac{e^{\mu_0} - 1}{\mu_0} \right) \frac{1}{\mu} \left\{ \frac{1}{\hat{L}_1^2}\left[\left(\sum_{n=1}^{m} n \left| r_n^{(1)} \right| I_{R_1}^{n-1} e^{(n-1)\mu_0} \right)\!\!\left(\sum_{n=1}^{m} (n+1) \left| l_n^{(1)} \right| I_{R_1}^n e^{n\mu_0} \right) + \right.\right.$$

$$+\left(U_0 e^{\mu_0}+I_0 e^{\mu_0}+\sum_{n=1}^{m}\left|r_n^{(1)}\right|\left|I_{R_1}^n\right|e^{n\mu_0}\right)\left(\sum_{n=1}^{m}(n+1)n\left|l_n^{(1)}\right|\left|I_{R_1}^{n-1}\right|e^{(n-1)\mu_0}\right)\right]+\frac{1}{\hat{L}_1}\right\}<1;$$

$$\dot{K}_I=\left(1+\frac{e^{\mu_0}-1}{\mu_0}\right)\frac{1}{\mu\hat{C}_1}\left(2+\frac{1}{Z_0}+\frac{U_0+I_0+2Z_0 I_{R_1}}{2Z_0}e^{\mu_0}\frac{H_1}{\hat{C}_1}\right)<1.$$

For a transmission line with length $\Lambda=1\,m,\quad L=0,45\,\mu H/m,\ C=80\,pF/m,$

$$v=1/\sqrt{LC}=1/(6.10^{-9})=1,66.10^8;Z_0=\sqrt{L/C}=75\,\Omega.$$

Then $T=\Lambda\sqrt{LC}=6.10^{-9}\,\sec.$

Let us check the propagation of waves with $\lambda_0=(1/6)10^{-3}m$. We have

$$f_0=1/(\lambda_0\sqrt{LC})=10^{12}\,Hz\Rightarrow T_0=1/f_0=10^{-12}.$$

We choose $\mu=10^{12}$, then $\mu T_0=\mu_0=1$ and

$$T=6.10^{-9}.2.10^{12}T_0=12000.T_0\Rightarrow m=12000.$$

We choose resistive elements with the following $V\text{-}I$ characteristics

$$R_0(i)=R_1(i)=0,028i-0,125i^3 \text{ i.e. } r_1=0,028, r_2=0, r_3=0,125$$

and inductive elements with $L_0(i)=L_1(i)=3i-(1/12)i^3$. Then

$$\tilde{L}_0(i)=i(dL_0(i)/di)+L_0(i)=i(3-(1/4)i^2)+3i-(1/12)i^3=6i-(1/3)i^3$$

If we choose $i_0=1$ one obtains $6i-(1/3)i^3>6-(1/3)=17/3$ and consequently $\dfrac{1}{\hat{L}_0}=\dfrac{3}{17}$.

Let us take $C_0(u)=C_1(u)=c_0/\sqrt{1-(u/\Phi_0)}=c_0\sqrt{\Phi_0}/\sqrt{\Phi_0-u}$, where $h=2$. Let us choose $\phi_0=0,2$, $c_0=c_1=50\,pF=5.10^{-11}F$ and

$$\Phi_0=\Phi_1=0,4\,V\Rightarrow U_0<\phi_0<0,4;$$

$$\hat{C}_p=c_p\sqrt[h]{\Phi_p}\frac{\Phi_p+\dfrac{h-1}{h}\phi_0}{(\Phi_p+\phi_0)^{\frac{1}{h}+1}},(p=0,1);$$

$$\hat{C}_0 = \hat{C}_1 = c_0\sqrt{\Phi_0}\,\frac{\Phi_0 + 0,5\phi_0}{\left(\Phi_0 + \phi_0\right)^{\frac{3}{2}}} = 5.10^{-11}\sqrt{0,4}\,\frac{0,4 + 0,5.0,2}{\left(0,4 + 0,2\right)^{\frac{3}{2}}} = 3,4.10^{-11}$$

$$H_p = \frac{2c_p\sqrt[h]{\Phi_p}\left(h\Phi_p + \phi_0\right)}{h^2\ \sqrt[h]{\left(\Phi_p - \phi_0\right)^{1+2h}}};$$

$$H_0 = H_1 = \frac{2c_0\sqrt{\Phi_0}\left(2\Phi_0 + \phi_0\right)}{4\ \sqrt{\left(\Phi_0 - \phi_0\right)^5}} = \frac{5.10^{-11}\sqrt{0,4}\left(0,8 + 0,2\right)}{2\ \sqrt{\left(0,4 - 0,2\right)^5}};$$

$$\frac{H_0}{\hat{C}_0} = \frac{2c_0\sqrt{\Phi_0}\left(2\Phi_0 + \phi_0\right)}{4\ \sqrt{\left(\Phi_0 - \phi_0\right)^5}}\,\frac{\left(\Phi_0 + \phi_0\right)^{\frac{3}{2}}}{c_0\sqrt{\Phi_0}\left(\Phi_0 + 0,5\phi_0\right)} = \frac{\left(2\Phi_0 + \phi_0\right)\left(\Phi_0 + \phi_0\right)}{2\left(\Phi_0 - \phi_0\right)^2\left(\Phi_0 + 0,5\phi_0\right)}\sqrt{\frac{\Phi_0 + \phi_0}{\Phi_0 - \phi_0}} = 15.\sqrt{3}$$

The above inequalities for $h = 2$, $U_0 = I_0 = 10^{-2}$; $I_{R_0} = I_{R_1} = 0,1$;

$r_1^{(0)} = r_1^{(1)} = 0,028, r_2^{(0)} = r_2^{(1)} = 0, r_3^{(0)} = r_3^{(1)} = 0,125$ become

$$e\,\frac{2.10^{-2}}{2} \le 0,2;\qquad \frac{3e}{17.10^{12}}\left(\frac{2.10^{-2}}{2} + \frac{0,028}{10} + \frac{0,125}{10^3}e^2\right) \le 10^{-1}\,;$$

$$\frac{e\left(2.10^{-2} + 4.75.10^{-1}\right)}{10^{12}10^{-11}.3,4.75} \le 10^{-2} \Leftrightarrow \frac{e\left(2.10^{-1} + 30\right)}{255} \le 1\,;$$

$$\frac{3e}{17.10^{12}}\left(\frac{2.10^{-2}}{2} + \frac{0,028}{10} + \frac{0,125}{10^3}e^2\right) \le 10^{-1}\,;$$

$$e\,\frac{2.10^{-2} + 150.10^{-1}}{10^{12}10^{-11}.3,4.75} \le 10^{-2}\,;$$

$$eK_0 = \frac{3e^2}{17.10^{24}}\left\{\frac{1}{2} + \frac{3}{17}\left[\left(\sum_{n=1}^{3} n\left|r_n^{(0)}\right|\left(\frac{e}{10}\right)^{n-1}\right)\left(\sum_{n=1}^{3}(n+1)\left|l_n^{(0)}\right|\left(\frac{e}{10}\right)^{n}\right) + \right.\right.$$

$$\left.\left. + e\left(2.10^{-2} + \sum_{n=1}^{3}\frac{\left|r_n^{(0)}\right|}{10}\left(\frac{e}{10}\right)^{n-1}\right)\left(\sum_{n=1}^{3}(n+1)n\left|l_n^{(0)}\right|\left(\frac{e}{10}\right)^{n-1}\right)\right]\right\} < 1\,;$$

$$eK_U = \frac{e^2}{10^{24}.3,4.10^{-11}}\left(2 + \frac{1}{75} + \frac{2.10^{-2} + 150.10^{-1}}{2}5e.\sqrt{3}\right) < 1\,;$$

$$eK_1 = \frac{3e^2}{17.10^{24}}\left\{\frac{1}{2} + \frac{3}{17}\left[\left(\sum_{n=1}^{3} n\left|r_n^{(1)}\right|\left(\frac{e}{10}\right)^{n-1}\right)\left(\sum_{n=1}^{3}(n+1)\left|l_n^{(1)}\right|\left(\frac{e}{10}\right)^{n}\right) + \right.\right.$$

$$\left.\left. + e\left(2.10^{-2} + \sum_{n=1}^{3}\frac{\left|r_n^{(1)}\right|}{10}\left(\frac{e}{10}\right)^{n-1}\right)\left(\sum_{n=1}^{3}(n+1)n\left|l_n^{(1)}\right|\left(\frac{e}{10}\right)^{n-1}\right)\right]\right\} < 1;$$

$$eK_I = \frac{e^2}{10^{24}.3,4.10^{-11}}\left(2 + \frac{1}{75} + \frac{2.10^{-2} + 150.10^{-1}}{2} 5e.\sqrt{3}\right) < 1;$$

$$\dot{K}_0 = \frac{e}{10^{12}}\left\{\frac{9}{17^2}\left[\left(0,028 + 3.0,125\frac{e^2}{10^2}\right)\left(2.6\frac{e}{10} + 4.\frac{1}{3}\left(\frac{e}{10}\right)^3\right) + \right.\right.$$

$$\left.\left. + e\left(2.10^{-2} + \frac{0,028}{10} + \frac{0,125}{10}\frac{e^2}{10^2}\right)\left(2.6 + 12.\frac{1}{3}\left(\frac{e}{10}\right)^2\right)\right] + \frac{3}{17}\right\} < 1;$$

$$\dot{K}_U = \frac{e}{10^{12}.3,4.10^{-11}}\left[2 + \frac{1}{75} + \frac{2.10^{-2} + 150.10^{-1}}{2.75} 5e.\sqrt{3}\right] < 1;$$

$$\dot{K}_1 = \frac{e}{10^{12}}\left\{\frac{9}{17^2}\left[\left(0,028 + 3.0,125\frac{e^2}{10^2}\right)\left(2.6\frac{e}{10} + 4.\frac{1}{3}\left(\frac{e}{10}\right)^3\right) + \right.\right.$$

$$\left.\left. + e\left(2.10^{-2} + \frac{0,028}{10} + \frac{0,125}{10}\frac{e^2}{10^2}\right)\left(2.6 + 12.\frac{1}{3}\left(\frac{e}{10}\right)^2\right)\right] + \frac{3}{17}\right\} < 1;$$

$$\dot{K}_I = \frac{e}{10^{12}.3,4.10^{-11}}\left[2 + \frac{1}{75} + \frac{2.10^{-2} + 150.10^{-1}}{2.75} 5e.\sqrt{3}\right] < 1.$$

We compute just $\dot{K}_0, \dot{K}_U, \dot{K}_1, \dot{K}_I$ since K_0, K_U, K_1, K_I are of order $\dfrac{1}{\mu^2}$:

$$\dot{K}_0 = \frac{e}{10^{12}}\left(\frac{9}{17^2} + \frac{3}{17}\right) < 1; \quad \dot{K}_U = \frac{e}{34}4,3694 = 0,35 < 1; \quad \dot{K}_I = 0,35 < 1 \Rightarrow K \approx 0,35.$$

8.7. LOSSY TRANSMISSION LINES TERMINATED BY NONLINEAR RL-LOADS PARALLEL TO C-LOAD

Here we consider a lossy transmission line terminated by the same circuits.

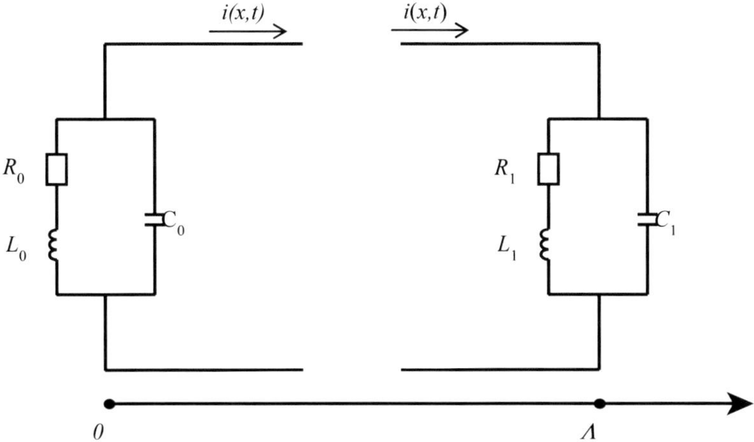

Figure 8.1.

We proceed from the system

$$C\frac{\partial u(x,t)}{\partial t} + \frac{\partial i(x,t)}{\partial x} + Gu(x,t) = 0,$$

$$L\frac{\partial i(x,t)}{\partial t} + \frac{\partial u(x,t)}{\partial x} + Ri(x,t) = 0 \tag{8.7.1}$$

$$(x,t) \in \Pi = \left\{(x,t) \in \Pi^2 : (x,t) \in [0,\Lambda] \times [0,\infty)\right\}$$

where $u(x,t)$ and $i(x,t)$ are the unknown voltage and current, while L, C, R and G are prescribed specific parameters of the line and $\Lambda > 0$ is its length. For the above system (8.7.1) can be formulated the following mixed problem: to find $u(x,t)$ and $i(x,t)$ in Π such that

$$u(x,0) = u_0(x), \ i(x,0) = i_0(x), \ x \in [0,\Lambda]. \tag{8.7.2}$$

The boundary conditions are already derived (cf.Fig. 8.1):

for $x = 0$

$$\left[i_{R_0L_0}(t)\frac{dL_0(i_{R_0L_0}(t))}{di_{R_0L_0}} + L_0(i_{R_0L_0}(t)) \right]\frac{di_{R_0L_0}(t)}{dt} = u(0,t) - R_0(i_{R_0L_0}(t)),$$

$$\left[\frac{dC_0(u(0,t))}{du}u(0,t) + C_0(u(0,t)) \right]\frac{du(0,t)}{dt} = -i(0,t) - \breve{I}_{in}(t) - i_{R_0L_0}(t) \qquad (8.7.3)$$

and for $x = \Lambda$:

$$\left[i_{R_1L_1}(t)\frac{dL_1(i_{R_1L_1}(t))}{di_{R_1L_1}} + L_1(i_{R_1L_1}(t)) \right]\frac{di_{R_1L_1}(t)}{dt} = u(\Lambda,t) - R_1(i_{R_1L_1}(t)),$$

$$\left[\frac{dC_1(u(\Lambda,t))}{du}u(\Lambda,t) + C_1(u(\Lambda,t)) \right]\frac{du(\Lambda,t)}{dt} = -i(\Lambda,t) - i_{R_1L_1}(t). \qquad (8.7.4)$$

8.8. REDUCING THE MIXED PROBLEM TO AN INITIAL VALUE PROBLEM ON THE BOUNDARY

First we present (8.7.1) in the form:

$$\frac{\partial u(x,t)}{\partial t} + \frac{1}{C}\frac{\partial i(x,t)}{\partial x} + \frac{G}{C}u(x,t) = 0,$$

$$\frac{\partial i(x,t)}{\partial t} + \frac{1}{L}\frac{\partial u(x,t)}{\partial x} + \frac{R}{L}i(x,t) = 0$$

and then in the matrix form

$$\frac{\partial U(x,t)}{\partial t} + A\frac{\partial U(x,t)}{\partial x} + BU(x,t) = 0 \qquad (8.8.1)$$

where $U(x,t) = \begin{bmatrix} u(x,t) \\ i(x,t) \end{bmatrix}$, $\dfrac{\partial U(x,t)}{\partial t} = \begin{bmatrix} \dfrac{\partial u(x,t)}{\partial t} \\ \dfrac{\partial i(x,t)}{\partial t} \end{bmatrix}$, $\dfrac{\partial U(x,t)}{\partial x} = \begin{bmatrix} \dfrac{\partial u(x,t)}{\partial x} \\ \dfrac{\partial i(x,t)}{\partial x} \end{bmatrix}$,

$$A = \begin{bmatrix} 0 & 1/C \\ 1/L & 0 \end{bmatrix}, \ B = \begin{bmatrix} G/C & 0 \\ 0 & R/L \end{bmatrix}.$$

To transform the matrix $A = \begin{bmatrix} 0 & 1/C \\ 1/L & 0 \end{bmatrix}$ in a diagonal form we solve the characteristic

equation $\begin{vmatrix} -\lambda & 1/C \\ 1/L & -\lambda \end{vmatrix} = 0$. The roots are $\lambda_1 = \dfrac{1}{\sqrt{LC}}$ and $\lambda_2 = -\dfrac{1}{\sqrt{LC}}$. Denote by H the

matrix formed by eigenvectors $H = \begin{bmatrix} \sqrt{C} & \sqrt{L} \\ -\sqrt{C} & \sqrt{L} \end{bmatrix}$ and by $H^{-1} = \begin{bmatrix} \dfrac{1}{2\sqrt{C}} & -\dfrac{1}{2\sqrt{C}} \\ \dfrac{1}{2\sqrt{L}} & \dfrac{1}{2\sqrt{L}} \end{bmatrix}$ its

inverse one.

Denote by $A^{can} = \begin{bmatrix} \dfrac{1}{\sqrt{LC}} & 0 \\ 0 & -\dfrac{1}{\sqrt{LC}} \end{bmatrix}$, then $A^{can} = HAH^{-1}$. Introduce a new variable

$Z = HU$ (or $U = H^{-1}Z$)

$$Z = \begin{bmatrix} V(x,t) \\ I(x,t) \end{bmatrix}, \quad H = \begin{bmatrix} \sqrt{C} & \sqrt{L} \\ -\sqrt{C} & \sqrt{L} \end{bmatrix}, \quad U = \begin{bmatrix} u(x,t) \\ i(x,t) \end{bmatrix}, \quad H^{-1} = \begin{bmatrix} \dfrac{1}{2\sqrt{C}} & -\dfrac{1}{2\sqrt{C}} \\ \dfrac{1}{2\sqrt{L}} & \dfrac{1}{2\sqrt{L}} \end{bmatrix}.$$

Then

$$\begin{vmatrix} V(x,t) = \sqrt{C}\, u(x,t) + \sqrt{L}\, i(x,t) \\ I(x,t) = -\sqrt{C}\, u(x,t) + \sqrt{L}\, i(x,t) \end{vmatrix} \tag{8.8.2}$$

and

$$\begin{vmatrix} u(x,t) = \dfrac{1}{2\sqrt{C}} V(x,t) - \dfrac{1}{2\sqrt{C}} I(x,t) \\ i(x,t) = \dfrac{1}{2\sqrt{L}} V(x,t) + \dfrac{1}{2\sqrt{L}} I(x,t). \end{vmatrix} \tag{8.8.3}$$

Replacing $U = H^{-1}Z$ in (8.8.1) and making transformations as in the previous chapters we obtain

$$\frac{\partial Z(x,t)}{\partial t} + A^{can} \frac{\partial Z(x,t)}{\partial x} + H\left(BH^{-1}\right)Z(x,t) = 0. \tag{8.8.4}$$

Taking into account Heaviside condition $\dfrac{R}{L} = \dfrac{G}{C}$ we have

$$HBH^{-1} = \begin{bmatrix} \dfrac{1}{2}\left(\dfrac{G}{C}+\dfrac{R}{L}\right) & \dfrac{1}{2}\left(-\dfrac{G}{C}+\dfrac{R}{L}\right) \\ \dfrac{1}{2}\left(-\dfrac{G}{C}+\dfrac{R}{L}\right) & \dfrac{1}{2}\left(\dfrac{G}{C}+\dfrac{R}{L}\right) \end{bmatrix} = \begin{bmatrix} \dfrac{R}{L} & 0 \\ 0 & \dfrac{R}{L} \end{bmatrix}.$$

and then (8.8.1) becomes:

$$\begin{bmatrix} \dfrac{\partial V(x,t)}{\partial t} \\ \dfrac{\partial I(x,t)}{\partial t} \end{bmatrix} + \begin{bmatrix} \dfrac{1}{\sqrt{LC}} & 0 \\ 0 & -\dfrac{1}{\sqrt{LC}} \end{bmatrix} \begin{bmatrix} \dfrac{\partial V(x,t)}{\partial x} \\ \dfrac{\partial I(x,t)}{\partial x} \end{bmatrix} + \begin{bmatrix} \dfrac{R}{L} & 0 \\ 0 & \dfrac{R}{L} \end{bmatrix} \begin{bmatrix} V(x,t) \\ I(x,t) \end{bmatrix} = \begin{bmatrix} 0 \\ 0 \end{bmatrix}. \tag{8.8.5}$$

The new initial conditions are

$$V(x,0) = \sqrt{C}\, u(x,0) + \sqrt{L}\, i(x,0) = \sqrt{C}\, u_0(x) + \sqrt{L}\, i_0(x) \equiv V_0(x)\,, \tag{8.8.6}$$

$$I(x,0) = -\sqrt{C}\, u(x,0) + \sqrt{L}\, i(x,0) = -\sqrt{C}\, u_0(x) + \sqrt{L}\, i_0(x) \equiv I_0(x)\,, x \in [0,\Lambda] \tag{8.8.7}$$

One can simplify (8.8.5) by the next substitution:

$$W(x,t) = e^{\frac{R}{L}t} V(x,t)\,, \quad J(x,t) = e^{\frac{R}{L}t} I(x,t)\,, \tag{8.8.8}$$

or

$$V(x,t) = e^{-\frac{R}{L}t} W(x,t)\,, \quad I(x,t) = e^{-\frac{R}{L}t} J(x,t)\,. \tag{8.8.9}$$

Replacing in (8.8.3) we obtain

$$\left| \begin{aligned} u(x,t) &= \frac{1}{2\sqrt{C}} e^{-\frac{R}{L}t} W(x,t) - \frac{1}{2\sqrt{C}} e^{-\frac{R}{L}t} J(x,t) \\ i(x,t) &= \frac{1}{2\sqrt{L}} e^{-\frac{R}{L}t} W(x,t) + \frac{1}{2\sqrt{L}} e^{-\frac{R}{L}t} J(x,t) \end{aligned} \right. \tag{8.8.10}$$

System (8.8.5) can be written in the form:

$$\left|\begin{array}{l} \dfrac{\partial V(x,t)}{\partial t} + \dfrac{1}{\sqrt{LC}}\dfrac{\partial V(x,t)}{\partial x} + \dfrac{R}{L}V(x,t) = 0, \\[2ex] \dfrac{\partial I(x,t)}{\partial t} - \dfrac{1}{\sqrt{LC}}\dfrac{\partial I(x,t)}{\partial x} + \dfrac{R}{L}I(x,t) = 0. \end{array}\right. \tag{8.8.11}$$

Then putting $V(x,t)$ and $I(x,t)$ from (8.8.9) into (8.8.11) we obtain

$$\left|\begin{array}{l} \dfrac{\partial W(x,t)}{\partial t} + \dfrac{1}{\sqrt{LC}}\dfrac{\partial W(x,t)}{\partial x} = 0, \\[2ex] \dfrac{\partial J(x,t)}{\partial t} - \dfrac{1}{\sqrt{LC}}\dfrac{\partial J(x,t)}{\partial x} = 0. \end{array}\right. \tag{8.8.12}$$

The mixed problem for (8.8.12) can be reduced to an initial value problem for a functional differential equation on the right boundary.

Let us replace (8.8.10) into the boundary conditions:

$$\frac{d\widetilde{L}_0(i_{R_0 L_0}(t))}{di_{R_0 L_0}}\frac{di_{R_0 L_0}(t)}{dt} = \frac{1}{2\sqrt{C}}e^{-\frac{R}{L}t}W(0,t) - \frac{1}{2\sqrt{C}}e^{\frac{R}{L}t}J(0,t) - R_0(i_{R_0 L_0}(t)),$$

$$\frac{1}{2\sqrt{C}}\frac{d\left(e^{-\frac{R}{L}t}W(0,t) - e^{\frac{R}{L}t}J(0,t)\right)}{dt} = -\frac{\dfrac{e^{-\frac{R}{L}t}}{2\sqrt{L}}W(0,t) + \dfrac{e^{\frac{R}{L}t}}{2\sqrt{L}}J(0,t) + \breve{I}_{in}(t) + i_{R_0 L_0}(t)}{d\widetilde{C}_0(u(0,t))/du}$$

$$\frac{d\widetilde{L}_1(i_{R_1 L_1}(t))}{di_{R_1 L_1}}\frac{di_{R_1 L_1}(t)}{dt} = \frac{1}{2\sqrt{C}}e^{-\frac{R}{L}t}W(\Lambda,t) - \frac{1}{2\sqrt{C}}e^{-\frac{R}{L}t}J(\Lambda,t) - R_1(i_{R_1 L_1}(t)),$$

$$\frac{1}{2\sqrt{C}}\frac{d\left(e^{-\frac{R}{L}(t-T)}W(t-T) - e^{-\frac{R}{L}t}J(t)\right)}{dt} = \frac{-\dfrac{1}{2\sqrt{L}}e^{-\frac{R}{L}t}W(\Lambda,t) - \dfrac{1}{2\sqrt{L}}e^{-\frac{R}{L}t}J(\Lambda,t) - i_{R_1 L_1}(t)}{d\widetilde{C}_1(u(\Lambda,t))/du}.$$

Then in view of $W(0,t) = W(\Lambda, t+T)$, $J(0,t+T) = J(\Lambda,t)$ we choose

$$W(0,t) = W(t), \; J(t) = J(\Lambda,t)$$

to be unknown functions. Then

$$\frac{di_{R_0 L_0}(t)}{dt} = \frac{\dfrac{1}{2\sqrt{C}}\left(e^{-\frac{R}{L}t}W(t) - e^{-\frac{R}{L}(t-T)}J(t-T)\right) - R_0(i_{R_0 L_0}(t))}{d\widetilde{L}_0(i_{R_0 L_0}(t))/di_{R_0 L_0}},$$

$$\frac{d\left(e^{-\frac{R}{L}t}W(t)-e^{-\frac{R}{L}(t-T)}J(t-T)\right)}{dt}=-2\sqrt{C}\,\frac{\frac{1}{2\sqrt{L}}\left(e^{-\frac{R}{L}t}W(t)-e^{-\frac{R}{L}(t-T)}J(t-T)\right)+\breve{I}_{in}(t)+i_{R_0L_0}(t)}{d\widetilde{C}_0(u(0,t))/du} \qquad (8.8.13)$$

$$\frac{di_{R_1L_1}(t)}{dt}=\frac{\frac{1}{2\sqrt{C}}e^{\frac{R}{L}(t-T)}W(t-T)-\frac{1}{2\sqrt{C}}e^{-\frac{R}{L}t}J(t)-R_1(i_{R_1L_1}(t))}{d\widetilde{L}_1(i_{R_1L_1}(t))/di_{R_1L_1}}\,;$$

$$\frac{d\left(e^{-\frac{R}{L}t}W(t)-e^{-\frac{R}{L}(t-T)}J(t-T)\right)}{dt}=2\sqrt{C}\,\frac{-\frac{1}{2\sqrt{L}}e^{-\frac{R}{L}t}W(t)-\frac{1}{2\sqrt{L}}e^{-\frac{R}{L}(t-T)}J(t-T)-i_{R_1L_1}(t)}{d\widetilde{C}_1(u(\Lambda,t))/du}\,.$$

We recall that

$$(RT)/L=\left(R\Lambda\sqrt{LC}\right)/L=(R\Lambda)/Z_0\,.$$

So our goal is to formulate conditions for the existence of periodic solutions of the above system. We notice, however, that the right-hand sides of the last system are not periodic functions. Therefore it is a natural to look for solutions of the type $e^{-\frac{R}{L}t}W(t)$, $e^{-\frac{R}{L}t}J(t)$, where $W(t)$, $J(t)$ are periodic functions.

Then $\widehat{W}(t)=e^{-\frac{R}{L}t}W(t)$, $\widehat{J}(t)=e^{-\frac{R}{L}t}J(t)$ become an oscillatory solution and (8.8.13) can be rewritten as

$$\frac{di_{R_0L_0}(t)}{dt}=\frac{\widehat{W}(t)-\widehat{J}(t-T)-2\sqrt{C}R_0(i_{R_0L_0}(t))}{2\sqrt{C}d\widetilde{L}_0(i_{R_0L_0}(t))/di_{R_0L_0}}\,,$$

$$\frac{d\left(\widehat{W}(t)-\widehat{J}(t-T)\right)}{dt}=-\frac{\widehat{W}(t)-\widehat{J}(t-T)+2Z_0\sqrt{C}\breve{I}_{in}(t)+2Z_0\sqrt{C}i_{R_0L_0}(t)}{Z_0d\widetilde{C}_0(u(0,t))/du} \qquad (8.8.14)$$

$$\frac{di_{R_1L_1}(t)}{dt}=\frac{\widehat{W}(t-T)-\widehat{J}(t)-2\sqrt{C}R_1(i_{R_1L_1}(t))}{2\sqrt{C}d\widetilde{L}_1(i_{R_1L_1}(t))/di_{R_1L_1}}\,,$$

$$\frac{d\left(\widehat{W}(t)-\widehat{J}(t-T)\right)}{dt}=-\frac{\widehat{W}(t)+\widehat{J}(t-T)+2Z_0\sqrt{C}i_{R_1L_1}(t)}{Z_0d\widetilde{C}_1(u(\Lambda,t))/du}$$

where in view of $W(0,t)=W(\Lambda,t+T)$, $J(0,t+T)=J(\Lambda,t)$ and $W(0,t)=W(t)$, $J(t)=J(\Lambda,t)$ we have:

$$u(0,t) = \frac{1}{2\sqrt{C}} e^{-\frac{R}{L}t} W(0,t) - \frac{1}{2\sqrt{C}} e^{-\frac{R}{L}t} J(0,t) = \frac{\widehat{W}(t) - \widehat{J}(t-T)}{2\sqrt{C}},$$

$$u(\Lambda,t) = \frac{1}{2\sqrt{C}} e^{-\frac{R}{L}t} W(\Lambda,t) - \frac{1}{2\sqrt{C}} e^{-\frac{R}{L}t} J(\Lambda,t) = \frac{\widehat{W}(t-T) - \widehat{J}(t)}{2\sqrt{C}}.$$

Now we are able to formulate the problem for the existence-uniqueness of an oscillatory solution of (8.8.14).

8.9. Operator Presentation of the Oscillatory Problem

Here we again denote $\widehat{W}(t)$, $\widehat{J}(t)$ by $W(t)$, $J(t)$. Now we are able to formulate the problem for the existence-uniqueness of an oscillatory solution of the following system:

$$\frac{di_{R_0 L_0}(t)}{dt} = \frac{W(t) - J(t-T) - 2\sqrt{C} R_0(i_{R_0 L_0}(t))}{2\sqrt{C} d\widetilde{L}_0(i_{R_0 L_0}(t))/di_{R_0 L_0}};$$

$$\frac{dW(t)}{dt} = \frac{dJ(t-T)}{dt} - \frac{W(t) - J(t-T) + 2\sqrt{L}\,\breve{I}_{in}(t) + 2\sqrt{L}\,i_{R_0 L_0}(t)}{Z_0 d\widetilde{C}_0\left(\dfrac{W(t) - J(t-T)}{2\sqrt{C}}\right)/du};$$

$$\text{(8.9.1)}$$

$$\frac{di_{R_1 L_1}(t)}{dt} = \frac{W(t-T) - J(t) - 2\sqrt{C} R_1(i_{R_1 L_1}(t))}{2\sqrt{C} d\widetilde{L}_1(i_{R_1 L_1}(t))/di_{R_1 L_1}};$$

$$\frac{dJ(t)}{dt} = \frac{dW(t-T)}{dt} + \frac{W(t-T) + J(t) + 2\sqrt{L}i_{R_1 L_1}(t)}{Z_0 d\widetilde{C}_1\left(\dfrac{W(t-T) - J(t)}{2\sqrt{C}}\right)/du}, \; t \in [T,\infty)$$

$$W(t) = \widetilde{W}_0(t), \;\; J(t) = \widetilde{J}_0(t), \, t \in [0,T], \, i_{R_0 L_0}(T) = i^{(0)}_{R_0 L_0} = 0, \, i_{R_1 L_1}(T) = i^{(0)}_{R_1 L_1} = 0.$$

We introduce an operator presentation of the oscillatory problem and by a fixed point theorem in uniform spaces [14] we establish an existence-uniqueness of oscillatory solution.

Now we are able to formulate the main problem: to find a solution of (8.9.1) with advanced prescribed zeros on $[t_0,\infty)$, $T \equiv t_0$, where $\widetilde{W}_0(t)$, $\widetilde{J}_0(t)$, $t \in [0,t_0]$ are prescribed initial oscillating functions on the interval $[0,t_0]$.

Let $S_T = \{\tau_k\}_{k=0}^n, n \in N$ be the set of zeros of the initial function, that is, $\tilde{W}_0(\tau_k) = 0$, $\tilde{J}_0(\tau_k) = 0$ such that $\tau_0 = 0$, $\tau_n = T \equiv t_0$. Besides $\max\{\tau_{k+1} - \tau_k : k = 0,1,...,n\} \le T_0$.

Let $S = \{t_k\}_{k=0}^\infty$ be a strictly increasing sequence of real numbers satisfying the following conditions **(C)**:

(C1) $\lim\limits_{k \to \infty} t_k = \infty$;

(C2) for every k there is $s < k$ such that $t_k - T = t_s$ where $t_s \in S_T \cup S$.

It follows

(C3) $0 < \inf\{t_{k+1} - t_k : k = 0,1,2,...\} \le \sup\{t_{k+1} - t_k : k = 0,1,2,...\} = T_0 < \infty$.

Introduce the set $C^1[t_0,\infty)$ consisting of all continuous and bounded functions differentiable with bounded derivatives on every interval $[t_k, t_{k+1}]$.

Remark 8.9.1. Let us note that the left and right derivative at t_k of any $W(.)$, $J(.) \in C^1[t_0,\infty)$ may not coincide. That is why we introduce a suitable topology for continuous functions with piece wise continuous derivatives. This is one reason to introduce uniform spaces (cf. Chapter I, also [14]).

Introduce the sets

$$M_0 = \left\{ i_{R_0L_0}(.) \in C^1[t_0,\infty) : i_{R_0L_0}(t_k) = 0 \wedge \left| i_{R_0L_0}(t) \right| \le I_{R_0} e^{-\frac{R}{L}t}, t \in [t_k, t_{k+1}]; i_{R_0L_0}(t_0) = 0 \right\},$$

$$M_W = \left\{ W(.) \in C^1[t_0,\infty) : W(t_k) = 0 \wedge \left| W(t) \right| \le W_0 e^{-\frac{R}{L}t}, t \in [t_k, t_{k+1}] \right\},$$

$$M_1 = \left\{ i_{R_1L_1}(.) \in C^1[t_0,\infty) : i_{R_1L_1}(t_k) = 0 \wedge \left| i_{R_1L_1}(t) \right| \le I_{R_1} e^{-\frac{R}{L}t}, t \in [t_k, t_{k+1}]; i_{R_1L_1}(t_0) = 0 \right\},$$

$$M_J = \left\{ J(.) \in C^1[t_0,\infty) : J(t_k) = 0 \wedge \left| J(t) \right| \le J_0 e^{-\frac{R}{L}t}, t \in [t_k, t_{k+1}] \right\}.$$

Further on the following assumptions will be hold:

Assumtion (**IN**): $\left|\widetilde{W}_0(t)\right| \le W_0 e^{-\beta} e^{-\frac{R}{L}(t+T)}$; $\left|\widetilde{J}_0(t)\right| \le J_0 e^{-\beta} e^{-\frac{R}{L}(t+T)}$, $t \in [0,T]$.

Assumption (Π): $\dfrac{W_0 + J_0}{2\sqrt{C}} \le \phi_0$, $I_{R_0}, I_{R_1} \le i_0$.

Remark 8.9.2. It follows that the functions from M_0, M_W, M_1 and M_J satisfy the inequalities

$$\left|i_{R_0 L_0}(t)\right| \le I_{R0} e^{\mu(t-t_k)}, t \in [t_k, t_{k+1}] , \left|W(t)\right| \le W_0 e^{\mu(t-t_k)}, t \in [t_k, t_{k+1}] ,$$

$$\left|i_{R_1 L_1}(t)\right| \le I_{R_1} e^{\mu(t-t_k)}, t \in [t_k, t_{k+1}] , \left|J(t)\right| \le J_0 e^{\mu(t-t_k)}, t \in [t_k, t_{k+1}], (k = 0,1,2,...\ )$$

where $I_{R_0}, W_0, I_{R_1}, J_0, \mu,\ \mu T_0 = \mu_0 = const.$ are positive constants.

Introduce the following family of pseudo-metrics

$$\rho^{(k)}(i_{R_0 L_0}, \bar{i}_{R_0 L_0}) = \max\left\{\left|i_{R_0 L_0}(t) - \bar{i}_{R_0 L_0}(t)\right| : t \in [t_k, t_{k+1}]\right\},$$
$$\rho^{(k)}(W, \overline{W}) = \max\left\{\left|W(t) - \overline{W}(t)\right| : t \in [t_k, t_{k+1}]\right\},$$

$$\rho^{(k)}(i_{R_1 L_1}, \bar{i}_{R_1 L_1}) = \max\left\{\left|i_{R_1 L_1}(t) - \bar{i}_{R_1 L_1}(t)\right| : t \in [t_k, t_{k+1}]\right\},$$
$$\rho^{(k)}(J, \overline{J}) = \max\left\{\left|J(t) - \overline{J}(t)\right| : t \in [t_k, t_{k+1}]\right\},$$

$$\hat{\rho}^{(k)}(i_{R_0 L_0}, \bar{i}_{R_0 L_0}) = \max\left\{\left|i_{R_0 L_0}(t) - \bar{i}_{R_0 L_0}(t)\right| : t \in [t_0, t_{k+1}]\right\},$$
$$\hat{\rho}^{(k)}(W, \overline{W}) = \max\left\{\left|W(t) - \overline{W}(t)\right| : t \in [t_0, t_{k+1}]\right\},$$

$$\hat{\rho}^{(k)}(i_{R_1 L_1}, \bar{i}_{R_1 L_1}) = \max\left\{\left|i_{R_1 L_1}(t) - \bar{i}_{R_1 L_1}(t)\right| : t \in [t_0, t_{k+1}]\right\},$$
$$\hat{\rho}^{(k)}(J, \overline{J}) = \max\left\{\left|J(t) - \overline{J}(t)\right| : t \in [t_0, t_{k+1}]\right\},$$

$$\rho_\mu^{(k)}(i_{R_0 L_0}, \bar{i}_{R_0 L_0}) = \max\left\{e^{-\mu(t-t_k)}\left|i_{R_0 L_0}(t) - \bar{i}_{R_0 L_0}(t)\right| : t \in [t_k, t_{k+1}]\right\},$$
$$\rho_\mu^{(k)}(W, \overline{W}) = \max\left\{e^{-\mu(t-t_k)}\left|W(t) - \overline{W}(t)\right| : t \in [t_k, t_{k+1}]\right\},$$

$$\rho_\mu^{(k)}(i_{R_1 L_1}, \bar{i}_{R_1 L_1}) = \max\left\{e^{-\mu(t-t_k)}\left|i_{R_1 L_1}(t) - \bar{i}_{R_1 L_1}(t)\right| : t \in [t_k, t_{k+1}]\right\},$$
$$\rho_\mu^{(k)}(J, \overline{J}) = \max\left\{e^{-\mu(t-t_k)}\left|J(t) - \overline{J}(t)\right| : t \in [t_k, t_{k+1}]\right\},$$

$$\hat{\rho}_\mu^{(k)}(i_{R_0 L_0}, \bar{i}_{R_0 L_0}) = \max\left\{\rho_\mu^{(0)}(i_{R_0 L_0}, \bar{i}_{R_0 L_0}), \rho_\mu^{(1)}(i_{R_0 L_0}, \bar{i}_{R_0 L_0}),..., \rho_\mu^{(k)}(i_{R_0 L_0}, \bar{i}_{R_0 L_0})\right\},$$

$$\hat{\rho}_{\mu}^{(k)}(W,\overline{W}) = \max\left\{\rho_{\mu}^{(0)}(W,\overline{W}),\rho_{\mu}^{(1)}(W,\overline{W}),...,\rho_{\mu}^{(k)}(W,\overline{W})\right\},$$

$$\hat{\rho}_{\mu}^{(k)}(i_{R_1L_1},\overline{i}_{R_1L_1}) = \max\left\{\rho_{\mu}^{(0)}(i_{R_1L_1},\overline{i}_{R_1L_1}),\rho_{\mu}^{(1)}(i_{R_1L_1},\overline{i}_{R_1L_1}),...,\rho_{\mu}^{(k)}(i_{R_1L_1},\overline{i}_{R_1L_1})\right\},$$

$$\hat{\rho}_{\mu}^{(k)}(J,\overline{J}) = \max\left\{\rho_{\mu}^{(0)}(J,\overline{J}),\rho_{\mu}^{(1)}(J,\overline{J}),...,\rho_{\mu}^{(k)}(J,\overline{J})\right\},$$

$$\rho_{\mu}^{(k)}(\dot{i}_{R_0L_0},\dot{\overline{i}}_{R_0L_0}) = \max\left\{e^{-\mu(t-t_k)}\left|\dot{i}_{R_0L_0}(t)-\dot{\overline{i}}_{R_0L_0}(t)\right| : t \in [t_k,t_{k+1}]\right\},$$

$$\rho_{\mu}^{(k)}(\dot{W},\dot{\overline{W}}) = \max\left\{e^{-\mu(t-t_k)}\left|\dot{W}(t)-\dot{\overline{W}}(t)\right| : t \in [t_k,t_{k+1}]\right\},$$

$$\rho_{\mu}^{(k)}(\dot{i}_{R_1L_1},\dot{\overline{i}}_{R_1L_1}) = \max\left\{e^{-\mu(t-t_k)}\left|\dot{i}_{R_1L_1}(t)-\dot{\overline{i}}_{R_1L_1}(t)\right| : t \in [t_k,t_{k+1}]\right\},$$

$$\rho_{\mu}^{(k)}(\dot{J},\dot{\overline{J}}) = \max\left\{e^{-\mu(t-t_k)}\left|\dot{J}(t)-\dot{\overline{J}}(t)\right| : t \in [t_k,t_{k+1}]\right\},$$

$$\hat{\rho}_{\mu}^{(k)}(\dot{i}_{R_0L_0},\dot{\overline{i}}_{R_0L_0}) = \max\left\{\rho_{\mu}^{(0)}(\dot{i}_{R_0L_0},\dot{\overline{i}}_{R_0L_0}),\rho_{\mu}^{(1)}(\dot{i}_{R_0L_0},\dot{\overline{i}}_{R_0L_0}),...,\rho_{\mu}^{(k)}(\dot{i}_{R_0L_0},\dot{\overline{i}}_{R_0L_0})\right\},$$

$$\hat{\rho}_{\mu}^{(k)}(\dot{W},\dot{\overline{W}}) = \max\left\{\rho_{\mu}^{(0)}(\dot{W},\dot{\overline{W}}),\rho_{\mu}^{(1)}(\dot{W},\dot{\overline{W}}),...,\rho_{\mu}^{(k)}(\dot{W},\dot{\overline{W}})\right\},$$

$$\hat{\rho}_{\mu}^{(k)}(\dot{i}_{R_1L_1},\dot{\overline{i}}_{R_1L_1}) = \max\left\{\rho_{\mu}^{(0)}(\dot{i}_{R_1L_1},\dot{\overline{i}}_{R_1L_1}),\rho_{\mu}^{(1)}(\dot{i}_{R_1L_1},\dot{\overline{i}}_{R_1L_1}),...,\rho_{\mu}^{(k)}(\dot{i}_{R_1L_1},\dot{\overline{i}}_{R_1L_1})\right\},$$

$$\hat{\rho}_{\mu}^{(k)}(\dot{J},\dot{\overline{J}}) = \max\left\{\rho_{\mu}^{(0)}(\dot{J},\dot{\overline{J}}),\rho_{\mu}^{(1)}(\dot{J},\dot{\overline{J}}),...,\rho_{\mu}^{(k)}(\dot{J},\dot{\overline{J}})\right\}.$$

The following inequalities imply the equivalence of the families of pseudo-metrics

$$\rho_{\mu}^{(k)}(i_{R_0L_0},\overline{i}_{R_0L_0}) \le \rho^{(k)}(i_{R_0L_0},\overline{i}_{R_0L_0}) \le e^{\mu T_0}\rho_{\mu}^{(k)}(i_{R_0L_0},\overline{i}_{R_0L_0}), \ (k=0,1,2,...),$$

$$\rho_{\mu}^{(k)}(W,\overline{W}) \le \rho^{(k)}(W,\overline{W}) \le e^{\mu T_0}\rho_{\mu}^{(k)}(W,\overline{W}), \ (k=0,1,2,...),$$

$$\rho_{\mu}^{(k)}(i_{R_1L_1},\overline{i}_{R_1L_1}) \le \rho^{(k)}(i_{R_1L_1},\overline{i}_{R_1L_1}) \le e^{\mu T_0}\rho_{\mu}^{(k)}(i_{R_1L_1},\overline{i}_{R_1L_1}), \ (k=0,1,2,...),$$

$$\rho_{\mu}^{(k)}(J,\overline{J}) \le \rho^{(k)}(J,\overline{J}) \le e^{\mu T_0}\rho_{\mu}^{(k)}(J,\overline{J}), \ (k=0,1,2,...).$$

It is easy to verify that

$$\begin{aligned}
\hat{\rho}^{(k)}(i_{R_0L_0},\overline{i}_{R_0L_0}) &= \max\left\{\rho^{(0)}(i_{R_0L_0},\overline{i}_{R_0L_0}),\rho^{(1)}(i_{R_0L_0},\overline{i}_{R_0L_0}),...,\rho^{(k)}(i_{R_0L_0},\overline{i}_{R_0L_0})\right\} \le \\
&\le e^{\mu T_0}\max\left\{\rho_{\mu}^{(0)}(i_{R_0L_0},\overline{i}_{R_0L_0}),\rho_{\mu}^{(1)}(i_{R_0L_0},\overline{i}_{R_0L_0}),...,\rho_{\mu}^{(k)}(i_{R_0L_0},\overline{i}_{R_0L_0})\right\} = e^{\mu_0}\hat{\rho}_{\mu}^{(k)}(i_{R_0L_0},\overline{i}_{R_0L_0}) \\
\hat{\rho}^{(k)}(W,\overline{W}) &= \max\left\{\rho^{(0)}(W,\overline{W}),\rho^{(1)}(W,\overline{W}),...,\rho^{(k)}(W,\overline{W})\right\} \le \\
&\le e^{\mu T_0}\max\left\{\rho_{\mu}^{(0)}(W,\overline{W}),\rho_{\mu}^{(1)}(W,\overline{W}),...,\rho_{\mu}^{(k)}(W,\overline{W})\right\} = e^{\mu_0}\hat{\rho}_{\mu}^{(k)}(W,\overline{W}), \\
\hat{\rho}^{(k)}(i_{R_1L_1},\overline{i}_{R_1L_1}) &= \max\left\{\rho^{(0)}(i_{R_1L_1},\overline{i}_{R_1L_1}),\rho^{(1)}(i_{R_1L_1},\overline{i}_{R_1L_1}),...,\rho^{(k)}(i_{R_1L_1},\overline{i}_{R_1L_1})\right\} \le \\
&\le e^{\mu_0}\max\left\{\rho_{\mu}^{(0)}(i_{R_1L_1},\overline{i}_{R_1L_1}),\rho_{\mu}^{(1)}(i_{R_1L_1},\overline{i}_{R_1L_1}),...,\rho_{\mu}^{(k)}(i_{R_1L_1},\overline{i}_{R_1L_1})\right\} = e^{\mu_0}\hat{\rho}_{\mu}^{(k)}(i_{R_1L_1},\overline{i}_{R_1L_1})
\end{aligned} \tag{8.9.2}$$

$$\hat{\rho}^{(k)}(J,\overline{J}) = \max\{\rho^{(0)}(J,\overline{J}),\rho^{(1)}(J,\overline{J}),...,\rho^{(k)}(J,\overline{J})\} \le$$
$$\le e^{\mu T_0}\max\{\rho_\mu^{(0)}(J,\overline{J}),\rho_\mu^{(1)}(J,\overline{J}),...,\rho_\mu^{(k)}(J,\overline{J})\} = e^{\mu_0}\hat{\rho}_\mu^{(k)}(J,\overline{J}).$$

The set $M_0 \times M_W \times M_1 \times M_J$ turns out into a complete uniform space with respect to the family of pseudo-metrics

$$\hat{\rho}_\mu^{(k)}((i_{R_0L_0},W,i_{R_1L_1},J,\dot{i}_{R_0L_0},\dot{W},\dot{i}_{R_1L_1},\dot{J}),(\bar{i}_{R_0L_0},\overline{W},\bar{i}_{R_1L_1},\overline{J},\dot{\bar{i}}_{R_0L_0},\dot{\overline{W}},\dot{\bar{i}}_{R_1L_1},\dot{\overline{J}})) =$$
$$= \max\{\hat{\rho}^{(k)}(i_{R_0L_0},\bar{i}_{R_0L_0}),\hat{\rho}^{(k)}(W,\overline{W}),\hat{\rho}^{(k)}(i_{R_1L_1},\bar{i}_{R_1L_1}),\hat{\rho}^{(k)}(J,\overline{J}),$$

$$\hat{\rho}_\mu^{(k)}(\dot{i}_{R_0L_0},\dot{\bar{i}}_{R_0L_0}),\hat{\rho}_\mu^{(k)}(\dot{W},\dot{\overline{W}}),\hat{\rho}_\mu^{(k)}(\dot{i}_{R_1L_1},\dot{\bar{i}}_{R_1L_1}),\hat{\rho}_\mu^{(k)}(\dot{J},\dot{\overline{J}})\}(k=0,1,2,...)..$$

Remark 8.9.3. Replacing $t = t_0$ in (8.9.1) we obtain conformity condition **(CC)**:

$$\frac{d\widetilde{C}_0(u(0,t_0))}{du}\frac{dW(t_0)}{dt} = \frac{d\widetilde{C}_0(u(0,t_0))}{du}\frac{dJ(0)}{dt} - \frac{W(t_0)-J(0)+2\sqrt{L}\,\breve{I}_{in}(t_0)+2\sqrt{L}\,i_{R_0L_0}(t_0)}{Z_0};$$

$$\frac{d\widetilde{C}_1(u(\Lambda,t_0))}{du}\frac{dJ(t_0)}{dt} = \frac{d\widetilde{C}_1(u(\Lambda,t_0))}{du}\frac{dW(0)}{dt} + \frac{W(0)+J(t_0)+2\sqrt{L}i_{R_1L_1}(t_0)}{Z_0}.$$

Since we can choose $t = t_0$ and $t = 0$ to be zero point of the initial functions, that is,

$$\frac{dW(0)}{dt} = \frac{dJ(0)}{dt} = \frac{dW(t_0)}{dt} = \frac{dJ(t_0)}{dt} = 0,\ W(0) = J(0) = W(t_0) = J(t_0) = 0,\ i_{R_0L_0}(t_0) = i_{R_1L_1}(t_0) = 0$$

it follows that **(CC)** is satisfied provided $\breve{I}_{in}(t_0) = 0$.

In order to avoid the problem with **(CC)** we define the operator

$$B = \left(B_0(i_{R_0L_0},W,i_{R_1L_1},J),B_W(i_{R_0L_0},W,i_{R_1L_1},J),B_1(i_{R_0L_0},W,i_{R_1L_1},J),B_J(i_{R_0L_0},W,i_{R_1L_1},J)\right)$$

by the formulas

$$B_0(i_{R_0L_0},W,i_{R_1L_1},J) = \int_{t_k}^{t} I_{R_0}(i_{R_0L_0},W,J)(s)ds - \frac{t-t_k}{t_{k+1}-t_k}\int_{t_k}^{t_{k+1}} I_{R_0}(i_{R_0L_0},W,J)(s)ds,$$

$$t \in [t_k,t_{k+1}],\ (k=0,1,2,...),$$

$$B_W(i_{R_0L_0},W,i_{R_1L_1},J)(t)=B_W^{(k)}(i_{R_0L_0},W,i_{R_1L_1},J)(t):=\int_{t_k}^{t}U(i_{R_0L_0},W,J)(s)ds-\frac{t-t_k}{t_{k+1}-t_k}\int_{t_k}^{t_{k+1}}U(i_{R_0L_0},W,J)(s)ds,$$

$$t\in[t_k,t_{k+1}], (k=0,1,2,...),$$

$$B_1(i_{R_0L_0},W,i_{R_1L_1},J)=\int_{t_k}^{t}I_{R_1}(W,i_{R_1L_1},J)(s)ds-\frac{t-t_k}{t_{k+1}-t_k}\int_{t_k}^{t_{k+1}}I_{R_1}(W,i_{R_1L_1},J)(s)ds,$$

$$t\in[t_k,t_{k+1}], (k=0,1,2,...),$$

$$B_J(i_{R_0L_0},W,i_{R_1L_1},J)(t)=B_J^{(k)}(i_{R_0L_0},W,i_{R_1L_1},J)(t):=\int_{t_k}^{t}I(W,i_{R_1L_1},J)(s)ds-\frac{t-t_k}{t_{k+1}-t_k}\int_{t_k}^{t_{k+1}}I(W,i_{R_1L_1},J)(s)ds$$

$$t\in[t_k,t_{k+1}], (k=0,1,2,...), \text{ where}$$

$$I_{R_0}(i_{R_0L_0},W,J)=\frac{W(t)-\vec{J}(t)-2\sqrt{C}R_0(i_{R_0L_0}(t))}{2\sqrt{C}d\widetilde{L}_0(i_{R_0L_0}(t))/di_{R_0L_0}},$$

$$U(i_{R_0L_0},W,J)=\frac{d\vec{J}(t)}{dt}-\frac{W(t)-\vec{J}(t)+2\sqrt{L}\,\breve{I}_{in}(t)+2\sqrt{L}\,i_{R_0L_0}(t)}{Z_0d\widetilde{C}_0\left(\dfrac{W(t)-\vec{J}(t)}{2\sqrt{C}}\right)/du}$$

$$I_{R_1}(W,i_{R_1L_1},J)=\frac{\vec{W}(t)-J(t)-2\sqrt{C}R_1(i_{R_1L_1}(t))}{2\sqrt{C}d\widetilde{L}_1(i_{R_1L_1}(t))/di_{R_1L_1}},$$

$$I(W,i_{R_1L_1},J)=\frac{d\vec{W}(t)}{dt}+\frac{\vec{W}(t)+J(t)+2\sqrt{L}i_{R_1L_1}(t)}{Z_0d\widetilde{C}_1\left(\dfrac{\vec{W}(t)-J(t)}{2\sqrt{C}}\right)/du}, \ t\in[T,\infty),$$

$$\vec{W}(t)=\begin{cases}\widetilde{W}_0(t-T),t\in[T,2T]\\W(t-T),t\in[2T,\infty)\end{cases}, \ \vec{J}(t)=\begin{cases}\widetilde{J}_0(t-T),t\in[T,2T]\\J(t-T),t\in[2T,\infty)\end{cases}.$$

Lemma 8.9.1. Problem (8.9.1) has a solution $(i_{R_0L_0},W,i_{R_1L_1},J)\in M_0\times M_W\times M_1\times M_J$ iff the operator B has a fixed point in $M_0\times M_W\times M_1\times M_J$, that is,

$$(i_{R_0L_0},W,i_{R_1L_1},J)=(B_0(i_{R_0L_0},W,i_{R_1L_1},J),B_W(i_{R_0L_0},W,i_{R_1L_1},J),B_1(i_{R_0L_0},W,i_{R_1L_1},J),B_J(i_{R_0L_0},W,i_{R_1L_1},J))$$

Proof: Let $(W,J) \in M_W \times M_J$ be a solution of (8.9.1). Then integrating every equation of (8.9.1) on every interval $[t_k,t] \subset [t_k,t_{k+1}]$ $(k = 0,1,2 \dots)$ we obtain

$$i_{R_0L_0}(t) - i_{R_0L_0}(t_k) = \int_{t_k}^{t} I_{R_0}(i_{R_0L_0},W,i_{R_1L_1},J)(s)ds \Leftrightarrow i_{R_0L_0}(t) = \int_{t_k}^{t} I_{R_0}(i_{R_0L_0},W,i_{R_1L_1},J)(s)ds \,,$$

$$W(t) - W(t_k) = \int_{t_k}^{t} U(i_{R_0L_0},W,i_{R_1L_1},J)(s)ds \Leftrightarrow W(t) = \int_{t_k}^{t} U(i_{R_0L_0},W,i_{R_1L_1},J)(s)ds \,,$$

$$i_{R_1L_1}(t) - i_{R_1L_1}(t_k) = \int_{t_k}^{t} I_{R_1}(i_{R_0L_0},W,i_{R_1L_1},J)(s)ds \Leftrightarrow i_{R_1L_1}(t) = \int_{t_k}^{t} I_{R_1}(i_{R_0L_0},W,i_{R_1L_1},J)(s)ds \,,$$

$$J(t) - J(t_k) = \int_{t_k}^{t} I(i_{R_0L_0},W,i_{R_1L_1},J)(s)ds \Leftrightarrow J(t) = \int_{t_k}^{t} I(i_{R_0L_0},W,i_{R_1L_1},J)(s)ds$$

and then

$$i_{R_0L_0}(t) = \int_{t_k}^{t} I_{R_0}(i_{R_0L_0},W,i_{R_1L_1},J)(s)ds \Rightarrow 0 = i_{R_0L_0}(t_{k+1}) = \int_{t_k}^{t_{k+1}} I_{R_0}(i_{R_0L_0},W,i_{R_1L_1},J)(s)ds \,,$$

$$W(t) = \int_{t_k}^{t} U(i_{R_0L_0},W,i_{R_1L_1},J)(s)ds \Rightarrow 0 = W(t_{k+1}) = \int_{t_k}^{t_{k+1}} U(i_{R_0L_0},W,i_{R_1L_1},J)(s)ds,$$

$$i_{R_1L_1}(t) = \int_{t_k}^{t} I_{R_1}(i_{R_0L_0},W,i_{R_1L_1},J)(s)ds \Rightarrow 0 = i_{R_1L_1}(t_{k+1}) = \int_{t_k}^{t_{k+1}} I_{R_1}(i_{R_0L_0},W,i_{R_1L_1},J)(s)ds \,,$$

$$J(t) = \int_{t_k}^{t} I(i_{R_0L_0},W,i_{R_1L_1},J)(s)ds \Rightarrow 0 = J(t_{k+1}) = \int_{t_k}^{t_{k+1}} I(i_{R_0L_0},W,i_{R_1L_1},J)(s)ds.$$

Therefore the ordered four $(i_{R_0L_0},W,i_{R_1L_1},J)$ satisfies

$$i_{R_0L_0}(t) = \int_{t_k}^{t} I_{R_0}(i_{R_0L_0},W,i_{R_1L_1},J)(s)ds - \frac{t-t_k}{t_{k+1}-t_k}\int_{t_k}^{t_{k+1}} I_{R_0}(i_{R_0L_0},W,i_{R_1L_1},J)(s)ds, \, t \in [t_k,t_{k+1}] \;;$$

$$W(t) = \int_{t_k}^{t} U(i_{R_0L_0},W,i_{R_1L_1},J)(s)ds - \frac{t-t_k}{t_{k+1}-t_k}\int_{t_k}^{t_{k+1}} U(i_{R_0L_0},W,i_{R_1L_1},J)(s)ds, \, t \in [t_k,t_{k+1}] \;; \quad (8.9.3)$$

$$i_{R_1L_1}(t) = \int_{t_k}^{t} I_{R_1}(i_{R_0L_0},W,i_{R_1L_1},J)(s)ds - \frac{t-t_k}{t_{k+1}-t_k}\int_{t_k}^{t_{k+1}} I_{R_1}(i_{R_0L_0},W,i_{R_1L_1},J)(s)ds, \, t \in [t_k,t_{k+1}] \;;$$

$$J(t) = \int\limits_{t_k}^{t} I(i_{R_0L_0}, W, i_{R_1L_1}, J)(s)ds - \frac{t - t_k}{t_{k+1} - t_k} \int\limits_{t_k}^{t_{k+1}} I(i_{R_0L_0}, W, i_{R_1L_1}, J)(s)ds, \; t \in [t_k, t_{k+1}]$$,

that is, $(i_{R_0L_0}, W, i_{R_1L_1}, J)$ is a fixed point of B.

Conversely, let $(i_{R_0L_0}, W, i_{R_1L_1}, J)$ be a fixed point of B, that is, $(i_{R_0L_0}, W, i_{R_1L_1}, J)$ satisfies (8.9.3). Then having in mind that $\mu_0 = \mu T_0 = \text{const.}$ we obtain

$$\left| \int\limits_{t_k}^{t_{k+1}} I_{R_0}(i_{R_0L_0}, W, i_{R_1L_1}, J)(s)ds \right| \leq \frac{1}{2\sqrt{C}} \left| \int\limits_{t_k}^{t_{k+1}} \frac{W(s) - J(s-T) - 2\sqrt{C} R_0(i_{R_0L_0}(s))}{d\widetilde{L}_0(i_{R_0L_0}(s))/di_{R_0L_0}} ds \right| \leq$$

$$\leq \frac{1}{2\hat{L}_0\sqrt{C}} \int\limits_{t_k}^{t_{k+1}} \left(|W(s)| + |J(s-T)| + 2\sqrt{C}|R_0(i_{R_0L_0}(s))| \right)ds \leq$$

$$\leq \frac{1}{2\hat{L}_0\sqrt{C}} \int\limits_{t_k}^{t_{k+1}} \left(W_0 e^{\mu(s-t_k)} + J_0 e^{\mu(s-t_k)} + 2\sqrt{C} \sum_{n=1}^{m} \left| r_n^{(0)} \right| |i_{R_0L_0}(s))|^n \right)ds \leq$$

$$\leq \frac{1}{2\hat{L}_0\sqrt{C}} \left(W_0 \int\limits_{t_k}^{t_{k+1}} e^{\mu(s-t_k)}ds + J_0 \int\limits_{t_k}^{t_{k+1}} e^{\mu(s-t_k)}ds + 2\sqrt{C} \sum_{n=1}^{m} \left| r_n^{(0)} \right| I_{R_0}^n \int\limits_{t_k}^{t_{k+1}} e^{n\mu(s-t_k)}ds \right) \leq$$

$$\leq \frac{1}{2\hat{L}_0\sqrt{C}} \left(W_0 \frac{e^{\mu(t_{k+1}-t_k)}-1}{\mu} + J_0 \frac{e^{\mu(t_{k+1}-t_k)}-1}{\mu} + 2\sqrt{C} \sum_{n=1}^{m} \left| r_n^{(0)} \right| I_{R_0}^n \frac{e^{n\mu(t_{k+1}-t_k)}-1}{n\mu} \right) \leq$$

$$\leq \frac{e^{\mu_0}-1}{\mu\hat{L}_0} \frac{1}{2\sqrt{C}} \left[W_0 + J_0 + 2\sqrt{C} \sum_{n=1}^{m} \left| r_n^{(0)} \right| I_{R_0}^n e^{(n-1)\mu_0} \right] \equiv M_0(\mu)$$;

$$\left| \int\limits_{t_k}^{t_{k+1}} U(i_{R_0L_0}, W, i_{R_1L_1}, J)(s)ds \right| \leq$$

$$\leq \left| \int\limits_{t_k}^{t_{k+1}} \frac{dJ(s-T)}{dt}ds \right| + \int\limits_{t_k}^{t_{k+1}} \left| \frac{W(t) - J(t-T) + 2\sqrt{L}\,\breve{I}_{in}(t) + 2\sqrt{L}\,i_{R_0L_0}(t)}{Z_0 d\widetilde{C}_0(u(0,t))/du} \right| dt \leq$$

$$\leq \frac{1}{Z_0\hat{C}_0} \int\limits_{t_k}^{t_{k+1}} \left(|W(s)| + |J(s-T)| + 2\sqrt{L}\left|\breve{I}_{in}(s)\right| + 2\sqrt{L}\left|i_{R_0L_0}(s)\right| \right)ds \leq$$

$$\leq \frac{1}{Z_0 \hat{C}_0} \left(W_0 + J_0 + 2\sqrt{L} I_{R_0} + 2\sqrt{L}\, I_{R_0} \right) \int_{t_k}^{t_{k+1}} e^{\mu(s-t_k)} ds \leq$$

$$\leq \frac{e^{\mu_0}-1}{\mu Z_0 \hat{C}_0} \left(W_0 + J_0 + 4\sqrt{L} I_{R_0} \right) \equiv M_W(\mu) \quad ;$$

$$\left| \int_{t_k}^{t_{k+1}} I_{R_1}(i_{R_0 L_0}, W, i_{R_1 L_1}, J)(s) ds \right| \leq \frac{1}{2\sqrt{C}} \int_{t_k}^{t_{k+1}} \left| \frac{W(s-T) - J(s) - 2\sqrt{C} R_1(i_{R_1 L_1}(s))}{d\widetilde{L}_1(i_{R_1 L_1}(s)) / di_{R_1 L_1}} \right| ds \leq$$

$$\leq \frac{1}{2\hat{L}_1 \sqrt{C}} \int_{t_k}^{t_{k+1}} \left(|W(s-T)| + |J(s)| + 2\sqrt{C}\left| R_1(i_{R_1 L_1}(s)) \right| \right) ds \leq$$

$$\leq \frac{1}{2\hat{L}_1 \sqrt{C}} \int_{t_k}^{t_{k+1}} \left(W_0 e^{\mu(s-t_k)} + J_0 e^{\mu(s-t_k)} + 2\sqrt{C} \sum_{n=1}^{m} \left| r_n^{(1)} \right| \left| i_{R_1 L_1}(s) \right|^n \right) ds \leq$$

$$\leq \frac{1}{2\hat{L}_1 \sqrt{C}} \left(W_0 \frac{e^{\mu(t_{k+1}-t_k)}-1}{\mu} + J_0 \frac{e^{\mu(t_{k+1}-t_k)}-1}{\mu} + 2\sqrt{C} \sum_{n=1}^{m} \left| r_n^{(1)} \right| I_{R_1}^n \frac{e^{n\mu(t_{k+1}-t_k)}-1}{n\mu} \right) \leq$$

$$\leq \frac{e^{\mu_0}-1}{\mu \hat{L}_1} \frac{1}{2\sqrt{C}} \left[W_0 + J_0 + 2\sqrt{C} \sum_{n=1}^{m} \left| r_n^{(1)} \right| I_{R_1}^n e^{(n-1)\mu_0} \right] \equiv M_1(\mu) ;$$

$$\left| \int_{t_k}^{t_{k+1}} I(i_{R_0 L_0}, W, i_{R_1 L_1}, J)(s) ds \right| \leq \left| \int_{t_k}^{t_{k+1}} \frac{dW(s-T)}{ds} ds \right| + \frac{1}{Z_0 \hat{C}_1} \int_{t_k}^{t_{k+1}} \left(|W(s-T)| + |J(s)| + 2\sqrt{L}\left| i_{R_1 L_1}(s) \right| \right) ds \leq$$

$$\leq \frac{e^{\mu_0}-1}{\mu Z_0 \hat{C}_1} \left(W_0 + J_0 + 4\sqrt{L}\, I_{R_1} \right) \equiv M_J(\mu) \quad .$$

As in Chapter II we conclude that

$$\int_{t_k}^{t_{k+1}} I_{R_0}(i_{R_0 L_0}, W, i_{R_1 L_1}, J)(s) ds = 0, \quad \int_{t_k}^{t_{k+1}} U(i_{R_0 L_0}, W, i_{R_1 L_1}, J)(s) ds = 0 \quad ,$$

$$\int_{t_k}^{t_{k+1}} I_{R_1}(i_{R_0 L_0}, W, i_{R_1 L_1}, J)(s) ds = 0, \quad \int_{t_k}^{t_{k+1}} I(i_{R_0 L_0}, W, i_{R_1 L_1}, J)(s) ds = 0 \quad .$$

Therefore

$$i_{R_0 L_0}(t) = \int_{t_k}^{t} I_{R_0}(i_{R_0 L_0}, W, i_{R_1 L_1}, J)(s)ds - \frac{t - t_k}{t_{k+1} - t_k}\int_{t_k}^{t_{k+1}} I_{R_0}(i_{R_0 L_0}, W, i_{R_1 L_1}, J)(s)ds \Leftrightarrow i_{R_0 L_0}(t) = \int_{t_k}^{t} I_{R_0}(i_{R_0 L_0}, W, i_{R_1 L_1}, J)(s)ds$$

$$W(t) = \int_{t_k}^{t} U(i_{R_0 L_0}, W, i_{R_1 L_1}, J)(s)ds - \frac{t - t_k}{t_{k+1} - t_k}\int_{t_k}^{t_{k+1}} U(i_{R_0 L_0}, W, i_{R_1 L_1}, J)(s)ds \Leftrightarrow W(t) = \int_{t_k}^{t} U(i_{R_0 L_0}, W, i_{R_1 L_1}, J)(s)ds$$

$$i_{R_1 L_1}(t) = \int_{t_k}^{t} I_{R_1}(i_{R_0 L_0}, W, i_{R_1 L_1}, J)(s)ds - \frac{t - t_k}{t_{k+1} - t_k}\int_{t_k}^{t_{k+1}} I_{R_1}(i_{R_0 L_0}, W, i_{R_1 L_1}, J)(s)ds \Leftrightarrow i_{R_1 L_1}(t) = \int_{t_k}^{t} I_{R_1}(i_{R_0 L_0}, W, i_{R_1 L_1}, J)(s)ds$$

$$J(t) = \int_{t_k}^{t} I(i_{R_0 L_0}, W, i_{R_1 L_1}, J)(s)ds - \frac{t - t_k}{t_{k+1} - t_k}\int_{t_k}^{t_{k+1}} I(i_{R_0 L_0}, W, i_{R_1 L_1}, J)(s)ds \Leftrightarrow J(t) = \int_{t_k}^{t} I(i_{R_0 L_0}, W, i_{R_1 L_1}, J)(s)ds$$

Differentiating the last integral equations we obtain (8.9.1).

Lemma 8.9.1 is thus proved.

Theorem 8.9.1. Let the following conditions be fulfilled:

(L) $|i| \le i_0 \Rightarrow \widetilde{L}_0(i) = \sum_{n=1}^{m}(n+1)l_n^{(p)}(i)^n \ge \hat{L}_p > 0, \; (p = 0,1);$

(IN) $\widetilde{W}_0(.), \widetilde{J}_0(.) \in C_{T_0}^1[0,T], \; T = mT_0, \; m \in \{2,3,...\},$

$$\left|\widetilde{W}_0(t)\right| \le e^{-\beta}e^{-\frac{R}{L}(t+T)}U_0, \left|\widetilde{J}_0(t)\right| \le e^{-\beta}e^{-\frac{R}{L}(t+T)}J_0, \left|\breve{I}_{in}(t)\right| \le I_{R_0}e^{-\frac{R}{L}t}; I_{R_0}, I_{R_1} \le i_0;$$

$$(\Pi): \frac{W_0 + J_0}{2\sqrt{C}} \le \phi_0, \; \frac{2L}{R\hat{L}_0\sqrt{C}}\sinh\left(\frac{RT_0}{L}\right)\left[W_0 + J_0 e^{-\beta} + 2\sqrt{C}\sum_{n=1}^{m}\left|r_n^{(0)}\right|I_{R_0}{}^{n}\right] \le I_{R_0};$$

$$J_0 e^{-\beta} + \frac{4L\left(W_0 + J_0 e^{-\beta} + 4\sqrt{L}\,I_{R_0}\right)}{Z_0\hat{C}_0 R}\sinh\left(\frac{RT_0}{L}\right) \le W_0;$$

$$\frac{2L}{R\hat{L}_1\sqrt{C}}\sinh\left(\frac{RT_0}{L}\right)\left(W_0 e^{-\beta} + J_0 + 2\sqrt{C}\sum_{n=1}^{m}\left|r_n^{(1)}\right|I_{R_1}{}^{n}\right) \le I_{R_1};$$

$$W_0 e^{-\beta} + \frac{4L}{R}\frac{W_0 e^{-\beta} + J_0 + 4\sqrt{L}I_{R_1}}{Z_0\hat{C}_1}\sinh\left(\frac{RT_0}{L}\right) \le J_0.$$

Then there exists a unique T0-periodic solution of (8.9.1).

Proof: As in Chapter VII one can show that the functions

$$B_0(i_{R_0 L_0}, W, i_{R_1 L_1}, J)(t), B_W(i_{R_0 L_0}, W, i_{R_1 L_1}, J)(t), B_1(i_{R_0 L_0}, W, i_{R_1 L_1}, J)(t), B_J(i_{R_0 L_0}, W, i_{R_1 L_1}, J)(t)$$

are continuous and differentiable on every $[t_k, t_{k+1}]$.

We prove the following inequalities:

$$\left|B_0^{(k)}(i_{R_0L_0}, W, i_{R_1L_1}, J)\right| \leq \int_{t_k}^{t}\left|I_{R_0}(i_{R_0L_0}, W, J)(s)\right|ds + \left|\int_{t_k}^{t_{k+1}}I_{R_0}(i_{R_0L_0}, W, J)(s)ds\right| \equiv J_1 + J_2 .$$

Since

$$J_1 \leq \frac{1}{2\sqrt{C}}\left|\int_{t_k}^{t}\frac{W(s) - \bar{J}(s) - 2\sqrt{C}R_0(i_{R_0L_0}(s))}{d\widetilde{L}_0(i_{R_0L_0}(s))/di_{R_0L_0}}ds\right| \leq$$

$$\leq \frac{1}{2\hat{L}_0\sqrt{C}}\int_{t_k}^{t}\left(|W(s)| + |\bar{J}(s)| + 2\sqrt{C}\left|R_0(i_{R_0L_0}(s))\right|\right)ds \leq$$

$$\leq \frac{1}{2\hat{L}_0\sqrt{C}}\int_{t_k}^{t}\left(W_0 e^{-\frac{R}{L}s} + J_0 e^{-\beta}e^{-\frac{R}{L}s} + 2\sqrt{C}\sum_{n=1}^{m}\left|r_n^{(0)}\right|\left|i_{R_0L_0}(s)\right|^n\right)ds \leq$$

$$\leq \frac{1}{2\hat{L}_0\sqrt{C}}\left(W_0\int_{t_k}^{t}e^{-\frac{R}{L}s}ds + J_0 e^{-\beta}\int_{t_k}^{t}e^{-\frac{R}{L}s}ds + 2\sqrt{C}\sum_{n=1}^{m}\left|r_n^{(0)}\right|\left(I_{R_0}\right)^n\int_{t_k}^{t}e^{-n\frac{R}{L}s}ds\right) \leq$$

$$\leq \frac{1}{2\hat{L}_0\sqrt{C}}\left[W_0\frac{L}{R}\left(e^{-\frac{R}{L}t_k} - e^{-\frac{R}{L}t}\right) + J_0 e^{-\beta}\frac{L}{R}\left(e^{-\frac{R}{L}t_k} - e^{-\frac{R}{L}t}\right) + 2\sqrt{C}\frac{L}{R}\left(e^{-\frac{R}{L}t_k} - e^{-\frac{R}{L}t}\right)\sum_{n=1}^{m}\left|r_n^{(0)}\right|I_{R_0}^{n}\right] \leq$$

$$\leq \frac{1}{2\hat{L}_0\sqrt{C}}\frac{L}{R}\left(e^{-\frac{R}{L}t_k} - e^{-\frac{R}{L}t}\right)\left[W_0 + J_0 e^{-\beta} + 2\sqrt{C}\sum_{n=1}^{m}\left|r_n^{(0)}\right|I_{R_0}^{n}\right] \leq$$

$$\leq \frac{1}{2\hat{L}_0\sqrt{C}}\frac{L}{R}e^{-\frac{R}{L}t}\left(e^{\frac{RT_0}{L}} - e^{-\frac{RT_0}{L}}\right)\left[W_0 + J_0 e^{-\beta} + 2\sqrt{C}\sum_{n=1}^{m}\left|r_n^{(0)}\right|I_{R_0}^{n}\right] \leq$$

$$\leq e^{-\frac{R}{L}t}\frac{1}{\hat{L}_0\sqrt{C}}\frac{L}{R}\sinh\left(\frac{RT_0}{L}\right)\left[W_0 + J_0 e^{-\beta} + 2\sqrt{C}\sum_{n=1}^{m}\left|r_n^{(0)}\right|I_{R_0}^{n}\right] .$$

In view of the inequalities

$$e^{-\frac{R}{L}t_k} - e^{-\frac{R}{L}t_{k+1}} \leq e^{-\frac{R}{L}t}\left(e^{\frac{RT_0}{L}} - e^{-\frac{RT_0}{L}}\right)$$

we have

$$J_2 \le \frac{1}{2\sqrt{C}} \left| \int_{t_k}^{t_{k+1}} \frac{W(s) - \bar{J}(s) - 2\sqrt{C}R_0(i_{R_0L_0}(s))}{d\tilde{L}_0(i_{R_0L_0}(s))/di_{R_0L_0}} ds \right| \le$$

$$\le \frac{1}{2\hat{L}_0\sqrt{C}} \left(W_0 \int_{t_k}^{t_{k+1}} e^{-\frac{R}{L}s} ds + J_0 e^{-\beta} \int_{t_k}^{t_{k+1}} e^{-\frac{R}{L}s} ds + 2\sqrt{C}\sum_{n=1}^{m} \left|r_n^{(0)}\right| \left|I_{R_0}\right|^n \int_{t_k}^{t_{k+1}} e^{-n\frac{R}{L}s} ds \right) \le$$

$$\le \frac{1}{2\hat{L}_0\sqrt{C}} \left[W_0 \frac{L}{R}\left(e^{-\frac{R}{L}t_k} - e^{-\frac{R}{L}t_{k+1}} \right) + J_0 e^{-\beta}\frac{L}{R}\left(e^{-\frac{R}{L}t_k} - e^{-\frac{R}{L}t_{k+1}} \right) + 2\sqrt{C}\sum_{n=1}^{m}\left|r_n^{(0)}\right|\left|I_{R_0}\right|^n \frac{L}{nR}\left(e^{-n\frac{R}{L}t_k} - e^{-n\frac{R}{L}t_{k+1}} \right) \right] \le$$

$$\le \frac{1}{2\hat{L}_0\sqrt{C}} \left[W_0 \frac{L}{R}\left(e^{-\frac{R}{L}t_k} - e^{-\frac{R}{L}t_{k+1}} \right) + J_0 e^{-\beta}\frac{L}{R}\left(e^{-\frac{R}{L}t_k} - e^{-\frac{R}{L}t_{k+1}} \right) + 2\sqrt{C}\sum_{n=1}^{m}\left|r_n^{(0)}\right|\left|I_{R_0}\right|^n \frac{L}{R}\left(e^{-\frac{R}{L}t_k} - e^{-\frac{R}{L}t_{k+1}} \right) \right] \le$$

$$\le \frac{L}{R}\left(e^{-\frac{R}{L}t_k} - e^{-\frac{R}{L}t_{k+1}} \right) \frac{1}{2\hat{L}_0\sqrt{C}} \left[W_0 + J_0 e^{-\beta} + 2\sqrt{C}\sum_{n=1}^{m}\left|r_n^{(0)}\right|\left|I_{R_0}\right|^n \right] \le$$

$$\le e^{-\frac{R}{L}t}\left(e^{\frac{RT_0}{L}} - e^{-\frac{RT_0}{L}} \right)\frac{L}{R}\frac{1}{2\hat{L}_0\sqrt{C}} \left(W_0 + J_0 e^{-\beta} + 2\sqrt{C}\sum_{n=1}^{m}\left|r_n^{(0)}\right|\left|I_{R_0}\right|^n \right) \le$$

$$\le e^{-\frac{R}{L}t} \sinh\left(\frac{RT_0}{L}\right) \frac{L}{\hat{L}_0 R\sqrt{C}} \left(W_0 + J_0 e^{-\beta} + 2\sqrt{C}\sum_{n=1}^{m}\left|r_n^{(0)}\right|\left|I_{R_0}\right|^n \right).$$

Then

$$\left| B_0^{(k)}(i_{R_0L_0}, W, i_{R_1L_1}, J) \right| \le$$

$$\le e^{-\frac{R}{L}t} \sinh\left(\frac{RT_0}{L}\right) \frac{L}{\hat{L}_0 R\sqrt{C}} \left[W_0 + J_0 e^{-\beta} + 2\sqrt{C}\sum_{n=1}^{m}\left|r_n^{(0)}\right|\left|I_{R_0}\right|^n \right] +$$

$$+ e^{-\frac{R}{L}t} \sinh\left(\frac{RT_0}{L}\right) \frac{L}{\hat{L}_0 R\sqrt{C}} \left(W_0 + J_0 e^{-\beta} + 2\sqrt{C}\sum_{n=1}^{m}\left|r_n^{(0)}\right|\left|I_{R_0}\right|^n \right) \le$$

$$\le e^{-\frac{R}{L}t} \sinh\left(\frac{RT_0}{L}\right) \frac{2L}{\hat{L}_0 R\sqrt{C}} \left[W_0 + J_0 e^{-\beta} + 2\sqrt{C}\sum_{n=1}^{m}\left|r_n^{(0)}\right|\left|I_{R_0}\right|^n \right] \le I_{R_0} e^{-\frac{R}{L}t}$$

Further on we have

$$\left| B_W^{(k)}(i_{R_0L_0},W,i_{R_1L_1},J)(t) \right| \le$$

$$\le \left| \int_{t_k}^{t} U(i_{R_0L_0},W,i_{R_1L_1},J)(s)\,ds \right| + \left| \int_{t_k}^{t_{k+1}} U(i_{R_0L_0},W,i_{R_1L_1},J)(s)\,ds \right| \equiv W_1 + W_2.$$

But

$$W_1 \le \left| \int_{t_k}^{t} \frac{d\vec{J}(s)}{dt}\,ds \right| + \frac{1}{Z_0} \int_{t_k}^{t} \left| \frac{W(s)-\vec{J}(s)+2\sqrt{L}\,\breve{I}_{in}(s)+2\sqrt{L}\,i_{R_0L_0}(s)}{d\widetilde{C}_0(u(0,s))/du} \right| ds \le$$

$$\le \left| \vec{J}(t) \right| + \frac{1}{Z_0\hat{C}_0} \int_{t_k}^{t} \left(|W(s)| + |\vec{J}(s)| + 2\sqrt{L}\,|\breve{I}_{in}(s)| + 2\sqrt{L}\,|i_{R_0L_0}(s)| \right) ds \le$$

$$\le J_0 e^{-\beta} e^{-\frac{R}{L}t} + \frac{1}{Z_0\hat{C}_0} \int_{t_k}^{t} \left(W_0 e^{-\frac{R}{L}s} + J_0 e^{-\frac{R}{L}s}e^{-\beta} + 2\sqrt{L}\,I_{R_0} e^{-\frac{R}{L}s} + 2\sqrt{L}\,I_{R_0} e^{-\frac{R}{L}s} \right) ds \le$$

$$\le J_0 e^{-\beta} e^{-\frac{R}{L}t} + \frac{1}{Z_0\hat{C}_0} \left(W_0 + J_0 e^{-\beta} + 4\sqrt{L}\,I_{R_0} \right) \int_{t_k}^{t} e^{-\frac{R}{L}s}\,ds \le$$

$$\le J_0 e^{-\beta} e^{-\frac{R}{L}t} + \frac{1}{Z_0\hat{C}_0} \left(W_0 + J_0 e^{-\beta} + 4\sqrt{L}\,I_{R_0} \right) \frac{L}{R} \left(e^{-\frac{R}{L}t_k} - e^{-\frac{R}{L}t} \right)$$

$$\le J_0 e^{-\beta} e^{-\frac{R}{L}t} + e^{-\frac{R}{L}t} \frac{L}{Z_0\hat{C}_0 R} \left(W_0 + J_0 e^{-\beta} + 4\sqrt{L}\,I_{R_0} \right) \left(e^{\frac{RT_0}{L}} - e^{-\frac{RT_0}{L}} \right) \le$$

$$\le e^{-\frac{R}{L}t} \left[J_0 e^{-\beta} + \frac{2L \left(W_0 + J_0 e^{-\beta} + 4\sqrt{L}\,I_{R_0} \right)}{Z_0\hat{C}_0 R} \sinh\left(\frac{RT_0}{L} \right) \right];$$

$$W_2 \le \left| \int_{t_k}^{t_{k+1}} \frac{d\vec{J}(s)}{dt}\,ds \right| + \frac{1}{Z_0} \int_{t_k}^{t_{k+1}} \left| \frac{W(s)-\vec{J}(s)+2\sqrt{L}\,\breve{I}_{in}(s)+2\sqrt{L}\,i_{R_0L_0}(s)}{d\widetilde{C}_0(u(0,s))/du} \right| ds \le$$

$$\le \frac{1}{Z_0\hat{C}_0}\int_{t_k}^{t_{k+1}}\left(|W(s)|+|\vec{J}(s)|+2\sqrt{L}\,|\breve{I}_{in}(s)|+2\sqrt{L}\,|i_{R_0L_0}(s)|\right)ds \le$$

$$\le \frac{1}{Z_0\hat{C}_0}\left(W_0+J_0e^{-\beta}+4\sqrt{L}\,I_{R0}\right)\frac{L}{R}\left(e^{-\frac{R}{L}t_k}-e^{-\frac{R}{L}t_{k+1}}\right)\le$$

$$\le e^{-\frac{R}{L}t}\left(e^{\frac{RT_0}{L}}-e^{-\frac{RT_0}{L}}\right)\frac{L}{Z_0\hat{C}_0R}\left(W_0+J_0e^{-\beta}+4\sqrt{L}\,I_{R0}\right)\le$$

$$\le e^{-\frac{R}{L}t}\frac{2L}{Z_0\hat{C}_0R}\left(W_0+J_0e^{-\beta}+4\sqrt{L}\,I_{R0}\right)\sinh\left(\frac{RT_0}{L}\right).$$

Then

$$\left|B_W^{(k)}(i_{R_0L_0},W,i_{R_1L_1},J)(t)\right|\le$$
$$e^{-\frac{R}{L}t}\left[J_0e^{-\beta}+\frac{4L\left(W_0+J_0e^{-\beta}+4\sqrt{L}\,I_{R0}\right)}{Z_0\hat{C}_0R}\sinh\left(\frac{RT_0}{L}\right)\right]\le e^{-\frac{R}{L}t}W_0.$$

Further on we have

$$\left|B_1^{(k)}(i_{R_0L_0},W,i_{R_1L_1},J)\right|\le \int_{t_k}^{t}\left|I_{R_1}(W,i_{R_1L_1},J)(s)\right|ds+\left|\int_{t_k}^{t_{k+1}}I_{R_1}(W,i_{R_1L_1},J)(s)ds\right|\equiv I_1+I_2.$$

Since

$$I_1\le \frac{1}{2\sqrt{C}}\int_{t_k}^{t}\left|\frac{\vec{W}(s)-J(s)-2\sqrt{C}\,R_1(i_{R_1L_1}(s))}{d\widetilde{L}_1(i_{R_1L_1}(s))/di_{R_1L_1}}\right|ds\le$$

$$\le \frac{1}{2\hat{L}_1\sqrt{C}}\left[W_0e^{-\beta}\int_{t_k}^{t}e^{-\frac{R}{L}s}ds+J_0\int_{t_k}^{t}e^{-\frac{R}{L}s}ds+2\sqrt{C}\sum_{n=1}^{m}\left|r_n^{(1)}\right|I_{R_1}^{\ n}\int_{t_k}^{t}e^{-n\frac{R}{L}s}ds\right]\le$$

$$\le \frac{1}{2\hat{L}_1\sqrt{C}}\frac{L}{R}\left(e^{-\frac{R}{L}t_k}-e^{-\frac{R}{L}t}\right)\left[W_0e^{-\beta}+J_0+2\sqrt{C}\sum_{n=1}^{m}\left|r_n^{(1)}\right|I_{R_1}^{\ n}\right]\le$$

$$\le e^{-\frac{R}{L}t}\frac{L}{\hat{L}_1\sqrt{C}R}\sinh\left(\frac{RT_0}{L}\right)\left(W_0e^{-\beta}+J_0+2\sqrt{C}\sum_{n=1}^{m}\left|r_n^{(1)}\right|I_{R_1}^{\ n}\right)$$

And

$$I_2 \le \frac{1}{2\sqrt{C}}\left|\int_{t_k}^{t_{k+1}} \frac{\vec{W}(s)-J(s)-2\sqrt{C}R_1(i_{R_1L_1}(s))}{d\widetilde{L}_1(i_{R_1L_1}(s))/di_{R_1L_1}}\,ds\right| \le$$

$$\le \frac{1}{2\hat{L}_1\sqrt{C}}\frac{L}{R}\left(e^{-\frac{R}{L}t_k}-e^{-\frac{R}{L}t_{k+1}}\right)\left[W_0 e^{-\beta}+J_0+2\sqrt{C}\sum_{n=1}^{m}\left|r_n^{(1)}\right|I_{R_1}{}^n\right] \le$$

$$\le e^{-\frac{R}{L}t}\frac{L}{\hat{L}_1\sqrt{C}R}\sinh\left(\frac{RT_0}{L}\right)\left(W_0 e^{-\beta}+J_0+2\sqrt{C}\sum_{n=1}^{m}\left|r_n^{(1)}\right|I_{R_1}{}^n\right)$$

then

$$\left|B_1^{(k)}(i_{R_0L_0},W,i_{R_1L_1},J)\right| \le$$

$$\le e^{-\frac{R}{L}t}\frac{2L}{\hat{L}_1\sqrt{C}R}\sinh\left(\frac{RT_0}{L}\right)\left(W_0 e^{-\beta}+J_0+2\sqrt{C}\sum_{n=1}^{m}\left|r_n^{(1)}\right|I_{R_1}{}^n\right) \le I_{R_1}e^{-\frac{R}{L}t}.$$

Finally

$$\left|B_J^{(k)}(i_{R_0L_0},W,i_{R_1L_1},J)(t)\right| \le \int_{t_k}^{t}\left|I(W,i_{R_1L_1},J)(s)\right|ds + \left|\int_{t_k}^{t_{k+1}}I(W,i_{R_1L_1},J)(s)ds\right| \equiv A_1 + A_2.$$

Since

$$A_1 \le \left|\int_{t_k}^{t}\frac{d\vec{W}(s)}{dt}ds\right| + \frac{1}{Z_0}\int_{t_k}^{t}\left|\frac{\vec{W}(s)-J(s)+2\sqrt{L}\,i_{R_1L_1}(s)}{d\widetilde{C}_1(u(\Lambda,s))/du}\right|ds \le$$

$$\le e^{-\frac{R}{L}t}\left[W_0 e^{-\beta}+\frac{2L}{R}\frac{W_0 e^{-\beta}+J_0+4\sqrt{L}I_{R_1}}{Z_0\hat{C}_1}\sinh\left(\frac{RT_0}{L}\right)\right]$$

and

$$A_2 \le \left|\int_{t_k}^{t_{k+1}}\frac{d\vec{W}(s)}{dt}ds\right| + \frac{1}{Z_0}\int_{t_k}^{t_{k+1}}\left|\frac{\vec{W}(s)-J(s)+2\sqrt{L}\,i_{R_1L_1}(s)}{d\widetilde{C}_1(u(\Lambda,s))/du}\right|ds \le$$

$$\le e^{-\frac{R}{L}t}\left[\frac{2L}{R}\frac{W_0 e^{-\beta}+J_0+4\sqrt{L}I_{R_1}}{Z_0\hat{C}_1}\sinh\left(\frac{RT_0}{L}\right)\right]$$

we have

$$\left| B_J^{(k)}(i_{R_0L_0}, W, i_{R_1L_1}, J)(t) \right| \le$$

$$e^{-\frac{R}{L}t}\left[W_0 e^{-\beta} + \frac{4L}{R}\frac{W_0 e^{-\beta} + J_0 + 4\sqrt{L}I_{R_1}}{Z_0 \hat{C}_1}\sinh\left(\frac{RT_0}{L}\right)\right] \le e^{-\frac{R}{L}t}J_0 .$$

Therefore B maps $M_0 \times M_W \times M_1 \times M_J$ into itself.

It remains to show the Lipschitz estimates for the elements of the operator B. Recall that

$$\widetilde{L}_p(i_{R_pLp})(t) = i_{R_pLp}(t)\frac{dL_p(i_{R_pLp}(t))}{di_{R_pLp}} + L_p\left(i_{R_pLp}(t)\right) = \sum_{n=1}^{m}(n+1)l_n^{(p)}\left(i_{R_pLp}(t)\right)^n \quad (p=0,1)$$

and

$$\frac{d\widetilde{L}_p(i_{R_pLp})(t)}{di_{R_pLp}} = \sum_{n=1}^{m}(n+1)nl_n^{(0)}\left(i_{R_pLp}(t)\right)^{n-1} .$$

We obtain

$$\frac{\partial I_{R_0}(i_{R_0L_0}, W, J)}{\partial i_{R_0L_0}} = \frac{\partial}{\partial i_{R_0L_0}}\left(\frac{W(t) - J(t-T) - 2\sqrt{C}R_0(i_{R_0L_0}(t))}{2\sqrt{C}\widetilde{L}_0(i_{R_0L_0})(t)}\right) =$$

$$= \frac{-2\sqrt{C}\left(dR_0(i_{R_0L_0}(t))/di_{R_0L_0}\right)\widetilde{L}_0(i_{R_0L_0})(t) - \left(W(t) - J(t-T) - 2\sqrt{C}R_0(i_{R_0L_0}(t))\right)\left(\widetilde{L}_0(i_{R_0L_0})(t)/di_{R_0L_0}\right)}{2\sqrt{C}\left(\widetilde{L}_0(i_{R_0L_0})(t)\right)^2} ;$$

$$\left|\frac{\partial I_{R_0L_0}(i_{R_0L_0}, W, J)}{\partial i_{R_0L_0}}\right| \le \frac{2\sqrt{C}\left|dR_0(i_{R_0L_0}(t))/di_{R_0L_0}\right|}{2\sqrt{C}\hat{L}_0} +$$

$$+ \frac{|W(t)| + |J(t-T)| + 2\sqrt{C}|R_0(i_{R_0L_0}(t))|}{2\sqrt{C}\hat{L}_0^2}\left|d\widetilde{L}_0(i_{R_0L_0})(t)/di_{R_0L_0}\right| \le$$

$$\le \frac{1}{\hat{L}_0}\sum_{n=1}^{m}n\left|r_n^{(0)}\right|\left|I_{R_0}\right|^{n-1} + \frac{W_0 + J_0 e^{-\beta} + 2\sqrt{C}\sum_{n=1}^{m}\left|r_n^{(0)}\right|I_{R_0}^n}{2\sqrt{C}\hat{L}_0^2}\sum_{n=1}^{m}(n+1)n\left|l_n^{(0)}\right|I_{R_0}^{n-1} \le$$

$$\leq \left(\frac{1}{\hat{L}_0} \sum_{n=1}^{m} n \left| r_n^{(0)} \right| \left(I_{R_0} \right)^{n-1} + \frac{W_0 + J_0 e^{-\beta} + 2\sqrt{C} \sum_{n=1}^{m} \left| r_n^{(0)} \right| I_{R_0}^n}{2\sqrt{C} \hat{L}_0^2} \sum_{n=1}^{m} (n+1) n \left| l_n^{(0)} \right| I_{R_0}^{n-1} \right);$$

$$\frac{\partial I_{R_0}(i_{R_0 L_0}, W, J)}{\partial W} = \frac{\partial}{\partial W} \left(\frac{W(t) - J(t-T) - 2\sqrt{C} R_0(i_{R_0 L_0}(t))}{2\sqrt{C} \widetilde{L}_0(i_{R_0 L_0})(t)} \right) = \frac{1}{2\sqrt{C} \, \widetilde{L}_0(i_{R_0 L_0})(t)};$$

$$\left| \frac{\partial I_{R_0}(i_{R_0 L_0}, W, J)}{\partial W} \right| \leq \frac{1}{2\sqrt{C} \, \hat{L}_0};$$

$$\frac{\partial I_{R_0}(i_{R_0 L_0}, W, J)}{\partial J} = \frac{\partial}{\partial J} \left(\frac{W(t) - J(t-T) - 2\sqrt{C} R_0(i_{R_0 L_0}(t))}{2\sqrt{C} \, \widetilde{L}_0(i_{R_0 L_0})(t)} \right) = -\frac{1}{2\sqrt{C} \, \widetilde{L}_0(i_{R_0 L_0})(t)};$$

$$\left| \frac{\partial I_{R_0}(i_{R_0 L_0}, W, J)}{\partial J} \right| \leq \frac{1}{2\sqrt{C} \hat{L}_0}.$$

Then we have

$$\left| B_0^{(k)}(i_{R_0 L_0}, W, J)(t) - B_0^{(k)}(\bar{i}_{R_0 L_0}, \overline{W}, \overline{J})(t) \right| \leq \int_{t_k}^{t} \left| I_{R_0}(i_{R_0 L_0}, W, J)(s) - I_{R_0 L_0}(\bar{i}_{R_0 L_0}, \overline{W}, \overline{J})(s) \right| ds +$$

$$+ \left| \int_{t_k}^{t_{k+1}} \left(I_{R_0}(i_{R_0 L_0}, W, J)(s) - I_{R_0 L_0}(\bar{i}_{R_0 L_0}, \overline{W}, \overline{J})(s) \right) ds \right| \equiv I_1 + I_2.$$

Since

$$I_1 \leq \frac{1}{2\sqrt{C} \, \hat{L}_0} \int_{t_k}^{t} \left| \bar{J}(s) - \bar{J}(s) \right| ds + \int_{t_k}^{t} e^{\mu(s-t_k)} \left| i_{R_0 L_0}(s) - \bar{i}_{R_0 L_0}(s) \right| ds \left(\frac{2\sqrt{C}}{\hat{L}_0} \sum_{n=1}^{m} n \left| r_n^{(0)} \right| I_{R_0}^{n-1} + \right.$$

$$+ \frac{W_0 + J_0 e^{-\beta} + 2\sqrt{C} \sum_{n=1}^{m} \left| r_n^{(0)} \right| I_{R_0}^n e^{(n-1)\mu_0}}{2\sqrt{C} \, \hat{L}_0^2} \sum_{n=1}^{m} (n+1) n \left| l_n^{(0)} \right| I_{R_0}^{n-1} \left. \right) + \frac{1}{2\sqrt{C} \, \hat{L}_0} \int_{t_k}^{t} \left| W(s) - \overline{W}(s) \right| ds \leq$$

$$\leq \rho^{(k)}(i_{R_0 L_0}, \bar{i}_{R_0 L_0}) \frac{e^{\mu(t-t_k)} - 1}{\mu} \left(\frac{2\sqrt{C}}{\hat{L}_0} \sum_{n=1}^{m} n \left| r_n^{(0)} \right| I_{R_0}^{n-1} + \right.$$

$$+\frac{W_0+J_0e^{-\beta}+2\sqrt{C}\sum_{n=1}^{m}\left|r_n^{(0)}\right|I_{R_0}^n}{2\sqrt{C}\,\hat{L}_0^2}\sum_{n=1}^{m}(n+1)n\left|l_n^{(0)}\right|I_{R_0}^{n-1}\right)+\frac{1}{2\sqrt{C}\hat{L}_0}\,\rho^{(k)}(W,\overline{W})\frac{e^{\mu(t-t_k)}-1}{\mu}\le$$

$$\le\frac{e^{\mu_0}\rho_\mu^{(k)}(i_{R_0L_0},\overline{i}_{R_0L_0})}{\mu}\frac{e^{\mu(t-t_k)}-1}{\mu}\left(\frac{2\sqrt{C}}{\hat{L}_0}\sum_{n=1}^{m}n\left|r_n^{(0)}\right|I_{R_0}^{n-1}+\right.$$

$$+\frac{W_0+J_0e^{-\beta}+2\sqrt{C}\sum_{n=1}^{m}\left|r_n^{(0)}\right|I_{R_0}^n}{2\sqrt{C}\,\hat{L}_0^2}\sum_{n=1}^{m}(n+1)n\left|l_n^{(0)}\right|I_{R_0}^{n-1}\right)+\frac{1}{2\sqrt{C}\,\hat{L}_0}\frac{e^{\mu_0}\rho_\mu^{(k)}(\dot{W},\dot{\overline{W}})}{\mu}\frac{e^{\mu(t-t_k)}-1}{\mu}+$$

$$\le e^{\mu(t-t_k)}\hat{\rho}_\mu^{(k)}((i_{R_0L_0},W,i_{R_1L_1},J,\dot{i}_{R_0L_0},\dot{W},\dot{i}_{R_1L_1},\dot{J}),(\overline{i}_{R_0L_0},\overline{W},\overline{i}_{R_1L_1},\overline{J},\dot{\overline{i}}_{R_0L_0},\dot{\overline{W}},\dot{\overline{i}}_{R_1L_1},\dot{\overline{J}}))\times$$

$$\times\frac{e^{\mu_0}}{\mu^2\hat{L}_0}\left(2\sqrt{C}\sum_{n=1}^{m}n\left|r_n^{(0)}\right|(I_{R_0})^{n-1}+\frac{W_0+J_0e^{-\beta}+2\sqrt{C}\sum_{n=1}^{m}\left|r_n^{(0)}\right|I_{R_0}^n}{2\sqrt{C}\,\hat{L}_0}\sum_{n=1}^{m}(n+1)n\left|l_n^{(0)}\right|I_{R_0}^{n-1}\right);$$

$$I_2\le\int_{t_k}^{t_{k+1}}e^{\mu(s-t_k)}\left|i_{R_0L_0}(s)-\overline{i}_{R_0L_0}(s)\right|ds\left(\frac{2\sqrt{C}}{\hat{L}_0}\sum_{n=1}^{m}n\left|r_n^{(0)}\right|I_{R_0}^{n-1}+\right.$$

$$+\frac{W_0+J_0e^{-\beta}+2\sqrt{C}\sum_{n=1}^{m}\left|r_n^{(0)}\right|I_{R_0}^n}{2\sqrt{C}\,\hat{L}_0^2}\sum_{n=1}^{m}(n+1)n\left|l_n^{(0)}\right|I_{R_0}^{n-1}\right)+\frac{1}{2\sqrt{C}\,\hat{L}_0}\int_{t_k}^{t_{k+1}}\left|W(s)-\overline{W}(s)\right|ds\le$$

$$\le e^{\mu_0}\rho_\mu^{(k)}(i_{R_0L_0},\overline{i}_{R_0L_0})\frac{e^{\mu(t_{k+1}-t_k)}-1}{\mu}\left(\frac{2\sqrt{C}}{\hat{L}_0}\sum_{n=1}^{m}n\left|r_n^{(0)}\right|I_{R_0}^{n-1}+\right.$$

$$+\frac{W_0+J_0e^{-\beta}+2\sqrt{C}\sum_{n=1}^{m}\left|r_n^{(0)}\right|I_{R_0}^n}{2\sqrt{C}\,\hat{L}_0^2}\sum_{n=1}^{m}(n+1)n\left|l_n^{(0)}\right|I_{R_0}^{n-1}\right)+\frac{1}{2\sqrt{C}\,\hat{L}_0}e^{\mu_0}\rho_\mu^{(k)}(W,\overline{W})\frac{e^{\mu(t_{k+1}-t_k)}-1}{\mu}\le$$

$$\le\frac{e^{\mu_0}\rho_\mu^{(k)}(\dot{i}_{R_0L_0},\dot{\overline{i}}_{R_0L_0})}{\mu}\frac{e^{\mu_0}-1}{\mu}\left(\frac{2\sqrt{C}}{\hat{L}_0}\sum_{n=1}^{m}n\left|r_n^{(0)}\right|I_{R_0}^{n-1}+\right.$$

$$+\frac{W_0+J_0e^{-\beta}+2\sqrt{C}\sum_{n=1}^{m}\left|r_n^{(0)}\right|I_{R_0}^n}{2\sqrt{C}\,\hat{L}_0^2}\sum_{n=1}^{m}(n+1)n\left|l_n^{(0)}\right|I_{R_0}^{n-1}\right)+\frac{1}{2\sqrt{C}\,\hat{L}_0}\frac{e^{\mu_0}\rho_\mu^{(k)}(\dot{W},\dot{\overline{W}})}{\mu}\frac{e^{\mu_0}-1}{\mu}\le$$

$$\leq \hat{\rho}_{\mu}^{(k)}((i_{R_0L_0},W,i_{R_1L_1},J,\dot{i}_{R_0L_0},\dot{W},\dot{i}_{R_1L_1},\dot{J}),(\bar{i}_{R_0L_0},\overline{W},\bar{i}_{R_1L_1},\overline{J},\dot{\bar{i}}_{R_0L_0},\dot{\overline{W}},\dot{\bar{i}}_{R_1L_1},\dot{\overline{J}}))\times$$

$$\times \frac{e^{\mu_0}-1}{\mu^2 \hat{L}_0}e^{\mu_0}\left(2\sqrt{C}\sum_{n=1}^{m}n\left|r_n^{(0)}\right|I_{R_0}^{n-1}+\frac{W_0+J_0 e^{-\beta}+2\sqrt{C}\sum_{n=1}^{m}\left|r_n^{(0)}\right|I_{R_0}^{n}}{2\sqrt{C}\,\hat{L}_0}\sum_{n=1}^{m}(n+1)n\left|l_n^{(0)}\right|I_{R_0}^{n-1}+\frac{1}{2\sqrt{C}}\right);$$

$$\left|B_0^{(k)}(i_{R_0L_0},W,J)(t)-B_0^{(k)}(\bar{i}_{R_0L_0},\overline{W},\overline{J})(t)\right|\leq$$

$$\leq e^{\mu(t-t_k)}\hat{\rho}_{\mu}^{(k)}((i_{R_0L_0},W,i_{R_1L_1},J,\dot{i}_{R_0L_0},\dot{W},\dot{i}_{R_1L_1},\dot{J}),(\bar{i}_{R_0L_0},\overline{W},\bar{i}_{R_1L_1},\overline{J},\dot{\bar{i}}_{R_0L_0},\dot{\overline{W}},\dot{\bar{i}}_{R_1L_1},\dot{\overline{J}}))\times$$

$$\times \frac{e^{2\mu_0}}{\mu^2 \hat{L}_0}\left(2\sqrt{C}\sum_{n=1}^{m}n\left|r_n^{(0)}\right|\left(I_{R_0}\right)^{n-1}+\frac{W_0+J_0 e^{-\beta}+2\sqrt{C}\sum_{n=1}^{m}\left|r_n^{(0)}\right|I_{R_0}^{n}}{2\sqrt{C}\hat{L}_0}\sum_{n=1}^{m}(n+1)n\left|l_n^{(0)}\right|I_{R_0}^{n-1}+\frac{1}{2\sqrt{C}}\right)\equiv$$

$$\equiv e^{\mu(t-t_k)}K_0\hat{\rho}_{\mu}^{(k)}((i_{R_0L_0},W,i_{R_1L_1},J,\dot{i}_{R_0L_0},\dot{W},\dot{i}_{R_1L_1},\dot{J}),(\bar{i}_{R_0L_0},\overline{W},\bar{i}_{R_1L_1},\overline{J},\dot{\bar{i}}_{R_0L_0},\dot{\overline{W}},\dot{\bar{i}}_{R_1L_1},\dot{\overline{J}}))\leq$$

$$\leq e^{\mu_0}K_0\hat{\rho}_{\mu}^{(k)}((i_{R_0L_0},W,i_{R_1L_1},J,\dot{i}_{R_0L_0},\dot{W},\dot{i}_{R_1L_1},\dot{J}),(\bar{i}_{R_0L_0},\overline{W},\bar{i}_{R_1L_1},\overline{J},\dot{\bar{i}}_{R_0L_0},\dot{\overline{W}},\dot{\bar{i}}_{R_1L_1},\dot{\overline{J}})).$$

It follows

$$\hat{\rho}^{(k)}(B_0(i_{R_0L_0},W,J),B_0(\bar{i}_{R_0L_0},\overline{W},\overline{J}))\leq$$

$$\leq e^{\mu_0}K_0\hat{\rho}_{\mu}^{(k)}((i_{R_0L_0},W,i_{R_1L_1},J,\dot{i}_{R_0L_0},\dot{W},\dot{i}_{R_1L_1},\dot{J}),(\bar{i}_{R_0L_0},\overline{W},\bar{i}_{R_1L_1},\overline{J},\dot{\bar{i}}_{R_0L_0},\dot{\overline{W}},\dot{\bar{i}}_{R_1L_1},\dot{\overline{J}}))\cdot$$

For the second component we get

$$\left|B_W^{(k)}(i_{R_0L_0},W,J)(t)-B_W^{(k)}(\bar{i}_{R_0L_0},\overline{W},\overline{J})(t)\right|\leq \int_{t_k}^{t}\left|U(s)-\overline{U}(s)\right|ds+\int_{t_k}^{t_{k+1}}\left|U(s)-\overline{U}(s)\right|ds \equiv W_1+W_2.$$

We need the following estimates of the partial derivatives:

$$\frac{\partial U(i_{R_0L_0},W,J)}{\partial i_{R_0L_0}}=\frac{\partial}{\partial i_{R_0L_0}}\left(\frac{dJ(t-T)}{dt}-\frac{W(t)-J(t-T)+2\sqrt{L}\,\breve{I}_{in}(t)+2\sqrt{L}\,i_{R_0L_0}(t)}{Z_0 d\widetilde{C}_0\left(\frac{W(t)-J(t-T)}{2\sqrt{C}}\right)/du}\right)=$$

$$=\frac{2\sqrt{L}}{Z_0 d\widetilde{C}_0(u(0,t))/du}\Rightarrow \left|\frac{\partial U(i_{R_0L_0},W,J)}{\partial i_{R_0L_0}}\right|\leq \frac{2\sqrt{L}}{Z_0 \hat{C}_0};$$

$$\frac{\partial U(i_{R_0L_0},W,J)}{\partial W}=\frac{\partial}{\partial W}\left(\frac{dJ(t-T)}{dt}-\frac{W(t)-J(t-T)+2\sqrt{L}\,\breve{I}_{in}(t)+2\sqrt{L}\,i_{R_0L_0}(t)}{Z_0 d\widetilde{C}_0\left(\frac{W(t)-J(t-T)}{2\sqrt{C}}\right)/du}\right)=$$

$$=\frac{d\widetilde{C}_0(u(0,t))/du-\left(W(t)-J(t-T)+2\sqrt{L}\,\breve{I}_{in}(t)+2\sqrt{L}\,i_{R_0L_0}(t)\right)\left(d^2\widetilde{C}_0(u(0,t))/du^2\right)\frac{1}{2\sqrt{C}}}{Z_0\left(d\widetilde{C}_0(u(0,t))/du\right)^2}\Rightarrow$$

$$\left|\frac{\partial U(i_{R_0L_0},W,J)}{\partial W}\right|\leq\frac{1}{Z_0\hat{C}_0}+\frac{W_0+J_0 e^{-\beta}+4\sqrt{L}\,I_{R_0}}{2Z_0\sqrt{C}\,\hat{C}_0^{\,2}}H_0;$$

$$\frac{\partial U(i_{R_0L_0},W,J)}{\partial J}=\frac{\partial}{\partial J}\left(\frac{dJ(t-T)}{dt}-\frac{W(t)-J(t-T)+2\sqrt{L}\,\breve{I}_{in}(t)+2\sqrt{L}\,i_{R_0L_0}(t)}{Z_0 d\widetilde{C}_0\left(\frac{W(t)-J(t-T)}{2\sqrt{C}}\right)/du}\right)=$$

$$=\frac{-d\widetilde{C}_0(u(0,t))/du-\left(W(t)-J(t-T)+2\sqrt{L}\,\breve{I}_{in}(t)+2\sqrt{L}\,i_{R_0L_0}(t)\right)d^2\widetilde{C}_0(u(0,t))/du^2\,\frac{1}{2\sqrt{C}}}{Z_0\left(d\widetilde{C}_0(u(0,t))/du\right)^2}\Rightarrow$$

$$\Rightarrow\left|\frac{\partial U(i_{R_0L_0},W,J)}{\partial J}\right|\leq\frac{1}{Z_0\hat{C}_0}+\frac{W_0+J_0 e^{-\beta}+4\sqrt{L}\,I_{R_0}}{2\sqrt{C}\hat{C}_0^{\,2}Z_0}H_0.$$

Then

$$W_1\leq\left|\int_{t_k}^{t}\left(\frac{dJ(t-T)}{dt}-\frac{d\bar{J}(t-T)}{dt}\right)ds\right|+\frac{2\sqrt{L}}{Z_0\hat{C}_0}\int_{t_k}^{t}\left|i_{R_0L_0}(s)-\bar{i}_{R_0L_0}(s)\right|ds+$$

$$+\left(\frac{1}{Z_0\hat{C}_0}+\frac{W_0+J_0 e^{-\beta}+4\sqrt{L}\,I_{R_0}}{2Z_0\sqrt{C}\,\hat{C}_0^{\,2}}H_0\right)\int_{t_k}^{t}\left|W(s)-\overline{W}(s)\right|ds+$$

$$+\left(\frac{1}{Z_0\hat{C}_0}+\frac{W_0+J_0 e^{-\beta}+4\sqrt{L}\,I_{R_0}}{2\sqrt{C}\hat{C}_0^{\,2}Z_0}H_0\right)\int_{t_k}^{t}\left|J(s-T)-\bar{J}(s-T)\right|ds\leq$$

$$\leq\frac{2\sqrt{L}}{Z_0\hat{C}_0}e^{\mu_0}\rho_\mu^{(k)}(i_{R_0L_0},\bar{i}_{R_0L_0})\int_{t_k}^{t}e^{\mu(s-t_k)}ds+\left(\frac{1}{Z_0\hat{C}_0}+\frac{W_0+J_0 e^{-\beta}+4\sqrt{L}\,I_{R_0}}{2Z_0\sqrt{C}\,\hat{C}_0^{\,2}}H_0\right)e^{\mu_0}\rho_\mu^{(k)}(W,\overline{W})\int_{t_k}^{t}e^{\mu(s-t_k)}ds\leq$$

$$\leq e^{\mu(t-t_k)}\frac{1}{\mu}\left[\frac{2\sqrt{L}}{Z_0\hat{C}_0}\frac{e^{\mu_0}\rho_\mu^{(k)}(\dot{i}_{R_0L_0},\bar{\dot{i}}_{R_0L_0})}{\mu}+\left(\frac{1}{Z_0\hat{C}_0}+\frac{W_0+J_0 e^{-\beta}+4\sqrt{L}\,I_{R_0}}{2Z_0\sqrt{C}\,\hat{C}_0^{\,2}}H_0\right)\frac{e^{\mu_0}\rho_\mu^{(k)}(\dot{W},\bar{\dot{W}})}{\mu}\right]\leq$$

$$\le e^{\mu(t-t_k)}\hat{\rho}_\mu^{(k)}((i_{R_0L_0},W,i_{R_1L_1},J,\dot{i}_{R_0L_0},\dot{W},\dot{i}_{R_1L_1},\dot{J}),(\bar{i}_{R_0L_0},\overline{W},\bar{i}_{R_1L_1},\overline{J},\dot{\bar{i}}_{R_0L_0},\dot{\overline{W}},\dot{\bar{i}}_{R_1L_1},\dot{\overline{J}}))\times$$

$$\times\frac{1}{\mu^2}\frac{e^{\mu_0}}{Z_0\hat{C}_0}\left[2\sqrt{L}+1+\frac{W_0+J_0e^{-\beta}+4\sqrt{L}\,I_{R_0}}{2\sqrt{C}\,\hat{C}_0}H_0\right];$$

$$W_2\le\left|\int_{t_k}^{t_{k+1}}\left(\frac{dJ(t-T)}{dt}-\frac{d\overline{J}(t-T)}{dt}\right)ds\right|+\frac{2\sqrt{L}}{Z_0\hat{C}_0}\int_{t_k}^{t_{k+1}}\left|i_{R_0L_0}(s)-\bar{i}_{R_0L_0}(s)\right|ds+\frac{1}{Z_0\hat{C}_0}\int_{t_k}^{t_{k+1}}\left|W(s)-\overline{W}(s)\right|ds+$$

$$+\frac{1}{Z_0\hat{C}_0}\int_{t_k}^{t_{k+1}}\left|J(s-T)-\overline{J}(s-T)\right|ds\le$$

$$\le e^{\mu(t-t_k)}\hat{\rho}_\mu^{(k)}((i_{R_0L_0},W,i_{R_1L_1},J,\dot{i}_{R_0L_0},\dot{W},\dot{i}_{R_1L_1},\dot{J}),(\bar{i}_{R_0L_0},\overline{W},\bar{i}_{R_1L_1},\overline{J},\dot{\bar{i}}_{R_0L_0},\dot{\overline{W}},\dot{\bar{i}}_{R_1L_1},\dot{\overline{J}}))\times$$

$$\times\frac{\left(e^{\mu_0}-1\right)}{\mu^2}\frac{e^{\mu_0}}{Z_0\hat{C}_0}\left(2\sqrt{L}+1+\frac{W_0+J_0e^{-\beta}+4\sqrt{L}\,I_{R_0}}{2\sqrt{C}\,\hat{C}_0}H_0\right);$$

$$\left|B_W^{(k)}U(i_{R_0L_0},W,J)(t)-B_W^{(k)}U(\bar{i}_{R_0L_0},\overline{W},\overline{J})(t)\right|\le$$

$$\le e^{\mu(t-t_k)}\hat{\rho}_\mu^{(k)}((i_{R_0L_0},W,i_{R_1L_1},J,\dot{i}_{R_0L_0},\dot{W},\dot{i}_{R_1L_1},\dot{J}),(\bar{i}_{R_0L_0},\overline{W},\bar{i}_{R_1L_1},\overline{J},\dot{\bar{i}}_{R_0L_0},\dot{\overline{W}},\dot{\bar{i}}_{R_1L_1},\dot{\overline{J}}))\times$$

$$\times\frac{e^{2\mu_0}}{\mu^2 Z_0\hat{C}_0}\left(2\sqrt{L}+1+\frac{W_0+J_0e^{-\beta}+4\sqrt{L}\,I_{R_0}}{2\sqrt{C}\,\hat{C}_0}H_0\right)\equiv$$

$$\equiv e^{\mu(t-t_k)}K_W\,\hat{\rho}_\mu^{(k)}((i_{R_0L_0},W,i_{R_1L_1},J,\dot{i}_{R_0L_0},\dot{W},\dot{i}_{R_1L_1},\dot{J}),(\bar{i}_{R_0L_0},\overline{W},\bar{i}_{R_1L_1},\overline{J},\dot{\bar{i}}_{R_0L_0},\dot{\overline{W}},\dot{\bar{i}}_{R_1L_1},\dot{\overline{J}})).$$

It follows

$$\hat{\rho}^{(k)}\left(B_W(i_{R_0L_0},W,J),B_W(\bar{i}_{R_0L_0},\overline{W},\overline{J})\right)\le$$
$$\le e^{\mu_0}K_W\,\hat{\rho}_\mu^{(k)}((i_{R_0L_0},W,i_{R_1L_1},J,\dot{i}_{R_0L_0},\dot{W},\dot{i}_{R_1L_1},\dot{J}),(\bar{i}_{R_0L_0},\overline{W},\bar{i}_{R_1L_1},\overline{J},\dot{\bar{i}}_{R_0L_0},\dot{\overline{W}},\dot{\bar{i}}_{R_1L_1},\dot{\overline{J}})).$$

For the third component we have

$$\frac{\partial I_{R_1}(W,i_{R_1L_1},J)}{\partial W}=\frac{\partial}{\partial W}\left(\frac{W(t-T)-J(t)-2\sqrt{C}R_1(i_{R_1L_1}(t))}{2\sqrt{C}\,\widetilde{L}_1(i_{R_1L_1}(t))}\right)=\frac{1}{2\sqrt{C}\,\widetilde{L}_1(i_{R_1L_1}(t))};$$

$$\frac{\partial I_{R_1}(W,i_{R_1L_1},J)}{\partial i_{R_1L_1}}=\frac{1}{2\sqrt{C}}\frac{\partial}{\partial i_{R_1L_1}}\left(\frac{W(t-T)-J(t)-2\sqrt{C}R_1(i_{R_1L_1}(t))}{\widetilde{L}_1(i_{R_1L_1}(t))}\right)=$$

$$= \frac{1}{2\sqrt{C}} \frac{-2\sqrt{C}\,dR_1(i_{R_1L_1}(t))/di_{R_1L_1}\cdot\widetilde{L}_1(i_{R_1L_1}(t)) - d\widetilde{L}_1(i_{R_1L_1}(t))/di_{R_1L_1}\left(W(t-T)-J(t)-2\sqrt{C}R_1(i_{R_1L_1}(t))\right)}{\widetilde{L}_1^{\,2}(i_{R_1L_1}(t))};$$

$$\frac{\partial I_{R_1}(W,i_{R_1L_1},J)}{\partial J} = \frac{\partial}{\partial J}\left(\frac{W(t-T)-J(t)-2\sqrt{C}R_1(i_{R_1L_1}(t))}{2\sqrt{C}\;\widetilde{L}_1(i_{R_1L_1}(t))}\right) = \frac{-1}{2\sqrt{C}\;\widetilde{L}_1(i_{R_1L_1}(t))};$$

$$\left|\frac{\partial I_{R_1}(W,i_{R_1L_1},J)}{\partial W}\right| \le \frac{1}{2\sqrt{C}\;\hat{L}_1};$$

$$\left|\frac{\partial I_{R_1}(W,i_{R_1L_1},J)}{\partial i_{R_1L_1}}\right| \le$$

$$\left[\frac{2\sqrt{C}}{\hat{L}_1}\sum_{n=1}^{m} n\left|r_n^{(1)}\right| I_{R_1}^{n-1} + \frac{W_0 + J_0 e^{-\beta} + 2\sqrt{C}\sum_{n=1}^{m}\left|r_n^{(1)}\right| I_{R_0}^{n}}{\hat{L}_1^{2}}\sum_{n=1}^{m}(n+1)n\left|l_n^{(1)}\right| I_{R_1}^{n-1}\right];$$

$$\left|\frac{\partial I_{R_1}(W,i_{R_1L_1},J)}{\partial J}\right| \le \frac{1}{2\sqrt{C}\;\hat{L}_1};$$

$$\left|B_1^{(k)}(W,i_{R_1L_1},J)(t) - B_1^{(k)}(\overline{W},\bar{i}_{R_1L_1},\overline{J})(t)\right| \le \int_{t_k}^{t}\left|I_{R_1}(W,i_{R_1L_1},J)(s) - I_{R_1}(\overline{W},\bar{i}_{R_1L_1},\overline{J})(s)\right|ds +$$

$$+ \int_{t_k}^{t_{k+1}}\left|I_{R_1}(W,i_{R_1L_1},J)(s) - I_{R_1}(\overline{W},\bar{i}_{R_1L_1},\overline{J})(s)\right|ds \equiv P_1 + P_2.$$

$$P_1 \le \frac{1}{2\sqrt{C}\;\hat{L}_1}\int_{t_k}^{t}\left|\vec{W}(s) - \vec{\overline{W}}(s)\right|ds +$$

$$+ \frac{W_0 + J_0 e^{-\beta} + 2\sqrt{C}\sum_{n=1}^{m}\left|r_n^{(1)}\right| I_{R_1}^{n} e^{(n-1)\mu_0}}{2\sqrt{C}\hat{L}_1^{2}}\sum_{n=1}^{m}(n+1)n\left|l_n^{(1)}\right| I_{R_1}^{n-1} e^{n\mu_0}\left.\right]\int_{t_k}^{t}\left|i_{R_1L_1}(s) - \bar{i}_{R_1L_1}(s)\right|ds +$$

$$+ \frac{1}{2\sqrt{C}\;\hat{L}_1}\int_{t_k}^{t}\left|J(s) - \overline{J}(s)\right|ds \le$$

$$\le \left[\frac{2\sqrt{C}}{\hat{L}_1}\sum_{n=1}^{m} n\left|r_n^{(1)}\right| I_{R_1}^{n-1} + \frac{W_0 + J_0 e^{-\beta} + 2\sqrt{C}\sum_{n=1}^{m}\left|r_n^{(1)}\right| I_{R_1}^{n}}{2\sqrt{C}\hat{L}_1^{2}}\sum_{n=1}^{m}(n+1)n\left|l_n^{(1)}\right| I_{R_1}^{n-1}\right]\frac{e^{\mu_0}\rho_\mu^{(k)}(i_{R_1L_1},\bar{i}_{R_1L_1})}{\mu}\frac{e^{\mu(t-t_k)}-1}{\mu} +$$

$$+\frac{e^{\mu_0}\rho_\mu^{(k)}(\dot{J},\dot{\bar{J}})}{2\mu\sqrt{C}\,\hat{L}_1}\frac{e^{\mu(t-t_k)}-1}{\mu}\le$$

$$\le e^{\mu(t-t_k)}e^{\mu_0}\frac{1}{\mu^2\hat{L}_1}\left[2\sqrt{C}\sum_{n=1}^{m}n\left|r_n^{(1)}\right|I_{R_1}^{n-1}+\frac{W_0+J_0e^{-\beta}+2\sqrt{C}\sum_{n=1}^{m}\left|r_n^{(1)}\right|I_{R_1}^n}{2\sqrt{C}\hat{L}_1}\sum_{n=1}^{m}(n+1)n\left|l_n^{(1)}\right|I_{R_1}^{n-1}+\frac{1}{2\sqrt{C}}\right]\times$$

$$\times\hat{\rho}_\mu^{(k)}((i_{R_0L_0},W,i_{R_1L_1},J,\dot{i}_{R_0L_0},\dot{W},\dot{i}_{R_1L_1},\dot{J}),(\bar{i}_{R_0L_0},\overline{W},\bar{i}_{R_1L_1},\bar{J},\dot{\bar{i}}_{R_0L_0},\dot{\overline{W}},\dot{\bar{i}}_{R_1L_1},\dot{\bar{J}}));$$

$$P_2\le\frac{1}{2\sqrt{C}\,\hat{L}_1}\int_{t_k}^{t_{k+1}}\left|\vec{W}(s)-\vec{W}(s)\right|ds+$$

$$+\left(\frac{2\sqrt{C}}{\hat{L}_1}\sum_{n=1}^{m}n\left|r_n^{(1)}\right|I_{R_1}^{n-1}+\frac{W_0+J_0e^{-\beta}+2\sqrt{C}\sum_{n=1}^{m}\left|r_n^{(1)}\right|I_{R_1}^n}{2\sqrt{C}\hat{L}_1^2}\sum_{n=1}^{m}(n+1)n\left|l_n^{(1)}\right|I_{R_1}^{n-1}\right)\int_{t_k}^{t_{k+1}}\left|i_{R_1L_1}(s)-\bar{i}_{R_1L_1}(s)\right|ds+$$

$$+\frac{1}{2\sqrt{C}\,\hat{L}_1}\int_{t_k}^{t_{k+1}}\left|J(s)-\bar{J}(s)\right|ds\le$$

$$\le\left(\frac{2\sqrt{C}}{\hat{L}_1}\sum_{n=1}^{m}n\left|r_n^{(1)}\right|I_{R_1}^{n-1}+\frac{W_0+J_0e^{-\beta}+2\sqrt{C}\sum_{n=1}^{m}\left|r_n^{(1)}\right|I_{R_1}^n}{2\sqrt{C}\hat{L}_1^2}\sum_{n=1}^{m}(n+1)n\left|l_n^{(1)}\right|I_{R_1}^{n-1}\right)\rho^{(k)}(i_{R_1L_1},\bar{i}_{R_1L_1})\int_{t_k}^{t_{k+1}}e^{\mu(s-t_k)}ds+$$

$$+\frac{\rho^{(k)}(J,\bar{J})}{2\sqrt{C}\,\hat{L}_1}\int_{t_k}^{t_{k+1}}e^{\mu(s-t_k)}ds\le$$

$$\le\left(\frac{2\sqrt{C}}{\hat{L}_1}\sum_{n=1}^{m}n\left|r_n^{(1)}\right|I_{R_1}^{n-1}+\frac{W_0+J_0e^{-\beta}+2\sqrt{C}\sum_{n=1}^{m}\left|r_n^{(1)}\right|I_{R_1}^n}{2\sqrt{C}\hat{L}_1^2}\sum_{n=1}^{m}(n+1)n\left|l_n^{(1)}\right|I_{R_1}^{n-1}\right)\frac{e^{\mu_0}\rho_\mu^{(k)}(i_{R_1L_1},\dot{\bar{i}}_{R_1L_1})}{\mu}\frac{e^{\mu T_0}-1}{\mu}+$$

$$+\frac{e^{\mu_0}\rho_\mu^{(k)}(\dot{J},\dot{\bar{J}})}{2\mu\sqrt{C}\,\hat{L}_1}\frac{e^{\mu T_0}-1}{\mu}\le$$

$$\le e^{\mu(t-t_k)}e^{\mu_0}\frac{e^{\mu_0}-1}{\mu^2\hat{L}_1}\left(2\sqrt{C}\sum_{n=1}^{m}n\left|r_n^{(1)}\right|I_{R_1}^{n-1}+\frac{W_0+J_0e^{-\beta}+2\sqrt{C}\sum_{n=1}^{m}\left|r_n^{(1)}\right|I_{R_1}^n}{2\sqrt{C}\hat{L}_1}\left(\sum_{n=1}^{m}(n+1)n\left|l_n^{(1)}\right|I_{R_1}^{n-1}\right)+\frac{1}{2\sqrt{C}}\right)\times$$

$$\times\hat{\rho}_\mu^{(k)}((i_{R_0L_0},W,i_{R_1L_1},J,\dot{i}_{R_0L_0},\dot{W},\dot{i}_{R_1L_1},\dot{J}),(\bar{i}_{R_0L_0},\overline{W},\bar{i}_{R_1L_1},\bar{J},\dot{\bar{i}}_{R_0L_0},\dot{\overline{W}},\dot{\bar{i}}_{R_1L_1},\dot{\bar{J}}));$$

$$\left|B_1^{(k)}(W,i_{R_1L_1},J)(t)-B_1^{(k)}(\overline{W},\bar{i}_{R_1L_1},\bar{J})(t)\right|\le$$

$$\le e^{\mu(t-t_k)}e^{\mu_0}\frac{1}{\mu^2\hat{L}_1}\left(2\sqrt{C}\sum_{n=1}^{m}n\left|r_n^{(1)}\right|I_{R_1}^{n-1}+\frac{W_0+J_0e^{-\beta}+2\sqrt{C}\sum_{n=1}^{m}\left|r_n^{(1)}\right|I_{R_1}^{n}}{2\sqrt{C}\hat{L}_1}\sum_{n=1}^{m}(n+1)n\left|l_n^{(1)}\right|I_{R_1}^{n-1}+\frac{1}{2\sqrt{C}}\right)\times$$

$$\times\hat{\rho}_\mu^{(k)}((i_{R_0L_0},W,i_{R_1L_1},J,\dot{i}_{R_0L_0},\dot{W},\dot{i}_{R_1L_1},\dot{J}),(\bar{i}_{R_0L_0},\overline{W},\bar{i}_{R_1L_1},\overline{J},\dot{\bar{i}}_{R_0L_0},\dot{\overline{W}},\dot{\bar{i}}_{R_1L_1},\dot{\overline{J}}))+$$

$$+e^{\mu(t-t_k)}e^{\mu_0}\frac{e^{\mu_0}-1}{\mu^2\hat{L}_1}\left(2\sqrt{C}\sum_{n=1}^{m}n\left|r_n^{(1)}\right|I_{R_1}^{n-1}+\frac{W_0+J_0e^{-\beta}+2\sqrt{C}\sum_{n=1}^{m}\left|r_n^{(1)}\right|I_{R_1}^{n}}{2\sqrt{C}\hat{L}_1}\sum_{n=1}^{m}(n+1)n\left|l_n^{(1)}\right|I_{R_1}^{n-1}+\frac{1}{2\sqrt{C}}\right)\times$$

$$\times\hat{\rho}_\mu^{(k)}((i_{R_0L_0},W,i_{R_1L_1},J,\dot{i}_{R_0L_0},\dot{W},\dot{i}_{R_1L_1},\dot{J}),(\bar{i}_{R_0L_0},\overline{W},\bar{i}_{R_1L_1},\overline{J},\dot{\bar{i}}_{R_0L_0},\dot{\overline{W}},\dot{\bar{i}}_{R_1L_1},\dot{\overline{J}}))\le$$

$$\le e^{\mu(t-t_k)}\frac{e^{2\mu_0}}{\mu^2\hat{L}_1}\left(2\sqrt{C}\sum_{n=1}^{m}n\left|r_n^{(1)}\right|I_{R_1}^{n-1}+\frac{W_0+J_0e^{-\beta}+2\sqrt{C}\sum_{n=1}^{m}\left|r_n^{(1)}\right|I_{R_1}^{n}}{2\sqrt{C}\hat{L}_1}\sum_{n=1}^{m}(n+1)n\left|l_n^{(1)}\right|I_{R_1}^{n-1}+\frac{1}{2\sqrt{C}}\right)\times$$

$$\times\hat{\rho}_\mu^{(k)}((i_{R_0L_0},W,i_{R_1L_1},J,\dot{i}_{R_0L_0},\dot{W},\dot{i}_{R_1L_1},\dot{J}),(\bar{i}_{R_0L_0},\overline{W},\bar{i}_{R_1L_1},\overline{J},\dot{\bar{i}}_{R_0L_0},\dot{\overline{W}},\dot{\bar{i}}_{R_1L_1},\dot{\overline{J}}))\equiv$$

$$\equiv e^{\mu_0}K_1\hat{\rho}_\mu^{(k)}((i_{R_0L_0},W,i_{R_1L_1},J,\dot{i}_{R_0L_0},\dot{W},\dot{i}_{R_1L_1},\dot{J}),(\bar{i}_{R_0L_0},\overline{W},\bar{i}_{R_1L_1},\overline{J},\dot{\bar{i}}_{R_0L_0},\dot{\overline{W}},\dot{\bar{i}}_{R_1L_1},\dot{\overline{J}})).$$

For the fourth component we obtain

$$\frac{\partial I(W,i_{R_1L_1},J)}{\partial W}=\frac{\partial}{\partial W}\left(\dot{W}(t-T)+\frac{1}{Z_0}\frac{W(t-T)+J(t)+2\sqrt{L}i_{R_1L_1}(t)}{d\widetilde{C}_1\left(\dfrac{W(t-T)-J(t)}{2\sqrt{C}}\right)/du}\right)=$$

$$=\frac{2\sqrt{C}d\widetilde{C}_1(u(\Lambda,t))/du-\left(W(t-T)+J(t)+2\sqrt{L}i_{R_1L_1}(t)\right)\left(d^2\widetilde{C}_1(u(\Lambda,t))/du^2\right)}{2\sqrt{C}Z_0\left(d\widetilde{C}_1(u(\Lambda,t))/du\right)^2}\Rightarrow$$

$$\Rightarrow\left|\frac{\partial I(W,i_{R_1L_1},J)}{\partial W}\right|\le\frac{2\sqrt{C}\left|d\widetilde{C}_1(u(\Lambda,t))/du\right|}{2\sqrt{C}Z_0\left(d\widetilde{C}_1(u(\Lambda,t))/du\right)^2}+\frac{\left(\left|W(t-T)\right|+\left|J(t)\right|+2\sqrt{L}\left|i_{R_1L_1}(t)\right|\right)\left|d^2\widetilde{C}_1(u(\Lambda,t))/du^2\right|}{2\sqrt{C}Z_0\left(d\widetilde{C}_1(u(\Lambda,t))/du\right)^2}\le$$

$$\le\frac{1}{Z_0\hat{C}_1}+\frac{W_0e^{-\beta}+J_0+2\sqrt{L}I_{R_1}}{2\sqrt{C}Z_0\hat{C}_1^2}H_1;$$

$$\frac{\partial I(W,i_{R_1L_1},J)}{\partial i_{R_1L_1}}=\frac{\partial}{\partial i_{R_1L_1}}\left(\frac{dW(t-T)}{dt}-\frac{W(t-T)-J(t)+2\sqrt{L}\,i_{R_1L_1}(t)}{Z_0d\widetilde{C}_1\left(\dfrac{W(t-T)-J(t)}{2\sqrt{C}}\right)/du}\right)=$$

$$=\frac{2\sqrt{L}}{Z_0d\widetilde{C}_1(u(\Lambda,t))/du}\Rightarrow\left|\frac{\partial I(W,i_{R_1L_1},J)}{\partial i_{R_1L_1}}\right|\le\frac{2\sqrt{L}}{Z_0\hat{C}_1};$$

$$\frac{\partial I(W, i_{R_1L_1}, J)}{\partial J} = \frac{\partial}{\partial J}\left(\frac{dJ(t-T)}{dt} - \frac{W(t-T)-J(t)+2\sqrt{L}\, i_{R_1L_1}(t)}{Z_0 d\widetilde{C}_1\left(\frac{W(t-T)-J(t)}{2\sqrt{C}}\right)/du} \right) =$$

$$= \frac{-d\widetilde{C}_1(u(\Lambda,t))/du - \left(W(t-T)-J(t)+2\sqrt{L}\, i_{R_1L_1}(t)\right)d^2\widetilde{C}_1(u(\Lambda,t))/du^2 \, \frac{1}{2\sqrt{C}}}{Z_0\left(d\widetilde{C}_1(u(\Lambda,t))/du\right)^2} \Rightarrow$$

$$\Rightarrow \left|\frac{\partial I(W, i_{R_1L_1}, J)}{\partial J}\right| \leq \frac{1}{\hat{C}_1 Z_0} + \frac{W_0 e^{-\beta} + J_0 + 2\sqrt{L}\, I_{R_1}}{2\sqrt{C}\,\hat{C}_1^{\,2} Z_0} H_1 .$$

We have

$$\left| B_J^{(k)}(W, i_{R_1L_1}, J)(t) - B_J^{(k)}(\overline{W}, \overline{i}_{R_1L_1}, \overline{J})(t) \right| \leq I_1 + I_2 .$$

Since

$$I_1 \leq \left| \int_{t_k}^{t}\left(\frac{dW(t-T)}{dt} - \frac{d\overline{W}(t-T)}{dt} \right)ds \right| + \frac{2\sqrt{L}}{Z_0 \hat{C}_1}\int_{t_k}^{t}\left| i_{R_1L_1}(s) - \overline{i}_{R_1L_1}(s) \right|ds +$$

$$+ \left(\frac{1}{\hat{C}_1 Z_0} + \frac{W_0 e^{-\beta} + J_0 + 2\sqrt{L}\, I_{R_1}}{2\sqrt{C}\,\hat{C}_1^{\,2} Z_0} H_1 \right)\int_{t_k}^{t}\left| J(s) - \overline{J}(s) \right|ds +$$

$$+ \left(\frac{1}{Z_0 \hat{C}_1} + \frac{W_0 e^{-\beta} + J_0 + 2\sqrt{L}I_{R_1}}{2\sqrt{C} Z_0 \hat{C}_1^{\,2}} H_1 \right)\int_{t_k}^{t}\left| W(s-T) - \overline{W}(s-T) \right|ds \leq$$

$$\leq \frac{e^{\mu(t-t_k)}-1}{\mu}\left(e^{\mu_0}\frac{2\sqrt{L}}{Z_0 \hat{C}_1}\frac{\rho_\mu^{(k)}(\dot{i}_{R_1L_1}, \dot{\overline{i}}_{R_1L_1})}{\mu} + e^{\mu_0}\left(\frac{1}{\hat{C}_1 Z_0} + \frac{W_0 e^{-\beta} + J_0 + 2\sqrt{L}\, I_{R_1}}{2\sqrt{C}\,\hat{C}_1^{\,2} Z_0} H_1 \right)\frac{\rho^{(k)}(\dot{J}, \dot{\overline{J}})}{\mu} \right) \leq$$

$$\leq e^{\mu(t-t_k)}\hat{\rho}_\mu^{(k)}\big((i_{R_0L_0}, W, i_{R_1L_1}, J, \dot{i}_{R_0L_0}, \dot{W}, \dot{i}_{R_1L_1}, \dot{J}), (\overline{i}_{R_0L_0}, \overline{W}, \overline{i}_{R_1L_1}, \overline{J}, \dot{\overline{i}}_{R_0L_0}, \dot{\overline{W}}, \dot{\overline{i}}_{R_1L_1}, \dot{\overline{J}}) \big) \times$$

$$\times \frac{e^{\mu_0}}{\mu^2 Z_0 \hat{C}_1}\left(2\sqrt{L} + 1 + \frac{W_0 e^{-\beta} + J_0 + 2\sqrt{L}\, I_{R_1}}{2\sqrt{C}\,\hat{C}_1} H_1 \right)$$

and

$$I_2 \leq \left| \int_{t_k}^{t_{k+1}}\left(\frac{dW(t-T)}{dt} - \frac{d\overline{W}(t-T)}{dt} \right)ds \right| + \frac{2\sqrt{L}}{Z_0 \hat{C}_1}\int_{t_k}^{t_{k+1}}\left| i_{R_1L_1}(s) - \overline{i}_{R_1L_1}(s) \right|ds + \frac{1}{Z_0 \hat{C}_1}\int_{t_k}^{t_{k+1}}\left| J(s) - \overline{J}(s) \right|ds +$$

$$+ \frac{1}{Z_0 \hat{C}_1}\int_{t_k}^{t_{k+1}}\left| W(s-T) - \overline{W}(s-T) \right|ds \leq$$

$$\leq e^{\mu(t-t_k)}\hat{\rho}_\mu^{(k)}((i_{R_0L_0},W,i_{R_1L_1},J,\dot{i}_{R_0L_0},\dot{W},\dot{i}_{R_1L_1},\dot{J}),(\bar{i}_{R_0L_0},\overline{W},\bar{i}_{R_1L_1},\overline{J},\dot{\bar{i}}_{R_0L_0},\dot{\overline{W}},\dot{\bar{i}}_{R_1L_1},\dot{\overline{J}}))\times$$

$$\times\left(e^{\mu_0}-1\right)\frac{e^{\mu_0}}{\mu^2 Z_0\hat{C}_1}\left(2\sqrt{L}+1+\frac{W_0 e^{-\beta}+J_0+2\sqrt{L}\,I_{R_1}}{2\sqrt{C}\hat{C}_1}H_1\right)$$

then

$$\left|B_J^{(k)}(W,i_{R_1L_1},J)(t)-B_J^{(k)}(\overline{W},\bar{i}_{R_1L_1},\overline{J})(t)\right|\leq$$

$$\leq e^{\mu(t-t_k)}\hat{\rho}_\mu^{(k)}((i_{R_0L_0},W,i_{R_1L_1},J,\dot{i}_{R_0L_0},\dot{W},\dot{i}_{R_1L_1},\dot{J}),(\bar{i}_{R_0L_0},\overline{W},\bar{i}_{R_1L_1},\overline{J},\dot{\bar{i}}_{R_0L_0},\dot{\overline{W}},\dot{\bar{i}}_{R_1L_1},\dot{\overline{J}}))\times$$

$$\times\frac{e^{2\mu_0}}{\mu^2 Z_0\hat{C}_1}\left(2\sqrt{L}+1+\frac{W_0 e^{-\beta}+J_0+2\sqrt{L}\,I_{R_1}}{2\sqrt{C}\hat{C}_1}H_1\right)\equiv$$

$$\equiv e^{\mu_0}K_J\hat{\rho}_\mu^{(k)}\left((i_{R_0L_0},W,i_{R_1L_1},J,\dot{i}_{R_0L_0},\dot{W},\dot{i}_{R_1L_1},\dot{J}),(\bar{i}_{R_0L_0},\overline{W},\bar{i}_{R_1L_1},\overline{J},\dot{\bar{i}}_{R_0L_0},\dot{\overline{W}},\dot{\bar{i}}_{R_1L_1},\dot{\overline{J}})\right).$$

It follows

$$\hat{\rho}^{(k)}\left(B_J(i_{R_0L_0},W,J),B_J(\bar{i}_{R_0L_0},\overline{W},\overline{J})\right)\leq$$
$$\leq e^{\mu_0}K_J\hat{\rho}_\mu^{(k)}((i_{R_0L_0},W,i_{R_1L_1},J,\dot{i}_{R_0L_0},\dot{W},\dot{i}_{R_1L_1},\dot{J}),(\bar{i}_{R_0L_0},\overline{W},\bar{i}_{R_1L_1},\overline{J},\dot{\bar{i}}_{R_0L_0},\dot{\overline{W}},\dot{\bar{i}}_{R_1L_1},\dot{\overline{J}}))$$

For the derivative of the first component we have

$$\left|\dot{B}_0^{(k)}(i_{R_0L_0},W,J)(t)-\dot{B}_0^{(k)}(\bar{i}_{R_0L_0},\overline{W},\overline{J})(t)\right|\leq\left|I_{R_0}(i_{R_0L_0},W,J)(t)-I_{R_0L_0}(\bar{i}_{R_0L_0},\overline{W},\overline{J})(t)\right|+$$

$$+\frac{1}{t_{k+1}-t_k}\left|\int_{t_k}^{t_{k+1}}\left(I_{R_0}(i_{R_0L_0},W,J)(s)-I_{R_0L_0}(\bar{i}_{R_0L_0},\overline{W},\overline{J})(s)\right)ds\right|\equiv \dot{I}_1+\dot{I}_2.$$

But

$$\dot{I}_1\leq\left|i_{R_0L_0}(t)-\bar{i}_{R_0L_0}(t)\right|\left(\frac{2\sqrt{C}}{\hat{L}_0}\sum_{n=1}^{m}n\left|r_n^{(0)}\right|I_{R_0}^{n-1}+\right.$$

$$+\frac{W_0+J_0 e^{-\beta}+2\sqrt{C}\sum_{n=1}^{m}\left|r_n^{(0)}\right|I_{R_0}^{n}}{2\sqrt{C}\hat{L}_0^2}\sum_{n=1}^{m}(n+1)n\left|l_n^{(0)}\right|I_{R_0}^{n-1}\right)+\frac{1}{2\sqrt{C}\hat{L}_0}\left|W(t)-\overline{W}(t)\right|+$$

$$+\frac{1}{2\sqrt{C}\hat{L}_0}\left|J(t-T)-\overline{J}(t-T)\right|\leq$$

$$\leq e^{\mu(t-t_k)}\rho_\mu^{(k)}(i_{R_0L_0},\bar{i}_{R_0L_0})\left(\frac{2\sqrt{C}}{\hat{L}_0}\sum_{n=1}^{m}n\left|r_n^{(0)}\right|I_{R_0}^{n-1}+\right.$$

$$\left.+\frac{W_0+J_0e^{-\beta}+2\sqrt{C}\sum_{n=1}^{m}\left|r_n^{(0)}\right|I_{R_0}^{n}}{2\sqrt{C}\hat{L}_0^2}\sum_{n=1}^{m}(n+1)n\left|l_n^{(0)}\right|I_{R_0}^{n-1}\right)+e^{\mu(t-t_k)}\frac{1}{2\sqrt{C}\hat{L}_0}\rho_\mu^{(k)}(W,\overline{W})\leq$$

$$\leq\frac{\rho_\mu^{(k)}(\dot{i}_{R_0L_0},\dot{\bar{i}}_{R_0L_0})}{\mu}\left(\frac{2\sqrt{C}}{\hat{L}_0}\sum_{n=1}^{m}n\left|r_n^{(0)}\right|I_{R_0}^{n-1}+\right.$$

$$\left.+\frac{W_0+J_0e^{-\beta}+2\sqrt{C}\sum_{n=1}^{m}\left|r_n^{(0)}\right|I_{R_0}^{n}}{2\sqrt{C}\hat{L}_0^2}\sum_{n=1}^{m}(n+1)n\left|l_n^{(0)}\right|I_{R_0}^{n-1}\right)+\frac{\rho_\mu^{(k)}(\dot{W},\dot{\overline{W}})}{2\sqrt{C}\mu\hat{L}_0}\leq$$

$$\leq\hat{\rho}_\mu^{(k)}((i_{R_0L_0},W,i_{R_1L_1},J,\dot{i}_{R_0L_0},\dot{W},\dot{i}_{R_1L_1},\dot{J}),(\bar{i}_{R_0L_0},\overline{W},\bar{i}_{R_1L_1},\overline{J},\dot{\bar{i}}_{R_0L_0},\dot{\overline{W}},\dot{\bar{i}}_{R_1L_1},\dot{\overline{J}}))\times$$

$$\times\frac{e^{\mu(t-t_k)}}{\mu\hat{L}_0}\left(2\sqrt{C}\sum_{n=1}^{m}n\left|r_n^{(0)}\right|I_{R_0}^{n-1}+\frac{W_0+J_0e^{-\beta}+2\sqrt{C}\sum_{n=1}^{m}\left|r_n^{(0)}\right|I_{R_0}^{n}}{2\sqrt{C}\hat{L}_0}\sum_{n=1}^{m}(n+1)n\left|l_n^{(0)}\right|I_{R_0}^{n-1}+\frac{1}{2\sqrt{C}}\right)$$

and

$$\dot{I}_2\leq\frac{1}{t_{k+1}-t_k}\left[\int_{t_k}^{t_{k+1}}e^{\mu(s-t_k)}\left|i_{R_0L_0}(s)-\bar{i}_{R_0L_0}(s)\right|ds\left(\frac{2\sqrt{C}}{\hat{L}_0}\sum_{n=1}^{m}n\left|r_n^{(0)}\right|I_{R_0}^{n-1}+\right.\right.$$

$$\left.+\frac{W_0+J_0e^{-\beta}+2\sqrt{C}\sum_{n=1}^{m}\left|r_n^{(0)}\right|I_{R_0}^{n}}{2\sqrt{C}\hat{L}_0^2}\sum_{n=1}^{m}(n+1)n\left|l_n^{(0)}\right|I_{R_0}^{n-1}\right)+\frac{1}{2\sqrt{C}\hat{L}_0}\int_{t_k}^{t_{k+1}}\left|W(s)-\overline{W}(s)\right|ds+$$

$$\left.+\frac{1}{2\sqrt{C}\hat{L}_0}\int_{t_k}^{t_{k+1}}\left|J(s-T)-\overline{J}(s-T)\right|ds\right]\leq$$

$$\leq\frac{e^{\mu(t_{k+1}-t_k)}-1}{\mu(t_{k+1}-t_k)}\left[\rho_\mu^{(k)}(i_{R_0L_0},\bar{i}_{R_0L_0})\left(\frac{2\sqrt{C}}{\hat{L}_0}\sum_{n=1}^{m}n\left|r_n^{(0)}\right|I_{R_0}^{n-1}+\right.\right.$$

$$\left.\left.+\frac{W_0+J_0e^{-\beta}+2\sqrt{C}\sum_{n=1}^{m}\left|r_n^{(0)}\right|I_{R_0}^{n}}{2\sqrt{C}\hat{L}_0^2}\sum_{n=1}^{m}(n+1)n\left|l_n^{(0)}\right|I_{R_0}^{n-1}\right)+\frac{1}{2\sqrt{C}\hat{L}_0}\rho_\mu^{(k)}(W,\overline{W})\right]\leq$$

$$\leq \frac{e^{\mu_0}-1}{\mu_0}\left[\frac{\rho^{(k)}(\dot{i}_{R_0 L_0},\dot{\bar{i}}_{R_0 L_0})}{\mu}\left(\frac{2\sqrt{C}}{\hat{L}_0}\sum_{n=1}^{m}n\left|r_n^{(0)}\right|I_{R_0}^{n-1}+\right.\right.$$

$$+\frac{W_0+J_0 e^{-\beta}+2\sqrt{C}\sum_{n=1}^{m}\left|r_n^{(0)}\right|I_{R_0}^{n}}{2\sqrt{C}\hat{L}_0^2}\sum_{n=1}^{m}(n+1)n\left|l_n^{(0)}\right|I_{R_0}^{n-1}\right)+\frac{1}{2\sqrt{C}\hat{L}_0}\frac{\rho_\mu^{(k)}(\dot{W},\dot{\bar{W}})}{\mu}\right]\leq$$

$$\leq \hat{\rho}_\mu^{(k)}((i_{R_0 L_0},W,i_{R_1 L_1},J,\dot{i}_{R_0 L_0},\dot{W},\dot{i}_{R_1 L_1},\dot{J}),(\bar{i}_{R_0 L_0},\overline{W},\bar{i}_{R_1 L_1},\overline{J},\dot{\bar{i}}_{R_0 L_0},\dot{\overline{W}},\dot{\bar{i}}_{R_1 L_1},\dot{\overline{J}}))\times$$

$$\times\frac{e^{\mu_0}-1}{\mu_0}\frac{1}{\mu\hat{L}_0}\left(2\sqrt{C}\sum_{n=1}^{m}n\left|r_n^{(0)}\right|I_{R_0}^{n-1}+\frac{W_0+J_0 e^{-\beta}+2\sqrt{C}\sum_{n=1}^{m}\left|r_n^{(0)}\right|I_{R_0}^{n}}{2\sqrt{C}\hat{L}_0}\sum_{n=1}^{m}(n+1)n\left|l_n^{(0)}\right|I_{R_0}^{n-1}+\frac{1}{2\sqrt{C}}\right).$$

Then

$$\left|\dot{B}_0^{(k)}(i_{R_0 L_0},W,J)(t)-\dot{B}_0^{(k)}(\bar{i}_{R_0 L_0},\overline{W},\overline{J})(t)\right|\leq$$

$$\leq e^{\mu(t-t_k)}\hat{\rho}_\mu^{(k)}((i_{R_0 L_0},W,i_{R_1 L_1},J,\dot{i}_{R_0 L_0},\dot{W},\dot{i}_{R_1 L_1},\dot{J}),(\bar{i}_{R_0 L_0},\overline{W},\bar{i}_{R_1 L_1},\overline{J},\dot{\bar{i}}_{R_0 L_0},\dot{\overline{W}},\dot{\bar{i}}_{R_1 L_1},\dot{\overline{J}}))\times$$

$$\times\left(1+\frac{e^{\mu_0}-1}{\mu_0}\right)\frac{1}{\mu\hat{L}_0}\left(2\sqrt{C}\sum_{n=1}^{m}n\left|r_n^{(0)}\right|I_{R_0}^{n-1}+\frac{W_0+J_0 e^{-\beta}+2\sqrt{C}\sum_{n=1}^{m}\left|r_n^{(0)}\right|I_{R_0}^{n}}{2\sqrt{C}\hat{L}_0}\sum_{n=1}^{m}(n+1)n\left|l_n^{(0)}\right|I_{R_0}^{n-1}+\frac{1}{2\sqrt{C}}\right)\equiv$$

$$\equiv e^{\mu(t-t_k)}\dot{K}_0\hat{\rho}_\mu^{(k)}((i_{R_0 L_0},W,i_{R_1 L_1},J,\dot{i}_{R_0 L_0},\dot{W},\dot{i}_{R_1 L_1},\dot{J}),(\bar{i}_{R_0 L_0},\overline{W},\bar{i}_{R_1 L_1},\overline{J},\dot{\bar{i}}_{R_0 L_0},\dot{\overline{W}},\dot{\bar{i}}_{R_1 L_1},\dot{\overline{J}})).$$

It follows

$$\rho_\mu^{(k)}\left(\dot{B}_0^{(k)}(i_{R_0 L_0},W,J),\dot{B}_0^{(k)}(\bar{i}_{R_0 L_0},\overline{W},\overline{J})\right)\leq$$
$$\leq \dot{K}_0\hat{\rho}_\mu^{(k)}((i_{R_0 L_0},W,i_{R_1 L_1},J,\dot{i}_{R_0 L_0},\dot{W},\dot{i}_{R_1 L_1},\dot{J}),(\bar{i}_{R_0 L_0},\overline{W},\bar{i}_{R_1 L_1},\overline{J},\dot{\bar{i}}_{R_0 L_0},\dot{\overline{W}},\dot{\bar{i}}_{R_1 L_1},\dot{\overline{J}})).$$

For the derivative of the second component on we have

$$\left|\dot{B}_W^{(k)}(i_{R_0 L_0},W,J)(t)-\dot{B}_W^{(k)}(\bar{i}_{R_0 L_0},\overline{W},\overline{J})(t)\right|\leq\left|U(t)-\overline{U}t\right|+\int_{t_k}^{t_{k+1}}\left|U(s)-\overline{U}(s)\right|ds\equiv \dot{W}_1+\dot{W}_2.$$

Since

$$\dot{W}_1\leq\left|\frac{dJ(t-T)}{dt}-\frac{d\overline{J}(t-T)}{dt}\right|+\frac{2\sqrt{L}}{Z_0\hat{C}_0}\left|i_{R_0 L_0}(t)-\bar{i}_{R_0 L_0}(t)\right|+\left(\frac{1}{Z_0\hat{C}_0}+\frac{W_0+J_0 e^{-\beta}+4\sqrt{L}I_{R_0}}{2Z_0\sqrt{C}\hat{C}_0^2}H_0\right)\left|W(t)-\overline{W}(t)\right|+$$

$$+\left(\frac{1}{Z_0\hat{C}_0}+\frac{W_0+J_0e^{-\beta}+4\sqrt{L}\,I_{R_0}}{2Z_0\sqrt{C}\,\hat{C}_0^{\,2}}H_0\right)\left|J(t-T)-\bar{J}(t-T)\right|\le$$

$$\le e^{\mu(t-t_k)}\hat{\rho}_\mu^{(k)}((i_{R_0L_0},W,i_{R_1L_1},J,\dot{i}_{R_0L_0},\dot{W},\dot{i}_{R_1L_1},\dot{J}),(\bar{i}_{R_0L_0},\overline{W},\bar{i}_{R_1L_1},\bar{J},\dot{\bar{i}}_{R_0L_0},\dot{\overline{W}},\dot{\bar{i}}_{R_1L_1},\dot{\bar{J}}))\times$$

$$\times\left[\frac{1}{\mu Z_0\hat{C}_0}\left(2\sqrt{L}+1+\frac{W_0+J_0e^{-\beta}+4\sqrt{L}\,I_{R_0}}{2\sqrt{C}\,\hat{C}_0}H_0\right)\right]$$

and

$$W_2\le\frac{1}{t_{k+1}-t_k}\left|\int_{t_k}^{t_{k+1}}\left(\frac{dJ(s-T)}{dt}-\frac{d\bar{J}(s-T)}{dt}\right)ds\right|+\frac{2\sqrt{L}}{Z_0\hat{C}_0}\int_{t_k}^{t_{k+1}}\left|i_{R_0L_0}(s)-\bar{i}_{R_0L_0}(s)\right|ds+\frac{1}{Z_0\hat{C}_0}\int_{t_k}^{t_{k+1}}\left|W(s)-\overline{W}(s)\right|ds+$$

$$+\frac{1}{Z_0\hat{C}_0}\int_{t_k}^{t_{k+1}}\left|J(s-T)-\bar{J}(s-T)\right|ds\le$$

$$\le e^{\mu(t-t_k)}\hat{\rho}_\mu^{(k)}((i_{R_0L_0},W,i_{R_1L_1},J,\dot{i}_{R_0L_0},\dot{W},\dot{i}_{R_1L_1},\dot{J}),(\bar{i}_{R_0L_0},\overline{W},\bar{i}_{R_1L_1},\bar{J},\dot{\bar{i}}_{R_0L_0},\dot{\overline{W}},\dot{\bar{i}}_{R_1L_1},\dot{\bar{J}}))\times$$

$$\times\frac{\left(e^{\mu_0}-1\right)}{\mu_0}\frac{1}{\mu Z_0\hat{C}_0}\left(2\sqrt{L}+1+\frac{W_0+J_0e^{-\beta}+4\sqrt{L}\,I_{R_0}}{2\sqrt{C}\,\hat{C}_0}H_0\right)$$

we have

$$\left|\dot{B}_W^{(k)}(i_{R_0L_0},W,J)(t)-\dot{B}_W^{(k)}(\bar{i}_{R_0L_0},\overline{W},\bar{J})(t)\right|\le$$

$$\le\left(1+\frac{e^{\mu_0}-1}{\mu_0}\right)\frac{1}{\mu Z_0\hat{C}_0}\left(2\sqrt{L}+1+\frac{W_0+J_0e^{-\beta}+4\sqrt{L}\,I_{R_0}}{2\sqrt{C}\,\hat{C}_0}H_0\right)\times$$

$$\times e^{\mu(t-t_k)}\hat{\rho}_\mu^{(k)}((i_{R_0L_0},W,i_{R_1L_1},J,\dot{i}_{R_0L_0},\dot{W},\dot{i}_{R_1L_1},\dot{J}),(\bar{i}_{R_0L_0},\overline{W},\bar{i}_{R_1L_1},\bar{J},\dot{\bar{i}}_{R_0L_0},\dot{\overline{W}},\dot{\bar{i}}_{R_1L_1},\dot{\bar{J}}))\equiv$$

$$\equiv e^{\mu(t-t_k)}\dot{K}_W\hat{\rho}_\mu^{(k)}((i_{R_0L_0},W,i_{R_1L_1},J,\dot{i}_{R_0L_0},\dot{W},\dot{i}_{R_1L_1},\dot{J}),(\bar{i}_{R_0L_0},\overline{W},\bar{i}_{R_1L_1},\bar{J},\dot{\bar{i}}_{R_0L_0},\dot{\overline{W}},\dot{\bar{i}}_{R_1L_1},\dot{\bar{J}})).$$

It follows

$$\hat{\rho}^{(k)}\left(\dot{B}_W(i_{R_0L_0},W,J)(t)-\dot{B}_W(\bar{i}_{R_0L_0},\overline{W},\bar{J})\right)\le$$

$$\le\dot{K}_W\hat{\rho}_\mu^{(k)}((i_{R_0L_0},W,i_{R_1L_1},J,\dot{i}_{R_0L_0},\dot{W},\dot{i}_{R_1L_1},\dot{J}),(\bar{i}_{R_0L_0},\overline{W},\bar{i}_{R_1L_1},\bar{J},\dot{\bar{i}}_{R_0L_0},\dot{\overline{W}},\dot{\bar{i}}_{R_1L_1},\dot{\bar{J}})).$$

For the derivative of the third component we obtain

$$\left|\dot{B}_1^{(k)}(W,i_{R_1L_1},J)(t)-\dot{B}_1^{(k)}(\overline{W},\bar{i}_{R_1L_1},\overline{J})(t)\right|\leq\left|I_{R_1}(W,i_{R_1L_1},J)(t)-I_{R_1}(\overline{W},\bar{i}_{R_1L_1},\overline{J})(t)\right|+$$

$$+\frac{1}{t_{k+1}-t_k}\int_{t_k}^{t_{k+1}}\left|I_{R_1}(W,i_{R_1L_1},J)(s)-I_{R_1}(\overline{W},\bar{i}_{R_1L_1},\overline{J})(s)\right|ds\equiv\dot{P}_1+\dot{P}_2.$$

Now

$$\dot{P}_1\leq\frac{1}{2\sqrt{C}\,\hat{L}_1}\left|W(t-T)-\overline{W}(t-T)\right|+\left[\frac{2\sqrt{C}}{\hat{L}_1}\sum_{n=1}^m n\left|r_n^{(1)}\right|I_{R_1}^{n-1}+\right.$$

$$+\frac{W_0+J_0e^{-\beta}+2\sqrt{C}\sum_{n=1}^m\left|r_n^{(1)}\right|I_{R_1}^n}{2\sqrt{C}\hat{L}_1^2}\left.\sum_{n=1}^m(n+1)n\left|l_n^{(1)}\right|I_{R_1}^{n-1}\right]\left|i_{R_1L_1}(t)-\bar{i}_{R_1L_1}(t)\right|+$$

$$+\frac{1}{2\sqrt{C}\,\hat{L}_1}\left|J(t)-\overline{J}(t)\right|\leq$$

$$\leq\left(\frac{2\sqrt{C}}{\hat{L}_1}\sum_{n=1}^m n\left|r_n^{(1)}\right|I_{R_1}^{n-1}+\frac{W_0+J_0e^{-\beta}+2\sqrt{C}\sum_{n=1}^m\left|r_n^{(1)}\right|I_{R_1}^n}{2\sqrt{C}\hat{L}_1^2}\sum_{n=1}^m(n+1)n\left|l_n^{(1)}\right|I_{R_1}^{n-1}\right)\frac{\rho_\mu^{(k)}(i_{R_1L_1},\bar{i}_{R_1L_1})}{\mu}+\frac{\rho_\mu^{(k)}(J,\overline{J})}{2\mu\sqrt{C}\,\hat{L}_1}\leq$$

$$\leq\frac{1}{\mu\hat{L}_1}\left(2\sqrt{C}\sum_{n=1}^m n\left|r_n^{(1)}\right|I_{R_1}^{n-1}+\frac{W_0+J_0e^{-\beta}+2\sqrt{C}\sum_{n=1}^m\left|r_n^{(1)}\right|I_{R_1}^n}{2\sqrt{C}\hat{L}_1}\sum_{n=1}^m(n+1)n\left|l_n^{(1)}\right|I_{R_1}^{n-1}+\frac{1}{2\sqrt{C}}\right)\times$$

$$\times e^{\mu(t-t_k)}\hat{\rho}_\mu^{(k)}((i_{R_0L_0},W,i_{R_1L_1},J,\dot{i}_{R_0L_0},\dot{W},\dot{i}_{R_1L_1},\dot{J}),(\bar{i}_{R_0L_0},\overline{W},\bar{i}_{R_1L_1},\overline{J},\dot{\bar{i}}_{R_0L_0},\dot{\overline{W}},\dot{\bar{i}}_{R_1L_1},\dot{\overline{J}}))$$

and

$$\dot{P}_2\leq\frac{1}{t_{k+1}-t_k}\left\{\frac{1}{2\sqrt{C}\,\hat{L}_1}\int_{t_k}^{t_{k+1}}\left|W(s-T)-\overline{W}(s-T)\right|ds+\left[\frac{2\sqrt{C}}{\hat{L}_1}\sum_{n=1}^m n\left|r_n^{(1)}\right|I_{R_1}^{n-1}+\right.\right.$$

$$+\frac{W_0+J_0e^{-\beta}+2\sqrt{C}\sum_{n=1}^m\left|r_n^{(1)}\right|I_{R_1}^n}{2\sqrt{C}\hat{L}_1^2}\left.\sum_{n=1}^m(n+1)n\left|l_n^{(1)}\right|I_{R_1}^{n-1}\right]\int_{t_k}^{t_{k+1}}\left|i_{R_1L_1}(s)-\bar{i}_{R_1L_1}(s)\right|ds+$$

$$+\frac{1}{2\sqrt{C}\,\hat{L}_1}\int_{t_k}^{t_{k+1}}\left|J(s)-\overline{J}(s)\right|ds\leq$$

$$\leq \left(\frac{2\sqrt{C}}{\hat{L}_1} \sum_{n=1}^{m} n \left| r_n^{(1)} \right| I_{R_1}^{n-1} + \frac{W_0 + J_0 e^{-\beta} + 2\sqrt{C} \sum_{n=1}^{m} \left| r_n^{(1)} \right| I_{R_1}^{n}}{2\sqrt{C} \hat{L}_1^2} \sum_{n=1}^{m} (n+1) n \left| l_n^{(1)} \right| I_{R_1}^{n-1} \right) \rho^{(k)}(i_{R_1 L_1}, \bar{i}_{R_1 L_1}) \int_{t_k}^{t_{k+1}} e^{\mu(s-t_k)} ds \ +$$

$$+ \frac{\rho^{(k)}(J, \bar{J})}{2\sqrt{C}\ \hat{L}_1} \int_{t_k}^{t_{k+1}} e^{\mu(s-t_k)} ds \leq$$

$$\leq \frac{e^{\mu(t_{k+1}-t_k)} - 1}{\mu(t_{k+1}-t_k)} \left\{ \left[\frac{2\sqrt{C}}{\hat{L}_1} \sum_{n=1}^{m} n \left| r_n^{(1)} \right| I_{R_1}^{n-1} + \right.\right.$$

$$\left.\left. + \frac{W_0 + J_0 e^{-\beta} + 2\sqrt{C} \sum_{n=1}^{m} \left| r_n^{(1)} \right| I_{R_1}^{n}}{2\sqrt{C}\hat{L}_1^2} \sum_{n=1}^{m} (n+1) n \left| l_n^{(1)} \right| I_{R_1}^{n-1} \right] \frac{\rho_\mu^{(k)}(i_{R_1 L_1}, \dot{\bar{i}}_{R_1 L_1})}{\mu} + \frac{\rho_\mu^{(k)}(\dot{J}, \dot{\bar{J}})}{2\mu\sqrt{C}\ \hat{L}_1} \right\} \leq$$

$$\leq e^{\mu(t-t_k)} \frac{e^{\mu_0} - 1}{\mu_0} \frac{1}{\mu} \left(\frac{2\sqrt{C}}{\hat{L}_1} \sum_{n=1}^{m} n \left| r_n^{(1)} \right| I_{R_1}^{n-1} + \frac{W_0 + J_0 e^{-\beta} + 2\sqrt{C} \sum_{n=1}^{m} \left| r_n^{(1)} \right| I_{R_1}^{n}}{2\sqrt{C}\hat{L}_1^2} \sum_{n=1}^{m} (n+1) n \left| l_n^{(1)} \right| I_{R_1}^{n-1} + \frac{1}{2\sqrt{C}\ \hat{L}_1} \right) \times$$

$$\times \hat{\rho}_\mu^{(k)}((i_{R_0 L_0}, W, i_{R_1 L_1}, J, \dot{i}_{R_0 L_0}, \dot{W}, \dot{i}_{R_1 L_1}, \dot{J}), (\bar{i}_{R_0 L_0}, \overline{W}, \bar{i}_{R_1 L_1}, \bar{J}, \dot{\bar{i}}_{R_0 L_0}, \dot{\overline{W}}, \dot{\bar{i}}_{R_1 L_1}, \dot{\bar{J}})).$$

Thus

$$\left| \dot{B}_1^{(k)}(W, i_{R_1 L_1}, J)(t) - \dot{B}_1^{(k)}(\overline{W}, \bar{i}_{R_1 L_1}, \bar{J})(t) \right| \leq$$

$$\leq \frac{1}{\mu\hat{L}_1} \left(2\sqrt{C} \sum_{n=1}^{m} n \left| r_n^{(1)} \right| I_{R_1}^{n-1} + \frac{W_0 + J_0 e^{-\beta} + 2\sqrt{C} \sum_{n=1}^{m} \left| r_n^{(1)} \right| I_{R_1}^{n}}{2\sqrt{C}\hat{L}_1} \sum_{n=1}^{m} (n+1) n \left| l_n^{(1)} \right| I_{R_1}^{n-1} + \frac{1}{2\sqrt{C}} \right) \times$$

$$\times e^{\mu(t-t_k)} \hat{\rho}_\mu^{(k)}((i_{R_0 L_0}, W, i_{R_1 L_1}, J, \dot{i}_{R_0 L_0}, \dot{W}, \dot{i}_{R_1 L_1}, \dot{J}), (\bar{i}_{R_0 L_0}, \overline{W}, \bar{i}_{R_1 L_1}, \bar{J}, \dot{\bar{i}}_{R_0 L_0}, \dot{\overline{W}}, \dot{\bar{i}}_{R_1 L_1}, \dot{\bar{J}})) \ +$$

$$+ e^{\mu(t-t_k)} \frac{e^{\mu_0} - 1}{\mu_0} \frac{1}{\mu\hat{L}_1} \left(2\sqrt{C} \sum_{n=1}^{m} n \left| r_n^{(1)} \right| I_{R_1}^{n-1} + \frac{W_0 + J_0 e^{-\beta} + 2\sqrt{C} \sum_{n=1}^{m} \left| r_n^{(1)} \right| I_{R_1}^{n}}{2\sqrt{C}\hat{L}_1} \sum_{n=1}^{m} (n+1) n \left| l_n^{(1)} \right| I_{R_1}^{n-1} + \frac{1}{2\sqrt{C}} \right) \times$$

$$\times \hat{\rho}_\mu^{(k)}((i_{R_0 L_0}, W, i_{R_1 L_1}, J, \dot{i}_{R_0 L_0}, \dot{W}, \dot{i}_{R_1 L_1}, \dot{J}), (\bar{i}_{R_0 L_0}, \overline{W}, \bar{i}_{R_1 L_1}, \bar{J}, \dot{\bar{i}}_{R_0 L_0}, \dot{\overline{W}}, \dot{\bar{i}}_{R_1 L_1}, \dot{\bar{J}})) \leq$$

$$\leq e^{\mu(t-t_k)} \left(1 + \frac{e^{\mu_0} - 1}{\mu_0} \right) \frac{1}{\mu\hat{L}_1} \left(2\sqrt{C} \sum_{n=1}^{m} n \left| r_n^{(1)} \right| I_{R_1}^{n-1} + \frac{W_0 + J_0 e^{-\beta} + 2\sqrt{C} \sum_{n=1}^{m} \left| r_n^{(1)} \right| I_{R_1}^{n}}{2\sqrt{C}\hat{L}_1} \sum_{n=1}^{m} (n+1) n \left| l_n^{(1)} \right| I_{R_1}^{n-1} + \frac{1}{2\sqrt{C}} \right)$$

$$\times \hat{\rho}_\mu^{(k)}((i_{R_0 L_0}, W, i_{R_1 L_1}, J, \dot{i}_{R_0 L_0}, \dot{W}, \dot{i}_{R_1 L_1}, \dot{J}), (\bar{i}_{R_0 L_0}, \overline{W}, \bar{i}_{R_1 L_1}, \bar{J}, \dot{\bar{i}}_{R_0 L_0}, \dot{\overline{W}}, \dot{\bar{i}}_{R_1 L_1}, \dot{\bar{J}})) \equiv$$

$$\equiv e^{\mu(t-t_k)}\dot{K}_1\hat{\rho}_\mu^{(k)}((i_{R_0L_0},W,i_{R_1L_1},J,\dot{i}_{R_0L_0},\dot{W},\dot{i}_{R_1L_1},\dot{J}),(\bar{i}_{R_0L_0},\overline{W},\bar{i}_{R_1L_1},\bar{J},\dot{\bar{i}}_{R_0L_0},\dot{\overline{W}},\dot{\bar{i}}_{R_1L_1},\dot{\bar{J}})).$$

Finally for the fourth component of the derivative we obtain

$$\left|\dot{B}_J^{(k)}(i_{R_0L_0},W,i_{R_1L_1},J)(t)-\dot{B}_J^{(k)}(\bar{i}_{R_0L_0},\overline{W},\bar{i}_{R_1L_1},\bar{J})(t)\right|\le \dot{I}_1+\dot{I}_2.$$

But

$$\dot{I}_1\le\left|\frac{dW(t-T)}{dt}-\frac{d\overline{W}(t-T)}{dt}\right|+\frac{2\sqrt{L}}{Z_0\hat{C}_1}\left|i_{R_1L_1}(t)-\bar{i}_{R_1L_1}(t)\right|+\frac{1}{Z_0\hat{C}_1}\left|J(t)-\bar{J}(t)\right|+$$

$$+\frac{1}{Z_0\hat{C}_1}\left|W(t-T)-\overline{W}(t-T)\right|\le$$

$$\le e^{\mu(t-t_k)}\hat{\rho}_\mu^{(k)}((i_{R_0L_0},W,i_{R_1L_1},J,\dot{i}_{R_0L_0},\dot{W},\dot{i}_{R_1L_1},\dot{J}),(\bar{i}_{R_0L_0},\overline{W},\bar{i}_{R_1L_1},\bar{J},\dot{\bar{i}}_{R_0L_0},\dot{\overline{W}},\dot{\bar{i}}_{R_1L_1},\dot{\bar{J}}))\times$$

$$\times\left[\frac{1}{\mu Z_0\hat{C}_1}\left(2\sqrt{L}+1+\frac{W_0e^{-\beta}+J_0+2\sqrt{L}\,I_{R_1}}{2\sqrt{C}\hat{C}_1}H_1\right)\right]$$

and

$$\dot{I}_2\le\frac{1}{t_{k+1}-t_k}\left[\left|\int_{t_k}^{t_{k+1}}\left(\frac{dW(t-T)}{dt}-\frac{d\overline{W}(t-T)}{dt}\right)ds\right|+\frac{2\sqrt{L}}{Z_0\hat{C}_1}\int_{t_k}^{t_{k+1}}\left|i_{R_1L_1}(s)-\bar{i}_{R_1L_1}(s)\right|ds+\frac{1}{Z_0\hat{C}_1}\int_{t_k}^{t_{k+1}}\left|J(s)-\bar{J}(s)\right|ds+\right.$$

$$\left.+\frac{1}{Z_0\hat{C}_1}\int_{t_k}^{t_{k+1}}\left|W(s-T)-\overline{W}(s-T)\right|ds\right]\le$$

$$\le e^{\mu(t-t_k)}\hat{\rho}_\mu^{(k)}((i_{R_0L_0},W,i_{R_1L_1},J,\dot{i}_{R_0L_0},\dot{W},\dot{i}_{R_1L_1},\dot{J}),(\bar{i}_{R_0L_0},\overline{W},\bar{i}_{R_1L_1},\bar{J},\dot{\bar{i}}_{R_0L_0},\dot{\overline{W}},\dot{\bar{i}}_{R_1L_1},\dot{\bar{J}}))\times$$

$$\times\frac{e^{\mu_0}-1}{\mu_0}\frac{1}{\mu Z_0\hat{C}_1}\left(2\sqrt{L}+1+\frac{W_0e^{-\beta}+J_0+2\sqrt{L}\,I_{R_1}}{2\sqrt{C}\hat{C}_1}H_1\right)$$

then

$$\left|\dot{B}_J^{(k)}U(W,i_{R_1L_1},J)(t)-\dot{B}_J^{(k)}U(\overline{W},\bar{i}_{R_1L_1},\bar{J})(t)\right|\le$$

$$\le e^{\mu(t-t_k)}\hat{\rho}_\mu^{(k)}((i_{R_0L_0},W,i_{R_1L_1},J,\dot{i}_{R_0L_0},\dot{W},\dot{i}_{R_1L_1},\dot{J}),(\bar{i}_{R_0L_0},\overline{W},\bar{i}_{R_1L_1},\bar{J},\dot{\bar{i}}_{R_0L_0},\dot{\overline{W}},\dot{\bar{i}}_{R_1L_1},\dot{\bar{J}}))\times$$

$$\times\left(1+\frac{e^{\mu_0}-1}{\mu_0}\right)\frac{1}{\mu Z_0\hat{C}_1}\left(2\sqrt{L}+1+\frac{W_0e^{-\beta}+J_0+2\sqrt{L}\,I_{R_1}}{2\sqrt{C}\hat{C}_1}H_1\right)\equiv$$

$$\equiv e^{\mu(t-t_k)}\dot{K}_J\hat{\rho}_\mu^{(k)}((i_{R_0L_0},W,i_{R_1L_1},J,\dot{i}_{R_0L_0},\dot{W},\dot{i}_{R_1L_1},\dot{J}),(\bar{i}_{R_0L_0},\overline{W},\bar{i}_{R_1L_1},\bar{J},\dot{\bar{i}}_{R_0L_0},\dot{\overline{W}},\dot{\bar{i}}_{R_1L_1},\dot{\bar{J}})).$$

It follows

$$\hat{\rho}_\mu^{(k)}\Big(\dot{B}_J(i_{R_0L_0},W,J),\dot{B}_J(\bar{i}_{R_0L_0},\overline{W},\overline{J})\Big) \le$$

$$\le \dot{K}_J\hat{\rho}_\mu^{(k)}((i_{R_0L_0},W,i_{R_1L_1},J,\dot{i}_{R_0L_0},\dot{W},\dot{i}_{R_1L_1},\dot{J}),(\bar{i}_{R_0L_0},\overline{W},\bar{i}_{R_1L_1},\overline{J},\dot{\bar{i}}_{R_0L_0},\dot{\overline{W}},\dot{\bar{i}}_{R_1L_1},\dot{\overline{J}})) \cdot$$

Let us denote by

$$K = \max\left\{e^{\mu_0}K_0,e^{\mu_0}K_W,e^{\mu_0}K_1,e^{\mu_0}K_J,\dot{K}_0,\dot{K}_W\dot{K}_1,\dot{K}_J\right\} < 1.$$

Then

$$\hat{\rho}_\mu^{(k)}((B_0,B_W,B_1,B_J,\dot{B}_0,\dot{B}_W,\dot{B}_1,\dot{B}_J),(\overline{B}_0,\overline{B}_W,\overline{B}_1,\overline{B}_J,\dot{\overline{B}}_0,\dot{\overline{B}}_W,\dot{\overline{B}}_1,\dot{\overline{B}}_J)) \le$$

$$\le K\hat{\rho}_\mu^{(k)}((i_{R_0L_0},W,i_{R_1L_1},J,\dot{i}_{R_0L_0},\dot{W},\dot{i}_{R_1L_1},\dot{J}),(\bar{i}_{R_0L_0},\overline{W},\bar{i}_{R_1L_1},\overline{J},\dot{\bar{i}}_{R_0L_0},\dot{\overline{W}},\dot{\bar{i}}_{R_1L_1},\dot{\overline{J}}))$$

The operator B has a unique fixed point. It is an oscillatory solution of the above problem on $[T,2T]$.

Theorem 8.9.1 is thus proved.

The obtained solution on $[T,2T]$ might be taken as an initial function and by the same way to obtain a unique solution on $[2T,3T]$ and so on.

8.9.1. Numerical Example

We collect all inequalities guaranteeing an existence-uniqueness of oscillatory solution:

$$\frac{W_0+J_0}{2\sqrt{C}} \le \phi_0, \ I_{R_0},I_{R_1} \le i_0,$$

$$\frac{2L}{R\hat{L}_0\sqrt{C}}\sinh\left(\frac{RT_0}{L}\right)\left[W_0+J_0e^{-\beta}+2\sqrt{C}\sum_{n=1}^{m}\left|r_n^{(0)}\right|I_{R_0}{}^n\right] \le I_{R_0};$$

$$J_0e^{-\beta}+\frac{4L\left(W_0+J_0e^{-\beta}+4\sqrt{L}\,I_{R_0}\right)}{Z_0\hat{C}_0R}\sinh\left(\frac{RT_0}{L}\right) \le W_0;$$

$$\frac{2L}{R\hat{L}_1\sqrt{C}}\sinh\left(\frac{RT_0}{L}\right)\left(W_0e^{-\beta}+J_0+2\sqrt{C}\sum_{n=1}^{m}\left|r_n^{(1)}\right|I_{R_1}{}^n\right) \le I_{R_1};$$

$$W_0e^{-\beta}+\frac{4L}{R}\frac{W_0e^{-\beta}+J_0+4\sqrt{L}I_{R_1}}{Z_0\hat{C}_1}\sinh\left(\frac{RT_0}{L}\right) \le J_0;$$

$$e^{\mu_0} K_0 = e^{\mu_0} \frac{e^{2\mu_0}}{\mu^2 \hat{L}_0} \left(2\sqrt{C} \sum_{n=1}^{m} n \left| r_n^{(0)} \right| I_{R_0}^{n-1} + \frac{W_0 + J_0 e^{-\beta} + 2\sqrt{C} \sum_{n=1}^{m} \left| r_n^{(0)} \right| I_{R_0}^{n}}{2\sqrt{C} \hat{L}_0} \sum_{n=1}^{m} (n+1)n \left| l_n^{(0)} \right| I_{R_0}^{n-1} + \frac{1}{2\sqrt{C}} \right) < 1;$$

$$e^{\mu_0} K_W = e^{\mu_0} \frac{e^{2\mu_0}}{\mu^2 Z_0 \hat{C}_0} \left(2\sqrt{L} + 1 + \frac{W_0 + J_0 e^{-\beta} + 4\sqrt{L} \, I_{R_0}}{2\sqrt{C} \, \hat{C}_0} H_0 \right) < 1;$$

$$e^{\mu_0} K_1 = e^{\mu_0} \frac{e^{2\mu_0}}{\mu^2 \hat{L}_1} \left(2\sqrt{C} \sum_{n=1}^{m} n \left| r_n^{(1)} \right| I_{R_1}^{n-1} + \frac{W_0 + J_0 e^{-\beta} + 2\sqrt{C} \sum_{n=1}^{m} \left| r_n^{(1)} \right| I_{R_1}^{n}}{2\sqrt{C} \hat{L}_1} \sum_{n=1}^{m} (n+1)n \left| l_n^{(1)} \right| I_{R_1}^{n-1} + \frac{1}{2\sqrt{C}} \right) < 1;$$

$$e^{\mu_0} K_J = e^{\mu_0} \frac{e^{2\mu_0}}{\mu^2 Z_0 \hat{C}_1} \left(2\sqrt{L} + 1 + \frac{W_0 e^{-\beta} + J_0 + 2\sqrt{L} \, I_{R_1}}{2\sqrt{C} \hat{C}_1} H_1 \right) < 1;$$

$$\dot{K}_0 = \left(1 + \frac{e^{\mu_0} - 1}{\mu_0} \right) \frac{1}{\mu \hat{L}_0} \left(2\sqrt{C} \sum_{n=1}^{m} n \left| r_n^{(0)} \right| I_{R_0}^{n-1} + \frac{W_0 + J_0 e^{-\beta} + 2\sqrt{C} \sum_{n=1}^{m} \left| r_n^{(0)} \right| I_{R_0}^{n}}{2\sqrt{C} \hat{L}_0} \sum_{n=1}^{m} (n+1)n \left| l_n^{(0)} \right| I_{R_0}^{n-1} + \frac{1}{2\sqrt{C}} \right) < 1;$$

$$\dot{K}_W = \left(1 + \frac{e^{\mu_0} - 1}{\mu_0} \right) \frac{1}{\mu Z_0 \hat{C}_0} \left(2\sqrt{L} + 1 + \frac{W_0 + J_0 e^{-\beta} + 4\sqrt{L} \, I_{R_0}}{2\sqrt{C}} \frac{H_0}{\hat{C}_0} \right) < 1;$$

$$\dot{K}_1 = \left(1 + \frac{e^{\mu_0} - 1}{\mu_0} \right) \frac{1}{\mu \hat{L}_1} \left(2\sqrt{C} \sum_{n=1}^{m} n \left| r_n^{(1)} \right| I_{R_1}^{n-1} + \frac{W_0 e^{-\beta} + J_0 + 2\sqrt{C} \sum_{n=1}^{m} \left| r_n^{(1)} \right| I_{R_1}^{n}}{2\sqrt{C} \hat{L}_1} \sum_{n=1}^{m} (n+1)n \left| l_n^{(1)} \right| I_{R_1}^{n-1} + \frac{1}{2\sqrt{C}} \right) < 1;$$

$$\dot{K}_J = \left(1 + \frac{e^{\mu_0} - 1}{\mu_0} \right) \frac{1}{\mu Z_0 \hat{C}_1} \left(2\sqrt{L} + 1 + \frac{W_0 e^{-\beta} + J_0 + 2\sqrt{L} \, I_{R_1}}{2\sqrt{C}} \frac{H_1}{\hat{C}_1} \right) < 1.$$

Consider a transmission line with length $\Lambda = 100 \, m$, cross-section area $S = 4 \, mm^2$, specific resistance for the copper $\rho_c = 0{,}0175$, the resistance per-unit length is

$$R = \frac{\rho_c \Lambda}{S} = \frac{0{,}0175 \cdot 100}{4} \approx 0{,}44 \, \Omega. \text{ Let } \quad L = 0{,}45 \, \mu H / m, \; C = 80 \, pF / m,$$

$$v = 1 / \sqrt{LC} = 1 / \left(6 \cdot 10^{-9} \right) = 1{,}66 \cdot 10^{8}; \; Z_0 = \sqrt{L / C} = 75 \, \Omega. \text{ Then}$$

$$T = \Lambda \sqrt{LC} = 10^2 \cdot 6 \cdot 10^{-9} = 6 \cdot 10^{-7}, \; \frac{R}{L} = \frac{0{,}22}{0{,}45 \cdot 10^{-6}} \approx 5 \cdot 10^{5}; \; \frac{RT}{L} = \frac{R\Lambda}{Z_0} \approx \frac{0{,}22 \cdot 100}{75} = 0{,}3; \; e^{0{,}3} \approx 1{,}35.$$

For waves with length $\lambda_0 = (1 / 6) 10^{-3} \, m$ we have

$$f_0 = 1/\left(\lambda_0 \sqrt{LC}\right) = 1/\left((1/6).10^{-3}.6.10^{-9}\right) = 10^{12}\, Hz \Rightarrow T_0 = 1/f_0 = 10^{-12}\ ;$$

$$\frac{RT_0}{L} = \frac{0,22.10^{-12}}{0,45.10^{-6}} \approx 5.10^{-7}.\ \text{We choose}\ \mu = 10^{12},\ \text{then}\ \mu T_0 = \mu_0 = 1,\ \text{and}$$

$$T = 6.10^{-5}7.10^{12}T_0 = 6.10^{5}T_0 \Rightarrow m = 6.10^{5}.\ \text{Consequently}$$

$$\mu T = 10^{12}.6.10^{-7} \approx 6.10^{5} \Rightarrow e^{-\mu T} = e^{-600000} \approx 0\ ;$$

$$\mu - \frac{R}{L} = 10^{12} - 5.10^{5} \approx 10^{12}\ ;\ \sinh\left(\frac{R}{L}T_0\right) \approx \frac{R}{L}T_0 = 5.10^{-7}.$$

We choose resistive elements with the following *V-I* characteristics

$$R_0(i) = R_1(i) = 0,028i - 0,125i^3\ \text{i.e.}\ r_1 = 0,028, r_2 = 0,\ r_3 = 0,125$$

and inductive elements with $L_0(i) = L_1(i) = 3i - (1/12)i^3$. Then

$$\widetilde{L}_p(i) = i\left(dL_p(i)/di\right) + L_p(i) = i(3 - (1/4)i^2) + 3i - (1/12)i^3 = 6i - (1/3)i^3;\ \ (p = 0,1)$$

If we choose $i_0 = 1$ one obtains $6i - (1/3)i^3 > 6 - (1/3) = 17/3$ and consequently

$$\frac{1}{\hat{L}_0} = \frac{1}{\hat{L}_1} = \frac{3}{17}.$$

Let us take $h = 2$ and then for $c_p = 50pF = 5.10^{-11}F$, $\Phi_p = 0,5\, V\ (p = 0,1)$ and $\phi_0 = 0,16$ we have

$$\hat{C}_0 = \hat{C}_1 = \frac{2c_0\sqrt[h]{\Phi_0}}{(\Phi_0 + \phi_0)^{\frac{1}{h}+2}} \frac{\Phi_0 - \dfrac{2h+1}{h}\phi_0}{h} \Rightarrow \hat{C}_0 = \frac{2c_0\sqrt{\Phi_0}}{\sqrt{(\Phi_0 + \phi_0)^5}} \frac{\Phi_0 - 2,5.\phi_0}{2} =$$

$$= \frac{2.50.10^{-12}\sqrt{0,5}}{\sqrt{(0,5 + 0,16)^5}} \frac{0,5 - 2,5.0,16}{2} \approx 1,45.10^{-11};$$

$$H_0 = H_1 = \frac{2c_0\sqrt{\Phi_0}\,(2\Phi_0 + \phi_0)}{4\sqrt{(\Phi_0 - \phi_0)^5}} = \frac{2.50.10^{-12}\sqrt{0,5}\,(1 + 0,16)}{4\sqrt{(0,5 - 0,16)^5}} \approx 3.10^{-10}.$$

Finally for $W_0 = J_0 = 10^{-5};\ I_{R0} = I_{R1} = 0,1;\ \beta > 0$ sufficiently large we obtain

$$\frac{W_0}{2\sqrt{C}} \le \phi_0 ; \quad \frac{J_0}{2\sqrt{C}} \le \phi_0 , \quad I_{R_0}, I_{R_1} \le 1 ;$$

$$\frac{3.2.10^{-12}}{17\sqrt{80.10^{-12}}} \left[W_0 + J_0 e^{-\beta} + 2\sqrt{80.10^{-12}} \sum_{n=1}^{3} \left| r_n^{(0)} \right| 10^{-n} \right] \le 0,1 ;$$

$$J_0 e^{-\beta} + 4.10^{-12} \frac{W_0 + J_0 e^{-\beta} + 4\sqrt{0,45.10^{-6}} \, I_{R_0}}{75.1,45.10^{-11}} \le W_0 ;$$

$$\frac{3.2.10^{-12}}{17\sqrt{80.10^{-12}}} \left(W_0 e^{-\beta} + J_0 + 2\sqrt{80.10^{-12}} \sum_{n=1}^{3} \left| r_n^{(1)} \right| I_{R_1}^{\ n} \right) \le 0,1 ;$$

$$W_0 e^{-\beta} + 4.10^{-12} \frac{W_0 e^{-\beta} + J_0 + 4\sqrt{0,45.10^{-6}} \, I_{R_1}}{75.1,45.10^{-11}} \le J_0 ;$$

$$\dot{K}_0 = \frac{0,45}{10^8}\left[0,3W_0 + 5,6\right] < 1 ; \quad \dot{K}_W = \frac{1}{10^2}\left(0,25 + 3,14.10^6.10^{-5}\right) \approx \frac{31,65}{100} < 1 ;$$

$$\dot{K}_0 = \frac{0,45}{10^8}\left[0,3J_0 + 5,6\right] < 1 ; \quad \dot{K}_J = \frac{1}{10^2}\left(0,25 + 3,14.10^6.10^{-5}\right) \approx 0,3165 < 1 ;$$

$$K = \max\left\{ e^{\mu_0} K_0, e^{\mu_0} K_W, e^{\mu_0} K_1, e^{\mu_0} K_J, \dot{K}_0, \dot{K}_W, \dot{K}_1, \dot{K}_J \right\} = \max\left\{ \dot{K}_0, \dot{K}_W, \dot{K}_1, \dot{K}_J \right\} = 0,3165 .$$

The successive approximations could be obtained as in the previous chapters.

CONCLUSION

We have noticed that the transmission lines loaded in the final configurations of *RLC* elements require as different way method of derivation of the boundary conditions, and respectively, leads to different types of differential equations. Here one final difficulty arises. One cannot exclude some transitional currents, and so we have to solve a system of 4 equations for 4 unknown functions. Our fixed-point method, however, is again applicable, and we obtain the existence-uniqueness of a periodic solution for lossless lines and an oscillatory ones solution (vanishing at infinity) for lossy lines.

We point out that the procedure for treating of time-varying specific parameters might may be applied to the present circuit (cf. § 5.7).

General Conclusion

- We investigate both lossless and lossy transmission lines terminated by various configurations of nonlinear RLC-loads. From a mathematical point of view the cases are different ones. More precisely, they need different types of transformations. They both lead, however, to differential equations (or system of equations) with retarded arguments of a neutral type on the boundary.
- The obtained equations (or systems) are nonlinear ones. Their nonlinearities depend on the characteristics of RLC-loads. In other words the nonlinearities of the loads are carried over the neutral equations or systems.
- In order to prove an existence-uniqueness theorem we introduce an operator (unknown in the literature up to now) whose fixed points are a periodic or oscillatory solution of the problem stated.
- Usually, the fixed-point method is applied to Lipschitz nonlinearities. We would like to point out that by applying the fixed-point method, we solve nonlinear equations with polynomial, exponential and transcendental ones. This is achieved due to the suitable selection of the appropriate functional spaces. An essential role plays the introducing of the family of pseudo-metrics of Bielecki type.
- It turns out that the space of oscillating functions does not form a metric space but rather a uniform one. This then requires applying the application of fixed-point theorems of operators acting on uniform spaces.
- By virtue of the theorems obtained in the book we show that attenuating oscillating modes are natural for the lossy transmission lines terminated by such configurations of the nonlinear loads.
- The numerical examples demonstrate a frame of applicability for the theory revealed (for instance to in the design of circuits) and show that the method could be applied by checking a few simple inequalities between the basic specific parameters of the lines and loads.
- Finally we note that a lot of research has been done where numerical (or other) methods are applied without assuring uniqueness. In these cases it is not clear to which solution being approached. Our fixed-point method guarantees solution uniqueness.
- The calculation of the successive approximations and the estimations of some terms (leading to them being disregarded) simplify the calculation of the next subsequent

approximations. This is an extremely important step for any program implementing this method.

- We consider transmission lines, taking into account the lossiness. This means there is attenuation of the signals. This natural physical fact is confirmed by the mathematical method we apply. Namely, the transformation (we have used to reduce the mixed problem for a hyperbolic system to a problem for a neutral system on the boundary) contains an exponential function that implies that signals (current and voltage) vanish exponentially. It reminds us that natural global solutions are not periodic ones. That is why we formulate the problem of the existence-uniqueness of an oscillatory solution (cf. Chapter V and Chapter VII).

- Some chapters contain an existence-uniqueness theorem for a periodic solution for linear *RLC*-loads. Our goal is to demonstrate a unified approach for linear and nonlinear problems.

- We show, that our methods are also applicable to the case of time-varying loads, provided they have strict positive lower bounds. At the very end of Chapters V, VI and VII we introduce a generalization of the Heaviside condition, which allows us to include this scenario in the general scheme.

- All results obtained may be applied to the transmission line analog of electron beams. In this case the role of the voltage and current are replaced by the series impedance and the shunt admittance:

SUGGESTIONS FOR FURTHER STUDIES

In all chapters of the present book we have considered homogenous systems:

$$\frac{\partial u(x,t)}{\partial t} + L\frac{\partial i(x,t)}{\partial x} = 0,$$

$$\frac{\partial i(x,t)}{\partial t} + C\frac{\partial u(x,t)}{\partial x} = 0$$

for lossless transmission lines and

$$C\frac{\partial u(x,t)}{\partial t} + \frac{\partial i(x,t)}{\partial x} + Gu(x,t) = 0,$$

$$L\frac{\partial i(x,t)}{\partial t} + \frac{\partial u(x,t)}{\partial x} + Ri(x,t) = 0$$

for lossy transmission lines.

If we consider the non-homogenous system from Chapter I

$$C\frac{\partial u(x,t)}{\partial t} + \frac{\partial i(x,t)}{\partial x} = j(t) \, ,$$

$$L\frac{\partial i(x,t)}{\partial t} + \frac{\partial u(x,t)}{\partial x} = e(t)$$

then by the transformation and its inverse

$$\left| \begin{aligned} U(x,t) &= \sqrt{C}\,u(x,t) + \sqrt{L}\,i(x,t) \\ I(x,t) &= -\sqrt{C}\,u(x,t) + \sqrt{L}\,i(x,t) \end{aligned} \right. ,$$

$$\left| \begin{aligned} u(x,t) &= \frac{1}{2\sqrt{C}}U(x,t) - \frac{1}{2\sqrt{C}}I(x,t) \\[2mm] i(x,t) &= \frac{1}{2\sqrt{L}}U(x,t) + \frac{1}{2\sqrt{L}}I(x,t). \end{aligned} \right.$$

the above system becomes

$$\frac{\partial U(x,t)}{\partial t} + \frac{1}{\sqrt{LC}}\frac{\partial U(x,t)}{\partial x} = \frac{1}{\sqrt{C}}j(t) + \frac{1}{\sqrt{L}}e(t),$$

$$\frac{\partial I(x,t)}{\partial t} - \frac{1}{\sqrt{LC}}\frac{\partial I(x,t)}{\partial x} = -\frac{1}{\sqrt{C}}j(t) + \frac{1}{\sqrt{L}}e(t).$$

We demonstrate on the line from Fig. 2.1 the reducing of the mixed problem to the neutral system on the boundary. Indeed, replace $u(0,t), i(0,t), u(\Lambda,t), i(\Lambda,t)$ into the boundary conditions

$$E(t) - u(0,t) = R_0 i(0,t), \quad t \geq 0, \, C_0\frac{du(\Lambda,t)}{dt} = i(\Lambda,t) - f(u(\Lambda,t)), \, t \geq 0$$

we have

$$E(t) - \frac{U(0,t)}{2\sqrt{C}} + \frac{I(0,t)}{2\sqrt{C}} = \frac{R_0}{2\sqrt{L}}U(0,t) + \frac{R_0}{2\sqrt{L}}I(0,t), \quad t \geq 0,$$

$$\frac{C_0}{2\sqrt{C}}\frac{d}{dt}\big(U(\Lambda,t) - I(\Lambda,t)\big) = \frac{U(\Lambda,t)}{2\sqrt{L}} + \frac{I(\Lambda,t)}{2\sqrt{L}} - f\left(\frac{U(\Lambda,t)}{2\sqrt{C}} - \frac{I(\Lambda,t)}{2\sqrt{C}}\right), \, t \geq 0.$$

Since

$$U(\Lambda, t+T) = U(0,t) + \frac{1}{\sqrt{L}} \int_t^{t+T} e(s)\,ds + \frac{1}{\sqrt{C}} \int_t^{t+T} j(s)\,ds \quad \Leftrightarrow$$

$$U(\Lambda, t) = U(0,t-T) + \frac{1}{\sqrt{L}} \int_{t-T}^{t} e(s)\,ds + \frac{1}{\sqrt{C}} \int_{t-T}^{t} j(s)\,ds$$

and then

$$\frac{dU(0,t)}{dt} = \frac{dU(\Lambda, t+T)}{\partial t} - \frac{e(t+T)-e(t)}{\sqrt{L}} - \frac{j(t+T)-j(t)}{\sqrt{C}}.$$

Analogously

$$I(\Lambda, t) = I(0,t+T) - \frac{1}{\sqrt{C}} \int_{t+T}^{t} j(s)\,ds + \frac{1}{\sqrt{L}} \int_{t+T}^{t} e(s)\,ds \Leftrightarrow$$

$$\Leftrightarrow I(\Lambda, t-T) = I(0,t) - \frac{1}{\sqrt{C}} \int_t^{t-T} j(s)\,ds + \frac{1}{\sqrt{L}} \int_t^{t-T} e(s)\,ds,$$

$$\frac{dI(\Lambda, t)}{dt} = \frac{dI(0,t+T)}{dt} - \frac{j(t)-j(t+T)}{\sqrt{C}} + \frac{e(t)-e(t+T)}{\sqrt{L}}.$$

From the above equations we obtain a system:

$$\frac{dU(\Lambda, t)}{dt} - \frac{dI(\Lambda, t)}{dt} = \frac{2\sqrt{C}}{C_0} \frac{U(\Lambda, t) + I(\Lambda, t)}{2\sqrt{L}} - f\left(\frac{U(\Lambda, t) - I(\Lambda, t)}{2\sqrt{C}} \right)$$

$$2\sqrt{C}E(t) - \frac{R_0 + Z_0}{Z_0}\left(U(\Lambda, t+T) - \frac{1}{\sqrt{C}} \int_t^{t+T} j(s)\,ds - \frac{1}{\sqrt{L}} \int_t^{t+T} e(s)\,ds \right) +$$

$$+ \frac{Z_0 - R_0}{Z_0}\left(I(\Lambda, t-T) - \frac{1}{\sqrt{C}} \int_t^{t-T} j(s)\,ds + \frac{1}{\sqrt{L}} \int_t^{t-T} e(s)\,ds \right) = 0.$$

Let us put in the first equation $t + T = t$:

$$2\sqrt{C}E(t-T) - \frac{R_0 + Z_0}{Z_0}\left(U(\Lambda, t) - \frac{1}{\sqrt{C}} \int_{t-T}^{t} j(s)\,ds - \frac{1}{\sqrt{L}} \int_{t-T}^{t} e(s)\,ds \right) +$$

$$+ \frac{Z_0 - R_0}{Z_0}\left(I(\Lambda, t-2T) - \frac{1}{\sqrt{C}} \int_{t-T}^{t-2T} j(s)\,ds + \frac{1}{\sqrt{L}} \int_{t-T}^{t-2T} e(s)\,ds \right) = 0$$

$$\frac{dI(\Lambda,t)}{dt} = \frac{dU(\Lambda,t)}{dt} - \frac{U(\Lambda,t)}{C_0 Z_0} - \frac{I(\Lambda,t)}{C_0 Z_0} + f\left(\frac{U(\Lambda,t)}{2\sqrt{C}} - \frac{I(\Lambda,t)}{2\sqrt{C}}\right)$$

solve it with respect to $U(\Lambda,t)$, and set $\delta = \dfrac{Z_0 - R_0}{Z_0 + R_0}$:

$$\frac{2Z_0\sqrt{C}E(t-T)}{R_0 + Z_0} + \frac{1}{\sqrt{C}}\int_{t-T}^{t} j(s)ds + \frac{1}{\sqrt{L}}\int_{t-T}^{t} e(s)ds +$$

$$+ \delta I(\Lambda, t-2T) - \frac{\delta}{\sqrt{C}}\int_{t-T}^{t-2T} j(s)ds + \frac{\delta}{\sqrt{L}}\int_{t-T}^{t-2T} e(s)ds = U(\Lambda,t)$$

Replacing $U(\Lambda,t)$ and $\dfrac{dU(\Lambda,t)}{dt}$ from the first equation into the second one we obtain the following neutral equation with respect to the unknown function $I(\Lambda,t)$:

$$\frac{dI(\Lambda,t)}{dt} = \frac{2\sqrt{C}\,Z_0}{R_0 + Z_0}\frac{dE(t-T)}{dt} + \frac{1}{\sqrt{C}}\frac{d}{dt}\left(\int_{t-T}^{t} j(s)ds\right) + \frac{1}{\sqrt{L}}\frac{d}{dt}\left(\int_{t-T}^{t} e(s)ds\right) +$$

$$+ \delta\frac{dI(\Lambda,t-2T)}{dt} - \frac{\delta}{\sqrt{C}}\frac{d}{dt}\left(\int_{t-T}^{t-2T} j(s)ds\right) + \frac{\delta}{\sqrt{L}}\frac{d}{dt}\left(\int_{t-T}^{t-2T} e(s)ds\right) -$$

$$- \frac{2Z_0\sqrt{C}E(t-T)}{C_0 Z_0(R_0 + Z_0)} - \frac{1}{C_0 Z_0 \sqrt{C}}\int_{t-T}^{t} j(s)ds - \frac{1}{C_0 Z_0 \sqrt{L}}\int_{t-T}^{t} e(s)ds -$$

$$- \frac{\delta I(\Lambda,t-2T)}{C_0 Z_0} + \frac{\delta}{C_0 Z_0 \sqrt{C}}\int_{t-T}^{t-2T} j(s)ds - \frac{\delta}{C_0 Z_0 \sqrt{L}}\int_{t-T}^{t-2T} e(s)ds +$$

$$+ f\left(\frac{Z_0 E(t-T)}{R_0 + Z_0} + \frac{1}{2C}\int_{t-T}^{t} j(s)ds + \frac{v}{2}\int_{t-T}^{t} e(s)ds + \frac{\delta}{2\sqrt{C}}I(\Lambda,t-2T) - \frac{\delta}{2C}\int_{t-T}^{t-2T} j(s)ds + \frac{\delta v}{2}\int_{t-T}^{t-2T} e(s)ds - \frac{I(\Lambda,t)}{2\sqrt{C}}\right).$$

Let us put $I(\Lambda,t) \equiv I(t)$. Then we can formulate a periodic problem: to find a T_0-periodic solution of the above equation for $t \in [T, 3T]$:

$$\frac{dI(t)}{dt} = \frac{2\sqrt{C}\,Z_0}{R_0 + Z_0}\frac{dE(t-T)}{dt} + \frac{j(t) - j(t-T)}{\sqrt{C}} + \frac{e(t) - e(t-T)}{\sqrt{L}} +$$

$$+ \delta\frac{dI(t-2T)}{dt} + \frac{\delta}{\sqrt{C}}\big(j(t-T) - j(t-2T)\big) - \frac{\delta}{\sqrt{L}}\big(e(t-T) - e(t-2T)\big) -$$

$$-\frac{2\sqrt{C}E(t-T)}{C_0(R_0+Z_0)}-\frac{1}{C_0Z_0\sqrt{C}}\int_{t-T}^{t}j(s)ds-\frac{1}{C_0Z_0\sqrt{L}}\int_{t-T}^{t}e(s)ds-$$

$$-\frac{\delta I(t-2T)}{C_0Z_0}+\frac{\delta}{C_0Z_0\sqrt{C}}\int_{t-T}^{t-2T}j(s)ds-\frac{\delta}{C_0Z_0\sqrt{L}}\int_{t-T}^{t-2T}e(s)ds+$$

$$+f\left(\frac{Z_0E(t-T)}{R_0+Z_0}+\frac{1}{2C}\int_{t-T}^{t}j(s)ds+\frac{v}{2}\int_{t-T}^{t}e(s)ds+\frac{\delta}{2\sqrt{C}}I(t-2T)-\frac{\delta}{2C}\int_{t-T}^{t-2T}j(s)ds+\frac{\delta v}{2}\int_{t-T}^{t-2T}e(s)ds-\frac{I(t)}{2\sqrt{C}}\right)$$

and

$$I(t)=I_0(t),\quad \frac{dI(t)}{dt}=\frac{dI_0(t)}{dt},\ t\in[-T,T],$$

where the initial function $I_0(t)$ is obtained as above.

In contrast to the equations from Chapter II the just obtained equation is more complicated one. It is easy to see that if $j(t),e(t)$ possess suitable properties they imply an existence-uniqueness of T_0-periodic solution. Otherwise one might exist unstable solutions or chaos.

REFERENCES

[1] Abolinya, W. E.; Myshkis, A. D. *Math. Sbornik*. 1960, *vol*. 50 (92), No 4.

[2] Andreev, V. S. *Theory of nonlinear electric curcuits.* Radio & Sviaz, Moscow, Russia, 1982, pp. 280.

[3] Angelov, V. G. *Czechoslovak Mathematical Journal*. 1987, *vol*. 37 (112).

[4] Angelov, V. G. *Acta Mathematica Hungarica*. 1992, *vol*. 59 (3-4).

[5] Angelov, V. G. *Annuare de l'Universite d'Architecture, de Genie Civil et de Geodesie, Sofia*. 2000-2001, *vol*. 41*, series 2* (*Mathematics and Mechanics*).

[6] Angelov, V. G. *Annual Univ. Mining & Geology "St. I. Rilski"*. 2003, *vol*. 46, part 3.

[7] Angelov, V. G. *Annual Univ. Mining & Geology "St. I. Rilski"*. 2003, *vol*. 46, part 3.

[8] Angelov, V. G. *Jubilee Scientific Conference '2003, VSU "L. Karavelov"*. 2003, *vol*. 1.

[9] Angelov, V. G.; Stefanov, S. A. *13-th International Symposium on Electrical Apparatus and Technologies, Siela, Plovdiv, Proceedings*. 2003, *vol*. 1.

[10] Angelov, V. G. *Fixed Point Theory*. 2006, *vol*. 7, No 2.

[11] Angelov, V. G. *J. Nonlinear Analysis RWA*. 2007, *vol*. 8, No 2.

[12] Angelov, V. G. *Applied Mathematics E-Notes*. 2007, *vol*. 7.

[13] Angelov, V. G. *Proceedings International Conference dedicated to 105 Anniversary of the Birth of J. Atanasoff & J. Von Neumann*. 2009, *vol*. 1.

[14] Angelov, V. G. *Fixed Points in Uniform Spaces and Applications*. Cluj University Press "Babes-Bolyia", Ciuj Napoca, Romania, 2009; pp. 230.

[15] Angelov, V. G. *Int. Conference of European Polytechnical University – Pernik*. 2011.

[16] Angelov, V. G.; Hristov, M. H. *Circuits and Systems*. 2011, *vol*. 2.

[17] Angelov, V. G.; Angelova, D. T. (2011). Oscillatory Solutions of Neutral Equations with Polynomial Nonlinearities. *Recent Advances in Oscillation Theory − special issue published in International Journal of Differential Equations* [Online serial] Volume number 2011 (Article ID 949547).

[18] Angelov, V. G., In *Transmission Lines: Theory, Types and Applications*; Welton, D. M. Ed.; Nova Science Publishers, Inc. NY, US, 2011; pp. 259-294.

[19] Angelov, V. G. *Int. J. Theoretical and Mathematical Physics*. 2012*, vol*. 2, No. 5.

[20] Angelov, V. G. *10^Th International Conference on Fixed Point Theory and its Applications. Cluj-Napoca, 2012*.

[21] Angelov, V. G. *Int. J. Theoretical and Mathematical Physics.* 2013, *vol*. 3, No 1.

[22] Bielecki, A. *Bull Acad. Polon. Sci*. 1956, *vol*. 4, No 5.

[23] Bellman, R. E; Cooke, K. L. *Differential Difference Equations*. Academic Press, New York/London, 1963; pp. 462.

[24] Bessonov, L. A. *Nonlinear Circuits*. High School, Moscow, Russia, 1977; pp. 342.

[25] Bloom, S.; Peter, R. W. *RCA Rev.* 1954, *vol.* 15.

[26] Bobrov, I. N.; Bobrov, S. *I. Proc. 5th Int. Symp. Recent Adv. Microwave Technol., IS RAMT 95, Kiev*, 11-16.09.1995, *vol.* 2.

[27] Branin, F. H. *Proc. IEEE.* November 1967, *vol.* 55.

[28] Brayton, R. K. *Quart. Appl. Math.* 1966, *vol.* 24, No 3.

[29] Brayton ,R. K. *Quart. Appl. Math.* 1967, *vol.* 24, No 4.

[30] Borisovich, J. G. *Uspekhi Mat. Nauk.* 1979, *vol.* 34, No 6 (210).

[31] Borisovich, J. G. In *Proceedings of Mathematical Faculty, vol.* 01.10, Voronez University Press, Russia, 1973; pp. 12-25.

[32] Brillouin, L. N.; Parodi, M. *Propagation des Ondes dans les Milieux Periodiques.* Masson at C-ie Editeurs, Dunod Editeurs, Paris, France, 1956; pp. 347.

[33] Burns, S. G.; Bond, P. R. *Principles of Electronic Circuits.* PWS Publ. Company, ITP, Boston, 1997.

[34] Centeno, A.; Conlay, J. *Asia-Pacific Conf. Applied Electromagnetics*, Shan Alam, Malaysia, 2003.

[35] Chan, A. F. *The Finite Difference Time Domain Method for Computational Electromagnetics.* Ph D Dissertation, 2006; pp. 282.

[36] Christoffersen, C. E. *Global Modelling of Nonlinear Microwave Circuits.* PhD Thesis, North Carolina State University, Electrical Engineering, Raleigh, 2000; pp. 151.

[37] Chua, L. O.; Desoer, C. A.; Kuh, E. S. *Linear and Nonlinear Circuits.* McGraw-Hill Book Company, New York, USA, 1987; pp. 839.

[38] Chua, L. O.; Pen-Min Lin. *Machine Analysis of Electronic Circuits.* Energy, Moscow, 1980; pp.700.

[39] Chua, L. O. *IEEE Transactions on Circuits and Systems.* 1984, *vol.* cas-31, No 1.

[40] Collatz, L. *Funktional Analysis und Numerische Mathematik*, Springer Verlag, 1964; pp. 371

[41] Cooke, K. L.; Krumme, D. W. *J. Math. Anal. Appl.* 1968, *vol.* 24.

[42] Corduneanu, C. C.; Poorkarimi, H. In *Differential Equations*, Knowles I. W.; Lewis R. T.; Ed.; North-Holland, New York, 1985; pp. 107-113.

[43] Corduneanu, C. C. *Integral Equations and Stability of Feedback Systems.* Academic Press, New York, USA, 1973; pp. 238.

[44] Damgov, V. N. *Nonlinear and Parametric Phenomena: Theory and Applications in Radio physical and Mechanical Systems.* World Scientific: New Jersey, London, Singapore, 2004; pp. 574.

[45] Danilov, L. V.; Mathanov, P. N.; Philipov, E. S. *The Theory of Nonlinear Electrical Circuits.* Published Energoatomizdat, Leningrad, Russia, 1990; pp. 256.

[46] Darlington, S. *IEEE Trans. on circuits and syst., I. Fundamental theory and appl.* 1999, *vol.* 46, No 1.

[47] Deutsch, A.; Kopsay, G. V.; Ranieri, V. A.; Cataldo, J. K.; Galligan, E. A.; Graham, W. S.; McGouey, R. P.; Nunes, S. L.; Paraszczak, J. R.; Ritsko, J. J.; Serino, R. J.; Shih, D. Y.; Wilczynski, J. S. *IBM J. Res. Development.* 1990, *vol.* 34, No 4.

[48] Dhaene, D.; Zutter, D. D. *IEEE Trans. Computer-Aided Design.* 1992, *vol.* 11, No 7.

[49] Dobrev, D. M.; Jordanova, L. T. *Radio-communications. vol.* I and II, Ciela, Soft and Publishing, Sofia, Bulgaria, 2001; pp. 376 and 360.

[50] Dodov, N. I. *Compatibility and measurement of parameters of antenna feeder devices.* Technique, Sofia, Bulgaria, 1984; pp. 282.

[51] Dunlop, J.; Smith, D. G. *Telecommunications Engineering.* Chapman&Hall, London, 1994; pp. 593.

[52] Elfadel, I. M.; Huang, H. M.; Rudehli, A. E.; Dounavis, A.; Nakhla, M. S. *IEEE Trans. Advanced Packag.* 2002, *vol.* 25, No. 2.

[53] Elsgolz, L. E.; Norkin, S. B. *Introduction to the Theory of Differential Equations with Deviating Arguments.* Nauka, Moscow, Russia, 1971; pp. 296.

[54] Flynn, M. P.; Kang, J. J. *Proc. ICCAD.* 2005.

[55] Gould, R. W. *IRE Transactions-Electron Devices.* 1955, *vol.* ED-2, No. 4.

[56] Gould, R. W. *IRE Transactions-Electron Devices.* 1958, *vol.* ED-5, No. 3.

[57] Grivet-Tagolia, S.; Huang, H. M.; Rudehli, A. E.; Canavero, F.; Elfadel, I. M. *IEEE Trans. Advanced Packag.* 2004, *vol.* 27, No. 1.

[58] Gruodis, A. J.; Chang, C. S. *IBM J. Res. Develop.* 1981, *vol.* 25.

[59] Haigh, D.G.; Webster, D. R.; Ataei, R.; Parker, T.E.; Scott, J. B. In *Proceedings of 2001 Workshop on Nonlinear Dynamics of Electronic Systems* (Arie van Staveren, Ed.), Delft, Netherlands, 2001; pp. 1-15.

[60] Halanay, A. *Differential Equations, Stability, Oscillations, Time Lags.* Academic Press, New York, USA, 1966; pp. 528.

[61] Halanay, A., Yorke, J. A. *SIAM Review.* 1971, *vol.* 13.

[62] Halmos, P. R. *Measure Theory*, Graduate Text in Mathematics, Springer Verlag, 1974; *vol.* 18, pp. 368.

[63] Hofeer, W. J. R. *IEEE Trans. Microwave Theory Tech.* 1985, *vol.* 33, No 10

[64] Janischewskyi, W. *IEEE Transactions on Power Apparatus and Systems.* 1972, *vol.* Pas-91, No. 6.

[65] Jankowski, T.; Kwapisz, M. *Annales Polonici Mathematici.* 1972, *vol.* 26.

[66] Jeong, J.; Nevels, R. *Antennas Propagat.* Int. Symp. 2005, *vol.* 3A.

[67] Jeong, J. *Analytical Time Domain Electromagnetic Field Propagators and Closed Form Solutions for Transmission Lines.* Ph D Dissertation, Texas A&M University, 2006; pp. 129.

[68] Jiang, L.; Nishio, Y.; Ushida, A. A. *IEEE Trans. on circuits and systems. I. Fundamental theory and appl.* 1998, *vol.* 45, No. 6.

[69] Kamenskii, G. A. *Proceedings of Moscow University* 181, Mathematics. 1956, *vol.* 8.

[70] Kamenskii G. A.; Skubachevskii, A. L. *Linear Boundary Value Problems for Differential-Difference Equations.* MAI, Moscow, Russia, 1992; pp. 192.

[71] Kelebekler, E.; Yener, N. *17th Telecommunications forum TELFOR* 2009, Belgrade.

[72] Kolesov, J. S.; Shvitra, D.I. *Self-Oscillation in Delay Systems.* Mokslas, Vilnjus, Litva, 1979; pp. 148.

[73] Kolesov, J. S. In *Differential Equations and their Applications*, Vilnius, Inst. of Physics and Mathematics of Litva Academy of Sciences, 1972, *vol.* 2.

[74] Krasnoselskii, M. A. *On Translating Operator along Trajectories of Differential Equations.* Moscow, Russia, 1966; pp. 332.

[75] Krasnoselskii, M. A.; Vainikko G. M.; Zabrejko, P. P.; Rutizkii, J. B.; Stecenko, V. J. *Approximative Solution of Operator Equations.* Nauka, Moscow, Russia, 1969; pp. 456.

[76] Komuro, T. *IEEE Trans. Circuits and Systems.* 1991, *vol.* 38.

[77] Kyu-Pyung Hwang, Jian-Ming Jin, *IEEE Transactions on Microwave Theory and Techniques.* 1998, *vol.* 46, No 8.

[78] Lee, S.Y.; Konard, A.; Saldanha, R. R. *IEEE Trans. Magnetics.* 1993, *vol.* 29.

[79] Lozenski, I. D. *Radioelectronics in Geophysics.* Ministry of Education Press, Sofia, Bulgaria, 1988; pp. 399.

[80] Lucic, R.; Jovic, V.; Kurtovich, M. *Int. Conf. and Exhibition on Electromagnetic Compatibility*, EMC York 99, 1999.

[81] Lu, K. *IEEE Trans. Microwave Theory Techniques.* 1997, *vol.* 45, No 1.

[82] Maffucci, A., Miano, G. *Int. Journal of Circuit Theory and Applications.* 1999, *vol.* 27, No 5.

[83] Magnusson, P. C.; Alrxander, G. C.; Tripathi, V. K. *Transmission Lines and Wave Propagation.* 3rd ed., CRC Press. Boka Raton, 1992; pp. 460.

[84] Matkanov, P. N., *Fundamentals of Analysis of Electrical Circuits. Nonlinear Circuits.* Visha schkola, Moscow, Russia, 1977; pp. 272.

[85] Melvin, W. R. *J. Differential Equations.* 1972, *vol.* 13.

[86] Miano, G.; Maffucci, A. *Transmission Lines and Lumped Circuits.* Academic Press, New York, USA, 2001; pp. 479.

[87] Myshkis, A. D. *Uspekhi Mat. Nauk,* 1949; *vol.* 4, 5 /33/. Additional bibliographical proceedings. *Uspekhi Mat. Nauk.* 1950, *vol.* 5, 2 /36/.

[88] Myshkis, A.D. *Uspekhi Mat. Nauk.* 1977, *vol.* 32, 2.

[89] Nagumo J., Shimura, M. *Proc. IRE.* 1961, *vol.* 49.

[90] Paul, C. R. *Introduction to Electromagnetic Compatibility.* A Wiley-Inter science Publication, J. Wiley &Sons, New York, USA, 2006; pp. 836.

[91] Pozar, D. M. *Microwave Engineering.* J. Wiley &Sons, New York, USA, 1998; pp. 736.

[92] Qinwei Xu, Zheng-Fan Li, Pinaki Mazumder, Jun-Fa Mao. *IEEE Transactions on Microwave Theory and Techniques.* 2000, *vol.* 48, No 2.

[93] Rasvan, V. B. *Tatra Mt. Math. Publ.* 2009, *vol.* 42.

[94] Romeo, F., Santomauro, M. *IEEE Trans. Microwave Theory and Techniques.* 1987, *vol.* 35, No 2.

[95] Rosenstark, S. *Transmission Lines in Computer Engineering*, Mc Grow-Hill, New York, USA, 1994; pp. 212.

[96] Rowe, J. E. *Nonlinear Electron Wave Interaction Phenomena.* Academic Press, New York, USA, 1965; pp. 616.

[97] Ramo, S.; Whinnery, J. R.; T. van Duzer, *Fields and Waves in Communication Electronics.* John Wiley & Sons, Inc. New York, USA, 1994; pp. 846.

[98] Rus, I. A. *Metrical Fixed Point Theorems.* Cluj University Press "Babes-Bolyia", Ciuj Napoca, Romania, 1979; pp. 111.

[99] Sadovskii, B.N. *Doklady Academii Nauk SSSR.* 1971, *vol.* 200, No 5.

[100] Schelkunoff, S. A. *Proceedings of the Institute of Radio Engineers.* 1937, *vol.* 25, No. 11.

[101] Schelkunoff, S. A. *Bell System Technical Journal.* 1952, *vol.* 31.

[102] Schelkunoff, S.A. *Bell System Technical Journal.* 1955, *vol.* 34, No 5.

[103] Sekine, T.; Kobayashi, K.; Yokokawa, S. *Electronics and Communications in Japan.* 2002, *vol.* 85, No 8.

[104] Shimura, M. E. *IEEE Transactions on Circuit Theory.* 1967, *vol.* 14, No 1.

[105] Sobolev, S. L. *Selected Issues of the Function Spaces Theory and Generalized Functions.* Nauka, Moscow, Russia, 1989; pp. 256.

[106] Sun Tao; Hou Shi-Ying; Wu Xiao-Bing. *Journal of Electrical and Electronics.* 2008-03.

[107] Smirnov, W. A. *Fundamentals of Microwave Radiocommunications.* Communication Press, Moscow, Russia, 1957; pp. 820.

[108] Szimoni, K. *Theoretische Elektrotechnik*, Mir, Moscow, Russia, 1964; pp. 775.

[109] Taflove, A.; Hagness, S. C. *Computational Electromagnetics: The Finite-Difference Time-Domain Method*, 2nd edition, Artech House, Boston London, 2000; pp. 997.

[110] Tamir, T. *Guided-Wave Optoelectronics.* Ed.; Mir, Moscow, Russia, 1991; pp. 401.

[111] Vizmuller, P. *RF Design Guide Systems, Circuits and Equations.* Artech House, Inc., Boston London, 1995; pp. 281.

[112] Vladimirov, V. S. *Generalized Functions in Mathematical Physics.* Nauka, Moscow, Russia, 1979; pp. 320.

[113] Volterra, V. *Sulle Equazioni Integrodifferenziali della Teorie dell'Elasticita*, Atti reale Accad. Lincei. 1909, *vol.* 18, pp. 295.

[114] Wang, L.; Wang, Z.; Xingfu Zou. *J. London Math. Soc.* 2002, *vol* (2), 65.

[115] Whitham, G. B. *Linear and Nonlinear Waves*, J. Wiley&Sons, New York, USA, 1974; pp. 141.

[116] Wu, J. *Theory and Applications of Partial Functional Differential Equations.* Springer Verlag, 1996; pp. 427.

[117] Wu, J.; Xia, H. *J. Differential Equations.* 1996, *vol.* 124.

[118] Xu, Q.; Li, Z. F.; Mazumder, P.; Mao, J. F. *IEEE Transactions on Microwave Theory and Techniques.* 2000, *vol.* 48, No 2.

[119] Xu, Q.; Mazumder, P.; Li, Z. F. *IEEE 14th Int. Conf. VLSI Design*, Jan. 2001.

[120] Xu, Q.; Mazumder, P. *IEEE Trans. Microwave Theory and Techniques.* 2002, *vol.* 50, No 10.

[121] You, S. H.; Kuester, E. F. *IEEE Transactions on Microwave Theory and Techniques.* 2005, *vol.* 53, No 9.

[122] Zaezdnii, A. M. *Foundation of Analysis of Nonlinear and Parametric Circuits.* Sviazi, Moscow, Russia, 1973; pp. 448.

[123] Zemanian, A. H. *Int. Journal of Circuit Theory and Applications.* 2005, *vol.* 33, No 3.

[124] Zhang, X.; Mei, K. K. *IEEE Trans. Microwave Theory and Techniques.* 1989, *vol.* 36, No 12.

[125] Zhong, X.; Liu, Y.; Mei, K. K. *Microwave and Optical Tech. Letters.* 2002, *vol.* 32, No 1.

[126] Zima, K. *Annales Polonici Mathematici.* 1973, *vol.* 27.

[127] Zverkin, A. M. In *Proceedings of the seminar on the theory of differential equations with deviating arguments.* Moscow, Russia, 1967; *vol.* 4, pp. 278-283.

Index

Φ

Φ-contractive functions, 29

B

Banach contraction condition, 27
Banach contraction mapping principle, 27
base, xvi, 25
beams, xii, xvi, xvii, 158, 648
boundary conditions, ix, xii, xiv, xv, xvii, xviii, 4, 5, 8, 12, 42, 57, 101, 140, 157, 160, 162, 163, 243, 244, 321, 323, 325, 327, 347, 401, 403, 405, 464, 465, 494, 497, 498, 500, 505, 553, 557, 559, 560, 605, 609, 646, 649
boundary value problem, ix, 4, 159
bounds, 274, 396, 489, 516, 549, 648
Bulgaria, 655, 656

C

cables, 321
Cartesian product, 33
Cauchy sequence, 25, 43
chaos, ix, 652
classification, 37
closure, 26
complement, 24
composition, 80, 83
computation, xiii, xiv
conductance, xii, xiii, 2
conduction, 2
conductivity, 19, 415, 493, 553
conductor(s), xi, xii, xv, 1, 2, 151
configuration, xv, xvi, xviii, 42, 321, 403, 557, 558

conformity, 39, 41, 42, 62, 63, 76, 79, 87, 89, 103, 114, 129, 141, 174, 209, 234, 253, 311, 333, 351, 385, 386, 454, 471, 472, 518, 615
connected RLC-loads, xvi, xvii, xviii, 157, 497
contradiction, 67, 91, 131
convergence, x, 24, 26, 27, 31, 33, 44, 50, 78, 87, 88, 113, 151, 155, 233, 241, 274, 304, 310, 319, 332, 351, 453, 469, 497
copper, 345, 376, 447, 544, 644

D

derivatives, 36, 37, 38, 39, 40, 42, 44, 50, 63, 68, 79, 85, 87, 102, 107, 115, 126, 142, 157, 165, 167, 172, 179, 213, 214, 223, 233, 240, 251, 276, 310, 332, 351, 364, 374, 384, 417, 439, 450, 451, 453, 469, 496, 509, 516, 518, 538, 566, 594, 612, 629
deviation, 41
differential equations, ix, xv, xvi, 1, 12, 26, 33, 36, 37, 38, 39, 40, 41, 42, 58, 409, 416, 506, 567, 646, 647, 657
diffusion, 113
diodes, 323
discontinuity, 39, 47, 48
discretization, 123
dispersion, xiv, 323
displacement, 2

E

electric field, 19
electromagnetic, ix, xi, 2, 4, 15, 228
electromagnetic fields, xi, 4, 15
electromagnetic waves, ix
electron, xiii, xvi, xvii, 152, 157, 229, 304, 648
electronic circuits, 321
energy, xi
engineering, 543

equality, 10, 31, 398, 405, 491, 550
Euclidean space, 25
existence-uniqueness result, xvii, 154, 157, 259, 403, 497, 600
exponential functions, x
extraction, xiv

F

families, x, 8, 13, 33, 35, 54, 57, 88, 333, 352, 365, 385, 471, 519, 562, 614
FEM, xiv, xv
finite element method, xiv
fixed-point method, xvi, xvii, xviii, 45, 53, 157, 321, 404, 497, 498, 646, 647
force, xiii, 152
formula, 48, 61, 169
France, 654

H

Hausdorff sequentially complete uniform space, 29
hybrid, xiv
hyperbolic systems, ix, xii

I

ideal, xiii, 4, 244, 321
identity, 35, 98, 138, 344, 376
inductor, 159
inequality, 25, 26, 27, 35, 37, 67, 71, 91, 106, 131, 216, 274, 343, 357, 416, 459, 574
initial value problem, ix, xv, xvi, xvii, xviii, 1, 5, 36, 37, 38, 39, 40, 42, 54, 55, 58, 60, 66, 83, 104, 113, 115, 143, 153, 157, 163, 229, 237, 243, 246, 251, 254, 272, 280, 323, 336, 368, 389, 404, 409, 421, 457, 464, 497, 498, 506, 557, 558, 609
integration, 11, 12, 40, 41, 66, 81, 101, 180, 254, 255, 260, 347, 420, 421, 426, 465, 512, 574
iteration, xvii

J

j-locally compact space, 35

L

laws, ix, xii, 19
lead, ix, xiv, xv, 157, 159, 300, 647
line voltage, xv, xvi, xvii

linear systems, 449
lossless transmission line(s), ix, xiii, xv, xvi, xvii, xviii, 4, 53, 55, 157, 160, 228, 229, 243, 321, 403, 409, 500, 506, 557, 648
lossy, vii, viii, 15, 321, 322, 403, 489, 497, 557, 604
lossy transmission lines, ix, xiv, xv, xvi, xvii, xviii, 321, 322, 323, 403, 404, 408, 497, 557, 647, 648

M

magnetic field, xii, 19, 20
MAI, 655
mapping, 27, 29, 30, 32, 34, 35, 36, 98, 138, 344, 376
mathematical methods, 24
mathematics, 52
matrix, xiv, 5, 6, 7, 8, 16, 17, 21, 22, 23, 323, 324, 396, 397, 406, 407, 489, 490, 503, 549, 551, 606
measurement, 655
media, 19
memory, xiv
metric, v, 24, 33
metric spaces, 27
mixed problem for transmission line equations, xvii, 52, 157, 243
modelling, xiv
models, xiii
multiplication, 17, 23, 25, 407
multiplier, 497

N

neutral equation(s), ix, xv, xvi, xvii, 1, 5, 36, 41, 42, 45, 52, 53, 54, 55, 128, 153, 157, 229, 243, 319, 321, 330, 331, 364, 647, 651
nodes, xiv
nonlinear loads, ix, xvi, xvii, xviii, 157, 159, 497, 558, 647
nonlinear neutral system, xvii, xviii, 157, 243, 403, 498

O

operations, 25
ordinary differential equations, 36

P

parallel, xi, xii, xv, xvi, xvii, xviii, 19, 53, 54, 243, 244, 319, 403, 497, 557, 558, 559
parallel connected RLC-loads, xvi, xvii, 244, 403

partial differential equations, ix, xi, xv, 3, 4, 42, 53, 321

periodic initial value problem, xvi, xvii, xviii, 54, 321, 557

periodic problem, xvii, xviii, 45, 47, 48, 63, 64, 81, 100, 102, 141, 157, 179, 183, 243, 321, 403, 557, 565, 574, 651

periodic solutions, xv, xvi, xvii, 1, 40, 45, 47, 49, 55, 60, 63, 114, 157, 227, 230, 243, 610

periodicity, 44, 45, 62, 64, 174, 495, 555

permeability, 19, 152, 230

permittivity, 19

plane waves, xvi, 1, 19

power lines, 321

propagation, ix, xii, xviii, 19, 54, 56, 75, 99, 122, 138, 154, 208, 231, 273, 300, 345, 377, 409, 447, 602

R

radio, 557

radius, 25

real numbers, 25, 44, 49, 87, 128, 232, 310, 331, 350, 364, 384, 452, 469, 518, 612

reasoning, xvi, 1, 212, 246, 325

recall, 2, 25, 56, 60, 62, 68, 90, 97, 123, 124, 129, 143, 236, 333, 411, 417, 435, 560, 574, 610

recalling, 66, 251, 252, 264, 279, 358, 371, 383, 417, 468

R-element, 53

relevance, xv

researchers, ix, xix

resistance, xii, xiii, xvi, 2, 53, 54, 152, 323, 345, 376, 447, 544, 644

resonator, 304

response, xiii, xv

retardation, 41

roots, 6, 16, 21, 324, 406, 503, 606

S

scattering, xiv

science, 656

self-control, 36

showing, xv, xvi, 259

signals, xi, xiv, 79, 80, 123, 648

smoothing, 40, 41

smoothness, 41

specific numeric data, xvii

stability, 36

state, ix

structural dimension, 2

structure, xi, xiv, 2, 4, 15, 244

substitution, 18, 325, 399, 408, 492, 552, 608

successive approximations, x, xvi, 26, 346, 377, 497, 646, 647

T

techniques, xiv, 1

TEM, xii, 19

TEM propagation, 19

TEM waves, xii, 19

topology, xiv, 24, 25, 28, 35, 44, 50, 88, 233, 310, 332, 351, 453, 469, 518, 612

transformation(s), xiii, 19, 52, 59, 100, 139, 162, 229, 246, 326, 346, 397, 400, 403, 409, 464, 492, 550, 552, 560, 607, 647, 648, 649

translation, 215, 251

traveling waves, 54

treatment, x, 496

U

uniform space, 27

V

vacuum, 152

variables, 6, 8, 17, 22, 101, 162, 164, 306, 324, 397, 407, 468, 490, 504, 549

vector, 19, 25, 27, 28, 37

velocity, ix, 151

V-I characteristic, xvi, xvii, 53, 54, 61, 75, 99, 112, 113, 152, 208, 231, 321, 322, 331, 345, 347, 363, 447, 544, 569, 602, 645

W

wave, 151, 656, 657

wave propagation, 151, 228

wires, 321

Y

yield, 161, 401, 559